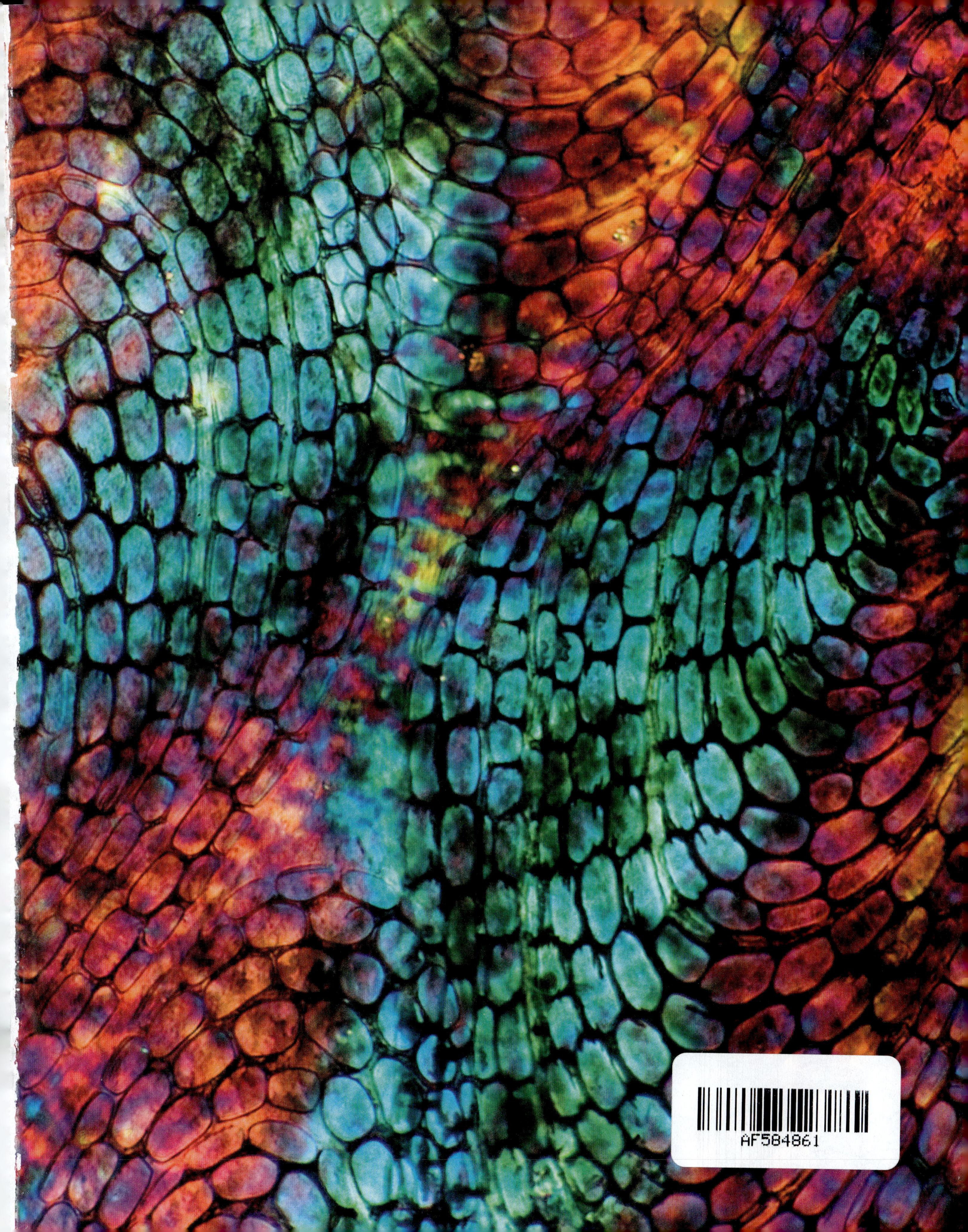

Pearson Australia
(a division of Pearson Australia Group Pty Ltd)
707 Collins Street, Melbourne, Victoria 3008
PO Box 23360, Melbourne, Victoria 8012
www.pearson.com.au

First published 2019 by Pearson Australia
2022 2021 2020 2019
10 9 8 7 6 5 4 3 2 1

Lead Publisher: Misal Belvedere and Malcolm Parsons
Project Manager: Michelle Thomas
Production Editor: Laura Pietrobon
Lead Development Editor: Amy Sparkes
Content Developer: Rebecca Wood
Development Editor: Rima Cilmi, Naomi Campanale, Haeyean Lee
Lead Editor: Leanne Poll
Editors: Julie Cooney, Pip Dundon
Editorial work: Fiona Maplestone
Designer: Anne Donald
Rights & Permissions Editor: Samantha Russell-Tulip
Senior Publishing Services Analyst: Rob Curulli
Typesetter: iEnergizer Aptara Pty Ltd
Proofreader: Trudi Ryan
Indexer: Max McMasters
Illustrator: DiacriTech
Printed in China by Golden Cup

ISBN 978 1 4886 1926 7
Pearson Australia Group Pty Ltd ABN 40 004 245 943

A catalogue record for this book is available from the National Library of Australia

Indigenous Australians
Some of the images used in *Pearson Biology 12 New South Wales Student Book* might have associations with deceased Indigenous Australians. Please be aware that these images might cause sadness or distress in Aboriginal or Torres Strait Islander communities.

We respectfully acknowledge the traditional custodians of the lands upon which the many schools throughout Australia are located. We acknowledge traditional Indigenous Knowledge systems are founded upon a logic that developed from Indigenous experience with the natural environment over thousands of years. This is in contrast to Western science, where the production of knowledge often takes place within specific disciplines. As a result, there are many cases where such disciplinary knowledge does not reflect Indigenous worldviews, and can be considered offensive to Indigenous peoples.

Practical activities
All practical activities, including the illustrations, are provided as a guide only and the accuracy of such information cannot be guaranteed. Teachers must assess the appropriateness of an activity and take into account the experience of their students and the facilities available. Additionally, all practical activities should be trialled before they are attempted with students and a risk assessment must be completed. All care should be taken whenever any practical activity is conducted: appropriate protective clothing should be worn, the correct equipment used, and the appropriate preparation and clean-up procedures followed. Although all practical activities have been written with safety in mind, Pearson Australia and the authors do not accept any responsibility for the information contained in or relating to the practical activities, and are not liable for any loss and/or injury arising from or sustained as a result of conducting any of the practical activities described in this book.

PEARSON
BIOLOGY
NEW SOUTH WALES

Writing and development team

We are grateful to the following people for their time and expertise in contributing to the *Pearson Biology 12 New South Wales* project.

Rebecca Wood
Content Developer
Subject Lead

Zoë Armstrong
Scientist
Author

Donna Chapman
Laboratory Technician
Safety Consultant

Wayne Deeker
Science Writer
Author

Anna Madden
Teacher
Author

Katherine McMahon
Teacher
Author

Kate Naughton
Scientist
Author

Yvonne Sanders
Teacher
Skills and Assessment Author

Sue Siwinski
Teacher
Author and Reviewer

Ana Sofia Wheeler
Teacher
Author

Krista Bayliss
Teacher
Contributing Author

Ian Bentley
Educator
Contributing Author

Sally Cash
Teacher
Contributing Author

Caroline Cotton
Teacher
Contributing Author

Sarah Edwards
Teacher
Contributing Author

Neil van Herk
Teacher
Contributing Author

Samantha Hopley
Educator
Contributing Author

Catherine Litchfield
Teacher
Contributing Author

Troy Potter
Teacher
Contributing Author

Helen Silvester
Teacher
Contributing Author

Reuben Bolt
Director of the Nura Gili Indigenous Programs Unit, UNSW
Reviewer

Judith Brotherton
Teacher
Reviewer

Jacqueline Kerr
Teacher
Reviewer

Sylvia Persis
Teacher
Reviewer

Timothy Sloane
Teacher
Reviewer

Alastair Walker
Teacher
Reviewer

Trish Weekes
Science Literacy Consultant

Christina Adams
Teacher
Answer Checker

Elaine Georges
Teacher
Answer Checker

Caryn Hellberg
Teacher
Answer Checker

Jacoba Kooy
Scientist
Answer Checker

Karen Malysiak
Educator
Answer Checker

Kelly Merrin
Scientist
Answer Checker

The Publisher wishes to thank and acknowledge Pauline Ladiges and Barbara Evans for their contribution in creating the original works of the series and longstanding dedicated work with Pearson and Heinemann.

Contents

Working scientifically

Module 5 Heredity

Module 6 Genetic change

Module 7 Infectious disease

Module 8 Non-infectious disease and disorders

How to use this book

Pearson Biology 12 New South Wales

Pearson Biology 12 New South Wales has been written to fully align with the new Stage 6 Syllabus for New South Wales Biology. The book covers Modules 5 to 8 in an easy-to-use resource. Explore how to use this book below.

Chapter opener

The chapter opening page links the syllabus to the chapter content. Key content addressed in the chapter is clearly listed.

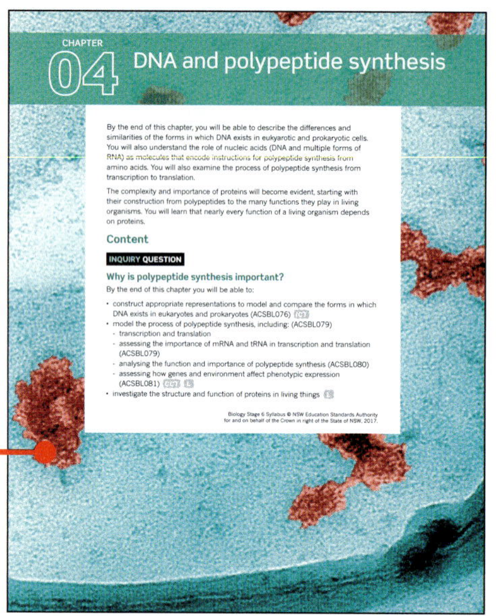

Section

Each chapter is clearly divided into manageable sections of work. Best-practice literacy and instructional design are combined with high-quality, relevant photos and illustrations to help students better understand the ideas or concepts being developed.

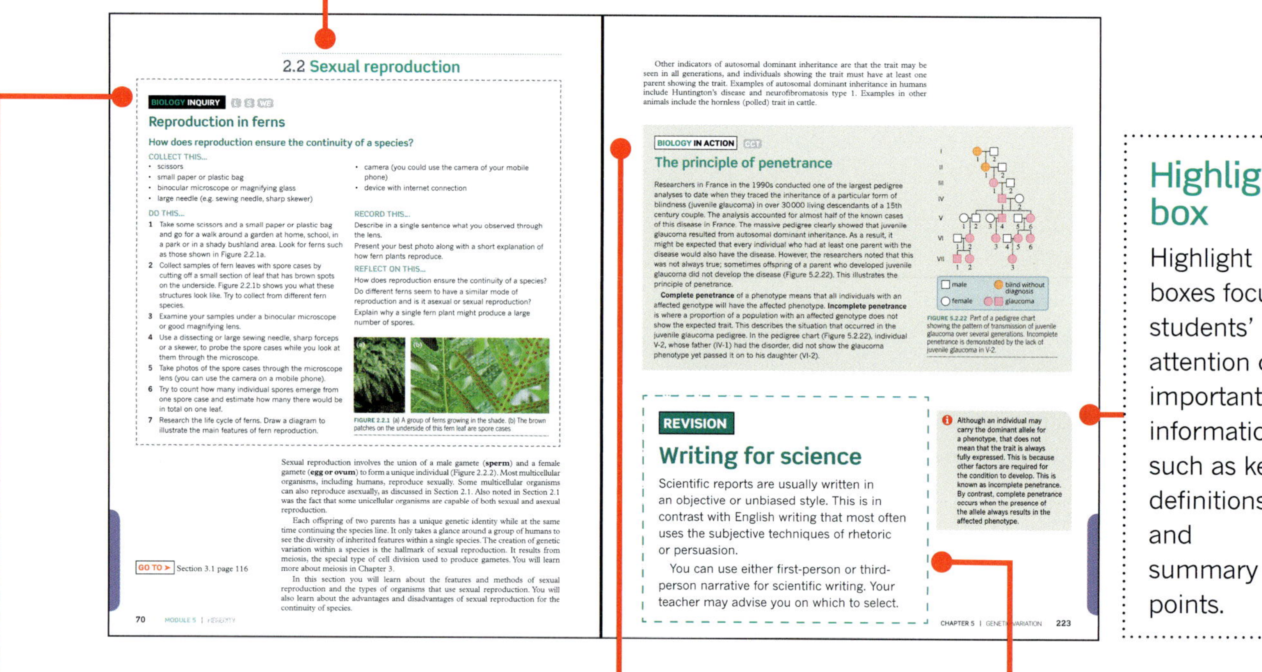

Highlight box

Highlight boxes focus students' attention on important information such as key definitions and summary points.

Biology Inquiry

Biology Inquiry features are inquiry-based activities that assist students to discover concepts before learning about them. They encourage students to think about what happens in the world and how science can provide explanations.

Biology in Action

Biology in Action boxes place biology in an applied situation or a relevant context. These refer to the nature and practice of biology, its applications and associated issues, and the historical development of its concepts and ideas.

Revision box

Revision boxes are used to remind students of vital concepts previously covered that are required for current learning.

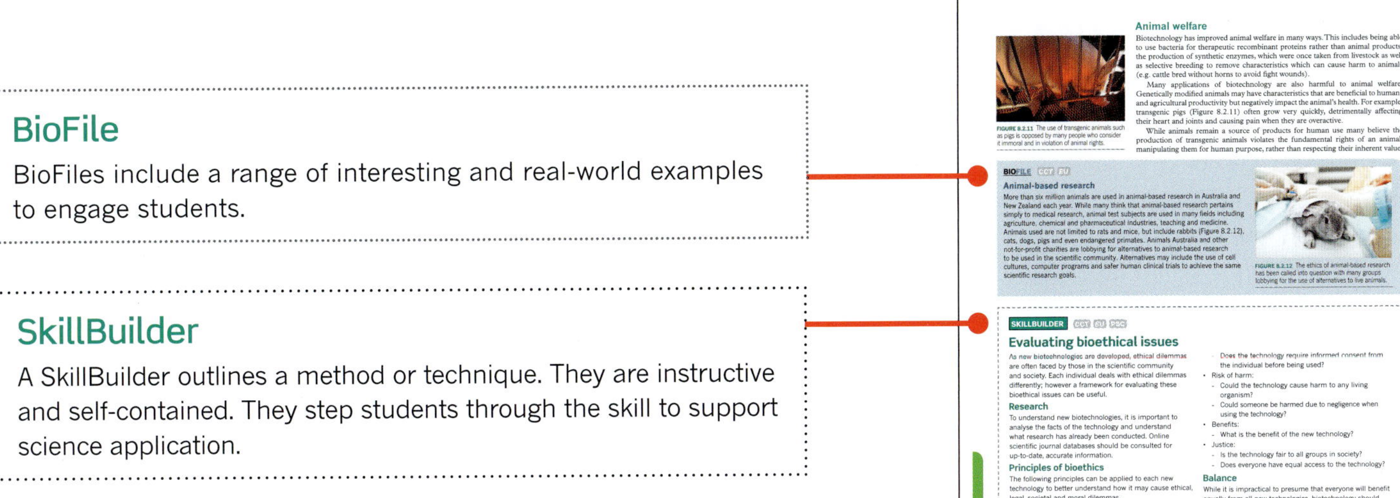

BioFile

BioFiles include a range of interesting and real-world examples to engage students.

SkillBuilder

A SkillBuilder outlines a method or technique. They are instructive and self-contained. They step students through the skill to support science application.

Worked examples

Worked examples are set out in steps that show thinking and working. This format greatly enhances student understanding by clearly linking underlying logic to the relevant calculations. Each Worked example is followed by a Try yourself activity. This mirror problem allows students to immediately test their understanding.

Additional content

Additional content includes material that goes beyond the core content of the syllabus. They are intended for students who wish to expand their depth of understanding in a particular area.

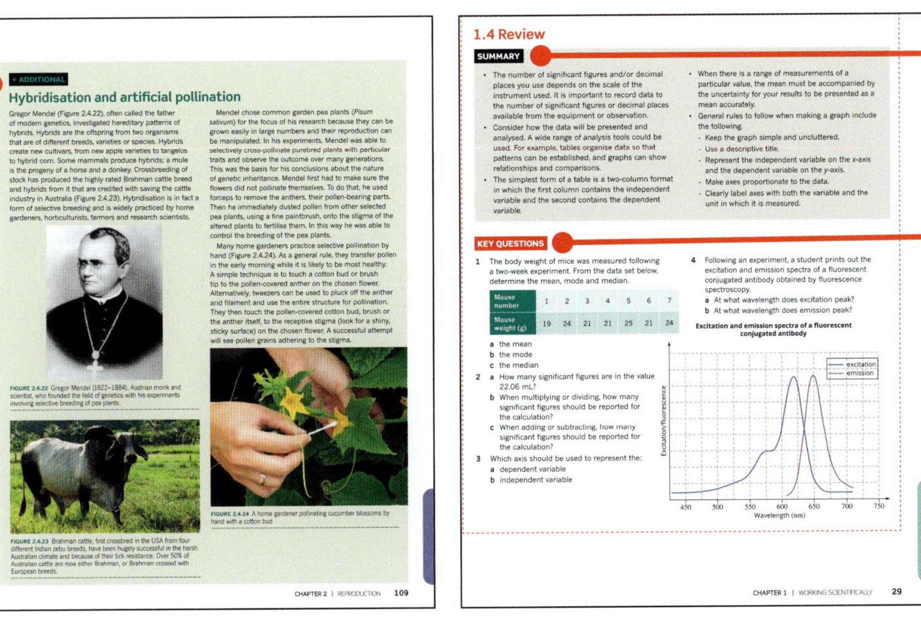

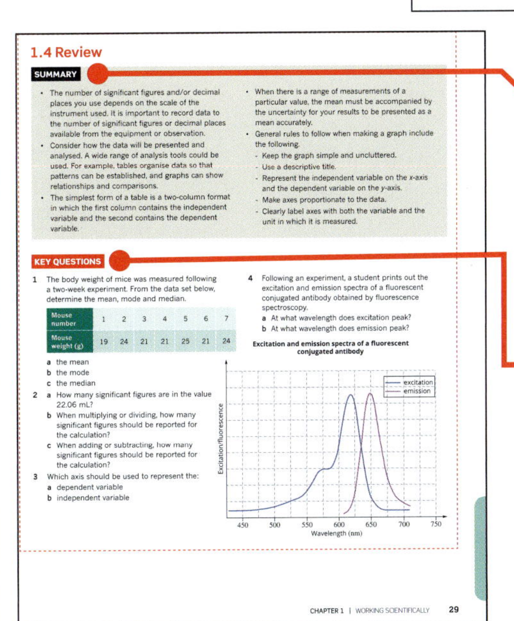

Section summary

Each section has a summary to help students consolidate the key points and concepts.

Section review questions

Each section finishes with key questions to test students' understanding and ability to recall the key concepts of the section.

How to use this book

Chapter review

Each chapter finishes with a list of key terms covered in the chapter and a set of questions to test students' ability to apply the knowledge gained from the chapter.

Module review

Each module finishes with a comprehensive set of questions, including multiple choice and short answer. These assist students in drawing together their knowledge and understanding, and applying it to these types of questions.

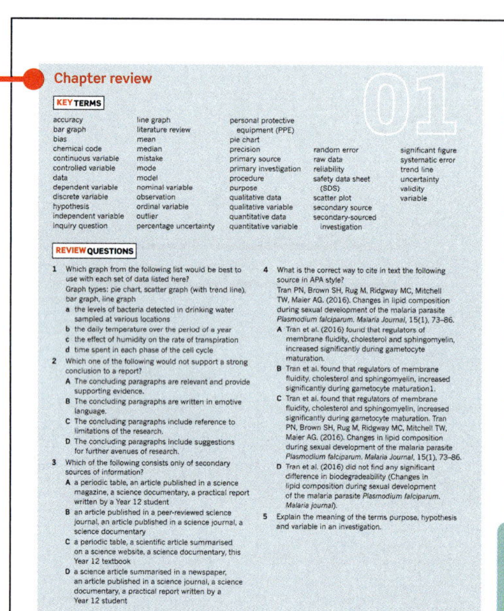

Chapter review

01

KEY TERMS

accuracy
bar graph
bias
chemical code
continuous variable
controlled variable
data
dependent variable
discrete variable
hypothesis
independent variable
inquiry question
line graph
literature review
mean
median
mistake
mode
model
nominal variable
observation
ordinal variable
outlier
percentage uncertainty
personal protective equipment (PPE)
pie chart
precision
primary source
primary investigation
procedure
purpose
qualitative data
qualitative variable
quantitative data
quantitative variable
random error
raw data
reliability
safety data sheet (SDS)
scatter plot
secondary source
secondary-sourced investigation
significant figure
systematic error
trend line
uncertainty
validity
variable

REVIEW QUESTIONS

1 Which graph from the following list would be best to use with each set of data listed here?
Graph types: pie chart, scatter graph (with trend line), bar graph, line graph
a the levels of bacteria detected in drinking water sampled at various locations
b the daily temperature over the period of a year
c the effect of humidity on the rate of transpiration
d time spent in each phase of the cell cycle

2 Which one of the following would not support a strong conclusion to a report?
A The concluding paragraphs are relevant and provide supporting evidence.
B The concluding paragraphs are written in emotive language.
C The concluding paragraphs include reference to limitations of the research.
D The concluding paragraphs include suggestions for further avenues of research.

3 Which of the following consists only of secondary sources of information?
A a periodic table, an article published in a science magazine, a science documentary, a practical report written by a Year 12 student
B an article published in a peer-reviewed science journal, an article published in a science journal, a science documentary
C a periodic table, a scientific article summarised on a science website, a science documentary, this Year 12 textbook
D a science article summarised in a newspaper, an article published in a science journal, a science documentary, a practical report written by a Year 12 student

4 What is the correct way to cite in text the following source in APA style?
Tran PN, Brown SH, Rug M, Ridgway MC, Mitchell TW, Maier AG. (2016). Changes in lipid composition during sexual development of the malaria parasite *Plasmodium falciparum*. *Malaria Journal*, 15(1), 73–86.
A Tran et al. (2016) found that regulators of membrane fluidity, cholesterol and sphingomyelin, increased significantly during gametocyte maturation.
B Tran et al. found that regulators of membrane fluidity, cholesterol and sphingomyelin, increased significantly during gametocyte maturation1.
C Tran et al. found that regulators of membrane fluidity, cholesterol and sphingomyelin, increased significantly during gametocyte maturation. Tran PN, Brown SH, Rug M, Ridgway MC, Mitchell TW, Maier AG. (2016). Changes in lipid composition during sexual development of the malaria parasite *Plasmodium falciparum*. *Malaria Journal*, 15(1), 73–86.
D Tran et al. (2016) did not find any significant difference in biodegradeability (Changes in lipid composition during sexual development of the malaria parasite *Plasmodium falciparum*. *Malaria journal*).

5 Explain the meaning of the terms purpose, hypothesis and variable in an investigation.

CHAPTER 1 | WORKING SCIENTIFICALLY 49

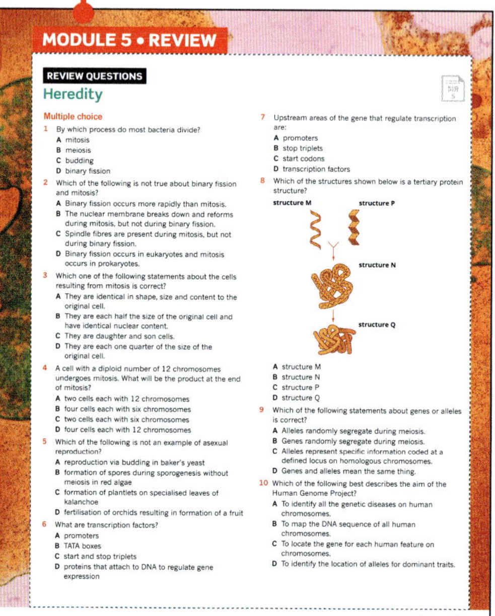

MODULE 5 • REVIEW

REVIEW QUESTIONS

Heredity

Multiple choice

1 By which process do most bacteria divide?
A mitosis
B meiosis
C budding
D binary fission

2 Which of the following is not true about binary fission and mitosis?
A Binary fission occurs more rapidly than mitosis.
B The nuclear membrane breaks down and reforms during mitosis, but not during binary fission.
C Spindle fibres are present during mitosis, but not during binary fission.
D Binary fission occurs in eukaryotes and mitosis occurs in prokaryotes.

3 Which one of the following statements about the cells resulting from mitosis is correct?
A They are identical in shape, size and content to the original cell.
B They are each half the size of the original cell and have identical nuclear content.
C They are daughter and son cells.
D They are each one quarter of the size of the original cell.

4 A cell with a diploid number of 12 chromosomes undergoes mitosis. What will be the product at the end of mitosis?
A two cells each with 12 chromosomes
B four cells each with six chromosomes
C two cells each with six chromosomes
D four cells each with 12 chromosomes

5 Which of the following is not an example of asexual reproduction?
A reproduction via budding in baker's yeast
B formation of spores during sporogenesis without meiosis in red algae
C formation of plantlets on specialised leaves of kalanchoe
D fertilisation of orchids resulting in formation of a fruit

6 What are transcription factors?
A promoters
B TATA boxes
C start and stop triplets
D proteins that attach to DNA to regulate gene expression

7 Upstream areas of the gene that regulate transcription are:
A promoters
B stop triplets
C start codons
D transcription factors

8 Which of the structures shown below is a tertiary protein structure?
structure M
structure P
structure N
structure Q
A structure M
B structure N
C structure P
D structure Q

9 Which of the following statements about genes or alleles is correct?
A Alleles randomly segregate during meiosis.
B Genes randomly segregate during meiosis.
C Alleles represent specific information coded at a defined locus on homologous chromosomes.
D Genes and alleles mean the same thing.

10 Which of the following best describes the aim of the Human Genome Project?
A To identify all the genetic diseases on human chromosomes.
B To map the DNA sequence of all human chromosomes.
C To locate the gene for each human feature on chromosomes.
D To identify the location of alleles for dominant traits.

REVIEW QUESTIONS 277

Icons

The New South Wales Stage 6 syllabus 'Learning across the curriculum' and 'General capabilities' content are addressed throughout the series and are identified using the following icons.

'Go to' icons are used to make important links to relevant content within the same Student Book.

This icon indicates the best time to engage with a worksheet (WS), a practical activity (PA), a depth study (DS) or module review (MR) questions in *Pearson Biology 12 New South Wales Skills and Assessment* book.

This icon indicates the best time to engage with a practical activity on *Pearson Biology 12 New South Wales* Reader+.

Glossary

Key terms are shown in **bold** in sections and listed at the end of each chapter. A comprehensive glossary at the end of the book includes and defines all the key terms.

Answers

Comprehensive answers and fully worked solutions for all section review questions, Worked example: Try yourself features, chapter review questions and module review questions are provided via *Pearson Biology 12 New South Wales* Reader+.

Pearson Biology 12 New South Wales

Student Book

Pearson Biology 12 New South Wales has been written to fully align with the new Stage 6 syllabus for New South Wales. The Student Book includes the very latest developments and applications of biology and incorporates best-practice literacy and instructional design to ensure the content and concepts are fully accessible to all students.

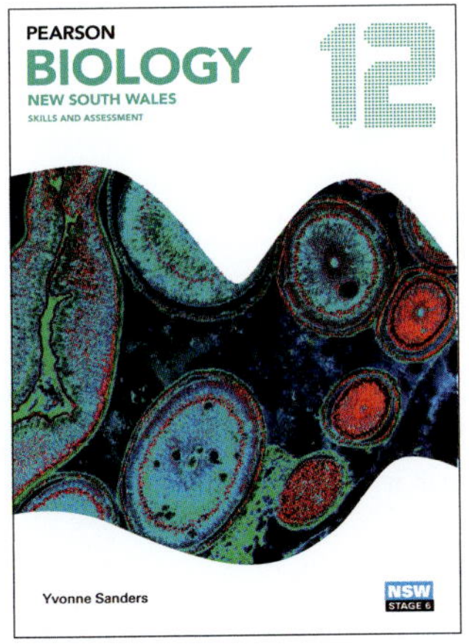

Skills and Assessment Book

Pearson Biology 12 New South Wales Skills and Assessment book gives students the edge in preparing for all forms of assessment. Key features include a toolkit, key knowledge summaries, worksheets, practical activities, suggested depth studies and module review questions. It provides guidance, assessment practice and opportunities to develop key skills.

Reader+ the next generation eBook

Pearson Reader+ lets you use your Student Book online or offline on any device. Pearson Reader+ retains the look and integrity of the printed book. Practical activities, interactives and videos are available on Pearson Reader+ along with fully worked solutions for the Student Book questions.

Teacher Support

Online teacher support for the series includes syllabus grids, a scope and sequence plan, and three practice exams per year level. Fully worked solutions to all Student Book questions are provided, as well as teacher notes for the chapter inquiry tasks. Skills and Assessment book resources include solutions to all worksheets, practical activities, depth studies and module review questions; teacher notes, safety notes, risk assessments and lab technician's checklists and recipes for all practical activities; and assessment rubrics and exemplar answers for the depth studies.

PEARSON
BIOLOGY 12
NEW SOUTH WALES

Access your digital resources at **pearsonplaces.com.au**
Browse and buy at **pearson.com.au**

CHAPTER 01

Working scientifically

This chapter covers the skills needed to successfully plan and conduct primary- and secondary-sourced investigations.

1.1 Questioning and predicting describes how to develop, propose and evaluate inquiry questions and hypotheses. When creating a hypothesis, a consideration of the variables must be included.

1.2 Planning investigations will help you learn to identify risks in your investigation and to consider all ethical concerns. It is important to choose appropriate materials and technology to carry out your investigation. You will also need to confirm that your choice of variables allows for a reliable collection of data.

1.3 Conducting investigations will help you develop and lead your investigations. It describes methods for accurately collecting and recording data to reduce errors. You will need to follow appropriate procedures when disposing of waste.

1.4 Processing data and information will help you learn how best to represent your information and how to identify trends and patterns in your data.

1.5 Analysing data and information explains how best to analyse your results. It explains error and uncertainty and how to construct mathematical models to better understand the scientific principles of your research.

1.6 Problem solving will help you use critical thinking to demonstrate an understanding of the scientific principles underlying the solution to your inquiry question.

1.7 Communicating explains how to communicate an investigation clearly and accurately using appropriate scientific language, nomenclature and scientific notation.

Outcomes

By the end of this chapter you will be able to:

- develop and evaluate questions and hypotheses for scientific investigation (BIO12-1)
- design and evaluate investigations in order to obtain primary and secondary data and information (BIO12-2)
- conduct investigations to collect valid and reliable primary and secondary data and information (BIO12-3)
- select and process appropriate qualitative and quantitative data and information using a range of appropriate media (BIO12-4)
- analyse and evaluate primary and secondary data and information (BIO12-5)
- solve scientific problems using primary and secondary data, critical thinking skills and scientific processes (BIO12-6)
- communicate scientific understanding using suitable language and terminology for a specific audience or purpose (BIO12-7)

Content

By the end of this chapter you will be able to:

- develop and evaluate inquiry questions and hypotheses to identify a concept that can be investigated scientifically, involving primary and secondary data (ACSBL001, ACSBL061, ACSBL096) L
- modify questions and hypotheses to reflect new evidence CCT
- assess risks, consider ethical issues and select appropriate materials and technologies when designing and planning an investigation (ACSBL031, ACSBL097) EU PSC
- justify and evaluate the use of variables and experimental controls to ensure that a valid procedure is developed that allows for the reliable collection of data (ACSBL002)
- evaluate and modify an investigation in response to new evidence CCT
- employ and evaluate safe work practices and manage risks (ACSBL031) PSC WE
- use appropriate technologies to ensure and evaluate accuracy ICT N
- select and extract information from a wide range of reliable secondary sources, and acknowledge them using an accepted referencing style L
- select qualitative and quantitative data and information, and represent them using a range of formats, digital technologies and appropriate media (ACSBL004, ACSBL007, ACSBL064, ACSBL101) L N
- apply quantitative processes where appropriate N
- evaluate and improve the quality of data CCT N
- derive trends, patterns and relationships in data and information
- assess error, uncertainty and limitations in data (ACSBL004, ACSBL005, ACSBL033, ACSBL099) CCT
- assess the relevance, accuracy, validity and reliability of primary and secondary data and suggest improvements to investigations (ACSBL005) CCT N
- use modelling (including mathematical examples) to explain phenomena, make predictions and solve problems using evidence from primary and secondary sources (ACSBL006, ACSBL010) CCT
- use scientific evidence and critical thinking skills to solve problems CCT
- select and use suitable forms of digital, visual, written and/or oral communication L N
- select and apply appropriate scientific notations, nomenclature and scientific language to communicate in a variety of contexts (ACSBL008, ACSBL036, ACSBL067, ACSBL102) L N
- construct evidence-based arguments and engage in peer feedback to evaluate an argument or conclusion (ACSBL034, ACSBL036) CC DD

1.1 Questioning and predicting

This section will guide you through some of the key steps you will need to take when first developing your **inquiry question** and **hypothesis**. Before you can start your investigation, you first need to understand the theory behind it.

REVISION

In Year 11, you learnt that scientific investigations are broken down into **primary investigations** (such as experiments in a laboratory, fieldwork or designing a **model**) and **secondary-sourced investigations** (such as a **literature review**).

WHAT INITIATES AN INVESTIGATION?

There are many starting points for an investigation. Curiosity can be triggered through observation and advances in technology.

Observation

Observation includes using all your senses, as well as available instruments, to allow closer inspection of things that the human eye cannot see. Through careful observation, during fieldwork or in the laboratory, you can find relationships and patterns that lead to a greater understanding of how organisms function under various conditions, and their interactions with each other and their environment.

The idea for a primary investigation of a complex problem may be sparked by a simple observation. For example, after observing differences in bacterial cells grown under different conditions, questions about the proteins involved in the growth of the bacteria and the biochemical pathways leading to the synthesis of these proteins can be investigated. Such investigations could lead to the development of new antimicrobial medicines or discoveries about the prevention of infectious diseases.

How observations are interpreted depends on past experiences and knowledge. But to inquiring minds, observations will usually provoke further questions, such as:

- How does antibiotic resistance transfer from a resistant bacterial strain to a non-resistant strain?
- How can a person be identified by their DNA from a mixture of blood samples?
- What factors are needed for a stem cell to differentiate into a neural cell?
- How do monoclonal antibodies fight cancerous cells?
- How does the human immune system respond to exposure to a pathogen?
- Does artificial manipulation of DNA have the potential to change populations forever?

While many of these questions cannot be answered by observation alone, they can be answered through scientific investigations. Many great discoveries have been made when a scientist has been busy investigating another problem. Good scientists have acute powers of observation and inquiring minds, and they make the most of these chance opportunities.

BIOLOGY IN ACTION CC CCT

Curiosity, observation and discovery

Scottish physician Alexander Fleming was growing cultures of *Staphylococcus* bacteria in his laboratory in the 1920s (Figure 1.1.1a) when he noticed that some of the agar plates he was growing the bacteria on became contaminated with a fungus called *Penicillium notatum*. From his observation that the bacteria were unable to grow in the region around the contaminating fungus, Fleming inferred that the fungus was releasing a substance that killed the bacteria. His curiosity led to further experiments that used extracts from the fungus, and when a paper disc was soaked in this extract and applied to an agar plate culture of *Staphylococcus*, a clear zone appeared around the disc (Figure 1.1.1b). The bacteria could not grow in this area, demonstrating the antibacterial properties of this substance. Fleming named it penicillin after the type of fungus producing the chemical.

After Fleming made the initial key observation that led to the discovery of naturally occurring antibiotics, the Australian scientist Howard Florey (then working at Oxford, England) and his colleagues further developed the procedures for extracting penicillin on a large scale, and showed it was effective against staphylococcal and pneumococcal infections. Following the success of penicillin, pharmaceutical companies searched for other naturally occurring antibiotics, many of which were found in fungi (Figure 1.1.2).

FIGURE 1.1.2 An agar plate with fungal colonies. Many naturally occurring antibiotics now used as medications were discovered by studying fungi.

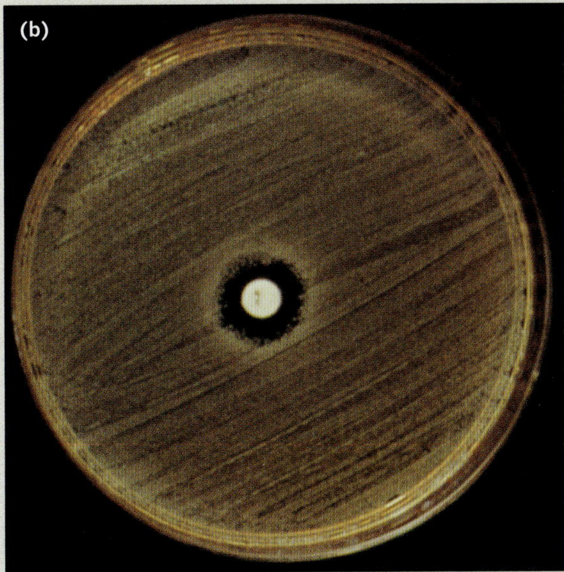

FIGURE 1.1.1 (a) Scottish biologist Alexander Fleming. (b) A culture of *Staphylococcus aureus* bacteria with a white disc containing penicillin placed at the centre. *Staphylococcus aureus* has not been able to grow near the penicillin disc.

BIOLOGY IN ACTION CC CCT

Australian discovery—peptic ulcers

In 2005, Australian scientists Professor Barry Marshall and J. Robin Warren (Figure 1.1.3) were awarded the Nobel Prize in Physiology or Medicine with their discovery of the bacteria species *Helicobacter pylori* and its role in gastritis and peptic ulcer disease.

Through careful observations and questioning, the two scientists showed that more than 90% of peptic ulcer cases that led to gastritis were not caused by stress, as previously thought, but caused by infection with *H. pylori*. *H. pylori* is able to withstand the extreme acidic conditions of the stomach, pass through the protective mucous layer and attach to and colonise the wall of the stomach. The bacteria release several compounds that lead to the damage of the stomach lining (Figure 1.1.4). The research of Marshall and Warren led to effective treatment of peptic ulcers with antibiotics.

An unexpected outcome of Marshall and Warren's research was the discovery that gastritis is a precursor to stomach cancer. Their research not only led to effective treatment for peptic ulcers but has also drastically reduced the rate of stomach cancer.

FIGURE 1.1.3 Australian scientists Professor Barry Marshall and Dr Robin Warren discovered *Helicobacter pylori* as the cause of peptic ulcers.

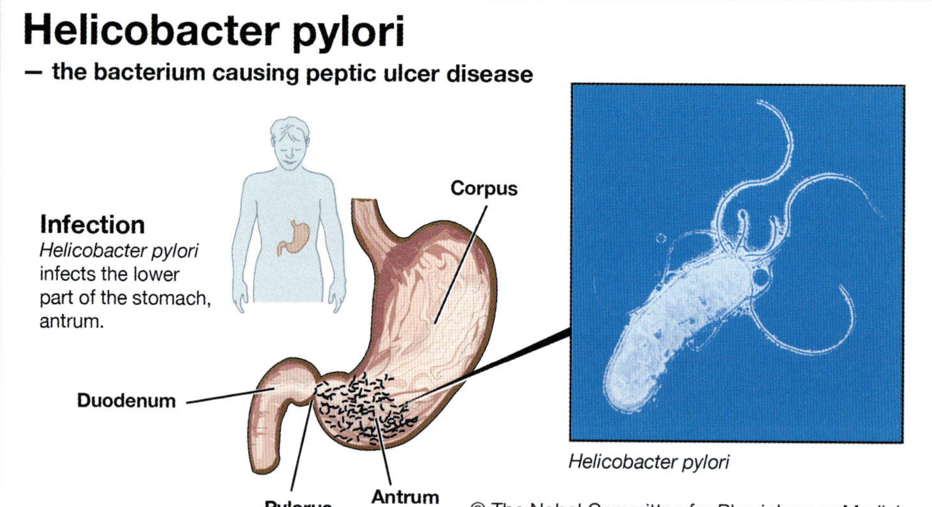

FIGURE 1.1.4 The majority of peptic ulcers are caused by the bacteria *Helicobacter pylori*.

Advances in technology

Technology plays an important role in science. Developments in technology have allowed evidence for scientific theories, laws and models to be accumulated. An example is the development of the polymerase chain reaction (PCR) to amplify targeted DNA so that it can be analysed or manipulated. The ability to manipulate DNA has propelled further technological advances in research, including the discovery of green fluorescent protein (GFP) from the jellyfish, *Aequorea victoria*, and incorporating this into proteins of interest. The use of GFP and its derivatives, which can be detected and tracked by imaging, flow cytometry or PCR, has given scientists a greater understanding of how cells function within a biological system.

The development of new scientific theories, laws and models also drives the need for new, improved technologies. Examples include developing robotics for easier and faster high throughput screening, information management systems to handle large volumes of **data**, and the advancements in imaging techniques such as the ability to visualise protein-to-protein interactions.

BIOLOGY IN ACTION 

The rapid rise of bioinformatics

Bioinformatics is the use of mathematics, statistics and computer science to analyse and understand biological data. Bioinformatics computer programs can be used to organise raw biological data to visualise patterns, identify genes, model protein structures (Figure 1.1.5), compare DNA sequences (Figure 1.1.6), understand evolutionary relationships and discover and design drugs, along with many other applications.

It is one of the fastest growing areas of biological science and is now integral to most research and development in biology. The global bioinformatics market is predicted to grow from AUD$5.2 billion in 2014 to AUD$15.8 billion by 2020. This rapid growth is primarily driven by the medical biotechnology sector, with its demand for more comprehensive and efficient storage and access to personal medical data, and the development of personalised medicine.

One of the most well-known applications of bioinformatics is whole genome sequencing. The Human Genome Project is still known as the world's largest collaborative biological project. It began in 1990, and by 2003 the three billion nucleotide bases of the human genome had been sequenced. With the technology available at the time, this was an enormous undertaking, costing approximately AUD$3.8 billion dollars. With rapid advances in sequencing technology, the output of genome sequencing has skyrocketed, while the cost to sequence a genome has plummeted. It now costs just over AUD$4000 to have your entire genome sequenced and analysed, making it affordable for many people.

Although sequencing technology is becoming more accessible, the pool of raw data for analysis is growing, requiring efficient data storage and management systems and the processing power to analyse it. The potential applications of whole genome data are vast but are limited by the bioinformatics tools, computational power and specialised knowledge of bioinformatics currently available to most biologists. As the demand for and capabilities of this technology grow, the scope of biological research is also shifting. Biologists are increasingly required to hone interdisciplinary skills in computer science, mathematics and statistics to keep pace with the rapid rise of bioinformatics.

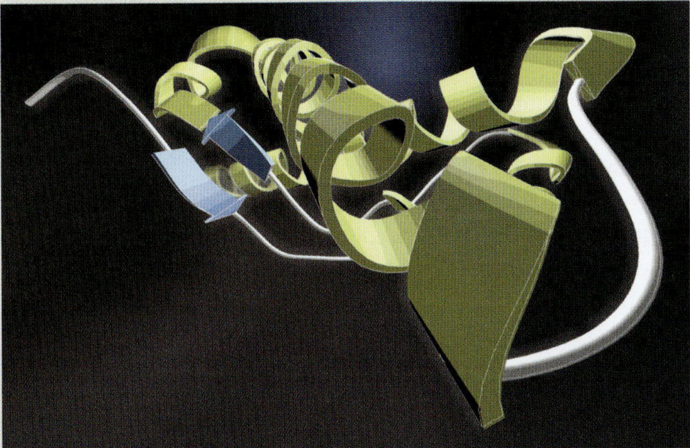

FIGURE 1.1.5 Bioinformatics tools enable you to view the secondary structure of proteins such as bovine prion protein, which misfolds and forms clumps in the brain, causing 'mad cow disease'.

```
A5ASC3.1   14 SIKLWPPSQTTRLLLVERMANNLST..PSIFTRK..YGSLSKEEARENAKQIEEVACSTANQ.....HYEKEPDGDGGSAVQLYAKECSKLILEVLK 101
B4F917.1   13 SIKLWPPSESTRIMLVDRMTNNLST..ESIFSRK..YRLLGKQEAHENAKTIEELCFALADE.....HFREEPDGDGSSAVQLYAKETSKMMLEVLK 100
A9S1V2.1   23 VFKLWPPSQGTREAVRQKMALKLSS..ACFESQS..FARIELADAQEHARAIEEVAFGAAQE......ADSGGDKTGSAVVMVYAKHASKLMLETLR 109
B9GSN7.1   13 SVKLWPPGQSTRLMLVERMTKNFIT..PSFISRK..YGLLSKEEAEEDAKKIEEVAFAAANQ.....HYEKQPDGDGSSAVQIYAKESSRLMLEVLK 100
Q8H056.1   30 SFSIWPPTQRTRDAVVRRLVDTLGG..DTILCKR..YGAVPAADAEPAARGIEAEAFDAAAA..SGEAAATASVEEGIKALQLYSKEVSRRLLDFVK 120
Q0D4Z3.2   44 SLSIWPPSQRTRDAVVRRLVQTLVA..PSILSQR..YGAVPEAEAGRAAAAVEAEAYAAVTES.SSAAAAPASVEDGIEVLQAYSKEVSRRLLELAK 135
B9MVW8.1   56 SFSIWPPTQRTRDAIISRLIETLST..TSVLSKR..YGTIPKEEASEASRRIEEEAFSGAST.......VASSEKDGLEVLQLYSKEISKRMLETVK 141
Q0IYC5.1   29 SFAVWPPTRRTRDAVVRRLVAVLSGDTTTALRKRYRYGAVPAADAERAARAVEAQAFDAASA....SSSSSSSVEDGIETLQLYSREVSNRLLAFVR 121
A9NW46.1   13 SIKLWPPSESTRLMLVERMTDNLSS..VSFFSRK..YGLLSKEEAAENAKRIEETAFLAAND.....HEAKEPNLDDSSVVQFYAREASKLMLEALK 100
Q9C500.1   57 SLRIWPPTQKTRDAVLNRLIETLST..ESILSKR..YGTLKSDDATTVAKLIEEEAYGVASN.......AVSSDDDGIKILELYSKEISKRMLESVK 142
Q2HRI7.1   25 NYSIWPPKQRTRDAVKNRLIETLST..PSVLTKR..YGTMSADEASAAAIQIEDEAFSVANA.......SSSTSNDNVTILEVYSKEISKRMIETVK 110
Q9M7N3.1   28 SFKIWPPTQRTREAVVRRLVETLTS..QSVLSKR..YGVIPEEDATSAARIIEEEAFSVASV.ASAASTGGRPEDEWIEVLHIYSQEIXQRVVESAK 119
Q9M7N6.1   25 SFSIWPPTQRTRDAVINRLIESLST..PSILSKR..YGTLPQDEASETARLIEEEAFAAAGS.......TASDADDGIEILQVYSKEISKRMIDTVK 110
Q9LE82.1   14 SVKMWPPSKSTRLMLVERMTKNITT..PSIFSRK..YGLLSVEEAEQDAKRIEDLAFATANK.....HFQNEPDGDGTSAVHVYAKESSKLMLDVIK 101
Q9M651.2   13 SIKLWPPSLPTRKALIERITNNFSS..KTIFTEK..YGSLTKDQATENAKRIEDIAFSTANQ.....QFEREPDGDGGSAVQLYAKECSKLILEVLK 100
B9R748.1   48 SLSIWPPTQRTRDAVITRLIETLSS..PSVLSKR..YGTISHDEAESAARRIEDEAFGVANT.......ATSAEDDGLEILQLYSKEISRRMLDTVK 133
```

FIGURE 1.1.6 Bioinformatics tools allow comparison of many gene or protein sequences at once. This is an amino acid sequence alignment.

REVISION

GO TO ➤ Year 11 Section 1.1

Inquiry question, hypothesis and purpose

The inquiry question, purpose (aim) and hypothesis are linked. You can refine these as you continue to plan your investigation.

An inquiry question defines what is being investigated. You need to be able to clearly interpret what an inquiry question is asking you to do. To formulate your inquiry question, compile a list of topic ideas and start a literature review. Once you have decided on a topic, you will need to evaluate and refine your inquiry question. Remember to consider the resources available to you when deciding on your inquiry question.

A hypothesis is a prediction that is based on evidence and prior knowledge. It often takes the form of a proposed relationship between two or more **variables** in a cause-and-effect relationship; in other words, 'If *x* is true and this is tested, then *y* will occur'. A good hypothesis is a statement that contains the **independent variable** and **dependent variable**, and is testable and falsifiable (can be proven false). You may need to adjust your hypothesis as you conduct further research into your chosen topic.

A **purpose** is a statement describing in detail what will be investigated. It is also known as the aim of your investigation. The purpose includes the key steps required to test the hypothesis. You should directly relate each purpose to the variables in the hypothesis, and describe how each will be measured. An experiment or investigation determines the relationship between variables and measures the results.

There are three categories of variables: independent, dependent and controlled and these are either **qualitative variables** (includes **nominal variables** and **ordinal variables**) or **quantitative variables** (includes **discrete variables** and **continuous variables**). Reliable primary and secondary sources should be used when researching your topic and during your investigation.

BIOFILE CCT WE

Detecting antibiotic resistance

Alexander Fleming's research involving bacteria (see Biology in Action p. 5) led to the development of standard operating procedures for detecting antibiotic sensitivity and resistance in bacteria. If a bacterium is susceptible to an antibiotic, its growth around a disc containing the antibiotic is inhibited and observed as a clear zone on the agar plate, called the zone of inhibition. The greater this zone, the more sensitive the bacteria are to the antibiotic. If a bacterium has developed resistance to an antibiotic, it can grow around its antibiotic disc (Figure 1.1.7). The spread of antibiotic resistance by gene transfer between different species of bacteria is one of the most significant threats to global health in the 21st century.

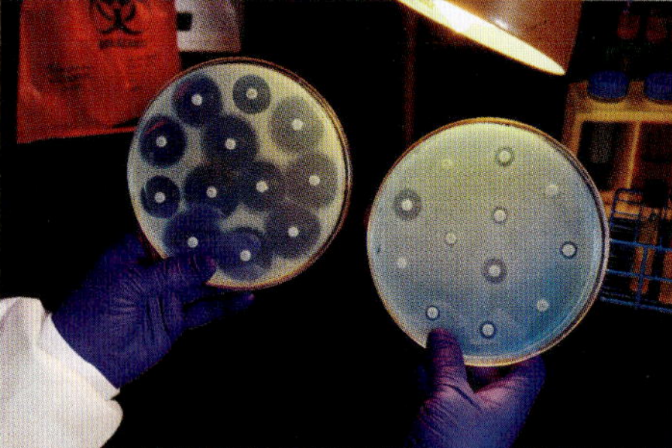

FIGURE 1.1.7 A microbiologist holding two Petri dish culture plates growing bacteria in the presence of discs containing various antibiotics. In the left plate, bacteria are not growing around the discs because they are susceptible to the antibiotics on the discs. The plate on the right was inoculated with an antibiotic-resistant bacterium that proved to be resistant to, and therefore able to grow well around, all the antibiotics tested. Photographed at the Centers for Disease Control, USA.

PEER REVIEW

Scientists often publish their findings in peer-reviewed journals. Examples of peer-reviewed Biology journals include:

- Nature
- Cell
- The Journal of Immunology
- PLOS Biology
- Journal of Animal Ecology
- The Journal of Biological Chemistry.

Peer-reviewed journals have an editorial board, made up of experts in a particular field, who read a draft article and ask questions of the author before agreeing to publish the article.

Your teacher may suggest that you partner with a student in your class, to provide each other with constructive feedback regarding your inquiry question and/or hypothesis. There are many benefits resulting from collaborating with others, including building on ideas and considering alternative perspectives. For example, in 2015, the World Health Organization endorsed a global action plan to tackle antibacterial resistance because of its detrimental health impact on the world's population.

When collaborating with others consider the following:

- Is the inquiry question clear?
- Can the inquiry question be answered in the time available?
- Is the purpose clear?
- Is the hypothesis written so that it can be tested and falsified?
- Are the independent, dependent and controlled variables clearly defined?
- What are the strengths of the inquiry question?

BIOLOGY IN ACTION CCT EU ICT

Printing body parts

Three-dimensional (3D) printing, or bioprinting, combines the areas of quantitative image analysis, computer-aided design and manufacturing. Its goal is to make functional human tissue, which is made up of a composite of cells and scaffolding material that mimics tissues of the human body.

There are already scientists investigating bioprinting a range of human tissues such as skin, capillaries/blood vessels, heart, bone and liver (see Figures 1.1.8a and 1.1.8b). Bioprinting is not only for medical research; a major cosmetic company is using bioprinting to make skin to test their products on before releasing them to the market rather than using animals in their trials.

(a)

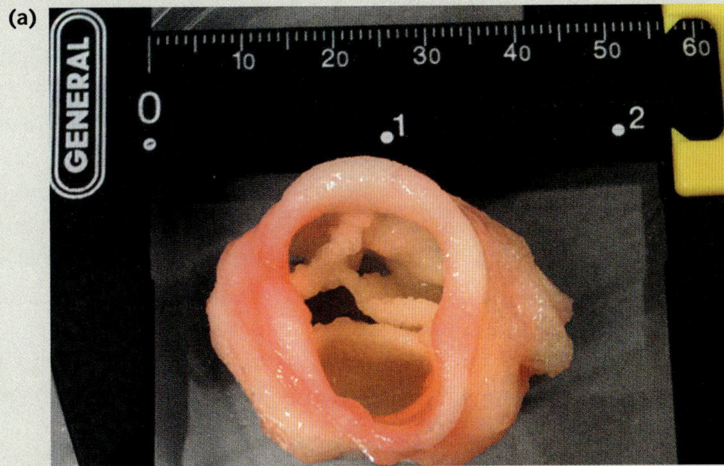

(b)

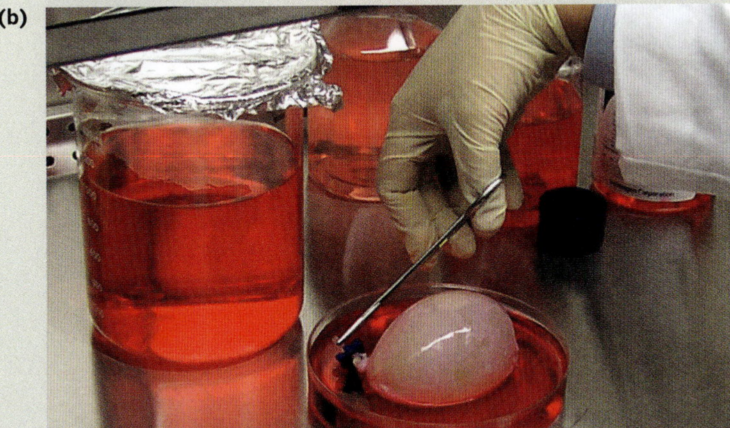

FIGURE 1.1.8 A 3D print of (a) a heart valve and (b) a bladder

1.1 Review

SUMMARY

- Before you begin your research, you need to conduct a literature review. Using data from primary and/or secondary sources will help you better understand the context of your investigation so you can create an informed inquiry question.
- The purpose (aim) is a statement describing in detail what will be investigated.
- A hypothesis is a tentative explanation for an observation that is based on evidence and prior knowledge. A hypothesis must be testable and falsifiable and define a proposed relationship between two variables.
- Once an inquiry question has been chosen, stop to evaluate the question before progressing. The question may need further refinement or even further investigation before it is suitable as a basis for an achievable and worthwhile investigation. It is better to not attempt something that you will not be able to complete in the time available or with the resources on hand.
- There are three categories of variables: independent, dependent and controlled.

KEY QUESTIONS

1. Distinguish between the terms 'inquiry question', 'hypothesis' and 'purpose'.
2. Which of the following is an inquiry question?
   A What is the sodium content of baby foods?
   B The sodium content of baby food will be less than 100 mg per 100 g.
   C Babies need to have a low sodium diet.
   D Sodium can be measured using gravimetric analysis and atomic absorption spectroscopy.
3. A scientific investigation is a multistep process. Which two of the following are important parts of the scientific method?
   A observations made by eye and with instrumentation
   B subjective decisions based on data collected
   C careful manipulation of results to fit your ideas
   D the use of prior knowledge to help objectively interpret new data
4. In an experiment testing enzyme activity, a student records the pH when the solution changes colour. What type of variable is 'colour'?
5. Which of the following is the most specific inquiry question?
   A What defence mechanisms do eucalypts use in response to fungal pathogens?
   B Do plants get infected by fungi?
   C How does a plant respond to infections?
   D What causes infections in Australian native plants?
6. Which of the following inquiry questions is objective and specific?
   A How does temperature affect the colour of a solution?
   B How does temperature affect the activity of the enzyme lactase?
   C Do enzymes behave differently in heated solutions?
   D Do enzymes work more efficiently when they are in hot or cold solutions?
7. Write a hypothesis for each of the following scenarios:
   a A student investigating algal blooms wondered whether *Chlorella*, a unicellular eukaryotic alga, carries out photosynthesis faster than *Anabaena*, a cyanobacterium.
   b A student on work placement at a dairy research station wondered whether dairy cows with mastitis (a bacterial infection of the udder) would have more white blood cells to fight the infection.

1.2 Planning investigations

After you have formulated your hypothesis, defined the purpose of your investigation and determined your variables, you will need to plan and design your investigation. Take the time to carefully plan and design a practical investigation before beginning. This will help you maintain a clear and concise focus throughout. Preparation is essential. This section will guide you through some of the key steps in planning and designing a practical investigation.

CHOOSING AN APPROPRIATE TYPE OF INVESTIGATION

Once you have drafted your inquiry question, purpose, hypothesis and variables, you will need to consider an appropriate type of investigation. Examples of various investigations are listed in Table 1.2.1. When selecting the type of investigation you will use, remember to consider how much time you have available, and whether you will need to work in a group or individually.

TABLE 1.2.1 Examples of different types of investigations

Type of investigation	Example
Primary data	
Design and conduct experiments	investigate the effect of temperature on the activity of the enzyme lipase
Test a claim	use bioinformatics to determine the evolutionary relationships of microbats
Test a device	test the accuracy and precision of laboratory equipment
Fieldwork	conduct a biodiversity survey in an urban environment
Data analysis	process and analyse raw data collected during fieldwork to construct graphs and tables, and understand trends in the data
Make a documentary or media report	investigate the concentration of lead in water passed through a water filter and compare your data with the data reported from tests commissioned by the manufacturer
Secondary data	
Conduct a literature review	gather data from secondary sources and write a literature review about the global decline of frog populations
Develop an evidence-based argument	find data reporting the concentrations of lead in soil in Sydney and propose strategies for residents to avoid lead contamination of home-grown vegetables
Write a journal article	look at the style of a peer-reviewed journal and write a literature review in that style
Write an essay—historical or theoretical	research an Australian scientist or their discovery and write an essay
Develop an environmental management plan	gather secondary-sourced data about a region of interest and write a management plan to increase the native vegetation on river banks and reduce erosion
Analyse a work of fiction or film for scientific relevance	book: *Ammonite* by Nicola Griffith (2002); film: *Gattaca*, director Andrew Niccol (1997) film: *An Inconvenient Truth*, director Davis Guggenheim (2006)
Create a visual presentation	create a flow chart, poster presentation or a model to show the spread of the Ebola virus
Investigate emerging technologies	investigate the prototype of the CLARITY light-sheet microscope developed by researchers from the University of Newcastle
Design and create	build your own potometer to measure the rate of water uptake and transpiration in a plant
Create a physical model	create a model using a 3D printer (e.g. hominid skull, cross section of an organ/tissue, an insect)
Create a portfolio	compile various resources to investigate the outcomes of engagement in STEM education in childhood for scientific literacy in adulthood
Data analysis (secondary-sourced data)	gather and analyse data from the scientific literature to predict the effects of climatic variation on the reproductive success of yellow-bellied three-toed skinks (*Saiphos equalis*) throughout New South Wales and Queensland

Organising information

You need to efficiently organise the information that you collect during your investigation. This is particularly important if your investigation is an in-depth literature review. Table 1.2.2 shows how you might be able to summarise information from primary and secondary sources.

TABLE 1.2.2 Example of categories that help you keep track of information as you conduct a literature review

Source of information	*Malaria Journal*, volume 15, https://malariajournal.biomedcentral.com/articles/10.1186/s12936-016-1130-z
Author	Tran P.N., Brown S.H., Rug M., Ridgway M.C., Mitchell T., Maier A.G.
Title	Changes in lipid composition during sexual development of the malaria parasite, *Plasmodium falciparum*.
Year of publication	2016
Procedure	parasite culture lipid labelling with fluorescent molecules mass spectrometry deconvolution microscopy
Summary of conclusion	The lipid composition of red blood cells infected with the malaria parasite, *Plasmodium falciparum*, is characteristic of particular life cycle stages of the parasite. This information is important for understanding lipid metabolism in *P. falciparum* and for developing drugs to interfere with life cycle stages of this parasite.
Key ideas	The researchers found significant differences in the morphology and lipid composition of red blood cells infected with *P. falciparum* in comparison to uninfected red blood cells. Knowledge of parasite lipid metabolism is a promising area of research for the discovery of novel biological mechanisms and the development of new drugs.

REVISION

Methodology elements

Validity refers to whether an experiment or investigation is in fact testing the set hypothesis and purpose. A valid investigation is designed so that only one variable is being changed at a time. To ensure validity, you will need to carefully determine the independent variable and how it will change, the dependent variable, as well as the controlled variables and how they will be maintained.

Reliability refers to the notion that the experiment can be repeated many times and the average of the results from all the repeated experiments will be consistent. You can maintain this by defining the control and ensuring the experiment is sufficiently replicated.

Accuracy is the ability to obtain the correct measurement. To obtain accurate results, you must minimise systematic errors.

Precision is the ability to consistently obtain the same measurement. To obtain precise results, you must minimise random errors.

In biology, it is important to know that the precision of glassware used in laboratory experiments varies. Table 1.2.3 shows the typical **uncertainty** of laboratory glassware.

Build some testing into your investigation to confirm the equipment is accurate and reliable and that you can read the information obtained. Reasonable steps to ensure the investigation is accurate include considering the unit in which the independent and dependent variables will be measured, and the instrument that will be used to measure the variables.

Select and use appropriate equipment, materials and procedures. For example, select equipment that measures to smaller degrees, which will reduce uncertainty, and repeat the measurements to confirm them.

Describe the materials and procedure in appropriate detail. This should ensure that every measurement can be repeated and the same result obtained within reasonable margins of experimental error (less than 5% is reasonable). **Percentage uncertainty**, sometimes referred to as percentage error, is a way to quantify how accurate a measurement is. This will be discussed in Section 1.4.

TABLE 1.2.3 Typical uncertainty of various laboratory glassware

Type of laboratory glassware	Typical uncertainty
250 mL beaker	250 ± 10 mL
10 mL measuring cylinder	10 ± 0.1 mL
20 mL pipette	20 ± 0.03 mL
5 mL pipette	5 ± 0.01 mL

WRITING THE PROCEDURE

The **procedure** (also known as the method) of your investigation is a step-by-step description of how the hypothesis will be tested. Consider using a diagram of your equipment set-up such as the one shown in Figure 1.2.1.

Procedures must be described clearly and in sufficient detail to allow other scientists to repeat the experiment. If other scientists cannot obtain similar results when an experiment is repeated and the results averaged, then the experiment is considered unreliable. It is also important to avoid personal **bias** that might affect the collection of data or the analysis of results.

The procedure must ensure that the results from an investigation are valid, reliable and accurate.

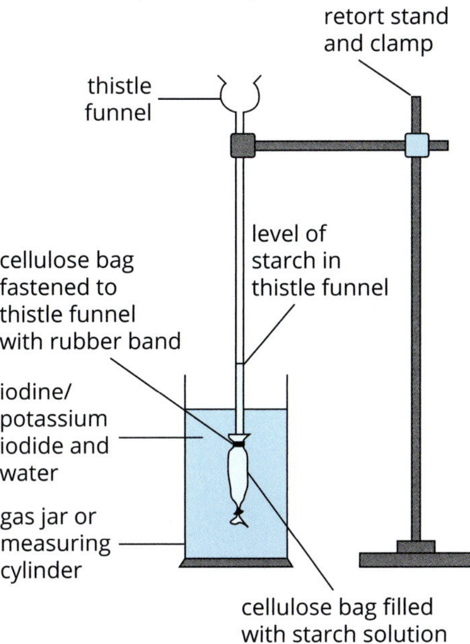

FIGURE 1.2.1 An equipment set up for dialysis tubing arrangement

Recording numerical data

When using measuring instruments, the number of **significant figures** (or digits) and decimal places you use is determined by how precise your measurements are. You will learn more about how to identify and use an appropriate number of significant figures and decimal places in Section 1.4.

GO TO ➤ Section 1.4 page 22

Data analysis

Data analysis is part of the procedure. Consider how the data will be presented and analysed (e.g. in graphs so that patterns can be seen clearly). Graphs can show relationships and enable comparisons. Preparing an empty table showing the data that needs to be obtained will help in planning the investigation. You should tailor the layout of your data table to suit your experiment. Table 1.2.4 is an example of a **raw data** table. It contains data from an experiment on the effect of temperature on the activity of enzyme X. A reaction between the enzyme and substrate was carried out for 10 minutes and the reaction product was measured. Three trials were performed. The same data is also presented in a processed format in Table 1.2.5. The replicate values from Table 1.2.4 have been averaged to calculate the mean product released (μg) and the mean reaction rate per minute (μg/min) (Table 1.2.5).

GO TO ➤ SkillBuilder page 24

TABLE 1.2.4 The effect of temperature on the reaction rate of enzyme X. This is raw data collected from three trials in which the reaction rate was estimated from the amount of reaction product released (μg).

	Product released (μg)		
Temperature (°C)	Trial 1	Trial 2	Trial 3
10	100	120	120
20	850	790	820
40	1350	1420	1390
60	1250	1210	1150
80	200	220	230

TABLE 1.2.5 The effect of temperature on the reaction rate of enzyme X. The raw data has been processed to find the mean product released (μg) and mean reaction rate (μg/min).

Temperature (°C)	Mean product released (μg)	Mean reaction rate (μg/min)
10	113.3	11.3
20	820.0	82.0
40	1386.7	138.7
60	1203.3	120.3
80	216.7	21.7

The nature of the data being collected, such as whether the variables are qualitative or quantitative, influences the type of procedure or tool that you can use to analyse the data. The purpose of the investigation and hypothesis will also influence the choice of analysis tool. Table 1.2.6 provides examples of variable types and the ways in which these variables can be measured.

TABLE 1.2.6 Types of variables and the ways in which they could be measured and used within an investigation

Inquiry question	If yeast cells are grown in higher amounts of glucose, will they carry out cellular respiration faster?
List the independent variable. Is the variable quantitative or qualitative?	glucose concentration, quantitative
List the dependent variable(s).	the rate of cellular respiration (measured as CO_2 release)
What equipment will you use to measure the dependent variable(s)?	This will be measured by the amount of carbon dioxide produced within a measuring cylinder.
List the variables that you will control.	yeast culture volume, temperature, light conditions, nutrients in the agar
What will you do to control these variables?	• temperature of a room will be measured with a thermometer; the uncertainty of the thermometer is ±0.1°C • electric balance will be used to weigh the glucose, which has the uncertainty of ±0.1 g • yeast culture volume will be measured in a 25 mL measuring cylinder, which has an uncertainty of ±0.3 mL • provide all yeast samples with the same nutrient levels (except glucose).

Sourcing appropriate materials and technology

When designing your investigation, you will need to decide on the materials, technology and instrumentation for your research. It is important to find the right balance between easily accessible items and those that will give you accurate results. You will also need to consider any costs or risks associated with using the technology, how familiar you are with using it, and any limitations the technology has that will impact your investigation. As you move onto conducting your investigation, it will be important to take note of the precision of your chosen instrumentation and how this affects the accuracy and validity of your results. This will be discussed in greater detail in Section 1.3.

Modifying the procedure

You may need to modify your procedure as you carry out the investigation. The following actions will help determine any issues in the procedure and how to modify them.

- Record everything.
- Be prepared to make changes to the approach.
- Note any difficulties encountered and how they were overcome. What were the failures and successes? Every test carried out can contribute to the understanding of the investigation, no matter how much of a disaster it may first appear.
- Do not panic. Go over the theory again, and talk to the teacher and other students. A different perspective can lead to a solution.

If the expected data is not obtained, do not worry. As long as limitations of the investigation are identified, the data can be critically evaluated, and further investigations are proposed, the investigation is worthwhile.

COMPLYING WITH ETHICAL AND SAFETY GUIDELINES

With rapid advances in biological research, we need to ensure that it fits in with our society's ethical and safety standards. Governing bodies implement ethical and safety regulations based on discussions between bioethicists, the public, health and law sectors. New South Wales and Australia have governing bodies that regulate ethical and safety guidelines.

Ethical considerations

Some investigations require ethics approval, so you will need to consult with your teacher. When deciding on an investigation, identify all possible ethical considerations and evaluate their necessity. You may be able to change the methodology to reduce or eliminate any possible ethical issues.

GO TO ➤ Year 11 Section 1.2

BIOLOGY IN ACTION EU ICT

Ethical issues of using CRISPR/Cas9 technology and gene editing

CRISPR/Cas9 is a genome editing tool that allows scientists to make precise targeted changes to the genome of living cells. There are two parts to the technique: CRISPR (Clustered Regularly Interspaced Short Palindromic Repeats) that target the DNA sequence that is to be modified and Cas9 (CRISPR-associated protein 9) that cuts the targeted DNA (Figure 1.2.2).

Some bacteria use a similar gene-editing system as a self-defence mechanism against pathogens. The bacteria cut out parts of the pathogen's DNA and use it to recognise the pathogen next time it encounters it. Professor Emmanuelle Charpentier and Professor Jennifer Doudna adapted this bacterial process so that it could cut any DNA sequence and edit genomes with high efficiency, ease of use and at a low cost.

Ethical concerns have been raised over the use of this technology. The most controversial use of the technique is to develop designer babies, now that gene manipulation can be performed in live sperm cells, eggs and embryos. These types of modifications are known as germ line editing and are passed down to subsequent generations.

Another concern that has been raised is the uncontrolled proliferation of modified genes in wild populations, known as gene drive. Gene drive occurs when modified genes have a greater than 50% chance of being inherited (due to molecular mechanisms) or confer a selective advantage over unmodified genes. These, and other molecular mechanisms, can result in the rapid spread of a modified gene throughout a population (Figure 1.2.3). The rapid spread of modified genes

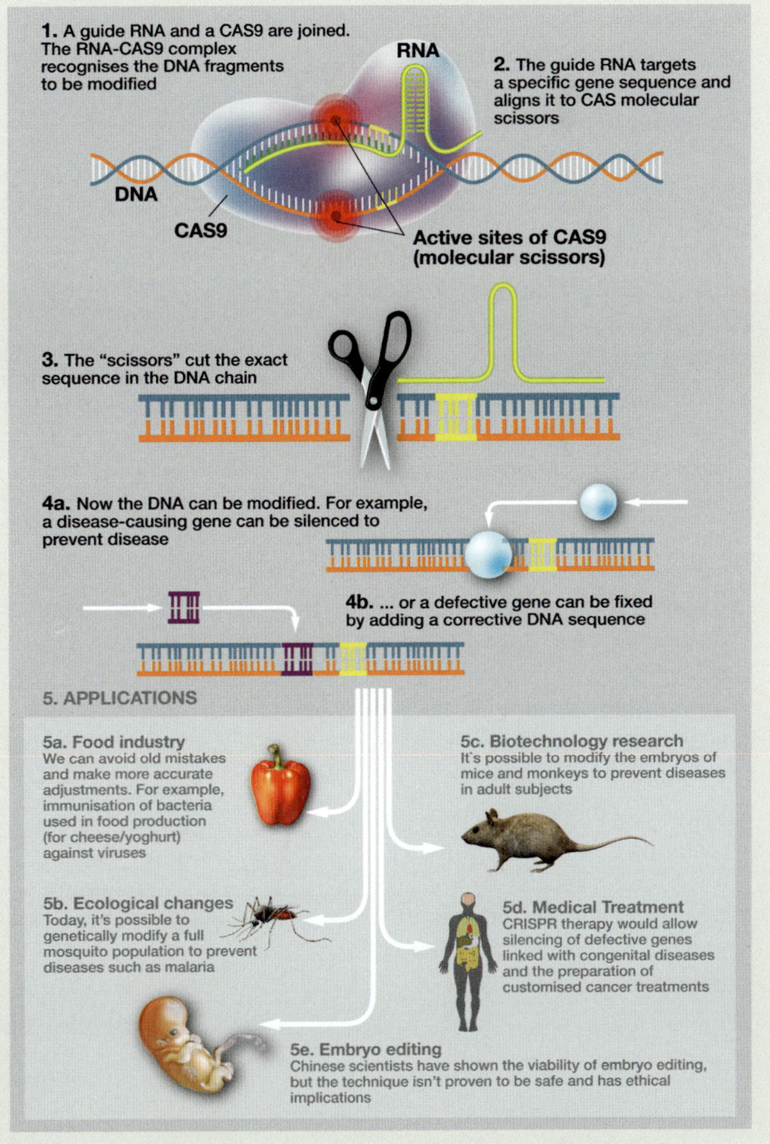

FIGURE 1.2.2 The steps involved in genome editing using CRISPR/Cas9

(continued over to next page)

BIOLOGY IN ACTION *continued*

in wild populations poses potential risks to biodiversity and may have unpredictable consequences for species and ecosystems.

While there is potential for negative outcomes, CRISPR/Cas9 technology and gene drives can also have favourable outcomes in wild populations, including reducing the reproductive fitness of an invasive species, disease vector or pathogen. One example is using CRISPR/Cas9 technology to introduce alleles that confer female sterility in the mosquito, *Anopheles gambiae*, one of the main vectors for malaria. Researchers have found a strong gene drive for these alleles, with an inheritance rate of over 90%. This technology has the potential to rapidly alter wild *A. gambiae* populations and eradicate the mosquito-borne disease, malaria.

The CRISPR/Cas9 system has many potential applications for the ability to cut DNA of any genome and therefore can be useful to society, such as:

- investigating models of diseases
- developing pharmaceutical products
- developing gene- and cell-based therapies
- eradicating pest species and disease such as the mosquito-borne disease, malaria; and schistosomiasis, a disease caused by the trematode worm
- improving the tolerance of crops and livestock to various environmental conditions
- increasing the yield of food crops
- restoring populations of endangered species by improving their reproductive success or controlling introduced predators or competitors.

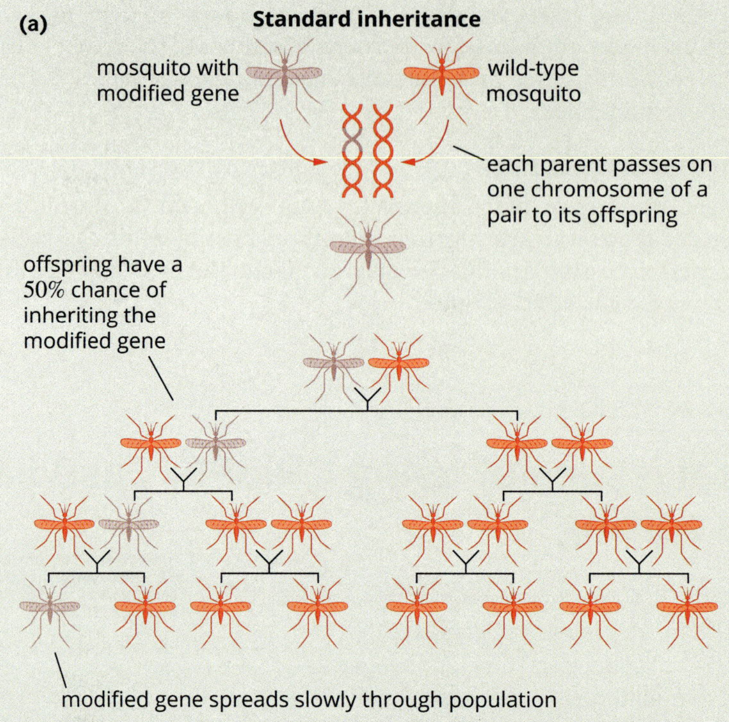

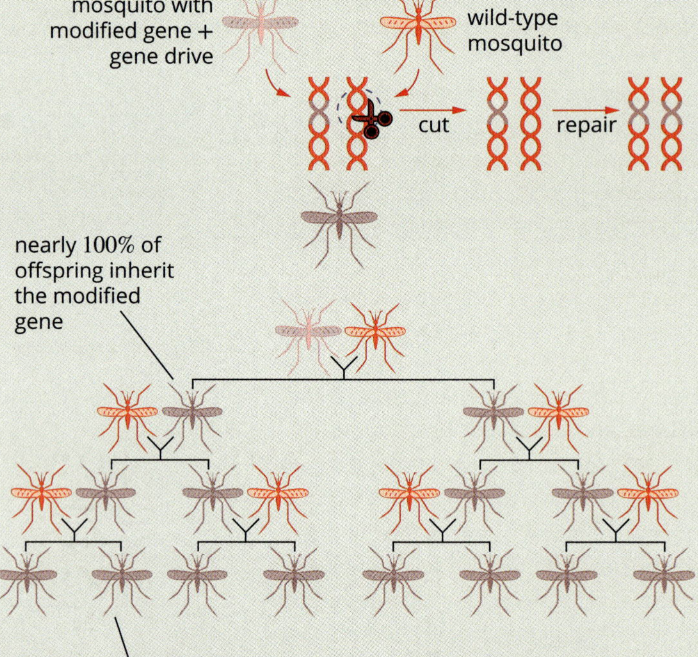

FIGURE 1.2.3 The difference between (a) standard inheritance and (b) gene drive inheritance. Gene drive inheritance has the potential to rapidly spread modified genes throughout a population.

REVISION

GO TO ➤ Year 11 Section 1.2

Risk assessments

Risk assessments identify, assess and control hazards. They should be performed for any situation, in the laboratory or outside in the field. Always identify the risks and control them to keep everyone as safe as possible.

Ways to reduce risk are, in order from most to least effective: elimination, substitution, engineering (modifying equipment), administrative control and **personal protective equipment (PPE)**. You must take special care to minimise the risks that come with working outdoors, such as wearing sunscreen, insect repellent, appropriate clothing, and gloves as necessary. Someone with first aid training should always be present, and you must immediately report any injuries or accidents to your teacher or lab technician.

REVISION

Chemical hazard codes

The chemicals at school or at the hardware shop have a warning symbol on the label. These are **chemical codes** (HAZCHEM). Some common codes and their meanings are shown in Figure 1.2.4. These codes may be displayed in hazard symbols on trucks or containers carrying chemicals.

From 1 January 2017, the Globally Harmonised System of Classification and Labelling of Chemicals (GHS) pictograms were introduced into Australia. This system is used for labelling containers and in safety data sheets. Some of the pictograms that you may see denote chemicals that a) are corrosive, b) pose a health hazard or c) are flammable.

Safety data sheets

Each chemical substance has an accompanying document called a **safety data sheet (SDS)**, previously a material safety data sheet. An SDS contains important safety and first aid information about each chemical you commonly use in the laboratory. If the products of a chemical reaction are toxic to the environment, you must pour your waste into a special container (not down the sink).

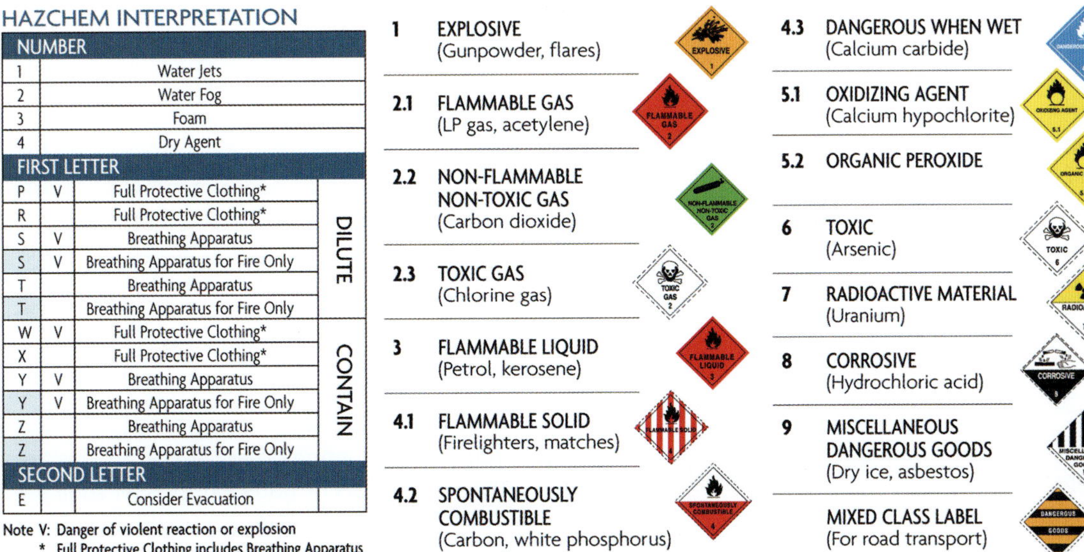

HAZCHEM INTERPRETATION

NUMBER			
1		Water Jets	
2		Water Fog	
3		Foam	
4		Dry Agent	
FIRST LETTER			
P	V	Full Protective Clothing*	DILUTE
R		Full Protective Clothing*	DILUTE
S	V	Breathing Apparatus	DILUTE
S	V	Breathing Apparatus for Fire Only	DILUTE
T		Breathing Apparatus	DILUTE
T		Breathing Apparatus for Fire Only	DILUTE
W	V	Full Protective Clothing*	CONTAIN
X		Full Protective Clothing*	CONTAIN
Y	V	Breathing Apparatus	CONTAIN
Y	V	Breathing Apparatus for Fire Only	CONTAIN
Z		Breathing Apparatus	CONTAIN
Z		Breathing Apparatus for Fire Only	CONTAIN
SECOND LETTER			
E		Consider Evacuation	

Note V: Danger of violent reaction or explosion
* Full Protective Clothing includes Breathing Apparatus

FIGURE 1.2.4 HAZCHEM and the Globally Harmonised System of Classification and Labelling of Chemicals (GHS) codes for identifying chemical hazards

PEER REVIEW

Your teacher may suggest that you partner with a student in your class, to provide each other with constructive feedback regarding your planning.

Consider the following questions:

- Is the type of investigation appropriate and relevant to the inquiry question?
- Is the procedure written in easy-to-understand logical steps?
- Is the procedure valid and reliable?
- Have you done a thorough and complete risk assessment before beginning the investigation?
- What are the strengths of the type of investigation and procedure?
- What questions do you have about the type of investigation and procedure?

1.2 Review

SUMMARY

- The procedure of your investigation is a step-by-step plan. When detailing the procedure, ensure it complies as a valid, reliable and accurate investigation.
- You also need to determine how many times to replicate the experiment. Many scientific investigations lack sufficient repetition to ensure that the results can be considered reliable.
- You must carry out risk assessments before starting an investigation. This will ensure the safety of yourself and others. If you have elements of your investigation that are too high risk, you will need to reevaluate your design.
- It is important to choose appropriate equipment for your experiment. Not just personal protective equipment (PPE) that will help keep you safe, but also selecting instrumentation that will give you accurate results.

KEY QUESTIONS

1 Give the correct term that describes an experiment with each of the following conditions.
  - **a** The experiment addresses the hypothesis and purpose.
  - **b** The experiment is repeatable and consistent results are obtained.
  - **c** Appropriate equipment is chosen for the required measurements.

2 A student wants to determine algae levels of water samples. Potential sites for water sampling in the field could include:
  - **A** rivers
  - **B** creeks
  - **C** lakes
  - **D** stormwater
  - **E** all of the above

3 A student wants to collect water samples from a local stream. Which of the following PPE should be included in the risk assessment for the collection of water samples in the field? More than one response may be correct.
  - **A** use of a fume cupboard
  - **B** ear muffs
  - **C** gloves
  - **D** non-slip rubber boots

4 A journal article reported the materials and procedure used to conduct an experiment. The experiment was repeated three times, and all values were reported in the results section of the article. Repeating an experiment and reporting results supports:
  - **A** precision
  - **B** reliability
  - **C** accuracy
  - **D** systematic errors

5 You are conducting an experiment to determine the effect of pH of water on mussel shell mass. Identify:
  - **a** the independent variable
  - **b** the dependent variable
  - **c** at least one controlled variable

1.3 Conducting investigations

Once you have completed planning and designing a practical investigation, the next step is to begin the investigation and record the results. As with the planning stages, you will need to keep in mind key steps and skills to maintain high standards and minimise potential error throughout the investigation (Figure 1.3.1).

This section will focus on the best procedures for conducting a practical investigation, as well as systematically generating, recording and processing data.

FIGURE 1.3.1 It is important to read the bottom of the meniscus at eye level to avoid parallax error. This student is showing how you can use a piece of white card (or a tile) to improve the contrast between the solution and the scale.

REVISION

GO TO ➤ Year 11 Section 1.3

Collecting and recording data

For an investigation to be scientific, it must be objective and systematic. Always record all **quantitative data** and **qualitative data** collected, as well as the procedures used to collect the data in your logbook. Also record any incident, feature or unexpected event that may have affected the quality or validity of the data. The recorded data is known as raw data. Usually this data needs to be processed in some manner before it can be presented.

Safe work practices

Remember to always employ safe work practices while conducting your experiment. See Section 1.2 for how to conduct risk assessments.

You will also need to keep in mind safe procedures to follow when disposing of waste. Consult your teacher or education and government websites for information.

REVISION

GO TO ➤ Year 11 Section 1.3

Identifying errors

Errors can happen for a variety of reasons. Being aware of potential errors helps you to avoid or minimise them. As shown in Figure 1.3.2 there are three types of errors that can occur in an experiment.

Mistakes are avoidable errors. A measurement that involves a mistake must be rejected and not included in any calculations, or averaged with other measurements.

A **systematic error** is an error that is consistent and will occur again if the investigation is repeated in the same way. Systematic errors are usually a result of instruments that are not calibrated correctly or procedures that are flawed. Whatever the cause, the resulting error is in the same direction for every measurement and the average will be either too high or too low as a result. These lead to bias. Examples of bias are shown in Figure 1.3.3.

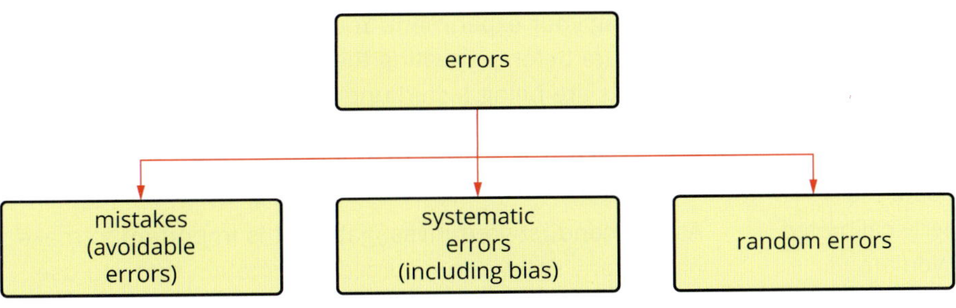

FIGURE 1.3.2 Types of errors that can be made in an experiment

(continued over to next page)

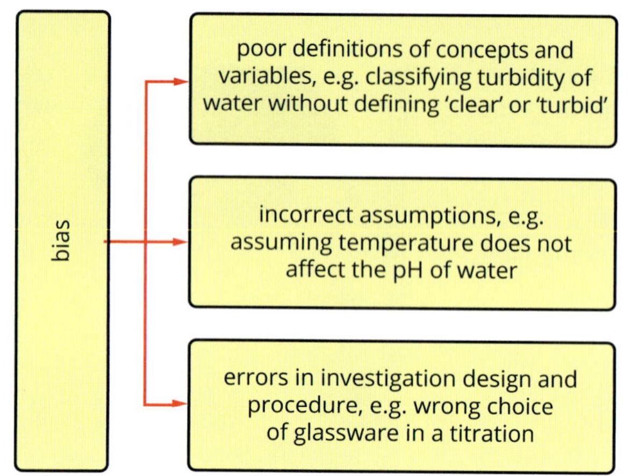

FIGURE 1.3.3 Types and examples of bias in an analysis of water quality

Random errors occur in an unpredictable manner and therefore follow no regular pattern. The measurement is sometimes too large and sometimes too small. The effects of random errors can be reduced by taking multiple measurements of the same quantity, then calculating an average.

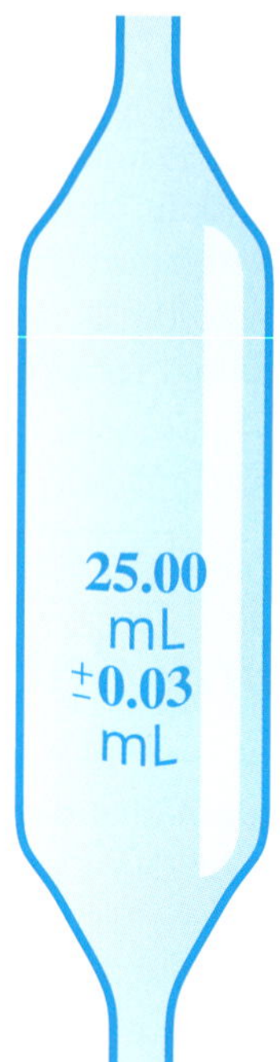

FIGURE 1.3.4 Record the uncertainty for glassware and instruments. This pipette can dispense a volume (aliquot) of 25.00 ± 0.03 mL. When used correctly the volume dispensed will be between 24.97 mL and 25.03 mL.

Techniques to reduce error

Use appropriate equipment

Use the equipment that is best suited to the data that needs to be collected. Determining the units of the data being collected and at what scale will help to select the correct equipment. Using the right unit and scale will ensure that measurements are more accurate and precise (with smaller systematic errors).

To minimise errors, check the precision of the equipment that you intend to use (Table 1.3.1). Pipettes, burettes and measuring cylinders have greater precision than using a beaker to measure volumes of solutions. However, you must still use all equipment correctly to reduce error.

TABLE 1.3.1 Precision of water-testing equipment and glassware

Equipment	Typical precision
pH meter	±0.1
50 mL burette	±0.02 mL
20 mL pipette	±0.03 mL

Often glassware and equipment have information that indicates their precision. Figure 1.3.4 shows where this can be found on a pipette.

Use calibrated equipment

Before carrying out the investigation, make sure the instruments or measuring devices are properly calibrated and are, in general, functioning correctly. If you are preparing a solution of known concentration you might have access to a measuring cylinder that has less uncertainty associated with measurements compared to using a beaker.

Use equipment correctly

Use the equipment properly. Ensure any necessary training has been done and that you have had an opportunity to practice using the equipment before beginning the investigation. Improper use of equipment can result in inaccurate, imprecise data with large errors, and the validity of the data can be compromised.

Incorrectly reading measurements is a common mistake. Make sure you can use all the equipment needed in the investigation correctly. Record the instructions in detail, so they can be referred to if the data doesn't appear correct.

Increase sampling size

In general, the larger the sample taken for analysis, the more precise the measured values will be. However, you will be limited by the size of the container and number of containers you can transport back to school, if collecting samples in the field.

Repeat the investigation

As discussed in Section 1.2, you can ensure reliability by repeating your experiment. You may need to modify your procedure before repeating the investigation to ensure all variables are being tested under the same conditions.

Referencing secondary-sourced information

As you conduct your investigation, it is important to make note of any secondary-sourced information that you use. Include this in your written report. This is discussed in further detail in Section 1.7.

PEER REVIEW

Your teacher may suggest that you partner with a student in your class, to provide each other with constructive feedback regarding your data collection.

Consider the following:

- Is the data presented clearly in tables and/or graphs?
- Are all entries in the logbook dated?
- Are appropriate headings and units included?
- If anything unexpected occurred during data collection, was this recorded?
- What are the strengths of the data collected?
- What questions do you have about the data collected?

1.3 Review

SUMMARY

- It is essential that you record the following in your logbook during the investigation:
  - all the quantitative and qualitative data you collect
  - the procedures you used to collect the data
  - any incident, feature or unexpected event that may have affected the quality or validity of the data.
- Mistakes are avoidable errors, and measurements affected by mistakes should be discarded.
- A systematic error is an error that is consistent and will occur again if the investigation is repeated in the same way. They are usually a result of instruments that are not calibrated correctly or procedures that are flawed.
- Random errors occur in an unpredictable manner and are generally small. A random error could be the result of a researcher reading the same result correctly one time and incorrectly another time.

KEY QUESTIONS

1 Where should observations and measurements be recorded?

2 Identify the type of error that is described in each scenario below, and how it could be avoided.
- **a** The electronic scale was not zeroed/tared before samples were weighed.
- **b** 1.0 mol L^{-1} hydrochloric acid is used rather than 0.1 mol L^{-1} hydrochloric acid.
- **c** One result is significantly lower than all the rest.
- **d** One student judged a water sample as 'cloudy' while another student recorded it as 'very cloudy'.

3 A student investigating the biomass of leaf litter from two different field sites recorded the following measurements in g/m^2:

Field site 1: 11.4, 10.9, 11.8, 10.6, 1.5, 11.1

Field site 2: 25, 27, 22, 26, 28, 23, 25, 27

Both sets of data contain errors.
- **a** Identify which set contains a random error and explain why.
- **b** Identify which set contains a systematic error and explain why.

4 Identify whether each of the following errors is a mistake, a systematic error or a random error.
- **a** A pipette that should have dispensed volumes of 25.00 ± 0.03 mL actually dispensed volumes of 24.92 ± 0.03 mL.
- **b** A student miscounts the number of cells across the field of view of a microscope.
- **c** A sample of glucose powder was weighed three times, and on the third weigh there was a fluctuation in power supply giving an unexpected value.

1.4 Processing data and information

Once you have conducted your investigation and collected data, you will need to find the best way to collate and present your data. This section is a guide to the different forms of representation that will help you to better understand your data.

REVISION

GO TO ➤ SkillBuilder page 24

Significant figures

Always avoid stating results to a higher number of significant figures or decimal places than the data justifies. Do not report results to a false level of precision. In some cases you may need to exercise your own judgment, but the following are good rules of thumb.

Significant figures are the numbers that convey meaning and precision. The number of significant figures used depends on the scale of the instrument. A significant figure is an integer or a zero that follows an integer. A zero that comes before any integers is not significant.

When you record raw data and report processed data, use the number of significant figures available from your equipment or observation. Using either a greater or smaller number of significant figures can be misleading. For example, Table 1.4.1 shows measurements of five samples taken using an electronic balance accurate to two decimal places. The data was entered into a spreadsheet to calculate the mean, which was displayed with four decimal places. You would record the mean as 5.69 g, not 5.6940 g, because two decimal places is the precision limit of the instrument. Recording 5.6940 g would be an example of false precision.

TABLE 1.4.1 An example of false precision in a data calculation

Sample	1	2	3	4	5	Mean
Mass (g)	5.67	5.75	5.62	5.71	5.72	5.6940

When multiplying or dividing measurements, report the final value to the least number of significant figures used within that calculation. For example, to calculate the amount, in mol, of a substance using concentration and volume, the formula is $n = cV$. If $c = 0.997\ \text{mol}\ \text{L}^{-1}$ (three significant figures), and the volume is 21.62 mL (four significant figures), the amount of substance, in mol, should be reported to the least number of significant figures (i.e. to three significant figures [$n = cV = 0.997 \times 0.02162 = 0.0216$ mol to three significant figures]).

Although digital scales can measure to many more than two figures and calculators can give 12 figures, be sensible and follow the significant figure rules.

Adding or subtracting

When adding or subtracting measurements, report the calculated value to the least number of decimal places.

For example, the following measurements were recorded.

Mass of leaf matter collected from forest floor: 5.68 g

Mass of same leaf matter after drying: 3.19602 g

Difference between initial and final mass = 5.68 − 3.19602 = 2.48398 = 2.48 g. The original answer of 2.48398 g is an example of false precision as one value in the data was not measured to five decimal places. The final answer must be stated to the smallest number of decimal places seen in the data (i.e. 2.48 g).

Multiplying or dividing

When multiplying or dividing values, the calculated answer should have no more significant figures than the value with the least number of significant figures.

Decimal places

Like with significant figures, you must be careful to record your measurements to the precision of the equipment used. If a weighing balance can report a mass measurement to two decimal places, you should record your value to two decimal places.

When adding or subtracting measurements, report the calculated value to the least number of decimal places used in the calculation. For example, the following measurements were recorded in a pH analysis.

Sample A 4.93

Sample B 5.54

Sample C 4.82

The average of these results is

$$\frac{4.93 + 5.54 + 4.82}{3} = 5.0966666667$$

As the sample results are all to two decimal places this average also needs to be rounded to two decimal places (e.g. 5.10). It would be misleading to report the average to a greater number of decimal places.

RECORDING AND PRESENTING QUANTITATIVE DATA

Raw data is unlikely to be used directly to test the hypothesis. However, raw data is essential to the investigation and plans for collecting the raw data should be made carefully. Consider the formulas or graphs that will be used to analyse the data at the end of the investigation. This will help to determine the type of raw data that needs to be collected to test the hypothesis.

Once you have determined the data that needs to be collected, prepare a table in which to record the data.

ANALYSING AND PRESENTING DATA

The raw data that has been obtained needs to be presented clearly, concisely and accurately.

There are many ways to present data, including tables (e.g. Table 1.4.1), graphs, flow charts (Figure 1.4.1) and diagrams. The best way of visualising the data depends on its nature. Try several formats before making a final decision, to create the best possible presentation.

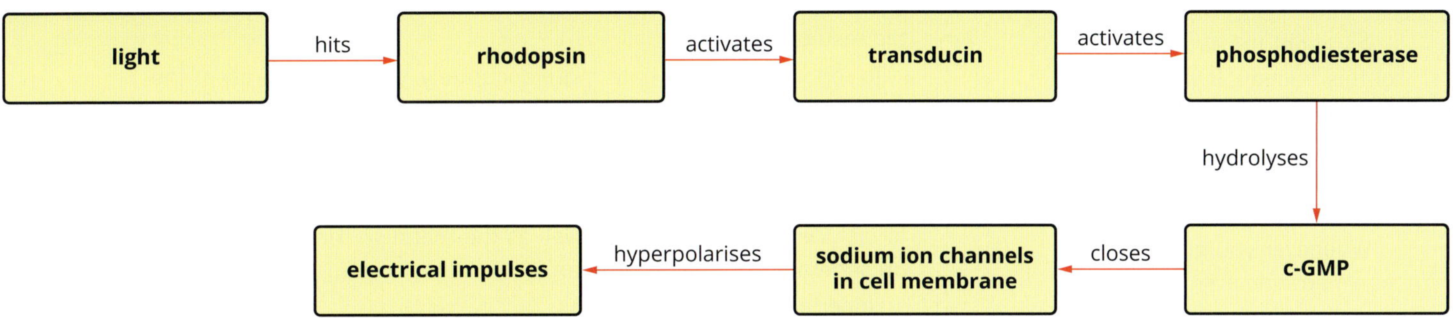

FIGURE 1.4.1 A flow chart summarises the processes that take place inside a rod cell in the retina of the human eye when it is exposed to light.

SKILLBUILDER N

Calculating percentages

Scientists use percentages to express a ratio or fraction of a quantity.

To express one quantity as a percentage of another, use the second quantity to represent 100%.

For example, expressing six as a percentage of 24 is like saying '6 is to 24 as x is to 100':

$$\frac{6}{24} = \frac{x}{100}$$
$$x = \frac{6}{24} \times 100$$
$$= 25\%$$

To calculate a percentage of a quantity, the percentage is expressed as a decimal then multiplied by the quantity.

For example, to calculate 40% of 20:

$$x = \frac{40}{100} \times 20$$
$$= 0.4 \times 20$$
$$= 8$$

SKILLBUILDER 

Converting between percentages and fractions

To write a percentage as a fraction, divide the percentage by 100.

For example:

$$25\% = \frac{25}{100} = \frac{1}{4}$$

$\frac{25}{100}$ is not the simplest form of this fraction. If you divide both the numerator and the denominator by 25 (their highest common factor) then the fraction simplifies to $\frac{1}{4}$.

Whenever you give a fraction as an answer, always try and simplify it by dividing the numerator and denominator by the highest common factor.

To write a fraction as a percentage, multiply the fraction by 100%. In many cases it is easier to convert the fraction to a decimal number first.

For example:

$$\frac{1}{4} = 0.25 \times 100 = 25\%$$

The value of the fraction or percentage has not changed. It is just being represented in a different way.

REVISION

Presenting raw and processed data in tables

Tables organise data into rows and columns and can vary in complexity according to the nature of the data. You can use tables to organise raw and processed data or to summarise results.

The simplest form of a table is a two-column format. In a two-column table, the first column should contain the independent variable and the second should contain the dependent variable.

Tables should have a descriptive title, column headings (including the unit), and numbers aligned to the decimal point.

A table of processed data usually presents the average values of data, the **mean**. However, the mean on its own does not provide an accurate picture of the results.

To report processed data more accurately, you should present the uncertainty as well.

Uncertainty

When there is a range of measurements of a particular value, you will need to include the uncertainty as well as the mean so that your results can be presented accurately as a mean.

Uncertainty is calculated as:

uncertainty = ± maximum variance from the mean

Percentage error

Percentage uncertainty (also known as percentage error) is a way to quantify how accurate a measurement is. To calculate the percentage uncertainty of your data, take the uncertainty and divide it by the measurement, then multiply by 100.

Other descriptive statistical measures

Other statistical measures that can be used, depending on the data obtained, are:

- **mode**—the value that appears most often in a data set, this measure is useful to describe qualitative or discrete data
- **median**—the 'middle' value of an ordered list of values, it is used when the data range is spread (e.g. because of unusual results, making the mean unreliable).

Table 1.4.2 is a guide to choosing the most appropriate measure of central tendency for a data set.

TABLE 1.4.2 The most appropriate measures of central tendency to use in descriptive statistics depends on the type of data.

Type of data	Mode	Median	Mean
Nominal (qualitative)	✓	✗	✗
Ordinal (qualitative)	✓	✓	possibly
Discrete or continuous (quantitative)	✓	✓	✓

Graphs

It is important that you choose an appropriate graph type to suit the data that you have collected. Table 1.4.3 summarises suitable graphs for qualitative and quantitative data.

TABLE 1.4.3 Suitable graph types for discrete and continuous data

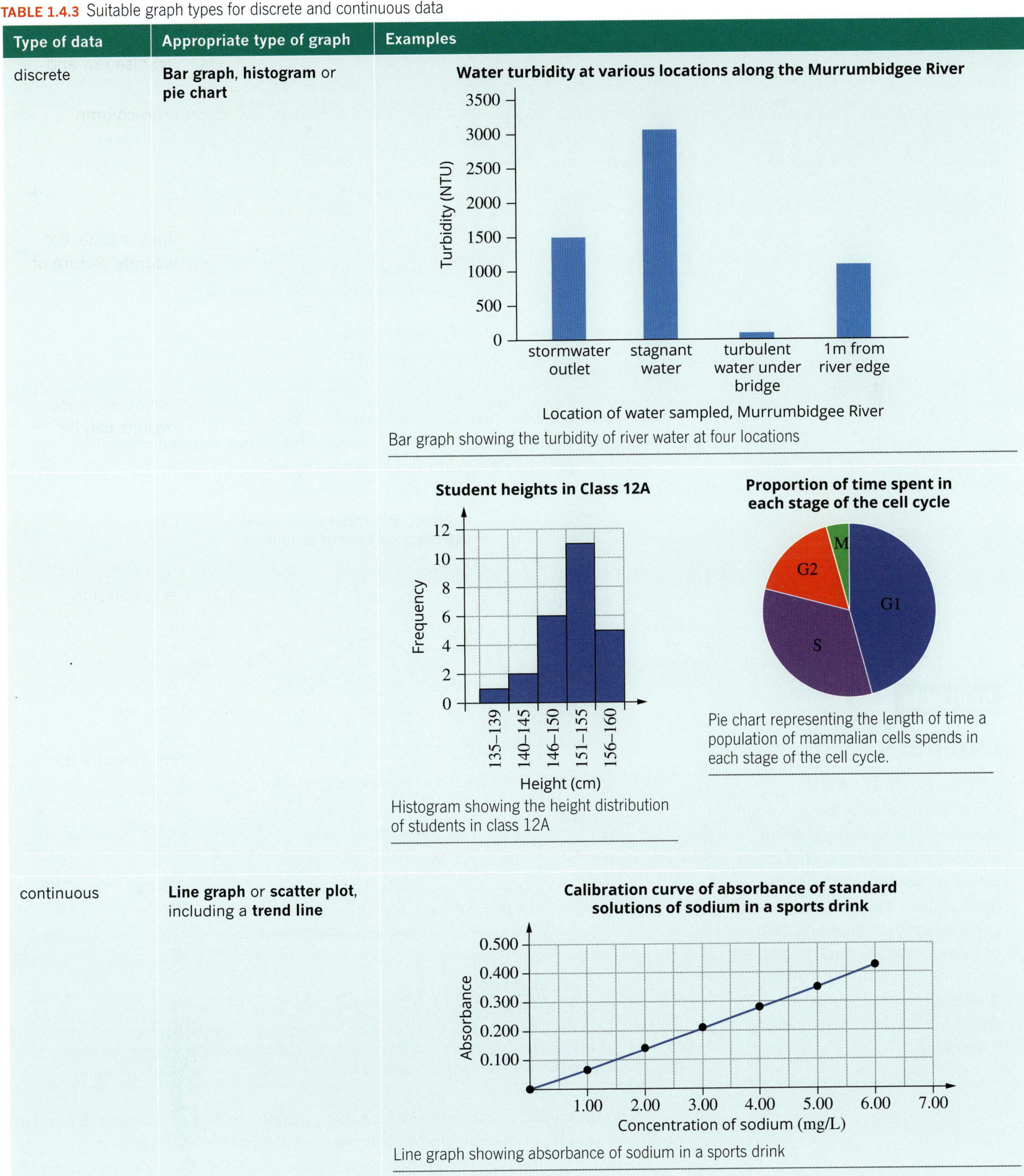

Type of data	Appropriate type of graph	Examples
discrete	**Bar graph**, **histogram** or **pie chart**	Bar graph showing the turbidity of river water at four locations Histogram showing the height distribution of students in class 12A Pie chart representing the length of time a population of mammalian cells spends in each stage of the cell cycle.
continuous	**Line graph** or **scatter plot**, including a **trend line**	Line graph showing absorbance of sodium in a sports drink

TABLE 1.4.3 *cont.*

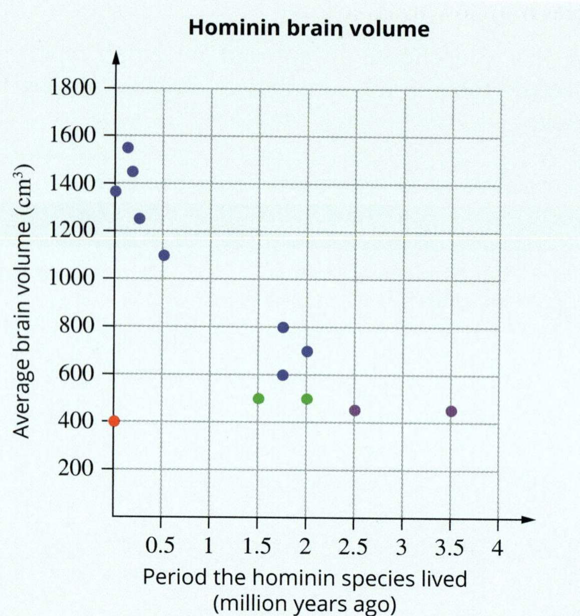

Scatterplot of brain volume in a range of hominin species—colour code for different genera: *Homo*, blue; *Paranthropus*, green; *Australopithecus*, pink; a modern chimpanzee included for comparison, red

In general, tables provide more detailed data than graphs, but it is easier to observe data trends and patterns in graphical form than in tabular form.

For example, the data from the enzyme experiment in Table 1.2.5, on page 13, is presented in Figure 1.4.2.

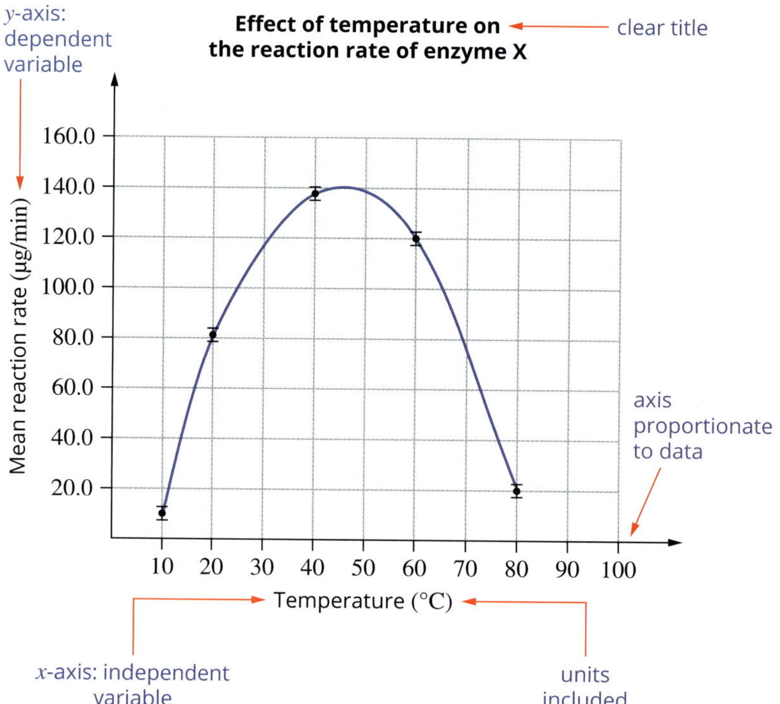

FIGURE 1.4.2 A line graph showing the relationship between two variables: temperature (independent variable) and reaction rate (dependent variable). The uncertainty values are represented as vertical bars above and below the mean at each data point.

REVISION

Outliers

Sometimes when data is collected, there may be one point that does not fit with the trend and is clearly an error. This is called an **outlier**. An outlier is often caused by a mistake made in measuring or recording data, or from a random error in the measuring equipment. If there is an outlier, include it on the graph but ignore it when adding a trend line.

Use graphs when you are considering two variables, and one variable is dependent on the other. The graph shows the relationship between the variables.

There are several types of graphs you can use, including line graphs, bar graphs and pie charts. The best one to use will depend on the nature of the data.

General rules to follow when making a graph include the following.

- Keep the graph simple and uncluttered.
- Use a descriptive title.
- Represent the independent variable on the *x*-axis and the dependent variable on the *y*-axis.
- Make axes proportionate to the data.
- Clearly label axes with both the variable and the unit in which it is measured.

REVISION

Distorting the truth

Poorly constructed graphs can alter your perception of the data. For example, in Figure 1.4.3, you can see two graphs that show the same data—the test results of two groups of students. One group of students did not eat breakfast before doing the test and scored an average of 42 marks out of 50. The other group did eat breakfast and scored an average of 48 marks out of 50. One graph exaggerates the difference in marks between the two groups by using a scale of 40 to 50 marks on the *y*-axis. You need to make sure the graphs you create do not misrepresent your data in any way. You should also be wary of distorted data when interpreting graphs in other publications.

(a)
Graph 1
Test score
0
5
10
15
20
25
30
35
40
45
50
Students who did not eat breakfast
Students who did eat breakfast

(b)
Graph 2
Test score
40
41
42
43
44
45
46
47
48
49
50
Students who did not eat breakfast
Students who did eat breakfast

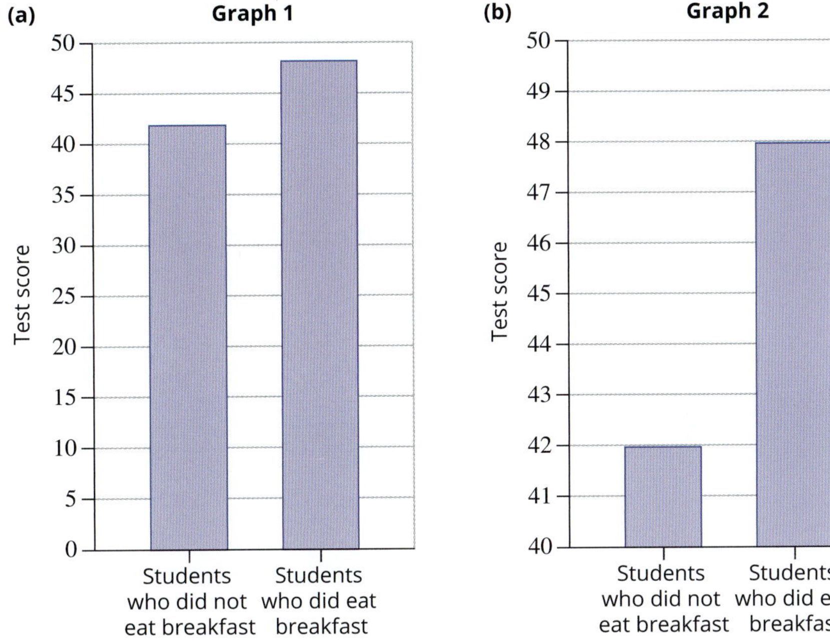

FIGURE 1.4.3 (a) Graph 1 shows the difference in test scores between two groups of students out of the total 50 marks on the *y*-axis. (b) Graph 2 shows the same difference but within only a narrow range of marks on the *y*-axis, which exaggerates the difference between the groups, making it appear larger than it really is.

Preparing diagrams

It is important to learn how to draw and label diagrams of equipment and biological specimens in your studies of biology. There are certain rules you should follow to produce a diagram that will be acceptable in your reports and exams.

When drawing scientific equipment and biological structures, diagrams should:

- be large, simple, two-dimensional pencil drawings
- have ruled lines where possible
- keep proportions realistic (Figure 1.4.4).

When drawing biological specimens follow these guidelines.

- Draw the whole diagram (including labels, lines, magnification, heading and scale if possible) in pencil. Do not shade the diagram.
- A diagram of microscopic objects does not require a circle representing the field of view.
- Draw one or a few cells to represent a sample; there is no need to try to draw every cell in a field of view.
- Draw your diagram with simple and clear lines (do not sketch).
- Use stippling rather than shading to indicate depth.
- Make your diagram as large as possible (at least 10 × 10 cm).

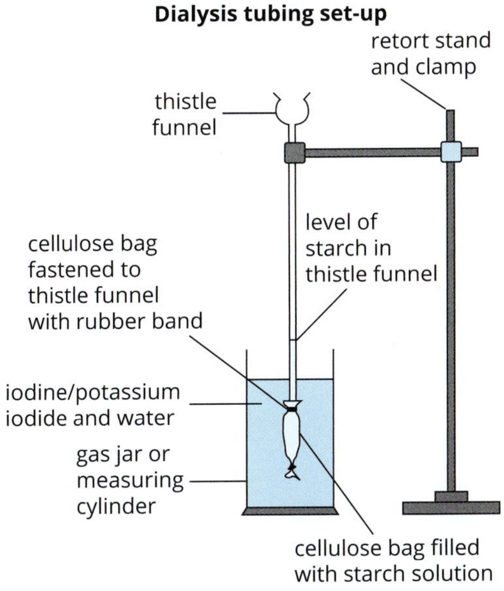

Dialysis tubing set-up
retort stand and clamp
thistle funnel
level of starch in thistle funnel
cellulose bag fastened to thistle funnel with rubber band
iodine/potassium iodide and water
gas jar or measuring cylinder
cellulose bag filled with starch solution

FIGURE 1.4.4 Diagram showing a dialysis tubing arrangement. Note the straight lines for labels that are horizontal where possible, and the realistic proportions of different parts in relation to each other.

- If there are many features to show, it is useful to pair a photo with a supporting diagram that shows cellular details (Figure 1.4.5).
- Draw only the structures that you see, not things you think you should see.
- Include clear labels for the features you want to highlight.
- Place labels outside the drawing.
- Make sure label pointers do not cross over each other.
- Labels should line up on either side of the diagram where possible.
- Use straight lines without arrowheads that meet the features being labelled.
- Include a scale bar or scale (e.g. 1:100) in the diagram, or state the magnification (e.g. ×400) in the caption.

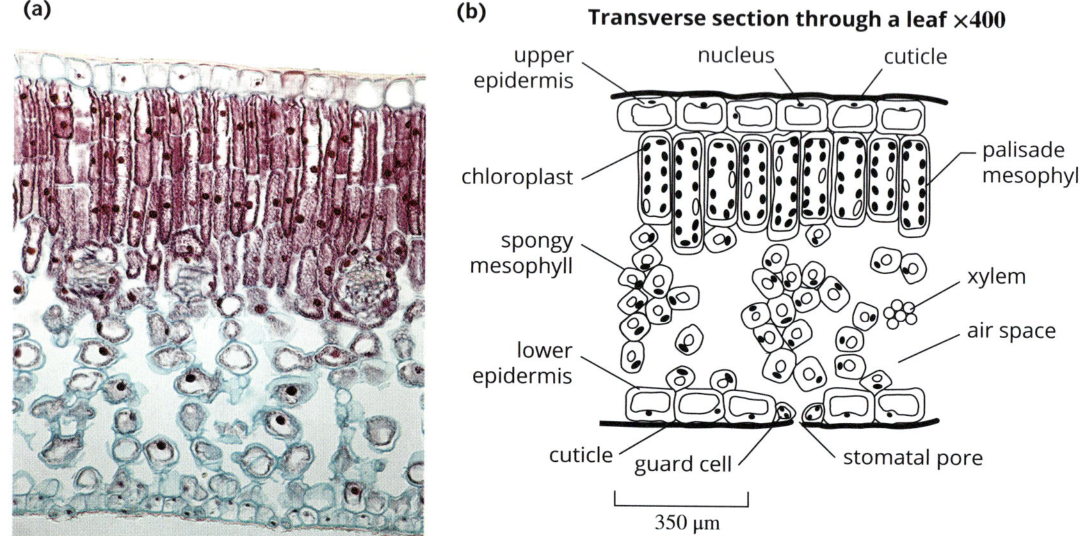

FIGURE 1.4.5 (a) A photomicrograph and (b) a diagram of a transverse section through a leaf. The diagram provides clear information about the size and structure of the leaf.

PEER REVIEW

Your teacher may suggest that you partner with a student in your class, to provide each other with constructive feedback regarding the processing of your data.

Consider the following:

- Is the processed data presented clearly in tables and/or graphs?
- Have any outliers been identified?
- Has the appropriate type of graph been selected to display the processed data?
- Are appropriate headings and units included?
- What are the strengths of the processed data collected?
- What questions do you have about the processed data collected?

1.4 Review

SUMMARY

- The number of significant figures and/or decimal places you use depends on the scale of the instrument used. It is important to record data to the number of significant figures or decimal places available from the equipment or observation.
- Consider how the data will be presented and analysed. A wide range of analysis tools could be used. For example, tables organise data so that patterns can be established, and graphs can show relationships and comparisons.
- The simplest form of a table is a two-column format in which the first column contains the independent variable and the second contains the dependent variable.
- When there is a range of measurements of a particular value, the mean must be accompanied by the uncertainty for your results to be presented as a mean accurately.
- General rules to follow when making a graph include the following.
  - Keep the graph simple and uncluttered.
  - Use a descriptive title.
  - Represent the independent variable on the *x*-axis and the dependent variable on the *y*-axis.
  - Make axes proportionate to the data.
  - Clearly label axes with both the variable and the unit in which it is measured.

KEY QUESTIONS

1 The body weight of mice was measured following a two-week experiment. From the data set below, determine the mean, mode and median.

Mouse number	1	2	3	4	5	6	7
Mouse weight (g)	19	24	21	21	25	21	24

a the mean
b the mode
c the median

2 **a** How many significant figures are in the value 22.06 mL?
b When multiplying or dividing, how many significant figures should be reported for the calculation?
c When adding or subtracting, how many significant figures should be reported for the calculation?

3 Which axis should be used to represent the:
a dependent variable
b independent variable

4 Following an experiment, a student prints out the excitation and emission spectra of a fluorescent conjugated antibody obtained by fluorescence spectroscopy.
a At what wavelength does excitation peak?
b At what wavelength does emission peak?

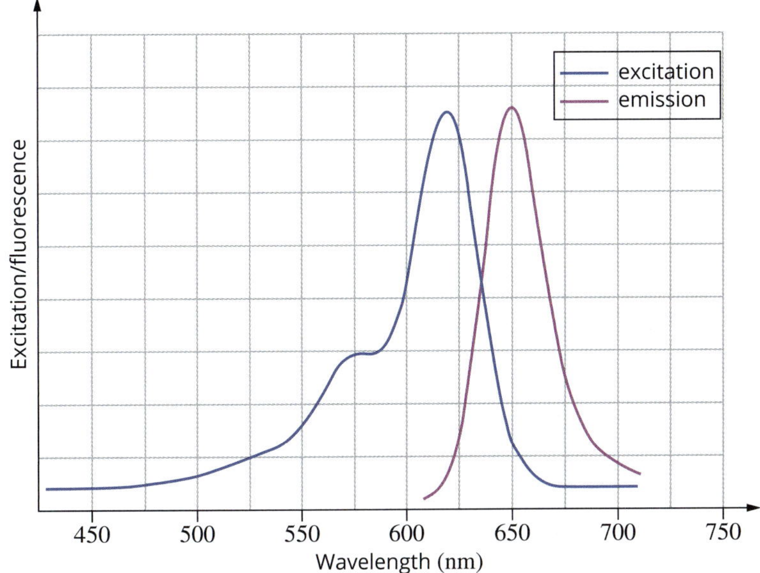

1.4 Review *continued*

5 Describe the trend in the following graph.

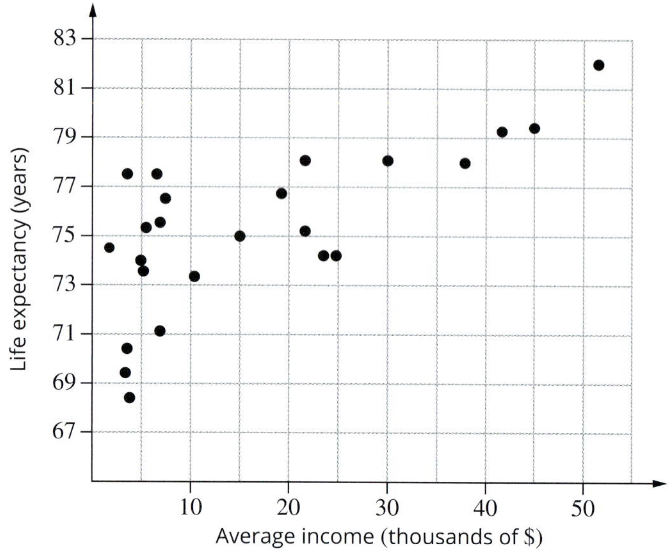

6 A student wants an estimate of the protein concentration in a sample that she isolated. The student prepares a standard curve of known protein concentrations. The following data was obtained.

Protein standard (mg/mL)	Absorbance (at 590 nm)
0	0
0.1	0.2
0.2	0.18
0.4	0.34
0.5	0.45
0.6	0.52
0.75	0.58
0.9	0.72
1	0.86
0	0
0.1	0.2
unknown	0.7

a Plot the data on a scatter plot, using graph paper or a spreadsheet program.

b Define the term outlier and describe what effect it could have on a line of best fit if left in the analysis.

c Identify any outliers in this set of data.

d Draw a trend line.

e Use the graph to determine the protein concentration of the unknown solution.

7 Describe at least four ways the graph below could be improved.

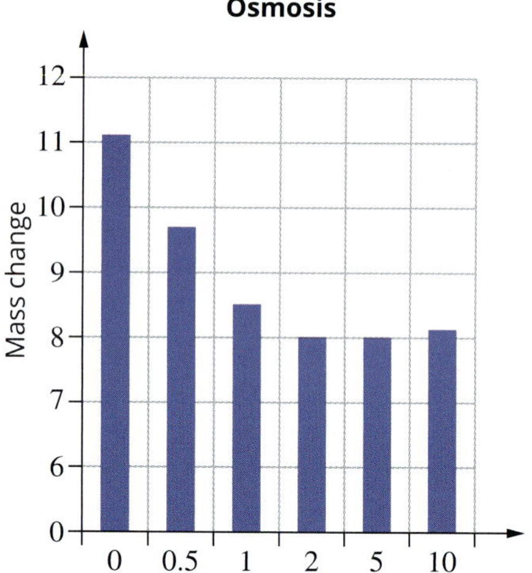

8 You are measuring the volumes of solutions in an experiment. Discuss the accuracy of your results if you are:

a using a 50 mL beaker with 5 mL increments

b using a 10 mL measuring cylinder with 0.5 mL increments

c using a 20 μl pipette

1.5 Analysing data and information

Now that the chosen topic has been thoroughly researched, the investigation has been conducted and data collected, it is time to draw it all together. You will now need to analyse your results to better understand the biological processes behind them (Figure 1.5.1).

FIGURE 1.5.1 To discuss and conclude your investigation, use the raw and processed data.

FACTORS THAT CAN AFFECT THE INTERPRETATION, ANALYSIS AND UNDERSTANDING OF DATA

Correlation and causation

You need to be careful to distinguish between a correlation between two variables and a cause-and-effect relationship. For example, Figure 1.5.2 shows a correlation between the number of worldwide non-commercial space launches and the number of sociology doctorates awarded in the USA between 1997 and 2009. Although the data shows a strong correlation, it cannot be concluded that there is a relationship between the number of worldwide non-commercial space launches and the number of sociology doctorates awarded. In other words, two sets of data might be correlated but have no relationship to each other (there is no causation).

It is important to consider other variables that may explain similar patterns in data sets before concluding that there is a correlation. For example, it is reported that sales of both ice-cream and sunscreen are greater in Sydney in February compared to in July. This is most likely a consequence of the hotter weather, instead of any relationship between the sales of each product.

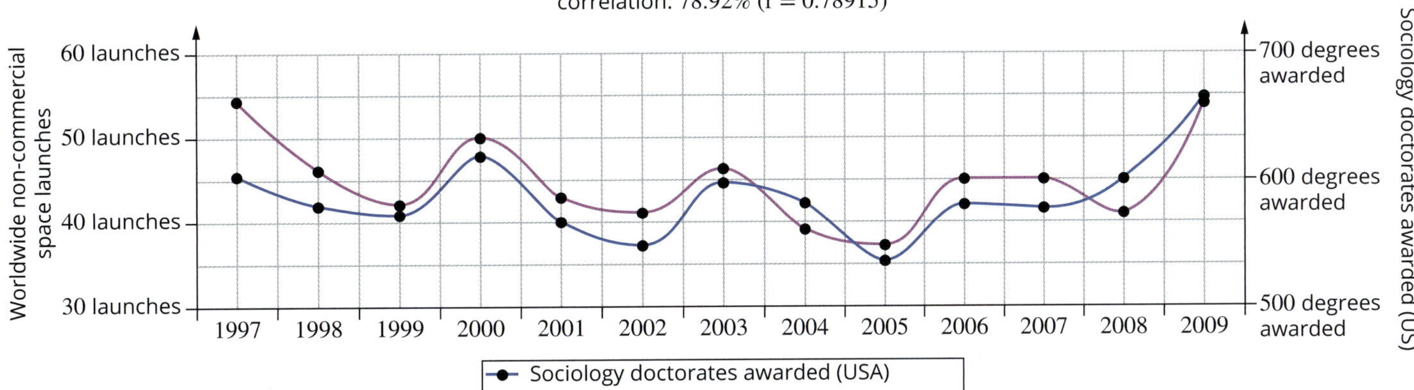

FIGURE 1.5.2 The number of worldwide non-commercial space launches and the number of Sociology doctorates awarded in the USA between 1997 and 2009 show a strong correlation but there is no relationship between the two data sets. Remember that correlation does not equal causation and you need to look at other variables that may explain similar patterns in data sets.

Evaluating the data

Some useful questions to consider when interpreting data include:

- Has the original question been answered?
- Do the results meet expectations? Do they make sense?
- What are the main conclusions? Are there other interpretations?
- Is the supporting data of sufficient quality? How current is it? How was it collected?
- Can the results be supported statistically? That is, are they statistically significant?

GO TO ➤ Year 11 Section 1.5

REVISION

Explaining results in the discussion

The key sections of the discussion are:

- analysing and evaluating data
- evaluating the investigative procedure
- explaining the link between the investigation findings and the relevant biological concepts.

When writing the discussion, you need to consider what message you want to convey. This will include clearly stating the context, results and implications of the investigation.

Analysing and evaluating data

This is where you need to analyse and interpret the findings of the investigation.

- State whether you observed a pattern, trend or relationship between the independent and dependent variables. Describe what kind of pattern it was and specify under what conditions it was observed.
- Were there discrepancies, deviations or anomalies in the data? If so, acknowledge and explain these.
- Identify any limitations in your data. Perhaps a larger sample or further variations in the independent variable would lead to a stronger conclusion.

Remember that the results may be unexpected. This does not make the investigation a failure. However, you will need to discuss and link the findings back to the hypothesis, purpose and procedure.

Describing more complex trends

You may find that you need to describe and explain a more complex trend in data than a simple linear or exponential relationship. The example shown in Figure 1.5.3 represents the energy changes during a chemical reaction with and without an enzyme. K is the activation energy of the reaction without the enzyme, L is the activation energy needed to start the reaction with the enzyme, M is the extra energy needed to start the reaction if an enzyme is not present.

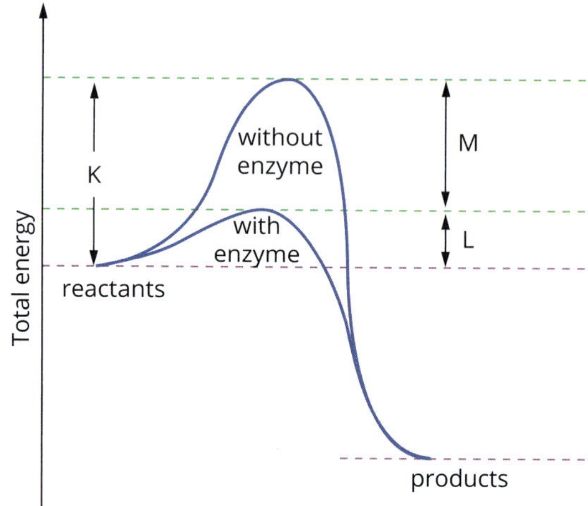

FIGURE 1.5.3 Energy changes during a chemical reaction with and without an enzyme

Reading information from a graph

Being able to extract information from a graph is an important skill. Figure 1.5.4 shows the results of a clinical trial testing the efficacy of a drug candidate. Both treatment groups (low and high dose) had an effect on the treatment group compared to the placebo group. The effect of the drug treatment is represented by the change from the baseline (%).

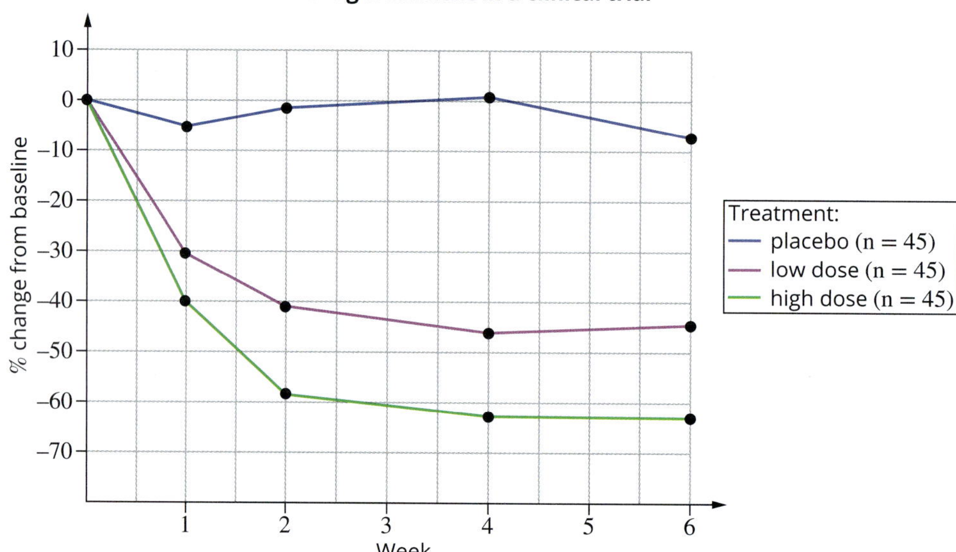

FIGURE 1.5.4 A clinical trial testing the efficacy of a drug candidate over six weeks (n = the number of people in each group)

SKILLBUILDER

Standard curves

Using standard curve graphs for quantification

You can use a standard curve (or calibration curve) to gain quantitative results for samples assayed with a biochemical colour-based assay. For example, you may be experimenting with an enzyme that breaks down protein (starting with a 1 mg/mL solution).

You can use the biuret test to detect changes in protein concentration at the end of a 24-hour reaction. This involves the use of biuret reagent and a protein such as bovine serum albumin (BSA). For the standard curve you would prepare a set of standard protein solutions of a known concentration (e.g. a range from 0.1 to 1.0 mg/mL), then perform the assay on the standards. Numeric values are then obtained for the standards by reading the absorbance (or optical density) with a colorimeter or spectrophotometer.

The absorbance values can be used to plot a graph and draw a line of best fit. This is the standard curve. Keep in mind that a standard curve may not be linear if you are not working in the correct concentration range for the substance being measured.

Once you have your standard curve, you can use it to determine the protein concentrations at different optical densities. For example, for a protein assay on a sample that gave an optical density of 0.32, from the protein standard curve you find that the protein concentration is 0.3 mg/mL (red lines on Figure 1.5.5).

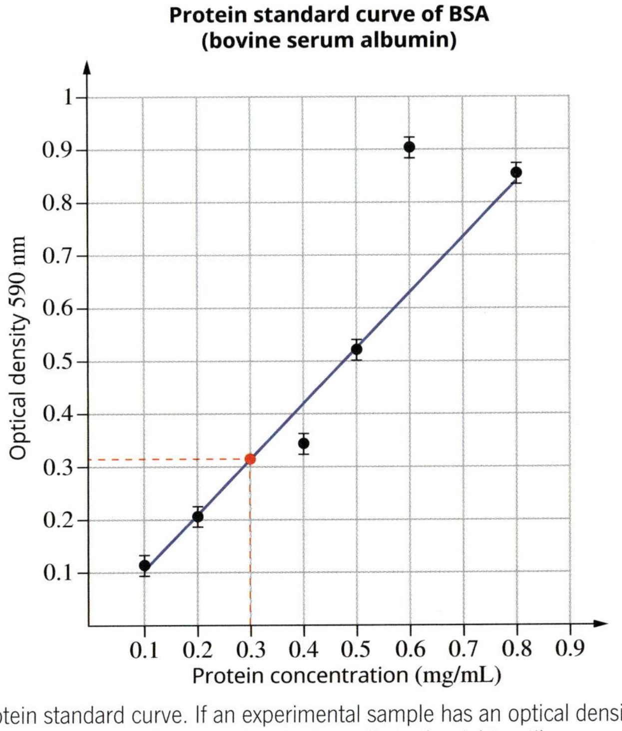

FIGURE 1.5.5 A protein standard curve. If an experimental sample has an optical density of 0.32, then go up the *y*-axis until you reach 0.32, trace a line horizontally to the right until you meet the line of best fit, then trace a line down to meet the *x*-axis. From the standard curve, you can estimate that the protein concentration of the sample is 0.3 mg/mL.

If you do not have an instrument to read absorbance, you can make a visual grading scheme with colour photographs (for identification) and relative values (for quantification) assigned to every colour intensity (e.g. –, +, ++, +++ or 0, 1, 2, 3). This provides a semi-quantitative result.

REVISION

Evaluating the procedure

You need to discuss the limitations of the investigation procedure. Evaluate the procedure and identify any issues that could have affected the validity, accuracy, precision or reliability of the data. You also need to state sources of errors and uncertainty in the discussion.

Once you have identified any limitations or problems in the procedure, recommend how the investigation could be improved if repeated.

Bias

Bias may occur in any part of the investigation procedure, including sampling and measurements. Bias is a form of systematic error resulting from the researcher's personal preferences or motivations. There are many types of bias, including poor definitions of both concepts and variables, incorrect assumptions and errors in the investigation design and procedure. You cannot eliminate some biases, but you should at least address these in the discussion.

Accuracy

In the discussion, evaluate the degree of accuracy and precision of the measurements for each variable in the hypothesis. Comment on the uncertainties obtained.

When relevant, compare the chosen procedure with any other procedures that might have been selected, evaluating the advantages and disadvantages of the selected procedure and the effect on the results.

Validity

Validity refers to whether an experiment or investigation is testing the set hypothesis and purpose. Factors influencing validity include:

- whether your experiment measures what it claims to measure (i.e. your experiment should test your hypothesis)
- whether the independent variable influenced the dependent variable in the way you thought it would (i.e. the certainty that something observed in your experiment was the result of your experimental conditions, and not some other cause that you did not consider)
- the degree to which your findings can be generalised.

Reliability

When discussing the results, indicate the range of the data obtained from replicates. Explain how you selected the sample size. Larger samples are usually more reliable, but time and resources might have been scarce. Discuss whether the sample size may have limited the results of the investigation.

The control group is important to the reliability of the investigation, as it helps determine if a variable that should have been controlled has been overlooked and may explain any unexpected results.

Error

Discuss any source of systematic or random error. When you have identified limitations of the procedure and results, suggest ways of improving the investigation.

CRITICALLY EVALUATING RESOURCES

Not all sources are credible. You must critically evaluate the content and its origin. Questions you should always ask when evaluating a source include:

- Who created this message? What are the qualifications, expertise, reputation and affiliation of the authors?
- Why was it written?
- Where was the information published?
- When was the information published?
- How often is the information referred to by other researchers?
- Are conclusions supported by data or evidence?
- What is implied?
- What is omitted?
- Are any opinions or values being presented in the piece?
- Is the writing objectively and accurately describing a scientific concept or phenomenon?
- How might other people understand or interpret this message differently from me?

When evaluating the validity or bias of websites consider its domain extension:

- .gov stands for government
- .edu stands for education
- .org stands for non-profit organisations
- .com stands for commercial/business.

When conducting a literature review, it can be useful to summarise your findings in a table, such as that shown in Table 1.5.1.

TABLE 1.5.1 You can use a table to help summarise information while researching. This example centres on the techniques used by Indigenous Australians to remove toxins from food.

Type of bush food	Notes	Reference	Reliability of source
bush tomato (*Solanum centrale*)	Purplish-green colour when unripe. Traditionally eaten raw when greenish-white to yellow–brown in colour. Dried on plant, then shaken from plant, rubbed into sand, followed by pounding or grinding into a paste with water, formed into balls or cakes (cited Peterson 1979).	Hegarty, M.P., Hegarty, E.E. & Wills, R. (2001). *Food Safety of Australian Plant Bushfoods*: Publication No. 01/28	Credible—Published by Rural Industries Research and Development Corporation and University of Newcastle, review article with credible primary sources
Triunia robusta (plant with toxic nuts)	Nuts are treated by soaking and baking to remove toxins.	Hegarty, M.P., Hegarty, E.E. & Wills, R. (2001). *Food Safety of Australian Plant Bushfoods*: Publication No. 01/28	Credible—Published by Rural Industries Research and Development Corporation and University of Newcastle, review article with credible primary sources
lemon-scented tea tree	Removal of toxins by grinding and baking.	Hegarty, M.P., Hegarty, E.E. & Wills, R. (2001). *Food Safety of Australian Plant Bushfoods:* Publication No. 01/28	Credible—Published by Rural Industries Research and Development Corporation and University of Newcastle, review article with credible primary sources
cycad seed (*Macrozamia* sp.)	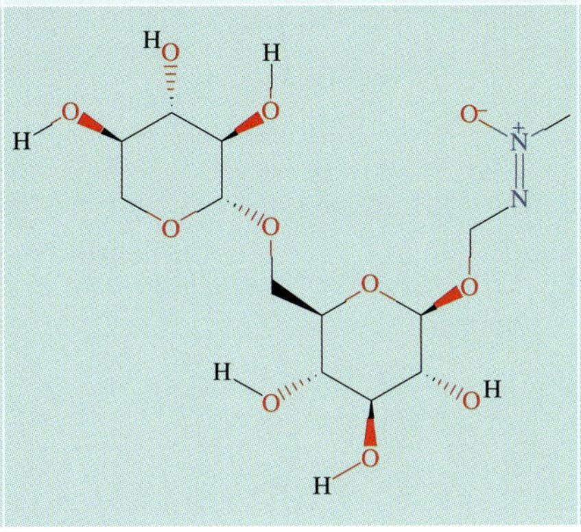	source: https://pubchem.ncbi.nlm.nih.gov/compound/Macrozamin#section=Top	Credible source

PEER REVIEW

Your teacher may suggest that you partner with a student in your class, to provide each other with constructive feedback regarding the analysis of your data.

Consider the following:

- Is the data accurately discussed?
- Have you identified any limitations of the procedure, data collection or data analysis?
- Have you made any recommendations to improve the investigation?
- If time permitted, did you repeat the investigation with the suggested improvements?
- What are the strengths of the data analysis?
- What questions do you have about the data analysis?

1.5 Review

SUMMARY

- After completing your investigation, you will need to analyse and interpret your data. A discussion of your results is required where the findings of the investigation need to be analysed and interpreted.
  - State whether a pattern, trend or relationship was observed between the independent and dependent variables. Describe what kind of pattern it was and specify under what conditions it was observed.
  - If possible, create a mathematical model to describe your data.
  - Were there discrepancies, deviations or anomalies in the data? If so, acknowledge and explain these.
  - Identify any limitations in the data collected. Perhaps a larger sample or further variations in the independent variable would lead to a stronger conclusion.
- It is important to discuss the limitations of the investigation procedure. Evaluate the procedure and identify any issues that could have affected the validity, accuracy, precision or reliability of the data. You also need to state sources of errors and uncertainty in the discussion.
- When discussing the results, indicate the range of the data obtained from replicates. Explain how you selected the sample size. Larger samples are usually more reliable, but time and resources are likely to have been scarce. Discuss whether the results of the investigation have been limited by the sample size.

KEY QUESTIONS

1 What types of graphs would be suitable for displaying discrete data?

2 Describe the trend in the following graph.

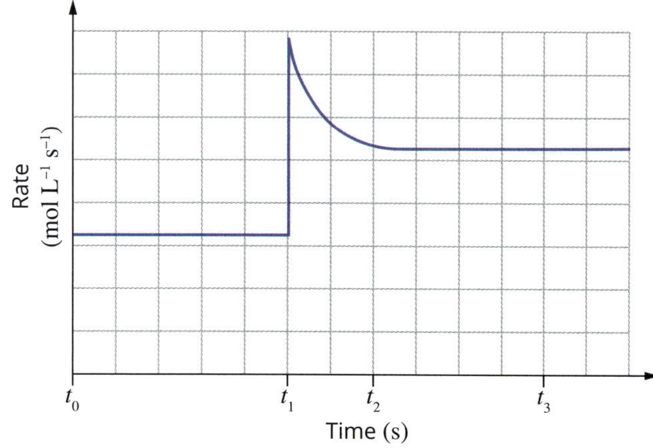

3 Which of the following graphs represents an immune response after vaccination (V) followed by an infection (I) with the pathogen that the vaccine protects against?

A

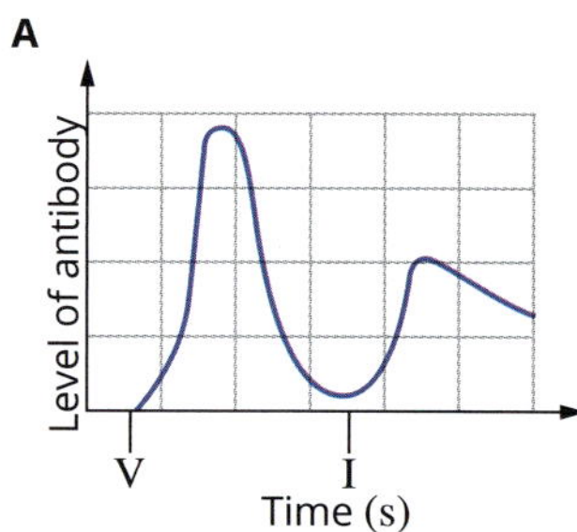

B

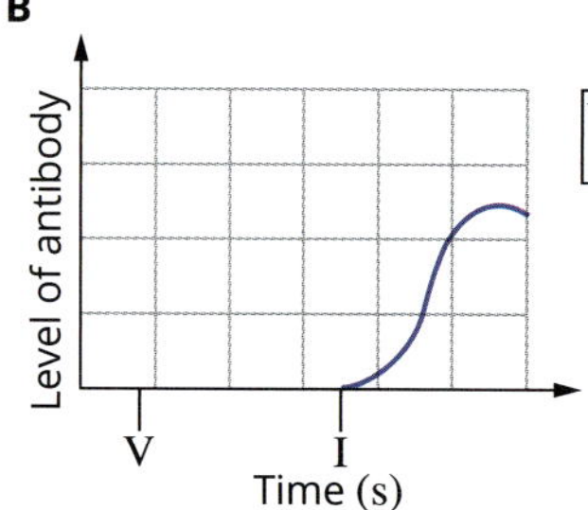

V = vacccination
I = infection

C

Level of antibody
V
I
Time (s)

D

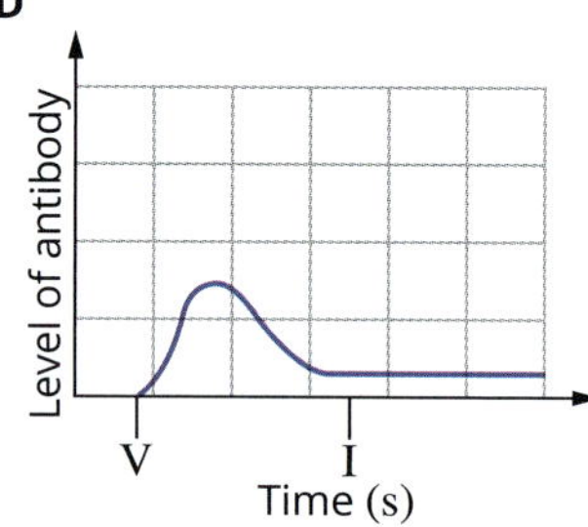

4 List features of a credible primary or secondary source.

1.6 Problem solving

Having analysed your results, you can then apply them to biological concepts to evaluate your conclusions. In this section you will learn how analysing your investigation leads to a better understanding of the underlying scientific principles of your research.

MODELLING

Table 1.6.1 lists examples of models that you can use to explain phenomena, visualise concepts, make predictions and solve problems.

TABLE 1.6.1 Examples of models that can be used in investigations

Model	Applications	Example
Mathematical model	Epidemiological modelling to predict the incidence, prevalence and spread of disease.	The SIR model is used in epidemiology to compute the theoretical number of infected people from a contagious disease. The mathematical model, used with diseases such as measles and rubella, uses several equations with the categories of susceptible (S) infected (I) and recovered (R). 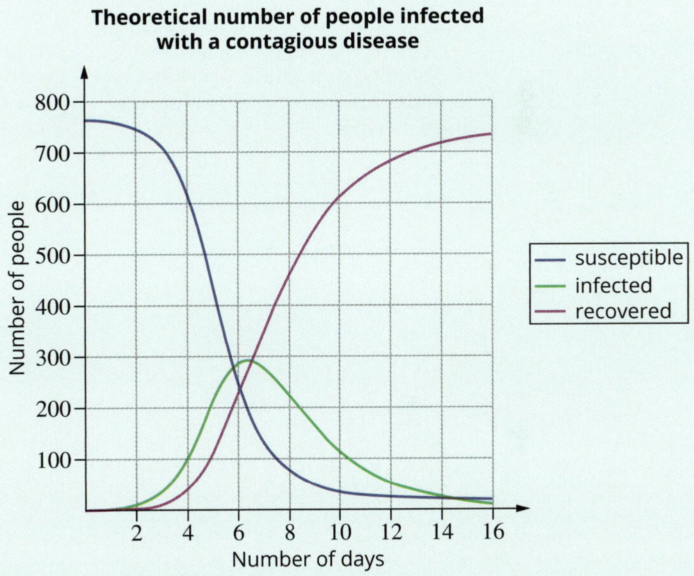
Molecular and protein models	Molecular and protein modelling can be used to understand biological functions, inheritance and disease and develop pharmaceuticals.	Modelling the quaternary structure of the haemoglobin protein to understand its function and the consequences of misfolded polypeptides. 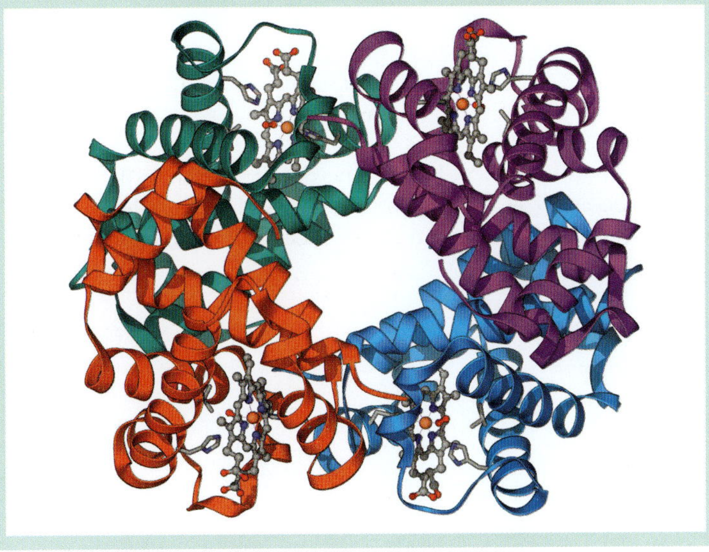

TABLE 1.6.1 *cont.*

Model	Applications	Example
Computer model	Computer models used to simulate biochemical reactions, cellular functions, body systems, ecological systems and evolutionary relationships.	Representative Concentration Pathways (RCPs) used to model atmospheric greenhouse gas concentrations under different emissions scenarios. **Predicted CO_2 emissions under four Representative Concentration Pathway (RCP) scenarios** 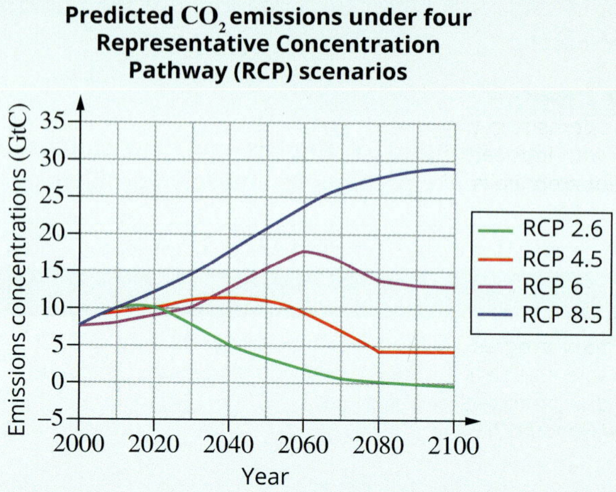
Model organism	Organisms that can be easily bred and studied in laboratory conditions to understand biological structures and functions.	*Drosophila melanogaster* is an important model organism in the study of genetics, development and disease. Much of what we understand about gene function and developmental genetics has been learnt from research on *D. melanogaster*. 
Physical and digital models	Physical and digital representations of biological structures and organisms	An artificial heart model made from metal (red) and plastic tubes (blue) is useful for showing the movement of blood into and out of the heart. However, it cannot replicate the rhythmic contractions of the heart muscles, or show the internal structure of the heart. 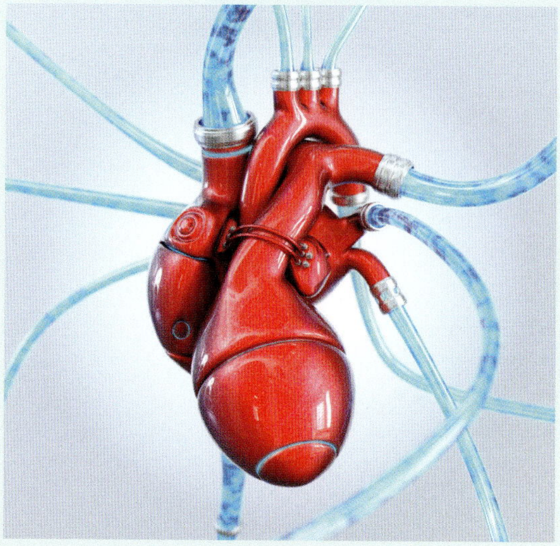

LITERATURE REVIEW

Table 1.6.2 shows an example of how summarising information in a table can help you identify patterns in a literature review.

TABLE 1.6.2 An in-progress summary of information obtained through a literature review of the use of iPS (induced pluripotent stem) cells in multiple sclerosis research

Source	Author	Year of publication	Type	Key ideas	Credible?
Multiple sclerosis: getting personal with induced pluripotent stem cells. *Cell Death Dis., 6,* e1806	Di Ruscio A, Patti F, Welner RS, Tenen DG, Amabile G	July 2015	review	potential impact of using iPS cell-derived models in multiple sclerosis (MS) research	yes, author qualified scientist, credible science journal
Efficient generation of myelinating oligodendrocytes from primary progressive multiple sclerosis patients by induced pluripotent stem cells. *Stem Cell Reports, 3*(2), 250–9.	Douvaras P, Wang J, Zimmer M, Hanchuk S, O'Bara MA, Sadiq S, Sim FJ, Goldman J, Fossati V	August 2014	primary source	skin cells from patients were cultured and transformed into functional oligodendrocytes	yes, primary source, credible authors, journal, recent article
Fine structure of neurally differentiated iPS cells generated from a multiple sclerosis (MS) patient: a case study. *Microsc Microanal., 20*(6), 1869–75.	Herszfeld D, Payne NL, Sylvain A, Sun G, Bernard CC, Clark J, Sathananthan H	December 2014	primary source	the presence of oligodendrocytes and analysed by transmission electron microscopy (TEM)	yes, primary source, credible authors, journal, recent article
Ultimate human stem cells created in the lab. *New Scientist.* Retrieved 13 June 2018 from https://www.newscientist.com/article/dn26209-ultimate-human-stem-cells-created-in-the-lab/	Coghlan, A	September 2014	secondary source	interview of a scientist researching into the production of iPS cells	yes, respected science magazine, gives the primary source
www.msaustralia.org.au	www.msaustralia.org.au	-	secondary source	about, symptoms, current treatments	yes, Australian Association for MS

DISCUSSING RELEVANT BIOLOGICAL CONCEPTS

To make the investigation more meaningful, you need to explain it within the right context, meaning the related biological concepts, theories and models. Within this context, explain the basis for the hypothesis.

Relating your findings to a biological concept

During the analysing stage of your investigation (Section 1.5), you were able to find trends, patterns and mathematical models of your results. This is the framework needed for you to discuss whether the data supported or refuted the hypothesis. Ask questions such as:

- Was the hypothesis supported?
- Has the hypothesis been fully answered? (If not, explain why this is so and suggest how you could improve or complement the investigation.)
- Do the results contradict the hypothesis? If so, why? (The explanation must be plausible and you must base it on the results and previous evidence.)

Providing a theoretical context also means that you can compare the results with existing relevant research and knowledge. After identifying the major findings of the investigation, ask questions such as:

- How does the data fit with the literature?
- Does the data contradict the literature?
- Do the findings fill a gap in the literature?
- Do the findings lead to further questions?
- Can the findings be extended to another situation?

Be sure to discuss the broader implications of the findings. Implications are the bigger picture. Outlining them for the audience is an important part of the investigation. Ask questions such as:

- Do the findings contribute to or impact on the existing literature and knowledge of the topic?
- Are there any practical applications for the findings?

REVISION

Interpreting scientific and media texts

Sometimes you may be required to investigate claims and conclusions made by other sources, such as scientific and media texts. As discussed in Section 1.4, some sources are more credible than others. Once you have analysed the validity of the primary or secondary source, you will be able to follow the same steps described above in evaluating their conclusions to solve scientific problems.

DRAWING EVIDENCE-BASED CONCLUSIONS

A conclusion is usually a paragraph that links the collected evidence to the hypothesis and provides a justified response to the research question. Indicate whether the hypothesis was supported or refuted and the evidence on which this is based (that is, the results). Do not provide irrelevant information. Only refer to the specifics of the hypothesis and the research question, and do not make generalisations.

What type of evidence is needed to draw valid conclusions?

You can draw a valid conclusion if:

- you designed the investigation to obtain data to address the purpose and hypothesis
- you only changed one independent variable
- you measured or observed the dependent variable
- you controlled all other variables
- you replicated the experiment
- another person could follow the procedure
- you obtained accurate and precise data
- there were no significant limitations with the procedure and data obtained
- you made any links between the data obtained and biological theory.

Read the examples of weak and strong conclusions in Table 1.6.3 and Table 1.6.4 for the hypothesis and inquiry question shown.

TABLE 1.6.3 Examples of weak and strong conclusions to the hypothesis

Hypothesis: An increase in the temperature of pond water will result in a decrease in the measured pH of the water sample	
Weak conclusion	Strong conclusion
The pH of water decreased as temperature increased.	An increase in temperature from 5°C to 40°C resulted in a decrease in the pH of the water from 7.4 to 6.8.

TABLE 1.6.4 Examples of strong and weak conclusions in response to the inquiry question

Research question: Does temperature affect the pH of water?	
Weak conclusion	Strong conclusion
The results show that temperature does affect the pH of water.	Analysis of data from an investigation on the effect of water temperature on pH showed an inverse relationship in which the pH of water decreased from 7.4 to 6.8. These results support the current knowledge that an increase in water temperature results in a decrease in its pH.

PEER REVIEW

Your teacher may suggest that you partner with a student in your class, to provide each other with constructive feedback regarding the analysis of your data.

Consider the following:

- Are the findings accurately discussed in relation to relevant biological concepts?
- Is the conclusion strong?
- Is the conclusion directed to the purpose, hypothesis and inquiry question?
- What are the strengths of the conclusion?
- What questions do you have about the conclusion?

1.6 Review

SUMMARY

- To make the investigation more meaningful, you need to explain it within the right context, meaning the related biological concepts, theories and models. Within this context, explain the basis for the hypothesis.
- Indicate whether you supported or refuted the hypothesis and on what evidence you based this (that is, the results). Do not provide irrelevant information. Only refer to the specifics of the hypothesis and the research question and do not make generalisations.

KEY QUESTIONS

1 Consider the experiment conducted by Jan Baptist van Helmont in the 17th century and then answer the following questions. Van Helmont planted a 2.25 kg willow plant in a pot containing 90 kg of dry soil. Over the next five years he watered the tree. After five years he weighed the tree and soil separately. The tree had a mass of approximately 76 kg and the soil mass had decreased by less than 1 kg. Van Helmont concluded that the plant's increased mass was due to water.

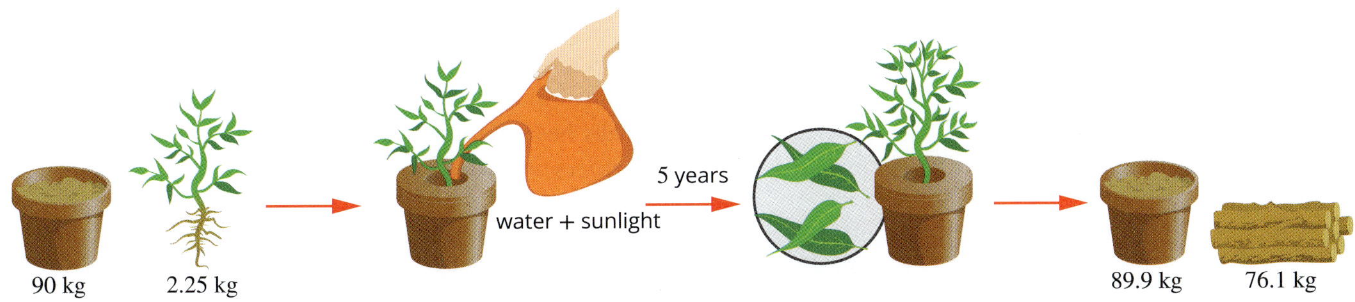

  a Does the experiment support the accepted thought at the time that plants 'eat' soil to get their nutrients? Justify your answer.
  b Suggest a control for the experiment.
  c Suggest another experiment to test van Helmont's hypothesis that plants only require water to grow.
  d Offer another explanation on how plants obtain their nutrients.

2 a Explain what is represented by the visual model to the right.
  b How does this model support your understanding of the biological concept that is represented?

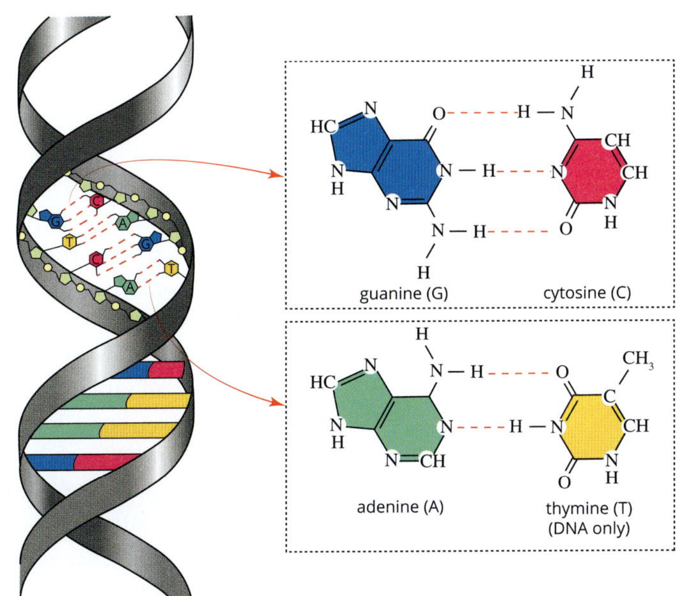

3 Outline the benefits and limitations of using models (physical and digital) as an alternative to animals in scientific research.

4 A procedure was repeated five times. How should the following statement be rewritten? To minimise the potential of random errors, many repeats of the procedure were conducted.

1.7 Communicating

The way you communicate your results will depend on your audience. If you are communicating with a general audience, you may want to present it as a news article or blog post. In this case, you cannot assume that your audience has a science background and you will need to make sure you do not use too much scientific language.

Throughout this course you will need to present your research using appropriate scientific language. There are many different presentation formats that you are used to, such as posters, oral presentations and reports. This section covers the main characteristics of effective science communication and report writing, including objectivity, clarity, conciseness and coherence.

PRESENTING YOUR WORK

Your teacher will specify how you should present your work. Two common formats are a scientific report and a scientific poster.

First identify your audience. This might be your teacher, your peers or the general public.

Secondly, consider how you could use figures such as labelled diagrams, graphs or flow charts to aid communication (e.g. to show how equipment was set up, or to summarise a biological pathway).

Scientific report

You can use scientific reports to describe the findings of an experiment or a literature review. A scientific report must have a clear, logical structure.

A scientific report usually has the following components/subheadings:

- and name(s) of lab partners and collaborator(s)
- date
- title
- Introduction
  - summary of relevant background information and definitions of key terms
  - purpose
  - hypothesis
  - definition of independent, dependent and controlled variables
- Materials
  - chemicals (including quantities and concentrations) and equipment
  - risk assessment (from safety data sheets)
- Procedure
  - step-by-step instructions that the reader can follow to replicate the experiment if required
  - can include two-dimensional scientific diagrams
- Results
  - summary data presented in tables and/or figures
  - all tables and figures are numbered and have an appropriate title
- Discussion
  - brief explanation of the significance of the results
  - direct reference to data in tables and/or figures; use of evidence to support statements
  - relevance to biology theory
  - limitations of the investigation and proposed avenues for further research
- Conclusion
  - summary of the investigation findings using evidence and the relevance of the findings to the field of research
  - relates to purpose and hypothesis
- References
  - formatted in a consistent style (e.g. American Psychological Association [APA] referencing style).

Scientific poster

Scientists often present their research at conferences in the form of a scientific poster. This is a summary of their research that includes visual support such as tables, diagrams, graphs and flow charts (Figure 1.7.1).

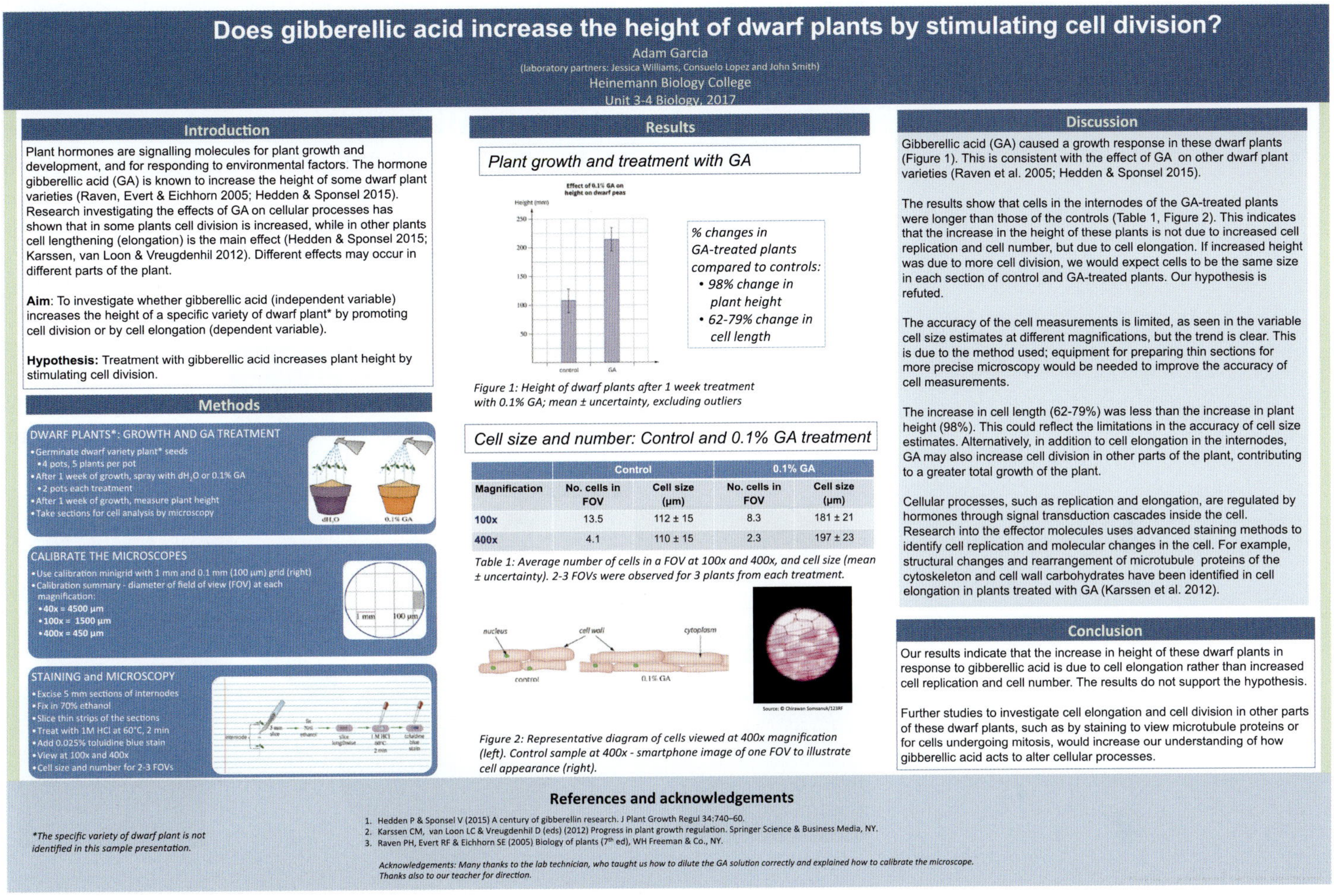

Does gibberellic acid increase the height of dwarf plants by stimulating cell division?

Adam Garcia
(laboratory partners: Jessica Williams, Consuelo Lopez and John Smith)
Heinemann Biology College
Unit 3-4 Biology, 2017

Introduction

Plant hormones are signalling molecules for plant growth and development, and for responding to environmental factors. The hormone gibberellic acid (GA) is known to increase the height of some dwarf plant varieties (Raven, Evert & Eichhorn 2005; Hedden & Sponsel 2015). Research investigating the effects of GA on cellular processes has shown that in some plants cell division is increased, while in other plants cell lengthening (elongation) is the main effect (Hedden & Sponsel 2015; Karssen, van Loon & Vreugdenhil 2012). Different effects may occur in different parts of the plant.

Aim: To investigate whether gibberellic acid (independent variable) increases the height of a specific variety of dwarf plant* by promoting cell division or by cell elongation (dependent variable).

Hypothesis: Treatment with gibberellic acid increases plant height by stimulating cell division.

Methods

DWARF PLANTS*: GROWTH AND GA TREATMENT

- Germinate dwarf variety plant* seeds
  - 4 pots, 5 plants per pot
- After 1 week of growth, spray with dH_2O or 0.1% GA
  - 2 pots each treatment
- After 1 week of growth, measure plant height
- Take sections for cell analysis by microscopy

CALIBRATE THE MICROSCOPES

- Use calibration minigrid with 1 mm and 0.1 mm (100 µm) grid (right)
- Calibration summary - diameter of field of view (FOV) at each magnification:
  - 40x = 4500 µm
  - 100x = 1500 µm
  - 400x = 450 µm

STAINING and MICROSCOPY

- Excise 5 mm sections of internodes
- Fix in 70% ethanol
- Slice thin strips of the sections
- Treat with 1M HCl at 60°C, 2 min
- Add 0.025% toluidine blue stain
- View at 100x and 400x
- Cell size and number for 2-3 FOVs

Results

Plant growth and treatment with GA

Effect of 0.1% GA on height on dwarf peas

% changes in GA-treated plants compared to controls:
- 98% change in plant height
- 62-79% change in cell length

Figure 1: Height of dwarf plants after 1 week treatment with 0.1% GA; mean ± uncertainty, excluding outliers

Cell size and number: Control and 0.1% GA treatment

	Control		0.1% GA	
Magnification	**No. cells in FOV**	**Cell size (µm)**	**No. cells in FOV**	**Cell size (µm)**
100x	13.5	112 ± 15	8.3	181 ± 21
400x	4.1	110 ± 15	2.3	197 ± 23

Table 1: Average number of cells in a FOV at 100x and 400x, and cell size (mean ± uncertainty). 2-3 FOVs were observed for 3 plants from each treatment.

Source: © Chirawan Somsanuk/123RF

Figure 2: Representative diagram of cells viewed at 400x magnification (left). Control sample at 400x - smartphone image of one FOV to illustrate cell appearance (right).

Discussion

Gibberellic acid (GA) caused a growth response in these dwarf plants (Figure 1). This is consistent with the effect of GA on other dwarf plant varieties (Raven et al. 2005; Hedden & Sponsel 2015).

The results show that cells in the internodes of the GA-treated plants were longer than those of the controls (Table 1, Figure 2). This indicates that the increase in the height of these plants is not due to increased cell replication and cell number, but due to cell elongation. If increased height was due to more cell division, we would expect cells to be the same size in each section of control and GA-treated plants. Our hypothesis is refuted.

The accuracy of the cell measurements is limited, as seen in the variable cell size estimates at different magnifications, but the trend is clear. This is due to the method used; equipment for preparing thin sections for more precise microscopy would be needed to improve the accuracy of cell measurements.

The increase in cell length (62-79%) was less than the increase in plant height (98%). This could reflect the limitations in the accuracy of cell size estimates. Alternatively, in addition to cell elongation in the internodes, GA may also increase cell division in other parts of the plant, contributing to a greater total growth of the plant.

Cellular processes, such as replication and elongation, are regulated by hormones through signal transduction cascades inside the cell. Research into the effector molecules uses advanced staining methods to identify cell replication and molecular changes in the cell. For example, structural changes and rearrangement of microtubule proteins of the cytoskeleton and cell wall carbohydrates have been identified in cell elongation in plants treated with GA (Karssen et al. 2012).

Conclusion

Our results indicate that the increase in height of these dwarf plants in response to gibberellic acid is due to cell elongation rather than increased cell replication and cell number. The results do not support the hypothesis.

Further studies to investigate cell elongation and cell division in other parts of these dwarf plants, such as by staining to view microtubule proteins or for cells undergoing mitosis, would increase our understanding of how gibberellic acid acts to alter cellular processes.

References and acknowledgements

**The specific variety of dwarf plant is not identified in this sample presentation.*

1. Hedden P & Sponsel V (2015) A century of gibberellin research. J Plant Growth Regul 34:740–60.
2. Karssen CM, van Loon LC & Vreugdenhil D (eds) (2012) Progress in plant growth regulation. Springer Science & Business Media, NY.
3. Raven PH, Evert RF & Eichhorn SE (2005) Biology of plants (7th ed), WH Freeman & Co., NY.

Acknowledgements: Many thanks to the lab technician, who taught us how to dilute the GA solution correctly and explained how to calibrate the microscope. Thanks also to our teacher for direction.

FIGURE 1.7.1 An example of a scientific poster

Regardless of the style of your presentation, it is a good idea to help the reader navigate the work by numbering tables and figures and referring to these in the body of the report.

Historical or theoretical essay

An essay contains the following elements/characteristics:

- a formal structure including introduction, body paragraphs and conclusion
- the introduction states the focus of the essay
- each paragraph makes a new point supported by evidence
- each paragraph has a link back to the previous paragraph
- the concluding paragraph draws all the information together but does not introduce any new information
- any visuals are included at the end of the essay, in an appendix.

Oral presentation

Consider the following elements when you prepare an oral presentation:

- must be engaging
- use cue cards but do not read directly from them
- look at the audience as you speak
- smile and appear confident (Figure 1.7.2)
- try not to fidget.

FIGURE 1.7.2 When giving an oral presentation, use your body language to engage the audience. Make eye contact, look around the audience and smile.

Other types of presentation

There are many other ways to present your findings, including:

- making a documentary or media report
- developing an evidence-based argument
- developing an environmental management plan
- analysing a work of fiction or film for scientific relevance
- creating a visual presentation.

Analysing relevant information

Scientific research should always be objective and neutral. Any premise presented must be supported with facts and evidence to allow the audience to make its own decision. Identify the evidence supporting or contradicting each point you want to make. It is also important to explain connections between ideas, concepts, theories and models. Figure 1.7.3 shows the questions you need to consider when writing your scientific report.

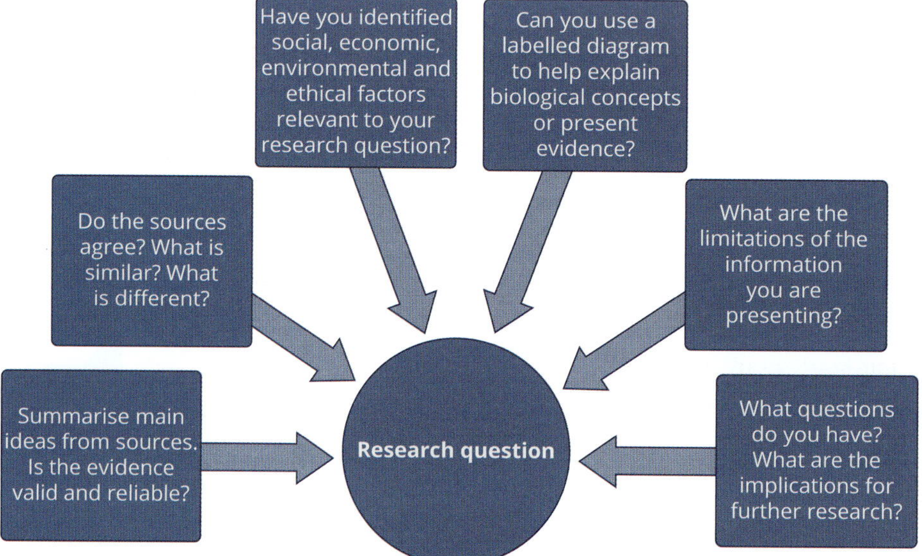

FIGURE 1.7.3 Discuss relevant information, ideas, concepts, implications and make sure your discussion is relevant to the question under investigation.

Once you have analysed your sources, annotate your outline, indicating where you will use evidence and what the source of that evidence is. Try to introduce only one idea per sentence and one theme per paragraph.

For example, for a report on 'Experimental research into biodegradability of plastics', the third paragraph might contain information from:

- Selke et al. (2015)—no significant effect
- Chiellini et al. (2007)—significant effect.

However when reviewing your summary table, you notice that different experimental procedures were used, and that Chiellini received funding from the manufacturer.

A report should include an analysis and synthesis of your sources. Therefore, you need to explicitly connect the information from different sources and make it clear where sources agree and disagree. In this example, the final sentences could be:

Selke et al. (2015) reported that tests of plastic polymers treated with biodegradation additives resulted in no significant biodegradation after three years. This finding contrasts with that of Chiellini et al. (2007), who reported significant biodegradability of additive-treated polymers.

The different results can be explained by differences in the studies. The 2007 study tested biodegradation in natural river water, whereas the 2015 study tested degradation under ultraviolet light, aerobic soil burial and anaerobic aqueous conditions (Chiellini et al. 2007; Selke et al. 2015). As well as using different additives and different experimental techniques, Selke et al. (2015) used additive rates of 1–5% and tested polyethylene terephthalate (PET) as well as polyethylene, whereas Chiellini et al. (2007) used additive rates of 10–15% and tested only polyethylene.

Both studies were conducted under laboratory conditions, so they may not reflect what happens in the natural environment.

REVISION

GO TO ➤ Year 11 Section 1.7

Writing for science

Scientific reports are usually written in an objective or unbiased style. This is in contrast with English writing that most often uses the subjective techniques of rhetoric or persuasion.

You can use either first-person or third-person narrative for scientific writing. Your teacher may advise you on which to select. In either case, ensure that you keep the narrative consistent.

Be careful of words that are absolute, such as *always*, *never*, *shall*, *will* and *proven*. Sometimes it may be more accurate and appropriate to use qualifying words, such as *may*, *might*, *possible*, *probably*, *likely*, *suggests*, *indicates*, *appears*, *tends*, *can* and *could*.

You need to write concisely, using shorter sentences, particularly if you want to engage and maintain your audience's interest.

Identify concepts that you can explain using visual models, as well as information that you can present in graphs or diagrams. This will both reduce the word count and also make it more accessible for your audience.

EDITING YOUR REPORT

Once your report is written, you will need to edit it. This is an important part of the process. After editing your report, save new drafts with a different file name and always back up your files in another location.

When editing, pretend you are reading your report for the first time. Once you have completed a draft, it is a good idea to put it aside and return to it with 'fresh eyes' a day later. This will help you find areas that need further work and give you the opportunity to improve them. Look for content that is:

- ambiguous or unclear
- repetitive
- awkwardly phrased
- too lengthy
- not relevant to your research question
- poorly structured
- lacking evidence
- lacking a reference (if it is another researcher's work).

Use a spellchecker tool to help you identify typographical errors, but first, check that your spellchecker is set to Australian English. Also be wary of words that are commonly misused, for example:

- where/were
- their/they're/there
- affect/effect
- which/that.

REVISION

References and acknowledgements

You will need to acknowledge all the quotations, documents, publications and ideas used in the investigation in the 'references and acknowledgments' section. This will help you avoid possible instances of plagiarism, as well as ensure authors are credited for their work. References and acknowledgements also give credibility to the study and allow the audience to locate information sources should they wish to study it further. The standard referencing style used is the APA academic referencing style.

For example, a book would be referenced as: Rickard, G., et al. (2016), *Pearson Science 9 Student Book* (2nd ed.), Pearson Education, Melbourne, Australia. An example of a website reference is: Adamson, D. H. (2017). 'Plastics' in *World Book Advanced*, accessed 26 February 2018, from http://www.worldbookonline.com/advanced/article?id=ar434080.

In-text citations

Each time you write about the findings of other people or organisations, you need to provide an in-text citation and provide full details of the source in a reference list. In the APA style, in-text citations include the first author's last name and date in brackets (author, date). List the full details in your list of references.

The following examples show the use of in-text citation.

Ben-Issa et al. (2017) concluded that the use of companion plants within a crop system has to be combined with other strategies to effectively control against pests.

Or

It was reported that the use of companion plants within a crop system has to be combined with other strategies to effectively control against pests (Ben-Issa et al., 2017).

The bibliographic details of the example above would be:

Ben-Issa, R., Gomez, L., & Gautier, H. (2017) Companion plants for aphid pest management. *Insects*, 8(4), 112-121.

There are many online guides to help you format your reference list and in-text citations. Some websites and journals suggest how to quote a source in a particular referencing style, such as APA.

PEER REVIEW

Your teacher may suggest that you show your work to a peer in order to seek constructive feedback. Part of the skill of a working scientist is to be open to receiving feedback, and to give constructive feedback to others.

Consider the following:

- Is the purpose clear?
- Is the format organised logically?
- Could you repeat the investigation without any additional information?
- Are tables and figures numbered?
- Does the body of the text guide readers as to when to refer to tables and figures?
- Are in-text citations used to acknowledge all sourced images and content?
- Are scientific conventions/nomenclature consistently followed?
- What are the strengths of the work?
- What questions do you have about the work?

MEASUREMENT AND UNITS

In biology, measurements are used for the very small, that cannot be seen by the naked eye, to the extremely large. In practical demonstrations and investigations we generally make measurements and process those measurements in order to draw conclusions. Scientists have a number of conventional ways of interpreting and analysing data from their investigations. There are also conventional ways of writing numerical measurements and their units.

REVISION

Prefixes and conversion factors

Use conversion factors carefully. You should be familiar with the prefixes and conversion factors in Table 1.7.1. The most common mistake made with conversion factors is multiplying rather than dividing. Note that the table gives all conversions as a multiplying factor.

TABLE 1.7.1 Prefixes and conversion factors

Multiplying factor		Prefix	Symbol
1 000 000 000 000	10^{12}	tera	T
1 000 000 000	10^{9}	giga	G
1 000 000	10^{6}	mega	M
1 000	10^{3}	kilo	k
0.01	10^{-2}	centi	c
0.001	10^{-3}	milli	m
0.000 001	10^{-6}	micro	μ
0.000 000 001	10^{-9}	nano	n
0.000 000 000 001	10^{-12}	pico	p

Do not put spaces between prefixes and unit symbols.

It is important to give the symbol the correct case (upper or lower case). There is a big difference between 1 mm and 1 mM.

PEER REVIEW

Your teacher may suggest that you partner with a student in your class, to provide each other with constructive feedback regarding the presentation of your investigation.

Consider the following:

- Are the findings accurately communicated using scientific language?
- Is the presentation written towards the appropriate audience?
- Are tables, images, diagrams, graphs and flow charts referenced in the text?
- Are appropriate in-text citations included for theoretical concepts and images that are not created by the student?
- Are all in-text citations and references listed in a consistent style, such as APA referencing style?
- What are the strengths of the presentation?
- What questions do you have about the presentation?

REVISION

Correct use of unit symbols

The correct use of unit symbols removes ambiguity, as symbols are recognised internationally. The symbols for units are not abbreviations and should not be followed by a full stop unless they are at the end of a sentence.

The product of a number of units is shown by separating the symbol for each unit with a dot or a space. The division or ratio of two or more units can be shown in fraction form, using a slash, or using negative indices. Prefixes should not be separated by a space.

Numbers and symbols should not be mixed with words for units and numbers. For example, thirty grams and 30 g are correct, while 30 grams and thirty g are incorrect.

Scientific notation

To overcome confusion or ambiguity, measurements are often written in scientific notation. Quantities are written as a number between one and ten, and then multiplied by an appropriate power of ten.

You should be routinely using scientific notation to express numbers.

1.7 Review

SUMMARY

- A scientific report must include an introduction, materials, procedures, results, discussion and conclusion.
- Evidence must be used to support statements in your scientific report. The evidence may be from your own results or results published by other researchers.
- The conclusion should include a summary of the main findings, limitations of the research, implications and applications of the research, and potential future research.
- Scientific writing uses unbiased, objective, accurate, formal language. Ensure your writing is concise and qualified.
- Visual support can assist in conveying scientific concepts and processes efficiently.
- Ensure you edit your final report.
- Use scientific notation when communicating your results.
- Every time you write about the research of other people or organisations (secondary sources), you need to provide an in-text citation and provide full details of the source in a reference list.

KEY QUESTIONS

1 Which of the following statements is written in scientific style?

A The results seemed to be very good ...
B The experiment took a long time ...
C The data in Figure 2 indicates ...
D The researchers felt ...

2 Which of the following statements is written in third-person narrative? (More than one response can be selected.)

A The researchers reported ...
B Samples were analysed using ...
C The experiment was repeated three times ...
D I reported ...

3 Complete the following information on scientific notation:

a convert 235 000 to scientific notation
b convert 0.000 000 655 to scientific notation
c explain why scientific notation is used

4 A scientist designed and conducted an experiment to test the following hypothesis: If eating fast food decreases liver function, then people who eat fast food more than three times per week will have lower liver function than people who do not eat fast food.

a The discussion section of the scientist's report included comments on the accuracy, precision, reliability and validity of the investigation. Read each of the following statements and determine whether they relate to precision, reliability or validity.

i only teenage boys were tested
ii six boys were tested.

b The scientist then conducted the fast-food study with 50 people in the experimental group and 50 people in the control group. In the experimental group, all 50 people gained weight. The scientist concluded all the subjects gained weight as a result of the experiment. Is this conclusion valid? Explain why or why not.

c What recommendations would you make to the scientist to improve the investigation?

Chapter review

01

KEY TERMS

accuracy
bar graph
bias
chemical code
continuous variable
controlled variable
data
dependent variable
discrete variable
hypothesis
independent variable
inquiry question
line graph
literature review
mean
median
mistake
mode
model
nominal variable
observation
ordinal variable
outlier
percentage uncertainty
personal protective equipment (PPE)
pie chart
precision
primary source
primary investigation
procedure
purpose
qualitative data
qualitative variable
quantitative data
quantitative variable
random error
raw data
reliability
safety data sheet (SDS)
scatter plot
secondary source
secondary-sourced investigation
significant figure
systematic error
trend line
uncertainty
validity
variable

REVIEW QUESTIONS

1 Which graph from the following list would be best to use with each set of data listed here?
Graph types: pie chart, scatter graph (with trend line), bar graph, line graph
- **a** the levels of bacteria detected in drinking water sampled at various locations
- **b** the daily temperature over the period of a year
- **c** the effect of humidity on the rate of transpiration
- **d** time spent in each phase of the cell cycle

2 Which one of the following would not support a strong conclusion to a report?
- **A** The concluding paragraphs are relevant and provide supporting evidence.
- **B** The concluding paragraphs are written in emotive language.
- **C** The concluding paragraphs include reference to limitations of the research.
- **D** The concluding paragraphs include suggestions for further avenues of research.

3 Which of the following consists only of secondary sources of information?
- **A** a periodic table, an article published in a science magazine, a science documentary, a practical report written by a Year 12 student
- **B** an article published in a peer-reviewed science journal, an article published in a science journal, a science documentary
- **C** a periodic table, a scientific article summarised on a science website, a science documentary, this Year 12 textbook
- **D** a science article summarised in a newspaper, an article published in a science journal, a science documentary, a practical report written by a Year 12 student

4 What is the correct way to cite in text the following source in APA style?
Tran PN, Brown SH, Rug M, Ridgway MC, Mitchell TW, Maier AG. (2016). Changes in lipid composition during sexual development of the malaria parasite *Plasmodium falciparum*. *Malaria Journal*, 15(1), 73–86.
- **A** Tran et al. (2016) found that regulators of membrane fluidity, cholesterol and sphingomyelin, increased significantly during gametocyte maturation.
- **B** Tran et al. found that regulators of membrane fluidity, cholesterol and sphingomyelin, increased significantly during gametocyte maturation1.
- **C** Tran et al. found that regulators of membrane fluidity, cholesterol and sphingomyelin, increased significantly during gametocyte maturation. Tran PN, Brown SH, Rug M, Ridgway MC, Mitchell TW, Maier AG. (2016). Changes in lipid composition during sexual development of the malaria parasite *Plasmodium falciparum*. *Malaria Journal*, 15(1), 73–86.
- **D** Tran et al. (2016) did not find any significant difference in biodegradeability (Changes in lipid composition during sexual development of the malaria parasite *Plasmodium falciparum*. *Malaria journal*).

5 Explain the meaning of the terms purpose, hypothesis and variable in an investigation.

6 Consider the planning process for the following investigation of an enzyme that breaks down cellulose:

Purpose

Procedure:
1. Set up 6 equal-sized test-tubes in a test-tube rack.
2. Label 2 test-tubes pH 5. Add 5 mL of pH 5 buffer to each tube.
3. Label 2 tubes pH 7. Add 5 mL of pH 7 buffer to each tube.
4. Label 2 tubes pH 9. Add 5 mL of pH 9 buffer to each tube.
5. Add 0.1 mL of cellulase enzyme solution to one tube at each pH.
6. Add 0.1 mL of the appropriate buffer (pH 5, 7 or 9) to the other tube at each pH.
7. Place the test-tube rack, with all tubes, in a 37°C water bath.
8. Add 0.1 g shredded cellulose paper to each of the test-tubes.
9. Incubate for 24 h.
10. Take 1 mL of solution from each tube and test for presence of glucose.

Experimental design

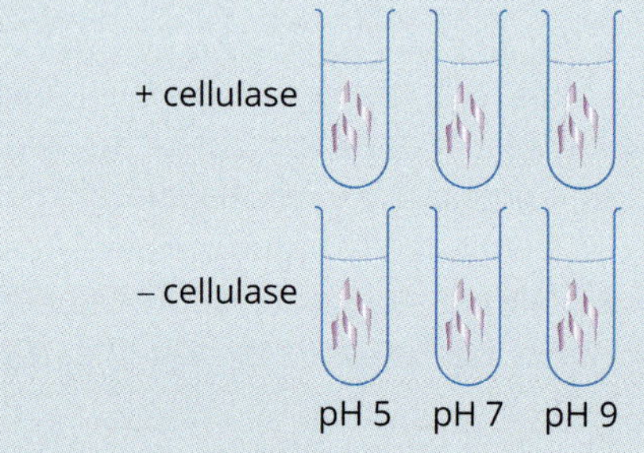

a Before conducting this experiment, what information would be researched as background for the introduction of the practical report?
b What information would you need to conduct the experiment effectively?
c Identify the independent variable for the experiment.
d Identify the dependent variable for the experiment.
e List the controlled variables stated in the procedure.
f State the purpose of this experiment.
g Why was it important to use the set of test-tubes without cellulase?
h Suggest improvements to the design of this experiment.

7 Consider the following experiment:
A student sets up an ELISA experiment to test if a sample of blood serum has any antibodies that will react to an antigen found on housedust mites. Identify the type of variable that each of the following represents in the experiment:
a sample of blood serum
b analytical technique, reagents, temperature
c positive and negative samples

8 Dwarf pea seeds were germinated and transferred into three pots with potting mix, five plants per pot. After one week, plants were sprayed with either dH_2O, 0.01% GA (a plant hormone) or 0.1% GA. At week three, a 5 mm section of internode was cut from a plant stem, placed on a microscope slide and sliced lengthwise. Three drops of 1 M HCl were added to the tissue and the slide placed on a 60 °C hotplate for two minutes. Two drops of toluidine blue stain were added for two minutes, then a coverslip was placed on the tissue and gently pressed down. The slide was viewed under the microscope at ×100 magnification. Two stems from each pot were stained and viewed in this manner. The procedure and the results are shown below:

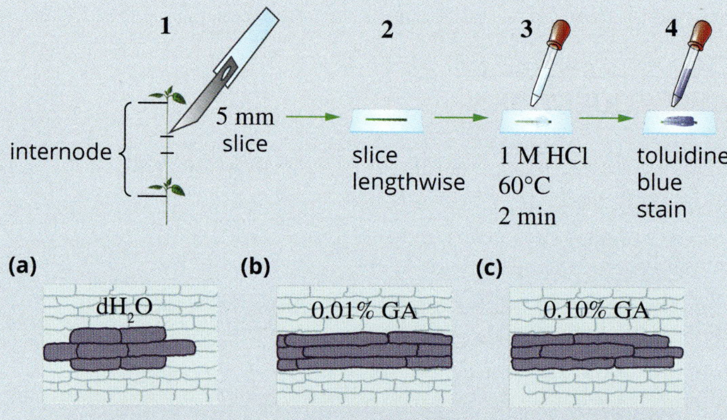

a Discuss whether the experimental design, materials and procedures were described clearly enough. For example, are there any missing experimental details needed to repeat the experiment?
b How would you interpret the results?
c Write a conclusion for the experiment.

9 What are the meanings of the following chemical codes?

a **b**

c

10 Convert the following units:

- **a** 70.60 mL into L
- **b** 0.015 g to mg
- **c** 10 μL to mL

11 Define the following terms:

- **a** mean
- **b** mode
- **c** median

12 The measurements of mass (g) of a compound were 16.05, 15.98, 16.80, 15.92, 16.20 and 17.01. What is the uncertainty of the average of these values?

13 Which of the statistical measurements of mean, mode and median is most affected by an outlier?

14 Identify whether the following are mistakes, systematic errors or random errors.

- **a** A student miscounts the number of cells in the haemocytometer.
- **b** The reported measurements are above and below the true value.
- **c** An electronic scale consistently reads 2 g more for each sample.

15 A student is confused about rules for significant figures. Summarise the following for the student:

- **a** Give examples of values with 2, 3, 4 and 5 significant figures.
- **b** Summarise the rules for reporting a calculated value that involved multiplication or division.
- **c** Summarise the rules for reporting a calculated value that involved addition or subtraction.

16 Answer the following questions.

- **a** Explain the terms 'accuracy' and 'precision'.
- **b** When might an investigation be invalid?
- **c** All investigations have limitations. Use an example to explain the meaning of 'limitations' of an investigation procedure.

17 A scientist designed and completed an experiment to test the following hypothesis: 'Increasing the temperature of water would result in a decrease in the measured pH of water.'
The discussion section of the scientist's report included comments on the reliability, validity, accuracy and precision of the investigation.
Determine which of the following sentences comment on the reliability, validity, accuracy or precision.

- **a** Three water samples from the same source were examined at each temperature. Each water sample was analysed and the measurements were recorded.
- **b** The temperature and pH of the water samples were recorded using data-logging equipment. The temperature of some of the water samples was measured using a glass thermometer.
- **c** The data logging equipment was calibrated for pH before use. The equipment was calibrated before measurements were taken.
- **d** The temperature probe (data logger) measured temperature to the nearest 0.1°C. The glass thermometer measured temperature to the nearest 1°C.

18 What is the purpose of referencing and acknowledging documents, ideas, images and quotations in your investigation?

19 A scientist designed and completed an experiment to test the following hypothesis: 'If there is a negative correlation between water temperature and pH, then water that is heated to 100°C will have a lower pH than water that is cooled to 5°C'

- **a** Write a possible purpose for this scientist's experiment.
- **b** What would be the independent, dependent and controlled variables in this investigation?
- **c** What kind of data would be collected? Would it be qualitative or quantitative?
- **d** List the equipment that could be used and the type of precision expected for each item.
- **e** What would you expect the graph of the results to look like if the scientist's hypothesis was correct?

20 What can you conclude from the visual model of a stained peripheral blood smear, shown below.

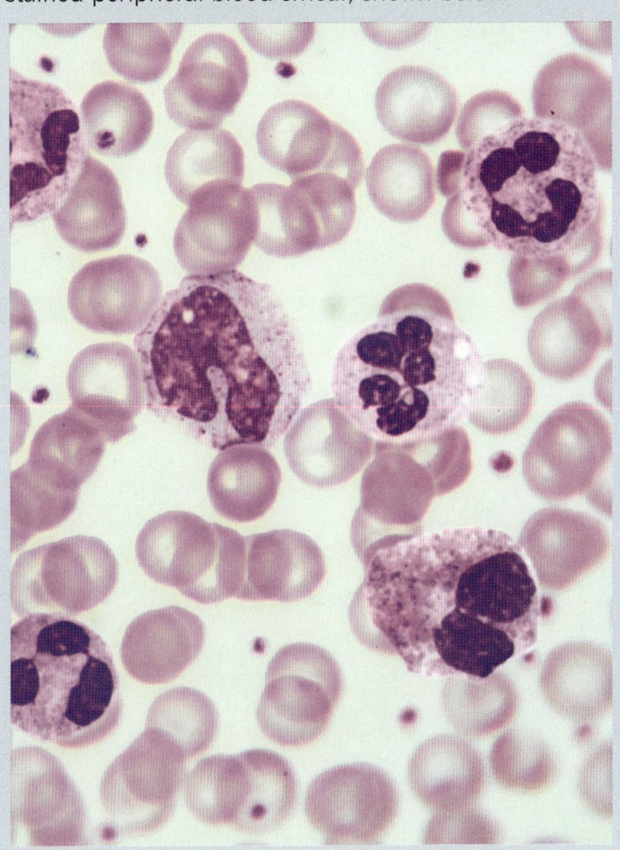

21 A student wants to draw some conclusions for the evolutionary history of bears. The student found an article published in a scientific journal with the phylogenetic tree, shown below.

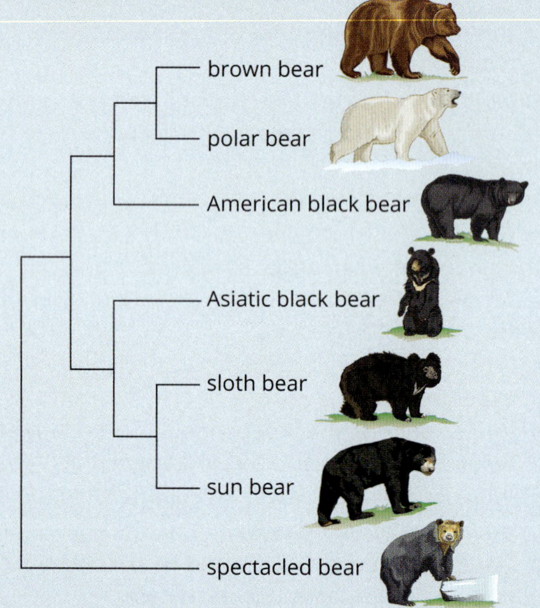

a Is the source of the article credible, and explain why/why not?

b Which conclusion can the student draw about the Asiatic black bear?

i The Asiatic black bear is more closely related to the American black bear than the spectacled bear.

ii There is no common ancestor for all the bears.

iii The sun bear diverged from the polar bear.

22 Look at the following graph below to answer the following questions about the photosynthetic pigments.

(a)

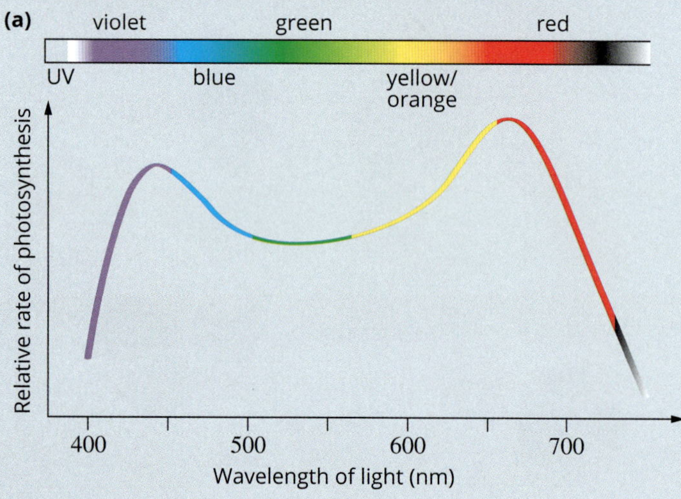

(b)

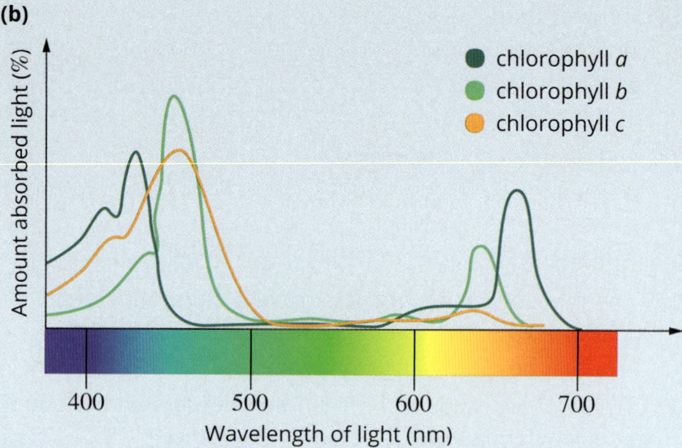

a Select the correct response for carotenoid:

i Carotenoid absorbs light above 550 nm.

ii Carotenoid absorbs light below 550 nm.

iii Carotenoid absorbs all light in the visible light spectrum.

b Select the correct response for chlorophyll *b*:

i Chlorophyll *b* only absorbs light below 550 nm.

ii Chlorophyll *b* absorption peak is above 550 nm.

iii Chlorophyll *b* absorption peak is below 550 nm.

23 A student wanted to test the effects of temperature on fermentation/cellular respiration in yeast. Look at the experimental set up shown below:

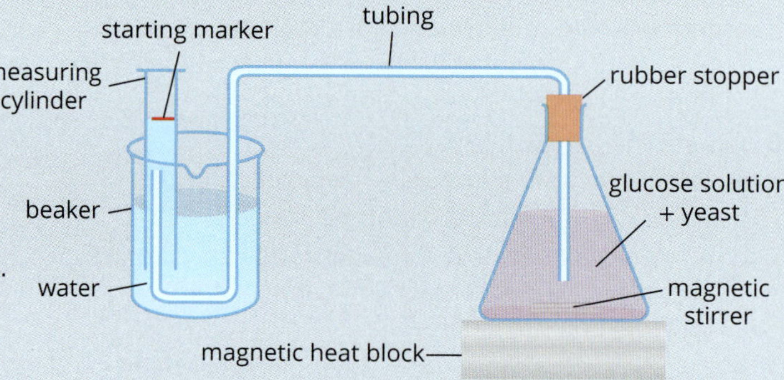

a What is the research question for this experiment?

b What is the product of respiration that can be tested?

c What would the height of space in the cylinder in the test samples 25°C, 35°C, 45°C and 55°C tell us about the rate of respiration in yeast?

24 A student designed an experiment to observe what would happen to the reaction rate of an enzyme if he added in different types of inhibitors. His set up included:

1 tube: enzyme + substrate

1 tube: enzyme + substrate + competitive inhibitor

1 tube: enzyme + substrate + non-competitive inhibitor

The results are presented in the below graph. Recall that competitive inhibitors compete for the same site as the substrate whereas non-competitive inhibitors allow the substrate to bind but don't allow the reaction to take place. V_{max} represents the maximum velocity of the enzyme.

Enzyme reaction rate in the presence of different inhibitors

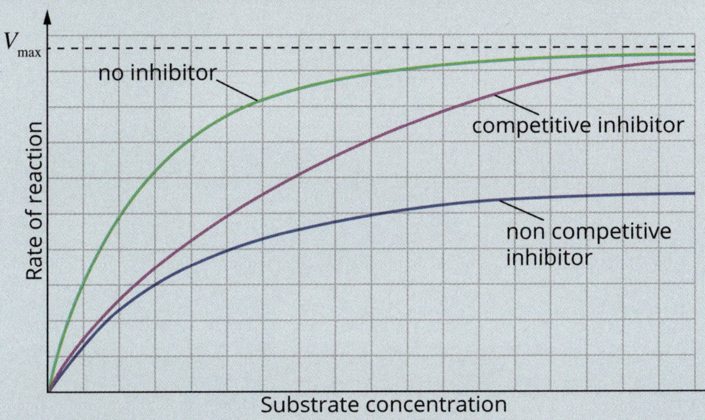

a Describe what happens when a competitive inhibitor is added to the enzyme.

b Describe what happens when a non-competitive inhibitor is added to the enzyme.

25 This flow chart provides an overview of the stages of aerobic cellular respiration. List the limitations of this flow chart.

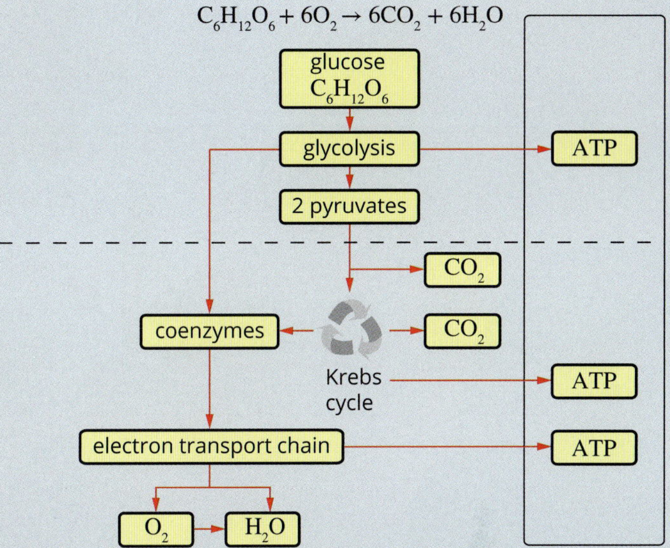

26 Research the effect that increasing global temperatures are having on the acidity of seawater and coral bleaching in the Great Barrier Reef. Ensure that you include a list of references and acknowledgements for your sources. Present your research in digital form.

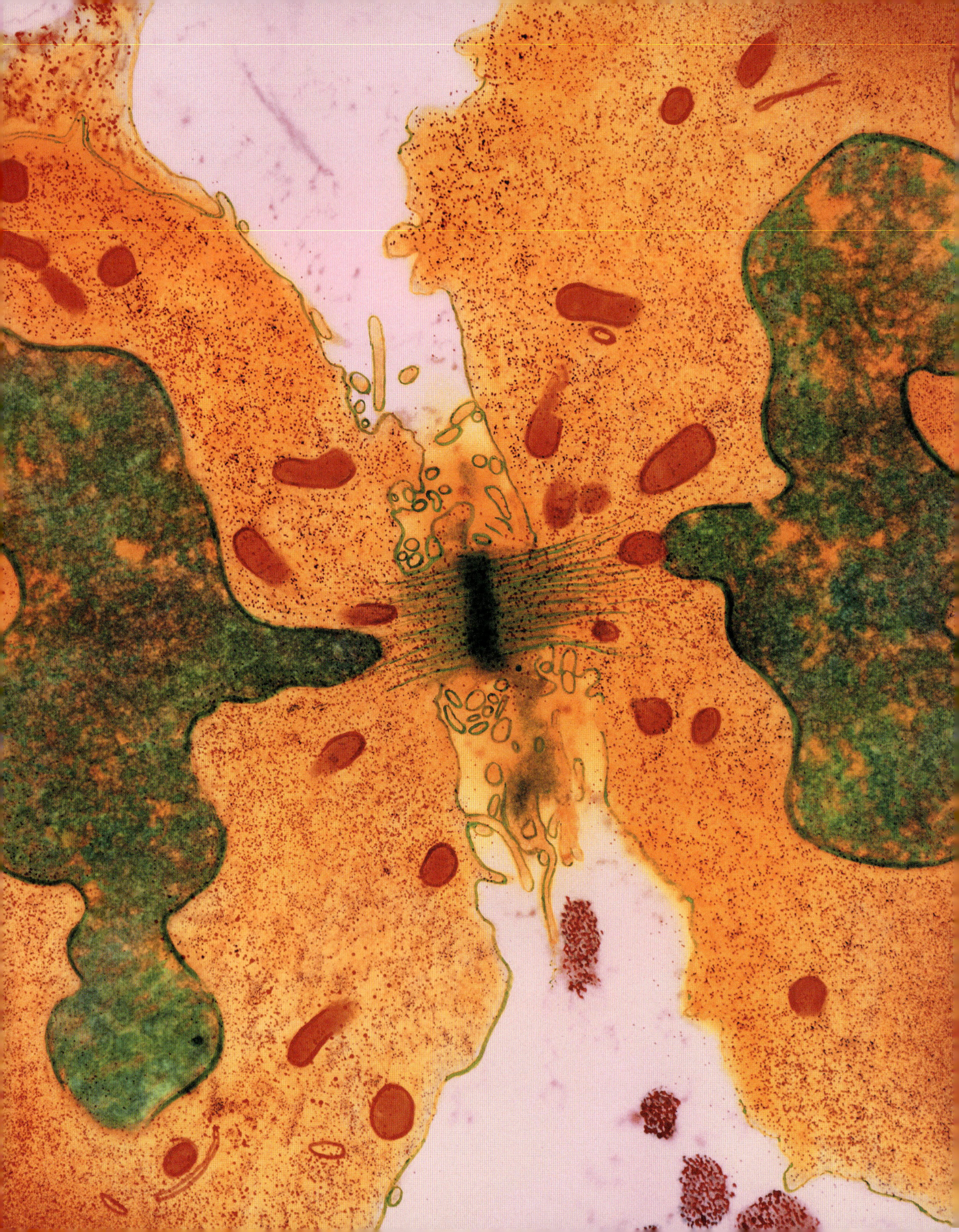

5 Heredity

Life continues through the processes of reproduction and heredity. You will expand your knowledge of evolution by understanding the cellular processes involved in increasing genetic diversity. You will investigate reproduction and inheritance patterns in both plants and animals as well as the role of DNA in polypeptide synthesis and the uses of technologies in the study of inheritance patterns.

You will also learn about contemporary research and the work of geneticists across a variety of industries, including medical applications and agriculture. You will explore the effects on society and the environment through the application of genetic research.

Outcomes

By the end of this module you will be able to:

- select and process appropriate qualitative and quantitative data and information using a range of appropriate media BIO12-4
- analyse and evaluate primary and secondary data and information BIO12-5
- solve scientific problems using primary and secondary data, critical thinking skills and scientific processes BIO12-6
- explain the structures of DNA and analyse the mechanisms of inheritance and how processes of reproduction ensure continuity of species BIO12-12

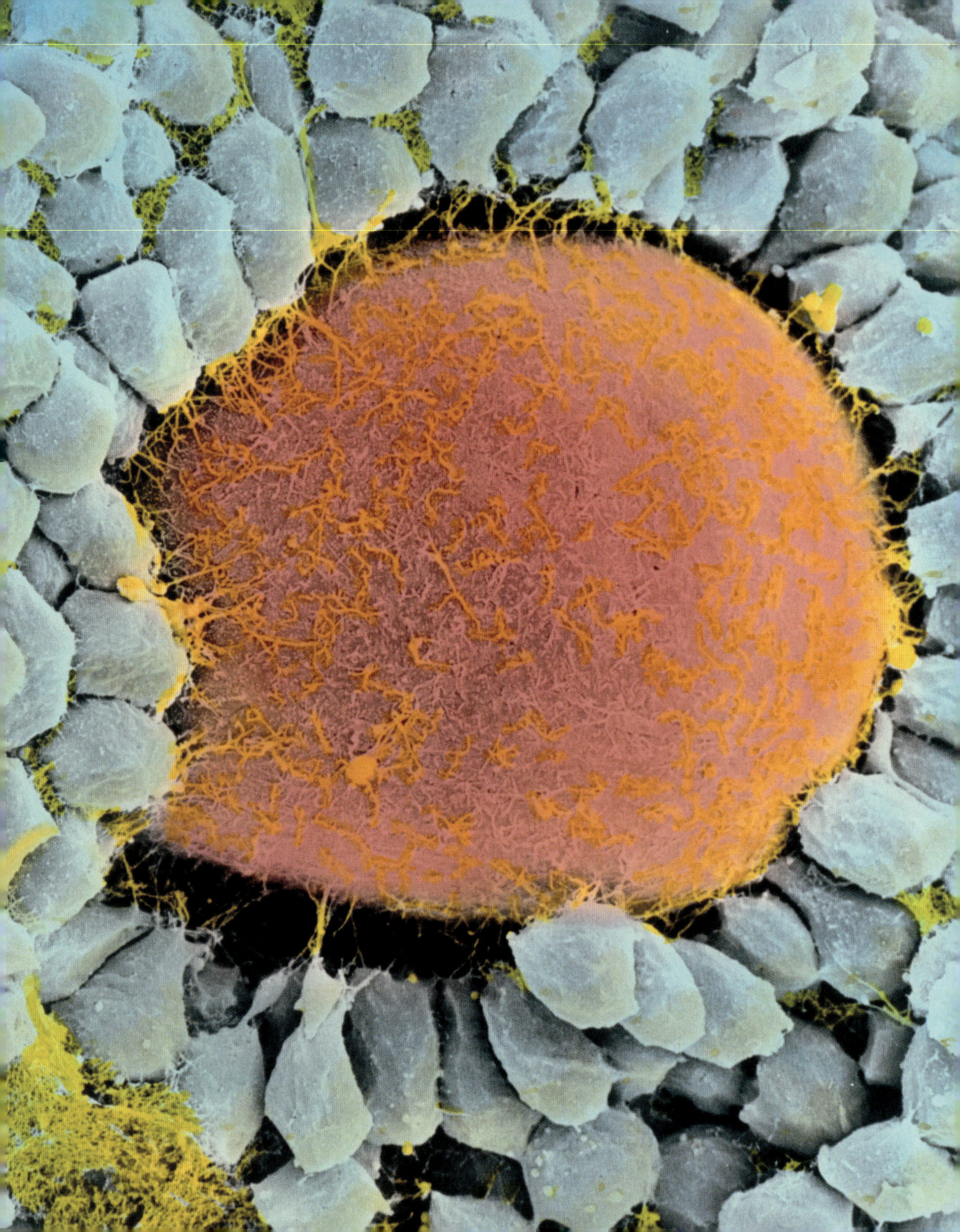

CHAPTER 02

Reproduction

This chapter examines how the continuity of a species is maintained through reproduction and the different mechanisms that organisms use to reproduce. You will learn about the advantages and disadvantages of different modes of reproduction, including internal and external fertilisation in animals and the various forms of asexual and sexual reproduction in animals, plants, fungi, bacteria and protists. You will analyse the features of reproduction in mammals, from fertilisation, implantation and pregnancy to the birth of their young. Lastly, you will critically examine the issues associated with the manipulation of plant and animal reproduction in agriculture.

Content

INQUIRY QUESTION

How does reproduction ensure the continuity of a species?

By the end of this chapter you will be able to:

- explain the mechanisms of reproduction that ensure the continuity of a species, by analysing sexual and asexual methods of reproduction in a variety of organisms, including but not limited to:
  - animals—advantages of external and internal fertilisation
  - plants—asexual and sexual reproduction
  - fungi—budding, spores
  - bacteria—binary fission (ACSBL075)
  - protists—binary fission, budding
- analyse the features of fertilisation, implantation and hormonal control of pregnancy and birth in mammals (ACSBL075) CCT EU
- evaluate the impact of scientific knowledge on the manipulation of plant and animal reproduction in agriculture (ACSBL074) EU L

2.1 Asexual reproduction

WS 5.1

Asexual means 'without sex'.

Clones are offspring that are genetically identical. Clones produced by asexual reproduction are genetically identical to their parent.

Individual organisms do not live forever. The continuity of life from generation to generation is the result of reproduction. Reproduction is one of the distinctive characteristics of living organisms.

The simplest way that organisms can reproduce is asexually. **Asexual reproduction** is the production of identical offspring from just one parent. It produces new individuals or offspring by **mitosis**, in which each daughter cell receives a copy of every **chromosome** of the parent cell. The offspring are therefore **clones**, that is, they are genetically identical to the parent unless genetic **mutations** occur.

Asexual reproduction occurs in every major group of life—Archaea, Bacteria, Protista and Fungi, as well as in many plants (Plantae) and some animals (Animalia) (Figure 2.1.1). Some species are capable of both asexual and **sexual reproduction**, either at different stages of their life cycle or under different environmental conditions. In this section you will learn about the different methods of asexual reproduction, the types of organisms that use these methods to reproduce and evaluate the advantages and disadvantages of asexual reproduction for the continuity of species.

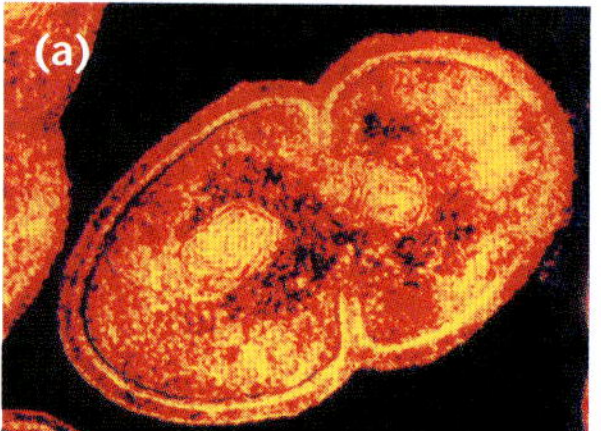

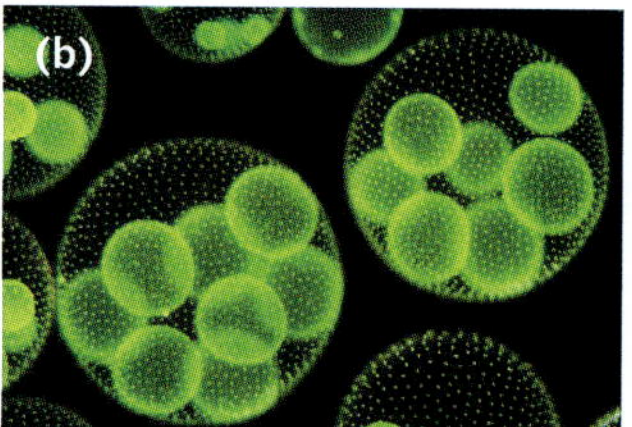

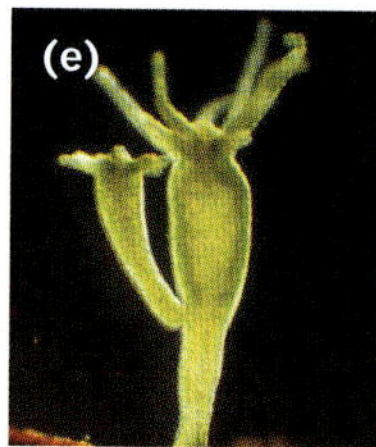

FIGURE 2.1.1 Asexual reproduction occurs in every kingdom of life in many different forms. (a) A coloured transmission electron microscope (TEM) image of a single bacterium cell dividing into two by the asexual process called binary fission. (b) The daughter colonies of cells in this protist (*Volvox aureus*) are copies that are reproduced asexually inside the main parent colony. (c) Pin moulds (*Mucor* species), like all fungi, reproduce asexually with spores released from capsules. (d) A single potato is a tuber with several growth points on it, known as eyes. Each eye can produce a sprout that grows into a whole new plant in an asexual process. (e) The *Hydra* is a small, multicellular animal common in freshwater. This magnified view shows one undergoing budding. The new individual is the bud on the left and will detach when fully grown.

ADVANTAGES AND DISADVANTAGES OF ASEXUAL REPRODUCTION

Asexual reproduction has short-term benefits, enabling rapid population expansion, but the lack of genetic variation in asexual populations limits their adaptability and evolutionary potential in the long term.

Mutation of DNA is the only mechanism that introduces genetic variation into an asexually reproducing population.

Asexual reproduction is an efficient and simple way to reproduce because individuals do not need to find a mate and can quickly reproduce large numbers of offspring in isolation. It uses the cell division process of mitosis, which is less demanding on an organism. The asexual reproductive strategy works well when environmental conditions are ideal and relatively static. For a species already well suited to a stable environment, there is no disadvantage in producing offspring with the same inherited traits. Asexually reproducing organisms are therefore commonly found in relatively stable and uniform environments to which they are well suited.

However, when environmental conditions are variable, asexually reproducing populations are at a disadvantage. Because asexual organisms are genetically the same, they will all respond to the environment in the same way. Lack of genetic variation in a population means that there will be no unusual individuals that may be able to tolerate new selection pressures in changed environmental conditions. As a group, they will either survive or die.

The advantages and disadvantages of asexual reproduction are summarised at the end of this section (Table 2.1.2, page 68).

METHODS OF ASEXUAL REPRODUCTION

Asexual reproduction is the most common method of reproduction for unicellular organisms. This is because there is no cell specialisation in unicellular organisms, so there are no reproductive organs or **germ cells** to produce **gametes**.

Many multicellular organisms also have the capacity to reproduce asexually. In multicellular organisms, the new individual arises from body cells, known as **somatic cells**.

Asexual reproduction takes place in a variety of organisms and can be by:

- fission—splitting of one cell into two (**binary fission**) or many (multiple fission), of equal size
- **budding**—outgrowths from a parent cell, each smaller than the parent
- **fragmentation** of body parts
- **spore** formation
- **vegetative reproduction** in plants
- **parthenogenesis**—in some female animals.

Germ cells produce gametes (e.g. sperm and eggs) by meiosis with half the chromosome number for sexual reproduction. Somatic cells are the normal body cells with a full set of paired chromosomes produced by mitosis and in asexual reproduction.

BACTERIA

Bacteria are unicellular, microscopic prokaryotes that reproduce asexually (Figure 2.1.2). You learnt about bacteria in Year 11. Fossil evidence has confirmed that bacteria were the first type of living organism on Earth, evolving at least 3.7 billion years ago. Bacteria have very diverse cell chemistry (metabolism) and can survive in a wide range of environments. They play an important role in ecosystems because they decompose many substances, including plant and animal remains and wastes. There are specialist bacteria that fix atmospheric nitrogen into a form that plants can use. Bacteria cause many diseases for plants and animals, such as tomato wilt, citrus canker, tetanus, cholera and toxic shock syndrome. Humans use specific bacteria in safe ways to manufacture foods such as cheese and yoghurt and medicines such as antibiotics, enzymes and prescribed drugs including human insulin for diabetics.

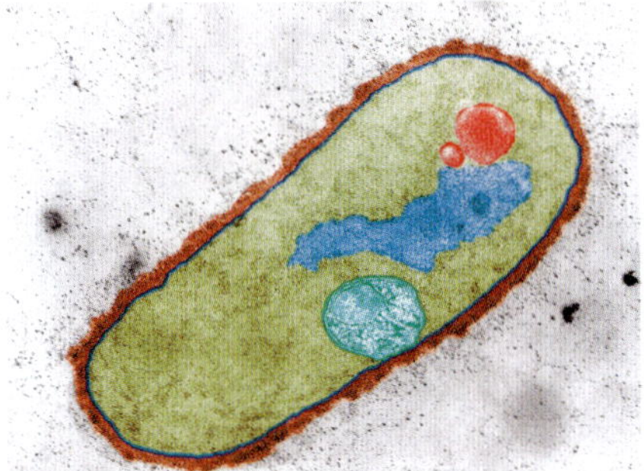

FIGURE 2.1.2 Coloured TEM of a single bacterium

Year 11 Chapter 2

Prokaryotic and eukaryotic cells

Prokaryotes are unicellular organisms without a membrane surrounding a nucleus and they lack most organelles inside their cells. They only have ribosomes and a circular loop of single-stranded DNA located in the nucleoid region of the cytoplasm. Some also have plasmids, which are small rings of double-stranded DNA. All Archaea and Bacteria are prokaryotic organisms.

In humans, all cells are of the alternative eukaryotic type. Protists, fungi and plants are also eukaryotes. A eukaryotic cell always has organelles bound by membranes, in particular the nucleus containing chromosomal DNA.

Reproduction by binary fission in bacteria

Due to the lack of organelles and smaller amount of DNA, cell replication in prokaryotes (Archaea and Bacteria) occurs more quickly than in eukaryotes. It is not the same process as mitosis because there is no nucleus in a prokaryotic cell. This asexual process is called binary fission because a single parent cell splits into two approximately equal daughter cells. Binary fission is a relatively rapid form of reproduction that produces new cells genetically identical to the parent (binary = two; fission = splitting) (Figure 2.1.3).

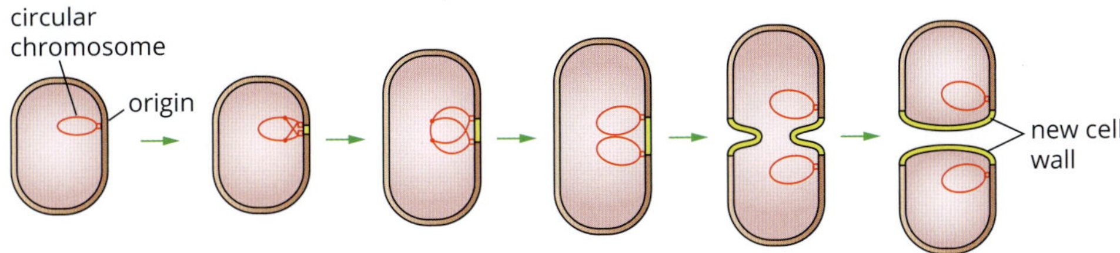

FIGURE 2.1.3 Division of a prokaryotic cell by binary fission. Replication of the single circular chromosome begins at a point called the origin. When the parent cell has grown large enough, a new cell wall and membrane are laid down to divide the cell into two genetically identical smaller daughter cells.

Exponential growth by cell replication:

$C = 2^n$

where

C = number of cells

n = number of cell replications that have occurred

Like mitotic division, binary fission is an exponential process because the population doubles after every cycle of division. In ideal conditions, some bacteria can undergo binary fission every 20 minutes, meaning that the number of cells can double every 20 minutes. This means that in six hours up to 18 cycles of binary fission could occur and in this time one bacterium could have produced a population of 262 144 individuals.

BIOFILE CCT

Cloning bacteria

One consequence of asexual reproduction is that, unless there are mutations, all organisms in a colony of bacteria are clones; that is, they are genetically identical (Figure 2.1.4). Scientists have used this knowledge to design drugs (antibiotics) that kill bacteria of a specific type. An antibiotic treatment will kill all or almost all members of a bacterial colony with which it comes into contact. But if even one bacterium has acquired antibiotic resistance by mutation, it can survive and produce a drug-resistant population.

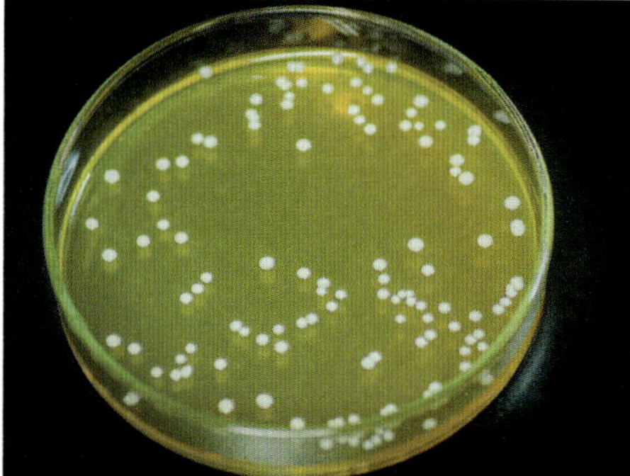

FIGURE 2.1.4 Colonies of bacteria growing on an agar plate. Each bacterium in a colony is formed from binary fission and is genetically identical to the others.

BIOLOGY IN ACTION CCT WE

Growing bacteria in pathology laboratories

Pathology laboratories put the exponential reproduction process of bacteria to good use when they culture them to diagnose a disease. A test sample from a patient is usually a swab from an infected site or body fluid, like blood or urine. Using sterile techniques, the sample from the patient is spread onto solid nutrient agar then incubated to encourage the bacterial cells to reproduce.

The nutrient agar, an extract from seaweed, is prepared using a specific recipe for encouraging rapid growth of the bacteria being tested and poured into Petri dishes where it sets like a firm jelly.

In the laboratory, ideal conditions are provided for rapid reproduction of the bacteria by providing suitable nutrients and growth factors in the agar, pH of 7.2–7.4 and incubation at the optimal temperature of 37°C in the dark. Under these conditions, colonies will rapidly reproduce by binary fission and grow large enough to allow diagnosis within 48 hours. Prior to incubation, the bacterial cells are too small to be seen with the naked eye. Under the right conditions, each bacterial cell will reproduce exponentially, forming colonies that can be easily seen and identified with the naked eye. Each colony is a clump of thousands of identical cells produced from one bacterial cell by binary fission.

To avoid contamination of samples and exposure to potentially pathogenic bacteria, strict safety precautions and sterile techniques need to be maintained in a pathology laboratory and in school laboratories when undertaking experiments to culture bacteria (Figure 2.1.5).

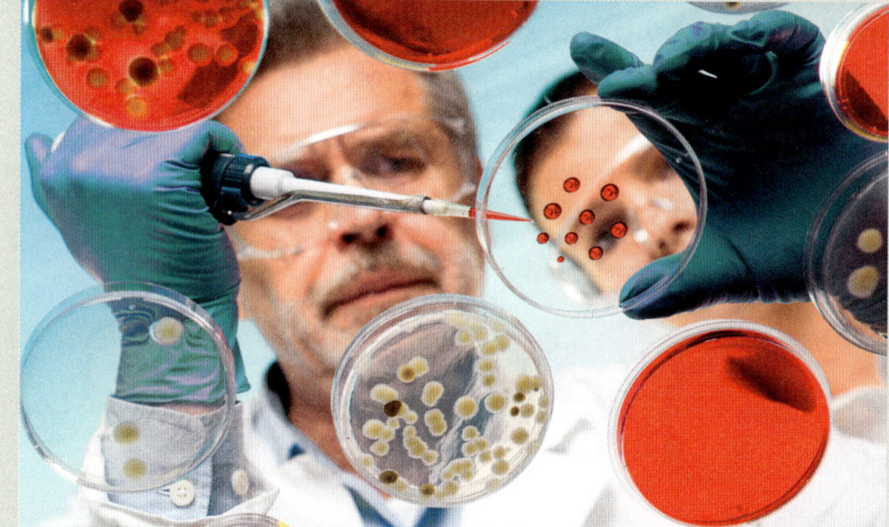

FIGURE 2.1.5 A microbiologist works with a variety of nutrient agar plates. When working with bacteria in a laboratory, it is important to wear protective equipment and use sterile techniques to avoid contamination and exposure to potentially pathogenic bacteria.

PROTISTS

All protists are eukaryotes that live in aquatic or moist environments. They can be unicellular or multicellular, but if multicellular they are not multi-tissued (i.e. their cells are not organised into functional tissues). Protists can reproduce both asexually and sexually. This extensive diversity within Protista makes it difficult to classify them as one group. Currently, protists are classified as any eukaryotic organism that is not a fungus, plant or animal. Because protists do not fit into any of the other kingdoms, they have been temporarily grouped together until there is more taxonomic knowledge available.

Protists include protozoans such as *Euglena* (Figure 2.1.6a), water moulds, slime moulds, dinoflagellates (Figure 2.1.6b), diatoms (Figure 2.1.6c) and some algae. Most protists are free-living but some are parasites (e.g. plasmodia that cause malaria and *Giardia* that causes intestinal disease with abdominal pain, diarrhoea and nausea when a person is infected from contaminated water; Figures 2.1.7b and 2.1.7d).

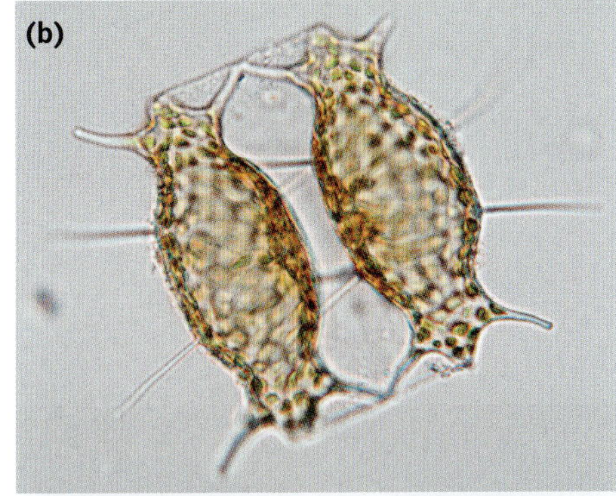

FIGURE 2.1.6 Protists are a very diverse group of organisms. (a) Freshwater *Euglena* (with flagella and chloroplasts) and some cilates (with covering of tiny cilia) swim together. (b) Microscopic view of two marine dinoflagellates (*Dinophysis* sp.). (c) Diatoms are single-celled algae that form an important part of the plankton at the base of marine and freshwater food chains. There are approximately 10 000 species of diatoms.

Reproduction by binary fission in protists

The typical mode of reproduction for most of the protists is asexual binary fission. The binary fission of protists is different to that of bacteria because protists have a membrane-bound nucleus that needs to be replicated. For protists without a cell wall, the body of an individual is simply pinched into two parts or halves; the parent cell disappears and is replaced by a pair of daughter nuclei in two new cells, although these may need to mature to be recognisable as members of the parental species. The length of time needed for binary fission varies among groups of organisms and with environmental conditions, ranging from a few hours in an optimal situation to many days. In protists the binary fission can be along a transverse or longitudinal axis or across a diameter, dividing the cytoplasm in half (Figure 2.1.7a–c).

Multiple fission is common for some parasitic protist species. The nucleus divides repeatedly to produce a number of daughter nuclei before division of the rest of the cell occurs. Eventually these daughter nuclei become the nuclei of multiple progeny after repeated cellular divisions (Figure 2.1.7d).

It is interesting to consider that after fission occurs, the parent cell no longer exists but neither has it died. It has redistributed itself to the daughter cells. In other forms of unicellular asexual reproduction (e.g. budding and spores) the parent cell continues with its life and death cycle.

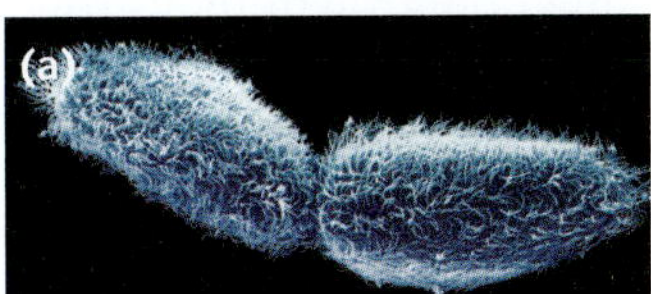

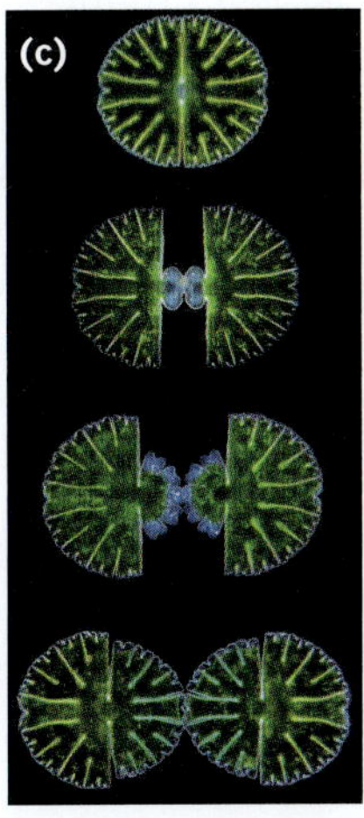

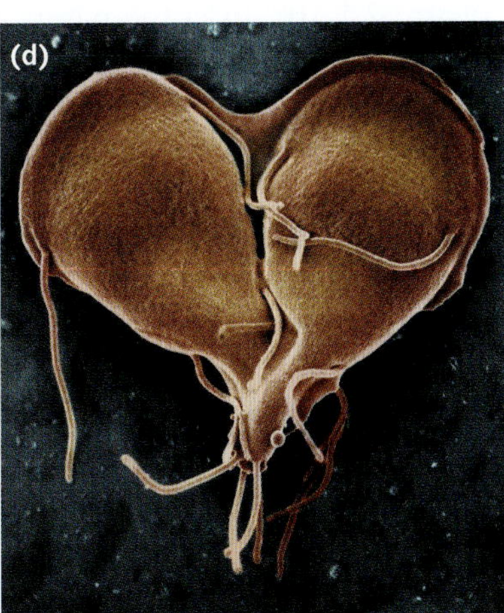

FIGURE 2.1.7 (a) Transverse fission in *Paramecium*, a free-swimming protist of the ciliate group. (b) Multiple fission of *Plasmodium falciparum*, the protist that causes malaria in humans (TEM). The parasite has divided multiple times inside a red blood cell. (c) Cell division of a desmid, a single-celled green algae. (d) Scanning electron microscope (SEM) image of longitudinal fission in *Giardia lamblia*. The protist is about to become two separate organisms during a late stage of cell division by binary fission. *Giardia lamblia* causes intestinal disease with abdominal pain, diarrhoea and nausea, spread through contaminated food and water.

Reproduction by budding in protists

Although fission is the most common form of asexual reproduction among protists, there are also other methods such as budding. Budding occurs when a new identical organism grows from the body of the parent. This usually occurs on the outside of the cell from which it detaches to live independently or sometimes remains in contact to form a colony. In some species a bud forms internally. The new nucleus is formed in a similar way to fission but, unlike fission, the division of the cytoplasm is unequal. At first the new organism is much smaller than the parent.

FUNGI

The kingdom Fungi includes species that are commonly known, like mould (Figure 2.1.8), mildew, mushrooms, yeasts, lichens (a symbiotic organism of fungi and algae) and truffles (Figure 2.1.9). Fungi are composed of eukaryotic cells that secrete enzymes over the surface of their food and absorb the breakdown products directly. Some fungi reproduce asexually by spores released from fruiting bodies. Mould, mushrooms and puffballs are examples of fungi that reproduce by spores (Figure 2.1.8). Other fungi, such as yeasts, reproduce by budding (Figure 2.1.10).

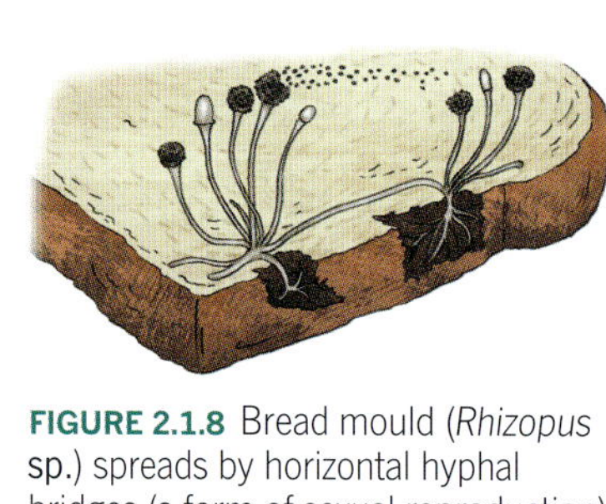

FIGURE 2.1.8 Bread mould (*Rhizopus* sp.) spreads by horizontal hyphal bridges (a form of sexual reproduction) and by asexual spores released from mature fruiting bodies (the black spore capsules). Bread moulds have dry spores that disperse through the air.

FIGURE 2.1.9 Truffles are the reproductive body of an underground fungus that grows in a symbiotic relationship with the roots of some trees. Small quantities are prized for their flavour in fine dining.

FIGURE 2.1.11 (a) Conidiophores of blue mould (*Penicillium expansum*), which produce spores called conidia by budding. (b) Blue mould on a rotting lemon. The spores are visible on the surface of the fruit; they are white at first but later turn blue-green.

Reproduction by budding in fungi

Most yeasts reproduce asexually by the asymmetric division process called budding (Figure 2.1.10). First the parent yeast cell produces a small outgrowth that grows larger and forms a bud. The nucleus of the parent cell splits off a smaller daughter nucleus, which migrates into the daughter cell. The bud detaches from the parent by pinching inwards at the base, although it may remain in contact with the parent cell. At this stage the bud is much smaller than the parent but is genetically identical. Repeated budding forms a chain of connected but independent cells.

The clouds of blue-green powder that come off the surface of a mouldy lemon are a type of asexually reproduced spore of blue mould (*Penicillium expansum*). These spores, produced by budding, are called conidia (Figure 2.1.11). In *Penicillium expansum* these spores are formed in long chains.

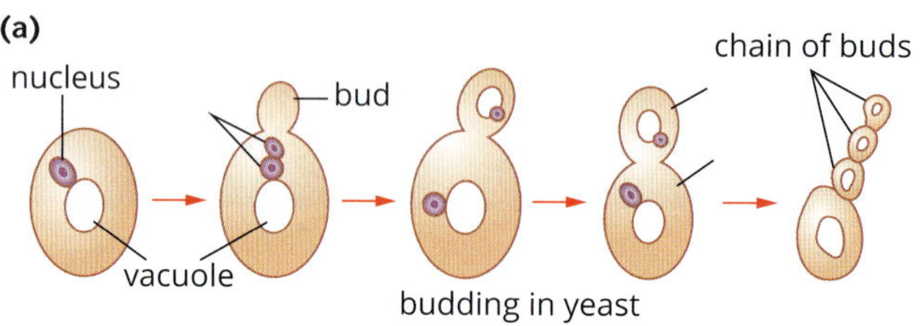

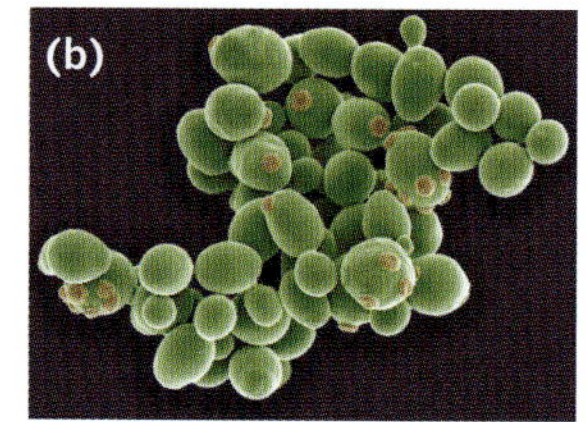

FIGURE 2.1.10 (a) A yeast cell forming a chain of identical but smaller cells by budding. The main nucleus produces a smaller daughter nucleus that migrates into the bud before pinching off. (b) SEM of brewer's yeast cells budding. The brown spots on the surface of some cells are buds that will grow and separate into identical, independent daughter cells.

Reproduction by spores

Spore formation is the method of reproduction most commonly associated with fungi. Spore formation can be either asexual or sexual. Spores that are produced asexually are called **mitospores** and are produced by mitosis. Spores produced sexually by **meiosis** are called **meiospores**. You will learn more about meiospores in Section 2.2.

Mitospores are **haploid** (one set of single chromosomes) reproductive cells capable of developing into an adult without fusion with a second cell. The daughter cells are genetically identical to the parent fungus and are usually produced in large numbers. These spores are cells that are encased in a protective coating that enables them to survive in unfavourable environments (Figure 2.1.12).

Some fungi produce a cluster of spores inside a structure called a **sporangium** (plural sporangia) (Figure 2.1.13). The spores are released when the sporangium wall disintegrates and are dispersed by wind or water. When a spore lands in a suitable environment it germinates, forming a new fungus. Spore formation and dispersal can rapidly increase identical fungal cells, expanding and spreading the population of a fungal species.

FIGURE 2.1.12 Puffball fungus releasing a cloud of microscopic mitospores

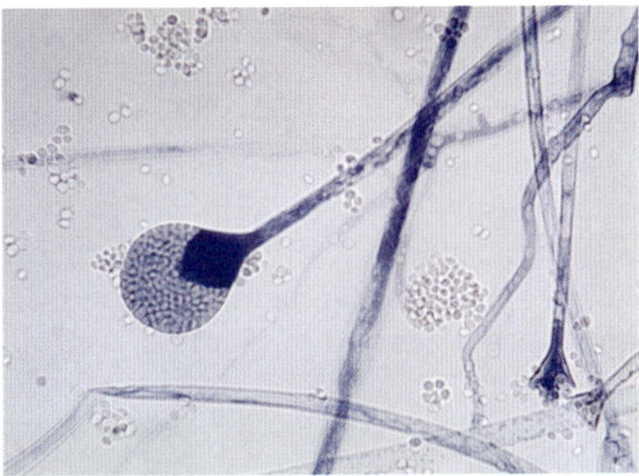

FIGURE 2.1.13 Sporangium of a mature *Mucor* fungus containing mitospores. When the wall of the sporangium disintegrates, the spores will be released. Broken sporangia can be seen in the lower right corner of the image, along with released spores in the background.

Mitosis is cell division where two daughter cells are produced that are identical to the parent cell. It is usually used for growth and repair.

Meiosis is cell division where four non-identical daughter cells are produced, each with half the chromosome number of the parent cell. It is used for production of gametes (sex cells).

Chromosomes that carry the genetic information in a somatic cell are matched in pairs. The number of chromosome pairs is called the diploid number ($2n$). Haploid is half ($1n$ or n) the diploid number because it is a set of single chromosomes, one from each pair.

BIOFILE CCT N S

Honey fungus—the largest living organism

The largest living organism in the world today is not a giant redwood tree or a blue whale. It is a type of mushroom, 3.8 km across, growing almost unseen in the Malheur National Forest in Oregon, USA. This parasitic fungus covers at least 965 hectares (close to 10 km^2), has an estimated biomass of 550 000 kg and is thought to be 2400 years old, but could be as ancient as 8650 years. It has been named Humongous Fungus and is classified as a species of honey fungus (*Armillaria solidipes*) (Figure 2.1.14). The *A. solidipes* colony is considered an individual organism because it is made up of genetically identical cells that communicate and have a common purpose.

Most of the time, the honey fungus reproduces sexually. The fungi begin their life as spores, released into the environment by a mature mushroom. When mycelia from genetically identical *A. solidipes* meet, they can fuse to form one individual. Their strange way of life has allowed them to endure through many centuries and thrive in such a way that one colony is now the largest single organism.

On the surface, the only evidence of the fungus is the fruiting bodies, clumps of golden mushrooms that pop up in the autumn with the rain. Underground there are mats of mycelium, which draw water and carbohydrates from the host tree to feed the fungus.

FIGURE 2.1.14 A group of fruiting bodies of the honey fungus (*Armillaria solidipes*) grows above ground on a fallen tree trunk.

PLANTS

The kingdom Plantae is the most conspicuous group of producer organisms on land. As with some protists and all the cyanobacteria, plants are autotrophs (self-feeders). Plants are multicellular and include small and structurally simple forms, such as mosses and liverworts, and large, more complex forms such as ferns and seed plants. Living seed plants include cycads, ginkgo, conifers and flowering plants, but in the current era flowering plants dominate the landscape. Plants use both asexual and sexual modes of reproduction. Flowering plants produce seeds by sexual reproduction. Other plants reproduce by a variety of asexual means and some species are capable of both.

Natural vegetative reproduction

Many members of the kingdom Plantae, including flowering plants, can reproduce asexually by vegetative reproduction. This does not involve the formation of seeds or spores. Instead it is the growth of specialised plant tissues that can grow into a new plant if it becomes separated from the parent plant. Naturally occurring vegetative reproduction may arise from many parts of the plant, such as the leaves and stems. An advantage of vegetative reproduction is that it can produce a rapid increase in the number of plants growing in a favourable area so that they outcompete, or displace, neighbouring species. In contrast, seeds produced by sexual reproduction may land in unfavourable conditions and fail to germinate. Potential disadvantages of vegetative reproduction are competition from sister and parent plants for resources, and lack of genetic variation to protect the population against disease or changing environmental conditions. Some types of vegetative reproduction in plants are described in Figure 2.1.15.

Rhizome

Underground stem that branches and gives rise to new shoots and roots. Well-known examples of plants with rhizomes are bracken fern, couch grass, irises and ginger.

Stolon/runner

Similar to rhizomes, but they grow above ground. Examples include strawberry plants, spider plants and spinifex grass. A spider plant with hanging stolons and new plantlets can be seen in the image.

Tuber

Swollen underground stems with buds ('eyes') that easily grow into new plants. Cassava, potatoes and *Begonia* are examples.

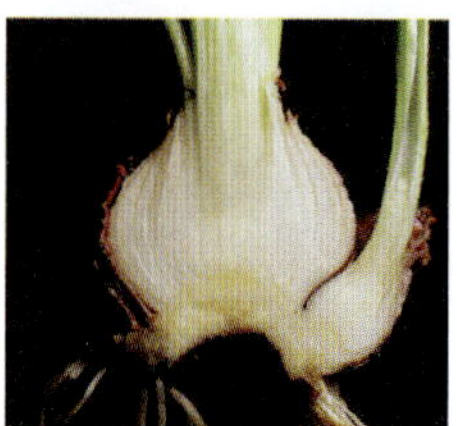

Bulb and corm

Produce lateral buds that develop into new plants. Examples include daffodil and hyacinth bulbs and gladioli corms. A hyacinth bulb (*Hyacinthus orientalis*) undergoing vegetative reproduction can be seen in the image.

Sucker

New shoots that arise from roots. Wattle trees, blackberries and bananas are examples.

Budding

An outgrowth (bud) of a new plant from the side of parent cells, much like the examples described earlier for protists and fungi. In plants, budding is unusual and can be seen in the *Kalanchoe* species and prickly pear.

Cutting/fragmentation

Broken pieces of branch regenerate into identical new plants.

Reproduction by means of cuttings falls under the general heading of fragmentation as a type of asexual reproduction. An example is *Salix babylonica* the weeping willow tree in Australia.

FIGURE 2.1.15 Common types of vegetative reproduction in plants

BIOFILE 

Weeds—rapid reproducers

A weed is often defined as a plant in the wrong place. Plants introduced to a new place without their usual ecological constraints keeping them in balance may become invasive. Three such examples in Australia are the weeping willow (*Salix babylonica*) (see Figure 2.1.15), prickly pear (*Opuntia* species) (Figure 2.1.16), and *Salvinia* (Figure 2.1.17). All have been able to spread unchecked because of their efficient methods of asexual reproduction.

Prickly pear (*Opuntia* species) is a type of prickly cactus that invaded large areas of Australia after it was introduced in 1788. It can reproduce asexually to grow new cloned plants directly from broken fleshy pads, as well as bear seeds in the red fruit from sexual reproduction.

FIGURE 2.1.16 The prickly pear (*Opuntia* species) grows new plants directly from the parent plant, enabling rapid reproduction.

Weeping willow (*Salix babylonica*) was first introduced into eastern Australia as a single female cloned specimen. From this specimen, the tree has become an invasive species because it reproduces asexually and disperses widely by means of broken fragments of branches floating along rivers and taking root on the banks. Pieces can travel many kilometres before establishing at a new site. Recreational fishers often break off twigs and stick them in the riverbank to hold their lines and these pieces will also grow.

Salvinia is a water plant capable of vegetative spread by breaking into daughter plants that grow when an abscission layer (where the leaf stem joins the plant stem) develops at each node. This process occurs very quickly in uncrowded, favourable growing conditions and many inland water areas in Australia have become choked with *Salvinia* (Figure 2.1.17).

FIGURE 2.1.17 *Salvinia*, an invasive waterweed in a Kakadu wetland, has spread rapidly by asexual production of daughter plants. Under favourable conditions, this plant can double its population size in two to five days by vegetative reproduction.

Artificial asexual propagation of plants

In the agricultural and horticultural fields propagation from fragments of plants, by grafting, cuttings or tissue culture, is advantageous for maintaining varieties that are genetically identical. It guarantees that the features of a desirable plant are preserved from generation to generation and can be produced in commercial quantities. Such techniques are also being used in research and recovery programs for endangered plant species, such as the round-leaved sundew (*Drosera rotundifolia*) (Figure 2.1.18) and the Wollemi pine (*Wollemia nobilis*) (Figure 2.1.19).

Methods of asexual propagation in plants are outlined in more detail in Section 2.4.

FIGURE 2.1.18 Round-leaved sundews (*Drosera rotundifolia*) growing from tissue cultures in a petri dish.

GO TO ➤ Section 2.4 page 99

ANIMALS

The kingdom Animalia includes an extremely diverse group of organisms, living in the sea, in freshwater and on land. All animals are multicellular. Most animals can move around to find food, nesting sites and mates. But many marine animals cannot, such as coral polyps, sponges and barnacles.

Asexual reproduction is less common in animals but it is not absent. The ability to move about means that animals are more likely to be able to find a mate and a suitable environment for sexual reproduction. When mates are not available or rapid identical reproduction is an advantage, asexual methods are used by some animals. Asexual methods of reproduction in animals include regeneration, fission, budding, fragmentation and parthenogenesis. Sometimes a species alternates asexual and sexual modes of reproduction, gaining the advantages of both while avoiding the disadvantages (Table 2.1.2, page 68).

BIOFILE N S

Wollemi pine—a living fossil

The Wollemi pine (*Wollemia nobilis*) (Figure 2.1.19) was thought to be extinct until 1994 when it was rediscovered in a remote region of the Blue Mountains in NSW. In nature this species reproduces sexually in a typical conifer reproductive mode of separate female and male cones with transfer of pollen through the air. Many people were interested in having one in their garden. Because it is so rare and its one known location was in need of protection, it was cloned for commercial sale using tissue culture and micro-propagation. The asexually cultivated pines have proved to be remarkably easy and fast to grow. Funds raised from the sale of the propagated specimens are used for protection of the rare pine in its natural habitat.

FIGURE 2.1.19 The Australian Wollemi pine (*Wollemia nobilis*) was thought to be extinct. After it was rediscovered in 1994, tissue culture was used to mass-produce pines for commercial sale. This kept the original location protected.

Regeneration

Regeneration occurs when a detached part of an individual grows into another individual. The echinoderms, which include sea stars, sea cucumbers and sea urchins, undergo tissue regeneration to grow replacement body parts. If the replacement is a large enough section with enough essential body tissues, the part could be regarded as reproduction of a new individual. However, the growing of a replacement arm for an otherwise intact sea star is not reproduction, but regeneration of a body part of the same adult. Likewise, the growing of a replacement tail for a lizard, a new leg for an axolotl, or replacement teeth for sharks, are not forms of reproduction.

Fission

As already described for unicellular organisms, longitudinal fission occurs when a cell splits along its longest axis and transverse fission occurs when the cell splits across its shortest axis. Fission can also occur in multicellular animals. A process similar to transverse fission is called **strobilation** in animals. A segment on the parent organism forms and when it matures it detaches to become a new individual (Figure 2.1.20).

Flatworms like planarians have been studied extensively for their ability to split and grow into new individuals. They can do this by longitudinal fission or fragmentation (Figure 2.1.21).

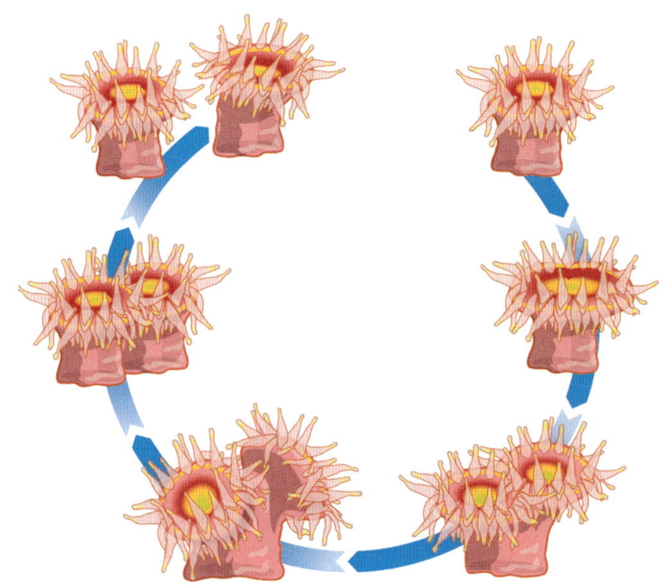

FIGURE 2.1.20 The jewel anemone (*Corynactis viridis*) undergoing longitudinal binary fission

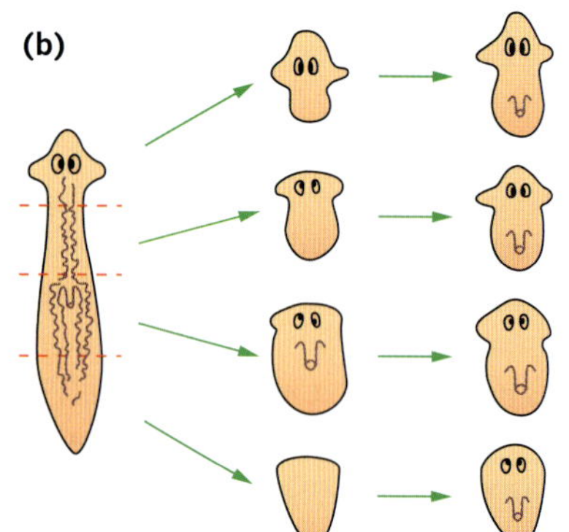

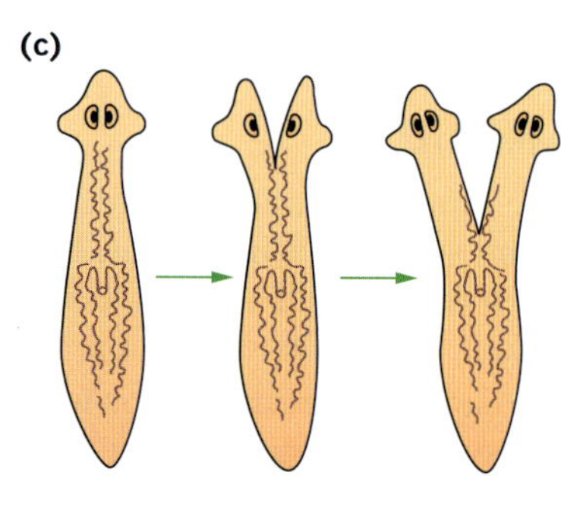

FIGURE 2.1.21 Planarians can regenerate completely after being fragmented or cut in half. (a) This marine planarian (flatworm) has a population of stem cells that enables it to reproduce or regenerate if body parts are severed. (b) One planarian that has been cut into fragments has reproduced four new *Planaria*. (c) A planarian that has been cut longitudinally through its anterior region can regenerate anterior regions, including a new head, faster than the posterior regions. If the planarian completely split, two new *Planaria* would form.

Budding

In a similar way to the fungi and unicellular protists, there are some animals that produce genetically identical but smaller buds as outgrowths from the parent body. The difference to fission is that there is an unequal division of cytoplasm with the buds being much smaller than the parent. Budding has been extensively documented in a small, multicellular animal called *Hydra* that is common in freshwater (Figure 2.1.22).

Sponges are aquatic animals that display asexual reproduction. In sponges this occurs by budding, which happens when a part of a sponge is broken off or one of its branch tips is constricted, and then this small piece grows into a new sponge. They may also reproduce asexually by producing multiple buds as packets of internal cells called **gemmules**, sometimes known as survival pods. If the parent sponge experiences damaging environmental conditions, such as drying out, the gemmules can remain dormant until conditions suit them to revive. Fragmentation and regeneration also occur in sponges (Figures 2.1.23, 2.1.24).

Like other simple organisms with long evolutionary histories, sponges exhibit sexual reproduction as well as asexual reproduction.

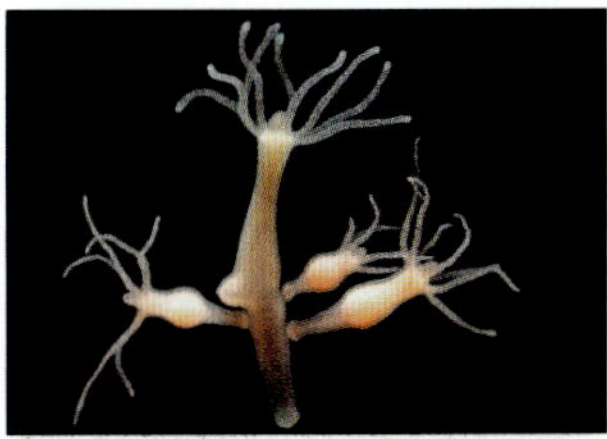

FIGURE 2.1.22 Parent *Hydra* with several buds destined to become new *Hydra*

FIGURE 2.1.23 A barrel sponge on a coral reef. The sponge reproduces asexually by budding and fragmentation.

BIOLOGY IN ACTION CCT S

A sponge experiment

An experiment with silicate-skeleton sponges, first performed by H.V. Wilson of the University of North Carolina, was published in a 1907 scientific journal. He separated the cells of a living marine sponge by forcing it through a fine silk cloth and catching the cells in a glass container of seawater. Most of the sponge separated into single cells. Wilson observed that these cells started behaving like individuals crawling around the bottom of the saucer. Then they began joining up to form small clumps of cells. Within days they grew to become whole new sponges (Figure 2.1.24).

His classic experiment has since been repeated many times by cutting sponges into sections, by sieving, by chemical dissociation of the cells (this happens when seawater has calcium and magnesium removed) and even by putting the sponge into a blender. Wilson's results proved reliable if the right conditions were provided and the separated cells included motile somatic cells called amoebocytes. The amoebocytes carry all the information needed to reassemble into new sponges.

Under natural conditions, when predators, currents or waves detach fragments of sponges, the pieces can reattach themselves to a suitable surface and rebuild over the course of several days (Figure 2.1.25). This provides great survival value for sponge species.

Knowledge of sponge regeneration has stimulated research in developmental biology and cytology, as well as continuing to fascinate people.

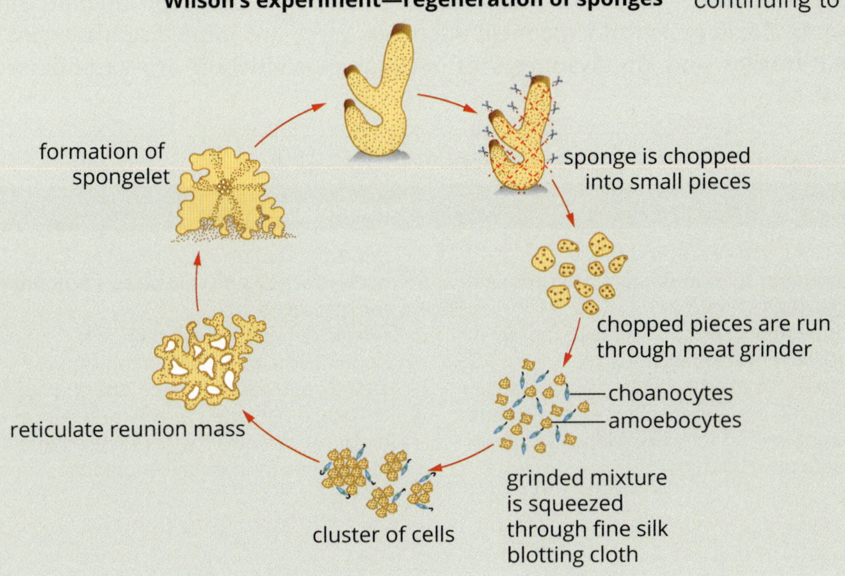

FIGURE 2.1.24 The basics of Wilson's experiment with fragmentation of sponges

FIGURE 2.1.25 Sponges and corals regenerate from fragments on mesh of an artificial reef.

Fragmentation

Fragmentation is similar to fission, but it happens in multicellular organisms. The body of the organism breaks into two or more parts, each of which regenerates the missing pieces to form a new, complete individual (Figure 2.1.24).

Fragmentation is common in some flatworms (Figure 2.1.21), marine worms and echinoderms. These organisms retain a population of **stem cells** throughout their life. The stem cells can develop into any cell type in the body, giving these organisms the remarkable ability to regenerate a body part that is completely lost through injury or even to reproduce new individuals.

Stem cell science is a prominent field of scientific and medical research. Stem cells are undifferentiated cells found in multicellular organisms. They have the potential to transform into specialised cells.

Many species of annelids (segmented worms) reproduce via fragmentation. For example, California blackworms (*Lumbriculus variegatus*) have 150–250 body segments and live in muddy sediment beside marshes and ponds, hence their other common name of mudworm. They are widely used for bait and aquarium fish food. Like earthworms, these worms are **hermaphrodites**, meaning they have both male and female reproductive parts in one body and can reproduce sexually. However, it is more common for them to reproduce using fragmentation. In this case, blackworms break apart and each segment can become a new worm. Hermaphrodites are covered in more detail in Section 2.2.

GO TO ➤ Section 2.2 page 70

Hermaphrodites are individuals that have both male and female reproductive organs.

Parthenogenesis

The development of an **egg** in the absence of **fertilisation** is an unusual form of asexual reproduction known as parthenogenesis, a Greek term meaning 'virgin birth'. Because it involves the development of an egg, parthenogenesis can only occur in females. Parthenogenesis is a normal part of the life cycle of some reptiles, birds, insects (bees, wasps and ants), rotifers and nematodes. There are at least 2000 species that are known to use parthenogenesis, even if it is not their usual means of reproduction.

BIOFILE CCT S

Virgin births of the Komodo dragon

The Komodo dragon (Figure 2.1.26) is native to three islands in Indonesia and is the world's largest lizard, growing 2–3 m in length. In 2006, in Chester Zoo in the UK, a captive female Komodo dragon called Flora laid 25 eggs. Flora had never been in contact with a male Komodo dragon. Genetic analysis of a collapsed egg confirmed that Flora was the only parent, with an extra egg having acted like a surrogate sperm to fertilise the main egg. Because the result was a combination of chromosomes from two cells, the offspring would not be exact clones of the parent, unlike other forms of asexual reproduction. Seven of Flora's eggs hatched and all were males. This reproductive adaptation would allow a single female to enter an isolated ecological niche, such as an island, produce male offspring by parthenogenesis and establish a genetically diverse population for the future that reproduces sexually.

FIGURE 2.1.26 The Komodo dragon is able to reproduce both by parthenogenesis and sexual reproduction.

SUMMARY OF THE ADVANTAGES AND DISADVANTAGES OF ASEXUAL REPRODUCTION

Given the abundance of organisms that reproduce by asexual means to ensure the continuity of their species, there must be significant advantages to asexual reproductive strategies. At the same time, because asexual reproduction almost always results in genetically identical offspring, there are some disadvantages. The key advantages and disadvantages of asexual reproduction are summarised in Table 2.1.2.

TABLE 2.1.2 Advantages and disadvantages of asexual reproduction

Advantages	Disadvantages
• efficient form of reproduction • amount of time and energy to produce offspring is minimal • population sizes can increase rapidly in optimal environments • there is no need to find a sexual partner • offspring are genetically identical to the parent cell, so they are well suited to a stable environment	• rapid population growth can lead to overcrowding and increased competition for resources • the lack of genetic variation in a population can cause death of the entire population if conditions change (e.g. a disease pathogen arrives or a severe drought) because they are not adaptable to new environmental conditions

2.1 Review

SUMMARY

- Asexual reproduction refers to a single parent producing a new individual from part of itself. It involves cell division by mitosis and produces offspring that are genetically identical (clones) to their parent.
- Asexual reproduction is rapid and is suited to organisms living in relatively stable and uniform environments. It is a disadvantage in changing environmental conditions where genetic variants are necessary for natural selection to act.
- Mechanisms of asexual reproduction include fission, budding, fragmentation, spore formation, vegetative reproduction and parthenogenesis.
- All groups of organisms (Bacteria, Archaea, Protista, Fungi, Plantae and Animalia) have at least some species that reproduce asexually.
- For unicellular organisms, the only purpose of cell replication is reproduction.
- Bacteria and Archaea are all unicellular and reproduce using binary fission.
- As cells reproduce by binary fission, their numbers increase exponentially with the formula $C = 2^n$, where C is the total number of cells and n is the number of cycles of replication that have occurred.
- Protists reproduce by binary fission and budding.
- Fungi reproduce both sexually and asexually. Their asexual reproduction is by budding or spores.
- Plants can reproduce by sexual and asexual means. Their asexual reproduction is called vegetative reproduction and includes rhizomes, stolons, tubers, bulbs, corms and cuttings.
- Animals can reproduce by sexual and asexual means, sometimes alternately in the same species. Sexual reproduction is the most common mechanism for animals. For the species that do use asexual reproduction, the methods include budding, fragmentation, strobilation (a type of fission in multicellular organisms) and parthenogenesis.
- Parthenogenesis is an unusual form of natural cloning in which an egg develops without fertilisation from another parent to form a new individual.

KEY QUESTIONS

1 Define 'asexual reproduction'.

2 Identify the sort of cell division that is involved in asexual reproduction.

3 **a** Outline the ideal environmental conditions for asexual reproduction.
  b Explain why these conditions are ideal for asexually reproducing organisms.

4 List the key events in binary fission for bacteria.

5 If a bacterial cell undergoes binary fission in ideal conditions every 30 minutes, calculate how many organisms there would be after four hours.

6 Match each type of asexual reproduction to its correct description.

budding	form of asexual reproduction of multicellular organisms in which an organism breaks into two or more parts, each of which regenerates the missing pieces to form a complete new organism.
fission	form of asexual reproduction in which the new organism arises as an outgrowth or bud from the parent
fragmentation	development of an egg in the absence of a fertilisation by sperm; a normal part of the life cycle of some insects and crustaceans
spore formation	form of asexual reproduction of unicellular organisms where the parent cell divides into two approximately equal parts
parthenogenesis	formation of structures that are resistant to adverse environmental conditions and can give rise to complete organisms when conditions become favourable

7 Identify each of the following as true or false.
  a All animals are multicellular.
  b Plants can only reproduce by sexual means.
  c Fungi can reproduce asexually and sexually.
  d Fission is only found in the bacteria.
  e Vegetative reproduction refers to the growing of vegetables.
  f Animals do not reproduce asexually.
  g Budding refers to the asexual formation of flower buds.

2.2 Sexual reproduction

BIOLOGY INQUIRY  

Reproduction in ferns

How does reproduction ensure the continuity of a species?

COLLECT THIS...

- scissors
- small paper or plastic bag
- binocular microscope or magnifying glass
- large needle (e.g. sewing needle, sharp skewer)
- camera (you could use the camera of your mobile phone)
- device with internet connection

DO THIS...

1. Take some scissors and a small paper or plastic bag and go for a walk around a garden at home, school, in a park or in a shady bushland area. Look for ferns such as those shown in Figure 2.2.1a.
2. Collect samples of fern leaves with spore cases by cutting off a small section of leaf that has brown spots on the underside. Figure 2.2.1b shows you what these structures look like. Try to collect from different fern species.
3. Examine your samples under a binocular microscope or good magnifying lens.
4. Use a dissecting or large sewing needle, sharp forceps or a skewer, to probe the spore cases while you look at them through the microscope.
5. Take photos of the spore cases through the microscope lens (you can use the camera on a mobile phone).
6. Try to count how many individual spores emerge from one spore case and estimate how many there would be in total on one leaf.
7. Research the life cycle of ferns. Draw a diagram to illustrate the main features of fern reproduction.

RECORD THIS...

Describe in a single sentence what you observed through the lens.

Present your best photo along with a short explanation of how fern plants reproduce.

REFLECT ON THIS...

How does reproduction ensure the continuity of a species?

Do different ferns seem to have a similar mode of reproduction and is it asexual or sexual reproduction?

Explain why a single fern plant might produce a large number of spores.

FIGURE 2.2.1 (a) A group of ferns growing in the shade. (b) The brown patches on the underside of this fern leaf are spore cases

Sexual reproduction involves the union of a male gamete (**sperm**) and a female gamete (**egg or ovum**) to form a unique individual (Figure 2.2.2). Most multicellular organisms, including humans, reproduce sexually. Some multicellular organisms can also reproduce asexually, as discussed in Section 2.1. Also noted in Section 2.1 was the fact that some unicellular organisms are capable of both sexual and asexual reproduction.

Each offspring of two parents has a unique genetic identity while at the same time continuing the species line. It only takes a glance around a group of humans to see the diversity of inherited features within a single species. The creation of genetic variation within a species is the hallmark of sexual reproduction. It results from meiosis, the special type of cell division used to produce gametes. You will learn more about meiosis in Chapter 3.

GO TO ➤ Section 3.1 page 116

In this section you will learn about the features and methods of sexual reproduction and the types of organisms that use sexual reproduction. You will also learn about the advantages and disadvantages of sexual reproduction for the continuity of species.

FEATURES OF SEXUAL REPRODUCTION

Life cycles of sexually reproducing organisms follow a pattern called **alternation of generations**. They alternate between haploid (n) and **diploid** ($2n$) stages. In most animals, including humans, the diploid stage is what we see as everyday body structure and function, and the haploid stage is the unseen internal production of sperm in males and eggs in females. For plants, there is a true alternation between a **gametophyte** stage (all cells haploid) and a **sporophyte** stage (all cells diploid) with the latter being the plant structure that we usually observe. The time spent in each stage varies greatly between different species.

Multicellular organisms are composed of two main types of cells: somatic cells and germ cells. Somatic cells are all the cells in the body of an organism apart from the sex cells (gametes). Examples of somatic cells in animals include skin cells, muscle cells and nerve cells. Germ cells are the cells that give rise to gametes, which are the specialised sex cells that combine in sexual reproduction.

Male gametes (sperm) and female gametes (eggs or ova) are often different in appearance (Figure 2.2.2). The formation of gametes occurs by meiosis in specialised reproductive organs, called **gonads**. In animals the gonads are the **testis** (plural testes) or the **ovary**. Although the male and female gonads look very different, they have the same function of producing haploid cells for reproduction. The sperm or eggs formed as a result of meiotic cell division are haploid, which means the number of chromosomes in the gametes is halved. A normal eukaryotic organism is composed of diploid cells (represented as $2n$) that carry one set of chromosomes (n) inherited from each parent. When the parental chromosomes pair up at fertilisation, they complete the diploid set for the new sexually reproduced offspring.

Female gametes (eggs or ova) are large, immobile cells. They contain the food stores needed for the development of the **embryo**. The male gametes (spermatozoa or sperm) contain limited food reserves and usually have a tail (flagellum) for motility, which enables them to move towards an egg (Figure 2.2.2).

After fertilisation the two haploid cells fuse to form a diploid **zygote**. The zygote then divides by mitosis to produce a large number of cells forming an embryo. The embryonic cells then differentiate to form the specialised tissues that make up a **fetus**. After birth for animals, or germination for plants, the organism continues to develop by mitotic divisions and becomes an adult. The reproductive cycle may then begin again. Fetal development in mammals will be covered in more detail in Section 2.3.

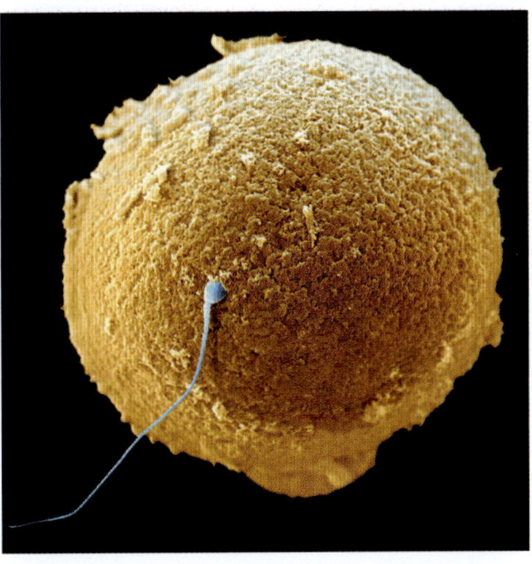

FIGURE 2.2.2 Human sperm (blue) fertilising an egg (orange) (coloured SEM). Only one sperm will penetrate the egg's thick outer layer (zona pellucida) and fertilise the central nucleus. Fertilisation occurs when the DNA of the sperm fuses with the DNA of the egg.

Alternation of generations is the fluctuation between haploid (n) and diploid ($2n$) stages in the life cycle of eukaryotes undergoing sexual reproduction. In plants the two stages grow as different structures but in animals the haploid stage only exists as gametes within the diploid body.

GO TO ➤ Section 2.3 page 87

INTRODUCING VARIATION—SEXUAL REPRODUCTION

Sexual reproduction requires the formation of gametes by the process of cell division called meiosis. Unlike mitosis it produces cells that are not genetically identical to each other or to the parent cell. There will be similarities in genetic content between parents and offspring, but the offspring are always genetically different from the parents and from each other (except for identical twins) (Figure 2.2.3). Here lies much of the advantage and disadvantage for any species using this form of reproduction. However, the widespread occurrence of sexual reproduction in almost all eukaryotic organisms shows that the long-term benefits to the species far outweigh any costs to the individuals (Table 2.2.3, page 85).

FIGURE 2.2.3 Sexual reproduction between two parents produces offspring to continue the species. Offspring will have some genetic differences to their parents. (a) Emperor penguin (*Aptenodytes forsteri*) family in Antarctica. (b) Orangutan and baby in Borneo (*Pongo pygmaeus*)

Advantages of sexual reproduction

The main advantage of sexual reproduction is the introduction of genetic variation that enables a species to survive and reproduce in varied and changing environments. In the long term, increased genetic variation allows greater adaptability and evolutionary potential in changing conditions. The pool of genetic variation in a population also facilitates the selection of beneficial traits and elimination of unfavourable traits, according to the survival and reproductive success of individuals. This process ultimately benefits the population, as those individuals that are most successful will reproduce, increasing beneficial genetic variants in the population.

Disadvantages of sexual reproduction

While there are advantages to sexual reproduction, there are also many disadvantages. In both plants and animals, energy must be used to produce gametes and ensure that mature gametes are brought together at the right time of year. In other words, not all the food the parent makes or eats is used to maintain its own body systems. For plants, relying on **pollination** by wind or an animal pollinator poses a risk, as does relying on environmental factors or another species for seed dispersal.

Finding and competing for a mate can be time consuming, energetically costly and risky. Some reproductive behaviours, such as calling or displays to attract mates, might also attract the attention of predators (Figure 2.2.4). In some animals, mating leads to considerable, and potentially harmful, competition between males. Providing parental care and protection for offspring is a substantial investment of time and resources, often shortening the lifespans of parents due to the excess expenditure of energy (Figure 2.2.5). Although individual adult animals are disadvantaged by the stress of parental care, the careful rearing of young has evolutionary advantages for the species.

FIGURE 2.2.4 Calling for a mate brings the risk of also attracting predator attention. This is a male African painted reed frog (*Hyperolius marmoratus*) displaying courting behaviour during a full moon.

FIGURE 2.2.5 Parenting young often requires a substantial investment of time and energy. (a) An Australian magpie lark (*Grallina cyanoleuca*) needs to work constantly to feed hungry chicks. (b) Parenting is especially challenging for the Emperor penguin (*Aptenodytes forsteri*) in the harsh environment of Antarctica.

FUNGI

The **mycelium** of a fungus is a mass of branching, threadlike hyphae (singular **hypha**), often growing underground, sometimes over a surface. The hyphae are the vegetative feeding state of the fungus and absorb the food digested by secreted enzymes. The nuclei of the mycelium cells are haploid (n).

In the reproductive state, a fungus produces a more conspicuous fruiting body or stalk with spores that result from either asexual or sexual reproduction. You learnt about asexual reproduction in fungi in Section 2.1.

GO TO ➤ Section 2.1 page 58

Although less common, many fungi can produce spores sexually with alternation of generations, as well as asexually. Two mating cell types, male and female, are formed for sexual reproduction.

There are many variations in fungal sexual reproduction but all include the following three stages: **plasmogamy** (cytoplasm union), **karyogamy** (nuclear union) and **gametangia** (haploid spores formed by meiosis).

Sexual reproduction introduces genetic variation into a population of fungi by combining genetic material from two fungal strains.

BIOFILE CCT S

Pilobolus crystallinus—specialised dung fungus

Pilobolus crystallinus is a specialised dung fungus that has evolved a way to shoot its spores away from the dung and onto nearby grasses. This helps to ensure the spores will be eaten by herbivores, allowing them to pass through the animals' digestive tracts and grow in fresh dung. Its 'shotgun' is a stalk swollen with cell sap, bearing a black mass of spores on the top (Figure 2.2.6). Below the swollen tip is a light-sensitive area. The light-sensitive area influences the growth of *P. crystallinus* by causing it to face towards sunlight. As the fungus matures, water pressure builds in the stalk until the tip explodes, shooting the spores into the air. The spores fly away at almost 40 km/h, reaching a height of 2 m and landing up to 2.5 m away from the parent fungus.

FIGURE 2.2.6 *Pilobolus crystallinus*, commonly known as the dung cannon or hat thrower

PROTISTA

Some unicellular organisms can reproduce sexually as well as asexually. This includes a group of protists known as ciliates. Ciliates have two types of nuclei: a **micronucleus**, which contains a normal diploid (2*n*) set of chromosomes; and one or more macronuclei (singular, **macronucleus**), which contain many sets of chromosomes (**polyploid**). You learnt about asexual reproduction in protists in Section 2.1.

Ciliates (known as paramecia) of the genus *Paramecium* usually reproduce asexually by mitosis of the micronucleus. The macronuclei simply pinch into two roughly equal pieces then fission occurs across the middle of the cell.

However, under stressful environmental conditions *Paramecium* can also reproduce sexually, by a method known as **conjugation** (Figure 2.2.7).

The benefit for a *Paramecium* species of switching to sexual reproduction and alternation of generations is that it introduces genetic variation, improving their ability to adapt to changing environments.

Polyploid cells are those that contain more than two sets of homologous chromosomes.

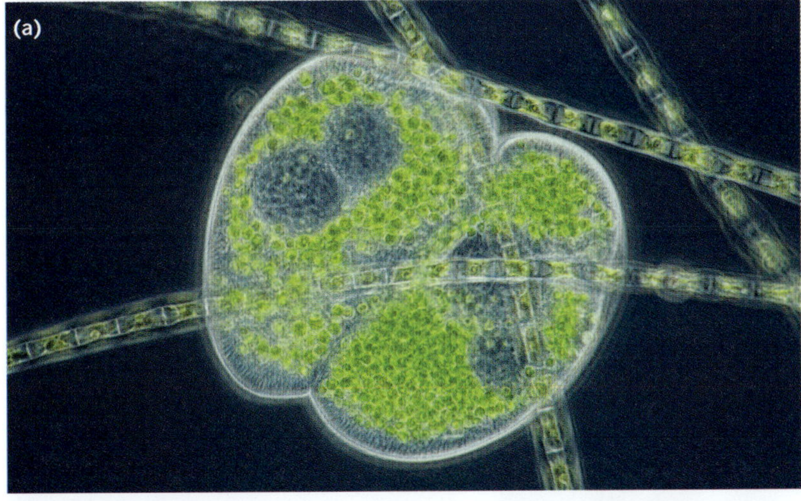

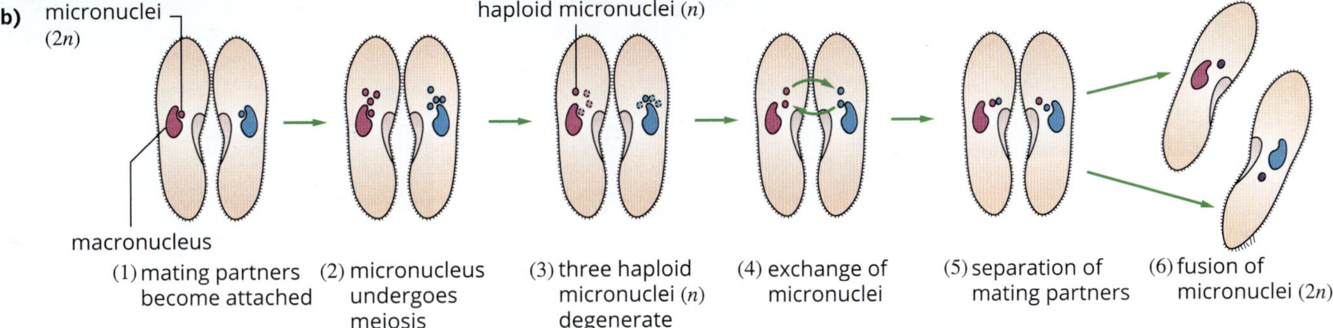

FIGURE 2.2.7 (a) Two paramecia *Paramecium bursaria* undergoing conjugation. This is a form of sexual reproduction in which two individuals fuse and exchange genetic material. (b) A diagram of the process of sexual reproduction (conjugation) in *Paramecium*.

PLANTS

GO TO ➤ Section 2.1 page 58

All groups in the kingdom Plantae reproduce sexually. Some also have asexual means of reproduction as detailed in Section 2.1. The two structures most commonly observed for sexual reproduction in seed-producing plants are seed cones of the **gymnosperm** plants and flowers of the **angiosperm** plants. The other plant types, grouped here as mosses and ferns, reproduce sexually with spores formed using less prominent structures.

Plants have a life cycle that alternates generations between haploid and diploid stages (Figure 2.2.8).

Gametophyte refers to the gamete-forming haploid stage and structures in a plant's life cycle.

Sporophyte refers to the spore-forming diploid stage and structures in a plant's life cycle.

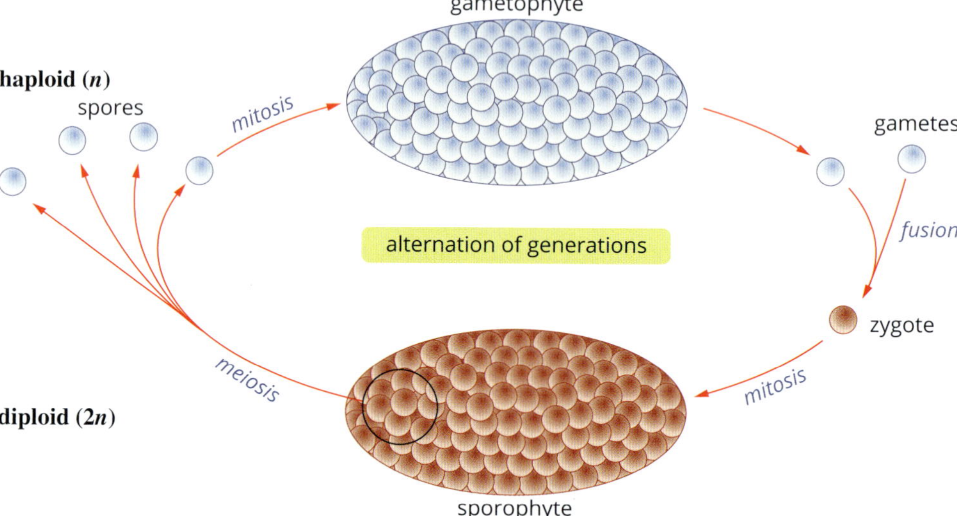

FIGURE 2.2.8 The alternation of generations between a diploid sporophyte (bottom) and a haploid gametophyte (top). This mechanism is common for sexually reproducing plants.

Ferns and mosses are the plant groups for which alternation of generations is most clearly displayed during their life cycles because both the sporophyte and gametophyte stages are free-living independent plants (Figures 2.2.9 and 2.2.10). In the seed-producing plants, gymnosperms and angiosperms, the diploid sporophyte stage is the dominant plant structure with the gametophyte stage present only as a small extension from the main plant, either a cone or a flower. The seeds are formed in or on the sporophyte after pollination. Seeds have great economic and survival value for humans as they are an important source of nutrition in the human diet.

FIGURE 2.2.9 Coloured SEM of the open mouth of a capsule (spore case) of the moss, *Homalothecium sericeum*, at the sporophyte stage. The triangular flaps that open and close the capsule to control release of spores can be clearly seen.

Mosses (Bryophyta)

Mosses and liverworts are the only plant groups without vascular systems, a state that restricts their size to low growing and their habitat to moist, shady places. Mosses reproduce sexually with spores and are characterised by small, flat, green leaf-like structures above ground (Figure 2.2.10) and root-like structures called **rhizoids** below ground. Mosses do not have true leaves or roots because they lack vascular tissue (xylem and phloem).

FIGURE 2.2.10 (a) Moss, *Leucobryum aduncum*, in the gametophyte stage (b) Moss-covered lava field in southern Iceland

The moss life cycle follows a sexual pattern of reproduction with alternation of generations and production of haploid meiospores by meiosis from a diploid plant (sporophyte). The meiospores grow into gametophytes that form male and female gametophores, which in turn fertilise to form a sporophyte again. Spores are released from capsules on the sporophyte to produce new gametophytes after meiosis and so on. Each spore capsule has a ring of tissue around the opening, made of triangular, close-fitting flaps that open and close to release spores when the moisture level in the surrounding environment is ideal (Figure 2.2.9). The haploid life stage of moss is dominant so we observe the gametophyte generation as the main plant (Figure 2.2.10a).

Ferns (Polypodiopsida)

Unlike mosses ferns have vascular tissues that transport water and soluble nutrients (Figure 2.2.11). They are a diverse group of plants characterised by the absence of flowers and fruit, the production of tiny spores instead of seeds, and by alternating generations of free living, spore producing plants (sporophytes, diploid) and gamete-producing plants (gametophyte, haploid) (Figures 2.2.12 and 2.2.13). Ferns are different from other land plants in that both the gametophyte and the sporophyte phases are free-living. The sporophyte is the dominant stage in the life cycle of a fern (Figures 2.2.11 and 2.2.12).

FIGURE 2.2.11 Kermadec tree-fern (*Cyathea kermadecensis*) one of the largest of the ferns

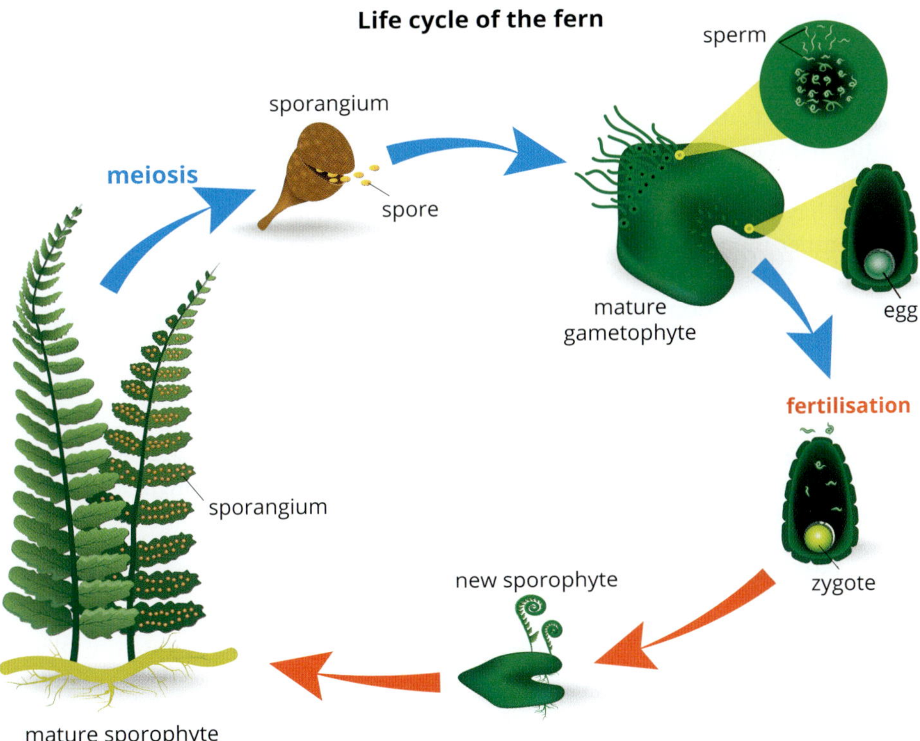

FIGURE 2.2.12 The life cycle of ferns is different from other land plants as both the gametophyte and the sporophyte phases are free living, with the sporophyte being the dominant stage. This diagram illustrates the alternation of generations in ferns.

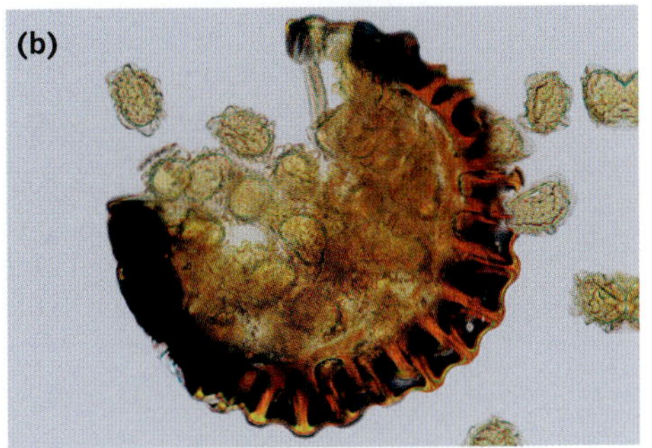

FIGURE 2.2.13 Spores are the reproductive cells of ferns. (a) Coloured SEM of a sporangium (spore case) from the leaf of a fern (*Dryopteris filix-mas*), which has split open to reveal the spores inside. (b) Sporangium of the western sword fern (*Polystichum munitum*) ruptured and spilling out fern spores.

Gymnosperms—cone-producing plants

Gymnosperms (meaning naked seed) are the vascular, non-flowering seed plants. They include the conifers (pine, spruce, fir, cedar and redwood trees), the cycads and ginkgo. The world's tallest, widest, heaviest and oldest living trees are all conifers of various types. The seeds of gymnosperms are produced by cones instead of flowers and when mature they are exposed rather than surrounded by a fruit (Figure 2.2.14). Pollination is always by wind because conifers do not have flowers to attract pollinators. Conifer seeds start their development as a haploid stage inside a protective cone that is woody for many species or, less commonly, fleshy in conifers like the *Podocarpus* (Figure 2.2.14c). Cones take from four months to three years to reach maturity, and vary in size from 2 mm to 60 cm long depending on the species.

Generally a conifer has a haploid stage with separate female and male cones, called the seed cone and pollen cone respectively. Small **pollen** grains (called **microspores**) develop in the male cones and when released they are transported by wind to the female cone which contains the **megaspores**. A **pollen tube** grows towards the **ovule** enclosed inside the female cone. In some species it may take months for the growing pollen tube to make its way to the megaspore inside the ovule where the mature haploid sperm can fertilise the haploid egg to form a new diploid cell. A conifer sperm has no tail and is not motile. It is carried on the tip of the growing pollen tube. Like the flowering plants (angiosperms), each pollen grain divides into two sperm (n) and only one will fertilise the egg cell and contribute its genetic information to the new seed ($2n$) that develops (Figure 2.2.15).

An unusual feature of conifer reproduction is that it can take two or more years from the haploid stage of pollination to the diploid stage of fertilisation and the release of the seed. Pollination is seasonal, once a year in spring, a fact noticed by the many hay fever sufferers (Figure 2.2.16). The female megaspore that has been formed earlier by meiosis remains in a dormant state until the pollen tube starts to grow, which may take months. Fertilisation is often a year or more after the pollen grains are deposited on the female cone. Subsequently the seed can take another year to mature and be released (Figure 2.2.15).

FIGURE 2.2.14 (a) Conifer seeds are protected by cones, which are usually woody structures such as the cone around this germinated pinecone seed or (b) the cones of the Wollemi pine (*Wollemia nobilis*). (c) In a few species, the scales of the cone have been modified to attract birds. The seeds of *Podocarpus* conifers are carried in fleshy colourful 'fruit' structures. Birds feed on the berry-like 'fruit' and disperse the seeds.

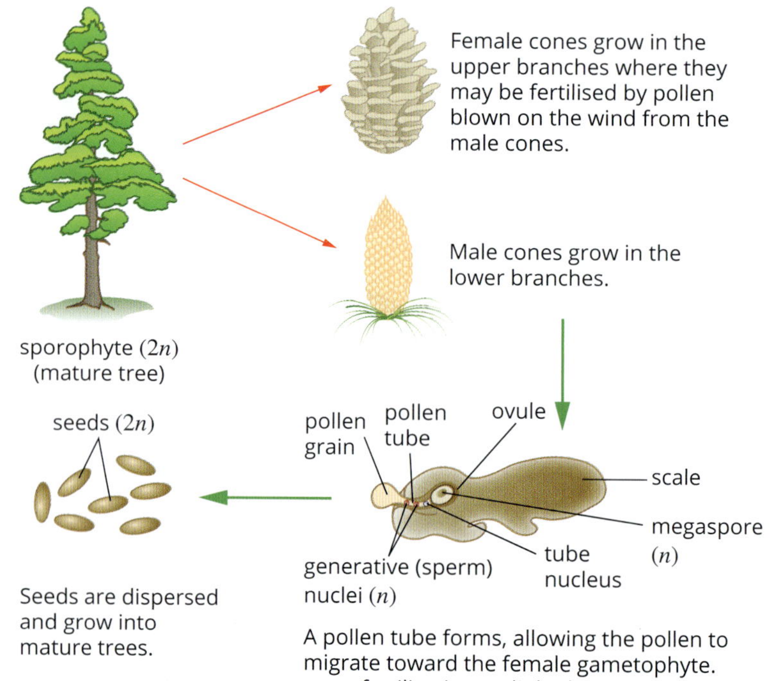

FIGURE 2.2.15 The life cycle of a conifer

BIOFILE N S

Conifers—the world's largest and oldest trees

On the western slopes of California's Sierra Nevada mountain range are the world's largest trees. The giant sequoias (*Sequoiadendron giganteum*) (Figure 2.2.17) are one of three members of a group that also includes the coastal redwoods (*Sequoia sempervirens*). One coastal redwood, named Hyperion, is the world's tallest tree at 115.7 m tall. A giant sequoia tree, named General Sherman, is the world's largest tree with a height of 84 m, a girth of 31 m at ground level and has an estimated weight of 2100 tonnes and a volume of 1487 m^3.

The giant sequoias only reproduce by seeds, which sometimes remain in the cone for 20 years. Forest fires help to open the cones and the seeds then grow from the burnt, bare soil. The reproductive success of these grand trees relies on their longevity. Each tree needs to produce just one maturing offspring over its lifespan of up to 4000 years for the species to persist. Nevertheless, they are closely protected.

Until 2013, the oldest known individual tree in the world was a 4845-year-old bristlecone pine (*Pinus longaeva*), in the White Mountains of California (Figure 2.2.18). Researchers at the Rocky Mountain Tree-Ring Research Group then estimated the age of another *P. longaeva* in the White Mountains to be 5062 years old. For this species, age is reliably calculated by tree ring counting. The exact locations of these two ancient, gnarled bristlecones are kept secret for their protection.

An Australian angiosperm tree comes close in height to Hyperion. Mountain ash are the tallest flowering trees, with one in Tasmania measured at 98.8 m tall.

FIGURE 2.2.17 The giant sequoia (*Sequoiadendron giganteum*) of California is thought to live for up to 4000 years. They grow to be the largest plants in the world, with heights of up to 100 m and volumes close to 1500 m^3.

FIGURE 2.2.18 Ancient bristlecone conifer trees (*Pinus longaeva*) in California. *P. longaeva* is the species that holds the record for the oldest known individual tree in the world.

FIGURE 2.2.16 Clouds of pollen leaving the male cones of an umbrella pine (*Pinus pinea*). Wind pollination of conifers is a trigger for hay fever in many people in spring.

BIOFILE CCT

Impossible bottle trick

Pine cones have long been used by people, particularly those in the Northern Hemisphere where conifer trees are prolific. Uses range from fire starters to Christmas decorations to toys, homemade hygrometers and a variation of the impossible bottle trick (ship in a bottle). Observation that some cones open out their scales and release seeds only in dry weather, led to creative uses like checking the moisture level of the air with a pine cone hanging outside as a weather predictor (hygrometer). A small green cone placed inside a bottle in wet weather will open into a much larger cone when it dries out, becoming impossible to remove and challenging the unsuspecting as to how it got into the bottle (Figure 2.2.19).

FIGURE 2.2.19 The impossible bottle trick with a pine cone

Angiosperms—flower-producing plants

Angiosperms are the flowering plants. Sexual reproduction for them involves meiosis, which produces haploid cells that then undergo several mitotic divisions and develop into haploid male or female gametophytes that are genetically different. The male gametophyte is the pollen grain, which contains sperm cells, held on the **anther** at the top of the **stamen**. The female gametophyte stage is the embryo sac, which contains the egg held inside an ovule in the plant's ovary at the base of the flower.

Flower structure

The reproductive organs of flowering plants are contained in the flowers (Figure 2.2.20). On the outside of a flower are sepals and petals. The sepals enclose and protect the other parts of the flower during the bud stage. Sepals are usually small and green, but in a few species they are large and brightly coloured.

The petals are arranged in a circle or cylinder around the reproductive organs. Inside the ring of petals are the stamens, which are the male reproductive organs. Each stamen usually has a long stalk called the **filament** with a small yellow pollen sac on the end, called the anther.

In the centre of the flower is the female reproductive organ, called a **pistil**. The term **carpel** can also be used to describe the female reproductive organ of a flower. The pistil can consist of one carpel or many, which may be fused or separate (Figure 2.2.21), combining to form the pistil in a flower. Each pistil consists of an ovary, which is a central swelling at the base, and a slender stalk called the **style** bearing the **stigma** at the top, which is the receptive surface for pollen. Inside the ovary are one or several ovules that contain the female gametophytes. The ovary is usually seen as a swelling and may be positioned above the petals and sepals or below (Figure 2.2.22).

Parts of a flower

male stamen
anther
filament
petal
ovule
receptacle
sepal
pedicel
stigma
style
ovary
female pistil

FIGURE 2.2.20 The sexual reproduction organs of a flower

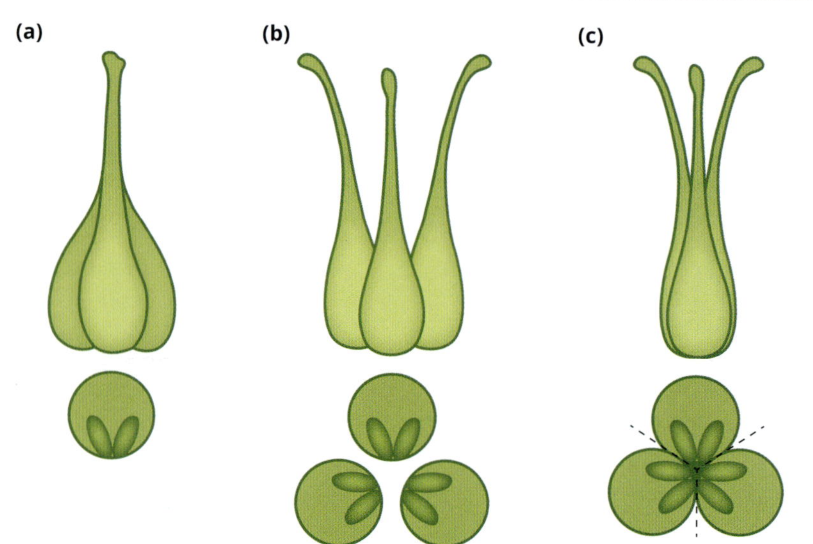

FIGURE 2.2.21 The female reproductive organ of a flower is the pistil, which consists of the stigma, style and ovary. The pistil may contain (a) a single carpel, (b) or many carpels, which may be separate or (c) fused.

There may be confusion with the names carpel and pistil used for female parts of a flower. A flower can have one or more carpels. Each carpel is made up of three structures: stigma, style and ovary. Often there are many carpels and they can be separate or fused together. Together, all the carpels are called a pistil, so the term pistil has been used here to mean the whole of the female reproductive organ.

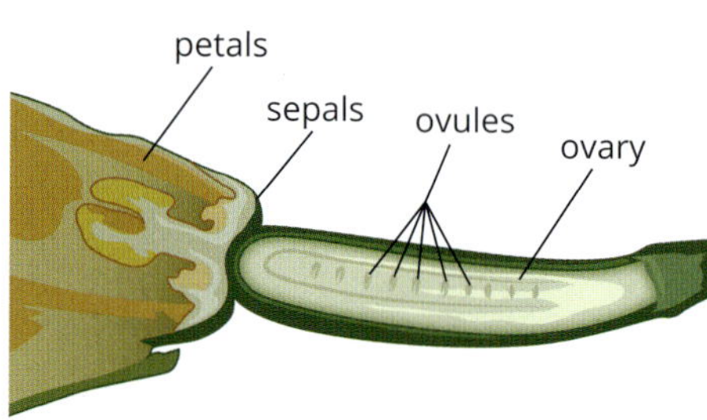

FIGURE 2.2.22 Longitudinal section of a female squash flower showing the ovules inside the ovary that has grown into a fruit. When petals and sepals are positioned above the ovary, as in the squash flower, the flower is said to have an inferior ovary.

Most flowers contain both male and female reproductive organs and are known as bisexual (e.g. rose, tea tree, mango). In some species, the flowers are unisexual, with the male and female organs in separate flowers on the same plant (e.g. maize, zucchini).

Pollination and fertilisation

Wind, insects or birds carry out pollination in most plants (Figure 2.2.23). Less commonly, the agents of pollination are bats, other animals or water.

Pollination occurs when a haploid male pollen grain lands on a receptive female stigma and begins to grow a tube. One of the cells in the pollen grain produces a tube that penetrates the surface of the stigma (Figure 2.2.24). The pollen tube carries two sperm cells and grows down through the style inside specialised nutritive tissues, towards the ovary, until it reaches an ovule. Fertilisation takes place in the ovule when the egg fuses with one of the two sperm cells, forming a diploid cell that grows into a seed. Other cells in the ovule combine with the second sperm cell and then divide rapidly to provide tissue called endosperm that nourishes the developing embryo inside the seed (Figure 2.2.25).

Although most flowers are bisexual, most of them do not self-pollinate because this would reduce the genetic variation in the offspring. Flowering plants have efficient mechanisms for preventing **self-pollination** and promoting **cross-pollination**. One mechanism includes the maturation of the anthers (male) and the stigma (female) at different times. For example, the stigma at the top of the pistil is sticky (for pollen grains to adhere) at different times to when pollen of the same flower is mature. Plants may also reject their own pollen, preventing the pollen tube from growing and therefore avoiding self-pollination.

Endosperm is a tissue produced inside the seeds of most flowering plants following fertilisation. It surrounds the embryo and provides nutrition in the form of starch, though it can also contain oils and protein.

FIGURE 2.2.23 Bees are one of the world's most important pollinators, carrying pollen between flowers as they collect nectar. Many food crops depend on pollination by bees.

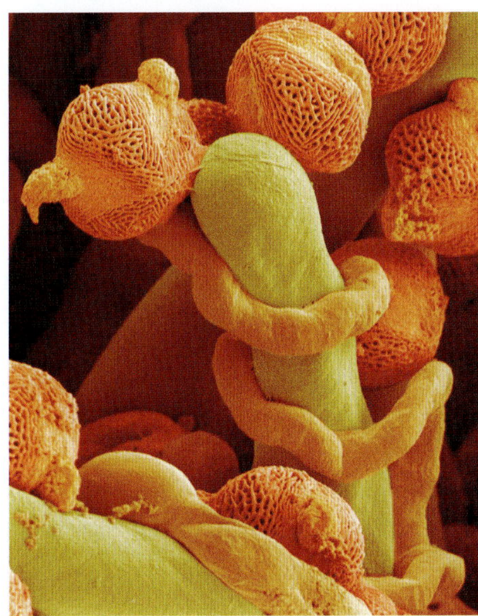

FIGURE 2.2.24 Coloured SEM of pollen tubes (orange) on the pistil of a prairie gentian flower (*Gentiana sp.*). Once the pollen grains, containing male gametes, land on the pistil (female parts) of a flower, they form tubes that burrow down through the stigma (tip) and style (shaft) of the pistil to reach the ovule, which holds the female sex cell. When an ovule is fertilised it forms a seed.

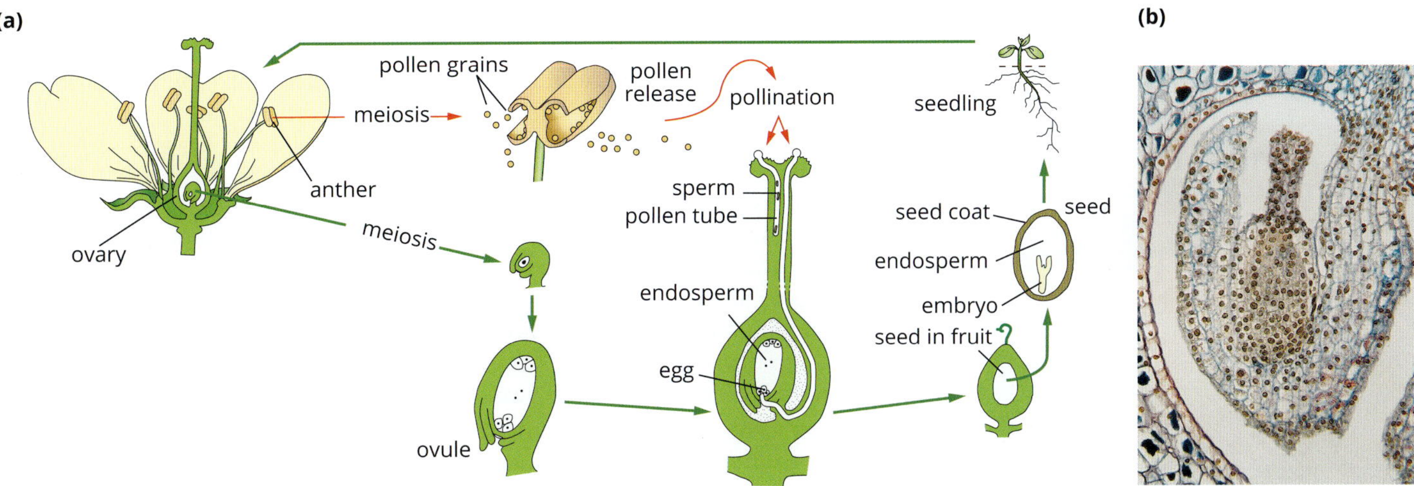

FIGURE 2.2.25 (a) Life cycle of a flowering plant showing fertilisation and seed formation (b). Cross-section through an ovule

BIOFILE S

Where are all the pollinators?

One of the many concerns about a reduction in biodiversity is the loss of insect species, given that they are prolific agents of pollination. In a 2016 report the independent organisation, the Intergovernmental Science-Policy Platform on Biodiversity and Ecosystem Services (IPBES), noted that many wild bees and butterflies have been declining in abundance, occurrence and diversity at local and regional scales in northwest Europe and North America. Declines have also been detected elsewhere in the world. In Europe, 9% of bee and butterfly species are threatened with extinction and populations are declining for 37% of bee species and 31% of butterfly species. The possible causes were identified as habitat loss, pesticides, pollution, invasive species, pathogens and climate change. The IPBES also identified 16% of bat and bird pollinator species worldwide as threatened with extinction.

The report stressed the importance of protecting pollinators to ensure stable fruit, nut and vegetable output, amid concern over the challenge of feeding the world's growing population in coming decades. The report also stated that animal pollination is directly responsible for 5–8% of global agricultural production and more than 75% of leading global food crops relies to some extent on animal pollination for yield and quality. Pollinator decline is therefore an issue to be taken seriously.

In an attempt to raise awareness about declining bee populations, a conservation group in Britain started The Great British Bee Count, with the public using an app to record numbers, identify species and post photos (Figure 2.2.27).

FIGURE 2.2.27 Two of the 26 species of bumblebees in Britain: common carder bee (*Bombus pascuorum*) and red-tailed bumblebee (*Bombus lapidarius*). Pollination by bees of fruit tree flowers is critical to the future of fruit production.

Successful fertilisation can only occur following acceptance of the pollen grain by the stigma and of the pollen tube by the style. Unlike animals, a plant cannot move around to select its partner, so plants have evolved mechanisms to choose gametes from appropriate partners and reject those that are inappropriate. Pollen of each species has a characteristic shape that allows recognition by the plant (Figure 2.2.26).

FIGURE 2.2.26 A 3D rendering of pollen grains from eight different plant species being transported by the wind

Seeds and fruit

After fertilisation the ovule develops into a seed protected by a tough outer seed coat. This process involves the ovule (in which the zygote develops) expanding, the endosperm forming, and the zygote undergoing a series of mitotic divisions to produce a multicellular embryo. All the cells are diploid. The embryo develops seed leaves (cotyledons) and a root tip, and epidermal and vascular tissues begin to form (Figure 2.2.25).

As the ovule changes into a seed, the ovary containing the ovule becomes a mature fruit. Nutrition for seed development and fruit growth is obtained through the phloem and xylem of the parent plant. Fruits are specialised structures that protect the seeds and may enhance seed dispersal. Some fruits contain a large store of nutrition to feed the seed after it germinates. Humans have taken advantage of this for their own food supply, and in some cases, have selected and bred fruit-bearing plants to increase or improve the flesh of the fruit. The many examples include apples, citrus, mango, watermelon and passionfruit, all of which carry seed inside fleshy nutritious fruit.

Germination and development

The embryos in seeds lie dormant until conditions are appropriate. Water, oxygen, temperature and day length are major environmental factors that influence seed germination. Many seeds can remain dormant and only germinate when conditions are favourable. Seed dormancy allows plants to disperse their progeny into the future, something that animals generally cannot do. Dormant seeds can wait months, years and even decades to continue propagation of their species. The oldest known germinating seed was almost 2000 years old.

Dormant seeds have a water content of around 10% compared to regular plant cells at 85% or more. When mature, a seed already contains within it a multicellular diploid embryo and one or two cotyledons (seed leaves) surrounded by the nutritious endosperm tissue and protected by a tough seed coat.

ANIMALS

Members of the kingdom Animalia have an amazing diversity of sexual reproductive strategies, often involving complex behavioural, physiological and structural adaptations for attracting mates, mating, and protecting and nurturing developing offspring.

As animals moved from protective aquatic environments to exposed terrestrial environments, there was a need to shift from external fertilisation to internal fertilisation to prevent dehydration of gametes. This evolution of reproductive strategies is evident in animals today, including the land-based mammals. It is understood that aquatic mammals like seals, dolphins and whales are species that returned to life in the water after an ancestral phase on land. These mammals still reproduce by internal fertilisation.

Most other aquatic animals reproduce by external fertilisation. The amphibians (frogs and toads) return to aquatic environments to lay and externally fertilise their eggs (Figure 2.2.28). The male frog appears to be mating internally with the female but he is actually depositing sperm onto her eggs as she lays them into the water. The fertilised eggs clump together until they hatch into tadpoles.

FIGURE 2.2.28 External fertilisation is practised by amphibians such as these common brown frogs (*Rana temporaria*). They are shown in position surrounded by floating eggs laid by the female ready for the male to deposit his sperm.

Most terrestrial animals use internal fertilisation. Reptiles and birds reproduce by internal fertilisation and then protect their developing offspring outside the body by laying shelled eggs. The eggs may have a hard outer shell (birds) or a tough membrane (reptiles). Mammals use internal fertilisation and most protect their developing offspring within the female's body until the fetus is fully developed. These are called the **placental** mammals because the fetus is nourished *in utero* through a **placenta** until it is born (Figure 2.2.29).

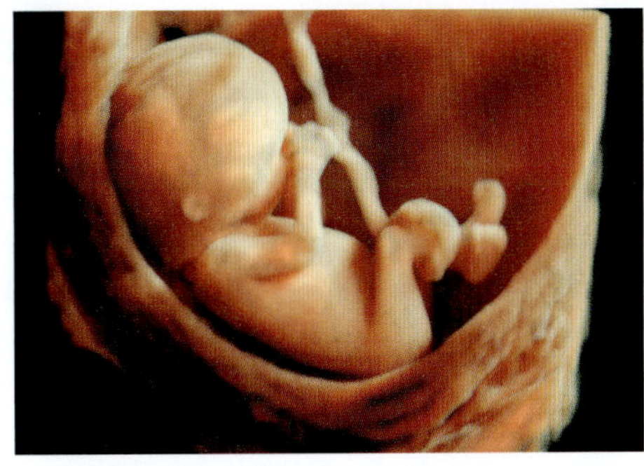

FIGURE 2.2.29 A 3D ultrasound image of a 12-week-old human fetus in its mother's uterus, clearly showing the umbilical cord that carries nourishment via the placenta. Humans are placental mammals.

Many native Australian animals are **marsupials**, unique amongst mammals in that they do not hold the developing offspring within the female's body for the full period of fetal development. The young are born at a very early stage and continue their development, nourished and protected, inside an external pouch (Figure 2.2.30). Even more unusual and iconic are the only two representatives of **monotremes**, the platypus (endemic to Australia) and the echidna (native to Australia and New Guinea) . The platypus and the echidna are classed as monotreme mammals, they practise internal fertilisation, then lay and protect eggs enclosed in tough flexible membranes (Figure 2.2.31).

The features of pregnancy and birth in mammals are covered in more detail in Section 2.3.

GO TO ➤ Section 2.3 page 87

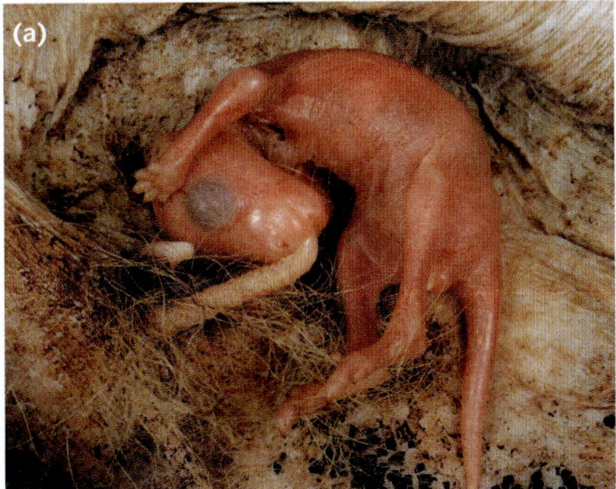

FIGURE 2.2.30 (a) A kangaroo joey 20 days after birth. In kangaroo reproduction a newborn finds its way to the pouch where it remains attached to a teat until fully developed. (b) A kangaroo joey (*Macropus sp.*) is protected in the mother's pouch for an extended period, even after it starts to feed on grass. A kangaroo is a marsupial mammal.

FIGURE 2.2.31 The egg of a short-beaked echidna (*Tachyglossus aculeatus*), measuring about 17 mm in diameter. An echidna is a monotreme mammal.

BIOLOGY IN ACTION CC S

Antechinus at North Head

The brown *Antechinus* (*Antechinus stuartii*) (Figure 2.2.32) is a carnivorous native marsupial still found in parts of Sydney but driven to extinction many years ago in Sydney's North Head area. Now they are being reintroduced to the headland under a program run by Australian Wildlife Conservancy, which is also expected to assist the pollination of the local banksia scrub.

A notable feature of all 12 species of *Antechinus* is that they only breed once in their life, something that is rare amongst mammals. Males live for approximately 11 months. They have a short and frenzied breeding cycle of about 14 days in winter, after which they die as a result of stress and exhaustion. During the mating period one female can breed with multiple males, which may result in pouch young from up to eight different fathers. This and the emergence of new males every year enhances the genetic variation in the *Antechinus* population.

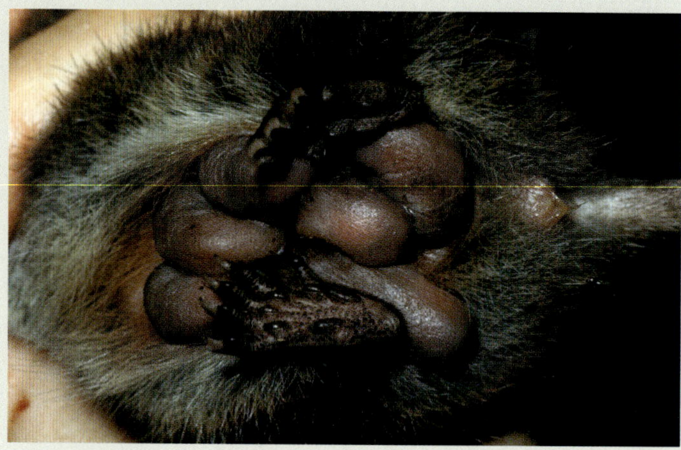

FIGURE 2.2.32 A brown *Antechinus* female (*Antechinus stuartii*) with six eight-week-old young attached to teats and overflowing from her pouch.

The mass death of mature males also protects the young from predation by the males.

Like all marsupials, females give birth to undeveloped naked young that latch onto teats in the pouch for up to 50 days.

External fertilisation

External fertilisation is when a male's sperm fertilises a female's egg outside of the female's body. Most aquatic animals have external fertilisation, indicating that the advantages outweigh the disadvantages (Table 2.2.1). Some examples of external fertilisation in animals are described below.

- In amphibians and bony fish the female usually lays her unfertilised eggs in water and the male waits nearby to deposit sperm onto the eggs. The mass release of eggs for fertilisation is called spawning and it can be spectacular to see. A notable example is the long distance, mass migration of Atlantic and Pacific salmon species from the ocean, upriver to the freshwater gravel beds, where they spawn in the same location that they hatched. (Figure 2.2.33). After the massive effort of annual spawning most adult salmon die but their species lives on.
- On tropical reefs once a year, coral colonies have a synchronised mass spawning event across the whole reef system. The invertebrate coral **polyps** in one area release eggs and sperm simultaneously forming a great floating cloud in the ocean, resembling a shaken snow dome with white, red, yellow and orange colours (Figure 2.2.34). Triggers for a coral spawning event and the way it is coordinated are still being researched. Most corals are hermaphrodites as they produce both male and female gametes inside one body. This means they could self-fertilise, but cross-fertilisation has the evolutionary advantage of creating genetic variation. By expelling the eggs and sperm at the same time, the coral increases the likelihood that cross-fertilisation will take place. When an egg is fertilised by a sperm it develops into a coral larva called a **planula** that floats around in the water for several days to weeks before settling on the ocean floor. After the planula has settled in a particular area it starts to bud (asexual reproduction) and the new coral colony develops.

FIGURE 2.2.33 Male and female red salmon (*Oncorhynchus nerka*) in the river gravel beds preparing to spawn. The female lays eggs and the male immediately deposits sperm onto the eggs externally.

FIGURE 2.2.34 Stony coral colony (*Acropora* sp.) releasing egg–sperm bundles during an annual mass coral spawning on the Great Barrier Reef, Queensland

Advantages and disadvantages of external fertilisation

The advantages and disadvantages of external fertilisation are outlined in Table 2.2.1.

TABLE 2.2.1 Advantages and disadvantages of external fertilisation

Advantages	Disadvantages
usually more rapid and prolific	more gametes need to be produced
female can continue to reproduce without pausing while the first young develop	no control over the gametes once released
parents do not expend energy for gestation and caring for the young	decreased chance of successful fertilisation. This can be improved by synchronised release of gametes (e.g. spawning events). young usually need to fend for themselves immediately
young are widely dispersed, reducing competition with the parent and with each other	must take place in an aquatic environment gametes and zygotes are exposed to predation and disease

Internal fertilisation

Internal fertilisation is when the male transfers his gametes directly into the female's body through a tube in his **penis**. This copulation process usually places his **semen**, containing the sperm, directly into the female's reproductive tract which greatly increases the chance of successful fertilisation with her egg (Figure 2.2.35). Internal fertilisation also overcomes the need for an environment with water because the reproductive tract is always moist. If a mature egg has been released from one of the female's ovaries and the swimming sperm meet it, only one sperm will be able to penetrate the protective layer (**zona pellucida**) surrounding the egg and fuse with the egg nucleus (Figure 2.2.36). Each sperm has a rounded head and a long tail, which it uses to both swim and burrow into the egg (Figure 2.2.36).

Fertilisation occurs when the chromosomes in the sperm's head pair up with those in the egg, forming the zygote. The egg then forms a barrier to other sperm. A diploid zygote cell is created with an equal mix of genetic information from male and female parents. After fertilisation, development of the zygote continues internally using mitotic division, either inside the female (mammals) or externally inside a shelled egg fed by the yolk (birds, reptiles and monotremes). Section 2.3 covers these reproductive features in more detail for mammals.

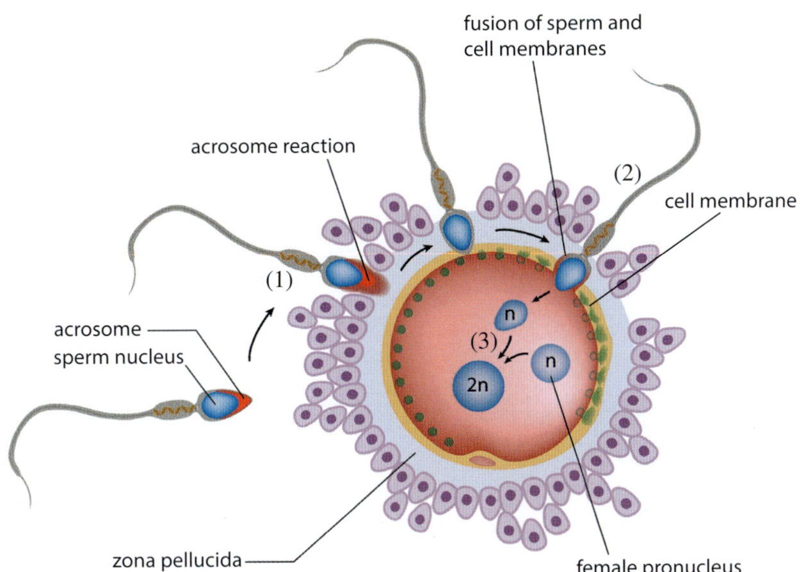

FIGURE 2.2.36 Illustration showing a sperm cell penetrating an egg cell. (1) The sperm cell comes into contact with the zona pellucida (the protective layer surrounding the egg) and an acrosome reaction breaks down the protective layer. (2) The sperm cell penetrates the egg's cell membrane. (3) The nucleus of the sperm enters the cytoplasm of the egg cell. When the haploid (*n*) set of chromosomes from the sperm and egg cells combine, fertilisation is complete and a diploid (2*n*) zygote forms.

FIGURE 2.2.35 Animals have a wide diversity of courtship and mating behaviours to achieve sexual reproduction. (a) Australian rainbow lorikeets mating (*Trichoglossus moluccanus*). (b) Blue ringtail damselflies (*Austrolestes annulosus*) form a 'wheel' when mating. The male (top) is holding the female's neck, while the female has moved her abdomen towards the male's genitalia to receive his sperm.

GO TO ➤ Section 2.3 page 87

Advantages and disadvantages of internal fertilisation

The advantages and disadvantages of internal fertilisation are outlined in Table 2.2.2.

TABLE 2.2.2 Advantages and disadvantages of internal fertilisation

Advantages	Disadvantages
can take place on dry land	usually slower with fewer progeny
less gametes have to be produced	mating rituals and practice are more complex to get to the point of copulation
more likely to be successful because gametes are positioned close together in female reproductive tract	potential for spread of sexually transmitted diseases throughout population
gametes and zygotes are protected from predation and disease	energetically costly; that is, requires a lot of ongoing energy input from the parent, especially the female
developing young are fed and protected increasing their chance of survival	parental care of young may be lengthy and demanding

BIOFILE CCT

Hermaphrodites

Sexual reproduction usually requires two parents for the production of offspring. However, this is not the case for all species. Many plants and some animals, such as tapeworms, snails, some fish and earthworms (Figure 2.2.38) have both male and female reproductive organs in the same individual—they are hermaphrodites. Some species are able to self-fertilise, while others require a partner. Even when hermaphrodites self-fertilise, the random assortment of chromosomes and recombination of genetic material that occurs during meiosis means that the offspring are always genetically unique.

FIGURE 2.2.38 Earthworms are hermaphrodites. They have both male and female reproductive systems in their bodies.

BIOFILE CCT

The quest for dominance

In Africa, fighting between male springboks (*Antidorcas marsupialis*) (Figure 2.2.37) establishes the stronger male when one of the competitors retreats before serious injury occurs. Despite its species name, the springbok is placental, not a true marsupial like the pouched Australian animals. The Latin word 'marsupium' means 'pocket'. In the the springbok it refers to the white flap above the tail that is lifted in mating displays.

FIGURE 2.2.37 Two male springboks (*Antidorcas marsupialis*) fighting in competition for a mate

SUMMARY OF ADVANTAGES AND DISADVANTAGES OF SEXUAL REPRODUCTION

Reproductive adaptations are the features that help an organism successfully reproduce, allowing their species to continue. The considerable benefit of sexual reproduction is evident from its widespread occurrence in almost all eukaryotic organisms. The most beneficial aspect of sexual reproduction is the genetic variation that is introduced through gamete production, genetic recombination and fertilisation. The benefits of sexual reproduction to the species outweigh the costs to the individual (Table 2.2.3).

TABLE 2.2.3 Advantages and disadvantages of sexual reproduction

Advantages	Disadvantages
fertilisation is less risky and the young are more likely to survive	slower reproductive rate—fewer offspring are produced over a longer time period
unfavourable (deleterious) genetic variation is eliminated from the population more efficiently	mates have to be found and accepted as suitable. Finding and competing for a mate can be risky and energetically costly
generates genetic variation through recombination during meiosis and selects for beneficial genetic variation more efficiently	recombination during meiosis can break apart beneficial genomic combinations and introduce deleterious variation to populations
populations are better able to adapt to and survive changing environmental conditions	potential for spread of sexually transmitted diseases throughout population
improves long-term evolutionary potential of populations	energetically costly; gamete production, mating, gestation and rearing young requires a lot of ongoing energy input from the parent

2.2 Review

SUMMARY

- Meiosis is a division of the nucleus that halves the normal number of chromosomes (diploid, $2n$) and produces different genetic combinations in the haploid ($1n$ or n) gametes.
- Germ cells are cells that give rise to gametes (egg and sperm cells) with half the chromosome number (haploid). Somatic cells are the normal body cells each with a full set of paired chromosomes (diploid).
- Sexual reproduction in multicellular organisms involves the fusion of gametes from two different individuals to form a zygote.
- The great advantage of sexual reproduction is that it produces variation between individuals of a population while still continuing the same species. However, there is often considerable cost to the parents.
- Sexual reproduction involves equal genetic contributions from male and female parents.
- Some unicellular organisms, like *Paramecium* protists, can undergo sexual reproduction, without gametes.
- Fungi, moss and ferns alternate between haploid and diploid generations with obvious gametophyte and sporophyte stages that produce spores.
- Gymnosperms (cone-producing plants) and angiosperms (flowering plants) are seed-bearing plants.
- Gymnosperms use wind pollination and form seeds on cones.
- Angiosperms use animal, wind and water pollination and their reproductive organ is the flower.
- In a flower, the male parts are stamens, composed of filaments and anthers. The anthers release pollen, which contains male gametes (sperm). The female part is the pistil (formed from one or more carpels), composed of the stigma, style and ovary, which contains ovules with egg cells.
- There are many adaptations in flowering plants that increase the chances of successful pollination.
- Pollination involves a specific interaction between pollen grains and stigma, growth of a pollen tube down the style, and fertilisation between sperm and egg in the ovule.
- After fertilisation, the ovule becomes a seed containing the embryo and endosperm, surrounded by a tough outer seed coat.
- The ovary containing the ovule(s) becomes the fruit, which may be dry or fleshy, and is often adapted for dispersal.
- Sexual reproduction in animals uses specialised reproductive organs called gonads. In female animals, the gonads are the ovaries, which produce eggs. In male animals, the gonads are the testes, which produce sperm.
- Mammals are classified into three groups based on their means of reproduction, which are placental, marsupial and monotreme.
- Animals living in water usually practice external fertilisation.
- Animals on land use internal fertilisation, primarily to avoid dehydration of the gametes and the zygote.
- There are advantages and disadvantages to both external and internal fertilisation with both being successful for for the continuation of species.

2.2 Review *continued*

KEY QUESTIONS

1. What is the difference between a somatic cell and a gamete? Give an example of each.
2. Explain the meaning of the term 'alternation of generations'.
3. Identify each of the following as true or false.
   - **a** In multicellular organisms, sexual reproduction is based on the fusion of a male and a female gamete.
   - **b** Fungi, like mushrooms, only reproduce asexually.
   - **c** Conjugation is a term used for the sexual reproduction method of some protists.
   - **d** Mosses are vascular plants with the gametophyte stage being dominant.
   - **e** Ferns are vascular plants that produce spores using alternation of generations.
   - **f** Seed dispersal relies on animals to carry the pollen.
   - **g** Spawning is a term used for spore production in mosses and ferns.
   - **h** Sexual reproduction creates genetic variation, allowing evolution by natural selection.
   - **i** Sexual reproduction is the only way to ensure continuity of a species.
4. Match each term to its correct description.

haploid	containing one set of paired chromosomes (2*n*)
diploid	plant classification group where seeds are formed on flowers
gametophyte	plant classification group where seeds are formed on cones
sporophyte	cell division of germ cells that produces four non-identical haploid gametes
meiosis	cell division of somatic cells that produces two identical new diploid cells
mitosis	gamete-forming haploid stage in a plant's life cycle
gymnosperm	containing one set of single chromosomes (*n*)
angiosperm	spore-forming diploid stage in a plant's life cycle

5. Compare pollination in gymnosperm and angiosperm plants.
6. Create a table with the following headings to distinguish between external and internal fertilisation in animals: definition, location, examples.
7. Classify the three groups of mammals based on reproductive strategies and give two examples for each group.
8. Use a table to outline three advantages and three disadvantages of sexual reproduction.

2.3 Pregnancy and birth in mammals

For all types of mammals, sexual reproduction produces genetically variable offspring, promoting the continuity of their species. The reproductive structures of female mammals are essential for creating a protective, watery environment for internal fertilisation. While all mammals reproduce using internal fertilisation, the physiology of **pregnancy** and birth varies widely. There are three types of mammals that are classified based on their reproductive strategies. These are placentals, marsupials and monotremes.

The differences between placental, marsupial and monotreme mammals are based on the extent of fetal development before birth and how the fetus is nourished during its growth period.

In this section you will learn about the stages of sexual reproduction and fetal development in mammals, particularly the placental mammals, of which humans are one. These stages include formation of gametes, fertilisation of gametes, formation and implantation of a zygote, the development of a zygote to embryo, the development of an embryo to fetus, and finally the birth of offspring. You will also gain an understanding of how these events are controlled by hormones.

FIGURE 2.3.1 A humpback whale (*Megaptera novaeangliae*) mother suckling her calf. Calves are 4–5 m long when they are born and they suckle for about five months. Whales are placental mammals.

MAMMALIAN REPRODUCTIVE SYSTEMS

In the placental mammals, a **uterus** provides nourishment and protection, via a placenta and **umbilical cord,** for the developing embryo and fetus until birth. After birth, placental babies are nourished with milk and develop a covering of fur. Some examples of placental mammals are humans, horses, dogs, mice, seals, elephants and whales (Figure 2.3.1).

For marsupial mammals, the under-developed joey is protected and nourished in the external pouch after an early birth, allowing another fertilisation to occur internally. Some examples of marsupials are kangaroo, brushtail possum, wombat and koala (Figure 2.3.2). Both placentals and marsupials are **viviparous**, that is they give birth to developed, live young.

FIGURE 2.3.2 A koala (*Phascolarctos cinereus*) with a six-week-old joey in her pouch. Koalas are marsupial mammals.

With the two monotreme mammals, the female lays eggs and each puggle (baby monotreme) develops inside a leathery eggshell, then hatches to be protected and fed milk by the mother. There are only two monotremes: the platypus and echidna (Figure 2.3.3). Monotremes are **oviparous**, that is they lay eggs in which their young develop.

For all the mammals, before fertilisation can occur, haploid gametes must be produced by specialised reproductive organs in each parent. And for internal fertilisation, there must be a way for the male to introduce sperm into the female's reproductive tract.

FIGURE 2.3.3 A short-beaked echidna (*Tachyglossus aculeatus*) puggle at 40 days old, still without its spines

The male reproductive system in mammals

The male reproductive system (Figure 2.3.4) consists of the following.

- Paired testes (testicles; singular, testis), held inside the scrotum, which produce and store mature sperm continuously during mating periods; the main structures are the **seminiferous tubules**, where sperm cells are formed, and the **epididymis** that stores the sperm cells.
- Accessory glands that produce secretions which make up about 95% of the volume of semen; these include prostate, seminal vesicles and cowper's glands.
- A paired system of ducts, called **vas deferens** (also known as ductus deferens and sperm duct), leading from the testes to the **urethra**.
- **Luteinising hormone** (**LH**) from the **pituitary gland** (in the brain) to stimulate the secretion of the male steroid hormone **testosterone** in the testes.
- A penis, the male organ that grows to full size during puberty and has both sexual and excretory functions. The urethra tube passes through the penis, delivering urine or semen out of the body but not at the same time.

In a male the penis becomes erect when ready for copulation. Erection results from increased blood flow into columns of spongy tissue until the organ is rigid.

Mitotic divisions of precursor germ cells in the testes produce diploid **spermatocytes**, each of which divides by meiosis to produce four haploid sperm cells. During mating, contractions of the vas deferens move sperm towards the urethra. Secretions of the accessory glands are added, forming the seminal fluid, which has two main functions: it causes the sperm to become motile, and it provides an alkaline nutritious medium that is rich in protein, ions, vitamins and fructose sugar. Mammalian sperm each have a single flagellum (plural: flagella) that is used to propel them through the female reproductive tract towards the egg after copulation. They literally swim through the liquid internal environment in a race to be the first to reach and fertilise an egg.

The head of a sperm contains the nucleus with a haploid set of chromosomes, and a cap called the **acrosome** that contains enzymes used for penetrating the outer layers of the female egg. Mitochondria in the midpiece produce adenosine triphosphate (ATP) for energy during the journey through the female reproductive tract. The tail makes lashing movements that propel the sperm on this journey (Figure 2.3.5).

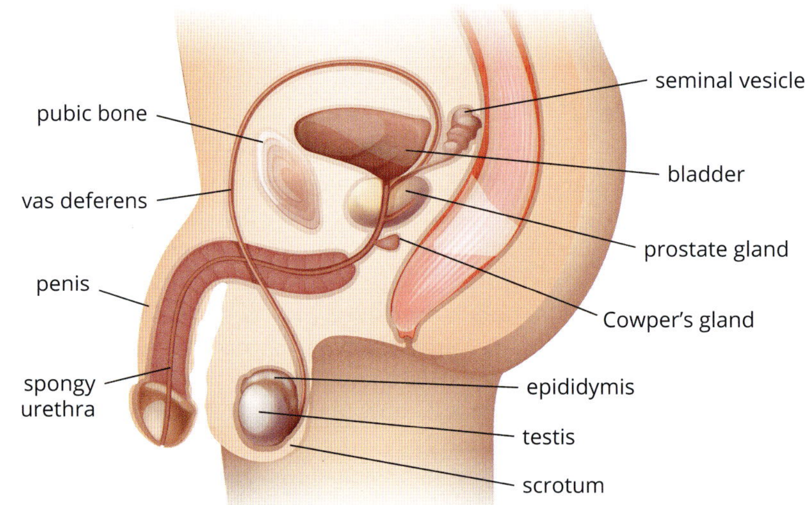

FIGURE 2.3.4 The human male reproductive system

BIOFILE CCT N

Human sperm—the facts

During sexual intercourse, about 5 mL (one teaspoon) of semen enters the vagina containing roughly 300–400 million sperm (Figure 2.3.5). Less than 100 000 sperm cells will pass through the cervix to begin their 15 cm journey to the egg. The fastest swimming sperm can reach the egg in half an hour, while others may take days. Sperm live for 48–72 hours inside the female reproductive tract but in the right conditions sperm can even live up to five days. An egg can be fertilised by sperm 12–24 hours after ovulation. In total the egg has a 24-hour lifespan during which it can be fertilised. A female's cervical mucous provides the sperm with nutrients to survive during their journey to the ovum. The typical lifespan of sperm in a female reproductive tract with fertile cervical mucous is three days (72 hours). So fertilisation can occur even if sperm entered the oviduct before an egg was released. Unlike females, males do not reach a period of infertility with age. However, the genetic quality and motility of sperm, as well as volume of semen all decrease with age.

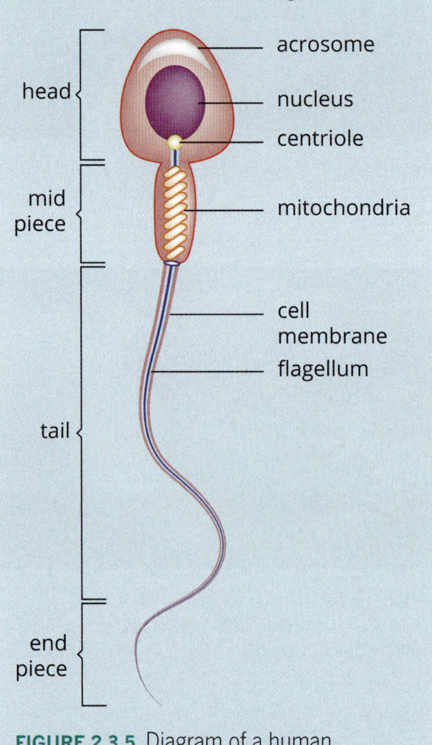

FIGURE 2.3.5 Diagram of a human sperm cell

The female reproductive system in mammals

The human female reproductive system (Figure 2.3.6) consists of the following.

- A single uterus where, if an egg is fertilised it implants in the uterine wall, a placenta forms and the fetus develops until the time of birth. The uterus undergoes changes that are controlled by **hormones**.
- Paired ovaries which hold the **oocytes** (immature egg cells) until puberty when monthly **ovulation** starts under hormonal control.
- Paired **fallopian tubes** (also called **oviducts** or uterine tubes) connecting each ovary to the uterus. The open end of each tube has fringe-like structures called fimbriae (singular **fimbria**) that surround the ovary to catch the eggs when released. Fertilisation usually takes place high in an oviduct.
- A **cervix**, a narrow muscular canal 2–3 cm length lined with mucous, that connects the uterus and **vagina**. The cervix dilates (stretches open) to at least 10 cm width for childbirth. During monthly **menstruation** it is controlled by the hormone **oestrogen** to become softer and more open.
- A vagina, a muscle-lined canal from the cervix to the genitals, which receives the male penis during sexual intercourse. Monthly menstrual blood flow from the uterus exits the body through the vagina. The vagina is also the birth canal for the baby to enter the outside world. Unlike in males, the opening for the excretion of urine (the urethra) is separate to the female reproductive tract.

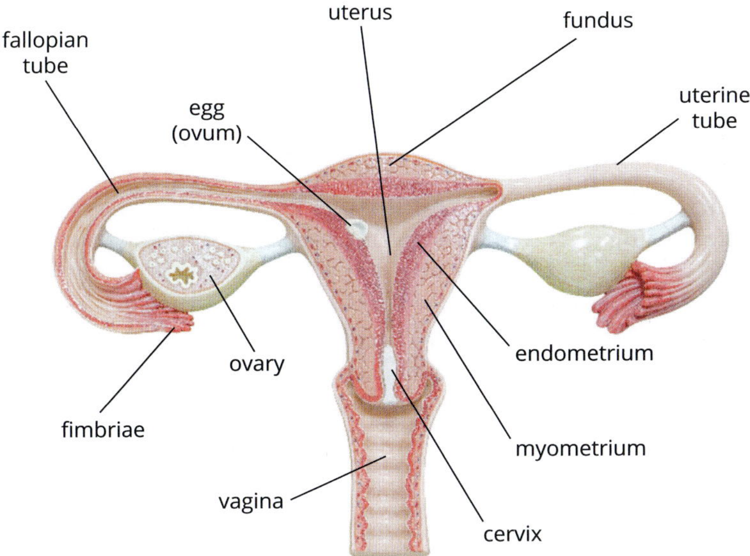

FIGURE 2.3.6 The human female reproductive system

Most female placental mammals have a single cervix. Humans and chimpanzees also have a single uterus, limiting the number of young the female can bear at one time. Some mammals such as cats, horses, deer, dogs and whales have two uteri, sometimes with the lower uteri joined into one. Rodents, rabbits and hares have a pair of uteri and cervices joined to a single vagina, allowing for reproduction of large litters. Marsupials also have paired reproductive tracts. An early birth at 4–5 weeks of age removes a developing marsupial from its mother's body much sooner than in placental mammals, thus marsupials have not developed a complex placenta. In monotremes, the uterus only functions to form a leathery eggshell around the embryo.

Ovulation

Unlike human males who continually produce sperm after puberty, a female is born with all the immature egg cells already in her ovaries. After reaching maturity (puberty), ovarian cycles commence. Later in life, at the time called **menopause**, females cease to ovulate. Thus before a female is born, meiosis has begun in all oocytes (immature egg cells) but is arrested at an early stage. Once she reaches puberty, pituitary hormones control the continuation and completion of meiosis I. Meiosis II begins, but again is paused until actual fertilisation occurs. Meiosis II only completes after fertilisation.

Meiosis only occurs in eukaryotes and only in germ cells to form gametes. There are two stages: meiosis I forms two cells and these two cells divide again in meiosis II to form four non-identical haploid gametes (either sperm or egg).

Under the influence of **follicle stimulating hormone** (FSH), one or more of the oocytes will resume its meiotic division up to metaphase II and matures within a group of nutritive cells called a **follicle** (Figure 2.3.7). Only one egg forms from each oocyte during meiosis. When the oocyte is maturing it grows much larger by adding nutrients and extra cell organelles. These are stored in its cytoplasm for use after fertilisation when **cleavage** begins and rapid mitotic division forms many new smaller cells.

Meiosis is explained in more detail in Chapter 3.

GO TO ➤ Section 3.1 page 116

Follicles containing a maturing egg release the hormone oestrogen, which causes changes to the lining of the uterus (**endometrium**) and also acts on the anterior pituitary gland. The uterine lining becomes thicker, softer and spongy, and richly supplied with blood vessels in readiness to receive a fertilised egg.

Follicles are a group of cells in the ovary that surround each immature egg to protect and nourish it. When an egg matures it ruptures the follicle and is released from the ovary into the oviduct. The follicle's full name is Graafian follicle.

Ovulation is the release of a mature egg and is triggered by a surge of luteinising hormone (LH) released from the anterior pituitary gland in the brain. The ovum (ripe egg) bursts out of the follicle and is drawn by fluid currents into the oviduct. Eggs, unlike sperm, cannot move by themselves. The fimbriae move to create a current that sweeps the egg into the start of the oviduct. Contractions of the oviduct and synchronised movement of cilia on its internal walls then help to propel the egg along towards the uterus. A human oviduct is about 10 cm long and 1 cm in diameter.

Left behind in the ovary, the burst follicle, now without its egg, is called the **corpus luteum**. The corpus luteum, stimulated by LH, secretes large amounts of both oestrogen and **progesterone**. These hormones cause a further thickening of the lining of the uterus during the latter part of the cycle. The actions on the endometrium are to prepare the uterus to receive an embryo, should fertilisation occur.

If it is not fertilised, the egg simply passes out of the reproductive tract. The corpus luteum slowly disintegrates and stops releasing its hormones. As a result, the thickened uterine lining breaks down and menstruation (monthly bleeding) occurs.

FERTILISATION

Fertilisation is the fusion of two haploid gametes (egg and sperm) to form a single diploid zygote cell. The zygote cell contains the genetic material of both the egg and the sperm (Figure 2.3.8). There are equal genetic contributions from the male and the female parents to the zygote and subsequent offspring. For humans, fertilisation is often called **conception**.

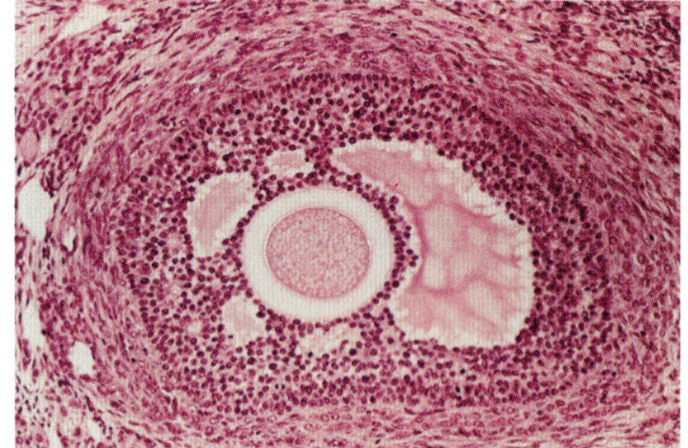

FIGURE 2.3.7 Light micrograph (LM) of a Graafian follicle in the human ovary. The follicle holds a round oocyte (centre) that is almost ready for release (ovulation). Around it (pale ring) is an accumulation of fluid then a layer called the zona granulosa (purple dots). To the outside of this dotted layer is the oestrogen-secreting tissue. Magnification: x200

The fertilisation event

In mammals, and most terrestrial animals, fertilisation occurs internally following mating and most often takes place in the upper part of the oviduct (Figure 2.3.10). The male inserts his penis into the female's vagina and a muscular contraction (ejaculation) pushes semen from his urethra into her vagina. From the vagina the sperm swim, using movement of their flagella, through the cervix into the uterus and into an oviduct until one sperm reaches and penetrates the egg. The other factor is timing. Ovulation to release a mature egg must have occurred and fertilisation of this egg has to take place. Sperm can survive for up to five days within the female's reproductive tract but three days is more typical.

Fertilisation occurs in four steps that are similar for all types of mammals.

1 The sperm uses enzymes from the acrosome to dissolve and penetrate the protective layer (zona pellucida) surrounding the egg to reach the cell membrane.
2 Molecules on the sperm surface bind to receptors (specialised proteins) on the egg's cell membrane to ensure that a sperm of the same species fertilises the egg, then the nucleus of the sperm enters the cytoplasm in the egg cell.
3 Changes at the surface of the egg occur to prevent the entry of multiple sperm nuclei into the egg.
4 Fusion of the haploid egg and sperm nuclei results in a diploid zygote cell (the fertilised egg).

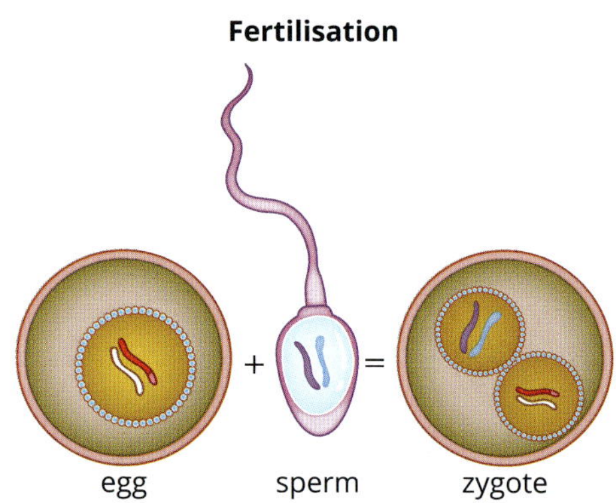

FIGURE 2.3.8 A basic model of fertilisation—fusion of two haploid gametes to form a diploid zygote

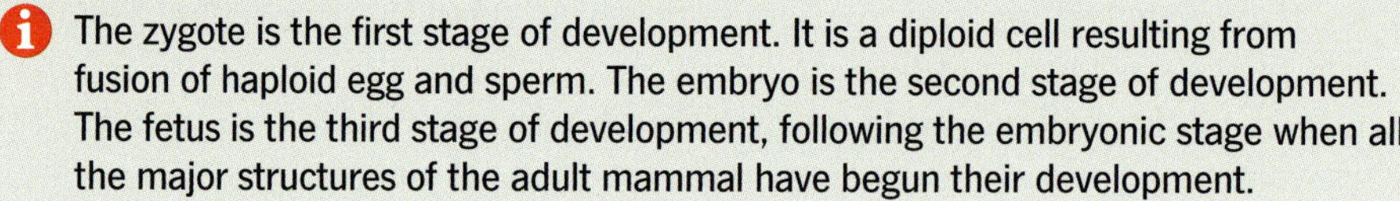

The zygote is the first stage of development. It is a diploid cell resulting from fusion of haploid egg and sperm. The embryo is the second stage of development. The fetus is the third stage of development, following the embryonic stage when all the major structures of the adult mammal have begun their development.

IMPLANTATION

After fertilisation, the zygote continues to travel down the oviduct until it reaches the uterus. Already the process of embryonic development has begun with a stage called cleavage. The development process continues as it passes down the oviduct. When the embryo reaches the uterus, ready for **implantation**, it is known as a **blastocyst** and looks like a ball of cells.

Cleavage

The first stage of development of the new zygote is cleavage, which commences following activation of the egg by sperm penetration. Cleavage is a period of rapid cell proliferation during which the single-celled zygote is divided into many hundreds of smaller cells by mitosis. You will learn more about mitosis in Chapter 3.

GO TO ➤ Section 3.1 page 116

Morula

The early embryo continues to divide until, three to four days later, it consists of 16 cells and then enters the uterus. At this stage, the embryo resembles a mulberry and is known as a **morula**. The morula is a ball of unspecialised embryonic stem cells.

Blastocyst

In the uterus, mitotic divisions continue, and the morula becomes a blastocyst as its cells begin to differentiate (Figure 2.3.9). By day eight to nine, the blastocyst is ready to attach to the wall of the uterus. The multicellular blastocyst consists of a single layer of surface cells and an inner cell mass that will later give rise to the embryo. The outer layer of cells sends out finger-like projections into a part of the wall of the uterus (endometrium) and this area develops into the placenta (Figure 2.3.10).

Gastrula

After the blastocyst is implanted, **gastrulation** occurs over approximately five days, and the blastocyst becomes a **gastrula**, which has three different layers of cells. Eventually the gastrula becomes an embryo then a fetus when it has formed all the basic adult features.

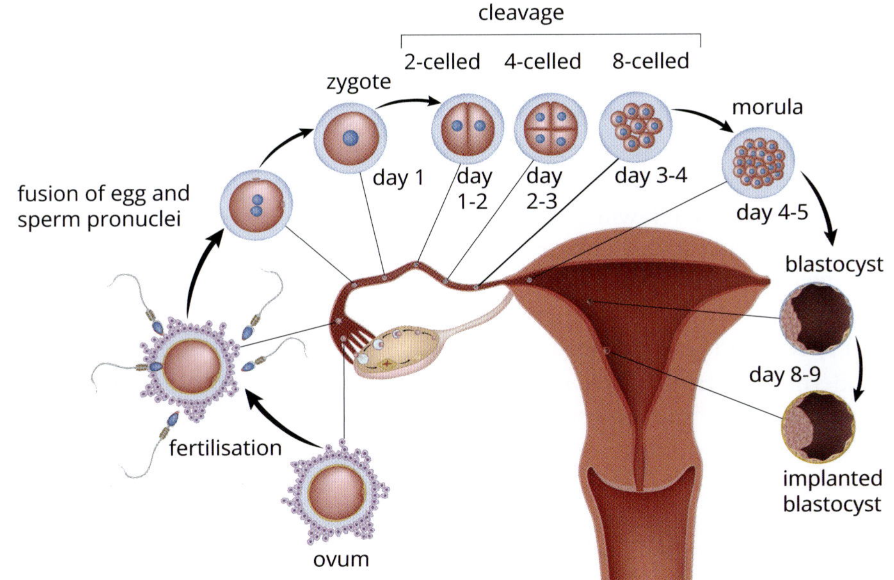

FIGURE 2.3.9 Diagram depicting ovulation to implantation of a human blastocyst in the uterine wall (endometrium)

The placenta and umbilical cord

The blastocyst adheres to the lining of the uterus (endometrium) and becomes implanted there. The outer layer of cells in the blastocyst initiates the formation of a placenta. Later an umbilical cord develops by the fifth week of the embryo stage from the remnants of the egg's yolk sac. It replaces the yolk sac as the source of nutrients for the embryo, acting as a conduit for embryonic blood vessels to reach the placenta. The umbilical cord remains attached to the fetus until after birth.

The placenta is an exchange organ bringing blood vessels of the fetus into close contact with maternal blood. There is no direct exchange of blood, rather nutrients and oxygen from the mother diffuse across into the blood of the umbilical vein and move to the fetus. The reverse happens for removal of waste products and circulation of depleted blood through the umbilical arteries from the fetus back to the placenta. After birth, the umbilical cord is cut (for humans) (Figure 2.3.11) or in other animals the mother often severs it by biting. A person's navel (belly button) is the scar where their umbilical cord dried up and dropped off after birth.

The placenta is also an important source of hormones to maintain the pregnancy.

The term 'morula' comes from the Latin word for 'mulberry'. This term originated because the early embryonic stage is a tight cluster of small cells bulging outwards into the zona pellucida which looks like a berry through a microscope.

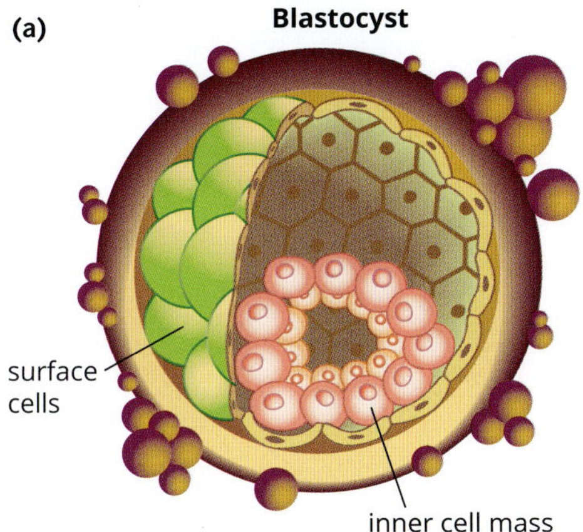

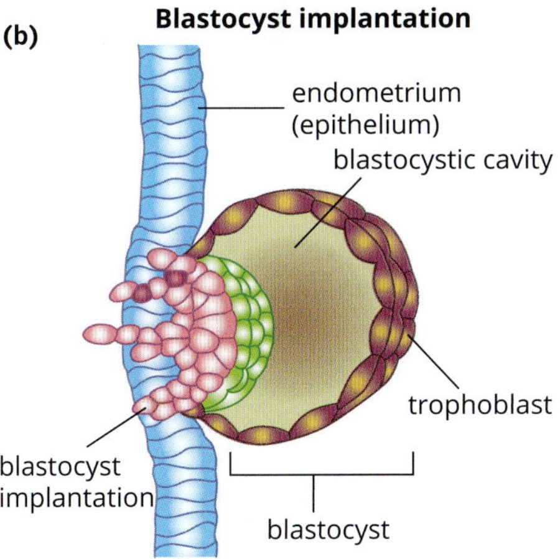

FIGURE 2.3.10 (a) The blastocyst is characterised by an inner and outer cell mass. The outer cells merge at implantation with the uterus wall to form a placenta and the umbilical cord grows much later from the inner yolk sac. (b) Blastocyst implantation takes place on the wall of the uterus (endometrium, blue). The outer layer of cells (brown) implants to form the placenta (pink). The inner cell mass (green) will form the embryo.

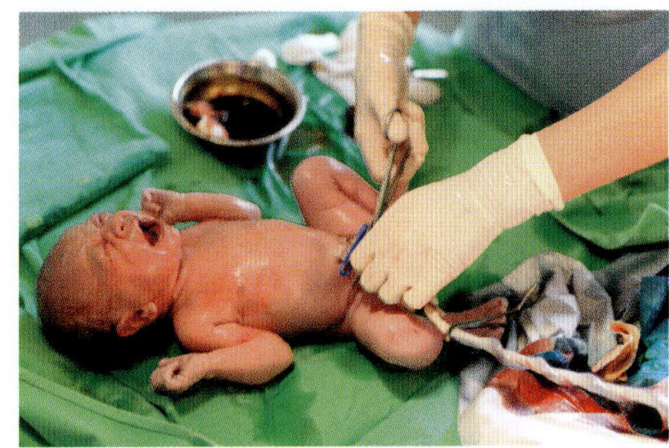

FIGURE 2.3.11 A doctor cutting the umbilical cord of a newborn baby

DEVELOPMENT OF THE EMBRYO

During the embryonic period of development, the major organs of the body are formed from the three primary layers of the gastrula. In humans this is completed about eight weeks after fertilisation (or 10 weeks after the last menstrual period). At the end of the embryonic stage, the developing organism has distinct features and is known as a fetus for the remainder of its development (Figure 2.3.12).

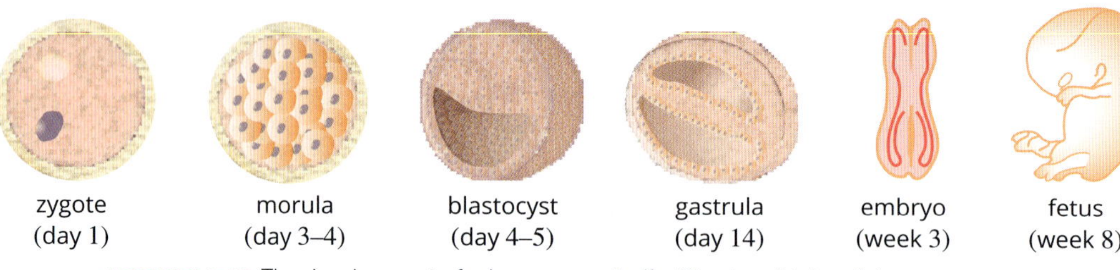

FIGURE 2.3.12 The development of a human zygote (fertilised egg) into a fetus

Embryonic germ layers and cell specialisation

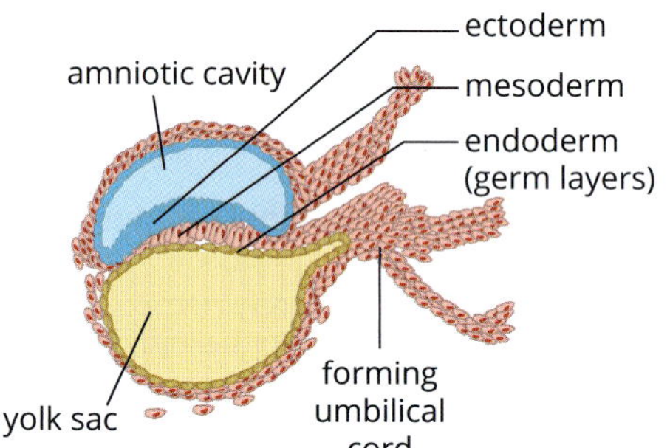

FIGURE 2.3.13 A gastrula approximately 16 days after fertilisation, showing the three primary germ layers: ectoderm, mesoderm and endoderm

After implantation in the uterus, the blastocyst undergoes gastrulation, folding in on itself to form a gastrula with three primary layers of cells: **ectoderm**, **mesoderm** and **endoderm**. These primary layers are known as **germ layers** (Figure 2.3.13) and are also supported by two membranes.

- The yolk sac which surrounds the egg yolk. It has a well-developed vascular system that transports nutrients from the egg yolk to the developing embryo.
- The **amniotic cavity**, which surrounds the developing embryo, is filled with fluid. Its main role is as a shock absorber to protect the embryo against any impacts or movements.

The three embryonic germ layers that form will eventually give rise to the different types of specialised cells that make up the tissues and organs in humans:

- ectoderm (outermost layer of the embryo) forms epidermis, hair, peripheral nervous system, brain and spinal cord cells
- mesoderm (middle layer of the embryo) forms muscle, cartilage, kidney and gonad cells
- endoderm (innermost layer of the embryo) forms the lungs, bladder and lining of the digestive system, including the stomach, colon, liver and pancreas.

DEVELOPMENT OF THE FETUS

As already mentioned, at the end of the embryonic stage the developing organism is known as a fetus for the remainder of its development, until birth at around 38 weeks after fertilisation. However, since it is almost impossible to determine exactly when fertilisation or implantation occurred, the period of development for a human baby is calculated from the first day of the mother's last menstruation. The time between the last menstrual period (LMP) and ovulation is approximately two weeks, so full term human pregnancy is usually 40 weeks (280 days but often estimated as nine months). The fetus grows in size and organs continue to develop for the rest of the pregnancy (this process is also known as prenatal development). Cells and tissues become specialised to carry out their particular functions. The fetus is protected in the amniotic cavity, which provides a fluid-filled environment in which it can move about.

BIOFILE CCT EU

Pregnancy by IVF

Artificial insemination (AI) and in vitro fertilisation (IVF) have made achieving pregnancy possible for those who cannot conceive via sexual intercourse. This approach may be undertaken due to infertility, as a voluntary choice or in agriculture to improve the breeding of livestock.

For IVF in humans, the prospective female first receives hormone treatment to stimulate ovulation. Then the mature eggs are drawn out from her and fertilised in vitro (outside the body) with sperm. The fertilised eggs (zygotes) are grown in vitro for three to five days. The growing blastocysts are closely monitored before implantation (Figure 2.3.14).

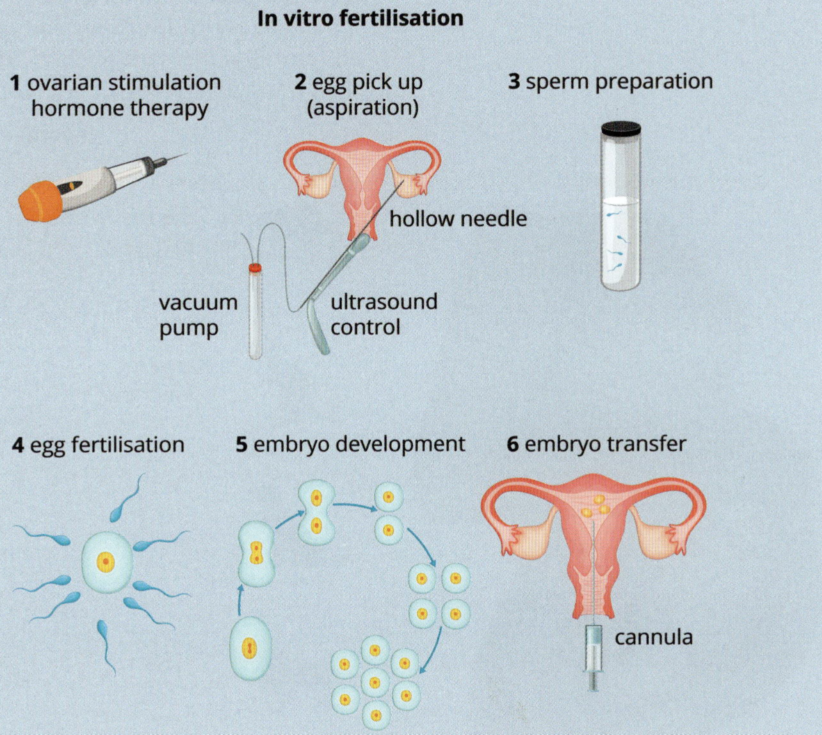

In vitro fertilisation

1 ovarian stimulation hormone therapy

2 egg pick up (aspiration)

hollow needle

vacuum pump

ultrasound control

3 sperm preparation

4 egg fertilisation

5 embryo development

6 embryo transfer

cannula

FIGURE 2.3.14 The stages of in vitro fertilisation (IVF) treatment

Gestation periods in mammals

In mammals, pregnancy is the period of reproduction during which a female carries one or more live offspring from implantation in the uterus through to birth. Pregnancy is also known as **gestation**. It begins when a zygote implants in the female's uterus and ends when the fetus leaves the uterus. Smaller species of mammals normally have a shorter gestation period than larger mammals. For example, gestation for the house mouse is 20 days (Figure 2.3.15), for the domestic cat it is 58–65 days, and 21 months for an elephant. Figure 2.3.16 shows a range of mammals and their gestation periods.

Human pregnancy can be divided roughly into three trimesters, each approximately three months long. The first trimester is from the last menstruation to week 13 of pregnancy, the second trimester is from week 13 to week 27, and the third trimester is from the week 28 to the week 42 (38 to 40 weeks growth plus two weeks from the start of the last menstruation). In humans, birth normally occurs at a gestational age of about 40 weeks (280 days) but births can occur between 37 and 42 weeks and still be considered a full term pregnancy (Figure 2.3.17).

FIGURE 2.3.15 A litter of baby mice (*Mus musculus*). Mice breed quickly and in large numbers compared to most other placental mammals.

The development of the fetus is monitored using ultrasound technology that is safe for both the fetus and mother. It uses very high frequency sound waves reflected back from structures inside a body to produce an image (Figure 2.3.20). The images produced in fetal ultrasound can be good enough to see the heart, with all four chambers and valves, at 20 weeks when the organ is only 15 mm long, less than the size of a five cent coin.

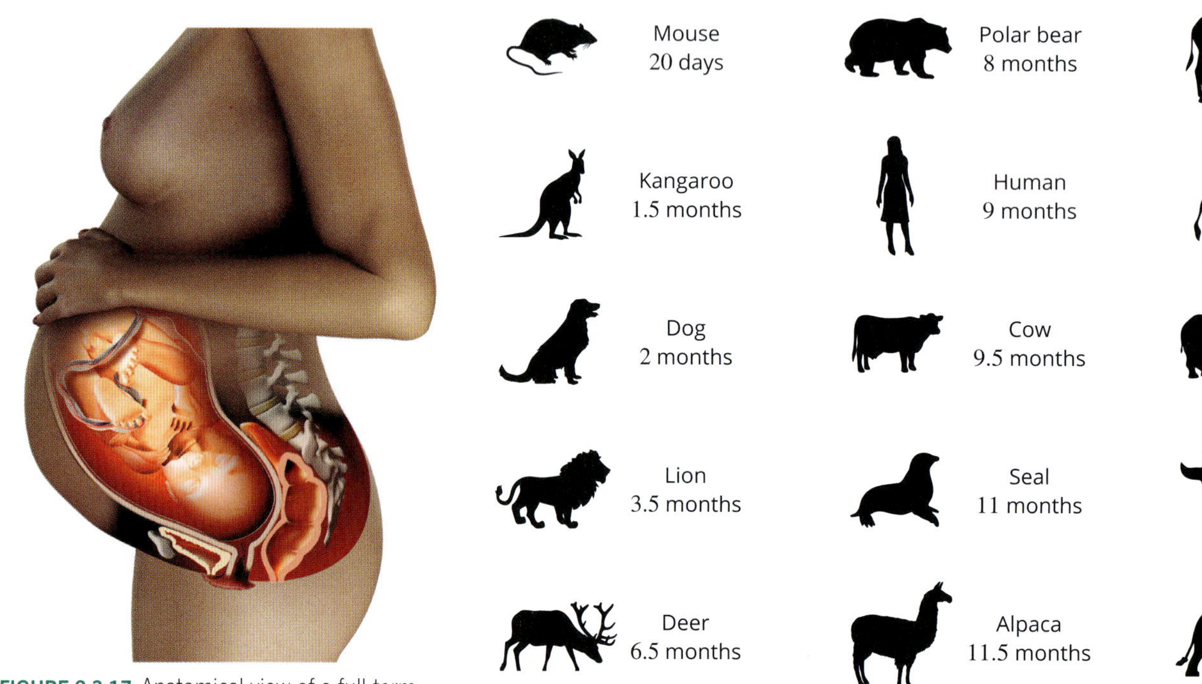

FIGURE 2.3.17 Anatomical view of a full-term (37–40 weeks) human pregnancy

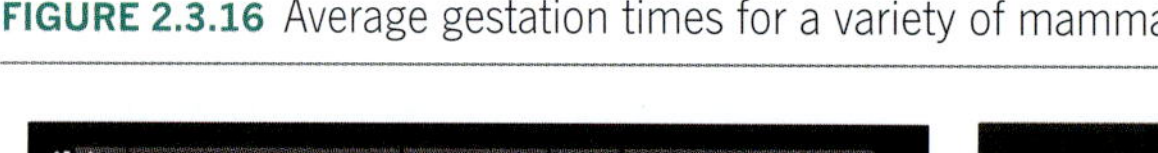

FIGURE 2.3.16 Average gestation times for a variety of mammals

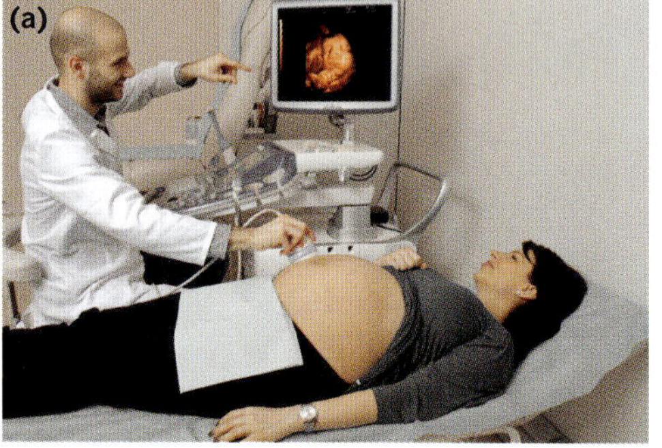

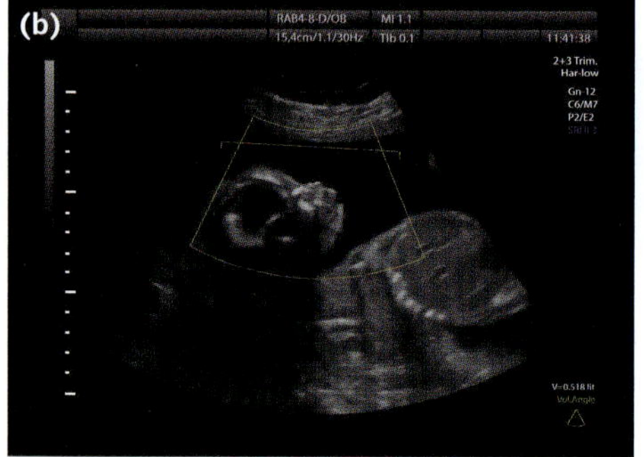

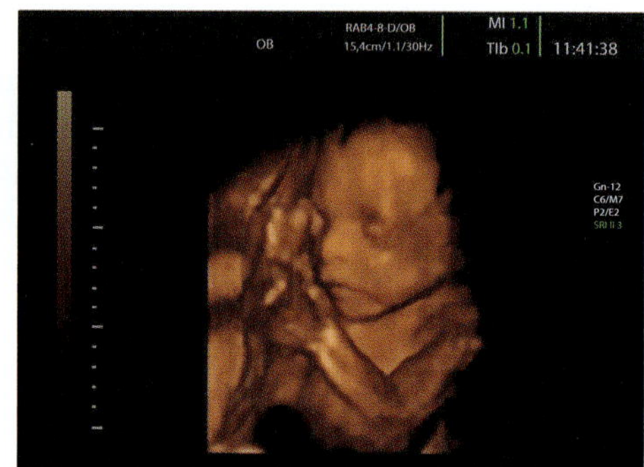

FIGURE 2.3.18 (a) A pregnant woman having an ultrasound scan to check the health and development of her baby. (b) Ultrasound images of a fetus *in utero*

HORMONAL CONTROL OF PREGNANCY AND BIRTH

Hormones are signalling molecules that are responsible for the communication between organs and tissues to regulate physiological and behavioural processes. In animals, hormones are synthesised by specialised cells (either in the endocrine glands or other tissues) and are transported to where they are needed via the circulatory and lymphatic systems or by diffusion through the extracellular fluid.

Hormones transmit signals to their target cells by altering specific biochemical reactions in these cells. The target cells have a matching surface receptor for a particular hormone. Hormones exert their effects either directly by passing through the cell membrane into the cell, or indirectly by interacting with a receptor on the surface of the cell. You will learn more about hormones in Chapter 14, but their role in pregnancy and birth in humans is examined here.

GO TO ➤ Section 14.2 page 510

The pituitary gland secretes hormones involved in the regulation of **lactation** and reproduction even though it is located in the brain (Figure 2.3.19 and Table 2.3.1). This gland lies immediately below, and is connected to, the **hypothalamus**, a region of the brain that acts as a master control centre. The hypothalamus receives information from all parts of the body and produces releasing hormones, which control the release of other hormones from the pituitary gland.

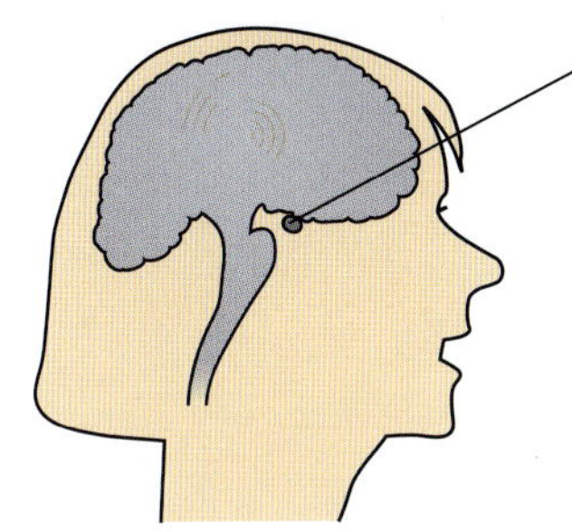

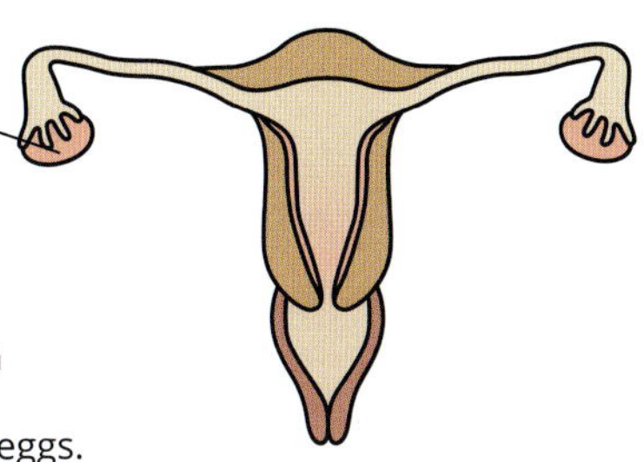

FIGURE 2.3.19 The connection between ovary and pituitary gland. A hormonal balance is necessary to establish and maintain a pregnancy.

TABLE 2.3.1 Hormones associated with reproduction in humans showing their target organs and functions

Gland	Hormone	Hormone type	Target	Function
anterior pituitary	• follicle stimulating hormone (FSH) • luteinising hormone (LH) • prolactin	• peptide • peptide • peptide	• ovaries • ovaries (female), testes (male) • mammary glands	• promotes development of follicle and secretion of oestrogen • promotes ovulation, development of corpus luteum, and secretion of progesterone (female); stimulates the secretion of the steroid hormone testosterone in the testes (male) • stimulates milk secretion
hypothalamus	• several releasing hormones	• peptides and amino acid derivatives	• anterior pituitary gland	• controls release of anterior pituitary hormones
ovary	• oestrogen • progesterone	• steroid • steroid	• reproductive tract, body generally • uterus	• promotes menstrual cycle, development of female features and behaviour • prepares uterus for, and maintains, pregnancy
posterior pituitary	• oxytocin	• peptide	• mammary glands	• causes release of milk; stretches cervix for birth
testis	• testosterone	• steroid	• reproductive tract, body generally	• development of masculine features and behaviour
placenta	• human chorionic gonadotropin (hCG)	• protein	• ovaries	• maintains corpus luteum for production of progesterone; stops ovulation

+ ADDITIONAL PSC

Menstrual cycle

After puberty, the balance of four different hormones controls a human female's menstrual cycle. The average cycle length is regarded as 28 days but individuals vary from 21 to 35 days and it may not be a regular pattern. Figure 2.3.20 shows the changes of hormone levels in conjunction with the ovarian cycle. It should also be noted that there are small changes in basal body temperature during the menstrual cycle. There is a drop just before and a rise just after ovulation but only by 0.2–0.5°C. These factors can be used for determining the most and least fertile times for a woman, although checking of cervical mucus and testing for hormone changes in the urine is more reliable. The oral contraceptive pill and the transdermal patch for birth control both use a combination of oestrogen and progesterone to prevent monthly ovulation and pregnancy.

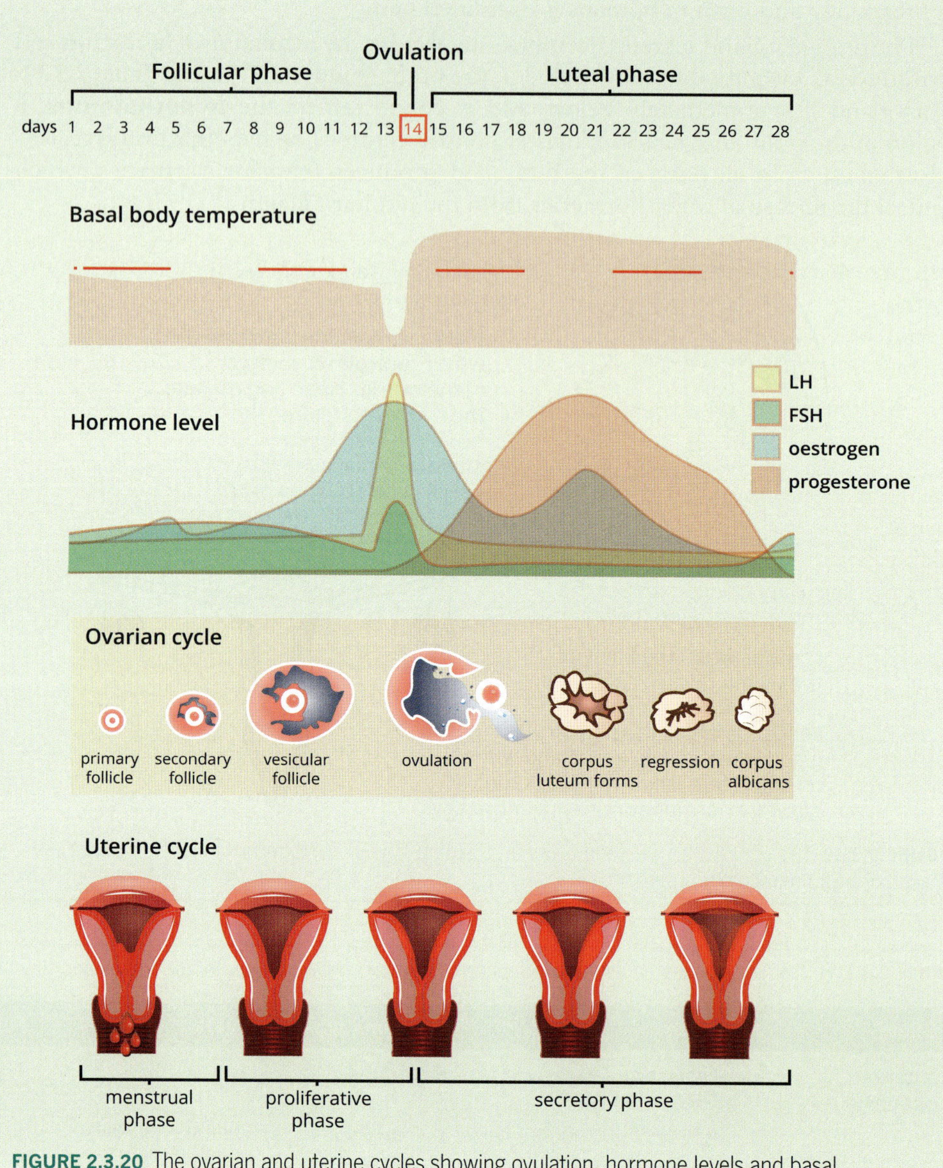

FIGURE 2.3.20 The ovarian and uterine cycles showing ovulation, hormone levels and basal temperature

The presence of human chorionic gonadotropin hormone (hCG) in urine is used as an indicator of pregnancy in over-the-counter pregnancy test kits. It can be detected as early as seven to nine days after fertilisation occurs.

Labour and birth in humans

A correct balance of hormones is essential to maintain the pregnancy, starting with **hCG (human chorionic gonadotropin)** from the placenta when the embryo has implanted. The hCG stimulates increased blood flow to the pelvic area and helps regulate the ovarian hormones.

Progesterone is required at high levels throughout pregnancy with levels steadily rising until the birth of the baby. Initially the progesterone comes from the corpus luteum then after six weeks the placenta produces it. Progesterone stimulates early preparation of the uterus for pregnancy and later it prevents lactation and uterine contraction until it is time for the birth.

Oestrogen levels also rise throughout the pregnancy to work in partnership with progesterone. Some organ development in the fetus, including liver, kidneys and lungs, requires oestrogen. This hormone promotes growth of breast tissue in preparation for maternal milk production.

There are several other human pregnancy hormones, not all of which have functions that are well understood. Some of the hormones cause side effects and emotional changes, including nausea (commonly known as morning sickness) and mood swings. These effects are mostly in the first trimester.

In the period just before a human birth the balance of two hormones, oestrogen and progesterone, changes. The natural level of **prostaglandins** increases which in turn increases the sensitivity of the cervix and uterus to **oxytocin**. Oxytocin is the hormone that causes uterine contractions. The hormonal changes create irregular uterine tightening or contraction. The fetus has usually moved with its head low in the pelvis, putting pressure on the cervix. This pressure stimulates further release of oxytocin and so labour begins. When the cervix reaches full dilation (10 cm or more in width), oxytocin and adrenaline hormones work together to start the final series of muscular contractions. After the baby is delivered the uterine contractions are maintained by oxytocin until the placenta is pushed out and the uterus starts shrinking back to normal size. Oxytocin also promotes the protective mothering instinct and works with **prolactin** to stimulate lactation for feeding the newborn baby (Figure 2.3.21).

Although the section above has specifically described hormonal controls for a human pregnancy, the hormones and features of pregnancy and birth are much the same in other mammals.

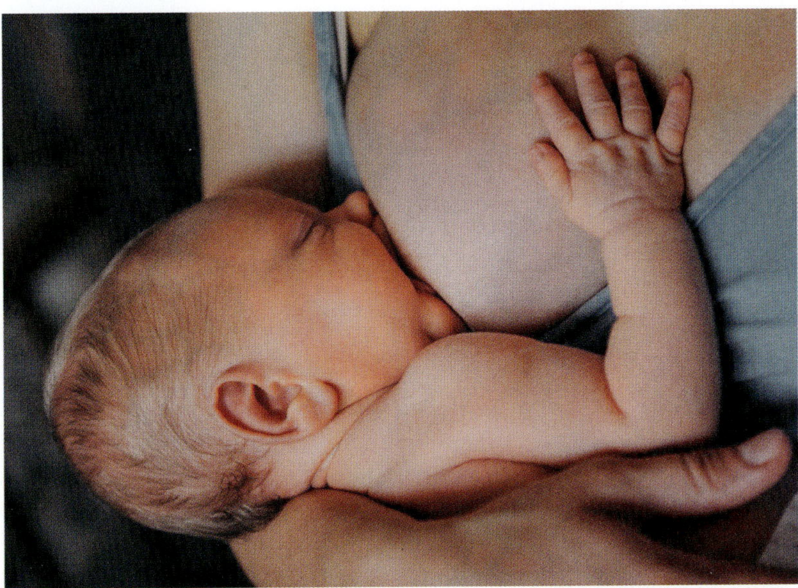

FIGURE 2.3.21 Newborn babies instinctively latch on to their mother's breast to feed soon after birth. Hormones such as oxytocin and prolactin play an important role in bonding between mother and baby and stimulating lactation.

> **i** Prostaglandins are a group of lipid chemicals found in most parts of the body. They play an important role in pregnancy and labour as they help control the induction of labour and uterine contractions.

2.3 Review

SUMMARY

- Sexual reproduction in multicellular organisms involves the fusion of gametes from two different individuals to form a zygote.
- Sexual reproduction involves equal genetic contributions from male and female parents and ensures the continuation of their species.
- The primary male sex organs are testes, which produce sperm.
- The primary female sex organs are ovaries, which produce eggs (ova).
- Secondary sex organs include other glands and organs involved in mating and reproduction.
- Sperm are made continuously after puberty and produced in the millions.
- Immature eggs (oocytes) are present in female ovaries before they are born.
- After puberty ova mature and are released every 21–35 days (average 28-day cycle) until menopause halts ovulation. It is normal to have some variations of the cycle.
- In mammals, the male ejaculates sperm into the female's reproductive tract for internal fertilisation in the oviduct.
- If fertilised, an egg provides a single set of chromosomes, nutrients for the growth of the embryo and regulatory factors that control early development.
- Within a week after fertilisation, a human zygote grows into a blastocyst before implanting in the wall of the uterus and forming a placenta that nourishes it until birth.
- In placental mammals, the blastocyst becomes an embryo then a fetus once it has developed major organs.
- Marsupials and monotremes follow a different course of reproduction relying on a pouch or an egg to protect their young as they develop.
- Each step of gamete formation, implantation, pregnancy and birth is controlled by a variety of hormones.
- Some hormones that are important in human reproduction are:
  - follicle stimulating hormone (FSH)
  - luteinising hormone (LH)
  - oestrogen
  - progesterone
  - testosterone
  - human chorionic gonadotropin (hCG).

KEY QUESTIONS

1. Distinguish between zygote, morula, blastocyst, gastrula, embryo and fetus.
2. Where would you find an acrosome and what is its function?
3. Where would you find a follicle and what is its function?
4. Identify each of the following as true or false.
   - **a** Human females produce new eggs throughout their lifetime.
   - **b** Human males and females both have separate openings for the urethra and reproductive tubes.
   - **c** A gastrula is the stage that implants and forms a placenta.
   - **d** A mature egg is much larger than a sperm.
   - **e** Multiple fertilisation of one egg is common.
   - **f** During cleavage mitosis produces many cells from a single zygote.
   - **g** The placenta mixes blood of the mother and fetus.
   - **h** Ultrasound scans are used for safe monitoring of a fetus.
5. Compare reproduction in placental and marsupial mammals.
6. Explain how two glands positioned in the brain, the hypothalamus and the pituitary, can control the reproductive organs.
7. Identify four hormones that are important in human reproduction and outline their roles.

2.4 Manipulation of plant and animal reproduction in agriculture

Humans have used **selective breeding** to produce animals and plants with more useful or more attractive characteristics for tens of thousands of years. This was done without using a high level of scientific or technological knowledge. They chose those animals or plants that expressed the characteristics they wanted to conserve and selectively bred them together, hoping that their offspring would show even more of these characteristics (Figure 2.4.1). In the past, selective breeding could only use characteristics that already existed in the genetic pool of a species (Figure 2.4.2). We now have the knowledge and skills to use cloning, recombinant DNA techniques, gene editing and transgenic technologies to transfer genes from one species to another and produce organisms with DNA combinations never seen before (Figure 2.4.3). This offers many benefits, but we also need to evaluate if the impacts of reproductive manipulation are environmentally, scientifically, socially and ethically acceptable.

In this section, you will learn about some examples of reproductive manipulation in agriculture and start to consider the advantages and disadvantages of such manipulations. Agricultural ecosystems are, by their very nature, low diversity systems dominated by only a few species that are protected by humans. You will gain an understanding of the importance of conserving species and genetic variation for the future.

Species diversity is the number of different species in a specific location or on the whole planet. Genetic diversity (also called genetic variation) is the diversity of genes within a species or population.

FIGURE 2.4.1 All of these animals that are farmed for human use have been selectively bred for desirable genetic traits. None of these animals are the same as their wild ancestors.

FIGURE 2.4.2 Reproduction of plants by cuttings is a simple, age-old form of cloning, still widely used in agriculture and horticulture. All the new plants will be identical genetic copies of the parent plant.

FIGURE 2.4.3 Trial plots of soybeans, including new genetically modified varieties, being grown in Wisconsin, USA

SELECTIVE BREEDING

Evolution by natural selection is an ongoing and, as the name implies, natural process. In addition, humans have been manipulating the gene pools of populations for thousands of years through deliberate selection of particular individuals for breeding. The process by which humans decide which individuals may breed and leave offspring to the next generation is called selective breeding or **artificial selection.**

All modern crops and livestock were developed by genetic manipulation of plant and animal species through the process of selective breeding (Figures 2.4.4 and 2.4.5). However, new molecular technologies are being used to alter the characteristics of organisms in a more targeted and specific way, and more quickly than by traditional breeding. These new methods, called **genetic engineering**, can also allow the exchange of genes between organisms that are sexually incompatible and normally cannot interbreed. New forms of plants and animals developed in this way are referred to as **genetically modified organisms** (**GMOs**). You will learn more about GMOs in Chapter 9.

GO TO ➤ Section 9.3 page 354

There are four basic steps that apply to all forms of selective breeding, whether it be with a plant or an animal.

1 Determine the desired trait.
2 Interbreed parents who show the desired trait.
3 Select the offspring with the best form of the trait and breed these offspring.
4 Continue this process until the population reliably reproduces the desired trait.

FIGURE 2.4.5 These varieties of maize, grown in Mexico, are among some of the first that were cultivated from the wild grass teosinte thousands of years ago.

FIGURE 2.4.4 Dairy cows, such as these Holstein Friesians, are selectively bred for their high milk yields.

Gene linkage

A common problem with selective breeding is **gene linkage**, meaning that it is not only the desired trait that is selected for. Other traits may be inadvertently selected because genes are carried on chromosomes with many other genes. Those genes that are located close together tend to travel linked together through cell division and into the offspring. For example, wheat that has been selected for the trait of high grain production may also carry a gene for tall stems that are not strong enough to support the ripe heads for harvesting (Figure 2.4.6). The success of selective breeding of both plants and animals may be limited by the presence of undesirable linked genes.

FIGURE 2.4.6 Plant geneticists select wheat varieties with high grain production, as well as strong, moderate-length stems. The danger in selective breeding is that high yield varieties may have gene linkage to tall stems that cannot support the weight of the grains.

Genes that are located close to each other on a chromosome are said to be linked. The genes usually stay linked together in cell division and are inherited as a group into the next generation.

Selective breeding in plants

Most selective breeding of plants is done to produce higher-quality food. Typically, seeds are collected from the individuals with the largest, most attractive or most numerous grains, fruits, nuts or other part of the plant that will be eaten. Those seeds are planted and the new generation of plants is cross-pollinated in a controlled way with other individuals having similar traits. The resulting plants produce larger, more nutritious or more aesthetically pleasing food products. Many food crops, such as cereals, tomatoes, potatoes and bread wheats have been modified by selective breeding to have higher yields, greater resistance to common diseases, to be more palatable or for improved nutritional value.

Once a desirable plant has been bred, artificial pollination or cloning methods, such as cuttings, grafts or tissue culture, may be used to mass produce identical plants by asexual means.

Examples of selective breeding in plants

- Maize or corn, (*Zea mays*)—maize is one of the most widely grown crops in the world. It is thought that maize was selectively bred from the wild grass, teosinte. Modern maize has significantly larger cobs with many more rows of much larger kernels compared to the ancestral teosinte. The higher-yielding modern maize provides more food for people than the ancestral form (Figure 2.4.5).
- Wheat (*Triticum aestivum*)—during 10 000 years of cultivation, numerous forms of wheat, many of them hybrids, have developed under a combination of artificial and natural selection. Modern wheat has become **polyploid** with strains that are tetraploid ($4n$, two sets of chromosomes) and hexaploid ($6n$, three sets). It has been selected for traits like high yield, high gluten content and heads that do not shed their seeds easily. The $4n$ variety called durum wheat is used for making semolina flour for pasta. The $6n$, called bread wheat, has several variants that are used for pastry, cakes and bread making. William Farrer (1845–1906) pioneered Australian wheat research when he systematically used cross-breeding (**hybridisation**) on his own property to improve bread wheats. Farrer was known to have cross-pollinated the plants using his wife's hairpins to transfer the fine grains of pollen until he acquired a pair of forceps (Figure 2.4.8).

FIGURE 2.4.8 William Farrer and his achievements with wheat hybrids were commemorated on the Australian two dollar note from 1966 until its replacement with a coin in 1988. The reverse side featured John Macarthur, who helped to establish the Australian wool industry.

BIOFILE EU S

Gene linkage in the silver fox

Silver foxes are a natural colour variant of the red fox (*Vulpes vulpes*) and range from blue-grey to black in colour (Figure 2.4.7). They were prized for their unusual colour and hunted for their fur. In 1959, Russian scientist Dmitri K. Belyaev began an experiment in which he selectively bred silver foxes for 'tameness'. He found that within eight to 10 generations the foxes showed clear signs of domestication, wagging their tails when people approached. However, they no longer resembled their wild ancestors. Instead, the domestic foxes had floppy ears, short or curly tails and their fur had changed considerably in colour and texture. The genes related to 'tameness' are carried on chromosomes linked with many other characteristics. The selection of this one trait affected the inheritance of other traits. By artificially selecting the tamest foxes, the gene frequencies for other linked traits changed, resulting in domesticated foxes with different phenotypes to the original wild population. Domesticated foxes are used in the agricultural industry for the commercial production of silver fox fur.

FIGURE 2.4.7 These foxes are both red foxes (*Vulpes vulpes*). (a) The silver colour is a natural colour variant of (b) the red fox.

FIGURE 2.4.9 Selective breeding has been used to biofortify the orange sweet potato for beta-carotene content, a precursor of vitamin A. They have been introduced recently to some African countries such as Uganda.

Biofortification is a term referring to improvement of nutritional quality of food crops through farming practices, conventional plant breeding or modern biotechnology as in the case of Golden Rice or the orange sweet potato. Often the goal is to prevent forms of micronutrient deficiency (e.g. vitamin A).

- Orange-fleshed sweet potatoes (OFSP)—the 2016 World Food Prize was awarded to a group of scientists who worked on breeding and introducing biofortified, orange-fleshed sweet potatoes to Africa (Figure 2.4.9). Their project recognised the importance of provitamin A, also known as beta-carotene, which is converted in the body to vitamin A. Deficiency of vitamin A is considered to be one of the most harmful forms of malnutrition in the developing world. It can cause blindness, limit growth, and weaken the body's immune system, especially in young children. An orange colour in vegetables indicates the presence of provitamin A. While orange-fleshed sweet potatoes are naturally occurring in South America, they were not grown in African countries where white-fleshed varieties are common and preferred (Figure 2.4.10). The biofortified varieties used for the OFSP project were specifically bred by plant breeders to be rich in provitamin A. A successful grassroots-education program has led to the introduction and acceptance of the orange variety in some African countries.

FIGURE 2.4.10 Ugandan women at a local market in Kampala selling the traditional white-flesh sweet potatoes and yams

Polyploidy

During the process of selective breeding of plants, it is quite common for polyploidy to emerge. This is the condition where the cell nucleus has more than two sets of chromosomes (e.g. $3n$, $4n$ or $6n$ rather than $2n$). Polyploidy can come about naturally through errors in meiosis where gametes may end up being diploid rather than haploid. In this case a fertilised egg becomes $3n$ or $4n$. In plants, a polyploid zygote often continues to develop into an adult plant that is sterile. Unlike vertebrate animals, many polyploid plant species can survive using asexual reproduction and continue to breed the polyploid variety into the future. Some banana varieties are triploid ($3n$); cultivated cotton and potatoes are examples of tetraploid ($4n$) organisms; bread wheats are hexaploid ($6n$); and strawberries are octoploid ($8n$) (Figure 2.4.11). In humans, polyploid zygotes do not survive. The condition is rare in animals but it is known to occur in some insects, earthworms and tree frogs.

Humans have selected for polyploidy in some crop plants because it can result in larger and more vigorous plants. It was discovered that polyploidy can be induced with a chemical called colchicine. When exposed to colchicine, the paired chromosomes are prevented from pulling apart during cell division, resulting in $2n$ gametes from meiosis or $4n$ cells from mitosis. As well as the examples listed above, polyploids are now found in a large number of agricultural crops such as turnips, spinach, apples, radishes, grapes and watermelons.

FIGURE 2.4.11 Strawberries are often polyploid. They can possess up to eight copies of each chromosome.

A drawback of inducing polyploidy in plants is that the seed crop produced by many polyploids is sterile or has lower fertility rates than their diploid types. In some situations, polyploid crops are preferred because they are sterile. For example, many seedless fruit varieties are seedless as a result of polyploidy, such as bananas, strawberries and seedless watermelon. To preserve the characteristics of the variety, such crops are propagated using asexual techniques, such as grafting, suckers, runners or tissue culture.

Hybridisation

The crossing of different varieties within one species to produce new varieties with different combinations of characteristics is one kind of hybridisation. In general, hybrid plants are more vigorous, higher yielding and may be more disease-resistant but the outcome of hybridisation is not always an improvement. It is also important to note that when hybrid offspring are produced artificially, they are designed to be cultivated or reared under controlled conditions of intensive agriculture, horticulture or farming and may not be suited to conditions in the wild.

When a hybrid has been deliberately bred, the induction of polyploidy is a common technique to overcome the sterility of a hybrid species. For example, triticale is the hybrid of a wheat (*Triticum turgidum*) and rye (*Secale cereale*) (Figure 2.4.12). It combines sought-after characteristics of both parent plants, but the initial hybrids are sterile. After polyploidisation using the chemical colchicine, the triticale hybrid becomes fertile and can continue to sexually reproduce itself, usually by self-fertilisation. At present, several types of triticale are grown for stock feed, either grain or forage, while research continues to try and improve it for human consumption.

FIGURE 2.4.12 Triticale, ripe ear and corns. Triticale is the hybrid of a wheat (*Triticum turgidum*) and rye (*Secale cereale*).

Heirloom plants

An heirloom plant, also known as a **heritage variety**, is a traditional cultivated plant that is maintained by small-scale gardeners and farmers. These may have been commonly grown during earlier periods in human history but are not used in modern large-scale agriculture. In modern agriculture most food crops are now grown using limited varieties in large, **monoculture** plots to keep the product of a consistent standard. These varieties have often been selectively bred for high productivity, ability to withstand mechanical picking and storage, and tolerance to drought, frost or pesticides. Fruit varieties such as apples have been propagated over the centuries through grafts and cuttings to maintain consistent traits such as size, colour and flavour. Many crop plants that are grown annually are from hybrid, or even genetically modified, seed purchased from a commercial supplier. The crop will be sterile as far as further reproduction goes and may have an intellectual property-right on it to prevent replanting. For example, in the USA (where it is legal), a grower of the patented GMO, Roundup Ready® soybean is prohibited by contract from saving and replanting the seed. This variety has been genetically modified to be resistant to herbicide spray used to kill weeds in the crop.

Heirloom gardening is a reaction against the limited varieties used in conventional agriculture and aims to preserve both species and genetic variation. The crops produced vary in output and quality, so are not acceptable to all consumers or economically viable for producers, especially in developed countries. However, maintaining species and genetic variation with heirloom varieties is a precaution against monopoly by a restricted number of companies. It also provides food security for the future in the face of climate change, new pests and diseases, salinity and other issues that may make current monoculture varieties no longer viable (Figure 2.4.13).

FIGURE 2.4.13 *Sorghum* is a genus of grass that includes many species and subspecies, some of which are native to Australia. Grain sorghum (*Sorghum bicolor bicolor*), which originated in Africa, is most commonly grown as a food source worldwide. It is important to conserve other wild species of *Sorghum* as sources of genetic variation that may be needed for crop breeding in the future.

BIOFILE EU S

Heirloom plants

It has been estimated that in 1880, farmers in the USA were growing more than 7000 named varieties of apple and since then 6800 have gone extinct. It is important to preserve genetic variation in agricultural populations to ensure the long-term evolutionary potential of crop varieties and food security for the future.

Many heirloom fruit and vegetables have kept their traits through open pollination, which preserves or increases genetic variation in the next generation. Carrots and tomatoes are examples. The original carrots were white or purple (Figure 2.4.14). Carrots of a rich orange colour are the product of a mutation selected by a Dutch horticulturist a few hundred years ago, because it was the colour of the Dutch Royal House of Orange-Nassau. Today in farmers' markets, the heirloom types are making a reappearance and heritage varieties can be appreciated as an interesting, colourful display (Figure 2.4.15).

FIGURE 2.4.14 Some varieties of heirloom Chantenay carrots

FIGURE 2.4.15 A collection of heritage tomatoes

BIOLOGY IN ACTION AHC CCT S

Selective breeding of edible Australian plants

There is renewed interest in the traditional bush plant foods of Australia's Indigenous peoples. Some have become widely used for cooking, often as flavourings such as lemon myrtle and finger lime. The macadamia nut, native to Queensland and selectively bred to improve nut size and flavour, has become popular all over the world. Bush tucker foods are much more varied than these examples and scientific research into their potential has only just begun.

Bush tucker

One staple food of desert Indigenous peoples is the maloga bean (*Vigna lanceolata*). It has an edible root and small beans that can be eaten raw. Another related species is *Vigna radiata*, the wild mung bean. Indigenous Australians did not traditionally make use of the bean's small black seeds, but in more recent times the wild mung bean has been selectively bred with other plants to produce the cultivated mung bean. The cultivated mung bean has a seed that can be green or black, and is more than double the size of the wild form (Figure 2.4.16). The cultivated plant also grows upright, instead of being a wiry creeper.

FIGURE 2.4.16 Seeds of wild mung bean (top) and the larger green seeds of the cultivated variety (bottom)

Conservation and use of wild genes

Indigenous wild plants are of interest to modern plant breeders. Wild plants are a source of genes that may be used for crop improvement. For example, crosses between wild and cultivated mung beans may produce hybrid forms best adapted to Australian soils and climates.

To conserve the genetic variation in wild populations for future use, native species that are the wild relatives of agricultural crops are collected and stored in a seed bank.

The CSIRO Centre for Plant Biodiversity Research has established a significant collection of another type of native pea, *Glycine*, which is related to the cultivated soybean. An Australian species of *Glycine* contains resistance to the leaf rust fungus, a trait needed for protecting soybean.

Functional genomics

Functional genomics aims to identify genes that determine particular functions. For example, it could be used to determine the genes that allow plants to grow under drier conditions, in saline soils or with resistance to fungal diseases. Once a gene is identified, collections of wild plants can be screened for natural variants of the desired phenotype (e.g. drought resistance). Any new and useful gene variants (alleles) discovered by genetic screening can be incorporated into existing agricultural variants through plant breeding by **cross-pollination**. Also, once identified, unwanted genes can be switched off and desired genes from other species can be introduced through genetic engineering.

Selective breeding in animals

Just as crops have been selectively bred for desired traits, so too have many animal species. In agriculture, sheep have been selected for the quality and quantity of the wool they grow, dairy cows have been selected for the milk they produce, beef cattle for their muscle mass and poultry for body weight and egg-laying. Aquaculture may involve selective breeding and is Australia's fastest growing primary industry producing fish, prawns, and shellfish including oysters and pearls. There are niche industries for sale of pedigree pets that involve intensive breeding programs.

When a selected species has a variety of traits, different traits may be useful in different situations. A single wild species can be the original source of a great variety of different breeds with an obvious example being domestic dogs bred from a single wolf species.

Examples of selective breeding in animals

- Domestic dogs—It is widely accepted that all today's domestic dogs were selectively bred from a wolf species. Today there are hundreds of domesticated dog breeds, some of which would be unlikely to survive in the wild. Examples include soft-mouthed, strong swimming dogs such as Labradors, which were bred for duck hunting, and sheepdogs bred for their intelligence and agility (Figure 2.4.17a). Humans have also artificially selected for a wide variety of traits in chickens, horses and many other domestic animals to produce gene pools that consistently produce the desired traits. Sometimes this has led to inbreeding with questionable outcomes for the health and welfare of the animal, such as the huge head and narrow hips of the bulldog that usually means their pups must be born by Caesarean section (Figure 2.4.17b). Some other issues with dog breeds are hip and arthritis problems, abnormally small skulls that cause pressure on the brain and breathing difficulty, and overgrowth of hair that covers and irritates their eyes (Figure 2.4.17c).
- Poultry—In the Australian poultry industry, scientific research is fundamental to their breeding programs. Geneticists select birds with the best characteristics for egg or meat production. Today's meat chicken looks quite different from its wild ancestor and is quite different from modern laying chickens, which are bred specifically for egg production (Figure 2.4.18). New genetic strains are imported into Australia as fertilised eggs, then hatched and reared to nine weeks under quarantine. Birds released from quarantine become the great-grandparents of the chicken you purchase at your local shop or butcher. These valuable great-grandparent eggs are imported from a handful of companies based in the USA and the UK that specialise in selective breeding of meat chickens.
- Prawns—The Australian black tiger prawn was developed through collaboration between CSIRO and industry partners. CSIRO scientists undertook a 10-year program of selective breeding in order to develop the prawn now being used by prawn farmers in Australia. The new breed has improved growth and survival rates, boosting pond yields by more than 50%. The average industry productivity for farmed prawns is five tonnes per hectare. The average yield of the new breed in 2010 was 17.5 tonnes per hectare.

FIGURE 2.4.18 (a) Wyandotte chickens are a dual purpose breed kept for its brown-shelled eggs and yellow-skinned meat. (b) Today's wild red jungle fowl of southeast Asia is the ancestral type of the modern domestic chicken.

FIGURE 2.4.17 Domestic dogs have been (and continue to be) selectively bred for a wide variety of characteristics. (a) Sheepdogs are bred for energy, agility and their hardworking nature. Other dogs are bred for their attractiveness or prize winning qualities, such as (b) the English bulldog and (c) the Lhasa Apso.

BIOFILE EU S

Musk ox

The musk ox (*Ovibos moschatus*) (Figure 2.4.19) is a large Arctic mammal, prized for its thick wool-like fleece called qiviut. Qiviut is a prized luxury item in North America and musk ox meat is considered a lean alternative to beef. Unregulated hunting of the musk ox led to the near extinction of the species in the late 1800s. Conservation efforts have allowed the species to survive. Hunting restrictions were introduced and musk ox from the surviving populations were relocated to repopulate regions where the animals had died out.

In the 1950s, the Musk Ox Farm Project was set up in Alaska in an attempt to domesticate the animals. Thirty-three individuals were captured from wild populations and selectively bred for domestication. Several other musk ox farms appeared after this time, greatly reducing the reliance on hunting. The domesticated individuals have been kept as livestock to create sustainable farms, in which qiviut is combed out of the living adults.

In many regions of the Arctic Circle, domesticated musk ox were released into the wild where the native populations had been hunted to extinction. In the 1970s, one farm in Northern Quebec closed due to poor profits and 54 musk ox were released into the wild. Slowly, the native population of musk ox has increased to over 1000 adults, all descending from domesticated individuals.

FIGURE 2.4.19 The musk ox (*Ovibos moschatus*) nearly became extinct in North America before captive breeding started for its fleece and meat. Now some wild populations are starting to establish again.

GENETIC MODIFICATION AND CLONING OF PLANTS AND ANIMALS

Over the last few decades, scientists have developed techniques that allow for the alteration of an organism's genome and for the transfer of genes from one organism to another. Genetic technologies are used in plant agriculture to increase crop productivity, provide crop resistance to insects and prevent disease. Genetic modification (GM) of animals has also been used to improve fertility, and the quality and yield of meat, milk, eggs and wool.

Cloning plants to retain desirable traits has been used for centuries in horticulture and agriculture. However, the direct manipulation and cloning of cells and embryos is a more recent development in agriculture. Cloning is the production of new individuals that contain the same genetic information as the parent organism. Natural clones are produced by asexual reproduction when a single parent cell divides to produce two new identical daughter cells. You learnt about asexual reproduction in Section 2.1. The term cloning is also used to refer to artificial methods of producing genetically identical organisms. The cloning techniques used in agriculture are artificial methods carried out by humans and include cuttings and grafts, tissue culture (Figure 2.4.20), embryo splitting (or artificial embryo twinning) (Figure 2.4.21) and somatic cell nuclear transfer (SCNT).

Genetic technologies in agriculture will be examined in more detail in Chapter 9.

ℹ Cloning results in the production of an organism that is a genetically identical copy of the original. Its DNA has not been altered at all, in fact the aim is to keep the new organism exactly the same.

GO TO ➤ Section 9.2 page 349

FIGURE 2.4.20 (a) Scientists working in a tissue culture laboratory. (b) Round-leaved sundews (*Drosera rotundifolia*) growing from tissue cultures in a petri dish. This carnivorous plant, endemic to North America, is endangered. Tissue culture allows large numbers of these plants to be produced rapidly for research and revegetation.

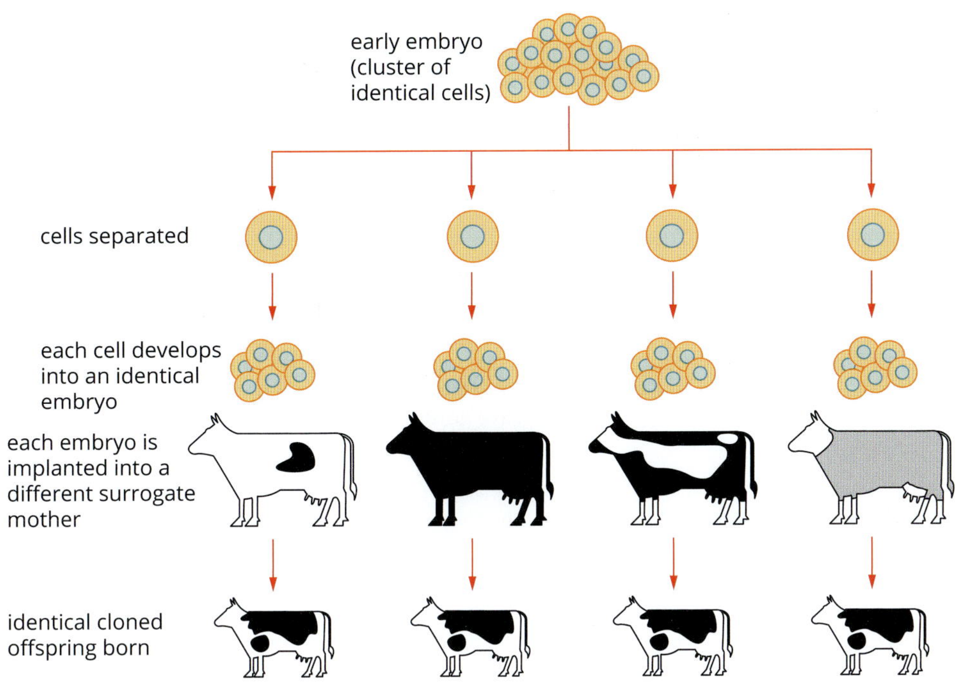

FIGURE 2.4.21 Cloning by embryo splitting is used in livestock breeding to increase the number of offspring produced each season.

BIOFILE EU S

Regulating GMOs

The European Food Safety Authority has taken a strong lead on regulation of GMOs. They have provided clear definitions and guidelines for applications to introduce any GMO to countries in the European Union. Their criteria provide a good summary of the issues involved.

GMO applications must address seven areas of potential risk for GM animals:

1. persistence and invasiveness of the GM animal, including vertical gene transfer
2. horizontal gene transfer
3. interactions of the GM animal with target organisms
4. interactions of the GM animal with non-target organisms
5. environmental impacts of the specific techniques used for the management of the GM animal
6. impacts of the GM animal on biogeochemical processes
7. impacts of the GM animal on human and animal health.

IMPACT OF MANIPULATING PLANT AND ANIMAL REPRODUCTION IN AGRICULTURE

By looking at the examples outlined in this section, it is clear that selective breeding and genetic modification has already brought considerable benefits to humankind. The quality and quantity of our food has been improved, sometimes costs have decreased, and plants and animals have become more suitable for human purposes. The improvements have not come without costs and controversy, especially in relation to animal welfare and long-term food security. Given the rapid development of genetic technology and its many potential applications in agriculture, the impacts of this technology must be evaluated. Some of the potential problems that have been raised concerning the manipulation of plants and animals in agriculture are outlined in Table 2.4.1.

Genetic modification (GM) involves deliberate altering of the genetic material of an organism. DNA sequences will have been inserted, removed or modified in order to introduce a new trait or change a characteristic such as the disease resistance of a plant.

Transgenic organisms are organisms that have genes from another species inserted into their DNA.

TABLE 2.4.1 Concerns associated with the manipulation of plant and animal reproduction in agriculture and their potential impact

Issue or concern	Impact
Health of the animal or plant	Crops approved for commercial growth appear relatively unaffected by a transgene (apart from the new trait). Transgenic animals may experience adverse effects from transgenes that affect growth rates.
Danger of altering our own DNA by eating GMOs	One argument against GMOs is the claim that eating these foods will change our own DNA in unpredictable ways. There is no reason to think this is the case as animals have been eating the DNA of plants and other animals for millions of years and this DNA does not result in any change to the genome of the consumer.
Uncontrollable pest plant species	Crops that have been modified for herbicide or insecticide resistance may breed with other plants, producing a hybrid pest species that farmers may not be able to control. Plants do not usually pollinate different species, so this is only a potential concern for cross-pollination with wild relatives, such as between crop canola and wild canola (see below). Uncontrollable growth would be a concern if the transgenes promote rapid growth and there was a selective advantage to having the genes in a wild environment.
Cross-pollination between GM crops and non-GM crops	Pollen distributed by wind and insects can transfer genes between neighbouring plots of the same species. Cross-pollination may happen in both directions and so will have an impact on both types of crop.
GM animals may interbreed or compete with natural populations	This may be a potential concern if there are wild relatives of the GM animals (animals usually don't breed with unrelated species) and there is a selective advantage to the presence of the transgene in a wild population (e.g. GM salmon escaping from aquaculture enclosures adjacent to the ocean and breeding with wild salmon. For this reason the adult GM salmon are made sterile).
Loss of biodiversity	As more farmers use GM crops or animals, there will be fewer crops grown that vary from these, reducing the genetic variation of some species. If disease or environmental change occurs, there could be widespread and catastrophic effects on food production. However, this is already an issue in modern agriculture where monocultures and commercial practices have reduced genetic variation due to selective breeding over many centuries. GM technology may accelerate the trend or it could be used to re-introduce characteristics lost from wild relatives. One solution is the use of seed banks to store samples of plants so that they are available if needed in the future.
The evolution of new species by artificial selection	Selective breeding, hybridisation, polyploidy and genetic modification can all lead to speciation (i.e. the evolution of new species). In cases of selective breeding, speciation may be induced artificially by human pressures, rather than by natural selection under environmental pressures. This could be regarded as an increase in biodiversity but there is debate over the value of such speciation and whether it destabilises natural ecosystems.
Reduced genetic variation	Selective breeding within a small population can further reduce genetic variation increasing the likelihood of detrimental homozygous recessive traits in offspring (i.e. one recessive allele is inherited from both parents). Genetic abnormalities that would be selected against in natural conditions may be maintained in selectively bred populations. For example, an eye disorder, known as progressive retinal atrophy, is common among many breeds of pure bred dogs.
Gene linkage	A complicating factor when selecting for a specific trait is gene linkage. This means that selecting for one trait will unavoidably carry other linked traits with it during cell division, some of which may be undesirable. The new techniques of editing individual genes have the potential to remedy issues associated with gene linkage.

The applications of genetic and reproductive technologies and the associated social and ethical concerns will be examined in more detail in Chapters 8 and 9.

+ ADDITIONAL

Hybridisation and artificial pollination

Gregor Mendel (Figure 2.4.22), often called the father of modern genetics, investigated hereditary patterns of hybrids. Hybrids are the offspring from two organisms that are of different breeds, varieties or species. Hybrids create new cultivars, from new apple varieties to tangelos to hybrid corn. Some mammals produce hybrids; a mule is the progeny of a horse and a donkey. Crossbreeding of stock has produced the highly rated Brahman cattle breed and hybrids from it that are credited with saving the cattle industry in Australia (Figure 2.4.23). Hybridisation is in fact a form of selective breeding and is widely practiced by home gardeners, horticulturists, farmers and research scientists.

FIGURE 2.4.22 Gregor Mendel (1822–1884), Austrian monk and scientist, who founded the field of genetics with his experiments involving selective breeding of pea plants.

FIGURE 2.4.23 Brahman cattle, first crossbred in the USA from four different Indian zebu breeds, have been hugely successful in the harsh Australian climate and because of their tick resistance. Over 50% of Australian cattle are now either Brahman, or Brahman crossed with European breeds.

Mendel chose common garden pea plants (*Pisum sativum*) for the focus of his research because they can be grown easily in large numbers and their reproduction can be manipulated. In his experiments, Mendel was able to selectively cross-pollinate purebred plants with particular traits and observe the outcome over many generations. This was the basis for his conclusions about the nature of genetic inheritance. Mendel first had to make sure the flowers did not pollinate themselves. To do that, he used forceps to remove the anthers, their pollen-bearing parts. Then he immediately dusted pollen from other selected pea plants, using a fine paintbrush, onto the stigma of the altered plants to fertilise them. In this way he was able to control the breeding of the pea plants.

Many home gardeners practice selective pollination by hand (Figure 2.4.24). As a general rule, they transfer pollen in the early morning while it is likely to be most healthy. A simple technique is to touch a cotton bud or brush tip to the pollen-covered anther on the chosen flower. Alternatively, tweezers can be used to pluck off the anther and filament and use the entire structure for pollination. They then touch the pollen-covered cotton bud, brush or the anther itself, to the receptive stigma (look for a shiny, sticky surface) on the chosen flower. A successful attempt will see pollen grains adhering to the stigma.

FIGURE 2.4.24 A home gardener pollinating cucumber blossoms by hand with a cotton bud

2.4 Review

SUMMARY

- Evolution of organisms can occur through natural selection or artificial selection.
- Selective breeding is the traditional form of artificial selection in agriculture. In selective breeding, humans select desired traits and interbreed plants or animals with these traits.
- Agricultural plants are typically bred for high yield and high resistance to common diseases. Animals are often bred for high quality traits and products (such as wool and milk), or for personality traits (such as loyalty in pets).
- There are four basic steps that apply to all forms of selective breeding:
  - determine the desired trait
  - interbreed parents who show the desired trait
  - select the offspring with the best form of the trait and interbreed these offspring
  - continue this process until the population reliably reproduces the desired trait.
- More recent molecular technologies have allowed for faster development of genetically modified organisms and for the transfer of DNA between species that cannot interbreed normally to form transgenic species.
- Selectively bred populations tend to have low genetic variation meaning that:
  - they are more susceptible to environmental change
  - biodiversity may be reduced if selectively bred populations replace wild populations and varieties
  - an increase in the incidence of genetic abnormalities can occur.
- Humans have been cloning organisms for hundreds of years, particularly in plant horticulture. Modern science has seen the development of much more advanced cloning techniques.
- Cloning techniques include cuttings and grafting, plant tissue culture, embryo splitting and somatic cell nuclear transfer (SCNT).
- Genetically modified organisms (GMOs) are organisms with modifications made to one or more of their natural genes.
- Transgenic organisms carry one or more genes inserted into their genome from a different species.
- Reproductive manipulation involving selective breeding and genetic modification has benefited humans by improving the quality, quantity and appearance of our food, improving the efficiency of food production and decreasing the cost of food.
- A range of social, ethical and scientific issues arise from the application of reproductive and genetic technologies in agriculture, including animal welfare concerns, reduced genetic variation in species used in agriculture, and cross-pollination between GM and non-GM crops.

KEY QUESTIONS

1. Identify the steps of artificial selection that would lead to the production of large corn cobs.
2. Outline how selective breeding has changed with the development of new technology.
3. Identify two common methods of plant propagation used in selective breeding.
4. Outline four methods of cloning currently used in horticulture and agriculture.
5. Recall a situation where plant tissue culture is of benefit.
6. Selective breeding of animals has been used with great success and resulted in healthy animals. In other cases it may not be in the best interests of the animals' welfare. Outline an example of each.
7. Make a list of at least 10 examples where the reproduction of plants and animals has been manipulated by humans.
8. Discuss the case for and against the manipulation of plant and animal breeding in agriculture. Use some specific examples in your answer.

Chapter review

02

KEY TERMS

acrosome
alternation of generation
amniotic cavity
angiosperm
anther
artificial selection
asexual reproduction
binary fission
blastocyst
budding
carpel
cervix
chromosome
cleavage
clone
conception
conjugation
corpus luteum
cross-pollination
diploid
ectoderm
egg
embryo
endoderm
endometrium
epididymis
fallopian tube
fertilisation
fetus
filament
fimbria (pl. fimbriae)
follicle
follicle stimulating hormone (FSH)
fragmentation
gametangia
gamete
gametophyte
gastrula
gastrulation
gemmule
gene linkage
genetic engineering
genetically modified organism (GMO)
germ cell
germ layer
gestation
gonad
gymnosperm
haploid
hCG (human chorionic gonadotropin)
hermaphrodite
hormone
hybridisation
hypha (pl. hyphae)
hypothalamus
implantation
karyogamy
lactation
luteinising hormone (LH)
macronucleus
marsupial
megaspore
meiosis
meiospore
menopause
menstruation
mesoderm
micronucleus
microspore
mitosis
mitospore
monoculture
monotreme
morula
mutation
mycelium
oestrogen
oocyte
ovary
oviduct
oviparous
ovulation
ovule
ovum (pl. ova)
oxytocin
parthenogenesis
penis
pistil
pituitary gland
placenta
placental
planula
plasmogamy
pollen
pollen tube
pollination
polyp
polyploid
pregnancy
progesterone
prolactin
prostaglandin
regeneration
rhizoid
selective breeding
self-pollination
semen
seminiferous tubule
sexual reproduction
somatic cell
sperm
spermatocyte
sporangium (pl. sporangia)
spore
sporophyte
stamen
stem cell
stigma
strobilation
style
testis (pl. testes)
testosterone
umbilical cord
urethra
uterus
vagina
vas deferens
vegetative reproduction
viviparous
zona pellucida
zygote

REVIEW QUESTIONS

1 A horticulturist grows a range of plants for sale. The table below outlines the reproductive strategies of the plants.

Plant	Type of reproduction
tulip	asexual
poppy	sexual
lily	asexual
strawberry	asexual and sexual

If a virus infects all the plants, which plants are most likely to survive?

A poppy and strawberry
B tulip and lily
C tulip, lily and strawberry
D poppy only

2 What type of reproduction and reproductive structures are responsible for the mouldy bread in your pantry?

3 Why is meiosis a necessary process in living organisms that reproduce sexually?
  A It happens in the reproductive organs.
  B It is necessary for the growth of an organism.
  C It produces new cells to replace dead or dying cells.
  D It enables each parent to contribute genetic information to the offspring.

4 Identify the classification category for each organism below and if they use asexual or sexual (or both) methods to reproduce. The first example has been completed for you.

Organism	Classification	Reproduction
yeast	Fungi	asexual (budding)
mushroom		
bacteria		
malaria parasite (*Plasmodium* sp.)		
potato cultivars		
strawberry cultivars		
bracken		
cypress pine tree		
Eucalyptus (gum) tree		
echidna		
kangaroo		
elephant		

5 Complete the missing labels for reproductive organs of the flower.

6 Clarify the difference between pollination, fertilisation and seed dispersal in flowering plants. Use examples to support your answer.

7 *Paramecium* protists can reproduce sexually. Name the process and list the following events for sexual reproduction in *Paramecium* in order, from first to last.
- three haploid micronuclei (n) degenerate
- separation of mating partners
- micronucleus undergoes meiosis
- mating partners become attached
- exchange of micronuclei
- fusion of micronuclei ($2n$)

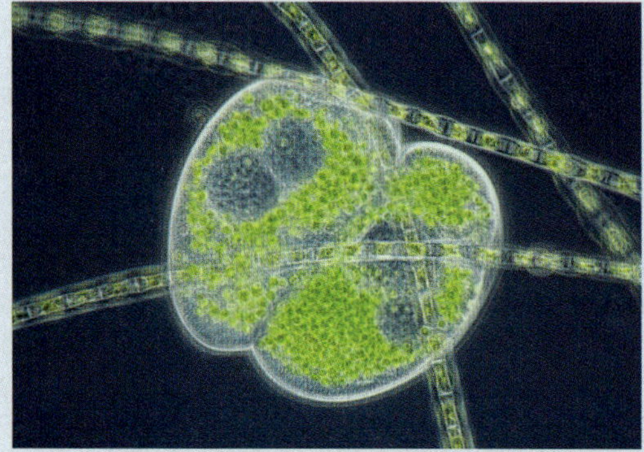

8 It is common to see red colours used for humming bird feeders in North and South America.
  a Suggest why this colour is chosen.
  b Identify the flower's natural process that is modelled by the use of an artificial feeder.
  c Does the use of a feeder pose any threat to this natural process?

9 The diploid number of chromosomes in horses is 64. State the number of chromosomes you would expect in a horse's:
  a fertilised ovum (zygote)
  b sperm cell
  c somatic cell

10 Do hermaphrodites reproduce via asexual or sexual reproduction? Explain your answer.

11 Identify each of the following statements as true or false.
  a Sexual reproduction produces offspring identical to their parents.
  b If a hybrid offspring is reproduced between a red kangaroo ($n = 10$) and a grey kangaroo ($2n = 16$), the diploid number of the offspring is 18.
  c Bony fish and frogs both use internal fertilisation.
  d All mammals use internal fertilisation.
  e Conifers are trees that reproduce with spores formed on cones.
  f Spores can be produced asexually or sexually.

12 a Which type of reproduction is common in many invasive species?
  b Discuss why this strategy makes organisms successful invaders of a new habitat and what impact this has on the native species in that environment.
  c Why is asexual reproduction more likely to be successful in the short term rather than the long term?

13 Name the developmental stages of an embryo shown below and number them to indicate the order in which they occur.

 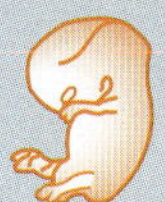 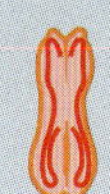 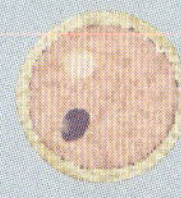 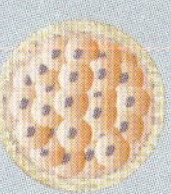 

14 Identify three hormones that are important in fertilisation, pregnancy and birth in mammals. Outline the role of each.

15 What does 'genetic modification' of an organism mean?

16 Selective breeding:
  A reduces biodiversity
  B reduces resistance to environmental change
  C increases genetic abnormalities
  D all of the above

17 A cloned organism:
  A increases biodiversity
  B is an identical copy of its parent
  C can only be a plant
  D all of the above

18 A transgenic organism:
  A is genetically modified
  B is a copy of its parent
  C is a common type of hybrid
  D reduces biodiversity

19 Wild bananas of south-east Asian countries have seeds embedded in their fruit.
  a Account for the lack of seeds in the bananas grown in Australia.
  b Assess the food security issues associated with growing seedless bananas.

20 A typical example of a breeding program in agriculture is for increased egg production in chickens. At the start of a particular breeding program, the average number of eggs per hen per year in a flock was 125. Hens that produced the most eggs per year were chosen as the female parents of the next generation. Roosters used in the program were the offspring of high-yielding hens. The average number of eggs per hen per year increased from 125 to 230 over 15 years, but the rate of increase was slower in subsequent years.
  a Outline how the breeding program is an example of artificial selection by humans.
  b What possible reasons might account for the slower rate of the increase in egg production for the subsequent years?
  c What are the possible negative effects on chickens of increasing egg production?

21 After completing the Biology Inquiry on page 70, reflect on the inquiry question: How does reproduction ensure the continuity of a species? Compare asexual and sexual reproduction and describe the short-term and long-term outcomes for offspring, populations and species produced via each method of reproduction.

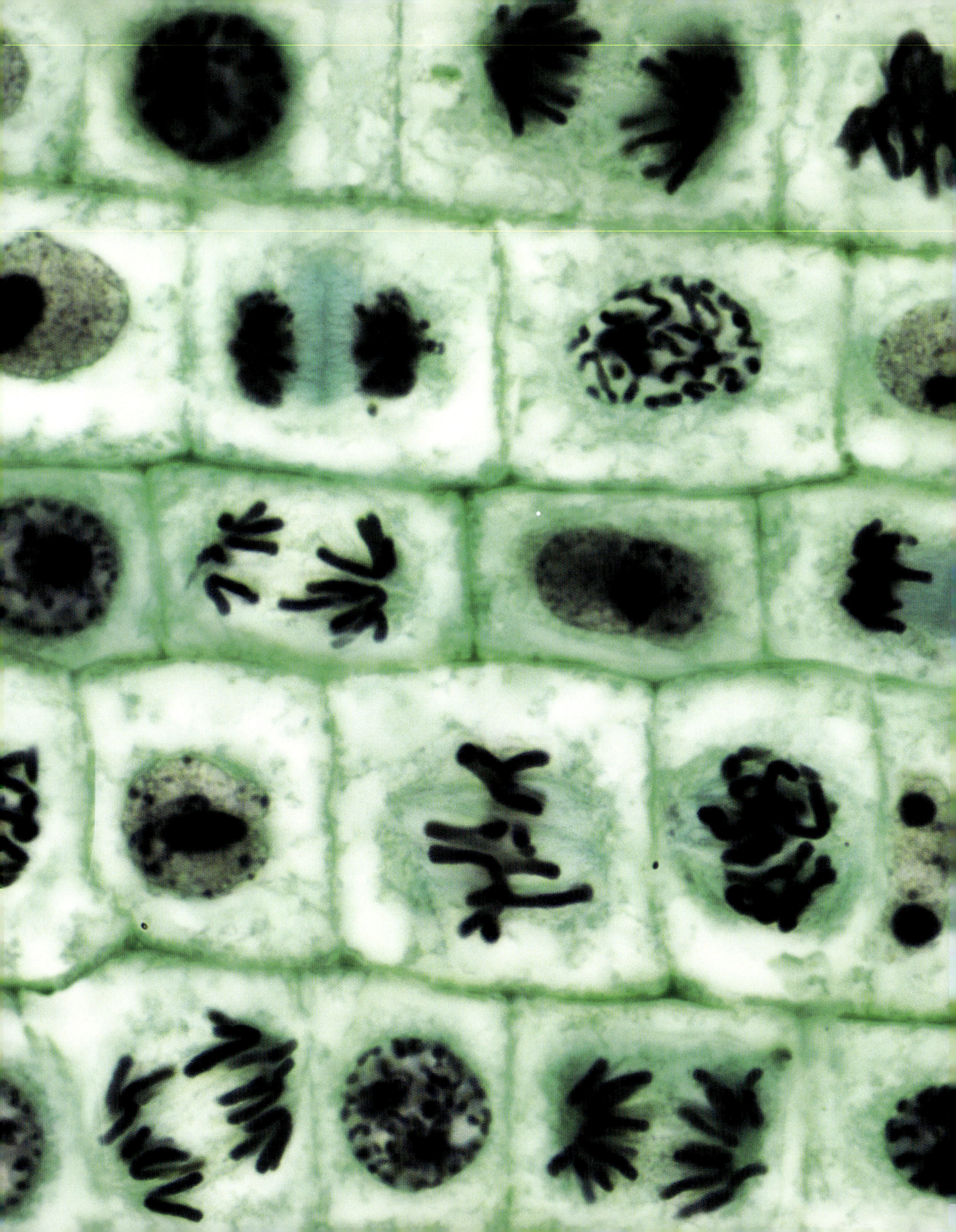

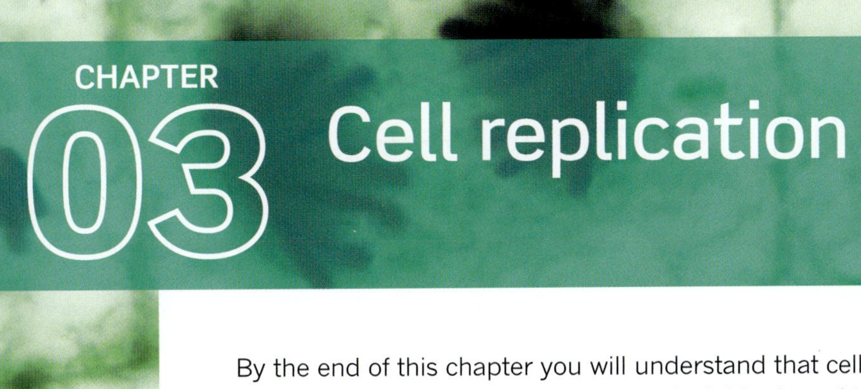

Cell replication

By the end of this chapter you will understand that cells are able to replicate in order to maintain continuity of life for individuals and their species. You will learn about the two processes of cell division—mitosis and meiosis—as well as the different outcomes for these two processes. You will examine the composition of nucleotides and the pairing and bonding between them to form the double-stranded helical structure of DNA.

Both cell division and replication are essential processes for the continuity of species. In some situations it is vital for cell replication to result in daughter cells with identical genetic information for the growth and health of individual organisms. In other contexts, genetic variation is a desirable outcome of cell division and may lead to the evolution of new species. The issues associated with defining the concept of a species are discussed as well as genetic changes within species.

Content

INQUIRY QUESTION

How important is it for genetic material to be replicated exactly?

By the end of this chapter you will be able to:

- model the processes involved in cell replication, including but not limited to:
  - mitosis and meiosis (ACSBL075) CCT ICT
  - DNA replication using the Watson and Crick DNA model, including nucleotide composition, pairing and bonding (ACSBL076, ACSBL077)
- assess the effect of the cell replication processes on the continuity of species (ACSBL084) ICT

3.1 Mitosis and meiosis

Cell theory states that all **cells** arise from pre-existing cells. In order for this to occur, cells must be able to replicate. This process is essential to the life of all organisms. Once you were a single cell—a fertilised egg known as a zygote. Now, your body is made up of about 37 trillion cells with many different specialisations (Figure 3.1.1). In order to start producing the millions of cells that make you, that first single-celled zygote had to replicate itself (Figure 3.1.2). Since then, as the cells in your body wear out and die or are damaged, more cells are replicated to replace them throughout your lifetime.

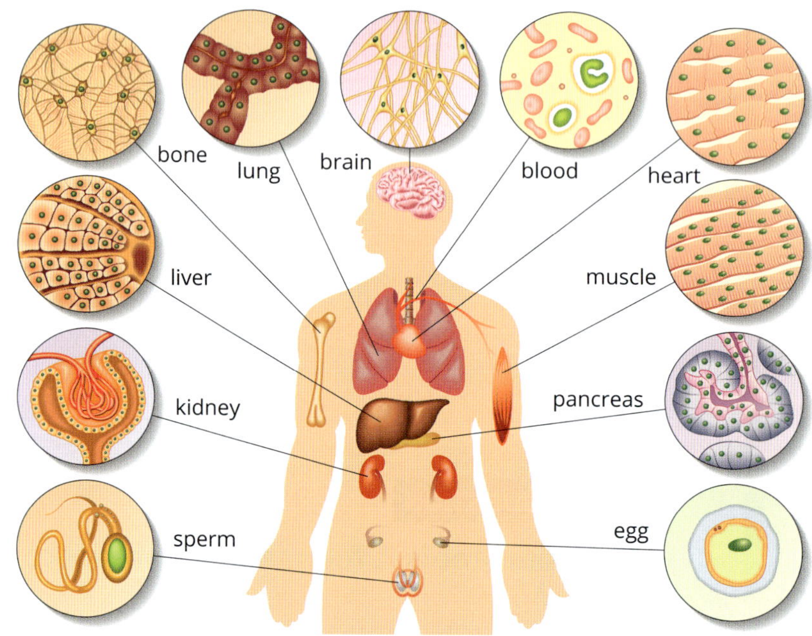

FIGURE 3.1.1 Some of the many cells that make up a human body. All cells originate from replication of one fertilised egg cell (zygote).

FIGURE 3.1.2 Early development of a human embryo by cell replication after the egg is fertilised by one sperm

In this section, you will learn about the process of identical **cell replication** called **mitosis** and the cell division process to form **gametes** called **meiosis** (Figure 3.1.3). You will compare the two processes and come to appreciate the complexity and importance of both in living things.

Replication and reproduction in multicellular organisms are different processes. Replication produces two genetically identical cells from one parent cell. Reproduction produces a new organism from one or two parent organisms.

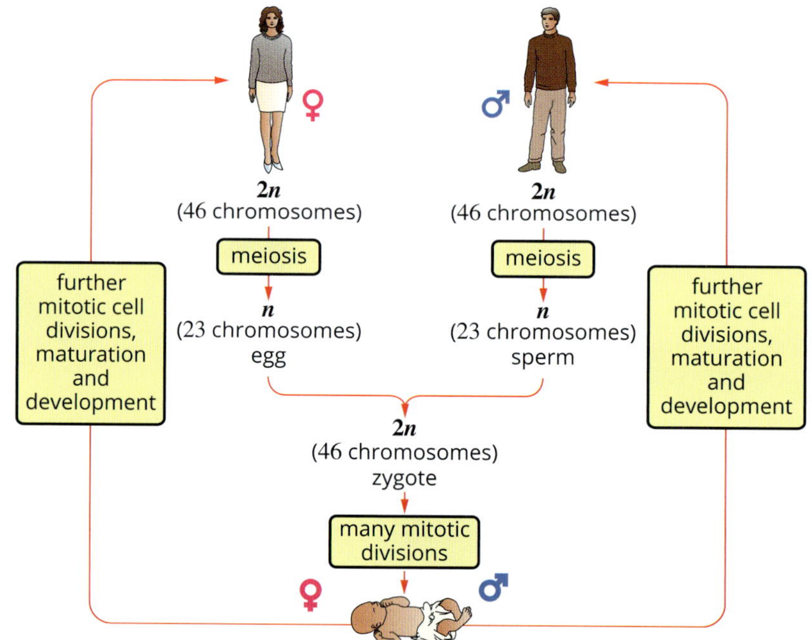

FIGURE 3.1.3 The cell division processes of mitosis and meiosis ensure growth and reproduction.

WHY DO CELLS REPLICATE?

Cell replication is the form of cell division in which a parent cell divides to produce two **daughter cells**. In eukaryotes (protists, fungi, plants and animals), cells replicate by mitosis. For a multicellular organism, cells replicate for the following reasons.

- Restoring the nucleus-to-cytoplasm ratio—egg cells have an unusually large volume of cytoplasm, too much for the nucleus to control. After fertilisation of an egg, the early rounds of mitosis proceed without pausing for cell growth, a process called **cleavage** that decreases the volume of cytoplasm per cell (see page 121 for more detail about cleavage).
- Growth and development—multicellular organisms grow in size by increasing the number of their cells through repeated cell replications. The new cells then grow in size, increasing the size of the organism. As the new individual continues to develop, new cells become specialised for different purposes, such as muscle cells in animals and phloem cells in plants. More replications follow and the specialised cells become organised into tissues, which form the body of the organism.
- Maintenance and repair—cells become damaged or die as a result of normal functioning, and also as a result of injury, such as sunburn in the case of human skin cells. Maintaining and repairing tissues requires the production of identical new cells to replace those that die. The new cells are produced by the cell replication process of mitosis.

Unicellular organisms do not need to replicate for these purposes because they remain a single cell throughout their entire life cycle. Instead, cell replication in unicellular organisms (whether prokaryotes or eukaryotes) is a simple form of reproduction and creates new, genetically identical individuals.

THE CELL CYCLE

The eukaryotic **cell cycle** is the life cycle of a cell, involving growth, replication of **DNA (deoxyribonucleic acid)** and division to produce two identical daughter cells (Figure 3.1.5). The cell cycle has three main phases:

- interphase
- mitosis
- cytokinesis.

These phases always occur in this order, with a cell spending most of its time in **interphase** (phases G_1, G_0, S, G_2) (Figure 3.1.5). During interphase the cell doubles its mass and duplicates its entire components. During mitosis the nucleus divides, and during **cytokinesis** the cytoplasm divides (Figure 3.1.6). The cell cycle is the period between one cytokinesis and the next. The time for a cell cycle is called generation time (T) and varies considerably between different cell types.

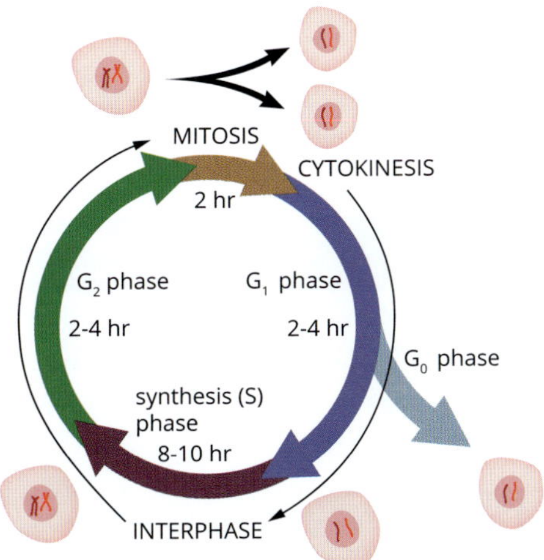

FIGURE 3.1.5 The cell cycle. Mitosis and cytokinesis (top) occupy only a small part of the whole cycle. A cell spends the majority of its time in interphase (G_1, G_0, S and G_2). The cell cycle shown here takes approximately 14–20 hours to complete but other cell cycles can take much longer.

BIOFILE CCT

Cell cycle 'gaps'

In 1879, German physician and professor of anatomy Walther Flemming (Figure 3.1.4) was the first to observe the behaviour of the chromosomes during cell division. He achieved this by using a stain he had developed that highlighted the nucleus of the cell during cell division. Because of the limited power of the microscopes available at the time, all the activity of the cell during interphase was not evident. For this reason the phases of interphase were incorrectly named Gap 1 (G_1) and Gap 2 (G_2).

We know now that interphase is a period of growth and activity in the cell cycle but have retained the G_1 and G_2 codes. You could usefully remember them as Growth 1 and Growth 2.

FIGURE 3.1.4 Walther Flemming (1843–1905), a pioneer in research on cell division

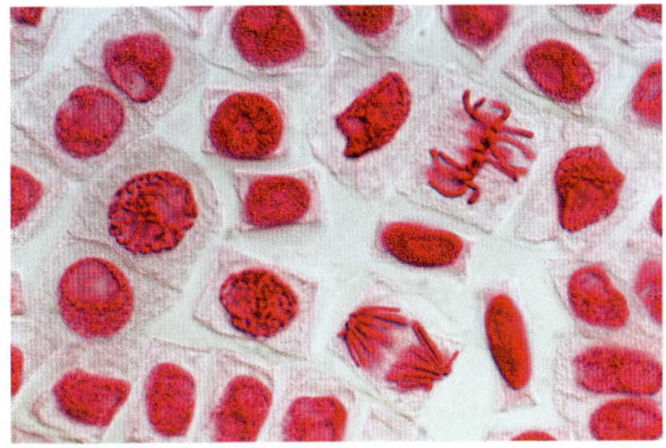

FIGURE 3.1.6 Light micrograph (LM) of hyacinth root cells undergoing mitosis, the phase of the cell cycle when the nucleus divides. The thick threads visible in two cells are chromosomes, at two different stages of the mitotic process.

BIOLOGY IN ACTION CCT N

The life span of cells

It has only been in the last decade that the life span of human cells could be determined accurately. The difficulty lay in being able to measure back in time to when each mature cell was born. Cell cycles measured in cells grown in laboratory cultures cannot be used to accurately represent those in living organisms.

In 2005, a team of Swedish researchers developed a dating technique for cells by applying carbon-14 (^{14}C) dating to the DNA of cells. A ^{14}C dating technique had long been used to date fossils but the decay rate was too slow to be used for short time spans.

The breakthrough came when the research team took advantage of the increased levels of ^{14}C in the atmosphere during the Cold War due to nuclear testing. By the time above-ground nuclear testing ended in 1963, the levels of atmospheric ^{14}C had doubled beyond natural background levels. Since then, it has halved every 11 years. By taking this into account, there were detectable changes in levels of ^{14}C in modern DNA over short time spans. A scale for converting the ^{14}C enrichment into calendar dates was calculated from ^{14}C measurements of tree rings in Swedish pine trees exposed to the atmospheric changes (Figure 3.1.7).

Since nuclear DNA fixes carbon atoms at the time of cell division it serves as a time capsule for measuring a cell's age.

Not surprisingly, the human cell cycle lengths were found to correlate in part to the function and position of each cell type. Cells that are worn away regularly need to be replaced regularly, meaning they have a shorter cell cycle (Figure 3.1.8). Some examples from the study are shown in Table 3.1.1.

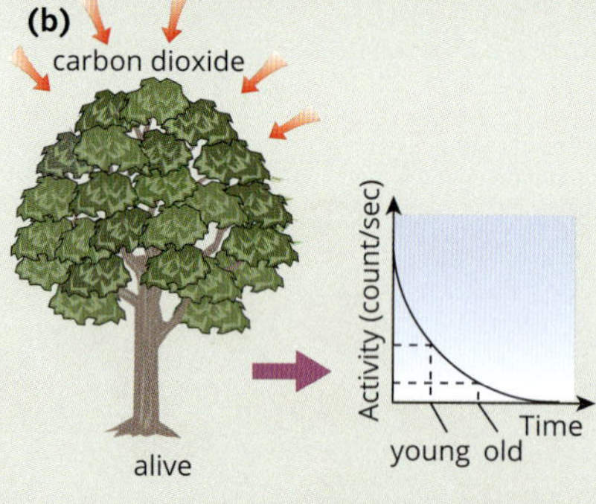

FIGURE 3.1.7 (a) The annual growth rings of pine trees containing carbon-14 from the atmosphere provided scientists with (b) a timeline for determining the age of human cells exposed to the same level of carbon-14.

FIGURE 3.1.8 Coloured scanning electron microscope (SEM) image of overlapping cells on the outer layer of the human skin. These cells have a short life span as they are located at sites where they are continually worn away and need to be replaced rapidly. Cells to be shed undergo programmed cell death (apoptosis).

The evidence supports that some of our cells continually replace themselves and others do not replicate again. Nerve cells of the visual cortex in the brain and the transparent cells of the eye lens are formed in the embryo as permanent cells to last the individual's lifetime. Most cells in the human body were found to be less than 10 years old and only a small minority of cell types last from birth to death without replication.

TABLE 3.1.1 Approximate cell cycle lengths of different cell types in the human body

Cell type	Approximate age of cell (cell cycle length)
gut lining	2–5 days
epidermis	2–4 weeks
red blood cell	3–4 months
liver	10–17 months
bone	10–30 years
rib muscle	15 years
visual cortex	same age as the person
eye lens	same age as the person

Interphase

A cell spends most of its time in interphase, carrying out cellular functions and preparing for cell division. There are three stages in interphase:

- G_1 (Gap 1)—during G_1 the cell produces more organelles and the cytoplasm increases in volume, doubling the size of the cell. If a cell is not going to divide again it will enter the G_0 phase. Human red blood cells, nerve cells and surface skin cells enter this phase from early in G_1, meaning that they cannot replicate again. Other cells enter the G_0 phase temporarily as a resting phase when they carry out cell functions but do not grow or replicate. These cells re-enter G_1 and continue with the cell cycle. That some cells spend temporary periods in G_0 explains the wide variations in interphase length. For example, adult human liver cells have generation times of 300–500 days, most of which will be spent in G_0.
- S (Synthesis)—DNA replication occurs during the S phase of interphase. Prior to the division of the cell's nuclei (mitosis), the DNA content must be replicated so that each new daughter cell receives a full set of the DNA-carrying chromosomes. It is essential that the genetic information is passed on accurately, because the activities of cells are ultimately controlled by the genetic information in the nucleus (in eukaryotes) or nucleoid (in prokaryotes). You will learn more about DNA replication in Section 3.2.
- G_2 (Gap 2)—during G_2 the cell undergoes a secondary stage of growth, metabolism and energy acquisition. It prepares for mitosis by synthesising the materials, such as **proteins**, needed for division.

  Interphase and the stages of mitosis are shown in the light micrographs (LM) in Figure 3.1.9.

Interphase, mitosis and cytokinesis are separate stages of the cell cycle.

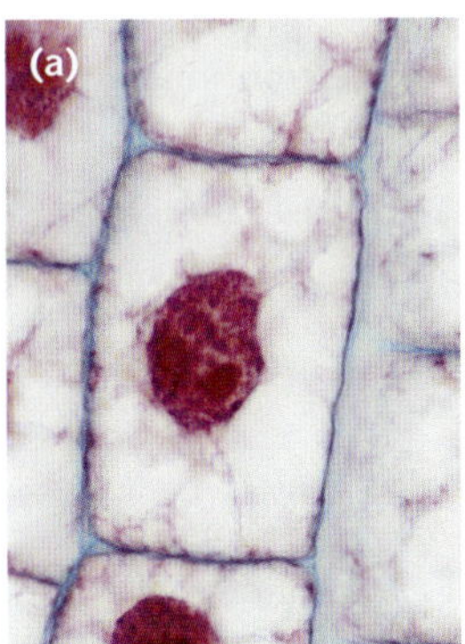

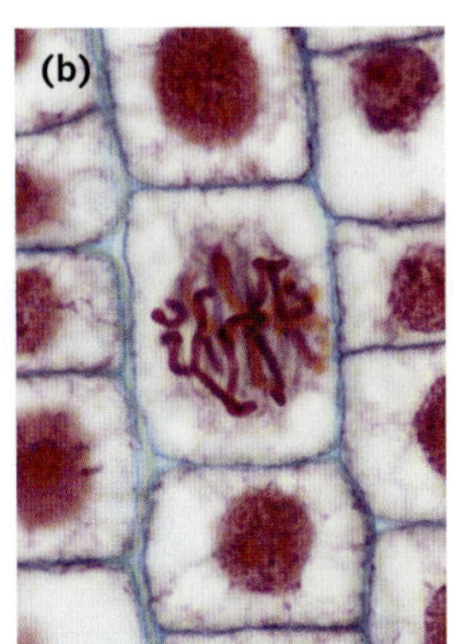

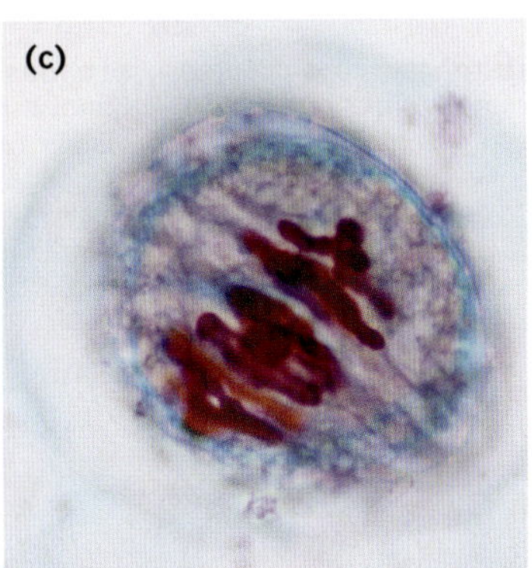

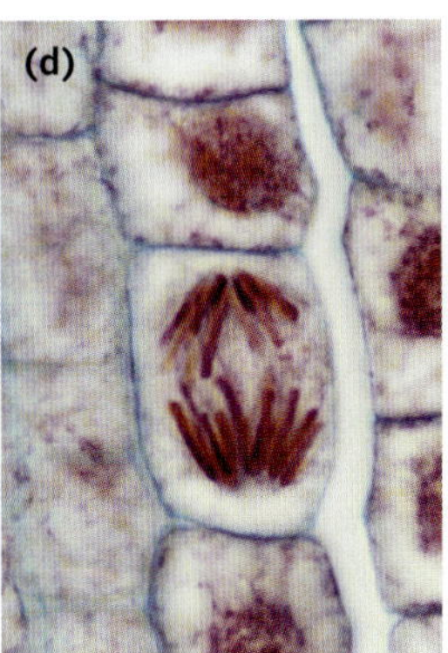

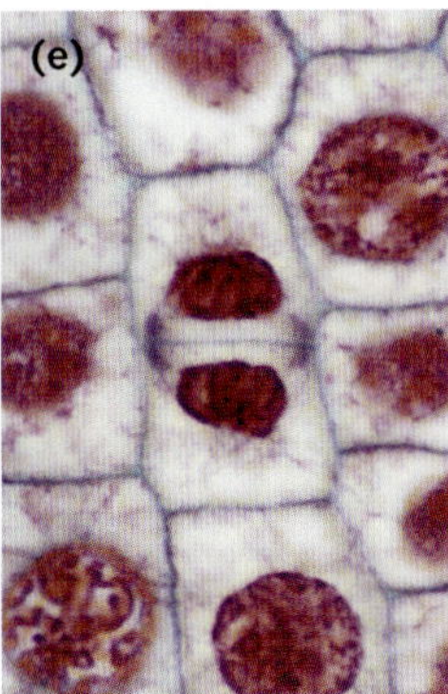

FIGURE 3.1.9 LMs of mitotic division in a garlic root tip cell: (a) interphase, (b) prophase, (c) metaphase, (d) anaphase and (e) telophase

Mitosis

The term mitosis refers to the division of the nucleus into two genetically identical daughter nuclei.

Mitosis is a continuous process but has four sub-phases (Table 3.1.2):

- prophase
- metaphase
- anaphase
- telophase.

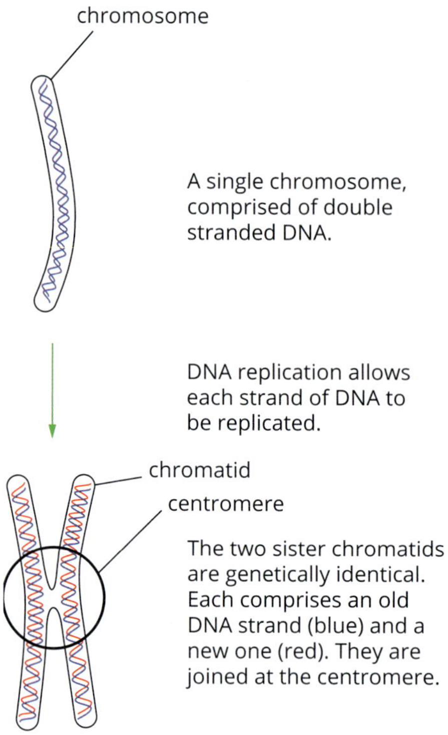

chromosome

A single chromosome, comprised of double stranded DNA.

DNA replication allows each strand of DNA to be replicated.

chromatid

centromere

The two sister chromatids are genetically identical. Each comprises an old DNA strand (blue) and a new one (red). They are joined at the centromere.

FIGURE 3.1.10 When a single chromosome replicates during the S phase (DNA synthesis), it forms two sister chromatids, joined at the centromere. This new structure is also known as a chromosome.

During interphase chromosomes are unfolded and exist like a tangled ball of thin strands in the nucleus. During prophase the strands organise themselves into tightly coiled, thick structures that are visible with a light microscope, a process called condensation.

WS 5.4

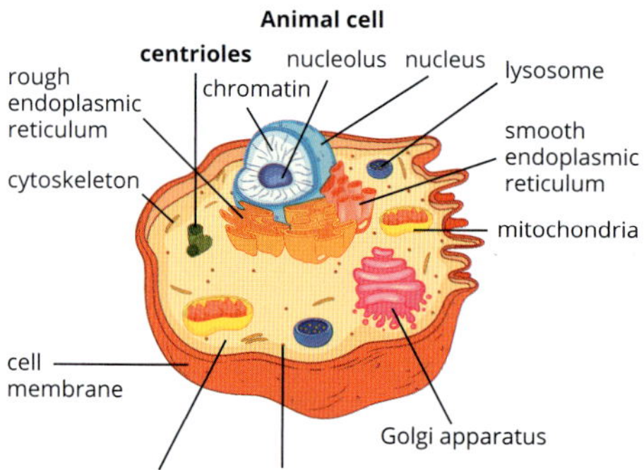

FIGURE 3.1.12 The centrioles are an organelle used in cell division to form and anchor the spindle which has the role of separating chromosomes.

TABLE 3.1.2 The characteristics of the sub-phases of mitosis

Mitosis	prophase	• chromosomes condense and become visible • centrioles move to opposite sides of the nucleus and form poles • nuclear membrane breaks down • centrioles form spindle fibres between the two poles
	metaphase	• chromosomes align at equatorial plane of cell • spindle fibres attach to centromeres of chromosomes
	anaphase	• spindle fibres contract, splitting the centromeres and separating the sister chromatids • the separated chromosomes are pulled to opposite poles
	telophase	• nuclear membrane reforms around the two sets of chromosomes • spindle fibres disappear • chromosomes become longer and thinner

Each sub-phase can be distinguished by the appearance and the position of the chromosomes in the cell. During interphase S in the cell cycle, each chromosome is duplicated forming a pair of sister **chromatids** (Figure 3.1.10). However, they are not visible under a microscope because they have not condensed. The structure and function of chromosomes and DNA will be explained in more detail in Section 3.2. For now it is sufficient to understand that DNA molecules carry coded genetic information and they are coiled into chromosome structures that reside in the nucleus of a cell. The movement of the double-stranded chromosomes during mitosis is shown in Figure 3.1.11.

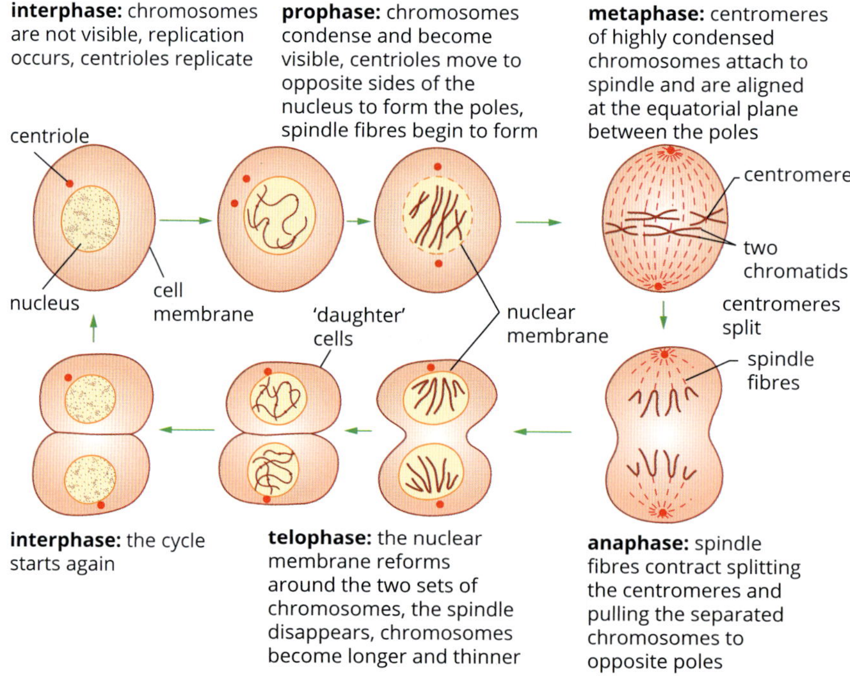

FIGURE 3.1.11 The major stages of mitosis

Prophase

Early in **prophase**, chromosomes begin to condense (shorten and thicken) and become increasingly visible under the microscope. As they condense further, each chromosome can be seen as two chromatids held together at the **centromere** (Figure 3.1.10). At the same time the **centrioles**, which were replicated during interphase, move to opposite ends of the cell to form the poles (Figure 3.1.12).

Later in prophase the nuclear membrane breaks down. The centrioles begin to form a network of fibres, called the **spindle**, which extends between the two poles of the cell. The centromere of each individual chromosome attaches to spindle fibres extending from each of the poles (Figure 3.1.11). Plant cells do not usually have centrioles; they use a different mechanism to produce the mitotic spindle.

Metaphase

During **metaphase** the centromeres continue to be drawn by the spindle fibres so that the chromosomes are aligned along the equator in the middle of the cell. Chromosomes are most easily observed at this stage because they are highly condensed.

A centromere is a constricted point along a chromosome where sister chromatids are joined and where the chromosome attaches to the spindle fibre.

Anaphase

In **anaphase** the spindle fibres contract, pulling the centromere in two directions. The centromere splits, separating the two chromatids. Contraction of the spindle fibres continues and the separated chromatids are pulled to opposite poles. Thus, daughter cells receive the same genetic information—one copy of every chromosome that was in the original nucleus at interphase.

Centrioles are a pair of tiny cylindrical organelles composed of a bundle of microtubules made from the protein tubulin. They assemble the spindle that stretches from pole to pole in a dividing cell.

Telophase

The final stage of mitosis is called **telophase**. It is rather like prophase in reverse. A nuclear membrane re-forms around the chromosomes at each pole. The spindle is dismantled and disappears. The chromosomes become longer and thinner, and therefore less visible under the microscope.

+ ADDITIONAL

Embryonic cell replication

The rapid mitotic cell replication that starts a new embryo is called cleavage. Cells spend little or no time in G_1 and G_2 during cleavage and, as a result, the cells have little growth time and halve in size with each division. They form the morula, which remains the size of the original zygote and stays inside the jelly-like layer of the zona pellucida (Figure 3.1.13). Cleavage ends when a manageable nucleus-to-cytoplasm ratio is restored in the cells and the blastula forms. Its embryonic cells will enter growth phases after each mitotic division. In humans, mitotic cleavage stops around day five after 16–20 cells have formed (Figure 3.1.14). These cells have used up all the yolk nutrients from the original egg and they break through the zona pellucida ready to implant on the wall of the uterus and form a placenta to receive further nutrients.

The pattern of cell division in cleavage is affected by the amount of yolk in the zygote. The process of embryonic cleavage and the development of the zygote was examined in detail in Chapter 2.

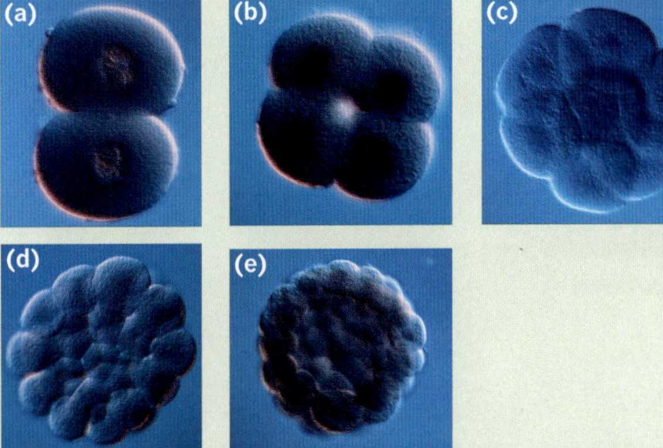

FIGURE 3.1.13 A series of LMs showing the first five divisions of a zygote by mitotic cell division from (a) two cells to (b) four cells, (c) eight cells, (d) 16 cells and (e) 32 cells. Between (d) the 16 cell morula stage and (e) the early blastocyst stage, cleavage will have finished and normal cell cycles started.

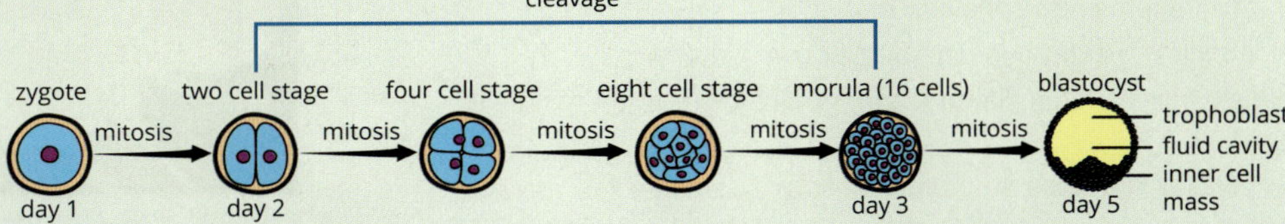

FIGURE 3.1.14 The early stages of cell replication in an embryo are known as cleavage and occur very rapidly without time for cells to increase in size. The cluster of cells fills the same volume as the original zygote.

Cytokinesis

At the end of mitosis the cytoplasm divides, separating the two nuclei and other organelles into two complete and identical daughter cells. The division of the cytoplasm is called cytokinesis and it finalises the cell division stage. Cytokinesis in animal cells occurs in a different way to cytokinesis in plant and fungi cells. In animal cells the cell membrane moves inwards, pinching the two daughter cells apart (Figure 3.1.15a).

Plant and fungi cells lay down a new cell membrane and cell wall between the two daughter nuclei to separate the daughter cells. Components of the new cell wall, called the cell plate, are initially deposited in the centre of the cell. The growth of the cell plate extends outwards until the two daughter cells are completely separated (Figure 3.1.15b).

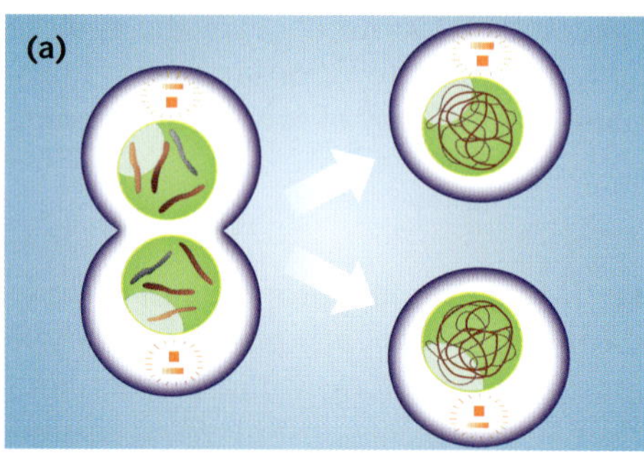

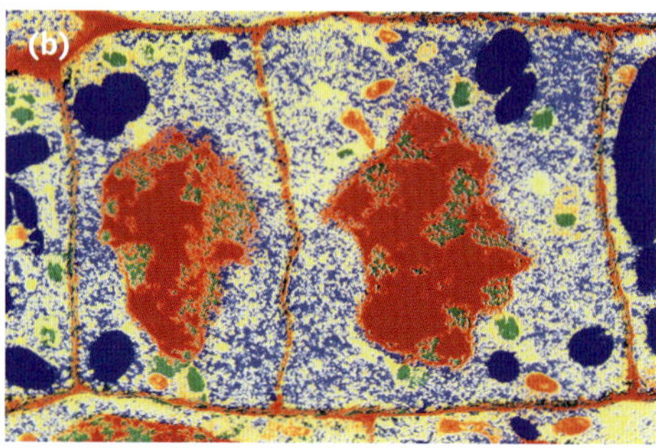

FIGURE 3.1.15 (a) Cytokinesis forms two new animal cells. The cytoplasm of the parent cell must be divided into two by the cell membrane pinching inwards. (b) Coloured TEM of a late stage of plant cell division. The daughter nuclei (red and green bodies) are reforming into membrane bound organelles. Between them is the new cell wall growing out from the middle (cytokinesis).

MEIOSIS

The other form of cell division in eukaryotic cells is meiosis. Meiosis is not identical cell replication because the nature of the process produces daughter cells that are different from each other and also from the parent cell. Meiosis is an important cell division process that is required for sexual reproduction and creating **genetic variation**; it produces four daughter cells (gametes) that are genetically unique. Meiosis occurs only in eukaryotes and only to form the gametes (Figure 3.1.17). The formation of gametes from **germ cells** occurs by meiosis in the specialised reproductive organs of sexually reproducing animals and plants.

Germ cells are the precursors to gametes.

BIOFILE CCT

Cancer cells that never die

Cancer cells (Figure 3.1.16) differ from normal cells in many ways. Cancer cells:

- divide at a faster rate than normal cells of the same type. The rate at which they divide varies
- are not affected by the normal signals that control the cell cycle, such as contact inhibition
- look different and may become less specialised
- release factors that stimulate the development of their own blood supply
- have DNA that mutates, making them different and sometimes resistant to earlier successful treatments
- can 'colonise' new parts of the body and continue to grow unchecked
- can continue dividing endlessly, whereas normal cells undergo a limited number of cell cycles
- avoid proceeding to death by apoptosis.

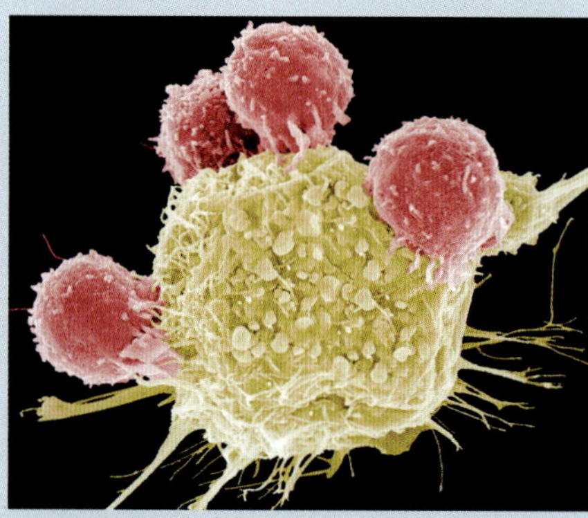

FIGURE 3.1.16 Cancerous cells usually look very different from normal cells. This coloured SEM shows T lymphocytes (pink) attached to a cancer cell (yellow). Magnification: x2300

Meiosis is called a **reduction division** because, unlike mitosis, it reduces the number of chromosomes in gametes (daughter cells) to half ($1n$ or n) of that in **somatic cells** ($2n$). Cells with n chromosomes are called **haploid** cells and cells with $2n$ chromosomes are **diploid** cells. Gametes receive only one copy of each pair of **homologous chromosomes** ($n = 23$ chromosomes in human gametes) (Figure 3.1.17). Compare this to mitosis, where each daughter cell receives a copy of every chromosome and they are genetically identical with $2n$ chromosomes.

Like mitosis, meiosis is a form of cell division that involves prophase, metaphase, anaphase, telophase and cytokinesis. Unlike mitosis, there are two sequential rounds of division in meiosis, called **meiosis I** and **meiosis II** each with these sub-phases (Figure 3.1.17)

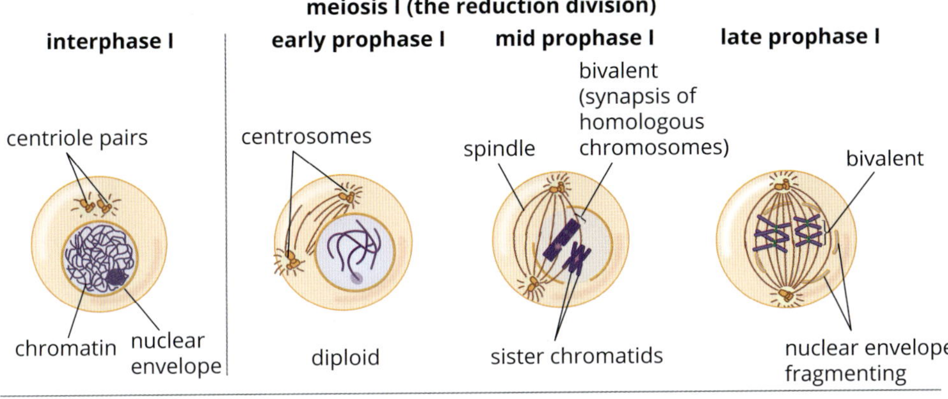

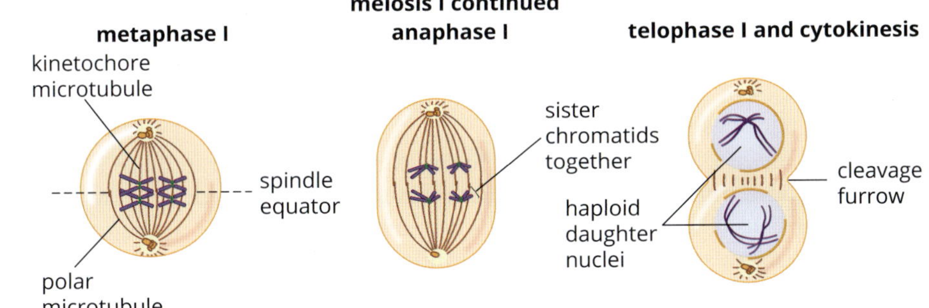

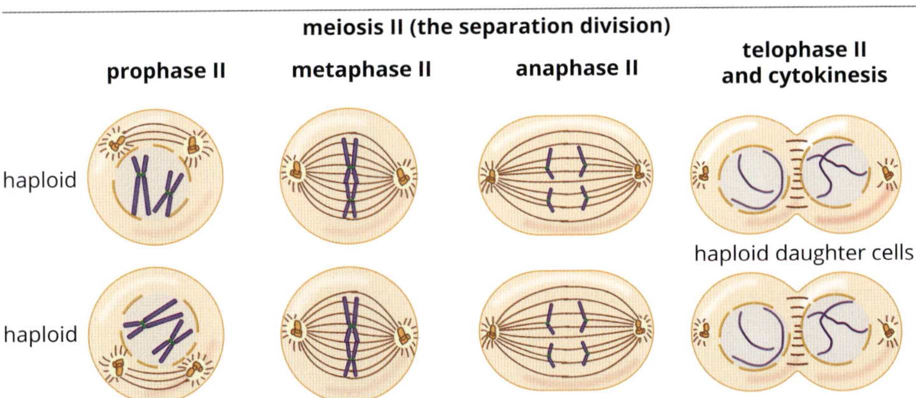

FIGURE 3.1.17 Stages of meiosis in an animal cell with a diploid number of chromosomes ($2n$). Reduction in the chromosome number (from diploid to haploid), crossing over and genetic recombination occur in the first meiotic division. By the end of the second division, four haploid nuclei (n) have been produced from the one original parent cell.

Somatic cells are all the cells in an organism other than the gametes. Somatic cells are diploid ($2n$) while gametes are haploid (n).

BIOFILE CCT EU

The HeLa cell line

A population of cells grown continuously by mitosis in a cell culture is called a 'cell line'. There are now many cell lines that can be isolated and grown in culture indefinitely, especially cells derived from tumours. 'HeLa' cells are a particular line of cultured cells that are used worldwide in experimental studies of cell functions. These cells were isolated from a human cervical carcinoma in 1951 and have been grown continuously ever since (Figure 3.1.18). They are named after the person from whom they were obtained, Henrietta Lacks. You will learn more about Henrietta Lacks in Chapter 8.

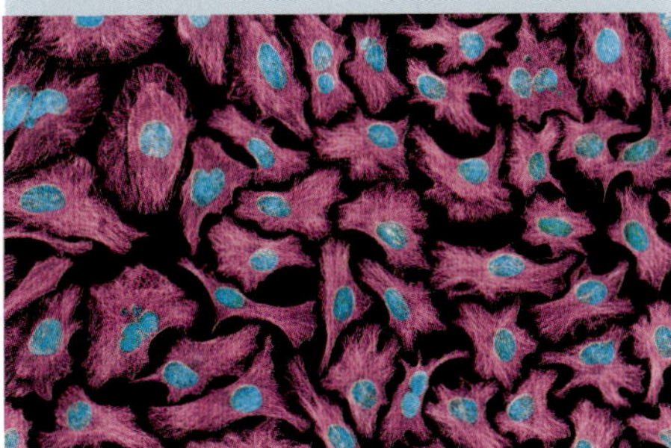

FIGURE 3.1.18 Fluorescence LM of a group of cultured HeLa cells, showing the cell nuclei (stained blue).

BIOFILE N

Meiosis produces haploid gametes

In order to maintain the chromosomal number of a species, the number of sets of chromosomes in somatic cells and gametes differs. For example, in humans, somatic cells have 46 chromosomes (23 pairs) whereas human gametes have 23 single chromosomes. This means that when the haploid gametes combine during fertilisation, the resulting zygote will have a full complement of chromosomes (23 chromosomes from the sperm plus 23 chromosomes from the ovum equals 46 chromosomes (23 pairs) in human somatic cells).

Somatic cells in most animals are diploid ($2n$) because they contain matching pairs of chromosomes called homologous chromosomes. In a homologous pair, one chromosome has been inherited from each parent. Geneticists use pictures called karyotypes that show a full diploid set of homologous chromosome pairs from an individual cell (Figure 3.1.19). These usually come from the metaphase stage of cell division.

In each gamete, there is a haploid (n) number of chromosomes because the homologous pairs are separated during meiosis of the germ cells, and only single chromosomes pass into the daughter cells.

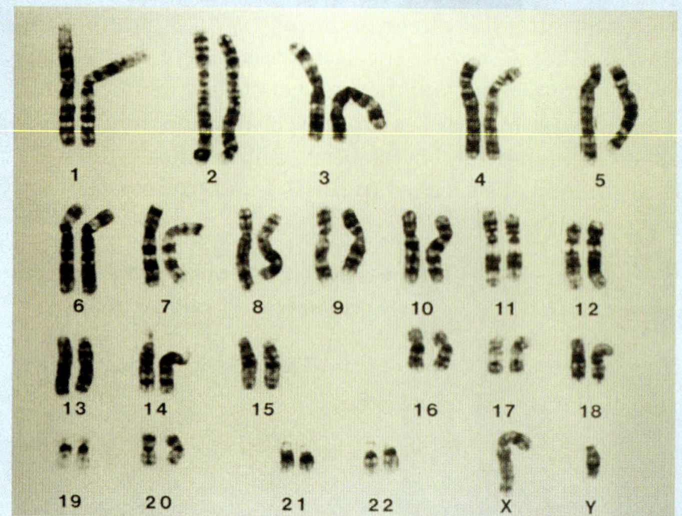

FIGURE 3.1.19 A normal human karyotype with 22 homologous pairs of chromosomes plus the XY sex chromosome pair, in this case a male. A female would have XX, which would be seen as two equal-sized X chromosomes.

The first division of meiosis: meiosis I

The sub-phases of meiosis I occur in the following order: prophase I, metaphase I, anaphase I, telophase I (Figure 3.1.17).

- During meiosis I, homologous chromosomes are separated, reducing the chromosome number by half (reduction division) and producing two haploid daughter cells. The sister chromatids remain joined together at the centromere so each chromosome is still double-stranded.
- Each chromosome pairs up precisely along its length with its matching (homologous) chromosome. This pairing is called **synapsis**. (This is very different from mitosis, where homologous chromosomes are completely independent of one another.) Because each chromosome has already replicated, each chromosome consists of two copies, called sister chromatids. So a pair of homologous chromosomes has a total of four chromatids.
- **Crossing over** may occur between homologous chromosomes.

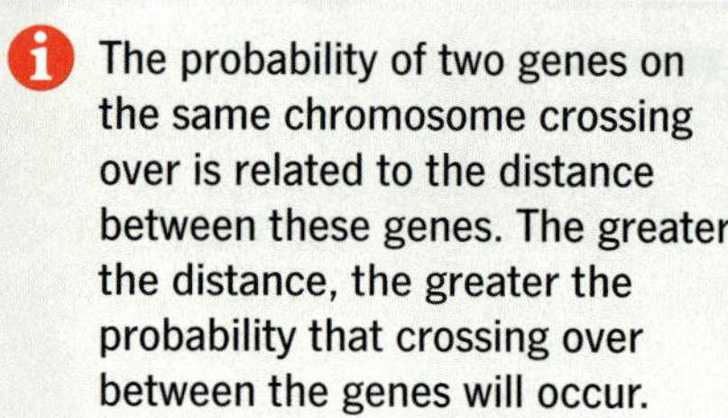
The probability of two genes on the same chromosome crossing over is related to the distance between these genes. The greater the distance, the greater the probability that crossing over between the genes will occur.

Crossing over and recombination

A key event now occurs—chromatids of homologous chromosomes may exchange portions of their genetic information in a process called crossing over. Crossing over is a natural genetic process that occurs between homologous chromosomes and leads to the switching of genetic material between the chromosomes. DNA strands from the chromatids of two homologous chromosomes are cut at the equivalent point, a segment is exchanged, and the strands are reconnected.

The point where crossing over occurs is called a **chiasma** (plural chiasmata). It consists of a temporary molecular scaffold that disappears later. A long chromosome may have several chiasmata (Figure 3.1.20).

The significance of crossing over is that it produces chromosomes with new combinations of genetic information. This process is called **recombination** and is essential to the production of genetic variation.

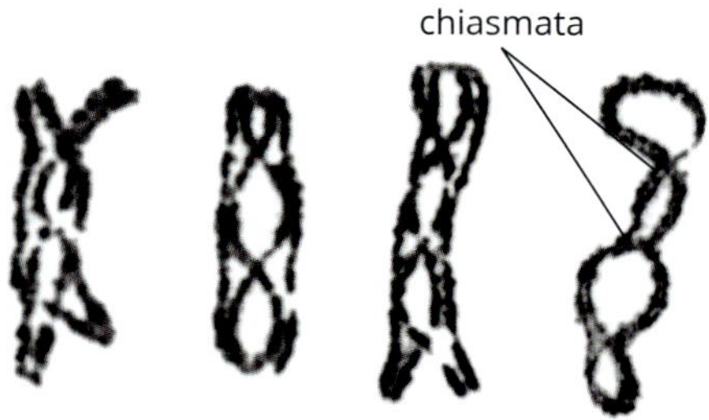

FIGURE 3.1.20 LM showing the formation of chiasmata between homologous chromosomes. These are the points where crossing over to swap genetic information can occur.

- When crossing over is finished, the homologous chromosome pairs align along the midline of the cell; they do this randomly, meaning the maternal and paternal chromosomes do not line up on the same side of the midline.
- The homologous chromosomes then separate and move to opposite poles. These two steps result in the **random assortment** of maternal and paternal chromosomes and their **alleles** (gene variants) in the gametes. The centromeres do not split. It is the chromosomes of a pair that separate, not the chromatids.
- The spindle breaks down and the nuclear membrane reforms.

At the end of this first division of meiosis, there are two daughter cells with the chromosome number halved—they contain only one set (n) of chromosomes (i.e. they are haploid). Each chromosome is still made up of two chromatids.

The second division of meiosis: meiosis II

The sub-phases of meiosis II occur in the following order: prophase II, metaphase II, anaphase II, telophase II (Figure 3.1.17). The second division of meiosis does not involve chromosome duplication. Meiosis II is similar to mitosis in that sister chromatids are separated by splitting the centromere. Each of the two haploid cells from meiosis I divide into two, producing four haploid daughter cells. This occurs in the following sequence:

- Chromosomes align on the spindle equator, the centromeres split and the chromatids separate.
- A chromatid from each chromosome moves to each pole.
- The final nuclei from the two divisions are each haploid (n).
- The cytoplasm divides by cytokinesis, and four daughter cells (non-identical) are formed from one original parent cell.

In a male, meiosis II results in four viable, haploid sperm (Figure 3.1.21a). However, in a female only one haploid ovum results and the other three haploid cells degenerate. This occurs because of the uneven distribution of cytoplasm in cytokinesis (Figure 3.1.21b). One daughter cell is very large, containing most of the cytoplasm and organelles, ready for the cleavage process if fertilised. The other three are called **polar bodies** and they usually enter **apoptosis** (programmed cell death). The polar bodies of human oocytes apoptose by 24 hours after formation and the resulting fragments remain within the zona pellucida of the large oocyte. That explains why the three polar bodies disappear from diagrams of meiosis II.

Polar bodies are being used as a DNA source for genetic testing of the health of IVF embryos rather than doing a biopsy on the embryo itself.

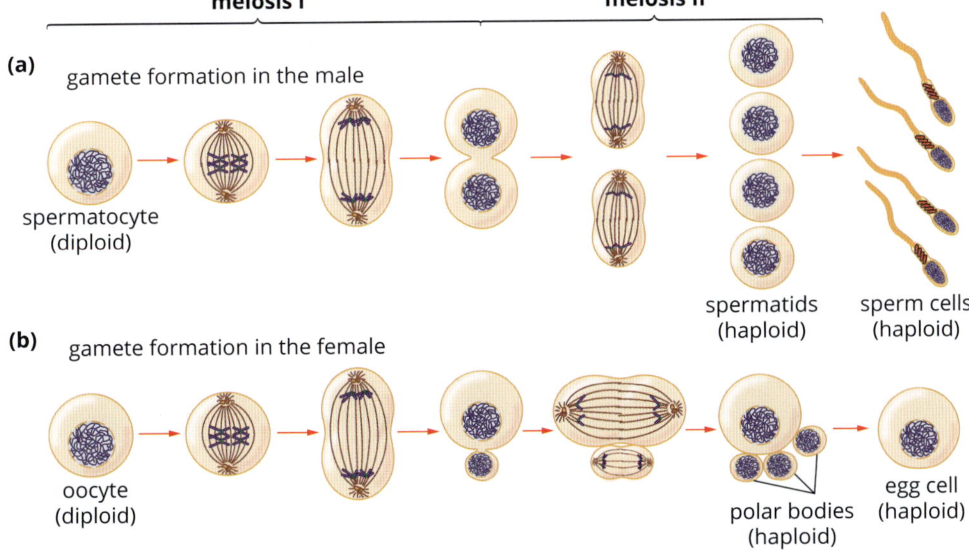

FIGURE 3.1.21 (a) Haploid male and (b) haploid female gametes are produced by meiosis in two sequential cell division processes.

Mitosis produces diploid somatic cells and meiosis produces haploid gametes.

COMPARISON OF MITOSIS AND MEIOSIS

The common feature between mitosis and meiosis is that both are processes of cell division to form new, additional cells (Figure 3.1.22). However, they have very different purposes, there are some different steps within the processes and different outcomes, which are outlined in Table 3.1.3. The distinction between somatic cells and germ cells is important to understanding mitosis and meiosis. Somatic cells are all the body cells except for the germ cells that produce gametes (sperm and egg in animals; pollen and egg in plants). Mitosis produces new somatic cells (e.g. skin cells) and meiosis produces sex cells (gametes, e.g. sperm and egg).

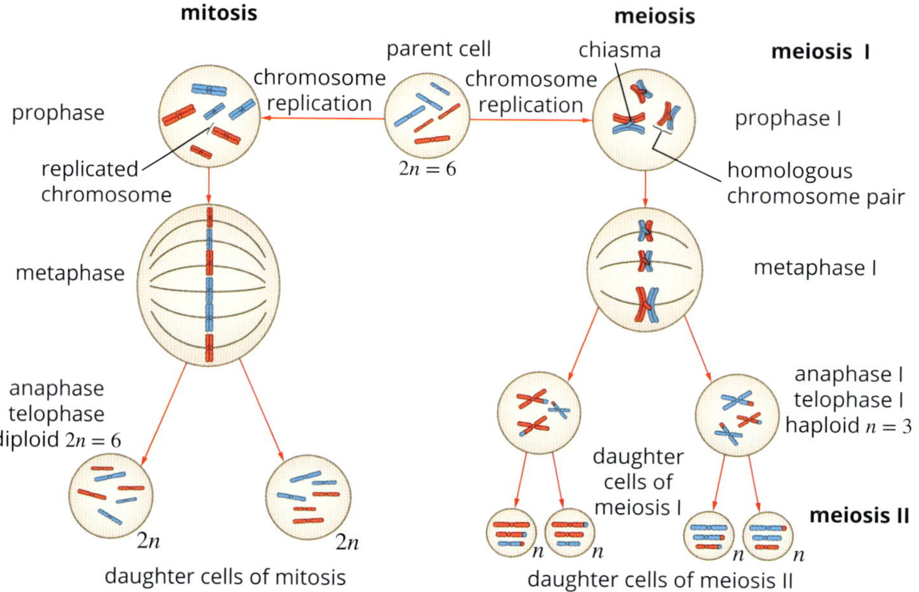

FIGURE 3.1.22 A comparison of mitosis and meiosis. They are both processes of cell division, but are different in a number of important ways

TABLE 3.1.3 Key differences between mitosis and meiosis

	Mitosis	Meiosis
Genetic recombination	Mitosis does not change the genetic information (chromosomes do not cross-over).	Meiosis rearranges genetic information between chromosome pairs, creating unique genetic variation (chromosomes cross-over).
Number of cells	Mitosis produces two genetically identical daughter cells.	Meiosis produces four genetically unique daughter cells.
Number of chromosomes	The daughter cells produced from mitosis have the same number of chromosomes (diploid, $2n$) as the parent.	The daughter cells produced from meiosis have half the number of chromosomes (haploid, n) of the parent.
Location	Occurs in all parts of the body to replicate somatic cells.	Only occurs in the gonads to produce gametes from germ cells

3.1 Review

SUMMARY

- The cell replication cycle has three main stages: interphase, mitosis and cytokinesis.
- Cytokinesis marks the beginning of two new cells, and the cell cycle is the period between one cytokinesis and the next.
- Interphase involves three phases: G_1 (growth, pre-DNA synthesis), S (DNA synthesis) and G_2 (growth, post-DNA synthesis).
- Some cells enter a G_0 phase (temporary rest from growth, or to start differentiation, or a permanent halt to cell cycle).
- Apoptosis is the natural and highly regulated cell death pathway when cells are targeted for destruction.
- Cellular activities that usually occur during interphase include DNA synthesis (replication of chromosomes), synthesis of organelles and normal cell biochemical processes, specialisation, increase in size, and preparation for mitosis.
- Mitosis is the form of cell replication used by all living things for growth and replacement of cells.
- Mitosis is a continuous process that divides the nucleus and is described in sub-phases: prophase, metaphase, anaphase and telophase.
- During mitosis, identical copies of each chromosome are passed from the parent cell to two daughter cells.
- Cytokinesis after mitosis is the division of the cytoplasm including molecules and organelles, and the separation of the new nuclei, to form two new daughter cells.
- Meiosis is another type of cell division that does not replicate the parent cell.
- Meiosis is a division of the nucleus over two stages that halves the normal number of chromosomes and produces different genetic combinations in haploid gametes. It follows the same sequence of steps as mitosis but with two divisions.
- The first stage meiosis I produces two haploid cells where each chromosome still has a sister chromatid.
- The second stage meiosis II follows immediately and separates the chromatid pairs in each cell into single chromosomes in four haploid cells.
- Production of haploid gametes only occurs in sexually reproducing eukaryotes, both plant and animal.
- Variation in gametes arises from the random assortment of chromosomes and exchange of alleles (gene variants) through recombination (crossing over) during meiosis.

KEY QUESTIONS

1. Within a cell cycle, cell replication involves three phases. Identify and summarise the three phases.
2. Classify the following examples of cell replication as forms of reproduction; repair and maintenance; growth and development; or restoration of nucleus-to-cytoplasm ratio.
   - toddler's height increasing by 2 cm
   - healing of a cut
   - bacteria cell dividing
   - cleavage of an embryonic cell
   - seed germinating
   - unicellular Protista organism dividing
3. Identify the two types of cell division, the number of daughter cells produced and the number of chromosomes in these daughter cells.
4. Recall the events that occur during interphase of the cell cycle.
5. Distinguish between a chromatid and a chromosome. When are they visible?
6. Explain how cell division in plants is different from cell division in animals.
7. Define 'apoptosis'. Explain the reasons for this process in a healthy organism, providing examples.
8. Summarise the two stages of meiosis using a flow chart.

3.2 DNA structure and replication

BIOLOGY INQUIRY    WE

DNA model

How important is it for genetic material to be replicated exactly?

COLLECT THIS...

- tablet or computer to access the internet
- a retort stand
- retort clamps, rings and bossheads (as many as possible)
- pencils or wooden skewers
- paper clips
- cardboard
- scissors
- four markers of different colours (or spot stickers in four colours)

DO THIS...

1 Find images of a DNA molecule in this chapter or online.
2 Make a cardboard pentagon template with each side approximately 2 cm.
3 Cut out at least 20 cardboard pentagons.
4 Label and colour-code these pentagons for the four nitrogenous bases in DNA: adenine (A), thymine (T), guanine (G) and cytosine (C).
5 Get creative with constructing a 3D model of DNA using your equipment and the pentagons.
6 Refer to the model made by Watson and Crick to help you construct your model (Figure 3.2.1).
7 Take a photo of your model and discuss the construction of your model with other students.

RECORD THIS...

Describe the structure of DNA in three or four sentences.

Present your photo and description of DNA structure in your notes for this section of work.

REFLECT ON THIS...

Can you see how the structure of DNA forms a repeating pattern that lends itself to self-replication?

How useful for your understanding was it to construct a 3D model of DNA rather than draw a 2D diagram on the page?

How important is it for genetic material to be replicated exactly?

FIGURE 3.2.1 This reconstruction of the double helix model of DNA contains some of the original metal plates used by Francis Crick and James Watson to determine the molecular structure of DNA. It is constructed out of metal plates and rods, arranged helically around a retort stand, and shows one complete turn of the famous double helix.

Throughout history, people have observed that children resemble their parents more than they resemble unrelated individuals. Today we know that many characteristics are inherited, such as the colour of our hair, eyes and skin. However, children are not identical to their mother or father, and they are not identical to their sisters or brothers (except in the case of identical twins).

We also know that the hereditary information is carried in coded form (genes) on complex molecules called deoxyribonucleic acid (DNA). During the formation of gametes by meiosis, the DNA is contained in structures called chromosomes in the nuclei of cells (Figure 3.2.2). Chromosomes are passed from parent to child after a sperm cell fertilises an egg cell. The DNA in the resultant zygote is a unique and equal mix of chromosomes from the mother and father. After fertilisation, the cell cycle at the S stage of DNA replication and the mitosis stage of cell division is responsible for accurately replicating cells including copying the inherited genetic code.

In this section you will learn about the structure of DNA and how this structure supports its function of carrying genetic information from one generation to the next. You will also come to understand the packaging of lengthy DNA molecules into compact chromosomes in a cell nucleus and how exact copies of the DNA are replicated prior to each cell division.

> DNA (deoxyribonucleic acid) is a complex molecule that contains all the genetic information necessary to build and maintain an organism.

> Macro-molecules like DNA and RNA are referred to as polymers made from repeated smaller monomer units, in this case the nucleotides. Proteins are also polymers built from amino acid monomers.

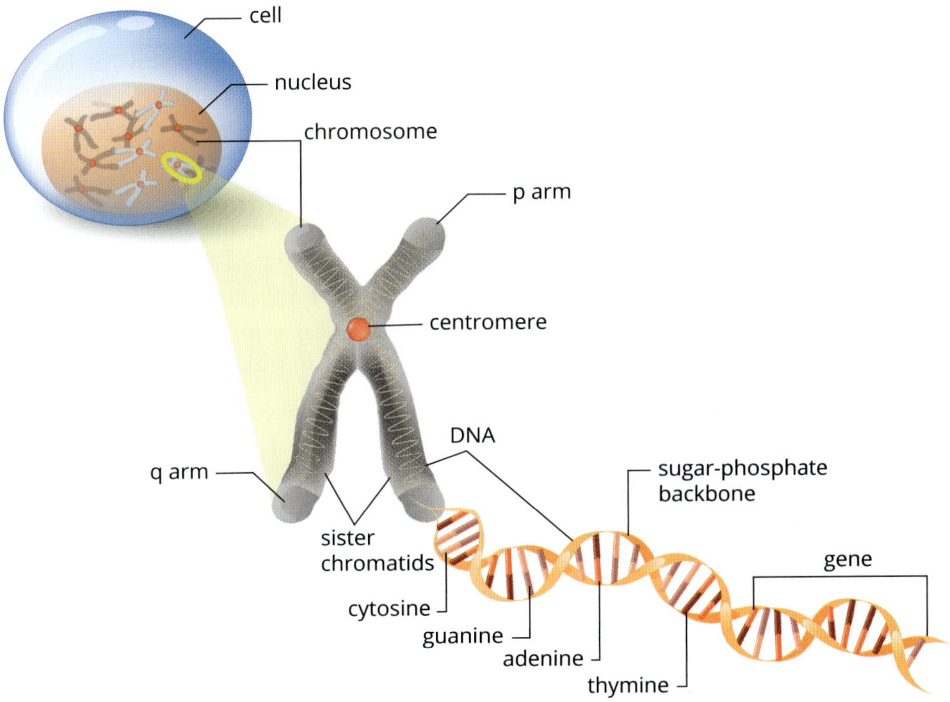

FIGURE 3.2.2 The relationship between a cell, chromosomes and the DNA molecule. The double helix structure and chemical sub-units of DNA are also depicted.

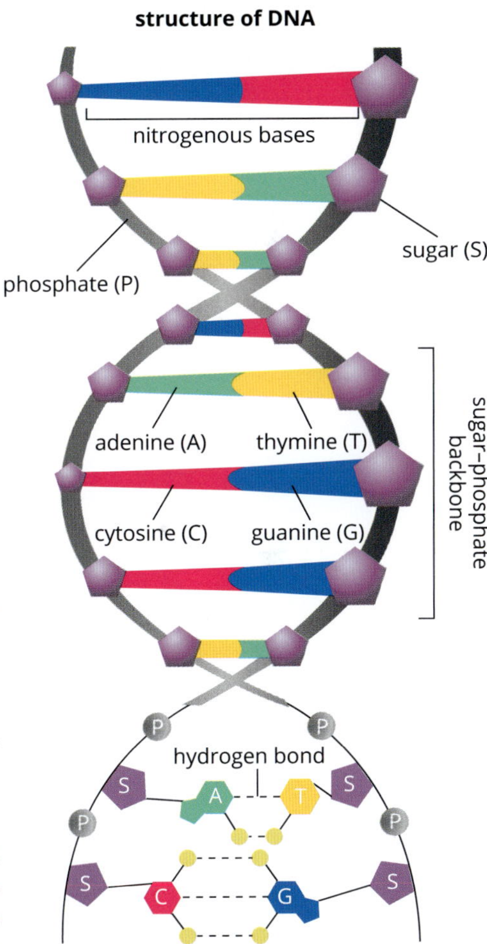

FIGURE 3.2.3 The polymer structure of a DNA double helix is formed from nucleotide monomer units. Complementary nitrogenous bases, joined across by hydrogen bonds, pull the DNA into a spiral. Each side of the helix is comprised of a chain of deoxyribose sugar and phosphate molecules, known as the sugar–phosphate backbone.

DNA STRUCTURE

DNA is one of the two types of **nucleic acids**: DNA (deoxyribonucleic acid) and **RNA (ribonucleic acid)**. Both DNA and RNA are made up of a repeating series of **nucleotide** units formed from **nitrogenous bases** and a **sugar–phosphate backbone**. RNA is single-stranded and DNA has a double-stranded helix (spiral) structure with **complementary base pairing** across the helix for its nitrogenous bases (Figure 3.2.3). DNA carries the genes (genetic information) needed to assemble functional protein molecules from amino acid sub-units. Identical copies of the original parental DNA are passed from a parent cell to each daughter cell during every mitotic cell division.

RNA works with DNA to play a role in the synthesis of the proteins within cells. Proteins are many and varied with a wide range of important structural and biochemical functions in all organisms. For example, haemoglobin protein carries oxygen, enzymes are proteins that control all the chemical reactions of an organism, antibodies and most hormones are protein molecules. Sections 4.2 and 4.3 cover more information about proteins and how they are made inside cells to the DNA specifications.

GO TO ➤ Section 4.2 page 165

GO TO ➤ Section 4.3 page 181

A molecule is made up of two or more atoms that are held together by chemical bonds. Any molecule that is found in a living organism is called a biomolecule. Large biomolecules, called biomacromolecules, can be made up of thousands of atoms and include proteins and nucleic acids.

The DNA **double helix** has become a familiar and iconic symbol since James Watson and Francis Crick announced the discovery of its structure in 1953. Like much of scientific discovery, the final understanding of DNA structure may seem like a eureka moment, but it was built on many years of painstaking earlier investigations and technological developments by other scientists. Some of the important discoveries that led to our understanding of DNA structure are listed below.

- 1866—Mendel and others had shown that information is inherited from parents to offspring. At first it was thought to be carried by protein molecules because of their complexity.
- 1869—Miescher isolated an unknown chemical from the nuclei of white blood cells and called it nuclein. The name was later changed to nucleic acid, then to deoxyribonucleic acid (DNA).
- Early 1900s—Sutton using grasshoppers, Bovery with sea urchins and Morgan with fruit fly all proved that chromosomes carry the inherited information between generations.
- 1905–39—Levene identified sugar, phosphate and base components of DNA, gave nucleotides their name and distinguished RNA from DNA.
- 1943—Avery showed that DNA carries the genetic code, after much dispute if it was carried by protein (with its complex structure) or nucleic acid (with a simpler structure).
- 1940s—Chargaff expanded on Levene's research and made three important discoveries that laid the foundation for fully understanding DNA structure:
  - different **species** have the same nucleotides but arranged in different orders
  - the amounts of nitrogenous bases A and T are always similar; amounts G and C are similar, but A-T and G-C may be present in different amounts—now known as Chargaff's rule
  - A + G always equals T + C.
- Early 1950s—Rosalind Franklin (Figure 3.2.4a) and Maurice Wilkins used X-ray crystallography to produce photos of DNA structure, after working out how to get DNA into crystal form (Figure 3.2.4b). This technology began with the Bragg father and son team in 1913–14. Franklin was skilful at improving it. Sadly, working with X-rays may have contributed to her early death in 1958 when she was 38 years old from cancer, making her ineligible for the 1965 Nobel Prize shared by Watson, Crick and Wilkins.

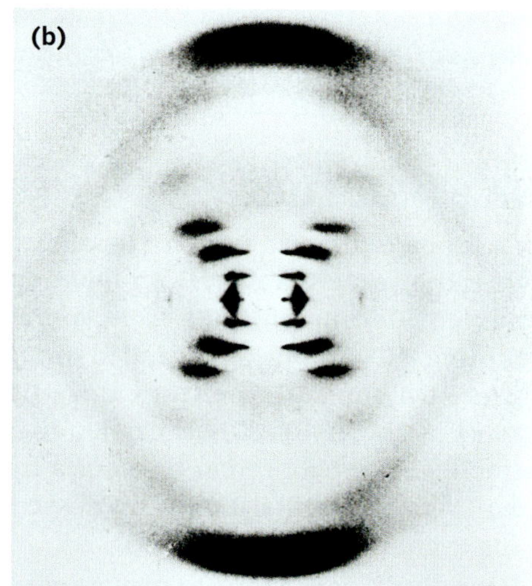

FIGURE 3.2.4 (a) Rosalind Franklin, the British X-ray crystallographer. Her images of DNA crystals were of the highest quality and she recognised important features of the structure of DNA. Watson and Crick finalised the structure of DNA using information gleaned from Franklin's X-ray images. (b) The famous Photo 51, an X-ray diffraction photograph of DNA obtained by Rosalind Franklin in 1953. The image results from a beam of X-rays being scattered onto a photographic plate by a DNA crystal. Features of the structure of DNA can be determined from the pattern of spots and bands. The cross of bands indicates the helical nature of DNA and position of bases inside the helix.

- 1953—James Watson and Francis Crick, using Chargaff's rule plus Franklin and Wilkins' information, made adjustments to their 3D modelling of DNA and announced the successful discovery of DNA structure amidst great excitement at their local pub (Figure 3.2.5). Positioning the bases inside the helix was the key to finally determining the structure.
- Genetics research has advanced rapidly since the 1950s discoveries, including understanding the universal genetic code in all species—in essence, there is a set of 64 codons (sets of three bases) corresponding to the 20 amino acids used for protein synthesis and as the signals for starting and stopping protein synthesis. Section 4.2 will explain the universal code of DNA and its all-important role in protein synthesis.

Hereditary information refers to genetic material that is passed on from parent to offspring; from one generation to the next.

A gene is a unit of heredity that determines the characteristics of an organism. At the molecular level, a gene is a section of DNA with a unique sequence of the bases (A, T, G, C). For example, GGGTTACGAACGT ... and so on.

GO TO ➤ Section 4.2 page 165

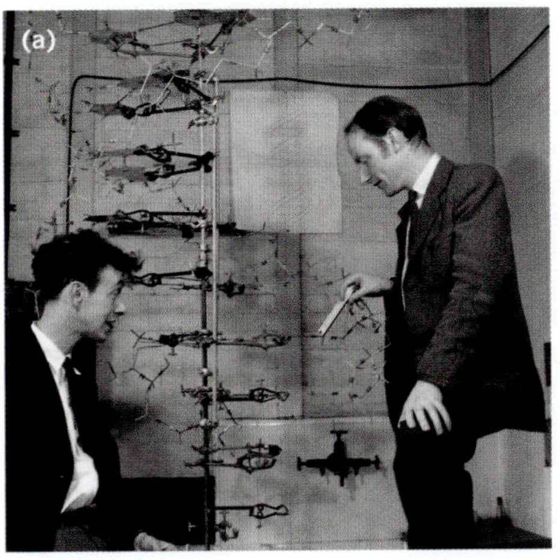

FIGURE 3.2.5 (a) The DNA model constructed in the Cavendish Laboratory at Cambridge, 1953, by Watson (left) and Crick (right). (b) Watson and Crick first announced their DNA discovery at their local pub, The Eagle.

Nucleotides—building blocks of DNA and RNA

Nucleotides are the chemical building blocks of DNA and RNA. Each nucleotide consists of:

- a phosphate group (Figure 3.2.6a)
- a five-carbon sugar (deoxyribose; Figure 3.2.6b). The five carbon atoms are numbered 1′–5′. In an individual nucleotide, a phosphate is attached to the 5′ carbon, and a base is attached to the 1′ carbon (Figure 3.2.9).
- one of four nitrogen-containing bases: adenine (A), guanine (G), thymine (T) and cytosine (C) (Figure 3.2.6c). There are two types: **purines**, with a double ring molecular structure, and **pyrimidines**, with a single ring. The purine bases are adenine and guanine, and the pyrimidine bases are thymine and cytosine (and uracil (U) in RNA where it replaces T).

All biological molecules are built from carbon, hydrogen and oxygen atoms in different combinations. Nitrogenous bases are so-called because, like protein molecules, they also contain nitrogen atoms.

Remember that one DNA nucleotide is built from one phosphate, one sugar and one nitrogenous base (either A or T or G or C).

DNA

DNA is a large (macro) molecule, which is made up of a series of nucleotides (Figures 3.2.6 and 3.2.7).

Figure 3.2.7 shows a single **polynucleotide** chain (strand) of DNA in which individual nucleotides are joined in a line by **phosphodiester bonds** (a type of strong covalent bond) (Figure 3.2.8). Note that in this diagram, the DNA has not yet formed into the typical double-stranded helix.

Covalent bonds are the force of attraction formed when one or more pairs of electrons are shared between two atomic nuclei.

The nitrogen-containing base distinguishes the nucleotides from one another (Figure 3.2.6d). The covalent bonds holding adjacent nucleotides together are between carbon, phosphorus and oxygen atoms (Figure 3.2.9). When many nucleotides are joined together, a single polynucleotide chain, which runs from 5′ to 3′, is formed (Figure 3.2.7). Different nucleotides can occur in any order within a strand—if a particular base is A, the next base in the sequence could be A, G, T or C.

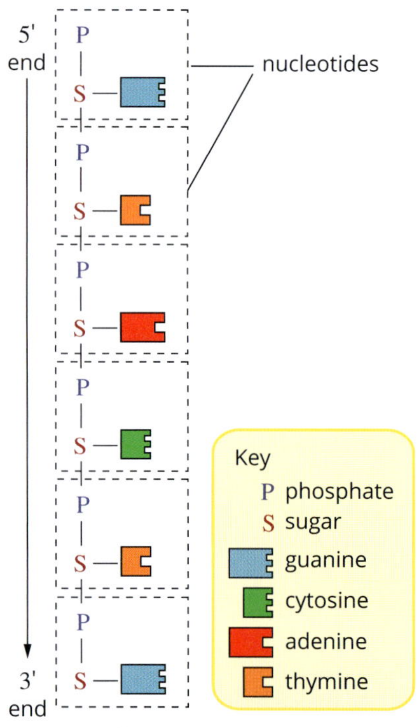

FIGURE 3.2.7 A single polynucleotide chain (strand) with individual nucleotides (shown in boxes) joined by phosphodiester covalent bonds. Two of these polynucleotide strands form the sugar-phosphate backbone of a DNA double helix.

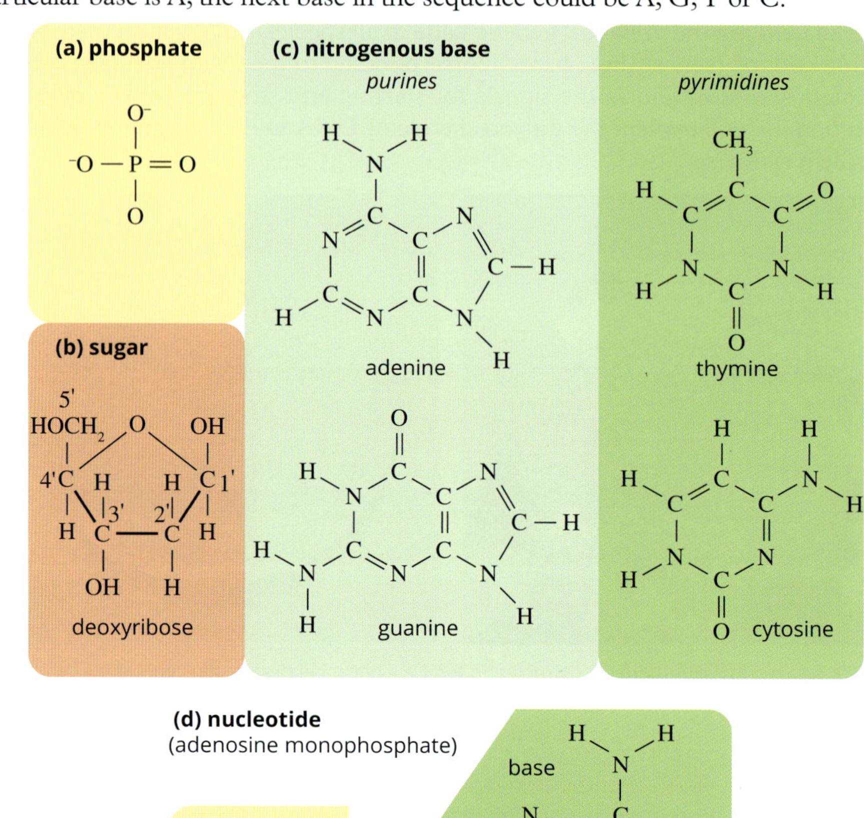

FIGURE 3.2.6 A nucleotide is composed of (a) a phosphate group, (b) the 5-carbon sugar, deoxyribose, and (c) one of four bases. (d) The nucleotide adenosine monophosphate includes the base adenine (A), a purine with a double ring molecular structure.

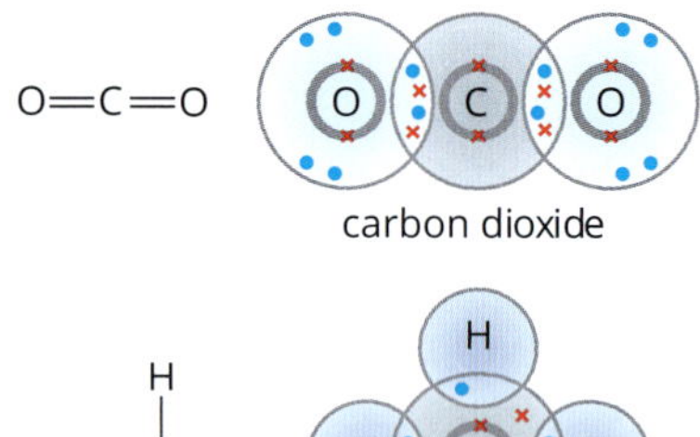

FIGURE 3.2.8 Two common examples of covalent bonding showing the sharing of electrons in carbon dioxide and also in methane.

In the structure of a DNA double helix, the bonds between nitrogenous bases are always a purine (double ring) to a pyrimidine (single ring) A-T and G-C; the bases always attach to the sugars in the backbone and there is always a phosphate between two sugars on the backbone.

The numbering convention of carbon atoms in the deoxyribose sugar (DNA) and ribose sugar (RNA) molecules is shown in Figure 3.2.6b. The numbers are referred to as 3 prime (3′), 5 prime (5′) and so on. The order of numbering is a complex system determined by functional groups attached to the carbon ring.

RNA

RNA is the other nucleic acid molecule and its role in the cell for **polypeptide** synthesis to produce functional proteins is explained in Chapter 4.

The differences between RNA and DNA are (Figure 3.2.10):

- RNA only exists as a single strand
- RNA polynucleotide strands are usually much shorter than DNA
- the sugar–phosphate backbone has ribose sugar not deoxyribose
- the nitrogenous bases are G, C, A and U (uracil) which in RNA replaces T (thymine)
- there are three main forms of RNA: messenger RNA (mRNA), ribosomal RNA (rRNA), transfer RNA (tRNA).

GO TO ➤ Section 4.2 page 165

Hydrogen bonds are an attractive force created between molecules when the hydrogen atom is already covalently bonded within a molecule to a highly electronegative atom such as oxygen or nitrogen. Due to the difference in electronegativity between the two atoms, the hydrogen develops a partial positive charge that bonds it to lone pairs of negative electrons on the atoms of neighbouring molecules. Hydrogen bonding holds bases together inside a DNA helix.

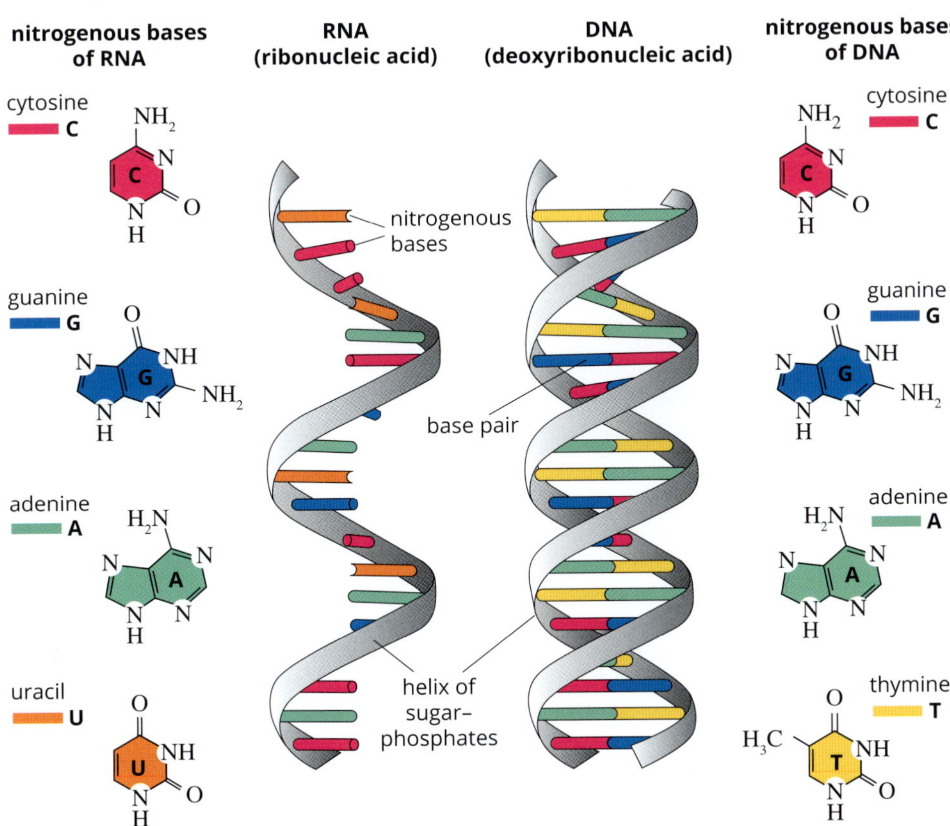

FIGURE 3.2.10 Comparison of the structures of RNA and DNA. Both molecules are made up of nitrogenous bases and a sugar–phosphate backbone, but in RNA ribose sugar replaces deoxyribose sugar and uracil replaces thymine. RNA is also single stranded and shorter in length; DNA is double stranded.

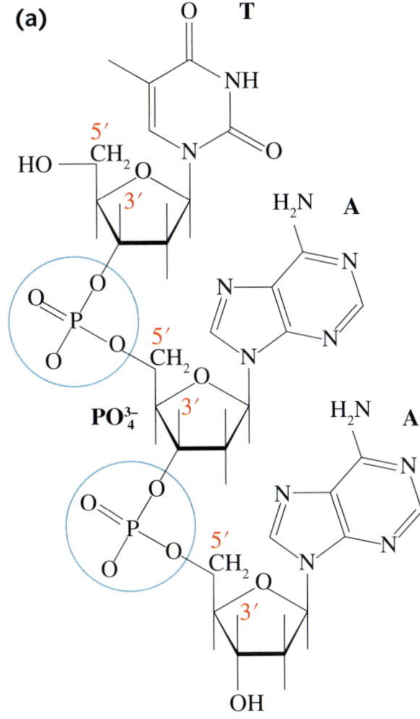

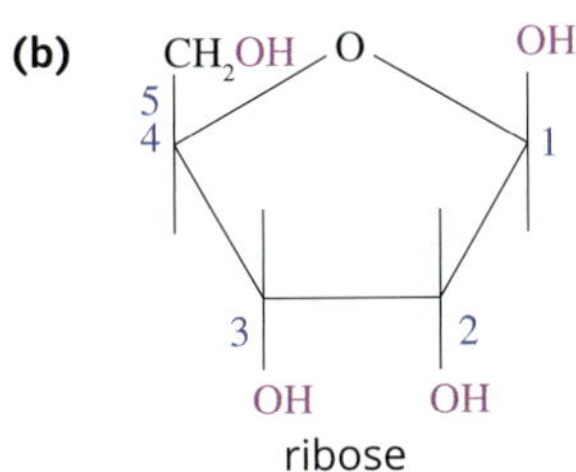

FIGURE 3.2.9 (a) Phosphodiester (PO_4^{3-}) covalent bonds (blue circles) formed between three nucleotides (T, A, A). (b) The numbering convention of carbon atoms in the deoxyribose sugar (DNA) and ribose sugar (RNA) molecules is shown in this diagram.

Phosphodiester bonds in DNA and RNA are the linkages between the 3′ carbon atom of one sugar molecule through a phosphate to the 5′ carbon atom of another sugar (deoxyribose in DNA or ribose in RNA). Strong covalent bonds form between the phosphate group and the two carbon ring sugar molecules.

Bonding and pairing of nucleotides

The way that free nucleotides bond together to form long strands is through a **condensation polymerisation reaction**. This reaction occurs initially between two nucleotides, enabling them to join to form a **dinucleotide**, and releasing a water molecule as described below.

- The hydroxyl group (OH) on the 3′ carbon atom of the sugar of one nucleotide joins with the phosphate (PO_4) on the 5′ carbon of the sugar of the other nucleotide to form water (H_2O), which is released.

Condensation polymerisation is when smaller chemical units (monomers) are bonded together by the removal of water molecules to create a large molecule of repeating units (polymer). Many biomolecules form this way including DNA, RNA and proteins.

In the DNA double helix, hydrogen bonds hold the pairs of polynucleotides together. Note that there are two hydrogen bonds between adenine (A) and thymine (T) and three hydrogen bonds between cytosine (C) and guanine (G) due to their structure as a pyrimidine or a purine.

- Free nucleotides can then be continuously added to the 3′ carbons in this way, forming a long sugar–phosphate–sugar–phosphate backbone strand known as a polynucleotide.
- The nucleotides in the sugar–phosphate chain are joined by phosphodiester bonds (Figure 3.2.9).
- In polynucleotide strands, one end has a free phosphate group on the 5′ carbon; this is called the 5′ end (five prime). The other end of the strand has a free hydroxyl on the 3′ carbon; this is called the 3′ end (three prime). The 5′ and 3′ ends are significant when DNA copies itself for cell division.

Both DNA and RNA are polynucleotides, formed through condensation polymerisation reactions (Figure 3.2.11).

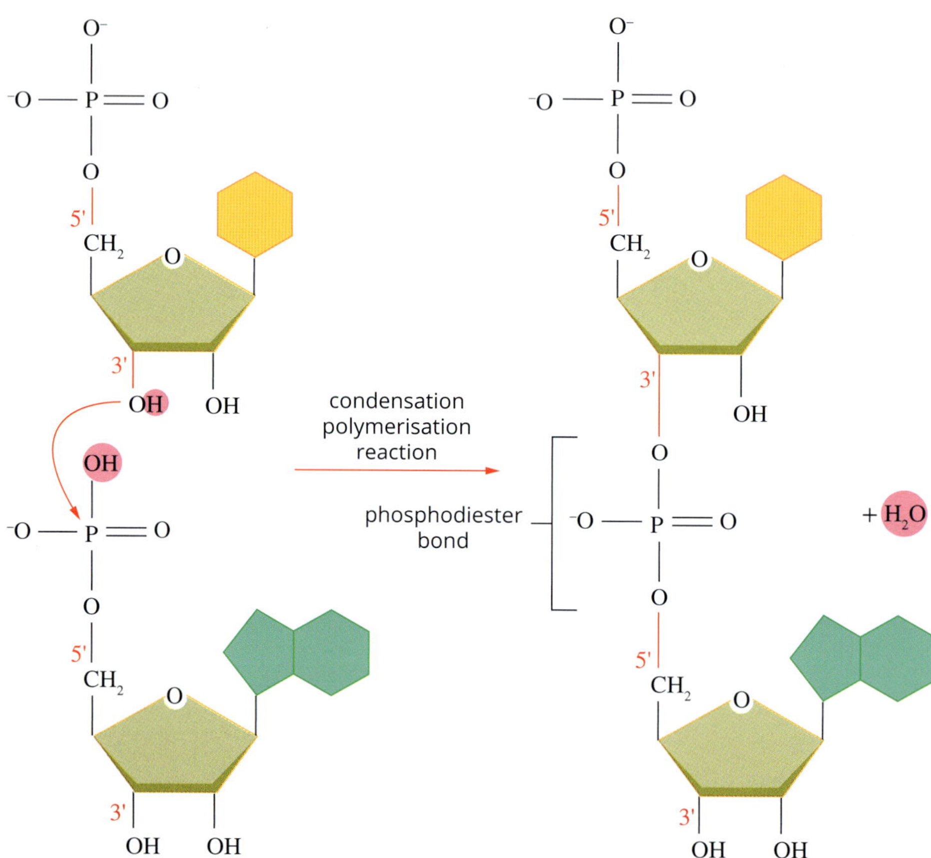

FIGURE 3.2.11 The condensation polymerisation reaction of two DNA nucleotides, forming an initial dinucleotide by the removal of water. More nucleotides are added to form the polynucleotide DNA backbone, made from alternate sugar and phosphate units joined by a type of covalent bond called phosphodiester bonds.

Adenine (A) always bonds with thymine (T).
Guanine (G) always bonds with cytosine (C).

DNA is a double-stranded helix

A full double-stranded DNA molecule is made up of two polynucleotide chains. The two polynucleotide chains of DNA are held together by **hydrogen bonds** between complementary base pairs, rather like steps on a ladder (Figures 3.2.12 and 3.2.13). The sides of the ladder are two sugar–phosphate backbones. The rungs of the ladder are the paired nitrogenous bases of each nucleotide. There is always direct pairing between A and T, and between G and C in the DNA molecule. This complementary base pairing results in the two polynucleotide strands joining together to form the double-stranded DNA molecule. Given the base sequence of one strand you can determine the sequence of the other by the complementary base pairing rule (Figure 3.2.12).

In complementary base pairing, the:

- purine adenine (A) always pairs with the pyrimidine thymine (T), held together with two weak hydrogen bonds

- purine guanine (G) always pairs with the pyrimidine cytosine (C), held together with three weak hydrogen bonds.

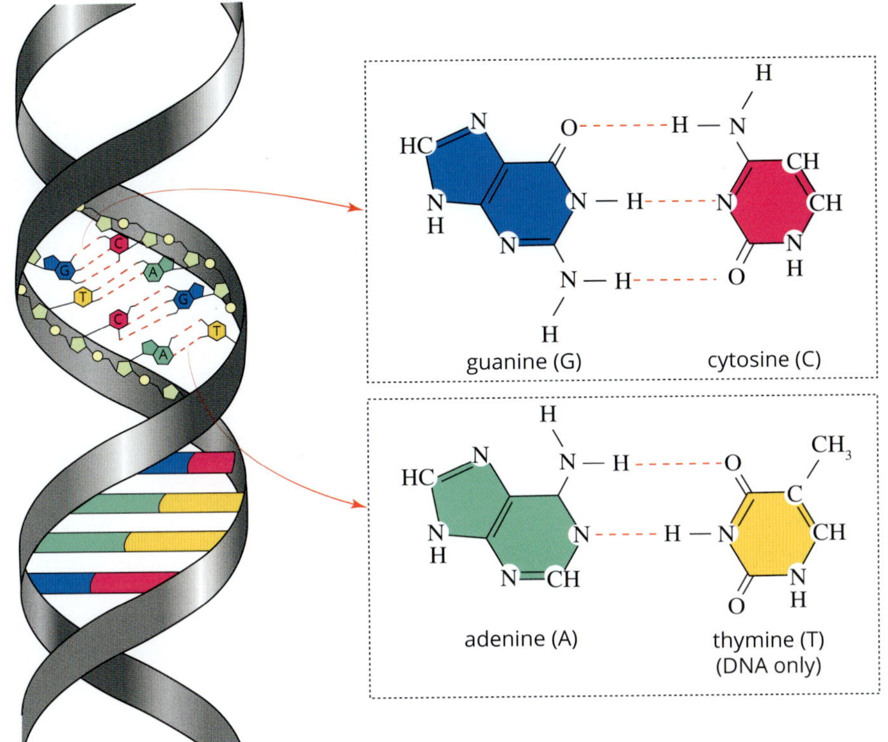

FIGURE 3.2.12 The helical structure of DNA. Two polynucleotide strands form a double helix joined by complementary base pairs: guanine (G) with cytosine (C) and adenine (A) with thymine (T). G-C has three hydrogen bonds and A-T has two.

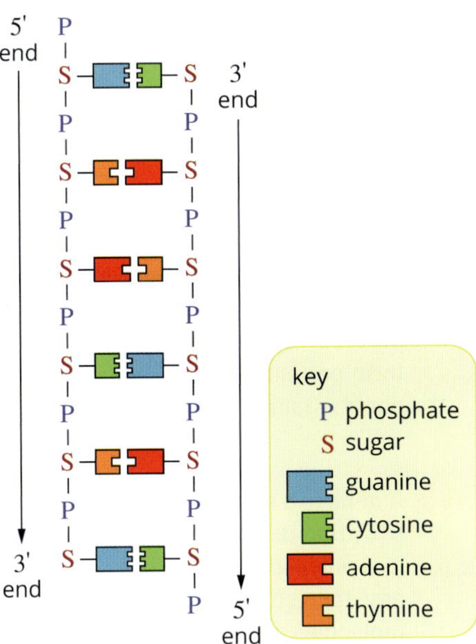

FIGURE 3.2.13 A two-dimensional representation of the DNA molecule. Hydrogen bonds between the complementary base pairs hold the two antiparallel polynucleotide strands together.

For the bases to face each other and form hydrogen bonds, one polynucleotide strand must run **antiparallel** to the other, meaning that one strand runs from 5′ to 3′ and the other runs from 3′ to 5′ (Figure 3.2.13). The two polynucleotide strands will then spiral around an imaginary axis in a right-hand turn, forming a double helix (Figure 3.2.12). It is like a rope ladder that is held at one end and twisted in a right-hand turn, or a spiral staircase. The 'steps' are only about 2 nm (nanometers) wide. The DNA double helix is a right-hand helix, meaning it curls in the direction of the right-hand fingers when holding the right thumb up. There are ten base pairs for each complete turn.

A nanometre (symbol nm) is one-billionth of a metre (10^{-9}m). This means there are 1 000 000 (1 million, 10^6) nanometres in a millimetre. The SI unit next largest to a nanometre is the micrometre (μm). There are 1000 nanometres in a micrometre and 1000 micrometres in a millimetre. A human hair is about 60 000–75 000 nm (60–75 μm or 0.06–0.075 mm) in diameter.

CHROMOSOMES

In a eukaryotic cell, chromosomes can have various shapes and sizes, and their appearance changes during the life of the cell. They remain inside the nucleus except during some stages of the cell division processes. Chromosomes carry genetic information as a unique set of genes inherited from parents. Eukaryotic organisms have sets of linear chromosomes and the number of chromosomes is constant in each species (e.g. 46 (23 pairs) in humans). These chromosomes are passed on to daughter cells during mitosis and to germ cells during meiosis.

Prokaryotes have a single circular chromosome. Prokaryotes may also contain smaller circular DNA molecules called plasmids, which can move between cells.

A chromosome is a structure containing a single DNA molecule and associated histone proteins.

A histone is a small protein that binds to DNA and plays a key role in the chromosome structure.

Structure

Common wisdom says that chromosomes are made from DNA but this is only partly correct. Eukaryotic chromosomes also contain many small structural proteins called histones that are roughly shaped like spheres. The long thin DNA molecules are wound around a series of these **histones** forming structures called **nucleosomes** (Figure 3.2.14).

A nucleosome is a structural unit that consists of histone proteins around which DNA is coiled.

> In eukaryotes, when the DNA is coiled around the small proteins called histones, they form particles (nucleosomes) about 10 nm in diameter.

> A gene is a specific length of DNA that directs the synthesis of a polypeptide with the assistance of enzymes, RNA molecules and ribosome organelles. Polypeptides then combine to form functional protein molecules. Each of the estimated 30 000 genes in the human genome makes an average of three proteins.

> Gene expression is the process by which the information in a gene is used to produce a functional product. The end products are usually proteins.

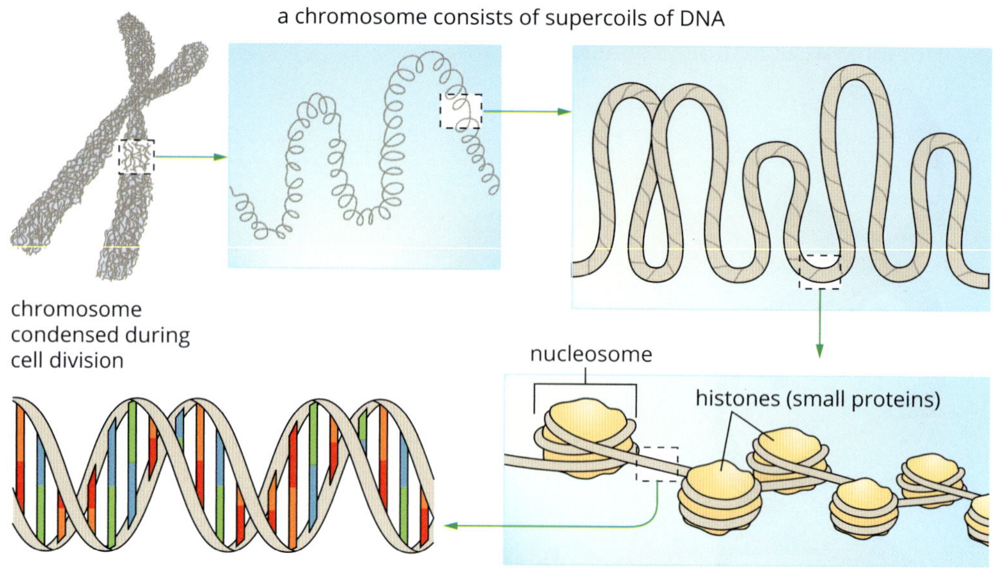

FIGURE 3.2.14 A chromosome unwound to reveal the complex substructure of supercoils of DNA. Each coil consists of histones that are structural proteins, which package and order the DNA into structural units called nucleosomes. Further unwinding of a nucleosome reveals the double-stranded DNA structure.

This complex structure is made into a **supercoil** that contracts tightly together to become the chromosome during condensation before the cell divides. At this stage in prophase, the chromosomes first become large enough to be seen under a light microscope if a stain for DNA is used (Figure 3.2.16).

Nucleosomes give the DNA strand the appearance of a string of beads. This arrangement of DNA wrapped around histones serves to package the long lengths of DNA efficiently and to protect it from enzymatic degradation. The end result is a fibre known as chromatin. The chromatin is looped and coiled again to produce a super-coiled thread-like structure called a chromosome.

BIOFILE N

DNA length

The 23 human haploid chromosomes contain about 3200 million base pairs. So, an 'average' human chromosome contains about 140 million base pairs of DNA. Ten base pairs are about 3.4 nm long, making the DNA in an average chromosome about 4.76 cm long (Figure 3.2.15). This is an extraordinary length considering that a DNA molecule has a very small diameter (2 nm) and that a full set of chromosomes fits into the nucleus of a microscopic cell. If you built a model of this one DNA molecule using a scale of 1 cm = 1 nm, the model would be 2 cm wide and 476 km long! Tight coiling of the skinny molecules enables them to compact their lengths inside a nucleus. A typical human somatic cell nucleus is around 6 μm in diameter and holds about 2 m of tightly coiled DNA with approximately 6.4 billion (6400 million) base pairs.

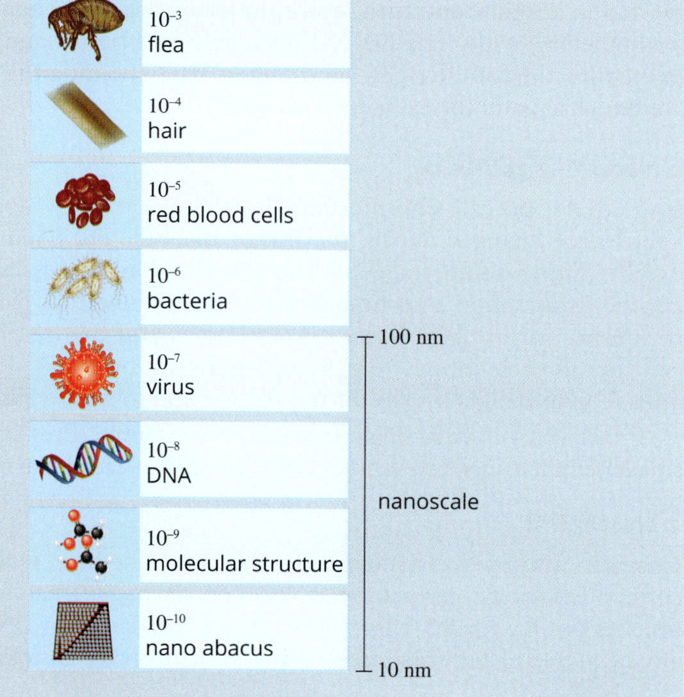

FIGURE 3.2.15 Size comparison on the nanometer scale

Function

Chromosomes carry genes encoded in the base sequence of the DNA. The process by which the information in the gene is decoded to make proteins is called **gene expression**. By definition, the information carried in each gene can be expressed to make one polypeptide chain of amino acids, the precursor of protein molecules.

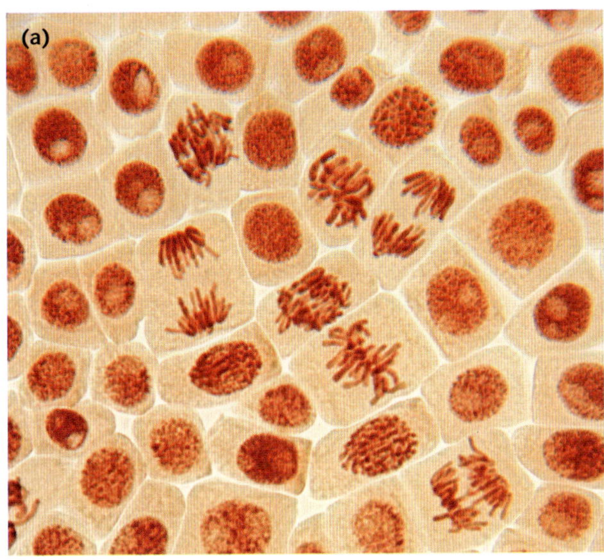

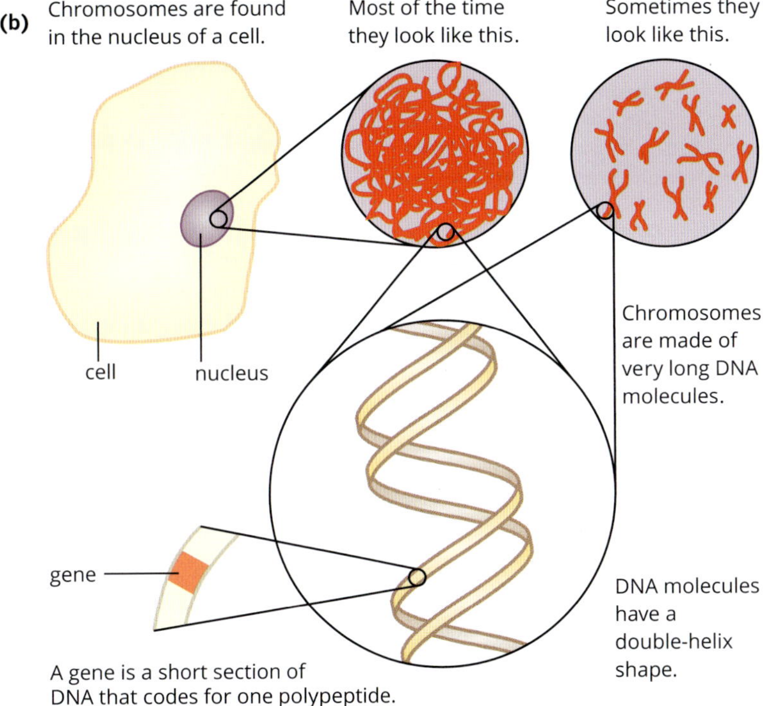

FIGURE 3.2.16 (a) Cells seen under the light microscope during the phases of mitosis have visible chromosomes because the nucleosomes (DNA and histones) are tightly condensed. (b) The packaging of DNA into chromosomes in a cell nucleus

Genomes

The term **genome** is often mentioned in connection with chromosomes, DNA and genes. A genome is the complete set of genetic information in an organism and is measured in the number of base pairs. It provides all the information the organism requires to function and is distinctive for each organism and species. In eukaryotes, each cell's genome is contained within the membrane-bound structure called the nucleus. Prokaryotes (bacteria), which contain no inner membranes, store their genome in a region of the cytoplasm called the nucleoid. The full range of RNA molecules expressed by a genome is known as its **transcriptome**, and the full assortment of proteins produced by the genome is called its **proteome**. The study and analysis of genomes is called **genomics**.

Chromosomes of organisms range in size from about 50 million to 300 million base pairs. Because both sets of chromosomes in a pair (diploid, $2n$) are almost identical, the genome of an organism is the total of an organism's DNA measured in the number of base pairs contained in a single (haploid, n) set of chromosomes.

Features of chromosomes

When a eukaryotic cell is preparing to divide, chromosomes become condensed and are visible when the cell is stained and viewed at high magnification under a light microscope. The nucleosomes fold in a regular manner, producing supercoils (Figure 3.2.16).

In comparison, the DNA in prokaryotic cells is usually present in a single circular chromosome that is located in the nucleoid region of the cytoplasm. Prokaryotic chromosomes are less condensed than their eukaryotic counterparts. The internal features of prokaryotic cells are not easily identifiable when viewed under a light microscope so more powerful electron microscopes are required (Figure 3.2.17). The differences between the DNA of prokaryotic and eukaryotic cells are covered more comprehensively in Chapter 4.

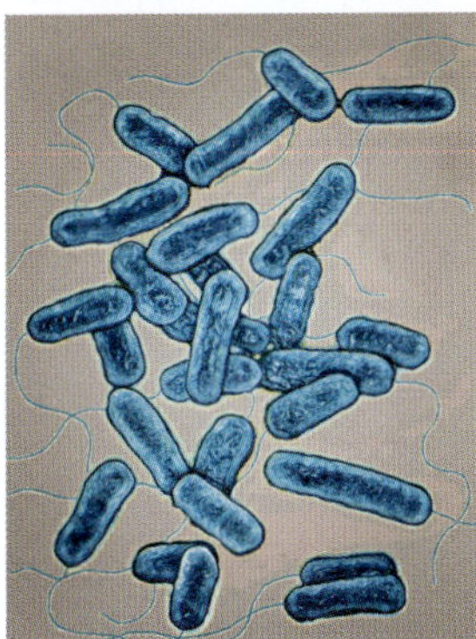

FIGURE 3.2.17 Coloured transmission electron microscope image (TEM) of the bacteria, *Legionella pneumophila*. The darker nucleoid region can be seen in the centre of these cells but a magnification of x10 000 or greater would be required to view chromosomes and plasmids in detail.

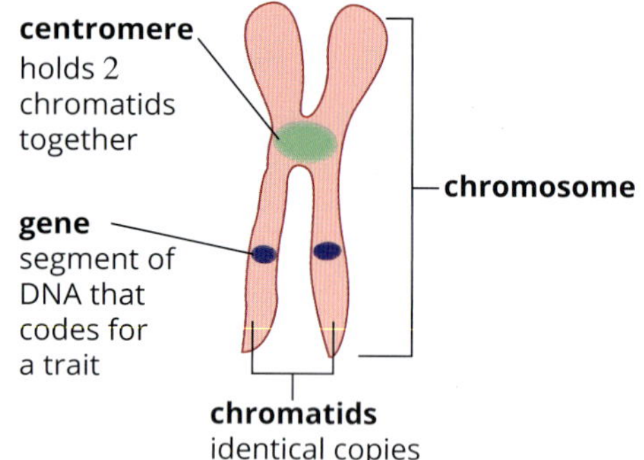

FIGURE 3.2.18 Each gene occupies a fixed position (locus) on a chromosome. The gene indicated here is on the q arm at a precise distance from the centromere.

The longest human chromosome (designated as chromosome number 1) has about 2000 genes. Each gene has a particular position, called a **locus** (plural loci), on a specific chromosome (Figure 3.2.18). Regions called spacer DNA separate the genes of each DNA molecule (Figure 3.2.19). Genes are highlighted in red and the spacer regions of DNA, separating the genes, are shown in blue. Spacer regions include DNA that does not encode a protein product. However, they may function in spacing genes far enough apart to enable enzymes or other molecules to interact easily with them. Chromosomes differ in length because of differences in the number of genes and the amount of spacer DNA between the genes.

Each chromosome has a constriction point called the centromere, which pinches the chromosomes into two sections. The regions on either side of the centromere are referred to as the chromosome arms. The shorter arm is called the 'p arm' (from the French word 'petite') and the long arm is referred to as the 'q arm' simply because it is the next letter after p in the alphabet. Photographs or diagrams of chromosomes are always arranged so that the p arm is at the top.

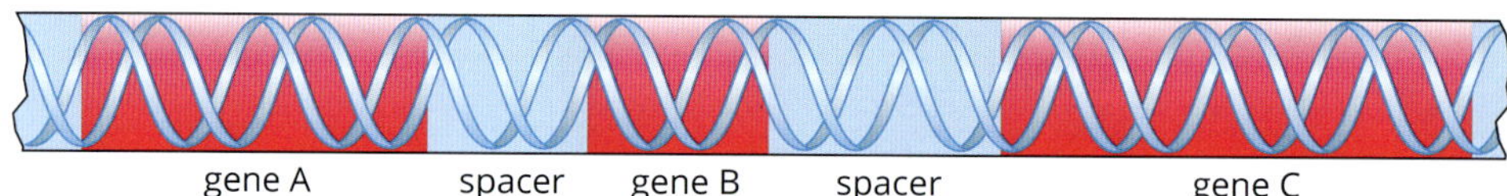

FIGURE 3.2.19 A short stretch of double-stranded DNA. Genes are highlighted in red and the spacer regions of DNA separating the genes are shown in blue.

The position of a gene on a chromosome is called a locus (plural loci).

Humans have 46 chromosomes, comprising 23 inherited from each parent. Forty-four of these chromosomes form 22 matching pairs. The same genes are found at the same locations (loci) on the two chromosomes in a matching pair. They are referred to as homologous chromosomes or homologues.

The sex chromosomes for all females are homologues because they have a matching pair of X chromosomes. The sex chromosomes for males are not homologous because they have an X and a Y chromosome, which contain different gene sets and are different lengths. Nevertheless, in most mammals the X and Y chromosomes behave as a homologous pair during meiosis because some small sections of these chromosomes match (i.e. are homologous).

Homologous chromosomes are the two chromosomes that make a matching pair. They carry the same genes at the same loci.

The number of sets of chromosomes in a cell is called the ploidy level. Haploid cells have one set, diploid cells have two sets, and polyploid cells have three or more sets.

Genes and alleles

Before continuing, it is worth revisiting some essential information from earlier in this section. DNA is the molecule of life that encodes the information from which organisms are built. The code carried by DNA consists of many genes, and these genes (genotype) determine the characteristics of an organism (phenotype). At a molecular level, a gene is a unique sequence of bases in the DNA. Each gene carries a particular instruction for a cell that is usually the production of a polypeptide chain. This process, by which the information in the gene is decoded to assemble the polypeptide, is called the gene expression. One or more polypeptides form proteins, which perform essential functions in an organism's cells. You will learn more about protein function in Chapter 4.

GO TO ➤ Section 4.3 page 181

Alleles

You can see, simply by looking at people around you, that there is variation in the human population. Observable physical characteristics or traits such as skin colour, eye colour and hair colour all vary within populations and even within families. Although an individual gene may be responsible for a specific trait, that gene can exist in different forms known as alleles (Figure 3.2.20). For example, there are genes that code for eye colour. Alternative forms of the gene (alleles) exist, including one for blue eye colour and another for brown eye colour. This means that the DNA sequence of bases for these two alleles is slightly different. Both alleles still code for eye colour and are found in the same place (locus) on the chromosome (DNA strand) (Figure 3.2.20a).

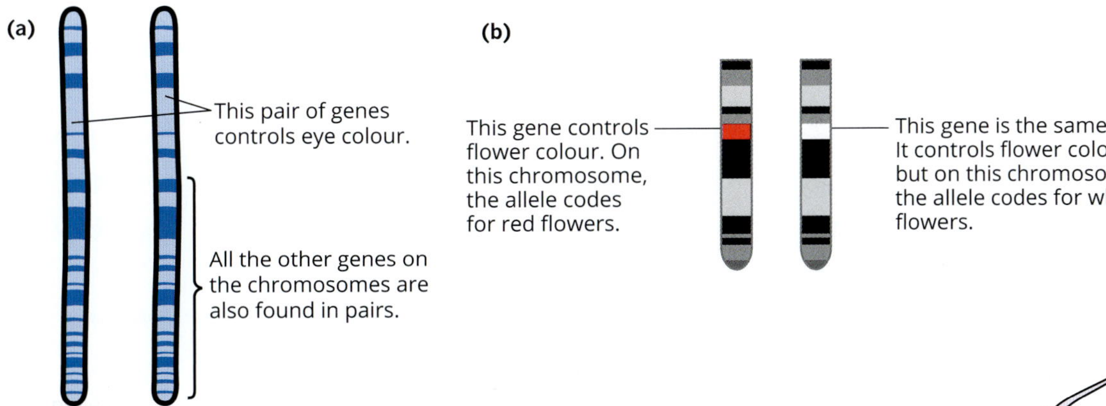

FIGURE 3.2.20 (a) The two alleles for human eye colour are a single gene with alternative phenotypic expressions for the eye colour pigment produced. They are on the same loci of matching (homologous) chromosomes, one inherited from each parent. (b) Alternative alleles for the flower colour gene in a plant's chromosomes.

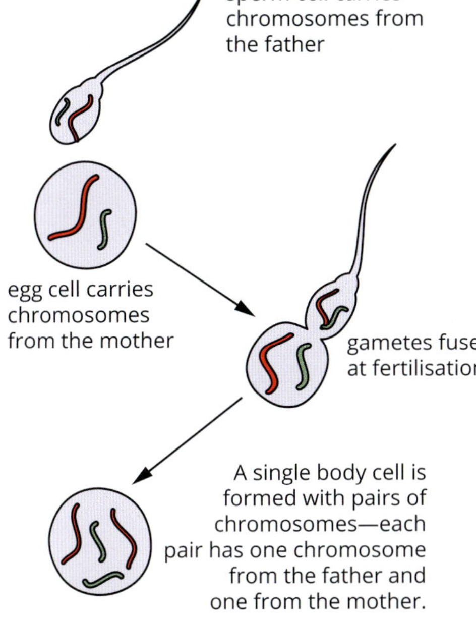

FIGURE 3.2.21 Each parent in sexual reproduction contributes one chromosome to each homologous pair of chromosomes. This means that one allele (of the two alternative forms of the same gene) has come from each parent.

Somatic cells of a diploid organism contain two alleles for every gene, with one allele of every gene in an organism inherited from each parent (Figure 3.2.21). In Chapter 5, the gene expression of alleles will be investigated in more detail.

GO TO ➤ Section 5.1 page 202

DNA REPLICATION

The DNA molecule is passed on from one cell to another when cells divide. In order to transmit an exact copy of the DNA molecule without losing any instructions, cells have a mechanism for synthesising and accurately copying new DNA in a process called self-replication. In the cell cycle before mitosis occurs, self-replication of DNA occurs in the S phase of interphase (see Figure 3.1.4 page 117).

During self-replication, each strand of a parental DNA molecule acts as a template strand on which a new strand is synthesised. This involves the double-stranded DNA unwinding or 'unzipping' under control of enzymes (helicase and topoisomerase) at the replication fork (Figure 3.2.22). There is a bank of free nucleotide molecules stored in the nucleus. An enzyme called primase initiates the first stage of synthesis. The free nucleotides start to pair up with each template strand, using the rules of complementary base pairing (A pairs with T, G pairs with C), to make new strands. The replication process proceeds from the 5′ towards the 3′ end of DNA. As each new complementary strand forms, the replication fork moves along and the completed pairs of DNA strands begin to rewind into doubles helices under control of yet more enzymes, DNA polymerases. Each double helix coils around histone proteins and reforms as a chromosome with twice as much DNA as a non-replicated chromosome. The structure now consists of sister chromatids joined at a centromere and ready to undergo cell division. The mitotic cell division process will separate the sister chromatids into single chromosomes again, each with a full set of identical genes.

In summary

DNA replication is described as semi-conservative replication because each daughter DNA molecule consists of one old and one newly synthesised strand. It is a very accurate process with three distinct phases (Figure 3.2.22) that can be summarised as:

- parent DNA helix starts to separate its strands from one end as enzymes break the weak hydrogen bonds joining base pairs
- free nucleotides attach to complementary partners on each strand, from the open end towards the replication fork
- two identical sets of DNA are formed that each coil back into a double helix.

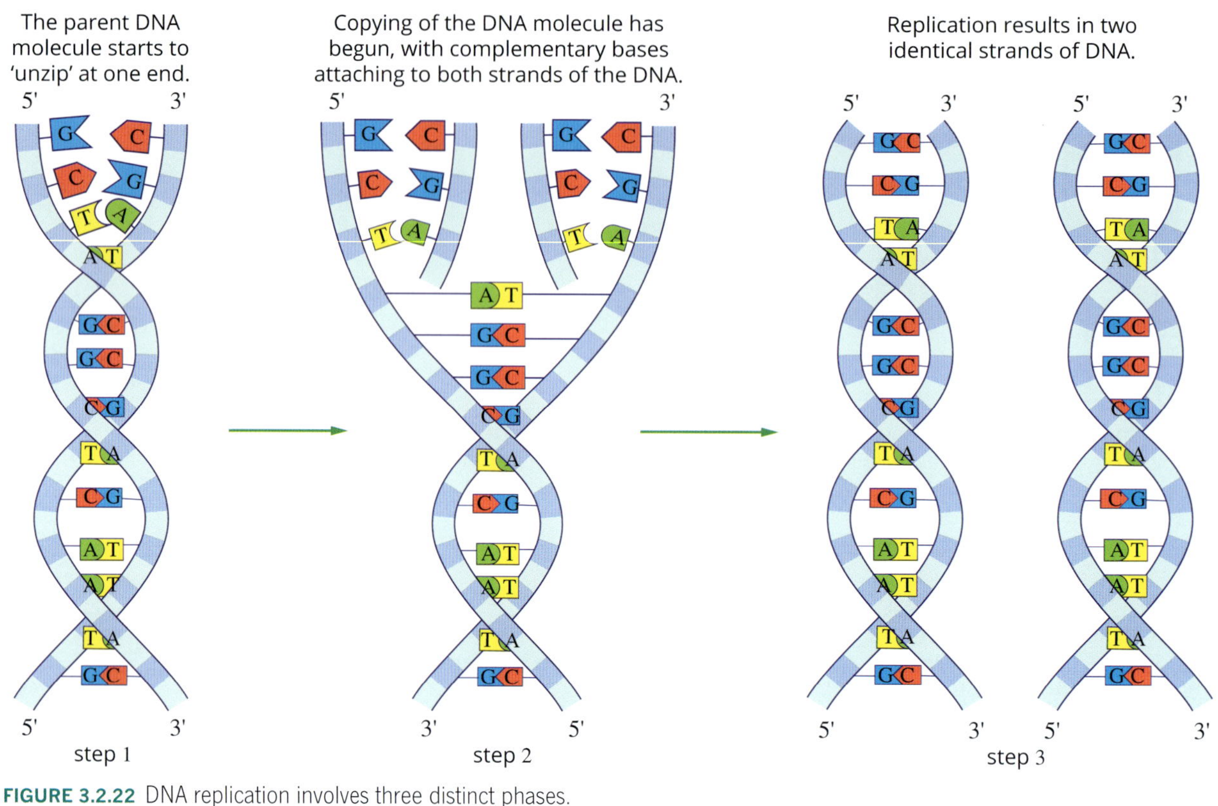

FIGURE 3.2.22 DNA replication involves three distinct phases.

The full story

The full story of the DNA replication process is complex and it requires many enzymes to work in a coordinated fashion (Figure 3.2.24). For example, DNA polymerase is actually a multi-protein complex, with as many as 10 different proteins, acting like a sliding clamp that circles the DNA. After primase initiates synthesis using a RNA primer piece, the polymerase complex slides along adding more complementary daughter bases in sequence. It also checks the bases already added, editing and replacing incorrect ones, before rewinding the double helix that now consists of one parent and one daughter strand. The precision of the initial assembly mechanism, plus the checking process, means that DNA replication has a very low number of errors even though it proceeds at an amazingly rapid rate. Up to 4000 free nucleotides can be added every second. The estimated error rate under normal conditions is only 1 in 10^9 (one in a billion).

As already mentioned, the daughter strands must be built from their 5′ to their 3′ end. This is because DNA polymerase can only add nucleotides to the 3′-OH of an existing nucleotide. The job of primase includes synthesis of a short RNA primer that is used as a starter nucleotide section for the DNA replication (Figure 3.2.24). This primer is removed before completion of the DNA replication process. Knowledge of primers from the natural process of replication has been usefully applied for the technology of PCR (polymerase chain reaction) now widely used to copy DNA in the laboratory.

From one of the parent strands, the synthesis of a daughter strand, called the leading strand, can proceed continuously from 5′ to 3′. Because the two parent strands ran anti-parallel, the second parent strand is running in the wrong direction. Synthesis of this daughter strand, called the lagging strand, is achieved by forming short lengths of DNA that are later joined together by the DNA ligase enzyme. These sections are named Ozaki fragments (Figure 3.2.24) in honour of the scientist who discovered them.

Thus, DNA replication is continuous on the leading strand and discontinuous on the lagging strand. On the lagging strand, DNA polymerase has to operate by moving away from the replication fork, which is why the copying must be done as discontinuous fragments that are joined later. Having a discontinuous process makes it even more remarkable that accuracy is so high.

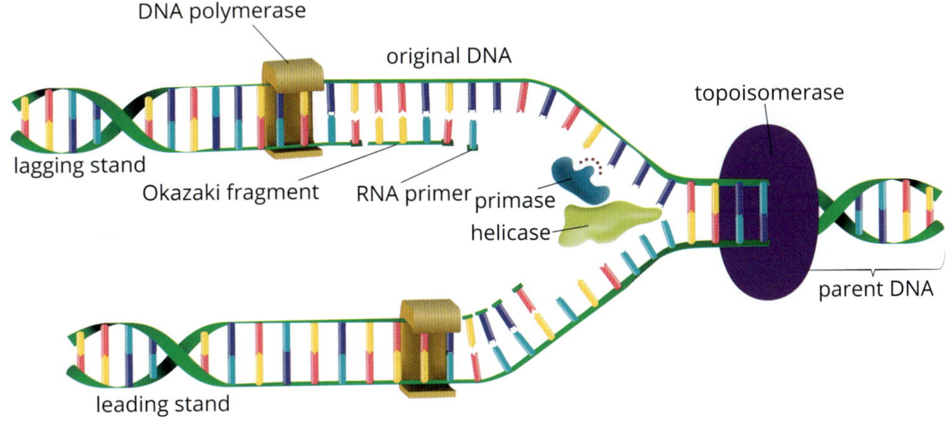

FIGURE 3.2.24 DNA replication under the control of enzymes

MUTATION AND OTHER DAMAGE TO DNA

Cells cannot function properly if the genetic material is changed, a situation usually classed as **mutation**. Mutations of DNA may affect a single gene or multiple genes. Despite the very high rate of accuracy, mistakes or damage to a cell's DNA can occur during DNA replication. It can also be caused by environmental factors. Checkpoints throughout the cell cycle monitor for damage to the DNA. If the damage cannot be repaired, the cell will undergo apoptosis (programmed cell death).

Somatic mutations occur in body cells and only affect that individual. **Germline mutations** are heritable because they affect gametes and can therefore be passed on to offspring. A germline mutation may bring a new allele into a **gene pool**, potentially influencing the allele frequencies. It is important to remember that mutation means change, not the introduction of a fault or disease. Mutations can have a beneficial or harmful effect, or no effect at all, on the organism. You will learn more about mutations in Chapter 7.

Repair

It has already been mentioned that cells have enzymes that can detect and repair damage to DNA. These enzymes run along strands of DNA like a zip, checking that the DNA is intact and has been replicated properly (Figure 3.2.25). If there is minor damage, such as DNA breakages, these will be corrected by enzymes before the cell continues its progress through the cell cycle. If the bases are not complementary matches, they can be replaced if detected before DNA replication is completed in the S phase of the cell cycle. Without repairs to any faulty DNA, the cell may not be able to make the correct proteins or it could become a cancerous cell that replicates itself without going through the complete cell cycle.

DNA is the extraordinary molecule that carries a genetic code universal to all living things on our planet, even the prokaryotic bacterial cells. The code is faithfully inherited from one generation to the next, no matter what the species or which method of reproduction they use. DNA's complex, but predictable and tightly-regulated structure was only elucidated fully in the 1950s. Since then, the developments and achievements of genomics have accumulated at an extraordinary pace.

To achieve its natural role, DNA self-replicates so that each time a cell divides there is a fresh set of identical DNA copied and ready to be inherited by a new cell. In both mitosis and meiosis of eukaryotic cells, the DNA has replicated before the start of nuclear and cell division.

BIOFILE CCT L

DNA replication enzymes

Like most enzymes, those that control replication of DNA have names ending in 'ase' and the word often gives a clue to their function:

- helicase unwinds the double helix (Figure 3.2.23)
- topoisomerase manages the twisting tension
- primase initiates the formation of each daughter strand starting with a short primer section
- polymerases synthesise the polymer strands from their monomer units (nucleotides)
- ligase controls the joining together (ligation) of DNA fragments
- uracil-DNA glycosylase, the base-excision repair (BER) enzyme, triggers the first step in the removal and replacement of damaged DNA bases.

Polymerases, ligase and BER act in the S phase and throughout the cell cycle to deal with mutations as they occur.

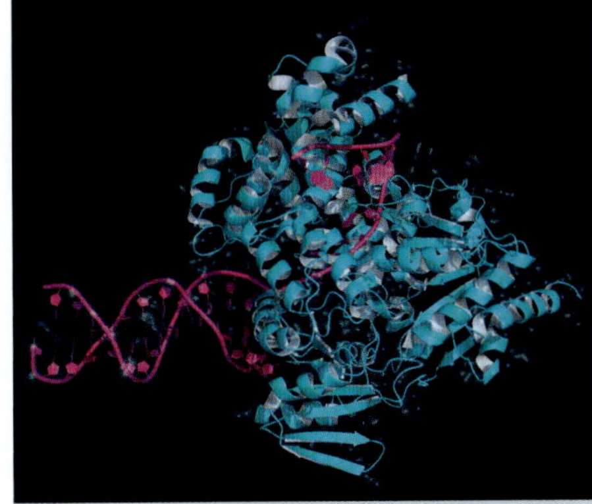

FIGURE 3.2.23 DNA helicase is the enzyme that separates the two DNA strands of the helix. In this model diagram the DNA molecule is coloured pink and the helicase enzyme's polypeptide chains are depicted by many curling blue ribbons.

GO TO ➤ Section 7.1 page 288

During most of the cell cycle, DNA is reliably carrying out its routine function of producing polypeptides to be made into the many proteins needed by even the simplest organism.

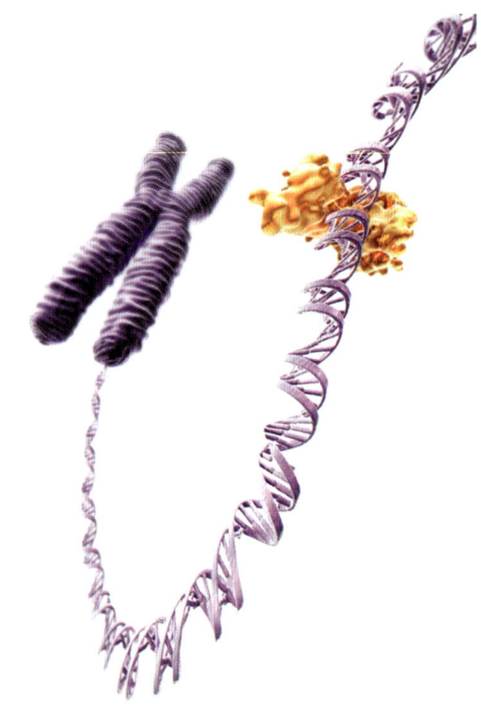

FIGURE 3.2.25 The enzyme, DNA ligase (shown in yellow), joining together a broken strand of DNA. Millions of DNA breaks occur during the normal course of a cell's life. Without molecules to connect the pieces, cells can malfunction, die or become cancerous.

3.2 Review

SUMMARY

- There are two types of nucleic acids: deoxyribonucleic acid (DNA) and ribonucleic acid (RNA).
- Nucleic acids are molecular polymers of monomer units called nucleotides.
- Nucleotides are each made up of a five-carbon sugar (pentose), a phosphate and a nitrogenous base.
- There are five nitrogenous bases: adenine (A), guanine (G), cytosine (C), thymine (T) and uracil (U). A always joins to T with two weak hydrogen bonds, while C always joins to G with three hydrogen bonds.
- Uracil (U) is only found in RNA where it replaces thymine (T).
- Adenine (A) and guanine (G) are purines; cytosine (C), thymine (T) and uracil (U) are pyrimidines.
- DNA is the genetic material that contains the instructions for cells to make proteins.
- RNA is usually a short, single-strand and works in partnership with DNA to synthesise proteins.
- DNA is a long, double-stranded nucleic acid molecule that forms a double helix.
  - nucleotides are linked to each other in a condensation polymerisation reaction to form chains of polynucleotides.
  - the two strands of DNA are antiparallel. One runs in the 5' to 3' direction, while the other runs in the opposite direction. This is significant for self-replication of the DNA molecule before a cell divides.
- Chromosomes are structures in a eukaryotic cell nucleus. Each chromosome contains a single DNA molecule and associated histone proteins organised into a chain of nucleosomes.
- In the prophase part of a cell cycle, the chromosomes condense into supercoils that are visible under a light microscope (if the DNA is stained).
- The genome of a eukaryote is the total of the organism's DNA, measured in the number of base pairs contained in a haploid (*n*) set of chromosomes.

- A gene is a unit of heredity made up of a unique and specific sequence of DNA bases that codes for a polypeptide or a functional RNA molecule. By so doing, it determines the characteristics of an organism. Polypeptides form into proteins.
- Genes have alternate forms for the same trait, known as alleles. One allele is inherited from each parent.
- DNA replication is necessary prior to cell division so that new cells will contain the same genome.
- DNA replication involves synthesis of new daughter strands copied from the template of parent strands when the DNA molecule unzips.
- The new DNA molecules consist of one parent strand and one newly synthesised daughter strand. For this reason it is referred to as semi-conservative replication.
- The replication process requires a team of enzymes and a pool of free nucleotides to be available in the cell nucleus.
- Replication of DNA is a complex, but rapid and accurate process, controlled by several enzymes. Despite this, there are occasional mutation mistakes during replication.

KEY QUESTIONS

1 a Outline the basic functions of DNA and RNA.
  b Explain how RNA differs from DNA, mentioning at least three features that differentiate them.

2 Identify the three basic units of a nucleotide, using a diagram to support your answer.

3 Complete the table by assigning 'purine or 'pyrimidine' to each nitrogenous base and filling in their complementary bases.

Base	Purine or pyrimidine	Complementary base	Purine or pyrimidine
adenine (A)			
guanine (G)			
cytosine (C)			
thymine (T)			
uracil (U)			

4 A strand of DNA has the sequence ATTCCGTA. Write this out, and under it write the sequence of the complementary strand.

5 Distinguish between these terms: chromosome, DNA, genome, gene and allele.

6 Using the terms chromosome, histone, DNA and nucleosome, describe how DNA is packaged into a cell nucleus.

7 How do cells ensure that the DNA is copied correctly to the daughter cells?

8 Assess the role of enzymes in DNA replication.

3.3 Cell replication and the continuity of species

It seems contradictory that, on the one hand, cell division can produce cells that are the exact replicas of the parent cell and, on the other hand, can also introduce enough genetic variation for the evolution of new species. Even a quick glance at Figure 3.3.1 shows the huge variety and complexity of species on Earth that have evolved by diverging from simple, unicellular bacteria. The answer to this puzzle lies in the different outcomes of mitosis and meiosis that you learnt about in Section 3.1.

According to the cell theory, all cells arise from pre-existing cells. You have learnt about the processes at the cellular and molecular level that allow cells to replicate themselves. It should be apparent by now that the DNA inside cells is the key to understanding cell replication and inheritance from one generation to the next. DNA is universal (found in all organisms) but also characteristic of each species. This ensures genetic continuity is maintained within species but also allows genetic change for adaptation and evolution.

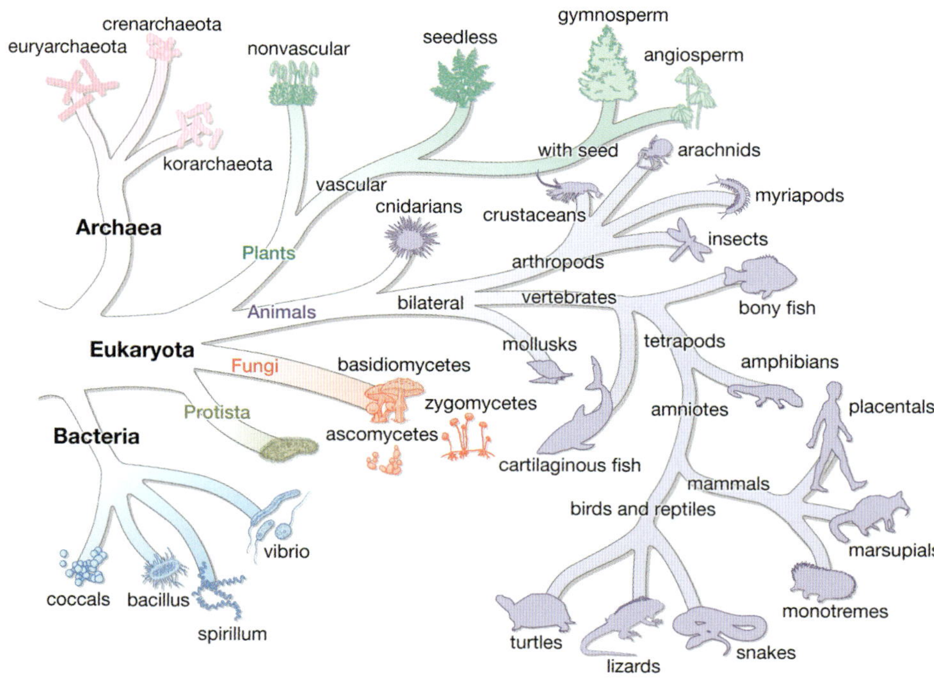

FIGURE 3.3.1 A phylogenetic tree representing the diversity and continuity of species since life first evolved on Earth

FIGURE 3.3.2 The process of passing genetic information from parent to offspring ensures the continuity of every species on Earth. (a) A queen honey bee (centre, with short wings) is cared for by worker bees. Her sole function in the colony is to reproduce the next generation, preserving the continuity of their species. (b) Spotted ladybugs procreate in spring to continue their species into the next generation. (c) Sporophytes of *Polytrichum* moss ensure the genetic continuity for this species.

You can think about genetic continuity in different ways. It refers to the identical replication of genetic information from a parent cell to two daughter cells. Another aspect is the continuance of parental traits in their offspring (Figure 3.3.2). At a broader level, you can look at the effects of evolution on the gene pool within a population or species. All of these perspectives depend on an understanding of DNA.

In this section you will learn about the different outcomes for cell replication processes and the effect this has on the continuity of species. You will also consider the meaning of the continuity of species in the context of evolution and come to understand how the boundaries between species are defined.

+ ADDITIONAL

The species concept

In the 17th and 18th centuries (1600–1800), there were two notable scholar scientists (John Ray and Carl Linnaeus) who contributed significantly to defining species in biological terms. Since then, and especially since the advent of knowledge about the molecular basis of inheritance, there have been various other ways to explain the concept of a species. Most notably and simply is the phylogenetic concept that a species is a group of organisms that share an ancestor. However, since we now understand that the same DNA molecule and code is common to all species, it raises the question of how many previous generations count as ancestors. In theory, all organisms share an ancestor right back to the first prokaryotic cells, but obviously not all can interbreed due to the evolutionary changes made since then (Figure 3.3.1).

A concept that is not strictly accurate, but useful and commonly used, is that individuals of the same species can interbreed successfully amongst themselves.

John Ray (1627–1705) studied plants and defined a species as a group that can only breed amongst itself.

Nearly 50 years later, Carl Linnaeus (1707–1778) made major contributions to biological classification with his hierarchical system and binomial naming of species that are still used today. Like Ray, Linnaeus studied plants but concentrated on their adult reproductive structures as the sole basis for species groupings, rather than a full range of morphological evidence.

By 1813, Augustin Pyramus de Candolle (1778–1841) had introduced the word taxonomy and defined species as *'... a collection of all the individuals which resemble each other more than they resemble anything else, which can by natural fecundation produce fertile individuals...'.*

In 1858, Darwin and Wallace jointly published their famous paper with the Linnean Society of London (Figure 3.3.3). This paper, sponsored by scientific supporters of Darwin, preceded his book in 1859, *The Origin of Species by Means of Natural Selection*. For these scientists, the species was a fundamental unit of evolution but difficult to pin down by a single definition.

In 1942, Ernst Mayr provided this widely-accepted definition for what is known as the biological species concept: *'groups of actually or potentially interbreeding natural populations which are reproductively isolated from other such groups'*. The concept has no reference to ancestry or morphology, instead it relies purely on the ability to reproduce. For example, if two plants are growing side by side and do not interbreed, they are two different species.

According to the biological species concept, members of a single species breed and produce offspring in the wild. Members of different species generally do not breed in

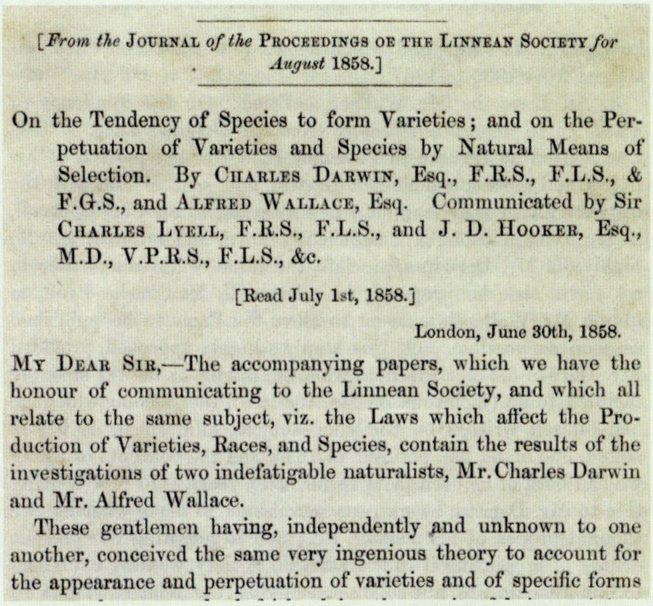

[*From the* JOURNAL *of the* PROCEEDINGS OF THE LINNEAN SOCIETY *for August* 1858.]

On the Tendency of Species to form Varieties; and on the Perpetuation of Varieties and Species by Natural Means of Selection. By CHARLES DARWIN, Esq., F.R.S., F.L.S., & F.G.S., and ALFRED WALLACE, Esq. Communicated by Sir CHARLES LYELL, F.R.S., F.L.S., and J. D. HOOKER, Esq., M.D., V.P.R.S., F.L.S., &c.

[Read July 1st, 1858.]

London, June 30th, 1858.

MY DEAR SIR,—The accompanying papers, which we have the honour of communicating to the Linnean Society, and which all relate to the same subject, viz. the Laws which affect the Production of Varieties, Races, and Species, contain the results of the investigations of two indefatigable naturalists, Mr. Charles Darwin and Mr. Alfred Wallace.

These gentlemen having, independently and unknown to one another, conceived the same very ingenious theory to account for the appearance and perpetuation of varieties and of specific forms

FIGURE 3.3.3 Title page of the Darwin-Wallace paper, *'On the Tendency of Species to form Varieties; and on the Perpetuation of Varieties and Species by Natural Means of Selection'*, published in the Journal of the Proceedings of the Linnean Society, 1858.

the wild. For example, house sparrows (*Passer domesticus*) breed with one another (Figure 3.3.4), but not with other species of the genus *Passer*, even if they live together. Different species do not generally interbreed freely in nature, even when they are members of the same genus, although there are exceptions to this.

However, species of many organisms do not reproduce sexually (e.g. bacteria, some insects and many protists undergo both sexual and asexual reproduction), so this criterion of a species cannot be used to classify organisms that reproduce asexually.

Traditionally, differences in the morphology (form and structure) of organisms have been used to divide organisms into species, however the development of the science of genetics throughout the 1900s presented a new way to define species. With the accelerating progress of genomic technology, quantifying genetic differences within and between groups of organisms is now increasingly used to determine species boundaries.

FIGURE 3.3.4 House sparrows (*Passer domesticus*) only breed within their own species. The genus *Passer* has at least 26 other species, including tree sparrow (*P. montanus*) cape sparrow (*P. melanurus*) and dead sea sparrow (*P. moabiticus*).

GENETIC CONTINUITY IN A SPECIES

In the classification hierarchy, only genus, species, subspecies and variety names are italicised or underlined. Names of families, orders, classes, phyla, kingdoms and domain are not. A genus name starts with an uppercase letter and a species name is all lower case. For example, if the genus is known but the species is not, or the topic refers to all species in that genus, it is written as '*Homo* sp.'.

Section 3.1 emphasised how the process of cell replication produces daughter cells that are genetically identical to the parent cell. Exact replicas of the DNA are made in the S phase of the cell cycle and distributed to the new cells. The cells need to make copies of themselves in order to maintain the continuity of life. This is the case for both unicellular and multicellular organisms.

In eukaryotes (protists, fungi, plants and animals), cells replicate by mitosis. Mitosis is part of the eukaryotic cell cycle described in Section 3.1. Cells replicate for a number of reasons, all of which must produce reliable copies. For a multicellular organism, cells replicate for:

- restoring the nucleus-to-cytoplasm ratio during the cleavage stage of earliest embryo development (see Section 2.3)
- growth and development of the fetus ready for birth, then of the infant to adult size
- maintenance and repair during an individual's lifetime.

Unicellular organisms do not need to replicate for these purposes because they remain a single cell throughout their entire life cycle. Instead, cell replication in unicellular organisms (whether prokaryotes or eukaryotes) is a simple form of reproduction and creates a new, genetically identical individual.

Cell replication therefore maintains the species continuity by processes that copy the same genetic make-up of cells for any one individual in its lifetime from fertilisation to death. Barring mutations and rare diseases, an individual organism has the same number of chromosomes with the same genes as the parent, even though they have a different set of alleles for those genes. For sexual reproduction, they are going to be compatible with other members of their species because their chromosomes will form homologous pairs at fertilisation. The species can therefore continue from one generation to the next.

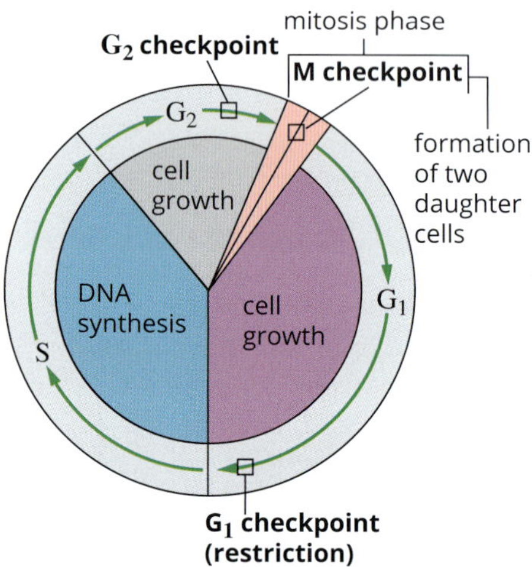

FIGURE 3.3.5 The cell cycle is controlled at three checkpoints: the G_1 checkpoint, G_2 checkpoint and M checkpoint.

Healthy cells replicate in a highly regulated way with a cell cycle that uses in-built checkpoints in a cell cycle control system (Figure 3.3.5). However, there are times when cells divide uncontrollably (Figure 3.3.6). If uncontrolled cell division occurs during embryo development, the embryo will be abnormal and, in most circumstances, will abort. If uncontrolled cell division occurs in a mature organism, a **neoplasm** may form.

A neoplasm is an abnormal growth of tissue that usually, but not always, forms a mass of abnormal cells. Neoplasms are more commonly referred to as tumours, but not all are cancerous. If a neoplasm develops, the genetic continuity has been lost because normal cell cycles of replication cannot continue. You will learn more about neoplasms in Chapter 7.

GO TO ➤ Section 7.2 page 295

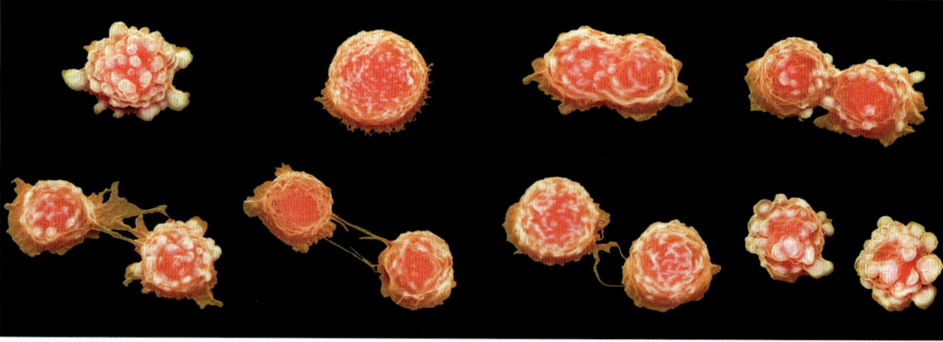

FIGURE 3.3.6 SEM of dividing bowel cancer cells. Cancerous cells ignore or override some of the factors that control cell division, often replicating a lot faster than the organism's own cells.

BIOFILE CCT S

Living fossils—staying the same through evolutionary time

Long-term continuity is obvious for some species that are called 'living fossils'. These are species that have shown little change over millions of years, because they are well adapted to an unchanging environment. There has been little or no natural selection pressure to cause changes, so they have continued with the same traits generation after generation. Fossil records of these species are used as evidence of their stability. Some examples are:

- Sea jellies—fossil evidence from 505 million years ago has placed them as existing in much the same form as today.
- Horseshoe crabs are one of the best-known living fossils, dated as existing from 445 million years ago to the present day with unchanged features.
- The coelacanth (Figure 3.3.7) is an ancient type of fish, plentiful in the Mesozoic (400–60 million years ago) and well-known from fossil evidence, but thought to be extinct until 1938 when one was caught off the coast of South Africa.
- The ant from Mars (*Martialis heureka*), discovered outside the Amazon jungle city of Manaus in 2008, has remained unchanged for 50 million years. It is regarded as a kind of proto-ant that has retained the genetic traits of blindness and forcep-like flexible mandibles. *M. heureka* remained living and breeding in the stable underground habitat, while other ant species emerged above ground and developed adaptive features for living with the newly evolved flowering plants.
- *Ginkgo biloba*, known as the ginkgo or maidenhair tree, is the only living gymnosperm species in the phylum Ginkgophyta (Figure 3.3.8). It is found in fossils dating back 270 million years with a worldwide distribution in the Triassic and Jurassic periods. Its natural distribution now is restricted to China but it is widely grown in gardens. The leaves of the survivor species are strikingly similar to those of Triassic gingko fossils.
- *Wollemia nobilis*, the Australian Wollemi pine, thought to be extinct until 1994 when a remnant population was discovered in a remote gorge of the Blue Mountains, only 200 km west of Sydney. The living leaves and cones appear the same as the oldest known Wollemi fossil from 90 million years ago. The living specimens feature no detectable genetic variation.

FIGURE 3.3.7 The coelacanth (*Latimeria chalumnae*), once thought to be extinct, inhabits steep rocky shores in the western Pacific and Indian Oceans, living at depths of 150–700 m. An adult coelacanth may reach a length of 2 m and has features unchanged from ancient fossil forms.

FIGURE 3.3.8 The foliage and pollen-bearing cones of a male ginkgo (*Ginkgo biloba*)

GENETIC CHANGE IN SPECIES

Genetic continuity should not be confused with a lack of variation. A sexually reproduced individual is unique, resembling both parents but identical to neither. This is largely due to the variation introduced by meiosis during the production of gametes.

Variation within a species

Over the course of two cell cycles (meiosis I and II), germ cells undergo meiosis and form gametes containing only one set of chromosomes (haploid, n). The change from diploid to haploid chromosomal number is why meiosis is referred to as a reduction cell division. The mixed set (n) of chromosomes in a gamete contains a single copy of each chromosome randomly supplied from either of the partner chromosomes in the diploid ($2n$) parental set. For example, human somatic cells contain 46 chromosomes (23 pairs) but the sperm and egg gametes have 23 single chromosomes drawn at random from the homologous partners, giving each gamete a different combination of alleles for the same genes.

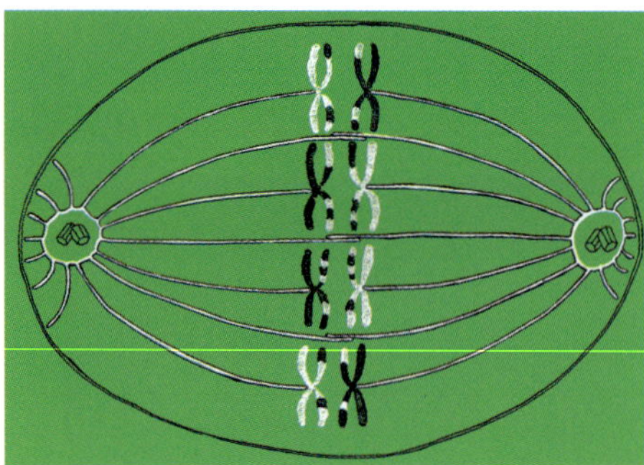

FIGURE 3.3.9 The process of meiosis I, showing the exchange of sections of genetic material (crossing over) between some homologous pairs of chromosomes

> Meiosis creates genetic variation during gamete formation and sexual reproduction combines the genetic variation of two individuals at fertilisation.

> Individuals within a sexually reproducing species are different due to the genetic variation introduced during meiosis and fertilisation. However, individuals within a species are still similar enough to interbreed, ensuring the continuity of their species.

During meiosis I even more variability is added by crossing over between some homologous chromosomes. Crossing over is a process found only in meiosis. It exchanges sections of DNA carrying some alleles, and creates new recombinant chromosomes with a set of alleles not found in either parent (Figure 3.3.9). At fertilisation, the union of egg and sperm (conception) restores the full diploid number of chromosomes ($2n$) to the zygote. Fertilisation itself is a random event, when only one sperm contributes its mixed set of chromosomes to form homologous pairs with the female chromosomes in the egg. The randomness of which specific sperm fertilises the egg introduces yet more variation to the offspring from one parental pair and helps explain why siblings are different (Figure 3.3.10).

The genetic continuity here is that a successful fertilisation event, leading to viable fertile offspring, can only occur with matching (homologous) chromosomes carrying the same genes at the same loci. Variation in individuals is present in the unique combinations of alleles at gene loci but the genes themselves must have continuity from generation to generation to maintain a species. The species continues generation after generation but with individuals in that species having some genetic variation, such as tall or short height, yellow or green pea pods, black or white fur, or blood group type. Such variation in traits does not prevent interbreeding for the next generation within the same species. In fact, populations of organisms that reproduce sexually have considerable genetic flexibility, which enables species to survive and reproduce in varied and changing environments.

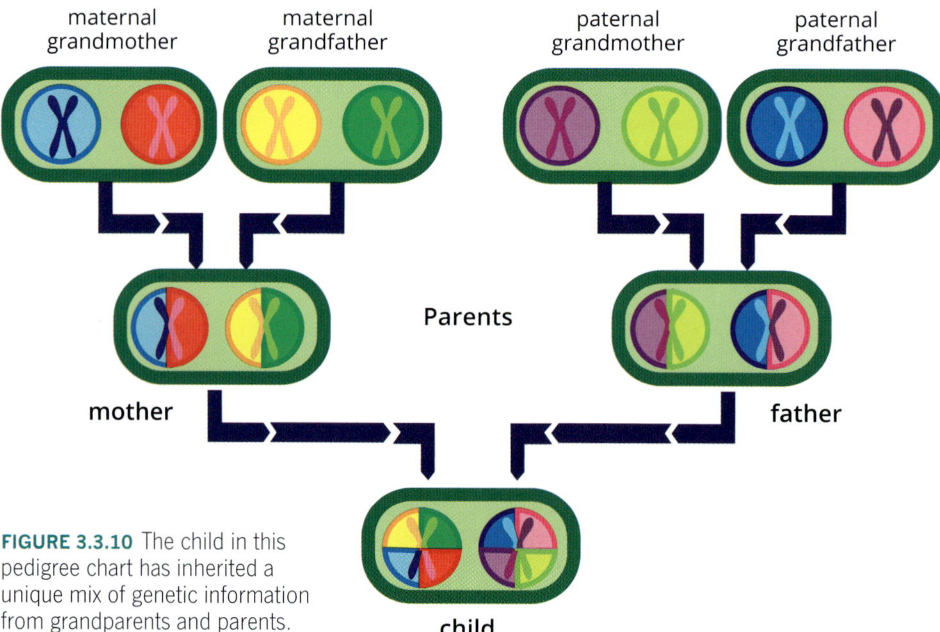

FIGURE 3.3.10 The child in this pedigree chart has inherited a unique mix of genetic information from grandparents and parents.

When meiosis goes wrong

Meiosis is usually an exact process, but sometimes errors occur. Missing, extra or malformed chromosomes can result from defective gametes, which may have serious consequences for offspring. One of the errors that can occur is if members of a pair of homologous chromosomes do not separate properly, resulting in abnormal development of the offspring. There are also changes in chromosome structure such as deletions, duplications, translocations and inversions of chromosome segments resulting in missing, duplicated or misplaced alleles. Deletions result in the embryo having a number of essential genes missing and are usually lethal. Duplications, translocations and inversions can result in changes in physical characteristics because a gene's expression can be influenced by its location among neighbouring genes. Another situation, much more common in plants, is polyploidy ($3n$, $4n$, and so on) that can arise through errors in meiosis. For example, gametes may end up being diploid rather than haploid.

> Crossing over (genetic recombination) is the process where homologous chromosomes pair up with each other and exchange different segments of DNA to form recombinant chromosomes. It occurs between prophase I and metaphase I of meiosis I.

Consequences of chromosomal errors are discussed more fully in Chapter 7. Polyploidy, and its occurrence in agricultural plants, is explained in Chapter 2.

GO TO ➤ Section 7.1 page 288

GO TO ➤ Section 2.4 page 99

Variation between species

There is a wide variation in the diploid number of chromosomes ($2n$) among species, ranging from 2 (1 chromosome pair) in some Australian ants to 78 (39 homologous pairs) for the dingo, and even higher for some plants. Difference in the number of chromosomes is the main distinction between species and the barrier to successful interbreeding between different species.

Hybrids between species were discussed in the Year 11 course. Even if their differing numbers of chromosomes did combine into a zygote at fertilisation, the resultant embryo is likely to be malformed and not develop into a fetus. The few species that do interbreed successfully usually have offspring that are not fertile and cannot continue to keep breeding. For example, the hybrid offspring of a female horse ($2n = 64$) and a male donkey ($2n = 62$) is a mule with a total of 63 chromosomes making it infertile (unable to produce viable offspring). The chromosomes of different species are incompatible because they have different genes at the specific loci on chromosomes, meaning that their chromosomes are not homologous and the alleles are not alternatives for the same genes. Without a full set of homologous pairs of chromosomes, meiosis cannot proceed normally to produce future generations. There is a discontinuity rather than continuity because the break down in the meiosis process blocks further cell replication.

GO TO ➤ Year 11 Section 9.2

Formation of new species

Given that both mitosis and meiosis tend to maintain genetic continuity, both in an individual and within a species, how then do new species emerge? The topic of **speciation** was covered in detail in the Year 11 course.

GO TO ➤ Year 11 Section 9.2

BIOFILE CCT L S

Naming new species

Every year, new species are listed, some that have evolved relatively recently and many that have been newly discovered. The discovery of new species is driven by new technologies, targeted surveys of little-studied ecosystems and a determined effort to identify plants and animals before their habitat is lost.

The International Commission on Zoological Nomenclature (ICZN) provides and regulates a uniform system of zoological nomenclature ensuring that every animal has a unique and universally accepted scientific name.

Examples of species discovered in the last 20 years include:

- the olive-backed forest robin from Gabon, Africa, *Stiphrornis pyrrholaemus*, which is Greek for 'stout bird that bears a flame-coloured throat'
- a new primate from Tanzania, *Rungwecebus kipunji*, meaning the monkey from around Mount Rungwe called kipunji in the local language
- a freshwater ray from a Brazilian river, *Potamotrygon rex*, given the name, 'King of Rays', for its king size proportions; up to 1.1 m in length with a weight of 20 kg.
- a tiny spider, 2 mm in length with a pointy shape, named *Eriovixia gryffindori*; the scientists who discovered the spider thought it resembled the Sorting Hat from the Harry Potter books and so named it after Godric Gryffindor, the original owner of the Sorting Hat (Figure 3.3.11).

FIGURE 3.3.11 The tiny spider, *Eriovixia gryffindori*, has an unusual body shape for camouflage amongst dried leaves.

BIOLOGY IN ACTION   

Seed banks

Continuity of species relies on natural processes of cell replication and the ability of DNA to self-replicate, as well as genetic variation. Human impact on the planet has reduced biodiversity and therefore the genetic variation available for natural evolutionary processes. Added to this is the impact of artificial genetic technologies at an ever-increasing rate. Concerns about reduction in biodiversity, especially for food crops, have led to the establishment of seed banks to preserve plant genetic material.

The Millennium Seed Bank Partnership aims to collect seeds from a quarter of the world's plants by 2020. The purpose of the seed bank is to ensure the continuity of plant species, especially plants that are most at risk from climate change and human activities, and plants most useful in the future (such as crop plants or plants with potential medicinal value). One potential use of the seed bank includes the ability to use the seeds for revegetation if a natural disaster destroys the native vegetation. The partnership is coordinated by the Royal Botanic Gardens in London, and involves organisations in 80 countries, including the Royal Botanic Gardens in Melbourne and Sydney. The seeds are stored in a vault in Sussex, England (Figure 3.3.12).

FIGURE 3.3.12 Examples of seeds kept as part of The Millennium Seed Bank Partnership

There is also a deep freeze seed bank in Svalbard, owned by the Norwegian government, where crop seeds are stored (Figure 3.3.13). More than 150 million seeds in 300 000 samples are now stored in the vault. The seeds are stored at −18°C in specially sealed compartments. If properly maintained, the seeds will remain viable for at least 1000 years.

Another seed gene bank is maintained by ICARDA (International Center for Agricultural Research in the Dry Areas), a global research-for-development organisation. Their storage includes barley, faba bean, durum wheat, chickpeas and lentils and holds a total of 151 000 accessions from the 'Fertile Crescent' in Western Asia, the Abyssinian highlands in Ethiopia and the Nile Valley, where the earliest known crop domestication practices occurred. Many of their plants are now extinct in their natural habitats. ICARDA aims to distribute germplasm to agricultural organisations around the world to enable development of improved climate-resilient crop varieties that are tolerant to drought, salinity, diseases and insects.

FIGURE 3.3.13 The Svalbard Global Seed Vault, constructed between 2006 and 2008, on the Norwegian island of Spitsbergen. Seeds held here are kept in sealed packets at −18°C and are expected to survive for hundreds and in some cases thousands of years. The Svalbard Global Seed Vault stores plant genetic material as an insurance against loss of crop species and has the capacity to store 4.5 million samples.

BIOLOGY IN ACTION  

Same, same but different—species convergence and divergence

Some groups of animals appear very similar and may have arisen from a common ancestral species, then diverged by a speciation process into separate species. In other situations, convergent evolution may have been at work and the species are actually unrelated but have evolved superficially similar features to suit life in the same habitat.

Sharks differ in origin from most other fish, even though when swimming side by side in the ocean they appear to be very similar (Figure 3.3.14). Most other fish have skeletons made of bone. A shark's skeleton is made of cartilage, a type of strong but flexible tissue. Most other fish are covered in smooth, flat scales. A shark is covered in sharp, tooth-like scales called denticles. In this case, convergent evolution, under selective pressures of living in the same marine habitat, has given both groups streamlined body shapes and fins for swimming.

On the other hand, the eight species of bear living today occur in some very different habitats, from polar (the Arctic region) to tropical (Southeast Asian forests) and have quite distinct physical differences (Figure 3.3.15). There are ongoing genomic studies that indicate a common origin for the different bear species and even that gene flow and extensive hybridisation is possible. The entire genomes of the eight bear species were sequenced by 2017. Each genome was found to be about 2.5 billion base pairs. It had previously been established that the polar and brown bears were related but the new studies showed evidence for gene flow among all the bear species. Genomic data showed a gene flow even between the polar and sun bears, despite the fact that the two species live in completely different geographical areas and never meet under natural circumstances. The researchers explained this apparent contradiction by suggesting that an intermediate host has passed the genes on in various directions, the brown bear being an ideal candidate with its geographic distribution overlapping with the other bear species and its genome containing polar bear genes.

The case of the bear species again raises the dilemma about the true definition of a species. Are the eight bear species in fact true species if they can hybridise and their genomes demonstrate gene flow? The lead researcher for the bear genome studies seems to be in agreement with Charles Darwin when he states that defining a species is less essential than preservation of genetic variation to allow evolutionary adaptation to future environmental changes.

FIGURE 3.3.15 The eight bear species of the world still appear to have gene flow between them all.

FIGURE 3.3.14 Despite their similar shapes and habitat, the shark has a very different ancestry to the bony fish it swims with in the ocean.

3.3 Review

SUMMARY

- Organisms need to reproduce in order for their species to survive. This is achieved by cell division in both mitosis and meiosis.
- The definition of the term 'species' is not fixed and has been approached differently over scientific history.
- The most useful concept of a species is a group of organisms that can interbreed successfully amongst themselves but not with others.
- The cell replication process of mitosis produces new cells with identical genetic content to their parent cell. This perpetuates the genetic information required for normal cell functions and ensures that individuals will survive to breed successfully and continue their species into the future.
- If uncontrolled cell division occurs, a neoplasm (abnormal growth) may form, disrupting the normal cell cycles of replication and genetic continuity.
- Inherited genetic variation is a feature of sexually reproducing organisms and meiosis.
- The reduction cell division process of meiosis produces gametes with a haploid set of chromosomes that are a unique mix from both parents.
- The different combinations of alleles have arisen from random mating, independent assortment and recombination during gamete formation, and also from mutations.
- The continuity of species ceases if numbers in a breeding population fall too low or the gene pool is so restricted that no individuals have adaptations to survive in times of unusual and severe environmental pressures.

KEY QUESTIONS

1. Recall the names of the two cell division processes and define their main features.
2. Identify the words used to describe a cell with:
   - **a** a set of paired chromosomes
   - **b** one copy of each chromosome.
3. Outline at least four benefits of classifying and naming organisms as species.
4. Write a definition of 'species'.
5. Examine the importance of cell division to a eukaryotic organism.
6. Consider a defect in a multicellular organism resulting in the inability to complete mitosis correctly all the time. Outline some of the possible consequences for the organism.
7. Predict the outcome for a species of lizard that becomes separated from the main population by a newly formed, wide and fast flowing river.

Chapter review

03

KEY TERMS

allele
anaphase
antiparallel
apoptosis
cell
cell cycle
cell replication
centriole
centromere
chiasma (pl. chiasmata)
chromatid
cleavage
complementary base pair
condensation
  polymerisation reaction
crossing over
cytokinesis
daughter cell
DNA
  (deoxyribonucleic
  acid)
dinucleotide
diploid
double helix
gamete
gene expression
gene pool
genetic variation
genome
genomics
germ cell
germline mutation
haploid
histone
homologous
  chromosome
hydrogen bond
interphase
locus (pl. loci)
meiosis
meiosis I
meiosis II
metaphase
mitosis
mutation
neoplasm
nitrogenous base
nucleic acid
nucleosome
nucleotide
phosphodiester bond
polynucleotide
polypeptide
prophase
protein
proteome
purine
pyrimidine
random assortment
recombination
reduction division
RNA (ribonucleic
  acid)
somatic cell
somatic mutation
speciation
species
spindle
sugar–phosphate
  backbone
supercoil
synapsis
telophase
transcriptome

REVIEW QUESTIONS

1 What does a nucleosome consist of?
A DNA and histones
B DNA and chromatid
C chromatid and nucleotides
D RNA and histones

2 Which of the following statements is not correct?
A Although it is divided into stages, mitosis is a continuous process.
B Cytokinesis marks the beginning of two new cells.
C DNA is replicated during interphase.
D Mitosis is the longest phase of the cell cycle.

3 Identify which one of the following processes does not occur in meiosis.
A cytokinesis
B DNA replication
C pairing of homologous chromosomes
D formation of two diploid daughter cells

4 Which of the following is not a purpose of cell replication by mitosis in multicellular organisms?
A growth
B repair
C reproduction
D restoring the nucleus-to-cytoplasm ratio

5 Identify which one of the following statements is true.
A Cytokinesis is also called binary fission.
B Cytokinesis involves the division of the nucleus.
C Cytokinesis occurs during meiosis.
D Cytokinesis occurs after mitosis.

6 Why is meiosis a necessary process in living organisms?
A It enables each parent to contribute genetic information to the offspring.
B It is necessary for the growth of an organism.
C It produces new cells to replace dead or dying cells.
D It happens in the reproductive organs.

7 What is the difference between the alleles of a gene?
A their locus on the chromosome
B the sequence of bases
C the type of sugar on the nucleotides
D their amino acid sequence

8 Describe where mitosis would be occurring in a pregnant woman.

9 The diagram below represents the stages of mitosis, but they are not in the correct order. Determine the correct sequence and describe what is happening at each stage of mitosis.

prophase

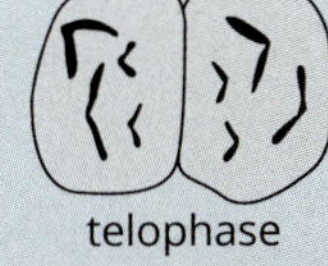
telophase

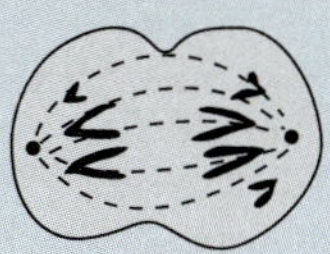
anaphase

metaphase

10 Examine the following diagram of a chromosome.

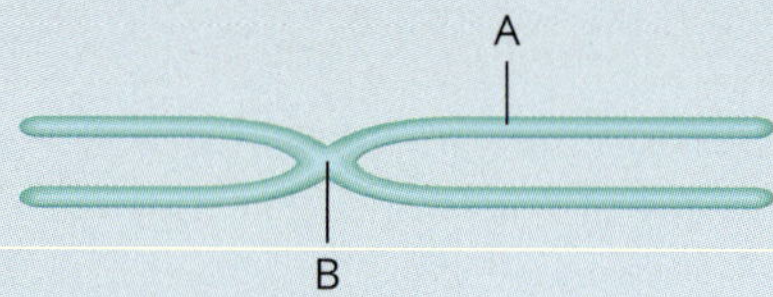

a Identify structures A and B.

b During which phase of mitosis does the chromosome first appear in this state? Explain what happens to cause this appearance.

c Chromosomes do not always look like the one shown in the diagram. Describe the changes in the appearance of chromosomes during the different phases of the cell cycle.

d Draw a typical representation of an animal cell in interphase.

11 Does every cell go through the G_0 phase? What happens when cells enter this phase?

12 a When a cell with chromosome number $2n = 24$ undergoes mitosis, how many daughter cells are produced, and what is their chromosome number?

b When a cell with chromosome number $2n = 24$ undergoes meiosis, how many daughter cells are produced, and what is their chromosome number?

13 The diploid number of chromosomes in horses is 64. How many chromosomes would you expect in a horse's:

a fertilised ovum (zygote)?

b sperm cell?

c somatic cell?

14 Arrange the following stages of meiosis in the correct order, from first to last.

- metaphase II
- telophase II
- prophase I
- anaphase I
- metaphase I
- anaphase II

15 Identify the process depicted at prophase I in this diagram and summarise what is happening.

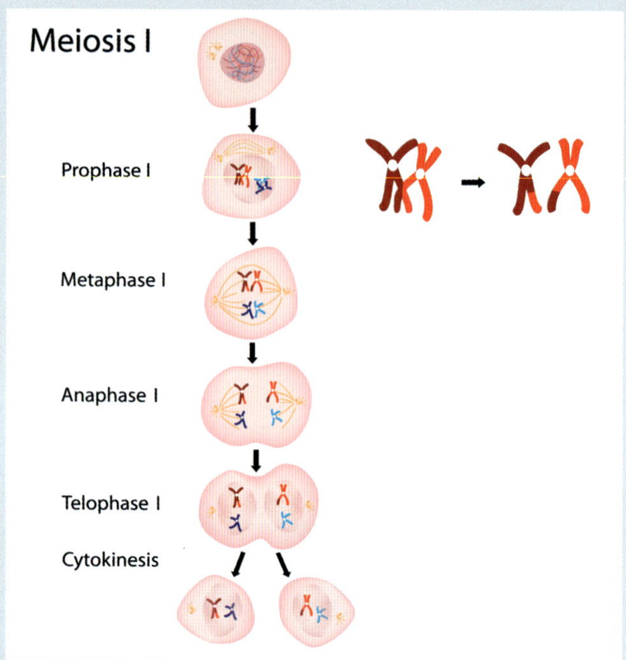

16 Assess the significance of crossing over in meiosis.

17 Explain how an error in meiosis can lead to Down syndrome (three copies of chromosome 21).

18 a Does this cell division diagram represent mitosis or meiosis? Justify your answer.

b Write a short description for each of the Stages 1, 2 and 3 shown on the diagram.

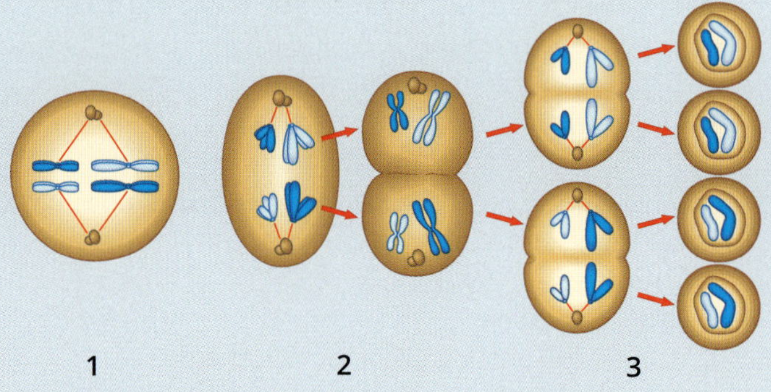

19 Outline the structure of the DNA molecule. In your answer, include:

- the name and components of the unit (monomer building block) of DNA
- the names of the four bases found in DNA
- how nucleotides are joined to build a single-stranded DNA molecule
- how a double-stranded DNA molecule is formed
- what is meant by 5′ and 3′ ends of a DNA strand, and the two strands of DNA being antiparallel
- details of the chemical bonding that holds the structure together.

20 Fill in the blanks with the most suitable term from the following list:
nucleotides, complementary, enzymes, hydrogen, semi-conservative, anti-parallel, covalent, Ozaki.
The replication of DNA is controlled by ______________. Because the replicated DNA consists of one original parent strand plus one new daughter strand it is known as ____________________ replication. The two strands of a DNA molecule run ____________________ to each other. As a result, the synthesis of daughter strands runs in different directions having one strand copied using ____________________ fragments. Free ____________________ are linked by ____________________ phosphodiester bonds and bases are joined in ____________________ fashion by ____________________ bonds.

21 Explain what is meant by DNA being coiled and supercoiled within a chromosome. Include a diagram in your answer. What benefit is there in DNA being packaged in this way in a cell?

22 A nucleic acid strand is under investigation. It has been found to contain 29% A, 32% G and 17% C.

a Is the fourth base uracil or thymine?

b How do you know?

c Calculate what percentage of the strand of nucleic acids is the fourth base.

23 In order to acquire the nucleotides needed for the production of DNA and RNA in cells, humans must have nucleic acids in their diet. These nucleic acids must be broken down into nucleotides so that they can be absorbed from the digestive tract and into the bloodstream so they can be carried to the cells that need them. The breakdown of the nucleic acids into nucleotides requires the input of water.

a Suggest why this is the case.

b A dinucleotide is shown in the molecular diagram. Draw the two nucleotides after digestion into single nucleotides.

c Predict what has happened to the water.

OH
$^{-}O-P=O$
$O-CH_2$ O thymine
H H H
O
$^{-}O-P=O$
$O-CH_2$ O adenine
H H H
OH

24 The diagram illustrates a proposed model to help students understand the concept of a gene and alleles as part of chromosome structure.

a Interpret how this model depicts the concept.

b Evaluate if such a model helps your own understanding of this area of the biology course.

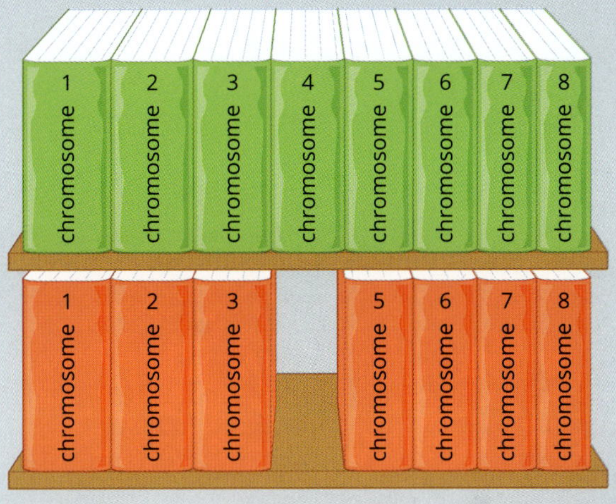

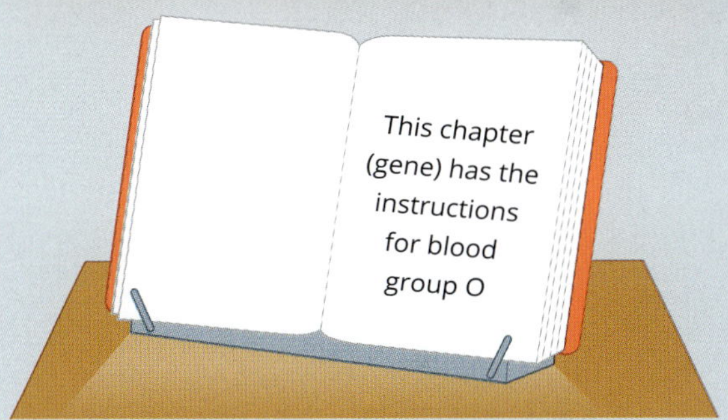

25 The LM shows cells at various stages of replication. Explain the importance of the genetic material in each cell being replicated exactly.

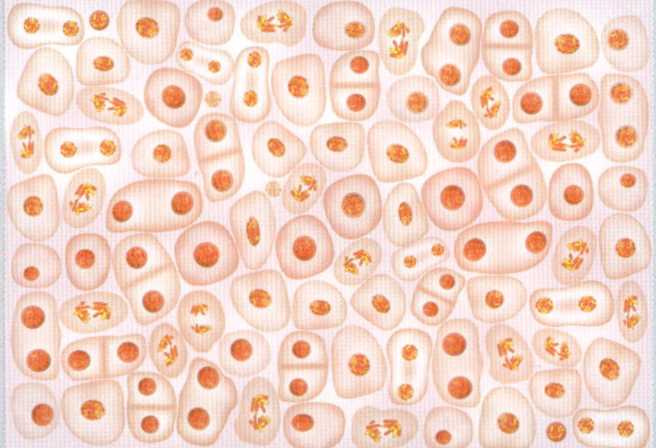

26 Explain the phrase 'continuity of species'.

27 After completing the Biology Inquiry on page 128, reflect on the inquiry question: How important is it for genetic material to be replicated exactly?

a Draw a diagram to show the replication of a DNA molecule.

b Make a list of the features of the DNA molecule and the cell processes that enable exact replication of a cell's genome.

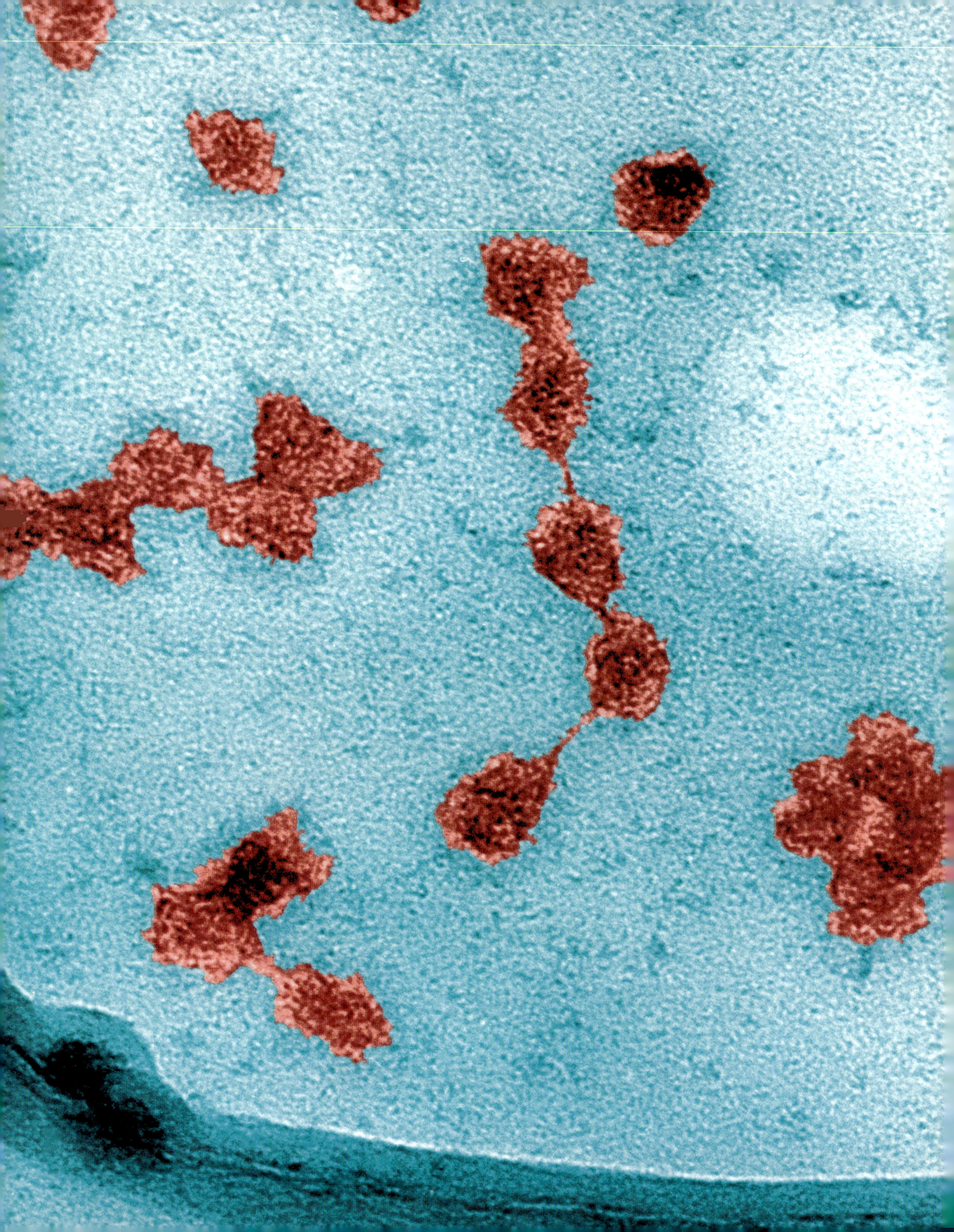

CHAPTER 04 DNA and polypeptide synthesis

By the end of this chapter, you will be able to describe the differences and similarities of the forms in which DNA exists in eukyarotic and prokaryotic cells. You will also understand the role of nucleic acids (DNA and multiple forms of RNA) as molecules that encode instructions for polypeptide synthesis from amino acids. You will also examine the process of polypeptide synthesis from transcription to translation.

The complexity and importance of proteins will become evident, starting with their construction from polypeptides to the many functions they play in living organisms. You will learn that nearly every function of a living organism depends on proteins.

Content

INQUIRY QUESTION

Why is polypeptide synthesis important?

By the end of this chapter you will be able to:

- construct appropriate representations to model and compare the forms in which DNA exists in eukaryotes and prokaryotes (ACSBL076) ICT
- model the process of polypeptide synthesis, including: (ACSBL079)
  - transcription and translation
  - assessing the importance of mRNA and tRNA in transcription and translation (ACSBL079)
  - analysing the function and importance of polypeptide synthesis (ACSBL080)
  - assessing how genes and environment affect phenotypic expression (ACSBL081) CCT L
- investigate the structure and function of proteins in living things L

4.1 DNA in eukaryotes and prokaryotes

GO TO ➤ Year 11 Section 2.1

There are two fundamentally different types of cells: prokaryotic and eukaryotic (Figure 4.1.1). Organisms are classified according to the type of cell that forms their structure. You learnt about the structure of typical prokaryotic and eukaryotic cells in Year 11.

Prokaryotes are usually unicellular, composed of a single prokaryotic cell, and include all species of Bacteria and Archaea. Prokaryotic cells are generally smaller and less complex than eukaryotic cells. The main distinction is that prokaryotic cells have chromosomal **DNA (deoxyribonucleic acid)** but do not hold it in a defined membrane-bound **nucleus** inside the cell.

Eukaryotes are composed of eukaryotic cells and include plants, animals, fungi and protists. They may be unicellular (protists, some fungi) or multicellular (plants, animals and some fungi). Eukaryotic cells contain membrane-bound organelles including the nucleus that contains the DNA inside a double **nuclear membrane**.

In this section, you will learn more about the way DNA is held in each type of cell and the differences and similarities between them.

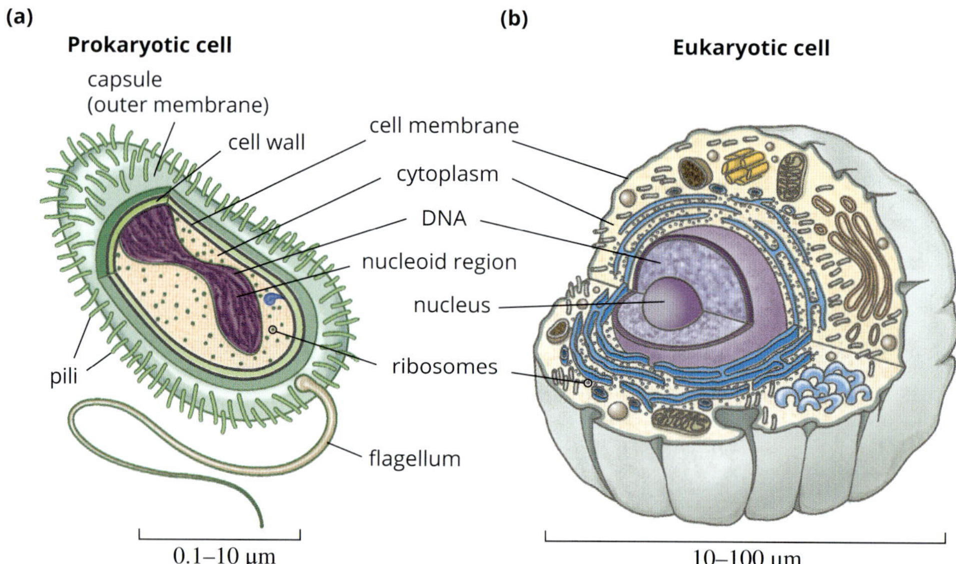

FIGURE 4.1.1 (a) A typical prokaryotic cell compared to (b) a typical eukaryotic cell. Prokaryotic cells and the processes involved in their DNA replication are generally much simpler than eukaryotic cells. Note the different membrane-bound organelles in the eukaryotic cell and the lack of such organelles in the prokaryotic cell.

BIOFILE N

Ribosomes in prokaryotic and eukaryotic cells

Ribosomes consist of two subunits joined together (Figure 4.1.2). The subunits in eukaryote ribosomes are different to those in prokaryote ribosomes. Ribosomes are either free in the cytoplasm or bound to rough endoplasmic reticulum (rough ER).

Ribosomes translate mRNA into polypeptides, which then combine to form functional proteins.

The prokaryotic bacteria and archaea have smaller ribosomes, called 70S ribosomes, which are composed of a small 30S subunit and a larger 50S subunit. The S stands for 'svedbergs', a unit used to measure sedimentation speed of molecules in a centrifuge. Eukaryotic ribosomes are larger and consist of a 40S subunit and a larger 60S subunit, which come together to form an 80S particle. The eukaryotic 80S ribosome has a mass of 4200 kd (kilodaltons), compared with 2700 kd for the prokaryotic 70S ribosome. The dalton is a non-SI unit used for mass on the molecular level. One dalton is approximately equal to the mass of one proton or one neutron, which is equivalent to 1 g/mol.

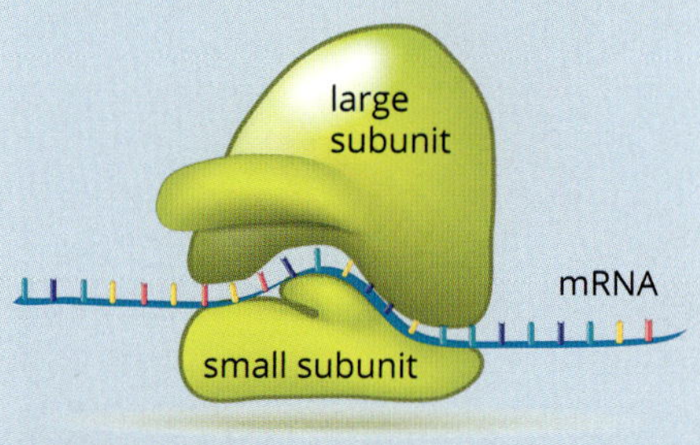

FIGURE 4.1.2 The 80S ribosome in eukaryotic cells is built of a 60S plus a 40S subunit. In prokaryotic cells, the 70S ribosomes are made from 50S plus 30S subunits.

DNA IN PROKARYOTIC CELLS

Prokaryotic cells are small and lack most membrane-bound organelles, including a distinct nucleus. The term prokaryote actually means 'before nucleus'. Their cytoplasm does contain scattered **ribosomes**, built of two subunits, which are not bound by membranes and are involved in the synthesis of **proteins**. The genetic material of prokaryotic cells is usually one double-stranded, circular DNA **chromosome**, which is contained in an irregularly shaped region called the **nucleoid**. Unlike the nucleus of eukaryotes, the nucleoid is a region of the cytoplasm, not an organelle, and it does not have a nuclear membrane surrounding it.

This circular chromosomal DNA is attached to the inside of the prokaryote's outer cell membrane by a region of the chromosome called the **origin**. Because the DNA is joined in a circle, unlike the linear chromosomes of eukaryotic cells, it does not have the protective end regions called **telomeres** that have a role in the ageing process of a eukaryotic cell.

In addition to the chromosomal DNA, many prokaryotic cells also contain small rings of double-stranded DNA called **plasmids**, which can move between cells (Figure 4.1.3). Replication of plasmids is independent of chromosomal replication during binary fission. Plasmids often provide bacteria with genetic advantages, such as antibiotic resistance. From the 1970s, seeing the possibilities in working with a complete small ring of self-replicating DNA, scientists began to use bacterial plasmids in genetic engineering as vectors to clone, transfer and manipulate **genes**. You will learn more about plasmids and their role in genetic technologies in Chapter 9.

GO TO ➤ Section 9.3 page 354

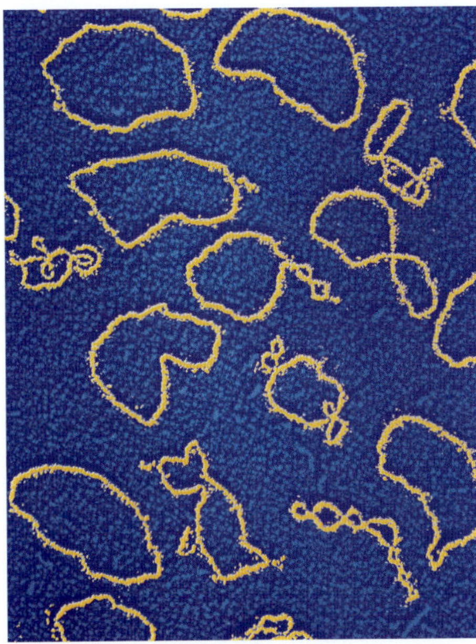

FIGURE 4.1.3 Coloured transmission electron microscope (TEM) of plasmids from the bacteria *E. coli*. Plasmids are independent units, commonly found in prokaryotes, made of small, circular DNA molecules that are much smaller than the single loop of the main chromosome.

Prokaryotic chromosomes are less condensed than their eukaryotic counterparts (Figure 4.1.4). Unlike eukaryotic chromosomes, the prokaryotic chromosomes do not have features that are easily identifiable when viewed under a light microscope. Usually a stained slide will just show the dark area of the nucleoid.

Most of what we know about the chromosomes of prokaryotes has been obtained from studies of the bacteria, *Escherichia coli*. This species is used as the model organism for much of the research on prokaryotes. It is known that *E. coli* and other bacteria do not have the histone proteins used to condense DNA in eukaryotes. However, some Archaea (the other domain of prokaryotes) do have histones, which is of significance in evolutionary history.

Many prokaryotes also have small hair-like projections called **pili**, which are involved in the transfer of plasmid DNA between organisms by conjugation and can also help generate movement. You learnt about asexual reproduction by conjugation in Chapter 2.

GO TO ➤ Section 2.1 page 58

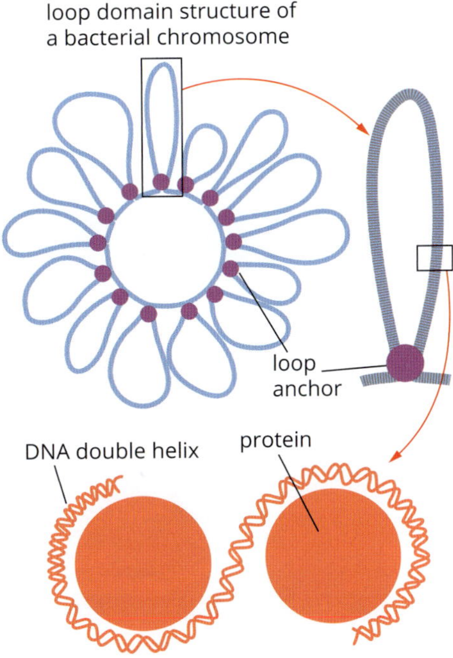

FIGURE 4.1.4 DNA is packaged into loop structures in prokaryotes. The DNA of prokaryotes does not have to be as tightly packaged as in eukaryotes because there is much less genetic material.

Gene expression in prokaryotes

In prokaryotic cells, **gene expression** consists only of **polypeptide synthesis** and occurs completely within the cytoplasm. Here the two processes of synthesis, **transcription** and **translation**, are closely coupled and occur at almost the same time. The action is between the DNA, **messenger RNA (mRNA)** and ribosomes in the cytoplasm of prokaryotic cells. The two stages involved in synthesis of polypeptide chains will be covered in detail in Section 4.2. Gene expression in prokaryotes is regulated during transcription, the first stage. Unlike eukaryotes, in prokaryotes there is very little processing of RNA, which is the level at which regulation of gene expression takes place for the eukaryotic cells.

DNA IN EUKARYOTIC CELLS

Eukaryotic cells are more complex than prokaryotes (Figure 4.1.1). Eukaryotic cells not only have a cell membrane around the cytoplasm, but also have internal membranes that form specialised membrane-bound compartments within the cell. This is known as cell compartmentalisation. The membrane-bound structures are called organelles. Organelles were covered in detail in Year 11.

REVISION

GO TO ➤ Year 11 Section 2.2

Organelles of eukaryotic cells

Organelles are subcellular structures that have a specific function. Because they have a specific function, their presence depends on the needs of the cell. In eukaryotes some organelles, like the nucleus, are membrane-bound and some are not. Prokaryotic cells possess some non-membrane bound organelles, such as ribosomes, flagella and a cell wall, although the structure and composition of these is usually different from that of equivalent eukaryotic organelles.

A eukaryotic cell has a nucleus which contains all of the chromosomal DNA and is wrapped in a double membrane (Figure 4.1.5). The outer nuclear membrane is continuous with the endoplasmic reticulum (ER) that usually holds ribosomes (known as rough endoplasmic reticulum), enabling unobstructed passage for mRNA from the nucleus and transport of the polypeptides after they are synthesised at the ribosomes. Nuclear pores in the membrane permit communication and selective transport between the nucleus and cytoplasm. Within the eukaryotic nucleus there is a smaller organelle called the nucleolus that has a role in forming ribosomes (Figure 4.1.6).

Ribosomes are the tiny organelles that construct polypeptide chains under instruction from the DNA code. In a eukaryotic cell, as in prokaryotes, the ribosome is also built from two subunits but they are both larger than those in prokaryotic cells.

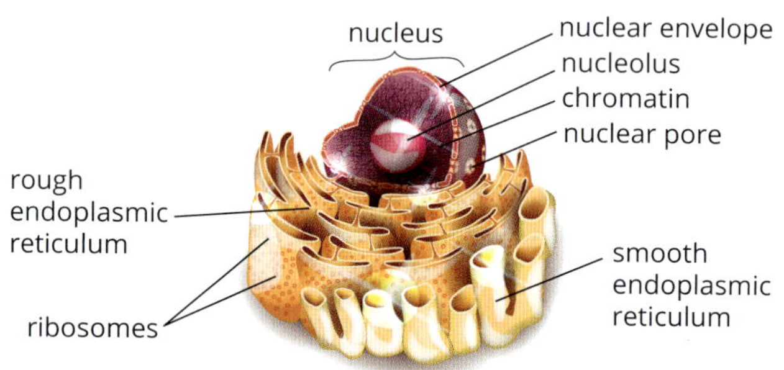

FIGURE 4.1.5 Diagram of the nucleus, nucleolus and rough endoplasmic reticulum (ER) inside a eukaryotic cell. The rough aspect of the ER is actually many ribosomes attached to the ER membrane, which is continuous with the outer nuclear membrane.

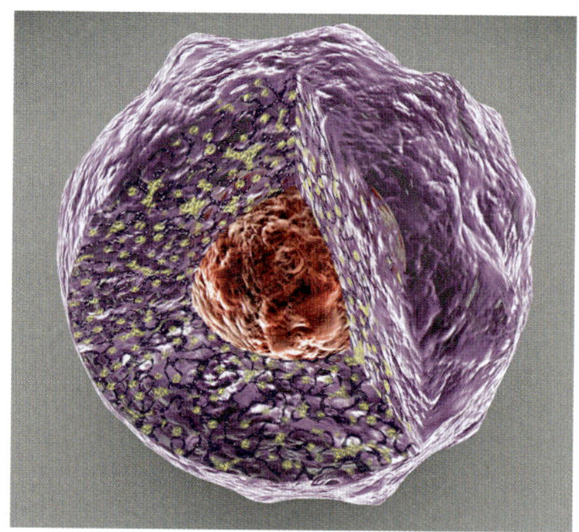

FIGURE 4.1.6 This image shows only the nucleus of a human body cell, with its double membrane dotted with ribosomes on the outside, the dark strands of uncondensed DNA and the nucleolus (red).

Packaging of DNA in eukaryotic cells

> ℹ Histones are proteins found in eukaryotic cells that tightly package DNA into structures called nucleosomes. Nucleosomes in turn form supercoils to become the chromosomes.

In eukaryotic cells, the double-stranded DNA is found in the nucleus where, at times, it is tightly coiled up (condensed) to form the linear structures called chromosomes (Figure 4.1.7). At other stages of the cell cycle, the uncondensed DNA strands form a tightly woven mass. Because eukaryotic cells are more complex than prokaryotes, the eukaryotes have an average of 25 times more DNA than prokaryotes. For it to fit into the nucleus, this amount of DNA has to be much more tightly packaged.

The processes involved in DNA replication are significantly faster in prokaryotic cells. Some bacterial cells take just 40 minutes to replicate their DNA, while in some animal cells this can take up to 10 hours before cell division.

The structure of DNA and its packaging into nucleosomes, then into linear pairs of homologous chromosomes for eukaryotic cells, has been discussed in Chapter 3. Figure 4.1.7 summarises this information in a diagram.

The tight packaging of chromosomes into the nucleus allows us to view a cell's genetic material under a light microscope.

GO TO ➤ Section 3.2 page 128

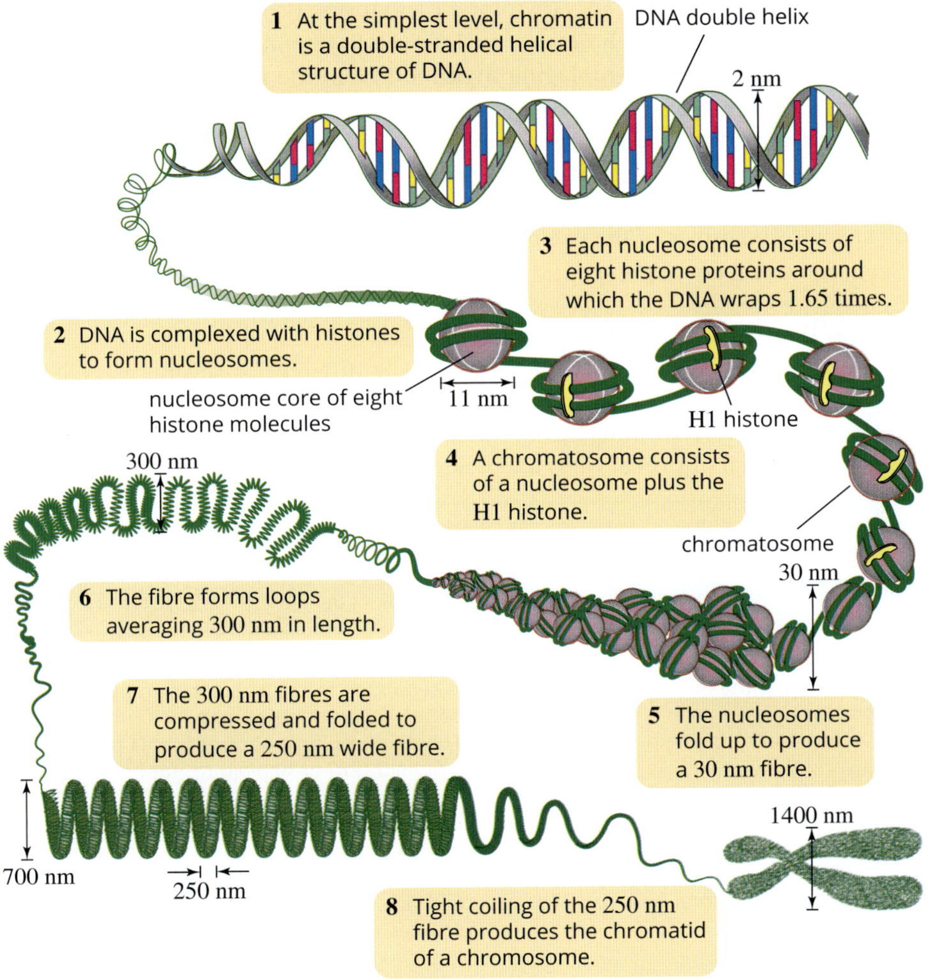

FIGURE 4.1.7 The packaging of DNA in eukaryotic cells. Because eukaryotes have large quantities of DNA to fit into a small space, the DNA needs to be tightly and efficiently packaged. Double-stranded DNA (1) is tightly coiled around histones to form nucleosomes (2 and 3). Nucleosomes and histones together form chromatosomes (4). The nucleosomes fold (5), loop (6) and compress (7) into chromatin. Tight coiling of the chromatin produces the chromatids of a chromosome (8).

A nanometre (nm) is one-billionth of a metre (10^{-9} m). This means there are 1 000 000 (1 million; 10^{6}) nanometres in a millimetre.

Gene expression in eukaryotes

Gene regulation is tightly controlled in both eukaryotes and prokaryotes. However, as the process of gene expression in eukaryotes is more complex, gene regulation checks occur at a greater number of stages.

In Section 4.2 you will learn that gene expression in eukaryotes comprises the processes of transcription, **RNA processing** and translation. Gene expression can be regulated at any of these stages. In eukaryotic cells, transcription and RNA processing occur within the nucleus, and translation occurs in the cytoplasm.

Gene expression in prokaryotes consists only of transcription and translation, whereas in eukaryotes, it involves transcription, RNA processing and translation.

COMPARISON OF PROKARYOTIC AND EUKARYOTIC DNA

Both types of cells contain double helix DNA as their inherited genetic material, making DNA universal in structure and function to all living things on our planet. An essential difference between prokaryotes and eukaryotes is the way the DNA molecule is packaged inside their cells.

Similarities

Although there are many differences between the gene structures and DNA packaging of prokaryotes and eukaryotes, shared evolutionary history means that there are also many fundamental similarities.

- Both prokaryotes and eukaryotes have double-stranded DNA, twisted into a double helix and built from the same type of nucleotides with the same nitrogenous base molecules—adenine (A), thymine (T), cytosine (C) and guanine (G).
- Both have mRNA, which acts as an intermediate code to building proteins, with the base uracil (U) replacing thymine (T) in the mRNA.
- Because the genetic information of prokaryotes and eukaryotes is composed of the same code, the way mRNA is translated into **amino acids** and proteins is also much the same. The universality of the DNA and mRNA coding has made the use of prokaryotic plasmids possible as vectors for genetic engineering across all species. Examples of the use of plasmid vectors can be found in Chapter 9.

GO TO ➤ Section 9.3 page 354

Section 4.2 will cover polypeptide synthesis in more detail.

Differences

Prokaryotic and eukaryotic genes are structurally different in several ways. These differences affect the way in which genetic information is transcribed, translated and expressed. Table 4.1.1 summarises the major differences between the structure of the genetic material of prokaryotes and eukaryotes.

TABLE 4.1.1 Summary of the major differences between the structure of the genetic material of prokaryotes and eukaryotes.

Prokaryote	Eukaryote
Chromosomal DNA is in a region of cytoplasm called the nucleoid, lacking a membrane. There is one chromosome per cell.	Chromosomal DNA is in the nucleus, which is separated from the cytoplasm by a double-layered membrane. There are multiple chromosomes per cell of a diploid ($2n$) or haploid (n) number.
A circular chromosome without ends (no telomeres).	Linear thread-like chromosomes with ends (telomeres).
Contains plasmids—small, circular DNA.	Contains no plasmids but there are other sources of DNA apart from chromosomes—mitochondrial DNA and chloroplast DNA.
There is much less DNA than in eukaryotes (thousands to millions of bases, depending on species).	There is much more DNA than in prokaryotes (millions to billions of bases, depending on species).
There are fewer genes than in eukaryotes (thousands).	There are more genes than in prokaryotes (tens of thousands).
There is less non-coding DNA (introns) than in eukaryotes (greater number of genes per number of bases).	There is more non-coding DNA (introns) than in prokaryotes (fewer genes per number of bases).
DNA is in a region called the nucleoid but is not packaged into an organelle (less DNA to fit into the cell).	DNA is tightly packaged—coiled around histones forming nucleosomes, which are condensed into chromatin and packed as chromosomes into the nucleus (a lot of DNA to fit into a small space).
Genes cluster into functional groups, known as operon regions (e.g. genes that code for enzymes in the same biochemical pathway are next to each other on the chromosome, so all the genes for the pathway can be transcribed and expressed at once).	Genes that code for functionally similar proteins can be physically far apart or located on different chromosomes. Eukaryotes have mechanisms to express these genes at the same time.

BIOFILE CCT

To code or not to code

The role of DNA in both prokaryote and eukaryote cells is to carry the inherited code (genes) for production of polypeptides that combine into the many essential protein molecules required by a cell. Prokaryotes need to carry all their coding genes in one relatively short chromosome, so there is not much space for non-coding sequences of DNA. In eukaryotes, the amount of non-coding lengths of DNA is about 98% compared to prokaryotes where it is only 12%. The coding segments of DNA are known as exons and the non-coding regions as introns (Figure 4.1.8).

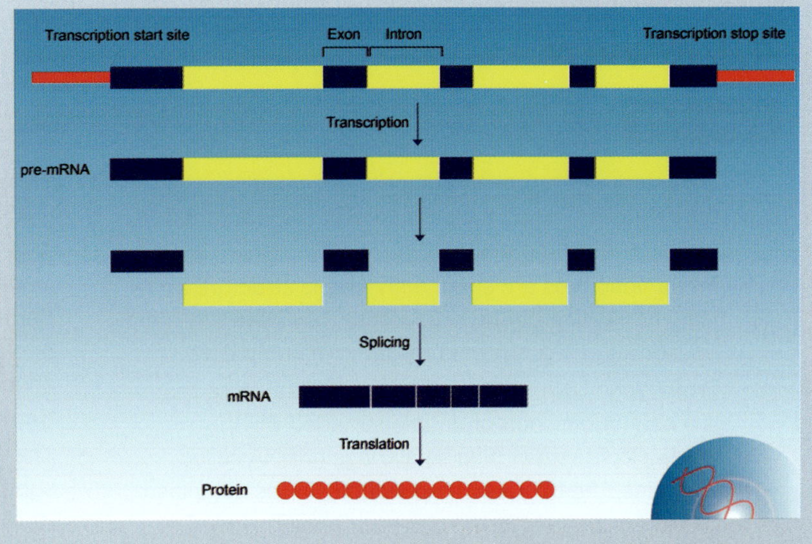

FIGURE 4.1.8 Gene codes for producing proteins are carried in the exon regions of a chromosome. Introns are the non-coding regions and there are far less of these in prokaryote cells. Some of the processes depicted in this diagram are explained in Section 4.2.

BIOLOGY IN ACTION 

The first genome

The first free-living organism to have its **genome** sequenced was a bacterium, *Haemophilus influenzae*. The work was completed in 1995 at TIGR (The Institute for Genomic Research, now called J. Craig Venter Institute) in the USA. The prokaryotic genome consists of about 1.8 million base pairs in a single circular chromosome. It was sequenced in 13 months at a cost of 50 cents (US) per base unit. At the time this was a record for speed and cost, now it is surpassed by vastly improved methods.

The rapid advancement of microchips and processors capable of managing and storing the vast amount of data, even from this relatively short, simple genome, was the key to completing the groundbreaking project. TIGR had developed new sequencing techniques and computational methods, as well as improved data storage and hardware. Software called the TIGR Assembler was used to assemble approximately 24 000 DNA fragments into the whole genome. At the time, the software required approximately 30 hours of processing time with half a gigabyte of RAM, an indication of the complex computation involved.

FIGURE 4.1.9 Zebrafish (*Danio rerio*) are freshwater aquarium fish that are commonly used as model organisms. Their eukaryotic genome was sequenced in 2013.

In 2003, the Human Genome Project was completed. This ushered in a new era, known as Next Generation Sequencing (NGS). Bioinformatics was born, a field where computer science and biology operate together.

Now, thanks to improved sequencing machines with increased accuracy and speed, the genomes of many eukaryotic species have also been sequenced. By 2013, the zebrafish (*Danio rerio*) (Figure 4.1.9) genome with 1.5 billion bases and 25 chromosomes, had been sequenced. More numerous and increasingly more complex genomes are being sequenced every month. Automation and digital technology with improved capacity are rapidly moving the field of genomics research forward, lowering costs and providing faster analysis.

The new era of bioinformatics has been likened to the establishment of the periodic table in the 19th century as a tool for chemists. Modern geneticists can now construct readily accessible databases of genes and DNA sequences (Figure 4.1.10).

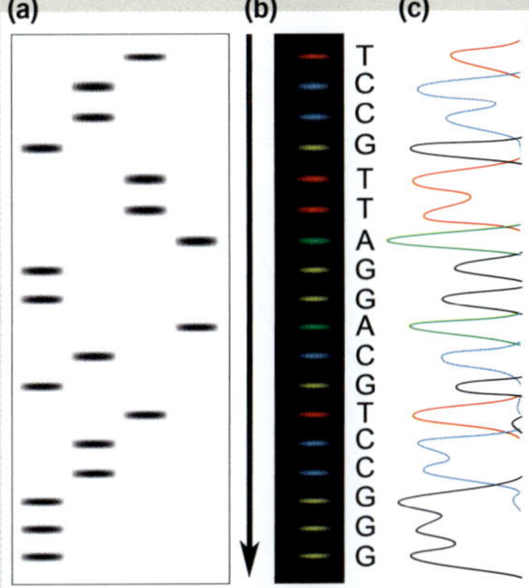

FIGURE 4.1.10 Illustration of the basic data from DNA sequencing. (a) capillary electrophoresis to separate nucleotide fragments, (b) colour fluorescence in a laser beam and (c) chromatogram generated by computer software.

4.1 Review

SUMMARY

- Organisms with prokaryotic cells are called prokaryotes. They are classified into two domains: Bacteria and Archaea.
- Prokaryotic cells are small, with simple structure and lack membrane-bound organelles. They have scattered ribosomes and a nucleoid region containing DNA.
- Prokaryotic DNA is carried as one chromosome that is a continuous loop of double-stranded DNA in the nucleoid region plus small rings of DNA called plasmids that operate independently in the cytoplasm.
- The single loop of chromosomal DNA in prokaryotic cells is formed mainly of coding sequences (exons).
- Organisms with eukaryotic cells are called eukaryotes. They are classified into the domain Eukarya, which is divided into four kingdoms: Protista, Fungi, Plantae and Animalia.
- Eukaryotic cells have a complex structure with a membrane-bound nucleus, many organelles in the cell cytoplasm, and DNA mainly in linear chromosomes in the nucleus.
- There are many differences in the structure of the genetic information of prokaryotes and eukaryotes (e.g. prokaryotes have much less DNA but more genes per number of bases than eukaryotes). Because eukaryotes have much more DNA, they package it more tightly. These structural differences affect the way genetic information is transcribed, translated and expressed into proteins.
- The ribosomes of both cell types are formed of two subunits with eukaryotes having the larger 80S ribosomes and prokaryotes having 70S ribosomes.
- Although there are many differences, the basic genetic structures of prokaryotes and eukaryotes share many similarities. For example, the same code of nitrogenous bases (adenine [A], thymine [T], cytosine [C], guanine [G] and uracil [U]) translates into amino acids, polypeptides and proteins in much the same way for both prokaryotes and eukaryotes.

KEY QUESTIONS

1 Use labelled diagrams to illustrate the main differences between prokaryotic and eukaryotic cells. Identify an example of each.

2 The figure below shows six chromosomes belonging to three homologous pairs in a eukaryotic cell.

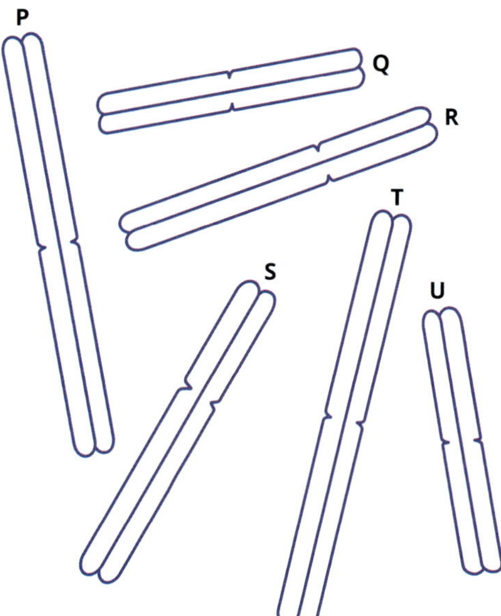

a Identify the three pairs of homologous chromosomes.

b Describe which features you used to match the chromosomes.

3 A human cell has approximately 25 000 genes and *E. coli* has approximately 4000 genes. Explain why the number of genes is not an indicator of the size of the genome.

4 Summarise five differences between prokaryotic and eukaryotic gene structure.

5 Examine some of the main differences in gene regulation between prokaryotic and eukaryotic cells.

6 Explain why the chromosomes of eukaryotic cells can be seen under a light microscope but those of prokaryotes cannot.

4.2 Polypeptide synthesis

The **genetic code** represents the inherited genetic information stored in DNA as triplet bases within sections called genes. This information is transcribed to RNA, then used to synthesise the amino acid sequences that form **polypeptides** through a process called gene expression. Chains of polypeptides combine to form proteins (Figure 4.2.1). Proteins are biological molecules that carry out most of the functions that are essential to life. Antibodies for immune response, collagen for tissue repair, enzymes that catalyse biochemical reactions, hormones for cellular signalling and haemoglobin for transport are all types of proteins.

Polypeptide means 'many peptide bonds'. A protein is formed by one or more polypeptides arranged in a biologically functional way.

In this section, you will learn about the roles of DNA and RNA in polypeptide synthesis, the different steps of gene expression and that functional proteins are formed from polypeptides. You will begin to appreciate the significance of protein molecules in a living organism, an area that will be further developed in Section 4.3. You may find it useful to revise the structure of nucleic acids from Chapter 3 before continuing with this section. The influence of the environment on phenotypic expression will also be explored.

GO TO ➤ Section 3.2 page 128

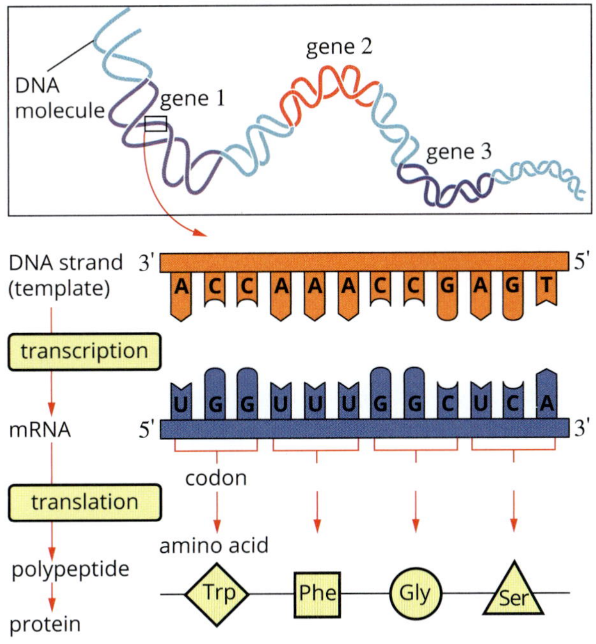

FIGURE 4.2.1 A model diagram for polypeptide and protein synthesis

THE GENETIC CODE

The genetic code is a set of rules that defines how the information in DNA and RNA is translated into polypeptides, which join to form proteins. The genetic code is universal—the rules are the same for all organisms on Earth.

As you learnt in Chapter 3, the information in DNA and RNA is stored as a three-letter code of nucleotides. RNA works in partnership with DNA to synthesise polypeptides as required by a eukaryotic cell. In this section you will learn about the roles of these nucleic acids in the synthesis of polypeptides.

Both DNA and RNA are made up of nitrogenous bases and a sugar–phosphate backbone. RNA is single-stranded and relatively short. DNA has a double-stranded helix (spiral) structure with complementary pairing of its nitrogenous bases holding the double strands together like rungs on a ladder.

RNA contains a pyrimidine base called uracil (U) in place of thymine (T) in DNA. Adenine (A) in DNA is transcribed into uracil (U) in mRNA.

Each DNA triplet or codon codes for one amino acid and may also provide specific instructions, such as 'start translation' and 'stop translation'.

Role of DNA in polypeptide synthesis

DNA provides the instructions, which are translated by RNA into polypeptides and proteins. DNA stores and transmits hereditary information as a sequence of nucleotides. The order of the nucleotides in DNA determines which polypeptides are synthesised. Groups of three nucleotides are called **triplets**. When a DNA triplet is transcribed into mature mRNA, the triplet is then called a **codon** (Figure 4.2.2). For polypeptide synthesis to occur, DNA and RNA must function together.

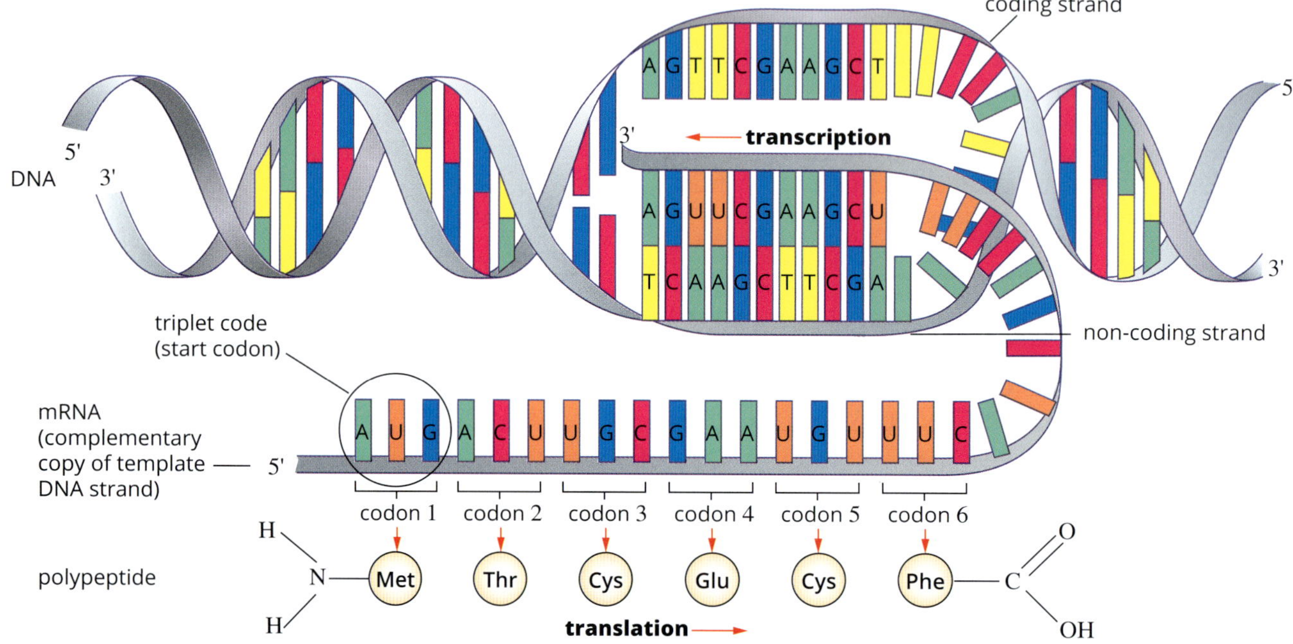

FIGURE 4.2.2 The DNA triplets are transcribed into messenger RNA (mRNA). mRNA is read as codons. Each codon codes for a particular amino acid with the exception of stop codons that end protein synthesis.

Role of RNA in polypeptide synthesis

RNA plays an important role in expressing the information contained in the DNA sequence to synthesise polypeptides. There are three main forms of RNA and each has a different role in polypeptide synthesis:

- Messenger RNA (mRNA) is formed in the nucleus by the process of transcription. The mRNA carries a complementary copy of the nucleotide sequence of DNA that specifies the amino acid sequence for a particular polypeptide (Figure 4.2.3). During transcription, the primary transcript (pre-mRNA) is first formed by the enzyme **RNA polymerase**. Pre-mRNA is then processed (post-transcriptional modification) to form mature mRNA, which is a single-stranded complementary copy of the coding DNA (gene or **exon**) (Figure 4.2.4). The mature mRNA travels from the nucleus to the cytosol where it binds to ribosomes ready for translation.
- **Ribosomal RNA (rRNA)** is synthesised in the nucleolus of the cell nucleus and is based on the nucleotide sequence of the DNA. Together with proteins, rRNA forms a small organelle called a ribosome. Ribosomes are the sites where the information in the mRNA is translated into a chain of amino acids (Figure 4.2.5).
- **Transfer RNA (tRNA)** molecules transfer amino acids from the cytoplasm to the ribosomes, where they are joined to form a polypeptide chain based on the sequence of nucleotides in the mRNA (Figures 4.2.5 and 4.2.6).

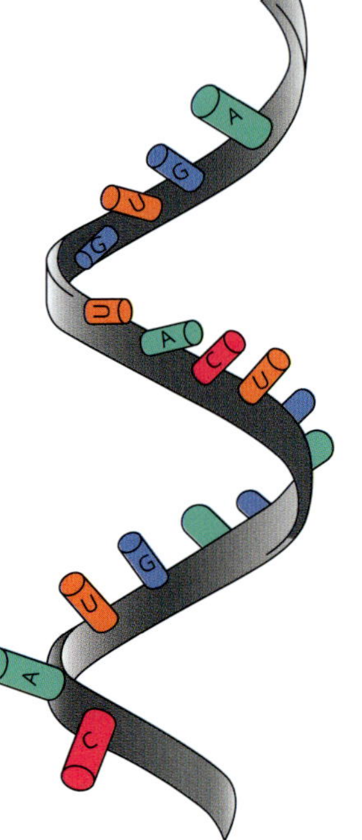

FIGURE 4.2.3 Messenger RNA (mRNA) carries a copy of the DNA's nucleotide sequence to be translated into proteins.

There are 61 different tRNA molecules, each of which combines with only one particular amino acid at one end of its molecule. (There are 64 codons that each represent an amino acid, three of which are **stop codons**. There are no tRNA molecules that recognise these stop codons and so translation is terminated.) There are three places for tRNA to bind to the ribosome: the exit site (E), the peptidyl site (P) and the aminoacyl (A) as shown in Figure 4.2.6b. At the other end of the tRNA molecule, there is a sequence of nucleotides known as the **anticodon**. The anticodon recognises a particular sequence of nucleotides in the mRNA. This enables an amino acid to be positioned in the correct place on a polypeptide chain.

> The primary transcript is the first RNA product synthesised during DNA transcription. The primary transcript undergoes processing to form mature, functional RNA products (e.g. mRNA, rRNA and tRNA). The primary transcript of mRNA is known as pre-mRNA.

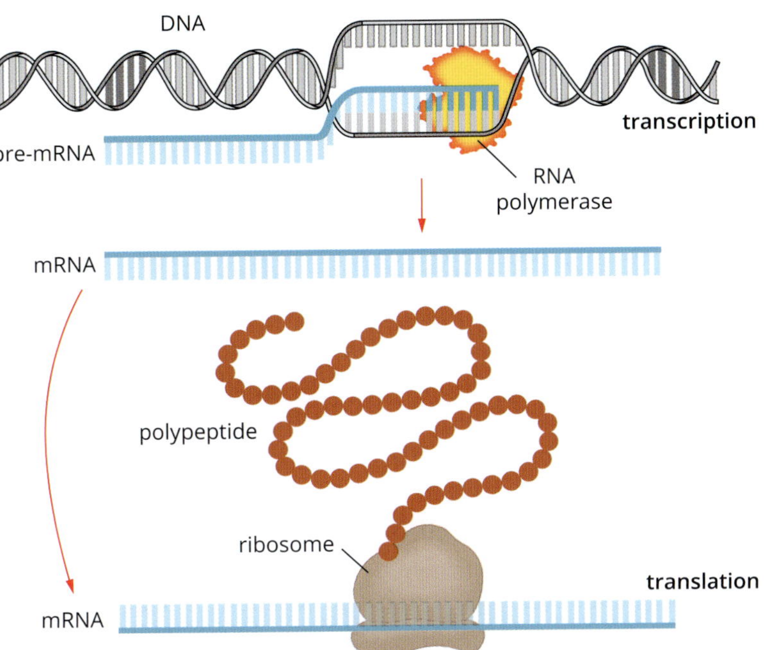

FIGURE 4.2.4 During transcription, RNA polymerase makes a complementary copy of the DNA, which first becomes pre-mRNA and then mature mRNA. The mRNA is then transported to the ribosomes where it is translated into a chain of amino acids, making a polypeptide.

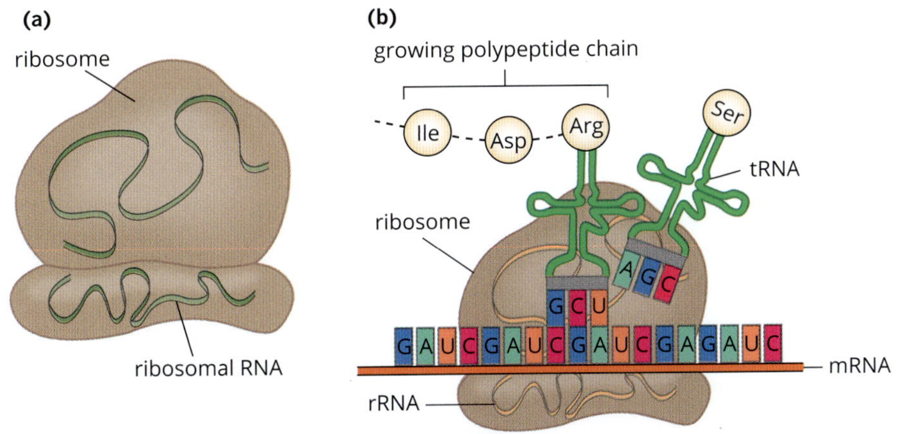

FIGURE 4.2.5 (a) Ribosomal RNA (rRNA), together with two protein subunits, forms ribosomes that are the site of translation of the mRNA into polypeptides. (b) The three different types of RNA (mRNA, rRNA and tRNA) work together to use the information contained in a DNA gene to synthesise a polypeptide.

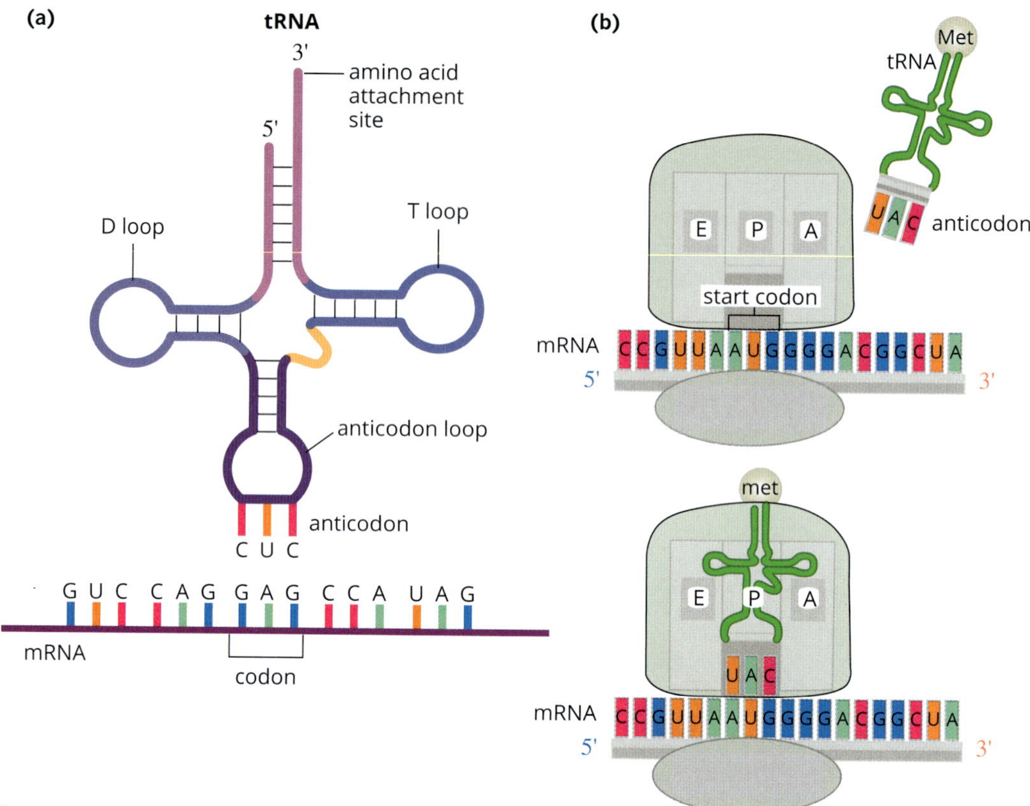

FIGURE 4.2.6 (a) Transfer RNA (tRNA) carries amino acids to the appropriate positions on the mRNA by matching its anticodon sequence to the complementary sequence in the mRNA. The amino acids transferred by tRNA build the polypeptide chain during translation. (b) In this figure, a tRNA molecule carrying the amino acid methionine (Met) recognises the codon on mRNA that is complementary to its anticodon sequence and transfers the amino acid in the correct position on the mRNA. Met is a start codon and the first amino acid to be incorporated into the polypeptide chain.

> The genetic code for determining amino acid sequences works in sets of three bases (nucleotides): on DNA, the set of three is called a triplet; on mRNA, it is called a codon; on tRNA, it is an anticodon. For example, the coding for lysine is: DNA triplet TTC, mRNA codon AAG, tRNA anticodon UUC (remembering that RNA has U instead of T).

Degeneracy of the genetic code

The genetic code is said to be degenerate because more than one RNA codon can code for the same amino acid (Figure 4.2.7). Differences in codons encoding the same amino acid usually occur at the second or third base. As the genetic code uses four nucleotides and three nucleotides code for an amino acid, the combinations of these nucleotides make a total of 64 possible codons ($4^3 = 64$), to code for the total 20 amino acids (Figure 4.2.7). The degeneracy of the code acts as a buffer against mutations in that a single change in one base may not necessarily lead to a change in the amino acid produced. Therefore, it may not necessarily change the structure of the protein produced.

First base of codon	Second base of codon: U		C		A		G		Third base of codon
U	UUU	phenylalanine (Phe)	UCU	serine (Ser)	UAU	tyrosine (Tyr)	UGU	cysteine (Cys)	U
	UUC		UCC		UAC		UGC		C
	UUA	leucine (Leu)	UCA		UAA	STOP	UGA	STOP	A
	UUG		UCG		UAG		UGG	tryptophan (Trp)	G
C	CUU	leucine (Leu)	CCU	proline (Pro)	CAU	histidine (His)	CGU	arginine (Arg)	U
	CUC		CCC		CAC		CGC		C
	CUA		CCA		CAA	glutamine (Gln)	CGA		A
	CUG		CCG		CAG		CGG		G
A	AUU	isoleucine (Ile)	ACU	threonine (Thr)	AAU	asparagine (Asn)	AGU	serine (Ser)	U
	AUC		ACC		AAC		AGC		C
	AUA		ACA		AAA	lysine (Lys)	AGA	arginine (Arg)	A
	AUG	methionine (Met) START	ACG		AAG		AGG		G
G	GUU	valine (Val)	GCU	alanine (Ala)	GAU	aspartic acid (Asp)	GGU	glycine (Gly)	U
	GUC		GCC		GAC		GGC		C
	GUA		GCA		GAA	glutamic acid (Glu)	GGA		A
	GUG		GCG		GAG		GGG		G

FIGURE 4.2.7 The genetic code for the 20 amino acids and stop codons. Remember that these are mRNA codons (not DNA triplets), a fact that should be apparent as soon as U for uracil is seen in the list of bases. The start codon methionine (Met) is highlighted yellow and the stop codons are highighted grey.

Worked example 4.2.1

UNLOCKING THE GENETIC CODE

Complete the table by entering the mRNA codons and the complementary DNA triplet bases that code for each amino acid listed.

Amino acid	mRNA codons	Complementary DNA triplets
Ala alanine		
Lys lysine		

Thinking	Working
Determine how to read the genetic code in Figure 4.2.7. A set of three bases in mRNA is a codon for one amino acid.	The codon is read first from the left column, then the top row, then the right column (e.g. alanine (Ala) is coded by G then C then U or C or A or G).
Remember that U replaces T in mRNA. The complementary DNA base for U is A.	One of the mRNA codons for alanine is GCU (not GCT) and the complementary DNA will be CGA.
Remember there are two to four repeat codons for each amino acid (64 codons and 20 amino acids). This is called degeneracy.	Alanine has four codons: GCU, GCC, GCA, GCG Lysine has two codons: AAA, AAG Enter the mRNA codons into the table:

Amino acid	mRNA codons	Complementary DNA triplets
Ala alanine	GCU, GCC, GCA, GCG	
Lys lysine	AAA, AAG	

Thinking	Working
Identify the complementary triplet codes (base sequence) of the DNA that these mRNA codons were transcribed from.	Alanine codons are: GCU, GCC, GCA, GCG Complementary DNA triplets are: CGA, CGG, CGT, CGC Lysine codons are: AAA, AAG Complementary DNA triplets are: TTT, TTC Enter the complementary DNA triplets into the table:

Amino acid	mRNA codons	Complementary DNA triplets
Ala alanine	GCU, GCC, GCA, GCG	CGA, CGG, CGT, CGC
Lys lysine	AAA, AAG	TTT, TTC

Worked example: Try yourself 4.2.1

Complete the table by entering the mRNA codons and the complementary DNA triplet bases that code for each amino acid listed.

Amino acid	mRNA codons	Complementary DNA triplets
Arg arginine		
Asn asparagine		
Asp aspartic acid		
Cys cysteine		
His histidine		
Leu leucine		
Tyr tyrosine		
Val valine		

THE STRUCTURE OF GENES

While the genetic code is universal, the structure of genes and chromosomes differs between prokaryotes and eukaryotes (see Section 4.1). Prokaryotes contain fewer introns than eukaryotes, which simplifies the process of polypeptide synthesis in prokaryotic cells.

Eukaryotic genes all have structural features in common, including:

- stop and start triplet sequences—regions where encoding DNA begins and ends for a specific gene
- **promoter**—sections of a gene that are found on the DNA before the start triplet (**start codon**), at the 5' end of the site where transcription will begin. The promoter is the location where the RNA polymerase (the enzyme that initiates transcription) attaches to the gene. In many eukaryotic genes, the promoter region is coded for by the sequence of bases TATAAA, which is sometimes called the TATA box
- exons—DNA regions that are the coding segments
- **introns** (or spacer DNA)—DNA regions that are non-coding segments (Figure 4.2.8).

The information on the coding strand of DNA is one side of the double helix. It is this information that is transcribed into mRNA ready for translation into proteins.

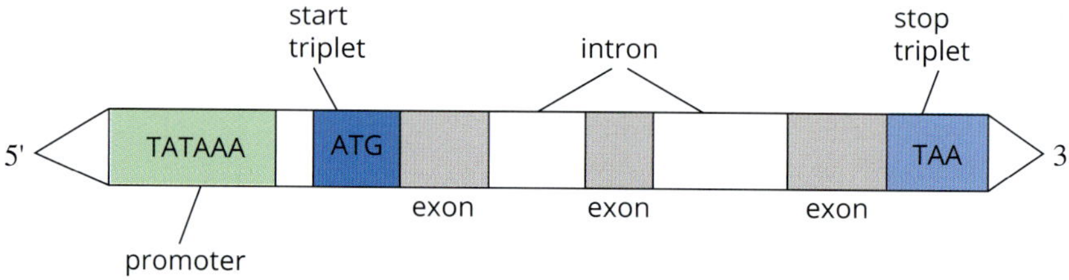

FIGURE 4.2.8 Eukaryotic genes have promoter regions, start and stop triplets, non-coding introns and coding exons.

+ ADDITIONAL

Start and stop instructions

A start codon indicates where the first stage of gene expression will begin. AUG (the amino acid methionine) is the most common start codon in mRNA. In DNA, the start triplet coding for AUG is written as ATG (Figure 4.2.8). This is because when writing a DNA sequence, the scientific standard is to show only the coding (non-template) strand of DNA and to write it in the 5' to 3' direction. The triplet that is complementary to AUG (TAC) is found on the non-coding (template) strand of DNA, which is read by RNA polymerase when building mRNA (see Figure 4.2.2). When transcribed into mRNA the DNA triplet ATG will become the start codon AUG. This codon initiates translation and codes for the amino acid methionine. Most functional proteins start with AUG, but there are some rare exceptions to this. For example, a protein in the fungus, *Candida albicans*, uses GUG as a start codon.

A stop triplet indicates where transcription will end. The stop triplet does not code for an amino acid. When the stop triplets are transcribed into mRNA they become the codons UAA, UAG and UGA which are not recognised by tRNA.

Figure 4.2.7 provides the full set of genetic codes for the 20 amino acids and start/stop codons. Remember that each amino acid corresponds to three bases of mRNA (a codon), not DNA triplets.

Introns and exons

In eukaryotes, not all DNA regions are translated.

- Exons are DNA regions that are usually 'expressed' as proteins or RNA. Exons come together to make up mRNA, which is then translated into proteins.
- Introns are non-coding, or intervening, regions of DNA. Introns are spliced out of the mRNA during RNA processing.

In models of nucleic acids, the nucleotides are simply referred to as bases and are identified by the base letter, A, T, C, G or U. This is because the sugar and phosphate units in all nucleotides are identical. The nitrogenous base is the unit that changes.

GENE EXPRESSION

Gene expression is the process by which the information stored in a gene synthesises a functional gene product (protein or RNA). This process is highly regulated so that proteins or RNA molecules are only produced if they are required by a cell. Multicellular organisms in particular have specialised cells that require a specific set of proteins. For example, in humans, the cells in connective tissue and bone require the protein fibrillin to form elastic fibres, and skin cells require the enzyme tyrosinase to produce melanin and other pigments. The ability to regulate gene expression conserves energy and materials (e.g. nucleotides and amino acids) in the cell.

Gene expression leading to polypeptide synthesis in eukaryotic cells occurs in three stages:

- transcription
- RNA processing
- translation.

In eukaryotic cells, transcription and RNA processing occur within the nucleus and translation occurs in the cytoplasm (Figure 4.2.9).

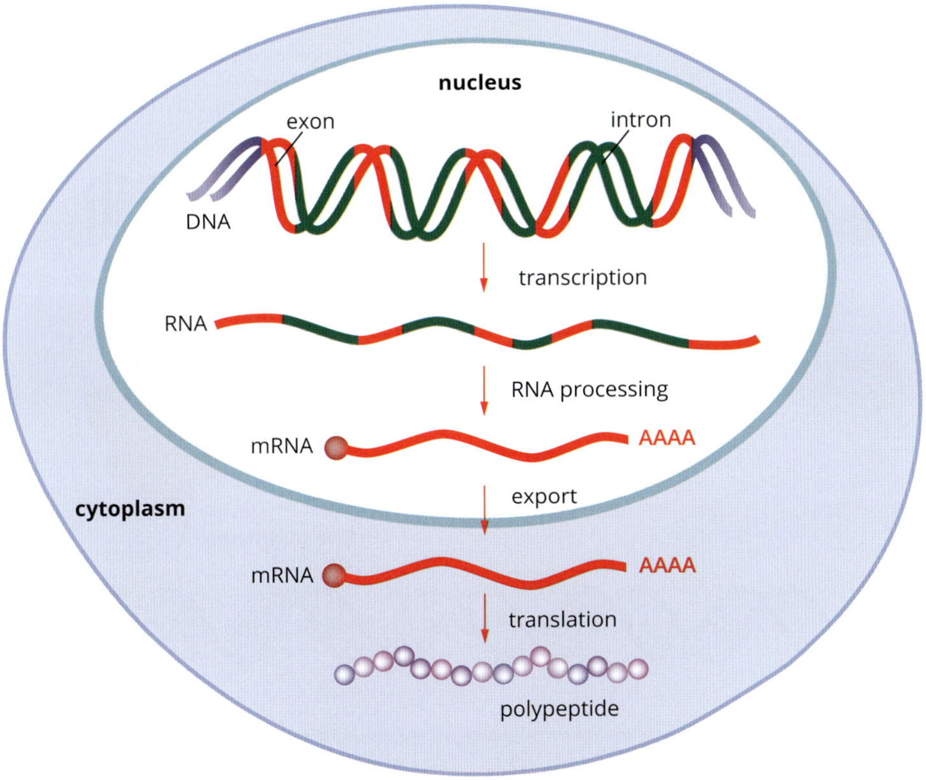

> The genetic code is always a three-letter code, read as groups of three bases. For example, AGU is an mRNA codon for the amino acid called serine. This codon was transcribed from the complementary DNA triplet of TCA. And the matching tRNA anticodon will be UCA.

> Messenger RNA (mRNA) is produced during transcription and then translated to produce an amino acid chain (polypeptide).

FIGURE 4.2.9 Transcription creates a primary transcript from DNA. The introns are then spliced (cut out) during RNA processing to create a mature strand of mRNA. The mRNA exits the nucleus via a nuclear pore. A ribosome translates the mRNA into a polypeptide chain during translation.

Transcription

Producing single-stranded mRNA from DNA is called transcription and occurs within the nucleus of eukaryotic cells. The DNA segment that undergoes transcription is known as the transcription unit.

Transcription occurs in three steps:

1. initiation
2. elongation
3. termination (Figure 4.2.10).

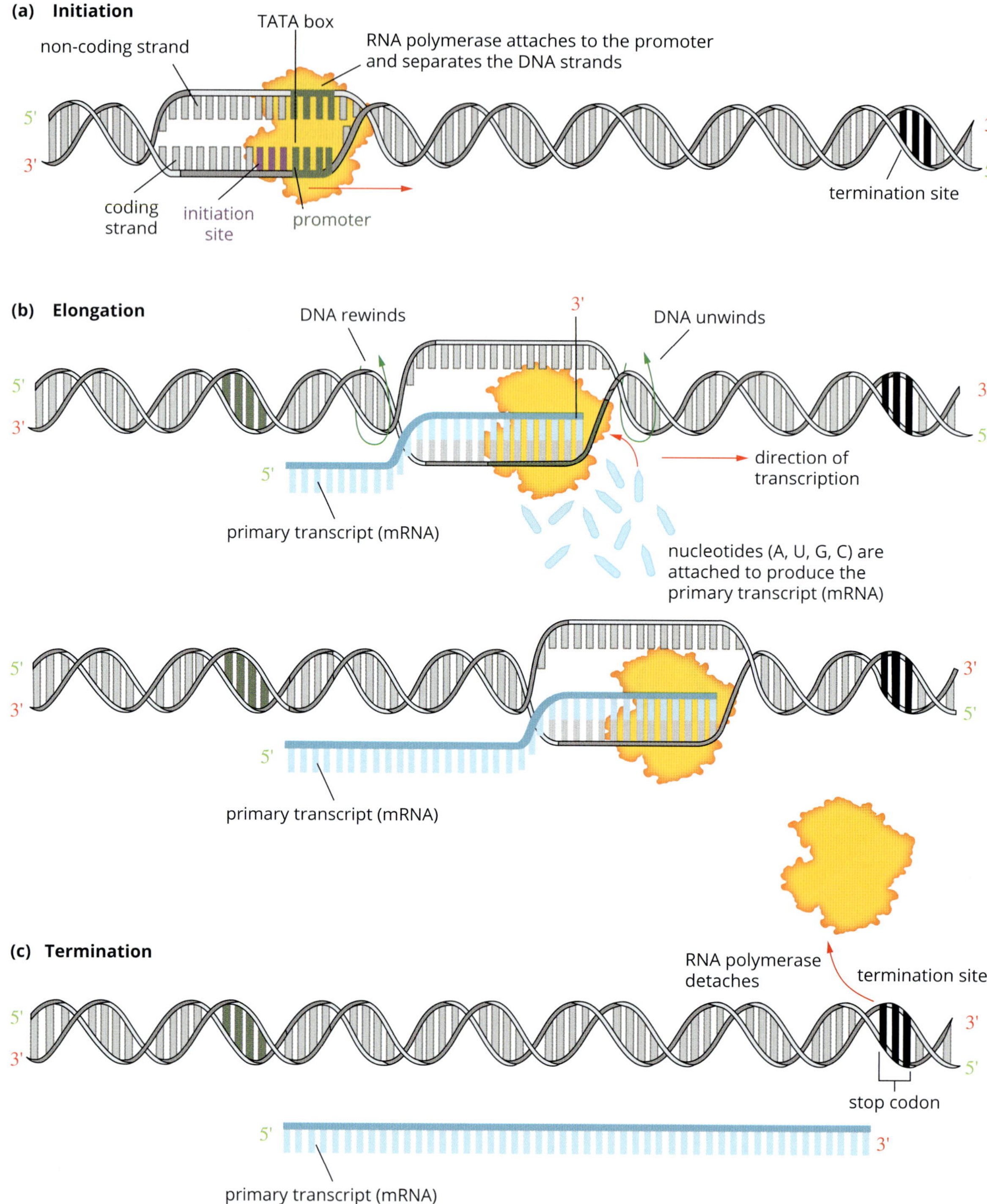

FIGURE 4.2.10 Transcription occurs in three stages. (a) Initiation of transcription, when RNA polymerase attaches to the promoter region of the DNA and unzips the two DNA strands by breaking their hydrogen bonds, exposing the nitrogenous bases (A, T, G, C). (b) During elongation, the RNA polymerase moves along the coding DNA strand and produces a strand of mRNA by attaching complementary nucleotides (A, U, G, C). The mRNA strand is known as the primary transcript (or pre-mRNA) at this stage. (c) Transcription is terminated when the RNA polymerase reaches the stop triplet code (stop codon) at the termination site. The RNA polymerase then detaches and the two DNA strands come together.

Many RNA polymerase molecules may attach to the gene being transcribed, producing many of the same mRNA molecules. The strand of DNA that is transcribed to the mRNA is known as the **non-coding strand** (or template strand) and the other complementary strand is known as the **coding strand** (or non-template strand). The mRNA carries the same base sequence as the coding strand, (except it contains uracil in place of thymine) because it has been copied from the non-coding strand into complementary bases.

RNA processing

After transcription, the primary RNA transcript is processed before being translated. RNA processing forms mature mRNA from pre-mRNA (the primary transcript of mRNA), after removing non-coding sequences (introns) so that only the coding sequences (exons) are carried to the ribosome for translation (Figure 4.2.11). RNA processing is present in both eukaryotic and prokaryotic cells but is much more complex in eukaryotes, as prokaryotes carry fewer introns that need to be spliced out after transcription. The RNA processing stage of gene expression in eukaryotes includes:

- the addition of a **5' cap**
- the addition of a **poly(A) tail** (The 5' cap and poly(A) tail make the mRNA more stable and prevent it from degrading.)
- **splicing** (removal) of the introns (mRNA maturation).

For prokaryotes, without a nuclear membrane, the transcription and translation processes are closely coupled in the cytoplasm, with the mRNA moving off the DNA and becoming attached to ribosomes even before transcription is complete.

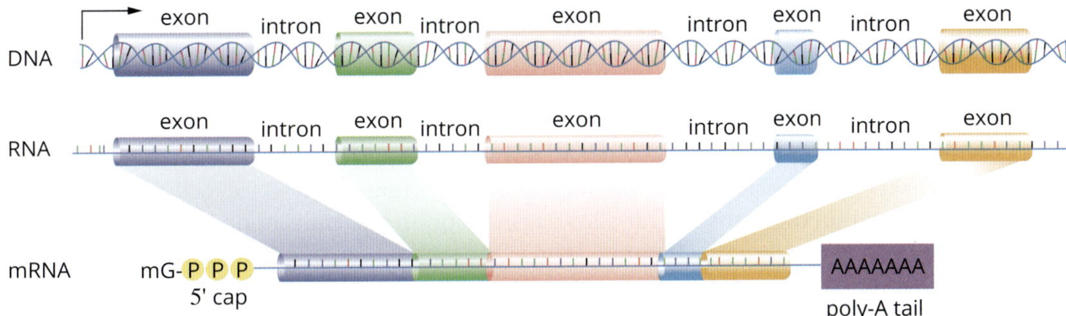

FIGURE 4.2.11 During RNA processing, the introns (non-protein-coding regions) are spliced from the primary pre-mRNA transcript, resulting in mature messenger RNA, which consists of only exons (protein-coding regions).

Translation

Translation is the process in which the codons on mRNA are translated into a sequence of amino acids resulting in a polypeptide. This process occurs on ribosomes in the cytoplasm. Ribosomes bind to an mRNA molecule and act as docking stations for the tRNAs to deposit their specific amino acids. A part of the tRNA, called an anticodon, recognises and binds to the codon on the mRNA by complementary base pairing. Each tRNA carries a specific amino acid related to the codon to which it binds.

Like transcription, translation also occurs in a series of three steps:

1. initiation
2. elongation
3. termination (Figure 4.2.12).

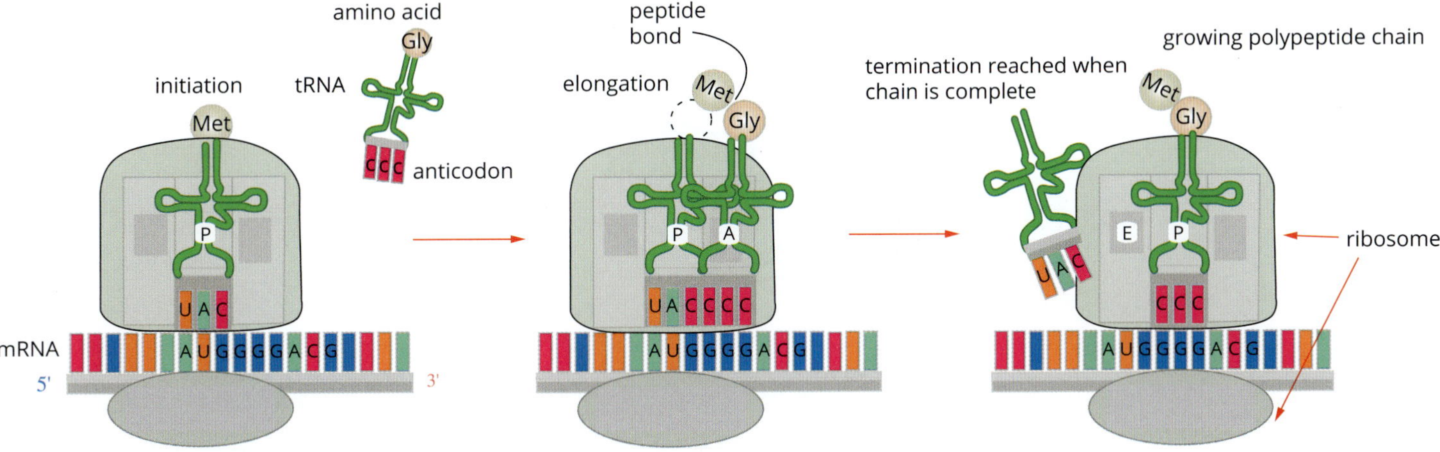

FIGURE 4.2.12 The process of translation on a ribosome. The ribosome moves along the mRNA one codon at a time and tRNA molecules bring their specific amino acids to their complementary mRNA codon. The amino acids join together by peptide bonds to form a polypeptide chain.

Many ribosomes can translate the same, single strand of mRNA, enabling many polypeptide chains to be produced at the same time (Figure 4.2.13). Once the polypeptides are fully functional, they either remain in the cell for use, or are exported from the cell by vesicles (via a process called exocytosis) for use elsewhere in the organism.

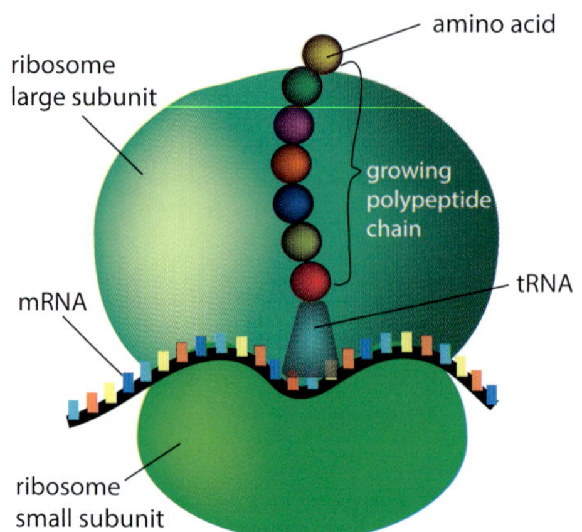

FIGURE 4.2.13 Eukaryotic ribosome translating mRNA into a polypeptide

Protein synthesis in prokaryotes

Prokaryotes do not have membrane-bound organelles, so all cellular processes occur within the cytosol. This allows transcription and translation to be a continuous process rather than two separate stages. Ribosomes can attach to the mRNA while it is being transcribed, so translation can occur at the same time. Prokaryotes mostly contain exons, so splicing rarely occurs before translation.

There are many differences between protein synthesis in prokaryotic and eukaryotic cells (Figure 4.2.14). Table 4.2.1 summarises the major differences, many of which have been used to develop drugs that target protein synthesis in prokaryotes only. For example, some antibiotics disrupt or inhibit the production of proteins in disease-causing bacteria.

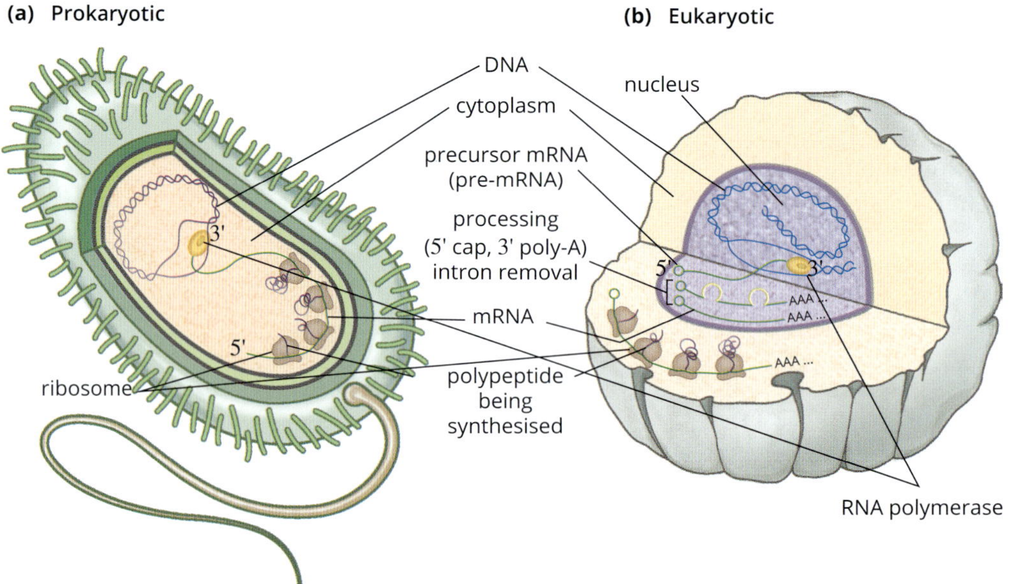

FIGURE 4.2.14 Comparison of protein synthesis in (a) prokaryotic and (b) eukaryotic cells. The structural differences between the cell types means that there are many differences in the way proteins are synthesised.

TABLE 4.2.1 A summary of the differences between prokaryotic and eukaryotic protein synthesis

Prokaryotic protein synthesis	Eukaryotic protein synthesis
30S and 50S ribosomal subunits (forms 70S ribosome)	40S and 60S ribosomal subunits (forms 80S ribosome)
Each mRNA may contain the coding sequences of several genes; this is known as polycistronic.	Each mRNA contains the coding sequence for one gene; this is known as monocistronic.
There is an overlap between transcription and translation, with protein synthesis beginning during transcription. This is known as coupled transcription–translation. This can occur because the DNA and ribosomes are in the cytosol together.	There is no overlap between transcription and translation. The transcription of DNA occurs in the nucleus, and translation and protein synthesis occur in the cytoplasm.
Prokaryotes mostly contain exons, with very few introns	Eukaryotes have introns and exons.
Because prokaryotes contain few introns, RNA processing is rarely required after transcription. RNA processing is much simpler than in eukaryotic cells.	RNA processing removes the introns (non-coding regions) so only the exons (coding regions) are translated.
Prokaryotes have about three different initiation factors.	Eukaryotes have around 10 different initiation factors.
No 5' cap is added to mRNA.	A methylguanosine triphosphate molecule, called a 5' cap, is added to the 5' end of the mRNA.
No poly(A) tail is added to mRNA.	A poly(A) tail is added to the 3' end of mRNA.

AMINO ACIDS, POLYPEPTIDES AND PROTEINS

Protein molecules are more complex than other **biomolecules** like carbohydrates or lipids. For example, the haemoglobin that carries oxygen in human blood has a chemical formula of $C_{2952}H_{4664}O_{832}N_{812}S_8Fe_4$. Proteins make up more than 50% of the dry weight of cells. There are thousands of different kinds of proteins, their functions vary widely and they are essential to the wellbeing of any organism. Although carbohydrates and lipids are similar in all plants and animals, organisms can have a variety of unique proteins that are specific to a species. This section explains the way proteins are synthesised inside cells from the inherited DNA code. Section 4.3 will discuss proteins in more detail.

Proteins are **biomacromolecules** made of chains of subunits called amino acids. Amino acids are linked by a chemical bond called a **peptide bond** and form polypeptides or polypeptide chains (Figure 4.2.16). Polypeptide means 'many peptide bonds'. A **peptide** is a linear sequence of fewer than 50 amino acids. A **dipeptide** is a molecule consisting of two amino acids joined with a peptide bond. A polypeptide is a molecule consisting of many (more than 50) amino acids joined together by peptide bonds. A protein is formed by one or more polypeptides arranged in a biologically functional way. In other words, 'protein' is the term used for a fully functioning molecule while 'polypeptide' refers to a non-functioning component.

There are 20 different amino acids commonly found in proteins. Nine of these are known as essential amino acids because they cannot be produced by humans. Humans must obtain these from their food. The properties of many proteins are determined by their shape, which is determined in turn by their amino acid sequence.

BIOFILE CCT PSC

Amino acids in the human diet

In the human diet, amino acids can be classified into three main groups.

- Essential amino acids: the body cannot synthesise these, we must obtain them from our diet (Figure 4.2.15). The nine essential amino acids are histidine, isoleucine, leucine, lysine, methionine, phenylalanine, threonine, tryptophan and valine.
- Non-essential amino acids: the body can produce these if not obtained from the diet. The five non-essential amino acids include alanine, asparagine, aspartic acid, glutamic acid and serine.
- Conditional amino acids: the body only requires these in times of illness or stress. The conditional amino acids include arginine, cysteine, glycine, glutamine, proline and tyrosine.

FIGURE 4.2.15 A collection of foods and supplements high in amino acids

Proteins

peptide bonds

amino acid subunits

FIGURE 4.2.16 The structure and bonding of proteins. Amino acid monomer units are bonded into peptides, then into polypeptides, then structured into functional proteins.

There are many steps involved in producing a functional protein. Although protein structure, size and function are quite diverse, all proteins are made up of amino acids. These smaller subunits (or monomers) are joined together in a particular order to form polypeptide chains (Figure 4.2.17). The polypeptide chains are then folded and coiled into proteins (Figure 4.2.18). Section 4.3 will explain the structure of proteins in greater detail.

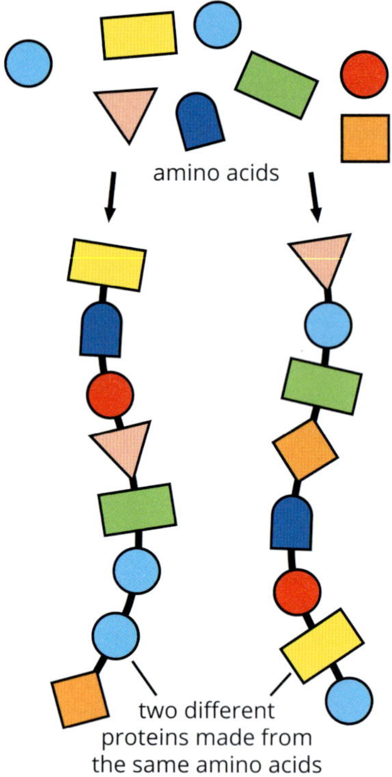

FIGURE 4.2.17 The diagram models how the same amino acids can be sequenced in a different order to make different polypeptide chains that, in turn, will form different proteins.

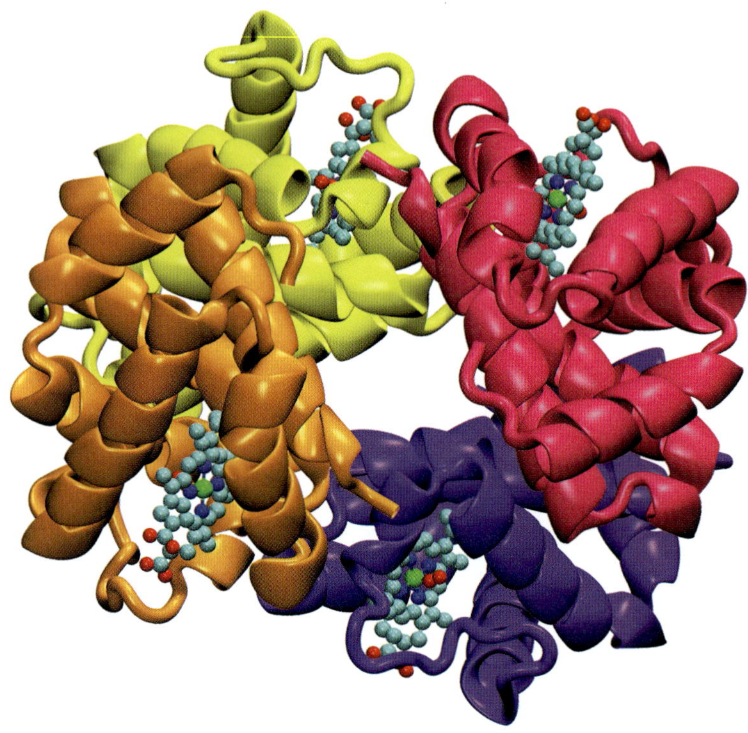

FIGURE 4.2.18 Haemoglobin is made up of four polypeptide subunits (coloured ribbon structures). Each has a haem group (turquoise), which is the oxygen-binding site (O_2=paired red spheres). Within each haem group there is one atom of iron (green). The chemical formula for an Hb molecule is $C_{2952}H_{4664}O_{832}N_{812}S_8Fe_4$.

BIOLOGY IN ACTION CCT ICT

Protein forensics and LOC

Because many proteins are unique to an organism, scientists can test a protein sample to determine which animal it came from. Some shops and restaurants have been caught selling shark as barramundi or putting horsemeat into hamburgers. The suspect protein sample can be analysed to determine its origin.

A technology referred to as lab-on-a-chip (LOC) miniaturises and compacts the different test processes that a researcher or a diagnostic lab technician uses. A tiny fluid sample is added to one end of the microchip-sized wafer of glass or plastic and the molecules of the suspect protein are channelled past circuits of nanometre-sized chemical and physical tests (Figure 4.2.19). The device is still being perfected and holds great promise for testing samples in the field or when the sample is very small in volume.

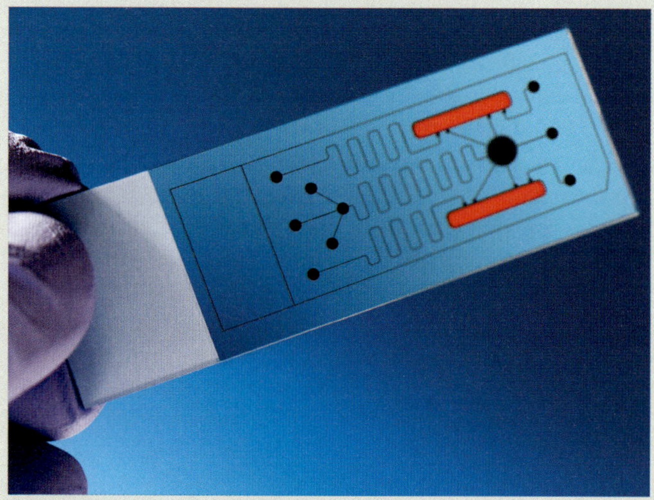

FIGURE 4.2.19 Biochip for identifying proteins using the LOC technology. Test systems for analysis of a sample are integrated into one microchip-sized thin glass plate.

The function and importance of polypeptide synthesis

The genome consists of many thousands of genes. A cell can express a selection of these genes at a given time. The genes that are expressed determine which proteins are produced, giving the cell its functionality and characteristics. Gene expression is the process through which information from a gene is used to synthesise a specific functional gene product—a polypeptide or length of RNA. Gene expression in eukaryotes is tightly regulated by multiple mechanisms at different points, usually in the transcription phase. The regulation determines that the correct polypeptides and proteins are produced when and where they are required. For example:

- growth hormone proteins are produced in infants, children and teenagers but switched off at other times
- plasma B cells of the immune system are regulated to only produce antibody proteins when they are required to fight an infection
- enzyme proteins are recycled but new ones will be synthesised to replace those that have reached the end of their useful life
- continual production of haemoglobin in the bone marrow of vertebrates.

BIOFILE L

Gene names

A gene name is always italicised, to distinguish it from the protein it encodes. For example, the disorder phenylketonuria (PKU) is caused by a mutation in the *PAH* gene that normally expresses the enzyme phenylalanine hydroxylase (PAH). The inherited disease of cystic fibrosis (CF) comes from mutation of the *CFTR* gene, which codes for CFTR protein, which regulates the movement of chloride ions across cell membranes.

BIOFILE CCT

Haemoglobin

Haemoglobin (Hb) is the oxygen transport protein carried on red blood cells of vertebrates. In humans, a functional haemoglobin protein molecule is built from four polypeptide chains. The polypeptides are constantly being synthesised by cells in the bone marrow so that newly differentiated red blood cells will carry haemoglobin (Figure 4.2.20).

Mature red blood cells have no nucleus and therefore cannot form new proteins while they circulate in the blood. This limits their lifetime to around 120 days, and means it is necessary for constant production of replacement red blood cells by the bone marrow—estimated at the extraordinary rate of around two million new red blood cells per second. Haemoglobin has the critical function of carrying oxygen to all parts of the body so that cells can produce energy using cellular respiration. Therefore, it is essential that gene expression in bone marrow cells is regulated to produce a constant supply of polypeptides for constructing the protein structure of haemoglobin.

When you donate blood, you lose red blood cells. Special cells in the kidneys, called peritubular cells, sense the decreased level of oxygen in the blood and start secreting a protein called erythropoietin. It passes through the bloodstream until it reaches the bone marrow where the erythropoietin triggers stem cells to develop into more red blood cells, rather than white blood cells or platelets.

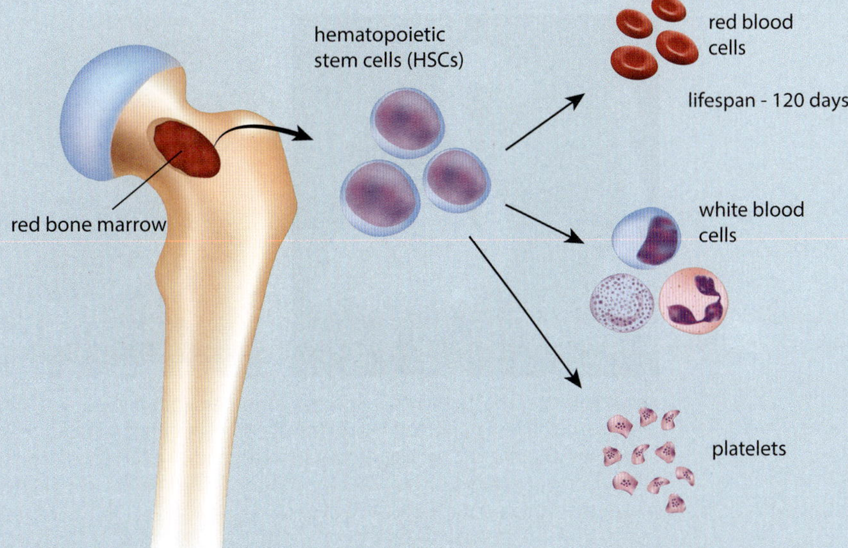

FIGURE 4.2.20 Blood cell formation from bone marrow. Haemoglobin protein must be available for red blood cells to fulfil the vital function of oxygen transport.

An organism's phenotype is all its observable characteristics. It is the result of inheritance and the effects of the organism's environment. The genotype is the set of alleles present in the DNA of an individual organism. The genotype is the result of inheritance and any mutations in the germ cells.

GENOTYPE, PHENOTYPE AND THE ENVIRONMENT

The **proteome** is the complete set of proteins expressed by the genome (the complete set of genes) of an individual cell or organism at a given time. The proteome varies between cell type, developmental stage and environmental conditions. Although a cell may contain the entire genome, only specific genes will be expressed, or 'switched on', at any given time. This ensures a cell produces only the proteins required for the specific functions it carries out.

The **phenotype** is the set of characteristics expressed from the genetic information of a cell. That is, the part of the **genotype** that has been switched on to allow transcription and translation to produce polypeptides, then proteins.

- An organism's genotype and environment determine its phenotype.
- If an individual with a given genotype develops in one environment, its phenotype may be different than if it had developed in a different environment.
  - For example, the average height of humans has gradually increased in the last few hundred years because of improved nutrition.
  - On the other hand, a human cannot grow to the full height potential that is coded in their genotype if they are severely undernourished as a child or if a bone disease affects their growth.
  - A tree will be stunted in height if it does not receive the water and minerals needed to reach its genetic potential or if it grows with continual strong wind, in constant shade or is grown in a small pot where it becomes root-bound.
  - A human's skin colour depends partly on the amount of sunlight exposure. In the short term, it can be altered by air temperature or exercise, such as being red-faced after strenuous exercise or pale-skinned on a very cold day.

Other examples of environmental influence on phenotype are:

- the development and severity of inherited disorders such as phenylketonuria (PKU)
- flower colour variation with soil pH
- seasonal variation in the fur colour of Arctic foxes
- feather colour variation in flamingos with different levels of carotenoids (organic pigments produced by algae, bacteria and fungi) in their diet (Figure 4.2.21).

You will learn more about the effect of the environment on phenotype in Chapter 5.

GO TO ➤ Section 5.1 page 202

FIGURE 4.2.21 The colour of a flamingo's feathers (phenotype) is determined by their dietary intake of carotenoids (environment). Carotenoids are pigments produced by some species of algae, bacteria and fungi. The flamingos' diet of brine shrimp and blue-green algae is a rich source of carotenoids. Feather colour varies from (a) light pink to (b) dark pink/red. The higher the intake of carotenoids, the pinker the flamingo's feathers will be.

4.2 Review

SUMMARY

- DNA stores hereditary information, carrying the instructions that code for the production of proteins and functional RNA molecules, in a specific sequence of nucleotides.
- The genetic code is the set of rules about how the instructions carried in nucleic acids are translated to synthesise proteins and functional RNA molecules. In DNA this information is stored as a three-letter code of nucleotides known as a triplet. When these triplets are transcribed into mature mRNA, they are then known as codons. On tRNA the group of three bases is called an anticodon because these bases are the opposite partners (complements) of the mRNA codons.
- The genetic code is universal and degenerate. There are 64 possible codons of three nucleotides each (e.g. UAC) for the 20 amino acids.
- A gene is a region of DNA that codes for a polypeptide in both prokaryotes and eukaryotes. Unique to eukaryotes, a gene may also code for a functional RNA molecule.
- Eukaryotic genes have many structural features in common:
  - Stop and start instructions—indicate where transcription of mRNA starts and stops. Stop codons do not code for amino acids.
  - Promoter region—the site at which the RNA polymerase attaches to the gene to begin transcription (sometimes called the TATA box).
  - Exons—coding DNA regions.
  - Introns—non-coding DNA regions.
- Gene expression is the process in which the information stored in a gene is used to synthesise a functional gene product (polypeptide or RNA). Gene expression is regulated so that it occurs if the particular protein or RNA is required by the cell.
- RNA is a short, usually single-stranded, nucleic acid.
- RNA contains nucleotides that are made up of ribose sugar, a phosphate and one of four nitrogenous bases (adenine, cytosine, guanine and uracil).
- The role of RNA is to express the information contained in the nucleotide sequence of a gene to synthesise polypeptides.
  - mRNA is produced by a process called transcription in the nucleus—a single-stranded nucleic acid that carries a copy of the genetic sequence in DNA, specifying the amino acid sequence for a polypeptide.
  - rRNA makes up part of a ribosome. Ribosomes are the sites where the information in the mRNA is translated into a chain of amino acids.
  - tRNAs carry specific amino acids to ribosomes to form polypeptide chains by the process called translation.
- Polypeptide synthesis in eukaryotes occurs in three stages: transcription, RNA processing and translation.
- Transcription occurs in the nucleus and involves RNA polymerase transcribing the DNA into a primary RNA transcript.
- The primary RNA transcript is spliced to remove the introns, and sometimes some exons, resulting in mature mRNA. The mature mRNA then leaves the nucleus.
- Translation occurs on a ribosome in the cytoplasm. The codons on mRNA are translated into a sequence of amino acids to form a polypeptide chain. The amino acids are delivered individually by their specific tRNA molecules each with an anticodon that complements the codon of the mRNA.
- After synthesis, polypeptide chains are folded and combined to form fully functional proteins. 'Protein' is the term used for a fully functioning molecule while 'polypeptide' refers to a non-functioning component.
- Polypeptide synthesis and protein formation is crucial to the functioning of all cells. Proteins are important because enzyme proteins control cell metabolism; other proteins are hormones and antibodies; proteins are important receptor, structural, storage and transport molecules; and they are important in movement.
- Changes in the amino acid sequence can lead to faults in proteins and cause health issues.
- The genotype plays an important role in determining the phenotype but gene expression can be affected by environmental influences such as temperature, soil pH, nutrition, availability of key chemicals and sunlight exposure.

4.2 Review *continued*

KEY QUESTIONS

1. Distinguish between peptides, dipeptides and polypeptides, using a diagram to support your answer.
2. Explain how RNA differs from DNA, mentioning at least three features that differentiate them.
3. Name the three types of RNA and outline their basic functions.
4. Identify the three stages of polypeptide synthesis in eukaryotes.
5. Transcription occurs in three stages: initiation, elongation and termination. Rearrange the table to match the transcription event with the correct stage at which it occurs.

Stage of transcription	Transcription event
initiation	• The RNA polymerase moves along the DNA molecule, producing a strand of mRNA. • The RNA polymerase detaches, releasing the mRNA and allowing the DNA molecule to reform.
elongation	• RNA polymerase uses a strand of DNA as a template, attaching nucleotides (A, U, G, C) by complementary base pairing. • Transcription factors combine with the region at the start of the gene, known as the promoter. • RNA polymerase reaches the termination site of the gene (stop codon) and translation ends.
termination	• RNA polymerase attaches to the promoter, unwinding and unzipping the DNA molecule by breaking the weak hydrogen bonds between the two strands to expose the bases.

6. Like transcription, translation occurs in three stages: initiation, elongation and termination. However, the events during each of these stages are different. Rearrange the table to match the translation event with the correct stage at which it occurs.

Stage of translation	Translation event
initiation	• Following the attachment of the amino acid methionine, another tRNA, with a complementary anticodon to the next codon on the mRNA, attaches and adds its specific amino acid to the growing polypeptide chain. • The tRNA reaches a stop codon.
elongation	• A small ribosomal subunit attaches to the 5' end of an mRNA strand. It then moves along the mRNA until it reaches a start codon (AUG). • The polypeptide chain is released from the ribosome into the cytoplasm or the endoplasmic reticulum.
termination	• The ribosome then releases the tRNA and moves further along the mRNA strand. At each codon, a new tRNA binds and adds another amino acid. • A tRNA molecule with an anticodon (UAC) brings the amino acid methionine to the mRNA. The tRNA molecule joins to the mRNA start codon, attaching by complementary base pairing between the codon and anticodon.

7. Outline the main structural features of eukaryotic genes and their functions.
8. Define what is meant by 'universal genetic code'.
9. There is a longstanding debate referred to as nature versus nurture. In the debate, nature refers to inherited information and nurture refers to environmental effects on phenotype. Use your scientific knowledge to assess how genes and the environment both influence the phenotype of an organism.
   Support your answer with examples.

4.3 Structure and function of proteins

BIOLOGY INQUIRY   L

Protein data banks

Why is polypeptide synthesis important?

COLLECT THIS...

- tablet or computer to access the internet
- note paper
- pen

DO THIS...

1. Search the internet for a protein data bank.
2. Explore the data bank and find a protein, such as a human enzyme, that you are interested in.
3. Find a 3D model of your selected protein and draw it or download the image.
4. Write notes about the structure and function of your selected protein such as how many polypeptide chains it has and if it is a tertiary or quaternary folded molecule (see pages 186–188 for information on levels of protein structure).

RECORD THIS...

Describe the value of the online protein data bank and identify how many molecules it currently holds in the database.

Present information about your chosen protein to your class and compare your information with that found by other students.

REFLECT ON THIS...

Why is polypeptide synthesis important?

Can you see how proteins are built from polypeptides in complex 3D configurations?

The Protein Data Bank (PDB) is a free international database that stores information about all known proteins and nucleic acids. How useful is it for scientists to have this data freely available to them?

Nearly every function of a living organism depends on proteins. Proteins speed up chemical reactions, and play a role in cell-to-cell recognition and cellular communication, movement, storage, cell membrane functions and even structural support. A human has tens of thousands of different proteins, and each protein has a specific sequence of amino acids, giving it a unique shape that enables it to carry out a particular function. Proteins have a large variety of functions, so they vary extensively in structure, with each type of protein having its own complex three-dimensional shape. Many proteins function with other chemical groups attached to them, such as the glycoproteins that are a combination of polypeptides and sugar molecules (Figure 4.3.1).

In this section, you will learn about the diversity in the functions of proteins and the nature of the proteome. First, the importance of the four hierarchal levels of protein structure will be investigated as molecular structure plays an essential role in effective functioning for each type of protein.

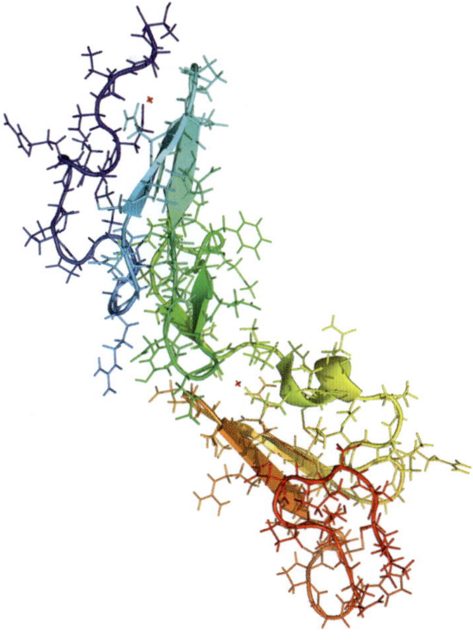

FIGURE 4.3.1 Computer-generated model of a fibrillin glycoprotein, a large, extracellular protein that binds calcium and is part of the human connective tissue. The protein component is represented by ribbons and the sugars are represented by the ring structures.

BIOFILE N

Protein size

Proteins come in vastly different sizes. The peptide hormones oxytocin, vasopressin and antidiuretic hormone are only nine amino acids long (Figure 4.3.2). Dystrophin is a large protein that is important for muscle cell structure and has 3600 amino acids (Figure 4.3.3). Defects in this protein lead to muscular dystrophy. Another muscle protein, titin, is involved in the elasticity of the muscle and is the largest known protein—it has around 30 000 amino acids.

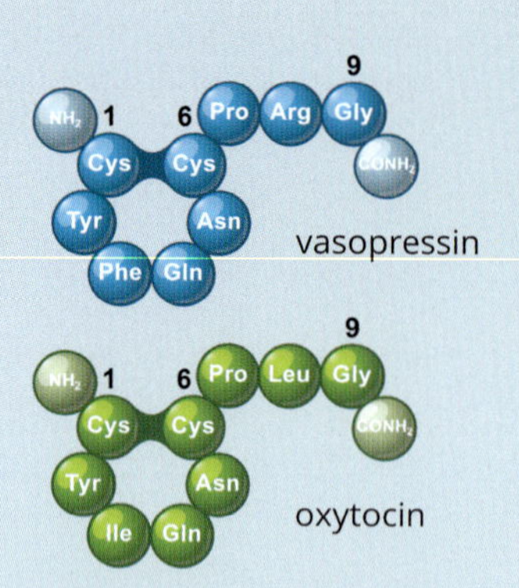

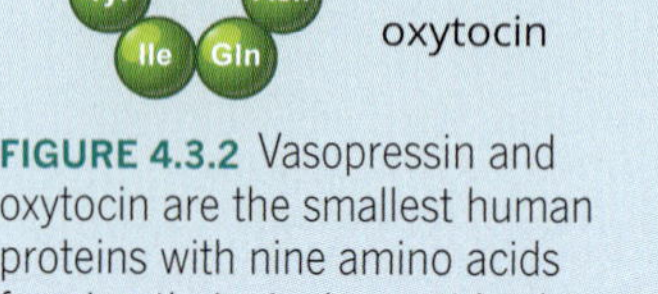

FIGURE 4.3.2 Vasopressin and oxytocin are the smallest human proteins with nine amino acids forming their single peptide chains.

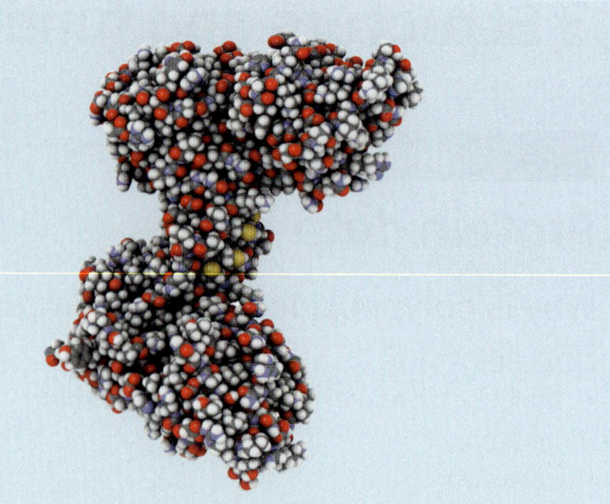

FIGURE 4.3.3 A model of the large protein dystrophin, which has 3600 amino acids. Defects in this protein cause muscle degeneration associated with some types of muscular dystrophy. Individual atoms are represented as spheres like those used in molecular model building kits.

PROTEIN STRUCTURE

Proteins are large biomolecules that vary in size, from less than 10 amino acids to thousands of amino acids, and may be synthesised as one or several polypeptide chains. These chains are folded and organised into specific shapes that are vital to the correct functioning of the protein. Most proteins are required to bind to other molecules to achieve functionality. A single change to one amino acid within the sequence can alter the shape, and consequently the function, of a protein. Likewise, if an essential partner atom or molecule is missing, the protein may lose or reduce its ability to function. For example, the haemoglobin protein that carries oxygen in the human bloodstream has iron as part of its quaternary structure to bind oxygen.

Synthesis of proteins

There are many steps involved in producing a functional protein. Although protein structure, size and function are quite diverse, all proteins are made up of amino acids. These 20 types of smaller subunits (or monomers) are joined together in a specific order to form polypeptide chains. The polypeptide chains are then folded and coiled into proteins. With 64 different triplet codes for gene expression, 20 different amino acids, more than one polypeptide per protein molecule and a variety of extra partner molecules, there are almost endless possibilities for different forms of proteins.

Amino acid structure

All amino acids have the same basic structure (Figure 4.3.4):

- an **amine group** (NH_2)
- a **carboxyl group** (COOH)
- an **R group** (or side chain).

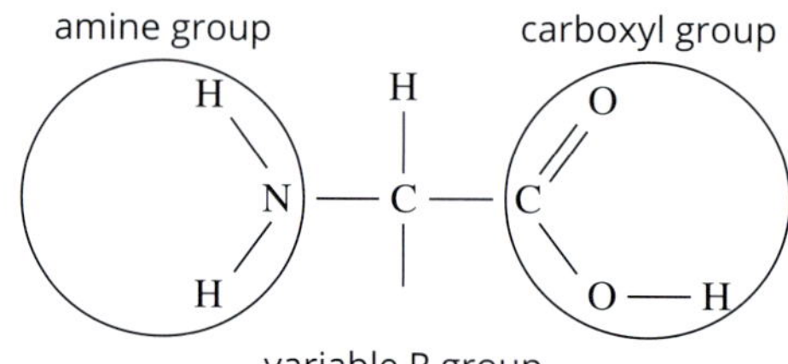

FIGURE 4.3.4 The basic structure of an amino acid, showing an amine, carboxyl and variable R group

In the synthesis of proteins in organisms, there are 20 different standard amino acids, and each has a different R group (Figure 4.3.5). The variable properties of the R group (e.g. charged or uncharged, polar or non-polar, hydrophobic or hydrophilic) determine the type of protein that the amino acid will form. R groups can be as simple as a hydrogen atom (as in the amino acid glycine) or more complex (e.g. $-CH(CH_3)^2$) as in the amino acid valine. The 20 amino acids with their variable R groups are shown in Figure 4.3.5.

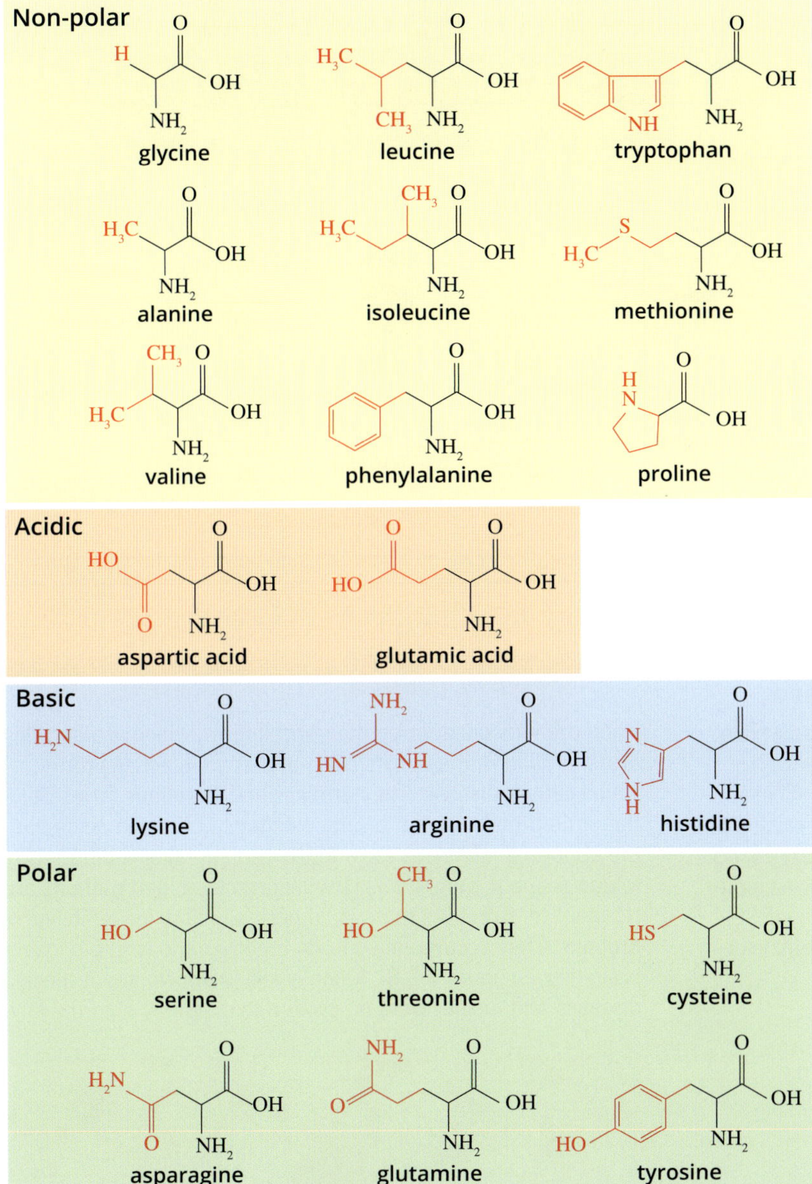

FIGURE 4.3.5 Chemical structure of the 20 standard amino acids. The variable R groups are coloured in red. Amino acids can be classified according to their chemical nature as non-polar, acidic, basic or polar.

Condensation polymerisation of amino acids

Amino acids are joined by peptide bonds in a **condensation polymerisation** reaction, which involves the removal of water (dehydration). A hydrogen atom and an oxygen atom from the carboxyl group of one amino acid join with a hydrogen atom from the amine group of another amino acid to produce water. The water is released and a dipeptide is synthesised, with a peptide bond holding the two amino acids together. A chain of amino acids joined by peptide bonds is known as a polypeptide chain. The backbone of the polypeptide chain is formed by the repeats of the carboxyl and amine groups, with the R groups forming the side chains of the polypeptide chain (Figure 4.3.6).

A polypeptide chain forms the primary structure of a protein. With further folding and modification, a fully functional protein can be formed.

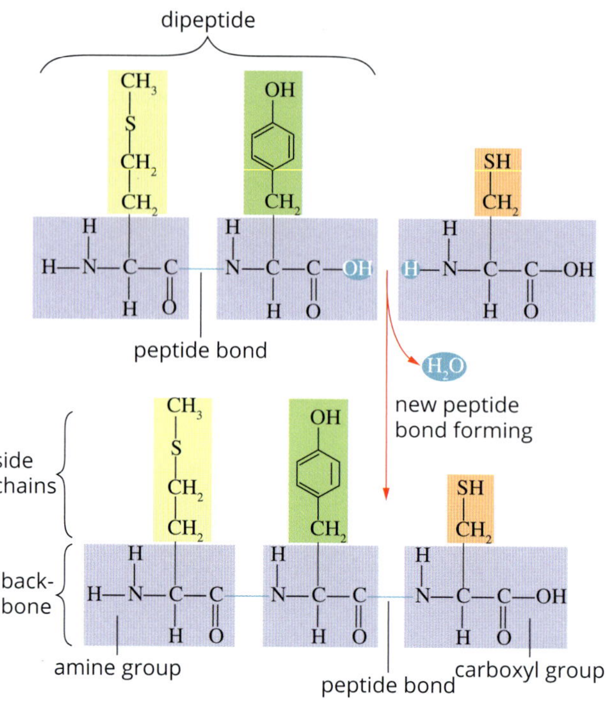

FIGURE 4.3.6 Formation of a polypeptide chain by a condensation polymerisation reaction

BIOFILE CCT

Protein folding and degenerative disease

A group of rare brain diseases in animals, called transmissible spongiform encephalopathies (TSEs), are associated with the accumulation of abnormal prion proteins in the brain. These diseases are fatal and characterised by spongy degeneration of the brain. At present, there are no reliable live animal tests, no treatments and no vaccines. They take various forms—scrapie in sheep and goats, bovine spongiform encephalopathy (BSE) in cattle (commonly called mad cow disease), and Creutzfeldt–Jakob disease in humans.

PrP^{C} is a normal protein

PrP^{Sc} the disease-causing form of the prion protein

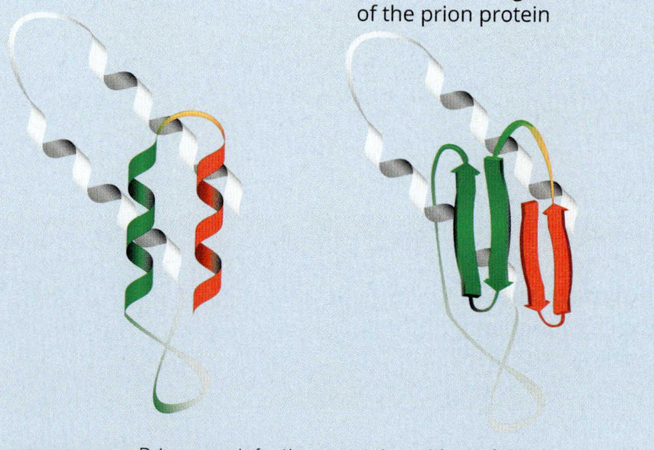

FIGURE 4.3.7 Prions are infectious proteins with misfolded structures. They are responsible for the transmissible mad cow disease and Creutzfeldt–Jakob disease in humans.

Prion proteins occur in both a normal form, which is a harmless protein found in the body's cells, and in an infectious form, which causes disease. The harmless and infectious forms of the prion protein have the same sequence of amino acids, but the abnormal infectious form of the protein takes a different folded shape (Figure 4.3.7). The incorrectly folded proteins are resistant to proteases that would normally destroy defective proteins. They accumulate, clump together and result in loss of brain tissue (Figure 4.3.8). Much is still unknown about these diseases and further work on protein folding may hold the key.

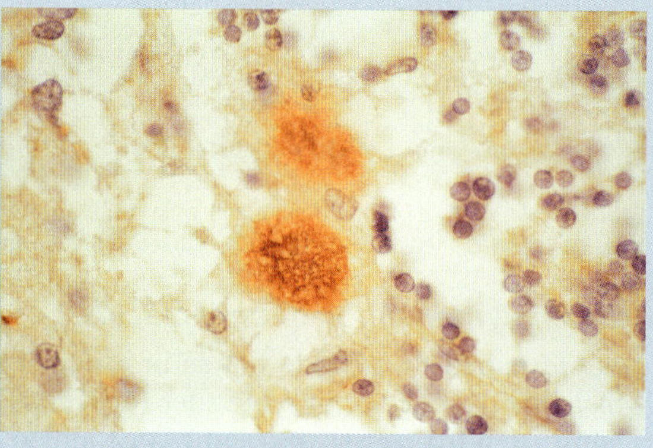

FIGURE 4.3.8 Light microscope image of a section through the cerebellum of a brain affected by Creutzfeldt–Jakob disease, showing two forms of amyloid plaques (dark yellow). Creutzfeldt–Jakob disease is the result of misfolded proteins (prions) within the brain that cause vacuoles and plaques to form, making the brain spongy and killing off the nerve tissue.

REVISION

GO TO ➤ Year 11 Section 3.4

Models of protein molecules

You learnt about the different representations of protein molecules in Year 11. Models of proteins allow for better understanding of a complex object, both by expert structural biologists and by other scientists, students and the general public. There are three models that are commonly used to represent protein structure:

- Ribbon diagrams are simple, yet powerful models for expressing the visual basics of a molecular structure (twist, fold and unfold). α-helices are shown as coiled ribbons or thick tubes, β-strands as arrows, and lines or thin tubes for non-repetitive coils or loops. The direction of the polypeptide chain is shown locally by the arrows, and may be indicated overall by a colour ramp along the length of the ribbon (Figure 4.3.9a).
- Wireframe diagrams have a line for each of the covalent bonds formed between the atoms. In many cases, small balls and sticks are used to make the three-dimensional shape easier to understand (Figure 4.3.9b).
- Space-filling diagrams have a sphere drawn around each atom to show their relative sizes (Figure 4.3.9c).

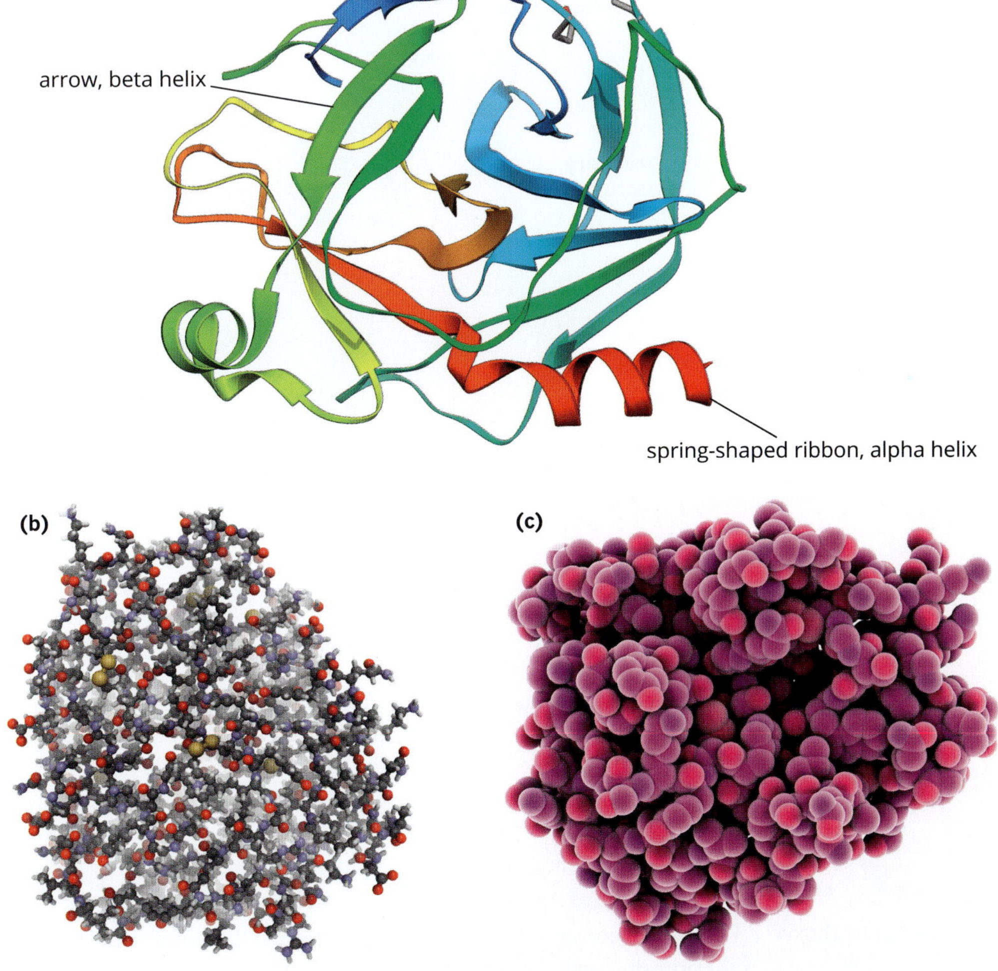

FIGURE 4.3.9 Three molecular models of the digestive enzyme trypsin, which is released by the pancreas and breaks down proteins into smaller amino acid chains: (a) ribbon diagram, (b) wireframe diagram filled with atomic spheres and (c) space-filling diagram.

Levels of protein structure

There are four different levels of organisation when describing protein structure (Figure 4.3.10):

- primary
- secondary
- tertiary
- quaternary.

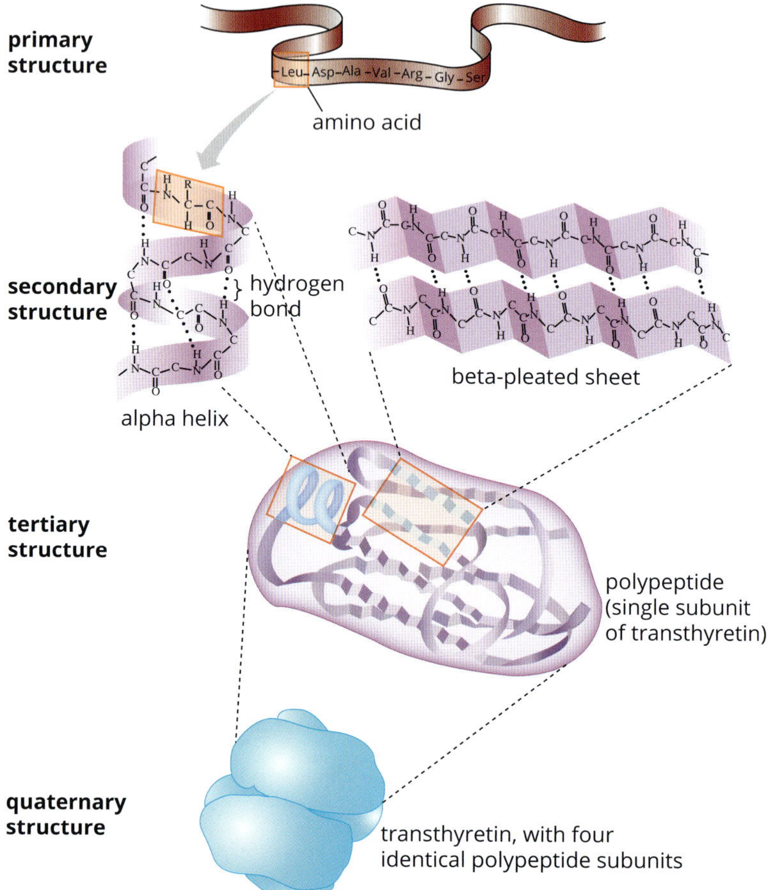

FIGURE 4.3.10 Overview of the four levels of protein structure: primary, secondary, tertiary and quaternary of the protein called transthyretin, which transports the thyroid hormone, thyroxine.

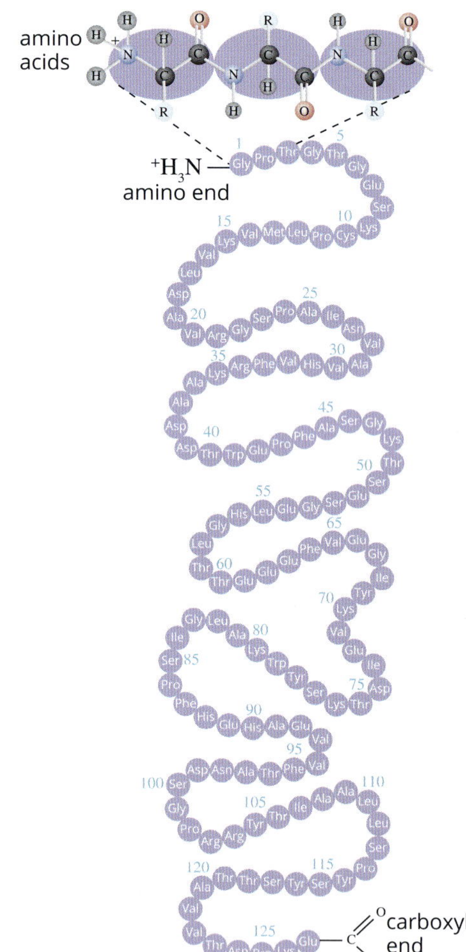

FIGURE 4.3.11 Primary structure of a protein showing the linear amino acid sequence of a polypeptide chain.

Primary structure

The linear sequence of amino acids in the polypeptide chain is referred to as the **primary structure** of a protein (Figure 4.3.11). It is unique to each protein. The primary structure of one polypeptide may consist of only 50 amino acids, whereas another may consist of 103 amino acids. Linear sequences shorter than 50 amino acids are known as peptides.

The linear sequence:

- provides information on how proteins will fold
- can be compared to identify differences between functional and non-functional proteins and determine the changes that render proteins non-functional
- can be compared between proteins to determine the evolutionary history of a protein.

Secondary structure

The next step in forming a functional protein is folding or coiling the polypeptide chain—its **secondary structure**. Folding or coiling occurs because of the formation of hydrogen bonds between the amine and carboxyl groups of amino acids within a polypeptide chain that have come in close proximity to each other. This results in the formation of secondary structures. There are three types of secondary structures.

Alpha helix—Hydrogen bonds form between adjacent amine and carboxyl groups within the polypeptide chain and the chain coils to form a helical shape (Figure 4.3.12a).

Beta-pleated sheets—Hydrogen bonds form between amine and carboxyl groups in different parts of adjacent polypeptide chains, causing the chains to fold back on each other (Figure 4.3.12b).

Random coil—Although parts of the polypeptide chain appear to have a random structure, the folding is not random and the same folding occurs in all molecules of the same protein. For example, all insulin molecules will have the same random coils within their structure (Figure 4.3.12c).

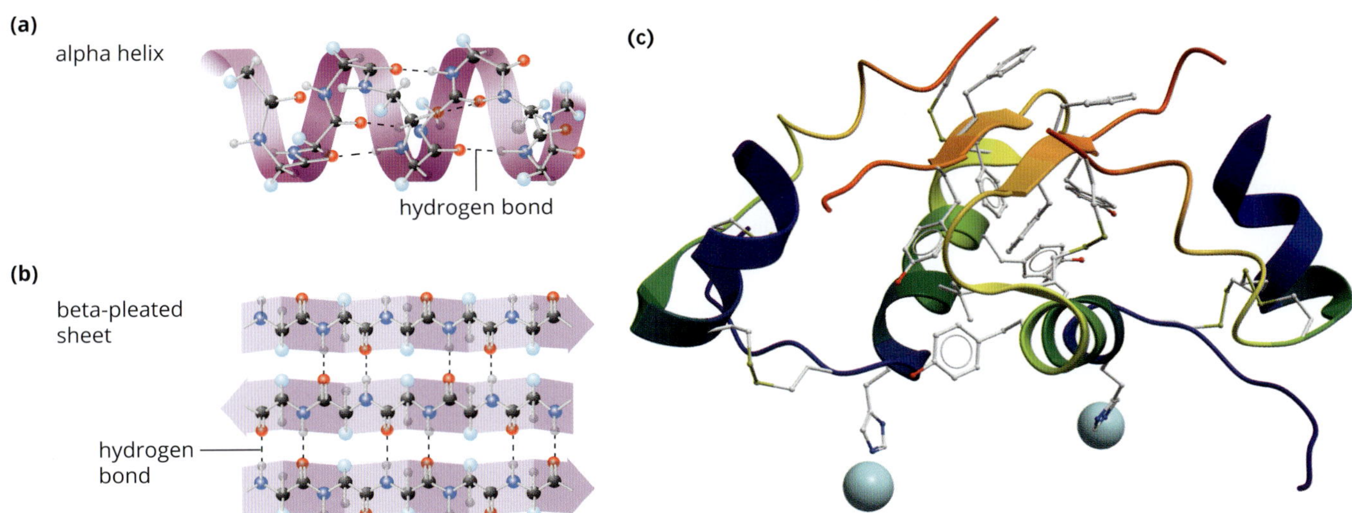

FIGURE 4.3.12 Regions stabilised by hydrogen bonds between atoms of the polypeptide chain result in the secondary structures (a) alpha helix and (b) beta-pleated sheets. (c) Ribbon model of human insulin protein showing apparently random coils that are consistent folding patterns for this protein.

Tertiary structure

Polypeptides also fold further, forming more compact and stable globular or fibrous three-dimensional shapes (Figure 4.3.13). This is known as the **tertiary structure**, and is usually the result of a combination of alpha helices and beta-pleated sheets along with other folded areas. The structure occurs because of different types of bonds, such as the disulfide bridge and the hydrogen bridge, between the R groups (side chains) of the amino acids (Figure 4.3.14).

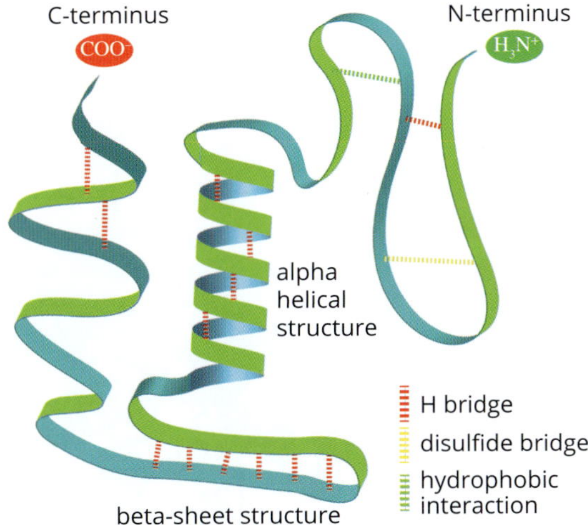

FIGURE 4.3.13 The tertiary structure of a protein is stabilised by the presence of different types of bonds.

The three-dimensional structure of a protein is critical to its function. In some smaller polypeptides, this folding process occurs spontaneously due to its chemical environment. However, larger, more complex proteins require specialised proteins (e.g. chaperonins) to help them fold correctly and, in some cases, to refold if they unravel and lose their natural shape (i.e. **denature**).

The tertiary structure is the final structure for some proteins.

> Van der Waals forces are weak individually but strong when present in large numbers in a biomacromolecule, and in combination with hydrogen bonds. They are caused by attractions between neutral molecules with temporary fluctuations of electron-rich regions on one molecule and electron-poor regions on another.

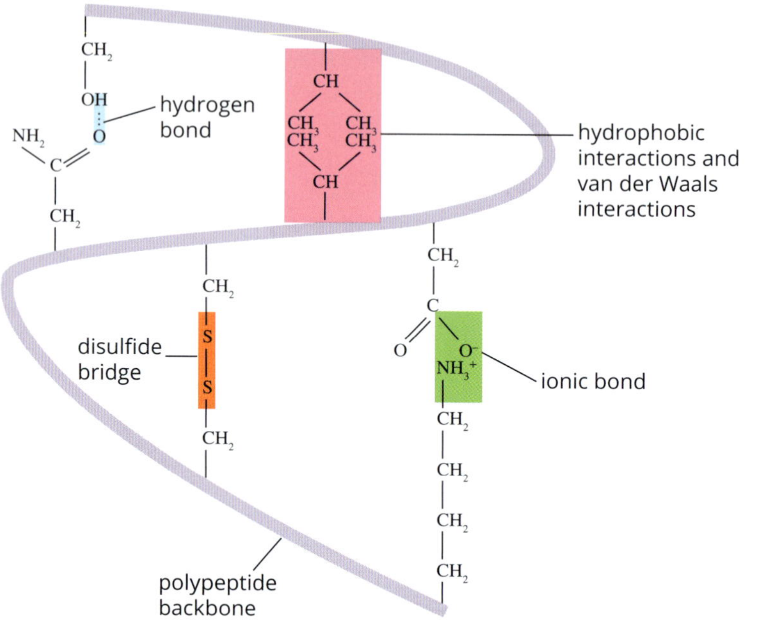

hydrogen bond
hydrophobic interactions and van der Waals interactions
disulfide bridge
ionic bond
polypeptide backbone

FIGURE 4.3.14 The different types of bonds between the R groups of the amino acids.

Quaternary structure

A **quaternary structure** is formed when two or more polypeptide chains or **prosthetic groups** (an inorganic compound that is involved in protein structure or function) join to create a single functional protein. The polypeptides may be identical or different. Some proteins will not become active until they achieve this structure. A protein with a prosthetic group is known as a **conjugated protein**. Haemoglobin is an example of a conjugated protein with a quaternary structure. It has two haem units containing iron (Fe) atoms that are essential to its oxygen transport function (Figure 4.3.15).

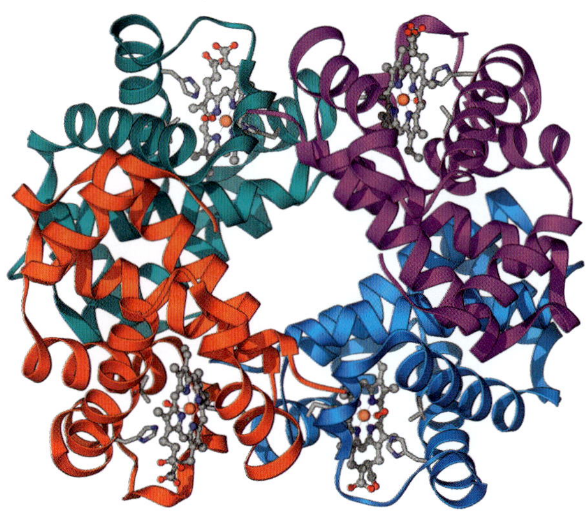

FIGURE 4.3.15 Quaternary structure of haemoglobin. Four polypeptides (two alpha (α) subunits of 141 amino acids and two beta (β) subunits of 146 amino acids) join with haem prosthetic groups to form the functional haemoglobin molecule. Haemoglobin is a conjugated protein.

BIOFILE CCT

Chaperonins

Crucial to the protein-folding process are **chaperonins** (or chaperone proteins). Chaperonins are protein molecules that assist in the proper folding of other proteins. Chaperonins do not specify the final structure of a polypeptide, but instead provide the polypeptides with an area to fold in without influences from the cytoplasmic environment (e.g. changes in pH) (Figure 4.3.16). Another function of chaperonins is to prevent newly synthesised polypeptide chains and assembled subunits from being denatured and becoming non-functional structures due to high temperatures.

It is now thought that some inherited diseases associated with the lack of function of a particular protein may be due to a fault in chaperonins rather than a mutation in the gene for the protein itself. The sequence of amino acids in the polypeptide may be correct, but the protein may not be correctly folded into its functional structure.

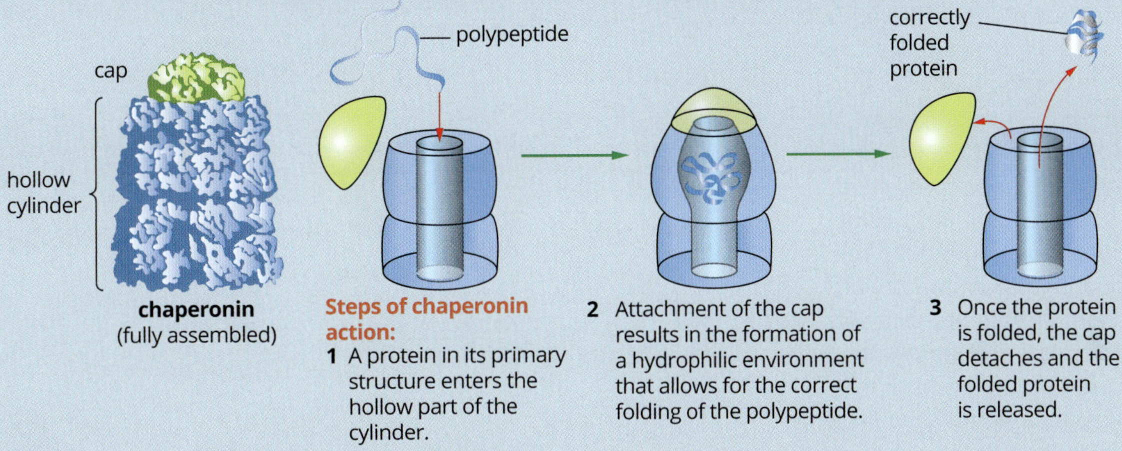

FIGURE 4.3.16 A computer graphic showing a large chaperonin protein complex from the bacteria, *E. coli*. It has an interior space that provides a shelter to allow the correct folding of newly made polypeptides.

PROTEIN CLASSIFICATION

Proteins can be classed as one of two types depending on their shapes.

Fibrous proteins are typically elongated and insoluble with little or no tertiary folding (Figure 4.3.17a). Many fibrous proteins have structural roles (e.g. collagen found in connective tissue and keratin found in hair and nails).

Globular proteins are compactly folded and coiled into spherical tertiary and quaternary structures (Figure 4.3.17b). They are generally soluble, having a core with hydrophobic properties and an outer hydrophilic region. Most enzymes and hormones are globular proteins.

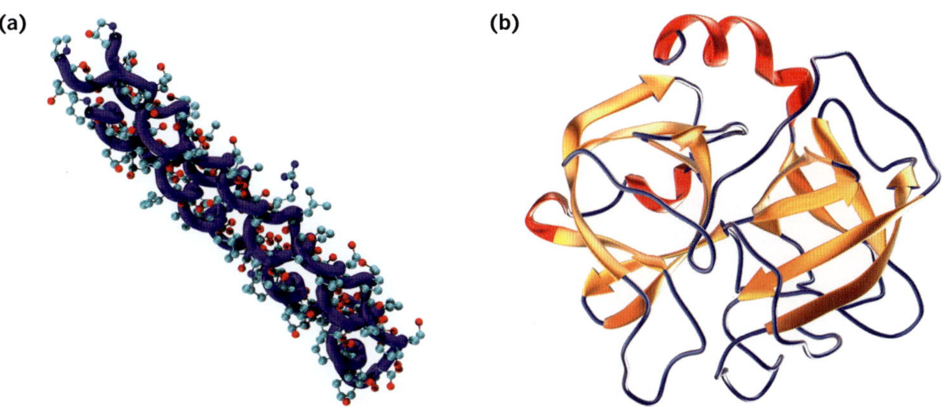

FIGURE 4.3.17 (a) The fibrous protein collagen, a major structural component of muscles, bones, blood vessels, tendons and skin is responsible for strength and elasticity throughout the human body. (b) The globular protein elastase is produced in the pancreas. It is an enzyme that catalyses the breakdown of elastin in human food, making it more digestible.

BIOLOGY IN ACTION ICT

Protein folding and the need for speed

The three-dimensional arrangement of proteins is vital to their biological functioning. Proteins display an extraordinary diversity in structural forms, with thousands of variations in size and folding configuration. Some examples are shown in Figure 4.3.18. It is not only the structure of the folds that is important for protein functioning, but also the speed at which they fold.

Researchers at the Heidelberg Institute for Theoretical Studies and the University of Illinois investigated the folding speed of all proteins known at that time, using computer analysis (bioinformatics). The researchers combined all known protein structures and genomes to obtain a dataset of 92 000 proteins and 989 genomes. By identifying protein sequences in the organisms' genomes that matched proteins for which the folding structure was known, the researchers were able to determine when different protein structures appeared in evolutionary history.

To determine the folding rates of different proteins, the researchers developed a mathematical model (algorithm) that used the known folding configurations of proteins. As proteins always fold at the same points, the speed and efficiency with which they fold is determined by how far apart these points are. The researchers found that, over the course of evolution from Archaea to complex multicellular organisms, protein-folding speed has increased. Amino acid chains that make up proteins have also become shorter over evolutionary time, contributing to the increased folding speeds. The researchers speculate that faster folding speeds of proteins may make them less susceptible to clumping and so increase their functional efficiency.

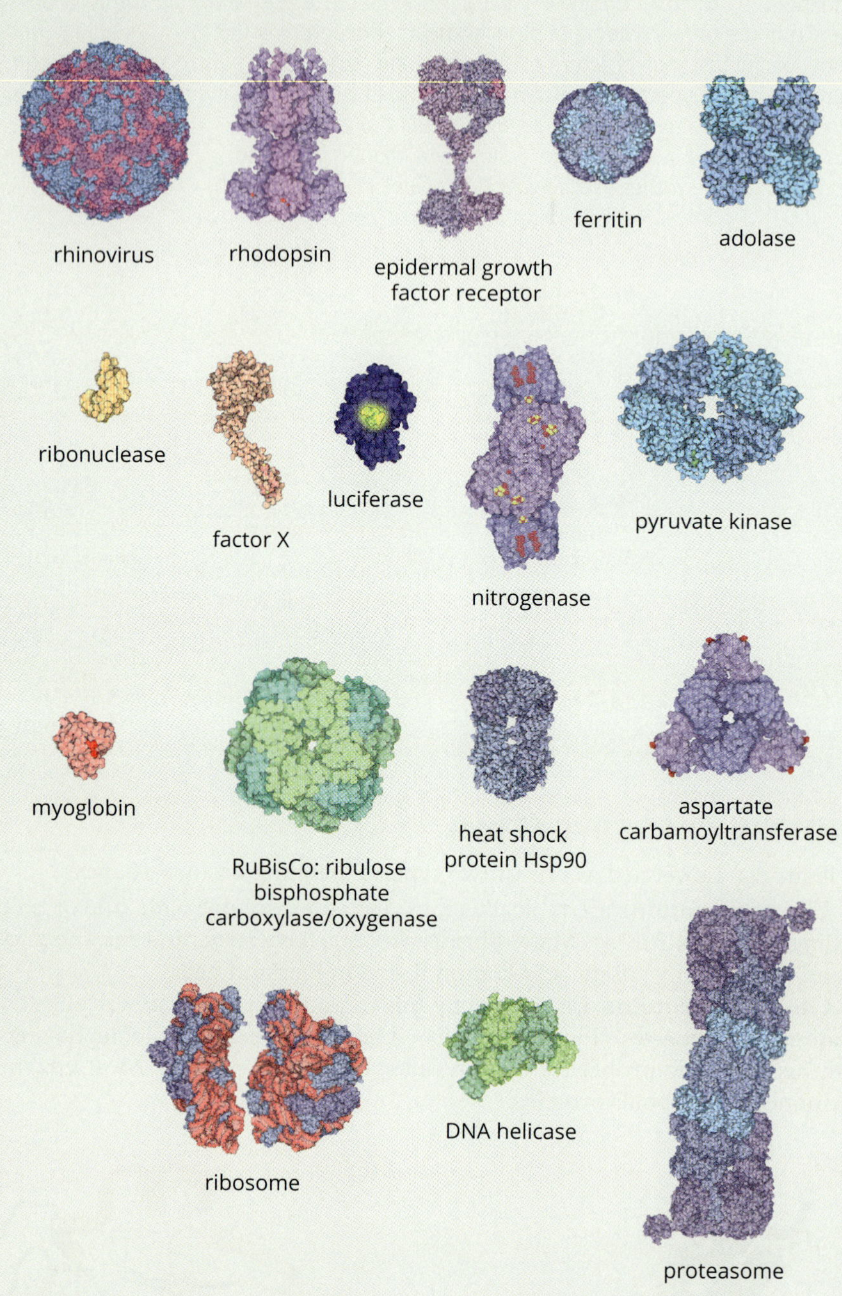

FIGURE 4.3.18 Examples of the diversity of structures and folding configurations of different proteins from the Worldwide Protein Data Bank

FUNCTIONAL DIVERSITY OF PROTEINS

There are many different types of proteins in organisms. Each protein has a different function, and each plays a vital role in the regulation, functioning and maintenance of both individual cells and entire organisms. In fact, almost every function of living organisms depends on proteins. The specific structure of each protein enables it to carry out its function. The main types of proteins can be grouped according to their function. Some of these functional types of proteins are described in Table 4.3.1.

TABLE 4.3.1 An overview of protein function

Function: enzymatic proteins	Function: hormonal proteins
Description: act as catalysts in cellular reactions (enzymes) Examples: catabolic enzymes, such as lipase and amylase, that catalyse the breakdown of bonds (also known as hydrolysis); anabolic enzymes, such as DNA polymerase, that catalyse the formation of bonds (also known as condensation)	Description: coordinate an organism's activities by triggering responses Examples: insulin, glucagon
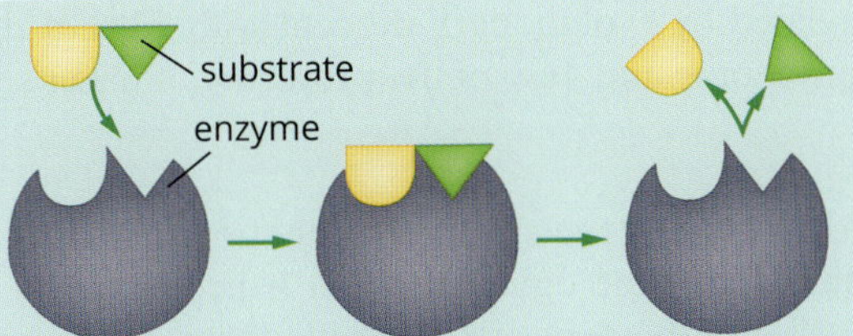	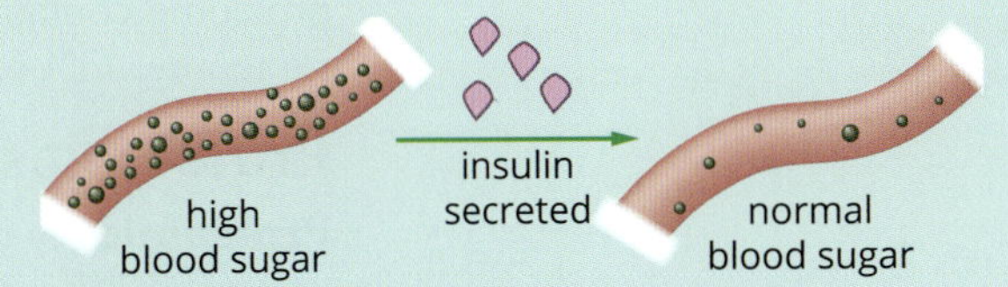
Function: immunological proteins (antibodies)	**Function: contractile and motor proteins**
Description: protect against disease by recognising foreign bodies and microbes; activate immune cells Examples: immunoglobulins (antibodies), complement, major histocompatibility complex proteins	Description: contractile proteins aid muscle contraction; motor proteins are responsible for the movement of cilia and flagella Examples: myosin, actin, kinesin, dynein, spindle fibres used in cell division
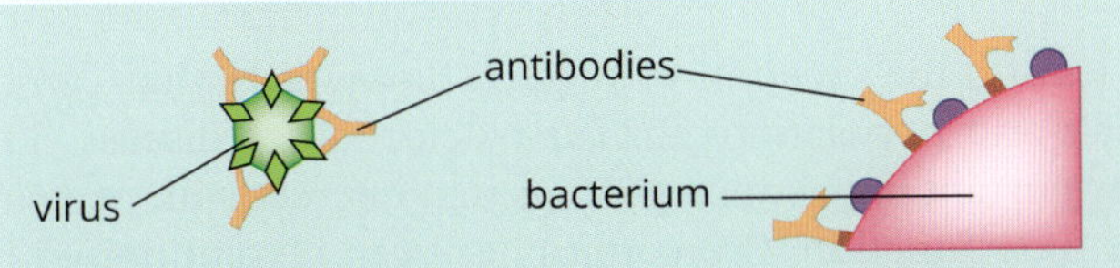	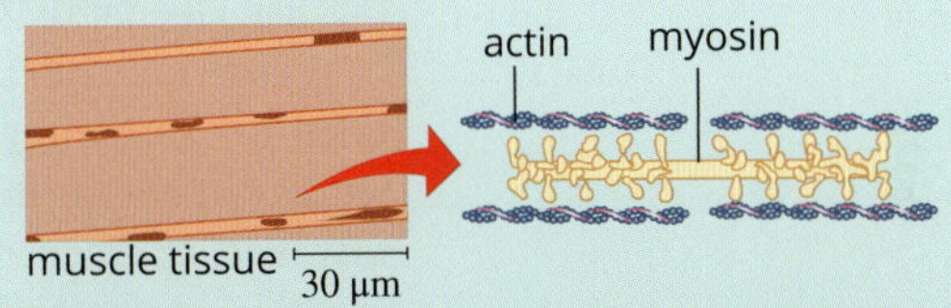
Function: structural proteins	**Function: transport proteins**
Description: provide support by forming the structural components of cells and organs; assist in contractile functions in tissue such as muscle Examples: collagen, keratin, actin, cytoskeleton, cell membranes	Description: transport of substances by acting as carrier molecules within or between cells; act as membrane channel proteins Examples: haemoglobin, sodium–potassium pump, calcium channel
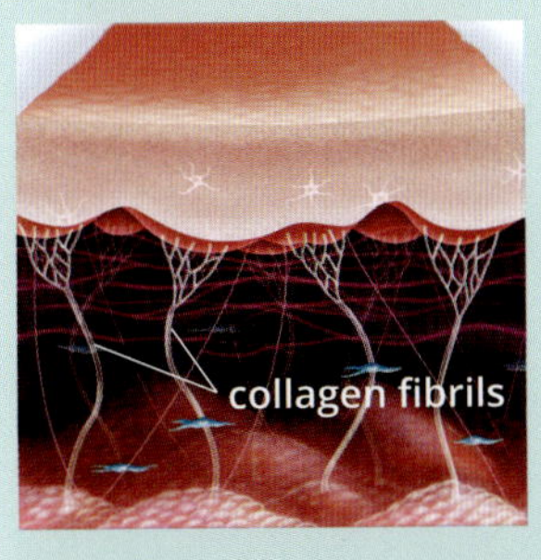	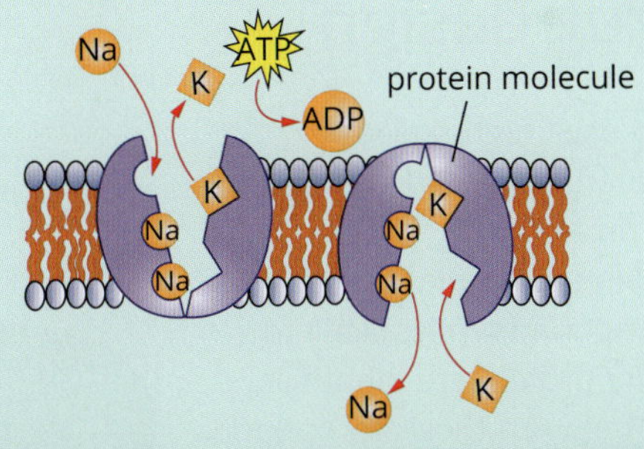
Function: receptor proteins	**Function: storage proteins**
Description: assist the cell in responding to chemical stimuli Examples: neurotransmitter receptors, hormone receptors	Description: storage of metal ions and amino acids Examples: ovalbumin and casein (to store amino acids), and ferritin (to store iron)
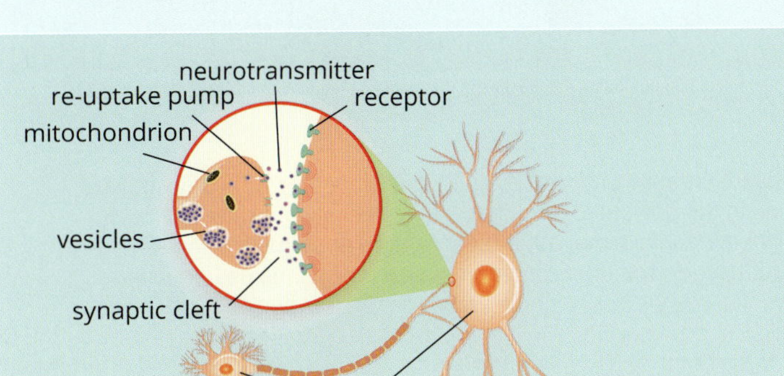	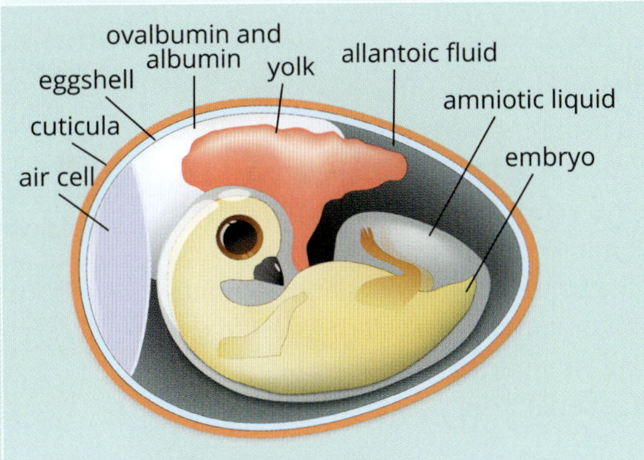

THE NATURE OF THE PROTEOME

The proteome is the entire set of proteins expressed by a genome, cell, tissue or organism at a given time.

The proteome varies between cell type, developmental stage and environmental conditions. Although a cell may contain the entire genome, only specific genes will be expressed, or 'switched on', at any given time. This ensures a cell produces only the proteins required for the specific functions it carries out.

Interestingly, there are many similarities between human and other proteomes, reflecting their common evolutionary origins and the universality of DNA. The human proteome contains proteins related by evolutionary descent with 61% of the fruit fly proteome, 43% of the worm proteome and 46% of the proteome of baker's yeast.

Proteomics

Proteomics is the large-scale study of the structure, function and interactions of proteins. Proteomics is a vital area of biology because it is proteins that carry out most of the activities of the cell, not the genes that encode them. By knowing when and where proteins are produced in an organism, as well as how proteins interact, we can better understand how cells and organisms function.

One of the ways to determine changes in the proteome is by comparing the proteomes of cells under different conditions. For example, by comparing the protein expression of a diseased cell and a healthy cell, the proteins affected by the disease can be determined.

Proteomic research can lead to the creation of protein biomarkers that can be used for screening individuals and populations for early detection of disease. The study of proteomics is also important in producing drugs that interact with proteins involved in disease and alter their function. Proteomics has been revolutionised by creation of the Protein Data Bank (PDB), a free international database that stores information about all known proteins and nucleic acids.

BIOLOGY IN ACTION  

Rational drug design

Rational drug design uses high-speed computers to compare the three-dimensional structure of a faulty protein with a database containing many different chemical compounds. The compounds most likely to interact with the faulty protein are identified, and the interactions between these compounds and the faulty protein can then be tested in the laboratory to design drugs.

The Australian CSIRO used rational drug design when developing the anti-viral drug Relenza® (Figure 4.3.19). Relenza works by targeting an enzyme that is found on the surface of all strains of the influenza virus. This enzyme helps the virus release new viral particles from infected host cells. Relenza attaches to the active site of the enzyme on the virus blocking the release of new viral particles, preventing the spread of the virus and allowing the body's natural immune system to destroy the remaining viral particles. You will learn more about rational drug design in Chapter 13.

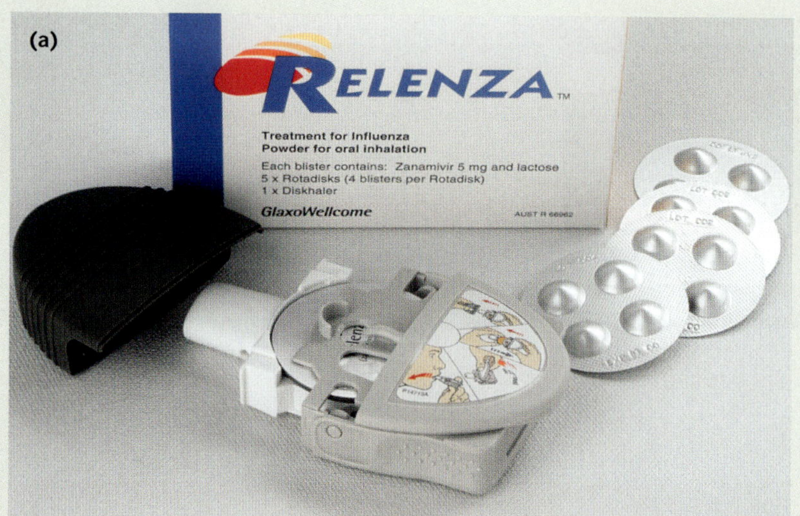

FIGURE 4.3.19 (a) Relenza, an anti-viral drug first developed by CSIRO Australia. Blister packs contain the powdered drug that is taken using an inhaler device. It cannot prevent influenza infections, but it reduces the symptoms and length of infection. Relenza contains the drug zanamivir, which inhibits the neuraminidase enzyme used by influenza viruses to break away from an infected cell and infect other healthy cells.

GO TO ➤ Section 13.3 page 468

FACTORS THAT AFFECT THE STRUCTURE AND FUNCTION OF A PROTEIN

The environment surrounding proteins plays an important role in maintaining its structure and function. Usually the protein's loss of function is due to denaturation of the protein. The factors in the environment that affect protein structure and function include:

- temperature
- pH
- concentration of ions or molecules that act as cofactors.

Denaturation and renaturation of proteins

A protein is said to have denatured when the hydrogen bonds, disulfide bridges, hydrophobic interactions and van der Waals forces that create the tertiary structure of the protein are broken and the shape of the protein is altered (Figure 4.3.20). As a result, the misshapen protein is biologically inactive and non-functional. If a protein becomes fully denatured, the reaction is non-reversible and the protein remains non-functional. In this case, a protease will usually destroy the defective molecule. However, a protein that is partially denatured may be able to fold again (i.e. **renature**) when the appropriate conditions are present.

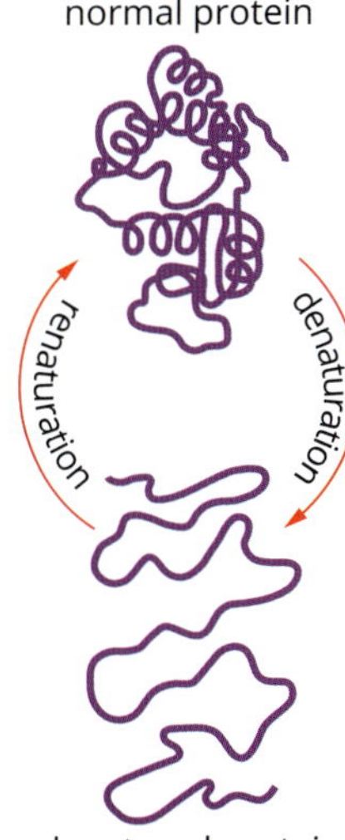

FIGURE 4.3.20 A denatured protein will lose its shape and hence its ability to function. Sometimes a protein can renature, when the chemical and physical aspects of its environment are restored to normal.

BIOLOGY IN ACTION CC CCT

How to unboil an egg

Professor Colin Raston from Flinders University, Australia, was honoured in 2015 with an Ig Nobel prize for creating a way to unboil an egg. The Ig Nobel is a humorous parody of the Nobel Prize that recognises unusual achievements in scientific research and is almost as well known as the actual Nobel Prize.

When an egg is boiled, the proteins unfold and refold into a more tangled, disordered form. The vortex fluidic device invented by Professor Colin Raston (Figure 4.3.21) is capable of unravelling proteins, a process required to make the white of a cooked egg runny again. The device only takes minutes to untangle denatured egg white protein molecules. A urea substance is used to liquefy the solid material then the vortex fluid device is employed to spin the material in a rapidly rotating tube. It applies shear stress within thin, microfluidic films to the proteins, forcing the tangled molecules apart and aligning them to allow folding back into their untangled, natural forms.

There is a serious side to this device. Other than being used to unboil an egg for demonstration purposes, the vortex fluid device has applications in the treatment of cancer, the manufacture of pharmaceuticals, the production of biofuels and in food processing. For example, drug companies often make cancer antibodies in expensive hamster ovary cells, because they rarely create misfolded proteins. If, instead, these companies could use the vortex fluid device and proteins from cheaper yeast or *E. coli* cells, it could make cancer treatments more affordable.

FIGURE 4.3.21 Professor Colin Raston and the vortex fluidic device that he invented. The vortex fluidic device unravels proteins and can be used to turn solid egg white to liquid. The device has potential applications in manufacturing products such as pharmaceuticals and biofuels.

The effect of temperature on protein function

Proteins can be denatured at high temperatures due to the breaking of bonds. For example, hydrogen bonds break at temperatures above 40°C. However, at temperatures below 35°C, the bonds are not flexible enough to allow the necessary conformational changes to break apart.

The optimal temperature for proteins varies with the organism and its environment. In humans, the optimum temperature for proteins is 37°C, but proteins found in organisms living in extreme environments, such as hot springs or icy environments, tend to have different optimal temperatures. Prokaryotes of the domain Archaea include cells that function at up to 120°C and in sub-zero temperatures where enzymes of other organisms would fail.

The effect of pH on protein function

A measure of the acidity level is pH, with the scale running from 1 to 14 (with 7 being neutral). Pure water has a pH of 7. Numbers less than pH 7 are acidic and those above are basic (alkaline). The scale is negative and logarithmic, based on powers of 10, so pH 5 is 10 times more acidic than pH 6.

Most proteins have a specific pH range in which their function is optimal, but this range can be quite different for each protein. The pH effect is unlike temperature, where the range of optimal temperatures is similar for most proteins within an organism. In humans the optimum pH for some enzymes of the digestive system are pH = 7 for amylase (starts digestion of starch in the mouth), pH = 2 for pepsin (digestion of protein food in the stomach), and pH = 8 for trypsin (digestive enzyme in small intestines) (Figure 4.3.22).

If the pH increases too far above or falls too far below the optimal pH, then the tertiary structure of a protein is affected. The interactions between the R groups of different amino acids are altered and the bonds between them are broken. As a result, the protein may be denatured, and in the case of enzymes the enzyme activity will decrease or halt.

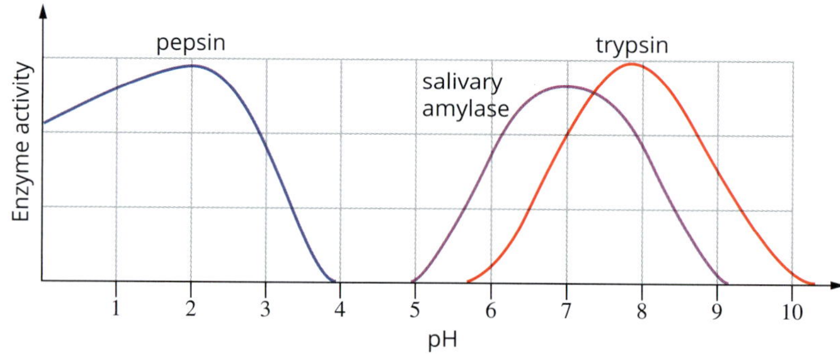

FIGURE 4.3.22 Graph showing the optimum pH for different digestive enzymes

The effect of cofactors on protein function

Some proteins require non-protein chemical compounds known as **cofactors** for their biological function. The presence and concentration of cofactors such as salts, specific elements such as iron, magnesium and calcium ions, or organic molecules such as vitamins can play a significant role in the folding and function of proteins.

Examples of proteins that require cofactors include:

- haemoglobin, which requires iron
- other respiratory proteins called haemocyanins use copper to attach and transport oxygen in a variety of arthropods and molluscs
- carboxypeptidase, an enzyme that requires zinc for splitting peptide bonds
- fibrillin, an important structural protein that uses calcium

- some molecular pump proteins in humans use copper and when these are defective there can be a toxic build-up of copper causing liver scarring or a shortage of copper that causes babies to have hair resembling steel wool (Menkes disease). The fleece of sheep grazing in copper-deficient areas show similar symptoms to humans with Menkes disease.
- the cytochrome group of proteins perform oxidation and reduction reactions in mitochondria of all living organisms, thus powering all the cellular processes that require energy. They require cofactors that include copper, iron and sulfur, without which the cell's metabolic rate reduces or stops (Figure 4.3.23).

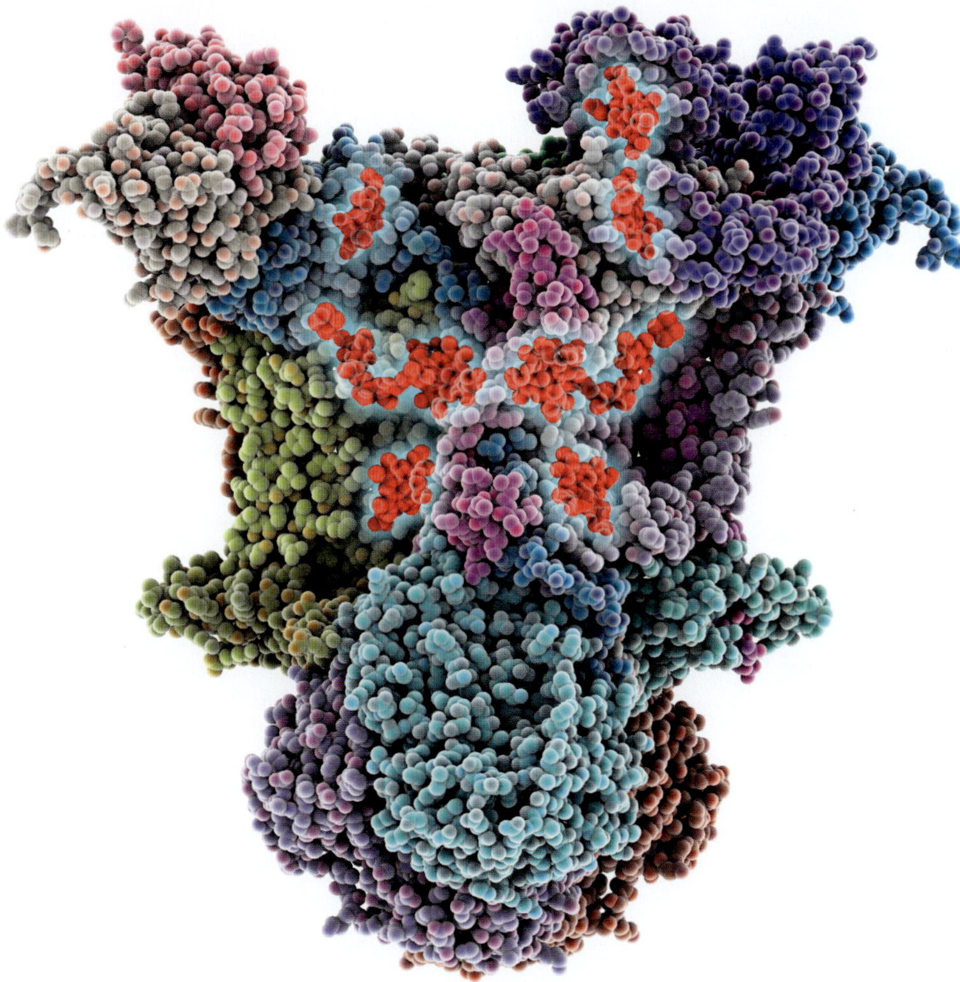

FIGURE 4.3.23 A model showing the structure of yeast cytochrome c. Inside the complex (in red) are cofactors: haems and iron-sulfur clusters. Cytochrome molecules perform a key step in the production of energy and are found in all living organisms.

4.3 Review

SUMMARY

- The proteome is the complete set of proteins expressed by a genome.
- Proteomics is the study of proteomes, including protein structure and function.
- Amino acids have an amine group, a carboxyl group and an R group. There are 20 standard amino acids. All have the same amine and carboxyl group, but differ in their R group.
- Amino acids join to form polypeptide chains in a condensation polymerisation reaction. This reaction removes water between two amino acids and forms a peptide bond between them.
- Proteins are made up of one or more polypeptide chains, which are folded and organised into specific shapes that relate to their specific function.
  - Primary structure of a protein is the linear sequence of amino acids in the polypeptide chain.
  - Secondary structure of a protein is achieved with the folding or coiling of the polypeptide chains due to hydrogen bonds. There are three forms: alpha helices, beta-pleated sheets and random coils.
  - Tertiary structure of a protein is achieved by further folding, which creates more stable shapes. This structure occurs because of bonds forming between the R groups of the amino acids.
  - Quaternary structure of a protein is achieved when two or more polypeptide chains join to create a single functional protein.
- Proteins can be either fibrous or globular:
  - Fibrous proteins are elongated and insoluble (e.g. keratin).
  - Globular proteins are spherical, compact and usually soluble for easy transport (e.g. most hormones and enzymes).
- Proteins have very diverse functions. The specific folding and final structure of proteins relate directly to their function.
- Functional types of proteins include:
  - enzymatic proteins (e.g. pepsin)
  - structural proteins (e.g. keratin)
  - transport proteins (e.g. haemoglobin)
  - hormonal proteins (e.g. insulin)
  - receptor proteins (e.g. hormone receptor on target cell)
  - immunological proteins (e.g. antibodies)
  - contractile and motor proteins (e.g. myosin)
  - storage proteins (e.g. ovalbumin).
- Factors within the environment can have an impact on the structure and function of a protein, and can also lead to denaturation. These factors include temperature, pH, concentration of ions and molecules that act as cofactors.
- Proteins have optimal temperature and pH ranges within which they function most effectively. Outside these ranges, they may become permanently denatured and would normally be dismantled by protease enzymes. Sometimes they can be returned to the correct configuration (i.e. renature).

KEY QUESTIONS

1. Proteins are key components of cells. Outline, with examples, at least five different roles carried out by proteins.
2. Distinguish between the proteome and proteomics.
3. Use a single sentence and a simple diagram to clarify what is meant by the following structures of a protein:
   a primary
   b secondary
   c tertiary
   d quaternary
4. Compare fibrous and globular proteins.
5. **a** Explain what is meant by a protein becoming 'denatured'.
   b Outline the factors that can cause a protein to become denatured.
6. **a** What are cofactors?
   b How does the presence of cofactors affect protein function?
7. Explain how chaperonins carry out their role in the folding process for proteins.
8. Identify two diseases or inherited conditions that are due to protein malfunctions in humans.
9. Demonstrate the relationship between amino acids, proteins and DNA.
10. Assess the value of a Protein Data Bank (PDB) to a scientist working in the field of proteomics.

Chapter review

04

KEY TERMS

5' cap (five prime cap)
alpha helix
amine group
amino acid
anticodon
beta-pleated sheet
biomacromolecule
biomolecule
carboxyl group
chaperonin (chaperone protein)
chromosome
coding strand
codon
cofactor
condensation polymerisation
conjugated protein
denature (n. denaturation)
dipeptide
DNA (deoxyribonucleic acid)
eukaryote
exon
fibrous protein
gene
gene expression
gene regulation
genetic code
genome
genotype (adj. genotypic)
globular protein
intron
messenger RNA (mRNA)
non-coding strand
nuclear membrane
nucleoid
nucleolus (pl. nucleoli)
nucleus
origin
peptide
peptide bond
phenotype (adj. phenotypic)
pili
plasmid
poly(A) tail
polypeptide
polypeptide synthesis
primary structure
prokaryote
promoter
prosthetic group
protein
proteome
proteomics
quaternary structure
R group
random coil
renature (n. renaturation)
ribosomal RNA (rRNA)
ribosome
RNA polymerase
RNA processing
secondary structure
splicing
start codon
stop codon
telomere
tertiary structure
transcription
transfer RNA (tRNA)
translation
triplet

REVIEW QUESTIONS

1 Select the statement that accurately describes eukaryotic cells.

A Eukaryotic cells have circular chromosomes and membrane-bound organelles, and some also have cell walls.

B Eukaryotic cells have linear chromosomes but not membrane-bound organelles, and some have cell walls.

C Eukaryotic cells have linear chromosomes and membrane-bound organelles, and some also have cell walls.

D Eukaryotic cells have linear chromosomes and membrane-bound organelles, but not cell walls.

2 Identify the sequences that are included in the final mRNA product of eukaryotic cells.

A introns **B** termons

C exons **D** spliced codons

3 In polypeptide synthesis, the function of the ribosome is to:

A synthesise the required amino acids

B ensure that the DNA base sequence is complete

C provide the energy needed for polypeptide synthesis

D provide the site for polypeptide synthesis

4 DNA provides the code for the synthesis of polypeptides. Which one of the following statements is true?

A Every codon codes for its own exclusive amino acid.

B The code is read as sets of three bases called triplets.

C Each triplet codes for at least two different amino acids.

D There are 20 different amino acids therefore there are 20 different codons.

5 Electron microscopy has greatly enhanced our understanding of cellular structure due to its ability to greatly magnify the internal structure of cells. Below are two cells observed under a TEM.

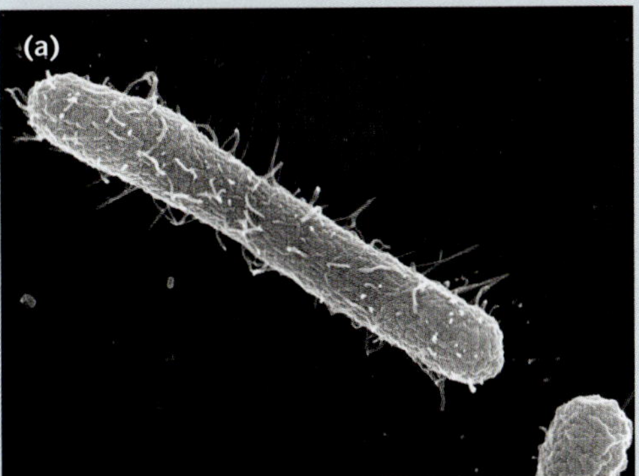

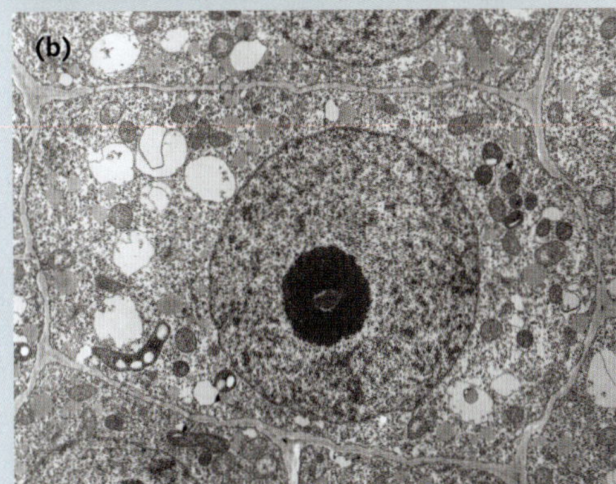

a One of the two cells is from a prokaryote. Explain which one.

b Is the eukaryotic cell from an animal or a plant? Give a reason for your choice.

6 A proteome is defined as:

A the sum of all of the functional proteins that an individual organism produces
B a primitive, simple form of protein
C the kinds of proteins produced by prokaryotic organisms
D the kinds of proteins produced by eukaryotic organisms

7 a Is the DNA packaging in this image typical of a prokaryotic or eukaryotic cell? Give at least one reason for your choice.

b Describe how DNA is packaged in the type of cell that was not your choice for (a).

c Identify the structures shown in red inside the DNA coils and state which class of biomolecules they belong to.

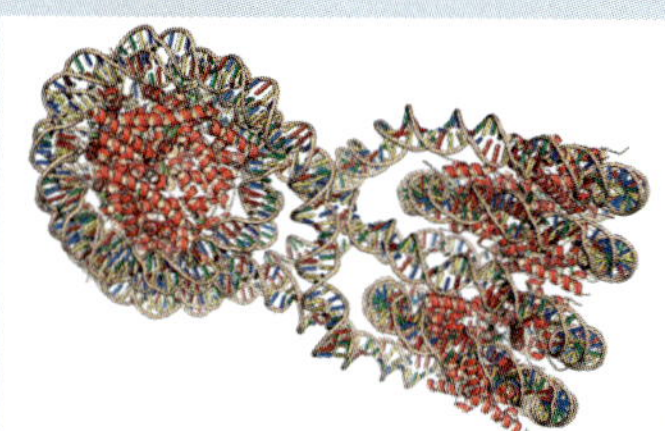

8 Match the terms with their correct functions.

Term	Function
mRNA	carries inherited genetic code
tRNA	carries DNA code to ribosomes
rRNA	catalyses transcription of RNA from DNA
DNA	forms part of ribosome structure
RNA polymerase	three base code on mRNA
triplet	carries amino acids to ribosomes
codon	three base code on tRNA
anti-codon	three base code on DNA

9 Fill in the blanks with the most suitable term from the word bank provided. You should only use a word once and there will be some words left.

> transcription, translation, replication, synthesis, complementary, nucleus, cytoplasm, peptide, dipeptide, polypeptide, amino acids, proteins

________ are important molecules for the functioning of an organism. They are usually composed of ________ chains folded into 3D configurations. In a eukaryotic cell, their ________ starts with a process called ________ in the ________ of the cell. A messenger molecule then moves into the ________ where it attaches to a ribosome. This is the start of the second stage called ________. Single ________ are carried to the ribosome and joined into a predetermined sequence by ________ bonds. The sequence is determined by matching ________ bases.

To answer Questions 10 to 12, refer to the genetic code in the table below. With a few rare exceptions, the genetic code is accepted as being universal.

First position (5' end)	Second position U	C	A	G	Third position (3' end)
U	Phe	Ser	Tyr	Cys	U
	Phe	Ser	Tyr	Cys	C
	Leu	Ser	STOP	STOP	A
	Leu	Ser	STOP	Trp	G
C	Leu	Pro	His	Arg	U
	Leu	Pro	His	Arg	C
	Leu	Pro	Gin	Arg	A
	Leu	Pro	Gin	Arg	G
A	Ile	Thr	Asn	Ser	U
	Ile	Thr	Asn	Ser	C
	Ile	Thr	Lys	Arg	A
	Met	Thr	Lys	Arg	G
G	Val	Ala	Asp	Gly	U
	Val	Ala	Asp	Gly	C
	Val	Ala	Glu	Gly	A
	Val	Ala	Glu	Gly	G

10 a The following flow chart represents the production of a polypeptide chain as directed by the DNA template.

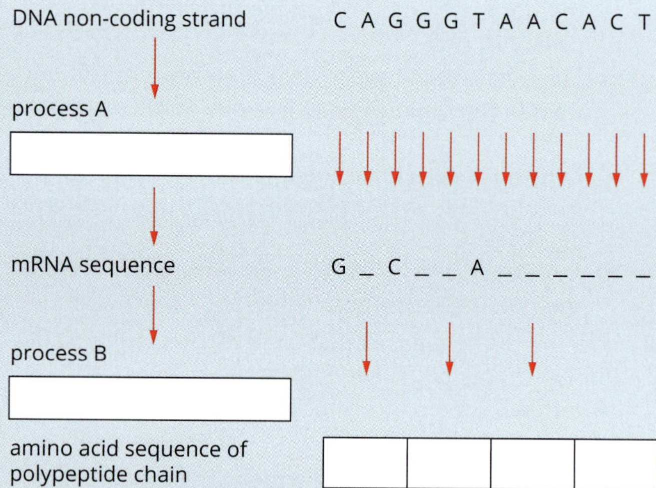

i Name processes A and B.
ii Complete the mRNA sequence.
iii Use the genetic code to determine the correct sequence of amino acids.

b i How is the fourth codon different from all the others in this sequence?
ii Outline the significance of this codon in terms of polypeptide production.

11 The DNA sequence of a particular eukaryotic gene is shown below.

TAC – GGA – TCT – AGA	– ATA – AAA –	CGG – AAT – GCT	– GGG –	ACA – CGG – GTA – ACA
exon 1		exon 2		exon 3

a What do the bright blue sections represent?

b Show the strand of mature mRNA that would be produced using this gene.

c Decode the mRNA.

d Identify the polypeptide that would be formed if exon two was skipped during splicing.

12 The DNA sequence of the promoter and first exon of a gene are shown below.

G G G C T C T A T A A A G G G T A C C A C T T C A A T G C T

a Identify the TATA box (the promoter region with a sequence of T and A bases).

b Explain where RNA polymerase will start transcription.

c The mRNA strand produced is shown below: A U G G U G A A G U U A C G A

i Explain whether the DNA strand shown is the coding strand or the non-coding strand.

ii Decode the strand.

d Predict the likely consequence for the organism of a mutation that changes the sequence of the TATA box.

13 a Use the following terms to label the structures A to I in the diagram:

amino acid, anticodon, tRNA loaded with an amino acid, mRNA, polypeptide, pre-mRNA (primary mRNA transcript), ribosome, RNA polymerase, unloaded tRNA

b Describe the role of structure B in protein synthesis.

c Describe the process occurring in structure H. Interpret the significance of structure I in your discussion.

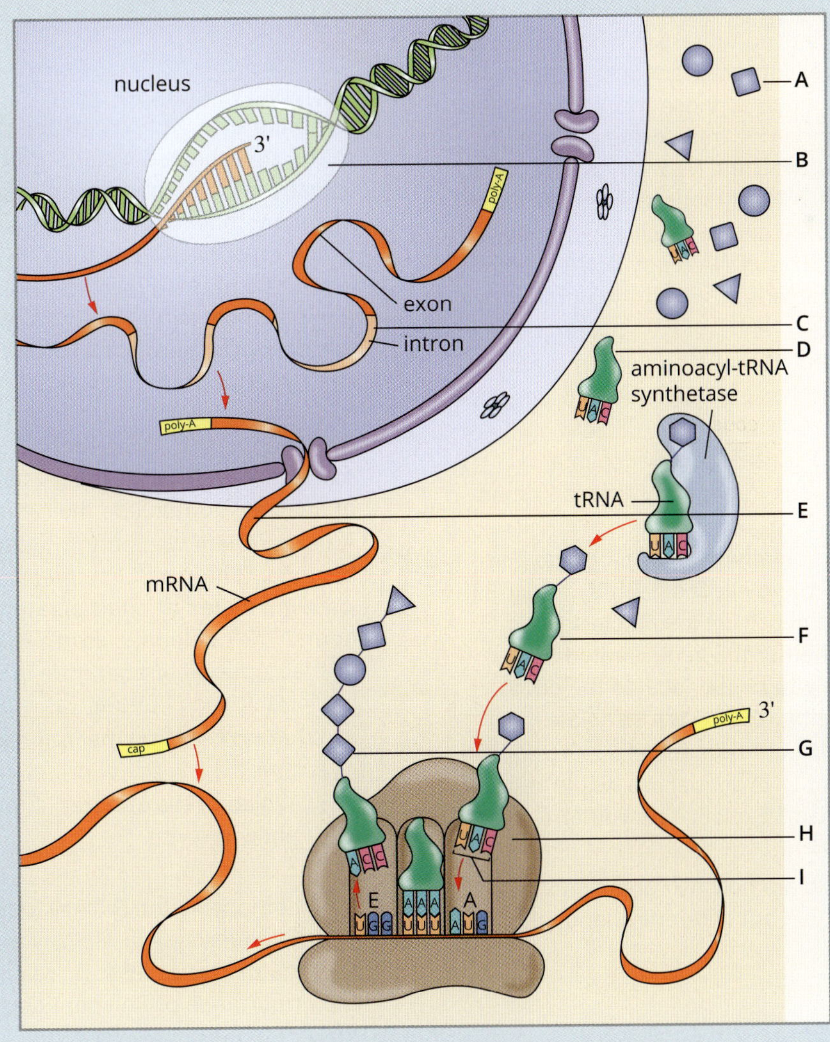

14 A strand of nucleic acid is shown below.
AUG AAU CCU UAU GGU GGC UUU UAA
The peptide produced because of the information encoded in the strand of nucleic acids is shown below.
Met–Asn–Pro–Phe

a Explain whether the strand given is DNA, pre-mRNA or mRNA.

b i During translation of the strand of nucleic acid shown, a tRNA having the anticodon UUA approached the ribosome. Which amino acid would the tRNA have been carrying?

ii Draw the tRNA molecule with its amino acid and anticodon.

15 The longest known gene is the dystrophin gene, which is 2.5 megabases long and is 99% introns.

a What is the maximum number of bases in the exons of the dystrophin gene?

b How many amino acids (approximately) make up the protein dystrophin?

c Even though the dystrophin gene is the longest, another muscle protein called titin has the longest sequence of amino acids. Explain how this comes about.

16 Genetic engineering is used to transform bacteria by inserting human genes into their genome to produce human polypeptides, such as those that form insulin. Before the bacterium can be transformed, a copy of the human gene is required. A common method of acquiring the gene is to extract the appropriate mRNA from human cells and to use it as a template to make a DNA copy. This cDNA (copy DNA) is then introduced into the bacterium, which then produces the required protein.

a Why can it be expected that a bacterium is able to decode a human gene and produce the correct protein?

b A gene made of cDNA is better for use in a bacterium than a gene cut directly from a human chromosome. Why?

c Another method of obtaining an appropriate gene is to analyse the protein needed, identify the amino acid sequence and construct a suitable section of DNA. Explain whether the gene created by this method is likely to be identical to the cDNA sequence made using mRNA as a template.

d Consider the amino acid sequence Leu–Pro–Val. How many different DNA sequences would result in this amino acid chain? Explain how you arrived at your answer.

17 Use an example to explain how two organisms can have the same genotype but different phenotypes.

18 In recent genetics research, scientists replaced the gene controlling eye development in *Drosophila* flies with the gene that controls eye development in mice. The transgenic *Drosophila* developed normal compound fly eyes. What does this observation suggest about:

a the gene controlling eye development in *Drosophila* and mice?

b the factors that control eye development in these two vastly different species of insect and mammal?

19 Consider the pictures of different levels of protein structure. Identify if each is primary, secondary, tertiary or quaternary and account for your choices.

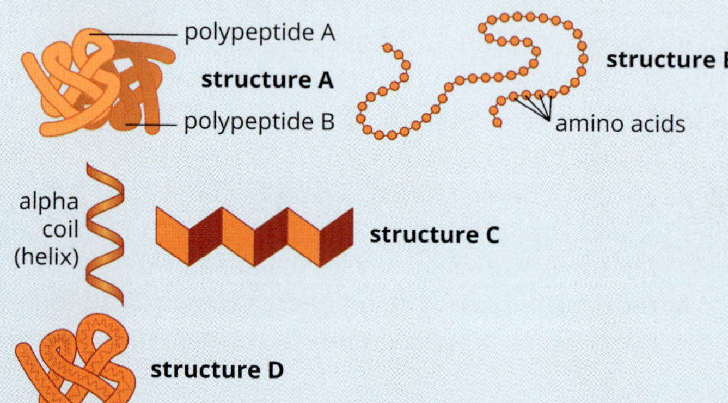

20 Bacteria live in a vast range of different environments. Conditions range from the ice sheets of Antarctica to the superheated water surrounding the undersea volcanoes of the mid-ocean ridges or the hot springs of Yellowstone. Bacteria living in the Antarctic ice are called cryophiles while those living in water at close to boiling point are hyperthermophiles. An example of a cryophilic bacterium is *Psychrobacter*, which thrives at temperatures between −10°C and 42°C. *Methanopyrus* is a hyperthermophilic bacterium. It has been shown to survive and reproduce at temperatures between 84°C and 110°C. Despite their extreme lifestyles these bacteria, like all living things, use proteins in the form of enzymes to regulate their metabolism.
An experiment was performed using both groups of bacteria. Cultures of *Psychrobacter* and *Methanopyrus* were incubated at a temperature of 60°C for three hours. The bacterial cultures were then returned to their optimal temperature and the growth of the bacteria in each culture was monitored.

a What is meant by the optimal temperature for a protein?

b i Which of the cultures, if any, would you expect to show growth?

ii Explain your reasoning.

21 After completing the Biology Inquiry on page 181, reflect on the inquiry question: Why is polypeptide synthesis important? Use at least three examples to account for the importance of polypeptide synthesis in cells.

CHAPTER

05 Genetic variation

Sexual reproduction results in offspring with a set of unique characteristics that are inherited from their parents. These characteristics vary among individual organisms. In this chapter you will learn how this genetic variation is generated and inherited. You will analyse patterns of inheritance, learn about the differences between independent and linked genes, interpret pedigree charts and Punnett squares, and predict the outcomes of genetic crosses. You will also learn how genetic variation between individuals can be used to identify trends, patterns and relationships in populations.

Content

INQUIRY QUESTION

How can the genetic similarities and differences within and between species be compared?

By the end of this chapter you will be able to:

- conduct practical investigations to predict variations in the genotype of offspring by modelling meiosis, including the crossing over of homologous chromosomes, fertilisation and mutations (ACSBL084)
- model the formation of new combinations of genotypes produced during meiosis, including but not limited to:
  - interpreting examples of autosomal, sex-linkage, co-dominance, incomplete dominance and multiple alleles (ACSBL085) CCT
  - constructing and interpreting information and data from pedigrees and Punnett squares
- collect, record and present data to represent frequencies of characteristics in a population, to identify trends, patterns, relationships and limitations in data, for example: ICT N
  - examining frequency data
  - analysing single nucleotide polymorphism (SNP)

5.1 Formation of genetic variation

> The genotype is the set of alleles present in the DNA of an individual organism. The genotype is the result of inheritance.

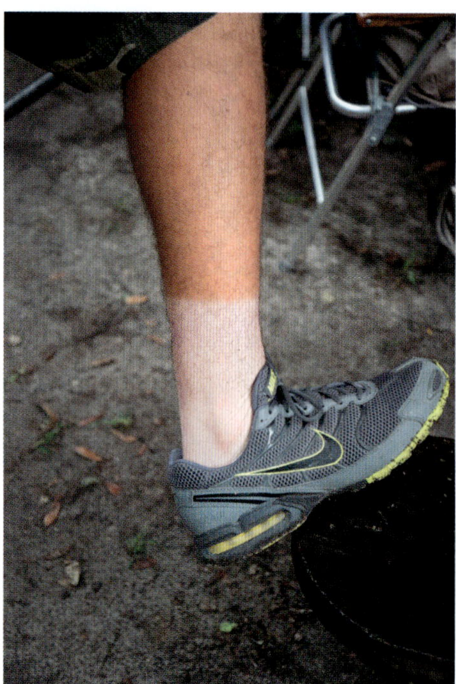

FIGURE 5.1.1 The effect of sun exposure on skin colour. The darker part of the skin has been exposed and has produced more melanin, causing it to darken. Unexposed skin (the ankle) does not produce extra melanin.

GO TO ➤ Section 3.1 page 116

> An organism's phenotype is all of its observable characteristics. It is the result of inheritance and the effects of the organism's environment.

> An organism that has two copies of the same allele of a gene is said to be homozygous for that gene; 'homo' means 'the same'. An organism that carries two different alleles of a gene is said to be heterozygous for that gene; 'hetero' means 'different'.

You inherited many of your physical features or **traits** from your parents. You share certain traits with your mother and others with your father. You might even appear to have a totally different version of a trait.

In this section you will look at what determines the traits and characteristics that humans have and how they are passed to children from parents. You will learn the difference between **genotype** and **phenotype**, the use of symbols for **alleles**, and the distinction between **dominant phenotypes** and **recessive phenotypes**. You will also explore how polygenic inheritance contributes to continuous variation in a **population**, through examples such as human skin colour and variation in height.

GENOTYPES AND PHENOTYPES

A genotype is the set of alleles present in the **DNA (deoxyribonucleic acid)** of an individual organism. It is the result of inheritance. An allele is an alternative form of a **gene**. Each individual usually only has two alleles for each trait: one inherited from their mother and one inherited from their father. But one gene may have many alleles, and this is what leads to variation in a population.

An organism's phenotype is all of its observable characteristics. It is the result of inheritance and the effects of the organism's environment. An example of a phenotype is skin colour. Your skin colour depends on how much skin pigment (melanin) is produced. But skin colour also depends on environmental factors such as exposure to sunlight, especially in pale-skinned people. The greater the exposure, the more melanin is produced (Figure 5.1.1).

Genotype

You will recall from Chapter 3 that a diploid cell has two sets of chromosomes (one set from each parent). Humans are described as **diploid** because we have two copies of every chromosome in all cells except gametes (which only have one copy and are therefore **haploid**). Chromosomes therefore exist in pairs. When both members of a pair of chromosomes are identical, that is, they carry the same genes, they are referred to as **homologous chromosomes**. In the human **genome**, only the *XY* pairing of **sex chromosomes** is non-homologous. All other chromosome pairs are homologous.

Consider a gene, which might be called gene *A*, that has two alleles. One allele can be represented by an upper case *A*, and the other by a lower case *a*. The names of genes and alleles are always italicised.

If gene *A* is in a homologous chromosome, there will be two copies of the gene. If you inherited the allele *A* from both parents, your genotype for gene *A* will be *AA*. On the other hand, if you inherited the *A* allele from one parent and the *a* allele from the other parent, you will have the genotype *Aa*. If you inherited the allele *a* from both parents, you will have the genotype *aa*.

Therefore, there are three different combinations or genotypes of gene *A*: *AA*, *Aa* or *aa* (Figure 5.1.2). Genotypes *AA* and *aa* contain only one type of allele, so the individual is said to be **homozygous** for that gene and is called a homozygote. Genotype *Aa* contains two different alleles (i.e. one copy of each allele), so the individual is said to be **heterozygous** for that gene and is called a heterozygote.

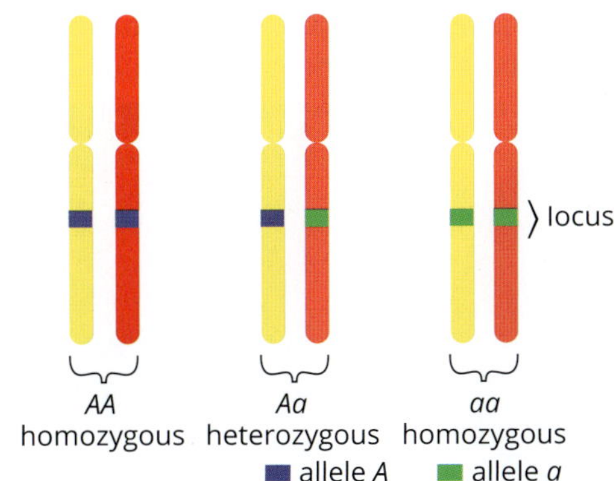

FIGURE 5.1.2 On homologous chromosomes, alleles of a gene occur at the same locus. If a gene has two alleles, there can be three different combinations or genotypes. Two of these combinations are homozygous, and one is heterozygous.

A single gene **locus** (position on a chromosome) may have more than two alleles. The human ABO blood group system is based on such alleles (Figure 5.1.3). In this case there are three alleles, represented as I^A, I^B and i. Each person carries two copies of these three possible alleles. There are therefore six possible genotypes and four phenotypes, as shown in Figure 5.1.3. Blood types are discussed further as **co-dominant phenotypes** on page 215.

A locus is the position of a gene or allele on a chromosome. The plural term for locus is loci.

ABO blood groups				
Blood type	**Type A**	**Type B**	**Type AB**	**Type O**
Possible allele combinations	$I^A I^A$ $I^A i$	$I^B I^B$ $I^B i$	$I^A I^B$	$i\ i$
Antigen (on RBC)	A antigen 	B antigen 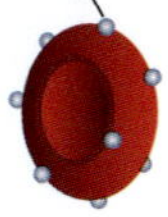	A and B antigens 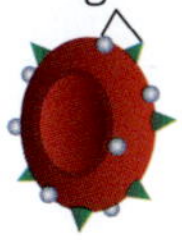	no antigens 

FIGURE 5.1.3 The ABO blood group system is based on three alleles. The production of antigens A and B depends on the combination of alleles present.

Phenotype

When studying inheritance it is important to know the genotypes of parents and offspring. However, it is equally important to know the characteristics that can result from a given genotype, known as the phenotype. The phenotype includes any distinct property of an organism: physical, chemical, physiological or behavioural.

GO TO ➤ Section 4.2 page 165

Importantly, environmental conditions can affect the phenotype of certain genotypes. In these cases, the genotype determines the possible range of phenotypes for a particular characteristic, and the environment influences where in that range the actual phenotype will be.

For example, Arctic foxes (*Vulpes lagopus*) have two genotypes that affect fur colour: 'white morph' and 'blue morph'. The fur of the blue morph remains dark blue-grey throughout the year, but the fur of the white morph varies from dark brown or grey to pure white. In summer the fur is dark, but as winter approaches the fur gradually changes to white in response to the increasing cold and shorter day length (Figure 5.1.4). At the end of winter the fur gradually returns to its summer colour. You will learn more about environmental influences on phenotypes on page 207.

Although the phenotype is relevant to the functioning of an individual organism, the genotype is what is passed on to the next generation.

FIGURE 5.1.4 The fur of the white morph genotype of Arctic fox (*Vulpes lagopus*) changes from dark brown or grey (a) to pure white (b) as winter approaches, in response to the increasing cold and shorter day length.

BIOFILE CCT L

Naming genes

There are internationally accepted names for genes and their abbreviated forms. The gene that codes for phenylalanine hydroxylase, an enzyme involved in the inherited disorder phenylketonuria, is abbreviated to *PAH*. The gene name is always italicised, to distinguish them from the proteins they encode. To provide another example, *BRCA1* is an enzyme expressed in the cells of breast and other tissue, where it helps repair damaged DNA or destroy cells if DNA cannot be repaired. The gene that codes for this enzyme is known as *BRCA1*.

FIGURE 5.1.5 A researcher sorting fruit flies (*Drosophila melanogaster*) in a breeding experiment

However, it is worth noting that some very odd gene names have been proposed, particularly for model organisms like mice and the common fruit fly (*Drosophila melanogaster*) that are closely studied on a genetic level. The fruit fly is an excellent organism for genetic study due to its very short generation time, so changes can be observed very quickly when researchers breed flies with particular traits (Figure 5.1.5). Because of the rapid rate of gene discovery for this organism, researchers focus more on memorable names rather than formality, which leads to some controversial names such as *INDY* (for 'I'm Not Dead Yet', referring to a gene that seemed to make flies live longer when a certain mutation occurred), *LUSH* (mutations in this gene resulted in flies being unusually attracted to alcohols such as ethanol, propanol and butanol), and *tinman* (mutations in this gene cause the fly embryo to fail to develop a heart). While these gene names are not so strictly formal as genes found in human studies, they are nevertheless maintained and used in future research.

DOMINANT AND RECESSIVE PHENOTYPES

The relationship between genotype and phenotype gives an insight into an important property of phenotypes known as dominance. For a given gene, you can determine whether a phenotype is completely dominant, co-dominant or recessive by examining the phenotype of the heterozygote compared with that of the homozygote. Dominant phenotypes are expressed if the individual carries at least one allele for the dominant trait. Recessive phenotypes are expressed only if the individual carries two alleles for the recessive trait. It is important to understand that dominance (and recessiveness) are properties of alleles, not genes. They are expressed as dominant or recessive phenotypes. Genes themselves are neither dominant nor recessive.

A dominant phenotype is one that is observed in the heterozygote and homozygote conditions.

A recessive phenotype is one that is observed only in the homozygous condition.

BIOLOGY IN ACTION CCT

Gregor Mendel, the founder of genetics

The birth of modern genetics began in an abbey garden, where a monk named Gregor Mendel proposed a mechanism for inheritance (Figure 5.1.6). Mendel developed his theory of inheritance before chromosomes were observed under the microscope and the significance of their behaviour was understood.

Mendel began studying inheritance by breeding garden peas. This was an excellent choice for a number of reasons. Firstly, garden peas have a range of easily observable phenotypes. Secondly, they are self-fertile but can also be outcrossed (that is, crosses could be carried out within and between pure-breeding lines). Finally, they had classic features of all genetic model organisms: large numbers of offspring can be counted for each cross, the generation time is very short, and the peas are easy to maintain.

Figure 5.1.6b shows some of the discrete traits that Mendel observed in his breeding experiments with peas: flower colour, pod colour, pod shape, seed colour and seed shape. Each of these traits was important in the choice of peas used for crossings. It was likely that Mendel devised his model of inheritance theoretically and used the data from peas to confirm the model.

Mendel presented his results to the Brunn Natural Science Society in 1865 and they were published by the society in 1866. But the results were ignored until three other scientists independently produced similar data in 1900. Only then was Mendel's scientific contribution recognised—16 years after his death.

FIGURE 5.1.6 (a) Gregor Mendel, the founder of genetics. (b) Some of the discrete traits that Mendel observed in his breeding experiments with peas are: flower colour, pod colour, pod shape, seed colour and seed shape.

Complete dominance

To understand **complete dominance**, it is useful to consider the white eye gene in blowflies. There are two alleles for the white eye gene, *W* and *w*. Individuals with genotype *WW* have red eyes, while individuals with genotype *ww* have white eyes (Figure 5.1.7). Individuals with genotype *Ww* do not show an in-between trait such as pink eyes but instead have red eyes, making them indistinguishable from those of the *WW* genotype.

Blowflies with genotype *Ww* display the red eye because the *W* allele makes enough membrane transporter protein to give the eye normal red pigment levels. The red eye colour phenotype is referred to as the dominant phenotype, because it only needs one *W* allele for that phenotype to be displayed. The white phenotype is referred to as the recessive phenotype because it is not observed in the heterozygote. It needs two copies of the *w* allele for it to be observed in the phenotype. This example shows that scientists can determine which phenotype is dominant by examining the heterozygote (assuming that a dominant phenotype exists, which is not always true).

By convention, the allele associated with the dominant phenotype is represented by an upper-case symbol (e.g. *W*). The allele associated with the recessive phenotype is represented by a lower-case symbol (e.g. *w*). The blowfly's white eye is an example of complete dominance. *Ww* individuals have the same eye colour as *WW* flies.

FIGURE 5.1.7 Blowflies with the *WW* and *ww* genotypes have a red-eye phenotype (top). Blowflies with the *ww* genotype have a white-eye phenotype (bottom).

> Dominance and recessiveness are properties of alleles, not genes. They are expressed as either dominant or recessive phenotypes.

> Continuous variation refers to phenotypes that occur on a continuous scale and cannot easily be placed into categories. Height and skin colour in humans are both examples of continuous variation. These tend to be polygenic traits. By contrast, **discontinuous variation** refers to a set of discrete phenotypes with no intermediate states. The ABO blood system is one example: you can have types A, B, AB or O, but there are no in-between states for blood types. Monogenic traits tend to be discontinuous (Figure 5.1.8b).

> Polygenic inheritance is the inheritance of an observable trait that is controlled by many genes.

POLYGENIC INHERITANCE

So far, we have been discussing phenotypes that are **monogenic**; that is, they are controlled by a single gene. For some traits, such as skin colour, eye colour and height in humans, more than one gene contributes to the phenotype of an individual. This is known as **polygenic inheritance** and results in a much greater range of phenotypes. Polygenic traits in non-human animals include wing shape and bristle count in *Drosophila*; birth weight, temperament and milking ability in cattle; and plumage and beak size in birds. When the phenotypes of a polygenic trait are shown on a graph, the result is often a bell-shaped curve (typical of **continuous variation**), which is referred to as a normal distribution.

Height in humans

It has been found that height in humans is controlled by about 50 genes or regions of the genome (Figure 5.1.8a). Some of these individual genes control secretion of thyroid gland hormones and human growth hormones. A deficiency in the production of these hormones during childhood and puberty can result in stunted growth. On the other hand, too much production can cause excessive growth resulting in exceptional height. The greater the number of genes that control a characteristic, the more possible gene combinations exist and as a result, the wider the range of possible phenotypes.

Human skin colour

The colour of human skin is determined by the amount of melanin (dark pigment) it contains, and there are three types of melanin: eumelanin, pheomelanin and neuromelanin. At least four genes are involved in melanin production. For each gene, one allele codes for melanin production while the other does not. For example, the melanocortin 1 receptor gene (*MC1R*) is primarily responsible for determining whether pheomelanin and eumelanin is produced in the human body. The agouti signalling peptide gene (*ASIP*) inhibits eumelanin production. The tyrosinase gene (*TYR*) is responsible for making an enzyme called tyrosinase. Tyrosinase converts tyrosine to dopaquinone. A series of additional chemical reactions convert dopaquinone to melanin in the skin. The presence or absence of various alleles of these genes affects the level of melanin production and hence dark skin colour.

Other genetic factors affect skin colour, including the formation of melanocytes (melanin producing cells). The KIT ligand gene (*KITLG*) is involved in the permanent survival, proliferation and migration of melanocytes to the skin surface.

The combination of melanin production (affected by *MC1R*, *ASIP* and *TYR*) and formation of melanocytes (affected by *KITLG*) together determines the degree of pigmentation. This results in a wide variation of observable skin tones.

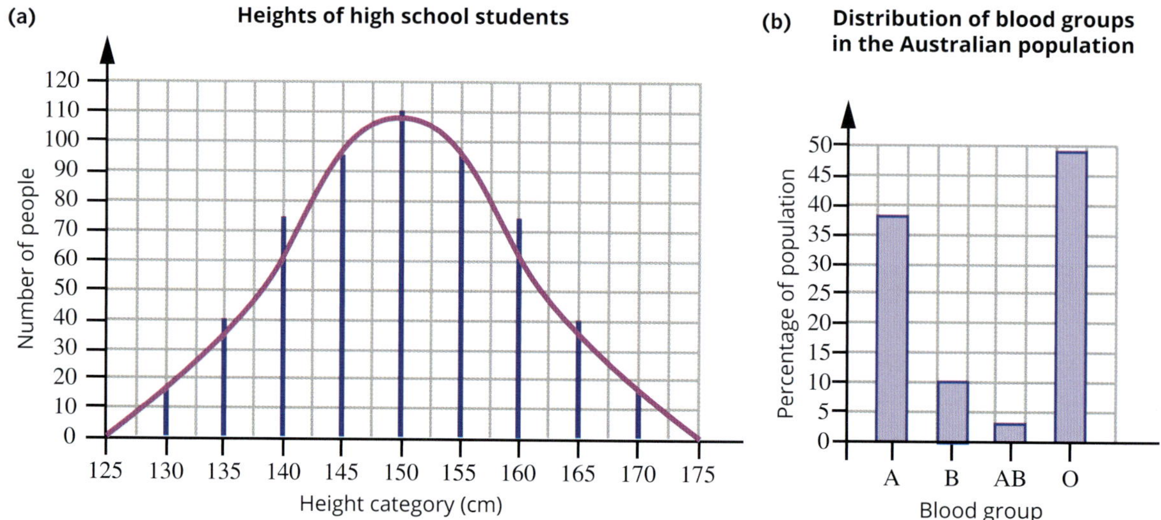

FIGURE 5.1.8 Examples of continuous and discontinuous variation. a) An example of continuous variation resulting from polygenic inheritance: height in humans. b) An example of discontinuous variation resulting from monogenic inheritance: blood groups in humans.

BIOFILE CCT

The redhead mutation

If you are a redhead, it's very likely that you have a mutation in the *MC1R* gene, responsible for the production of pheomelanin and eumelanin. The 'redhead mutation' means that the gene does not function normally, leading to the combination of pale skin, freckles, and red hair (Figure 5.1.9). You might expect that redheads will be more susceptible to skin cancer, due to the lower pigmentation and reduced protection from UV, but what you might not know is that *MC1R* is expressed in some parts of the brain as well as in the skin, and it affects pain tolerance and anaesthetic sensitivity. Redheads often need up to 20% more anaesthetic and pain relief to achieve the same effect as people with other skin tones and hair colours, and they tend to be much more sensitive to cold and tooth pain. On the other hand, they are much tougher when it comes to stinging pain such as that caused by needles, slaps and cuts; they are also unusually sensitive to opioid-based medications, meaning that they need less of the same medication to achieve the same effect.

We still do not fully understand the way in which *MC1R* modulates pain and thermal tolerance, but it is an excellent example of how the same gene, when expressed in different tissues of the body, can have profoundly different effects.

FIGURE 5.1.9 Mutations in *MC1R* affect the production of pheomelanin and eumelanin, resulting in a combination of red hair, pale skin and freckles. Since it is also expressed in the brain, the redhead allele also affects pain tolerance and response to anaesthetics.

ENVIRONMENTAL INFLUENCES ON PHENOTYPE

The examples discussed so far show that the phenotype is determined by the genotype but may also be affected by the environment. If an individual with a given genotype develops in one environment, its phenotype may be different than if it had developed in some other environment. For example, the average height of humans has gradually increased in the last few hundred years because improvements in nutrition have had a positive overall effect on growth.

Phenylketonuria

The inherited disorder phenylketonuria (PKU) is a consequence of the build up of an amino acid called phenylalanine in the blood. This is toxic to developing neurons, leading to abnormal development of the nervous system and intellectual disability. PKU is caused by a **mutation** in the *PAH* gene, which codes for an enzyme that converts phenylalanine into another amino acid, tyrosine. If an individual inherits two copies of the mutant allele (that is, they are homozygous for the gene), they will develop PKU. Fortunately, development of the symptoms can be prevented by modifying the diet (environment) of babies that test positive for PKU shortly after birth. If homozygous individuals reduce their intake of dietary phenylalanine, particularly during childhood, they show normal brain development. Newborns in Australia are routinely tested for genetic disorders like PKU.

Fur colour in Himalayan rabbits

Fur colour of Himalayan rabbits provides an example of the effect of environmental temperature on phenotype. The Himalayan rabbit is homozygous for a mutant allele that encodes a heat-sensitive tyrosinase. Tyrosinase is produced but it is inactivated at normal body temperature, meaning that no melanin is produced and the rabbit develops a white coat. At low temperatures, tyrosinase is activated and results in the formation of melanin, causing black fur to form. When a small section of fur is shaved from a white region on the back, the fur grows back black if the animal is kept at low temperatures, but white if the animal is kept at high temperatures (Figure 5.1.10).

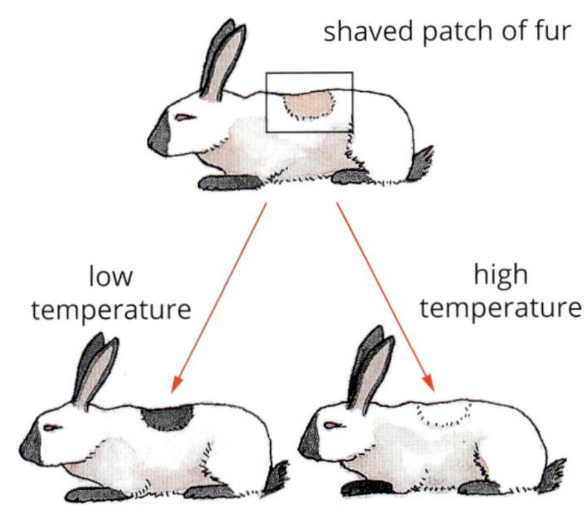

FIGURE 5.1.10 The relationship between temperature and fur colour in Himalayan rabbits

FIGURE 5.1.11 Flower colours of cuttings of the same hydrangea plant grown in (a) an acid soil and (b) an alkaline soil

The term 'epigenetics' means 'on top of genetics'. It is the study of changes in the phenotype due to gene expression (the switching on and off of genes), rather than changes in the genotype.

FIGURE 5.1.12 All the honeybees in a colony are genetically identical to each other. The queen bee (marked with blue paint on her head) looks different to the worker bees because of epigenetics.

A methyl group is a carbon with three hydrogen atoms bound to it. The attachment of a methyl group to DNA, referred to as DNA methylation, can affect gene expression; the gene might not be 'turned on' to produce the protein, or it might be silenced.

Flower colour in hydrangeas

Hydrangeas are a commonly seen example of environmental effects on phenotype. If cuttings of a single hydrangea plant are grown in very acidic soil (pH 5.5 or less), the flowers produced are blue (Figure 5.1.11a); if the cuttings are grown in weakly acidic or alkaline soil (pH 6.5 or more), the flowers are pink (Figure 5.1.11b). The cuttings are of identical genotype, so it must be the environment (the pH of the soil) that affects the phenotype of the hydrangea.

This effect is caused by the relationship between soil pH, a pigment called anthocyanin and the availability of aluminium in the soil for uptake by the plant. At a soil pH of 5.5 or less, aluminium is free to be taken into the plant. Anthocyanin is normally red, but it binds to aluminium in the plant to form a blue pigment called metalloanthocyanin, resulting in blue flowers. At a soil pH of 6 or more the aluminium binds to soil particles and is less available to the plants. This leaves most of the anthocyanin in the plant in its red form, resulting in pink flowers.

Epigenetics

Phenotypes can sometimes be affected by the interaction of DNA with other molecules. For example, the queen and worker honeybees (*Apis mellifera*) are genetically identical, but their behaviour, physiology and appearance are different (Figure 5.1.12). The phenotype differences are due to the differences in the diet of the bees. Queen bees are fed royal jelly while worker bees are fed nectar. Royal jelly contains ingredients that inhibit an enzyme called cytosine methyltransferase. This enzyme adds a methyl group ($-CH_3$) onto cytosine bases in honeybee DNA, preventing certain genes from being expressed (Figure 5.1.13). When the enzyme is inhibited, methylation does not occur and these genes are expressed. When scientists mimicked the effects of royal jelly on worker bees, worker bees exhibited characteristics of queen bees.

This effect of royal jelly in the honeybee is an example of **epigenetics**. Epigenetics is a term used to describe molecular events, such as adding methyl groups, which occur on DNA without altering the DNA sequence. Such modifications are called 'epigenetic marks' and result in changes in gene expression and variations in phenotype. Other forms of epigenetic modification include addition of methyl or phosphate groups to **histones**, which affect how DNA is coiled and whether particular genes are expressed. Another example of epigenetics is X-inactivation, where one of the two X chromosomes in female mammals is inactivated during embryonic development. In each cell, X-inactivation randomly switches off one X chromosome to ensure the correct number of genes are expressed and development proceeds as normal.

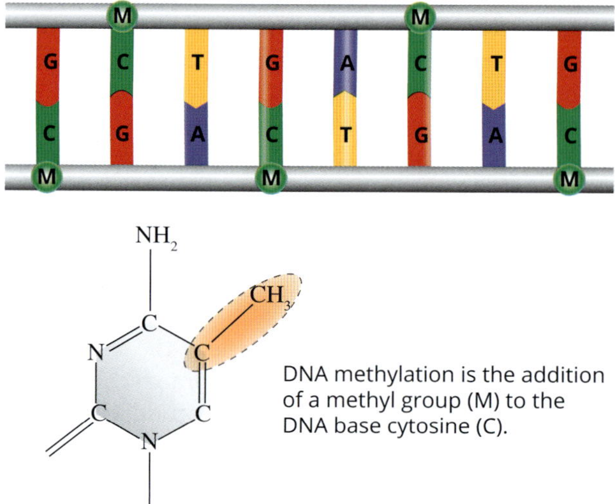

FIGURE 5.1.13 When a methyl group ($-CH_3$) is attached to cytosine bases, it prevents the expression of genes.

5.1 Review

SUMMARY

- Genotype is the combination of alleles at a particular locus.
- An organism that has two copies of the same allele is homozygous for that allele.
- An organism that carries two different alleles is heterozygous.
- Phenotype is an observable characteristic or trait that results from the genotype under the influence of the environment.
- Dominance and recessiveness are properties of alleles and are expressed as dominant or recessive phenotypes.
- A phenotype can be dominant or recessive depending on its appearance in the heterozygote.
  - A dominant phenotype is one that is visible in the heterozygote and one homozygote.
  - A recessive phenotype is only observed in the homozygous condition.
- An italic upper case letter is used to signify the allele for a dominant phenotype.
- An italic lower case letter is used to signify the allele for a recessive phenotype.
- Gene names and allele symbols are always italicised.
- When more than one gene influences a trait, it is called polygenic inheritance.
- Polygenic inheritance causes a wide variety of phenotypes. This is called continuous variation.
- Discontinuous variation occurs when a single gene determines a trait.
- Phenotype is influenced by:
  - genotype
  - interaction between genotype and the environment
  - interaction between DNA with other molecules (epigenetic factors).
- Fur colour in rabbits, flower colour in hydrangeas and the management of PKU are examples of how the environment can affect an organism's phenotype.
- Epigenetics refers to molecular events that affect the expression of genes without altering the DNA sequence. These events usually involve switching genes on or off. Examples of epigenetic events are the addition of methyl or phosphate groups to histones, which affect how DNA is coiled.

KEY QUESTIONS

1. Explain the difference between the genotype and phenotype of an individual.
2. What factors contribute to an individual's phenotype? Give an example.
3. *E* and *e* are alleles of a particular gene. Write down the possible combinations of these alleles and state whether each is homozygous or heterozygous.
4. Use an example to explain how two organisms can have the same phenotype but different genotypes.
5. Use an example to distinguish between dominant and recessive phenotypes.
6. Describe an example where an organism's phenotype can be affected by the environment in which it is raised.
7. Mutations are changes in a DNA sequence, and can result in new alleles for a gene.

   Outline the difference between epigenetic events affecting phenotypes and mutations affecting phenotypes.

5.2 Inheritance of genetic variation

FIGURE 5.2.1 After carefully studying the results of crossing different pea plants (*Pisum* species) in his garden over two years, Gregor Mendel deduced the basic principles of inheritance.

Much of what is now understood about natural variation and patterns of inheritance in sexually reproducing organisms was originally gained through the work of Gregor Mendel in the 1860s. Mendel accurately deduced the basic principles of inheritance by studying several inheritable traits in pea plants (*Pisum* species) (Figure 5.2.1), using precise experimentation and careful observations over many years.

In this section you will learn about the basic principles of inheritance, focusing on autosomal and sex-linked inheritance.

Autosomes are chromosomes that are not involved in sex determination. Humans have 22 pairs of autosomes and one pair of sex chromosomes.

MENDEL'S STUDY OF PATTERNS OF INHERITANCE

Mendel demonstrated that traits are passed from parents to offspring, and that these traits form specific patterns over generations of crossbreeding.

Mendel made several observations in his pea experiments: purple flowers were dominant over white flowers, a round seed shape was dominant over a wrinkled seed shape, and a green pod colour was dominant over a yellow pod colour. These different phenotypes occur because of variations in genes (alleles).

A cross of the traits being studied can be carried out to determine which trait is dominant. A **monohybrid cross** is a cross between two individuals with different alleles at a single locus. In the standard monohybrid cross, homozygous parents (P) with different phenotypes of the same trait (e.g. white fur and black fur) are crossed first to produce heterozygous offspring (the first filial generation, or **F1 generation**). These heterozygous offspring are then crossed with each other to produce the second filial generation (**F2 generation**). The **phenotypic ratios** in the offspring of F1 and F2 generations indicate which phenotypes are dominant or recessive. Mendel used monohybrid crosses to discover the dominance relationships of traits in pea plants.

parent generation
P homozygous dominant × homozygous recessive
↓
F1 first filial generation
heterozygous offspring
↓
F2 second filial generation

The term 'filial' refers to the offspring of a cross. The symbols for filial generations are sometimes written in the form F1 (first filial generation), F2 (second filial generation), etc.

PUNNETT SQUARES

In 1905, geneticist Reginald Punnett devised a simple method for showing the random combination of gametes and the genotypes of the resulting offspring. In a Punnett square the alleles of each parent are first written in the top and side cells. Then by going down each column and across each row, the alleles are combined and written into the remaining cells (Figure 5.2.2).

Punnett squares make it easy to establish all the possible combinations of alleles carried by the gametes and, therefore, all the possible genotypes of the offspring. This is useful in fields such as animal husbandry and horticulture because it allows breeders to select individuals to cross according to the desired traits of the offspring.

Genotypic and phenotypic ratios

Genotypic and phenotypic ratios are used to express the expected frequency of genotypes and phenotypes in the offspring from a genetic cross. Punnett squares are used to calculate the expected outcomes of a cross and the possible genotypes and phenotypes generated in the offspring.

The ratio of genotypes in the offspring is written in the following order: homozygous dominant : heterozygous : homozygous recessive.

The ratio of phenotypes observed in the offspring is written as: dominant phenotype : recessive phenotype.

The genotypic and phenotypic ratios sometimes differ because the dominant or recessive nature of traits means that different genotypes can result in the same phenotype (e.g. both *WW* and *Ww* result in red eyes in flies).

The **genotypic ratio** 1 : 2 : 1 of the F2 generation resulting from a monohybrid cross occurs because of the following reasons.

In meiosis, heterozygous (*Ww*) individuals (both male and female) produce two gametes (a *W* gamete and a *w* gamete). This is because of the separation of pairs of alleles during the formation of reproductive cells.

Fertilisation occurs at random. For example, a *W* sperm has equal chance of fertilising a *W* egg or a *w* egg, because these eggs are produced in equal frequency. A *w* sperm also has an equal chance of fertilising a *W* egg or a *w* egg. So the four equally possible genotypic outcomes are *WW*, *Ww*, *wW*, *ww*.

The Punnett square accounts for both of these factors in demonstrating the possible outcomes of the cross.

BIOLOGY IN ACTION CCT

The Law of Segregation

In the 1860s, Gregor Mendel conducted breeding experiments on 34 different varieties of pea plants. During this time he carefully collected data and made many observations that would later lay the foundations for modern genetics and the study of inheritance.

One of his most significant observations was that the offspring of the pea plants did not always have the same phenotype as the parents, and that offspring from the same parents were often different from one another. Mendel hypothesised that hereditary units or 'factors' (now called genes) must have different forms (now called alleles) that separate randomly during the production of gametes. These forms would then unite after fertilisation, with each parent contributing one allele to the offspring. Mendel's hypothesis became known as the **Law of Segregation**, or Mendel's first law.

With the advancement of cell biology, we now have a better understanding of the process of the Law of Segregation. During meiosis, each daughter cell (or gamete) receives one chromosome from each homologous pair. The alleles for each trait are separated into different gametes. Because gametes are haploid, they carry only one of the two alleles of a genotype. Offspring then receive one allele from each parent at fertilisation (Figure 5.2.2).

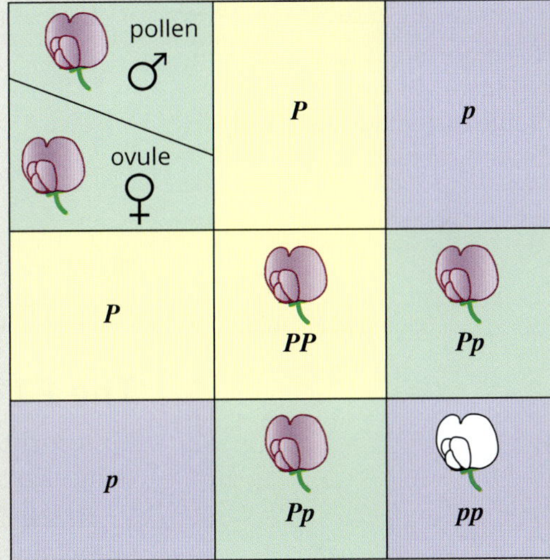

FIGURE 5.2.2 Punnett square for flower colour resulting from a cross between two heterozygous pea plants (*P* = purple flowers, *p* = white flowers). The resulting offspring have a 3 : 1 phenotypic ratio.

AUTOSOMAL DOMINANT INHERITANCE

Autosomal dominant inheritance (complete dominance) refers to a dominant trait that is passed on to offspring via an autosomal gene. An **autosome** is any chromosome that is not a sex chromosome. Only one copy of an autosomal allele (i.e. from one parent) is needed to express a dominant phenotype.

BIOFILE CCT N

Punnett squares versus experimental data

Punnett squares provide only the theoretical results of a cross; the actual results from an experiment may be different. Fertilisation can be compared to tossing a coin—for most genes there are two possible outcomes. If a coin is tossed, there is a 50% chance of getting heads and a 50% chance of getting tails. If the coin is tossed 10 times, you might not get five heads and five tails, but if it is tossed many times, a heads : tails ratio very close to 1 : 1 would be observed.

Similarly, the more fertilisation events (data) there are in a breeding experiment, the closer the results will be to the theoretical ratio.

Parent (P) generation

The inheritance of eye colour in the Australian sheep blowfly is an example of a single gene with two alleles (found on an autosomal chromosome) coding for the trait. In the blowfly, red eye colour is dominant over white eye colour. The homozygous genotypes are *WW* and *ww*. Homozygous genotypes produce only one type of gamete. *WW* individuals produce only *W* gametes and *ww* individuals produce only *w* gametes.

In a cross between two red-eye homozygous (*WW*) individuals, all the offspring would be homozygous *WW* (red eye). As long as *WW* individuals were crossed together, it would be a **true-breeding** strain. True-breeding (also called pure-breeding) crosses produce offspring with the same trait as the parents. Similarly, as you can see on the right side of Figure 5.2.3, crosses between two homozygous white-eye (*ww*) individuals would yield a true-breeding white eye (*ww*) strain.

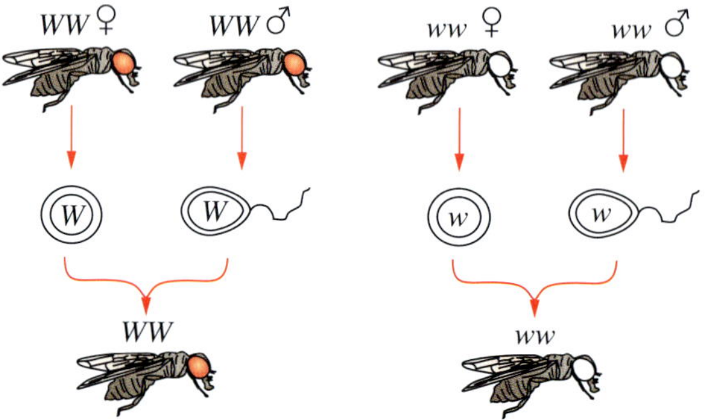

FIGURE 5.2.3 Homozygous genotypes produce only one type of gamete. By crossing homozygotes of the same genotype together, a true-breeding strain can be established.

The F1 generation

To test the principle of dominance, two true-breeding parents with two different traits can be crossed.

In the blowfly example, two true-breeding strains (one with red eyes, *WW*, and one with white eyes, *ww*) can be crossed to produce an F1 generation. The results of the cross can be shown in a Punnett square, as shown in Figure 5.2.4.

Each of the offspring in the F1 generation has the heterozygous genotype *Ww*. The phenotype resulting from this genotype is red eyes. From this, it can be deduced that the red-eye phenotype is dominant over the white-eye phenotype.

When choosing symbols for alleles, it is common practice to select one that relates to the dominant phenotype (e.g. *B* for dominant black fur and *b* for recessive white fur). Sometimes allele symbols are chosen based on the name of the gene (e.g. *w* for the 'white eye gene' in blowflies).

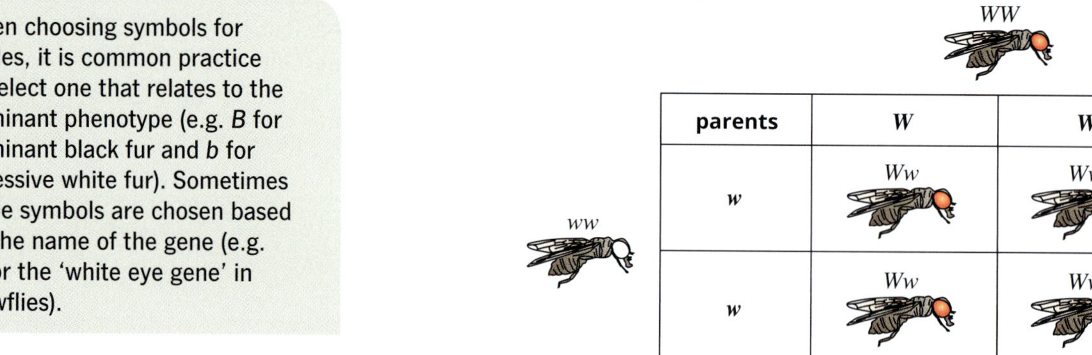

parents	W	W
w	Ww	Ww
w	Ww	Ww

FIGURE 5.2.4 A Punnett square for a cross between two homozygous parents to produce an F1 generation. All F1 individuals are heterozygous.

The F2 generation

The F2 generation is the result of crossing the individuals from the F1 generation. In this example, half of the gametes produced by an F1 individual (*Ww*) will be *W* and half will be *w*. So three different combinations of alleles are possible in the F2 generation: *WW*, *Ww* and *ww*, as shown by the Punnett square in Figure 5.2.5.

In the F2 generation the dominant phenotype is likely to occur in three out of four crossings, and the recessive phenotype only once.

> The wild type is the typical or standard form of an organism, gene or characteristic that is the most common in natural or normal populations. It is distinguished from forms that may result from selective breeding.

Ww

Ww

parents	W	w
W	*WW*	*Ww*
w	*Ww*	*ww*

FIGURE 5.2.5 Punnett square of a cross between two F1 individuals to produce the F2 generation

The Punnett square shows that the F2 genotypic ratio is:

WW : *Ww* : *Ww* : *ww* or 1*WW* : 2*Ww* : 1*ww*

Because red eye colour is dominant over white eye colour, the F2 phenotypic ratio is:

3 red-eyed flies (*WW* : *Wr* : *Wr*) : 1 white-eyed fly (*ww*)

From this information it can be determined that the **wild type** is red eyed. The wild type is the typical or most common form of a trait in a natural population.

TEST CROSSES

It is not immediately obvious whether an individual with a dominant phenotype is homozygous, because it might be either *AA* or *Aa*. Apart from sequencing the gene involved, the only way to determine this is to do a **test cross**. A test cross involves crossing an individual with the dominant phenotype (heterozygous or homozgous) with an individual with the recessive phenotype (homozygous). Homozygous individuals produce gametes with one type of allele, whereas heterozygous individuals can produce gametes with two types of alleles.

If the offspring from the test cross all have the dominant phenotype, then both the parents are likely to be homozygous (although it is not possible to be certain because of the random nature of fertilisation). If the offspring have both dominant and recessive phenotypes, then the parent with the dominant phenotype must also carry a recessive allele and is therefore heterozygous (Figure 5.2.6).

Coat colour in guinea pigs

The coat colour of guinea pigs is determined by the alleles of one gene, and black fur is the dominant phenotype. If a true-breeding white guinea pig (*bb*) is crossed with a true-breeding black guinea pig (*BB*), the resulting F1 has black fur (*Bb*). But if the genetic history of a black guinea pig is unknown, its genotype can be determined by crossing the black guinea pig with a white-coated guinea pig, which must be *bb* (homozygous recessive).

Figure 5.2.6 illustrates the test cross that would be carried out. Of the resulting offspring in this example, half are white and half are black. This ratio of about 1 : 1 is consistent with the results of a heterozygote crossed with a homozygote if the trait is determined by the alleles of one gene and one trait is dominant. So the black guinea pig is likely to be heterozygous. If all the offspring of the test cross were black-coated (all *Bb*), the F1 guinea pig in question would have been shown to be homozygous.

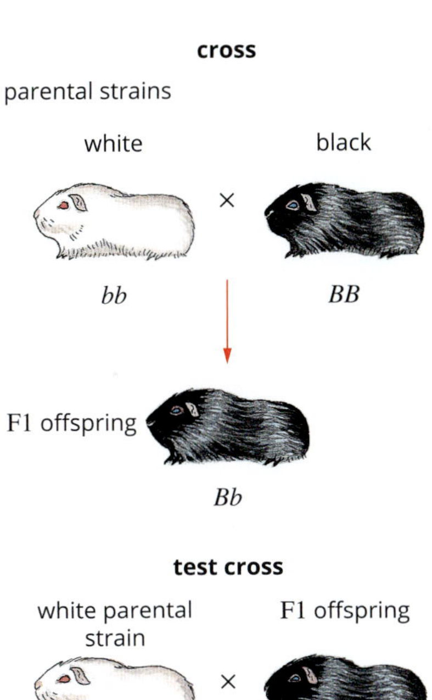

gametes from parental strain

		b	*b*	
gametes from F1	***B***	*Bb*	*Bb*	27 black
	b	*bb*	*bb*	23 white

FIGURE 5.2.6 A cross and test cross between true-breeding strains of guinea pigs and their offspring (F1)

The predicted outcome for a cross between heterozygote black guinea pigs is 1 black (*Bb*) : 1 white (*bb*). However, as the diagram shows, the resulting ratio of the test crosses was 27 black (*Bb*) : 23 white (*bb*) rather than, for example, 23 : 23 (which is equal to a 1 : 1 ratio). The difference between predicted and observed ratios is due to chance.

AUTOSOMAL CO-DOMINANT INHERITANCE

Some traits do not show simple dominance or recessiveness. These are instances in which both alleles are expressed to varying degrees in the phenotype of heterozygous individuals. This is called co-dominance. Sometimes neither phenotype is completely dominant, so that intermediate phenotypes occur. This is known as **incomplete dominance**.

Autosomal co-dominance and incomplete dominance

Flower colour in snapdragons is a trait showing incomplete dominance. For example, snapdragons (Figure 5.2.7) can have a red flower phenotype (R_1R_1) or a white flower phenotype (R_2R_2). In this case upper-case letters and subscripts are used to distinguish the alleles because neither phenotype is completely dominant. Plants of the R_1R_2 genotype have pink flowers. In R_1R_1 flowers, both copies of the R_1 allele produce an enzyme required to produce red pigment. In R_2R_2 flowers, no pigment is produced because the R_2 allele produces no enzyme or a defective enzyme.

FIGURE 5.2.7 Red, white and pink snapdragons. When pure-breeding red (R_1R_1) and white (R_2R_2) snapdragons are crossed, the resulting heterozygotes are pink (R_1R_2) because only half the amount of red pigment is produced.

Since the R_1R_2 flower has one R_1 allele (which produces the active enzyme) and one R_2 allele (which does not), it will produce half the amount of pigment as the R_1R_1 flower, and will be pink in colour.

Crossing red-flowering snapdragons (R_1R_1 genotype) with white-flowering snapdragons (R_2R_2 genotype) will yield an F1 generation in which all individuals have the genotype R_1R_2 and are pink-flowering (Table 5.2.1).

TABLE 5.2.1 Cross between red-flowered snapdragons and white-flowered snapdragons

Parents		Red flowers	
		R_1	R_1
White flowers	R_2	R_1 R_2	R_1 R_2
	R_2	R_1 R_2	R_1 R_2

If the F1 plants ($R_1R_2 \times R_1R_2$) are crossed, an F2 generation with a 1 : 2 : 1 genotypic ratio (1 R_1R_1 : 2 R_1R_2 : 1 R_2R_2) would be expected (Table 5.2.2).

TABLE 5.2.2 Cross between two pink-flowered snapdragons

Parents		Pink flowers	
		R_1	R_2
Pink flowers	R_1	R_1 R_1	R_1 R_2
	R_2	R_1 R_2	R_2 R_2

The heterozygote pink-flowering snapdragon (R_1R_2) can be distinguished from the two homozygotes, red R_1R_1 and white R_2R_2, due to the co-dominance of both the red and white alleles resulting in a pink-flowering phenotype.

Genotype ratio: 1 R_1R_1 : 2 R_1R_2 : 1 R_2R_2

Phenotype ratio: 1 white : 2 pink : 1 red

This phenotypic ratio of 1 : 2 : 1 (Figure 5.2.8) is different to the 3 : 1 ratio of two phenotypes observed in complete dominance (Figure 5.2.2 on page 211.)

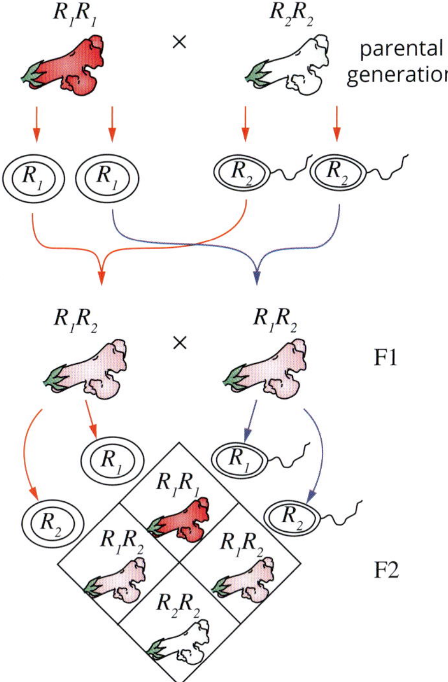

FIGURE 5.2.8 A cross between homozygous red and white snapdragons produces pink-flowering progeny in the F1 generation. If plants from the F1 are then crossed, a phenotypic ratio of 1 red : 2 pink : 1 white would be expected in the F2 generation.

ABO blood grouping

As described in Section 5.1 human blood type is an example of autosomal co-dominant inheritance. In this case, depending on the alleles inherited, the expression of the genotype differs. There are three alleles for blood type at the same locus, and individuals can have A, B, AB or O phenotypes. This is illustrated in Figure 5.1.3. Those with the less common AB blood group are heterozygotes carrying one allele that produces an A antigen and one allele that produces a B antigen.

Because both the A and B antigens are present on the surface of red blood cells, which can be detected using antibodies, neither phenotype is fully dominant. This is another example of co-dominance. So A and B phenotypes are co-dominant, while the O phenotype is recessive.

Multiple alleles at a single locus

Blood group systems, provide a demonstration of the effects of multiple alleles at the same locus. In this case, the three alleles are represented as I^A, I^B and i. I^A codes for the A antigen, I^B codes for the B antigen while i does not produce either antigen. Therefore the effects of I^A and I^B dominate over i. Each person carries copies of one or two of these three possible alleles.

The possible genotypes and phenotypes of the offspring of a parent with blood type O and a parent with blood type AB can be determined using a Punnett square (Table 5.2.3).

To represent co-dominance or incomplete dominance in a cross, the alleles are written as capital letters with subscript numbers (e.g. R_1 represents the red flowers in snapdragons and R_2 represents the white flowers). This is because neither phenotype is completely dominant or recessive. In the case of blood typing, the alleles are written as superscripts.

TABLE 5.2.3 Cross between an individual with blood type AB and an individual with blood type O

Parents		Blood type AB	
		I^A	I^B
Blood type O	i	I^Ai	I^Bi
	i	I^Ai	I^Bi

The F1 generation in this example would be either blood type A or B, but all would be heterozygous.

If a heterozygous individual for blood type A and a heterozygous individual for blood type B were to have children, all four combinations of blood type are possible: AB, A, B and O. (Table 5.2.4).

The important principle illustrated by this example is that phenotypes are not always dominant or recessive. The dominance of a phenotype is always in relation to another phenotype. Thus, phenotype A is co-dominant with B, but dominant to O.

TABLE 5.2.4 Cross between an individual with blood type B and an individual with blood type A

Parents		Blood type B	
		i	I^B
Blood type A	I^A	I^AI^B	I^Ai
	i	I^Bi	ii

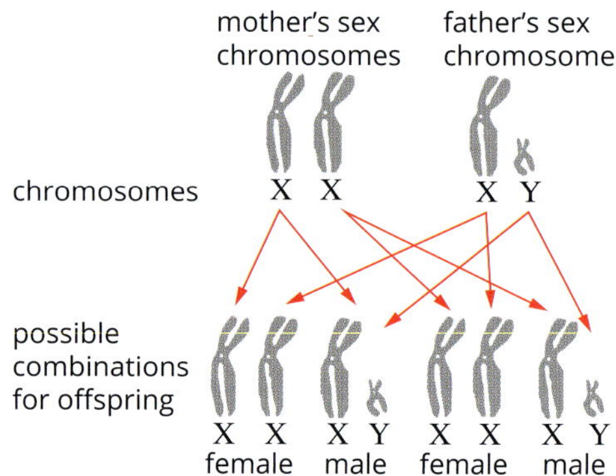

FIGURE 5.2.9 Inheritance of the sex chromosomes in the XY sex-determination system. The outcome of this inheritance is two possible arrangements—XX or XY with half the offspring being female and half being male.

SEX-LINKED INHERITANCE

So far you have examined the inheritance of genes located on autosomes. However, the patterns of inheritance are not the same for genes located on either of the two sex chromosomes. Phenotypes inherited through genes on sex chromosomes are said to be 'sex-linked' and they show **sex-linked inheritance**. It is important to remember that sex chromosomes also carry other genes which are not related to sex determination.

Figure 5.2.9 shows how sex chromosomes are transferred to the offspring, with an equal probability of the offspring being female or male. The XY system determines sex in humans, most other mammals, some insects and some plants. In this system, females are **homogametic** (XX) and males are **heterogametic** (XY). The female passes one X chromosome on to her offspring, while the male can pass on either an X or Y chromosome; an X chromosome produces female offspring and a Y chromosome produces male offspring. It is therefore the father's genetic contribution that determines the sex of the offspring. In birds, some fish, some insects and some reptiles, sex is determined by ZW chromosomes. In this system females are the heterogametic sex (ZW) and males are the homogametic sex (ZZ).

X-linked recessive inheritance

Males are most affected by X-linked recessive traits because they only carry one X chromosome. The second X chromosome in females can mask the recessive trait.

In humans, **X-linked** recessive traits are predominantly expressed in males, because males carry only one X chromosome. Females carrying an X-linked recessive allele might not express the trait, or show only mild expression. This is because the second X chromosome that females carry could mask the recessive trait. The probability in humans of a female carrying two X-linked recessive alleles is very low.

The pattern of sex-linked inheritance is evident when a **reciprocal cross** is performed. This is a cross used to investigate the role of parental sex on the inheritance of genotypes. Two crosses are performed: one crossing a male with the trait of interest with a female not expressing the trait (usually homozygous wild type), and another crossing a female with the trait of interest (homozygous) with a male that does not express the trait (usually wild type). If the trait is sex-linked (carried on the X chromosome), the phenotypic ratios of the male and female offspring will be different.

Paralysis in Drosophila

The temperature-sensitive paralytic gene, named after the mutant phenotype, is on the X chromosome of the fruit fly (*D. melanogaster*). A mutant phenotype is one that is different to the normal wild type and arises from a genetic mutation. Individuals with the mutant allele are paralysed when incubated to a temperature of 29°C, whereas wild type flies show normal behaviour at this temperature. The paralytic phenotype is recessive to wild type. The wild type (for this trait) is defined as normal movement at an ambient temperature of 29°C, that is, a lack of paralysis.

The alleles are defined differently for sex-linked traits. An *X* is used to indicate that the trait is carried on the X chromosome, and the allele is written in superscript next to the X. In the example of the fruit flies, the alleles can be written as:

X^P = wild type

X^p = paralysis

As the paralytic phenotype is recessive, females that are homozygous for the mutant allele (X^pX^p) express the mutant paralysis phenotype; females that are homozygous dominant (X^PX^P) and females that are heterozygous (X^PX^p) are both wild type phenotype.

As males have only one X chromosome there are only two male genotypes: X^pY males are paralytic and X^PY males are wild type.

If paralytic females (X^pX^p) are crossed with wild type males (X^PY), all of the F1 male offspring will be paralytic (X^pY) and all of the F1 female offspring will be wild type phenotype (X^PX^p) (Figure 5.2.10a). This pattern of transmission of the mutant phenotype from the female parent to male offspring is characteristic of X-linked recessive inheritance.

The reciprocal (reverse) cross, shown in Figure 5.2.10b, produces a different outcome. If a wild type homozygous female (X^PX^P) is crossed with a paralytic male (X^pY) all of the offspring (male and female) are wild type phenotype (X^PX^p and X^PY).

These different outcomes of the reciprocal crosses are characteristic of X-linked recessive inheritance. In contrast, reciprocal crosses give the same outcome for autosomally inherited genes.

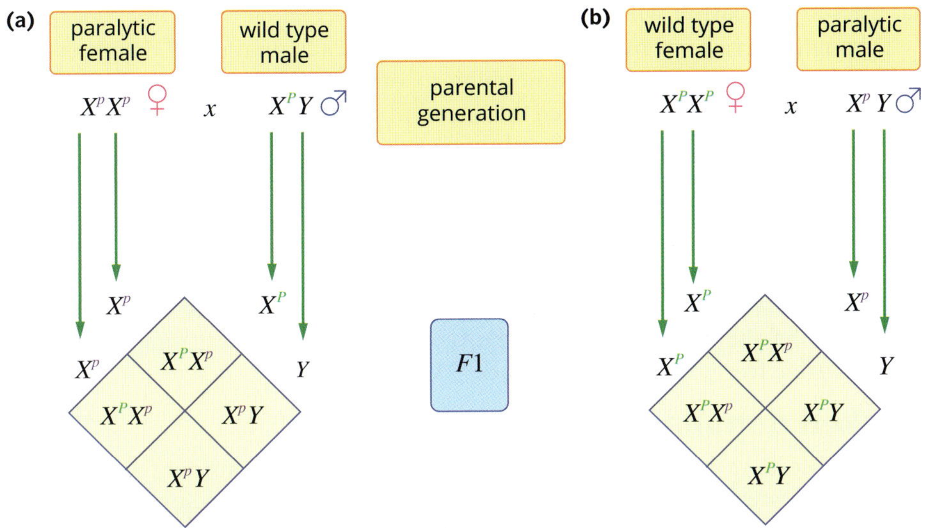

FIGURE 5.2.10 The characteristics of X-linked inheritance are evident in a reciprocal cross. (a) A male receives an X chromosome from the female parent, so males are paralytic (X^pY) and females are wild type (X^PX^p). (b) In the reciprocal cross, both the male and female offspring are wild type. The outcome of the reciprocal cross is always different for X-linked traits.

BIOFILE CCT

Spontaneous mutations

Spontaneous mutations are mutations in DNA that have not been inherited. In the case of Queen Victoria (Figure 5.2.12) there is no history of haemophilia in her ancestry, so it seems likely that she (or possibly her mother, Victoria, Duchess of Kent) was the source of the mutation. Spontaneous mutations are believed to cause about one-third of all haemophilia occurrences.

FIGURE 5.2.12 A portrait of Queen Victoria and her family, produced in the 1870s. Depicted are nine children, two daughters-in-law, four sons-in-law and twenty-three grandchildren. The spontaneous mutation for haemophilia may have arisen with Queen Victoria or her mother. Three of her children inherited the mutation.

Haemophilia in the British royal family

Figure 5.2.11 shows part of the family tree of the British royal family, including Queen Victoria, whose eighth child Leopold was born with haemophilia. Haemophilia is a blood disorder in which blood clotting is slow, resulting in excessive bleeding. It results from a mutation in a gene on the X chromosome that is involved in the production of a blood-clotting protein that controls bleeding.

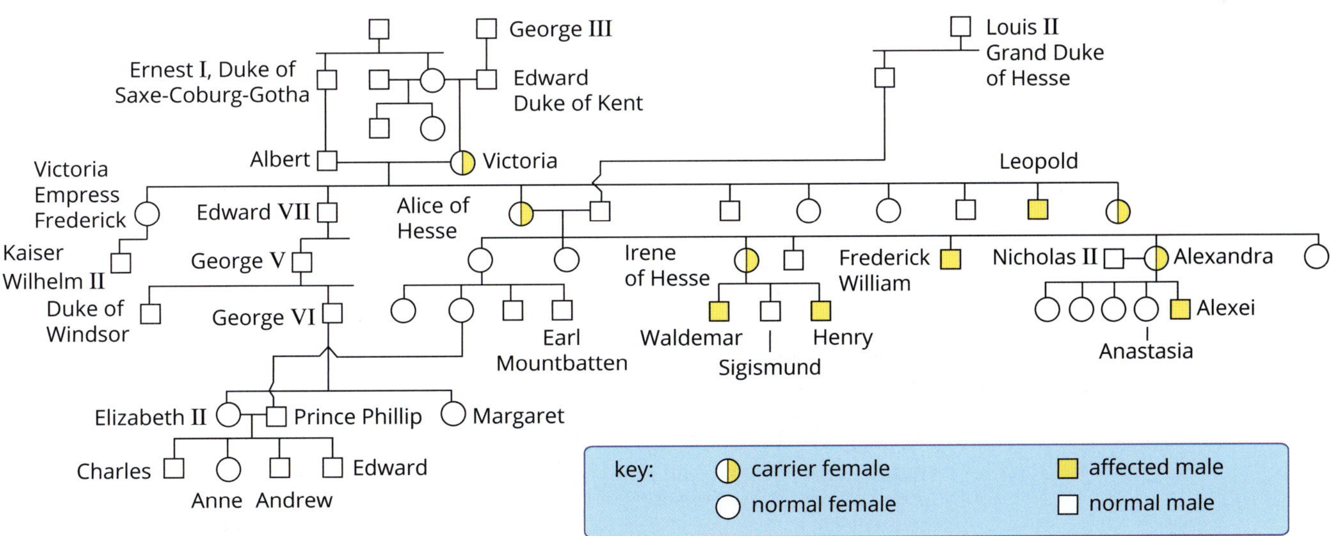

FIGURE 5.2.11 Queen Victoria was a carrier of a mutation that causes haemophilia. The affected gene codes for a blood-clotting protein. She had one copy of the normal allele and one copy of the mutant allele. Some of Queen Victoria's female descendants have been carriers, and some male descendants have had the disease. The bent line above Prince Philip's name indicates that he is a descendant of Alice of Hesse not George VI.

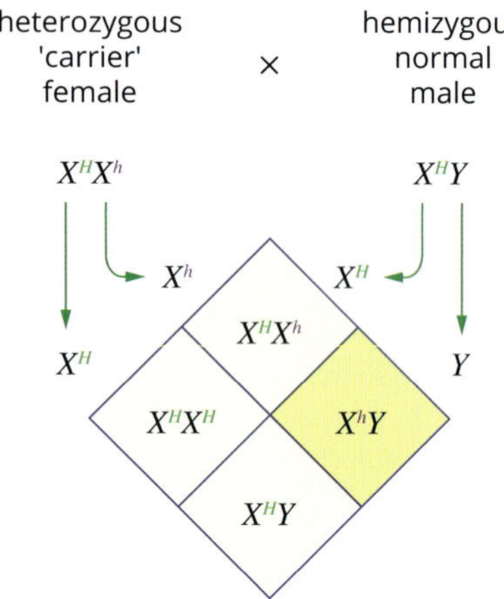

FIGURE 5.2.13 Diagram highlighting the typical situation for the inheritance of X-linked recessive diseases such as haemophilia. A female who is phenotypically normal carries one copy of the allele for the dominant normal phenotype X^H and one for the recessive (mutant) phenotype X^h. All of the female offspring will be normal, but there is a 50% chance that each male offspring will inherit the disease.

The incidence of haemophilia in the descendants of Queen Victoria shows the hallmarks of X-linked recessive inheritance. All of the haemophiliacs shown in the tree are male. The female **carriers** of the disease are heterozygous, carrying one haemophiliac allele and one normal allele. Given that the haemophilia phenotype is recessive, carrier females are phenotypically normal. However, because females produce eggs carrying the normal and haemophiliac alleles with equal frequency, and males receive their single X chromosome from the egg, there is a 50% chance that the son of a carrier will have haemophilia.

Through marriage, some of Victoria's phenotypically unaffected daughters who carried this mutation spread haemophilia to other royal families in Europe. For example, Irene of Hesse transmitted the haemophilia allele to her eldest and youngest sons Waldemar and Henry, and the normal allele to her other son, Sigismund. This form of haemophilia occurs at a frequency of one in 10000 males and one in 100 million females.

In general, X-linked recessive disorders occur at much higher frequencies in males than females because females need to inherit a copy of the allele from both parents to be affected (that is, the mother must be a carrier and the father must be affected by the disorder). Males, however, need only inherit one copy of the X-linked allele from their carrier mother (Figure 5.2.13).

X-linked dominant inheritance

X-linked disorders may also display a dominant phenotype. Consider the inheritance of vitamin D-resistant rickets disorder, which causes bone deformities. This is shown in the pedigree chart in Figure 5.2.14. The mother of the first generation is heterozygous and affected by the condition. Her children had a 50% chance of having vitamin D-resistant rickets, regardless of whether they were male or female.

When a father is affected and a mother is normal (as in the second generation), all female offspring will show the condition, and all male offspring will be normal. This is because female children all receive an X chromosome, which carries the allele for the disease, from their father (Figure 5.2.9 on page 216).

The alleles for this trait could be shown as:

X^D = vitamin D-resistant (rickets) allele

X^d = normal allele

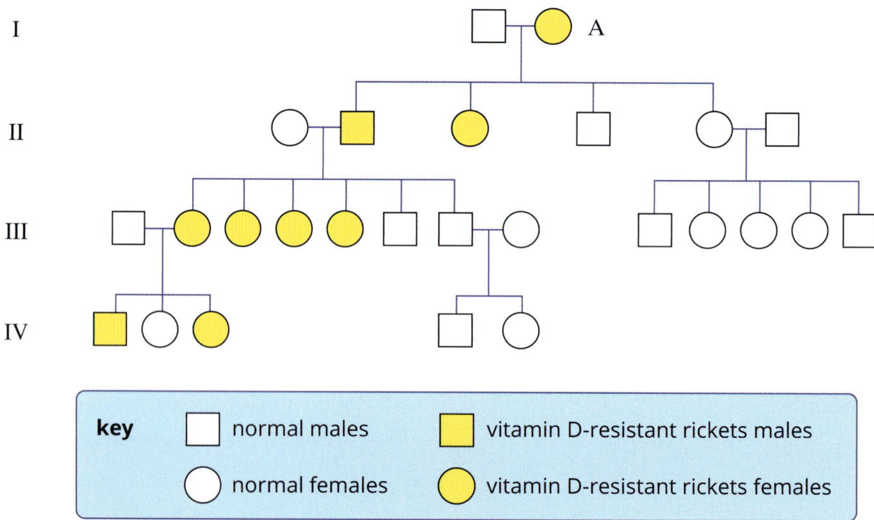

FIGURE 5.2.14 Pedigree chart showing the inheritance of the X-linked dominant condition vitamin D-resistant rickets. The mother (A) of the first generation is heterozygous for the gene that controls the condition.

The following Punnett square shows the pattern of inheritance of offspring of a heterozygous female affected by the vitamin D-resistant rickets allele and a male with the normal allele (Table 5.2.5).

TABLE 5.2.5 Cross between a female carrying an allele for vitamin D-resistant rickets and an unaffected male

Parents		Father	
		X^d	Y
Mother	X^D	X^DX^d	X^DY
	X^d	X^dX^d	X^dY

The possible genotypes of the offspring are:

$X^DX^d : X^dX^d : X^DY : X^dY$ (1 : 1 : 1 : 1)

The possible phenotypes are therefore:

female with vitamin D-resistant rickets : normal female : male with vitamin D-resistant rickets : normal male (1 : 1 : 1 : 1)

The following Punnett square shows the pattern of inheritance in the offspring of a homozygous unaffected female and a male with vitamin D-resistant rickets (Table 5.2.6).

TABLE 5.2.6 Cross between an unaffected female and a male carrying an allele for vitamin D-resistant rickets

Parents		Father	
		X^D	Y
Mother	X^d	X^DX^d	X^dY
	X^d	X^DX^d	X^dY

The possible genotypes of the offspring are:

$X^DX^d : X^dY$ (1 : 1)

The possible phenotypes are therefore:

All females have vitamin D-resistant rickets : all males are unaffected (1 : 1)

Y-linked inheritance

Compared to the X chromosome, the Y chromosome has few genes: it has only about 72 protein coding genes, compared to 800–900 on the X chromosome. Most of these genes are involved in male sex determination and fertility. Therefore there are far fewer **Y-linked** traits than X-linked traits.

If a trait is passed from father to son and never observed in females, it is likely to be Y-linked, meaning the gene for that trait is on the Y chromosome. Until recently it was thought that hairy ears were controlled by a Y-linked gene, but recent studies suggest there are also autosomal genes involved in the trait.

Sex-limited inheritance

The Y-linked pattern of inheritance is sometimes confused with **sex-limited inheritance**. Sex-limited traits can only occur in one sex because the feature affected is unique to that sex. Therefore, males and females have different phenotypes. For example, complete androgen insensitivity syndrome, in which the fetus is unresponsive to male hormones, can only occur in males, because only males carry the Y chromosome. This means that even if females have the genotype for this syndrome, they cannot express the condition.

PEDIGREE ANALYSIS

Studying the patterns of inheritance in humans has its challenges, but patterns of inheritance of alleles across generations of families (Figure 5.2.16) can be analysed using pedigree charts.

BIOFILE CCT

Male pattern baldness—sex-limited inheritance

It is estimated that 80% of hair loss is genetic, and though the causes are not yet well understood, it is known that several genes are involved (that is, it is a polygenic trait).

Male pattern baldness is the most common type of baldness and affects around 40% of men by the age of 40 and around 60% by the age of 60. Affected males gradually start losing their hair, until eventually they have hair only on the sides and back of the head (Figure 5.2.15).

You may have heard that baldness is inherited from your mother's father. This is because one of the key genes associated with balding is on the X chromosome. If your mother's father has male pattern baldness, then your mother will carry the allele for this characteristic on the X chromosome that she inherited from her father. Because you inherited one of your X chromosomes from your mother, there is a 50% chance that you will inherit the affected X chromosome. However, males are much more likely to express the balding phenotype because females have a second X chromosome to mask the expression of the gene.

Balding can also be passed from fathers to offspring, indicating that autosomal genes must be involved. Two genes on chromosome 20 have been found to contribute to balding. The effects of these genes are neither dominant nor recessive, but have an additive effect—the more copies of the alleles you have, the more likely you are to go bald. Even though these genes are found on autosomes, males are affected more than females. This is because some of the genes are associated with male hormone receptors. This is an example of sex-limited inheritance—males and females may have the same genotype but express different phenotypes.

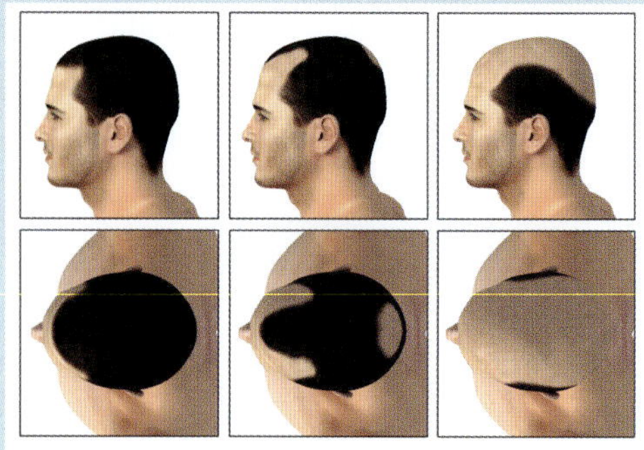

FIGURE 5.2.15 Head of a man showing the change over time of hairline with male pattern baldness. This phenotype can be caused by several genes located on autosomes and the X chromosome.

Pedigree analysis can be used to follow the inheritance of traits through a family over a number of generations. Given sufficient data, the likely mode of inheritance can be determined (e.g. dominance patterns and whether inheritance is autosomal or sex-linked). When it comes to studying the genetics of humans, pedigree analysis is often the only method available for the following reasons:

- Experimental crosses cannot be set up as required.
- The environment in which humans live cannot be controlled experimentally.
- There are strict legal and ethical laws concerning human experimentation.
- Humans tend not to have large families, so there are rarely large numbers of offspring to score.
- Each generation of humans takes many years to reach sexual maturity and produce offspring.

Some of the unique problems of human genetic analysis are being overcome as a result of the Human Genome Project (HGP). Knowledge of the mechanisms of inheritance in humans and other organisms continues to advance through research in model genetic organisms, such as the bacterium *Escherichia coli*, fruit fly *D. melanogaster*, yeast *Saccharomyces cerevisiae*, mouse *Mus musculus*, nematode *Caenorhabditis elegans* and the plant *Arabidopsis thaliana*.

In the meantime, studying existing families can assist in tracing inheritance of traits. Pedigree analysis is a technique involving studying a family tree for the occurrence of a particular character or trait in one family over a number of generations. In practice it may be necessary to combine information gained from the pedigree data of several families to determine the most likely mode of inheritance of a particular character.

Pedigrees can be used to determine the pattern of inheritance of particular alleles, as well as the presence of particular alleles within a family and the chances of the allele occurring in offspring.

FIGURE 5.2.16 Pedigree charts help to determine patterns of inheritance for different traits in families.

A pedigree is a record of the ancestry (also called the lineage) of an individual or a group of related individuals.

Pedigree analysis is the determination of the pattern of inheritance of a trait or condition (or disease) by reference to a family tree in which the presence or absence of the condition is recorded over generations.

The observable characteristic that is evident in the heterozygote is referred to as the dominant phenotype.

Pedigree charts

Pedigree analysis makes use of pedigree charts to track and organise data. When analysing a pedigree chart, key features in the pattern of inheritance can be used to distinguish between one type of inheritance and another.

Symbols and conventions used for pedigrees

Pedigree charts use a number of standard symbols and conventions (Figure 5.2.17). The main ones you need to know are described below.

- Circles represent females and squares represent males.
- Shapes are shaded or unshaded to represent the presence of a phenotype for a particular trait.
- A horizontal line represents a cross between the individuals.
- A vertical line represents a link from parents to offspring.
- Individuals are numbered from left to right (if required).
- Generations are represented with Roman numerals (if required), with the first generation in the pedigree being generation I.
- A carrier of an autosomal trait is shown with a half-filled symbol. A carrier of an X-linked recessive trait is shown with a dot in the centre of the symbol.

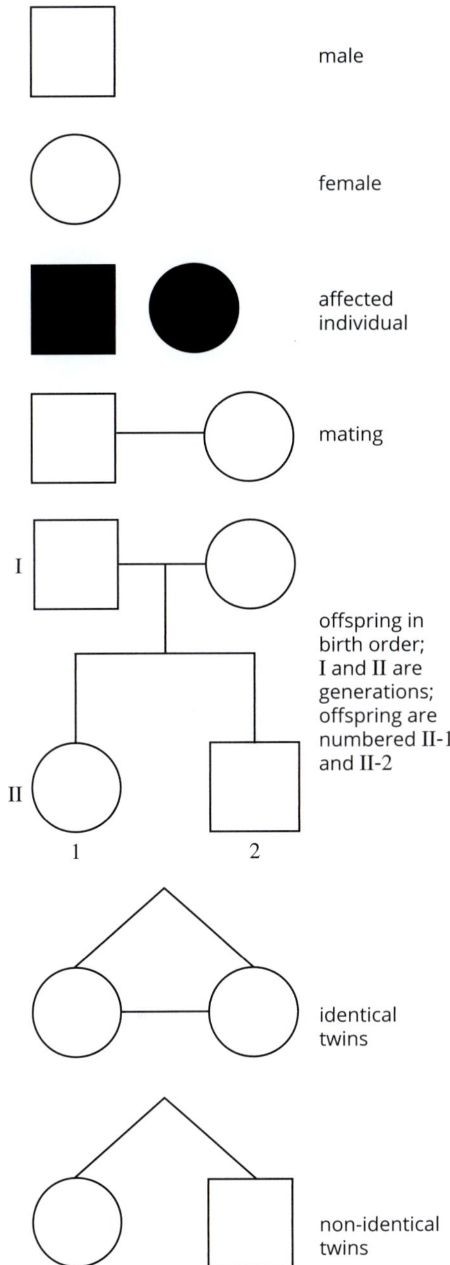

FIGURE 5.2.17 Conventions for pedigree charts

RECOGNISING INHERITANCE PATTERNS IN PEDIGREES

Pedigree analysis can be used to understand the way in which particular genetic traits are inherited, but as you will have learnt by now, there are many possible modes of inheritance. Fortunately, these can be recognised if you are aware of a few simple methods. Once you know these methods, you should be able to examine a comprehensive pedigree and then be able to work out the inheritance pattern for a trait.

Autosomal inheritance

When the inheritance of a trait is just as likely in males as in females, it is an autosomal trait—that is, the gene(s) associated with the trait are not located on the sex chromosomes, but on the autosomes. This is not always immediately obvious, particularly if a mating produces only male or only female offspring.

Autosomal recessive inheritance

Autosomal recessive inheritance of a trait is likely if two parents do not have a particular phenotype but one or more of their offspring does. Figure 5.2.18 shows the inheritance pattern of a particular trait in a population of unisexual plants (so plants are either male or female). Shaded individuals are affected (that is, they express the trait), and unshaded individuals are not affected.

For this exercise it may be assumed that the inheritance of the trait is not sex-linked. Figure 5.2.18 shows that:

- the cross between individuals II-3 and II-4 resulted in three offspring
- individuals II-3 and II-4 are unaffected
- offspring III-1 is female and unaffected, III-2 is male and unaffected, and III-3 is female and affected.

The parents both contribute one allele each for the trait to III-3, so both parents must carry the allele responsible for the trait. Since the parents are both unaffected, both must be heterozygous. The trait is therefore autosomal recessive.

Another indicator of autosomal recessive inheritance is that the trait skips generations (that is, it does not appear in every generation). However, not skipping a generation (as in Figure 5.2.18) does not rule out autosomal recessive inheritance.

Once the form of inheritance is determined, it is possible to work out the genotypes of the individuals in the pedigree. First, a symbol should be allocated for the two alleles. In this particular example:

- *P* can represent the allele for purple flower (dominant trait)
- *p* can represent the allele for white flower (recessive trait).

So III-3 (white flowers) must be *pp*, both parents must be heterozygous (*Pp*) and individuals III-1 and III-2 must be heterozygous (*Pp*) or homozygous (*PP*).

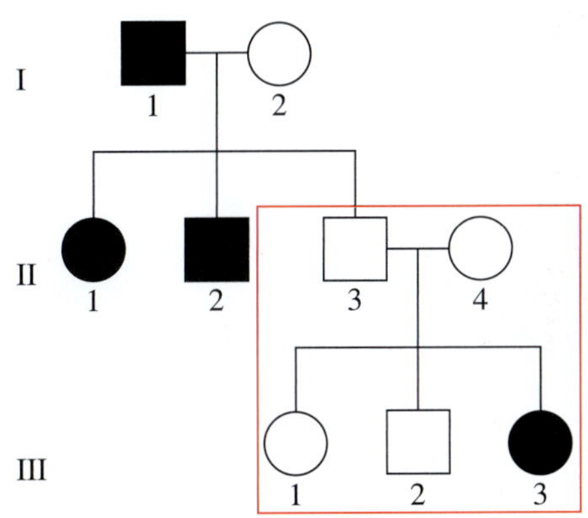

FIGURE 5.2.18 Example of autosomal recessive inheritance

BIOFILE S

Pedigree analysis and conservation

Pedigree analysis can be extended to a number of animals that are difficult to study genetically for similar reasons to those for humans. Studbooks are records of parent and offspring phenotypes kept for many years for animals for which ancestry is important. Such animals include recognised dog and cat breeds, thoroughbred horses, and a number of breeds of farm animals.

Studbooks and pedigree analyses are also crucial tools in conservation. Many zoos worldwide participate in global captive breeding programs for endangered animals such as the Sumatran tiger (*Panthera tigris sumatrae*) (Figure 5.2.19). The estimated population of Sumatran tigers in the world is less than seven hundred, a dangerously low number resulting in a profoundly restricted gene pool. The risk of damage to the species from inbreeding is very real when genetic variation is this low. For this reason, zoos that participate in such programs have to keep very careful breeding records to ensure that closely related animals do not breed.

FIGURE 5.2.19 Zoos that participate in global captive breeding programs must keep very careful records to prevent inbreeding and maintain genetic variation. This image shows Sumatran tiger cub Achilles and mother Melati at the London Zoo in 2017.

Autosomal dominant inheritance

Autosomal dominant inheritance of a trait is likely if both parents show the trait but one or more of their offspring do not show the trait. The pedigree in Figure 5.2.20 shows the inheritance pattern of round (smooth) and wrinkled pea shape in pea plants (Figure 5.2.21). The shaded individuals represent round peas and the unshaded individuals represent wrinkled peas.

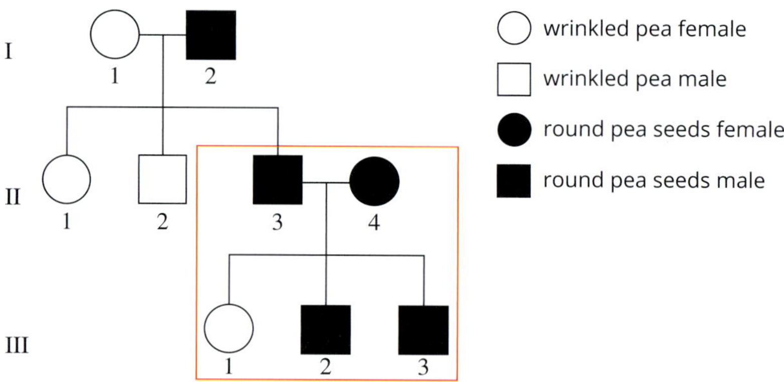

FIGURE 5.2.20 Example of autosomal dominant inheritance

FIGURE 5.2.21 Hands holding two types of pea, with wrinkled ones at left and round ones at right.

Again, it may be assumed for this example that the trait is not sex-linked. Consider the section of the chart highlighted by the square. This shows that:

- II-3 and II-4 produce three offspring
- both II-3 and II-4 are affected
- offspring III-1 is female and unaffected, and III-2 and III-3 are male and affected.

Because both II-3 and II-4 contribute one allele each to III-1, each must carry the allele for the unaffected phenotype. Since each parent carries the wrinkled allele, they must both be heterozygous. As they are heterozygous and show the round phenotype, their phenotype (round) must be dominant.

Other indicators of autosomal dominant inheritance are that the trait may be seen in all generations, and individuals showing the trait must have at least one parent showing the trait. Examples of autosomal dominant inheritance in humans include Huntington's disease and neurofibromatosis type 1. Examples in other animals include the hornless (polled) trait in cattle.

BIOLOGY IN ACTION 

The principle of penetrance

Researchers in France in the 1990s conducted one of the largest pedigree analyses to date when they traced the inheritance of a particular form of blindness (juvenile glaucoma) in over 30 000 living descendants of a 15th century couple. The analysis accounted for almost half of the known cases of this disease in France. The massive pedigree clearly showed that juvenile glaucoma resulted from autosomal dominant inheritance. As a result, it might be expected that every individual who had at least one parent with the disease would also have the disease. However, the researchers noted that this was not always true; sometimes offspring of a parent who developed juvenile glaucoma did not develop the disease (Figure 5.2.22). This illustrates the principle of penetrance.

Complete penetrance of a phenotype means that all individuals with an affected genotype will have the affected phenotype. **Incomplete penetrance** is where a proportion of a population with an affected genotype does not show the expected trait. This describes the situation that occurred in the juvenile glaucoma pedigree. In the pedigree chart (Figure 5.2.22), individual V-2, whose father (IV-1) had the disorder, did not show the glaucoma phenotype yet passed it on to his daughter (VI-2).

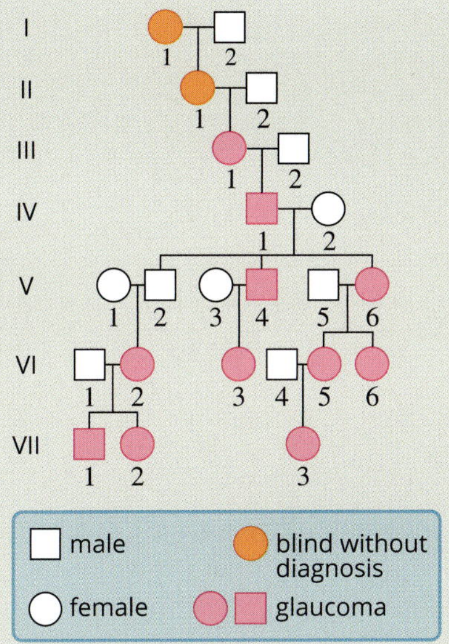

FIGURE 5.2.22 Part of a pedigree chart showing the pattern of transmission of juvenile glaucoma over several generations. Incomplete penetrance is demonstrated by the lack of juvenile glaucoma in V-2.

Sex-linked inheritance

When an inheritance pattern is 'sex-linked', that means the trait in question affects males and females differently. We therefore conclude that one or more of the gene(s) responsible for the trait is located on a sex chromosome. If sufficient male and female offspring are produced in each generation, sex-linked inheritance should be obvious when looking at the pedigree.

Distinguishing between autosomal and sex-linked inheritance

In the previous examples of autosomal inheritance it was assumed that inheritance was autosomal. However, when investigating inheritance it is important to check whether the inheritance might be sex-linked, because similar patterns can occur in both types.

The inheritance pattern of colour blindness in humans (Figure 5.2.23) can be examined by studying the pedigree of a family in which some individuals are colourblind. By convention, if a mating partner is not shown in the pedigree, he or she is not affected by the phenotype (in this case, not colourblind).

Differences in the incidence of the trait between males and females, and the frequency of the trait across generations, are two inheritance patterns that help determine whether a trait is sex-linked or autosomal.

ℹ Although an individual may carry the dominant allele for a phenotype, that does not mean that the trait is always fully expressed. This is because other factors are required for the condition to develop. This is known as incomplete penetrance. By contrast, complete penetrance occurs when the presence of the allele always results in the affected phenotype.

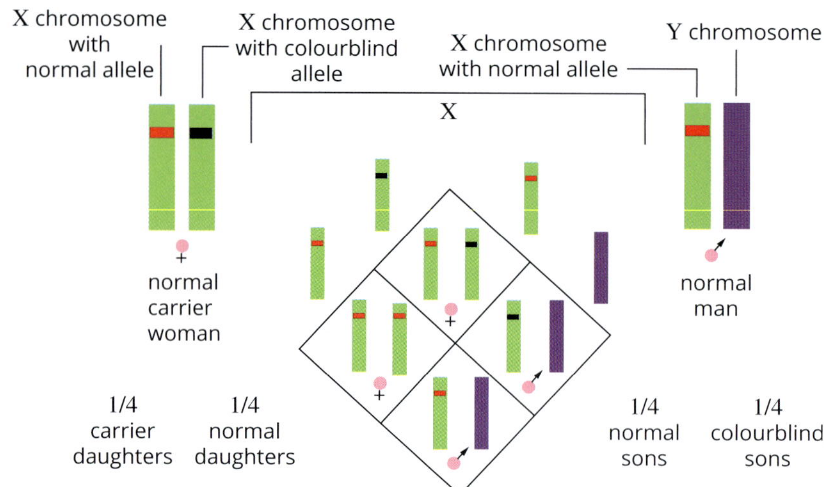

FIGURE 5.2.23 The inheritance of sex-linked colour-blindness: the probability is shown for the disorder in children from a mother carrying the recessive allele and an unaffected father. There is a 25% chance each for colourblind sons, unaffected sons, unaffected daughters and carrier daughters. This trait is classed as X-linked recessive.

In the pedigree shown in Figure 5.2.24 only the males are colourblind. This demonstrates that the colourblind trait is not dominant and is potentially sex-linked (X-linked). Y-linked alleles affect only males and are very rare, and the phenotype will be present in every male offspring. Because of these factors, colour blindness cannot be a Y-linked trait.

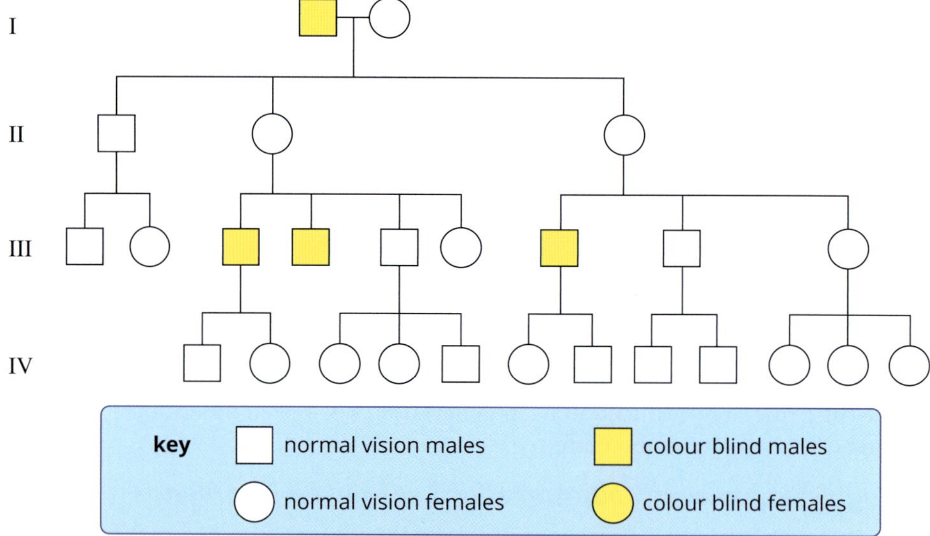

FIGURE 5.2.24 Pedigree of a family in which colour blindness is present. The inheritance of colour blindness is X-linked recessive and so males are affected more frequently than females. No colourblind females were observed in this particular pedigree.

X-linked recessive inheritance

The pedigree in Figure 5.2.24 also indicates that the inheritance of colour blindness is more likely to be X-linked recessive than autosomal recessive, because X-linked recessive traits affect more males than females. This is because males only inherit one X chromosome, while females inherit two. This second X chromosome has a masking effect on recessive alleles, resulting in females carrying the affected allele but not expressing the phenotype. In order for females to be affected by X-linked recessive traits, they must carry two copies of the allele, one on each X chromosome.

Given the low number of offspring, it is possible that all the colourblind individuals are males purely by chance. However, in reality, many such pedigrees have been studied and show similar patterns, confirming that the inheritance of colour blindness is X-linked recessive.

X-linked dominant inheritance

Traits that are X-linked dominant are rare and affect more females than males. This is because females inherit two X chromosomes and so have twice the chance of inheriting an affected X chromosome compared to males, who inherit only one.

Evidence of X-linked dominant inheritance is seen in a pedigree in which affected males have daughters who are all affected and sons who are not affected. This is because daughters inherit their father's only X chromosome, while sons inherit their father's Y chromosome. If the X chromosome carries an allele for a dominant trait, then the daughter will express its phenotype. X-linked dominant inheritance can be seen in Figure 5.2.25.

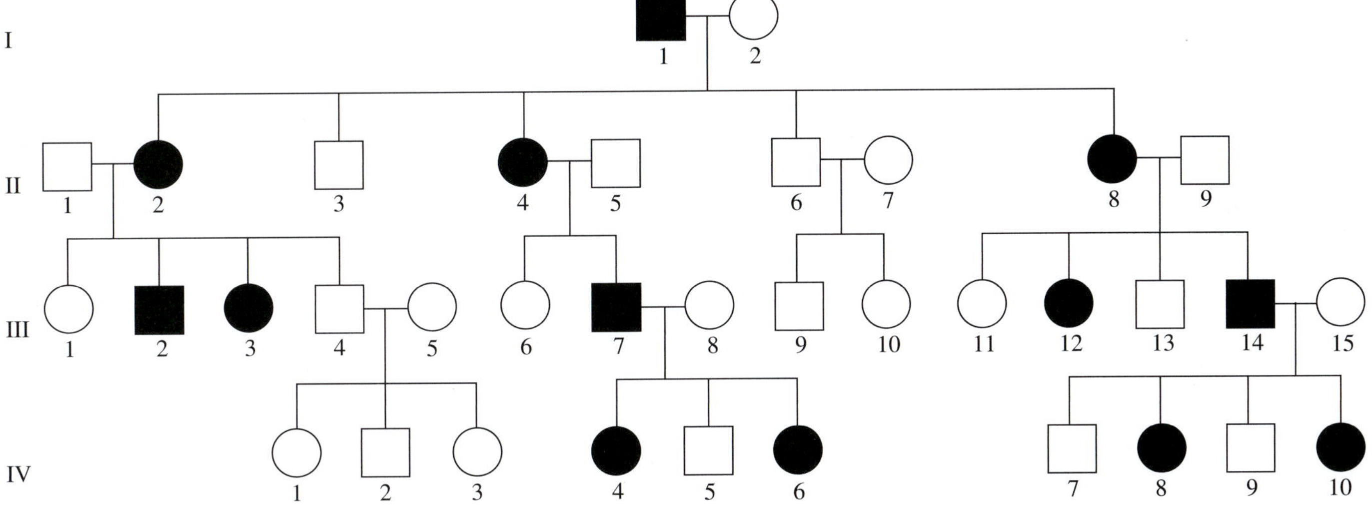

FIGURE 5.2.25 An example of X-linked dominant inheritance

Males I-1, III-7 and III-14 have daughters who are all affected and sons who are unaffected.

Because females pass on one of their two X chromosomes to both their daughters and sons, there is a 50% chance that they will pass on an X-linked dominant trait to their offspring. This pattern of inheritance is seen in female individuals II-2, II-4 and II-8, who have both affected and unaffected daughters and sons. An example of X-linked dominant inheritance in humans is Rett syndrome (RTT), a rare genetic neurological disorder of the grey matter of the brain. Another example is vitamin D-resistant rickets (Figures 5.2.14 and 5.2.26).

Y-linked inheritance

Male offspring inherit their father's Y chromosome, so any alleles carried on this chromosome will be passed on from father to son. The phenotype of Y-linked disorders is therefore only seen in fathers and all their sons. Females are never affected by Y-linked traits because they do not inherit a Y chromosome. As there is only one Y chromosome (**hemizygous**), and thus only one allele present, the general principles of dominant and recessive inheritance do not apply.

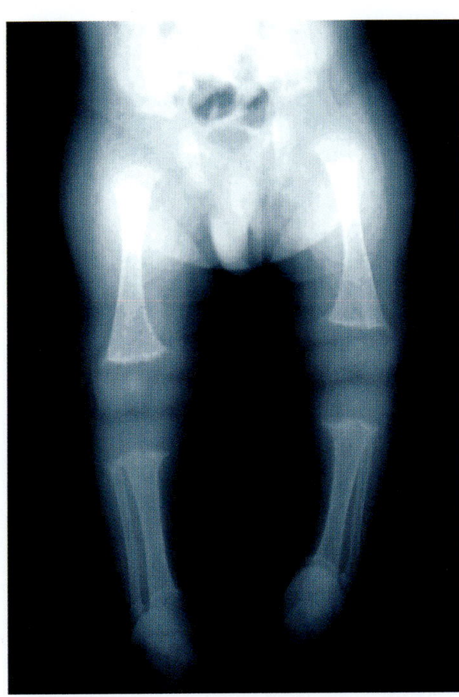

FIGURE 5.2.26 X-ray of the legs of a young child with vitamin D-resistant rickets

A Y-linked trait is likely if:

- only males are affected
- all male offspring are affected
- the trait is observed in every generation in which males are born.

This pattern of inheritance can be seen in Figure 5.2.27. Because the Y chromosome has far fewer genes that encode for proteins compared to the X chromosome, Y-linked inheritance of disorders is relatively rare.

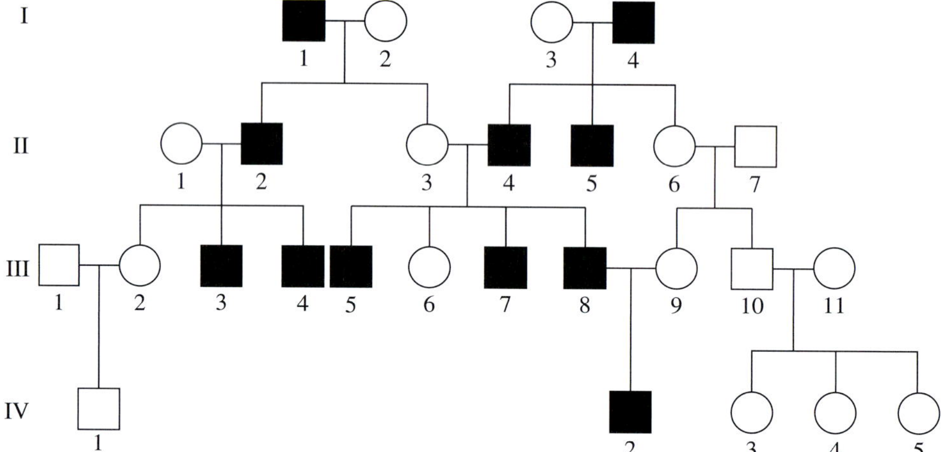

FIGURE 5.2.27 Pedigree chart for a trait with Y-linked inheritance. This pattern of inheritance is evident because only males are affected and the trait is present in males in every generation.

Ruling out sex-linked inheritance

Consider the following pedigree for a trait in humans. It is possible to rule out sex-linked inheritance in this pedigree (Figure 5.2.28) by looking for a pattern that will rule out each type of sex-linked inheritance in turn. (If no such pattern can be found, then that type of inheritance cannot be ruled out.)

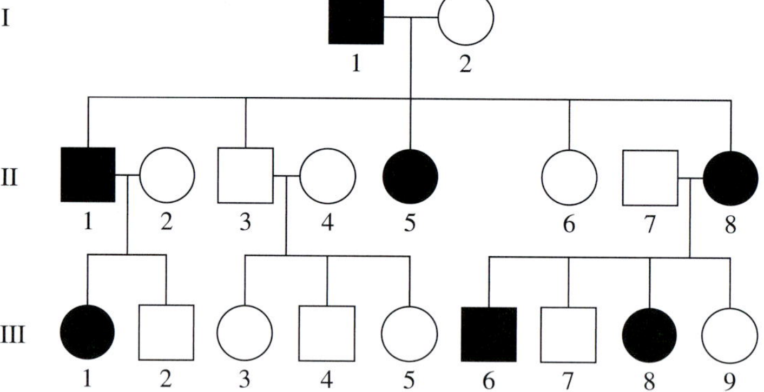

FIGURE 5.2.28 A pedigree chart for an inherited trait in humans. At this stage it is not known whether the inheritance is sex-linked or autosomal.

Ruling out X-linked recessive inheritance

In order for a trait to be X-linked recessive, affected mothers must have affected sons. In Figures 5.2.28 and 5.2.29, individual II-8 has two sons, one affected by the trait and the other unaffected. For the trait to be X-linked recessive, the mother would have to carry the allele on both her X chromosomes (X^hX^h) to be affected. The mother contributes one of these chromosomes to her offspring, so both sons would have to receive a chromosome with the affected allele. But one son does not show the trait (and therefore did not receive the affected allele), so X-linked recessive inheritance cannot be involved.

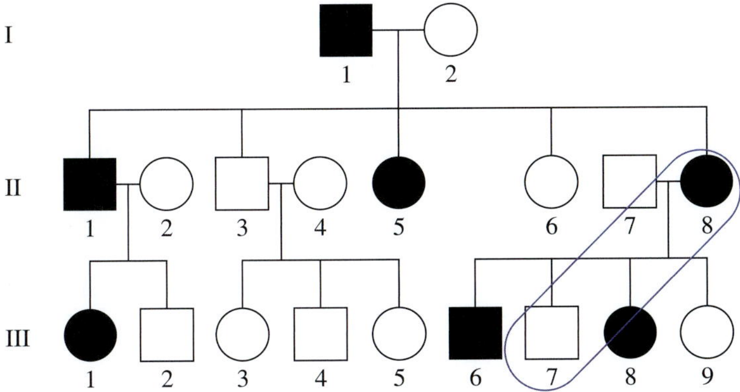

FIGURE 5.2.29 Ruling out X-linked recessive inheritance

Ruling out X-linked dominant inheritance

For X-linked dominant inheritance to be involved, every daughter of an affected male must be affected. This is because the daughters must receive an X chromosome from their father, and the father has only one X chromosome. In the pedigree shown in Figure 5.2.30, male I-1 has the trait but one of his daughters (II-6) does not. X-linked dominant inheritance therefore cannot be involved in this pedigree.

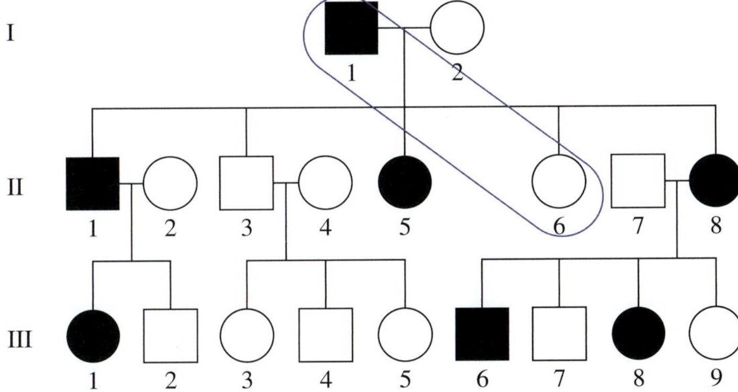

FIGURE 5.2.30 Ruling out X-linked dominance

Ruling out Y-linked inheritance

If a trait is Y-linked, only males are affected (because females lack a Y chromosome) and affected fathers pass the trait on to all their sons. In the pedigree shown in Figure 5.2.31, some females have the trait, and not all fathers with the trait passed it on to their sons (e.g. male I-1 and son II-3; also II-1 and III-2). Y-linked inheritance is therefore not involved in this pedigree.

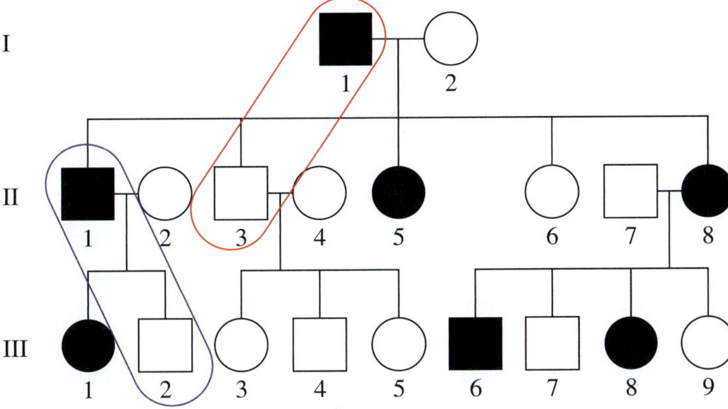

FIGURE 5.2.31 Ruling out Y-linked inheritance.

After ruling out X-linked recessive, X-linked dominant and Y-linked inheritance, it can be concluded that the trait shown in the previous figures must be the result of autosomal inheritance. Although the trait is observed in every generation, which often indicates dominant inheritance, in this case the trait could be autosomal dominant or autosomal recessive inheritance. Further information would be needed to determine which is involved in this pedigree.

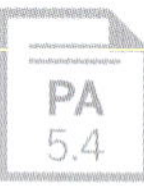

STEPS IN PEDIGREE ANALYSIS

A pedigree analysis should be carried out in a methodical series of steps to determine the pattern of inheritance. These steps are outlined below and illustrated in Figure 5.2.32.

Step 1: Determine whether the condition is sex-linked.

a Are mostly males affected? If only males are affected, and the trait passes from father to son in every generation, inheritance is most likely Y-linked.

b Do affected daughters have an affected father?

c Do affected mothers have affected sons but not affected daughters? If conditions in (b) or (c) are met, inheritance is most likely X-linked recessive.

d If you answered 'no' to either question in (b) and (c), go to step 2.

Step 2: Look for two affected/unaffected parents that have a child with a different phenotype.

a If two unaffected parents have an affected offspring, inheritance is autosomal recessive.

b If two affected parents have an unaffected offspring, inheritance is autosomal dominant.

WS 5.8

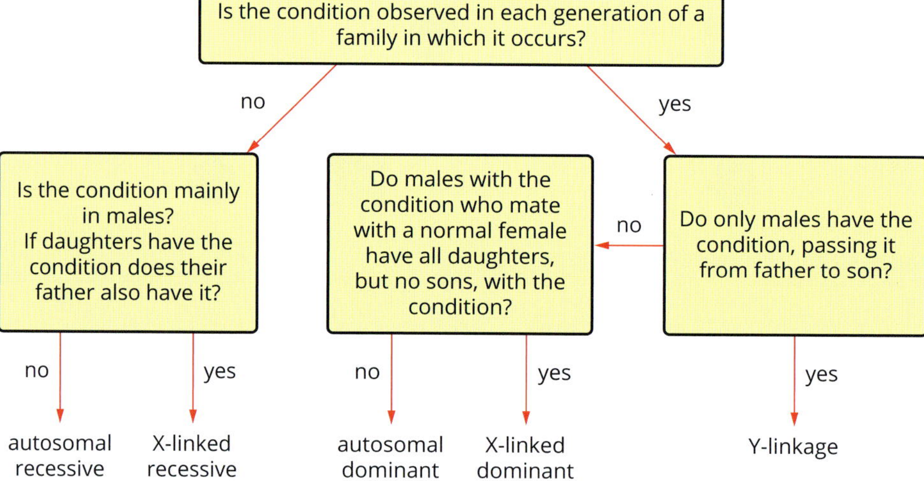

FIGURE 5.3.32 Flow diagram for pedigree analysis of simple modes of inheritance

INDEPENDENT ASSORTMENT

The crosses discussed so far relate to one trait, such as the colour of a flower. In this section, you will learn about the inheritance of two characteristics as either independent or linked, and the biological consequence of crossing over for linked genes.

Mendel's second law of inheritance, the **Law of Independent Assortment**, states that the alleles of a gene controlling one trait assort independently of alleles of another gene controlling a different trait. This can be illustrated by considering crosses involving two genes that affect two distinct traits.

You can cross true-breeding strains of fruit flies (*D. melanogaster*) that differ for two traits: eye colour and body colour (Figure 5.2.33) and then conduct a **dihybrid cross** ('di' meaning two). The two traits in this example are eye colour and body colour in *D. melanogaster*.

FIGURE 5.2.33 This fruit fly (*Drosophila melanogaster*) has red eyes and a brown body.

The eye-colour gene in this example is the yellow eye gene. (This is a different gene from the white eye-colour gene discussed earlier, and is located on a different chromosome). The 'yellow eye' gene is autosomal; the alleles are *Y* and *y*.

The wild type 'red eye' phenotype (genotypes *YY* and *Yy*) is dominant and the 'yellow eye' phenotype (genotype *yy*) is recessive.

The second trait is body colour and the gene in this example is called 'brown body'. It is an autosomal gene with two alleles *B* and *b*. The wild type 'green body' phenotype (genotypes *BB* and *Bb*) is dominant and the 'brown body' phenotype (genotype *bb*) is recessive.

The eye-colour and body-colour genes are on different chromosomes.

Wild-type phenotypes are not always dominant to mutant phenotypes.

The Law of Independent Assortment states that the alleles of genes that code for different traits are inherited independently from each other.

+ ADDITIONAL

Dihybrid crosses

A dihybrid cross is a cross between two individuals that differ in two phenotypes (i.e. have different alleles at two loci). This is different to a monohybrid cross which is a cross between two individuals with different alleles at a single locus.

An example of a dihybrid cross is shown below:

F1 generation

In a dihybrid cross, a cross is first set up between a true-breeding red eye, green body (*YY*, *BB*) strain and a yellow eye, brown body (*yy*, *bb*) strain. The homozygous *YY*, *BB* strain will produce gametes carrying only the *Y* and *B* alleles and the homozygous *yy*, *bb* strain will produce only *y* and *b* alleles. The gametes fuse to produce F1 progeny, which are heterozygous for both genes (*Yy*, *Bb*). Following meiosis, a gamete will end up with any of four possible combinations of alleles: *YB*, *yb*, *yB* or *Yb*.

During gamete formation in F1, the chance of a sperm or egg cell containing a *Y* allele is 0.5 (because half of the gametes contain a *Y* allele). The chance that a gamete will contain a *B* allele is also 0.5. Therefore, the chance of the gamete being *YB* is $0.5 \times 0.5 = 0.25$. The probability of each of the other three possible gametic combinations is also 0.25.

This is because the segregation of alleles of a gene on one chromosome in meiosis is independent of the segregation of alleles of a gene on another chromosome. The homologue (matching chromosome) carrying the *Y* allele can move to either pole, as can the homologue carrying the *y* allele. These homologues move independently of the chromosomes that carry the *B* and *b* alleles (Figure 5.2.34). This is the Law of Independent Assortment of genes.

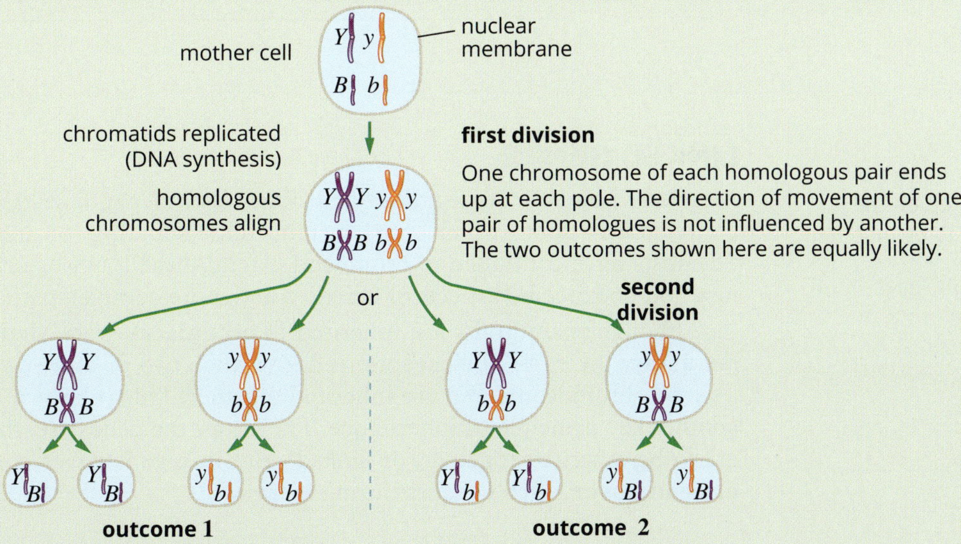

FIGURE 5.2.34 New combinations of alleles result from cells dividing by meiosis. Two genetic loci are shown, each on separate chromosomes within the cell nucleus. Daughter cells produced by meiosis may have any of four possible combinations of alleles.

F2 generation

The heterozygotes generated in the F1 can be crossed together (a dihybrid cross) to produce an F2 generation. Figure 5.2.35 shows that the expected ratio of phenotypes in the F2 generation is:

- 9 red eye, green body
- 3 red eye, brown body
- 3 yellow eye, green body
- 1 yellow eye, brown body.

If these crosses were actually performed, the phenotypic ratio in the F2 generation should be close to the 9 : 3 : 3 : 1 ratio. There would be some difference between the expected and observed phenotypic ratios due to sampling error. The larger the number of F2 progeny scored, the closer the result will be to the 9 : 3 : 3 : 1 ratio.

A phenotypic ratio approximating 9 : 3 : 3 : 1 will be observed in the F2 generation of a dihybrid heterozygous cross if the following five conditions apply:

- the two genes control two distinct traits
- there are two alleles for each of the genes
- one phenotype is dominant for each trait
- both genes are on autosomes
- the two genes assort independently.

In this example, independent assortment occurs because the two genes are on different chromosomes. However, you will learn later in this section that independent assortment can occur via another mechanism.

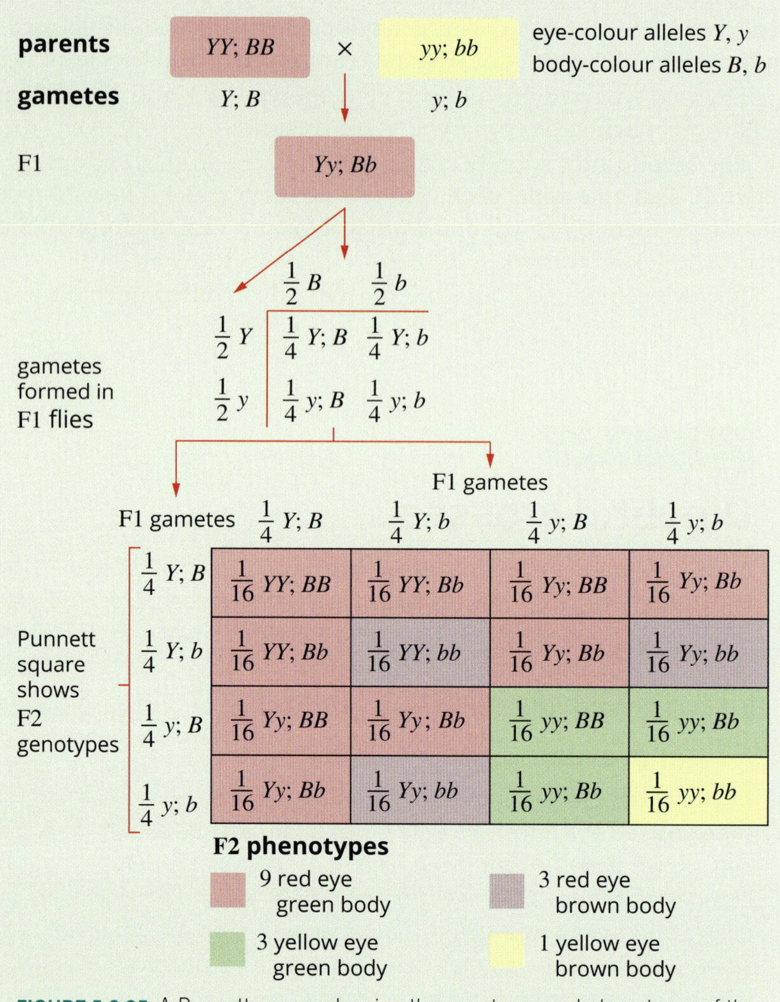

FIGURE 5.2.35 A Punnett square showing the genotypes and phenotypes of the F2 progeny. The F1 generation produces *YB*, *yB*, *yB* and *yb* gametes in equal frequency. When F1 individuals are crossed, the resulting F2 shows a 9 red eye, green body : 3 red eye, brown body : 3 yellow eye, green body : 1 yellow eye, brown body phenotypic ratio.

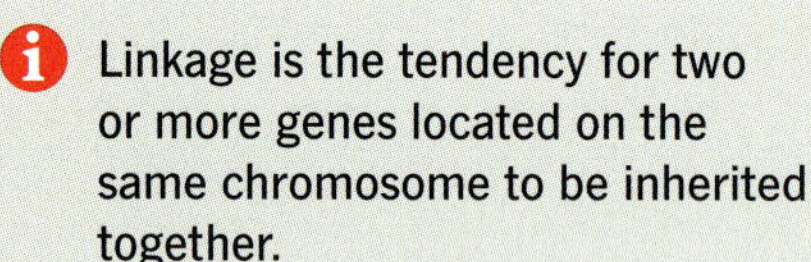

Linkage is the tendency for two or more genes located on the same chromosome to be inherited together.

LINKED GENES

Mendel's Law of Segregation and Law of Independent Assortment are the cornerstones upon which our current understanding of inheritance is built. Scientists have continued to refine and extend these Laws or principles to explain new and unexpected aspects of heredity, and more complex patterns of inheritance.

Although many traits are inherited in accordance with Mendel's Laws, this is not always the case. The exceptions occur when two or more genes are located on a single chromosome and are inherited together. This is known as **linkage**, and is another key principle of inheritance. The closer the genes are, the more likely they are to be inherited together, or 'linked'. But linkage is never complete because of **crossing over**, which occurs during meiosis.

Linkage and recombination

Crossing over is a normal event that results in genetic exchange between non-sister chromatids. It will occur in most germline cells going through meiosis. The probability of at least one cross-over event occurring somewhere on the chromosome is high because there is usually at least one **chiasma** (point of crossing over between chromosomes; plural: chiasmata) for each homologous pair in meiosis (Figure 5.2.36).

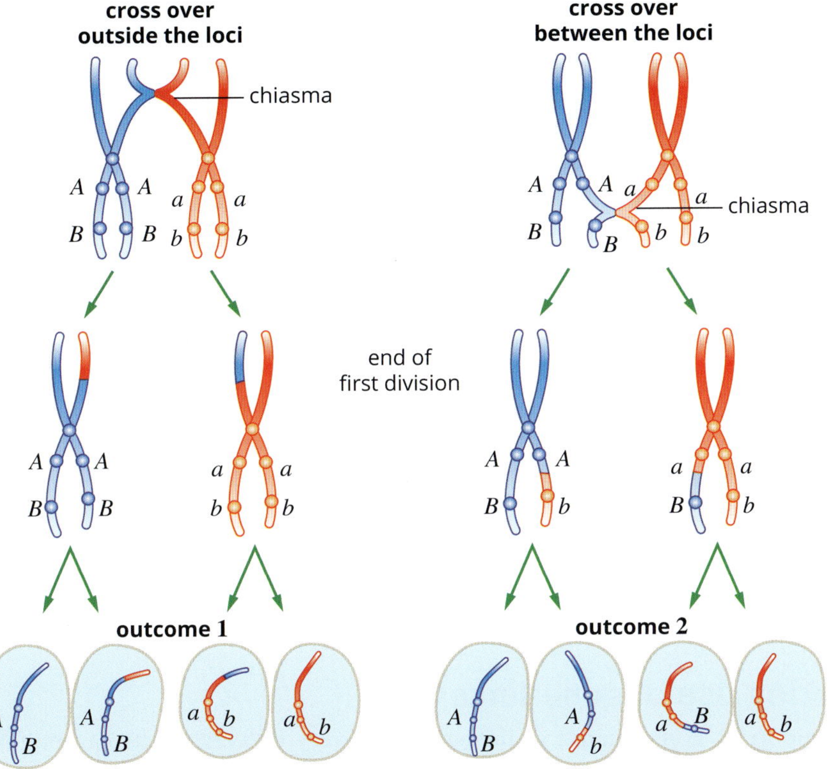

FIGURE 5.2.36 Illustration showing the consequences of crossing over. When two genes, *A* and *B*, are located on the same chromosome, crossing over may occur outside the loci (outcome 1) or between the two loci (outcome 2). The gametes produced in these two situations (outcome 1 and outcome 2) are very different.

If the A and B loci are very close together, the probability of a random cross-over event occurring between them (outcome 2) is very low.

If genes are close together, there will be fewer **recombinant gametes** and more parental gametes produced (outcome 1). The closer the two genes are together, the rarer the recombinant gametes will be.

If genes are so far apart (on the same chromosome) that close to 50% of the gametes are recombinants, then independent assortment is observed. If the percentage of recombinant gametes is less than 50%, the two genes are considered to be linked (Figure 5.2.37).

> Crossing over is the exchange of chromosomal material between members of a homologous pair of chromosomes during meiosis.

> Non-sister chromatids are chromatids of paired homologous chromosomes, one from each parent. Paired non-sister chromatids form chiasmata (crossing points) during prophase I of meiosis.

> Recombinant gametes are produced during crossing over and contain a combination of alleles that are not in either parent.

Recombination and distance between linked genes

By measuring the percentage of recombinant gametes produced by an F1 heterozygote when genes are linked, it is possible to estimate the distance between the two genes. The farther apart two genes are on the chromosome, the more frequently crossing over will occur and the higher the observed percentage of recombination. By repeating such measurements for different pairs of genes, the position of any identifiable gene on a particular chromosome can be found. This process is called **gene mapping**.

> Genes are said to be linked when the percentage of recombinant gametes falls below 50%.

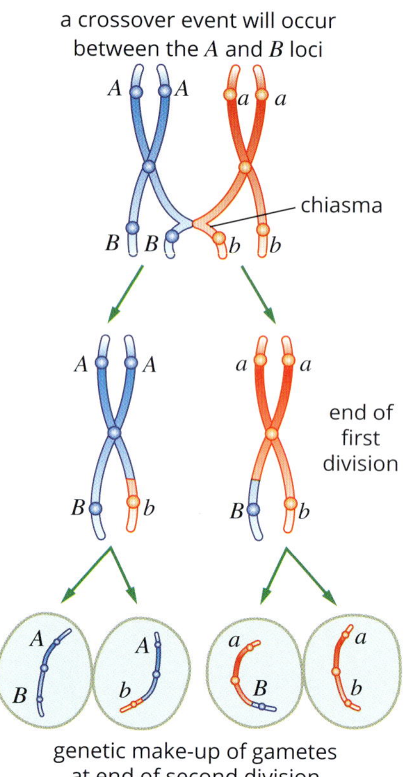

FIGURE 5.2.37 If the genes *A* and *B* are far enough apart on the same chromosome, there will be an average of one cross-over event between the genes in every cell, and 50% of the gametes will be recombinant. Therefore, the genes and their alleles assort independently.

Making use of gene linkage

An important consequence of linkage is that very different traits can be inherited together. This can be extremely important in terms of understanding human health, and in plant and animal breeding.

> The percentage of recombination between two linked genes is correlated with their physical distance apart along the length of the chromosome.

Information about one locus provides us with the likely genotype at the other locus. When discussing gene mapping, the term **genetic marker** refers to a sequence of DNA with a known location on a chromosome (it is important to note that in other contexts, such as population genetics, the location of a genetic marker might be unknown). If a gene of interest is closely linked to a genetic marker, then this can be used to determine if someone carries a mutation (a change in DNA sequence). The closer two loci are, the greater the linkage and the more precisely a genotype can be used at one locus to predict the other.

Molecular techniques and genetic markers play an important role in this area. Linkage relationships are becoming increasingly important in providing information about the likelihood that a child will have a particular disorder. A marker locus closely linked with the gene that causes a disorder can be extremely useful in determining the genotype of a parent who has the disorder. If a child has the same marker as the parent with the disease, the child is also likely to have the disorder.

BIOLOGY IN ACTION CCT DD

Cystic fibrosis and linkage mapping

Cystic fibrosis (CF) is an inherited disorder that affects the respiratory and digestive systems. It can significantly shorten the lifespan of people with the condition. In a person with cystic fibrosis, the mucous glands secrete thick, sticky mucus, which clogs the airways, leading to breathing difficulties, respiratory infections and lung damage (Figure 5.2.38). The mucus also affects the pancreas, inhibiting the release of important digestive enzymes, which causes a range of nutritional problems. There is currently no cure for the disorder, but modern treatments are continuing to improve life expectancy for those with cystic fibrosis.

The symptoms of cystic fibrosis were first identified in 1938. Finding the gene responsible was a difficult task because its protein product was not known at the time and the gene could have existed on any of the 23 human chromosomes.

In the 1980s, researchers conducting linkage analysis tracked and mapped five genes that were linked to the gene that causes CF (the *CFTR* gene). The data showed that the *CFTR* gene was located on the long arm of chromosome 7. The gene was subsequently cloned in 1989 and its gene product (protein) was identified as a membrane chloride channel protein in 1992. This protein regulates the movement of salt in and out of cells. Because the gene is faulty, the regulation of salt movement is inefficient and leads to a build up of salt in the cells, which causes the production of thick mucus.

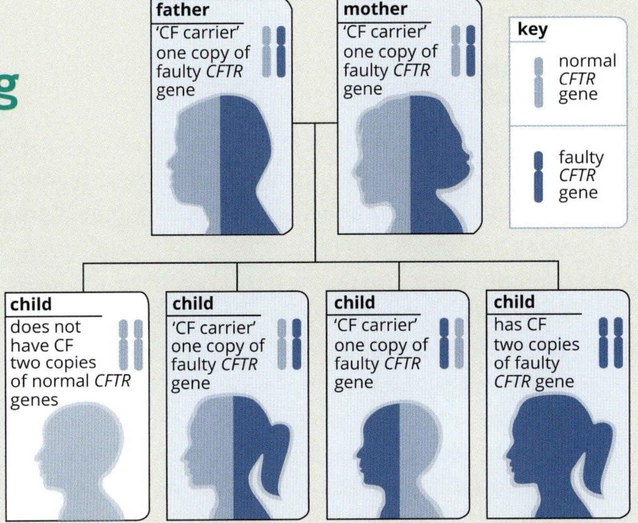

FIGURE 5.2.39 Autosomal recessive inheritance of cystic fibrosis. Regardless of their biological sex, an individual has a 25% chance of inheriting cystic fibrosis if both their parents are carriers of the *CFTR* gene.

It is now known that cystic fibrosis is an autosomal recessive disorder (Figure 5.2.39). It is the most common genetic life-threatening disorder in Australia; more than one million Australians carry a copy of the faulty *CFTR* gene.

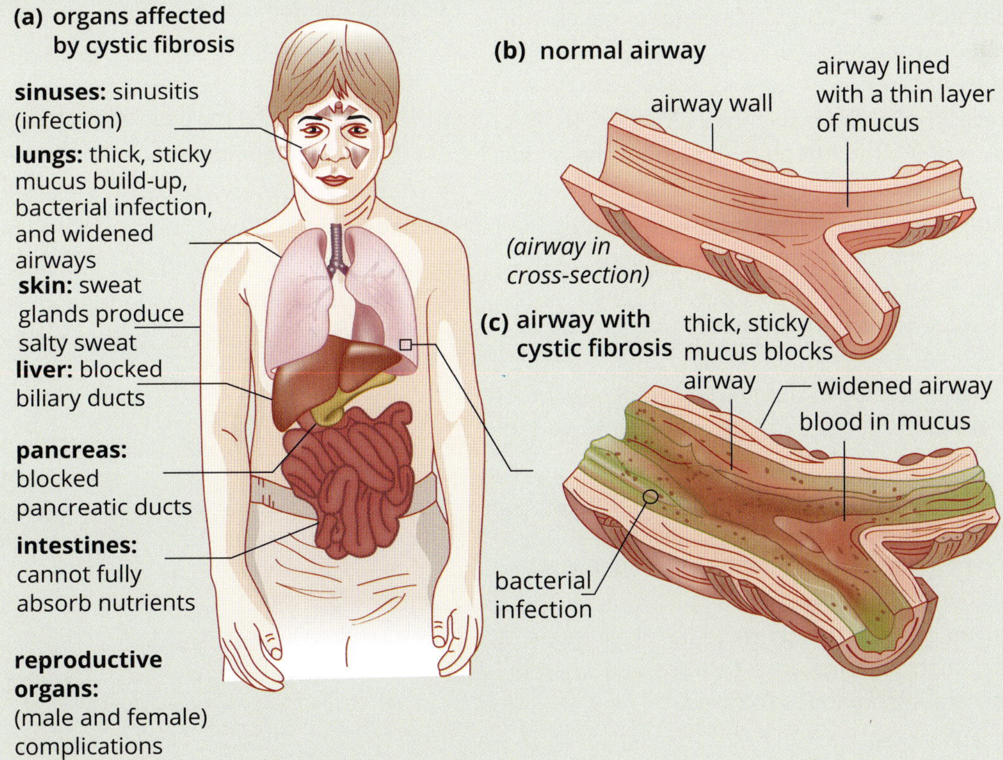

FIGURE 5.2.38 (a) The effects of cystic fibrosis on the systems of the human body and (b) a normal airway compared to (c) the airway of someone with cystic fibrosis.

5.2 Review

SUMMARY

- The Law of Segregation states that individuals carry pairs of alleles of each gene, which segregate into gametes during meiosis so that each gamete carries one allele of each gene.
- True-breeding strains are homozygous at the locus of interest and produce genetically identical progeny when crossed with each other.
- A phenotypic ratio of approximately 3 : 1 will be observed in the F1 generation of a monohybrid cross between two heterozygous individuals for any trait controlled by a single autosomal gene, with two different alleles (one dominant and one recessive).
- A test cross involves crossing an individual displaying the dominant phenotype but unknown genotype with an individual displaying the recessive phenotype(s).
- Test crosses are used to determine whether an individual of dominant phenotype is homozygous or heterozygous.
- A phenotypic ratio of approximately 1 : 2 : 1 will be observed in the F2 generation of a monohybrid cross for any trait controlled by a single autosomal gene, with two different alleles, displaying co-dominance.
- Co-dominant inheritance can be seen in ABO blood grouping.
- Phenotypes inherited through the action of genes located on either the X or Y chromosomes show sex-linked inheritance.
- X-linked recessive inheritance shows a pattern of transmission of the mutant phenotype from the female parent to male offspring.
- X-linked dominant inheritance shows a pattern of transmission of the dominant trait from an affected male parent to all female offspring and from an affected heterozygous female parent to 50% of all offspring.
- Y-linked inheritance shows a pattern of transmission of the trait from father to son, and it is never observed in females.
- Pedigree analysis is a technique of looking through a family tree (of humans or other organisms) for the occurrence of a particular characteristic in one family over a number of generations.
- Pedigrees can be used to determine the likely mode of inheritance, such as dominance patterns and whether inheritance is autosomal or sex-linked.

Autosomal recessive

- Both sexes display the trait in equal numbers in a pedigree.
- Offspring of unaffected parents have a 25% chance of being affected.
- Affected individuals are homozygous.

Autosomal dominant

- Both sexes display the trait in equal numbers in a pedigree.
- One parent must be affected to have an affected offspring.
- Two affected parents with an unaffected offspring indicate dominance: the parents need only one 'dominant' allele to express the trait, so their offspring may inherit their unaffected alleles.
- The trait is observed in each generation.

X-linked recessive

- The trait is rare within the pedigree, but males are more affected than females.
- Affected fathers do not pass the trait on to their sons, so the condition can skip generations.
- Females can be carriers and not show the condition; females pass the trait on to their sons.

X-linked dominant

- Males and females are affected (often more females than males).
- All affected sons have an affected mother.
- All affected daughters have an affected father.
- The trait is observed in each generation.

Y-linked

- Only males are affected, not females.
- Fathers pass the trait on to their sons.
- The trait is observed in each generation.

- Independent assortment occurs because the segregation of one pair of homologous chromosomes (and the alleles they carry) in meiosis, does not influence the segregation of other homologous pairs of chromosomes.
- Linkage is the tendency for two or more genes located on the same chromosome to be inherited together.
- Genes are 'linked' when the percentage of recombinant gametes falls below 50%.
- Recombinant gametes carry a combination of alleles not observed in the parents.
- The percentage of recombinant progeny can be used to estimate the distance between two genes on chromosomes.

KEY QUESTIONS

1 What is meant by the term 'monohybrid cross'?

2 Freckles are an inherited trait that results in the formation of spots on fair skin. The gene associated with freckle formation is found on chromosome 4 and the trait for freckles will be expressed if only one parent carries the allele.

a State the type of inheritance (autosomal, sex-linked, dominant and/or recessive).

b Using suitable symbols, draw a Punnett square to show how a mother and father, who have freckles, can have a child that does not have freckles. State the probability of the child not having freckles.

3 Why is pedigree analysis often the easiest way to investigate inheritance patterns in humans? Give three reasons.

4 Individual III-5 has a genetic condition but the rest of her family are unaffected. What are the genotypes of her parents (II-3 and II-4)? Choose from options A–D and draw a Punnett square to show your reasoning.

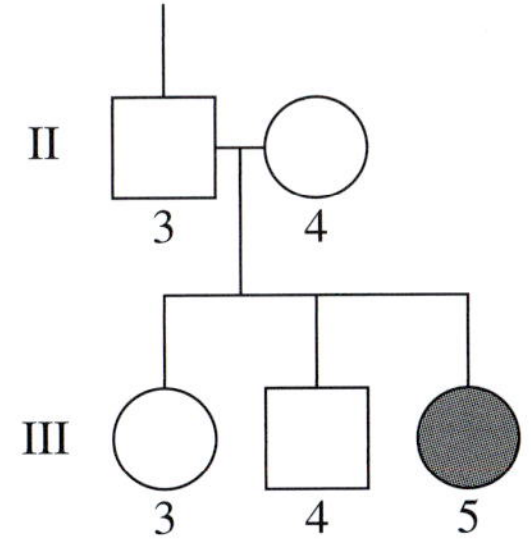

A *BB, BB*

B *Bb, Bb*

C X^BX^B, X^BY

D X^BX^B, X^bY

5 What type of inheritance is shown in the pedigree below? Give three reasons for your choice of inheritance pattern.

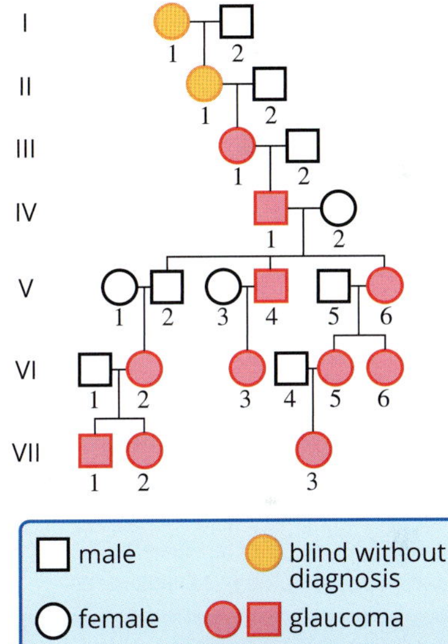

6 What is the Law of Independent Assortment, and what does it state?

7 What are the possible outcomes of meiosis in a heterozygote individual with the genotype *AaBb*? List all the gametes that this individual could produce.

8 For each of the following statements, state whether it is true or false.

a Genes are linked when the percentage of recombinant gametes falls below 50%.

b Recombinant gametes are observed in the offspring.

c If A and B loci are very close together, the probability of a random cross-over event is very low.

BIOLOGY INQUIRY  

Comparing big cat DNA

How can the genetic similarities and differences within and between species be compared?

COLLECT THIS...

- paper and pen
- a tablet or computer to access the internet

DO THIS...

1. Refer to Table 5.3.1, which shows amino acid sequences of big cat species and work out the number of amino acid differences between each individual.
2. Draw a table to compare the number of amino acid differences between each individual. Your teacher can provide an example table.
3. Refer to Table 5.3.2, which shows DNA nucleotide sequences of big cat species and work out the number of nucleotide differences between each individual.
4. Draw another table (like the table in step 2) to compare the number of DNA nucleotide differences between each individual.
5. Examine the tables carefully, noting which species are distinct and which are the same at each site.
6. Use the internet to research the species in the tables. Examine pictures of all species and look up where the big cats are found. Do they look as similar or different as their sequence data might suggest?
7. Try to find a phylogenetic tree showing the currently understood relationships between the species. Does this match the relationships you have worked out? You only have a small genetic data set. Would a larger data set show a different result?

RECORD THIS...

Describe the evolutionary relationships between the species and genera (plural of genus) of the big cats.

Present your findings to the class, using your tables.

Discuss the benefits of using amino acid and DNA nucleotide data when investigating evolutionary relationships.

REFLECT ON THIS...

How can the genetic similarities and differences between species be compared?

Why is it important to use more than one genetic marker and analyse more than one individual from a species?

Why might you use DNA sequences instead of amino acid sequences to understand evolutionary relationships?

TABLE 5.3.1 Five variable amino acid (AA) sites in a mitochondrial gene (*COI*), for nine big cat species

Species	Variable AA sites				
	1	2	3	4	5
Bengal tiger (*Panthera tigris tigris*)	L	V	W	M	Y
Siberian tiger (*Panthera tigris altaica*)	L	V	R	M	Y
South China tiger (*Panthera tigris amoyensis*)	L	V	W	M	Y
African lion (*Panthera leo*)	F	V	W	M	H
Leopard (*Panthera pardus*)	L	V	W	M	Y
Puma (*Puma concolor*)	L	V	W	M	Y
Eurasian lynx (*Lynx lynx*)	L	I	W	M	Y
Canadian lynx (*Lynx canadensis*)	L	I	W	M	Y
Bobcat (*Lynx rufus*)	L	I	W	M	Y

TABLE 5.3.2 Ten variable DNA nucelotide sites in a mitochondrial gene (*COI*), for nine big cat species

Species	Variable DNA nucleotide sites									
	1	2	3	4	5	6	7	8	9	10
Bengal tiger (*Panthera tigris tigris*)	T	G	A	G	A	A	C	G	C	A
Siberian tiger (*Panthera tigris altaica*)	T	A	G	G	A	A	C	G	C	A
South China tiger (*Panthera tigris amoyensis*)	T	G	A	G	G	A	C	G	C	A
African lion (*Panthera leo*)	T	G	A	A	A	A	C	A	C	A
Leopard (*Panthera pardus*)	C	G	A	A	A	A	C	G	C	A
Puma (*Puma concolor*)	T	G	A	A	A	T	C	C	C	A
Eurasian lynx (*Lynx lynx*)	T	G	A	A	A	G	C	C	C	G
Canadian lynx (*Lynx canadensis*)	T	G	A	A	A	G	T	C	C	G
Bobcat (*Lynx rufus*)	T	G	A	A	A	G	C	C	T	G

Now that you have learnt how genetic variation is inherited in related groups, you can begin to predict the patterns of genetic variation in large groups of unrelated individuals. In this section, you will learn how to model the behaviour of alleles in populations.

ALLELE FREQUENCIES IN A POPULATION

When a gene has different alleles in a population, it is referred to as **polymorphic**. These different alleles are more common in some populations than others. The rate at which alleles occur in a population is known as the **allele frequency**. An allele that occurs in 30% of people in a population would have therefore an allele frequency of 0.3. An allele that has a frequency of 1.0 in a population (that is, it occurs in every single individual in that population) is referred to as a **fixed allele**. In that population, the allele is considered **monomorphic**. All allele frequencies in a population must add up to 1.0.

The allele frequencies in a population can be estimated by examining phenotypes. When discussing genes with only two alleles, the alleles are represented as p and q.

Differences between populations

When studying the genetics of a population, it is important to understand what is meant by population. The term 'population' often refers to the inhabitants of a particular place, but in biology, it is used to refer to any group of interbreeding organisms. Genetics can be used to estimate where one population ends and another begins by determining if there is **gene flow** between the populations. Gene flow is the exchange of genes between populations (e.g. via migration and interbreeding).

One way to do this is by comparing allele frequencies of different populations. Imagine two populations of snapdragons. If the two populations are exchanging genes/interbreeding then their allele frequencies should be very similar (i.e. in equilibrium). Note that this does not mean that their phenotypic ratios will be the same. If the two populations are not exchanging genes/interbreeding the allele frequencies will be different (i.e. not in equilibrium).

If there is a high level of gene flow between populations they may be considered a single interbreeding population.

If there is not much gene flow between populations, they may be considered separate gene pools or populations.

In some cases, where multiple genes are examined, there may be enough evidence to conclude that the populations are even different species.

Of course, examining a single gene does not provide much information. The results could be due to random chance, as fertilisation success is subject to chance. To get a statistically significant result, you need samples with many individuals and many different genes. Then you can run a more robust statistical analysis, to see whether the allele frequencies in these populations are significantly different from one another. If they are, this would indicate that something is restricting the flow of genes between these populations. This could be distance, a physical barrier such as a mountain range, or even that the populations have overlapping distributions but can no longer interbreed.

FIGURE 5.3.1 Many genes for coat colour in horses show incomplete dominance. A cross between (a) a cremello horse and (b) a chestnut horse results in (c) a palomino offspring.

Worked example 5.3.1

HORSE COAT COLOUR

The iconic palomino horse (Figure 5.3.1) is the result of a cross between a chestnut and a cremello (cream) horse. Palomino horse coat colour is an example of incomplete dominance, meaning that phenotypic ratios can be estimated by examining the individuals.

In a paddock with 80 horses, 50 are chestnut, 20 are palominos and 10 are cremello. Use the symbols C_1 to represent the allele for a chestnut coat, C_2 to represent the allele for a cremello coat, and C_1C_2 to represent the genotype of a palomino horse. What are the allele frequencies of C_1 and C_2 in this population? Let $C_1 = p$ and $C_2 = q$.

Thinking	Working
Determine how many copies of the gene are in the population.	Horses are diploid, so each horse has two copies. Therefore: $80 \times 2 = 160$ copies.
Identify how many copies of each allele are in each group.	Each of the 50 chestnut horses has two copies of the C_1 allele, resulting in a total of 100 copies. Each of the 20 palomino horses has one copy of the C_1 allele and one copy of the C_2 allele, making for $20 \times C_1$ and $20 \times C_2$. Each of the 10 cremello horses has two copies of the C_2 allele, resulting in a total of 20 copies of the C_2 allele.
Calculate how many copies there are of each allele across the groups.	C_1 allele: 100 (chestnut) + 10 (palomino) + 0 (cremello) = $120 \times C_1$ C_2 allele: 0 (chestnut) + 20 (palomino) + 20 (cremello) = $40 \times C_2$
Check that these numbers add up to the total number of gene copies.	$120\ C_1 + 40\ C_2 = 160$
Calculate the allele frequencies by dividing the number of allele copies by the total number of gene copies.	C_1: 120 divided by 160 = 0.75 C_2: 40 divided by 160 = 0.25
Remember to use appropriate notation (i.e. p and q).	$p = 0.75$ $q = 0.25$
Check that all allele frequencies add up to 1.0.	$0.75 + 0.25 = 1.00$

Worked example: Try yourself 5.3.1

HORSE COAT COLOUR

In a paddock with 50 horses, 12 are chestnut, 18 are palominos and 20 are cremello. Use the symbols C_1 to represent the allele for a chestnut coat, C_2 to represent the allele for a cremello coat, and C_1C_2 to represent the genotype of a palomino horse.

What are the allele frequencies of C_1 and C_2 in this population? Let $C_1 = p$ and $C_2 = q$.

TABLE 5.3.3 Modified Punnett square showing the calculation of allele frequencies p and q for the two alleles (A and a) of the next generation

	Pollen	
	p A	q a
p A	p^2 AA	pq Aa
q a	pq aA	q^2 aa

TABLE 5.3.4 Modified Punnett square depicting the expected genotypic ratios of a second generation of snapdragons, based on allele frequencies from a parent population

Ovule	Pollen	
	0.65 R_1	0.35 R_2
0.65 R_1	p^2 R_1R_1 0.65×0.65 = 0.4225	pq R_1R_2 0.35×0.65 = 0.2275
0.35 R_2	pq R_2R_1 0.65×0.35 = 0.2275	q^2 R_2R_2 0.35×0.35 = 0.1225

+ ADDITIONAL

HARDY-WEINBERG EQUILIBRIUM

Now that you understand how genes are inherited between generations, you can use the allele frequencies of a population to predict what they should be in the next generation. This is assuming that the population is not undergoing selection; that is, the environment is not influencing which alleles survive and which do not. You can use a modified Punnett square to model this, using our snapdragons as an example.

In Table 5.3.3, the allele frequencies p and q for two alleles A and a are included in a modified Punnett square. This can tell you what gametes are produced (the usual function of a Punnett square) but it can also tell us how these will be arranged in terms of genotypes in the next generation.

You learnt in Section 5.1 that for a gene with an intermediate heterozygote phenotype (i.e. pink snapdragons), the expected phenotypic ratio of an F1 cross is 1 (red): 2 (pink): 1 (white). The logic with regard to calculating genotypic ratios at equilibrium is much the same, except you are now looking at the whole population instead of a single cross. For this reason, you must account for the allele frequencies of the parent population. To put it another way, if the allele frequency p for the parent generation is 0.65, that means that only 65% of the population can contribute that allele to the next generation.

Now substitute the numbers for our snapdragons into the modified Punnett square (Table 5.3.4) and multiply the frequency of each genotype together to determine the outcome.

If the population is not undergoing evolution, it is expected that the next generation of snapdragons will be approximately 42% red (R_1R_1), represented by p^2; 12% white (R_2R_2), represented by q^2; and 46% pink (R_1R_2), represented by $2pq$.

Note that, because you are looking at ratios, $p^2 + 2pq + q^2$ must equal 1. If you remember this equation, you do not have to draw a Punnett square every time, but it is a helpful way to derive the formula if you need to.

This model is referred to as **Hardy-Weinberg equilibrium**, because it calculates allele frequencies for a stable population at equilibrium, and was developed independently by two scientists: G. H. Hardy (a mathematician) and W. Weinberg (a German obstetric gynaecologist).

Limitations of Hardy-Weinberg equilibrium

Like many models of biological systems, Hardy-Weinberg equilibrium (HWE) is useful but imperfect. There are a number of assumptions in the model that are often not met in real life. HWE assumes an 'ideal population'. In an ideal population, the following statements are true.

1 All alleles are mutating at the same rate, so that inheritance is not affected.
2 There is no migration into or out of the population.
3 Generations do not overlap.
4 The population is not affected by selection (i.e. evolution is not taking place).
5 The population is large enough to not be affected by random events.
6 All mating is random (i.e. individuals do not show any preference or selection for mates).

You can probably already see that most populations will not meet all these requirements. This means that, if it turns out that a population is not in HWE, it might be because it is violating one of these assumptions. However, HWE is what we call a 'robust' theory, meaning that violations of the assumptions only affect allele frequencies in a minor way. For this reason, HWE is still a good starting point for understanding allele frequencies in populations and forms the basis for many other analyses. HWE can also be used to understand the ways in which a population is not an 'ideal' population. For example, if allele frequencies are particularly skewed, we might suspect that mating is not random, or that the population is small and is experiencing genetic drift.

GENETIC MARKERS

A genetic marker is any piece of DNA that can be reliably analysed using sequencing or genotyping. Genetic markers can be in coding regions (exons) or non-coding regions (introns). The location of a genetic marker in the genome is usually known but sometimes it is not. When the location of a genetic marker is not known, it is referred to as **anonymous DNA**. Although the location and function of anonymous DNA is unknown, sequences of this DNA can provide valuable information about the genetic variation within and between populations. Such genetic markers provide records of change and insight into evolutionary relationships.

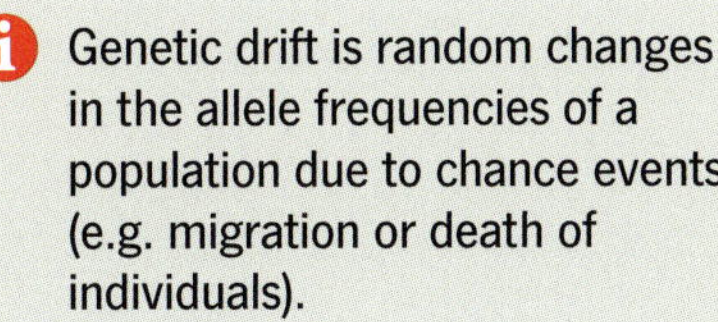
Genetic drift is random changes in the allele frequencies of a population due to chance events (e.g. migration or death of individuals).

Although you do not need to know where a genetic marker comes from in order for it to be useful, you cannot just use any genetic marker for population genetic research. An ideal genetic marker has the following properties.

1 It is polymorphic within the population of interest. That is, it must show variation. If a genetic marker is the same for all individuals, it will not provide any information.
2 It is usually a **neutral marker**, meaning that it is not currently under selection. This is so that probability-based models such as HWE can be applied. Probability assumes a level of random chance, whereas selection is non-random. There are situations where a marker that is under selection can be very useful in distinguishing populations, but they are difficult to analyse because of that selection pressure.

3 It is easy and affordable to work with (practical consideration). Some genetic markers are very expensive to isolate and analyse from the organism; some are more reliable while others are difficult to work with. Many population genetics projects have a focus on conservation and there is often a limited budget.

GO TO ➤ Section 6.1 page 248

Below, there is a quick overview of genetic markers and how they are used in population genetics. These markers will be covered in more detail in Chapter 6.

Sequence data

The most direct way to examine genetic variation is to examine the DNA sequence (Figure 5.3.2). It is common in population genetics to look at mitochondrial DNA (mtDNA), rather than DNA from the nuclear genome with its multiple chromosomes. Mitochondrial DNA is only inherited maternally (from the mother) in most species, and therefore each sequence will only have one allele. In a haploid genome such as the mitochondrial genome, an allele is referred to as a **haplotype**. MtDNA changes more rapidly than most nuclear DNA, meaning that differences in populations will be easier to see. A single DNA sequence from a useful genetic marker will have multiple points of difference in a population, rather than just one. We can then use these differences to work out which individuals are more closely related to one another, and how these populations have changed over time.

(a)

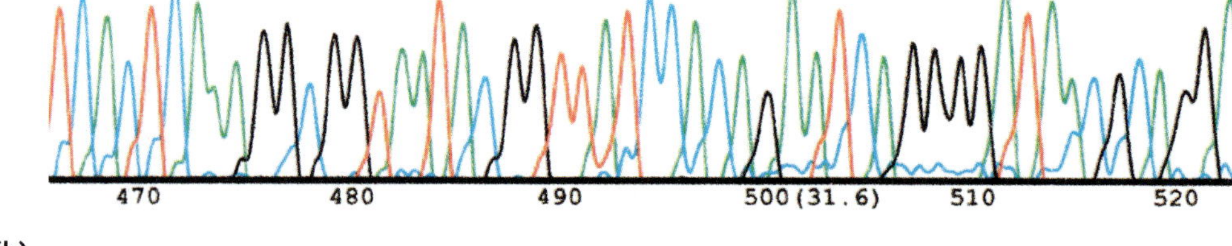

(b)

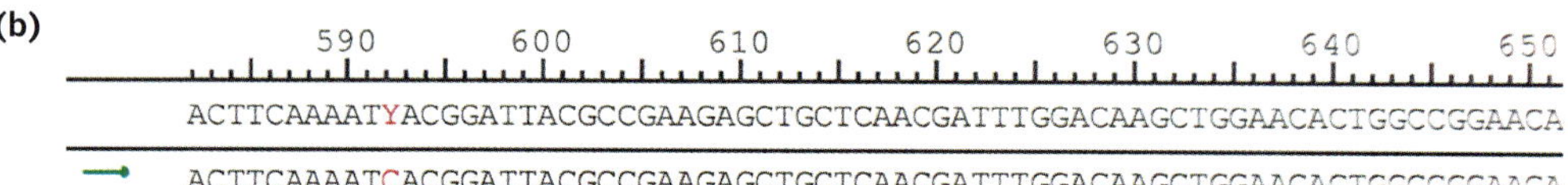

FIGURE 5.3.2 DNA sequence analysis is the most direct way of looking at genetic variation. (a) The raw result from an automated DNA sequencer is called a chromatogram. (b) When looking at variation, sequences from different individuals (but from the same genetic marker) are examined together and aligned. The sequence at the top is called the 'consensus' sequence. The consensus sequence is supposed to represent all individual sequences. Note that, in position 592, the consensus sequence uses the symbol Y, as some of the sequences have T and some have C in this position. Y means 'T' or 'C'. There are other symbols, called 'degenerate nucleotides', that refer to such variable positions in a sequence.

Single nucleotide polymorphisms (SNPs)

Single nucleotide polymorphisms (SNPs), pronounced 'snips', are another way of examining genetic variation. These are single base changes (as shown in Figure 5.3.2b), but unlike the examination of a single DNA sequence, these are scattered throughout the genome across multiple chromosomes. This means that many points of variation can be examined simultaneously, sometimes thousands at a time, providing a snapshot of genetic variation. Development of SNPs—that is, finding out what parts of the genome vary and how they can best be isolated—can be technically challenging and expensive (Figure 5.3.3). SNPs usually have two alleles per marker.

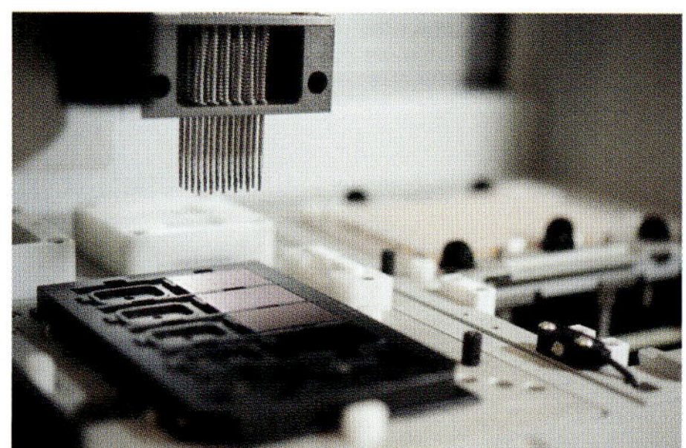

FIGURE 5.3.3 There are many technologies available to develop a SNP 'library' but this process can be expensive and challenging. Pictured here is a Sequenom DNA MassARRAY machine used for SNP genotyping.

Short tandem repeats (STRs)

Another way of looking at genetic variation is to use anonymous markers called **short tandem repeats (STRs)** (also known as **microsatellites**). Instead of directly sequencing the DNA, each STR marker is examined for size differences, which is much cheaper and easier than sequencing a large number of individuals.

STRs are usually non-coding pieces of DNA that contain a string of repeating nucleotides (e.g. 'AGAGAG' or 'TGATTGATTGAT'). STRs usually have repeats of two nucleotides (dinucleotide, e.g. AG), three nucleotides (trinucleotide, e.g. GCT) or four nucleotides (tetranucleotide, e.g. TGAT). An STR with three repeats of 'TGAT' will be shorter in length (have fewer nucleotides) than an STR with seven repeats of 'TGAT'. The number of repeats in an STR varies between individuals in a population. For example, one individual may have 10 repeats of ATG and another individual may have 14 repeats of ATG for the same STR (Figure 5.3.4).

The number of repeats in an STR can also vary within an individual. Individuals have two copies of every STR (one inherited from each parent) because the same DNA sequence occurs on each member of a homologous pair of chromosomes (Figure 5.3.4). If the number of repeats for the STR is the same on each chromosome, the individual is **homozygous** for that STR. If the number of repeats for the STR is different on each chromosome, the individual is **heterozygous** for that STR.

It is this variation within and between individuals that make STRs useful for population genetic studies and identifying individuals in paternity tests or crime investigations. Unlike SNPs, which usually only have two alleles, STRs are usually highly variable and have many different repeat variations (alleles) within a population. Therefore fewer STRs are needed (compared to SNPs) to get the same amount of information.

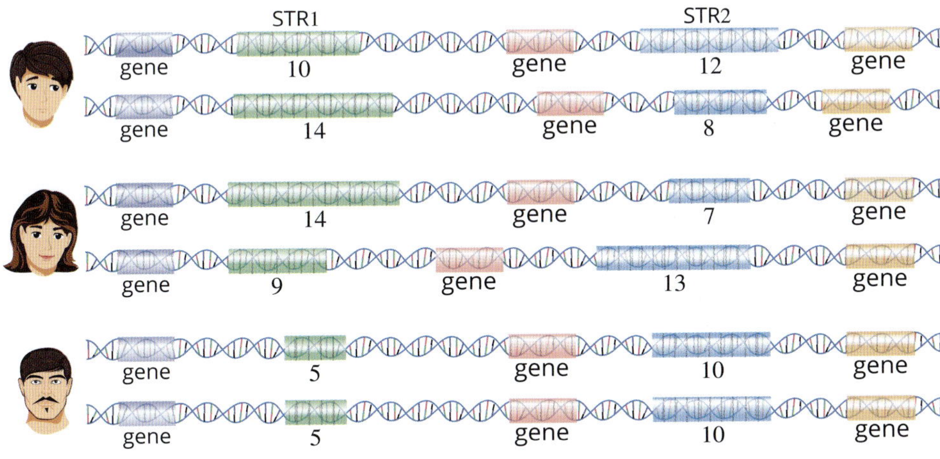

FIGURE 5.3.4 An example of the variation that can be seen in the short tandem repeats (STRs) in three individuals at two STR sites (STR1 and STR2). STR sites are also called loci or genetic markers. As chromosomes occur as homologous pairs, each person has two copies of each STR, which may vary in length (number of repeats, shown under each STR).

SKILLBUILDER

Analysing short tandem repeats (STRs)

STRs are anonymous pieces of DNA that are isolated from the nuclear genome and then analysed in a process known as 'genotyping'. The DNA pieces (STRs) are tagged with a fluorescent dye and run through a capillary, which estimates the size of the STR (length of the DNA pieces). This process is known as chromatography and the visual output is called a chromatogram (Figure 5.3.2a, Figure 5.3.5). A chromatogram shows peaks of different heights at different points along the *x*-axis. Each peak represents an STR allele in an individual. The height of the peaks indicates the fluorescence of the dye (i.e. how much dye was attached to the DNA) and the point along the *x*-axis indicates the number of nucleotides (i.e. the size of the STR) (Figure 5.3.5).

Estimating the size and number of alleles for an STR is known as 'genotyping'. Figure 5.3.5a shows a chromatogram with two peaks. These two peaks represent the genotype of an STR in an individual. Because there are two different sized peaks (alleles), the individual with this genotype is a heterozygote for this STR. This STR is called a 'dinucleotide repeat', meaning the alleles differ by two nucleotides in length; one microsatellite allele is 142 nucleotides long and the other is 144 nucleotides long. This genotype can be written as '142/144'. Figure 5.3.5b shows what this might look like if an individual is homozygous for a microsatellite marker; the two alleles have the same number of nucleotides so the peaks overlap and appear as one peak in the chromatogram. The genotype of this individual is written as '146/146'. The short peak in front of each tall peak is known as a 'stutter' and is not included in the genotyping.

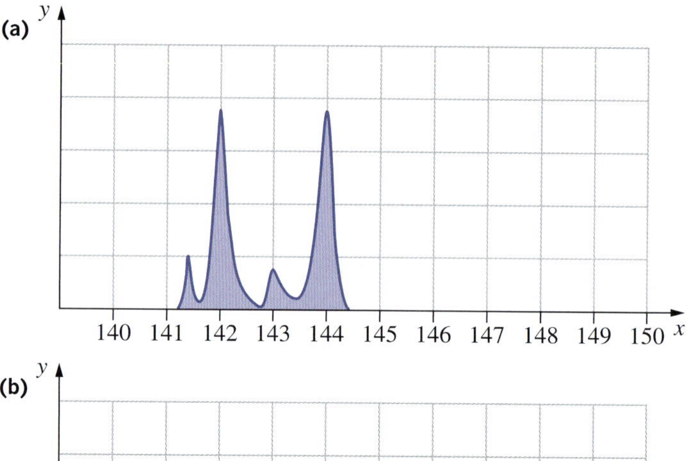

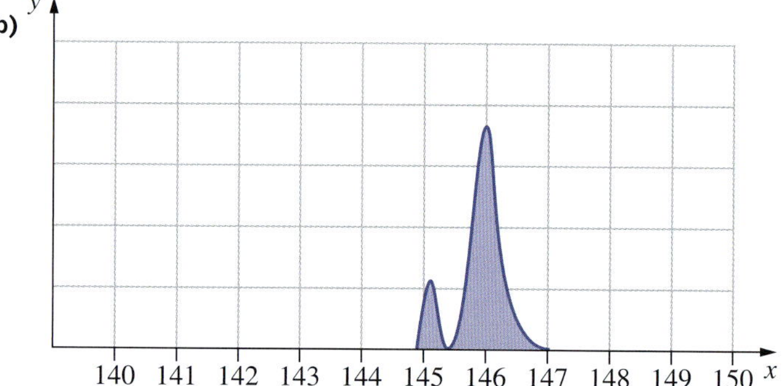

FIGURE 5.3.5 Chromatograms representing the genotypes of two different individuals at the same locus. (a) An individual with a heterozygous genotype '142/144' and (b) an individual with the homozygous genotype '146/146'. The numbers indicate the length of the DNA fragment under analysis (e.g. individual (a) has one allele that is 142 nucleotides long and one allele that is 144 nucleotides long).

5.3 Review

SUMMARY

- The rate at which an allele occurs in a population is known as the allele frequency.
- Allele frequencies are proportions, so the total frequencies of all available alleles for a gene must add up to 1.0.
- Allele frequencies can be easily calculated by examining genotypic ratios.
- When genotypic ratios can be determined from phenotypic ratios (e.g. in some cases of incomplete dominance and co-dominance), allele frequencies can be calculated from phenotypic ratios.
- Population genetics can be used to monitor the health of populations by examining genetic variation.
- Population genetics can also be used to determine whether two populations are actually distinct species.
- Genetic markers can refer to any section of DNA that can be examined for a purpose.
- Genetic markers are pieces of DNA that can be reliably analysed by sequencing or genotyping. They may be coding DNA, non-coding DNA or anonymous DNA.
- DNA sequences can be examined directly and checked for differences between individuals.
- Single nucleotide polymorphisms (SNPs) can show a snapshot of differences in the genome.
- Short tandem repeats (STRs) (also known as microsatellites) are scored by size instead of base sequence, and can also look at multiple parts of the genome at once.

KEY QUESTIONS

1. There is a population of 40 fruit flies. Ten of them have a green body, and 30 have a brown body. All have red eyes. The genes for body colour and eye colour are located on different chromosomes. Based on this information, and for this population only, which of the following statements is true?
   - **A** The gene for body colour is monomorphic in this population.
   - **B** The red eye allele frequency is 1 in this population.
   - **C** The gene for eye colour is polymorphic in this population.
   - **D** This is an example of incomplete dominance.
2. Red roan cattle have coats that contain both red (chestnut) hairs and white hairs. This is an example of co-dominance, rather than incomplete dominance, because both the red coat allele and the white coat allele are fully expressed. If it were incomplete dominance, all hairs would be a light reddish brown. In a population of 150 cattle, 90 are red, 30 are roan and 30 are white. Calculate the allele frequencies for the white coat allele ($p : W_1$) and the red/chestnut coat allele ($q : W_2$).
3. List one way in which population genetics is useful for conservation purposes.
4. List three characteristics of an ideal genetic marker and explain why they are necessary in population genetic analyses.

Chapter review

05

KEY TERMS

allele
allele frequency
anonymous DNA
autosome
carrier
chiasma (pl. chiasmata)
co-dominant phenotype
complete dominance
complete penetrance
continuous variation
cross
crossing over
dihybrid cross
diploid
discontinuous variation
DNA (deoxyribonucleic acid)
dominant phenotype
epigenetics
F1 generation
F2 generation
fixed allele
gene
gene flow
gene mapping
genetic marker
genome
genotype (adj. genotypic)
genotypic ratio
haploid
haplotype
Hardy-Weinberg equilibrium
hemizygote (adj. hemizygous)
heterogametic
heterozygote (adj. heterozygous)
histone
homogametic
homologous chromosome
homozygote (adj. homozygous)
incomplete dominance
incomplete penetrance
Law of Independent Assortment
Law of Segregation
linkage
locus (pl. loci)
microsatellite
monogenic inheritance
monohybrid cross
monomorphic
mutation
neutral marker
pedigree analysis
phenotype (adj. phenotypic)
phenotypic ratio
polygenic inheritance
polymorphism (adj. polymorphic)
population
recessive phenotype
reciprocal cross
recombinant gametes
sex chromosome
sex-limited inheritance
sex-linked inheritance
short tandem repeat (STR)
single nucleotide polymorphism (SNP)
test cross
trait
true-breeding
wild type
X-linked
Y-linked

REVIEW QUESTIONS

1 In mice, coat colour is controlled by a single gene. Assuming that black coat colour is dominant to white coat colour.

- **a** Assign allele symbols for the gene responsible.
- **b** How many genotypes are possible with respect to these alleles? State the genotypes and phenotypes.

2 The shape of a human earlobe is determined by a single autosomal gene. Free lobe is dominant to attached lobe.

- **a** Write appropriate allele symbols for this gene.
- **b** How many genotypes are possible with respect to these alleles? How many phenotypes are possible?
- **c** A homozygous man with free lobes married a heterozygous woman. Show the genotypes and phenotypes possible in their children.
- **d** Can two people with free lobes have a child with attached lobes? Explain your answer.
- **e** Two parents heterozygous for earlobe shape have a child. What is the probability that the child has attached lobes? Write your answer as a percentage, and as a ratio.

3 In mice, black coat colour is dominant to white coat colour. Calculate the expected genotypic and phenotypic ratio for a cross between two heterozygotes. Use appropriate notation.

4 A genetics student undertakes a study of inheritance patterns of feather colour in domestic chickens. The student observes the following matings between:

- black-feathered adults always result in black-feathered offspring
- white-feathered adults always result in white-feathered offspring
- black-feathered adults and white-feathered adults produce only blue/grey-feathered offspring
- blue/grey-feathered adults results in black, blue/grey and white offspring in a ratio of 1 : 2 : 1.

- **a** Describe the mode of inheritance of this trait. Outline the evidence that leads you to this conclusion.
- **b** How many genes and alleles control this trait? Outline the evidence that leads you to this conclusion.
- **c** Use appropriate notation to set up a model that explains the student's observations.

5 Using the key terms that relate to different sorts of crosses (testcross, monohybrid cross and other crosses), make a poster that distinguishes between the different crosses.

6 What is the most likely mode of inheritance for each of the diseases shown in the following pedigrees? Explain your choices.

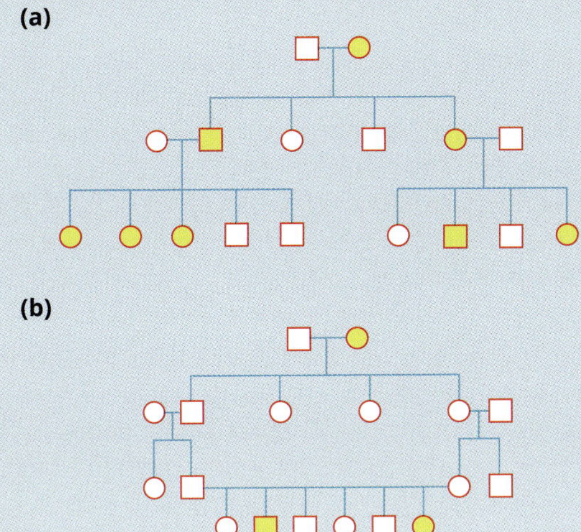

7 The following pedigree shows the inheritance of albinism.

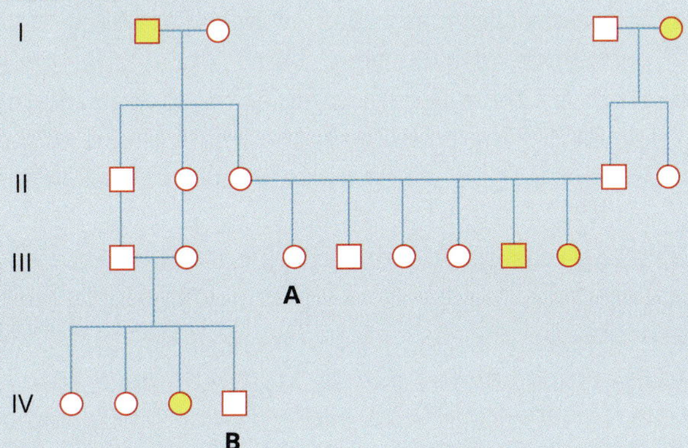

a What is the most likely mode of inheritance of the condition? Explain.

b If III-A and IV-B were to have offspring, what would you need to know about these individuals to calculate the chances of their offspring having albinism?

8 This diagram represents a linkage group on a chromosome from a common crop plant. R, S and T represent different loci on the chromosome. MU = map units (the distance between loci).

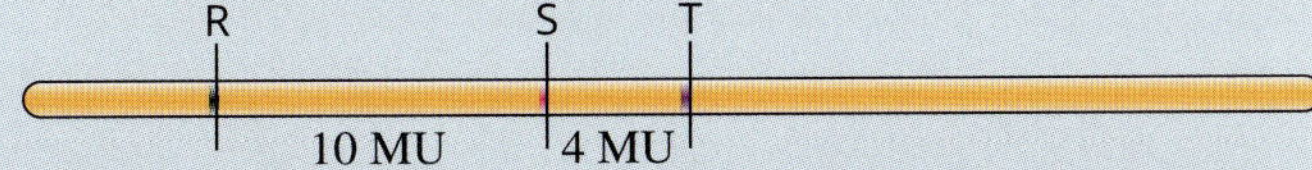

Would you expect the greatest percentage of recombination to occur between RS, ST or RT? Explain your reasoning.

9 Why might the results of a monohybrid cross differ from the expected ratio of 3 : 1? Select the correct answer.

A mutations

B alleles not segregating

C chance

D incomplete meiosis

10 Robert has blood type A and Lee has blood type B. Is it possible for them to have a baby of blood type O? What is the probability of this occurring? Draw a Punnett square to explain your answer.

11 The figure below shows the inheritance of haemophilia in a family. Haemophilia is a recessive X-linked inheritance.

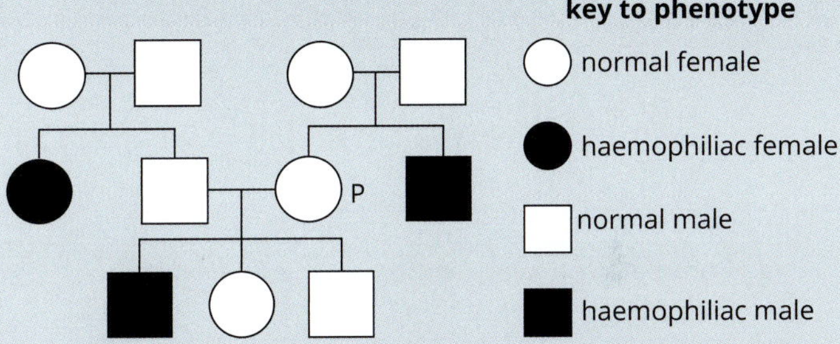

What is the genotype of P? Show your working using a Punnett square and appropriate symbols.

12 Why do sex-linked disorders affect males more than females?

13 Why are there fewer Y-linked disorders than X-linked disorders?

14 The figure below shows part of a family pedigree. If individual III-3 was shaded, which of the following best describes the trait?

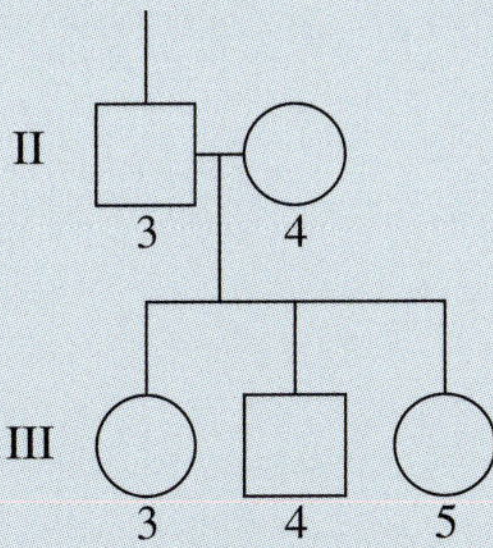

A dominant

B sex-linked

C recessive

D co-dominant

15 What does 'carried on the X chromosome' and 'occurs more in males than females' suggest?

A a monohybrid cross

B a dihybrid cross

C Mendel's experiments

D sex-linked inheritance

16 Complete the sentences below about the principle of linkage.
The tendency for two or more _______ located on the same ______ is that they are inherited ______.
The ______ the ______ are to each other, the more likely they are to be ______ together, and appear to be ______.

17 Sheep blowfly chromosome 5 carries genes for resistance to the insecticide dieldrin (gene *D*). The same chromosome carries a gene called furrowed eyes (*F*).

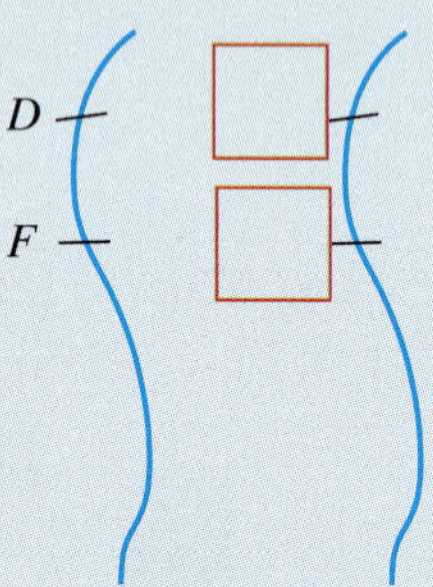

- **a** Complete the allele symbols for a fly that is heterozygous at both loci.
- **b** If no crossing over occurs in meiosis, the gametes will carry either DF or df alleles. What combinations of alleles will be present in gametes if crossing over does occur?
- **c** Construct a Punnett square for a cross between the fly shown above (genotype DdFf) (after recombination has occurred in meiosis) and a homozygous recessive fly (ddff).
- **d** If dieldrin resistance is a dominant trait and furrowed eye is recessive, what proportion of the offspring with normal (wild type) eyes are resistant to the chemical dieldrin?

18 Which of the following is not a violation of the HWE model?
- **A** Giant cuttlefish (*Sepia apama*) only breed once before dying.
- **B** The New Zealand kakapo (*Strigops habroptila*), a species of parrot, had a population of only 18 individuals in the 1970s. There are now an estimated 150 birds.
- **C** The Australian fur seal (*Arctocephalus pusillus*) has a mating system where one dominant male keeps a harem of females.
- **D** Human populations (*Homo sapiens*).

19 A single gene controls the flower colour of a plant. There are three alleles and all interactions show incomplete dominance. One allele (R_1) codes for red pigment; one (R_2) codes for yellow pigment; one (R_3) limits the production of any pigment. In a garden, there are 100 of these plants. Thirty have bright red flowers, 20 bright yellow, 20 pink, 15 white, 5 orange and 10 pale yellow.
- **a** There are six phenotypes. What are the likely genotypes for each phenotype?
- **b** What are the allele frequencies for each of the three alleles? (using *p* and *q* for two of the alleles, and *r* for the third) (Hint: make sure that all three values add up to 1).

20 Give two reasons why mitochondrial DNA is often preferable to nuclear DNA when looking at population genetics.

21 After completing the Biology Inquiry on page 236, reflect on the inquiry question: How can the genetic similarities and differences within and between species be compared? Discuss the value of understanding genetic variation within and between species. Provide at least three examples of areas where genetic information is useful.

CHAPTER

06 Inheritance patterns in a population

This chapter examines the inheritance patterns in populations and the technologies used to identify trends, patterns and relationships in population genetic data. In this chapter you will look at the nature and uses of DNA sequencing and profiling, and learn how this technology can be used to determine inheritance patterns in a population. You will learn about whole genome sequencing, population genetics and bioinformatics, as well as the application of these techniques in wildlife conservation and management, understanding the inheritance of disease and tracing the evolution of modern humans.

Content

INQUIRY QUESTION

Can population genetic patterns be predicted with any accuracy?

By the end of this chapter you will be able to:

- investigate the use of technologies to determine inheritance patterns in a population using, for example: (ACSBL064, ACSBL085) ICT
  - DNA sequencing and profiling (ACSBL086) EU
- investigate the use of data analysis from a large-scale collaborative project to identify trends, patterns and relationships, for example: (ACSBL064, ACSBL073) A CCT IU N
  - the use of population genetics data in conservation management S
  - population genetics studies used to determine the inheritance of a disease or disorder CCT ICT N
  - population genetics relating to human evolution IU

6.1 DNA sequencing and profiling

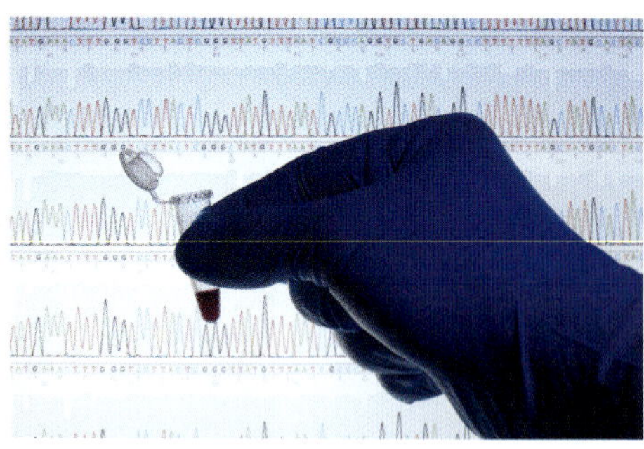

FIGURE 6.1.1 DNA can be extracted from tissue samples and sequenced. DNA sequencing has a wide range of applications, including understanding the inheritance of genetic disorders, determining parentage or tracing evolutionary relationships.

Most cells contain the entire complement of an organism's **DNA (deoxyribonucleic acid)**, known as its **genome**, which in humans includes thousands of **genes**. Today biologists can isolate fragments of DNA (Figure 6.1.1) and study a single gene or determine the base sequence of an entire genome. DNA sequencing and profiling are powerful technologies that use DNA manipulation techniques. In this section you will learn about these techniques and how they are used to determine inheritance patterns in a population.

GENETIC TESTING

Genetic testing is a medical test used to detect specific alleles, mutations, **genotypes** or karyotypes that are associated with **heritable traits**, diseases or predispositions to diseases. It may also be used to determine parentage or ancestry and has become an important part of forensic analysis.

There are many kinds of tests, but the methodologies fall into three main categories—molecular genetic testing, cytogenetic testing and biochemical genetic testing. Examples of methods used in genetic testing are listed in Table 6.1.1.

TABLE 6.1.1 Examples of methods used in genetic testing

Disease examples	Type of mutation	Methods used	Detection
• Huntington's disease	• CAG repeat mutation	• polymerase chain reaction (PCR) • gel electrophoresis	• different length of allele on gel
• phenylketonuria (PKU) • cystic fibrosis	• point mutation occurs in restriction enzyme recognition site	• PCR • restriction enzymes • gel electrophoresis	• specific fragments detected on gel
• Fragile X syndrome • Di George syndrome • Patau syndrome	• duplication mutation • deletion mutation • trisomy (three copies of chromosome)	• fluorescence in situ hybridisation (FISH) • fluorescence microscopy	• mutation located on chromosomes

Molecular genetic testing

Molecular genetic testing is used to identify single genes or short lengths of DNA. Through molecular testing, scientists can identify variations or **mutations** that lead to a genetic disorder. It is the most effective method if the gene sequence of interest is known and the function of the protein is unknown.

Molecular genetic testing can be performed on any tissue sample and requires only a small sample. There are a few different kinds of procedures used to analyse the sample. Two of the most common are the **polymerase chain reaction (PCR)** and **restriction fragment length polymorphism (RFLP)**.

Polymerase chain reaction (PCR)

PCR is a technique used to make many copies of a piece of DNA, a process referred to as **DNA amplification** (Figure 6.1.2). A DNA sample is mixed with **primers** (short lengths of DNA that target specific regions for PCR) and other chemicals in a small tube, and the mixture is heated and cooled in cycles. At each cycle of synthesis, the number of copies of the DNA fragment doubles—the DNA has been amplified. In this way a large amount of DNA can be produced in less than an hour. The DNA primers used in PCR target the genes associated with the disorder. The amplified DNA product is then sequenced using **DNA sequencing** to reveal the presence or absence of the disease-causing mutation.

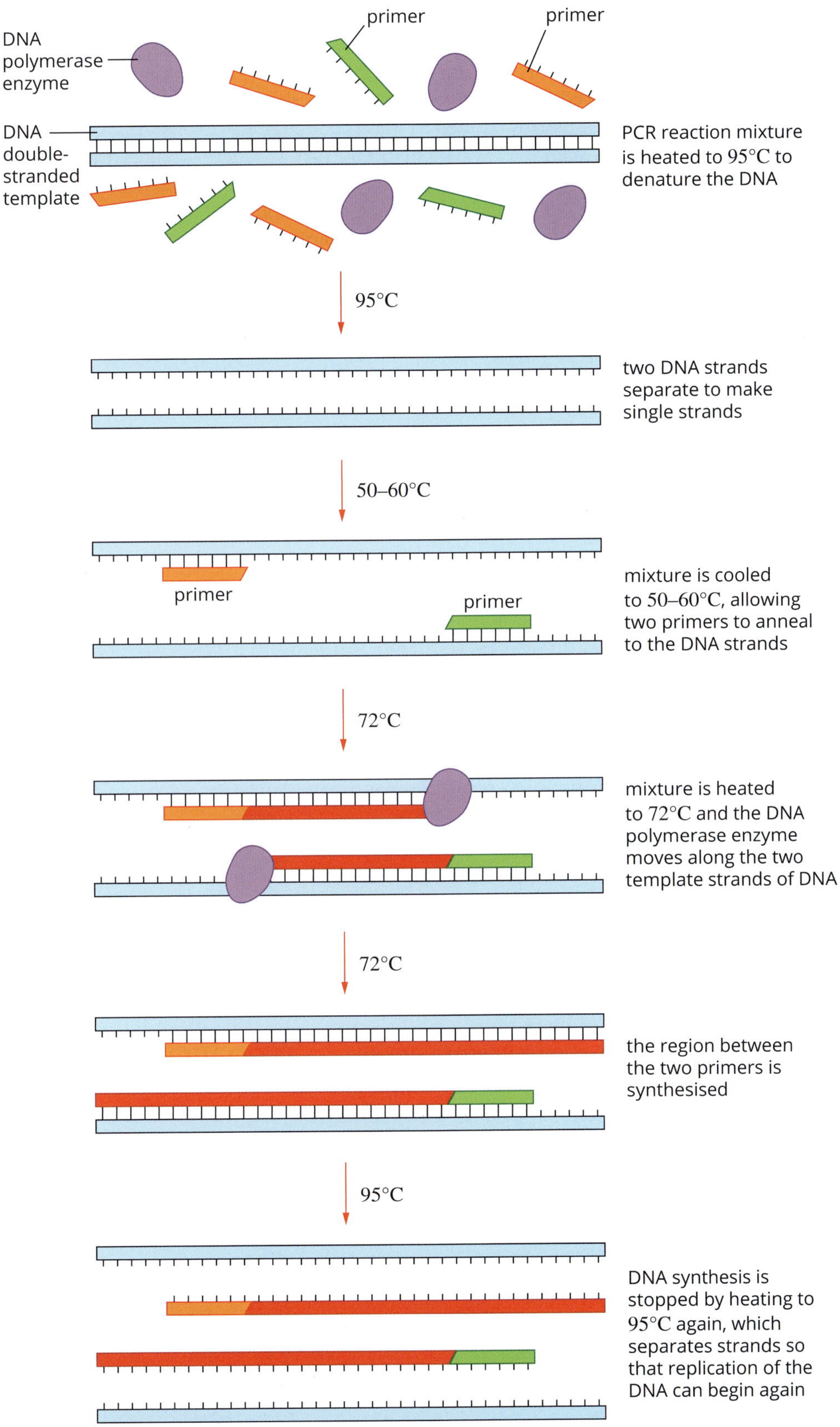

DNA polymerase enzyme

primer

primer

DNA double-stranded template

PCR reaction mixture is heated to 95°C to denature the DNA

95°C

two DNA strands separate to make single strands

50–60°C

primer

primer

mixture is cooled to 50–60°C, allowing two primers to anneal to the DNA strands

72°C

mixture is heated to 72°C and the DNA polymerase enzyme moves along the two template strands of DNA

72°C

the region between the two primers is synthesised

95°C

DNA synthesis is stopped by heating to 95°C again, which separates strands so that replication of the DNA can begin again

FIGURE 6.1.2 Amplifying DNA using the polymerase chain reaction (PCR). All the components of the PCR reaction mixture are present in the tube. (The four nucleotides, containing bases A, T, C and G are also in the mixture, although they are not shown here.)

BIOFILE CCT

Forensic applications of PCR

The polymerase chain reaction (PCR) is an extraordinarily sensitive technique for forensic investigations of crimes. A DNA sample of only a few cells can provide enough DNA for amplification. Scientists at the Victoria Forensic Science Centre have shown that merely touching an object deposits sufficient material for successful DNA amplification. In handling keys, opening a door or driving a car, the cellular material deposited by a criminal provides ample DNA for analysis.

After 40 cycles of amplification using the PCR technique (Figure 6.1.3), one target molecule of DNA will produce more than one million molecules.

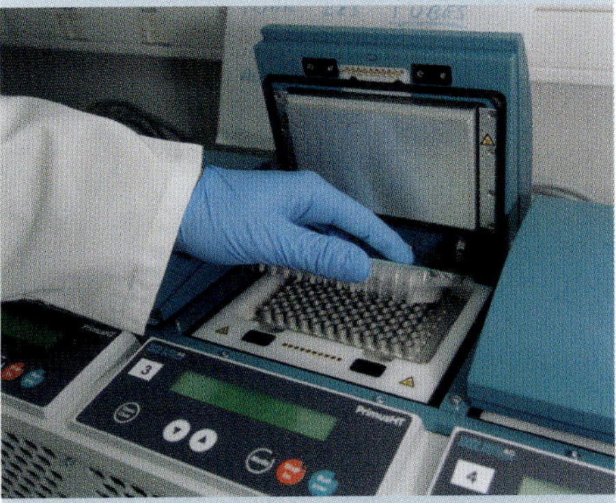

FIGURE 6.1.3 A scientist inserts a set of prepared samples into a thermocycler for PCR amplification.

PCR is a key tool for generating a **DNA profile** (also known as a DNA fingerprint) using DNA regions called **short tandem repeats (STRs)** (also known as microsatellites). STRs are short sequences of DNA that frequently appear throughout the **genome**, characterised by short motifs (e.g. AT) that repeat multiple times (e.g. ATATAT). These are not usually located in the coding regions (exons), but in the non-coding spacer DNA between the exons. The STRs vary in length—that is, the number of times that the motif repeats—in different people. These differences usually have no impact on the phenotype or the health of a person, but they are very useful for DNA comparisons. In forensic analysis, PCR is used to amplify 13 different STR regions (using different primers that target different STRs) to gain a DNA profile unique to each individual. STRs are covered in more detail in Chapter 5.

GO TO ➤ Section 5.3 page 236

In molecular genetic testing, the DNA primers used in PCR target the genes associated with the disorder. For example, specific primers can target the gene associated with cystic fibrosis (*CFTR* gene) when screening parents to identify carriers of the disease. Different primers target the *PAH* gene when screening newborns carrying an **allele** (gene variant) that causes phenylketonuria. The amplified DNA product is then sequenced using DNA sequencing to reveal the presence or absence of the disease-causing mutation. Alternatively, the amplified DNA is treated with special enzymes, known as **restriction enzymes** (also known as restriction endonucleases) to identify specific alleles.

BIOLOGY IN ACTION CCT

Discovery of heat-stable DNA polymerase

A field trip to Yellowstone National Park in the USA radically altered the course of molecular genetics research in the 1960s. Thomas Brock, a bacteriologist from the University of Wisconsin–Madison, found bacteria in water taken from a hot spring (Figure 6.1.5). He named the new species *Thermus aquaticus*.

Enzymes from most organisms are normally denatured (they unfold and stop working) if heated to temperatures of 95°C for more than a few seconds. For *T. aquaticus* to survive in the hot springs, its enzymes, including DNA polymerase, must be able to tolerate these high temperatures. Therefore, the DNA polymerase from *T. aquaticus* (called *Taq*) has proved to be an ideal enzyme for PCR.

FIGURE 6.1.5 Thomas Brock, USA, microbiologist and discoverer of the bacterium, *Thermus aquaticus*.

Sequencing DNA

The products of PCR—copies of a particular sequence of interest—can be used for STR analysis, but they can also be sequenced directly in a DNA sequencer. This allows individual sequences to be directly compared.

The first stage of DNA sequencing involves replicating a piece of purified DNA in a test-tube, just like the usual PCR process. However, in this case, the mixture contains an additional ingredient—terminating nucleotides that are tagged with coloured fluorescent dyes. Each type of terminating nucleotide (A, T, G or C) is tagged with a different coloured dye.

During PCR the DNA strand is split and one strand (called the template strand) is used to build the section of DNA required. When a primer binds to a DNA strand, the enzyme **DNA polymerase** begins to replicate the sequence by adding nucleotides. However, when it uses a terminating nucleotide, the replication stops. This can happen at any point during the replication, so the process results in many different lengths of DNA fragments (Figure 6.1.4).

DNA template to be sequenced: AGCTTGGATT
Sequence of the complementary strand: TCGAACCTAA
If the terminating nucleotide is for A, four fragments are produced:
TCGA
TCGAA
TCGAACCTA
TCGAACCTAA
If the terminating nucleotide is for T, two fragments are produced:
T
TCGAACCT

FIGURE 6.1.4 Example showing how different fragment sizes are produced when terminating nucleotides are added at different points in the sequence during a polymerase chain reaction.

After the PCR is completed, the DNA mixture is cleaned up to remove any unused nucleotides and other substances, so that it contains only fragments of the DNA sequence required.

In the next stage, the DNA fragments are injected into very fine capillaries filled with a polymer. A large voltage is then applied, causing the negatively charged DNA fragments to be drawn towards the positive electrode. This technique is called capillary electrophoresis. Smaller fragments move faster than larger ones. Just before reaching the positive electrode, the fragments pass through a laser beam that causes the terminating nucleotide to fluoresce. The colour and strength of the fluorescence is recorded and the data is assembled into a data file (Figure 6.1.6).

This file is then analysed by a computer program that arranges the detected bases in the correct sequence. The output from the software is the sequence of bases and a matching chromatogram showing the strength of the fluorescence and the base that produced it (Figure 6.1.6). Because the data from the fluorescence detector is sometimes misread (especially if the quality or volume of DNA is low), the sequence must be checked manually by the researcher. Sequences are usually obtained in both directions (5′ to 3′ and 3′ to 5′), and the two sequences are matched up by the program.

The order of bases in the original sample of DNA can be determined from the order of the peaks in the chromatogram (Figure 6.1.7).

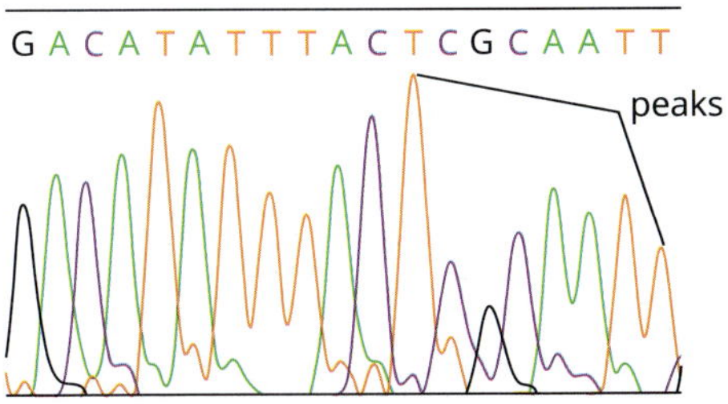

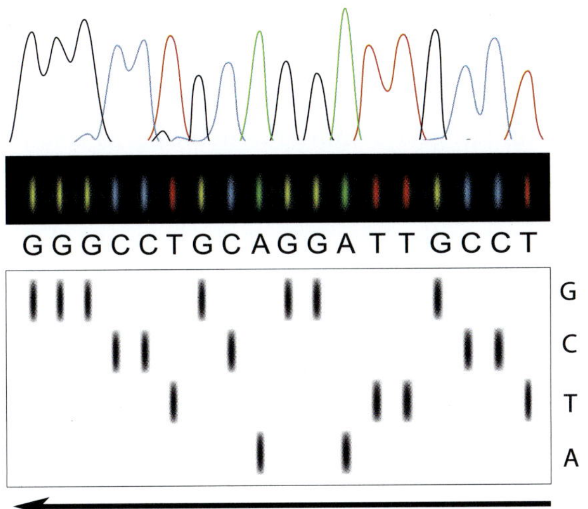

FIGURE 6.1.6 Capillary electrophoresis separates the DNA fragments according to size. The arrow indicates the direction in which the DNA fragments moved through the gel. The smallest fragments move furthest through the gel and are at the left and the largest fragments are at the right. The fragments pass through a laser beam, causing the terminating nucleotides to fluoresce. Computer software then transforms the data into a sequence and a matching chromatogram.

GACATATTTACTCGCAATT

peaks

FIGURE 6.1.7 The output from sequencing software is a sequence of bases (top row) and a matching chromatogram that shows the strength of the fluorescence detected (indicated by the height of the peaks). The bases are colour-coded to make them easy to identify in the chromatogram.

> **i** The term 'polymorphism' means different form. In regard to DNA, different forms are alleles or differences in a single nucleotide (e.g. single nucleotide polymorphisms (SNPs)) between individuals.

Restriction fragment length polymorphism (RFLP)

Restriction fragment length polymorphism (RFLP) is a technique that identifies individuals based on differences (**polymorphisms**) in their DNA sequences at restriction enzyme cut sites. RFLP analysis rapidly provides information about DNA sequences without having to sequence the DNA itself. There are two parts to this procedure—producing DNA fragments using restriction enzyme digestion (cutting) of a DNA molecule, then separation of the fragments using **gel electrophoresis**.

DNA can be cut into fragments by a set of enzymes called restriction enzymes. These enzymes recognise short sequences of bases (a **recognition site**) in a DNA molecule. The different sequences of bases in individuals result in restriction enzymes cutting the target DNA at different sites (Figure 6.1.8). Therefore, DNA samples from two individuals that have been mixed with the same restriction enzymes will form different DNA fragments, giving each person a unique DNA profile. This type of analysis is used in forensics and paternity testing.

Restriction enzymes are frequently used in combination with PCR to identify specific alleles, such as in screening for carriers of inherited disorders.

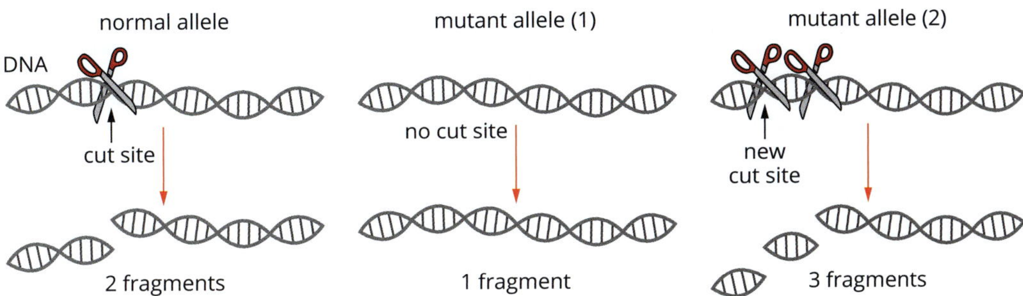

FIGURE 6.1.8 The ability of a restriction enzyme (indicated with scissors) to cut DNA can identify specific sequences in DNA (e.g. alleles).

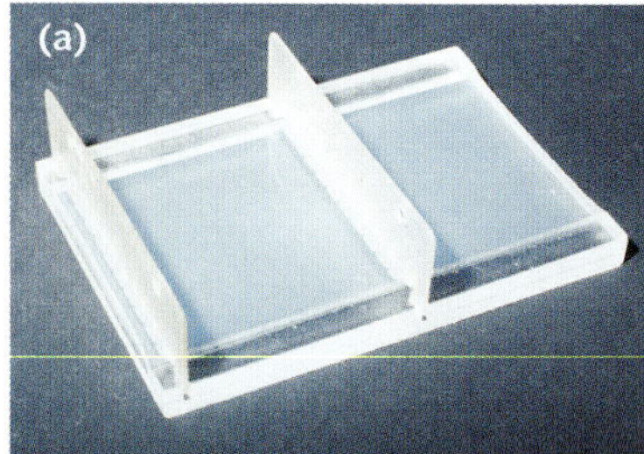

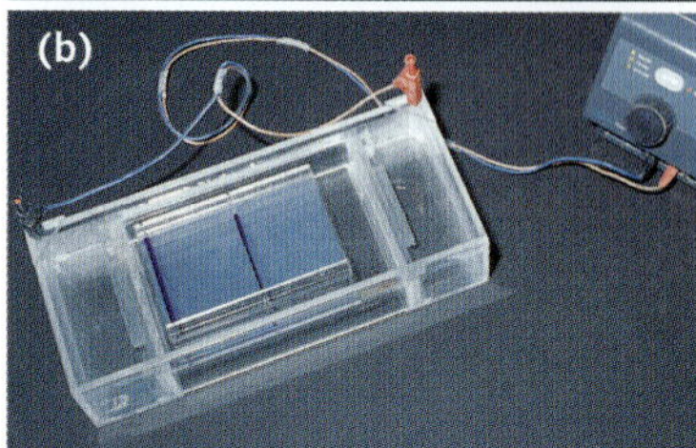

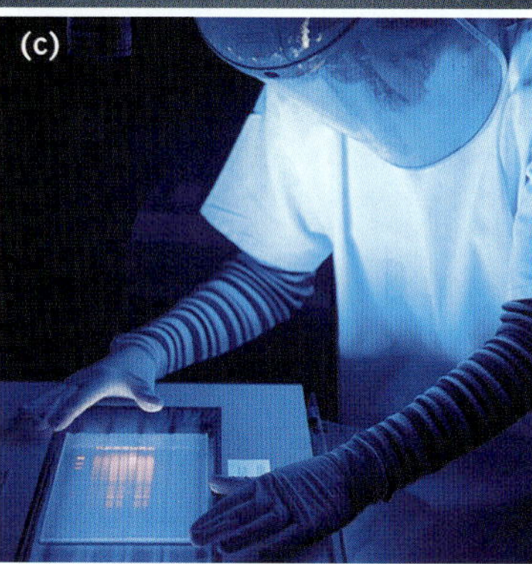

FIGURE 6.1.9 The process of gel electrophoresis. (a) The gel is made in a mould. The two combs insert rows of wells for samples and standards. (b) The gel is placed in an electrophoresis bath where it is covered with a buffer solution and an electric current is applied. (c) The DNA in the gel can be observed when the gel is placed under ultraviolet light.

Comparing DNA using gel electrophoresis

When the mixture of DNA fragments has been produced, gel electrophoresis is used for separating fragments of DNA. DNA molecules are separated in a slab of jelly-like substance called agarose.

The gel is immersed in a buffered salt solution to keep the pH suitable for DNA molecules; salts are needed to conduct the electric current through the gel. The DNA is mixed with a fluorescing dye that attaches to the DNA fragments. A small amount of this mixture is transferred into wells along the top of the gel, together with a standard mixture of DNA fragments of different known molecular weights (also known as a molecular weight standard or DNA ladder due to its appearance) (Figure 6.1.9). The standard is crucial to be able to compare the different sizes of the DNA fragments.

The phosphate groups in DNA fragments are negatively charged, so the DNA fragments move through the gel towards the positive terminal when an electric current is applied. Heavier (longer) fragments move more slowly than lighter (shorter) fragments, so the fragments become more and more separated with time. After a set time the current is turned off. The gel is removed from the buffer solution and placed on an ultraviolet (UV) light box. When the light is switched on, the dye attached to the DNA fragments fluoresces so the DNA fragments can be seen and their sizes estimated by comparing them to the fragments in the standard (DNA ladder).

This technique can be used to compare the DNA of individuals and identify relationships. Because individuals have different DNA sequences, restriction enzymes cut DNA at different sites, resulting in a unique banding pattern on a gel. This unique banding pattern is known as a DNA profile (Figure 6.1.10).

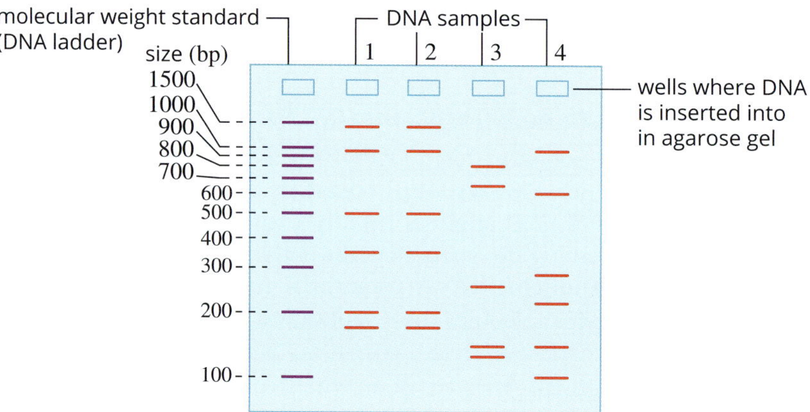

FIGURE 6.1.10 Diagram of a gel showing fragments of digested DNA from four different plant samples. The left lane contains a standard mixture of DNA fragments (DNA ladder), which is used to compare the sizes of DNA fragments in the test samples. Plant samples 1 and 2 have identical banding, so they are genetically identical, indicating the samples most likely came from the same individual.

Cytogenetic testing

There are two types of **cytogenetic testing**—conventional cytogenetic testing (also known as karyotyping) and molecular cytogenetic testing by **fluorescence in situ hybridisation (FISH)**.

Conventional cytogenetic testing (karyotyping)

Conventional cytogenetic testing (or karyotyping) is used to detect abnormalities in the number or structure of chromosomes in metaphase cells. This method is useful for detecting large changes in chromosomes using light microscopic methods at approximately ×1000 magnification. An image of an individual's set of chromosomes is known as a **karyotype** (Figure 6.1.11).

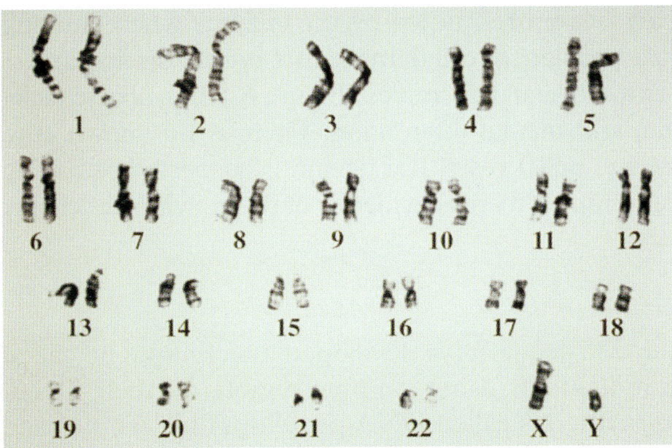

FIGURE 6.1.11 A normal male karyotype. Chromosomes in metaphase cells are stained and viewed under the light microscope. Chromosome photos are arranged in order, to construct the karyotype chart. Conventional cytogenetic testing allows for the determination of abnormal chromosome numbers (such as additional chromosome number 21 in Down syndrome) or large changes in chromosome structure (such as abnormally short or long chromosomes).

Molecular cytogenetic testing by FISH

Fluorescence in situ hybridisation (FISH) is a technique for detecting the presence or absence of specific DNA sequences. Molecular cytogenetic testing by FISH is a method in which fluorescently labelled DNA fragments are allowed to hybridise, or attach, to whole chromosomes (Figure 6.1.12). This allows scientists to assess the chromosomes for changes in their structure, such as the deletion or duplication of certain sections. For example, a chromosome that normally has sections in the order of A-B-C-D might have sections A-B-B-C-D (duplication of B), A-C-D (deletion of B) or A-B-C-P (replacement of D with section P from a different chromosome). FISH can detect these chromosomal alterations (Figure 6.1.13a). Chromosomes can also be assessed for the presence of certain DNA sequences of interest (Figure 6.1.13b).

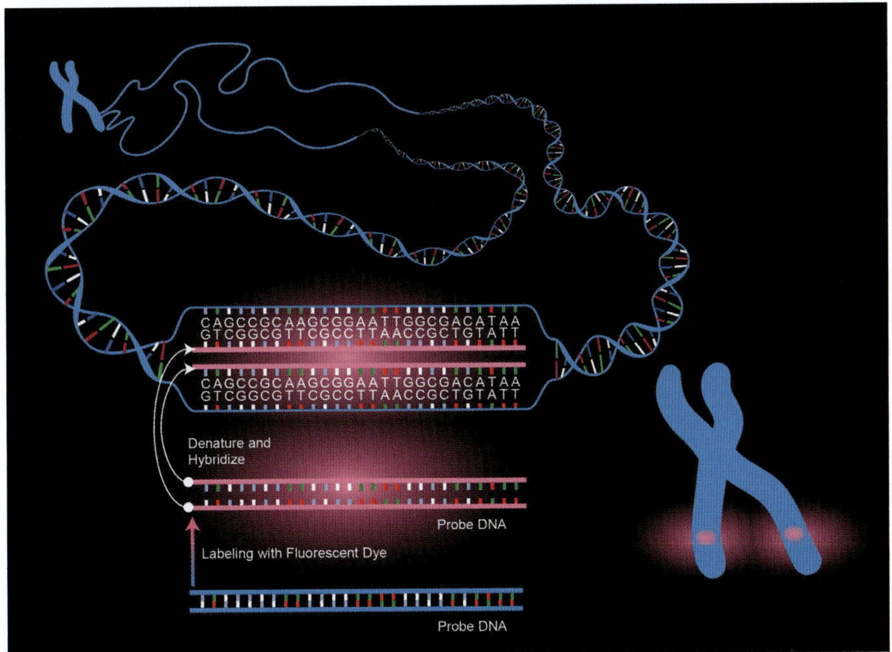

FIGURE 6.1.12 This diagram illustrates how a fluorescent DNA probe binds by complementary base pairing to the target sequence on a chromosome.

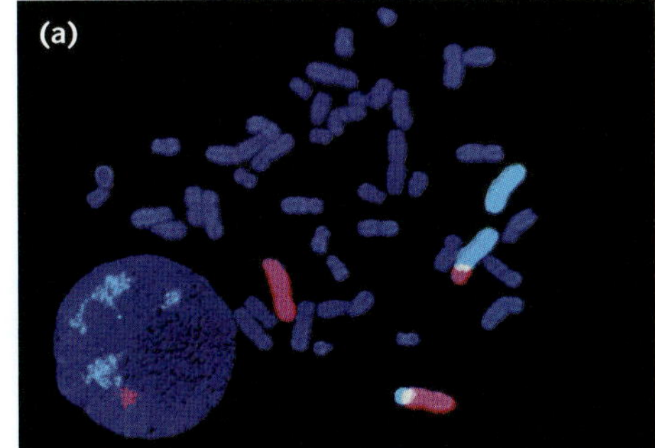

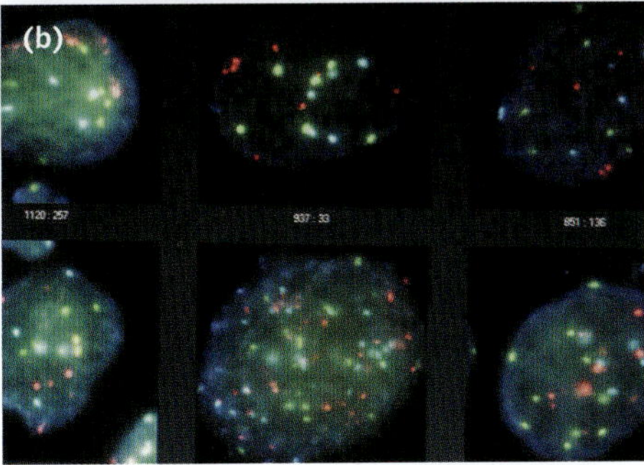

FIGURE 6.1.13 Fluorescence in situ hybridisation (FISH) is a technique for detecting the presence or absence of specific DNA sequences. (a) FISH micrograph of chromosomes. In this image, chromosomes two and three fluoresce pink and light blue. Segments of these chromosomes have undergone translocation (exchanged location) so that part of chromosome two is attached to three, and part of three is attached to two. (b) FISH is used as part of a breast cancer screening process to determine the presence of *HER-2* (human epidermal growth factor receptor 2) gene, which codes for a protein that increases aggressiveness in breast cancers.

Biochemical genetic testing

Biochemical genetic testing studies the amount or activity level of certain proteins. Abnormalities in protein activity levels can indicate mutations in the DNA sequence that are disrupting normal polypeptide synthesis. Scientists may use biochemical tests when the specific gene defect has not been identified, or when there are several possible causes of the disease, such as different mutations in one or more genes. For example, a genetic disease called mucopolysaccharidosis I (MPS I) is caused by an enzyme deficiency that results in an accumulation of molecules called glycosaminoglycans (GAGs) within

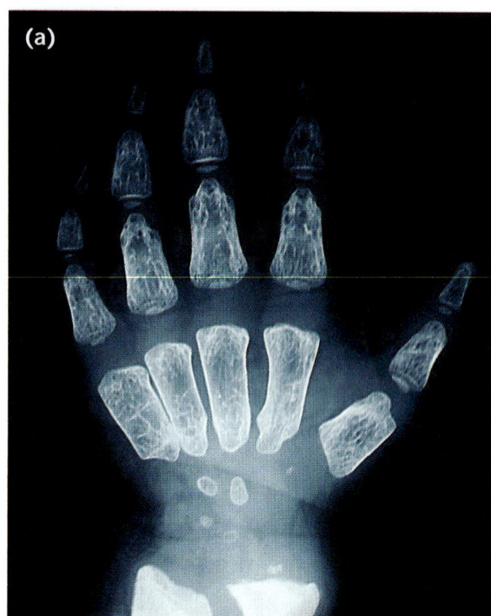

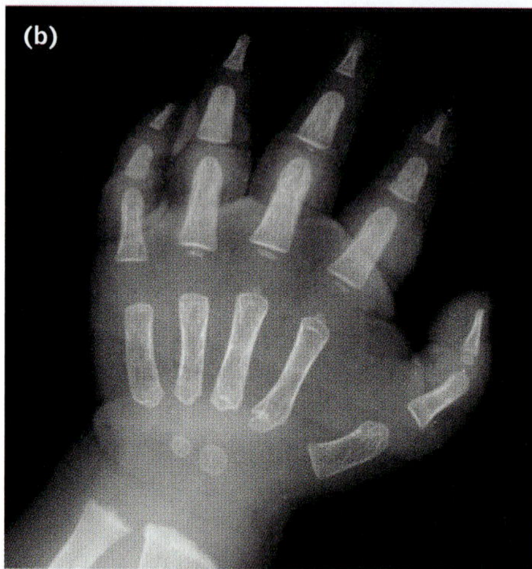

FIGURE 6.1.14 (a) X-ray of the hand of a 28-month-old child with mucopolysaccharidosis type I (MPS I). MPS I is caused by a defective enzyme that is unable to break down glycosaminoglycans (GAGs). GAGs are large molecules normally found in the fluid lubricating the joints. (b) X-ray of a normal child's hand.

GO TO ➤ Section 5.3 page 236

the lysosomes of cells. If GAGs are not broken down by the enzyme when entering the cells, they accumulate and cause permanent cell damage that progressively hinders physical development. MPS I leads to skeletal deformities (Figure 6.1.14), coarse facial features, enlarged liver and spleen, and mental retardation. There is no known cure and individuals rarely live past the age of 10 years. If a person is suspected of having MPS I, biochemical testing can determine if there is evidence of the enzyme deficiency associated with the disorder.

DNA PROFILING

In 1984, an English geneticist named Alec Jeffreys developed a technique he called DNA fingerprinting to identify individuals using variable regions of their DNA. Today, the more commonly used method is called DNA profiling. This can be used to distinguish one individual from another. It is often used in forensics to identify the perpetrator of a crime, identify bodies after disasters or to confirm if a child is genetically related to a parent.

DNA profiling relies on an individual's unique DNA (Figure 6.1.15). The non-coding sections of the DNA, those that do not code for proteins, can vary widely between individuals. DNA profiling uses multiple polymorphic (variable) DNA regions (called **genetic markers**) to identify individuals.

STRs are examples of commonly used genetic markers. They are short sections of DNA with repeating nucleotides. Variation in the number of repeating units within an STR is common between individuals, making them valuable for identification purposes. There are thousands of STRs throughout the human genome—using multiple STRs increases the variation available for analysis, which in turn improves the likelihood of identifying individuals. The inheritance pattern, structure and useage of STRs is described in more detail in Chapter 5.

DNA profiling is also used for genealogy, biogeographical population comparisons, historical population migration patterns and evolutionary relationships. For these purposes, the DNA sequences used for comparisons include STRs (different markers from those used for crime scene analysis), mitochondrial DNA, Y chromosome genes and **single nucleotide polymorphisms (SNPs)**.

Most recently, DNA profiles have been used to identify key features of an individual's appearance. The identification of alleles for eye, skin and hair colour allows investigators to narrow down the list of suspects based on these characteristics. Occasionally the suspect's ancestry can also be determined. This is a developing field of science and the analysis may not yet be reliable.

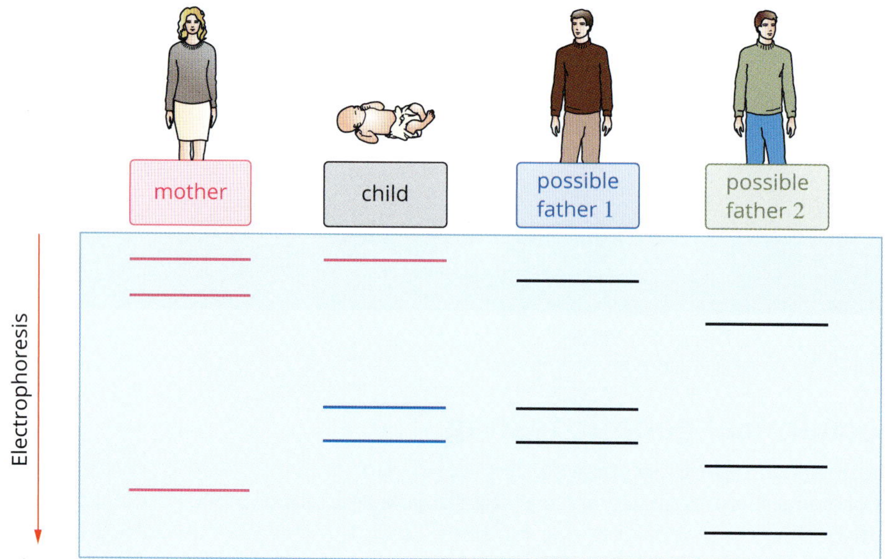

FIGURE 6.1.15 Family members share common bands in DNA profiles, although the combination of bands in individuals is unique. Each band represents an allele (DNA variant), which is inherited from your mother or father.

BIOFILE ICT

The FBI's CODIS

The CODIS system developed by the FBI in the USA uses 13 STR sites in DNA profiling (Figure 6.1.16). CODIS stands for Combined DNA Index System. Using 13 STRs gives an extremely high probability that the DNA profile is unique and that the only perfect match of all 13 DNA sites will be with DNA from the same person. DNA profiling also includes regions on the X and Y chromosome for determining sex.

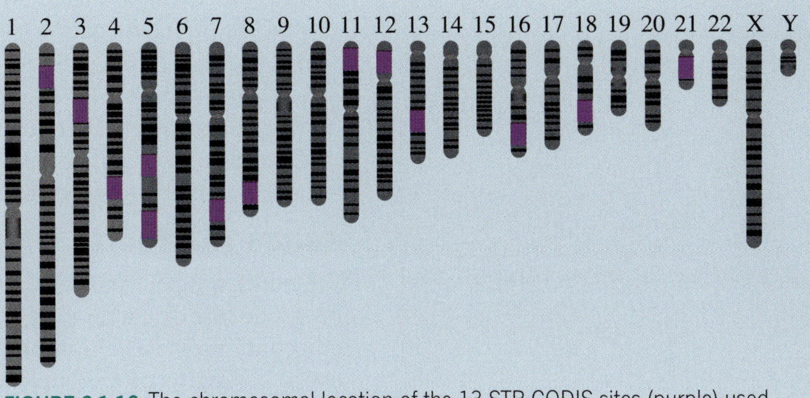

FIGURE 6.1.16 The chromosomal location of the 13 STR CODIS sites (purple) used for forensic DNA profiling

Techniques involved in DNA profiling

If a small amount of blood (white blood cells contain DNA, but red blood cells do not), semen or another sample containing DNA is found at a crime scene, the following steps will be carried out to determine the DNA profile and match it to a suspect.

- DNA is extracted from the sample.
- A small amount of this DNA is added to a PCR, which contains specific primers for each STR.
- The STRs are amplified using PCR, producing a much larger sample for testing, even from a very small amount of DNA.
- Differences in the size of the STRs can be detected by standard gel electrophoresis or by capillary electrophoresis, a rapid, automated method. In capillary electrophoresis, the DNA fragments move in a thin tube under the influence of an electric field. The smaller the size of the fragment, the faster it moves through the capillary tube. As each fragment moves through the tube a laser detector registers a peak on a graph (Figure 6.1.17).

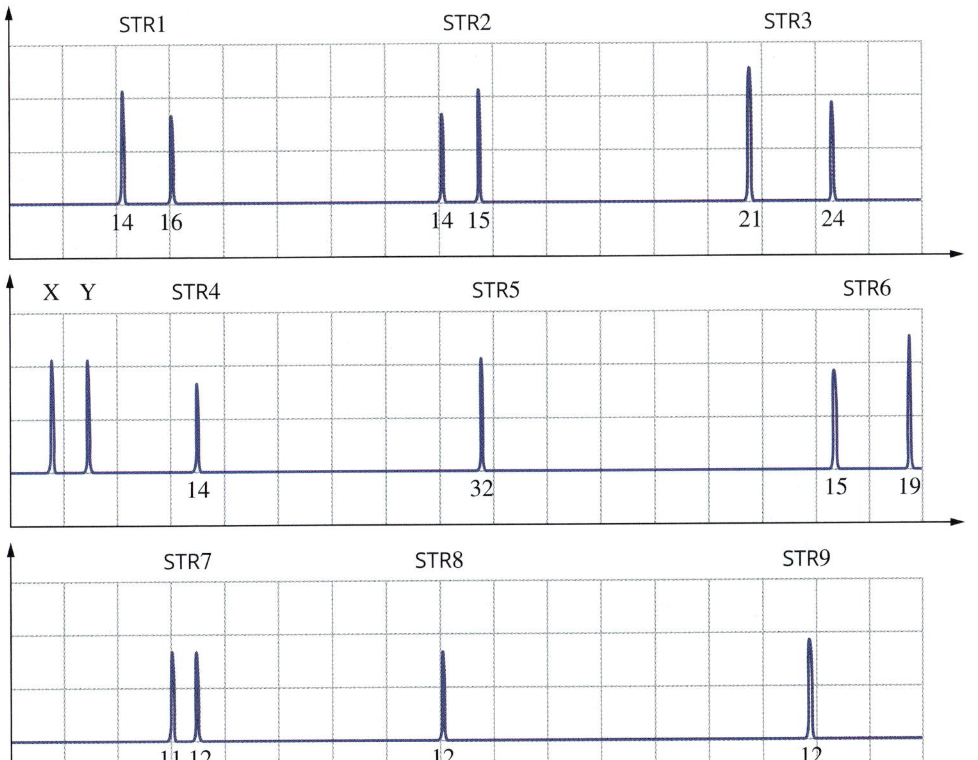

FIGURE 6.1.17 DNA profiling results of one individual at ten genetic markers (nine STRs and the sex chromosome markers). STRs either appear as a single peak or a pair of peaks on the graph. A single peak indicates the individual has the same number of repeats on each homologous (paired) chromosome and is homozygous for that STR. A pair of peaks indicates that the individual has a different number of repeats on each homologous chromosome and is heterozygous for that STR. The numbers below the peaks indicate the number of repeats for that STR. For example, the first two peaks (STR1) show this person has 14 repeats on one chromosome and 16 repeats on the other for this STR. The last peak (STR9) indicates this person has 12 repeats on both homologous chromosomes for this STR. The X and Y chromosome markers indicate this sample is from a male because there is an X and a Y chromosome present. A DNA profile from a female would have one peak for her two X chromosomes.

- The STR analysis of DNA from the crime scene is compared with the STR analysis of DNA from a suspect.
- If the repeats for all the STRs analysed are the same for the crime scene DNA and the suspect's DNA, then it is a match. The chance that two individuals have the same STR repeats at multiple markers is hundreds of billions to one. Therefore, a match means that it is highly probable that the DNA from the crime scene belongs to the suspect.

Issues related to DNA profiling

While it is unlikely that a person will be incorrectly identified by DNA profiling if the procedure is carried out accurately, it is possible. For example, foreign DNA may contaminate a sample at a crime scene or at the laboratory where the sample is tested. Failing to properly clean equipment could cause a sample from one suspect to contaminate a sample for another test.

When it comes to DNA profiling, privacy is a contentious issue. In Victoria, DNA samples cannot be obtained from a person unless they give permission. However, they can be ordered to provide a sample if there is strong evidence that they may have committed the crime and if the DNA profile could help to confirm or deny their guilt. These DNA samples must be destroyed if the person is not guilty or is not charged. However, in some countries, the DNA may be kept for up to 10 years. This has enabled the identification of criminals who have committed crimes in

BIOFILE CCT IU

Tsar rediscovered using DNA

In July 1918, at the end of the Russian revolution, Tsar Nicholas II of Russia, the Tsarina Alexandra, their five children, Olga, Tatiana, Maria, Anastasia and Alexei (Figure 6.1.18), three female servants and the royal physician were executed by a Bolshevik firing squad in the town of Ekaterinburg, Russia. In the chaotic rise of the Soviet Union and in the wake of civil war, the Tsar and his family were seen as dangerous symbols of the old regime. Historical accounts indicate that two of the children's bodies were burned, although others claim that Anastasia escaped execution. The remaining bodies were thrown into a shallow grave and sulfuric acid poured over them.

In 1991, two amateur historians, Gely Ryabov and Alexander Avdonin, discovered nine skeletons in a grave near Etkaterinburg. The remains were tested to find out whether they came from the Tsar and his family. DNA extracted from bone tissue samples was amplified by PCR. They identified the sex of the skeletons using PCR of a gene that is found on the Y chromosome. This indicated there were two males and seven females.

Using DNA profiling of the bone samples, it was possible to conclude with certainty that five of the skeletons were two parents and their three daughters. But these could have been the remains of any family. To establish the identity of the bones, comparison to the DNA from a related person was needed. The evidence came when the DNA profiles of the skeletons were compared with those generated from known relatives of the Tsar (George, brother of Nicholas II, whose remains were exhumed from a crypt in St Petersburg) and Tsarina (Prince Philip, husband of Queen Elizabeth II). The presence of common bands in the DNA profiles of the five bodies, Prince Philip and George indicated that all the individuals were related. The probability that the bones found in Etkaterinburg are the remains of the Tsar and his family is approximately 99 999 out of 100 000.

FIGURE 6.1.18 Tsar Nicholas II of Russia with his wife, Tsarina Alexandra, and their five children in 1913

unsolved cases that occurred before DNA profiling technology was developed. It has also resulted in the exoneration of wrongly accused people.

Storing DNA after a person has served their sentence for a crime may be seen as unethical. Others are in favour of the creation of a 'bank' of DNA samples, provided by everyone in the community, which could be used to solve crimes and perhaps trace the remains of unidentified missing persons. Opponents of a DNA bank argue the potential for these samples to be stolen or used unethically.

6.1 Review

SUMMARY

- Genetic testing is a medical test used to detect specific alleles, mutations, genotypes or karyotypes that are associated with heritable traits, diseases or predispositions to diseases.
- Genetic testing may also be used to determine parentage or ancestry.
- There are three main categories for genetic testing—molecular genetic testing, cytogenetic testing and biochemical testing.
- Molecular genetic testing is used to identify single genes or short lengths of DNA using techniques such as polymerase chain reaction (PCR), DNA sequencing and gel electrophoresis.
  - PCR amplifies target sections of DNA for analysis. Target DNA may be a gene of interest or polymorphic non-coding sites such as short tandem repeats (STRs).
  - Restriction fragment analysis uses restriction enzymes to cut amplified DNA at specific recognition sites. Because individuals have different DNA sequences, restriction enzymes cut their DNA sequences at different sites, resulting in unique DNA profiles.
  - Gel electrophoresis is a technique used to visualise and compare DNA fragments that have been amplified by PCR. DNA fragments are run through a gel using an electric current, and separated based on size.
  - DNA amplified by PCR can also be sequenced for analysis. DNA sequencing involves arranging the DNA nucleotides (A,T, C and G) in the correct order. DNA can be sequenced for a particular gene/genetic marker or the whole genome of an organism.
- Conventional cytogenetic testing, or karyotyping, analyses whole chromosomes for major genetic changes such as an extra copy of a chromosome.
- Molecular cytogenetic testing via fluorescent in situ hybridisation (FISH) allows for the chromosomes to be assessed for changes in their structure, such as the insertion, deletion or duplication of sections. Chromosomes can also be assessed for the presence of DNA sequences of interest.
- Biochemical testing studies the amount or activity level of certain proteins. Any abnormality in this is an indication of a genetic disorder, due to changes in the DNA.
- DNA profiling compares variable short tandem repeat (STR) regions of the genome to identify individuals.

KEY QUESTIONS

1 Define 'genetic testing'.

2 What are the three main methodologies used in genetic sequencing and profiling?

3 The following diagram represents STR regions of DNA used to identify a person who died in a natural disaster. Three people who were looking for a missing sibling submitted DNA for comparison.

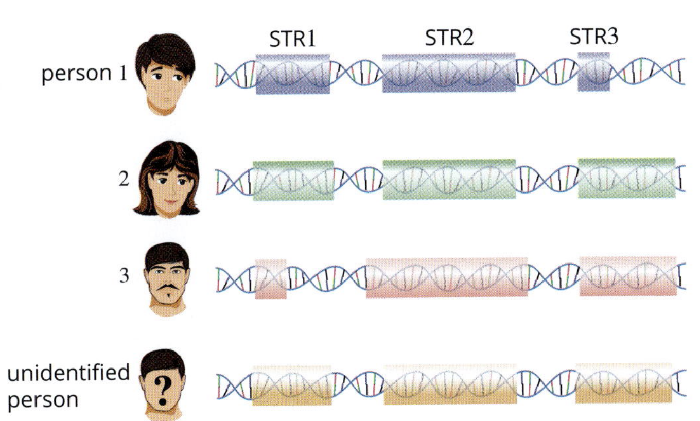

6.1 Review *continued*

a What is an STR? List the steps for determining a DNA profile by STR analysis.

b The following diagram represents gel electrophoresis of the STR analysis. Which person is most likely the sibling of the unidentified person? Explain your choice.

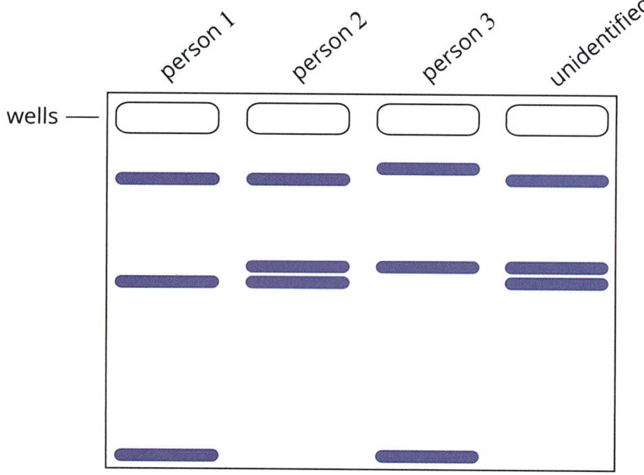

4 a Explain how you can use DNA profiling to help determine the guilt or innocence of a murder suspect found with blood stains on their clothes.

b Explain why you may need to test a sample in more than one laboratory.

5 Genetic testing may be performed on blood taken from a newborn baby whose family carries the inherited disease phenylketonuria (PKU).

a What would the genetic test be looking for?

b Describe the methods that may be used to perform the genetic test.

6 Explain why DNA profiling can or cannot discriminate between:

a a brother and sister

b a mother and daughter

c identical twins.

6.2 Population genetics and bioinformatics—identifying trends, patterns and relationships

BIOLOGY INQUIRY   

Counting cards

Can population genetic patterns be predicted with any accuracy?

COLLECT THIS...

- up to four decks of playing cards
- notebook
- calculator

DO THIS...

Population sample

1 Remove jokers from the decks of cards.
2 For a particular gene in a sexually reproducing species, there are two alleles:
   - '*A*' is represented by the black suits
   - '*a*' is represented by the red suits
   - Shuffle the decks of cards. Each student in the class takes two cards. These two cards represent your genotype. Note your genotype and then replace the cards in the deck before passing them to the next student.
3 Write down the number of each genotype (*AA*, *Aa* or *aa*) in the class.
4 Calculate the allele frequencies in this population.
5 For the next generation, divide into (breeding) pairs. You will produce two offspring. Work out the genotypes of your offspring. For each heterozygous parent, shuffle the deck and take one card to indicate which allele you pass on, then replace the card in the deck.
6 Each couple produces two offspring. Record the genotypes of the offspring.
7 Combine the class data for this new generation.
8 Calculate the new allele frequencies. Are they close to the starting frequencies?
9 Each parent takes on the genotype of one of their offspring. Repeat steps 5 to 8, breaking into new pairs.

Genetic drift and small populations

10 Remembering your original genotype, break the class into three smaller groups.
11 Work out the allele frequencies of your group.
12 Repeat the breeding step (steps 5 to 6) to generate another generation.
13 Calculate the allele frequencies of this new generation.
14 Repeat for two more generations, recording your data.
15 Pool your sub-population data with the class.

RECORD THIS...

Describe the difference between how allele frequencies behave in the full class data and the sub-population data.

Present your sub-population's allele frequencies, noting if there was a change from the starting frequencies.

REFLECT ON THIS...

Can you predict population genetic patterns with any accuracy?

How does the size of the sample affect the accuracy of the prediction?

Do allele frequencies shift more quickly in small populations than large ones?

Analysing large-scale genetic data is its own field of study and is called 'bioinformatics'.

In this section you will learn about the complex field of genetic data analysis, or **bioinformatics**. This usually involves analysing large amounts of genetic information. It uses powerful computers to detect patterns and build models that will allow researchers to understand more about the structure and function of genes and the genome. You will also learn about the **Human Genome Project** and the role of genomic research in sequencing the genes of many organisms, exploring the relatedness between species and determining gene function. You will also explore how population genetic techniques and our understanding of inheritance patterns in populations can be used in conservation management, in broad scale studies of disease inheritance and to highlight key aspects of human evolution.

THE HUMAN GENOME PROJECT

In the human genome there are 23 pairs of chromosomes. Between 1990 and 2003, all 23 chromosome pairs were fully sequenced through a massive international research effort known as the Human Genome Project (HGP) led by an American Institute. It was the world's largest collaborative biological project ever undertaken, involving biologists from at least 20 universities, one private company and thousands of people from around the world working towards one common goal—sequencing the human genome. The goal of HGP was to determine the sequence of the complete human genome—the precise order of nucleotides within a DNA molecule and the number of genes in one human individual. Through the Human Genome Project it was found that each haploid set of human chromosomes consists of approximately 3.2 billion DNA base pairs. Diploid cells contain twice as many base pairs—approximately 6.4 billion. The sequence of bases tells scientists where particular genes are located on the chromosomes. The position is known as the **locus** (plural loci). **Genomics** technology has improved rapidly since the HGP, private individuals can now pay to have their own genome analysed and many other species have been sequenced. Today's large-scale sequencing projects would be impossible without computer power for the data and automatic sequencing machines, which became commercially available in the late 1980s and have made DNA sequencing much quicker and more reliable. Previously it took up to a year for a scientist to complete a sequence of 20 000 to 50 000 bases, which is just a single chromosome of the shortest length. Today's automated machines can produce a rough draft of a DNA sequence of 50 000 bases in a few hours, ready for assembly and final checks.

BIOFILE CCT

Genomes of model organisms

The first genome to be fully sequenced was a bacteria species in 1995, the first eukaryote was baker's yeast in 1996 and since then genomes of many other species have been generated—rice, quinoa, watermelon, fruit fly, centipede, rainbow trout, saltwater crocodile, peregrine falcon, emperor penguin, platypus, tammar wallaby, gorilla and chimpanzee to name but a few of the wide range. Most are associated with specific research projects. Many are what biologists call model organisms; non-human species that breed well in captivity and can be used to represent a group of similar organisms. Model organisms and their DNA (Figure 6.2.1) are used in the laboratory to help scientists understand biological processes such as genetic inheritance and diseases.

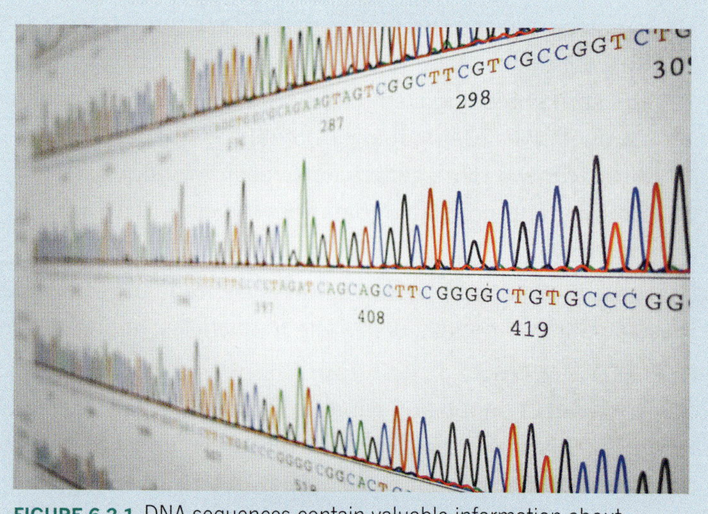

FIGURE 6.2.1 DNA sequences contain valuable information about biological processes and are an important tool in research of disease, inheritance and evolutionary relationships.

BIOLOGY IN ACTION  

The Human Genome Project—collaborative science

The Human Genome Project (HGP), completed in 2003, was a truly international cooperative research project led by the US National Human Genome Research Institute (NHGRI) (Figure 6.2.2). From the beginning, it was clear that expanding scientific knowledge of the genome would have a far-reaching impact upon humankind. Accordingly 5% of the annual budget was used for examining ethical, legal and social implications related to human genome research and to providing guidance for policy makers and the public. An example is their decision to sequence the DNA of a group of anonymous individuals, rather than a known individual, to protect privacy.

Biological research had traditionally been based on individual and independent efforts. However, for the HGP, the magnitude of the technological challenge and the large cost required the collaboration of interdisciplinary teams with specialties in biology, engineering and information technology. Procedures were automated wherever possible and the research concentrated in several major centres around the globe.

As a result, the era of team-oriented research in biology was ushered in. Research involving other genome-related projects is now characterised by large-scale, cooperative efforts involving many institutions, often from many different nations, working collaboratively. In addition to introducing large-scale approaches to biology, the HGP has produced new tools and technologies that can be used by individual scientists to carry out smaller scale research in a much more effective manner.

Every part of the genome sequenced by the HGP was made public immediately and new data continues to be shared in freely accessible public databases.

The human genome sequence produced by the HGP is not that of one person but a composite of several individuals, making it a kind of generic or representative sequence for ethical reasons. To ensure anonymity, blood samples were taken from nearly 100 donors, no names were attached and only a few were selected for analysis. Not even the donors knew if their sample was actually used for sequencing.

The HGP was designed to generate a resource that could be used for a broad range of biomedical studies. For example, to look for genetic variations that increase the risk of specific diseases such as breast cancer, or to look for the type of genetic mutations associated with cancerous cells.

Analysis of the vast amount of data processed in genomics is only possible with computing power and the expertise of a team of specialists across the STEM fields (science, technology, engineering, maths).

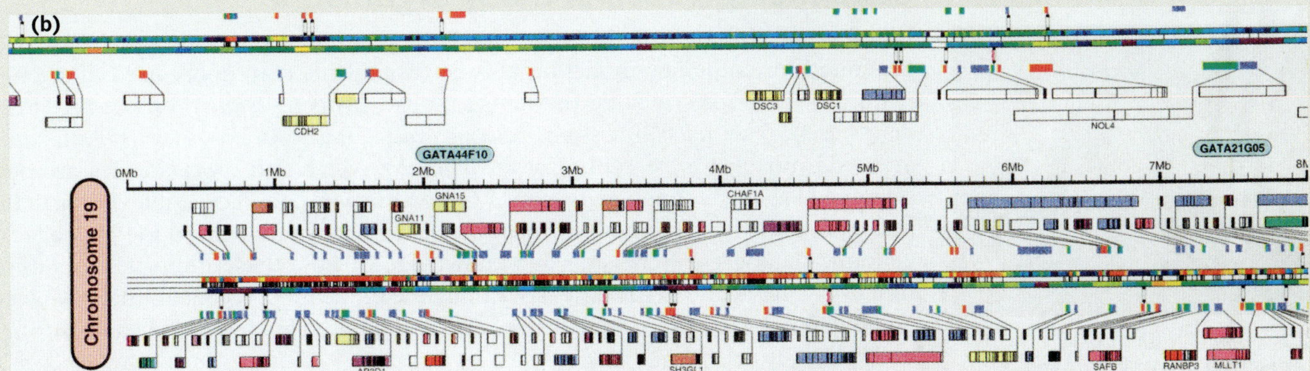

FIGURE 6.2.2 (a) In April 2003, the director of the USA's National Human Genome Research Institute (NHGRI), Francis Collins, announced that the first human genome had been successfully sequenced and mapped. The HGP was a collaboration between scientists from across the world. (b) A human genome map of chromosome 19.

THE ROLE OF GENOMIC RESEARCH

When the Human Genome Project began in 1990, a few laboratories helped sequence approximately 100 000 bases. Since then, technological improvements and automation have increased the speed of DNA sequencing enormously. New methods could reportedly sequence a human genome in 10 days, compared to more than 10 years for the first full human genome sequence. These methods are now being used to sequence many other organisms.

Determining gene function

Finding the sequence of the entire human genome does not tell us about the functions of the genes. The protein encoded by the gene must be identified in order to study its location and function. A gene sequence can be analysed and compared to other species where its function is already understood.

Molecular techniques based on gene sequences allow scientists to identify alleles and mutant forms of a gene that may cause a malfunction. The gene sequence can be modified and inserted into model organisms, and studying the effects of the modification helps scientists to understand the normal function. Alternatively, genes can be disabled in model organisms to provide direct evidence for the function of the gene and the protein it encodes. The complete human genome sequence provides the raw data for scientists to search for genes already identified in other organisms that may be useful for therapeutic purposes in humans.

Early detection and diagnosis of human diseases

By analysing DNA sequences, scientists can identify base differences in a gene suspected to be the cause of a genetic disorder. This is done by comparing the candidate gene in a DNA sequence from a healthy person with a DNA sequence from those with the disorder. For example, an inherited form of breast cancer is caused by mutations in the *BRCA1* gene on chromosome 17. The *BRCA1* gene is responsible for the production of enzymes that repair damaged DNA. When normal function is disrupted by mutations in this gene, the cells are more likely to develop additional genetic mutations that can lead to cancer. In such cases, determining exactly how a gene differs in affected and healthy individuals of a family is important for predicting who may be at risk of developing cancer, and in developing a cure.

A person with a family history of an inherited disease can have their DNA sequenced to determine if they carry the allele that caused the disease in other family members. In some cases, they can then take measures to delay the onset or reduce the severity of the disease.

Sequencing genes of many organisms

FIGURE 6.2.3 The fruit fly (*Drosophila melanogaster*) is a model organism often used in genetic research. The fly on the right is a normal fruit fly with red eyes and the fly on the left is a mutant with white eyes.

Besides sequencing the human genome, genome sequences have now been generated for some key model organisms (organisms commonly used in scientific research). Examples include the house mouse (*Mus musculus*), brown rat (*Rattus norvegicus*), fruit fly (*Drosophila melanogaster*) (Figure 6.2.3), rice (*Oryza sativa*) and wild mustard cress (*Arabidopsis thaliana*). More and more species are being sequenced as the technology becomes more affordable and accessible. In particular, the relatively small genomes of prokaryotes and viruses are being added to collective genomic databases, increasing our knowledge of how these important organisms function genetically. At the time of writing, one of the most recent animals to have an entire genome sequenced was the rare and endangered golden bat (*Myotis rufoniger*) in South Korea.

Scientists have used gene sequencing to understand the function of many genes used in agriculture. For example, researchers have found that fragrant varieties of rice, such as jasmine and basmati rice, have a change in the allele of a gene on chromosome 8. This different allele causes a fragrant protein to be produced. The difference does not affect any other part of the development of the plant.

BIOLOGY IN ACTION 

Unlocking the secrets of the wallaby genome

An international team of researchers have sequenced the genome of the tammar wallaby (*Macropus eugenii*) (Figure 6.2.4). Researchers from the National Human Genome Research Institute (NHGRI), National Institutes of Health (NIH), the Australian Genome Research Facility (AGRF), the Victorian State Government, and the Brockhoff Foundation collaborated on the project. The sequence of the tammar wallaby genome and its analysis was published in 2011.

The availability of the genome sequence provides important knowledge for understanding human and animal health. It helps unravel the biology of some interesting traits, such as the regulation of lactation (the formation and secretion of milk from the mammary glands) and embryonic diapause (a period of suspended embryonic development).

For example, researchers noted that the lactating female wallaby can feed newborn and older joeys at the same time. Remarkably, each joey is delivered milk that has the correct composition for its stage of development. Understanding the control mechanisms involved in lactation is of major interest to the dairy industry.

In diapause, wallaby embryos develop to a stage of about 100 cells and then go into a state of suspended animation for 9 to 10 months until an environmental change spurs the older joey to leave the pouch. When the older joey leaves the pouch and stops suckling, the development of the suspended early-stage embryo resumes. Understanding the mechanisms that control the embryonic development of the tammar wallaby could lead to new treatments for infertility or methods of contraception in humans.

FIGURE 6.2.4 A tammar wallaby (*Macropus eugenii*) mother and joey. The complete genome of the tammar wallaby was sequenced as part of an international collaboration between research institutions.

REVISION

GO TO ➤ Year 11 Section 10.1

Building phylogenetic trees

Phylogenetic trees are diagrams that show the evolutionary relationships between different groups of organisms. The groups compared can be different species, genera, phyla or any other level of Linnaean classification.

You can build a phylogenetic tree by placing taxa in a branching sequence, according to their shared biological characteristics (i.e. physical traits or DNA sequences). See Figure 6.2.5 for a simple example using morphological (physical) characters. By assessing the characters that different organisms share, the evolutionary relationships between taxa can be hypothesised. Starting with the most shared character, which is assumed to be the most ancestral, taxa are added to the tree sequentially, ending with the least shared character at the top of the tree (Figure 6.2.5). Each branch in the tree represents a change in character state (e.g. DNA mutation) from the last common ancestor. The greater the number of nucleotide differences between sequences or taxa, the greater the distance between them in the tree, reflecting their evolutionary relationships. Most phylogenetic trees are now built using software to generate more complex trees from large datasets.

(a) Character table

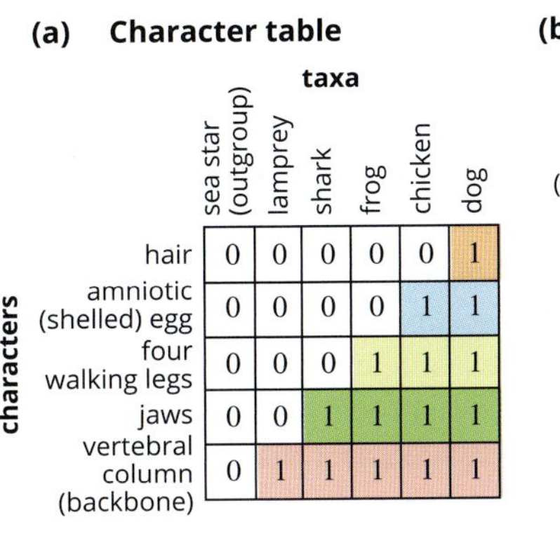

characters \ taxa	sea star (outgroup)	lamprey	shark	frog	chicken	dog
hair	0	0	0	0	0	1
amniotic (shelled) egg	0	0	0	0	1	1
four walking legs	0	0	0	1	1	1
jaws	0	0	1	1	1	1
vertebral column (backbone)	0	1	1	1	1	1

(b) Phylogenetic tree

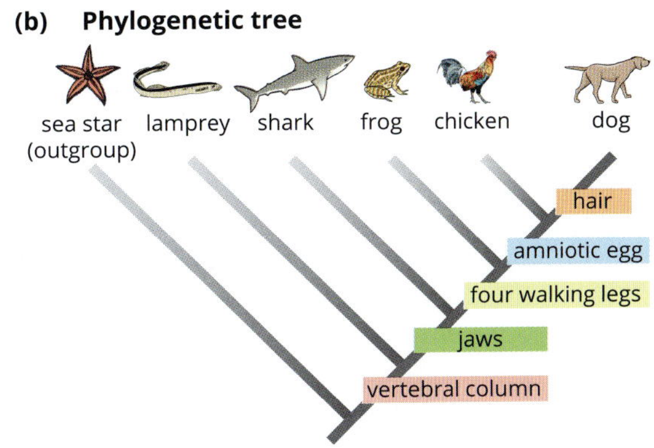

FIGURE 6.2.5 Building a simple phylogenetic tree using morphological characters. (a) A character table lists different morphological characters and their presence (1) or absence (0) in each taxon is indicated. (b) Each taxon is sequentially added to the tree according to the presence of shared characters, which are assumed to reflect when they appeared in evolutionary history. The more ancestral the character, the more likely it is to be present in a greater number and variety of organisms (e.g. all taxa except the most ancestral group (sea star) have a vertebral column, so this character is assumed to be the most ancestral and placed at the base of the tree).

BIOLOGY IN ACTION 

The origin and future of the Australian dingo

Most theories about how the dingo (*Canis lupus dingo*) (Figure 6.2.6), came to inhabit Australia have assumed Asian seafarers introduced it between 3500 and 5000 years ago, and that the dingo is therefore a kind of dog domesticated from Chinese wolves. These theories were supported by the fact that fragments of wolf DNA were found in the genomes of dogs. However, the results of a study that sequenced and compared the full genome of the dingo with the basenji (from central Africa), the domesticated boxer (from Europe) and grey wolves (from China, Croatia and Israel), suggests these theories are wrong.

The study found that the wolves were all more closely related to each other than to the dogs, and that the dogs (the dingo, basenji and boxer) were all more closely related to each other than to the wolves. From these results researchers can infer that dogs evolved earlier than previously thought, from a relative of grey wolves that has since become extinct, and that the fragments of wolf DNA in the genomes of dogs are the result of more recent interbreeding.

The study also suggested that dogs were probably domesticated by nomadic hunter-gatherers, rather than groups of people who had settled to farm the land, as previously thought. Unlike domestic dogs, most dingoes do not have the genes needed to digest starch, which is found in farmed grains. So, the dingo appears to have branched off from the other dogs before humans settled into farming societies. What this means is that the dingo is probably the closest relative of dogs that lived 10000 to 35000 years ago. However, interbreeding between dingoes and domestic dogs is changing this. Indeed, hybridisation between dingoes and domestic dogs may soon lead to the extinction of pure dingoes, which could have negative ecological consequences. For example, there is some evidence that the breeding season of dingoes has become longer because of domestic dog genes, leading to a large increase in their numbers.

FIGURE 6.2.6 An Australian dingo (*Canis lupus dingo*)

POPULATION GENETICS IN CONSERVATION AND MANAGEMENT

Population genetics is a diverse field, and has uses outside of animal and plant breeding. Examining the variation within and between populations can give us insight into the processes of evolutionary change—after all, evolution itself is merely the result of changes in allele frequencies in populations over time.

Genetic variation and conservation

For populations of species to remain viable in changing environments, there must be **genetic variation** (also known as genetic diversity).

One reason for this is that genetic variation is what allows populations to adapt to environmental change—genetic variation creates a greater variety of individuals that will respond to selection pressures differently. For example, populations with high genetic variation are more resistant to disease. Individuals with alleles that provide resistance to a disease are more likely to survive and reproduce than individuals that do not have these alleles. In populations with low genetic variation, alleles that provide disease resistance are less likely to occur, leaving the entire population vulnerable in the event of a disease outbreak.

Another reason that genetic variation is necessary for the long-term survival of a population or species is that, over time, **inbreeding** allows unhealthy mutations to accumulate. While one mutation in one gene in one individual might not be harmful (because that individual most likely has another healthy copy of that gene from the other parent), two copies of that mutation may be lethal. The more closely related individuals are, the more likely it is that they share copies of potentially lethal mutations. This is not a concern if they do not breed, but inbreeding concentrates those alleles, increasing the chance that offspring will suffer from genetic diseases.

Population genetics researchers can measure the genetic variation of a population by sampling multiple individuals and analysing multiple genetic markers. Population genetic data has many applications in conservation and the preservation of populations and species, such as captive breeding programs, taxonomy (identifying and classifying species) and prioritising areas for habitat protection.

Another way in which population genetics can assist in conservation is by identifying invasive species. When these appear in a new environment and begin to cause damage to the local ecosystem, population genetics can be used to find out where they came from. For example, the Northern Pacific sea star (*Asterias amurensis*) (Figure 6.2.7) was introduced to Tasmania in ballast water and has since spread to coastal areas in Victoria and in southern NSW. This highly successful invasive species has caused significant damage to local ecosystems. Using population genetics, researchers were able to pinpoint Japan as the most likely region of origin.

FIGURE 6.2.7 The Northern Pacific sea star (*Asterias amurensis*) was introduced into Tasmania through ballast water and has spread to Victoria and NSW. Population genetics revealed that the species was originally introduced from Japan.

BIOFILE S

Genetic variation in endangered species—giant pandas vs Siberian tigers

When it comes to saving endangered species, genetic variation is one of the most important factors to consider. The largest of all the tigers, the Amur tiger (*Panthera tigris altaica*) (Figure 6.2.8a), is estimated to have only around 500 individuals remaining in the wild. This is a very low population in terms of genetic variation, but unfortunately the story is more alarming than that—when the Amur tiger was first protected in 1947, there were only an estimated 50 individuals in the wild. The remaining 500 individuals are descended from those 50 individuals. Although the population size has increased, the genetic variation has not. This is referred to as a 'population bottleneck', and is often a cause of low of genetic variation in endangered species. This lack of genetic variation means that the Amur tiger population is extremely vulnerable to inbreeding, disease and environmental change.

Not all endangered species have such a low genetic variation. There are only an estimated 1500 giant pandas (*Ailuropoda melanoleuca*) (Figure 6.2.8b) still breeding in the wild, confined to mountain ranges in northern China. Although pandas have been difficult to breed in captivity scientists have discovered that they have a surprisingly high genetic variation in terms of their immune systems. This means that, while the giant pandas are still considered endangered, they are less susceptible to disease and inbreeding than the Amur tiger.

FIGURE 6.2.8 (a) The Amur tiger has experienced a population bottleneck and therefore has a much lower genetic variation than the current population size would suggest. (b) By contrast, the giant panda appears to have a higher genetic variation than expected, which indicates a larger and healthier gene pool.

Defining species

It is not unusual for distinct populations of a species to be quite genetically different from one another, simply because they are separated by geographic distance or other barriers that prevent **gene flow**. However, population genetic studies sometimes reveal that two populations that are thought to belong to one species are in fact two distinct species. These are called **cryptic species**. Such research is very important as it allows scientists to properly understand the species that make up an ecosystem—if a species is not described, it cannot be adequately monitored or protected. For example, the common octopus (*Octopus vulgaris*) was once thought to be a single species with a global distribution (Figure 6.2.9). However, genetic studies have shown that the common octopus could be at least six species, if not more. When there are many species together that are difficult to tell apart, this is called a species complex.

To monitor the populations of species, it is also important to be able to tell them apart without sequencing their DNA. Compared to simply looking at an animal, genetic analysis is time consuming and expensive, and conservation work often has limited funding. For this reason, finding a visible difference between the species in a complex is very important.

FIGURE 6.2.9 The common octopus (*Octopus vulgaris*) is a species complex—that is, it consists of many closely related species that were once thought to be one. This individual is from the Great Barrier Reef in Australia.

BIOLOGY IN ACTION 

The barcode of life

A DNA barcode is a short sequence of nucleotides that uniquely identifies a species. It is obtained by using PCR and DNA sequencing (Figure 6.2.10). DNA sequences are submitted to online databases such as the Barcode of Life Database (BOLD) or other similar databases. There are global DNA barcoding projects underway to catalogue all life, including bees, butterflies, mosquitoes, fungi, mammals and plants. Scientists and non-scientists alike can access these sequences for research.

To compare and identify species, you need a gene sequence that is present in all organisms but differs slightly between different groups or species. For eukaryotes, the mitochondrial gene *COI* (cytochrome oxidase subunit I) is often used. Genes in plant plastids help to identify plant species. For bacteria, a gene for ribosomal RNA can be used.

Examples of barcoding projects include:

- species identification and diversity (e.g. barcoding organisms of the Great Barrier Reef, CSIRO scientists barcoding Australian fish species)
- tracking pathogenic and non-pathogenic bacterial populations
- food authentication—Is it shark and chips? Is the beef burger really a horse burger?
- monitoring wildlife crime, such as illegal trade in protected and endangered species
- ecology and evolution (e.g. seed identification and seed banking of the Australian *Acacia* genus for conservation and restoration of biodiversity).

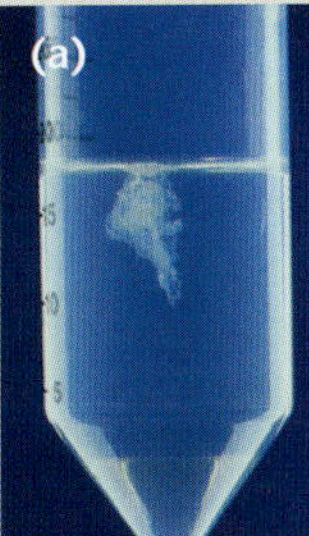

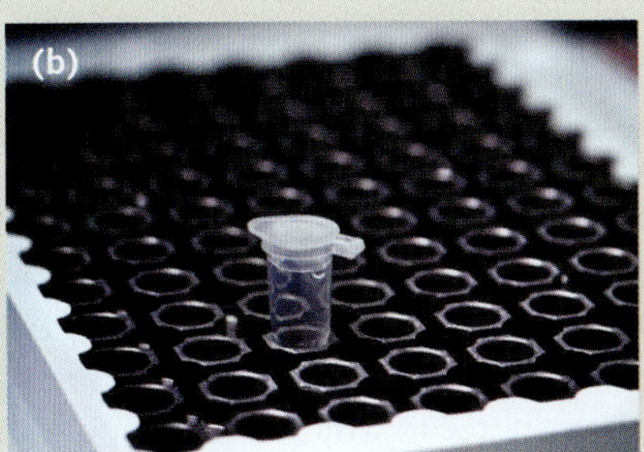

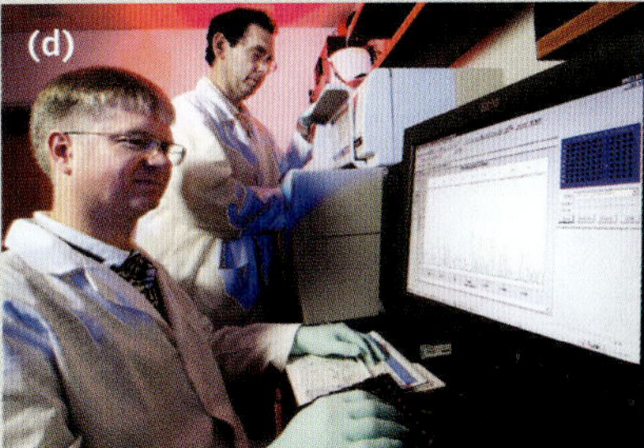

FIGURE 6.2.10 The process of obtaining a DNA barcode: (a) extract DNA, (b) amplify DNA using the polymerase chain reaction (PCR), (c) sequence DNA, and (d) analyse the DNA sequence data using bioinformatics.

POPULATION GENETICS TO DETERMINE DISEASE INHERITANCE

GO TO ➤ Section 5.2 page 210

While pedigree analyses can tell us a great deal about the way in which certain genetic diseases are inherited, they are often only useful in fairly straightforward situations where historical family information is readily available. You learnt about pedigrees in Chapter 5. In diseases where multiple genes are involved or are only partially expressed, inheritance patterns are more complicated and more data is required. Where time is of the essence and family history data is not available, population genetics can shed some light on the inheritance of diseases.

Isolated populations

Isolated populations are a particularly rich source of data when it comes to genetic disorders. This is because isolation often results in inbreeding and the genetic characteristics of the population become concentrated over multiple generations. The overall genetic variation in isolated populations is reduced, and traits that are rare in the wider population are more likely to occur. When these traits are more common, they are easier to study.

A famous isolated population is that of the Pitcairn Islanders. Pitcairn Island is a volcanic island in the South Pacific Ocean. The population on the island descends from the survivors of the famous mutiny on the *HMS Bounty* in 1789 (Figure 16.2.11),

FIGURE 6.2.11 1867 wood engraving of Pitcairn Island and the *HMS Bounty*. The island proved to be too small to support the population once it reached approximately 200, and the inhabitants migrated to Norfolk Island. The founders of the Pitcairn/Norfolk population were a small group, resulting in significant inbreeding and an increased occurrence of cardiovascular disease markers.

including a number of British sailors and Polynesians kidnapped from another island in Tahiti. When the founders of a population are very few in number, and there is no subsequent immigration, the population will experience a phenomenon known as the **founder effect**. Even if the founding population is genetically variable (in this case, a mix of Polynesian men and women, and the surviving mutineers from the British navy ship), inbreeding will rapidly concentrate alleles in subsequent generations. In 1856, the Pitcairn Islanders (having discovered that Pitcairn Island was too small to feed their growing population) migrated to Norfolk Island. However, 18 months later, 17 of the Pitcairn Islanders returned to Pitcairn Island and another 27 followed five years later. In 2014, only 57 individuals remained on Pitcairn Island.

Today, over half of the population of Norfolk Island can still be genetically traced to the survivors of the *Bounty*, which is a strong indicator of inbreeding. Studies of this population have discovered a much higher proportion of cardiovascular disease indicators in the Norfolk Island population than in the wider population. The incidence of cardiovascular disease is correlated with the proportion of Polynesian ancestry present in an individual. The founder effect has in this case resulted in these traits being concentrated in the population. Studying the genetic profiles of individuals descended from the *Bounty* survivors may shed light on the multiple genes associated with heart disease.

Twin studies

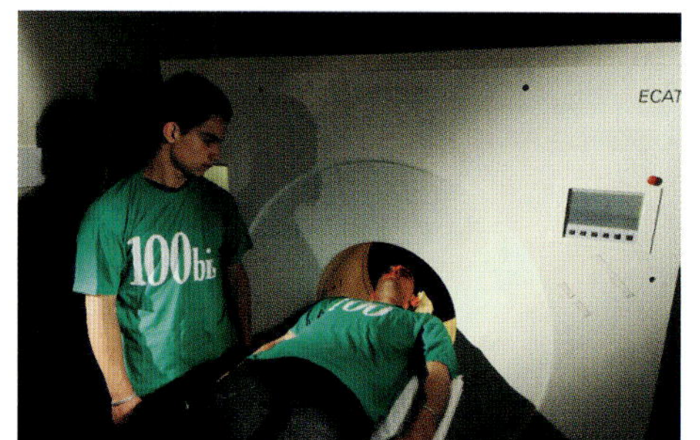

FIGURE 6.2.12 A pair of identical twins undergoing magnetic resonance imaging (MRI) of the brain to look for differences in adolescent development.

The study of genetically identical (or monozygotic) twins in a population is another way to monitor disease inheritance (Figure 6.2.12). One of the key questions facing medical researchers is how to distinguish between the effects of genetic and environmental influences on a disorder. Sometimes a disorder will not necessarily be expressed to the same degree in different individuals, and it is not clear whether this is due to differences in their DNA or in their lifestyle or environment. Identical twins, however, offer an opportunity to separate those factors due to their identical DNA. If a condition occurs in both identical twins less than 100% of the time, it has an environmental component. By comparison, fraternal (or dizygotic) twins are only related to the extent that most siblings are. Comparing the occurrence of a disorder between identical twins and fraternal twins can shed light on the degree to which a condition is genetic (or heritable).

Twin studies are particularly common in psychological research, because the presentation of mental illness is often highly variable and subject to environmental influence. For example, if one identical twin has a diagnosis of schizophrenia, the other twin will also have a diagnosis in 50% of cases. By contrast, if a fraternal twin has schizophrenia, the other twin will only have a diagnosis in 10 to 15% of cases. This difference suggests that genetics plays a very strong role in the expression of the disease. Similar numbers have also been found for bipolar disorder, anorexia, and other mental illnesses, suggesting that these also have a strong genetic component as well as being subject to environmental influences.

Genome-wide association studies

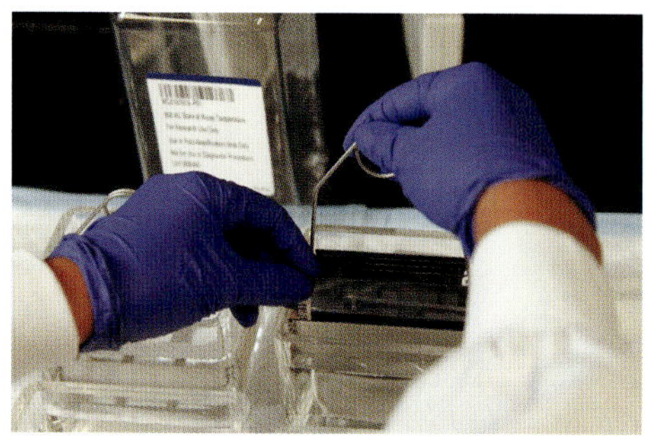

FIGURE 6.2.13 A scientist preparing a single nucleotide polymorphism (SNP) genotyping array. Such arrays are very commonly used in genotype-wide association studies (GWAS).

DNA sequence variation is being studied on a large scale in populations. Genome-wide association studies (GWAS) search for gene variants, often single base changes called single nucleotide polymorphisms (SNPs) that are associated with various traits (Figure 6.2.13). This sort of study requires very large samples, comparing the SNP arrays of a large group of people with a particular disorder, and another large group without the disorder (the control group). A GWAS will screen for thousands of SNPs at once. Because there is so much natural variation in the genome, it is often very difficult to find a true association between an SNP and a disorder.

The demographics of the sample also make these studies difficult. Biological sex and age need to be controlled for in the sample. In addition, genetic mutations leading to disorders are known to be much more common in the populations where they first occurred—therefore, samples need to be controlled for the geographical, historical and ethnic background of the individuals. This is called population stratification, and is the main reason that sample sizes for GWAS need to be very large. However, there have been some notable successes. SNPs associated with age-related macular degeneration were the first to be identified in a study in 2005. These SNPs can then be used for screening individuals to identify those at risk of developing the disorder.

POPULATION GENETICS AND THE EVOLUTION OF MODERN HUMANS

The quest for answers about human origins and the evolutionary paths that led modern humans to where we are today is more exciting than ever, with rapidly advancing DNA technology and the discovery of new fossils and archaeological artefacts. Our evolutionary story is a dynamic one, changing as new information comes to light or new ideas are found to have greater explanatory power.

Modern humans and evolutionary relationships

Homo sapiens is the Latin term for our species. The term translates as 'wise man' and refers to anatomically and behaviourally modern humans. *Homo sapiens* is one of the most widespread, adaptable and influential species to have ever existed. Under the Linnaean system of biological classification, *Homo sapiens* is a eukaryote and a member of the animal kingdom.

Among animals, humans are mammals, with the characteristics of body hair and the ability to suckle young. Humans are also classified as primates, having a grasping hand, bicuspid teeth, a short nose and well-developed eyes and brain. Primates are also characterised by their mobile hip and shoulder joints. Within the Primate order, humans belong to the same family (Hominidae) as the great apes, which include orangutans, gorillas, chimpanzees and bonobos. All members of the family Hominidae (hominids) lack a tail and have similar skeletal and skull features. In addition, hominids are bipedal (meaning they walk on two legs) and have very large brains. As well as sharing many anatomical and behavioural features with all the great apes, modern humans share 98.8% of their DNA with chimpanzees and bonobos, our closest living relatives. Although modern humans are very similar to chimpanzees and bonobos, humans did not directly evolve from chimpanzees or any other living primate. DNA and fossil evidence reveals that humans last shared a common ancestor with chimpanzees and bonobos approximately six to eight million years ago (Figure 6.2.14).

> Chimpanzees and bonobos are our closest living relatives, but we did not directly evolve from them. Our last shared common ancestor existed approximately six to eight million years ago.

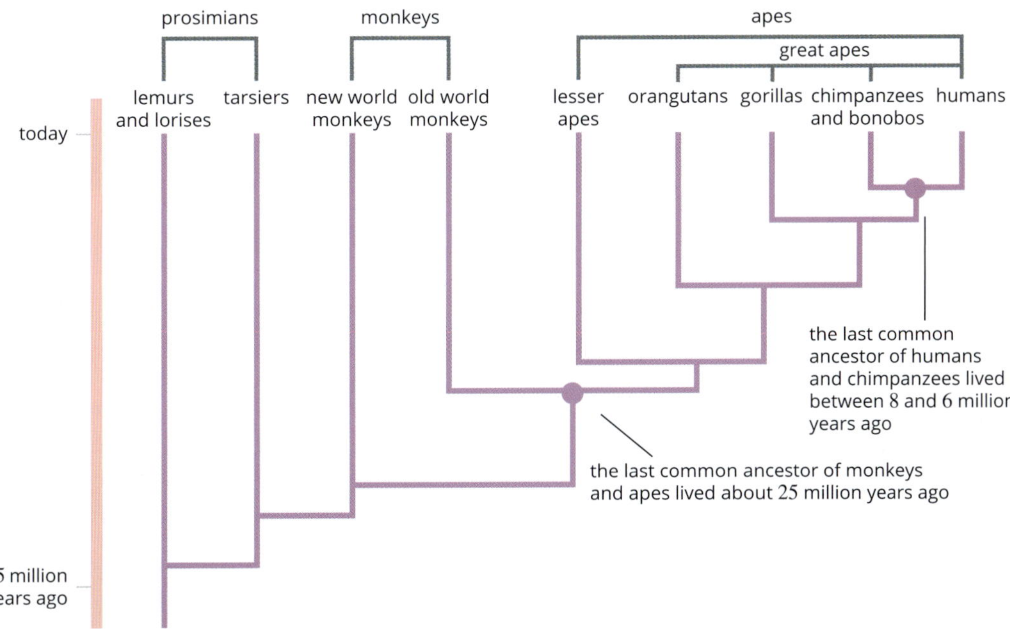

FIGURE 6.2.14 Evolutionary relationships of living primates. The closest living relatives of modern humans are chimpanzees and bonobos. Although humans share many similar characteristics with chimpanzees and bonobos we did not directly evolve from them or any other living primates. The last common ancestor humans shared with chimpanzees and bonobos lived about six to eight million years ago. The lineage that led to our species includes many fossils of human-like species that walked upright.

BIOFILE L

Hominins, hominids and hominoids

The taxonomic classification of humans has changed over time, with some confusion arising about the terms 'hominin', 'hominid' and 'hominoid.' Humans belong to the subfamily Homininae and hence are hominins. Homininae also includes chimpanzees and gorillas, as well as extinct *Homo* species, *Australopithecus* and *Paranthropus*. We are also part of a larger group, the family Hominidae (hence the term hominid), which includes orangutans. Finally, the superfamily Hominoidea includes humans, the great apes and the lesser apes (gibbons), hence the term hominoid.

There are various fossilised human-like species that have been discovered, all characterised by walking upright. *Sahelanthropus* is the oldest known ancestor of modern humans, with skulls found in central Africa that date to approximately seven million years ago. Some scientists think that *Sahelanthropus* may represent a common ancestor of humans and chimpanzees, but this is still debated in the scientific community. Species of the extinct genus *Australopithecus* are the closest relatives to our genus, *Homo*. *Australopithecus* had a forward-jutting face, a small brow and a pronounced ridge above the eyes, but hands and teeth that were similar to ours. *Homo* diverged from *Australopithecus* approximately 2.8 million years ago. The genus *Homo* is characterised by a larger brain and includes nine to 15 extinct species known to date (the identities of some fossil specimens are uncertain). Our species, *Homo sapiens*, is the only living member of the genus *Homo*. Of the extinct species, *Homo habilis*, found in Africa, is the oldest. A later species, *Homo erectus*, which had a larger brain than *Homo habilis*, spread into Southeast Asia about 1.6 million years ago, possibly surviving to about 300 000 years ago.

Fossil and DNA evidence is continuing to uncover new information and helping us to tell our story, but many questions remain concerning the origins of modern humans and fossil species.

BIOLOGY IN ACTION  

Omo Kibish

Our species, *Homo sapiens*, has been around for approximately 200 000 years, with fossil and genetic evidence placing the earliest of our species in Africa. The oldest fossil record of our species was found in a region called Omo Kibish in Ethiopia and has been dated to 195 000 years ago (Figure 6.2.15). Two specimens, known as Omo I (Figure 6.2.15b) and Omo II, were recovered from this region between 1967 and 1974. Omo I has anatomically modern human morphology with some primitive traits, representing a transition from archaic *Homo sapiens* to early modern *Homo sapiens* (Figure 6.2.15). Omo II has more primitive features than Omo I, such as a sloping forehead and more robust build, leading researchers to suggest that it belonged to a population that was transitional between *Homo heidelbergensis* and *Homo sapiens*. As both specimens have been dated to approximately the same time of 195 000 years ago, they may have co-existed. Omo Kibish and the surrounding regions are of great archaeological significance as this area is the source of many important findings regarding the origin of our species. For this reason, Ethiopia is currently thought to be the site where some of the first modern *Homo sapiens* lived.

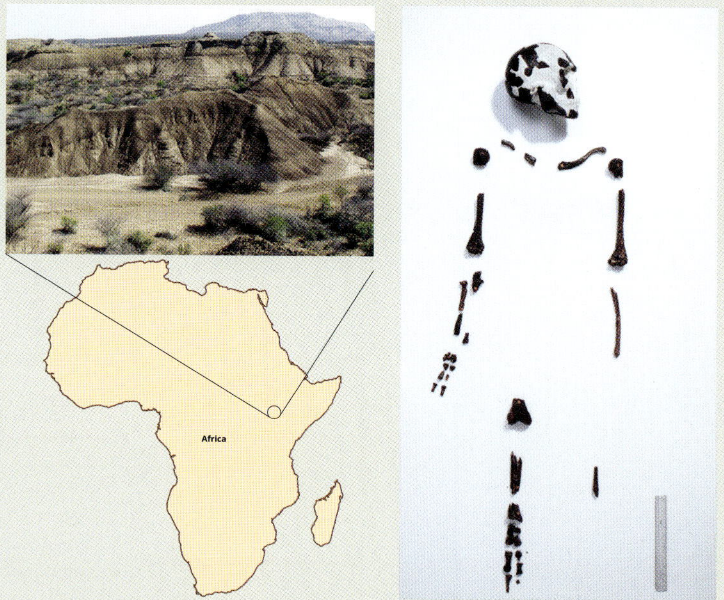

FIGURE 6.2.15 (a) The Omo Kibish formation in Ethiopia, Africa, is the site of the earliest known *Homo sapiens* fossils. These fossils date back to 195 000 years ago. (b) Omo I, the oldest anatomically modern *Homo sapiens* specimen discovered in Omo Kibish in 1967.

DNA evidence for early human origins

Fossil evidence is a rich source of information when it comes to the evolution of humans and other hominids, but in the last few decades, DNA evidence has shed light on more recent events. Fossil evidence indicates that humans almost certainly originated in Africa approximately 200 000 years ago, and that the first appearance of modern humans outside of Africa dates to around 92 000 years ago, but fossil evidence could not reliably tell us what happened afterwards. DNA evidence can help us answer a number of questions. When and how did humans emerge from Africa and spread across the globe? Did modern humans replace the other hominids, like the Neanderthals (*Homo neanderthalensis*) that lived at the same time, or did they interbreed with them?

One of the first ways that researchers used DNA to look at human origins was by using mitochondrial DNA (mtDNA). Since mtDNA does not recombine and is only maternally inherited (i.e. from mother to offspring), it is much simpler to analyse and offers a much more straightforward story than nuclear DNA. A wide sampling of the mitochondrial genome from humans currently living provided an important insight into human evolution. In 1987, researchers learnt that the most recent common ancestor—that is, the woman from whom all current human mitochondrial genomes were inherited—could be traced to Africa around 200 000 years ago. In the science media of the time, this woman was referred to as 'Eve', but it is important to remember that she is only the **mitochondrial Eve**—different genes have different histories, and this tells us only about the mitochondrial genome of lineages that have survived to the present day.

Other analyses of the mitochondrial genome can tell us about the migration pattern of early humans out of Africa approximately 100 000 years ago. By examining the differences in gene sequences between people from different regions with differing ancestries, we can reconstruct this path (Figure 6.2.16).

After the discovery of mitochondrial Eve, other research focused on the inheritance of the Y chromosome (Figure 6.2.16) and other nuclear genes that undergo very limited recombination. Like mitochondrial DNA, the Y chromosome can be used to trace the most recent common ancestor from whom all current human Y chromosomes were inherited. Because the Y chromosome is paternally inherited (i.e. from father to son) this ancestor is a male who lived in Africa approximately 90 000 years ago. In some ways, the migratory patterns shown by the analyses of nuclear genes are very similar to those of the mitochondrial analyses, but there are distinct differences. This does not mean that either pattern is incorrect, but that different genes experience different changes due to recombination, independent assortment, mutation and **genetic drift** (random changes in allele frequencies). The more genes included in a study, the more closely the results will represent the true history of the organism.

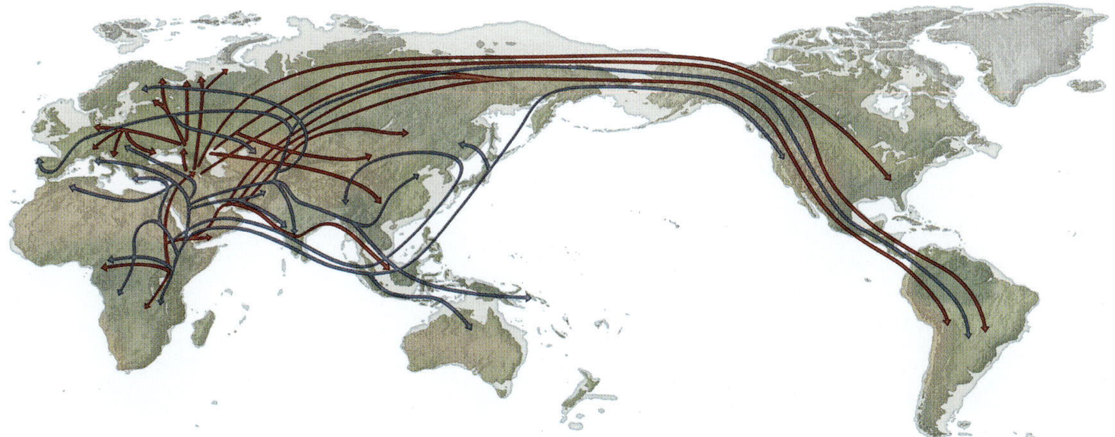

FIGURE 6.2.16 Scientists have employed population genetic techniques to study the migration of early humans out of Africa. On this map, the red arrows represent migration routes based on mitochondrial DNA (maternally inherited), while the blue arrows show the human migration routes based on evidence from Y-chromosome markers (paternally inherited).

Ancient DNA evidence, Neanderthals and the Denisovans

Fossil evidence has also shown that early *Homo* species co-existed with two other species of the genus *Homo*—the Neanderthals (Figure 6.2.17) and the Denisovans (*Homo denisovans*). While the Neanderthals have been well-studied, the Denisovans are known only from a single finger bone and two teeth. DNA has, however, been successfully recovered from both species. Isolation of DNA from very old material (such as the bones of a Neanderthal) is a difficult and complex process. The DNA is heavily degraded (broken into very small pieces) and requires special care. Techniques used to handle such old genetic material are known as 'ancient DNA techniques'. Fragmentary DNA, such as **ancient DNA**, must be sequenced in very small overlapping pieces. The process is also vulnerable to contamination (from other samples or even from researcher DNA), so this work is only done in specialised clean laboratories.

A small segment of mitochondrial DNA from a Neanderthal bone was first analysed in 1997, and showed that the Neanderthals were genetically distinct from modern humans. It was a further 11 years before an entire mitochondrial genome of Neanderthals was sequenced in 2008, and finally in 2014 an entire nuclear genome was sequenced. The genetic evidence from multiple Neanderthal samples now shows indisputable evidence of interbreeding between early modern humans and Neanderthals, but the presence of Neanderthal DNA in human genomes only occurs in non-African genomes, suggesting that this occurred after modern humans left Africa. This is not surprising, however, as Neanderthals were restricted to Europe and Asia.

It appears that while Neanderthals have contributed 1–4% of the genomes of non-African modern humans, this contribution is restricted to the nuclear genome. There is no evidence of Neanderthal contribution to the human mitochondrial genome. There are a number of possible explanations for this.

1. It is possible that human lineages existed that carried Neanderthal mitochondrial DNA, but they have since died out.
2. If only Neanderthal males were interbreeding successfully with female humans, the resulting offspring would have the mother's (human) mitochondrial DNA. Given that humans and Neanderthals were widely divergent genetically, they may have only been capable of fertilisation between a male Neanderthal and a female human, and not between a female Neanderthal and a male human.
3. There may have been social or cultural issues affecting the survival or production of any hybrid offspring.

Genetic research has also found that Denisovans were distinct from humans and Neanderthals (though this is based only on a single sample). There is also evidence that the Denisovans may have interbred with Neanderthals, with a 17% similarity in their genomes. Nuclear DNA studies have also shown that 3–5% of the DNA of present-day humans is shared with Denisovans, indicating that the Denisovans interacted closely with both early humans and Neanderthals.

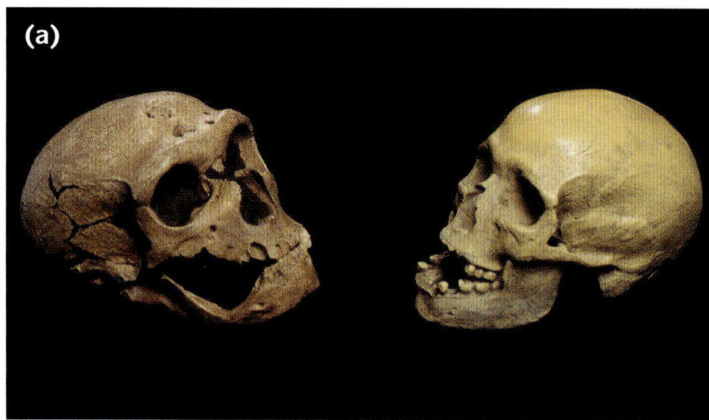

FIGURE 6.2.17 The Neanderthals (*Homo neanderthalis*) existed at the same time as early humans. (a) The skull of a Neanderthal is shown on the left, with a human skull on the right for comparison. The differences in the shape of the brow and cranium are immediately apparent. (b) Reconstruction of a Neanderthal family based on fossil evidence. It is thought that Neanderthals lived in small family groups and had complex social structures and language.

6.2 Review

SUMMARY

- The Human Genome Project determined the first full sequence of the human genome; that is the precise order of nucleotides within a DNA molecule and number of genes in a human.
- DNA technology from the Human Genome Project allowed for the DNA sequencing of other species.
- DNA sequencing of different species allows for comparison between species to determine evolutionary relationships and also determine which model organism to use for different diseases.
- DNA sequences of different species can be compared to detect genes and also determine gene function.
- The gene (or genes) responsible for a human disease can be determined by comparing the DNA sequence of a healthy individual with that of an individual suffering from the disease. Knowledge of the genes involved enables diseases to be detected early and some may even be prevented.
- Population genetics can be used to monitor the health of populations by examining genetic variation.
- Population genetics can also be used to determine whether two populations are actually distinct species.
- Genetic and fossil evidence has shed light on the evolutionary story of modern humans (*Homo sapiens*).
- Modern humans (*Homo sapiens*) are the only living species belonging to the genus *Homo*.
- The closest living relatives of humans are chimpanzees and bonobos, but we did not directly evolve from them or any other living primate. Our last shared common ancestor existed around six to eight million years ago.
- The oldest fossil record of our species was found in a region called Omo Kibish in Ethiopia and has been dated to 195 000 years ago.
- Our closest extinct relative is Neanderthal (*Homo neanderthalensis*). *Homo sapiens* coexisted with Neanderthals and there is evidence of interbreeding between the two species. DNA and fossil evidence also suggests that both these species coexisted and interbred with Denisovans (*Homo denisovans*).

KEY QUESTIONS

1 **a** What was the original goal of the Human Genome Project?

b Outline three outcomes of the Human Genome Project.

2 What major question does the Human Genome Project not answer?

3 List two reasons that genetic variation is important for the long-term survival of a gene pool.

4 Why are isolated populations so useful for studying disease inheritance?

5 When did humans last share a common ancestor with chimpanzees?

6 Why is it important to study multiple genes when looking at evolutionary history?

Chapter review

06

KEY TERMS

allele
ancient DNA
biochemical testing
bioinformatics
cryptic species
cytogenetic testing
DNA amplification
DNA (deoxyribonucleic acid)
DNA polymerase
DNA profiling
DNA sequencing
founder effect
fluorescence in situ hybridisation (FISH)
gel electrophoresis
gene
gene flow
genetic drift
genetic marker
genetic testing
genetic variation
genome
genomics
genotype (adj. genotypic)
heritable trait
Human Genome Project
inbreeding
karyotype
locus (pl. loci)
mitochondrial Eve
molecular genetic testing
mutation
population genetics
polymerase chain reaction (PCR)
polymorphism (adj. polymorphic)
primer
recognition site
restriction enzyme
restriction fragment length polymorphism (RFLP)
short tandem repeat (STR)
single nucleotide polymorphism (SNP)
species complex
twin study

REVIEW QUESTIONS

1 Which one of the following best describes restriction enzymes?

A enzymes that cut DNA at particular base sequences
B enzymes that replicate DNA
C enzymes involved in gene expression
D digestive enzymes involved in protein breakdown

2 Which one of the following lists components used in a polymerase chain reaction (PCR)?

A restriction enzymes and a primer to copy a DNA sequence
B DNA polymerase to produce a primer
C DNA polymerase and a primer to produce many copies of DNA
D restriction enzymes to produce a primer

3 A couple wishes to find out if their unborn child has sickle cell anaemia. The figure below shows the results from the gel electrophoresis of the restriction fragments of the sickle cell gene (located on chromosome 11) for the family. The mother carries the mutation, which results in sickle cell anaemia, while the father does not carry the mutation.

a How does restriction fragment analysis of the alleles of a gene result in different banding patterns?
b Does the child carry the mutation for sickle cell anaemia? Explain your answer.

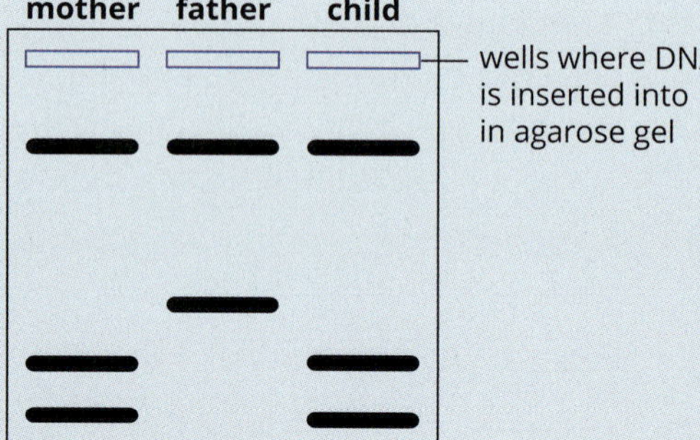

4 WAGR syndrome is a genetic condition caused by deletion of several genes from chromosome 11. It results in a number of complications, including cancer of the kidneys, vision loss, genital and urinary tract abnormalities and intellectual disability.

When WAGR syndrome is suspected, FISH is used to confirm the diagnosis.

a What does FISH stand for?
b Normally when FISH is used two gene probes are used—one for a sequence known to be on the same chromosome as the target DNA and the other for the target DNA. The first probe allows the researcher to clearly identify the correct chromosome. When the probe joins to its target it fluoresces. This fluorescence can be observed under a microscope.

What is a probe?

c **i** To perform FISH, what must the geneticist first know?
ii If the individual has normal chromosomes, how many spots would fluoresce on the chromosomes?
iii If the individual has WAGR, how many spots would fluoresce on the chromosomes?

5 *Drosophila melanogaster* is commonly used for genetic research. One particular mutation results in the deletion of a section of DNA, 200 nucleotides long, from one particular gene. The gene was extracted from a fly that is homozygous for the mutant gene and the same gene was extracted from a fly that is homozygous for the wild type (normal) version of the gene. Both versions of the gene were amplified using PCR. The PCR products were then separated using gel electrophoresis. Which gel most accurately shows the PCR products?

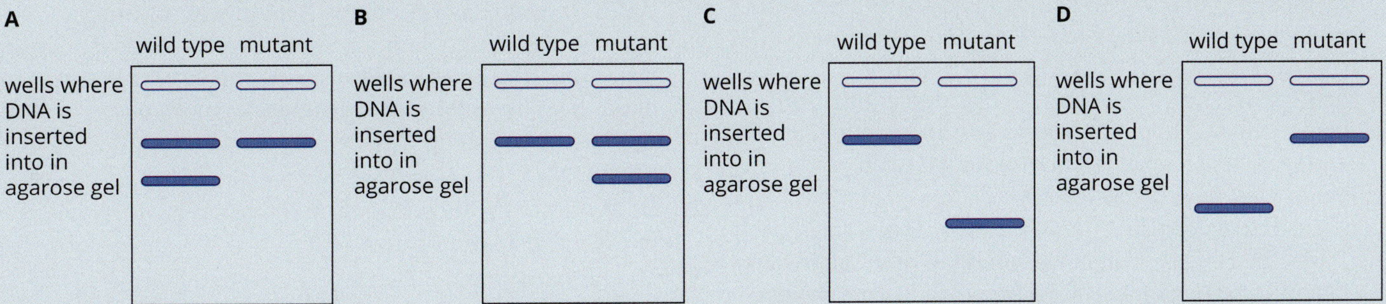

6 At a small country hospital three babies were born on one night. This stretched the resources of the hospital to such an extent that normal procedures failed and the babies were not labelled with their mother's name. To ensure the correct babies were taken home by the correct parents, the hospital performed DNA testing. Homologous chromosomes come in pairs. Each member of the pair has alleles for the same genes. STRs also come in allelic forms. The number of repeats on both members of a homologous pair can be different. An STR on chromosome 6 that has between seven and 20 ATTG repeats was investigated to match the parents with their babies. The number of STR repeats for the couples and the babies are shown below.

Couple one		Couple two		Couple three	
Mother	Father	Mother	Father	Mother	Father
11, 14	7, 12	14, 20	12, 18	18, 20	11, 18

Baby one	Baby two	Baby three
12, 20	11, 20	12, 14

a Match each baby with its correct parents.

b Explain how you matched the couples with their children.

c Figure 6.1.17 shows one way of analysing a series of STRs. It shows the analysis of 10 sites. Some sites have two peaks and others only one. Explain why this is the case.

7 Consider the evolutionary tree below, which includes the genera *Homo, Paranthropus, Australopithecus* and *Sahelanthropus*.

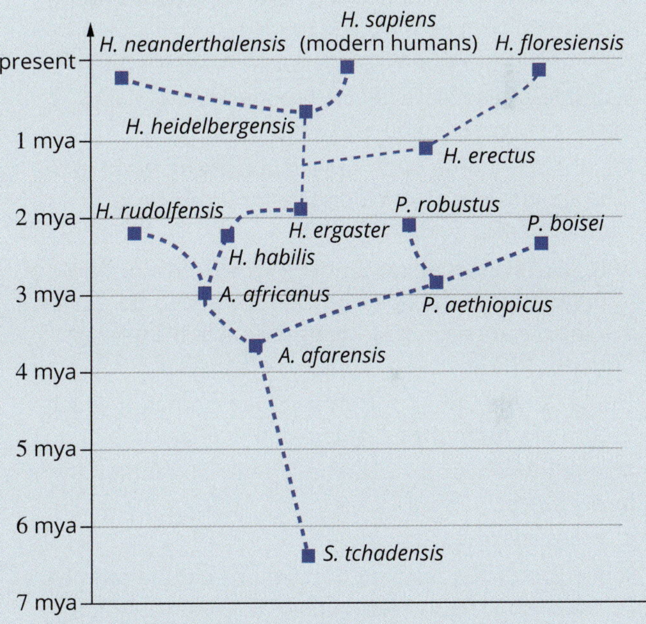

According to the evolutionary tree, which of the following is true?

A *Homo erectus* gave rise to modern humans.

B *Homo habilis* is a direct descendent of *Homo ergaster*.

C *Australopithecus afarensis* lived 4.5 million years ago.

D The most recent common ancestor of *Homo erectus* and *Homo heidelbergensis* lived about 1.5 million years ago.

8 Sections of DNA from two archaic species of *Homo*, Neanderthals and Denisovans, have been sequenced and compared to the genome of modern humans. This research has shown that Neanderthals in Europe and the Middle East and Denisovans in Asia interbred with modern humans. Only modern humans descended from European populations have Neanderthal DNA in their genomes and only populations descended from Indigenous Australians, Melanesians and some native Southeast Asian groups, such as the Manobo of the Philippines, have Denisovan DNA.

- **a** There is a significant number of scientists who argue that the classification of Neanderthals and Denisovans should be *Homo sapiens denisovan* and *Homo sapiens neanderthalensis*. Why might they suggest this?
- **b** Mitochondrial DNA from Neanderthals and modern humans have both been fully sequenced. Modern human mitochondrial DNA shows no relationship between Neanderthals and modern humans. How does this support the idea that they are separate species?

9 List at least two ways in which population genetics is useful for conservation purposes.

10 What is a Genome Wide Association Study (GWAS)? Why do these sorts of studies require very large sample sizes?

11 With some disorders, it is not clear how severe they are in different individuals or how they are inherited. What are the advantages of studying identical twins when it comes to these sorts of complex disorders?

12 Humans are hominids. Hominids are primates, which in turn are mammals. List two defining features of

- **a** mammals
- **b** primates
- **c** hominids.

13 While genetically isolated populations can be physically isolated (such as the residents of Pitcairn Island), they can also be culturally isolated. Cultures that practice consanguineous marriage—that is, marriage between blood relations—will also show signs of inbreeding over multiple generations. Why are such populations of interest to genetic researchers?

14 Describe two techniques that allow scientists to determine the function of genes.

15 At one point during the DNA sequencing process, fragments of DNA are injected into fine capillaries. When a voltage is applied, the DNA moves towards a positive electrode, and the small fragments move faster than the larger ones, allowing the correct order of bases to be determined. What other molecular technique covered in this chapter uses the same principle?

16 Briefly describe the technical and ethical issues related to DNA profiling.

17 The human genome has been sequenced in its entirety. Name another organism that has had its genome sequenced, and what might be learned from that.

18 How is the polymerase chain reaction (PCR) of use to researchers and profilers?

19 Genetic techniques allow researchers to determine whether similar-looking organisms are the same species or whether they are different. Why is this important for conservation purposes?

20 Why is the gene *cytochrome c* often used when comparing relatedness between species?

21 After completing the Biology Inquiry on page 259, reflect on the inquiry question—Can you predict population genetic patterns with any accuracy? Describe two methods for analysing population genetic patterns and explain how data collected using these methods can be applied to:

- **a** conservation of biodiversity
- **b** disease treatment or control
- **c** understanding human evolution.

MODULE 5 • REVIEW

REVIEW QUESTIONS

Heredity

Multiple choice

1 By which process do most bacteria divide?
  - **A** mitosis
  - **B** meiosis
  - **C** budding
  - **D** binary fission

2 Which of the following is not true about binary fission and mitosis?
  - **A** Binary fission occurs more rapidly than mitosis.
  - **B** The nuclear membrane breaks down and reforms during mitosis, but not during binary fission.
  - **C** Spindle fibres are present during mitosis, but not during binary fission.
  - **D** Binary fission occurs in eukaryotes and mitosis occurs in prokaryotes.

3 Which one of the following statements about the cells resulting from mitosis is correct?
  - **A** They are identical in shape, size and content to the original cell.
  - **B** They are each half the size of the original cell and have identical nuclear content.
  - **C** They are daughter and son cells.
  - **D** They are each one quarter of the size of the original cell.

4 A cell with a diploid number of 12 chromosomes undergoes mitosis. What will be the product at the end of mitosis?
  - **A** two cells each with 12 chromosomes
  - **B** four cells each with six chromosomes
  - **C** two cells each with six chromosomes
  - **D** four cells each with 12 chromosomes

5 Which of the following is not an example of asexual reproduction?
  - **A** reproduction via budding in baker's yeast
  - **B** formation of spores during sporogenesis without meiosis in red algae
  - **C** formation of plantlets on specialised leaves of kalanchoe
  - **D** fertilisation of orchids resulting in formation of a fruit

6 What are transcription factors?
  - **A** promoters
  - **B** TATA boxes
  - **C** start and stop triplets
  - **D** proteins that attach to DNA to regulate gene expression

7 Upstream areas of the gene that regulate transcription are:
  - **A** promoters
  - **B** stop triplets
  - **C** start codons
  - **D** transcription factors

8 Which of the structures shown below is a tertiary protein structure?

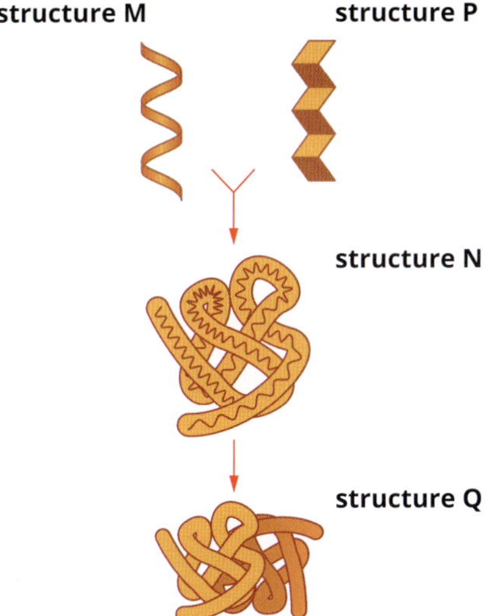

  - **A** structure M
  - **B** structure N
  - **C** structure P
  - **D** structure Q

9 Which of the following statements about genes or alleles is correct?
  - **A** Alleles randomly segregate during meiosis.
  - **B** Genes randomly segregate during meiosis.
  - **C** Alleles represent specific information coded at a defined locus on homologous chromosomes.
  - **D** Genes and alleles mean the same thing.

10 Which of the following best describes the aim of the Human Genome Project?
  - **A** To identify all the genetic diseases on human chromosomes.
  - **B** To map the DNA sequence of all human chromosomes.
  - **C** To locate the gene for each human feature on chromosomes.
  - **D** To identify the location of alleles for dominant traits.

11 Which of the following describes the composition of a eukaryotic chromosome?

A one DNA molecule and one large protein
B many DNA molecules and many proteins
C one DNA molecule and many proteins
D many DNA molecules and one large protein

12 How many autosomes are there in a human sperm?

A 1
B 22
C 23
D 44

13 What is the difference between the X chromosome and Y chromosome in humans?

A The X chromosome is much shorter.
B Many genes found on the X chromosome are absent from the Y chromosome.
C Both chromosomes carry the same genes, but the loci of the genes are different.
D Only the X chromosome determines biological sex.

14 Consider the following types of information:

i size of the chromosomes
ii gene mutations of the chromosomes
iii age of the individual

Which one or more of these are evident in a karyotype?

A i only
B ii only
C i and ii only
D i, ii and ii

15 To make a karyotype, which phase of cell division is photographed?

A anaphase of mitosis
B anaphase I of meiosis
C metaphase of mitosis
D metaphase I of meiosis

16 Which of the following genotypes shows alleles for a heterozygous trait?

A AA
B Bb
C CD
D Cd

17 'Carried on the X-chromosome' and 'occurs more commonly in males than in females' suggests:

A monohybrid cross
B dihybrid cross
C autosomal-linked inheritance
D sex-linked inheritance

18 The cross-over percentage between linked genes P and Q is 40%, between Q and R is 20%, between R and S is 10%, between P and R is 20%, and between Q and S is 10%. What is the sequence of genes on the chromosome?

A P, Q, R, S
B P, R, S, Q
C P, Q, S, R
D P, S, R, Q

19 Which step in the process of PCR best describes annealing?

A separating the DNA strands
B binding the primers
C adding the polymerase
D building the complementary DNA strands

20 The figure below is a DNA profiling printout obtained from one individual. Ten regions have been analysed, nine STRs and the sex chromosome markers. In the centre of the figure there is a peak labelled as 32.

Which of the following best describes why there is only one peak?

A The individual has one chromosome with that STR and the STR is 32 units in length.
B The individual has two chromosomes with that STR and the STR is 32 units in length.
C The individual has two chromosomes with that STR and the STR is 16 units in length.
D The peak represents an STR on the Y chromosome and the person tested was male.

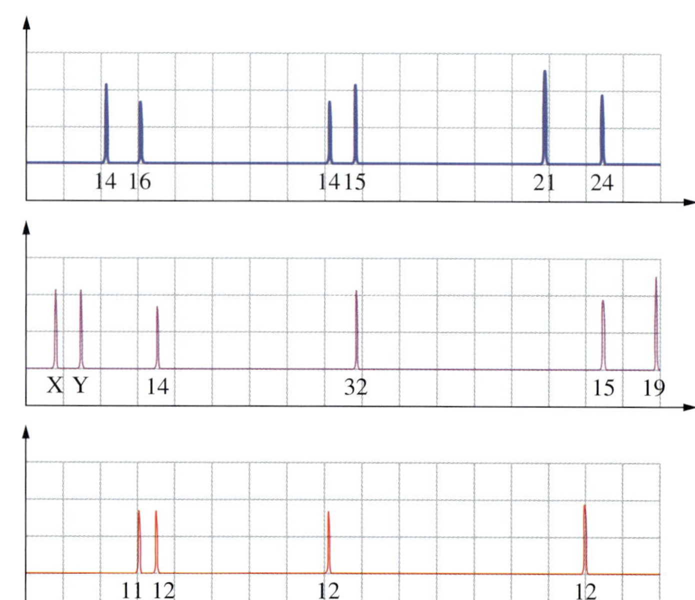

Short answer

21 Some species of aphids are capable of reproducing either asexually (via parthenogenesis) or sexually.

a What might be the advantage or disadvantage of:

i asexual reproduction when the environment is favourable?

ii switching to sexual reproduction when the environment becomes unfavourable?

b Examine the differences between parthenogenesis and fragmentation.

22 According to cell theory, all cells arise from pre-existing cells. This figure shows the cell cycle of a eukaryotic cell of a diploid organism.

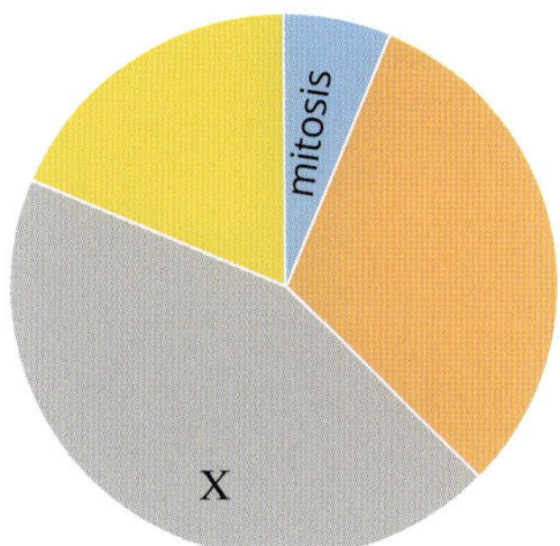

a Explain what is meant by the term 'cell cycle'.

b Identify the phase of the cell cycle labelled X.

c Describe the process that is occurring during the phase marked as X on this cell cycle diagram.

d For some cells, there is a fourth phase in the cell cycle known as G_0. Copy the cell cycle diagram shown, and mark on it where the G_0 phase occurs.

e What are some of the reasons why cells enter the G_0 phase?

f There are checkpoints in the cell cycle to ensure that two genetically identical daughter cells are produced at the end of the cell cycle.

i On your diagram, draw and label the three checkpoints in the cell cycle.

ii Outline what happens during each of the three checkpoints.

iii Cells can undergo cell death if damage to the cell is too great. State the two types of cell death and outline the conditions that will result in each type of cell death.

23 The following diagram shows a stage during meiosis. The circles represent the cell and the structures within represent a homologous pair of chromosomes.

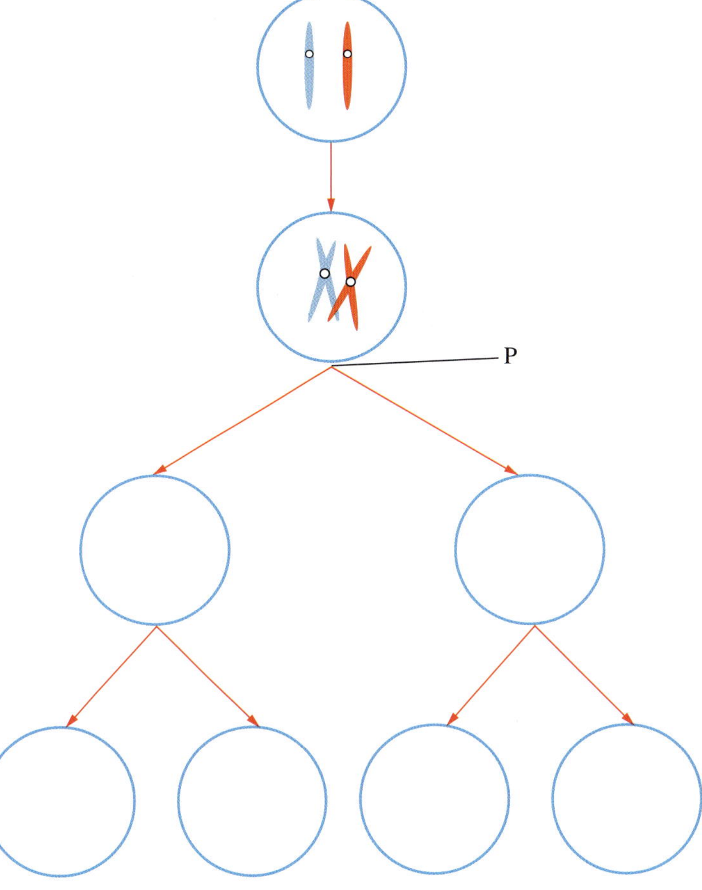

a Copy the diagram and then complete it by drawing the chromosomes inside the blank cells.

b Explain what is happening at P.

c Analyse how meiosis promotes variation in a species.

24 Meiosis is the type of cell division that produces gametes. Meiosis is divided into phases, but before it can start a special phase called interphase is required. The following diagram is a summary of interphase and meiosis.

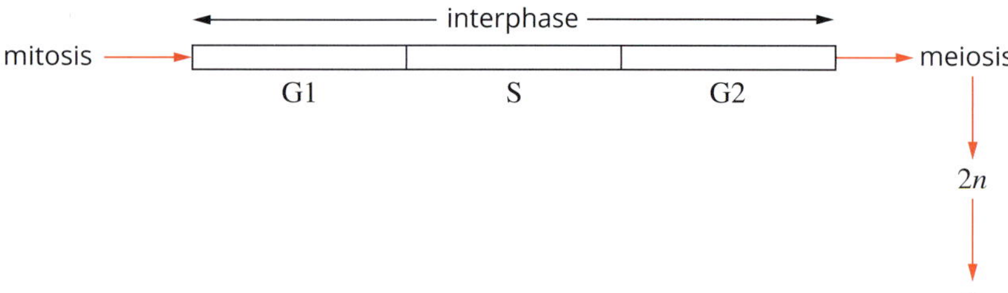

a The first cell division of meiosis is called the reduction division. Explain why.

b Which stage of meiosis is similar to mitosis? Explain your answer.

c The following graph represents the changes in the amount of DNA in a cell as it goes through meiosis.

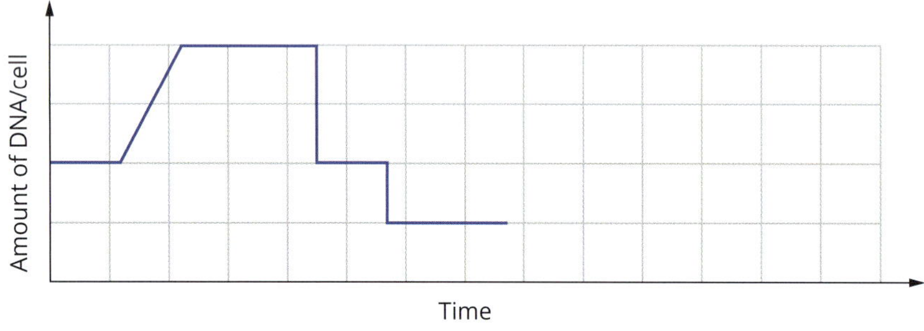

i On the graph, write the letter D where DNA replication is occurring, and the letter C where cytokinesis is occurring.

ii On the graph, continue the line to show what would happen to the amount of DNA if fertilisation occurred and the cell carried on with one mitotic cell division.

d Distinguish between mitosis and meiosis.

25 The following diagram shows a short section of a polynucleotide.

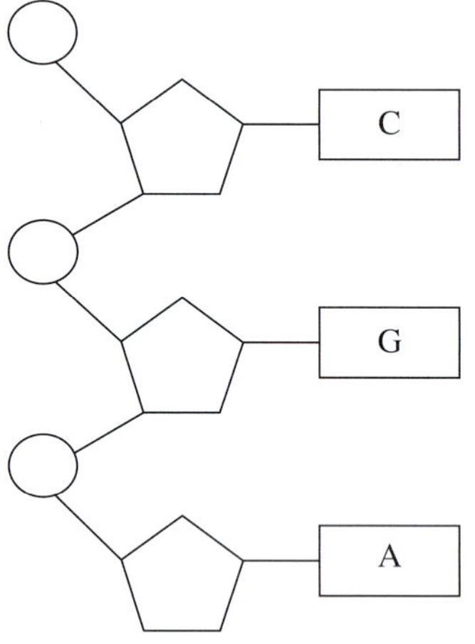

a On the diagram, label a nucleotide, a phosphate group, a deoxyribose sugar and a nitrogenous base.

b Draw on the diagram the complementary strand of polynucleotide.

c Describe how DNA is packaged in a chromosome.

26 The snowshoe hare's fur is brown in summer and white in winter. Explain how different factors influence the snowshoe hare's fur colour.

27 Myotonic dystrophy is a serious disease that causes wastage of muscles. It can affect cardiac muscle, resulting in heart problems. The most severe form of the disease is caused by a mutation in the DMPK gene, which is found on the long arm of chromosome 19. It is caused by a CTG trinucleotide repeat. In most people there are between 5 and 37 repeats, but in individuals with myotonic dystrophy the number of repeats exceeds 50. It is often an adult-onset disease and has an autosomal dominant pattern of inheritance. This means that if the allele is inherited it is certain that the disease will develop, but the person may not know until later in life.

a Explain how electrophoresis could be used to identify whether an individual has the mutated allele.

b Before a person can undergo genetic testing they must spend some time discussing associated issues with a counsellor. Propose some issues that could be associated with genetic testing for myotonic dystrophy.

28 Following the completion of the Human Genome Project in 2003, the DNA technologies developed were used to sequence the genomes of non-human organisms. Answer the following questions.

a Distinguish between the terms 'genome' and 'genes'.

b What is the purpose of sequencing the genomes of non-human organisms?

c Compare the use of the Human Genome Project with karyotypes as ways to identify the location of harmful genes in humans.

29 In humans, the adenomatous polyposis coli (APC) gene is located on chromosome 5. APC controls cell division and is also known as a tumour suppressor gene. Mutations of APC will cause a genetic disease called Familial Adenomatous Polyposis (FAP).

a Is FAP a sex-linked genetic disease? Explain your answer.

b Of the gametes produced by a person with FAP, 50% have an APC gene with the mutation. State whether FAP is a dominant or recessive phenotype. Explain your answer.

c A male who is heterozygous for FAP and an unaffected female are planning to have children. Predict the possible phenotypes and genotypes of the children, showing your working.

30 Boys can inherit the X-linked allele (Xc) that causes the recessive phenotype red–green colour blindness.

a Explain the following terms:

i X-linked allele

ii recessive phenotype.

b Write down the possible genotypes for red–green colour blindness in:

i men

ii women.

c A boy inherited red–green colour blindness from one of his grandfathers. Which of the boy's grandfathers (maternal or paternal) was also colour blind? Explain your reasoning

d A red–green colour-blind woman and an unaffected man had five children—three boys and two girls. The three boys and the elder girl did not have children. The younger girl married a man with normal colour vision, and they had four children—two boys and two girls. Draw a pedigree chart to illustrate the inheritance of the X-linked condition in this family. Use conventional symbols.

31 Consider the following cross-section of an embryo, showing the three primary germ layers.

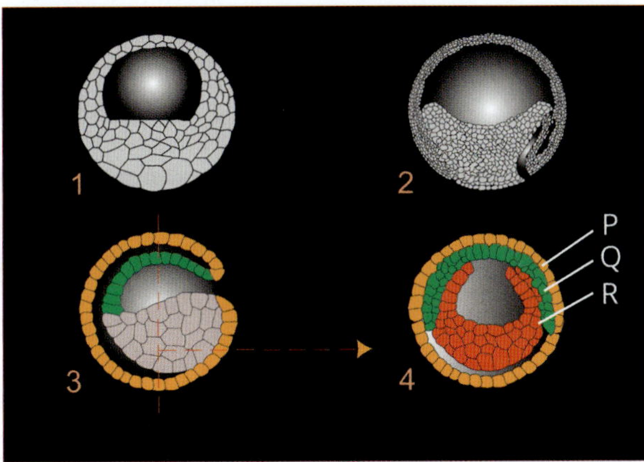

a Clarify the difference between an embryo and a fetus.

b Identify the germ layers P, Q and R.

c State one type of tissue that will arise from each of the germ layers.

d Can the cells from this embryo be used as embryonic stem cells? Explain your answer.

e Describe the different types of stem cells.

f Outline the advantages and disadvantages of using embryonic stem cells and adult stem cells.

32 Lectin is a type of glycoprotein produced by plants, and can have toxic effects on animal cells. After investigating whether plants would produce molecules that are also toxic to plant cells, a group of students conducted an experiment to determine whether lectin effects the rate of mitosis in onion root tips.

The roots of one onion bulb were immersed in distilled water for 48 hours, while the roots of another were immersed in distilled water containing lectin for 48 hours. Three onion root tips from each onion were then harvested, stained and viewed under a microscope (see below). For each root tip, the number of cells in the field of view was counted. These cells were in interphase and undergoing mitosis in the apical meristem.

The results were recorded in tables, as shown below. Table 1 shows the number of cells in interphase and mitosis for onion tips immersed in distilled water only. Table 2 shows the number of cells in interphase and mitosis for onion tips immersed in distilled water containing lectin.

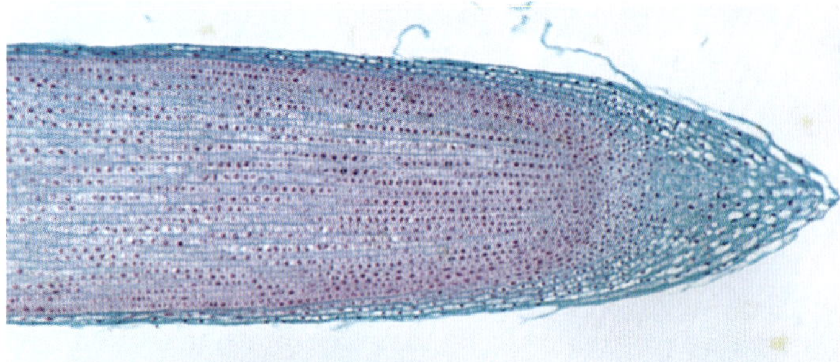

TABLE 1 Results for onion tips immersed in distilled water only

Onion root tip	Number of cells at:		
	Interphase	Undergoing mitosis	Total
1	47	34	81
2	36	29	65
3	37	30	67
Total	120	93	213

TABLE 2 Results for onion tips immersed in distilled water containing lectin

Onion root tip	Number of cells at:		
	Interphase	Undergoing mitosis	Total
1	52	44	96
2	83	25	108
3	90	54	144
Total	225	123	348

a Write a hypothesis for this experiment.

b Calculate the percentage of cells in interphase and undergoing mitosis for onion root tips treated with lectin and immersed in distilled water. Write down your answers in the following table.

Type of onion root tip	Percentage of cells (%)	
	Interphase	Undergoing mitosis
treated with lectin		
immersed in distilled water		

c Based on the results, what conclusion can you make about the effect of lectin on the rate of mitosis in onion root tips? Do the results support your hypothesis?

33 In the garden pea, *Pisum sativum*, the phenotype for tall plants (allele represented as T) is dominant over the phenotype for short plants (t). The phenotype for round seeds (R) is dominant over the phenotype for wrinkled seeds (r). The alleles are not linked.

Pure-breeding tall plants with round seeds were crossed with pure-breeding short plants with wrinkled seeds. The F1 plants were then crossed with plants that had the genotype *ttrr*. The table below shows the results obtained in the F2 generation.

Phenotype	Frequency
tall plants with round seeds	22%
short plants with round seeds	26%
tall plants with wrinkled seeds	25%
short plants with wrinkled seeds	27%

a State the genotype and the phenotype of the F1 individuals.

b Draw up a Punnett square to show the expected ratio of phenotypes in the F2 generation.

c Are the results listed in the table exactly as you expected? Suggest an explanation for why they are as expected or why they are different.

d Outline an experiment to investigate the genotype of a tall *Pisum sativum* whose genetic history is unknown.

34 In fruit flies (*Drosophila melanogaster*), the grey body phenotype (allele G) is dominant to black body (g), and normal wing shape (N) is dominant to vestigial (very small) wings (n). The genes for these traits are linked. Male flies heterozygous for both grey body and normal wings were mated with black-bodied and vestigial winged females. Five thousand offspring were examined for body colour and wing type. The following table shows the results obtained.

Offspring	Frequency
grey body, normal wings	36%
black body, vestigial wings	38%
grey body, vestigial wings	12%
black body, normal wings	14%

a Contrast linked and unlinked genes.

b The genotype for the female parental flies can be represented as ggnn. Suggest a possible genotype designation for the male parentals.

c Describe the process of recombination (crossing over).

d Does the experiment provide evidence for recombination of linked genes? Explain your answer.

e Identify which offspring are the result of recombination in this cross.

35 The 'Ever-Open Convenience Store' had experienced several robberies. The police were keen to catch the offender, who brandished a gun during each robbery. The police had five suspects, but were unable to gather sufficient evidence to clearly identify the perpetrator.

The robber wore rubber gloves, a mask, concealing clothing and a balaclava. After the fourth robbery the police found the little finger ripped from a pair of rubber gloves. The piece of glove was carefully collected and sent to the forensic science laboratory to be tested for DNA. Such material will contain a very small amount, if any, of DNA.

a What is the source of the DNA found inside the glove?

b Such small amounts of DNA are not suitable for preparing a DNA profile. How will the forensic scientists acquire enough DNA to create a profile? Draw a flow chart describing the process.

c DNA from the crime scene was collected and amplified. A DNA profile was made using the amplified DNA from the crime scene and DNA from each of the five suspects. The profile is shown below.

i The standards are 1000 kb, 2000 kb, 4000 kb, 5000 kb, 7000 bp and 10 000 bp. What is the size of the band indicated by the red arrow?

ii Why are standards needed?

iii Explain which suspect best matches the crime scene sample.

iv Does a match mean that the suspect performed the crime?

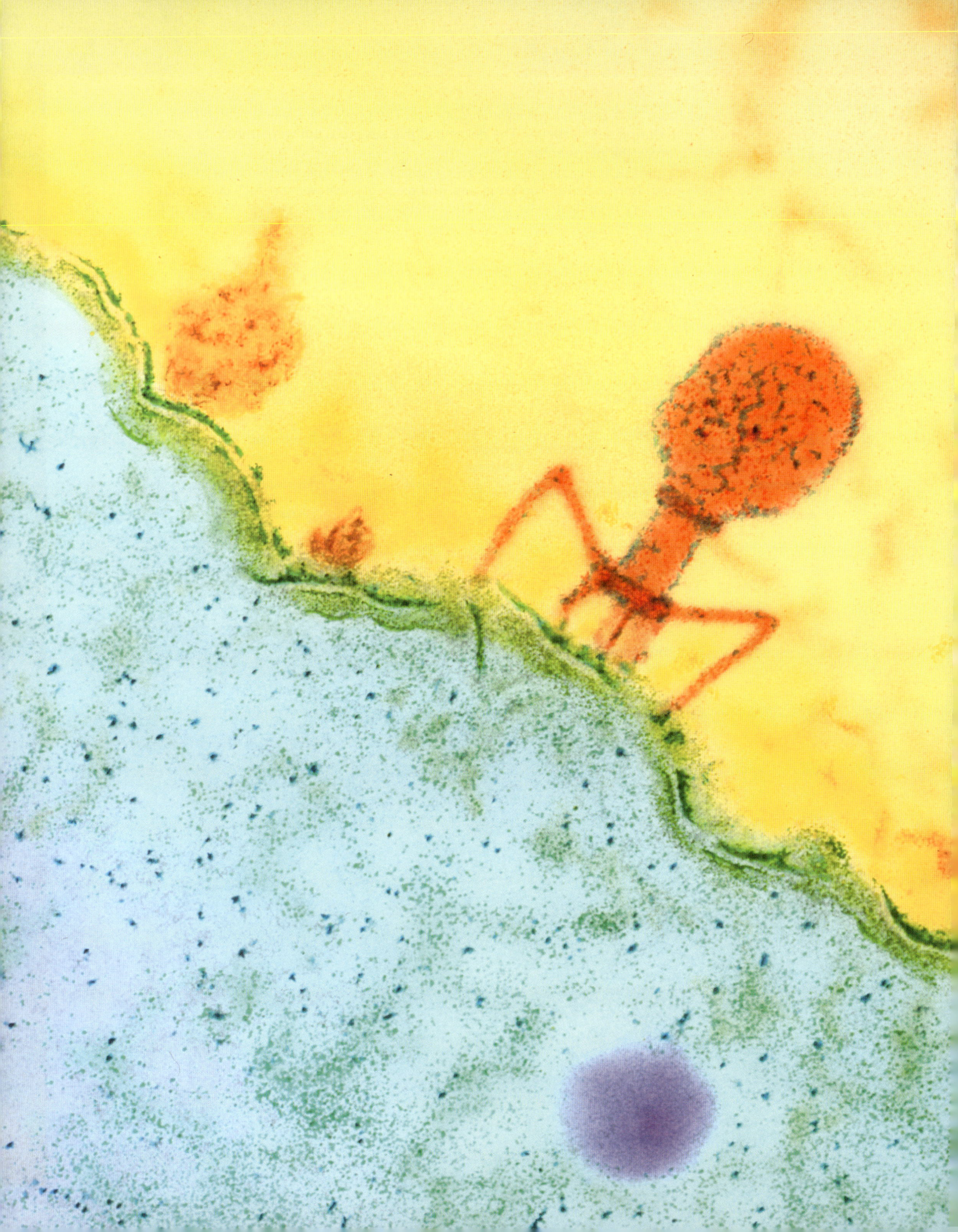

Genetic change

You will learn about natural and human-induced causes and effects of genetic change, including mutations, environmental pressure and uses of biotechnology. You will also learn how to investigate the processes of inheritance and evolution.

The work of scientists in various fields of work, including agriculture, industry and medicine, can be explored within the context of biotechnology. You will also explore the impact of biotechnology on biological diversity in this module.

Outcomes

By the end of this module you will be able to:

- solve scientific problems using primary and secondary data, critical thinking skills and scientific processes BIO12-6
- communicate scientific understanding using suitable language and terminology for a specific audience or purpose BIO12-7
- explain natural genetic change and the use of genetic technologies to induce genetic change BIO12-13

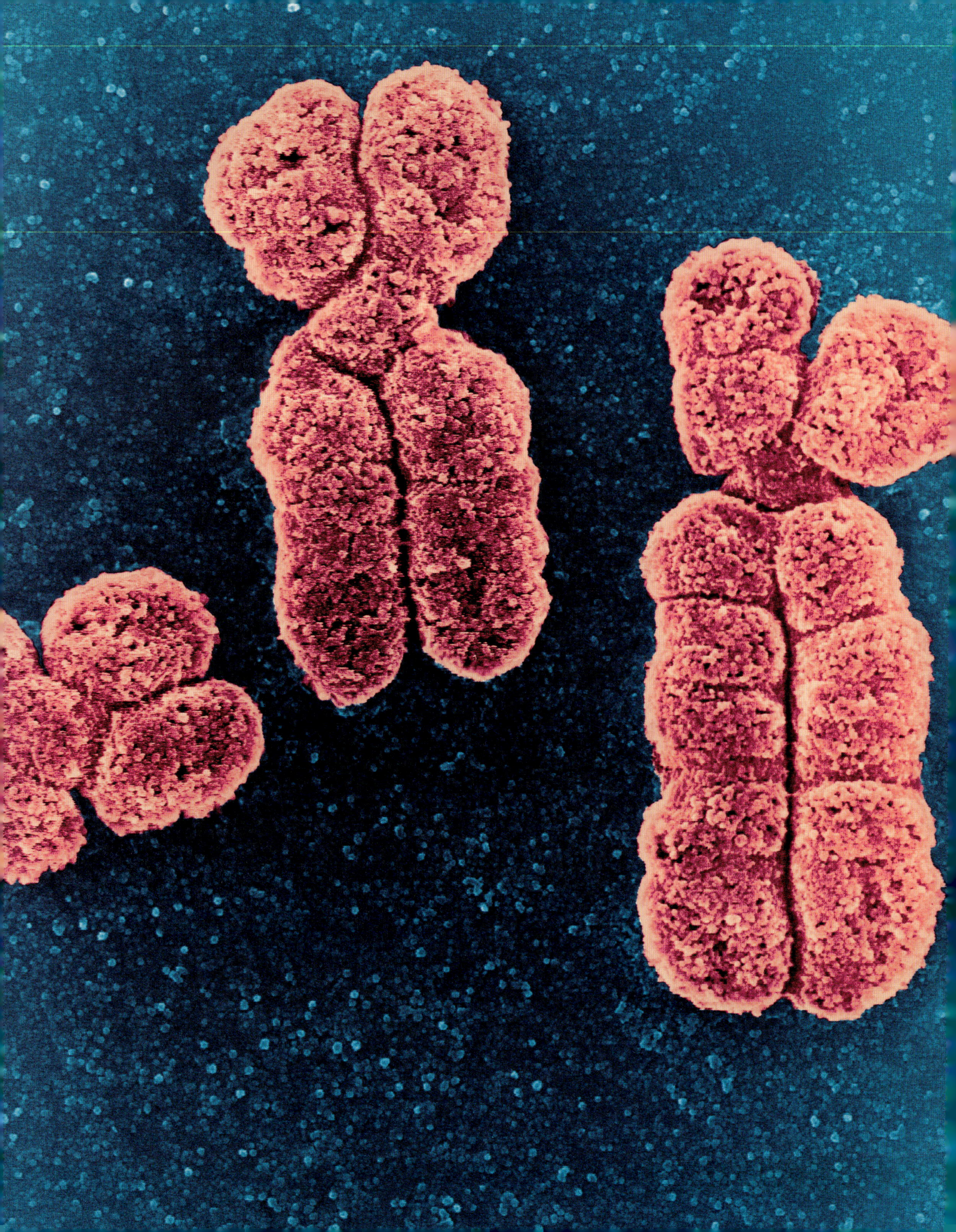

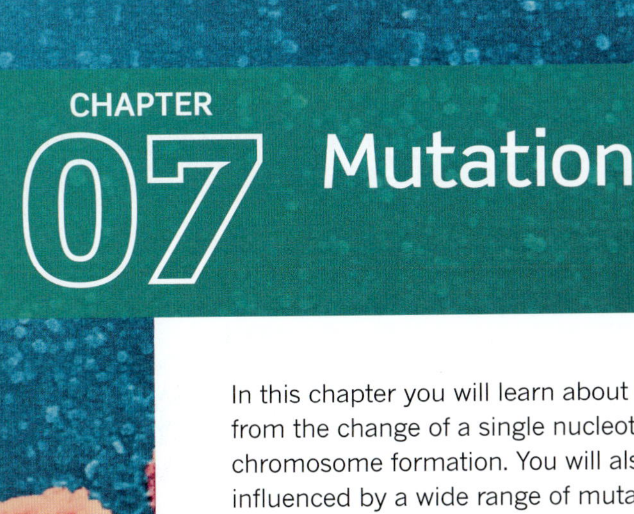

CHAPTER 07 Mutation

In this chapter you will learn about how mutation alters the structure of a genome, from the change of a single nucleotide in a DNA sequence to the change in chromosome formation. You will also learn how mutation occurs and how it is influenced by a wide range of mutagens, both artificial and naturally occurring. You will come to understand how different mutations are expressed in the phenotype of an individual and the long-term implications of mutation in somatic and germline cells.

Content

INQUIRY QUESTION

How does mutation introduce new alleles into a population?

By the end of this chapter you will be able to:

- explain how a range of mutagens operate, including but not limited to: ICT
  - electromagnetic radiation sources
  - chemicals
  - naturally occurring mutagens
- compare the causes, processes and effects of different types of mutation, including but not limited to: ICT N
  - point mutation
  - chromosomal mutation
- distinguish between somatic mutations and germline mutations and their effect on an organism (ACSBL082, ACSBL083) ICT
- assess the significance of 'coding' and 'non-coding' DNA segments in the process of mutation (ACSBL078) ICT N
- investigate the causes of genetic variation relating to the processes of fertilisation, meiosis and mutation (ACSBL078) N

7.1 Types of mutations

All mutations are changes in DNA, but this can take many forms and occur on many scales, from a large-scale change in chromosomal arrangement to a single change in one nucleotide. In this section, you will learn about the three major types of mutations and how these different processes affect the organism.

MUTATION

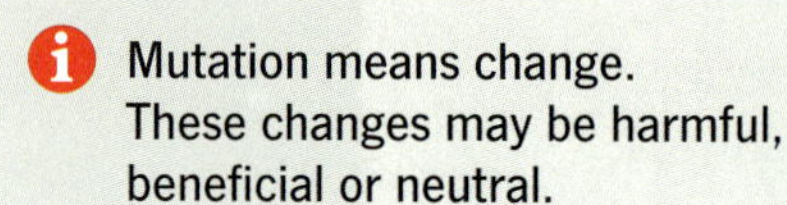

All **genetic variation** between species and between individuals of the same species is a result of **mutation**. Mutations are changes in **DNA (deoxyribonucleic acid)**. It is important to remember that mutation means change, not the introduction of a fault or disease. Mutations can have a beneficial or harmful effect or no effect at all on an organism.

Mutations can occur randomly as errors during cell replication and can affect a single gene, multiple genes or may involve entire **chromosomes**. They occur spontaneously or as a result of **mutagens**—factors that induce mutation. Most mutations are detected and repaired by enzymes. Those that cannot be repaired fall into one of three categories—neutral, beneficial or harmful:

- neutral mutations have no effect on survival
- beneficial mutations increase the likelihood of survival
- harmful mutations decrease the likelihood of survival.

POINT MUTATIONS

GO TO ➤ Section 4.2 page 165

Genetic sequences are read in sets of three **nucleotides**. These nucleotide 'triplets' are more formally known as **codons**. Codons code for specific **amino acids**. As you learnt in Chapter 4, amino acids begin as DNA triplets, which are then transcribed into mRNA codons. The mRNA codons (which now have the nucleotide U, uracil in place of T, thymine) directly code for specific amino acids. Using the four nucleotides, A, C, G and U, there are 64 possible codons but only 20 amino acids. This means that most of the amino acids are coded for by more than one codon. This is why the DNA code is referred to as **degenerate**.

A mutation that alters, adds or removes only one or very few nucleotides from a sequence of DNA or RNA is called a **point mutation** (Figure 7.1.1). Point mutations typically only affect a single gene. They include substitution and frameshift mutations.

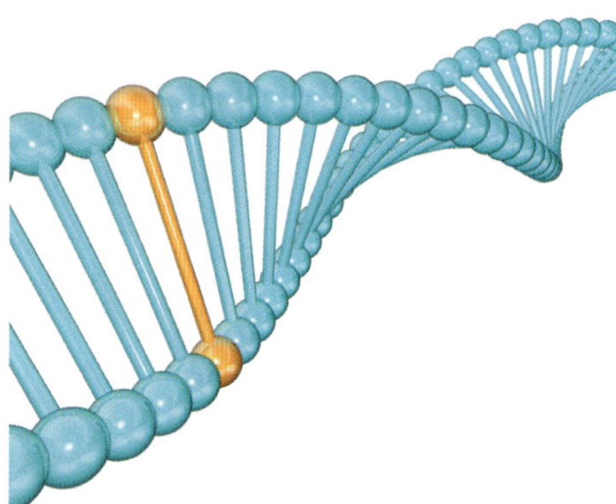

FIGURE 7.1.1 A point mutation involves a change in just one nucleotide in single-stranded RNA or nucleotide pair in double-stranded DNA.

Substitution mutations

A **substitution mutation** is a point mutation in which one nucleotide is replaced by another type of nucleotide. There are different types of substitution mutations including:

- silent mutations
- missense mutations
- nonsense mutations.

Silent mutations

A **silent mutation** occurs when a substitution results in a new codon that still codes for the same amino acid. For example, the DNA triplet CCA is transcribed into the **messenger RNA (mRNA)** codon GGU, which codes for the amino acid glycine (Gly). If the last nucleotide in the sequence is substituted for any other, the codon will still code for glycine and will not have any effect on the final **polypeptide**. The normal and mutated sequences of this silent mutation are shown below as an example (the mutation is indicated in red).

Original mRNA sequence:	AUG AAG GAG CGU UUC GGU AUU CAG
Amino acid sequence:	Met – Lys – Glu – Arg – Phe – Gly – Ile – Gln
Mutated mRNA sequence:	AUG AAG GAG CGU UUC GGG AUU CAG
Amino acid sequence:	Met – Lys – Glu – Arg – Phe – Gly – Ile – Gln

Missense mutations

Substitution mutations that result in an amino acid replacement are said to be **missense mutations**. Missense mutations still produce a protein. Whether this altered protein can function properly or not depends on the importance of the amino acid that was replaced.

Normal and sickle cell haemoglobins differ due to a missense mutation in which the amino acid glutamic acid (Glu) is replaced by valine (Val) in the beta chain of the haemoglobin protein. The distorted shape of the sickle-cell haemoglobin, called 'haemoglobin S', affects the overall shape of red blood cells, causing them to become sickle- or crescent-shaped (Figure 7.1.2).

Below are the normal and the mutated sickle-cell nucleotide sequences (the mutation is indicated in red).

Original mRNA sequence:	AUG AAG GAG CGU UUC GGU AUU CAG
Amino acid sequence:	Met – Lys – Glu – Arg – Phe – Gly – Ileu – Gln
Mutated mRNA sequence:	AUG AAG GUG CGU UUC GGU AUU CAG
Amino acid sequence:	Met – Lys – Val – Arg – Phe – Gly – Ileu – Gln

Nonsense mutations

When a substitution mutation results in the creation of a **stop codon** (UAA, UAG or UGA) it is classified as a **nonsense mutation**. Stop codons (also known as termination codons) signal **translation** to stop, resulting in the termination of protein synthesis. Stop codons are an important part of normal protein synthesis but if it appears too early in an amino acid sequence, it can have severe effects. Mutations that result in premature stop codons are called nonsense mutations because they result in incomplete and usually non-functional protein products.

Thalassaemia is an example of a nonsense mutation in which codon 17, of a 147-codon sequence, codes for 'stop' instead of lysine (Lys) (AAG is changed to UAG).

Below are the normal and the mutated thalassaemia nucleotide sequences (the mutation is indicated in red).

Original mRNA sequence:	AUG AAG GAG CGU UUC GGU AUU CAG
Amino acid sequence:	Met – Lys – Glu – Arg – Phe – Gly – Ileu – Gln
Mutated mRNA sequence:	AUG UAG GAG CGU UUC GGU AUU CAG
Amino acid sequence:	Met–stop

Frameshift mutations

Frameshift mutations involve one or two nucleotides being either added or removed from a nucleotide sequence, altering every codon in that sequence from that point onwards. These mutations can have significant effects on the polypeptide because as every codon is altered, so too is every amino acid they code for after the point of mutation. This results in the loss of functional protein, as it is likely that the resulting polypeptide would be completely different. In cases where the frameshift mutation creates a stop codon earlier in the sequence, the resulting polypeptide will be shorter.

BIOFILE CCT DD

Haemoglobin mutations

The altered shape of haemoglobin S reduces its ability to carry oxygen effectively. The fragile nature of sickle cells, their awkward shape (see Figure 7.1.2) and their limited oxygen-carrying abilities result in anaemia and oxygen-deficiency diseases in people who are homozygous for haemoglobin S. Interestingly, people who are heterozygous for haemoglobin S have an increased resistance to malaria—this is known as a heterozygote advantage.

This resistance has resulted in significant differences in allele frequencies of haemoglobin S in different populations. Sickle cell has much higher frequencies in populations in sub-Saharan Africa, and small areas of the Mediterranean, Middle East and India as a result of the higher incidence of malaria compared to northern European countries.

Thalassaemia is, like sickle cell, a mutation of haemoglobin. It results in oxygen-deficiency disease. The allele frequency of thalassaemia varies greatly among different populations. Like sickle cell, thalassaemia increases resistance to the parasite that causes malaria and so tends to have higher frequencies in populations with Mediterranean, Indian, Middle Eastern and Asian descent.

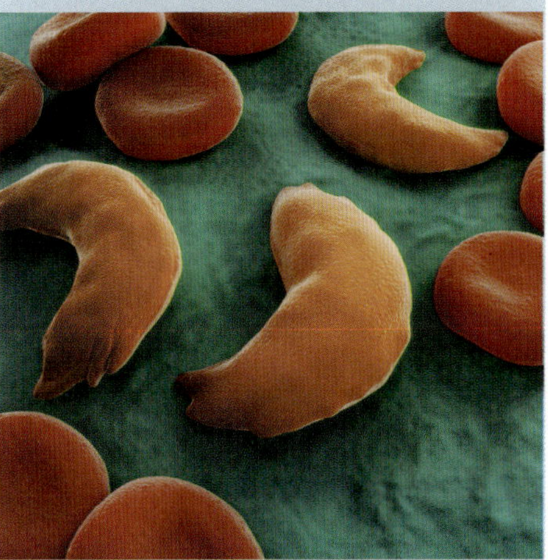

FIGURE 7.1.2 Sickle cells are characteristically crescent-shaped compared to normal round red blood cells. Sickle cells are caused by a missense mutation that changes the shape of the haemoglobin protein.

BIOFILE DD CCT

A beneficial mutation

An example of a beneficial substitution point mutation is one that involves the *ApoA-1* gene. This gene codes for a protein (apolipoprotein A-1) that is normally involved in the transport of cholesterol and phospholipids to the liver, where they are then redistributed or broken down and excreted. One of the mutated forms of the protein, ApoA-1 Milano, involves a substitution of the amino acid arginine (Arg) for cysteine (Cys). This mutated protein acts as an antioxidant, reducing cholesterol deposition in arteries, significantly decreasing the risk of cardiovascular disease.

The mutant form of the ApoA-1 protein (Figure 7.1.3) was first identified in Milan, and so the mutated gene was named after this city. Further investigation, including blood tests of an entire Italian village, traced the origin of the mutation to a single man. The 3.5% frequency of the gene in the population of the Italian village can be attributed to the descendants of this one man.

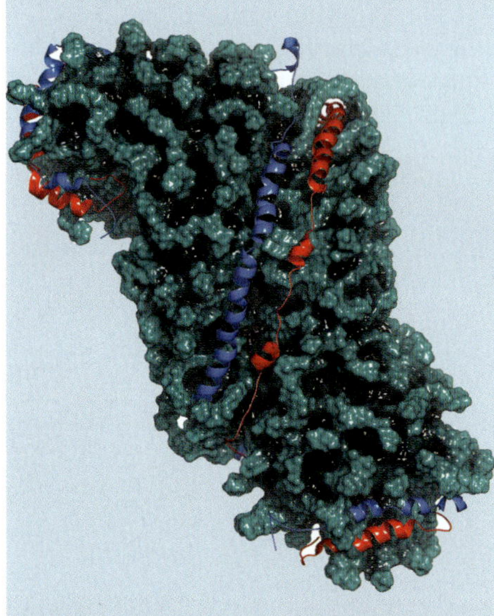

FIGURE 7.1.3 ApoA-1 Milano is a mutated form of a protein that can reduce cholesterol levels in the human blood stream. ApoA-1 Milano is caused by a beneficial point mutation.

There are two types of frameshift mutation:

- a nucleotide insertion adds one or two new nucleotides into the sequence and pushes the rest of the nucleotides back one or two places
- a nucleotide deletion removes one or two nucleotides and pulls all the following nucleotides forwards by one or two places.

Original mRNA sequence:	AUG AAG GAG CGU UUC GGU AUU CAG
Amino acid sequence:	Met – Lys – Glu – Arg – Phe – Gly – Ile – Gln
Insertion mRNA sequence:	AUG AAG GAG CGU AUU CGG UAU UCA G
Amino acid sequence:	Met – Lys – Glu – Arg – Asn – Arg – Tyr – Ser
Deletion mRNA sequence:	AUG AAG GAG CGU UU*G GUA UUC AG
Amino acid sequence:	Met – Lys – Glu – Arg – Leu – Val – Phe

CHROMOSOMAL MUTATIONS

Mutations that affect large sections of a chromosome, typically multiple genes, are called **chromosomal mutations** (or block mutations). These types of mutations usually occur during **meiosis** in eukaryotic cells. They can also be caused by mutagens such as **radiation**. When a gene is disrupted by the mutation, the effects are serious, even lethal. There are five main forms of block mutations:

- duplication
- inversion
- deletion
- insertion
- translocation.

Duplication mutations

Duplication mutations involve the replication of a section of a chromosome that results in multiple copies of the same genes on that chromosome (Figure 7.1.4a). There can be thousands of repeats. This often increases **gene expression**, which can be harmful or beneficial depending on the gene involved.

Inversion mutations

An **inversion mutation** involves a section of the sequence breaking off the chromosome, rotating 180° and reattaching to the same chromosome (Figure 7.1.4b). Inversions may involve as few as two bases or they may involve several genes.

Deletion mutations

Deletion mutations remove sections of a chromosome (Figure 7.1.4c). Deletions lead to disrupted or missing genes, which can have serious effects on growth and development. Chromosomal deletions are often fatal.

Insertion mutations

An **insertion mutation** occurs when a section of one chromosome breaks off and attaches to a different chromosome (Figure 7.1.4d). In eukaryotes, the effects of this type of mutation depend on whether the cell retains two copies of every gene. During meiosis, random assortment may separate the chromosome with the insertion from the chromosome that originally contained the inserted section. This can result in some **gametes** with two copies of the genes in the inserted section, while other gametes may be missing the genes entirely (Figure 7.1.5).

Translocation mutations

In **translocation mutations**, a whole chromosome or a segment of a chromosome becomes attached to or exchanged with another chromosome or segment. For example, sections from two **non-homologous chromosomes** (chromosomes that are not a pair) may break off at the same time. They may reattach to the other chromosome, swapping genetic material (Figure 7.1.4e). Translocations typically interrupt normal gene regulation and are the cause of some forms of **cancer**.

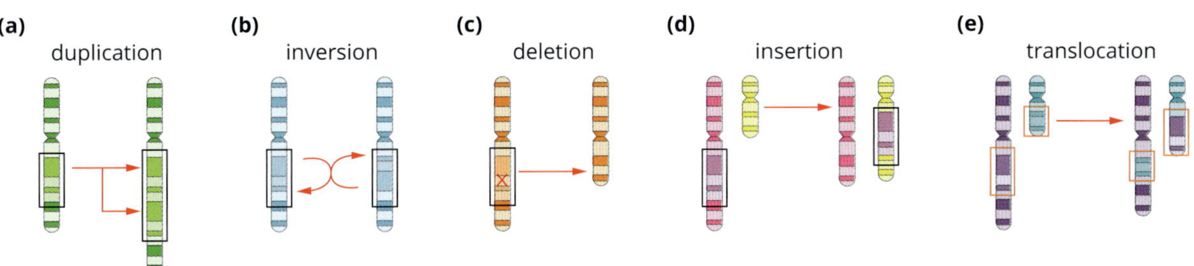

FIGURE 7.1.4 (a) Chromosomal duplication mutations result in multiple repetitions of a sequence of DNA. (b) Chromosomal inversion mutations involve a broken section of the sequence rotating 180° before reattaching. (c) Chromosomal deletion mutations involve the loss of large sequences of DNA (sometimes whole genes). (d) Chromosomal insertion mutations involve a sequence breaking off one chromosome and attaching to another. (e) Chromosomal translocation mutations can involve two different chromosomes exchanging segments.

PA 6.1

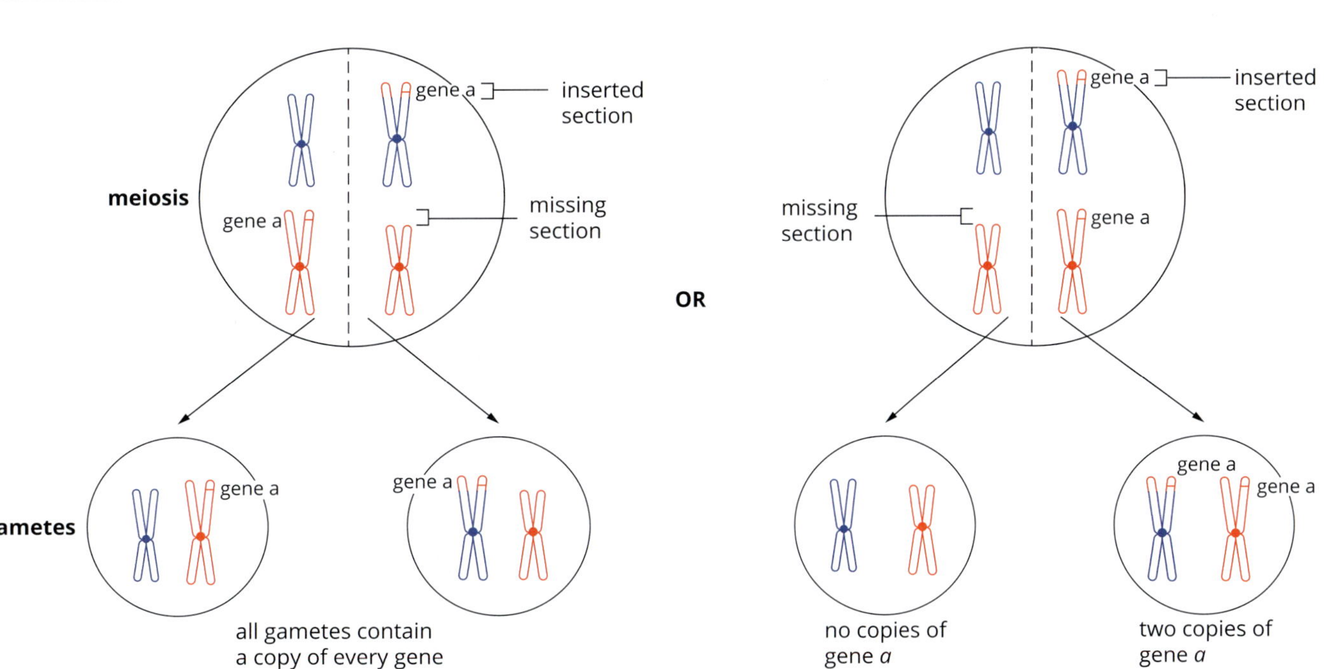

FIGURE 7.1.5 As a result of random assortment during meiosis, a chromosomal insertion mutation may lead to gametes with one, two or no copies of the inserted region.

CHROMOSOMAL ABNORMALITIES

When a mutation involves whole chromosomes, or the number of chromosomes, it is termed a **chromosomal abnormality**. This type of mutation is easily detected with a **karyotype**, which is a technique of staining and photographing chromosomes to help classify them and detect chromosomal mutations. There are two main forms of chromosomal abnormalities: **aneuploidy** and **polyploidy**.

> The haploid number (n) is the number of chromosomes an organism has when only one copy of the genome is present in a cell. The diploid number ($2n$) is the number of chromosomes present when an organism has two copies of the genome present. Most animals, including humans, are diploid. The usual diploid number for humans is 46, meaning that there are 23 pairs of chromosomes in all non-gamete cells. The usual haploid number for humans—and the number of chromosomes in gametes—is therefore 23. People with aneuploidy have less or more chromosomes than the usual 46.

Aneuploidy

Aneuploidy is the presence of an abnormal number of a particular chromosome, either an extra chromosome (called a **trisomy** when there are three copies of one chromosome) or a missing chromosome. Examples are Down syndrome (trisomy 21) and Patau syndrome (trisomy 13). Aneuploidy is usually caused by **non-disjunction** during meiosis. This is when two **homologous chromosomes** do not separate during the first division of meiosis. Aneuploidy results in gametes with an incorrect **haploid** number, and so results in an abnormal **diploid** number, which often leads to miscarriage of the embryo in humans. Aneuploidy can also occur in plants, such as maize, resulting in sterility. You will learn more about non-disjunction of chromosomes in Section 7.4.

Aneuploidy of sex chromosomes

Aneuploidy of the human sex chromosomes, X and Y, can result in a number of different conditions, some of which are summarised in Table 7.1.1.

TABLE 7.1.1 Types of aneuploidy in sex chromosomes in humans

Condition	Genotype	Gender	Incidence	Common characteristics
triple X syndrome	XXX (extra X chromosome)	female	1:1000	no observable characteristics, although typically very tall
Turner syndrome	X (missing X chromosome) (Figure 7.1.6b)	female	1:2500	short stature, low-set ears, webbed neck, typically infertile
Klinefelter syndrome	XXY (extra sex chromosome)	male	1:650	typically sterile, often of tall stature
48, XXXY syndrome	XXXY	male	1:50 000	tall stature, testicular dysfunction and intellectual impairment
48, XXYY syndrome	XXYY	male	1:18 000–40 000	sterility, intellectual impairment, tall stature and developmental delays
XYY syndrome	XYY	male	1:1000	rapid physical growth, but usually appear as 'normal' males

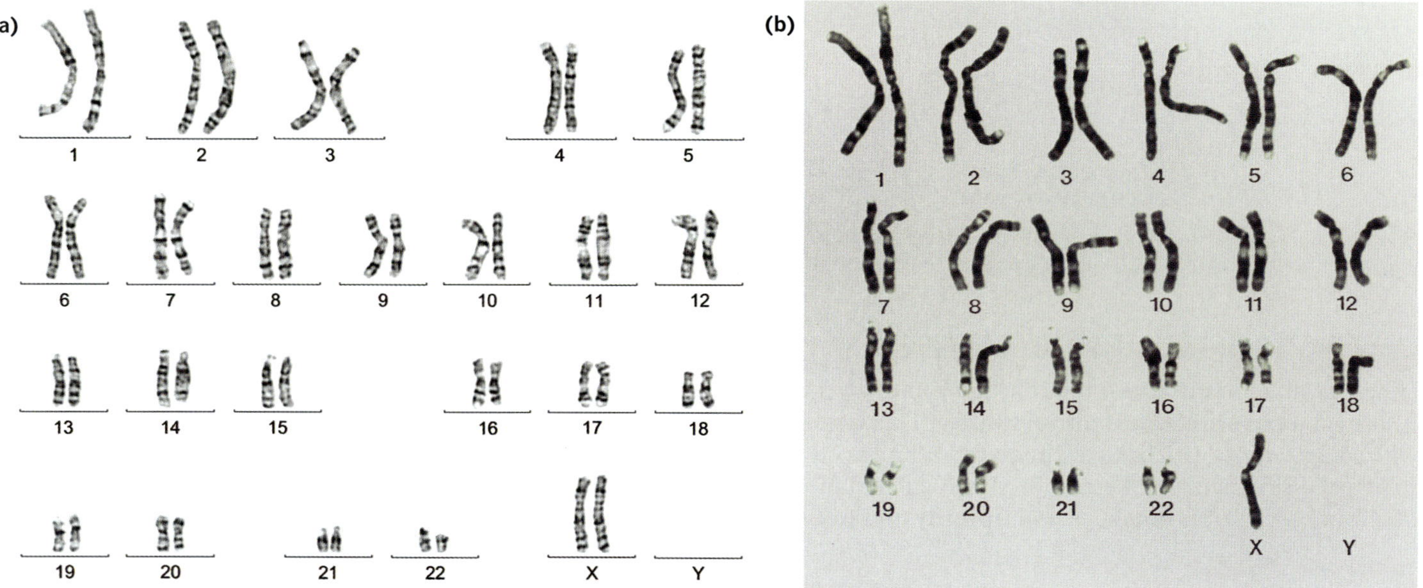

FIGURE 7.1.6 (a) A normal human karyotype with 23 pairs of homologous chromosomes. The presence of two X chromosomes indicates this person is a female. (b) A human karyotype that is missing a sex chromosome, with just one X chromosome. This karyotype indicates that this person is a female with Turner syndrome.

Polyploidy

Most eukaryotic cells are diploid, that is, they have two complete sets of chromosomes. For sexually reproducing organisms, the ovum provides one set of chromosomes and the sperm provides the other set. The gametes themselves are typically haploid, to ensure that fertilisation restores the correct number of chromosomes to the zygote.

However, errors during meiosis can result in diploid gametes. If one of these gametes is fertilised, it will result in a zygote with more than the usual two sets of chromosomes. Polyploidy is the condition that results from cells and organisms that contain more than two full sets of chromosomes; that is, more than two of every chromosome in the set (Figure 7.1.7).

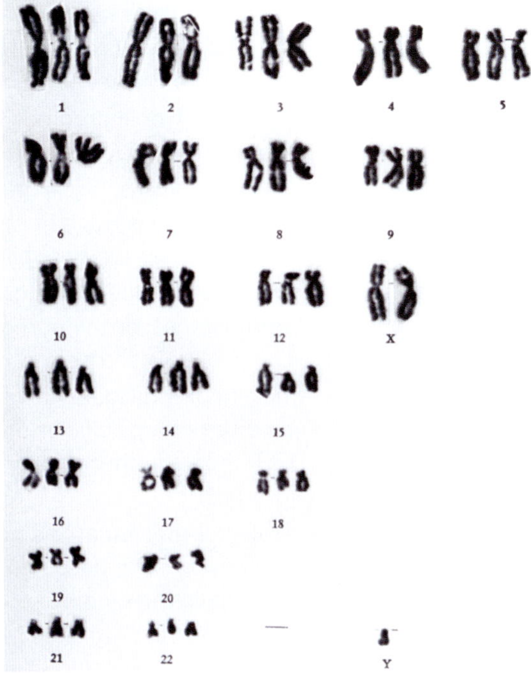

FIGURE 7.1.7 Three copies of all autosomal chromosomes plus two X and one Y chromosome shown in this karyotype indicates a male with triploidy.

In humans, polyploid zygotes do not survive. However, polyploidy can also arise from errors during mitosis and produce groups of polyploid **somatic cells**. Having a few cells with an abnormal chromosome number may not affect health. A liver, for example, may function normally but have patches of polyploid cells.

Polyploid animals include some insects, earthworms and tree frogs. However, polyploidy is more common in plants than animals because many plants can survive by asexual or vegetative reproduction. A triploid ($3n$) organism, resulting from the fusion of one haploid gamete and one diploid gamete, is typically sterile or has low fertility because of problems with chromosome pairing during meiosis and gamete formation. However, a triploid plant could survive by vegetative reproduction or by spontaneous duplication of all chromosomes to $6n$. Tetraploid organisms can result from two diploid gametes fusing. A tetraploid organism can produce viable offspring if it breeds with another tetraploid, but not with a diploid individual.

BIOFILE CCT

Polyploid plants

Some banana varieties are triploid ($3n$); cultivated cotton and potatoes are examples of tetraploid ($4n$) organisms; bread wheats are hexaploid ($6n$); and strawberries (Figure 7.1.8) vary according to their variety, some of which are diploid ($2n$), tetraploid ($4n$), hexaploid ($6n$), octoploid ($8n$) and even decaploid ($10n$). There are heptaploid ($7n$) and pentaploid ($5n$) varieties of strawberries, but these are hybrids and can only reproduce vegetatively.

FIGURE 7.1.8 Strawberries are often polyploid. Depending on the variety, they can possess up to 10 copies of each chromosome.

7.1 Review

SUMMARY

- A mutation is a change in the genetic structure of an organism.
- New alleles, genes and chromosomes are created through mutation.
- Mutations may have a beneficial effect, a harmful effect or no effect at all on the individual.
- Point mutations include:
  - substitution—the mutation replaces one nucleotide in the sequence for another nucleotide
  - frameshift—the addition or deletion of a nucleotide results in a frameshift.
- Substitution mutations appear as three types:
  - silent mutations—the new codon (nucleotide triplet) still codes for the same amino acid
  - missense mutations—the new codon codes for a different amino acid (the effects of which may vary depending on the new amino acid and the resulting functionality)
  - nonsense mutations—the new codon is a stop codon and shortens the amino acid chain (which may have severe effects).
- Addition or deletion mutations are also called frameshift mutations because they alter every triplet from that point onwards in that gene. They typically have more severe effects than substitution mutations.
- Chromosomal mutations (also called block mutations), typically involve multiple genes. Types include:
  - duplication mutations—a section of a chromosome is repeated multiple times on that chromosome
  - deletion mutations—entire genes are cut from the chromosome
  - inversion mutations—a large section of a chromosome is removed and rotated 180° before being reinserted, so that the sequence is reversed
  - insertion mutations—a whole chromosome or section of a chromosome is added to a different chromosome
  - translocation mutations—sections from two non-homologous chromosomes are swapped.
- Chromosomal abnormalities are mutations that involve whole chromosomes, or the number of chromosomes. The two main forms are:
  - aneuploidy—the cell or individual has more than or fewer than two copies of a particular chromosome (e.g. trisomy 13 [Patau syndrome] or trisomy 21 [Down syndrome])
  - polyploidy—the cell or individual has more than two copies of every chromosome (e.g. triploid organisms have three copies of every chromosome).

KEY QUESTIONS

1. Why are frameshift mutations more significant than substitution point mutations?
2. Explain why silent mutations are called 'silent'.
3. For each of the conditions listed below, state which chromosome is affected and whether the chromosome is in excess or missing.
   a. Down syndrome
   b. Turner syndrome
   c. Klinefelter syndrome
4. Choose two forms of chromosomal mutations. Draw a diagram explaining each.
5. The figure opposite shows a human karyotype. Is there any evidence of chromosomal abnormality? Explain your answer.

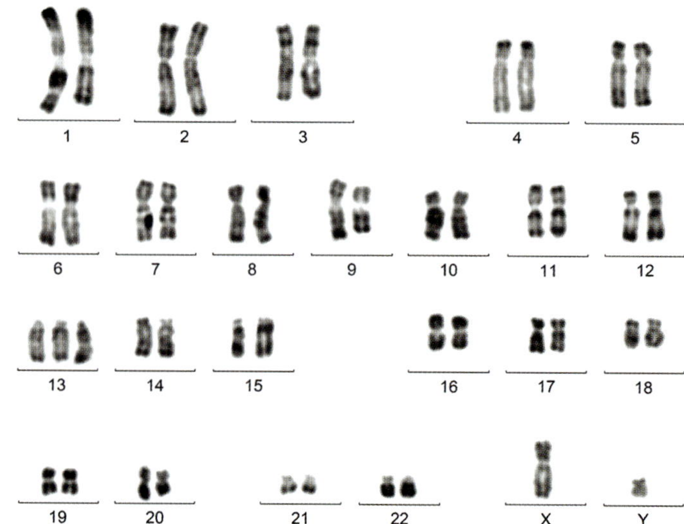

7.2 Mutagens and their impacts

Mutagens are substances or processes that can dramatically increase the rate of mutations, often beyond what an organism can tolerate. While we often think of mutagens as artificial, many occur naturally. In this section, you will learn about a wide range of mutagens and how they function. Mutagens can also function as **carcinogens**, or cancer-causing agents; you will also learn about how this process takes place.

HOW CARCINOGENS FUNCTION

You will be familiar with the process of cell division and the steps of the cell cycle from Chapter 3, and the processes of DNA **transcription** and translation from Chapter 4. While the cell cycle is highly regulated, the regulatory mechanisms can be disrupted by a number of factors, such as damage to DNA caused by exposure to mutagens. If uncontrolled cell division occurs during embryonic development, the embryo will be abnormal and, in most circumstances, will abort. If this occurs in a mature organism, uncontrolled cell division may lead to **neoplasms**, which can eventually transform into cancer.

GO TO ➤ Section 3.1 page 116

GO TO ➤ Section 4.2 page 165

A neoplasm is an abnormal growth of tissue that usually, but not always, forms a mass. Neoplasms are more commonly referred to as tumours, but not all are cancerous. There are three types of neoplasm:

- benign—these form localised masses but do not transform into cancer
- potentially malignant—these form localised masses that will eventually invade other tissues and transform into cancer
- malignant—these form masses that invade other tissues and transform into cancer (Figure 7.2.1).

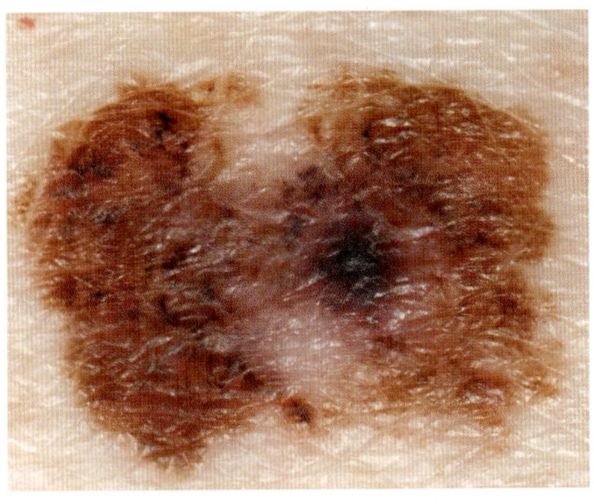

FIGURE 7.2.1 A melanoma is a malignant neoplasm that will continue to grow as a skin cancer unless removed.

Comparing benign and malignant neoplasms

Benign and malignant neoplasms can appear similar. Table 7.2.1 outlines the main differences between them.

TABLE 7.2.1 Comparison of benign and malignant neoplasms

Benign neoplasms	Malignant neoplasms
Cells divide uncontrollably, yet not as rapidly as those of a malignant neoplasm.	Cells divide uncontrollably.
The organism controls the growth of the neoplasm to a certain extent by encapsulation. Cells are contained and do not penetrate the blood and lymph vessels.	Growth of the neoplasm: uncontrolled cell growth breaks out of capsule. Neoplastic cells can spread to other tissues (i.e. they metastasise).
Because the neoplasm grows inside a capsule, it does not destroy the surrounding tissues.	The growing neoplasm destroys the surrounding tissues.

Disruptions to the control of cell division

While the cell cycle is highly regulated, the regulatory mechanisms can be disrupted by a number of factors. These interruptions can result in uncontrolled cell divisions; this is the process by which neoplasms are produced, which can eventually develop into cancer.

Genetic factors

Genes code for the enzymes that regulate cell division or regulate **apoptosis** (programmed cell death). When mutations to these genes occur, regulation of cell division or apoptosis can cease.

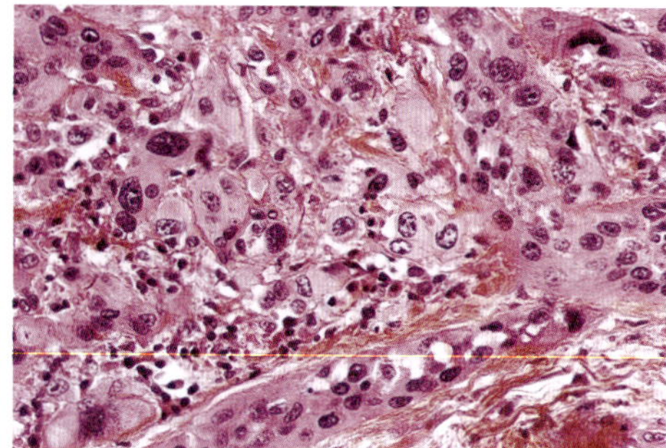

FIGURE 7.2.2 Light micrograph (LM) of thyroid cancer tissue

Oncogenes are cancer-causing genes. Oncology is the medical field that studies and treats cancer.

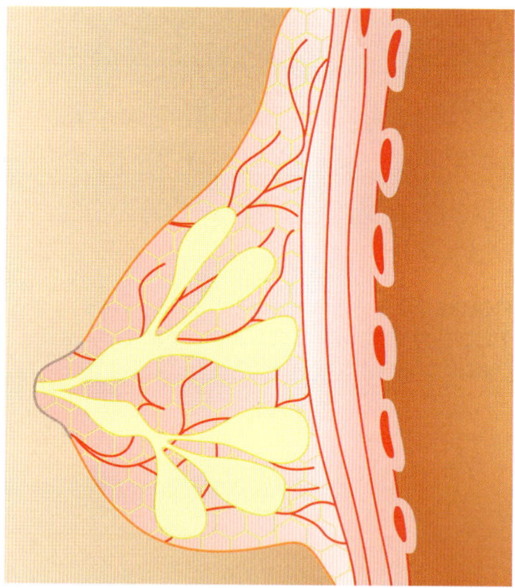

FIGURE 7.2.3 The mammary glands of the breast (highlighted in yellow) contain tissues that depend on oestrogen for normal functioning. This tissue is vulnerable to neoplasms caused by changes in the *BRCA1* or *BRCA2* genes.

Proto-oncogenes are a group of normal genes involved in the regulation of cell division. One of their functions is to stimulate cell growth. They are required for the normal growth and development of cells. However, mutations of these genes can change them into **oncogenes**, which induce uncontrolled cell division leading to the development of neoplasms.

One example is an important regulator of cellular proliferation called platelet derived growth factor (PDGF). Mutations of the *PDGF* gene are connected with the development of brain cancer. Mutations of the *PDGF* receptor gene are associated with thyroid cancer (Figure 7.2.2).

Genes that play a role in the development of cancer include:

- **Tumour-suppressor genes**—the group of genes that code for proteins involved in the slowing down of cell division, the repair of DNA or apoptosis are called tumour-suppressor genes. If these genes are damaged, cell division may go unchecked.
- Inherited oncogenes or mutations—some people also have a genetic predisposition to certain forms of cancer. An individual could receive an oncogene or mutated tumour-suppressor gene from one or both parents. This does not necessarily mean that all family members with that gene will develop cancer as sometimes more than one mutation needs to occur before cancer develops. Environmental factors also play a significant role in cancer development. A family history of breast cancer, for example, is associated with mutations in the *BRCA1* or *BRCA2* genes, which are located on chromosome 17. A protein coded by these genes takes part in the repair of damaged DNA, but changes in its structure lead to uncontrolled growth. The growth occurs mainly in tissues whose functioning depends on oestrogen (e.g. tissues in the mammary glands (Figure 7.2.3) and ovaries).

Loss of immunity

The immune system is usually able to detect and destroy abnormal cells, including those that are replicating in an uncontrolled manner. If an individual's immune system is weakened, cells that are dividing in an uncontrolled manner may not be detected and may continue to divide to form tumours.

TYPES OF MUTAGENS

While there are internal genetic factors that are linked to naturally occurring mutations in an organism, the external environment also contains numerous mutagens that can damage DNA and, in some cases, disrupt cell division. This disruption will occur if the damage occurs to proto-oncogenes or tumour-suppressor genes. Mutagens can be divided into three categories: chemical, physical, and biological.

Chemical mutagens

Chemical mutagens can be divided into three categories in terms of function:

1. **Intercalating agents** are chemicals that insert themselves into the bonds between base pairs and alter the shape of the DNA, leading to subsequent errors in replication. A common laboratory stain used to visualise DNA on agarose gels, ethidium bromide, falls into this category.
2. **Base analogues** are chemicals that are structurally similar enough to that of the nitrogenous bases in DNA that they are incorporated into the DNA sequence during replication instead of the usual bases, meaning that the DNA no longer functions.
3. **DNA reactive chemicals** are chemicals that react directly with DNA, such as **reactive oxygen species (ROS)** (highly reactive molecules containing oxygen), which can cause breakages and cross-links in DNA strands. While ROS are produced during normal cellular function, excessive amounts are thought to contribute to the development of cancer. There are numerous other chemicals that fall under this heading. Many of these are not mutagenic on their own, but they do interact with metabolic processes and produce mutagenic compounds.

In many cases, we do not know why some chemicals are mutagenic and what category they fall into. The first step in identifying a mutagenic or carcinogenic chemical is noting a clear correlation between exposure and obvious signs of cancer or mutation. A dose-dependent effect must be shown; that is, the higher the exposure, the greater the risk of a mutagenic effect. In studies of human health, this is very difficult, as human lifestyles are enormously complex and variable and only observational studies are possible. Fortunately, experimental work on model organisms such as fruit flies (*Drosophila melanogaster*) and mice have allowed us to draw some more solid conclusions.

Numerous chemical carcinogens occur in the world. The first chemical mutagen to be demonstrated was mustard gas, which was shown to cause mutations in fruit flies in the 1940s. Mustard gas was already known to be toxic and deployed as a chemical weapon in World War I, but its long-term mutagenic effects were unknown at the time. This is not uncommon when it comes to chemical mutagens, as often long-term exposure is required to cause obvious mutagenic effects (such as cancer).

FIGURE 7.2.4 Tobacco contains mutagenic and carcinogenic compounds. Smokers are much more likely to develop malignant neoplasms of the respiratory system, pharynx, larynx and lungs than non-smokers.

Tobacco contains mutagenic compounds, some of which are carcinogenic (Figure 7.2.4). Prior to the widespread smoking of tobacco, lung cancer was an uncommon form of cancer. Now we know that smokers are much more likely to develop malignant neoplasms of the respiratory system, pharynx, larynx and lungs than non-smokers are. In Australia, smoking is responsible for 84% of new lung cancers in men and 77% in women.

Outdoor air pollution is another significant cause of cancer (Figure 7.2.5). Air pollution contains a number of known or suspected carcinogens, such as diesel engine exhaust and solvents.

FIGURE 7.2.5 Outdoor air pollution can contain chemicals that increase the risk of developing a neoplasm.

Not all chemical mutagens are artificial. Some processed foods have been linked to an increased risk of cancer, such as refined sugar and processed meat (Figure 7.2.6). These substances are generally present in such small quantities that they are extremely unlikely to do you any harm. To experience an increased risk of neoplasm (or other form of damage to your DNA), exposure would have to be significantly higher than that of the average diet. This is not the case with all naturally occurring mutagens. Alcohol, for example, is a Group 1 carcinogen, meaning that the evidence linking it to cancer is extremely strong. It is worth noting that tobacco itself is a plant containing mutagenic compounds and therefore is a naturally occurring mutagen.

FIGURE 7.2.6 Carcinogens and mutagens occur in many foods, either naturally, or as a result of production, preservation, cooking and other processing. The levels of such carcinogens in our foods are generally very low.

Other naturally occurring chemical mutagens include metals such as cadmium and arsenic. While the mechanism by which mutagenic metals damage DNA is not known in all cases, it is thought that some affect the processes of DNA repair, that is, prevent DNA from fixing random errors in replication that would normally be corrected. Other metals cause the production of reactive oxygen species (ROS).

FIGURE 7.2.7 A nuclear power plant in France. Nuclear fission and fusion are forms of particle radiation.

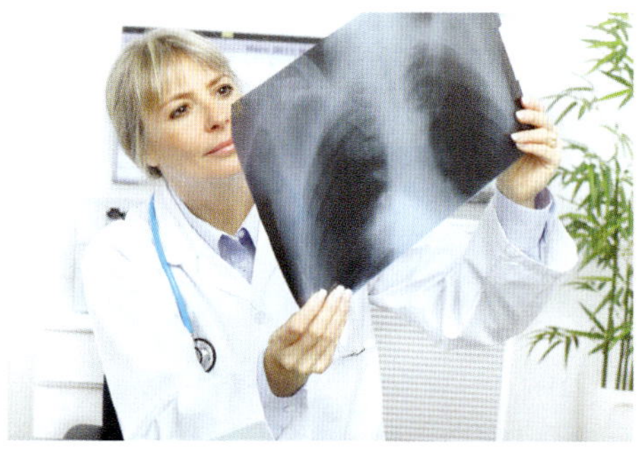

FIGURE 7.2.8 The ionising radiation in X-rays can have dangerous effects on cells because it can cause cancerous mutations. However, the low exposure from a few X-rays over a lifetime is highly unlikely to cause cancer.

FIGURE 7.2.9 Ultraviolet (UV) light is a mutagen. Overexposure to sunlight or tanning beds, and therefore ultraviolet radiation, increases the risk of skin cancer or melanoma.

Ionising radiation is a physical mutagen and can cause cancer, but non-ionising radiation is not associated with damage to DNA.

Physical mutagens

Physical mutagens include **particle radiation** and **electromagnetic radiation (EMR)**. Radiation is simply the emission or movement of energy through space or materials. In some cases, this energy can interrupt cellular processes and ionise molecules, damaging DNA; however, not all radiation is strong enough to do this. For this reason, when considering biological effect, it is more useful to split radiation into two types: **non-ionising radiation** and **ionising radiation**.

Particle radiation is the energy that is emitted by fast-moving subatomic particles. This is the radiation produced by nuclear fission and fusion, such as that which occurs in nuclear power plants (Figure 7.2.7) and nuclear weaponry, as well as in solar flares. Particle radiation is an extremely powerful mutagen.

Electromagnetic radiation is the radiant energy emitted by the electromagnetic field, and it exists on a spectrum. It includes radio waves, microwaves, infrared, (visible) light, ultraviolet (UV), X-ray and gamma radiation. Unlike particle radiation, EMR does not have mass. EMR is emitted by electrical devices of all kinds and is usually encountered at relatively low levels. The electromagnetic spectrum is split into low frequency and high frequency radiation (see Figure 7.2.10a on page 295).

Ionising radiation

Ionising radiation includes both particle radiation and high-frequency electromagnetic radiation (X-rays, gamma rays and the high-frequency end of the UV light spectrum). On an atomic and chemical level, such radiation contains enough energy to push an electron out of its orbit, thus creating an ion. This alters electric charges, thus it can break chemical bonds causing damage to DNA and basic cellular function. Repeated long-term exposure to ionising radiation can cause cancer. The effect is dependent on dosage and frequency; a few X-rays over a lifespan is unlikely to result in neoplasm (Figure 7.2.8) while continuous exposure to UV light is responsible for the high levels of skin cancer in the Australian population (Figure 7.2.9). It is estimated that around two in three Australians will be diagnosed with some form of skin cancer by the time they are 70, although this figure includes non-malignant neoplasm (see Figure 7.2.1).

In the short term, high levels of exposure can result in **acute radiation syndrome (ARS)**, commonly known as radiation sickness or radiation poisoning. While this is a very serious condition (and often fatal), it is extremely rare. Historically, ARS is most commonly associated with large-scale nuclear incidents such as the bombing of Hiroshima and Nagasaki, and the containment failure of nuclear power plants such as Chernobyl in Russia in 1986. In the present day, ARS is largely caused by industrial workplace accidents at nuclear power plants and other worksites related to nuclear energy and waste containment.

Non-ionising radiation

Non-ionising radiation is that which exists at the low frequency end of the electromagnetic spectrum: infrared light, microwaves, radio waves, visible light and the near and middle portions of the ultraviolet spectrum. Non-ionising radiation is not associated with damage to DNA or with acute radiation syndrome. The level of energy emitted is simply too low to have an effect on the chemical bonds of DNA or to cause cellular damage. This is why energy emitted by mobile phone usage and by wireless Internet signal is not dangerous or carcinogenic. However, non-ionising radiation can cause susceptible materials, particularly those containing water (such as food, fluids, and living tissue) to heat up. This is why you can use microwaves to heat food as they cause the food materials to warm, without altering the chemical structure of the food.

BIOLOGY IN ACTION ICT PSC N

Non-ionising radiation: weighing-up the evidence

Concern about the effects of using mobile phones, microwaves and wireless Internet are waning, but they have been widespread in the past. The radiation emitted by mobile phones and by wireless internet signals is low frequency and low energy, occupying the non-ionising range of the electromagnetic spectrum. They do not emit sufficient energy to ionise atoms or molecules, so it is not possible for them to cause damage to DNA or to interrupt cellular processes in the way that ionising radiation such as X-rays and gamma rays can do.

In spite of this, public concern was high enough that numerous independent and government-funded scientific studies have been funded to study the effects of mobile phone usage, wireless internet and other sources of non-ionising radiation. No conclusive evidence of mutagenic or carcinogenic effects has ever been found from non-ionising radiation from these sources.

Non-ionising radiation does cause tissues to heat up, and can cause thermal damage (burns). The low and mid-level frequencies of the UV light spectrum are technically defined as non-ionising radiation, while the high frequencies are considered ionising radiation, placing UV light in a controversial category. These low and mid-level frequencies can still cause sunburn, and long-term, repeated damage to the skin can indirectly result in cancer, thus mimicking the effect of ionising radiation. It is important to remember that non-ionising radiation is not completely harmless.

However, mobile phones, microwaves, wireless internet and other items of concern emit radiation well below the UV portion of the spectrum and below even visible light (Figure 7.2.10a). There is no doubt about their status as ionising or non-ionising radiation, and the only way in which they could cause damage to DNA would be via repeated burns from continued thermal damage (Figure 7.2.10b). Most people would stop using their phone if it were to get that hot.

Ultimately, the lower end of the electromagnetic spectrum has been shown to be reasonably safe, which is good news for people who use mobile phones, wireless internet, microwaves and radios of any kind.

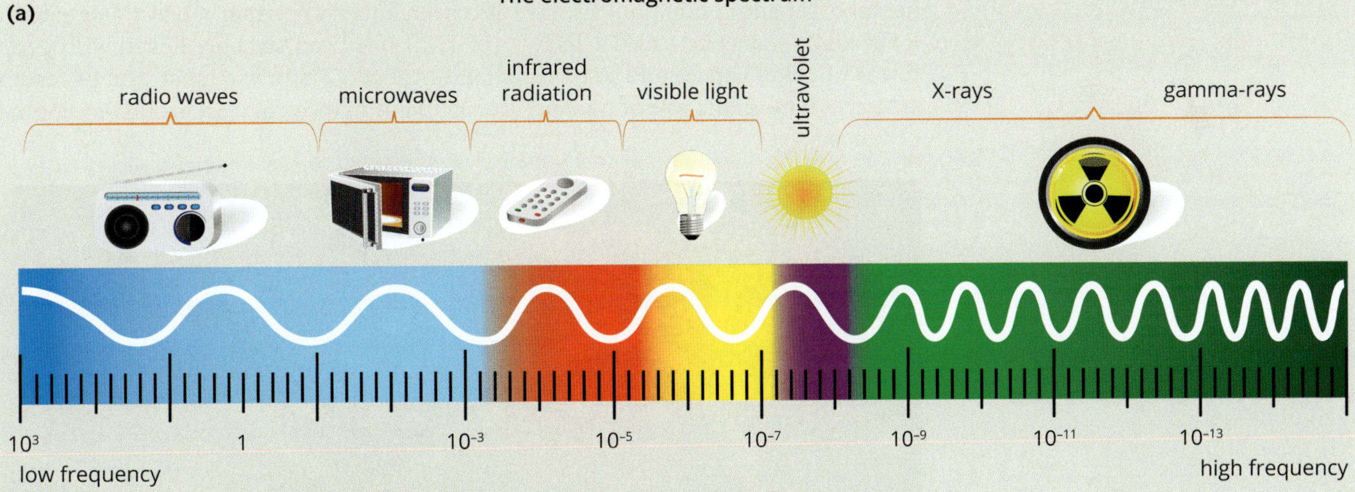

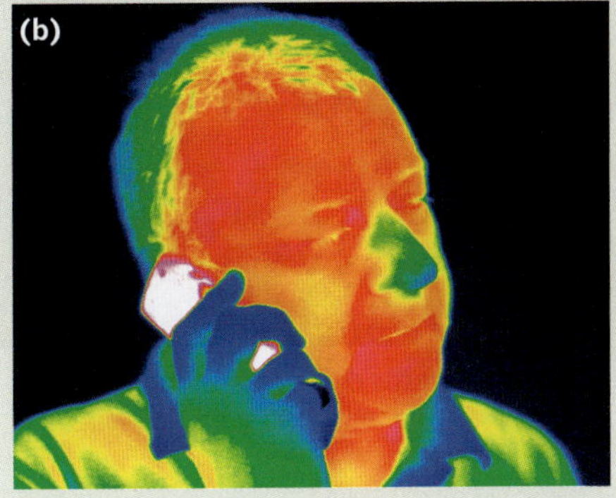

FIGURE 7.2.10 (a) Image of the electromagnetic spectrum showing the frequencies at which common household objects emit radiation. (b) Thermal image of a man using a mobile phone. The temperature scale runs from white (warmest) through pink, red, orange, yellow, green and blue to black (coldest). We can see that the phone emits heat, which is expected due to the presence of non-ionising radiation. However, after many years of independent and government-funded research, no conclusive evidence linking mobile phone use with cancer has been found.

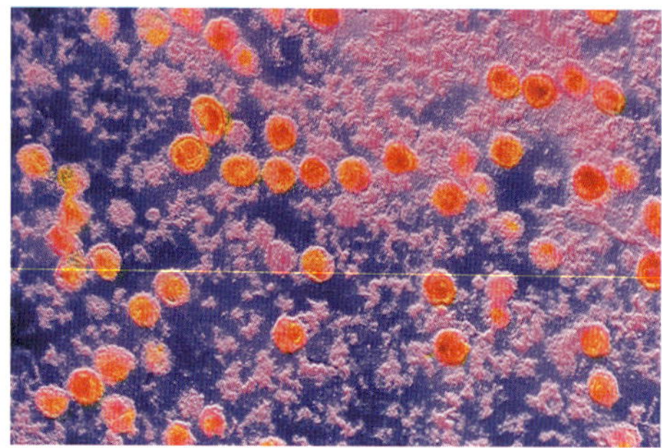

FIGURE 7.2.11 Some viruses are oncogenic (cancer-causing), such as the HTLV-1 virus (red spheres), which is responsible for certain types of leukaemia or blood neoplasms. HTLV-1 is a retrovirus, a type of virus that inserts its genetic information into the host cell's chromosomes. Such viruses either carry oncogenes themselves or enhance the host cell's proto-oncogenes. This image is a false-colour transmission electron micrograph (TEM).

Biological mutagens

Biological mutagens are living molecules that can cause mutations and cancers by interfering with the functions of oncogenes and tumour-suppressor genes. These fall into two categories: certain viruses (termed **oncogenic viruses**) and a category of DNA known as a **transposable element (TE)** (also called a transposon).

Viruses

Retroviruses, such as HTLV-1, insert their genetic information into the chromosomes of the host cell (Figure 7.2.11). HTLV-1 is responsible for certain types of leukaemia or blood neoplasms and is considered to be an oncogenic virus. Oncogenic viruses can either be direct or indirect. Direct oncogenic viruses either carry oncogenes themselves or enhance the host cell's proto-oncogenes, while indirect oncogenic viruses cause cancer through decades of chronic inflammation. HTLV-1 is a direct oncogenic virus, while hepatitis C is an indirect oncogenic virus. Another well-known oncogenic virus includes human papillomavirus (HPV), also known as genital warts, which is responsible for the majority of cervical cancers.

Transposable elements

Transposable elements (also known as TE or transposons) are short DNA sequences that move around the genome (Figure 7.2.12), giving them the name 'jumping genes'. These are extremely common in eukaryotic genomes and are thought to comprise over 45% of the human genome. These can also cause errors in replication and interfere in the functioning of certain genes. Multiple copies of the same sequence can change chromosome crossover points and cause errors in chromosome duplication. TEs may also insert themselves into the middle of a functional gene, interrupting its function. If these errors are not corrected, this can result in mutations that are passed to the next generation. The original mutation for one form of haemophilia was caused by the insertion of a TE into a gene necessary for effective blood clotting. When this error occurred during meiosis, it was passed on to subsequent generations, resulting in a heritable genetic condition. TEs have also been linked to certain cancers. For example, a virus-like sequence called LINE-1 (or L-1) has been strongly linked with certain forms of colon cancer and also oesophageal cancer. There is a great deal of research yet to be done on the role of transposable elements in the progression and development of cancers.

> Transposable elements are genes that copy and insert themselves into different places in the genome. This gives transposable elements the name 'jumping genes'.

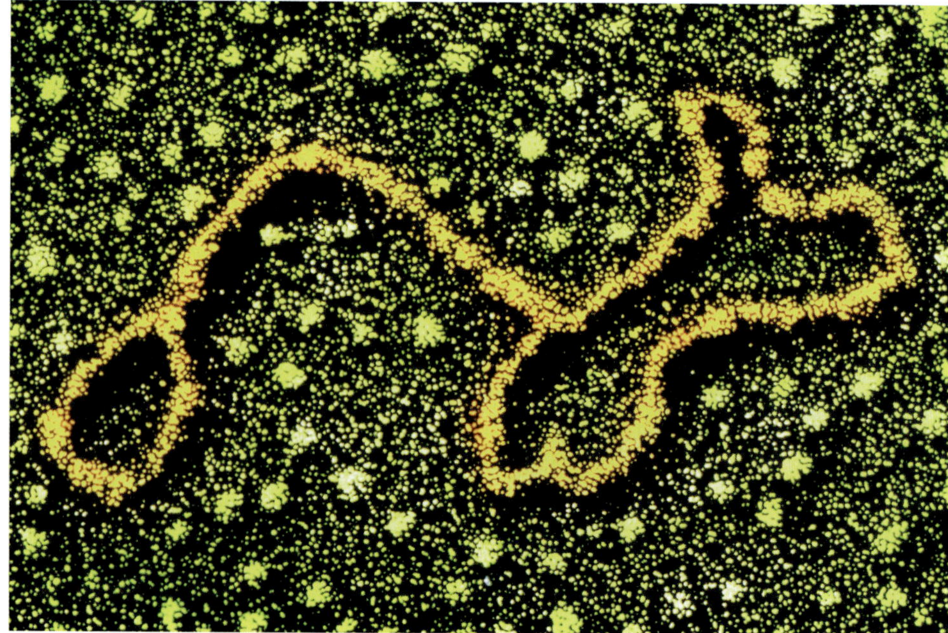

FIGURE 7.2.12 Transposable elements or 'jumping genes' are segments of DNA that move around in the genome. This false-colour TEM of a transposon, shows their characteristic stem and loop structure.

BIOLOGY IN ACTION CCT

Mutagens in research and medicine

Mutagens should be treated with caution, due to the obvious danger they represent, but there are many uses for them in biological research and in medical therapies.

Genetic research

One of the ways in which geneticists explore the functions of genes is to deliberately damage them (often using radiation) and find out what biochemical pathways are interrupted. Comparing the phenotype of the wild-type organism with the mutant organism and then comparing their DNA sequence or gene expression can shed light on which genes have been affected and how they have resulted in the mutant phenotype (Figure 7.2.13).

FIGURE 7.2.13 Genetic studies often begin with comparing the phenotypes of a mutant and a wild-type organism before comparing the DNA sequence or gene expression. This is a LM of a normal and a mutant fruit fly (*Drosophila melanogaster*). The fly on the right is the natural, or wild type. The fly on the left is a mutant type with almost no wings.

Cancer research

Scientists can study the development and progression of different cancers by examining cancer cells in culture. This helps them understand how the cancers formed in the first place. Inducing cancer in cell cultures is usually achieved by use of a chemical mutagen (Figure 7.2.14).

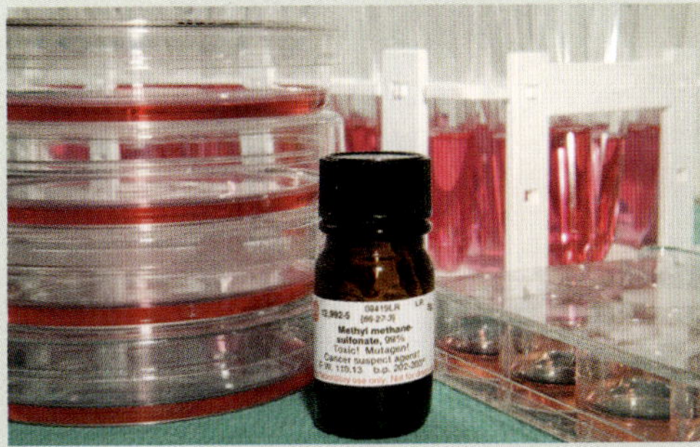

FIGURE 7.2.14 Methyl methane sulphonate (MMS) is a chemical mutagen that is used to induce cancer in cell cultures so that the effects on the cells can be studied.

Mutagens in medicine

In radiotherapy, radiation is used to damage the DNA of rapidly dividing cancer cells, causing cell death and preventing them from dividing (Figure 7.2.15). If effective, radiotherapy causes the tumour to shrink. Non-cancer cells are also damaged, but they are more robust than cancer cells and recover more easily. This damage is the reason that radiotherapy side effects can be so severe.

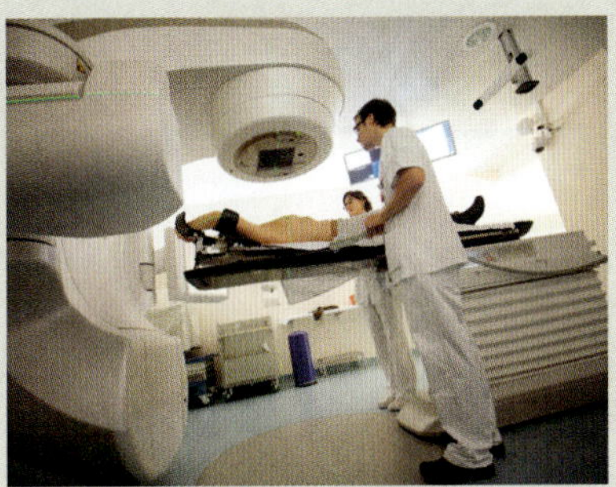

FIGURE 7.2.15 Radiotherapy is a therapeutic use of ionising radiation. Here, it is being used to treat breast cancer.

Other medical therapies involve exposure to mutagens, but in these cases the hope is that damage to DNA can be avoided by keeping the dose low. This includes X-rays, as well as computed tomography (CT or CAT) scans, which also use X-rays. X-rays work by passing through soft tissues such as skin and being absorbed by more dense tissues such as bones and organs. Normal X-rays involve a simple straightforward beam (Figure 7.2.16a). In CT scans the X-ray beam moves in a circle forming a more detailed image (Figure 7.2.16b). These are crucial diagnostic technologies and are staples of modern medicine. X-rays are a form of ionising radiation, which means they can cause damage to DNA. Exposure to X-rays is very brief and comprises an extremely low dose of radiation.

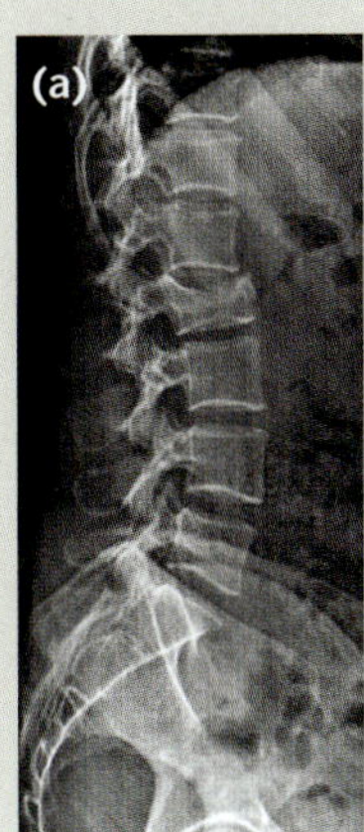

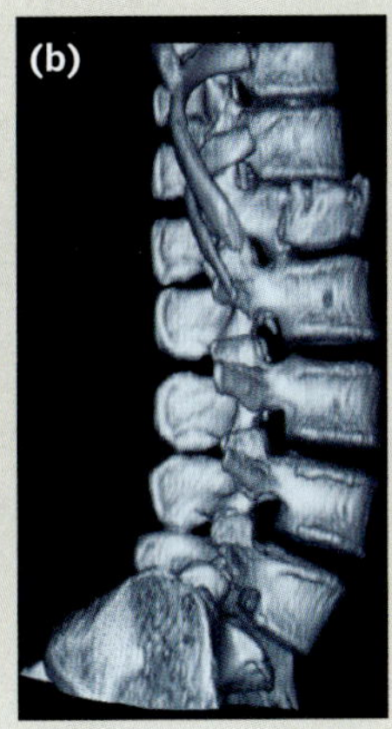

FIGURE 7.2.16 (a) X-ray image of a spinal injury (fractured vertebra). X-rays work because they pass through soft tissues and are absorbed by more dense tissues. (b) A computed 3D tomography (CT) scan of a similar spinal injury (although a different vertebra is fractured). In this case, the X-ray moves around the target in a circle, providing a more detailed and useful image than an X-ray.

7.2 Review

SUMMARY

- Mutagens are substances or processes that can dramatically increase the rate of mutations. Mutagens may be artificial or natural in origin.
- Mutagens can function as carcinogens (cancer-causing agents).
- Disruption to the cell cycle can result in uncontrolled cell division:
  - in a mature organism, uncontrolled cell division may lead to the formation of a neoplasm and eventually cancer
  - in an embyro, uncontrolled cell division will result in an abnormal embryo, which will usually abort.
- Neoplasms may be benign or malignant:
  - benign neoplasms are encapsulated and do not invade surrounding tissues
  - malignant neoplasms divide more rapidly, break out of the capsule, and invade and destroy surrounding tissues.
- Genetic factors can influence the development of mutations and cancers:
  - mutations can change normal proto-oncogenes into oncogenes, which result in neoplasms
  - damage to tumour-suppressor genes may lead to uncontrolled cell division
  - damaged proto-oncogenes or tumour-suppressor genes can be inherited, leading to a family predisposition to certain forms of cancer.
- Damage to the immune system may also promote the development of tumours.
- Environmental mutagens can be chemical, physical or biological.
- Chemical mutagens include intercalating agents, base analogues and DNA reactive chemicals.
- Physical mutagens include particle radiation and high energy electromagnetic radiation, but it must be ionising radiation.
  - non-ionising radiation cannot damage DNA directly.
- Biological mutagens include oncogenic viruses (which may be direct or indirect) and transposable elements ('jumping genes').

KEY QUESTIONS

1 What is a neoplasm?

2 Give two examples of each of the following:
  a sources of chemical mutagens
  b physical mutagens
  c biological mutagens

3 Explain how proto-oncogenes and tumour-suppressor genes can cause cancer.

4 What is the difference between ionising and non-ionising radiation? Briefly explain the effects of each.

5 How do transposable elements cause mutations?

7.3 Mutations in non-coding regions

So far you have learnt about mutations mostly as they affect the coding regions of genes, but most of the DNA in a genome is non-coding. Non-coding DNA does not code for amino acids and polypeptides. As you have already learnt in Chapter 4, this does not mean that this DNA is non-functional. **Introns** (non-coding DNA), non-coding RNAs (genes that form ribosomal RNA and transfer RNA), promoter and terminator regions are essential to the functioning of the cell. In this section, you will learn about the effects of mutations in non-coding DNA and how they can be just as significant as alterations to coding sequences. A large proportion of known human genetic disorders can be traced to non-coding regions of DNA.

GO TO ➤ Section 4.2 page 165

INTRON MUTATIONS

Introns are the non-coding sequences of DNA separating one **exon** (coding region) of a gene from another. Often, a protein-coding gene is several exons long. During the transcription process, the introns (which do not code for proteins) must be removed to create functional mRNA. The removal is known as **splicing** and requires a protein complex known as a **spliceosome**. The spliceosome is made up of **small nuclear ribonucleoprotein particles (snRNPs)**. This spliceosome binds to three recognised sequences along the intron: one at the 5' end (known as the 5' junction) one at the 3' end (the 3' junction) and one near the 3' end (the branch point) that allows the intron to fold into a loop before it is removed (Figure 7.3.1). After the intron is removed, the exons can join together to create an mRNA molecule with a neat, continuous coding sequence.

Because introns are non-coding, they can often tolerate mutations without having any effect on an organism. However, a mutation at any one of these spliceosome binding sites (either the 3' or 5' junctions, or the branch point) may affect the splicing of the intron. Mutations can even have significant effects if they are more than 100 nucleotides from the binding sites if the sequence starts to resemble a splice site. These are known as **deep intronic mutations**.

If the mutations occur at the binding site, splicing may fail to occur and the intron will be retained in the mRNA. This means that the intron sequence will be read as a coding sequence when the mRNA codes for a polypeptide, which could cause any of the mutations detailed in Section 7.1, including nonsense mutations and unexpected stop codons. This results in what is known as a **pseudo-exon**. Pseudo-exons are the cause of most of the well-known genetic disorders resulting from intron mutations, such as beta thalassaemia, some of the mutations that cause cystic fibrosis, Duchenne muscular dystrophy, Becker muscular dystrophy, retinitis pigmentosa (a condition causing tunnel vision and ultimately blindness) and Pompe disease (a severe disorder of glucose metabolism).

If the mutations within the intron resemble a splice site, splicing may sometimes occur at the wrong point. This means that the expression of the gene will be reduced, as the correct protein product will be produced only when the correct splicing site is recognised. Therefore, intron mutations that create new splicing sites can reduce gene expression.

Introns may also contain sequences that regulate transcription and translation. Mutations in these sequences can either reduce or increase the amount of gene product produced.

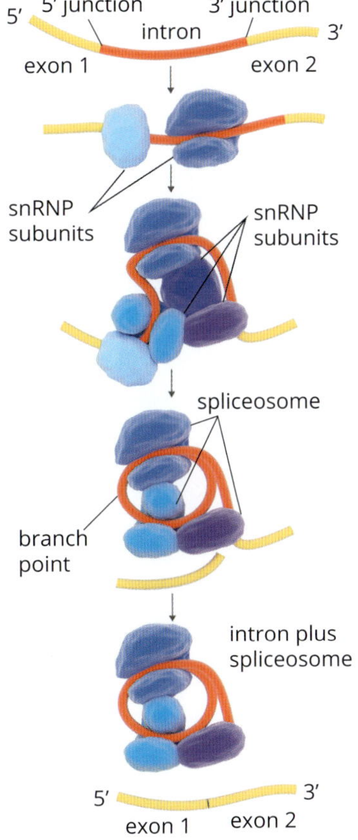

FIGURE 7.3.1 The process by which an intron is removed from pre-mRNA. The intron is shown in red and the neighbouring exons in yellow. The enzyme complex (spliceosome) binds to various sites on the intron and cuts out the intron, before binding the two exons together to form the mature messenger RNA (mRNA). The spliceosome consists of small nuclear ribonucleoprotein particle (snRNP) subunits (blue and purple). Any mutations to the junctions or the branch point sequence may interfere with the binding of the spliceosome and prevent the intron being completely removed.

PROMOTER AND TERMINATOR MUTATIONS

Promoter and terminator sequences are neither introns nor exons. Mutations in these regions will mostly result in either over- or under-expression of the gene, leading to either an excess or shortage of the gene product. There are many genetic diseases that could be the result of changes to promoter or terminator sequences, but these are rarely monogenic diseases (diseases caused by a single gene). Some forms of beta thalassaemia are caused by mutations to promoter regions in the globin B gene (*HBB*).

BIOFILE DD

Intronic mutations, promoter mutations and disease: beta thalassaemia

Beta thalassaemia is an inherited blood disorder, characterised by an inability of the body to produce sufficient quantities of beta globin (HBB), resulting in small, fragile and easily damaged red blood cells (Figure 7.3.2). Beta globin is necessary for fully functioning haemoglobin A.

Beta thalassaemia is caused by mutations of the *HBB* gene on chromosome 11. The mutation can occur in several places within the gene. The severity of the disease, which ranges from mild (beta thalassaemia minor) to moderate (beta thalassaemia intermedia), to severe (beta thalassaemia major) and even lethal, depends on which mutation is present, and whether the person is heterozygous or homozygous for a particular mutation. The mutations causing beta thalassaemia are sorted into two groups:

- deletion—there are nucleotides missing from the coding sequence in the *HBB* gene. These missing fragments can be of different sizes
- non-deletion—the mutations occur in the non-coding portions of the *HBB* gene, either the intron or the promoter region. This leads to the production of abnormal and sometimes non-functional mRNA. Many mutations have been identified in the promoter region, but in some cases there appear to be deep intronic mutations that lead to splicing errors during mRNA production. Pseudo-exons are then included in the mRNA and the *HBB* gene is not correctly transcribed.

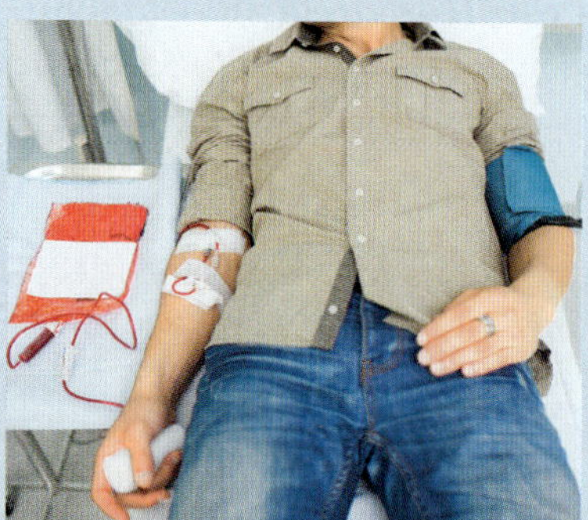

FIGURE 7.3.2 Thalassaemia results in the production of malformed red blood cells, which are smaller, more fragile and less functional than normal red blood cells. They often break up as they move through the body. Many people with beta thalassaemia intermedia will require occasional or regular blood transfusions, while those with beta thalassaemia major will need lifelong transfusion.

NON-CODING RNA MUTATIONS

Genes that code for RNA proteins such as ribosomal RNAs (rRNAs) or transfer RNAs (tRNAs) are known as **non-coding RNAs**, because these genes are not based on a sequence of codons. The product of a non-coding ribosomal RNA gene is not a chain of amino acids, but a highly structured piece of RNA that binds to ribosomal proteins to form a ribosome, while transfer RNA genes produce a structure necessary for the attachment of amino acids to mRNA molecules. These are therefore known as **structural genes**. The correct structure and binding of ribosomal and transfer RNAs are crucial to the functioning of the cell, and therefore mutations are not well tolerated. Minor changes to certain parts of the sequence will not have an effect, but changes in other sites will be lethal, resulting in a failure of the cell to function. The effect of the mutation will depend exactly where in the sequence it occurs and how it affects the folding of the resulting RNA molecule. For this reason, there are currently no known human genetic diseases based on functional mutations to non-coding RNA in the nuclear genome. Mutations in the genes that form ribosomal proteins can cause disease, but these are changes to coding genes, rather than non-coding RNA.

However, there is some evidence of disruptive mutations in non-coding RNA sequences in the mitochondrial genome. There are many different forms of mitochondrial disease, many of which affect protein coding genes, but two of the more common mutations occur in a tRNA gene. This means that the particular amino acid that binds to this tRNA (leucine) cannot be effectively added to mitochondrial proteins. Overall, this has a negative effect on cellular respiration. One of these mutations is associated with a cluster of symptoms known as Melas syndrome (myopathy [muscle weakness] encephalopathy [brain disease] lactic acidosis [a build-up of lactic acid in the body and stroke-like symptoms]) and the other is associated with myopathy alone. The effects of Melas syndrome on the brain are shown in Figure 7.3.3. There are also other mitochondrial diseases associated with mutations to ribosomal RNA genes.

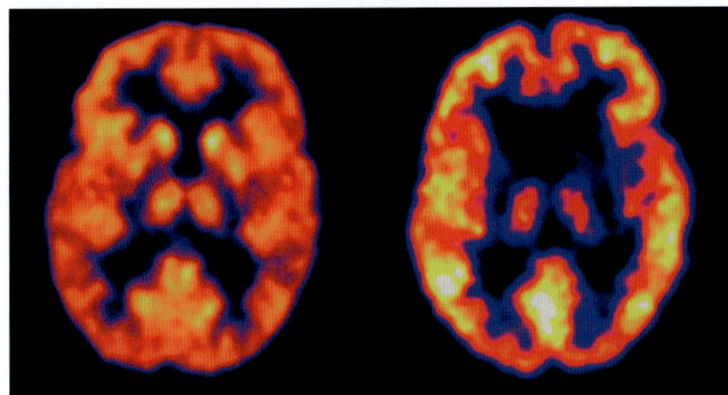

FIGURE 7.3.3 Positron emission tomography (PET) brain scans showing a normal brain (left) and the brain of a patient with Melas syndrome (right). Melas syndrome is caused by mitochondrial mutations. One of the most common causes of Melas is a mutation to the tRNA(Leu) gene. The Melas brain has an absence of a part of the brain called the striatum. This absence is visible in the upper central portion of the image, and results in dementia, seizures and stroke-like symptoms.

7.3 Review

SUMMARY

- Even though introns do not code for proteins, intron mutations can have significant effects.
- To be successfully spliced from a pre-mRNA molecule, an intron needs three recognisable sequence sites for a spliceosome to bind to:
  - 5' junction
  - 3' junction
  - branch point.
- Mutations to any of these points will lead to incomplete removal of the intron.
- Mutations deep within the intron may result in false splicing sites, which may lead to incorrect splicing and competition for natural splicing sites.
- Failure to remove the intron may result in the inclusion of pseudo-exons.
- Intronic mutations are associated with numerous genetic disorders.
- Mutations to promoter regions can also profoundly affect transcription, mostly by affecting how much of a gene product is produced.
- Mutations to non-coding RNA in the nuclear genome are usually lethal (i.e. the cell does not survive).
- Mutations to non-coding RNA in the mitochondrial genome are associated with Melas syndrome and other forms of mitochondrial disease.

KEY QUESTIONS

1 Is the following statement true or false?
Mutations to non-coding regions of DNA will not have an effect on the organism. Explain your answer.

2 Describe the process by which introns are removed from the pre-mRNA molecule during transcription. Use a diagram to illustrate this process.

3 Referring to your diagram in Question 2, explain why mutations to introns can cause errors in transcription.

4 What is a deep intronic mutation?

5 **a** What is non-coding RNA?
b What are the effects of mutations to tRNA genes?

7.4 Mutation and genetic variation

BIOLOGY INQUIRY   

Mutation—the genetic wild card

How does mutation introduce new alleles into a population?

COLLECT THIS...

- one or two decks of playing cards (depending on class size)
- coin (for coin tossing)
- dice
- notebook
- calculator or tablet

DO THIS...

Neutral mutations

1 Keep the jokers in the deck of cards. Shuffle one deck of cards and distribute two cards to each student without replacement. If all cards are used (i.e. if there are more than 28 students), begin from a second deck. These determine genotypes as in the Chapter 6 Biology Inquiry activity (black cards represent '*A*' alleles, red cards represent '*a*' alleles). Students with the joker have a mutated form of the allele of the same suit colour. Indicate the black joker with *A** and the red joker with *a**. Note your genotype (generation 1) and note starting allele frequencies for this population.

2 Divide into breeding pairs and draw cards for offspring. Those who are heterozygous for a mutation should toss a coin to see which allele is passed on: the mutated allele or one of the wild-type alleles. Note your genotype (generation 2).

3 If one or more of the mutations are passed on, divide into new pairs and repeat step 2 (generation 3).

4 Repeat steps 2–3 again (generation 4). Note the genotypes.

Beneficial mutations

5 Many generations after step 4, one of the mutant alleles has become more common (*A**) while the other (*a**) has died out. Distribute two cards to each student to determine starting genotypes.

6 Each student with the genotype *AA* (two black cards) rolls a dice:

- 1—homozygous for the mutant allele (*A*A**)
- 2 or 3—heterozygous for the mutant allele (*A*A*)
- 4, 5 or 6—homozygous for the wild-type allele (*AA*).

7 Calculate the allele frequencies for the class, including the mutant alleles.

8 A disease sweeps through the population. The mutation turns out to offer resistance to this disease. All individuals carrying the mutation live. All individuals without a mutation have a 50% chance of dying (decided by coin toss).

9 If you survived, you produce more offspring as in step 2. Note the genotypes (generation 5).

10 Note the new allele frequencies, including the mutated alleles.

RECORD THIS...

Describe the changes in the mutant allele frequency in both the neutral and beneficial models.

Present your prediction of what would happen to the mutant allele(s) if this continued for several more generations after the disease passes through the population.

REFLECT ON THIS...

How does mutation introduce new alleles into a population?

How do selection pressures affect the survival of mutant alleles in a population?

How might this activity be relevant to understanding the problems of antibiotic or pesticide resistance?

If the mutation had not occurred (or had died out) prior to the arrival of the disease, what would have happened to this population?

Somatic mutations occur in body cells and only affect the individual. **Germline mutations** occur in **germ cells** (cells that differentiate into gametes). They are heritable because they affect gametes and can therefore be passed on to offspring. Mutations in germline cells are not just a matter of concern for individual offspring: these mutations are the source of new alleles in the population. While some of these mutations can cause problems, others can be advantageous. For **natural selection** to take place, there must be variation in a population. You can think of mutation as the raw material of evolution. In this section, you will learn about both somatic and germline mutations and how they affect an individual organism, the next generation, populations and even species.

Germline mutations can be passed from parent to offspring but somatic mutations cannot.

SOMATIC MUTATIONS

Somatic mutations are genetic changes in somatic cells (body cells other than gametes). Any mutation that occurs during the normal process of mitotic cell division and growth in non-reproductive tissue is a somatic mutation. Somatic mutations can also be referred to as postzygotic mutations, since they are mutations that occur after fertilisation.

Mutations in somatic cells will not be passed onto the next generation and therefore do not affect the gene pool of a population.

Mutation and the ageing process

Like all mutations, somatic mutations may occur as a result of exposure to mutagens (as explained in Section 7.2), but they also occur naturally as a result of errors in DNA transcription. The rate of natural mutation in human cell division is much higher in somatic cells than in germline cells. While somatic mutations may result in neoplasm and possible cancer if the mutation occurs in certain genes, this is not the most common result. We are all experiencing a low level of somatic mutation as we grow, develop and age. In fact, the ageing process is largely due to an accumulation of errors during DNA replication over time (Figure 7.4.1). To put it another way, we get old because we gather somatic mutations. This is another reason that humans become more prone to develop certain cancers as we age; the more errors we accumulate in our DNA, the more likely it is that those errors will trigger the formation of a neoplasm.

The more that cells divide, the more opportunity there is for replication error (and therefore mutation). Therefore, we would expect a higher rate of cancers in children who are growing rapidly—but we do not see that. In fact, for most cancers, the rate of occurrence in children is very low. This is because evolution has selected for the increased activity of DNA repair enzymes in children and young people, but this protective effect does not last.

FIGURE 7.4.1 The symptoms of ageing are linked to a build-up of somatic mutations as we age, due to our reduced ability to correct errors during DNA replication and cell division.

Mosaicism

Most somatic mutations do not survive to produce daughter cells, either because of destruction by the immune system, or because of regulatory mechanisms that cause the cell to self-destruct. If the mutations do survive, the altered cells will continue to divide as normal, populating the relevant tissues. This means that an individual may carry different genomes: the normal, inherited genome, which occurs throughout most of the body; and the mutated genome, which only occurs in a select few cells. This condition is called **mosaicism**. Somatic mutation is common enough that most likely all multicellular organisms experience a certain amount of mosaicism. Most of the time, the effects of genetic mosaicism are not immediately obvious but there are some situations where it is easy to see. The port-wine birthmark is an example of this. This birthmark is caused by a single somatic mutation that occurs early in development, before birth. The mutated cell divides and proliferates, causing expanded blood vessels just beneath the skin in the affected tissues (Figure 7.4.2).

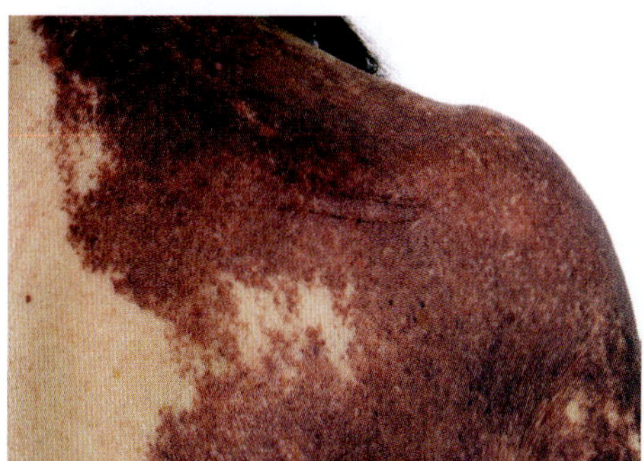

FIGURE 7.4.2 The port-wine stain is an example of genetic mosaicism. The cells in these areas carry a non-cancerous somatic mutation that occurred before birth, allowing the cells time to rapidly spread throughout the tissue during development. This mutation causes the blood cells to expand just under the skin.

BIOFILE CCT

Somatic mutations in horticulture

You might think that somatic mutations are not very useful, given that they can only affect a single individual. The truth is quite the opposite. Because plants can grow asexually via vegetative reproduction, somatic mutations can be of great use to the agricultural industry.

The Golden Delicious apple (Figure 7.4.3a) is derived from a single mutation in the flower of a Red Delicious apple (Figure 7.4.3b), affecting the colour of the ovary wall. Although the flower itself is a reproductive structure, it is not germline tissue. Meiosis does not occur in the ovary wall because it is somatic tissue. When the fertilised flower developed into a fruit, the mutated cell produced daughter cells that populated the flesh of the apple. Because the mutation does not exist in the germ cells of the plant, Golden Delicious apples do not reproduce themselves. Any seeds that germinate from the Golden Delicious will grow into trees that will produce Red Delicious apples. Effectively, the Golden Delicious is a dead end, single event mutation—or it would be, if it were not for the fact that plant tissues can be grafted onto one another (Figure 7.4.3c). Horticulturists can graft branch tissue from one plant onto tissue from another, thus combining somatic characteristics from different plants.

FIGURE 7.4.3 (a) Golden Delicious apples are the result of a mutation in the somatic cells of a (b) Red Delicious apple. (c) Grafting maintains Golden Delicious apples in the horticultural industry. Here, an apple is being grafted onto the rootstock of a different plant. Grafting is a common practice in horticulture, which allows leaves, stems and flowers of one plant to be grown on the base of another plant, which might be more robust or appropriate.

GERMLINE MUTATIONS

Germline mutations are mutations in reproductive tissue (germ cells) that give rise to gametes. Germline mutations are significantly rarer than mutations in somatic cells. This is because of the elevated levels of DNA repair enzymes in these tissues.

If a germline cell (gamete) undergoes mutation and survives to form a zygote, the offspring will be heterozygous for that mutation in all cells; however, the parent organism will not experience any changes to their own phenotype. These mutations will only affect the next generation.

Like all mutations, germline mutations may occur due to exposure to mutagens that damage DNA, or occur naturally due to errors in meiosis.

Non-disjunction in meiosis

Meiosis is usually an exact process but sometimes errors occur. Missing, extra or malformed chromosomes can result from defective gametes, which may have serious consequences for offspring. Most of these have been discussed in Section 7.1, but there is one error that is specific to meiosis—non-disjunction. This is the usual cause of aneuploidy.

Sometimes during meiosis a chromosome does not separate. This is known as a non-disjunction. Non-disjunction can occur during meiosis I if members of a pair of homologous chromosomes do not move apart properly, or during meiosis II if sister chromatids fail to separate. As a result, one gamete receives two of the same type of chromosome and another gamete receives no copy. If either of the gametes unites with a normal gamete at fertilisation, the zygote will also have an abnormal number of a particular chromosome. An example is non-familial Down syndrome. In this disorder, an individual has three copies of chromosome 21. Therefore, non-familial Down syndrome is often referred to as trisomy 21 (Figure 7.4.4).

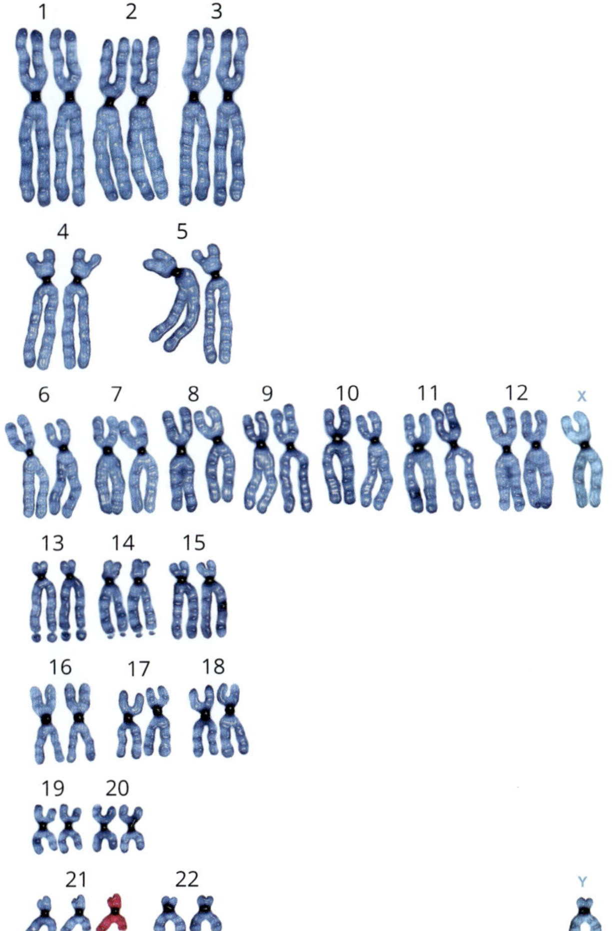

FIGURE 7.4.4 The karyotype of a male with Down syndrome. This condition is caused by trisomy 21, which is the presence of an additional copy of chromosome 21 (highlighted in red).

In people with non-familial Down syndrome, the extra copy of chromosome 21 usually results from an error in meiosis in one of the parents. Chromosome 21 undergoes non-disjunction so that cell division produces gametes with either an extra or missing chromosome 21. If a gamete with $(n + 1)$ chromosomes unites with a normal gamete (n) the zygote will be trisomic $(2n + 1)$ (Figure 7.4.5).

The additional chromosome 21 results in slower and limited physical development, so people with non-familial Down syndrome tend to be shorter and have less muscle development than people without the syndrome. Intellectual development is also slowed, but well-designed support programs for families, communities and educators can help individuals with Down syndrome to lead healthy, productive lives. The physical and intellectual expressions of Down syndrome also vary greatly; some individuals are more severely affected than others.

The use of karyotypes to identify other chromosomal abnormalities is covered in Section 7.1.

 Year 11 Chapter 9

Natural selection

Natural selection favours some phenotypes over others and results in evolution. When there is variation in a population, **natural selection** can take place. The natural world places pressure on a population that cause some phenotypes to be more successful (or 'fit') than others. When it comes to genetics, fitness is best understood in terms of reproductive success. To put it another way, individuals that survive to pass on their genes are the most fit.

The pressures that cause some individuals to die off while others survive are known as selection pressures. Some phenotypes might be favoured due to competition for resources such as food and habitat, either within species or between species; others might be favoured due to mate choice. Some phenotypes may survive climate effects more readily than others, being more robust. These phenotypes are considered to have higher adaptive value. The response of a population to selection pressures will result in evolution.

Evolution is most easily understood as the change in allele frequencies in a population over time. Natural selection; however, is only one of the ways in which allele frequencies can be changed. Mutation, gene flow and genetic drift all interact to change allele frequencies over time and their relative influences will dictate the level of genetic change in the population.

Mutation is the source of new alleles in a population and ultimately for all the variation in a species.

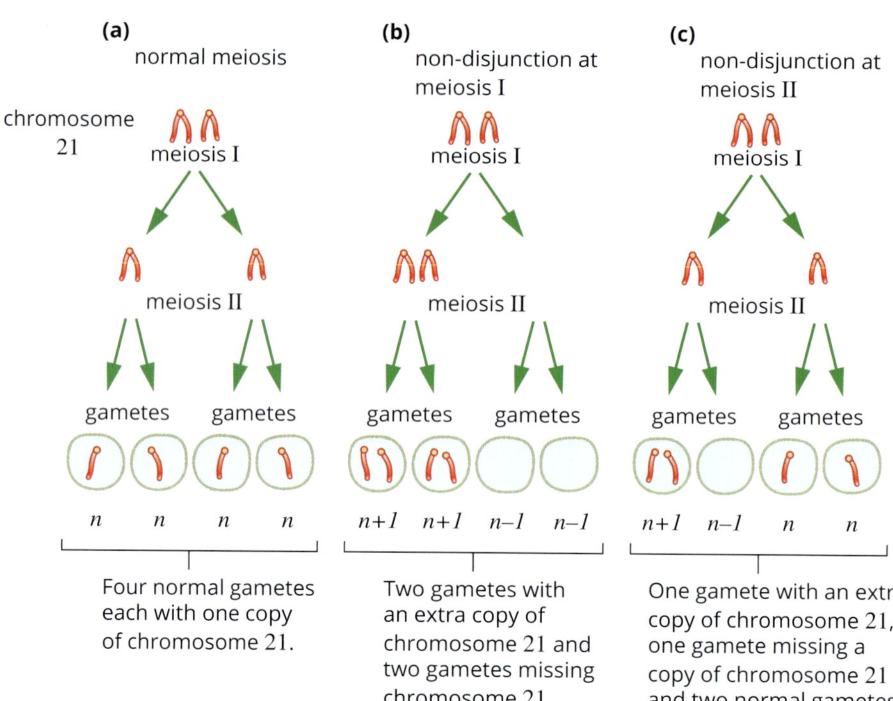

FIGURE 7.4.5 Gametes resulting from non-disjunction of chromosome 21 at the first or second division of meiosis in a parent who has a child with Down syndrome. Normal meiosis is shown for comparison. All other chromosomes segregate normally in each case.

MUTATION AS A SOURCE OF NEW ALLELES

When you look at the people around you, you will notice that they all look quite different (Figure 7.4.6). They have different heights, different shapes, different skin tones and different facial features. They have different eye colours and hair textures. These phenotypic variations are mostly genetic in origin (although environmental factors will play a role). Most of these variations have their origin in the distant past as mutations in germline cells. Transcription and translation are precisely regulated processes and any variation in those processes can cause problems. This is why mutations are usually harmful, or—depending on when and how they occur—neutral. However, if that were always the case then all humans would be identical. Mutation is ultimately the source of genetic variation and necessary for evolution to occur.

Variation in one population can be a source of new alleles for another population. The movement of alleles between populations is known as **gene flow**. When gene flow between populations is high, their **gene pools** (total genetic variation) will be similar. When gene flow is low or absent between populations, their gene pools will become increasingly different. If two populations are isolated for long enough, they may eventually become different species. Allele frequencies in a gene pool may also change randomly over time as a result of chance events (e.g. birth and death). This is called **genetic drift**. In small, isolated populations, the random loss of alleles due to genetic drift can have a significant effect, resulting in loss of genetic variation.

FIGURE 7.4.6 Humans are an extremely variable species. Every genetic variant (allele) originated in a germline mutation somewhere in the distant past.

7.4 Review

SUMMARY

- Somatic mutations occur in body cells and only affect the individual.
- Germline mutations are heritable because they affect gametes and can therefore be passed on to offspring.
- The ageing process is linked to the accumulation of somatic mutations.
- Mutation is more common in somatic tissues than in reproductive tissues.
- Mutation is also more common in adults than in children and infants.
- Somatic mutations also result in mosaicism, a condition where an individual can have multiple genomes within the same tissues.
- If a germline cell (gamete) undergoes mutation and survives to form a zygote, the offspring will be heterozygous for that mutation in all cells.
- Chromosomal abnormalities, such as non-disjunction, sometimes occur during meiosis, and an extra or missing chromosome in a gamete can have severe effects in offspring.
- Down syndrome is the effect of having three copies of chromosome 21 (trisomy 21).
- If whole sets of chromosomes fail to separate, a gamete may end up with a number of sets of chromosomes, leading to polyploidy.
- Mutation creates genetic variation and is required for evolution by natural selection.
- Natural selection is the influence of environmental pressures on allele frequencies of a population, which occurs because of genetic variation between individuals, and the survival and reproduction of those individuals with favourable phenotypes (traits).

KEY QUESTIONS

1. What is the difference between a somatic cell and a germ cell? Give an example of each.
2. Explain how an error in meiosis can lead to Down syndrome.
3. Why is polyploidy more common in plants than animals?
4. Explain how somatic mutations give rise to mosaicism.
5. Somatic mutations are responsible for which of the following?
   A the symptoms of ageing
   B port-wine birthmarks
   C neoplasm
   D all of the above
6. Evolution is the change in allele frequencies over time. List four major factors that affect allele frequencies in a population.

Chapter review

07

KEY TERMS

acute radiation syndrome (ARS)
amino acid
aneuploidy
apoptosis
base analogue
biological mutagen
cancer
carcinogen
chemical mutagen
chromosomal abnormality
chromosomal mutation
chromosome
codon
deep intronic mutation
degenerate
deletion mutation
diploid
DNA (deoxyribonucleic acid)
DNA reactive chemical
duplication mutation
electromagnetic radiation (EMR)
exon
frameshift mutation
gamete
gene flow
genetic drift
gene expression
gene pool
genetic variation
germ cell
germline mutation
haploid
homologous chromosome
insertion mutation
intercalating agent
intron
inversion mutation
ionising radiation
karyotype
meiosis
messenger RNA (mRNA)
missense mutation
mosaicism
mutagen
mutation
natural selection
neoplasm
non-coding RNA
non-disjunction
non-homologous chromosome
non-ionising radiation
nonsense mutation
nucleotide
oncogene
oncogenic virus
particle radiation
physical mutagen
point mutation
polypeptide
polyploid (n. polyploidy)
proto-oncogene
pseudo-exon
radiation
reactive oxygen species (ROS)
retrovirus
silent mutation
small nuclear ribonucleoprotein particle (snRNP)
somatic cell
somatic mutation
spliceosome
splicing
stop codon
structural gene
substitution mutation
transcription
translation
translocation mutation
transposable element (TE or transposon)
trisomy
tumour-suppressor gene

REVIEW QUESTIONS

1 A sequence of DNA is GGA TTA CCG TCT. It undergoes a mutation that changes the sequence to GGA CTA CCG TCT. What type of mutation is this?

A a frameshift mutation
B a block mutation
C a point mutation
D a deletion mutation

2 Sickle-cell anaemia is a hereditary disease in which the haemoglobin of red blood cells is abnormally formed. The condition results in rapid destruction of red blood cells and severe anaemia. Individuals suffering from sickle cell disease usually do not survive childhood. The disease is controlled by a single gene with the alleles *HbS* (sickle cell disease) and *HbA* (normal haemoglobin). The varying degrees of the disease are determined by the combinations of alleles in individuals.

Normal haemoglobin	Sickle cell trait	Sickle cell disease
HbAHbA	*HbAHbS*	*HbSHbS*

Heterozygotes show greater resistance to the mosquito-borne parasite, *Plasmodium falciparum*, which causes malaria, than do normal individuals.

a Sickle cell disease and sickle cell trait are relatively common in Africa, India and the Mediterranean, where malaria is prevalent. Given that the homozygous normal genotype develops into a generally healthier individual, explain the steps that would have occurred in human populations originally colonising one of these regions that resulted in a change in the frequency of the *HbS* allele.

b Explain how the variation in these alleles is maintained in human populations in the Mediterranean.

c Based on the above information, what trend do you expect to see in *HbS* allele frequencies in Australia?

3 Which sort of mutation involves the movement of DNA between non-homologous chromosomes?

A duplication
B deletion
C inversion
D translocation

4 A child was born with some abnormalities. A chromosomal analysis was performed as a part of the testing to identify the cause of these abnormalities. There were some changes identified to two chromosome pairs. The normal chromosomes are shown on the left in the diagram below. Each coloured section is one gene and each letter represents an exon of that gene.

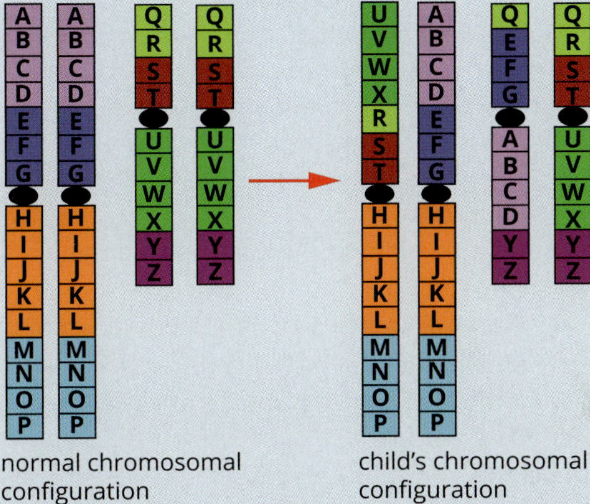

The analysis showed that two translocations had occurred between the chromosomes of these pairs. The first swapped ABCD with UVWX and the second swapped EFG with RST as shown on the right in the diagram above.

Despite the abnormalities the child grew up, met a partner, and they went on to try to start a family of their own. The couple wish to know if the abnormal chromosomal arrangement means that they will be unable to have children. The diagram below shows the possible gametes of the partner with the translocation. The other partner has a normal chromosomal arrangement. Use this information to advise the couple.

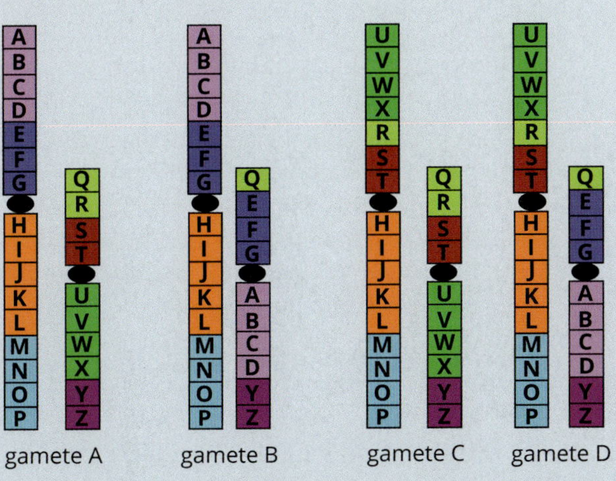

5 Below are the karyotypes of a normal individual and one with a chromosomal mutation.

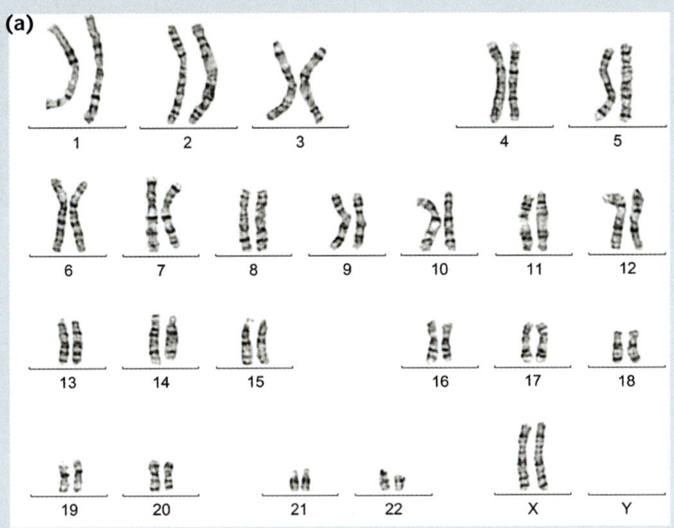

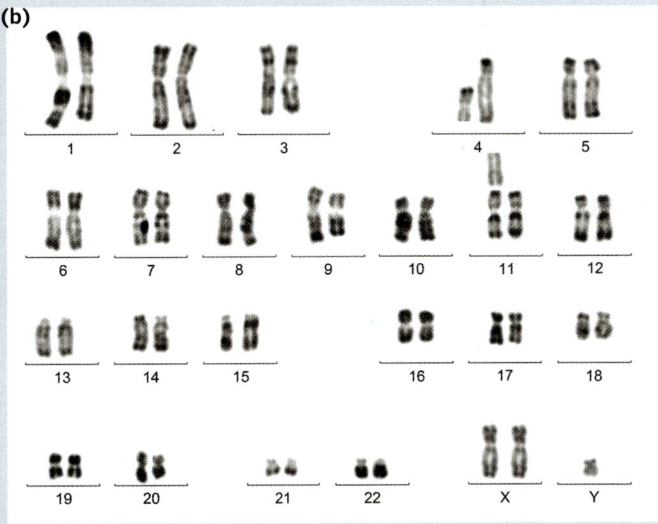

a Identify the sex of the individuals in the karyotypes.

b Identify the abnormalities in karyotype B.

6 Why are triploid organisms usually sterile?

7 List three differences between malignant and benign neoplasms.

8 Some cancers are at least partially hereditary; the genetic predisposition can be passed down in a family line.

a Give one example of a hereditary predisposition to cancer.

b If one family member has a particular type of hereditary cancer, do all other relatives develop the cancer? Explain why or why not.

c What other factors can influence the development of a neoplasm?

9 Which of the following factors can contribute to a person developing cancer?
- **i** genetic factors
- **ii** oncogenes
- **iii** exposure to a carcinogen
- **iv** infection by human papillomavirus (HPV)

A i only
B i and iii only
C ii and iv only
D i, ii, iii and iv

10 Can exposure to each of the following forms of electromagnetic radiation (EMR) result in damage to DNA? Note whether each involves ionising or non-ionising radiation. Comment on the likelihood of experiencing harm as a result.
- **a** getting an X-ray for a possible broken arm
- **b** regular usage of wireless internet (Wi-Fi)
- **c** regular trips to the beach on days of high UV exposure

11 The Gardasil® vaccine works by immunising against four strains of human papilloma virus (HPV). HPV has been implicated in cervical cancer due to its interference with tumour suppressor proteins and is spread through sexual contact. Blocking the activity of these four strains can prevent up to 70% of cervical cancers and 90% of genital warts. Which of the following terms does not describe HPV? Explain your answer.

A biological mutagen
B indirect oncovirus
C sexually transmitted disease
D physical mutagen

12 Although a substance may be known to be toxic in the short-term, it often takes a long time to identify whether or not a chemical is a mutagen. Why is this the case?

13 Although it is extremely rare in the present day, acute radiation syndrome (ARS) can be caused by high doses of which of the following?
- **i** particle radiation
- **ii** ionising radiation
- **iii** non-ionising radiation
- **iv** electromagnetic radiation

A i, iii, iv
B iii and iv
C i and iv
D i, ii and iv

14 What is meant by the term 'deep intronic mutation'? How can a deep intronic mutation interfere with the function of a gene?

15 In spite of the rapid rate of cell division during early growth and development, most cancers are rare in children and infants. This is because evolution has selected for elevated levels of repair enzymes, so most errors in DNA replication are corrected before they have a chance to result in neoplasm. However, after middle age, the risk of cancer greatly increases. Why do you think this is the case?

16 Port-wine birthmarks are caused by a mutation in the *GNAQ* gene that occurs after fertilisation. However, if this mutation occurs very early in embryonic development it will result in a much more severe condition called Sturge-Weber syndrome, which is linked to seizures, stroke, blood vessel deformities and other symptoms.
- **a** Based on this information, is this a somatic or germline mutation?
- **b** Can Sturge-Weber syndrome be passed onto the next generation of offspring?

17 Which of the following is a source of new genetic material?

A gene flow
B natural selection
C genetic drift
D mutation

18 What is meant by the statement 'Mutation is the raw material of evolution'?

19 A rare species of butterfly is found in an archipelago (a group of islands) in the Pacific. There are over 1000 butterflies on each island, except for the smallest island. On that island, the population is quite small (only 100 individuals) and a recessive mutation has arisen in this population. Individuals on the small island that are homozygous for this mutation are camouflaged in their surroundings and avoid predators. These camouflaged butterflies have an advantage and the mutation is spreading rapidly. The camouflage phenotype has not been observed in the other butterfly populations. A tropical storm blows through the Pacific and wipes out two-thirds of the small island's butterfly population, including all the camouflaged butterflies.
- **a** How did the allele frequencies change in the small population before and after the tropical storm?
- **b** Is there likely to be gene flow between butterfly populations on the different islands? Explain your answer.

20 After completing the Biology Inquiry on page 306, reflect on the inquiry question: How does mutation introduce new alleles into a population? Explain how mutation can be both beneficial and detrimental to the evolutionary potential of a population. Use at least two examples to support your answer.

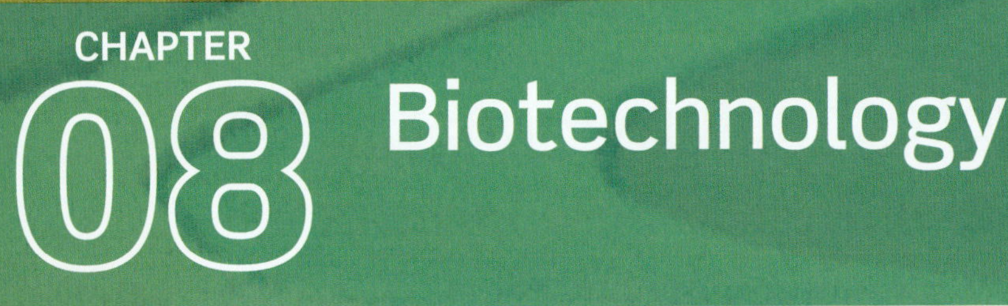

Biotechnology

This chapter will explore a range of historic and modern biotechnology applications, including DNA techniques and genetically modified organisms. You will examine the social and ethical dilemmas that can arise with biotechnology use and identify the potential benefits to society as a result of genetic research. You will evaluate the effect of biotechnology on biodiversity, with reference to the impact of genetically modified crops on biodiversity and the use of molecular techniques in wildlife conservation.

Content

INQUIRY QUESTION

How do genetic techniques affect Earth's biodiversity?

By the end of this chapter you will be able to:

- investigate the uses and applications of biotechnology (past, present and future), including: (ACSBL087)
  - analysing the social implications and ethical uses of biotechnology, including plant and animal examples S EU ICT CC
  - researching future directions of the use of biotechnology CCT ICT
  - evaluating the potential benefits for society of research using genetic technologies S EU PSC
  - evaluating the changes to the Earth's biodiversity due to genetic techniques S EU PSC

8.1 Biotechnology—past, present and future

Biotechnology uses organisms and biological processes to create products and develop technologies.

At its simplest, **biotechnology** is the merging of 'biology' and 'technology', and at its broadest, it is an interdisciplinary technology which encompasses many fields of science and technologies along with biology including chemistry, physics, mathematics and engineering, to help improve the future of humans and the planet. Biotechnology uses biological processes and organisms to create new products and develop new technologies. These new products and technologies are used in many fields including medicine, agriculture, industry and the environment.

HISTORIC BIOTECHNOLOGY

Humans have been using biological processes to improve their lifestyles long before **genes** and **DNA (deoxyribonucleic acid)** were understood or the term 'biotechnology' was first used. Most historic biotechnology was discovered by accident or through simple observation of nature, and was then used to improve everyday life. For example, agriculture developed in ancient times when humans needed to grow their food closer to home to avoid travelling long distances for supplies (Figure 8.1.1). Plant seeds for crops were tended to by experimenting with different amounts of water, light and nutrients, and crop improvement soon developed where the most successful plants were used to obtain seeds for future crops. The manipulation of biological processes for human use in early agriculture was the basis of biotechnology.

FIGURE 8.1.1 An artwork depicts common agricultural practices in ancient Egypt including tending and harvesting crops

Humans have used biological processes such as fermentation and selective breeding for thousands of years.

The deliberate selection and breeding of individuals with favourable characteristics is known as **selective breeding** or **artificial selection**. Selective breeding of plants and animals is carried out with the intention of retaining favourable characteristics in future generations. For example, corn existed in 5000 BC with small cobs and very few kernels. By 1500 AD, selective breeding had produced corn cobs that were almost five times the size of ancestral corn cobs and full of large, juicy kernels. **Cross-breeding** of individuals with different favourable characteristics led to a vast array of local varieties of common crops such as rice, corn and wheat that had higher productivity than their wild ancestors.

Early agriculture also involved the domestication of wild animals by using selective breeding to produce animals suitable for food and transport plus fur and hides for clothing. A classic example of selective breeding is the mule, which is bred from a female horse and a male donkey. Mules were commonly used in ancient times for the transport of heavy loads. Dogs have also been selectively bred for a variety of purposes including rounding-up livestock, hunting small animals and companionship. As a result of this selective breeding there are now hundreds of dog breeds, all with different appearances and behaviours.

One of the most popular dog breeds in the world, the Labrador Retriever was originally bred to retrieve fishing nets and catch from the freezing Newfoundland waters of Canada. Known in the 1700s as the St John's dog, or Newfoundland dog, the breed went on to become a hunting companion, retrieving game from land and water. The breed became extremely popular in England and America with black coats (Figure 8.1.2a) preferred over the recessive yellow and chocolate coloured coats (Figure 8.1.2b).

FIGURE 8.1.2 (a) Historically, the black Labrador Retriever coat (dominant trait) was favoured in selective breeding, while (b) in modern times the recessive traits, expressed in yellow and chocolate coat colours, have been selected for.

Fermentation has been dated back to 6000 BC when Sumerians and Babylonians used fermentation to make beer. Fermentation occurs when microorganisms (such as yeast and bacteria) breakdown sugars during anaerobic respiration (when there is no oxygen present). This biological process was commonly used in other ancient civilisations to make bread, wine, cheese, yoghurt, tofu and sake (Figure 8.1.3). Vinegar was another product of historic biotechnology, produced for consumption with food and used to prevent food spoilage. We now know that it is the acidity of vinegar which prevents the growth of microorganisms, however in ancient times it was simply used like drying, salting and freezing to help food last longer.

FIGURE 8.1.3 Ancient painting (2500–2350 BC) depicting evidence of Egyptian people making bread batter

Many traditional medicines also used biological processes and organisms. For example, ancient Egyptians used honey to treat wounds and infections, Ukrainians used mouldy cheese to heal infections, and the Chinese used mouldy soybean curds to treat boils. These industrious uses of biological products to improve living conditions and human health demonstrate the important role biotechnology has played in past societies, even before people understood the biological processes they were using or manipulating.

BIOFILE IU

Ancient antibiotics

The first primitive antibiotic was believed to have been used in 500 BC when the Chinese spread the curd of mouldy soybeans on boils, carbuncles and other skin infections. Many other ancient cultures used moulds and soil to treat infections. It is unknown how effective this treatment was, however it is likely that trace amounts of antibiotic-producing bacteria or fungi that are used to cultivate modern antibiotics (e.g. penicillin from the *Penicillium* fungi) may have helped to heal surface wounds and prevent infection. Ancient Serbs and Greeks used mouldy bread pressed against wounds to treat infection (Figure 8.1.4), while civilisations of Central Asia used mouldy barley and apple paste. Primitive antibiotics were not just used topically (on the skin surface), the Sudanese Nubia civilisation (350 AD) were found to have levels of antibiotics in their bones, which was believed to have come from microorganisms contaminating their stores of wheat, barley and millet which was kept in mud bins. The Sudanese Nubia civilisation was found to have naturally low rates of infection. This trend was also seen in the Herculaneum civilisation who stored pomegranates and figs in beds of straw, contaminating them with bacteria.

FIGURE 8.1.4 Ancient civilisations frequently applied mouldy food, such as bread, to a wound to prevent infection. The sporadic success of this treatment was likely due to trace amounts of antibiotic-producing bacteria or fungi, such as *Penicillium*, which are used to make modern antibiotics.

> Modern biotechnology uses molecular techniques such as polymerase chain reaction (PCR) and DNA sequencing.

MODERN BIOTECHNOLOGY

While Gregor Mendel identified genes and their role in inheritance in 1865, it was the discovery of DNA in biological processes during the 1940s that marked the beginning of modern biotechnology. It has led to more sophisticated production methods including pasteurisation to remove **pathogens** (disease-causing bacteria), as well as using purified cultures of bacteria. Agriculture has also advanced by using modern techniques to improve crop yields, decrease the use of chemicals and water as well as develop healthier food alternatives.

Modern biotechnology, while continuing to use classic biological processes such as fermentation, also uses a range of molecular techniques such as polymerase chain reactions (PCR) and DNA sequencing. Some common techniques and applications of biotechnology include DNA profiling, gene cloning, genetically modifying organisms, genetic screening, gene therapy and recombinant DNA technologies. You learnt about some of these technologies in Chapter 6 and others are examined in more detail in Chapter 9.

GO TO ➤ Section 6.1 page 248

Molecular techniques are used in modern medicine where genetic technology is being used to decrease rates of infectious and chronic diseases, provide improved methods of disease detection and personalise patient treatment. Modern biotechnology is also used in modern industries, for streamlining manufacturing processes, developing clean energy sources, increasing the use of recyclables, providing more accurate forensic data and decreasing water waste.

FUTURE OF BIOTECHNOLOGY

The interdisciplinary science of biotechnology is driving progress across many fields, from medicine and industry, to agriculture and environmental sustainability. Biotechnology research and innovation has the potential to bring about advances on a global scale, such as sustainable agriculture methods to meet the food demands of future world populations, as well as improvements to the lives of individuals (e.g. improved disease detection). While biotechnology may raise social, ethical and environmental concerns, the field continues to bring about new scientific advances worldwide.

BIOLOGY IN ACTION CC

Say 'Cheese!'—the modern cheese industry

Bacteria and fermentation have always been used to make cheese, however the modern cheese industry uses pure microbial starter cultures (in ancient times bacteria from the environment was used). Modern large-scale cheese production involves pasteurisation, which kills pathogenic bacteria, as well as genetically engineered rennet enzymes responsible for clotting the milk (the basis of making cheese). The type of bacteria added to the cheese, as well as fermentation times, determines the flavour and texture of the cheese produced. Blue-veined cheeses are produced by the *Penicillium* fungi, and production involves spiking the cheese with stainless steel rods to allow oxygen into the cheese encouraging mould (fungal) growth; this gives the cheese its characteristic green, grey, blue or black veins (Figure 8.1.5). Cheese production plays a major role in Australia's dairy industry, with domestic sales estimated at approximately $1.5 billion per year, and exports to countries such as Japan bringing in over $715 million per annum.

FIGURE 8.1.5 Modern cheese production uses pasteurisation and other food safety technologies to develop a wide variety of cheese products.

8.1 Review

SUMMARY

- Biotechnology is an interdisciplinary science used to create new products and develop new technologies using living organisms and biological processes.
- Simple forms of biotechnology were used in ancient civilisations to improve everyday life.
- Modern biotechnology arose in the 1940s with the discovery of DNA.
- Processes that arose in ancient civilisations such as fermentation and selective breeding are used today in more sophisticated biotechnology applications.
- The future of biotechnology can be seen in the current research and development of technologies in fields such as medicine, agriculture and the environment.

KEY QUESTIONS

1 Define the term 'biotechnology'.

2 Name and describe a biological process that was used by ancient civilisations to create products for human use.

3 Define the term 'selective breeding' and provide an example of its application in ancient times.

4 Identify the significant scientific event that marked the commencement of modern biotechnology.

5 Provide an example of a biotechnology that was used in ancient times and also in modern biotechnology. Describe ways this ancient biotechnology has become more sophisticated in modern applications.

8.2 Biotechnology and society

Scientific knowledge can bring great benefits to society; but it can also bring great challenges, as societies consider if or how new technologies should be implemented, and the possible biological, social, economic and ethical implications of their use.

SOCIAL IMPLICATIONS OF BIOTECHNOLOGY

Technology has the power to change people's lives and social structures. Biotechnology is no exception to this; it has the potential to influence the values of a society and change its demographic and economic conditions. In turn, the financial position, lifestyle and social profile of an individual or community can determine whether they have access to new products and services created by biotechnology.

Social equality, accessibility and cost

Biotechnology has allowed people greater access to goods, services and opportunities. For example, **recombinant DNA technology** can be applied to the development of safer vaccines. One of the current techniques is to remove selected genes from a virus so that it cannot replicate. It can then be safely injected into humans to trigger an immune response (Figure 8.2.1).

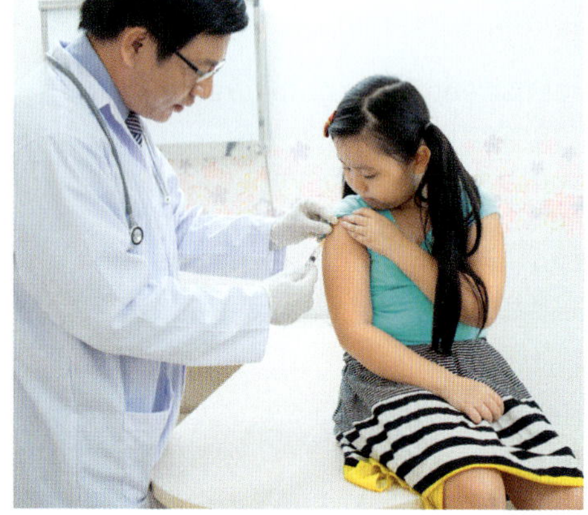

FIGURE 8.2.1 Modern biotechnology has allowed for increased patient access to safe, affordable preventative medicine such as vaccines.

Biotechnology also has the potential to improve the nutrition and yield of crops grown by farmers. This could lead to improved health and result in greater availability of food to some of the poorest people in the world. Golden Rice (Figure 8.2.2) is an example of a **genetically modified organism (GMO)** developed to improve health and nutrition in some of the poorest communities. Rice is a staple food for millions of people around the world; traditional white rice varieties grown in countries such as Vietnam, Bangladesh and the Philippines are low in vitamin A. Vitamin A is a coenzyme that is essential for healthy eyes and the immune system. Diets poor in vitamin A contribute to the deaths of many children in developing countries each year. Vitamin A deficiency can lead to preventable blindness in children and increases the risk of common infections such as measles and diarrhoea.

FIGURE 8.2.2 Golden Rice is yellow due to the presence of pro-vitamin A, also known as beta-carotene.

Golden Rice is a **transgenic** rice crop that contains pro-vitamin A. Golden Rice is produced when two plant genes and one bacterial gene are inserted into the white rice **genome**. These genes switch on a biochemical pathway in the rice plant that sends vitamin A to the rice grains rather than the leaves, as would occur in non-modified rice. Australian researchers are exploring other genetic modifications to improve the ability of rice plants to use nitrogen for improved growth and reduce the need for nitrogen fertiliser. They also aim to make rice grains richer in elements such as iron, zinc and phosphorus through a process called **biofortification**. While the success of some GMOs such as Golden Rice is documented, some suggest that the best way to end hunger and malnourishment is a political solution to inequitable global food distribution rather than reliance on any agricultural technology, including GMOs.

Patenting

Biotechnology companies often control access to new technologies due to patenting of products. **Patents** are available for any invention and allow the patent-holder to be protected by law so that no one else can make, use or sell the invention for a certain period of time. This raises an issue in biotechnology as new biological material is generally classed as a 'discovery' and non-patentable. Some argue that biotechnology products are human-made 'inventions' and therefore should be patentable.

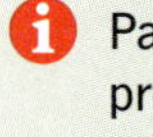

Patents are a form of intellectual property that grant the inventor exclusive commercial rights to their invention.

Patenting of biotechnology has raised concerns regarding the affordability of the technology. If companies control the rights to the genome of genetically modified (GM) crops, they also control the prices of the seeds. Farmers that plant GM crops need to spend money each year to buy GM seeds. For example, Bt cotton is a transgenic crop that has been modified to contain genes from the soil bacterium, *Bacillus thuringiensis*. Expression of these genes produces resistance to common crop pests such as certain caterpillars. Bt cotton seeds lose their efficacy after one generation and the farmer is therefore forced to purchase new seeds each year, limiting the farmer's income.

Patenting of GM crops also raises legal concerns for farmers. Cross-pollination between GM and non-GM crops occurs naturally as pollen and seeds are carried by the wind or animals. If a farmer's GM crop is 'contaminated' with a non-GM crop, the company that own the patent for the GM crop has legal grounds to sue the farmer for breaching commercial agreements. The opposite can also happen when organic crops are contaminated with GM crops and can no longer be certified as organic. For example, a Western Australian farmer lost his organic certification after GM canola crops were planted at a neighbouring farm and cross-pollinated with the organic GM canola. As a result the organic farmer sued the GM canola farmer for losses of over $85 000.

New technologies are not equally accessible to all communities because countries vary in terms of needs, priorities, legislation and economic and social stability. This raises the question of whether biotechnology creates **social inequality**. Many of these technologies are expensive but also have the potential to bring about great benefits. For example, GM crops that require less insecticides allow money to be channelled into other needs such as health and education. This has caused debate around who should be able to benefit from these technologies and if limiting access through patents and pricing brings about greater social inequality. These concerns do not appear to have limited the uptake of GM technologies, and many developing countries have embraced GM food and fibre crops while some developed countries oppose their introduction.

Privacy

The ability to choose which personal information is made available to others, and when, is becoming increasingly difficult in today's society. Privacy is a contentious issue in biotechnology, with applications such as **DNA profiling** requiring the storage of people's genetic information in databases. In NSW, DNA samples (Figure 8.2.4) cannot be obtained from a person unless they give their permission. They can, however, be ordered to do so if there is evidence that they may have committed a crime currently under investigation. These DNA samples must be destroyed if the person is not guilty or is not charged. However, in some countries, the DNA may be kept for up to 10 years. This has enabled the identification of criminals who have committed crimes in unsolved cases that occurred before DNA profiling technology was developed. It has also resulted in the exoneration of people who have been wrongly accused.

Collecting blood samples from newborn babies for **genetic screening** for abnormalities is another common reason for storing people's genetic information. In NSW, genetic information is retained for 15 years after the individual reaches 18 years old. However, this varies from state to state, with Victoria retaining the information indefinitely. Access to this information is also a contentious issue; parents may choose to give permission for the samples to be used in research projects for which individual identification is removed. Access to the samples may be legally protected by legislation in some states because they contain personal, identifiable, genetic information. However, the genetic information in the samples is the property of the child who cannot, at the time of testing, give consent for it to be used.

Many people argue that there is a potential for misuse of this stored genetic information in the future. For example, unless there is legislation to prevent discrimination, an insurance company may refuse life insurance to a person carrying disease-associated alleles detected by genetic screening. The legislation relating to DNA data varies across states and countries. Laws may change in response to new understanding of the technology and improved techniques, experience in particular legal cases, or changes in political or social attitudes.

BIOFILE EU

Giants of the seed world

An agricultural multinational company in the United States that patented the production and sale of its seeds, prevented the farmers who bought seeds from replanting them the following season, forcing the famers to buy new seeds each planting season (Figure 8.2.3). The multinational company sued hundreds of small farmers who had replanted seeds in a bid to protect its assets. The company won over $23 million dollars in damages and financially devastated many farms.

FIGURE 8.2.3 Farmers who replanted patented GM crops were sued by biotechnology companies. Farmers are forced to buy new seeds each planting season.

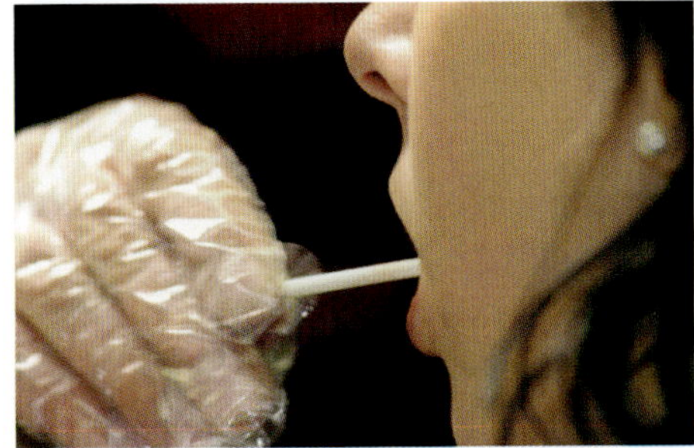

FIGURE 8.2.4 Privacy of genetic information has become a debated issue in society with data, such as DNA samples, being stored for purposes including criminal evidence and genetic screening.

Health

While the products and services of biotechnology become more accessible, some concerns have been raised about the safety of these new technologies for humans. GM foods have been introduced to modern diets and are readily available in grocery stores (Figure 8.2.5a) however, some groups have raised concerns about the safety of these foods for human consumption. While some claim that GM foods cause toxic effects on organs and have the potential to change consumers' DNA, most studies have found no evidence of this. Research from independent groups worldwide has concluded that GM foods are safe for human consumption, with no adverse effects on health, reproduction or DNA evident. However, consumer groups are lobbying to have all GM foods labelled so people can choose whether or not they consume them. In Australia, manufacturers are required by law to label foods that contain GM ingredients (Figure 8.2.5b); however, these laws do not apply to foods prepared by restaurants and other food vendors.

Concerns for human health have also been raised in response to **gene therapy**. The death of a participant in an early clinical trial prevented further gene therapy trials for some time while scientists investigated the cause. It was found that the participant had experienced an unexpected adverse immune reaction to the viral vector that was used to transport the therapeutic gene. In another gene therapy trial, the target gene was inserted into a vital gene, disrupting its function and causing additional health problems in the patient. There have also been cases of contamination with an infectious virus when viral vectors have been used. As a result of the rise in concerns over the safety of biotechnology, risk assessments and evaluations are conducted to determine any potential threats to the health and safety of the community. The risks of all known and future biotechnologies are evaluated with regard to their durability, exact purpose, social relevance, safety and security. In Australia, biotechnology is regulated by the following governing bodies:

- Therapeutic Goods Administration (TGA)—medicines, medical devices, blood and tissues
- Food Standards Australia New Zealand—foods
- Office of the Gene Technology Regulator (OGTR)—gene technologies
- Australian Pesticides and Veterinary Medicines Authority—chemicals used in agriculture and veterinary medicine
- National Chemicals Notification and Assessment Scheme—industrial chemicals
- Department of Agriculture (Biosecurity)—imported GM products.

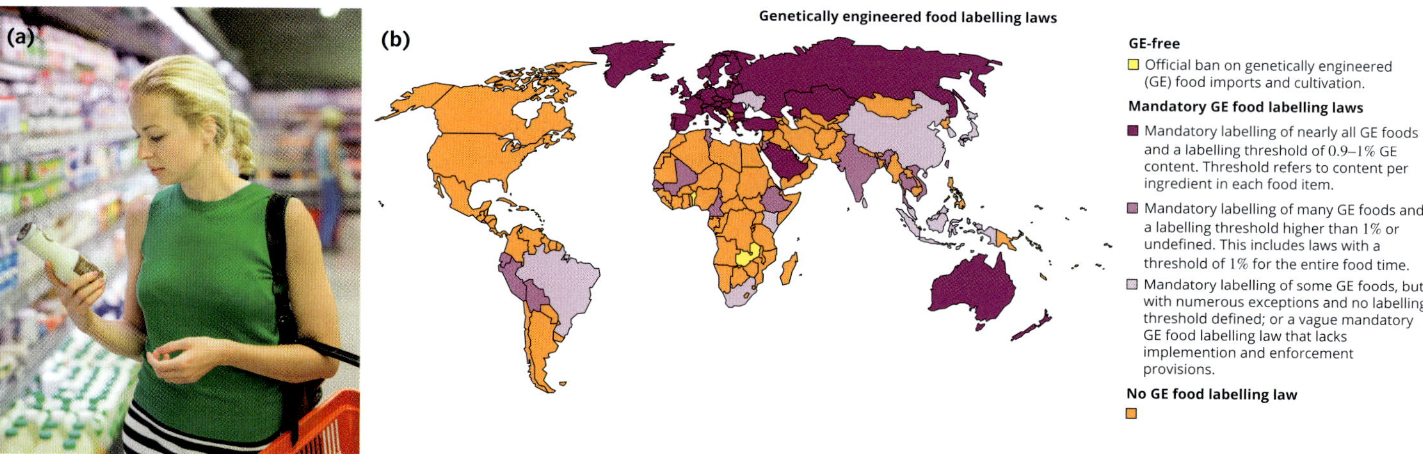

FIGURE 8.2.5 (a) Concerns over the health and safety of GM foods has led to stricter guidelines on food labelling so consumers can make informed choices. (b) Australia has some of the toughest food labelling laws. Purple—mandatory Genetic Engineering (GE) food labelling laws. Dark purple shows the the strictest countries, medium and light purple show less strict food labelling laws. Yellow—GE-free. Orange—no GE food labelling laws.

Society's view of biotechnology

The rapid advancement of biotechnology represents one of the great technological revolutions of modern times. It has introduced unprecedented changes in how we deal with the challenges faced in human health, agriculture and the environment. The public's perception of biotechnology is complex, and there is no clear correlation between people's knowledge of biotechnology and their acceptance of new technologies. In general, society is more accepting of biotechnology's application to the human health sector, and more opposed to its application in agriculture and industry. For example, the development of genetic testing for the *BRCA1* and *BRCA2* genes that can be used to assess a woman's genetic predisposition to breast and ovarian cancers has been embraced by society, while the consumption of produce from GM crops has been strongly opposed. Biotechnology is now also impacting society's expectations and values. For example, gene therapy and genetic screening have changed people's expectations of medical intervention. However, like many other procedures, genetic testing can have errors. Parents may sue individual doctors or companies for the cost of raising their child because the doctors failed to identify that the child had a genetic defect in utero.

ETHICAL USE OF BIOTECHNOLOGY

New technologies have had a profound impact on society, and biotechnology can be celebrated for the many benefits it has brought to humanity. However, biotechnology tends to produce products and technologies that lead to polarised views. As a result, ethical, philosophical and religious issues are also part of the biotechnology debate.

Philosophical, cultural and religious views

The philosophy, culture and religion of many in society shape what they know of the world. Biology, religion and philosophy all attempt to explain the world, and life on Earth, each in different ways. While biotechnology has the potential to bring about benefit, many people believe biotechnology practices and products overstep ethical boundaries. Many biotechnological developments, such as **cloning** and gene therapy, are raising issues and generating questions that traditionally fall within the domain of religion and philosophy, such as the definition of human life, DNA as central to the individual identity and the control of nature. Many cultures and religions have opposed biotechnologies that are seen to intervene with nature, with the aim of maintaining respect for all life and protecting what many people view as central to the human species and identity.

The religious, cultural and philosophical views of society, and an individual, become particularly relevant when dealing with the areas of genetic screening and reproductive decisions. For example, if a fetus is found to have a genetic abnormality, the parent(s) must decide if they want to continue the pregnancy or terminate it. The difficult decision requires balancing the loss of the pregnancy with the knowledge or prediction of the suffering and challenges for the parents and child if it were born. The decision will be informed by the severity of the genetic disorder and the capacity of the family to accommodate any special needs of the child. Genetic counsellors (Figure 8.2.7) may assist by providing information on the disorder, discuss issues related to passing on the genetic defect to other potential children and help the parents come to a decision consistent with their values. Genetic abnormalities can also be identified via **preimplantation genetic diagnosis**, which identifies genetic abnormalities before the embryo is implanted. At this stage, the embryo is a collection of undifferentiated cells. Many people feel that choosing not to implant an affected embryo is a choice that reduces potential suffering for both the potential child and its parents. Others believe that at fertilisation the embryo is uniquely human and has the same right as any other child. Some argue that removing embryos that are less than perfect will bring about a society that lacks compassion and acceptance of difference.

BIOFILE EU

Couple sues Royal Children's Hospital

In 2015 a couple sued a Melbourne hospital because it had failed to identify that their first child had a condition called Fragile X syndrome before they had their second child. Fragile X syndrome is characterised by behavioural, emotional, developmental and physical features such as anxiety, intellectual disabilities and hyperflexible joints.

The hospital performed genetic testing on the first child but did not test for Fragile X, even though the child displayed characteristics for this. Following the birth of their second child, who also displayed these characteristics, it was discovered that both children had the genetic disorder. Both children would need care and medical treatment for the rest of their lives.

The couple is also considering bringing legal action against a relative who knew they were a carrier of the condition but did not tell the rest of their family. The couple said that if they had known they were carriers for this condition they would not have chosen to have children, or would have chosen to use IVF to select unaffected embryos or to have used donor eggs.

The mutation responsible for Fragile X syndrome is a duplication of 3 bases, CGG, more than 200 times in the *FMR1* gene on the X chromosome (Figure 8.2.6).

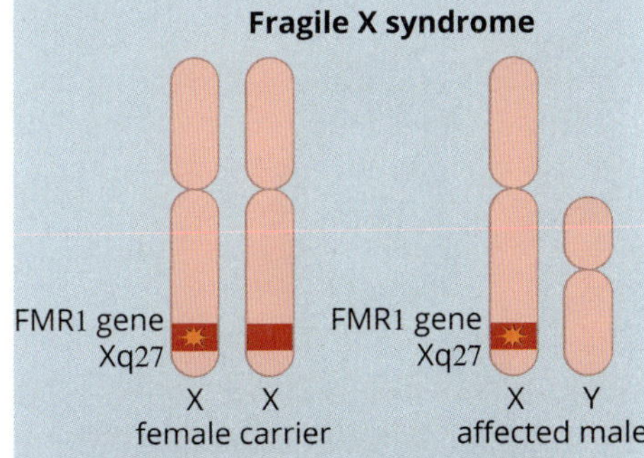

FIGURE 8.2.6 Fragile X is a sex-linked mutation causing a number of health and behavioural issues. As females carry two X chromosomes they are less often affected than males who only carry one X chromosome.

FIGURE 8.2.7 Trained counsellors help patients make decisions about medical treatments consistent with their religious, cultural and moral values.

Religious, philosophical and cultural views also influence a person's view of transgenic animals and genetically modified foods. For example, some religious and cultural traditions have dietary practices that restrict the use of transgenic animals for food. This is because while a transgenic animal may appear to be of one species, it may contain some genetic elements of a species that members of the religion choose not to eat. Many cultures also place high value on all life, which raises animal rights issues in regard to biotechnology.

Medical intervention and consent

With technologies such as cloning and genetic screening, it is easy to see how relevant biotechnology becomes to the field of medicine. However, with the use of this new technology come ethical issues such as the values of dignity, bodily integrity and independent choice. Regulations are in place in Australia to ensure that patients must give informed consent for all forms of medical and genetic testing, screening and therapeutic treatment. Informed consent for prenatal genetic testing of a fetus, for example, would include medical staff informing the pregnant woman whether the disorder being tested for can be prevented or treated, whether she will be faced with a decision to terminate the pregnancy and any potential risk of testing to her or the fetus. In Australia, strict regulations also govern medical research. Guidelines are in place to instruct researchers in the selection of clinical trial participants, the handling and storage of biological material, and the storage, analysis and reporting of data.

BIOFILE EU

Henrietta Lacks

In 1951, Henrietta Lacks (Figure 8.2.8a)—an African-American tobacco farmer—was diagnosed at John Hopkins Hospital with cervical cancer. During her treatment, doctors removed two cervical tissue samples without her knowledge. Lacks died in 1951 aged 31 years old. The cells taken from the tumour were sent to researcher, Dr George Otto Gey. Dr Gey was surprised by the high quality and durability of the cells and created a cell line, naming it HeLa (after Henrietta Lacks). The HeLa strain (Figure 8.2.8b) revolutionised medical research and modern medicine, helping to develop the polio vaccine, as well as cloning, gene mapping and in vitro fertilisation technologies.

Since the cloning of the HeLa cells, tens of thousands of patents involving the cells have been bought and sold for billions of dollars. The Lacks family only learnt of the HeLa cells in the 1970s, and have not benefited in any way from the contribution of the HeLa cells to the medical industry, with the Lacks family unable to even afford health insurance. In 2010 Johns Hopkins Hospital publicly admitted that the cells were taken from Henrietta Lacks without her knowledge or consent and acknowledged the contribution the HeLa cells had made to medical research. In 2017 a film based on the book 'The immortal life of Henrietta Lacks' was released detailing her life.

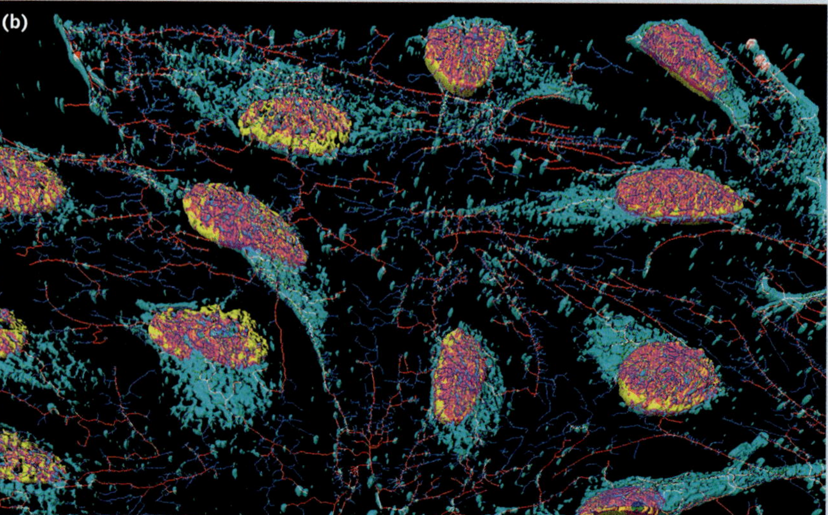

FIGURE 8.2.8 (a) Henrietta Lacks had two cell samples taken without her knowledge during cancer treatment. These cells went on to become the HeLa strain of cells. (b) HeLa cells have revolutionised modern medicine leading to developments such as the polio vaccine.

Legal implications

Legislation in Australia tightly regulates biotechnology, with laws to prohibit cloning humans, as well as regulating research conducted on human embryos. However, legal 'grey areas' can arise such as the legal impacts of storing genetic information. For example, if a four-year-old child tests positive for a genetic disorder that will affect them as an adult, issues regarding who should have legal rights to access this information arises (e.g. future employers, insurers and financial institutions). Another issue is whether medical staff have the legal right to inform the child's blood relatives that they tested positive for a genetic disorder if the condition is likely to be heritable.

Legislation in Australia regulates the use of biotechnology.

DNA profiling raises significant legal implications as it is regularly used in forensics to help personnel in the legal system determine a person's guilt or innocence of a crime (Figure 8.2.9). These days it is unlikely that a person will be incorrectly identified by DNA profiling if the procedure is carried out accurately. However, there is the possibility of incorrectly identifying an individual through DNA profiling. For example, foreign DNA may contaminate a sample at a crime scene or at the laboratory where the sample is tested. Failure to properly clean equipment could cause a sample from one suspect to contaminate a sample for another test. Legal implications from these potential errors include the wrongful conviction of suspects of crimes. Further, storing DNA after a person has served their sentence for a crime may be seen as unethical. Others are in favour of the creation of a 'bank' of DNA samples, provided by everyone in the community, which could be used to solve crimes and perhaps trace the remains of unidentified missing persons. Opponents of a DNA bank argue that there would be potential for these samples to be stolen or used unethically.

FIGURE 8.2.9 DNA profiling is often used in forensic evidence during criminal trials to help determine the guilt or innocence of an accused offender.

BIOLOGY IN ACTION 

Freeing the innocent

Greg Hampikian, a biology and criminal justice expert at Boise State University, founded the Idaho Innocence Project. Hampikian works with police and defence lawyers around the world to free innocent people wrongly convicted of crimes by using new DNA profiling technology or exposing errors with existing technology. One such case Hampikian worked on involved an error in police DNA sampling (Figure 8.2.10) that caused two innocent people to be jailed. Hampikian reviewed the DNA sampling procedures and found many errors, including one where an item of clothing containing a DNA sample was not collected until 46 days after the crime, and was passed around to several police investigators before being placed in a different position at the crime scene for photographing, allowing ample time for contamination of the sample. Hampikian also found samples from the same crime scene that were not tested by forensics teams because the quantity of the sample was smaller than what the FBI deem 'valid for testing'. In fact, as Hampikian discovered, the samples were sufficient in quantity and quality to provide valid DNA evidence.

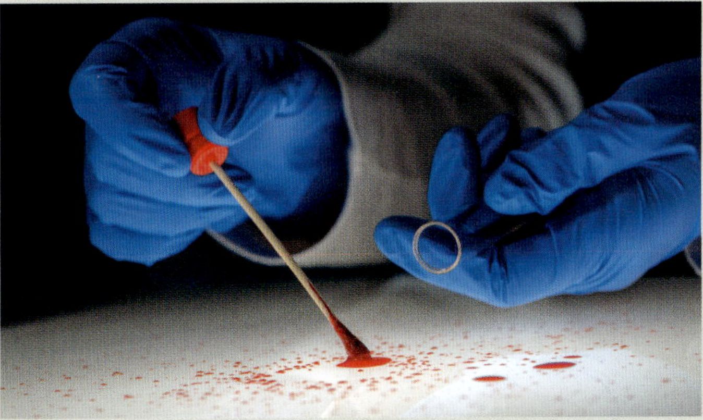
FIGURE 8.2.10 The Idaho Innocence Project uses experts in biology and criminal justice to analyse potential flaws in evidence, such as DNA sampling, to appeal cases.

Animal welfare

FIGURE 8.2.11 The use of transgenic animals such as pigs is opposed by many people who consider it immoral and in violation of animal rights.

Biotechnology has improved animal welfare in many ways. This includes being able to use bacteria for therapeutic recombinant proteins rather than animal products, the production of synthetic enzymes, which were once taken from livestock as well as selective breeding to remove characteristics which can cause harm to animals (e.g. cattle bred without horns to avoid fight wounds).

Many applications of biotechnology are also harmful to animal welfare. Genetically modified animals may have characteristics that are beneficial to humans and agricultural productivity but negatively impact the animal's health. For example, transgenic pigs (Figure 8.2.11) often grow very quickly, detrimentally affecting their heart and joints and causing pain when they are overactive.

While animals remain a source of products for human use many believe the production of transgenic animals violates the fundamental rights of an animal, manipulating them for human purpose, rather than respecting their inherent value.

BIOFILE CCT EU

Animal-based research

More than six million animals are used in animal-based research in Australia and New Zealand each year. While many think that animal-based research pertains simply to medical research, animal test subjects are used in many fields including agriculture, chemical and pharmaceutical industries, teaching and medicine. Animals used are not limited to rats and mice, but include rabbits (Figure 8.2.12), cats, dogs, pigs and even endangered primates. Animals Australia and other not-for-profit charities are lobbying for alternatives to animal-based research to be used in the scientific community. Alternatives may include the use of cell cultures, computer programs and safer human clinical trials to achieve the same scientific research goals.

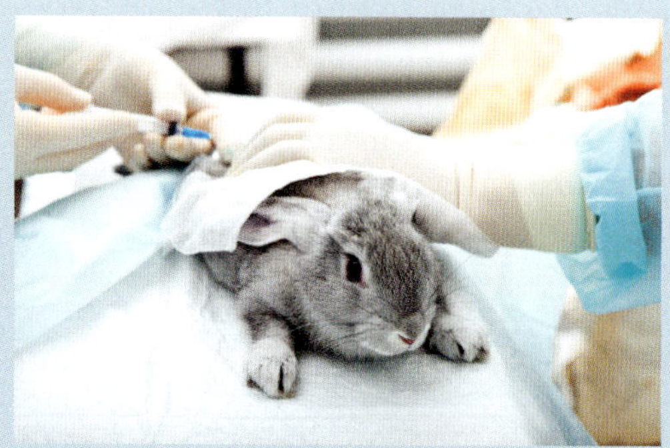

FIGURE 8.2.12 The ethics of animal-based research has been called into question with many groups lobbying for the use of alternatives to live animals.

SKILLBUILDER CCT EU PSC

Evaluating bioethical issues

As new biotechnologies are developed, ethical dilemmas are often faced by those in the scientific community and society. Each individual deals with ethical dilemmas differently; however a framework for evaluating these bioethical issues can be useful.

Research

To understand new biotechnologies, it is important to analyse the facts of the technology and understand what research has already been conducted. Online scientific journal databases should be consulted for up-to-date, accurate information.

Principles of bioethics

The following principles can be applied to each new technology to better understand how it may cause ethical, legal, societal and moral dilemmas.

- Respect for autonomy:
  - Does the technology interfere with an organism's free will?
  - Does the technology require informed consent from the individual before being used?
- Risk of harm:
  - Could the technology cause harm to any living organism?
  - Could someone be harmed due to negligence when using the technology?
- Benefits:
  - What is the benefit of the new technology?
- Justice:
  - Is the technology fair to all groups in society?
  - Does everyone have equal access to the technology?

Balance

While it is impractical to presume that everyone will benefit equally from all new technologies, biotechnology should strive to provide more benefits than disadvantages as well as adhere to the principle of 'do no harm'.

THE POTENTIAL BENEFITS OF BIOTECHNOLOGY RESEARCH TO SOCIETY

While biotechnology prompts many ethical, moral and legal discussions, it has great potential to benefit society. Biotechnology research can be applied to fields including medicine and agriculture, and may include studies on gene expression, natural genetic variation and other biological processes (Figure 8.2.13). In Australia, biotechnology research is undertaken by scientific institutions like the CSIRO and universities. Research institutions can also be affiliated with hospitals, such as the Westmead Institute for Medical Research. Research is supported by funding from various sources including government agencies, not-for-profit organisations and private donors.

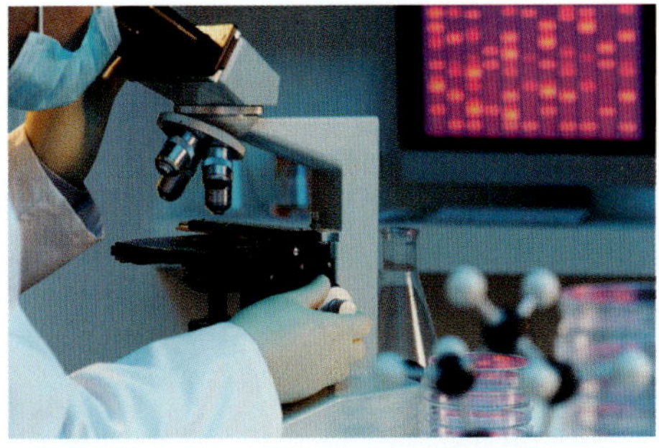

FIGURE 8.2.13 Biotechnology research has the potential to bring many benefits to society including advances in the medical, agricultural and environmental fields.

Pioneering medicine

Potential advancements in the medical field as a result of biotechnology research include improved diagnosis of disease and predisposition to medical conditions, designer drugs individualised to match a patient's specific genetic profile, new vaccines and new treatments for chronic diseases. For example, current research is looking to develop a technique to switch off a gene in muscle cells that causes a certain form of muscular dystrophy. These medical advancements have benefits for the individual patient and the wider community. Preventative medicine and early intervention help alleviate pressures on the medical system and improve the economy as people are better able to maintain health and remain employed.

Pharmacogenomics is the study of how genes affect a person's response to drugs. Traditional pharmacology has used a 'one size fits all' approach to medicine, leaving patients at risk of adverse side effects ranging from minor to severe to fatal. The mapping of the human genome allowed researchers to identify how a person's genes can affect their response to certain drugs. This has the potential to develop effective and safe medications tailored to an individual's specific needs. Pharmacogenomics holds the potential to create safe and more effective treatments for cancer, HIV/AIDS, asthma, epilepsy, depression, cardiovascular disease and Alzheimer's disease.

Gene therapy continues to be a field transforming into the future. The success of new technologies such as vector engineering, CRISPR (Clustered Regularly Interspaced Short Palindromic Repeats) technology, and the discovery of new structures such as microRNA hold incredible promise for new clinical trials. This sophisticated gene technology has the potential to correct inheritable (germline) genetic disorders, meaning families with genetic defects could have access to targeted treatment. These advances in gene technology may also provide improved treatments for diseases caused by somatic mutations. For example, viral vectors that directly deliver dopamine-synthesising DNA are being trialled for the treatment of Parkinson's disease, a condition in which the brain cannot produce enough dopamine for normal function.

Nanomedicine is another emerging field of medical biotechnology that involves the construction of technologies on a nanoscale (1 nanometre = one-billionth of a metre). The use of nanoparticles allows new technologies to be relevant in size to biological structures such as antibodies, which are approximately 10 nanoparticles in size. Nanomedicine will allow targeted interaction (Figure 8.2.14) and manipulation of biological structures to improve health care. For example, nanoparticles are being trialled to deliver anti-cancer drugs directly to cancer cells without damaging healthy cells. Other applications of nanomedicine include the use of nanoparticles for biological devices, biosensors and biological machines.

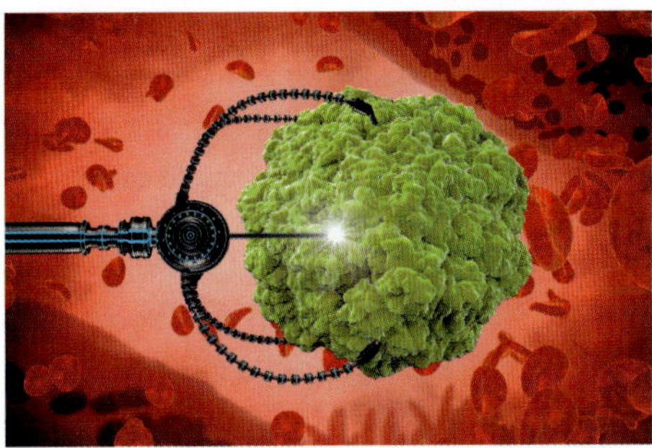

FIGURE 8.2.14 An artist's impression of medical nanotechnology—a nanorobot interacting with a bacterium.

BIOFILE ICT

Nanomedicine for disease detection

Researchers are successfully developing nanomachines made from compartments of DNA enzyme molecules and substrates. The tiny disease detectors are being trialled to locate specific microRNA sequences found in breast cancer cells (Figure 8.2.15). When the nanomachine comes into contact with the target cells, a DNA motor is turned on producing fluorescence. This allows fast detection of diseases before they progress.

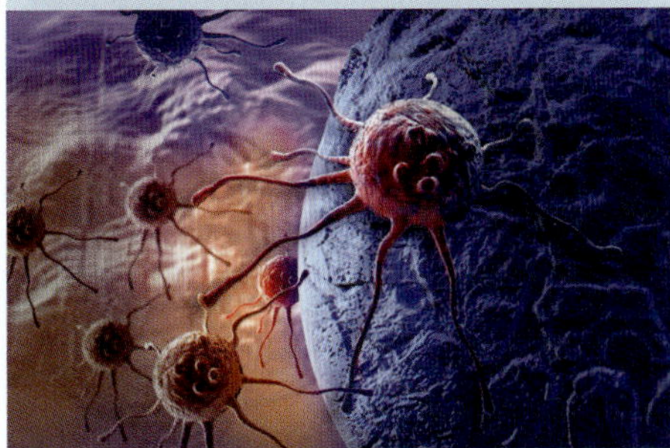

FIGURE 8.2.15 Cancer cells are being targeted by nanomachines to enable earlier diagnosis

BIOLOGY IN ACTION DD ICT

Three-dimensional printed prosthetics

Biotechnology is also moving us towards 'the bionic age', in which technologies such as mind-controlled bionic limbs are becoming a reality. Such applications are a clear example of the interdisciplinary nature of biotechnology—incorporating elements of biology with mechatronics, chemistry and exercise science. Mind-controlled bionic limbs allow prosthetics, such as artificial limbs, made out of synthetic materials to be integrated with body tissues including the nervous system. This would allow near-normal movement and functionality for recipients of the prosthetics.

While mind-controlled bionic limbs remain costly and require expertise to produce and fit, three-dimensional (3D) printing is making medical biotechnology accessible to many in the poorest of countries where prosthetics are in demand due to high rates of amputation as a result of disease, violence and accidents. Industries making automobiles and clothing use 3D printing and it is frequently used in medicine and dentistry for routine products such as new tooth crowns. The use of 3D printing for prosthetics (Figure 8.2.16) is cost-effective and allows prosthetics to be easily produced and fitted by almost anyone, anywhere. The prosthetics are fully-customised for the wearer and allow motility and function to amputees. Even in more developed countries, 3D printing offers advantages for paediatric medicine. With children outgrowing expensive prosthetics quickly, 3D printing offers temporary prosthetics, which can be affordably replaced as needed.

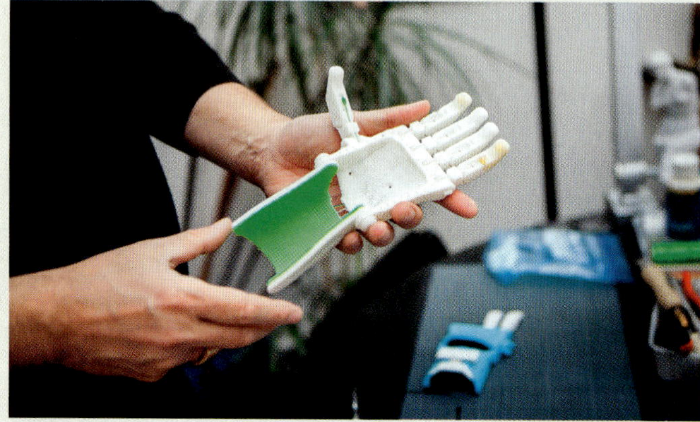

FIGURE 8.2.16 A hand prosthesis produced by a 3D medical printer promises better access to medical prosthetics.

BIOLOGY IN ACTION CCT EU

Plant a vaccine

Research is being conducted to produce edible vaccines in plants such as bananas (Figure 8.2.17) and potatoes. The vaccine antigens are protected by the cell wall of the plant cells and, once ingested, are released in the intestines and absorbed into the circulatory system. The plant vaccines would enable higher protection by stimulating the immune system and would provide a low-cost alternative to regular vaccines. Additionally, as they do not require refrigeration, they would be ideal for medicine in rural communities and less developed countries where access to electricity and suitable transport is not always reliable.

FIGURE 8.2.17 Genetic research is trialling the use of bananas and other edible plants to deliver vaccine antigens into patients through ingestion.

Food production

Biotechnology research also holds potential benefits for society through developments in agriculture and environment. Research is continually developing GM crops, making them more disease and pest resistant and tolerant to a range of environmental conditions. For farmers, this means increased productivity and lower input of labour and costs. This technology has the potential to benefit society by producing more nutritious food, grown with fewer chemicals.

Biotechnology can be used to improve the yield and nutritional value of food plants.

Feeding the hungry

In 2017 the global human population reached 7.6 billion and this number is expected to climb to 9.5 billion by 2050. With this steep rise in the population the demand for food also continues to rise. Along with the escalating demand for food is the ongoing trend of food distribution inequality, with world hunger and food insecurity a recurring problem. Malnutrition and poor diet is the leading driver of disease in the world, with malnutrition most prevalent in Asia and Africa. To simply feed the projected global population by 2050 will require a 70% increase in food production.

Biotechnologists are hoping to meet the demand for increased food sources, as well as improve food accessibility for the world's poorest people. However, biotechnology in agriculture has been a debated topic in society. GM crops may have adverse effects on non-target species and soil ecosystems, reduce biodiversity and potentially transfer genes to wild species creating 'superweeds'. Further, they may exacerbate social inequality and financial loss. However many biotechnologists believe that these issues can be mitigated, and that the future of genetic research and technology to improve the global food supply remains optimistic.

GM crop technology is focused on producing more yield on less land, with benefits for society and the environment. Further, new crops with enhanced vitamin and nutrient levels are aiming to lower malnutrition levels globally.

BIOFILE EU S

Genetically modified salmon

Recently the USA government cleared the way for genetically modified Atlantic salmon to be used for human consumption. A gene from another salmon species, along with a promoter sequence from a pout fish. means that the transgenic salmon eat all through the year, not only when the water temperature is warm. This increases the growth rates of these fish dramatically and means they are ready for harvest much sooner than non-modified Atlantic salmon (Figure 8.2.19). The eggs of the GM salmon are treated to create infertile adult fish (99% of the adults are reported to be sterile), thus reducing the chances of interbreeding with wild salmon if they escape from their pens. This will be the first genetically modified animal of any type to be cleared for human consumption in the USA.

FIGURE 8.2.19 Transgenic salmon have an increased growth rate compared to non-modified salmon.

BIOLOGY IN ACTION 

Banning bovine belching

Cows (Figure 8.2.18) are herbivores with a four-chamber stomach and special gut microbes that help breakdown plant matter. The gut microbes also produce methane, which is released by cow burps or flatulence. With over one billion cows on Earth, the amount of methane they produce accounts for over 14% of all greenhouse gases produced each year. Like carbon dioxide, methane traps solar heat, leading to global warming and climate change.

Australian scientists have discovered that an intake of red seaweed (just 2% of a cow's food intake) can reduce the amount of methane produced by 80 to 90%. In the USA, researchers discovered that by adding 3-nitrooxypropanol into cow feed, enzymes prevented the gut bacteria emitting methane by up to 30%. The most promising research appears to be conducted by Danish scientists who are researching the DNA of grass (cows' staple food) to hopefully engineer a 'super-grass' which holds all the nutrients the cows require while yielding less gas.

FIGURE 8.2.18 Researchers are discovering ways to alter cow feed and reduce bovine production of methane, a gas which when released through cow burping or flatulence adds to greenhouse gas accumulation in the atmosphere.

Using secondary sources

Secondary sources describe or explain data collected by others (primary data). Secondary sources may include textbooks, websites or scientific journals. When searching for secondary sources, the reliability and accuracy of information is important and some secondary sources are more reliable than others. Some sources of reliable scientific information are:

- scientific journal or magazine articles (e.g. *Journal of Biotechnology*)
- textbooks
- government or organisation websites (these will have '.gov' or '.org' in their web address).

The above secondary sources are generally more reliable than commercial or special interest websites, which may contain biased information. The date of the secondary source should always be checked to ensure data is current. For example, a scientific journal article from 2016 providing data on breast cancer research would contain more up to date information than an article from 1999.

Triangulating data

Secondary sources should always be compared to other sources to ensure accurate and reliable information. Secondary data should be compared to at least two other sources so that the information has been verified in a total of three sources (representing the three points on a triangle).

Referencing secondary sources

There are many different referencing styles (e.g. APA and Harvard referencing) and it is important to apply the style consistently throughout the reference section of your report. Secondary sources are referenced in reports and other bodies of work to validate statements made, and so readers can find more information when required.

There are two types of referencing.

- In-text citations:

  Referencing in the main wording of the report, for example: Scientists believe they may have found a cure for white-nose syndrome that is wiping out bats in the United States (Lee 2015).

  This shows to the reader that the statement is validated by information found in a secondary source published by the author, Lee, in the year 2015.

- End of text referencing:

  At the end of an essay or other body of work, a reference list or bibliography of all secondary sources used to research the work must be listed in alphabetical order. Each referencing system has strict rules about how to format the reference. For example, to reference the following book:

  Book title: *The Diversity of Life*

  Author: Edward Osborne Wilson

  Date of publication: 1992

  Publisher: Harvard University Press, Cambridge.

  The reference using the APA referencing style would be:

  Wilson E.O. (1992). *The Diversity of Life*. Cambridge: Harvard University Press.

8.2 Review

SUMMARY

- The social issues surrounding biotechnology include cost and accessibility, privacy, health and safety.
- Due to factors such as economy and social status, biotechnology may not be equally accessible to all groups in society.
- Issues of privacy exist in regard to genetic information being stored in databases.
- Health and safety concerns exist for some biotechnology applications such as GM foods.
- People's religious, cultural and philosophical perspectives impact their view of new genetic technologies.
- Informed consent is required for all medical treatments and clinical research in Australia.
- Some biotechnology applications, such as DNA profiling, have legal implications.
- Some believe biotechnology, such as the use of transgenic organisms, violates animal rights and welfare.
- Biotechnology research has the potential to make revolutionary advances in the fields of medicine, agriculture and the environment, with applications such as new vaccines and pest-resistant crops.
- Pharmacogenomics, nanotechnology and gene therapy are providing potential for improved prosthetics, disease detection, treatment of chronic illness and equal access to health care.
- GM crop technology aims to increase the nutritional value of grains, increase vitamins and micronutrients in seeds, eliminate allergens and improve crop productivity.

KEY QUESTIONS

1. Describe the positive social impacts of the GM crop Golden Rice.
2. **a** What are patents?
   b Provide an example of how patents may affect access to new biotechnology.
3. Explain why a religion or culture may oppose the use of transgenic animals.
4. Explain how DNA profiling can have legal implications.
5. Create a table, listing the advantages and disadvantages of biotechnology for animal welfare.
6. List some of the social and ethical issues surrounding GM crops.
7. Identify some potential benefits of biotechnology research in the medical field.
8. Explain how the use of technology on a nanoscale is an advantage for medicine and health care.

8.3 Biotechnology and biodiversity

Genetically transforming the wild

How do genetic techniques affect Earth's biodiversity?

COLLECT THIS...

- paper
- pen
- tablet or computer to access the internet

DO THIS...

1 Working in small groups or pairs, choose a hypothetical scenario from the following (each group should have a different scenario).
   - Peach (*Prunus persica*) trees naturally require approximately one month of temperatures below 10°C to produce fruit the following growing season. They are found in cool climates with moderate rainfall. Scientists have used genetic techniques to produce a peach tree that can grow in hot, dry climates.
   - Sardines (*Sardina pilchardus*) are small fish that travel in large schools. With demand for this fish species growing worldwide, scientists have genetically engineered a breed of sardine that grows as large as a medium shark (approximately 600 kg).
   - Poaching threatens the African elephant (*Loxodonta africana*) due to the demand for their ivory tusks. Scientists have discovered a genetic mutation that prevents elephants growing tusks. They have enabled this mutation to be transferred into many elephant genomes so the next generation of elephants will be tuskless.
   - With polar bears (*Ursus maritimus*) losing their Arctic habitat to human-driven pressures such as climate change and urban development, scientists have developed a genetic technique which would give polar bears genes that allow them to survive in hot, humid climates.
   - Scientists have discovered a genetic technique to rebuild the genome of the saber-tooth cat (*Smilidon* sp.), allowing it to be brought back from extinction.

2 Read the scenario carefully.

3 Draw a table with two columns—one titled 'Advantages' and one titled 'Disadvantages'.

4 As a group, take 10 minutes to discuss the possible advantages and disadvantages of this genetic technique. Think about how the change to this organism would affect individual organisms, other species, the environment, humans and society. You can use the internet to research ideas.

5 As a class, discuss the advantages and disadvantages of the genetic technique in each scenario.

RECORD THIS...

Describe how genetic techniques may have an impact on society and the environment.

Present the advantages and disadvantages of the genetic techniques from each scenario in a table.

REFLECT ON THIS...

How do genetic techniques affect Earth's biodiversity?

How might scientists decide which genetic techniques to use to ensure a balance between advantages and disadvantages?

Could certain groups in society oppose the use of these genetic techniques?

Biodiversity is the variety of life on Earth and is valued by many simply for its natural, cultural and aesthetic values. However, biodiversity also has many practical benefits. Biodiversity provides **ecosystem services** such as food and fuel production, shelter, air and water purification, soil generation and nutrient cycling. Biodiversity naturally pollinates plants, controls pests and disease, and stabilises climate. Further, biodiversity provides resilience to changes whether this be at a genetic, organism or ecological level.

LOSS OF BIODIVERSITY

The biotechnology feared to be posing one of the most significant threats to global biodiversity is the use of GM crops in agriculture. It is believed that as more farmers use GM crops, there will be fewer crop varieties grown, reducing the gene pool of some species. If disease or environmental change occurs, there could be widespread and catastrophic effects on food production. This is already an issue in modern agriculture where monocultures and commercial practices have reduced genetic variation. Genetic variation is not only essential for evolution, it is also important for biotechnology—if it is lost, so is the potential for new cultivars of plants and breeds of animals. There is a modern movement to revive heritage varieties of crops and breeds of livestock such as Berkshire pigs (Figure 8.3.1a) and heirloom tomatoes (Figure 8.3.1b). Maintaining the biodiversity of organisms used in agricultural practices ensures the health of future generations of crops and livestock and improves their ability to adapt to changing environmental conditions.

FIGURE 8.3.1 Movements to promote the use of heritage breeds of crops and livestock such as (a) Berkshire pigs and (b) heirloom tomatoes are important for the preservation of genetic variation in agriculture.

Cross-pollination between GM crops and non-GM crops can further exacerbate the risk to genetic variation; pollen distributed by wind and insects can transfer genes between neighbouring plots of the same species. Cross-pollination may happen in both directions, potentially impacting both types of crop. Uncontrollable pest plant species can then be a threat, with crops that have been modified for herbicide or insect resistance having the potential to breed with wild plants, producing a hybridised pest species that farmers may not be able to control. Plants do not usually pollinate different species, so this is only a potential concern for cross-pollination with wild relatives, such as between crop canola and wild canola. Uncontrollable growth would be a concern if there was a selective advantage to having the genes in a wild environment. A loss of biodiversity can also occur due to non-target effects of GM plants on beneficial and native organisms such as bees, beetles and moths. Some GM crops have also been known to affect soil ecosystems including microbial populations, decomposition rates and carbon and nitrogen levels.

Another potential threat to biodiversity is the risk of GM animals escaping and interbreeding or competing with natural populations with unforeseen consequences. This may be a potential concern if there are wild relatives of the GM animals and there is a selective advantage to the presence of the **transgene** (the gene transferred to the organism) in a wild population (e.g. if GM salmon escaped from aquaculture enclosures adjacent to the ocean and bred with wild salmon). Biotechnology also poses an indirect threat to biodiversity by increasing the value of certain species. For example, if a successful biotechnology that uses a wild species of plant is developed, the value of this plant may increase and put the species at risk of overexploitation.

BIOFILE 

Diversity vs devastation: Ireland's potato famine

The importance of biodiversity was demonstrated in devastating fashion in 1845 during Ireland's Great Famine, which left over one million people dead and forced the emigration of nearly two million more. While potatoes were a crop favoured throughout the Northern Hemisphere during the 19th century, the potato was a sole subsistence food for one-third of Ireland's population—particularly impoverished farmers who could grow the plant in poor soil and feed it to their family as well as livestock. The difference in Ireland's potato consumption compared to other countries was that Ireland depended on only one type of potato—the Irish Lumper (Figure 8.3.2a). In 1845 a plant pathogen known as *Phytophthora infestans* caused widespread loss of Irish Lumper crops, causing famine and further poverty in the country (Figure 8.3.2b).

FIGURE 8.3.2 (a) The Irish Lumper potato, the staple crop wiped out by the plant pathogen *Phytophthora infestans* in the 19th century, causing widespread famine in Ireland. (b) An artwork depicting the poverty of many peasants during Ireland's potato famine.

BIOFILE S

GM crops linked to collapse of bee colonies

While it has been hypothesised that GM crops will reduce global pesticide use in the agricultural industry, the use of pesticides has actually risen since the introduction of GMOs. Coinciding with this rise has been the sudden decline of bee colonies (Figure 8.3.3) since the 1990s, with numbers of some bee species so low that they have now been listed as endangered. Bees are key to ecosystem functioning and one-third of human food production depends on their pollination services. Recent studies have found that a group of pesticides known as neonicotinoids (NNIs) are negatively effecting bee populations. NNIs are largely used on GM crops and are produced by most multinational biotechnology corporations who produce and sell GM seeds. Pollen containing NNIs was found in treated crops that naturally attract bees, as well as non-target plants such as wildflowers. Research on honeybees, bumblebees and solitary bees found that NNIs had adverse effects on bee reproduction with fewer egg cells found in nests, and fewer queen bees in hives. Other effects include a decrease in colony size and shorter life spans.

FIGURE 8.3.3 Recent research has linked bee colony collapse to neonicotinoid pesticides used on GM crops.

CONSERVATION OF BIODIVERSITY

While there is concern that some biotechnologies are leading to a loss of biodiversity, many molecular methods and gene technologies are being used to conserve biodiversity. Today's ecosystems are faced with several key threats including habitat loss, climate change and invasive species. Molecular tools are now more easily accessible due to improved ease-of-use and cost-efficiency, and this, combined with a new generation of scientists who are more comfortable using biotechnology, has led to a range of promising new **conservation** methods.

Traditional conservation methods focus on **in situ** practices, that is, strategies focusing on maintaining organisms in their natural environment (Figure 8.3.4a). In situ methods are favoured for their ability to retain the wild, natural state of organisms, along with genetic variation. However, conservationists are facing increasing challenges of habitat loss and fragmentation in ecosystems. As a result, biologists are looking toward **ex situ** conservation methods (Figure 8.3.4b) that conserve organisms outside of their natural habitat (e.g. in zoos or seed banks). Ex situ conservation methods are being revolutionised by biotechnology and the new molecular techniques being offered.

FIGURE 8.3.4 (a) In situ conservation methods maintain the organism in its natural habitat, while (b) ex situ methods remove the organism from its natural habitat for conservation.

Studies that are in situ maintain the organism in its natural environment, such as when plants are studied in a forest.

Studies that are ex situ remove the organism from it natural habitat, such as when animals are captured from the wild and entered into captive breeding programs at zoos.

BIOLOGY IN ACTION 

Wildlife trafficking forensics

Conservationists are fighting an ongoing battle to prevent the trafficking of animals and animal parts on the global black market, and genetic technologies have an important role to play. DNA profiling on seized ivory (Figure 8.3.5) has been used to trace the tusks back to the geographic location where the elephants once roamed. The technique is proving highly successful in tracking and locating the source of the ivory. For example, after a 6.5 tonne shipment of ivory arrived in Singapore, DNA profiling revealed it had come from elephant populations in Zambia. As a result of this data, Zambia has enforced harsher sentences for wildlife traffickers and poachers in the region with the hopes of stopping the trade at its source.

FIGURE 8.3.5 Seized elephant tusks are just some of the trafficked wildlife parts that are being DNA profiled to trace poachers and smugglers.

Plant propagation

As discussed earlier in the chapter, with the human population continuing to grow, GM crops are being increasingly utilised to ensure future food security. Increasing crop productivity, especially in less developed countries, often leads to massive habitat destruction when land is cleared for agriculture. However, biotechnology offers avenues to increase the yields of crops, allowing more productivity on smaller parcels of land. This is particularly promising for biodiverse tropical regions, where land clearing for agriculture poses a significant threat to global biodiversity. GM crops can also be used to reintroduce characteristics from wild crops that have been lost through modern agricultural practices. Ex situ conservation methods are being used to help preserve genetic variation and endangered native plant species, with molecular methods being used to establish seed and gene banks, as well as plant micro-propagation and cloning.

BIOFILE CC S

Australia's botanic seed banks

The Australian National Botanic Gardens National Seed Bank is an ex situ conservation project with functions including research, propagation and seed supply. The seed bank holds over 5000 individual seed collections from more than 3000 plant taxonomic groups. The National Seed Bank works to protect the biodiversity of Australian native plants (Figure 8.3.6), especially Endangered species and those with a key ecological role. For example, *Sphagnum* moss species are crucial to the survival of the Critically Endangered southern corroboree frog (*Pseudophryne corroboree*) in the alpine bogs and fens.

FIGURE 8.3.6 Native plant species, such as the coastal wattle (*Acacia sophorae*), are researched and propagated at the Australian National Botanic Gardens National Seed Bank.

Monitoring genetic variation

FIGURE 8.3.7 The endangered northern quoll (*Dasyurus hallucatus*) is one species in Australian captive breeding programs benefiting from new genetic technologies.

With global biodiversity declining at unprecedented rates, conservation of wild animal species requires a targeted approach. Biotechnology provides tools which allow conservationists a wealth of genetic information to better utilise both ex situ and in situ conservation programs. For example, **next-generation sequencing (NGS)** is being used to determine the pedigree of individual organisms, allowing more informed choices of which individuals to use in breeding programs. The technology also allows biologists to monitor the genetic variation of captive populations, ensuring it is reflecting (or improving) the variation in wild populations. This technique is being used in the captive breeding of endangered northern quolls (*Dasyurus hallucatus*) (Figure 8.3.7) to provide greater options for reintroducing quolls to the wild, and to assist setting up a more robust captive population as an insurance policy in case the species becomes extinct in the wild.

Biotechnology can be used to monitor the genetic variation of populations and improve captive breeding programs.

DNA sequencing techniques and software are also allowing conservationists to study both functional and neutral genetic variation. Functional genetic variation can provide insight into the adaptive potential of organisms to new conditions such as the ability of captive-bred organisms to survive in the wild, the adaptability of wild organisms to environmental change or the susceptibility of populations to disease. For example, scientists are closely monitoring genes associated with immune response in the Tasmanian devil (*Sarcophilus harrisii*) to determine if there are individuals with natural immunity to devil facial tumour disease (DFTD). This knowledge could then help researchers to develop a vaccine.

BIOLOGY IN ACTION

De-extinction

New genetic technologies such as CRISPR and targeted gene editing are also creating potential for the conservation method termed 'de-extinction', a controversial concept of bringing extinct species back to life. While some studies have focused on a Jurassic Park-style method of bringing to life species from the geologic past such as the woolly mammoth (*Mammuthus primigenius*) (Figure 8.3.8), modern conservation strategies are hoping to focus on using well-preserved tissues of recently extinct species to give a second chance to keystone species and help restore the natural balance of at-risk ecosystems. Other applications of these genetic technologies would be to use genes from the extinct species to diversify inbred endangered populations. Many conservationists are using cryopreservation of gametes, embryos and tissues of threatened species for these future options.

With de-extinction conservation attempts, often scientists only have small or damaged fragments of DNA to work with. Biotechnology has allowed the development of genomic techniques, which can extract information from contaminated or degraded DNA samples. This technology has allowed DNA to be studied from the single leaf of a 140-year-old herbarium specimen and 100-year-old insects.

FIGURE 8.3.8 New genetic technologies are providing potential for the 'de-extinction' of species such as the woolly mammoth (*Mammuthus primigenius*).

Heterozygosity

Genetic variation in populations is measured in terms of **heterozygosity**. This is a measure of the variation (number of alleles) at a locus (region in a DNA sequence). Scientists use computer software to analyse data from a population and provide estimates of genetic variation.

High heterozygosity indicates that the population contains many different alleles (i.e. has high genetic variation), while low heterozygosity indicates that the population contains very few different alleles (i.e. has low genetic variation).

Low heterozygosity is a concern for the evolutionary potential and long-term viability of populations and may signal issues such as genetic bottlenecks and inbreeding. High heterozygosity is generally a positive sign for the evolutionary potential and stability of populations but it may also indicate a sub-division in the population. Sub-divided populations may require taxonomic assessment to determine if the populations are in fact separate sub-species or species.

BIOLOGY IN ACTION  

Facilitated adaptation

Researchers are currently studying an idea known as 'facilitated adaptation', which involves the genetic modification of plants and animals to deal with climate change (Figure 8.3.9). Introducing gene variants to organisms could allow the organisms to survive in climates different to their natural range and fill different ecological niches. The technique can be done in one of three ways:

1 animals from the threatened population are hybridised with individuals from the same species who are better adapted to the new conditions
2 scientists identify genes that make one population more suited to the new conditions, and then insert these genes directly into the less-suited populations or individuals of the same species
3 scientists take genes from the well-adapted species and insert them into the genome of a different species threatened by the new conditions.

The technique is seen as less challenging than moving whole populations or species to new geographical locations, and also lessens the threat of producing invasive species or unbalancing existing ecosystems. However, others see this technique as intervening in natural evolutionary processes, raising ethical and moral debates.

FIGURE 8.3.9 New genetic research and technologies have the potential to make possible the facilitated adaptation of endangered species such as the ringed seal (*Pusa hispida*) and other species that are severely threatened by climate change.

A SUSTAINABLE FUTURE

Biotechnology is helping to develop more efficient biofuels to decrease the carbon emissions produced from traditional fossil fuel-derived petroleum. Biofuels are renewable alternatives to the fossil fuels used for petroleum and fuel used in transport and industry. Australia has embraced the first-generation biofuels made from waste starch, sorghum, cooking oil and other plant and animal waste products. Biofuels have proven to reduce carbon emissions by 85%. Their sustainable nature provides fuel as well as economic security, and as it is waste products that are being used, it does not require more land and does not impact food security.

Current studies are looking at third generation biofuels, which use microalgae to produce an oil which can be refined to gasoline and even diesel fuel. The microalgae have the added benefit of being carbon dioxide consumers, meaning they decrease atmospheric carbon through the production process. Fourth generation biofuels go one step further, looking to develop an engineered biological process to create oil directly from carbon dioxide through the use of an embryonic cell culture. The future of biofuels is also looking to diversify its use and scale, with studies being conducted to produce biofuel jet fuel for military and commercial aviation (Figure 8.3.10).

FIGURE 8.3.10 A fighter jet is successfully flight tested using a 50/50 biofuel blend. Future military and commercial jet biofuel is being researched to decrease carbon emissions.

Biotechnology is also providing options to help clean up environmental pollution, through processes of phytoremediation and bioremediation. Phytoremediation is the use of transgenic plants to remove heavy metal pollution from contaminated soils. For example, mercury pollution, which is an ongoing problem has been found to be reducible by producing plants that accumulate the heavy metal in their tissues and are then harvested, allowing the waste to be extracted and disposed of safely. Bioremediation on the other hand uses microorganisms to destroy hazardous contaminants, or convert them to a safer form. Research is looking at new methods to achieve better contact between microorganisms and contaminants and better understand microbial processes to use bioremediation for previously untreatable contaminants.

8.3 Review

SUMMARY

- Biodiversity is essential for the functioning and stability of ecosystems.
- Biodiversity exists at a range of scales including genetic, species and ecosystem.
- Biotechnology in agriculture can cause biodiversity loss through favouring a small number of varieties and losing the characteristics of wild species.
- Genetically modified (GM) plants can adversely affect organisms such as bees as well as soil ecosystems.
- GM organisms pose a risk through breeding or cross-pollination, with the potential for invasive species to develop in the wild with the GM characteristics (e.g. herbicide resistance).
- A range of biodiversity conservation methods have become possible with biotechnology.
- The genetic variation of wild and captive populations can be measured and monitored using genetic technologies.
- 'De-extinction' techniques could be used in the future to return keystone species to at-risk ecosystems.
- Ethical questions have been raised regarding using genetic technologies to interfere with evolutionary processes.
- Biofuels have the potential to lower atmospheric carbon dioxide by providing cleaner energy sources.
- Bioremediation and phytoremediation have the potential to clean up environmental pollution such as mercury deposits in soil.

KEY QUESTIONS

1. Describe how genetic variation can benefit agricultural biotechnology.
2. Provide an example of the negative impacts of low genetic variation.
3. GM crops have been criticised for causing biodiversity loss. Identify some of the key reasons for this criticism.
4. Explain how the increased use of a species for biotechnology could cause biodiversity loss.
5. Identify two factors that have facilitated the integration of biotechnology in conservation practices.
6. Differentiate between in situ and ex situ conservation methods.
7. Biotechnology can help ecologists monitor the genetic variation in a population. Explain one way this data could be advantageous to an ecologist.
8. Name a genetic technique and explain how it is being used to conserve biodiversity.

Chapter review

08

KEY TERMS

artificial selection
biodiversity
biofortification
biotechnology
cloning
conservation
cross-breeding
DNA (deoxyribonucleic acid)
DNA profiling
ecosystem service
ex situ
fermentation
gene
genetically modified organism (GMO)
genetic screening
gene therapy
genome
heterozygosity
in situ
next generation sequencing (NGS)
patent
pathogen
pharmacogenomics
preimplantation genetic diagnosis
recombinant DNA technology
selective breeding
social inequality
transgene
transgenic

REVIEW QUESTIONS

1 Provide an example of how biotechnology has been used in past civilisations.

2 Describe an application of biotechnology which began primitively in ancient civilisations and has evolved in modern biotechnology.

3 Identify two molecular techniques used in modern biotechnology that were not present in ancient civilisations

4 a Describe the disadvantage of the current 'one-size fits all' approach to medications.
  b Explain how pharmacogenomics is aiming to improve this current state.

5 Describe an application of gene therapy in medicine.

6 Provide an example of a medical biotechnology that allows greater access to improved health care. In your answer explain its advantage.

7 Draw up a table to list a positive and a negative aspect of genetic screening. Include social and ethical issues.

8 Huntington's disease is a genetic condition. It results in significant and continuing neurological decline leading to severe problems with cognition, emotional control and movement. Many sufferers develop severe depression. This condition is caused by a mutation in a single gene that codes for a protein called huntingtin. Most people that have the mutated allele do not show symptoms until they are in their 40s or later.
  There is no treatment for Huntington's disease. It will result in serious mental deterioration. It is ultimately fatal. It is a dominant condition so if an individual has one mutated allele it is certain that they will develop Huntington's disease. If a child tests positive for the Huntington's disease allele, then one of their parents must also have the mutated allele and will develop Huntington's disease. This raises issues associated with testing for the condition.
  a Describe an ethical issue associated with testing for Huntington's disease.
  b Describe a social or financial issue associated with testing for Huntington's disease.

9 In your notebook draw up a table to identify what you consider to be the key issues surrounding the use of genetically modified organisms. List the points in categories that you consider to be positive or negative aspects of the technology.

10 Australian regulations prohibit genetic or medical testing without consent. Why is consent necessary prior to using these technologies?

11 Explain how advances in the medical field may benefit society.

12 Describe two examples of how nanotechnology is being applied to medicine.

13 It has been said that medicine is moving into the 'bionic age'. Explain this statement, providing examples.

14 Explain one method by which GM crops aim to curb rates of global malnutrition.

15 Explain why some communities have opposed the introduction of GM crops.

16 List three benefits of biodiversity.

17 Describe two ways that GM crops are causing biodiversity loss.

18 Taronga Zoo has a captive breeding program for the endangered southern and northern corroboree frogs (*Pseudophryne corroboree* and *Pseudophryne pengilleyi*). Is this an in situ or ex situ conservation method?

19 List at least two ways in which biotechnology is promoting the conservation of plant biodiversity.

20 Describe the two main conservation aims of 'de-extinction' genetic technologies.

21 Construct a table to list the ways in which biotechnology can reduce and increase biodiversity.

22 Explain how third generation biofuels provide an advantage over previous biofuels.

23 **a** Differentiate between phytoremediation and bioremediation.

b Describe how bioremediation methods are being improved.

24 List one positive and one negative for the future of GM crops.

25 From the cases described in this chapter or your own knowledge, identify one example of the following:

a a genetic modification that leads to a reduced environmental impact

b a genetic modification that increases productivity

c a genetic modification that impacts the viability of an animal but is of benefit to human health

26 **a** Give an example of how genetic modification may be used to change a characteristic in:

i an animal

ii a crop plant

b For the above examples, list any advantages or disadvantages for the organism, the environment and/or society.

27 After completing the Biology Inquiry on page 332, reflect on the inquiry question: How do genetic techniques affect Earth's biodiversity? Given that many scientists think we are currently experiencing the sixth major extinction event in Earth's history, evaluate the positive and negative consequences of using genetic techniques for biodiversity. Propose ways in which genetic techniques can be used (or not used) for the greatest good in terms of promoting biodiversity and the long-term evolutionary potential of species and ecosystems.

CHAPTER 09 Genetic technologies

In this chapter you will explore a range of current genetic technologies that induce genetic change, including reproductive technologies, cloning and recombinant DNA technology. You will evaluate the benefits of these technologies to the fields of agriculture, medicine and industry as well as the potential impacts of genetic technologies on biodiversity.

Content

INQUIRY QUESTION

Does artificial manipulation of DNA have the potential to change populations forever?

By the end of this chapter you will be able to:

- investigate the uses and advantages of current genetic technologies that induce genetic change
- compare the processes and outcomes of reproductive technologies, including but not limited to: S
  - artificial insemination
  - artificial pollination
- investigate and assess the effectiveness of cloning, including but not limited to: EU ICT
  - whole organism cloning
  - gene cloning
- describe techniques and applications used in recombinant DNA technology, for example: EU CCT
  - the development of transgenic organisms in agricultural and medical applications (ACSBL087)
- evaluate the benefits of using genetic technologies in agricultural, medical and industrial applications (ACSBL086) S EU
- evaluate the effect on biodiversity of using biotechnology in agriculture S
- interpret a range of secondary sources to assess the influence of social, economic and cultural contexts on a range of biotechnologies EU ICT IU DD

9.1 Reproductive technologies

BIOLOGY INQUIRY   

Propagating a population of plants

Does artificial manipulation of DNA have the potential to change populations forever?

COLLECT THIS...

- small pots
- gardening secateurs (or small scissors)
- potting soil
- safety equipment per student (safety goggles, face mask, gloves, apron, closed shoes)
- existing garden plants or a small pot plant
- honey
- water

DO THIS...

1. Dress in safety equipment before proceeding.
2. Working in small groups or pairs, fill a small pot with potting soil.
3. Take a cutting from a plant, using Figure 9.1.1 as a guide. The whole class should take cuttings from the same plant.
4. Dip the end of the stem (cut end) into the honey.
5. Make a hole in the soil and place ¾ of the stem below the soil.
6. Compress the soil around the stem.
7. Wet the soil slightly and place the pot in a sheltered sunny position.

RECORD THIS...

Describe this method of plant propagation and the expected outcome for the gene pool of this population of plants.

Present a diagram depicting the change in the gene pool of a population of propagated plants over four generations.

REFLECT ON THIS...

How does the manipulation of one generation of plants change the genetic variation in subsequent generations?

Does the artificial manipulation of DNA have the potential to change populations forever?

FIGURE 9.1.1 Taking a cutting from a plant

Reproductive technologies are used to artificially intervene in processes of sexual and asexual reproduction. Imitating asexual reproduction may include the genetic cloning of a cell or propagating a plant through cuttings. While reproductive technologies such as artificial insemination and artificial pollination are used to improve the chances of pregnancy they are also used to deliberately select the phenotype of offspring. Reproductive technologies can also assist in the screening of disease and genetic conditions that can be transmitted to offspring.

ARTIFICIAL INSEMINATION

Artificial insemination is a reproductive technology that involves the deliberate introduction of male sperm into the female reproductive tract by a method other than sexual intercourse. This reproductive technology is used in animal breeding as well as to help humans conceive a child.

Whether in wildlife conservation or animal husbandry, artificial insemination allows for the expansion of the genetic pool, allowing inbreeding to be avoided and increasing resilience through **genetic variation**. It also allows for semen samples to be screened for disease and quality, improving reproductive health of the population. However, there are some disadvantages, such as breeders having to detect the best time for female conception, as well as selecting the most valuable male sire for the **gene pool**.

Animal husbandry

Artificial insemination is a beneficial reproductive technique in **animal husbandry** (breeding and caring for farm animals) as it allows desirable characteristics from a specific male animal to be passed on to future generations. The process involves the collection of semen from the male and the insertion of fresh or stored sperm into the female reproductive tract under hygienic conditions (Figure 9.1.2a). The sperm can also be frozen using a technique known as **cryopreservation**, where the biological samples are cooled to very low temperatures (usually −135°C) (Figure 9.1.2b). This allows for the long-term storage and transportation of semen without losing viability. Artificial insemination also allows for the impregnation of multiple female animals in different geographic locations, meaning that more offspring can be produced with desirable characteristics than would be possible with normal mating. For example, one semen specimen from a sire (male parent of an animal) can produce thousands of offspring. Further, artificial insemination is a viable breeding option for farmers who often cannot afford or do not want to maintain a large male animal, such as a bull, on their property.

FIGURE 9.1.2 (a) A vet artificially inseminates a ewe using an endoscope. (b) A farmer removes cryopreserved bull semen from its liquid nitrogen storage

BIOFILE CC

Prized pedigree

Livestock auctions are a regular part of Australian farming. Farmers and breeders will gather at saleyards to purchase animals (Figure 9.1.3) including prized sires for breeding. It is not unusual for a single prized bull, such as an Angus bull, to be auctioned for over $100 000.

FIGURE 9.1.3 A Santa Gertrudis bull goes up for auction in Baryugil, NSW

Wildlife conservation

Artificial insemination is also widely used in wildlife conservation to improve the reproductive success of endangered species. Semen samples that have been cryopreserved are often transported between zoos and wildlife parks to introduce new genetic variation to captive breeding programs. This is advantageous as animals do not have to be transported long distances between wildlife centres, which can place a strain on their health and requires teams of specialist animal handlers. Introduction of wild animals to a new potential mate can also result in a lack of natural mating behaviour or even aggression, leading to injury.

BIOLOGY IN ACTION S

Breeding the Giant Panda

In 2016 the Giant Panda (*Ailuropoda melanoleuca*) (Figure 9.1.4) had its conservation status downgraded from Endangered to Vulnerable by the International Union for Conservation of Nature (IUCN) following a population growth of nearly 17% over the last ten years. The giant panda is threatened in their native China by habitat fragmentation and destruction as well as hunting and tourism. While protection of wildlife corridors and national reserves has been important to conservation efforts, captive breeding programs and successful reintroductions to the wild have played a large role in the species' new status. This is surprising given the difficulties biologists have faced breeding the giant panda in captivity. Attempts to breed the species in captivity began in 1955, but it wasn't until 1963 that the first captive-bred giant panda, 'Ming Ming', was born in Beijing. Giant pandas in captivity rarely mate, with individuals either losing interest or not knowing how to. Further adding to the complexities of captive breeding was the fact that female pandas only have one oestrus cycle (when they are in heat, or ready to mate) per year for two to seven days, and will only be fertile for 24–36 hours. Biologists have found artificial insemination to be invaluable to the success of the breeding programs, with scientists able to perform the procedure at the precise time of optimum female fertility (keepers carefully monitor hormone levels and note changes in behaviour). Further, fresh semen from a male panda at the same zoo can be used, or frozen semen from a panda at another facility can be introduced to increase the genetic variation of the captive population.

FIGURE 9.1.4 A giant panda (*Ailuropoda melanoleuca*) adult and cub in China. Artificial insemination has played a key role in the recovery of populations, helping to overcome the reproductive challenges of individuals.

Human reproduction

Artificial insemination in humans is referred to as **intrauterine insemination (IUI)**, or **assisted reproductive technology**. This reproductive technology is often used when a male is unable to engage in sexual intercourse, has low sperm count or when donor sperm is used. However, female fertility issues such as endometriosis can also require the use of this reproductive technology. The semen may be fresh or from a cryopreserved sample. The semen is inserted directly into the woman's vagina, cervix or uterus close to the ovulation phase of her menstrual cycle (Figure 9.1.5). The treatment may require medication to stimulate ovulation prior to treatment. Artificial insemination is a less invasive form of fertility treatment than other reproductive technologies and requires little to no recovery time for the female patient and it has very few side effects. Cryopreservation also allows semen to be stored and used in the future; this often allows a couple to conceive a child if the male partner becomes sterile from treatment such as chemotherapy or radiotherapy. Artificial insemination differs from **in vitro fertilisation (IVF)** which is a reproductive technology which takes the egg and sperm from patients and fertilises them in a specialised laboratory, before transferring the fertilised egg back into the female uterus.

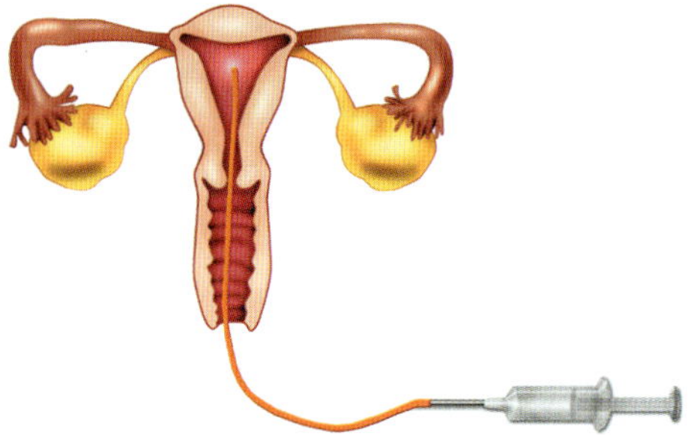

FIGURE 9.1.5 Intrauterine insemination is a relatively safe and effective reproductive technique to deliver sperm directly to the female uterus when sexual intercourse between partners is not possible.

GENETIC SCREENING

Genetic screening is used to detect abnormalities or changes to **genes**. Screening is used to confirm or rule out suspected genetic conditions, to determine if a person carries a faulty gene that can be passed on to their offspring or whether a person is at risk of developing a disease. Genetic screening can be undertaken from the embryo stage, right up to adulthood.

BIOFILE EU

Designer babies

The term 'designer baby' refers to a baby whose phenotype has been specifically selected during the IVF process (Figure 9.1.6). Preimplantation genetic diagnosis, originally used to detect genetic disorders, could potentially be used to select traits such as appearance, intelligence and biological sex. This raises a lot of ethical issues including the termination of unwanted embryos, the loss of individuality, and social implications of parents 'designing' a child. It is also an issue that needs to be addressed internationally; some countries have laws to ban genetic editing in viable embryos, whilst others have only guidelines or no rules at all.

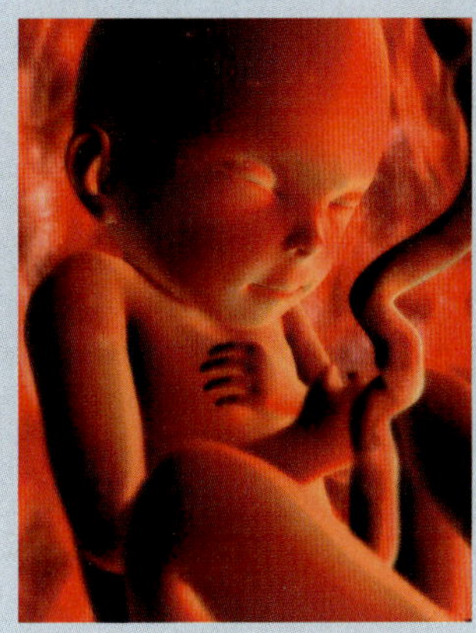

FIGURE 9.1.6 Preimplantation genetic diagnosis is being used to 'design' babies, allowing parents to select traits such as appearance, intelligence and biological sex.

SKILLBUILDER   

Assessing the influences of human reproductive technology

As new reproductive technologies continue to develop, the ways in which individuals can become parents continue to expand in range and diversity. However, awareness of the social influences and implications of reproductive technologies are just as important to be aware of as the scientific merits of such biotechnology.

Research

Many studies have been conducted to understand the social, economic and cultural influences of reproductive technologies on society and of attitudes toward reproductive technologies. Online sociological journal databases should be consulted for up-to-date, accurate information.

Examples of social, economic and cultural contexts

Reproductive technology information in the science community often does not address the societal issues that will influence its application. Examples of factors influencing the use of reproductive technologies may include the following.

- Laws of the country or state:
  - What are the laws governing surrogacy, egg donation, biological sex selection, access to donor sperm?
- Cultural influences and attitudes:
  - Does each country and culture view reproductive technology the same way?
  - What influence do genetic relationships have on child-rearing and child development?
  - What are the differing attitudes to infertility?
- Economics:
  - Can everyone afford reproductive technology treatment?
  - Are there ongoing costs for some reproductive technologies?

ARTIFICIAL POLLINATION

Artificial pollination occurs when humans manually facilitate the natural plant pollination process. The pollen is taken from the stamen (male part) of one flower and dusted over the stigma (female part) of another flower (Figure 9.1.7) resulting in fertilisation and the developments of seeds. These seeds may develop into fruits or new plants. Artificial pollination allows humans to choose pollen from plants with desired characteristics and to spread these genetic traits quickly through a population.

FIGURE 9.1.7 The stamens (male structures) and stigma (female structures; yellow, centre of flower) of an oriental lily (*Lilium orientalis*)

Traditionally, artificial pollination was used for small-scale plant breeding and study. For example, Gregor Mendel used artificial cross-pollination of pea plants to study heredity and genetics in the 1800s. Plant breeders used the reproductive technique to develop new varieties of plants and select for specific characteristics (e.g. pollinating from a specific plant population to ensure the genetic traits for larger fruit are passed on to a new crop). Farmers may use artificial pollination to speed-up the fertilisation process and improve crop yields. This is common in orchard farms where fruit trees such as apples and pears need to be pollinated by a different tree of the same species. This would naturally require the pollen to be carried by the wind or a pollinating animal; however, artificial pollination can quickly achieve the same result. The need for artificial pollination has also risen with the decline in natural pollinators such as honeybees. Some countries such as China have so few natural pollinators that some crops can only continue to be produced through artificial pollination.

BIOFILE S

China's pollinator crisis

The use of excessive pesticides and the degradation of natural habitat in south-west China has eradicated native bee populations. The native bees once pollinated the apple and pear orchards of this region. Farmers now have to hand-pollinate their trees (Figure 9.1.8) a laborious task requiring every flower on the tree to be dusted with pollen using a small paintbrush. Farmers often climb to the top of trees to hand-pollinate high value crops.

FIGURE 9.1.8 A farmer in Sichuan, China, hand-pollinates fruit trees due to the demise of natural pollinators in the region.

Mechanical pollination

Artificial pollination can occur mechanically or by hand. Mechanical pollination involves the mass-dusting of pollen onto plants from small aircraft or large blowers. Although this can be a quick process the accuracy of the pollen reaching plants is limited. Scientists have discovered that natural pollination (bees and other pollinators distributing the pollen directly to the female organs of the flower) is far more accurate than mechanical pollination.

Hand pollination

Hand pollination generally involves the gardener or farmer using a small brush to transfer the pollen from one plant to the stigma of the other plant (Figure 9.1.9a). This method is more accurate than mechanical pollination, but does take more time and labour. Passionfruit (*Passiflora edulis*) (Figure 9.1.9b) is a commonly grown home-garden fruit. However, many gardeners complain that the vine never produces any fruit. The reason for this is that the male and female parts of this plant are very far apart and require a pollinator to carry pollen to the pistil (female organs). Home-gardeners will often use a small paint brush to take some pollen from the anthers (male) and dust it onto the female stigma. This fertilises the plant and artificially induces the development of fruit. However, in Brazil, farmers hand-pollinate passionfruit crops due to the decline of the natural bee population from pesticide use in the region. This expensive labour is pushing up the price of fruits and vegetables and causing Brazilians to look to more affordable, but unhealthy, options such as fast-food and fatty meats.

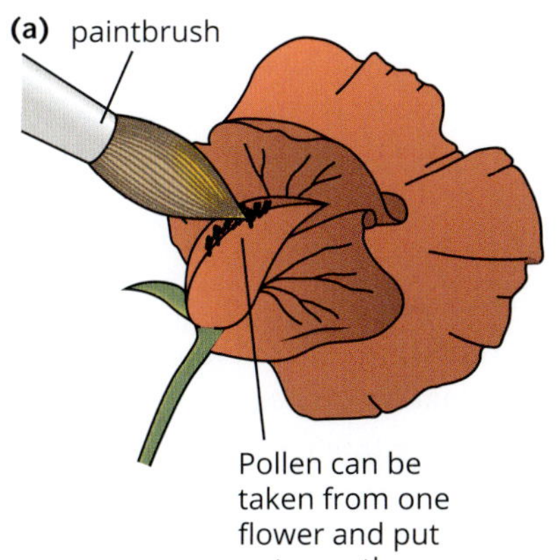

FIGURE 9.1.9 (a) Hand pollination involves the transfer of pollen from one flower to another usually using a simple tool like a paintbrush. (b) Passionfruit (*Passiflora edulis*) plants often require hand pollination due to the large distances between the male and female parts of the flowers.

Conservation of plant biodiversity

Artificial pollination is also being used in conservation efforts to save endangered species of plants. Global plant **biodiversity** is at-risk globally from human-induced threats of habitat destruction and degradation, climate change and invasive species. However, the key ecological interaction of plants and pollinators (Figure 9.1.10) is also under threat, with pollinator species such as bees globally in decline. The extinction of a pollinator species may cause the coextinction of a plant, and the extinction of a plant may lead to the coextinction of a pollinator species. This is known as **mutualism disruption**, where the mutualistic relationship between two species is disturbed, which can have far-reaching consequences to ecosystem functioning. Studies show that Africa, Asia and the Caribbean are at particular risk of mutualism disruptions. As a result, scientists are turning to artificial pollination to help fertilise endangered plants and increase their populations both in the wild and in botanic gardens.

FIGURE 9.1.10 The key ecological interaction between pollinators and plants is being threatened with the extinction of either species leading to mutualism disruption.

BIOLOGY IN ACTION 

Trekking to save Hawaiian plants

The Plant Extinction Prevention Program, based in Hawaii's Kauai National Tropical Botanical Garden, is currently working to save Critically Endangered Hawaiian plant species—often with the assistance of artificial pollination. Some species, such as the native Hawaiian alula (*Brighamia insignis*), are so close to extinction that it is believed there is only one living specimen left in the wild. The native Hawaiian alula is found on steep cliffs in Kauai and is pollinated by the green sphinx moth, which is rarely seen in the area. Biologists in Kauai rappel down cliffs, hike through forests and traverse waterfalls to reach the 1000 metre-high cliffs of the native Hawaiian alula and hand-pollinate its flowers (Figure 9.1.11). The extent to which scientists are willing to go for conservation work is understandable as there are more than 1300 Hawaiian plant species already extinct.

FIGURE 9.1.11 Biologists go to extreme measures such as (a) repelling and crossing waterfalls to pollinate (b) the endangered native Hawaiian alula (*Brighamia insignis*).

9.1 Review

SUMMARY

- Reproductive technologies are used to artificially intervene in processes of sexual reproduction.
- Artificial insemination is a reproductive technology that involves the deliberate introduction of male sperm into the female reproductive tract by a method other than sexual intercourse.
- Artificial insemination is used to help humans conceive a child and is also used in agriculture and wildlife conservation.
- Genetic screening is used to detect abnormalities or changes in genes.
- Genetic screening identifies disease-associated mutations in fetuses, newborns or adults.
- Artificial pollination occurs when humans manually facilitate the natural plant pollination process.
- Artificial pollination allows humans to choose pollen from plants with desired characteristics.
- The need for artificial pollination in agriculture and conservation has risen with the global decline in natural pollinators.

KEY QUESTIONS

1. Define the term 'artificial insemination'.
2. What advantages does artificial insemination bring to agriculture?
3. How does artificial insemination differ from in vitro fertilisation (IVF)?
4. Describe the two main methods of artificial pollination.
5. What is driving the increased need for artificial pollination in agriculture and conservation?
6. Explain the term 'mutualism disruption'.

9.2 Cloning

A **clone** is a genetically identical copy of a gene, cell or organism. It may occur naturally, as in the case of bacteria and plants which reproduce asexually, or identical twins, but clones can also be created artificially. There are three types of cloning: whole organism cloning, gene cloning and therapeutic cloning.

WHOLE ORGANISM CLONING

Whole organism cloning (also known as reproductive cloning) has many applications in the areas of medicine and agriculture. Whole organism cloning is used in many areas of scientific research where genetically identical test subjects are desired for reliable research outcomes, in animal husbandry and in agriculture where cloning enables desirable traits to be passed onto many offspring in a method faster than traditional breeding. The cloning of plants through techniques such as cutting and grafting has been used both commercially and in home-gardens for centuries to create identical plants (Figure 9.2.1). **Tissue culture** techniques have also developed to clone individual plants.

FIGURE 9.2.1 Taking a cutting from a mature plant and propagating a genetically identical plant is a traditional form of cloning that has been used both commercially and in home gardens.

Whole organism cloning of animals has been a more challenging process. To date, it has been limited to mammals such as cattle, chickens, sheep and dogs. While whole organism cloning has many potential uses, the disadvantages of this reproductive technology include the high cost of breeding, consumer resistance to cloned foods and the higher rate of mortality and health problems in cloned animals. Cloning also raises social, moral and ethical concerns. Human cloning is illegal in most countries, including Australia.

The two methods for whole organism cloning are **artificial embryo twinning** and **somatic cell nuclear transfer (SCNT)**.

Artificial embryo twinning

Artificial embryo twinning is a technique that mimics the natural process which leads to identical twins where the embryo splits in two very early in the development process (Figure 9.2.2) and both halves of the embryo continue to divide leading to two separate individuals. As both individuals developed from the same fertilised egg they are genetically identical. Artificial embryo twinning is carried out in a laboratory petri dish. The early embryo is separated into individual cells and allowed to continue dividing. The embryos are then transferred into a surrogate mother.

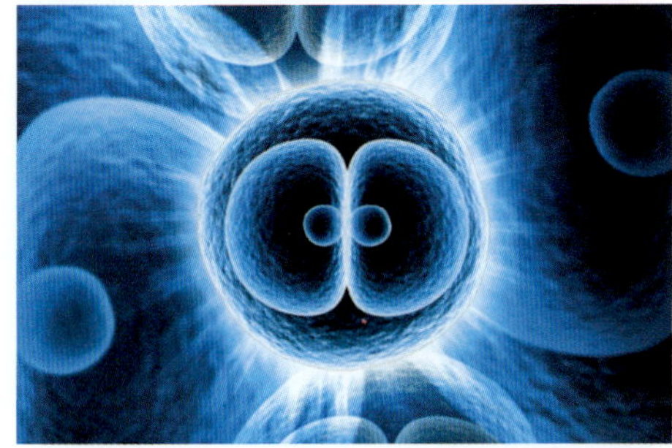

FIGURE 9.2.2 Artificial embryo twinning promotes the early separation of an embryo into individual cells which continue to divide.

Somatic cell nuclear transfer (SCNT)

Somatic cell nuclear transfer (SCNT) removes the single set of chromosomes from an egg cell and replaces them with the nucleus of a somatic cell (which has two sets of chromosomes) taken from the organism being cloned. The egg cell is then induced to divide as though under the natural fertilisation process. Once the embryo has developed, it is transferred to the uterus of a surrogate female organism. The individual that is born following gestation will be an identical genetic copy of the animal from which the original somatic cell was taken.

Cloning in agriculture

Somatic cell nuclear transfer was the technique used to successfully breed the first large cloned animal in 1997. The cloned sheep, named Dolly (Figure 9.2.3a), was bred in Scotland by researchers at the Roslin Institute in Edinburgh. Cells from the udder of an adult sheep were collected and transferred to an egg cell whose **DNA (deoxyribonucleic acid)** had been removed. The egg was then implanted into a female sheep and, following gestation, a live sheep was born (Figure 9.2.3b). The lamb was genetically identical to the sheep whose tissues were sampled. Dolly was euthanised in 2003 when a large tumour, due to a viral infection, was found in her lungs. Australia bred its first cloned animals in 2000—Suzi, a Holstein cow, and Matilda, a merino sheep. While Australia has some 100 clones of beef and dairy cattle, as well as sheep clones, products from cloned animals have not been approved as safe for consumption.

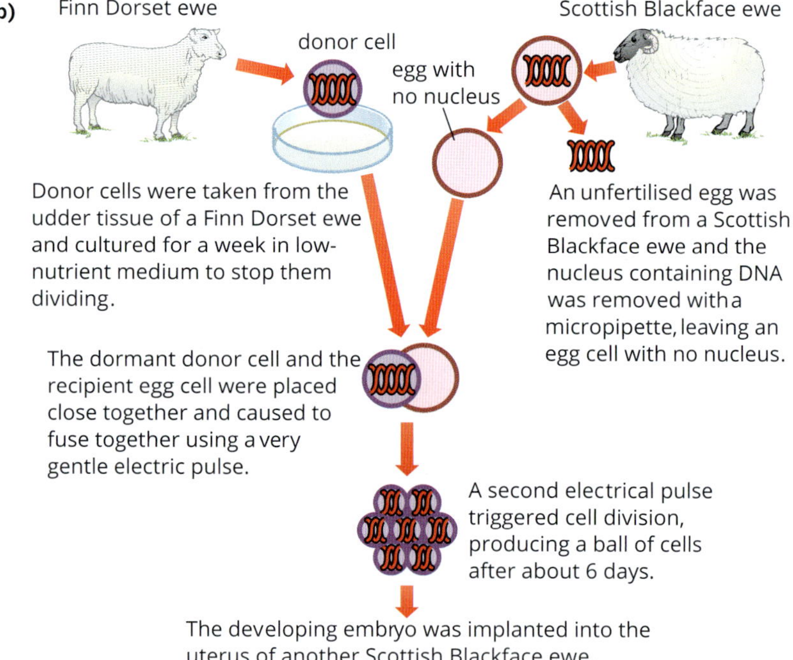

FIGURE 9.2.3 (a) Dolly the sheep was born in 1997, the first large cloned animal. (b) The cloning process that produced Dolly.

Cloning for conservation

Cloning has also been suggested for use in conservation practices to save Critically Endangered species and as a method of **de-extinction** of species. For example, the northern white rhinoceros (*Ceratotherium simum cottoni*) stands on the brink of extinction with only two females remaining. The last male northern white rhinoceros died in March 2018. The heavily-poached species remain under armed guard in Kenya's Ol Pejeta Conservancy (Figure 9.2.4). Scientists have harvested the last eggs from the two remaining females with the plan to use stem cell technology and IVF to create embryos to be surrogated by another rhinoceros sub-species.

FIGURE 9.2.4 Whole organism cloning may offer a reprieve from extinction for the northern white rhinoceros (*Ceratotherium simum cottoni*) whose only two remaining individuals are under 24-hour guard in Kenya's Ol Pejeta Conservancy to protect them from poachers.

BIOLOGY IN ACTION  

Waking the woolly mammoth

The woolly mammoth (*Mammuthus primigenius*) (Figure 9.2.5a) which became extinct approximately 4000 years ago, could be on the verge of de-extinction (discussed in Section 8.3 on page 332). The woolly mammoth lived during the Pleistocene epoch and lived along what is now the northern parts of Europe, Asia and America. It was approximately the size of a large elephant, was covered in fur with long outer hairs and had two large curved tusks. The species has been well-studied due to the discovery of frozen carcasses in Alaska and Siberia. The woolly mammoth's closest living relative is the Asian elephant (*Elephas maximus*) (Figure 9.2.5b). More than 70% of the woolly mammoth's genome was decoded in 2008. Since then scientists have been working to clone an individual woolly mammoth using cells from frozen carcasses. The cloning process would involve:

- the nucleus of a viable mammoth cell being extracted from a frozen carcass
- removing the DNA from the egg cell of an Asian elephant and replacing it with the nucleus from the mammoth cell
- scientists stimulating the division of the egg cell
- implantation of the embryo into the uterus of a female Asian elephant
- following gestation, a live baby mammoth would be born as an exact clone of the original frozen mammoth.

Suggestions of cloning a woolly mammoth have already raised moral and ethical arguments, including challenges of how the mammoth would learn behaviours without role models and how it would survive in a modern world facing climate change, amongst other threats. Many have suggested that cloning the woolly mammoth would simply result in a biological orphan as a curio for exhibits.

FIGURE 9.2.5 (a) Scientists are currently working to clone a woolly mammoth (*Mammuthus primigenius*) by removing the nucleus from a preserved cell in a frozen carcass and transferring the nucleus into an embryo of (b) an Asian elephant (*Elephas maximus*) for gestation and birth.

GENE CLONING

Gene cloning allows scientists to produce exact copies of a gene of interest. Gene cloning is very different from the cloning of whole organisms. In gene cloning the end product is many copies of a specific gene. Two techniques can be used for gene cloning: **in vivo** and **in vitro** methods.

The in vivo technique involves the use of **restriction enzymes**, **ligases** and **vectors** to incorporate the desired gene into the DNA of a living organism, where this gene will replicate. Because the genetic code is universal—the same nucleotides, adenine (A), guanine (G), cytosine (C) and thymine (T), are found in all species—the original gene taken from one organism will express the same protein in the host organism.

In the in vitro gene cloning technique, the **polymerase chain reaction (PCR)** is used to produce multiple copies of the specific gene. You learnt about PCR in Chapter 6.

Gene cloning has many potential benefits for individuals and society and contributes to important areas in biological research, medicine, food production and industry. Applications of gene cloning include **whole genome sequencing**, characterising genes (e.g. where they are located, how they are expressed and their biological function), the production of proteins, which are important in medicine and research, the development of **transgenic organisms** and **gene therapy**.

In molecular biology, a vector is a vehicle used to transfer foreign DNA into a cell. Plasmids and viruses are common vectors for transporting genes.

GO TO ➤ Section 6.1 page 248

Studies that are in vivo are 'within the living', such as when cells are studied in a living organism. Studies that are in vitro are 'in glass' or in a dish or test tube, such as when cells are removed from the organism and studied in a culture dish.

DNA and genome sequencing

Gene cloning techniques allow scientists to obtain DNA sequences of different species. The DNA sequences can then be compared to understand evolutionary relationships and the changes in DNA sequences that have led to functional changes over evolutionary time. Using software, DNA sequences of an organism can be pieced together and genes mapped to their location within the **genome**. DNA sequencing is covered in more detail in Chapter 6.

GO TO ➤ Section 6.1 page 248

Characterising genes

Gene cloning and sequencing allows scientists to examine how **gene expression** may be regulated by other nearby genes or the environment and how mutations can disrupt gene expression and lead to disease. Researchers can also manipulate the expression of cloned genes to understand their functional roles in organisms. For example, researchers have found that mutations in the tumour suppressor gene *BRCA1* disrupt normal tumour suppressive activity increasing a person's risk of developing breast cancer. Information about specific genes can be used to develop methods for earlier disease detection and treatments.

Gene therapy

Gene therapy is an application of cloning technology that is used in medicine. Gene therapy refers to the insertion of a gene into an individual's cells to correct or replace defective gene function that leads to disease. Gene therapy is a promising area of research and gives hope to many people living with disease. Gene therapy for disease will be examined in detail in Chapter 17.

GO TO ➤ Section 17.1 page 582

THERAPEUTIC CLONING

The somatic cell nuclear transfer (SCNT) cloning technique used to create Dolly the sheep can be used to produce human embryos for therapeutic purposes (i.e. the treatment of disease) (Figure 9.2.6). Cloning embryonic stem cells for the research and treatment of disease is known as **therapeutic cloning** (sometimes simply called somatic cell nuclear transfer). Much scientific and ethical debate has surrounded the use of somatic cell nuclear transfer on human cells, both for reproductive purposes and also as a means to extract embryonic stem cells for treatment of certain diseases. Currently, the reproductive cloning of humans is banned due to ethical and scientific objections. Stem cells are unspecialised cells and have the potential to develop into specialised cells such as nerve and muscle cells as well as being used for a variety of functions including growth and repair. Stem cells found in the embryo stage are the most plentiful and versatile. Therapeutic cloning has the potential to overcome problems with immunological rejection of cells and tissues if a patient's own stem cells are used to create the therapeutic cells. Research is currently underway to reprogram cloned adult stem cells. These reprogrammed cells are called induced pluripotent stem (iPS) cells. This would allow patient-specific stem cells to be produced without the need to use human eggs or embryos.

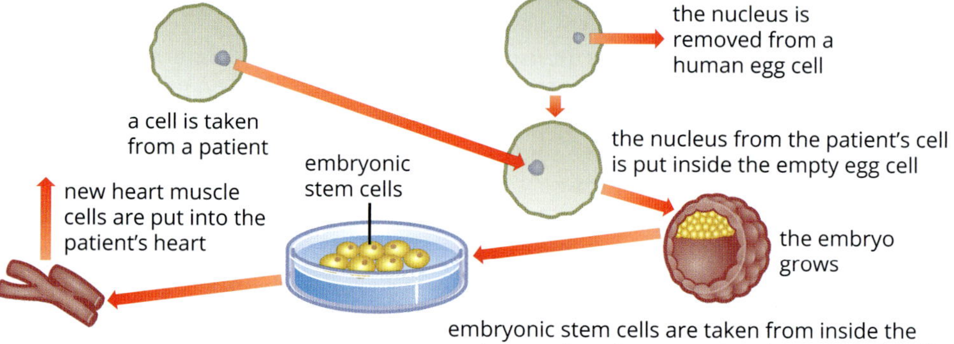

FIGURE 9.2.6 The cloning and harvesting of embryonic stem cells is known as therapeutic cloning and is used for research and the potential treatment of disease. Stem cells are incredibly versatile and can be used for growth and repair of many cells and tissues in the human body, such as heart muscle cells.

9.2 Review

SUMMARY

- A clone is a genetically identical copy of a gene, cell or organism.
- There are three types of cloning: whole organism (or reproductive) cloning, gene cloning and therapeutic cloning.
- The two methods for whole organism cloning are artificial embryo twinning and somatic cell nuclear transfer.
- Suggested applications for whole organism cloning include scientific research, wildlife conservation and agriculture.
- Gene cloning allows scientists to make exact copies of a gene of interest.
- Gene cloning can occur via in vivo or in vitro methods.
- Cloned genes can be used for sequencing, research and the production of recombinant proteins.
- Gene therapy is an application of gene cloning in medicine; this technique replaces a defective gene that causes disease with a normal gene.
- Cloning techniques often raise social, moral and ethical concerns in society.

KEY QUESTIONS

1. What is artificial twinning? How does it differ from the birth of natural identical twins?
2. Explain the process of somatic cell nuclear transfer.
3. What are some of the potential advantages and applications of whole organism cloning?
4. Explain why the cloning and 'de-extinction' of a woolly mammoth remains a controversial issue.
5. Describe two reasons why scientists clone genes.
6. Differentiate between the 'in vivo' and 'in vitro' methods of gene cloning.
7. Describe the term 'gene therapy'.

9.3 Recombinant DNA technologies

When DNA from two different sources is joined together, the resulting molecule is called **recombinant DNA**. Scientists create recombinant DNA to clone a particular gene. Following this step, they may also produce large quantities of the protein expressed by the cloned gene (Figure 9.3.1).

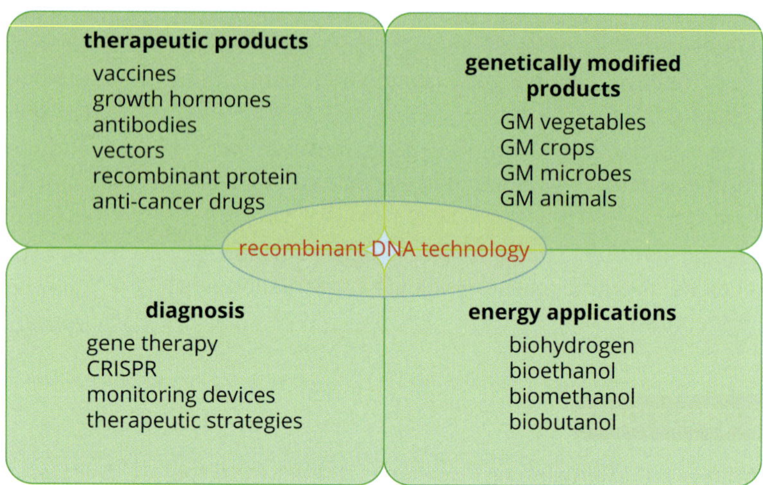

FIGURE 9.3.1 Potential applications of recombinant DNA technologies

Plasmids are small circular DNA molecules found in bacteria.

USING PLASMIDS AS VECTORS

To introduce foreign genetic material into a cell, scientists often use **plasmids** as the vector. They insert target DNA into a double-stranded circular plasmid, producing a **recombinant plasmid**. The plasmid is inserted in a cell, where the self-replicating system of the plasmid and cell replicates the plasmid genes. Each cell containing the plasmid will produce the protein products of the genes in the plasmid, including those of the target DNA. For example, if a recombinant plasmid containing the insulin-coding gene is placed in a bacterial cell, the cell will produce large amounts of the protein insulin which can be used in the treatment of diabetes.

A reporter gene is a gene that allows detection of gene expression in genetic engineering, such as the genes for *lacZ* and fluorescent proteins.

Plasmids are used as vectors as they are self-replicating, stable and easy to manipulate in a laboratory. Plasmids vary in size and functionality but contain several common features:

- origin of replication (also called ori)—so that the plasmid can continue to replicate independently within the host cell
- at least one restriction enzyme recognition site—so the target DNA can be inserted into the plasmid
- a selectable marker or a reporter gene—most commonly confers antibiotic-resistance or produces a coloured or fluorescent protein so that cells containing the plasmid are selected.

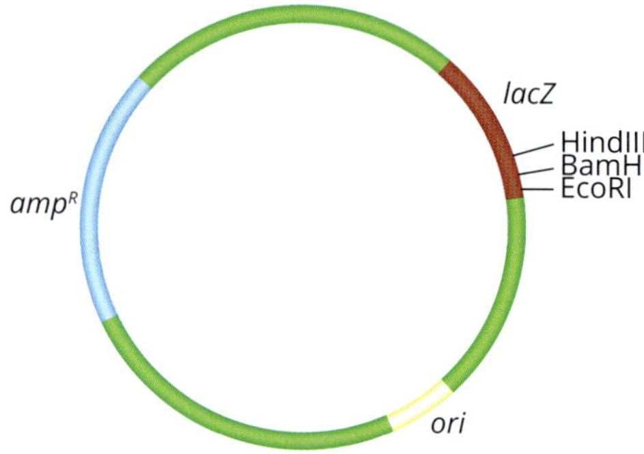

FIGURE 9.3.2 An example of a plasmid used in recombinant DNA technology showing the gene for ampicillin resistance (amp^R), the origin of replication (*ori*) and the *lacZ* gene, which creates blue colonies when grown on agar containing an indicator called X-gal. Sites for three restriction enzymes, HindIII, BamHI and EcoRI, lie within the *lacZ* gene.

A plasmid containing the ***lacZ* gene** used in bacteria is shown in Figure 9.3.2. Restriction enzyme sites for inserting the gene of interest are located within the *lacZ* gene. If the gene insertion is successful it will disrupt the *lacZ* gene and therefore protein expression of this gene. You will learn more about how plasmids are used in a bacterial system later in this section. The plasmid also contains the antibiotic resistance gene amp^R, which encodes resistance to ampicillin.

Creating recombinant plasmids

The process of creating a recombinant plasmid is outlined below and in Figure 9.3.3.

1 DNA is isolated from the cell and the target DNA is cut out using a restriction enzyme that will create either sticky or blunt ends. In the example in Figure 9.3.3 sticky ends are produced.

2 The bacterial plasmid is cut by the same restriction enzyme used to cut out the target DNA. The plasmid and the target DNA now have the same sticky ends with exposed bases that are complementary to each other.

3 The target DNA and plasmids are placed together. Some plasmids will simply close back up (known as non-recombinant plasmids), while other plasmids will incorporate the target DNA by complementary base pairing (known as recombinant plasmids).

4 The enzyme DNA ligase is added to rejoin the sugar–phosphate backbone of the DNA.

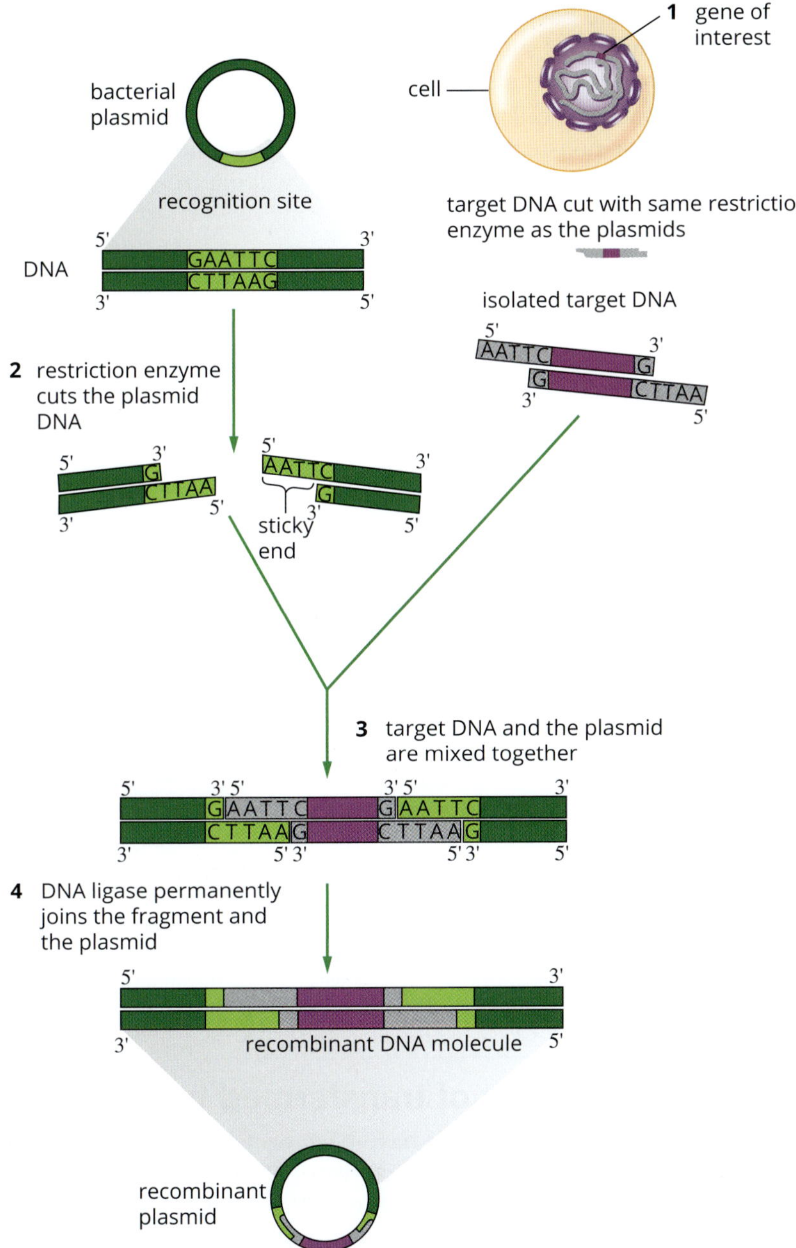

FIGURE 9.3.3 A recombinant plasmid is created by joining a target DNA fragment and a plasmid that have both been cut with the same restriction enzyme, which in this instance has created sticky ends. They are joined using the enzyme DNA ligase.

Transforming bacterial cells

Bacterial cells that have had foreign DNA incorporated into them are said to be transformed because the cells can express a new gene and therefore has a new characteristic.

Two methods of artificial **bacterial transformation** are used: heat shock and electroporation.

- Heat shock involves placing bacterial cells and a mixture of recombinant and non-recombinant plasmids in an ice-cold solution containing calcium ions and then rapidly increasing the temperature to disrupt the cell membranes of the bacteria. The plasmids can then penetrate the membrane and enter the bacteria.
- In electroporation, the bacterial cells and a mixture of recombinant and non-recombinant plasmids are subjected to an electrical current that alters the cell membrane. Again, the plasmids are then able to enter the bacteria.

Very few of the bacterial cells will be transformed with recombinant plasmids. Some will take up the non-recombinant plasmids (plasmids without the target DNA) and others will not be transformed at all or will die from the heat shock or electroporation treatment (Figure 9.3.4).

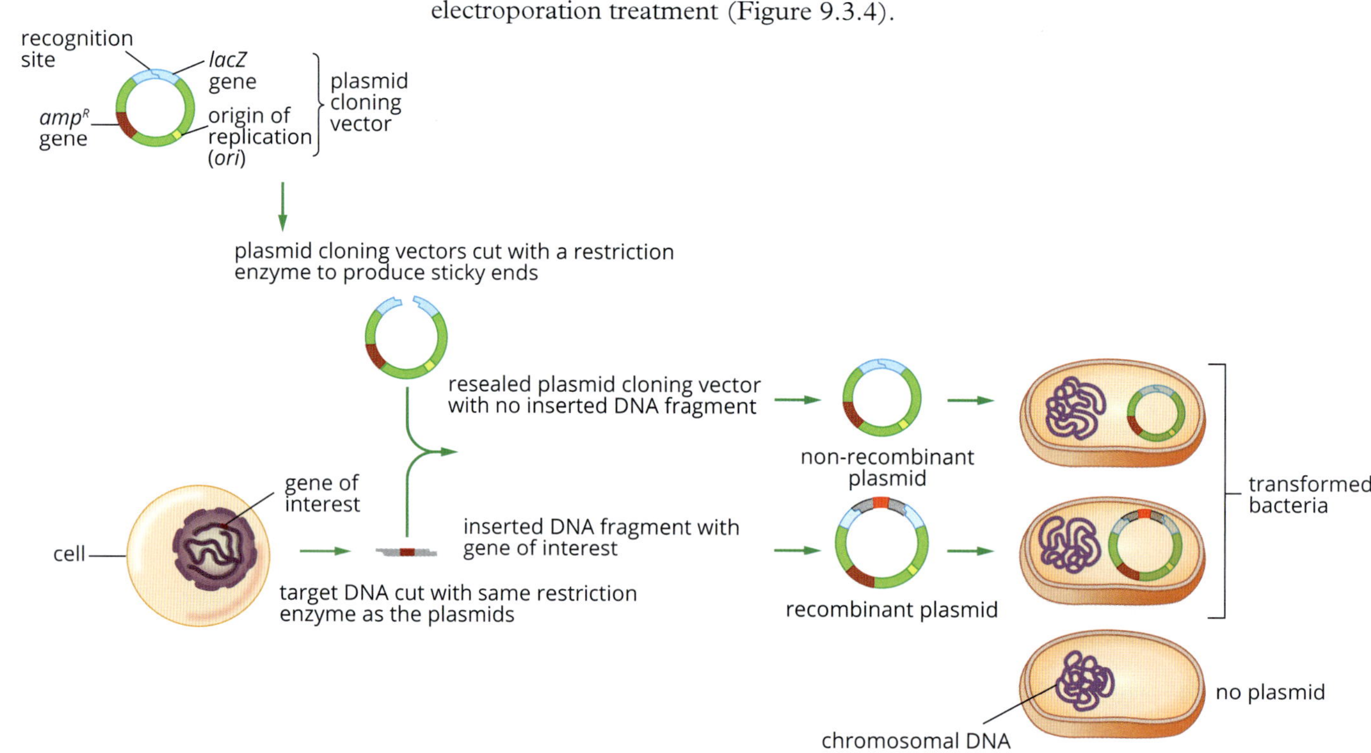

FIGURE 9.3.4 The process of bacterial transformation involves creating recombinant plasmids and then inserting them into bacterial cells. Some will successfully take up the recombinant plasmid but others will either contain a non-recombinant plasmid or no plasmid at all.

Selection and screening of transformed bacteria

When determining which bacterial cells have been transformed with recombinant plasmids containing target DNA, the characteristics of the plasmid vectors become important. Recall from the earlier example (Figure 9.3.2 page 354) that the plasmid vector contains other genes, including a gene for antibiotic resistance (this example has ampicillin resistance) and a gene that displays a particular phenotype, such as a coloured product.

To determine which of the bacterial cells have been transformed, the cells are incubated at 37°C on agar plates that contain the antibiotic ampicillin (Figure 9.3.5). The only bacteria to survive will be those that have taken up the plasmid, whether it is a recombinant or non-recombinant plasmid. These bacteria have the ampicillin resistance gene. Bacteria that have not taken up the plasmid will not have the ampicillin resistance gene and will be killed.

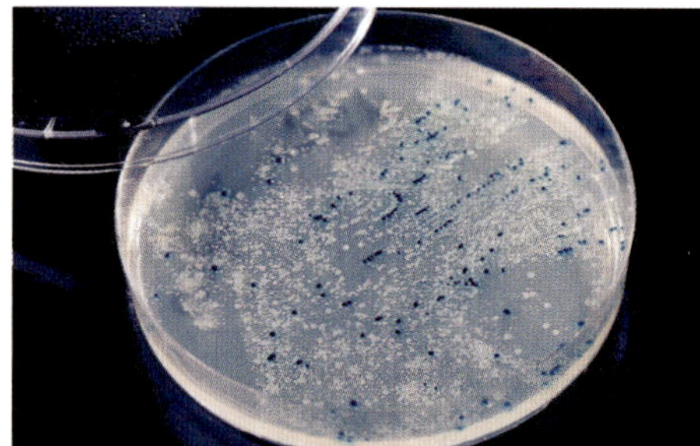

FIGURE 9.3.5 Growth of transformed bacteria on agar plates containing ampicillin and X-gal. Bacteria with the non-recombinant plasmid appear blue because the *lacZ* gene is expressed. Bacteria with the recombinant plasmid appear white.

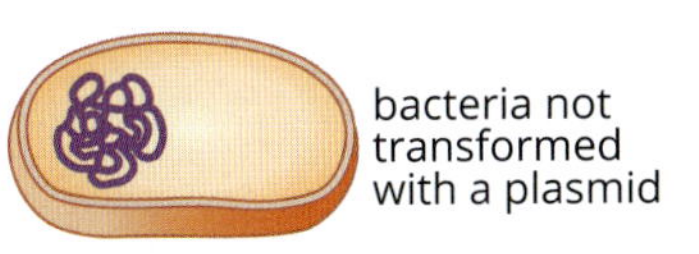

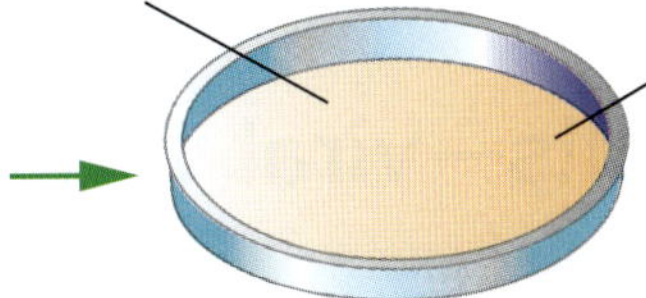

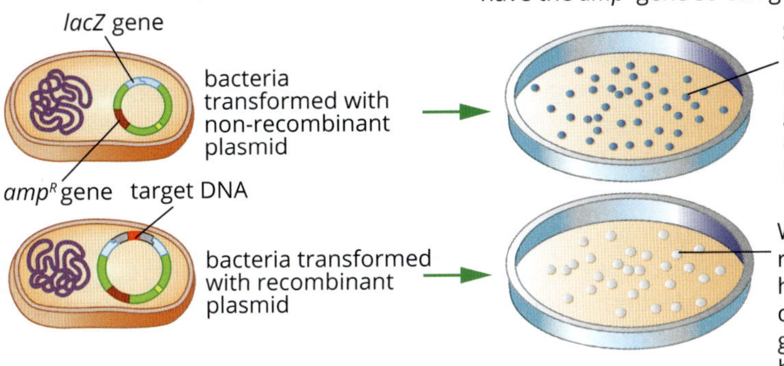

FIGURE 9.3.6 Selection and screening of bacterial cells to identify which cells have been transformed (contain a plasmid), which colonies contain recombinant plasmids with the target DNA and which contain non-recombinant plasmids

The next step is to determine which bacterial colonies contain the recombinant plasmid. In our example, the plasmid carries the *lacZ* gene, which codes for an enzyme that breaks down an indicator called X-gal, resulting in a blue product. Bacteria carrying the non-recombinant plasmid with an intact and functioning *lacZ* gene produce blue colonies on agar plates. If the target DNA has been successfully inserted within the *lacZ* gene, expression is disrupted and the enzyme coded by this gene is not produced. Therefore, bacteria transformed with recombinant plasmids appear as white colonies (Figures 9.3.5 and 9.3.6).

Bacteria transformed with the recombinant plasmids are then cultured with nutrients in order for them to replicate and produce the protein encoded by the target DNA.

The *lacZ* gene is one of the genes in the bacterial lac operon. It codes for the enzyme β-galactosidase, which breaks down lactose into glucose and galactose. The *lacZ* gene is removed from the lac operon and inserted into plasmids to act as a reporter gene in recombinant DNA technology.

X-gal is a colourless synthetic compound with a very similar structure to lactose, so it fits in the active site of β-galactosidase and is broken down, releasing a blue reaction product. Using agar plates containing X-gal is an easy way to see whether cells produce β-galactosidase.

BIOFILE CCT

Fluorescent proteins

Green fluorescent protein (GFP) is a protein first isolated from jellyfish. The GFP gene and its many coloured variants can be inserted into plasmids to be used in bacterial transformation. Bacterial cells that have incorporated the plasmid vector fluoresce under UV light (Figure 9.3.7).

Genes that code for fluorescent proteins are often attached to another gene that is being investigated. When this gene of interest is expressed, the fluorescent protein is also produced, so it sends an obvious 'visual report' that the gene is being expressed. Hence they are referred to as reporter genes. Other reporter genes and proteins include luciferase, an enzyme from fireflies that produces a yellow fluorescent product, and a red fluorescent protein from a coral.

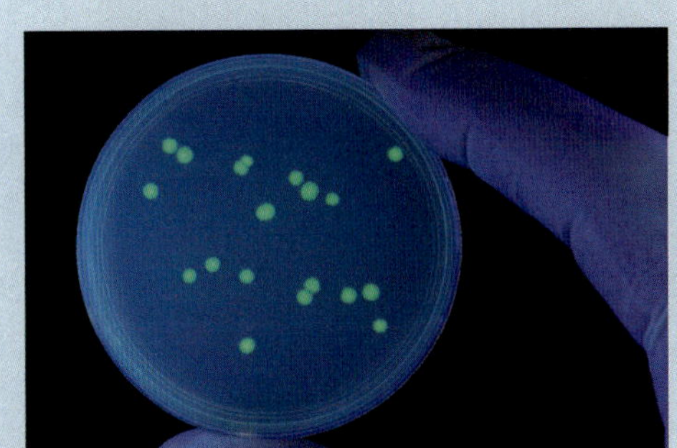

FIGURE 9.3.7 Transformed bacterial colonies can be identified as they contain the GFP gene and fluoresce under UV light.

BIOLOGY IN ACTION CCT

Getting genes into eukaryotic cells—viral vectors

Bacteria are limited in their ability to produce functioning eukaryotic proteins. Many biologically active proteins are glycoproteins, which have sugars attached to the protein chain. Sugars are added in the endoplasmic reticulum and Golgi apparatus of eukaryotic cells. Because bacteria lack these organelles and the necessary enzymes they are unable to produce glycoproteins. Therefore some recombinant proteins are mass produced in eukaryotic cell cultures, such as yeast or mammalian fibroblasts. Examples are Gardasil®, the vaccine that provides protection against the human papilloma virus, made from recombinant DNA in yeast, and erythropoietin, produced from recombinant DNA in cultured mammalian cells.

Recombinant DNA technology is also used to produce genetically modified plants and animals. Plasmids are not always effective for getting genes into eukaryotic cells, so other methods and vectors are used.

Viruses can be effective vectors for delivering genes (Figure 9.3.8a) as viruses naturally insert genetic information into their host cells. Artificial chromosomes can carry large genes into cells treated by electroporation to open pores in the cell membrane. A gene gun (Figure 9.3.8b) has been used to shoot beads coated with DNA into plant cells. DNA can also be microinjected directly into cells (Figure 9.3.8c). The process of introducing genes into mammalian cells is referred to as 'transfection' (the term 'transduction' is also used when a viral vector delivers the gene). You will learn more about viral vectors and their use in gene therapy in Chapter 17.

GO TO ➤ Section 17.1 page 582

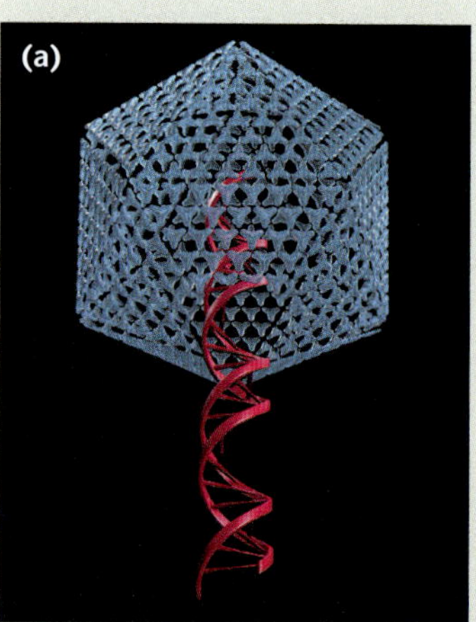

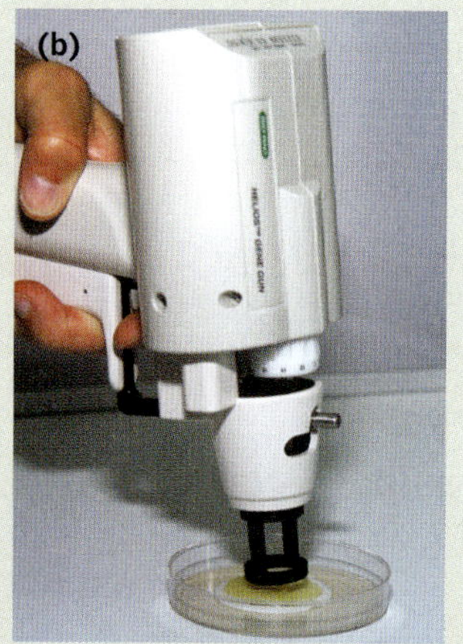

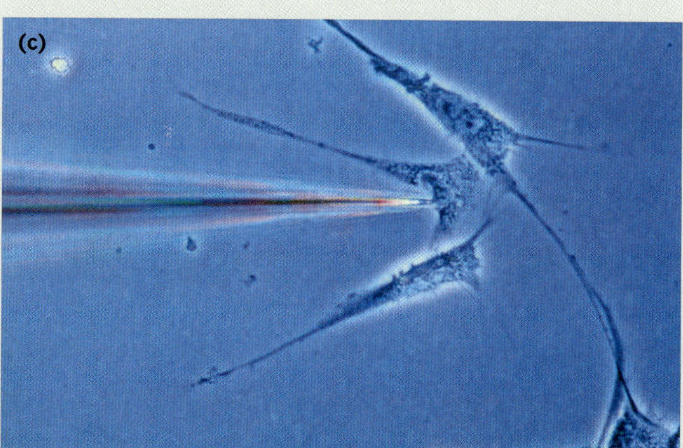

FIGURE 9.3.8 (a) Viral vectors are used to transfer genes into animal cells. (b) The gene gun, a biolistic device, shoots DNA into plant cells. (c) Microinjection of DNA into mammalian cells in a culture dish

FIGURE 9.3.9 Genetically modified cotton on the left shows its insect resistance compared to the non-modified cotton on the right, which has been ravaged by insects

GO TO ➤ Section 17.1 page 582

GENETICALLY MODIFIED AND TRANSGENIC ORGANISMS

As discussed in Chapter 8, humans have used selective breeding to produce animals and plants with more useful or more attractive characteristics since ancient times. They chose those animals or plants that expressed the characteristics they wanted to conserve and selectively bred them together, so that their offspring inherited those characteristics. In the past, selective breeding could only use characteristics that already existed in the genetic pool of a species. We now have the knowledge and skills to use recombinant DNA techniques and techniques to transfer genes from one species to another to produce organisms with DNA combinations never seen before (Figure 9.3.9).

You will learn about genetically modified cells for the treatment and prevention of disease in Chapter 17.

Genetically modified organisms (GMOs)

Over the last few decades techniques have been developed that allowed for the alteration of an organism's genome and for the transfer of genes from one organism to another. Because the DNA code is universal, almost any gene transferred from one organism to another will express the protein that it expressed in the original organism. This means that a desirable characteristic seen in one animal or plant could be transferred to another organism lacking this characteristic.

An organism that has had its genome altered in this way is considered to be a **genetically modified organism (GMO)**. An organism that has had genes from another species inserted is called a transgenic organism. The gene that came from another organism is called a **transgene**. Organisms may have their genome modified by directed mutation (mutagenesis) or by newer technologies called **gene editing**. These methods may change the genes of an individual without introducing a new gene, so it would be genetically modified but not a transgenic organism.

Transgenic crops

Transgenic crops are used in agriculture to increase crop productivity, provide resistance to insect herbivory and prevent disease. Several genetically modified crops have been developed or are grown in Australia. For example, insect-resistant GM cotton has been grown since 1996, and herbicide-tolerant GM canola was approved for commercial production in Victoria in 2008. In Australia, the Office of the Gene Technology Regulator (OGTR) assesses all GM animals and plants before research, agricultural and commercial use.

Techniques for producing transgenic plants

Transferring a gene into plant cells can be limited by the presence of the cell wall. The introduction of foreign genes into plants is usually done using a biological vector. One method uses *Agrobacterium tumefaciens*, a soil bacterium that is able to naturally transfer a plasmid into plant cells (Figure 9.3.10). *A. tumefaciens* normally causes crown gall disease because it carries a plasmid with genes that cause the growth of a tumour. A recombinant plasmid (the vector) carrying a desired gene from a different species, but lacking the tumour-inducing genes, is introduced into *A. tumefaciens* cells. When the transformed *A. tumefaciens* is cultured with plant cells, the recombinant plasmid can move from the bacterium into plant cells and be directed into the chromosomes of plants. These transformed plant cells are then grown in tissue culture into new plants for transplanting into the field as a transgenic crop (Figure 9.3.11).

FIGURE 9.3.11 (a) Plant material is exposed to *Agrobacterium* to allow transfer of recombinant plasmids from bacteria to plant cells. (b) The plants carrying the new gene are cultured and selected in the laboratory before release for field testing.

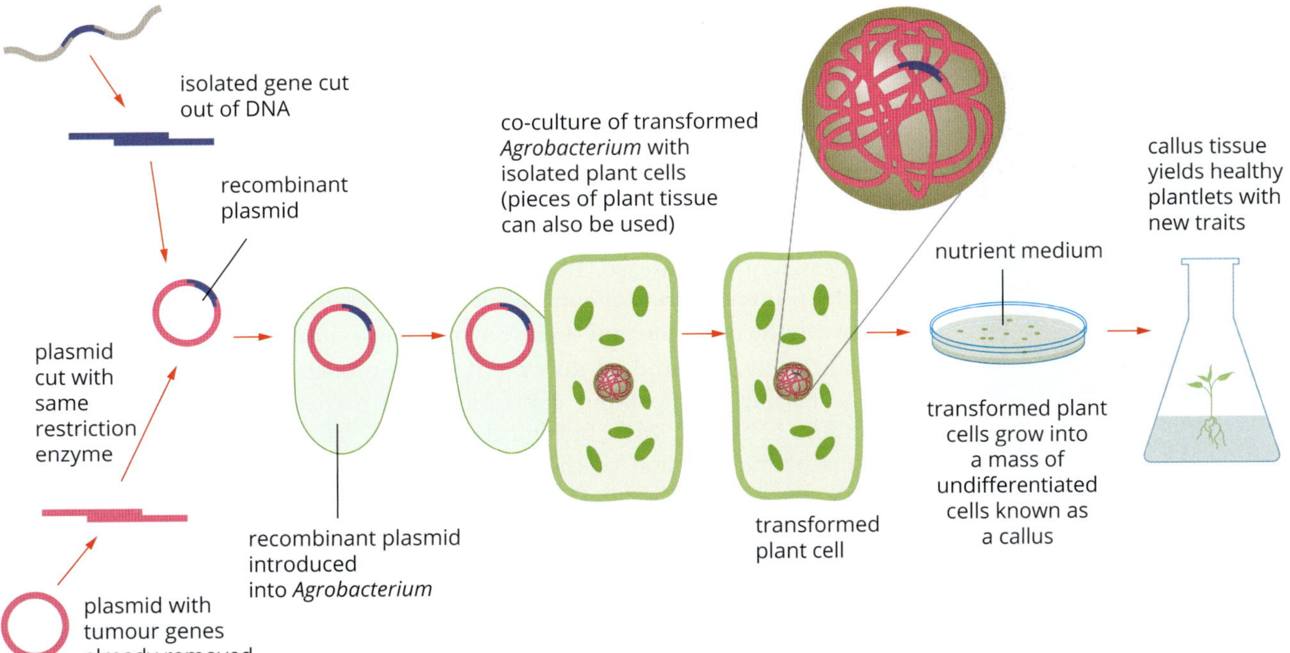

FIGURE 9.3.10 The use of *Agrobacterium tumefaciens* in gene cloning and the production of genetically modified and transgenic plants. Relative sizes of the plasmid, bacterium and plant cell are not to scale

Salt-tolerant wheat

FIGURE 9.3.12 Australian scientists have produced wheat plants that can grow in salty soil.

Soil salinity is a major problem for Australian agriculture. A high level of sodium salts in the soil leads to osmotic water loss from roots and other tissues in which salt accumulates. Cells are stressed due to the altered ratios of sodium and potassium ions in cells. Salt-tolerant plants protect themselves from the effects of salinity by preventing sodium entry into cells, storing the salt in the vacuole or pumping the sodium out of the cells. Scientists have found the genes that control these features of salt-tolerant plants.

To increase crop productivity, Australian scientists from the University of Adelaide introduced a gene from a salt-tolerant Australian native plant into wheat plants. This greatly improved the grain yield of wheat grown on salty soils without affecting grain yield in normal soil (Figure 9.3.12). The salt-tolerant gene codes for a protein that removes sodium from the leaves, allowing water to move normally from the roots to the leaves. This increases the geographical range that can be used for wheat production in Australia and other countries facing salinity problems. This is becoming increasingly important as the global population grows.

Bt cotton

Cotton is a plant that attracts many insect pests. To protect the cotton crops, they are sprayed with insecticides up to four times before the crop is harvested. This high use of insecticides impacts the populations of both harmful and beneficial insects as well as the animals that feed on them. Insecticides may also have an impact on human health. In addition, insecticides are expensive.

Bt cotton is a transgenic crop that has been modified to contain two genes from the soil bacterium, *Bacillus thuringiensis*. Expression of these genes produces proteins in the cotton plant that kill the main caterpillar pest of cotton by disrupting its digestive system. In Australia, almost all cotton grown is Bt cotton and this has reduced the use of pesticides dramatically. This decreases the environmental impacts of pesticides and saves farmers money. Australian regulators have reported no adverse effects over 15 years of Bt cotton use in Australia. Cotton seed oil extracted from Bt cotton can be sold without GM labelling as the extraction processes separate the oil from the plant's proteins and nucleic acids, therefore the oil does not have any GM components.

Transgenic animals

FIGURE 9.3.13 This transgenic mouse has been genetically modified so that it carries the gene for green fluorescent protein (GFP). This mouse will be used in scientific research. The GFP gene is linked to the gene being studied, and so acts as a reporter gene for the expression and location of the protein of interest.

Transgenic animals have been used in scientific research (Figure 9.3.13), medicine, pharmaceutical production and agriculture. In research, mice, rats and rabbits are used as living models of biological processes in healthy and diseased states. Scientists can use transgenic animals to study disease progression and potential treatments. For example, scientists studying motor neuron disease have identified a mutation in the gene for the enzyme superoxide dismutase 1 (SOD1) that is associated with some inherited cases of this disease. To study the disease in an animal model, scientists produced a transgenic mouse expressing the SOD1 enzyme mutation to establish links between the mutation and disease symptoms to investigate treatments.

Scientists may also produce knock-out mouse models in which gene technology is used to disable or knock-out expression of a particular gene. The phenotype (appearance, behaviour and biological function) of the animal gives the first indications of the function of the protein encoded by the gene. Structural and regulatory genes are studied in this way.

In agriculture, transgenic sheep and cows are used for improved fertility, meat production, milk quality and yield, and wool quality and yield. The use of genetically modified farm animals has not expanded to the extent it has for GM plants, perhaps because of detrimental effects of some modifications in animals. For example, genes that promote growth may also cause altered skeletal growth, arthritis, heart and kidney problems.

To date, GM animals are not approved for human consumption in Australia. Transgenic fish have been approved in the USA, but they are unlikely to be approved for sale in Australia in the near future.

Spider silk protein (Figure 9.3.14) is an example of a potentially useful product made in transgenic goats. The gene for spider 'dragline' silk has been put into the genome of goats, along with regulatory genes so that it is expressed in the milk. Spider silk is of great interest for its extraordinary strength and flexibility. Potential applications include a biopolymer for artificial ligaments and tendons, bandages, biodegradable bottles and tough bulletproof clothing.

FIGURE 9.3.14 Researchers are untangling the mystery of what makes spider silk so strong. The key to the silk's strength, which exceeds that of steel, is its cross-linked beta-pleated sheet structure.

Some animals are being used for the production of therapeutic proteins such as antibodies that are difficult to make in bacteria and cultured cells. This process has been referred to as 'pharming', combining the words farming and pharmaceutical. The products are released into blood or milk from where they can be readily extracted.

Medical researchers are also exploring ways to modify genes in insect vectors of disease and modify pig embryos to make their organs suitable for transplantation into humans, a procedure called **xenotransplantation**. You will learn more about xenotransplantation in Section 9.4.

BIOFILE CCT EU

Camel pharmacy

Researchers in Dubai are attempting to produce transgenic camels (*Camelus* sp.) (Figure 9.3.15) that will synthesise pharmaceutically-active proteins in their milk for use in medicine. It is hoped that the camels will provide a cost-effective way to produce the medicinal molecules which are unaffordable in many parts of the world. Camels were chosen over other species due to their high resistance to disease and ability to easily convert molecules from food. However, research is still in the formative stages, with regulators unsure if the GM stock would be effective for manufacturing and what impact the production would have on animal health.

FIGURE 9.3.15 Transgenic camels (*Camelus* sp.) could soon be used to synthesise proteins for modern medicine in their milk.

9.3 Review

SUMMARY

- When DNA from two different sources is joined together, the resulting molecule is called recombinant DNA.
- Therapeutic recombinant human proteins, made from recombinant DNA in bacteria, yeast or animal cell cultures, include hormones, cytokines, enzymes and vaccines.
- Plasmids are small, circular pieces of double-stranded DNA found in bacterial cells. They replicate independently of the bacteria's chromosomal DNA.
- Recombinant plasmids are plasmids that have had target DNA inserted into them. The same sticky-end or blunt-end restriction enzyme is used to cut both the targeted gene and the plasmid, then DNA ligase is used to permanently join the two together.
- Plasmids with antibiotic resistance are generally used to enable identification of bacterial transformation, as only bacterial cells containing these plasmids will survive when grown in cultures containing the antibiotic.
- Genetically modified organisms (GMOs) are organisms with modifications made to one or more of their genes.
- Transgenic organisms carry a gene from a different organism.
- Genetically modified animals are used in research, disease control, medicine and biomolecule production.
- *Agrobacterium tumefaciens* and plasmid transfer is a well established method of transferring genes into plant cells.
- Transgenic plants are used in agriculture, providing varieties that resist insect attack, are herbicide resistant or have improved yield or nutritional content.

KEY QUESTIONS

1. **a** What types of human proteins are commonly produced by recombinant DNA technology?
   b Suggest an advantage of this method of production compared to a traditional approach.
2. The following diagram illustrates an application of DNA technology. Name and describe the method illustrated. Give an example of its application.

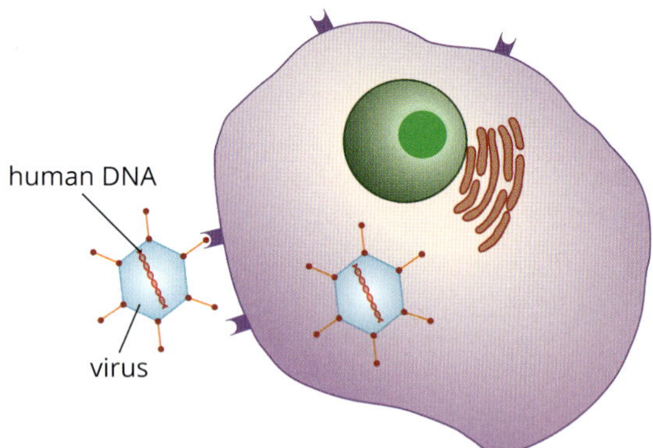

3. Recombinant human proteins may be produced in eukaryotic cells, such as cultured mammalian cells or yeast, rather than bacteria. Suggest a reason for this method of producing recombinant proteins.
4. What is a plasmid? Describe the role played by plasmids in gene cloning.
5. **a** What does 'genetic modification' of an organism mean? Include an example in your answer.
   b Describe a successful application of genetic modification in agriculture in recent years.
6. Describe a genetically modified organism and compare it to a transgenic organism.
7. The following diagram illustrates two model organisms used in research and the molecular procedures being used to alter a genetic characteristic. State whether the resulting organism is genetically modified, transgenic or both.

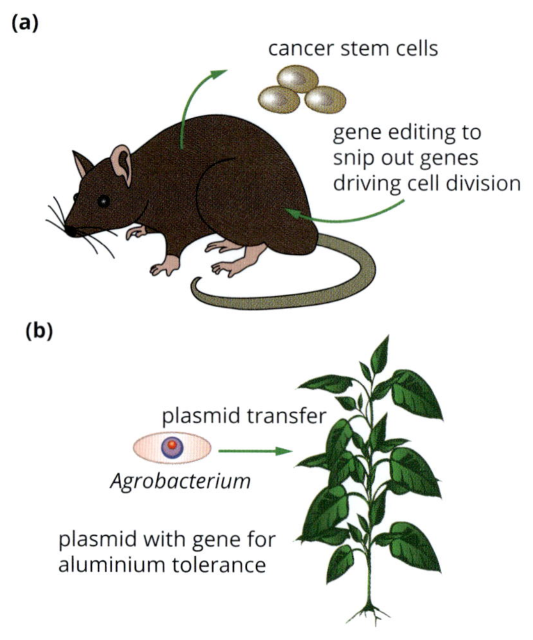

9.4 Benefits of genetic technologies

Reproductive technologies, cloning and recombinant DNA technology have a range of applications across fields including wildlife conservation, agriculture and medicine, many of which have the potential to benefit society in the future.

Medicine

Artificial insemination, which is mostly used via intrauterine insemination, allows humans to conceive a child despite challenges including fertility problems or the absence of a sexual partner. It is a relatively inexpensive and non-invasive procedure, allowing many to access reproductive technology.

Cloning holds many potential applications for medicine. One biomedical application is xenotransplantation, which involves the transplantation of cells, tissues or organs from one donor species to humans. Scientists are currently working on the genetic modification of animal sample grafts, allowing them to be disguised to reduce the likelihood of rejection by the human body (Figure 9.4.1). This would be a medical breakthrough in the need for organ donors.

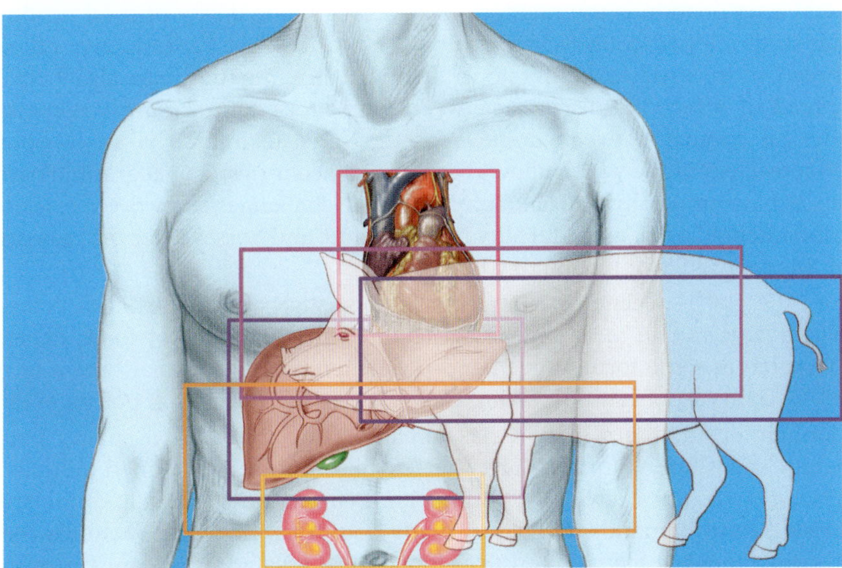

FIGURE 9.4.1 Xenotransplantation using genetically modified tissues could provide hope for the many patients awaiting organ donation. Scientists are researching the genetic modification of cloned pig organs so they do not trigger an immune rejection in humans.

Cloning can produce genetically altered cells that can be used in the treatment of diseases, such as Parkinson's and Alzheimer's, in which brain cells degenerate. Researchers aim to replace these cells with cloned and modified cells from other animals such as pigs. Cloning can also be used to develop genetically modified animals, which more closely mimic the human genome to better test new drugs. For example, many viruses such as HIV do not affect animals the same way as humans, a genetically modified animal would allow scientists to better develop drugs if they are able to monitor human-like responses.

A range of proteins for therapeutic purposes are produced using recombinant DNA technology (Figure 9.4.3). Examples include epidermal growth factor used in the treatment of burns to improve the survival of skin grafts, interleukin-2 used in cancer treatment, antibodies for immunotherapy and vaccines against a number of viruses.

Recombinant protein production

Recombinant proteins are produced by introducing recombinant DNA into bacteria or eukaryotic cells and allowing them to synthesise the protein. The main types of proteins produced by this technology are hormones, cytokines, enzymes and vaccines for human therapeutic purposes. This is much safer and more effective than using proteins purified from other organisms, such as the purification of insulin from pigs and growth hormone from human pituitary glands, as was done in the past. Examples of human proteins that are now produced in bacteria, yeast or animal cell culture are viral coat proteins for hepatitis B vaccines, viral capsid protein for human papilloma virus vaccines, cytokines (e.g. interleukin-2) for the treatment of cancer and insulin for the treatment of diabetes.

Agriculture

Artificial insemination is the method for breeding the majority of food production animals in developed countries; it can be used to improve reproductive efficiency and breed more productive livestock. Artificial insemination can also help limit the spread of infectious diseases between animals as samples are screened, and no physical contact between animals occurs. This technique is an inexpensive reproductive technique which allows breeding to occur despite geographic barriers, as well behavioural, physical and physiological challenges. However, researchers have documented a decline in fertility in dairy cattle and horses since the mainstream use of this reproductive technology began.

With the global decline of natural pollinators, such as bees, artificial pollination will play a large role in the future of agriculture, with over 90% of foods dependent on pollination in some form. While hand-pollination is still a popular option in many sectors, mechanical pollination is taking a leap forward with the help of pollinator drones.

Cloning for food production has already been approved in many parts of the world, and some countries have started mass-producing cloned animals. For example, in China, upwards of 500 pigs are being cloned annually at facilities to meet rising food demands. However, many countries including Australia, believe there is not enough research to ensure cloned meat is safe for human consumption.

Industry

Recombinant DNA technology is widely used in biotechnology from its use in the food industry with the production of GM foods and synthetic enzymes such as amylase and lipase, to the development of pharmaceutical products, textile treatments and detergent additives. There is also continued advances in the fields of bioremediation and biofuels. You learnt about these technologies in Chapter 8.

GO TO ➤ Section 8.2 page 320

BIOLOGY IN ACTION 

Recombinant human erythropoietin

Red blood cell production is essential for maintaining oxygen homeostasis. A drop in oxygen supply to tissues (hypoxia) normally triggers the release of the protein erythropoietin (also known as EPO) from the kidneys. EPO promotes red blood cell production in the bone marrow to restore the oxygen-carrying capacity of blood and its delivery to tissues (Figure 9.4.2).

In chronic kidney disease, not enough EPO is made by the kidneys, resulting in low red blood cell counts and anaemia. Recombinant human EPO for the medical treatment of this disease is produced in cultured mammalian cells. A copy of the human EPO gene is inserted into a plasmid which is introduced into mammalian host cells.

EPO is a glycoprotein and must have the correct carbohydrates attached to the protein chain to function properly. Bacteria cannot do this therefore mammalian cells must be used for making the recombinant protein (Figure 9.4.3). Recombinant EPO has also been developed for veterinary use, such as recombinant feline EPO for cats with chronic kidney disease.

Because EPO promotes red blood cell production and oxygen-carrying capacity, it has been used by athletes seeking an advantage. EPO has been at the centre of sports doping scandals in recent years, particularly in endurance sports such as cycling, long-distance running and the triathlon. It has also been used in horse racing. The World Anti-doping Agency (WADA) works with drug testing laboratories to develop and validate tests that can distinguish between EPO produced naturally in the athlete and pharmaceutical EPO.

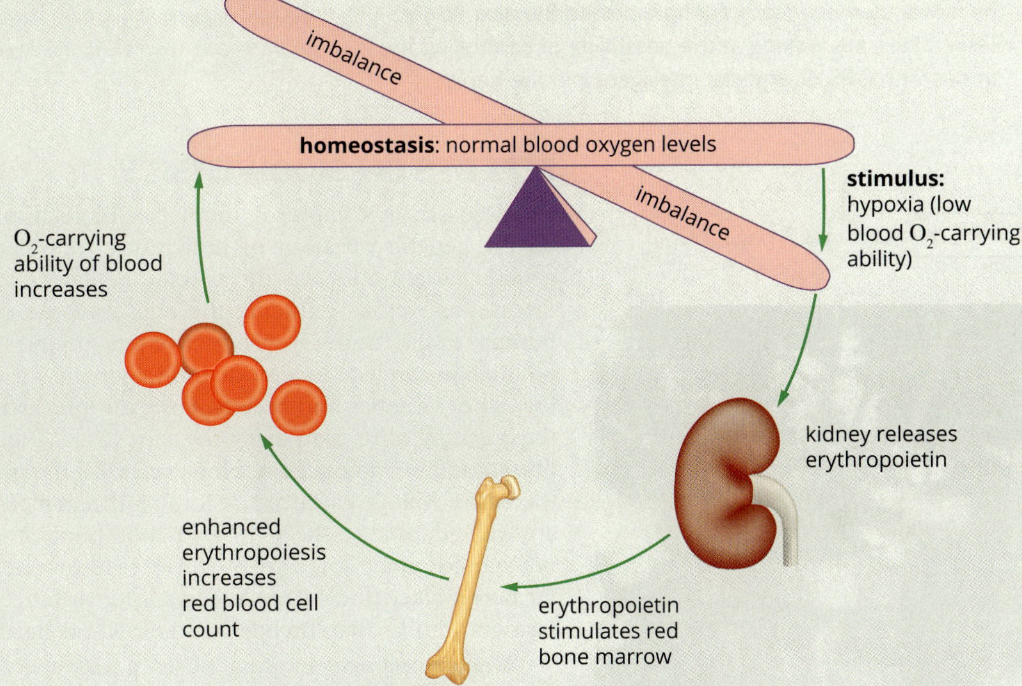

FIGURE 9.4.2 Erythropoietin is a protein released mainly by the kidney to maintain oxygen homeostasis. It promotes red blood cell production (erythropoiesis) in the bone marrow.

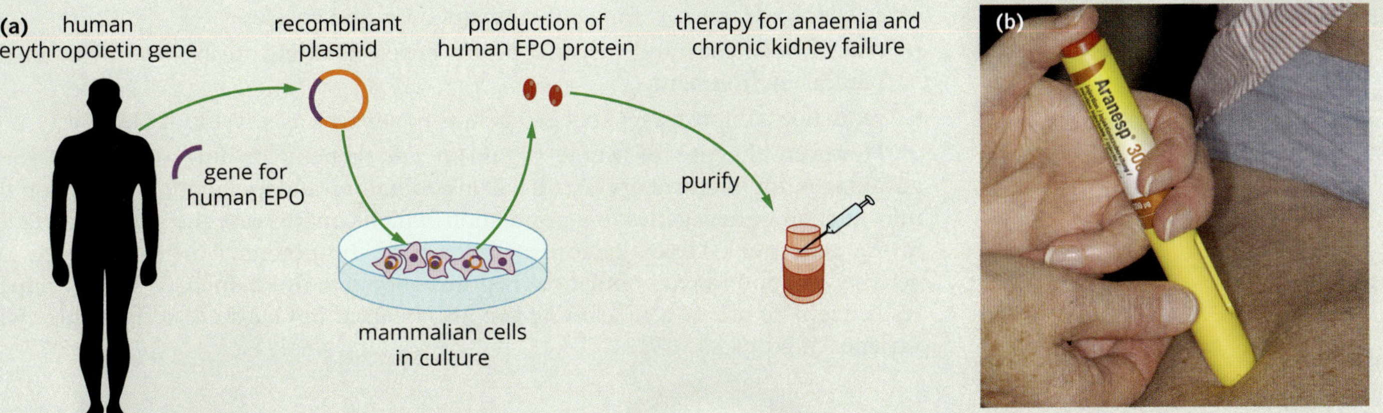

FIGURE 9.4.3 Recombinant human erythropoietin is produced in cultured mammalian cells. (a) It is used to treat anaemia and chronic kidney disease. (b) The brand Aranesp®, for example, is used to treat severe anaemia caused by chemotherapy and chronic kidney failure.

BIOLOGY IN ACTION 

Pollinator drones

With approximately 40% of pollinator species currently facing extinction globally, and with 75–90% of food crops that humans depend on for world food supply dependent on pollinators, scientists are looking toward mechanical artificial pollination methods to secure future food security. Enter the pollinator drone: a robotic artificial pollinator the size of a hummingbird (Figure 9.4.4). With four rotating blades the drone can travel from flower to flower via human-driven controls. Attached to the drone are bristles covered in a camouflaged ionic liquid gel which collect the pollen off the flower stamens ready for transport to the next flower. Researchers are looking at the possibility of controlling the drones with GPS or artificial intelligence in the future.

FIGURE 9.4.4 The pollinator drone, driven by human controls, is able to transport pollen from one flower to another.

EFFECTS ON BIODIVERSITY

GO TO ➤ Section 8.3 page 332

As discussed in Chapter 8, the use of biotechnology can have a significant impact on the genetic variation of individuals, populations and species. A reduction in genetic variation lessens the species' resilience to changes in the environment and disease, as well as reducing the ability of the environment to meet the needs of humans in the future. Reproductive technologies such as artificial insemination and pollination can lead to a reduction in genetic variation by focusing on, and selecting for, fewer varieties and breeds. For example, artificial insemination has allowed for the breeding of Fresian dairy cows in Africa (Figure 9.4.5), resulting in greater milk output and economic gain. However, with the introduction of this new population, the native Ankole cattle are becoming threatened. This is a concern as Ankole cattle are well-adapted to the semi-arid conditions of countries such as Uganda, and do not require great amounts of water, feed or veterinary attention. During droughts farmers walked their Ankole cows long distances to water sources, whereas those farmers with Fresian breeds lost their whole herd.

FIGURE 9.4.5 A Ugandan veterinarian prepares to artificial inseminate a Fresian cow, one of the most productive breeds of dairy cattle

Whole organism cloning poses a risk to biodiversity, with the potential for multiple identical genetic copies becoming widespread in a population resulting in a loss of natural vigour. As discussed in Chapter 8, transgenic and genetically modified species pose a significant threat to genetic variation, with threats to agriculture and the natural environment as a result of:

- selection of a narrow range of cultivars over wild species
- risk of pesticide- and insect-resistant species leading to invasive pests in the natural environment
- reduction of non-target species such as honeybees as a result of GM.

However, the use of whole organism and genetic cloning does hold some advantages for biodiversity. Artificial insemination and pollination allow for the introduction of new genes to a population, which can increase the genetic variation in the gene pool. De-extinction methods hold the potential to re-introduce lost genetic variation into a population, restore ecosystem functioning as well as genetic resilience. The use of gene cloning can also be used in GMOs to re-introduce wild varieties of crops.

DS 6.1 | WS 6.7 | WS 6.8

9.4 Review

SUMMARY

- Reproductive technologies, cloning and recombinant DNA technology have a range of applications in wildlife conservation, agriculture, medicine and industry.
- A biomedical application of cloning is xenotransplantation, which involves the transplantation of cells, tissues or organs from a donor species to a different species (e.g. a non-human donor to a human).
- A range of proteins for therapeutic purposes are produced using recombinant DNA technology.
- With the global decline of natural pollinators, such as bees, artificial pollination will play a large role in the future of agriculture, with more than 90% of foods dependent on pollination in some form.
- Reproductive technologies such as artificial insemination and pollination can lead to a reduction in genetic variation by focusing on, and selecting for, fewer varieties and breeds.
- Transgenic and genetically modified species pose a significant threat to genetic variation, with threats to agriculture and the natural environment.

KEY QUESTIONS

1. Identify two advantages of artificial insemination for human fertility treatment.
2. Define the term 'xenotransplantation'.
3. Explain why stem cells are desirable for scientific research into therapeutic applications.
4. Provide an example of a therapeutic application of recombinant DNA technology in medicine.
5. Explain how reproductive technologies such as artificial insemination and artificial pollination can lead to a decrease in biodiversity.
6. Identify one advantage and one disadvantage of the artificial insemination and breeding of Fresian cattle in Uganda.

Chapter review

09

KEY TERMS

animal husbandry
artificial embryo twinning
artificial insemination
artificial pollination
assisted reproductive technology
bacterial transformation
biodiversity
clone
cryopreservation
de-extinction
DNA (deoxyribonucleic acid)
gene
gene cloning
gene editing
gene expression
gene pool
gene therapy
genetic screening
genetic variation
genetically modified organism (GMO)
genome
intrauterine insemination (IUI)
in vitro
in vitro fertilisation (IVF)
in vivo
lacZ gene
ligase
mutualism disruption
plasmid
polymerase chain reaction (PCR)
recombinant DNA
recombinant plasmid
recombinant protein
restriction enzyme
somatic cell nuclear transfer (SCNT)
therapeutic cloning
tissue culture
transgenic organism
transgene
vector
whole genome sequencing
whole organism cloning
xenotransplantation

REVIEW QUESTIONS

1 **a** Describe cryopreservation.
b What is an advantage of cryopreservation to breeding?

2 Explain why artificial insemination can be a low-risk option for breeding wild animals in captivity.

3 What is artificial pollination?

4 The native bee population on Truebark Peninsula has become threatened due to pesticide-reliant local agriculture and habitat destruction. The native Truebark blossom plant relies on the pollination of its flowers by the native bee population.
a Is it likely that the Truebark blossom plant will be affected by the decline in native bees?
b If the Truebark blossom plant became extinct due to the lack of pollination, what would this break down in ecological interaction be called?

5 Define the following terms and give an example of each of the terms below.
a gene cloning
b recombinant DNA

6 Construct a flow diagram showing the process involved to clone Dolly the sheep.

7 Differentiate between whole organism cloning and gene cloning.

8 Provide an example of a genetic technology being used in wildlife conservation.

9 Describe how genetic modification can be useful as a tool to fight vector-borne disease, such as a disease carried and transmitted by an insect vector.

10 **a** Restriction enzymes are a basic molecular tool in gene technology. What is a restriction enzyme and what can it do?
b Describe the difference between sticky ends and blunt ends produced by restriction enzymes.
c Outline the purpose of DNA ligase in recombinant DNA technology.

11 Draw a flow diagram outlining the different techniques and processes involved in insulin production. Use diagrams where possible to assist your explanations.

12 **a** Explain what is meant by 'genetic transformation'.
b List three protein products manufactured using genetic transformation and outline their importance in medicine or agriculture.

13 Some students are doing an experiment involving bacterial transformation. Bacteria were incubated with plasmids containing resistance to the antibiotic ampicillin and then grown on agar plates. A plate that will have only transformed bacteria growing will have which of the following?
A nutrient agar only
B nutrient agar and ampicillin
C plain agar with ampicillin
D nutrient agar, ampicillin and penicillin

14 Genes such as the *lacZ* gene can be used as reporter genes. Reporter genes can be used to determine:
A that a bacterium has taken up a plasmid
B that a plasmid has accepted the gene of interest
C that a bacterium has taken up a plasmid containing the gene of interest
D none of the above

15 One result of a genetic application was the Flavr Savr tomato. Tomatoes have a short shelf life due the effects of an enzyme called polygalacturonase. This enzyme catalyses the breakdown of the cell walls of the tomato, causing the tomatoes to become soft and unappetising. To slow down this process, the sequence of the polygalacturonase gene was determined and an antisense gene was produced. The antisense gene has a complementary nucleotide sequence to the polygalacturonase gene. The antisense gene was inserted into the tomatoes.

When the antisense gene is transcribed, the mRNA produced is complementary to the mRNA for the polygalacturonase gene, so the two mRNAs join to form double-stranded mRNA. Double-stranded mRNA cannot be translated, so the enzyme is not formed and the cell walls are not broken down.

The Flavr Savr tomatoes can be considered to be:

A only transgenic

B only genetically modified

C both genetically modified and transgenic

D none of the above

---

16 Arsenic contamination of soil is a serious problem in some countries. The arsenic contaminates groundwater and drinking wells. The following flow chart illustrates a process used to insert a gene that enables plants to absorb arsenic from the soil. In your notebook write the terms that describe steps A–E.

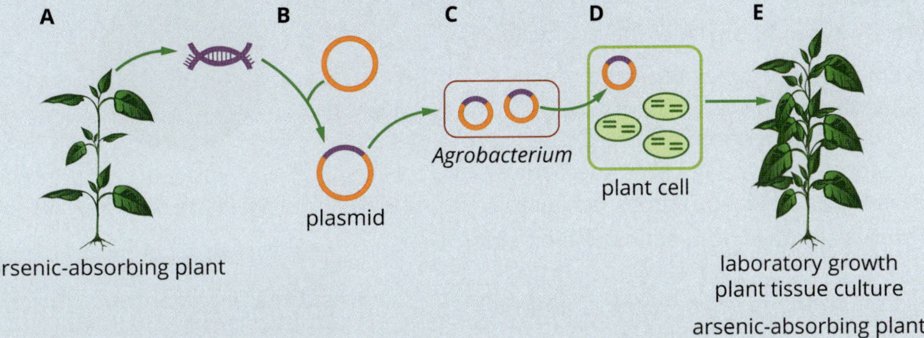

---

17 Strawberries are a very fragile fruit. If they are exposed to freezing temperatures the fruit becomes soft and unappealing. This results in considerable economic loss to strawberry farmers. Scientists have been searching for a way to make strawberries more resistant to frost. One approach that is still in laboratory trials is to genetically engineer strawberries with a gene from Arctic flounder (*Liopsetta glacialis*). Arctic flounder live in near-freezing waters but their blood does not freeze because they make a protein that acts as an antifreeze. Scientists have cut this gene from the genome of the Arctic flounder and inserted it into a plasmid. They have also inserted a gene that makes epidermal cells produce a blue pigment into the same plasmid. A species of bacteria, *Agrobacterium tumefaciens*, is used to modify the strawberry cells. When the strawberry cells have been successfully modified they produce both the antifreeze protein and the blue pigment. Blue frost-resistant strawberries are produced.

a The resulting strawberry plants are genetically modified. Are they also transgenic? Explain.

b Blue strawberries are unlikely to have large-scale consumer appeal so when the strawberries finally move to field trials this gene will not be in the modified strawberries. Why is it being used in this early stage of research?

c Draw a flow chart of the steps needed to make the strawberries frost resistant.

d If the frost-resistant strawberries ever get to the stage of field trials, applications will need to be made to the government department that oversees the control of genetically modified organisms. Why can't scientists just plant genetically modified crops without any oversight?

18 Describe two applications of genetic technologies to the field of medicine.

19 Identify one advantage and one disadvantage of artificial insemination use in agriculture.

20 Explain why the loss of genetic variation is detrimental.

21 Describe the threats whole organism cloning poses to global biodiversity.

22 What are some of the effects (positive and negative) of genetically modified and transgenic organisms on biodiversity?

23 After completing the Biology Inquiry on page 342, reflect on the inquiry question: Does artificial manipulation of DNA have the potential to change populations forever? Consider examples of GMOs that are already in use in Australia. Use secondary sources to evaluate the impact these GMOs have had on the populations of these organisms in the short and long term.

MODULE 6 • REVIEW

REVIEW QUESTIONS

Genetic change

MR 6

Multiple choice

1 Identify which feature of cancer cells makes them different from normal cells.

A Cancer cells are unable to synthesise DNA.
B Cancer cells are arrested in S phase of cell cycle.
C Cancer cells continue to divide even when they are tightly packed together.
D Cancer cells are always in M phase of the cell cycle.

2 The use of embryonic stem cells has attracted much attention in the scientific world and in the media. Pluripotent stem cells are taken from embryos. They can be stimulated to become any type of cell in the body. This technology has led to ethical questions because:

A differentiated stem cells have no practical use outside the laboratory
B stem cells must be taken from two-week-old embryos that have been removed from the uterus
C there is no source of embryonic stem cells other than from aborted fetuses
D stem cells are from excess embryos produced in the IVF process

3 Scientists perform an experiment using a bacterial culture in which the cells have a mutation in both the *lacI* and *lacA* genes, causing these genes to be non-functional. The cells are exposed to both lactose and glucose. Which result shows the expected activity of the *lac* genes?

	lacZ	*lacY*	*lacA*
A	low	low	low
B	high	high	high
C	high	high	low
D	low	low	high

4 Students were performing an experiment to transform *E. coli* bacteria and make them resistant to the antibiotic kanamycin. The experiment involved four plates—two with plain nutrient agar and two with nutrient agar and kanamycin. One plain nutrient plate and one of the kanamycin plates were exposed to untreated *E. coli*. A second batch of *E. coli* was incubated with a plasmid containing the kanamycin-resistance gene and then heat shocked. These bacteria were then divided between the remaining plates—one nutrient only and the other with kanamycin. The plates were then incubated at 37°C for 24 hours. When the students came to check their plates they realised that they had forgotten to label them. Using your understanding of the processes involved, determine which of the following alternatives correctly identifies each of the plates.

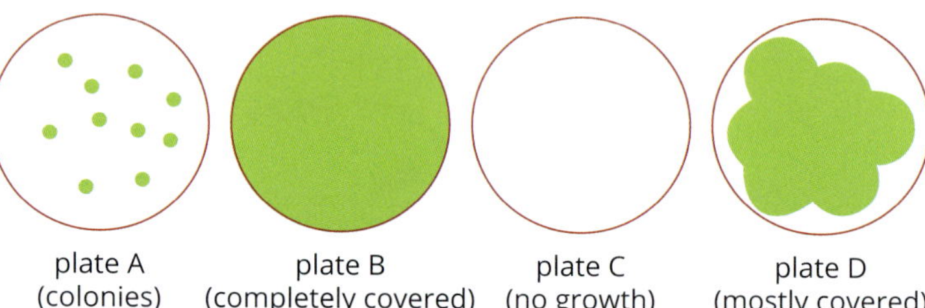

	Untreated *E. coli*		Transformed *E. coli*	
	Nutrient agar only	Nutrient agar and kanamycin	Nutrient agar only	Nutrient agar and kanamycin
A	plate B	plate C	plate D	plate A
B	plate A	plate B	plate C	plate D
C	plate D	plate A	plate B	plate C
D	plate C	plate D	plate A	plate B

5 Select the most accurate description for restriction enzymes:

A enzymes that replicate DNA
B enzymes that cut DNA at particular base sequences
C enzymes involved in gene expression
D digestive enzymes involved in protein breakdown

6 Which list displays chemicals that are used in polymerase chain reactions (PCR)?

A restriction enzymes and a primer to copy a DNA sequence
B DNA polymerase to produce a primer
C DNA polymerase and a primer to produce many copies of DNA
D restriction enzymes to produce a primer

7 Identify the mutation that would be most likely to affect a species.

A mutations in the DNA of somatic cells
B mutations in DNA during mitosis
C mutations in DNA during polypeptide synthesis
D mutations in DNA during meiosis

8 Select the answer that would be most likely to result in polyploidy:
- **A** gamete with n number of chromosomes
- **B** a gametic cell with $2n$ number of chromosomes
- **C** a sperm or ovum with n number of chromosomes
- **D** a somatic cell with $2n$ number of chromosomes

9 Germline mutations:
- **A** are generally lethal
- **B** are located in some somatic cells
- **C** will be found in all cells of the offspring
- **D** affect the individual but not the offspring

10 Which answer distinguishes between point mutation and chromosomal mutation?
- **A** Point mutation codes for one new protein; chromosomal mutation codes for many new proteins.
- **B** Point mutation affects a single gene; chromosomal mutation affects a whole chromosome.
- **C** Point mutation affects somatic cells; chromosomal mutation affects germline cells.
- **D** Point mutations are also called block mutations; chromosomal mutations are also called trisomy.

11 A frameshift mutation is:
- **A** adding or deleting a base pair
- **B** substituting one base with another base
- **C** translocating a part of a chromosome onto another chromosome
- **D** when the mRNA cannot read the code

12 Select the statement that explains the use of genetically engineered organisms most accurately.
- **A** Genetic engineering produces better species.
- **B** Genetic engineering will help humans to survive better.
- **C** Genetically engineered organisms can improve agricultural production.
- **D** Genetically engineered organisms cannot be used in medicine.

13 Which is the most accurate definition of a clone?
- **A** a genetically identical organism
- **B** an exact copy of a different species
- **C** an organism produced by genetic engineering
- **D** a genetically modified organism

14 Identify which series best summarises processes involved in producing a transgenic species:
- **A** induction of ovulation, artificial insemination, normal intrauterine development
- **B** extracting target gene from DNA of one species, copying the gene, inserting it into a recipient organism
- **C** gene shearing to obtain target gene, gene replication, microinjection of gene into cell of another species
- **D** extracting nucleus from a parental somatic cell, transferring nucleus to enucleated ovum, implanting ovum in utero

15 Which description describes a recombinant protein?
- **A** has a structure that has been denatured then restored
- **B** contains polypeptides from different types of proteins
- **C** copied multiple times using PCR technology
- **D** expressed from DNA that has a gene inserted from a different species

16 Heterozygosity can be used as a measure of genetic diversity in a population. It refers to:
- **A** genotypes with different alleles at matching chromosome loci, resulting in greater population diversity
- **B** genotypes with the same alleles at matching chromosome loci, resulting in greater population diversity
- **C** genotypes with different alleles at matching chromosome loci, resulting in lower population diversity
- **D** genotypes with mutated alleles at chromosome loci, resulting in lower population diversity

17 Reporter gene is the term used for a:
- **A** mutated gene that signals its presence with expression of a coloured pigment
- **B** gene tag attached to a target gene so the target can be tracked in genetic engineering
- **C** a transgene created by genetic engineering to improve nutritional value of a food crop
- **D** recombinant gene used for cloning

18 A plasmid is a:
- **A** structure in prokaryotic cells used in asexual reproduction by binary fission
- **B** short synthetic segment of DNA used in the polymerase chain reaction to replicate DNA in a laboratory
- **C** circular section of DNA in bacteria, which is separate and smaller than the chromosomal DNA
- **D** virus vector used in genetic engineering

19 Identify the two methods commonly used for inserting transgenic plasmids in the bacterial transformation process:
- **A** vector transfer and microinjection
- **B** PCR and co-culture
- **C** heat shock and electrolysis
- **D** heat shock and electroporation

20 Artificial embryo twinning is a technique used for:
- **A** whole organism cloning by in vitro separation of an early stage embryo into two
- **B** human IVF programs to provide twins for the parents
- **C** tissue culture of endangered plant species
- **D** somatic cell nuclear transfer research to create identical clones of valuable livestock

MODULE 6 • REVIEW

Short answer

21 Match the terms in the left column with the correct description in the right column.

term	description
intron	sequence of chromosomal DNA in prokaryotic cells with a cluster of genes
exon	incorrect diploid number of chromosomes
operon	sequence of DNA that does not code for proteins
aneuploidy	changes to DNA in body cells that are not gametes
polyploidy	sequence of DNA that codes for proteins
genetic drift	changes to DNA in gamete-forming cells
genetic isolation	more than one diploid set of chromosomes
somatic mutation	random changes to allele frequencies in a gene pool
germline mutation	gene flow between populations of the same species is blocked

22 Identify if each statement is true or false and justify each of your answers.

a Mutations always harm an organism.

b Mutagens can be chemicals, nuclear radiation or naturally occurring.

c Neoplasms result from uncontrolled cell division and may be benign or malignant.

d Promoter and terminator mutations prevent synthesis of any polypeptides.

e The bottleneck and founder effects both lead to lower genetic variation in populations.

f Sticky ends are single sections of DNA left exposed after use of a restriction enzyme.

23 a Define the term 'mutation'.

b Account for how mutations may lead to new alleles and influence the process of evolution.

c Describe the circumstances in which mutations can affect:

i individuals in the next generation

ii only the organism in which the mutation occurs.

24 a Define the term 'mutagen'.

b Outline the evidence for the statement 'some types of electromagnetic radiation are mutagenic'.

c Assess the potential impact of mutagens released into the environment by human activity on the evolution of living organisms.

25 Biodiversity provides resilience to change. Interpret this statement, using examples to provide substance in your answer.

26 Complete a table to distinguish between DNA, mtDNA and mRNA.

27 Match the terms in the left column with the correct description in the right column.

term	description
biotechnology	use of mechanical systems that function like part of a living organism
biodiversity	genetically modified by transfer of specific genes from another species
bioremediation	use of technology on a very small (molecular) scale
bionics	the number, variety and variability amongst living organisms
gene editing	using living organisms and biological processes to develop new products
gene therapy	changing a gene at a specific site on the DNA
nanotechnology	use of microbes to destroy or lessen the risk from contaminants
transgenic	introduction of healthy genes to replace defective ones in a patient

28 Glycoproteins are common in eukaryotic organisms. Many hormones are glycoproteins (e.g. EPO for red blood cell production). They are part of each cell's plasma membrane, they create mucous linings, hold skin cells tightly together and help the immune system. Glycoproteins have sugars attached to the protein's polypeptide chains and they are formed by specialised organelles in the eukaryotic cells.

Explain why recombinant bacterial cells cannot be used to produce human glycoproteins for therapeutic use.

29 Scientists conducted an experiment to grow transformed bacteria with the *lac* operon on nutrient agar plates containing ampicillin and X-gal. The results are shown below. Explain why bacterial colonies that have been successfully transformed with recombinant plasmids appear white and the non-recombinant colonies are blue due to expression of the *lacZ* gene.

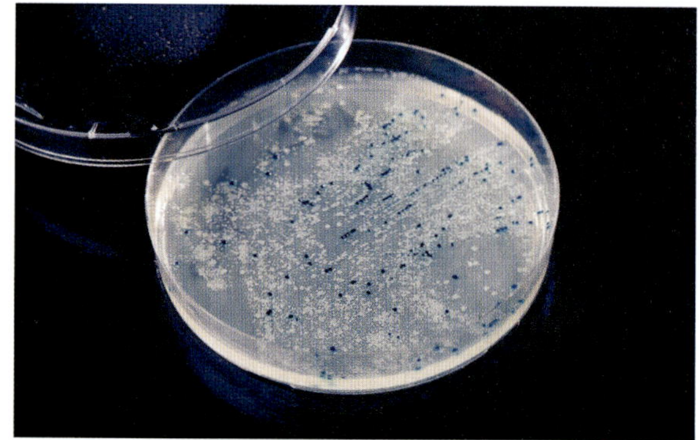

30 The figure below shows a karyotype of a human baby.

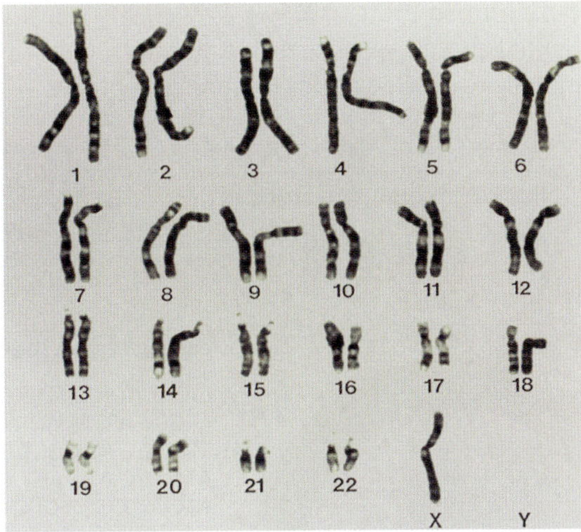

a The chromosomes are arranged in pairs. State the name given to chromosomes that contain the same gene loci.

b Explain the term 'diploid' and why all human somatic cells are diploid.

c Is there any chromosome abnormality with this baby? Explain your answer.

d Describe the types of variation seen in chromosome structure and number between species.

31 a Give three examples of different types of mutation with a brief outline to explain each.

b Using the sequence shown in the figure below, draw up separate flow charts that identify the consequences of the following changes in the mRNA sequence:

i substituting the first guanine (G) base with adenine (A)
ii inverting the bases AAAUCC
iii duplicating the sequence GCU
iv inserting a cytosine following GCU
v deleting the first uracil base.

Note: You are working with the mRNA sequence in the middle NOT the DNA at the top.

Refer to Figure 4.2.7 on page 168 to check the genetic codes for the 20 different amino acids.

G C T A A A T C C T A G
base sequence in DNA template strand C G A T T T A G G A T C

TRANSCRIPTION (copying)

base sequence in mRNA G C U A A A U C C U A G

TRANSLATION (change language)

amino acid sequence in polypeptide Ala — Lys — Ser stop translation

32 Today, scientists are using genetic technologies that artificially manipulate DNA and its inheritance for many organisms.

a Reproductive technologies include artificial insemination, artificial pollination and cloning. Outline the methods used in each of these techniques.

b Predict how each of these three techniques would alter the genetic composition of a population.

c Evaluate the benefits of using genetic technologies like these.

33 Research has been conducted to produce an improved form of a cereal crop plant that is genetically modified (GM) but is not a transgenic plant.

a Clarify the terms, genetically modified and transgenic.

b Critically evaluate each type of genetic technology. Use examples to support your answer.

34 Antithrombin is a plasma protein that stops blood clots forming in inappropriate places. People with a mutation in the gene for antithrombin production will easily develop a thrombosis (blood clot) and will generally require hospitalisation. Blood clots in the brain and heart can cause death. Like many blood-clotting diseases, antithrombin deficiency is treated by injecting required amounts of the protein. The challenge for doctors is the supply of the protein. To create a steady supply of antithrombin, goats have been engineered to produce the protein in their milk. The process is summarised below:

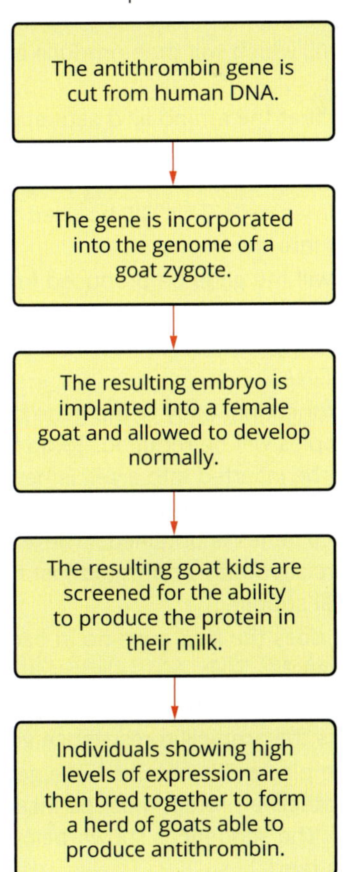

Note that the goats are both genetically modified and transgenic.

a Explain how it is possible that a goat can be genetically modified to express the human antithrombin protein.

b Are there any possible drawbacks to using animals to make human proteins?

c Identify an ethical issue associated with the use of animals to produce human pharmaceuticals.

35 Kuru is a disease that was once common in the highlands of New Guinea. It has been established that it is caused by a prion. Like mad cow disease in cattle and Creutzfeldt-Jakob disease (vCJD) in humans, this prion builds up in neurons causing plaques that eventually destroy the cells and compromise neurological function.

Researchers studying the problem in New Guinea have discovered that some individuals are highly resistant to the misfolding of their proteins into the prion form. Further study has established these individuals possess a mutated protein.

Further research of this protein is needed as it may lead to a treatment or cure for both Kuru and vCJD. To study this protein, a large and readily available pure supply is needed, so the scientists wish to introduce the gene for the protein into bacteria, which will then produce a constant supply for research.

a Assume that the amino acid sequence of only part of the protective protein has been identified. Using this information, mRNA for the protein has been extracted from human cells. This will be used to make the gene to insert into the bacterium.

  i How will the gene be produced from the mRNA?

  ii Why is it better to use mRNA in this case rather than DNA?

b Once a functional copy of the gene has been created, many copies will be required. To do this a plasmid that can be inserted into a cell is needed. A plasmid containing the *lacZ* and ampicillin (an antibiotic) resistance genes is obtained. (These are made commercially today and would just be bought from a biological supplier.)

  i What does the *lacZ* gene do in bacteria normally?

  ii What does it do in the transformed bacteria?

  iii Why is the ampicillin resistance gene included?

c Once the plasmid is obtained, the gene for the Kuru-protective protein must be inserted into the plasmid. The sequences shown below are for the coding strand.

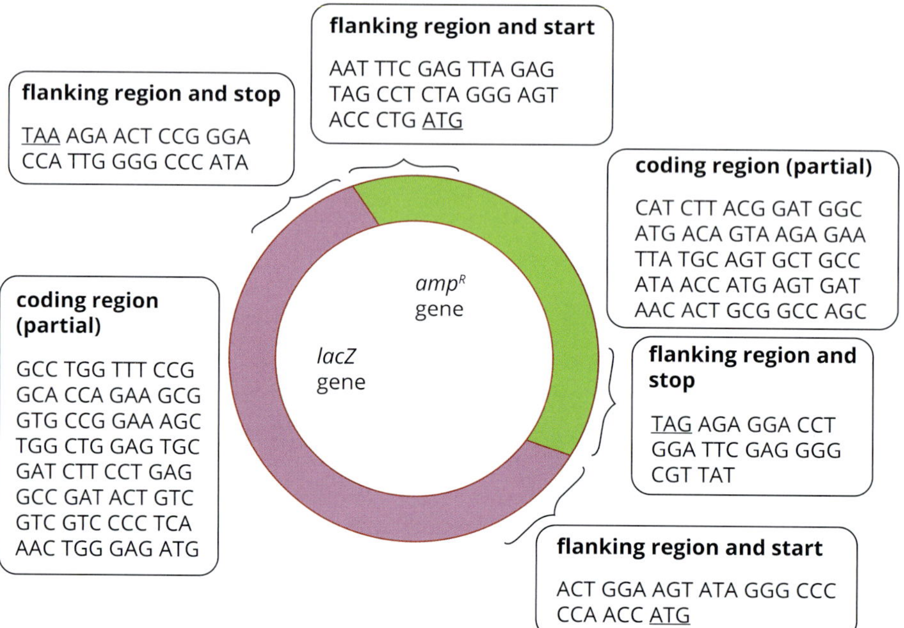

Many plasmids are cut open with a restriction enzyme and are incubated with the gene of interest. You have four restriction enzymes you could use. The enzymes have cutting sites as shown. The slash (/) indicates the cutting site. Enzymes 1 and 2 create sticky ends and enzymes 3 and 4 create blunt ends.

enzyme	1	2	3	4
cutting site	GGG/CCC	CTT/CCT	GA/TACT	GAA/AGC

  i What is the difference between sticky ends and blunt ends?

  ii Explain which enzyme should be used to cut the plasmid.

d Once the plasmid has been cut, it should be incubated with the gene to allow the gene to be incorporated into it. The bacteria will then be mixed with the plasmids and a proportion of the bacteria will be transformed.

  i Name the enzyme needed to incorporate the gene for the protein of interest into the plasmid.

  ii How will the proportion of bacteria that is transformed be increased?

7 Infectious disease

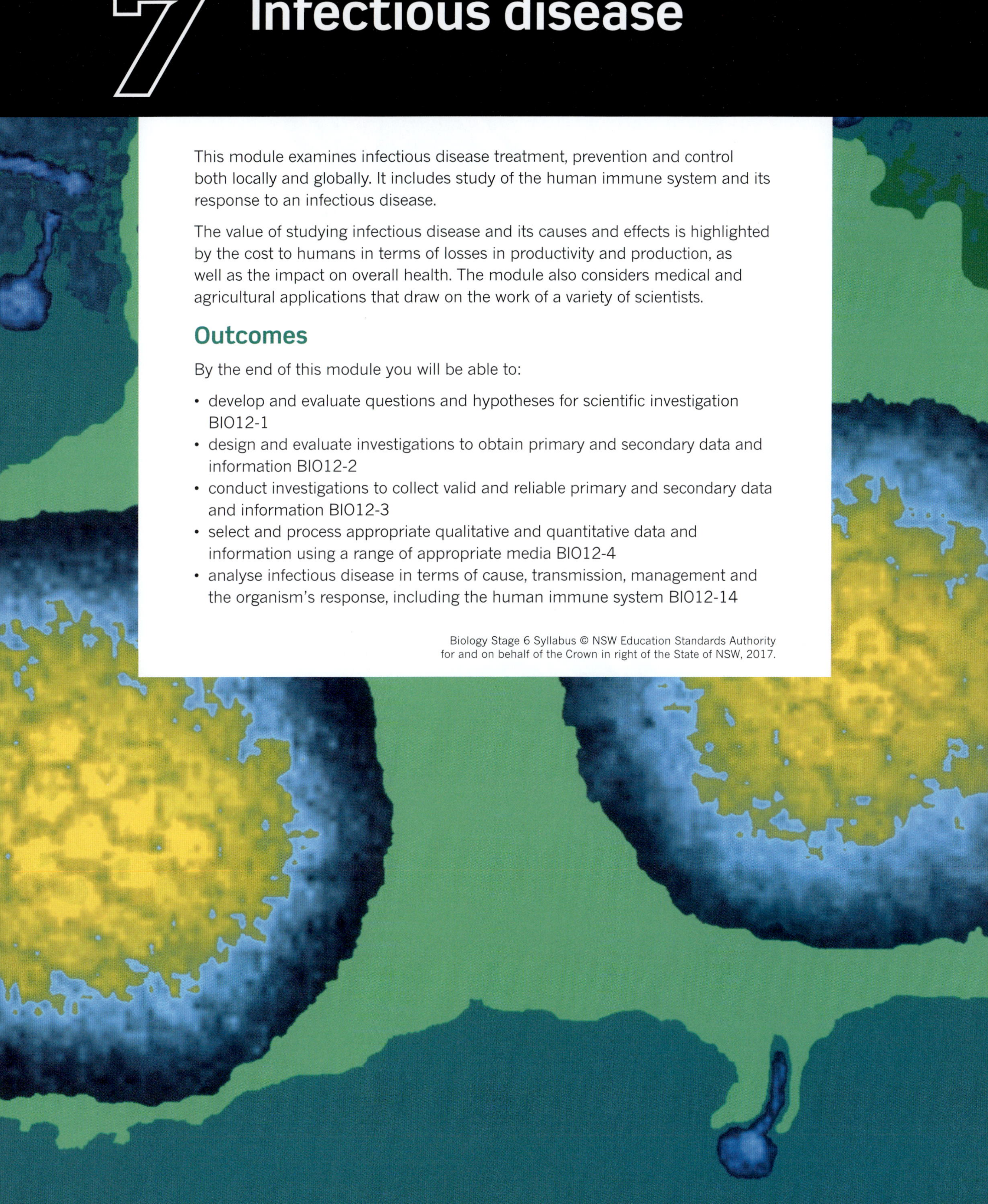

This module examines infectious disease treatment, prevention and control both locally and globally. It includes study of the human immune system and its response to an infectious disease.

The value of studying infectious disease and its causes and effects is highlighted by the cost to humans in terms of losses in productivity and production, as well as the impact on overall health. The module also considers medical and agricultural applications that draw on the work of a variety of scientists.

Outcomes

By the end of this module you will be able to:

- develop and evaluate questions and hypotheses for scientific investigation BIO12-1
- design and evaluate investigations to obtain primary and secondary data and information BIO12-2
- conduct investigations to collect valid and reliable primary and secondary data and information BIO12-3
- select and process appropriate qualitative and quantitative data and information using a range of appropriate media BIO12-4
- analyse infectious disease in terms of cause, transmission, management and the organism's response, including the human immune system BIO12-14

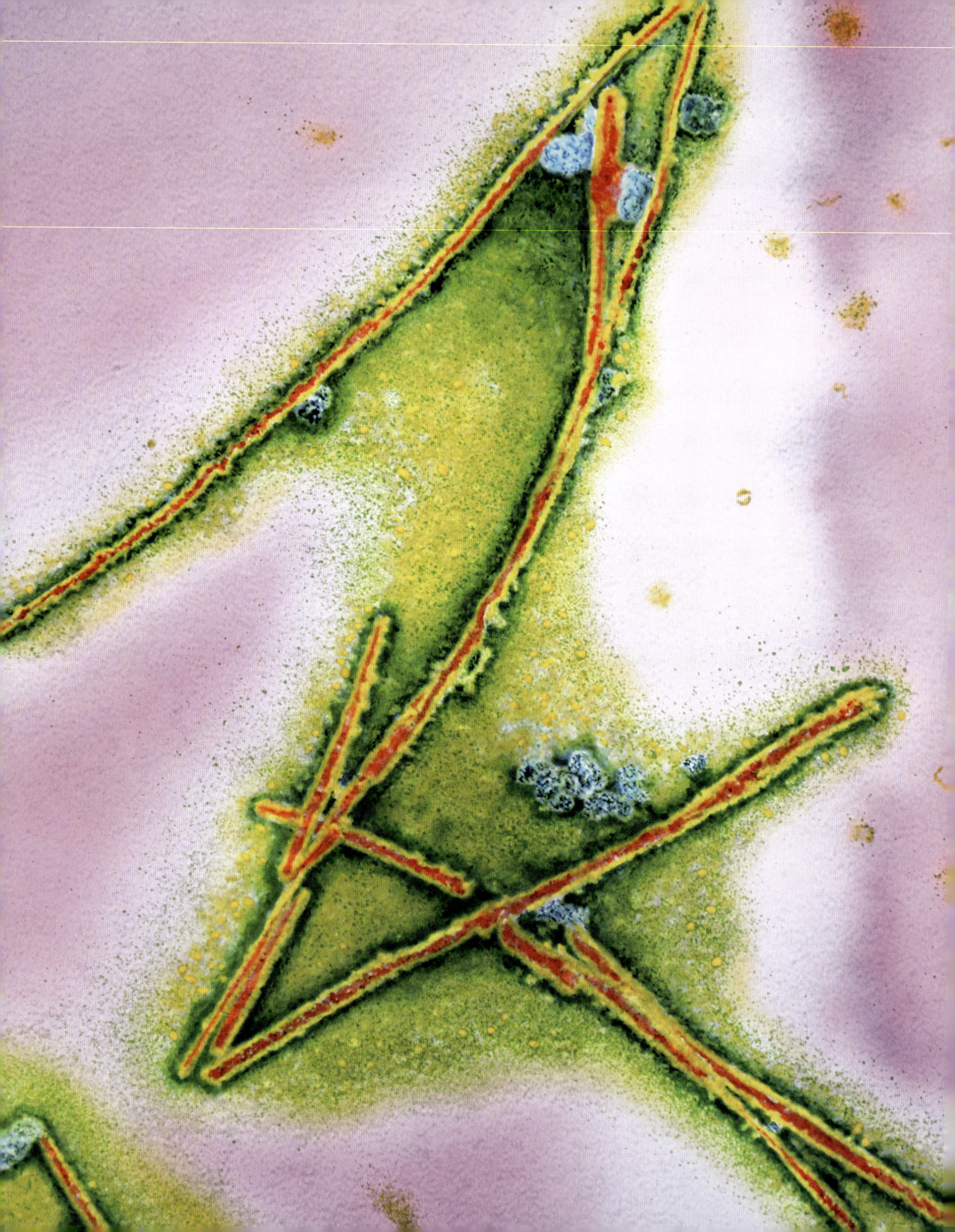

CHAPTER 10

Causes of infectious disease

This chapter examines pathogens as the infectious agents of disease and how disease is spread. You will become familiar with a range of different pathogens and the diseases they cause in humans, other animals and plants. You will investigate early research of infections in microbiology and medicine and how it has informed our present understanding of pathogens.

You will learn how pathogens influence agricultural industries and the consequence of disease in productive plants and animals. You will then examine the different adaptations that pathogens have evolved in order to combat the challenges of host colonisation, survival within the host and transmission between hosts.

Content

INQUIRY QUESTION

How are diseases transmitted?

By the end of this chapter you will be able to:

- describe a variety of infectious diseases caused by pathogens, including microorganisms, macroorganisms and non-cellular pathogens, and collect primary and secondary-sourced data and information relating to disease transmission, including: (ACSBL097, ACSBL098, ACSBL116, ACSBL117)
  - classifying different pathogens that cause disease in plants and animals (ACSBL117)
  - investigating the transmission of a disease during an epidemic
  - design and conduct a practical investigation relating to the microbial testing of water or food samples ICT
  - investigate modes of transmission of infectious diseases, including direct contact, indirect contact and vector transmission
- investigate the work of Robert Koch and Louis Pasteur, to explain the causes and transmission of infectious diseases, including: L WE
  - Koch's postulates
  - Pasteur's experiments on microbial contamination
- assess the causes and effects of diseases on agricultural production, including but not limited to: S WE
  - plant diseases
  - animal diseases
- compare the adaptations of different pathogens that facilitate their entry into and transmission between hosts (ACSBL118)

10.1 Pathogens—agents of disease

BIOLOGY INQUIRY   

The journey of germs

How are diseases transmitted?

COLLECT THIS…

- large sheet of paper
- pens or pencils
- tablet or computer to access the internet

DO THIS…

1 Working in groups, choose one of the following infectious diseases.
  - **a** chicken pox
  - **b** athlete's foot
  - **c** whooping cough
  - **d** giardiasis

2 Research your chosen infectious disease and identify:
  - **a** the infectious agent or pathogen that causes the disease
  - **b** the symptoms and physiological response of an infected person
  - **c** direct and indirect ways the disease is transmitted from one host to another.

RECORD THIS…

Describe the effects of the disease and how it is passed on to others.

Present this information in a poster and explain to your class how the disease affects its host and how the disease is transmitted.

REFLECT ON THIS…

How are diseases transmitted?

What adaptations does the pathogen have that assist it in infecting hosts?

How does an infected host respond to the pathogen?

AGENTS OF DISEASE

A **pathogen** is a biological agent that causes disease or illness in a **host** organism. Pathogens only cause **infectious diseases**—diseases that can be spread from one host to another. Non-infectious diseases are not caused by an infectious agent and cannot be passed on from one person (or host) to another. Non-infectious disease may be caused by environmental or genetic factors. Some examples of non-infectious diseases are most cancers, heart disease, diabetes and genetic disorders. You will learn more about non-infectious diseases in Chapters 14–18.

Pathogens live and reproduce at the expense of the host organism and cause disease by releasing toxins, damaging tissue or competing for nutrients. The presence of a pathogen in a host organism does not always result in **infection**. Infection requires the pathogen to invade, reproduce and elicit a host response. Some organisms that are normally non-pathogenic can cause disease when a host is unhealthy or under stress or, if the pathogens colonise a part of the body where they are not normally found. Organisms that cause disease in plants and animals can be **cellular pathogens** or **non-cellular pathogens**.

Microorganisms

Most pathogens are very small and cannot be seen with the naked eye; they are **microscopic**. Microscopic cellular pathogens include bacteria, fungi, oomycetes and protozoa.

Bacteria

Bacteria are **prokaryotes** that exist almost everywhere including on the inside and outside of our bodies. Most bacteria are not pathogenic; the human body supports and relies on a range of bacteria to stay healthy. Only about 100 species of bacteria have been identified as pathogenic to humans out of an estimated 1 trillion different bacteria species on earth.

Those bacteria that are pathogenic cause many diseases, from common illnesses such as food poisoning to life-threatening disease like meningitis, whooping cough and serious pandemics (e.g. cholera). Bacterial pathogens can also have a detrimental effect on plants.

FIGURE 10.1.1 The wildflower, large-flowered trillium (*Trillium grandiflorum*), sometimes has green stripes on its petals. This is the result of an infection by a mycoplasma. As the infection spreads it eventually impairs the reproductive ability of the infected plant.

Bacteria can be transferred to plants via soil, water, and through animals and insects. Many plant diseases are also transferred mechanically through wounds caused by horticultural and agricultural tools. Propagation, grafting or pruning with bacterial infected tools is one of the main ways that infections can be spread in agricultural or horticultural crops. Table 10.1.1 highlights some of the diseases in plants caused by bacteria.

TABLE 10.1.1 Bacterial pathogens of plants and the common diseases they cause

Type of bacteria	Typical host	Common disease	
Necrotrophic bacteria			
Pectobacterium carotovorum	affects a wide range of plants: • potato • tomato seedlings • cauliflower • onion • carrot • cabbage	soft rot disease	
Biotrophic bacteria			
Erwinia amylovora	• roses • apple tree and fruit • pear tree and fruit	fire blight	
Pantoea stewartii	• corn	Stewart's wilt	

Mycoplasma is a genus of bacteria that lacks a cell wall. They are non-rigid and non-motile organisms that are approximately 0.1 to 0.3 μm in size. These bacteria are transferred by insect pests travelling through plant phloem tissue. Mycoplasmas are responsible for more than 50 plant diseases collectively known as the 'yellows' disease (Figure 10.1.1).

Fungi

Fungi are **eukaryotes** and include yeasts, moulds and mushrooms. Most pathogenic fungi are microorganisms which cause disease in humans, plants and other animals. Fungi secrete digestive enzymes and other chemicals into their environment to break down organic matter, which can then be absorbed into the fungus. These secreted substances are usually responsible for causing disease and symptoms in the host organism.

In humans, fungi most commonly cause skin infections such as tinea, and a variety of diseases including athlete's foot and ringworm. These skin infections are caused by fungal species belonging to three genera: *Microsporum*, *Epidermophyton* and *Trichophyton*, which live on the outside layer of skin and break down keratin tissue, producing by-products that cause itchiness and inflammation (Figure 10.1.2).

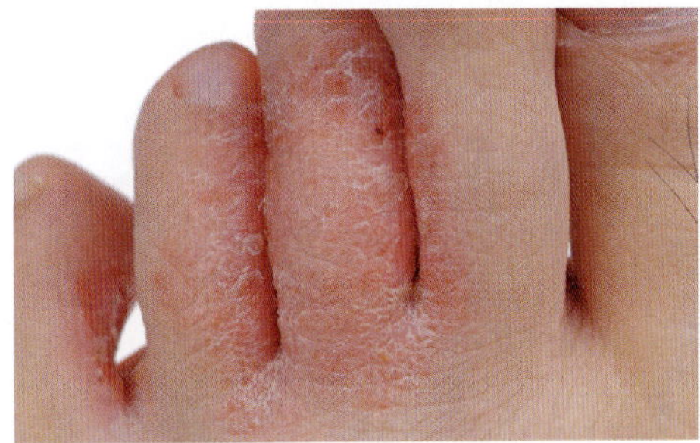

FIGURE 10.1.2 Fungal skin infection Athlete's foot (medical term, tinea pedis) causing itchiness, redness and flaking skin

Some fungi are **zoonotic** meaning they can be transferred between humans and animals (e.g. ringworm is often spread from pets to their owners and vice versa).

Overgrowth of normal fungal flora or invasion of fungi into areas of the body where it is not normally found can also cause disease. *Candida* is a genus of yeasts that are normally found in mucous membranes of the intestinal tract, respiratory tract and vagina. Yeast infections can occur when the balance of natural flora is disturbed and *Candida* is able to flourish causing diseases such as thrush. If *Candida* enters the bloodstream it can also cause serious illness (invasive candidiasis).

Fungal diseases are the most common diseases among plants, as outlined in Table 10.1.2.

TABLE 10.1.2 Common plant diseases, associated pathogens and symptoms

Disease	Pathogen	Disease classification and symptoms	
Blight	Fungi	This disease is classified through the extensive death of plant tissue. Symptoms include: • chlorosis (loss of green pigment) • tissue browning • leaf, branch, floral organ death.	
Powdery mildew	Fungi	A common disease that causes minimal long-term effects for most plants. Symptoms include: • white powdery spots or white fungal blooms on leaf and other plant surfaces • blackening leaves if infection is severe • reduced photosynthesis • increased transpiration.	
Rust	Fungi	Disease identified through orange coloured areas or spots (pustules) on the underside of leaves and which may also appear on stems of plants. Symptoms include: • chlorosis of leaves • plant death in severe cases • leaves falling prematurely.	

Oomycetes

Oomycetes (class Oomycota) include organisms that cause blight and downy mildew on plants and life-threatening infections in animals. Originally thought of as fungi, the oomycetes are now classified as protists. Oomycetes have motile cells (with flagella), walls of cellulose, and many cellular processes that are not found in fungi. Oomyctes reproduce using two types of **spores**: **zoospores**, which are asexual, motile spores and **oospores**, which are sexual spores that can survive in extreme environmental conditions. Spores may be dispersed via water, soil or wind. When spores of oomycetes are released on a leaf they can be carried in water droplets to other leaves, germinate directly, or swim to another germination site, sending out a **hypha** (fungal thread) that branches and invades plant tissue (Figure 10.1.3).

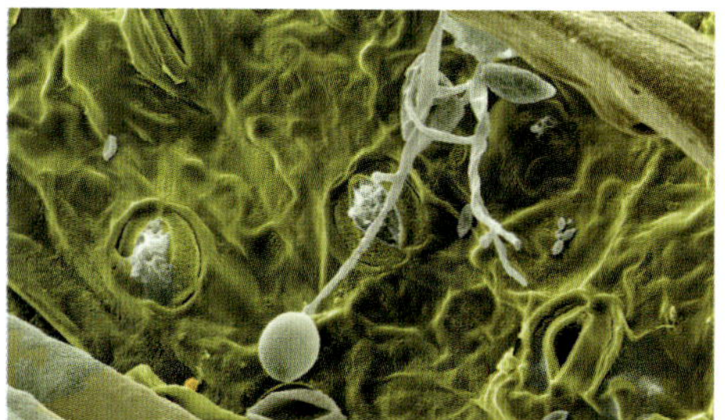

FIGURE 10.1.3 Leaf of a potato plant infected with *Phytophthora infestans*, the oomycetes that causes potato blight. A hyphal thread (white) with a sporangium (round) containing spores can be seen in this scanning electron micrograph (SEM)

The oomycetes includes the genus *Phytophthora* (meaning 'plant destroyer'), which is one of the most invasive plant pathogens worldwide. It infects many economically important crops such as potato, tomato, apple, tobacco and citrus and can also devastate natural ecosystems. In Australia, *Phytophthora cinnamomi* has destroyed tens of thousands of hectares of native forest. *Banksia*, *Eucalyptus*, *Grevillea* and grass trees (*Xanthorrhoea*) are some of the most susceptible plants (Figure 10.1.4).

The spores of *P. cinnamomi* can survive for years in moist soil and are spread through the movement of soil on animals, people's shoes, vehicle tyres and via root-to-root contact between plants (Figure 10.1.4). *Pythium insidiosum* causes disease in mammals, with humans, dogs and horses becoming infected by *Pythium* after exposure to stagnant water where they come in contact with the zoospores of the pathogen.

FIGURE 10.1.4 Eucalypt dieback caused by *Phytophthora cinnamomi*

Protozoa

Protozoa is a diverse group of unicellular eukaryotes that cause disease in plants and animals. *Giardia lamblia* (Figure 10.1.5) is common pathogenic protozoan which infects the small intestine of many animal species including humans. *Giardia* is often transmitted through water that has been contaminated with faeces.

The life cycles of some protozoans include multiple stages in different hosts. For example, *Trypanosoma brucei* causes African sleeping sickness and is transmitted by the biting tsetse fly. Malaria is also an insect-borne infectious disease caused by protozoans of the *Plasmodium* genus. Malaria is transmitted from person to person via an infected female *Anopheles* mosquito.

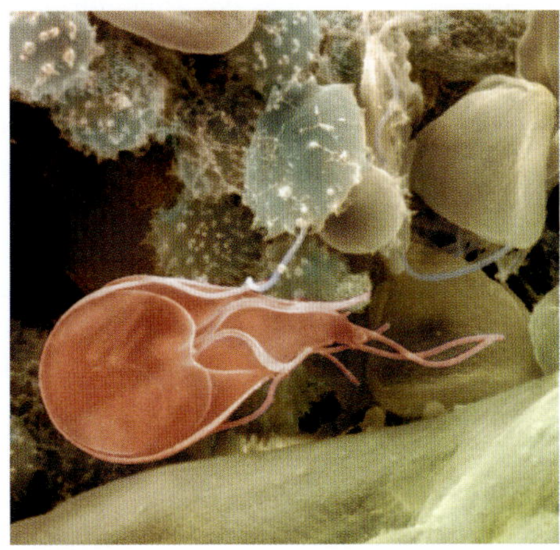

FIGURE 10.1.5 Coloured SEM of a *Giardia lamblia* protozoan (pink), which undergoes asexual reproduction in the small intestine and causes diarrhoea in humans

Macroorganisms

Macroscopic cellular pathogens are disease-causing agents that are visible to the naked eye. These pathogens are large, multicellular organisms that live on the inside (**endoparasites**) or outside (**ectoparasites**) of another organism. Endoparasites generally have a longer association with their host organism and can complete multiple life stages or their mature life stage inside the host. Ectoparasites usually have shorter associations with a single host organism.

Helminths

Helminths are a group of endoparasitic worms that live and feed on a live host, which can be a plant or animal. In plants, roundworms, or nematodes, infect roots and are major pests of orchard trees and crops (Figure 10.1.6). Helminths include intestinal worms such as flukes, tapeworms, flatworms and nematodes (Figure 10.1.7) which reside in the intestinal tract of an animal host and prevent the host from absorbing nutrients properly, which can lead to weight loss, anaemia and nutrient deficiencies. Helminths are also capable of inhibiting the immune systems of animals so that the immune response against them is suppressed. Some helminths are able to survive within their host for many years.

FIGURE 10.1.6 Rotting of fruit caused by nematode infection.

FIGURE 10.1.7 A light micrograph (LM) of female and juvenile roundworms (or nematodes)

BIOLOGY IN ACTION S

Transmissible cancer and the Tasmanian devil

Cancer is generally not transmissible. Most cases of cancer are caused by environmental factors and, although an individual's genetics may increase their risk of getting cancer, cancers are not hereditary. Some cases are caused by viruses such as cervical cancer, which is caused by human papillomavirus (the cancer itself is not transmissible). However, in recent years a massive population decline of the Tasmanian devil has been attributed to a transmissible facial tumour (Figure 10.1.8). The devil facial tumour disease is spread by biting, a frequent interaction among devils competing for food or mates. Living cancer cells from an infected individual are transmitted to another individual where they form lesions and tumours around the face which can spread throughout the body. The mortality rate among infected devils is very high and the disease usually causes death within a few months. Fears that the disease could cause Tasmanian devils to become extinct in the wild have led to the establishment of healthy captive populations and substantial research into the disease. The Tasmanian devil was listed as Endangered in 2009 by the International Union for Conservation of Nature (IUCN) due to population declines and potential future impact of the disease.

FIGURE 10.1.8 A transmissible cancer has infected the wild population of Tasmanian devils and causes facial lesions and tumours. Living cancer cells can spread through biting.

Non-cellular pathogens

Some infectious diseases are caused by non-cellular, non-living agents that require host cells to reproduce. Non-cellular pathogens have no metabolism and use the reproductive mechanisms of the host cell to replicate themselves. These non-living pathogens are simply made up of nucleic acids, proteins or both.

Viruses

Viruses are very small (around 20–300 nm) agents of disease. Viruses can infect all kinds of organisms and require the cells of their host for reproduction. When outside a host's cells, viruses exist as independent particles known as **virions** (Figure 10.1.9). Virions have an outer protein coat (known as a **capsid**) and inner core of nucleic acid (DNA or RNA). Some virions also have an outer envelope of lipids that surrounds the protein coat. The virion is the vector stage which is transmitted from host to host. Once the virus penetrates the organism's defenses, the virion sheds its capsid (and envelope if it is present) releasing its RNA or DNA into the host's cells. Once inside the cell, the virus multiplies by incorporating its own RNA or DNA into the host's genetic material and replicating during the cell's normal DNA replication processes.

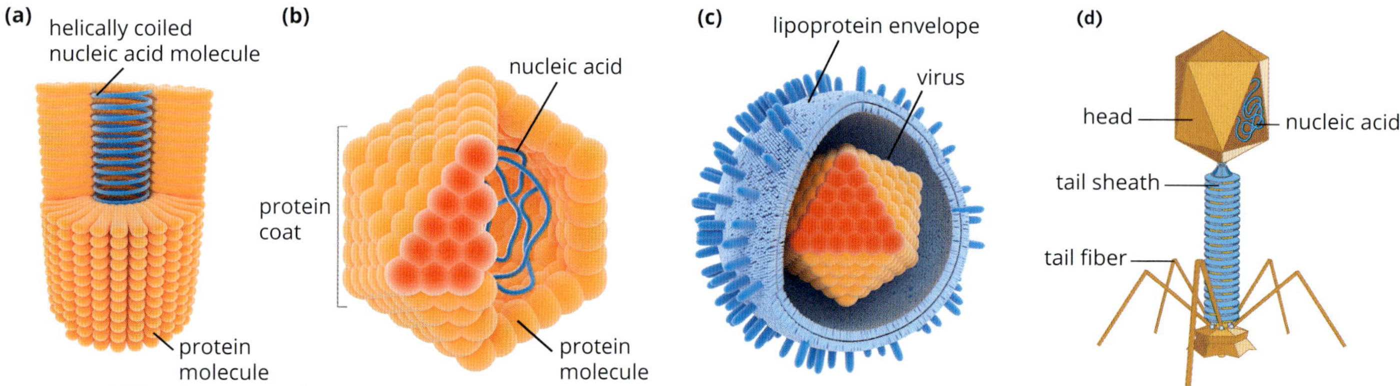

FIGURE 10.1.9 Different structures of virions. (a) A rod-shaped virion with proteins (orange) surrounding a helically coiled nucleic acid molecule (blue). (b) An isometric virion with an icosahedral protein coat surrounding a nucleic acid core (blue). (c) An icosahedral virion (orange) enclosed by a lipoprotein envelope (blue) (d) a bacteriophage composed of a capsid, tail sheath and fibres.

Viruses infect and multiply in the following way (Figure 10.1.10):

1. attachment—the attachment site of the virion binds to a specific receptor site of the host cell
2. penetration—viral proteins are released into the cell's cytoplasm
3. uncoating—the viral capsid is removed and genetic material of the virus is released into the host cell
4. biosynthesis—viruses will begin to replicate within the host cell. DNA viruses use host enzymes for transcription and replication; RNA viruses use their own enzymes and replicate in the cytoplasm
5. maturation and release—viral particles separate from the host cell and assemble into new virions. These will ultimately be released by the host cell, usually through cell lysis.

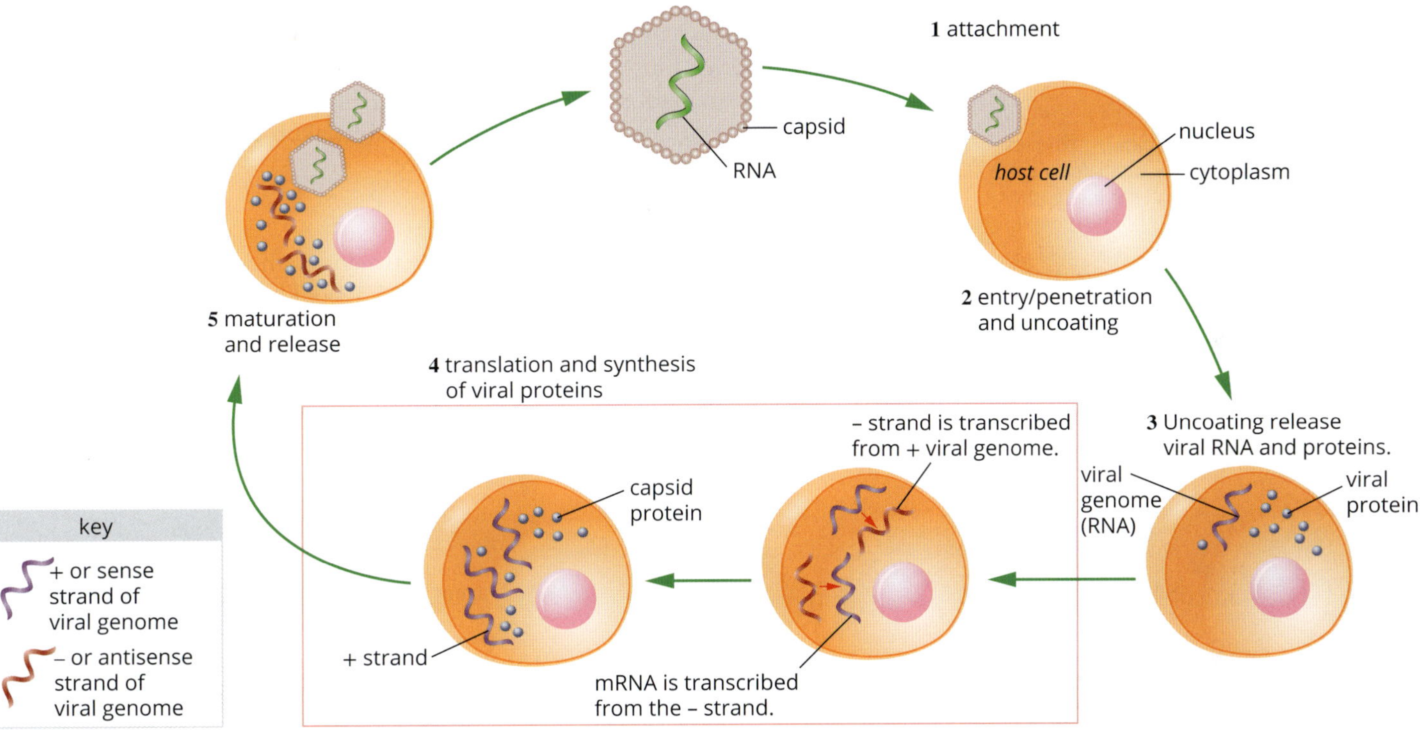

FIGURE 10.1.10 A diagram demonstrating an RNA virus attachment, penetration, uncoating, replication, maturation and release.

The wheat streak mosaic virus (WSMV) caused severe damage to many crops in NSW in previous years with affected plants either dying prematurely, failing to grow or stunted. The leaves developed yellow streaks and blotching creating a mosaic pattern (Figure 10.1.11). Viruses are responsible for many diseases ranging from mild to life threatening, including common colds, influenza and HIV/AIDs, which has killed over 35 million people since the 1980s.

Vaccinations have been created to stop the spread of many viruses that cause diseases such as influenza, smallpox, measles, mumps, genital warts and cervical cancer caused by the human papillomavirus.

Plant and animal viruses can be transmitted via mechanical means and through **vectors**. Plant viruses may also be transmitted into an ovule via a pollen tube of an infected pollen grain. Although there are many modes of transmission, plants are typically infected with viruses by insect vectors with piercing, sucking mouthparts. Insects that have fed on diseased plants will often carry viruses and infect healthy plants as they continue to feed on other plants. Plants may also be infected with a virus via wounds inflicted through horticulture, or plant propagation (grafting and budding) by humans.

FIGURE 10.1.11 Wheat affected with a mosaic virus creating a mottled appearance and stunted growth

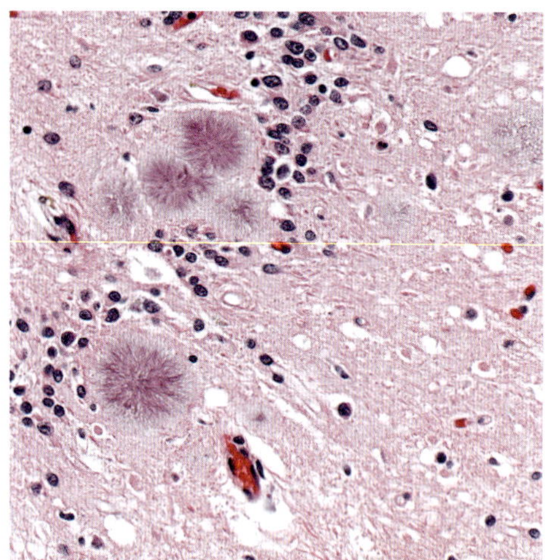

FIGURE 10.1.12 LM section showing plaques in a human brain effected by Creutzfeldt-Jakob disease (CJD). White areas on the image show vacuoles where prions have destroyed neurons and tissue, making the brain appear 'spongy'.

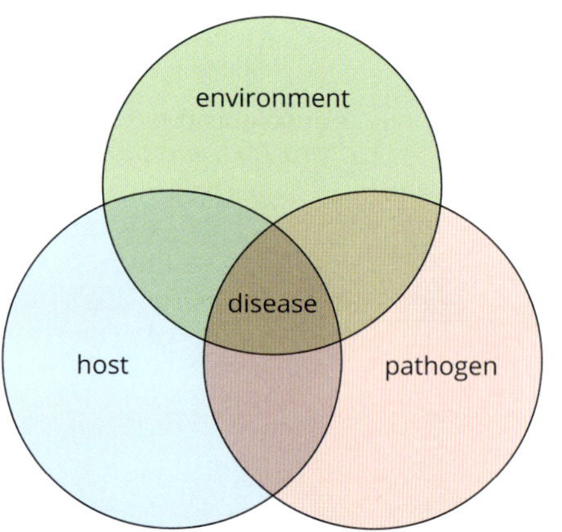

FIGURE 10.1.13 A Venn diagram illustrating the relationship between pathogen, host and environment that leads to disease.

Viroids

Viroids are similar to viruses but are even smaller and less complex. Viroids lack a protein coat and are essentially just a circular, single strand of RNA. Viroids are only known to be pathogens of plants. They damage plants by competing for nucleotides and forming viroid bundles, which mechanically interfere with the internal structures of plants, much like a tumour.

Prions

Prions are the smallest known agents of disease comprising solely of protein and no genetic material. Prions cause neurodegenerative diseases in humans and other animals by promoting abnormal folding of proteins in the host's central nervous system.

In humans, prions cause Creutzfeldt-Jakob disease (CJD) in which vacuoles and misfolded proteins (or plaques) form in the brain, killing neurons and making the brain appear 'spongy' under a microscope (Figure 10.1.12). Symptoms include dementia and sudden muscle contractions, leading to death. The equivalent disease in cattle, bovine spongiform encephalopathy (BSE), commonly known as mad cow disease, has been linked to variant Creutzfeldt-Jakob disease (vCJD) through human consumption of BSE-contaminated beef.

THE CHAIN OF INFECTION

Infectious diseases require interactions between pathogens, hosts and the environment in order to spread (Figure 10.1.13). These interactions can be described as the chain of infection (Figure 10.1.14). Breaking a link in the chain will stop the spread of disease.

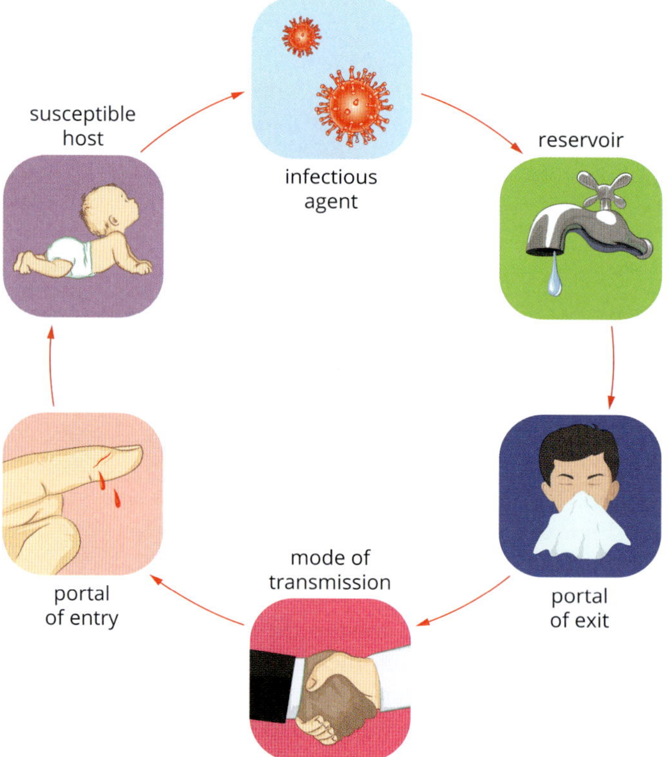

FIGURE 10.1.14 The chain of infection decribes the spread of infectious disease.

Infectious agent

The **infectious agent** is the pathogen that is capable of infection and causing disease. This includes the cellular and non-cellular pathogens previously described. The ability of the pathogen to infect and cause harm to its host is described as **virulence**. The more virulent a pathogen is the greater effect it will have on the health of the host. Less virulent pathogens cause mild symptoms while highly virulent pathogens may cause death.

Reservoir

The reservoir is where the pathogen normally lives and reproduces. Reservoirs can either be living organisms or the environment. Soil, water, food, faeces, and surfaces such as door handles and toilet seats are environmental reservoirs for diseases such as tetanus and gastroenteritis.

Portal of exit

The portal of exit is the pathway that a pathogen uses to exit a host and be transmitted to another. Pathogens can cause symptoms in a host that assist its exit and spread.

Respiratory

The respiratory tract includes the mouth, nose, throat and lungs. Pathogens can cause irritations and increased mucus production to exit the body through the respiratory tract via coughing, sneezing and exhaling.

Gastrointestinal

The gastrointestinal tract includes the stomach, and small and large intestines. The oral cavity and rectum are portals of exit and a pathogen may cause symptoms such as vomiting and diarrhoea to aid its spread.

Genitourinary

The genitourinary tract includes the reproductive system and the urinary system. Microorganisms can exit the tract via the urethra (in both male and females) and the vagina. Diseases can be spread through sexual contact and infected body fluids such as sperm, blood and vaginal fluids.

Skin

The skin normally provides a barrier to infection. If the skin is broken it can act as a portal of exit. Wounds or lesions on the skin from an infectious disease can spread the pathogen through blood or secretions, like pus, from the wound.

Mode of transmission

Transmission is the way a pathogen spreads disease from one host to the next. Transmission of a pathogen or disease can be direct or indirect. Direct transmission includes contact between the reservoir or infected host and a new host. Sexual intercourse, skin-to-skin contact or biting in animals allow for direct transmission of disease. Infection through sneezing, coughing and breathing is also considered direct transmission as droplets of body fluid are spread directly from one host to another. Indirect transmission requires an intermediate between one host and the next. Indirect transmission includes the spread of disease through the air and dust, in food and water or on surfaces and objects like bedding, clothes and medical equipment. Vectors such as mosquitoes, fleas, aphids and nematodes also spread disease indirectly (Table 10.1.3). Some pathogens require a vector to complete their life cycle and reach maturity in an intermediate host.

Portal of entry

The portal of entry into a new host is often the same as the portal of exit from the previous host. The pathogen must be able to access the type of tissues where it can grow and reproduce (e.g. influenza viruses may exit a host's respiratory tract through sneezing and the virus can be transported in mucus droplets to a new host where the pathogen enters the respiratory tract through the nose or mouth).

Susceptible host

Exposure to a pathogen does not necessarily result in infection. The last link in the chain of infection is a susceptible host, an organism that will become infected. The susceptibility of a host depends on many factors including their genetics, immunity to the pathogen, and overall health. If the host is suffering from malnutrition or other diseases they are likely to be more susceptible to further infections.

TABLE 10.1.3 Bacterial pathogens of plants and the common diseases they cause

Invertebrate vector	Description and characteristics	
aphids	• a common insect pest that is approximately 1–3 mm, may have a green, grey, or black colouration • cluster on the tips of the shoots, sucking the sap from the plant reducing plant strength and health • may be vectors for other plant pathogens (virus, fungus)	
borers	• Includes beetles, wood moths, weevils and termites, which tunnel holes into stems, trunks and branches of plant hosts • chewing mouthparts may transmit viral and bacterial pathogens • most healthy plants are able to withstand borer attacks and new growth will outgrow borer attacks; however, in weakened plants, holes caused by borers results in weakening of plants and in severe cases may cause plant death	
lerps, mealybugs and scales	• sap sucking insects covered in a protective waxy coating • attack the leaves of a variety of native plants, particularly eucalypts feeding on new shoots and leaves causing severe distortion, chlorosis of leaves and death of plant tissue • lerps produce honeydew, which encourages the growth of sooty mould. Soon after a lerp infestation, leaves turn brown and are shed prematurely • severe infestations of scale insects can also result in defoliation, stunted plant growth and plant death	
nematodes	• microscopic soil-borne organisms affecting the roots of many plants creating enlarged galls on infected plants • may also cause plant wilting • some nematodes affect the stem or the leaves and may also be vectors for other pathogenic organisms	
caterpillars, snails, slugs and slaters	• common pests that feed off herbaceous plants often chewing holes in leaves and stems • seedling death caused by feeding • in severe cases may cause leaf and plant death • may be vectors for other plant pathogens (virus, fungal and bacterial pathogens)	

EPIDEMICS

Epidemics occur when the number of people affected by a particular disease is higher than usual. More specifically, an epidemic is a sudden increase in the occurrence of a particular disease among the population of a given area (Figure 10.1.15). The usual level of disease found within a population is known as the endemic level, where the level of a particular disease is steady and its occurrence is not increasing (or decreasing) within a population. When a disease reaches above endemic levels on a global scale and epidemics are occurring across multiple countries or continents it is called a **pandemic**.

Epidemics of infectious disease occur through the chain of infection where the pathogen and susceptible hosts are present in high numbers and transmission can easily occur. Epidemics are the result of a change in one or more of the following factors: virulence of the pathogen, migration of the pathogen and host exposure and susceptibility.

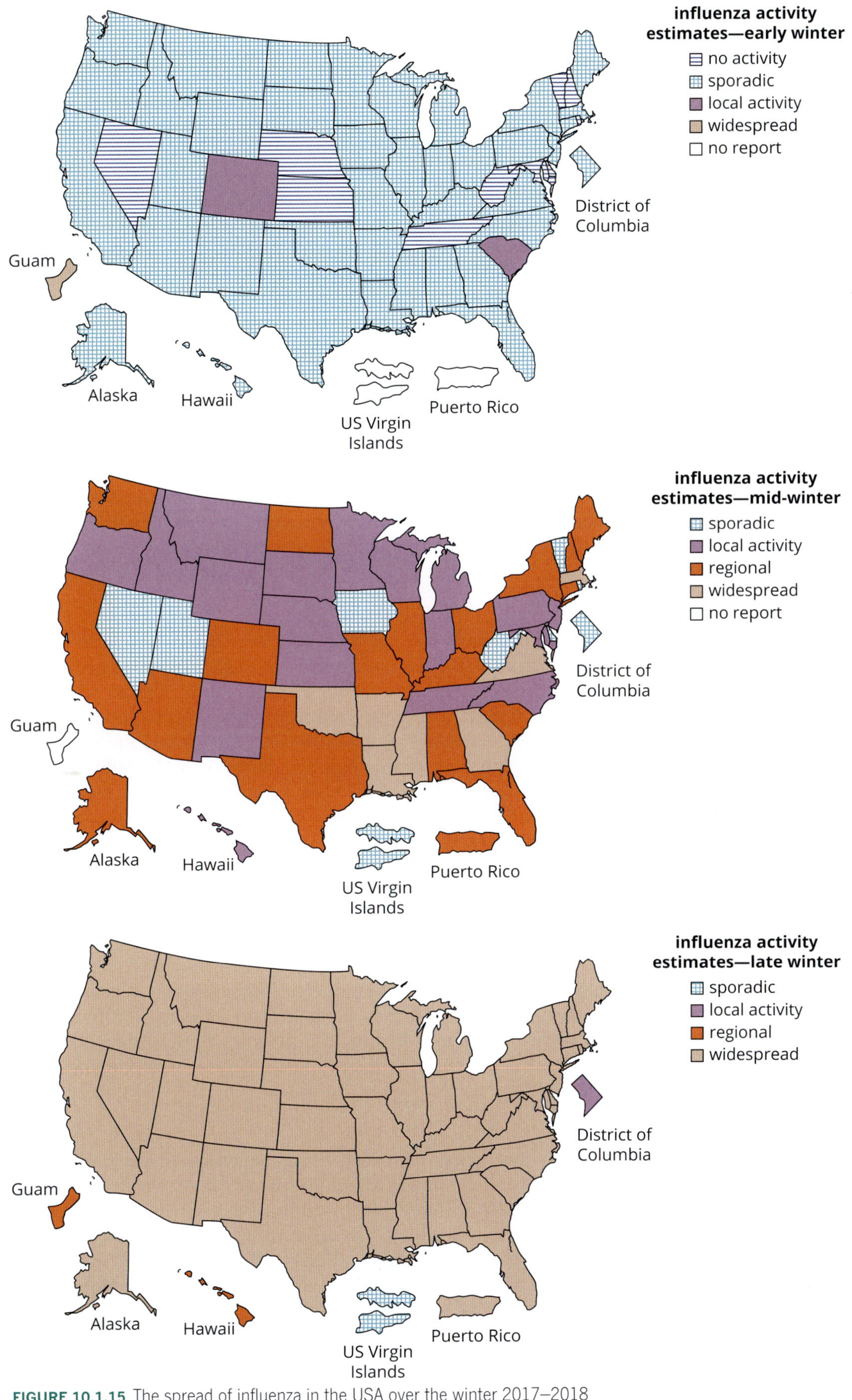

FIGURE 10.1.15 The spread of influenza in the USA over the winter 2017–2018

BIOFILE IU

Identification of the Ebola virus in 1976

In 1976, approximately 280 people died in a remote village in The Democratic Republic of Congo. The deaths were initially believed to be due to an outbreak of yellow fever. Two vials containing blood samples from an infected patient were sent from the village to a laboratory in Belgium for identification. The vials were packaged in a regular insulated flask and carried in the hand luggage of a passenger. On arrival, it was discovered that one of the vials had broken in transit and its contents were mixed through the ice. It was extremely fortunate that the pathogen did not infect the passengers or the scientists, as it could have rapidly spread, possibly causing a pandemic.

The scientists in Belgium used transmission electron microscopy (TEM) to examine the blood samples and discovered a large and unusual virus that was unknown to science. It was named 'Ebola' after the Ebola River, near the village where the first known outbreak of the disease occurred (Figure 10.1.16).

Today the Ebola virus is identified by an assay (called an enzyme-linked immunosorbent assay or ELISA) that detects the viral antigen or antibodies present in the blood of people who have been exposed to the virus. The virus is also detected by viral RNA genome sequencing.

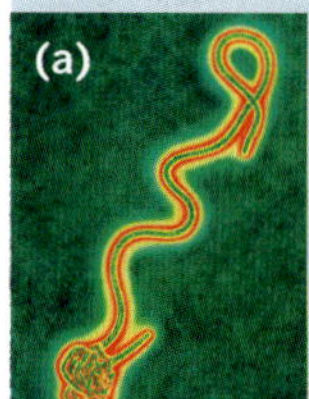

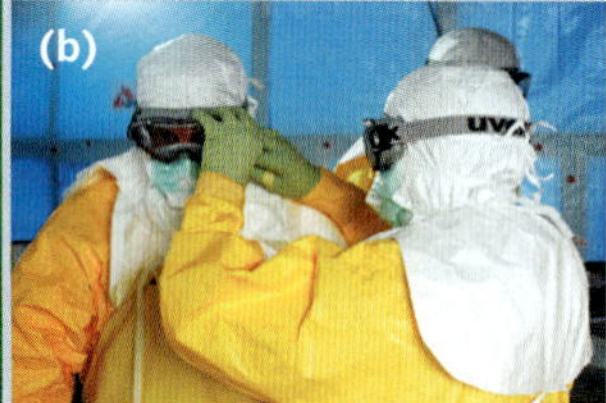

FIGURE 10.1.16 (a) TEM of the Ebola virus. (b) The personal protective equipment that healthcare workers were issued during the 2014 Ebola outbreak in Sierra Leone.

Virulence of the pathogen

Mutations can occur that increase the virulence of a disease. New strains of a disease may emerge in a population where hosts had built up immunity to the previous strains. A host shift may also cause virulence of a pathogen to be extremely high. The virulence of a pathogen can dramatically increase after a shift to infect a new host species.

Ebola

Ebola is a virus that is largely asymptomatic when found in fruit bats. However, the virus entered into the human population when people began consuming bat meat and became highly virulent and often fatal when it crossed the species barrier. Ebola is spread through bodily fluids and has a relatively short incubation period of as little as two days. In 2014, an epidemic of the Ebola virus occurred in western Africa, affecting over 11 000 people in multiple countries. Ebola was first identified in 1976 and, although this was the largest and longest outbreak, Ebola had affected small numbers of people previously at least 26 times, mainly in equatorial African countries.

Pathogen migration

Migration of individuals into new populations can result in an epidemic. The migrating individuals may bring with them a disease the new population has never been in contact with before and expose hosts to a new pathogen that they are not immune to.

The Spanish flu

From 1918 to 1920, approximately 50 million people were killed worldwide during an influenza (flu) pandemic. World War I was occurring during the same period and infected soldiers spread the disease throughout Europe, the USSR and the USA in three deadly waves. Increased travel and improved transportation technology allowed infected travellers to reach new countries and introduced the disease to new populations. The influenza virus was highly contagious and deadly, even to healthy young adults. The disease was spread by air as the pathogen was breathed or coughed out of the infected person's lungs. It was dubbed the Spanish flu, although it did not originate in Spain; the origins of the flu are still debated.

Host exposure and susceptibility

A host population may become more susceptible to disease if the population's genetics change. Loss of genetic variation or selection pressures can cause a population to lose genetic resistance to certain pathogens. Increased exposure to pathogens can also result in epidemics. This is often caused by poor sanitation or hygiene and dense populations. Pathogen exposure can be increased by lack of infrastructure for reliable sanitation amenities such as clean running water, effective sewerage systems and adequate health services. This can occur in developing areas, or places recently affected by natural disaster or war.

Cholera

The cholera epidemic that occurred in Haiti in the wake of the devastating 2010 earthquake is the worst cholera epidemic in recent history, with more than 700 000 cases and almost 9000 deaths. It was a surprise, as Haiti had not experienced a cholera epidemic in more than a century. The epidemic was caused by the toxigenic strain of bacteria, *Vibrio cholerae* (strain O1, Figure 10.1.17). Cholera affects the gastrointestinal tract, causing watery diarrhoea and vomiting, dehydration and in some cases death.

The origin of the Haitian epidemic was traced to a camp of Nepalese army personnel who came to the country to aid the relief effort. Epidemiology and scientific methods were used to trace the source and identify the infectious agent.

DNA fingerprinting and sequencing matched the epidemic strain to a *V. cholerae* strain previously identified in Bangladesh, rather than strains that already existed at low levels in Latin America. The *V. cholerae* bacterium was most likely introduced by the Nepalese soldiers, who had recently been in a region experiencing a cholera outbreak. Contaminated sewage from the camp was released into the local river. The communities downstream used this water for washing and cooking, because the water supply network had been damaged and was still being repaired.

Despite improved sanitation and repairs to water infrastructure new infections continue to occur. Reasons why cholera continues to be a problem in Haiti sometime after the initial cause of the epidemic include:

- limited sanitation infrastructure
- limited access to adequate medical services
- gaps in the water quality systems, limited chlorination
- limits to the alert and coordination systems
- displacement of people.

A different strain of *V. cholerae*, which emerged in 1992–93, was responsible for a cholera epidemic in a large area along the Bay of Bengal. Scientific studies of this *V. cholerae* strain (O139) are providing clues as to how bacteria evolve as new variants with the potential to cause epidemics.

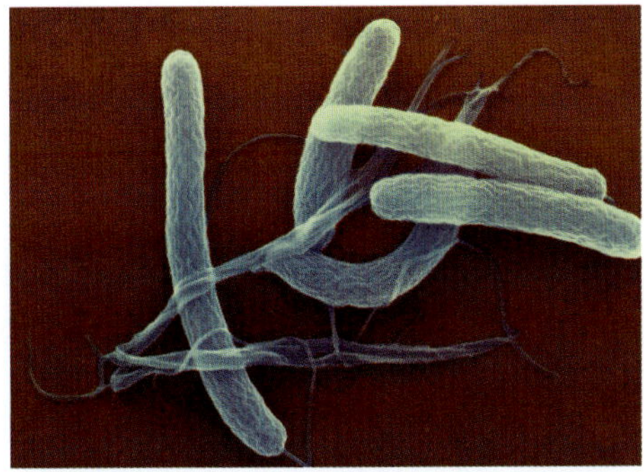

FIGURE 10.1.17 Coloured SEM of *Vibrio cholerae*

BIOLOGY IN ACTION 

IU

Reconstructing the Spanish flu virus

In an attempt to understand more about the Spanish flu virus and why it was so deadly, the body of a person killed by the virus was exhumed from permafrost in Alaska. The Spanish flu had killed the majority of the people living in the village at the time. Genome sequencing of the virus was completed in 2005 and it was identified as H1N1 influenza A, originating in birds. Researchers extracted the viral RNA, grew active virus in cultured cells, and infected mice and monkeys with the reconstructed virus (Figure 10.1.18). The virus was highly virulent, killing the first experimental mice in six days.

In a normal response to the flu, the immune system wanes over a period of time. However, when monkeys were infected with the reactivated Spanish flu virus, their immune systems went into overdrive and did not switch off. Large quantities of cytokines were produced, resulting in uncontrolled inflammation. Similar responses were observed in the H5N1 avian flu, an epidemic that appeared in Asia in 1997. Researchers are using this information to investigate possible treatments.

The reactivated virus may help to identify the mutations that produce a virulent virus so that they can be identified in new variants of seasonal flu, and help prevent another pandemic.

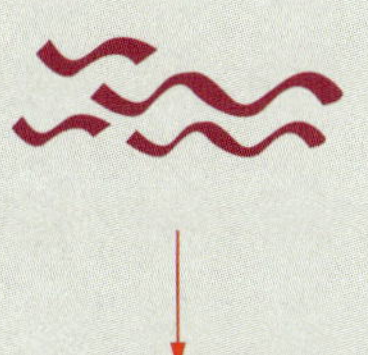

FIGURE 10.1.18 How the Spanish flu virus was reconstructed in the laboratory and tested in animal models

10.1 Review

SUMMARY

- Pathogens are the agents of infectious disease. There are many kinds of pathogens including, prokaryotes, eukaryotes, microscopic and macroscopic organisms, and cellular and non-cellular agents.
- For infectious disease to spread interaction is required between pathogens, hosts and the environment.
- The chain of infection describes the process of spreading disease, it must include: an infectious agent, reservoir, portal of exit, mode of transmission, portal of entry and a susceptible host.
- A pathogen often uses the same portal of exit and portal of entry to pass from one host to another. In humans they include the respiratory tract, gastrointestinal tract, genitourinary tract and skin.
- An epidemic is the increase in the occurrence of a particular disease within the population of a defined area. A pandemic is when a disease reaches above epidemic levels on a global scale across multiple countries or continents.
- Epidemics can be caused by:
  - increased pathogen virulence, mutations occur in the disease causing organism that allow it to affect its host more seriously or it may be able to move to a new host
  - introduction of a pathogen to a previously unexposed population which has no immunity to the new pathogen
  - increased exposure and susceptibility of host organisms through loss of genetic resistance or increased pathogen exposure.

KEY QUESTIONS

1. Give an example of a disease caused by each of the following organisms.
   - a fungi
   - b bacteria
   - c viruses
   - d helminths
2. Outline the chain of infection.
3. Name three portals of exit in human hosts.
4. Distinguish between an epidemic and a pandemic.
5. What is meant by the term 'virulence'?
6. Using an example explain how an epidemic is caused by increased exposure to disease.

10.2 Pasteur and Koch

Up until the 1700s and early 1800s it was thought that 'germs' which caused decay and disease in non-living matter, like rotting flesh or food, would spontaneously generate. People believed that microbes and other organisms like maggots and flies were created from the decaying matter. One of the most important discoveries in medical history was the evidence for the **germ theory**. Pathogens and other microorganisms were found to exist in the environment, rejecting the previous theory that they spontaneously generated from non-living matter. The germ theory states that all living matter comes from existing living matter.

GERM THEORY AND LOUIS PASTEUR

In the mid 19th century French chemist, Louis Pasteur (1825–1892), carried out laboratory research that would provide evidence for the germ theory of disease. Pasteur worked for the wine industry and recognised that some wine became sour over a period of time while other bottles of wine would not. He discovered that the sour wine had been contaminated and contained many smaller, microbial cells than just yeast cells, which are important for fermentation. These small cells were bacteria that produce lactic acid, rather than alcohol, when undergoing respiration, giving the wine a sour taste. Pasteur discovered that heating wine to about 55°C soon after it was made would kill most microbes responsible for spoiling the product and the wine could subsequently be left to age. This process is known as **pasteurisation**, after Pasteur himself, and is now used widely in the food and beverage processing industry.

Other important research conducted by Pasteur provided additional evidence of the germ theory—most famously his experiment with bone broth and different types of flasks (Figure 10.2.1).

Pasteur prepared a broth and emptied equal amounts into two glass flasks. He heated the long necks of the flasks and bent them into an S shape—this allowed air to enter the flask but any 'dust' or 'spores' (which Pasteur proposed would cause contamination) would settle in the bend and not reach the broth. Pasteur heated the flasks to kill any microbes in the broth. He then removed the S-bend neck from one flask, exposing the broth to the air, and left the flasks to sit at room temperature for several weeks.

Pasteur observed that the broth in open-necked flasks had become cloudy, indicating contamination from germs. The broth in the S-bend flask was unchanged and free of germs. This experiment demonstrated that microbial germs are airborne and that life cannot be generated from a sterilised medium. Pasteur's work identified microbes as the agents of decay and contamination. He was able to ascertain that microbial pathogens are free living in the environment and capable of infecting certain substrates and organisms but cannot appear on sterile substrates that haven't been exposed to microbes.

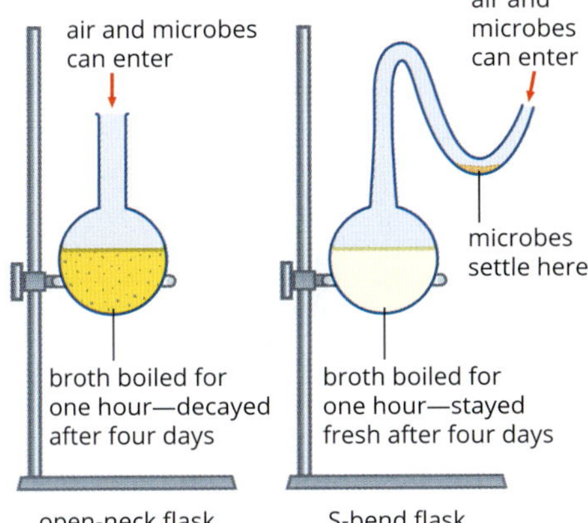

FIGURE 10.2.1 Pasteur's experiment disproving spontaneous generation and supporting germ theory

KOCH'S POSTULATES

Other research important for medicine and microbiology was also being carried out in Germany in the 19th century. A physician named Robert Koch (Figure 10.2.2) was able to establish the relationship between microbial pathogens and disease through his experiments.

Koch's experimental method involved examining blood samples taken from patients with different diseases, then growing microbes from the blood on nutrient plates. When he injected specific microbes into mice he found that they developed diseases similar to those of the original patient. As a result of these studies, specific microbes became recognised as the cause of particular diseases.

Koch formulated a set of criteria, known as **Koch's postulates**, which were used to establish whether a specific microorganism was the cause of a particular disease.

FIGURE 10.2.2 Dr Robert Koch (1843–1910), German bacteriologist, in his laboratory in South Africa

Koch's postulates are as follows.

1 The microorganism must be present in the tissues of the infected organism and not in a healthy organism.
2 The microorganism must be able to be cultivated in isolation from the infected organism.
3 When an uninfected organism is then inoculated with the culture, it should develop symptoms of the disease.
4 Samples from the second infected organism should be able to be isolated and found to be the same as the microorganism from the first infected organism.

Koch's postulates were very significant for microbiology and medicine but they did have some limitations, which Koch himself recognised. Koch isolated *Vibrio cholerae* bacteria as the pathogenic agent causing cholera but found that the pathogen did not meet his first postulate. He was able to isolate *V. cholerae* from healthy individuals, not just from those suffering symptoms. Koch's postulates did not take into account the existence of asymptomatic carriers. Only some individuals (5–10% of cases) that are infected with *V. cholerae* will display symptoms associated with the disease, such as diarrhoea and vomiting. Others that are infected will not display any symptoms but are still able to pass on the disease.

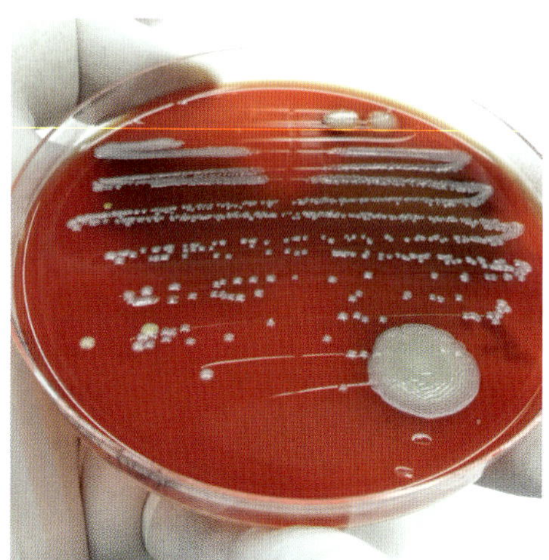

FIGURE 10.2.3 Bacteria grown on blood agar. Viruses cannot be cultured in this way and required a new set of postulates.

Koch's postulates were formulated before the discovery of viral diseases. Many viruses are not able to be cultivated in cell culture so do not meet postulates two and three (Figure 10.2.3).

BIOFILE CCT IU L

Rivalry between Koch and Pasteur

Louis Pasteur and Robert Koch (Figure 10.2.4) were both conducting significant research and had risen to fame in a difficult political climate of European nationalism. Each scientist represented a nation on one side of the Franco-Prussian war, between the French Empire and German states, which began in 1870. Pasteur and Koch were working toward similar goals in microbiology and were at first civil and respectful of each other's work despite being on opposite sides of war and neither speaking the other's language. However the language barrier sparked an intense rivalry at a conference in 1882 when a phrase from a lecture Pasteur was presenting was mistranslated from French to German. Pasteur was referring to some of Koch's published works as 'recueil allemand' meaning a 'collection of German works' in French. Pasteur's translator misheard and translated 'orgeuil allemande' meaning 'German arrogance' from French to German for the audience. Koch was in the audience and protested the insult, which Pasteur did not realise had been given. This led to an ongoing feud, beginning with Koch publishing criticisms of Pasteur's entire body of work. In the aftermath of war, tensions were still high between France and Germany and many rejected one scientist's findings in favour of the other instead of viewing their work as complementary.

FIGURE 10.2.4 (a) Louis Pasteur, French microbiologist and chemist (b) Robert Koch, German bacteriologist

+ ADDITIONAL

Fredericks and Relman: 21st century postulates

Fredricks and Relman proposed a new set of postulates for the 21st century based on DNA or RNA identification of pathogens so that viruses and viroids are included. These postulates still have limitations as prion diseases are not included because they have no DNA or RNA sequence.

1. A DNA or RNA sequence belonging to a pathogen should be present in most cases of an infectious disease. Pathogen-associated DNA or RNA should be preferentially found in diseased tissue and not in healthy tissue.
2. Hosts or tissues without disease should contain fewer, or no, copies of pathogen-associated DNA or RNA.
3. As the disease resolves, the copy number of pathogen-associated DNA or RNA sequences in the host should decrease. If the disease reoccurs, the copy number should increase.
4. If the detection of pathogen-associated DNA or RNA occurs before disease symptoms or the copy number of pathogen-associated DNA or RNA sequences correlates with disease severity, then the sequence-disease relationship is likely to be causal.
5. The characteristics of the pathogen detected from the DNA or RNA sequence data should be consistent with the characteristics of the group of organisms to which it belongs.
6. Correlation between the diseased tissue and pathogen-associated sequences should be sought at the cellular level. Efforts should be made to locate and visualise microorganisms in affected tissue.
7. The sequence-based evidence for the disease-microorganism relationship should be reproducible.

10.2 Review

SUMMARY

- Up until the end of the 18th century it was believed that disease was caused by spontaneous generation, the idea that life could arise from nothing.
- Germ theory replaced the idea of spontaneous generation. Germ theory states that germs already exist in the environment and that these life forms do not spontaneously generate.
- A French microbiologist, Louis Pasteur, and a German physician, Robert Koch, were instrumental in providing evidence for the germ theory of disease.
- Louis Pasteur's laboratory research led to the development of pasteurisation to avoid microbial contamination of food and beverages.
- Pasteur's flask experiments demonstrated that microbial germs are airborne and life cannot be generated from a sterilised medium.
- Robert Koch's experiments established relationships between particular pathogens and specific disease.
- Koch developed four postulates as criteria to determine the relationship between pathogen and disease. These postulates have some limitations.

KEY QUESTIONS

1. Compare germ theory to the theory of spontaneous generation.
2. Outline the steps in Pasteur's bone broth flask experiments.
3. What are Koch's four postulates?
4. Explain one limitation of Koch's postulates.

10.3 Diseases in agriculture

FIGURE 10.3.1 Potatoes infected by the fungus, *Phytophthora infestans*

Pathogens and disease have a huge effect on the agricultural industry. Disease management and prevention is a major consideration and expense for famers and producers. When disease outbreaks occur among crops and livestock the effects can be far reaching. Disease can result in reduced production and shortages that drive up prices for consumers.

DISEASES AND GENETIC VARIATION

Plants and domestic or stock animals have been bred selectively for particular traits and characteristics over many years. Plants are susceptible to many types of disease, which affect the yield and quality of crops. Populations of plants used in agriculture can be particularly vulnerable to disease as they often lack genetic variation.

Potato blight

In the 1800s the Irish imported a variety of potato from South America to feed their growing population in Ireland. The potatoes were grown asexually through vegetative propagation, so all the potato plants were genetically identical to one another. When the potato crops were infected by the fungal pathogen, *Phytophthora infestans*, in the 1840s, the entire crop was destroyed because the lack of genetic variation meant every potato was susceptible to the disease (Figure 10.3.1). The crop failure occurred across Europe but Ireland was the country worst affected. At that time, an Irish working-class family grew their own crop and an adult labourer would each eat up to five kilograms of potatoes a day as their main food.

The Irish potato famine is a reminder of the importance of maintaining biodiversity in crops to minimise the transmission of disease. The famine was responsible for the death of over one million Irish people between 1845 and 1851. Another one million emigrated to other countries.

Panama disease

Panama disease is another well-known example of a disease causing devastation among genetically similar crops. Up until the 1950s one variant of banana, Gros Michel, dominated the banana trade. Gros Michel bananas were all almost genetically identical and eventually fell victim to a fungal pathogen (*Fusarium oxysporum*) which was first recognised in Panama. The pathogen spread out of Panama to most banana plantations and caused a worldwide banana crisis. A new banana variety, the Cavendish, was thought to be immune to Panama disease and replaced the Gros Michel as the most internationally traded banana variety. However, Cavendish bananas are also genetic clones and are not immune to a newly discovered strain of *F. oxysporum* called tropical race 4 (TR4). TR4 reached the Northern Territory of Australia in the 1990s and wiped out the local banana industry. TR4 threatens to invade other banana-producing regions of Australia and the rest of the world.

INTENSIFICATION AND CHANGE IN FARMING PRACTICES

Changes in farming practices can result in emergence of new diseases and facilitate transmission of existing diseases. To meet increasing global demands for high value protein the livestock industry is pressured to increase production of meat, eggs and dairy products. Intensive farming means animals are stocked at higher densities which increases direct contact between animals and the risk of infectious disease transmission. To combat these disease-promoting conditions, antibiotics and vaccines must be administered to maintain the health of the livestock. Some pathogens, such as *Staphylococcus aureus*, which also infects humans, are becoming resistant to antibiotics. Consumption of and contact with infected animals can lead to the spread of zoonotic diseases (diseases that are transmitted from non-human animals to humans). The beef industry in the UK was severely affected when farmers began to change practices by feeding cattle protein supplements.

BIOFILE S

Phytophthora in Australia

There are about thirty-five species of *Phytophthora*, which infect many crops including potato, tomato, apple, tobacco plants and citrus trees. *Phytophthora* species cause diseases in plants that greatly affect the Australian timber and fruit industries. *Phytophthora cinnamomi* (Figure 10.3.2) penetrates the roots and stem to get nutrients, clogging the vascular tissue and thus starving the plant. Other species affect commercially important fruit crops. Although it has been in the country since the 1930s, it has recently spread to regions that were previously unaffected, and poses a significant threat to Australian biodiversity. In Australia, *P. cinnamomi* has destroyed tens of thousands of hectares of valuable eucalypt woodland and forest. The spores of *P. cinnamomi*, some of which can survive for years in moist soil, are attracted to the roots of the plants they infect by a chemical released from the roots.

FIGURE 10.3.2 *Phytophthora cinnamomi* causes root rot or dieback disease in Jarrah forests in Western Australia.

Mad cow disease

Bovine spongiform encephalopathy (BSE) or mad cow disease is caused by an infectious prion protein and is a fatal neurological disease in cows. The first documented case of BSE in the UK was recorded in 1986 and the incidence grew rapidly. At the height of the epidemic in 1993, up to 1000 new cases were reported weekly. Over 180 000 cattle were affected.

In the 1990s, a new human disease appeared, variant Creutzfeldt-Jakob disease (vCJD). It was acquired by eating products from BSE-infected cattle. Transmission occurred because the abnormal prion proteins are not denatured or destroyed when cooked, as occurs with most proteins.

A farming practice that contributed to the rapid and extensive spread of BSE in the UK was feeding meat and bone meal to farm animals. It is possible that BSE started when this meal contained tissue from sheep with scrapie (a prion disease). Controlling the spread initially involved animal culls. The practice of feeding mammalian protein to cattle was banned and stringent slaughter practices, testing (Figure 10.3.3) and surveillance were introduced. Australia's thorough screening and tight importation regulations on animal products has kept the country free of BSE.

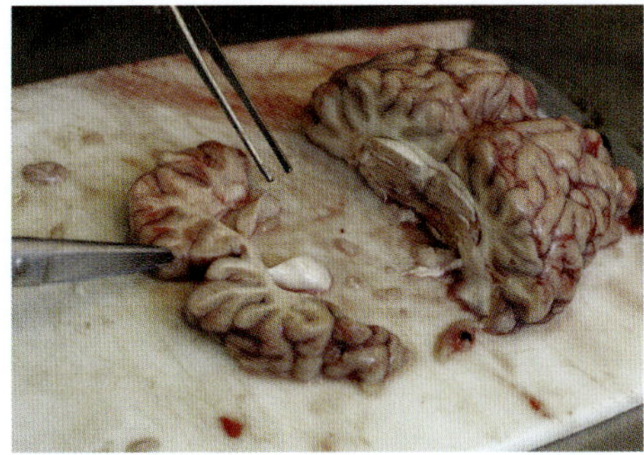

FIGURE 10.3.3 A vet slicing a cow brain to look for signs of BSE and other diseases. Tissue showing signs of damage must not enter the food chain.

BIOSECURITY AND BORDER CONTROL

Australia is fortunate to be free of many of the diseases that have damaged agricultural industries in other countries around the globe. As an isolated continent Australia has a natural border of water that stops many pathogens from reaching our shores. However, as travel, imports, exports and global connection increases so do the chances of introducing new diseases.

As an example, infectious fungal diseases are emerging globally as major threats to whole ecosystems and in particular plants and animals (Figure 10.3.4). Human activity, such as travel within and between countries, bushwalking and mountain biking through affected areas, and the import and export of agricultural products, increases the speed at which these pathogens move around the globe.

In Australia, governments and agriculture industries spend about $650 million on biosecurity each year (2014–15). Australia has strict regulations at sea and airports to prevent the introduction of new diseases. If you have ever travelled overseas you will have noticed strong screening and inspection of people, luggage and cargo coming into the country. X-ray machines and detection dogs are used for inspection and severe consequences apply to those who do not declare items that could import disease. Australia also implements strict quarantine conditions for any pets and other animals arriving in the country, which must reach clearance before and after they are imported.

It is also important for Australia to have interstate biosecurity measures in place as it is such a large country with diverse environments. Some devastating agricultural diseases have entered Australia but have been contained to a particular region. Restrictions apply to each state and territory for the movement of fruit, vegetables, plants, soils, flowers, plant products, agricultural machinery, animals or animal products and recreational equipment.

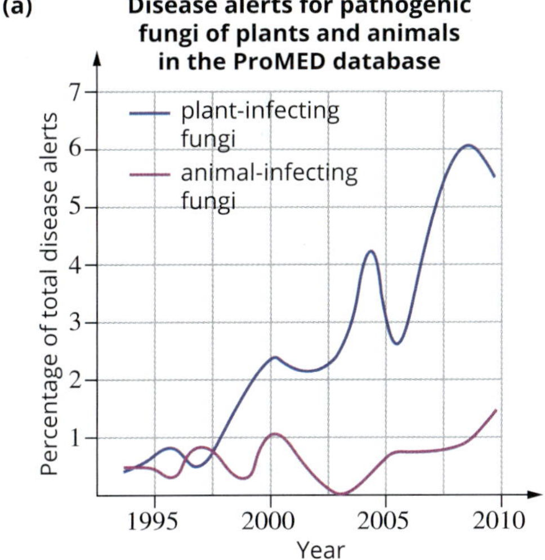

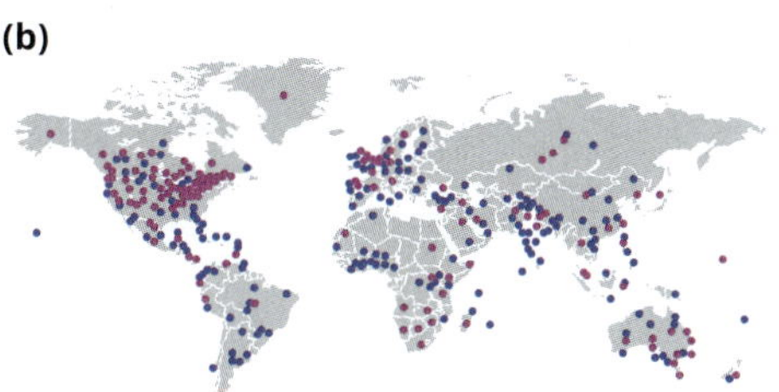

FIGURE 10.3.4 (a) Data showing the global rise in emerging fungal infections of animals and plants and (b) the location of the reported diseases

BIOFILE WE

Small scale, large consequences—biosecurity on the farm

Biosecurity is incredibly important for individual farmers to keep their crops and livestock free from disease and to prevent the spread of infection on one property to another. Most producers have an on farm biosecurity plan to identify and minimise risks of disease and implement an appropriate response if an outbreak does occur (Figure 10.3.5). Management involves recording all incomings and outgoings on the farm and monitoring the livestock, crops and environmental conditions on the farm itself. A biosecurity plan is unique to each property and distinctive features of the farm must be taken into account:

- size
- species present: livestock, crops, pests
- farm inputs: new stock, feed, water, fertiliser, seed
- farm outputs: agistment, sale of stock, crops, grains and wool
- people, vehicles, equipment: visitors, employees, rental or new machinery
- production practices: irrigation, effluent and waste management, pesticide use, vaccinations and drenching
- emergency planning: disease identification, quarantine, veterinarian and local authority contacts.

FIGURE 10.3.5 Chicken farm workers wear protective clothing following an outbreak of bird flu, which resulted in the death of 15 000 chickens.

10.3 Review

SUMMARY

- Disease in agriculture can result in reduced production and shortages; prevention and management is a huge cost to producers, industry and government.
- Selective breeding among plants and animals used for production has resulted in loss of genetic variation making livestock and crop populations very susceptible to disease.
- Intensification of production provides conditions that allow for emergence of new diseases and facilitate transmission of existing diseases.
- Changes in farming practices can result in emergence of new diseases that can lead to industry crashes and crises.
- Zoonotic diseases can be borne from human contact and consumption of infected animal products.
- Biosecurity is extremely important to prevent epidemics and the spread of disease in agriculture. It is especially important for isolated nations like Australia that have avoided exposure to many catastrophic diseases.

KEY QUESTIONS

1. Give two examples of pathogens that cause disease in plants.
2. Describe the impact of the pathogens from question 1 on the agricultural industry.
3. What is a zoonotic disease? Give an example of this type of disease.
4. How do human activities influence disease among plants and animals in agriculture?

10.4 Adaptations of pathogens

Pathogens possess many adaptations to penetrate the defence barriers of their plant and animal hosts, enhance their transmission and the spread of disease. Once pathogens have caused infection, many are able to hide within the host or combat the host's immune response. Host symptoms often aid in pathogen reproduction and increase the chance of future infection in a new host.

ENTRY INTO HOST AND TRANSMISSION BETWEEN HOSTS

In plants cuticle layers, bark, lignin and other structural defences protect against infection. Some fungi have developed a more complex and sophisticated mechanism to penetrate the cuticle of host plants. Some pathogenic fungi form a specialised infection cell, called an **appressorium**. These specialised cells are typical of many fungal plant pathogens and are used to infect host plants. The pathogenic fungi firmly attach the appressorium to the plant surface with extracellular adhesives. With the assistance of hydrolytic (water-splitting) enzymes they can infect the plant host. Hydrolytic enzymes split molecules in two pieces using water. They break the bonds between proteins, nucleic acids, starch and other macromolecules.

Vectors

An important adaption for many animal pathogens is the use of vectors to overcome the skin barrier. The outer layer of skin is impenetrable to most microorganisms unless the skin in broken. Even when the skin is broken, blood clotting occurs quickly to create a scab for protection. Pathogens that use biting insects, such fleas, ticks and flies, as vectors are able to penetrate the skin.

Malaria

Plasmodium falciparum, a protozoan that causes malaria, has developed an intricate relationship with two hosts, the *Anopheles* mosquito (Figure 10.4.1) and primates, including humans. *P. falciparum* parasite is transmitted between humans when they are bitten by the infected mosquito. A mosquito becomes infected through a blood meal of an infected person. When the mosquito feeds again on another person the pathogen is injected into the person's blood.

The antimalarial drug chloroquine was developed in 1934 and used in the 1940s as it was a safe and highly effective drug. However, in the late 1950s chloroquine-resistant parasites emerged in Asia. A mine along the border of Thailand and Cambodia attracted many workers from neighbouring regions. As well as the environmental changes that occurred during the construction of the mine conditions favouring the breeding of mosquitoes and workers were given inadequate doses of chloroquine. Under these pressures a mutation in the DNA of the parasite allowed it to survive in the presence of chloroquine, spread to infect the many workers in the region, and then further afield as the workers migrated (Figure 10.4.2).

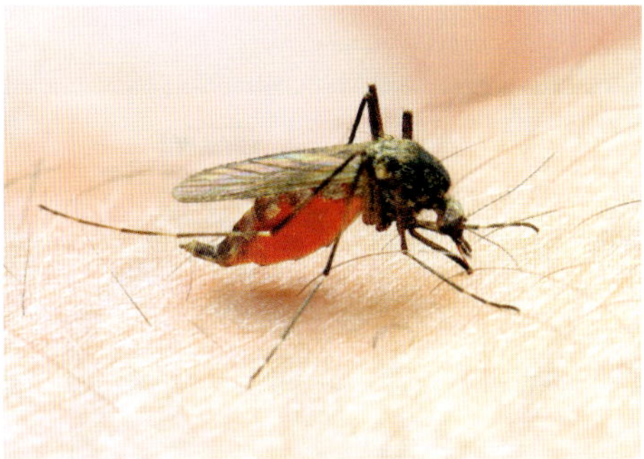

FIGURE 10.4.1 *Plasmodium falciparum*, a pathogenic protozoan, causes malaria in humans. The pathogen is transmitted by female *Anopheles* mosquitoes that can penetrate the skin of the host, a defensive barrier to infection.

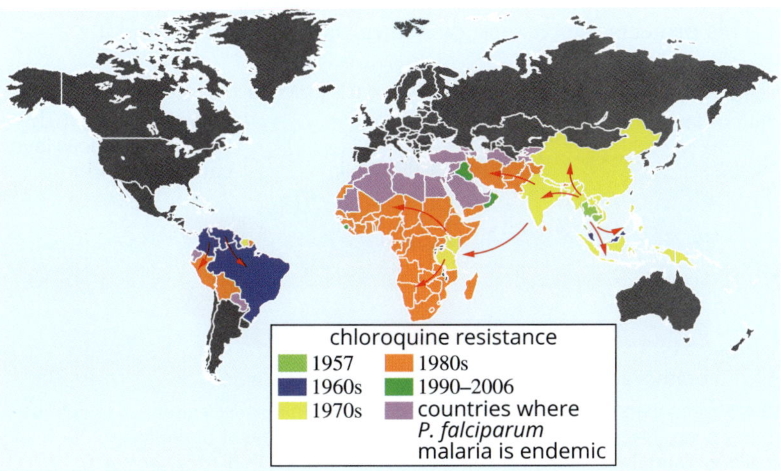

FIGURE 10.4.2 The history of chloroquine-resistant *P. falciparum*

Bubonic plague

The bubonic plague historically caused devastation to human populations and remains a re-emerging disease in several parts of the world. Sometimes simply called the plague, it is caused by the bacterium *Yersinia pestis* which uses fleas (Figure 10.4.3) as a vector to penetrate the skin of its host and spread disease.

FIGURE 10.4.3 (a) LM of the rat flea (*Xenopsylla cheopis*). The rat flea plays a major role in the spread of *Yersinia pestis*, the bacterium that causes bubonic plague (the Black Death of the Middle Ages). The bacterium is carried by the rat flea and infects domestic and brown rats that live in close association with humans. (b) In Sydney in 1900, tens of thousands of rats were killed to prevent further spread of the bubonic plague.

This bacterium infects rodents and is transmitted from rodent to rodent by infected fleas. It is transmitted to humans when a flea carrying the bacterium bites a susceptible person. Once bacteria reach the lungs, they become airborne and highly contagious (Figure 10.4.4). Symptoms appear 7–10 days after infection. The first pandemic plague recorded occurred in the 6th century, and is believed to have been brought to Europe from Africa by the fleas on rats in trade ships. There have been several epidemics and pandemics of the plague throughout the centuries, which have claimed the lives of millions of people.

FIGURE 10.4.4 (a) A mask used by 16th-century Venetian doctors to protect themselves against the highly contagious bubonic plague. The beak was filled with herbs to prevent the 'bad' air from being inhaled. (b) An illustration demonstrating the full outfit worn by doctors attending to patients suffering from the bubonic plague. The mask and outfit acted as protective wear so the doctor did not become infected.

Mucous membranes and chemical barriers

Pathogens have adaptations to help them penetrate the mucus barrier through physical propulsion or the chemical breakdown of mucus. Mucus contains **antibodies** that bind to pathogens preventing them from invading. Lysozyme is also present in mucus and is an enzyme that can break down bacterial cell walls.

Helicobacter pylori

Helicobacter pylori is a spiral-shaped bacteria that causes stomach infections and ulcers in humans (Figure 10.4.5). *H. pylori* is motile with special flagella that help it move in a screw-like motion through the mucus lining of the stomach. A pH gradient exists between the acidic mucus and the less acidic epithelial cells underneath. The bacteria use **chemotaxis** to move through the mucus and bind to the epithelial cells. *H. pylori* is also able to change the pH of the micro environment directly surrounding the bacterial cell. *H. pylori* can only survive between a pH of 4 and 8.2, the pH of gastric acid is around 1.5 to 3.5. *H. pylori* produces urease, an enzyme which catalyses the conversion of urea, from the surrounding mucus, to ammonia which has a very high pH. This protects the bacteria by neutralising the acid around it.

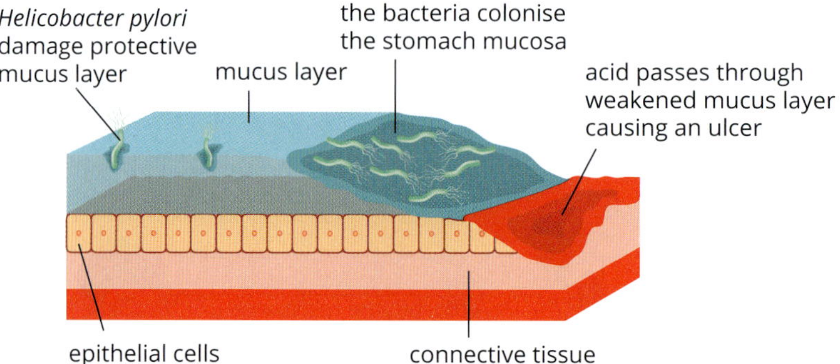

FIGURE 10.4.5 *Helicobacter pylori* is a pathogenic bacteria that causes stomach infections in humans. It has special adaptations to combat the harsh environment of the gastrointestinal tract. It uses flagella to swim through mucus and produces chemicals to neutralise acid.

Antigenic variation

GO TO ➤ Section 12.2 page 432

Antigens are molecules present on pathogens that induce an immune response from a host. You will learn more about antigens in Chapter 12. The host is able to recognise a pathogen from its antigen and create targeted antibodies to fight the specific pathogen. During the **primary immune response** the host will create memory cells and subsequent colonisation from the same pathogen will stimulate a faster immune response. Some pathogens are able to change their antigens so that the host does not recognise it upon secondary exposure. This is known as **antigenic variation** and gives the pathogen more time to colonise its host.

Influenza

Flu viruses circulate in populations of many species of animals and are constantly changing to produce new antigens. In humans, a new variant emerges almost every year and infects large numbers of people; this is called seasonal flu. New strains evolve from the genetic mixing or re-assortment of genes. This occurs when two or more different viruses infect the same human or other host animal. New strains resulting from re-assortment have new combinations of the surface antigens haemagglutinin (H) and neuraminidase (N) (e.g. H1N1 is the strain know to cause Spanish flu and Swine flu, H5N1 caused bird flu in 2004). New strains may be more virulent and are unrecognisable to a host previously infected with a different influenza strain.

White blood cells

White blood cells (T and B **lymphocytes**) are involved in the host's immune response and are responsible for finding and recognising antigens and destroying the associated pathogen. Some pathogens have adaptations that allow them to survive within the white blood cells sent to destroy them. Other pathogens may infect and deplete the cells responsible for coordinating the immune response.

HIV

Human immunodeficiency virus (HIV) is the virus that causes acquired immune deficiency syndrome (AIDS) by infecting and destroying helper T lymphocytes (Figure 10.4.7), a specialised white blood cell. Helper T lymphocytes normally coordinate an immune response and call on other types of white blood cells to fight infection. HIV has a protein on its surface that binds to receptors on T lymphocytes and hijacks those cells to make replicates of the virus. HIV depletes the amount of T lymphocytes in the body leaving the host susceptible to further infection from other pathogens. As HIV progresses the number of helper T lymphocytes declines and once the cell count drops below 200 the infected person is diagnosed with AIDS. AIDS is the most common cause of death from infectious disease. More than 35 million people have died of AIDS-related diseases since the beginning of the epidemic in 1981.

Biofilms and antibiotic resistance

Some bacteria live and grow together as an aggregate group connected to each other in a sticky extracellular matrix. These associated microorganisms and matrix are known as a **biofilm**. Biofilms can cause chronic infection by enabling bacteria to increase tolerance to antibiotics and host immune system responses.

Biofilms offer more protection to the aggregated microorganisms than to free living cells of the same species (e.g. chlorine can be used to effectively kill planktonic *Staphylococcus aureus* bacteria but a chlorine concentration of 600 times is needed to kill biofilm cells).

New spontaneous mutations may occur in the bacterial DNA, making the bacteria resistant to an antibiotic. These bacteria are able to multiply rapidly without competition from sensitive strains, and produce cloned offspring that carry the resistance gene. Antibiotics target specific biological pathways that disrupt cellular function and kill or slow the growth of the bacteria. There is the chance that a population of bacteria will become resistant to that particular antibiotic when treated with it. This can occur because some bacteria in a population already carry resistance genes; they are not killed by the antibiotic and pass on the resistance genes.

BIOFILE CCT

The changing face of malaria

The malarial protozoan *Plasmodium* infects red blood cells, where it resides to evade recognition by circulating immune cells (Figure 10.4.6). *Plasmodium* produces adhesion proteins, which it presents on the surface of the red blood cell. These proteins interfere with the cell's activities within capillaries.

The immune system recognises the adhesion proteins as non-self antigens, but before it can mount an effective immune response, the parasite replaces the adhesion protein with a different adhesion protein. *Plasmodium* has approximately 60 different adhesion proteins that it can continually interchange to remain a step ahead of the immune system. Scientists have recently discovered that an enzyme known as ribonuclease (RNase) causes this process of antigenic variation.

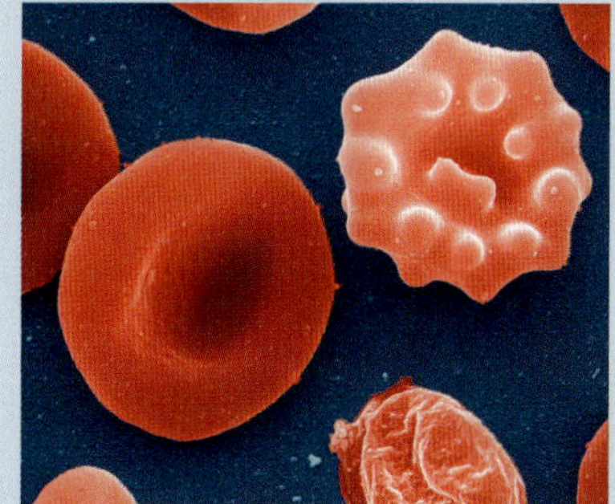

FIGURE 10.4.6 SEM of red blood cells infected with *Plasmodium*

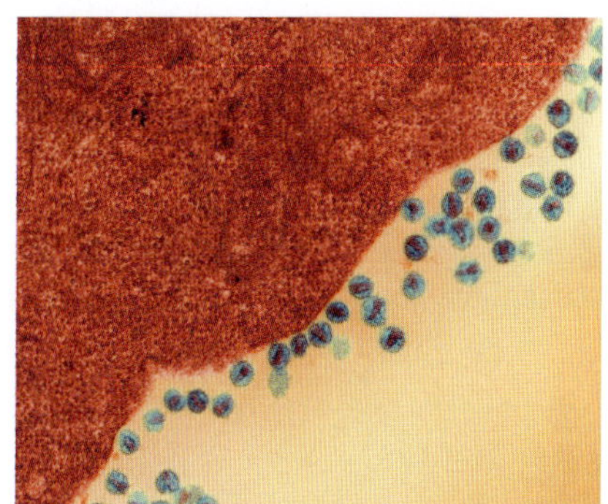

FIGURE 10.4.7 Human immunodeficiency virus (HIV) infecting a cell. TEM showing HIV particles (virions, round) on the surface of a cultured cell. HIV attacks helper T lymphocytes (specialised white blood cells), which are crucial in the body's immune system.

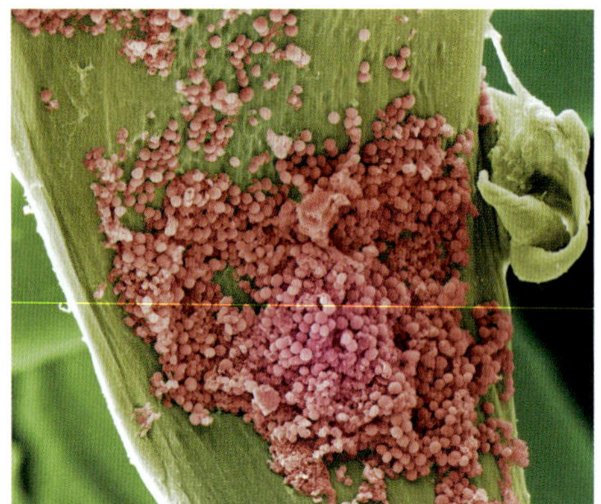

FIGURE 10.4.8 Methicillin-resistant *Staphylococcus aureus* (MRSA, red) shown in an SEM. It is resistant to many commonly prescribed antibiotics and can grow as a biofilm.

Biofilm growth also promotes the transfer of antibiotic resistance between bacterial cells in a process known as **horizontal gene transfer**. Plasmids carrying a drug resistance gene can be transferred between bacteria in this way. Transmission of resistant bacteria can be indirect, through food and water, or through direct contact with an infected person or animal. Methicillin-resistant *Staphylococcus aureus* (MRSA) (Figure 10.4.8) and vancomycin-resistant *Enterococcus* (VRE) are examples of bacteria that are now resistant to top-line antibiotics, making them almost impossible to treat.

SPREADING DISEASE

Once a pathogen has managed to colonise and reproduce inside a host it must complete the chain of infection through a portal of exit and into another host. Some pathogens have adaptations that cause symptoms within the host to aid in the spread of disease. Other pathogens can modify the behaviour of their host to assist pathogen transmission.

Host symptoms

In animals, symptoms of infection such as diarrhoea, coughing and sneezing are an attempt to remove pathogens and flush disease from the body. These host responses help to rid the body of pathogenic cells but also enhance disease transmission by potentially contaminating water and food, intermediate surfaces or through direct contact.

Rhinovirus

Rhinoviruses are a major cause of the common cold, which affects the nose, throat and sinuses. The portal of entry is usually through the respiratory tract and primarily spread through respiratory aerosols, particles of secretion from the nose or mouth. When the rhinovirus enters a new host it reaches the back of the throat and attaches to cell surface receptors. An immune response is stimulated and the host's mucous membranes will secrete a large amount of fluid. The immune system response results in the release of inflammatory cytokines, dilation of blood vessels and nerve fibre stimulation that can cause pain in the host and stimulate coughing and sneezing.

BIOFILE PSC

Sneezing

The function of sneezing is to expel mucus, which may contain pollutants, pathogens or any other foreign particles, from the nasal cavity. A sneeze can send over 40 000 droplets of mucus up to 8 m at speeds of over 150 km an hour. The size of the mucus droplets and the environmental air conditions influence the speed and distance that the droplets can travel (Figure 10.4.9). Small droplets may stay suspended in the air for 10 minutes. Causing a host to sneeze is a very effective way for pathogens to spread disease. Droplets from an uncovered sneeze could certainly reach susceptible hosts on the other side of a room.

FIGURE 10.4.9 High-speed video still of a person sneezing. After a sneeze, large droplets of saliva and mucus (green) shoot out of the mouth, but fall relatively quickly. A turbulent cloud carries smaller droplets (red) and allows them to drift for up to 8 m.

Change in host behaviour

Some pathogens are capable of altering the behaviour of their host to assist pathogen transmission and spread of disease. In animal hosts these pathogens affect the central nervous system or neurochemical communication.

Toxoplasma gondii

Toxoplasma gondii is a protozoan that can cause toxoplasmosis in many mammals. Infection does not normally result in any symptoms or health problems in adult humans except among individuals with compromised immune systems. *T. gondii* does have a marked effect on the behaviour of its intermediate hosts, rodents. Infected rats and mice are seen to have a reduced fear of cats (Figure 10.4.10). Causing this behavioural change in the intermediate host is a useful adaptation for *T. gondii* because cats are the definitive host where the protozoan completes its life cycle and sexually reproduces. Increasing the likelihood that infected rodents will be preyed on by cats increases the reproductive success of the pathogen.

FIGURE 10.4.10 *Toxoplasma gondii* is a pathogen that reduces infected rodents' innate fear of cats. Infected rodents are more likely to be preyed upon by cats, the definitive host of *T. gondii*.

10.4 Review

SUMMARY

- Pathogens have many adaptations to spread disease and defeat their hosts' defence mechanisms.
- Colonisation is the first stage of infection. Infection occurs if a pathogen is able to reproduce and elicit a host response.
- Host organisms have adaptations to combat infection by pathogens. These include physical barriers and immune system responses.
- In order to colonise an animal, pathogens must overcome many host barriers or defence mechanisms including the skin, mucous membranes and harsh chemicals.
- Plant pathogens have adaptations to enter cuticle layers, bark, lignin and other structural defences.
- Once a host is colonised, pathogens need to avoid the immune system with adaptations that make them unrecognisable or resistant to their hosts.
- Host response and symptoms support transmission of pathogens and help the spread of disease into a new susceptible host.

KEY QUESTIONS

1. Describe antigenic variation. What is the purpose of this adaptation in pathogens?
2. Give examples of two adaptations of *Helicobacter pylori* that help it to colonise the human body.
3. Briefly describe how HIV overcomes the host's immune response.
4. How does sneezing advance transmission of pathogens?
5. a Which animals are the definitive and intermediate hosts of *Toxoplasma gondii*?
   b What behavioural change does toxoplasmosis cause in its intermediate host and how does this aid transmission?

Chapter review

10

KEY TERMS

antibody
antigen
antigenic variation
appressorium (pl. appressoria)
bacteria
biofilm
capsid
cellular pathogen
chemotaxis
ectoparasite
endoparasite
epidemic
eukaryote
fungi
germ theory
helminth
horizontal gene transfer
host
hypha (pl. hyphae)
infection
infectious agent
infectious disease
Koch's postulates
lymphocyte
macroscopic
microscopic
non-cellular pathogen
oomycetes
oospore
pandemic
pasteurisation
pathogen
primary immune response
prion
prokaryote
protozoa
spore
transmission
vector
virion
viroid
virulence
virus
zoonotic
zoospore

REVIEW QUESTIONS

1 What is meant by the term 'pathogen'?

2 Give an example for each:
  a macroscopic pathogen
  b non-cellular pathogen
  c microscopic cellular pathogen.

3 Briefly describe the components of a virus and how a virus is able to reproduce.

4 Give an example of a disease epidemic caused by a virus.

5 Explain how the disease in question 4 managed to reach epidemic levels.

6 What is a prion? Give an example of a disease caused by a prion and its host.

7 How does the disease in question 6 affect its host?

8 What is a vector? Give an example.

9 Explain the difference between direct and indirect transmission.

10 Describe how increased pathogen exposure can contribute to an epidemic. Give examples.

11 Describe pasteurisation.

12 Describe the limitations of Koch's postulates.

13 Fredericks and Relman's 21st century postulates based on nucleic acids also have limitations. What kind of pathogens do these postulates not account for?

14 What did Pasteur's flask experiments help to demonstrate?

15 How does depleted crop variation impact global food security?

16 Give an example of a crop disease that has caused a food shortage.

17 Why is biosecurity and border control important for Australia's agricultural industry?

18 Explain one adaptation of a pathogen to penetrate the skin of animal hosts.

19 What is a biofilm and how does it aid infection?

20 After completing the Biology Inquiry on page 378, reflect on the inquiry question: How are diseases transmitted? Choose an infectious disease and use secondary sources to learn more about how this disease is transmitted. Use the chain of infection to represent the transmission of this disease.

CHAPTER 11 Responses to pathogens

In this chapter you will learn about the effects that a variety of pathogenic organisms have on plant and animal health and the ways in which plants and animals respond to infection. We will examine diseases caused by fungal and viral plant pathogens and the responses of Australian plants to these pathogens. We will also look at the physical and chemical changes that occur in the cells and tissues of animals in response to the presence of pathogens. You will learn about the different pathogens that infiltrate animal defences and how animals combat these pathogens with innate and adaptive immune systems.

Content

INQUIRY QUESTION

How does a plant or animal respond to infection?

By the end of this chapter you will be able to:

- investigate the response of a named Australian plant to a named pathogen through practical and/or secondary-sourced investigation, for example:
  - fungal pathogens
  - viral pathogens
- analyse responses to the presence of pathogens by assessing the physical and chemical changes that occur in the host animal's cells and tissues (ACSBL119, ACSBL120, ACSBL121, ACSBL122) CCT ICT

11.1 Plant responses to pathogens

BIOLOGY INQUIRY   

Fighting infection: Australian plants and animals

How does a plant or animal respond to infection?

COLLECT THIS...

- tablet or computer to access the internet
- large sheet of paper
- coloured pens, pencils or craft supplies

DO THIS...

1. Working in groups of two to four, choose one of the following Australian plants or animals:
   - Wollemi pine
   - eucalyptus tree
   - banksia tree
   - koala
   - Tasmanian devil
   - sulphur-crested cockatoo
2. Use the internet to find a pathogen that affects your chosen plant or animal.
3. Record information about the pathogen and how the plant or animal responds to infection.
4. As a group, create a flow chart or model to represent the response of your chosen plant or animal to infection by this pathogen.

RECORD THIS...

Describe the adaptations of your chosen plant or animal that help it resist infection by the pathogen.

Present your flow chart or model to the class.

REFLECT ON THIS...

How does a plant or animal respond to infection?

How has the plant or animal adapted to the presence of the pathogen over evolutionary time?

How could this disease impact the biodiversity of the ecosystem the plant or animal belongs to?

GO TO ➤ Section 10.1 page 378

Plants have evolved over millions of years to adapt to the variety of environments they occupy and the many environmental challenges they face, such as nutrient and light availability, temperature change, drought and **disease**. **Pathogens** are disease-causing organisms that have detrimental effects on the organisms they infect. To fight off pathogens, plants respond in several ways. In Chapter 10 you learnt about some of the common pathogens that cause disease in plants. In this section you will learn about the mechanisms plants use to resist **infection** by these pathogens and survive disease.

PLANT DISEASES

Plants can be affected by several types of pathogens, these include fungi, bacteria, nematodes and viruses. Symptoms of disease in plants include **senescence**, **wilt**, abnormal or stunted growth, **chlorosis** and vascular destruction. Plant health can also be negatively affected by fluctuations in environmental factors, such as rainfall and sunlight.

Some pathogens—**biotrophs**—invade a plant host but do not kill it. They depend on living plant tissue for nutrients, growth and reproduction. Examples of biotrophic pathogens include certain fungi (e.g. powdery mildew, *Blumeria graminis*) and some bacterial pathogens (e.g. rice pathogen, *Xanthomonas oryzae*).

Other pathogens—**necrotrophs**—invade their plant host through excretions of **enzymes** and toxic chemicals that deteriorate plant tissue, ultimately overcoming plant defences and causing **necrosis** (cell and tissue death). Examples of necrotrophs are the grey mould fungus (*Botrytis cinerea*) and the bacterial soft-rot pathogen (*Pectobacterium carotovorum*). They are parasitic organisms that obtain nutrients from their plant host and secrete enzymes that kill the plant. Plants infected with necrotrophs often exhibit necrotic **lesions** that may be either confined or so extensive they cause the plant to die. Most bacteria and viruses of plants are considered biotrophs, while most fungal plant pathogens are necrotrophs.

Plants wilt because of inadequate water and nutrient uptake, usually as a result of interference with plant vascular tissue (**xylem** and **phloem**) and water conduction within the plant. **Stomata** play a vital role in gas exchange and water regulation in plants. These allow the plant to absorb carbon dioxide from the air and remove excess oxygen and water. Plants may also wilt because of ineffective **transpiration** due to loss of control of the stomata (pores in the leaf surface).

Galls or **cankers** are often the result of hormonal imbalance in the presence of parasitic pathogens causing abnormal and distorted growth in plant tissue. A range of pathogens can cause plant chlorosis and tissue discolouration (mosaic effect) on plant tissue. Any pathogenic organism that interferes with chlorophyll production in plants will ultimately cause discolouration in plant leaves, turning them yellow or brown.

> Stomata (singular stoma) are tiny openings in the epidermis of the plant usually found on the underside of leaves.

Viral pathogens

Viruses lack cell membranes, cytoplasm, ribosomes and enzymes necessary for protein synthesis and energy production and so the viral envelope and capsid play an important role in enabling viruses to penetrate host cells.

Once established within the host the virus can become a localised infection and remain in a confined area of the plant or may move throughout the plant body, resulting in a systemic infection. Viruses may move through the plant in several ways:

- cell-to-cell movement via **plasmodesmata**
- parenchyma cells (these cells make up most of the plant's soft tissue and mesophyll tissue—viruses are able to move through the **symplastic pathway**, between the mesophyll cells and through to vascular tissue (Figure 11.1.1) where they continue to infiltrate the plant and cause disease)
- transported in conducting channels (phloem tissue and **sieve tubes**).

> Plants have two main pathways for water and solute movement: the apoplastic pathway and the symplastic pathway. The apoplastic pathway is through the intercellular spaces, along the cell walls, and the symplastic pathway is through the cytoplasm.

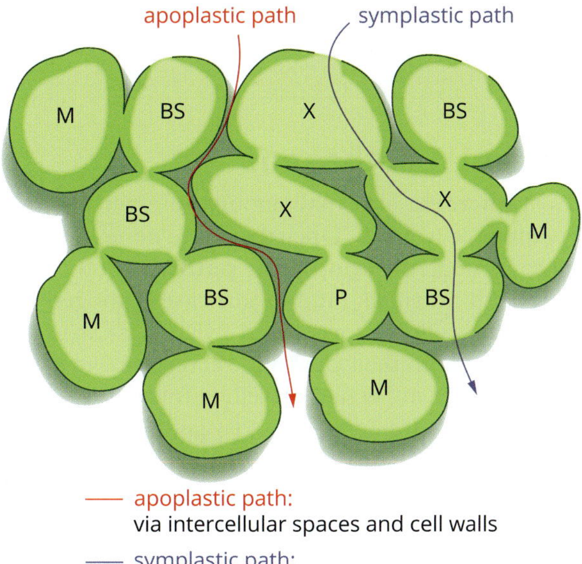

FIGURE 11.1.1 A schematic diagram of a cross-section of vascular tissue showing the apoplastic (orange line) and symplastic (blue line) pathways that viruses can move through in plant tissue. Mesophyll (M) and bundle sheath (BS) cells surround xylem (X) and phloem (P).

Fungal pathogens

Fungi are eukaryotic and heterotrophic organisms that inhabit almost every environment on Earth, from hot, dry deserts to deep sea hydrothermal vents. The Fungi kingdom is divided into taxonomic groups based on their reproductive strategies and structures. Fungal pathogens can inhibit cellular processes in plants, such as **photosynthesis** and the uptake of nutrients, and cause wilting, leaf and seedling chlorosis and inhibit the functioning of stomata. Some common plant diseases caused by fungi are listed in Table 11.1.1.

TABLE 11.1.1 Common plant diseases caused by fungi

Type of Fungi (phylum and genus)	Common plant diseases
Chytridiomycota *Synchytrium*	black wart of potato
Blastocladiomycota *Physoderma*	brown spot of corn, crown wart of alfalfa
Zygomycota *Rhizopus*	soft rot of several plant parts
Ascomycota *Neurospora* *Monilinia* *Cryphonectria*	powdery mildew, brown rot of stone fruits, chestnut blight, Dutch elm disease
Basidiomycota *Puccinia* *Ustilago* *Cronartium*	black stem rust of wheat, common corn smut, blister rusts

Hyphae are thread-like filaments that protrude from fungi and secrete enzymes to break down and absorb nutrients.

Some fungal pathogens infect plants through the secretion of a variety of enzymes that disrupt the physical defence mechanisms of plants, such as the **cuticle** and **cell wall**. Some other fungal species (some rusts) gain entry to the plant via the stomata. These fungi produce a specialised infection cell called an **appressorium** (plural appressoria) over the stomatal opening. A fungal **hypha** (plural hyphae) then penetrates the leaf and forms a sub-stomatal vesicle (Figure 11.1.2). Hyphae emerge from the vesicle and form **haustoria** (a projection from a hyphae) enabling the pathogenic fungus to penetrate the tissues of its host and absorb nutrients from it.

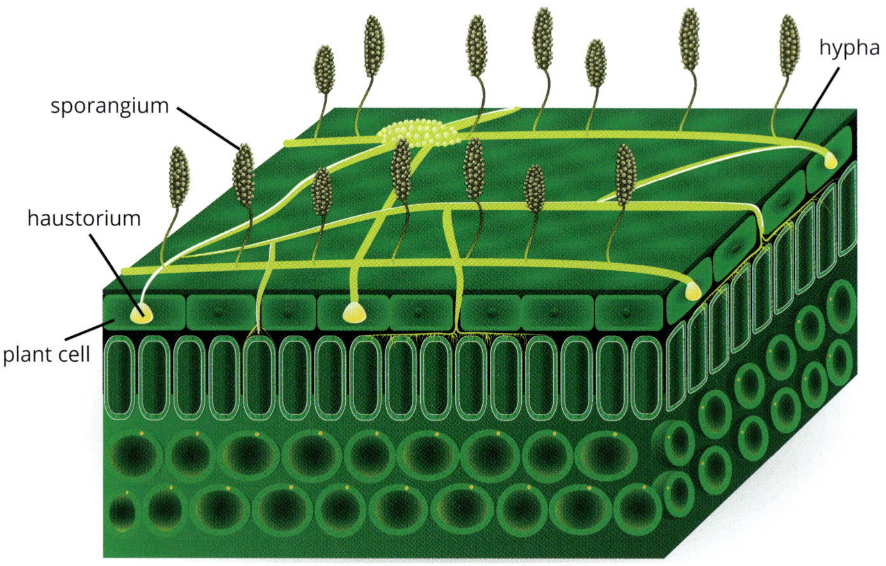

Fungi-parasites on plants

sporangium
hypha
haustorium
plant cell

FIGURE 11.1.2 Pathogenic fungi absorb their food material from the living tissues of the hosts on which they parasitise.

Bacterial pathogens

Most **bacteria** are small enough to enter plants through stomata and other small openings or wounds. Once they have penetrated the plant's defences they are able to travel through the **apoplastic pathway** of the plant. The apoplastic pathway allows water and nutrients (and in this case pathogenic bacteria) to travel around the cell walls, intercellular spaces and xylem vessels of the plant (Figure 11.1.3). Bacteria multiply within their host and secrete enzymes that breakdown plant cell walls and destroy other molecules travelling through the apoplastic system.

Common diseases caused by apoplastic bacteria include diseases such as scabs, cankers, galls and other overgrowths, wilts, leaf spots, blights and soft rots. Some pathogenic bacteria that affect plants may also produce toxins or enzymes that breakdown plant structures, such as cell walls.

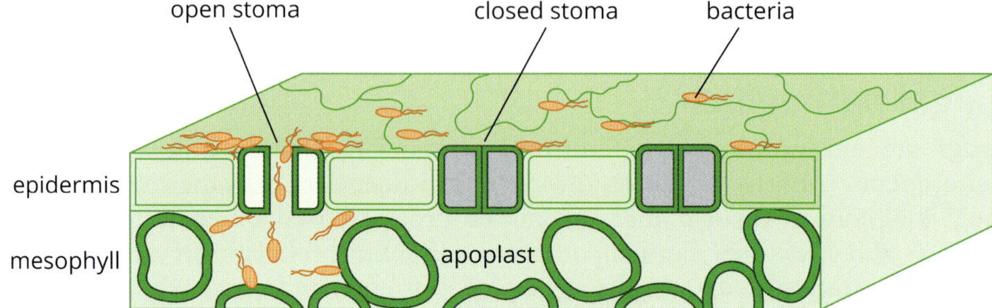

FIGURE 11.1.3 A schematic diagram of a cross-section of a leaf with open and closed stomata in the epidermis. This diagram shows bacterial invasion through open stomata into the leaf apoplast (mesophyll).

BIOFILE S

Antibiotics for plants

Plants, like animals, are susceptible to bacterial, fungal and viral infections. Although plants have an arsenal of antimicrobial compounds to defend them against pathogens, sometimes pathogens overcome these barriers. Spraying with antimicrobial chemicals, such as copper compounds, may be sufficient to control an infection. In some places, when agricultural commercially valuable crops are infected, antibiotics may be used.

Apple crops in New Zealand are affected by the highly contagious disease fire blight (Figure 11.1.4), which is caused by the bacterium *Erwinia amylovora*. Affected parts of the plant may appear as though they have been scorched by fire; fruits and leaves may be blackened, dried, shrivelled and cracked. Some orchard owners in New Zealand treat crops by spraying with the antibiotic streptomycin. In other countries fire blight is treated by injecting the antibiotic oxytetracycline into trees. This is a costly exercise, so is less common than chemical sprays. Antibiotics are not approved for use as a crop spray in Australia. There is also concern that the widespread use of antibiotics as sprays would speed the development of resistant bacteria in the environment.

Australia does not have fire blight, and orchardists want the restriction on the importation of New Zealand apples and pears to remain in place because of the high biosecurity risk. This remains a major concern for Australian fruit growers.

FIGURE 11.1.4 Fruit tree damaged by fire blight disease, which is caused by the bacterium *Erwinia amylovora*

PLANT-PATHOGEN INTERACTIONS

Recognition of pathogens by plants involves several processes and includes proteins, sugars and lipopolysaccharides (large molecules made up of a lipid and polysaccharide components), which allow the plant to initiate defence responses. These molecules are released to counter chemicals secreted by pathogenic organisms that have penetrated the plant in some way. Initial interactions between plant defences and pathogens take place in the extracellular spaces (apoplast) of the plant. Materials can move freely in the plant through the spaces outside of the cell membrane. In the apoplast pathogens with specific microbial **elicitors** (foreign molecules), otherwise known as **microbe-associated molecular patterns (MAMPs)**, are identified by membrane-localised **pattern recognition receptors (PRRs)** of plants.

If recognition of microbial elicitors occurs PRRs activate the first line of defence in the host plant. If the host cannot overcome the pathogen the host will kill both pathogen and the infected plant cell.

PLANT DEFENCE MECHANISMS

Plants have developed structural, chemical and protein based defence mechanisms in order to combat pathogenic organisms. Some physical barriers involved in plant defences are cell walls, waxy epidermal cuticles and bark. These substances are composed of **lignin**, **cutin** and other macromolecules, which inhibit the entrance of many pathogens and protect the plant. In addition, many chemical barriers are utilised by plants in response to infection with pathogenic organisms. These include the production of toxic chemicals, proteins (enzymes) which breakdown pathogens and programmed cell death (**apoptosis**) in order to prevent further infection of the plant from the point of entrance.

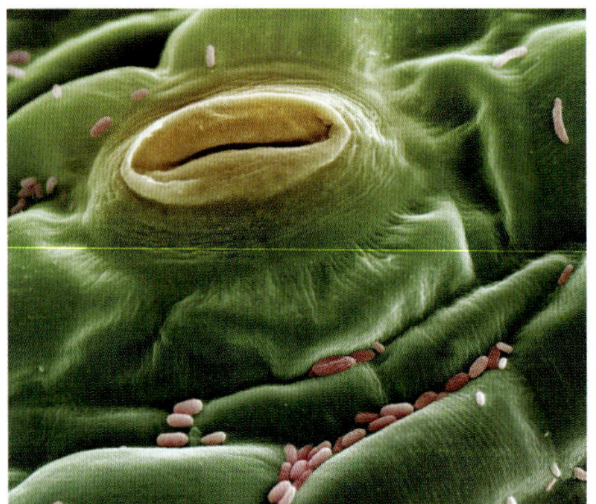

FIGURE 11.1.5 Scanning electron micrograph (SEM) of a single stomata on the surface of the leaf of a tomato plant. Stomata are able to close to prevent bacteria (rod-shaped, pink) entering and infecting the plant.

Physical barriers in plants

Physical barriers in plants largely involve cell walls that provide strength and flexibility. Cutin and waxes are fatty substances that make up the cuticle, which is found on the outer cell wall. A thicker cuticle generally prevents more pathogens from infecting the plant than a thinner cuticle. Likewise, in trees, a thicker layer of bark is better able to prevent pathogens from entering the plant. Stomata create openings in the physical barriers of plants, providing an entry point for pathogens, but these openings can be closed when signalled (Figure 11.1.5). In addition to the presence of physical barriers, the orientation of leaves can also play a role in defence. By positioning leaves vertically, water is unable to collect on the surface of leaves. This prevents infection by pathogens that are reliant on water for motility. This adaptation is important for plants that live in environments with high rainfall.

Chemical barriers in plants

Plants have developed a vast array of chemical defences. Table 11.1.2 lists some of the chemicals produced by plants to defend against pathogens.

TABLE 11.1.2 Chemical barriers in plants

Chemical	Source/plant	Function
saponin	wheat	disrupts cell membranes of fungi
caffeine	coffee, tea, cocoa plants	toxic to insects and fungi
tannins	tea and grapes	toxic to insects
citronella	essential oil from lemongrass	repels insects
defensins	barley and wheat	toxic to microbes
chitinases	barley, tomato, banana	enzymes that disrupt cell membranes of fungi

The immune system of plants

FIGURE 11.1.6 Bread wheat (*Triticum aestivum*) contains small cysteine-rich proteins that act as plant defensins to inhibit the growth of bacteria and fungi.

Unlike mammals, plants don't have an **adaptive immune response**. They also lack mobile immune cells that can travel to the site of infection, so every plant cell has to respond to pathogens independently. Plants have an **innate immune response** and secrete a number of **defensive molecules** (Figure 11.1.6). As an example, certain plants produce considerable amounts of the insect moulting hormone, ecdysone. When parasitic insect larvae eat the leaves, their hormonal balance can be fatally disrupted, with obvious advantages to the plant. Eucalypts are under heavy and continuous attack from a vast array of organisms that suck, chew or nibble. Eucalypt leaves contain high levels of toxic compounds, but some insect attackers have adapted to become resistant to these chemicals. Some species of *Acacia* deter insect grazers by producing cyanogenic glycosides (cyanide-generating chemicals that are cytotoxic), killing cells by blocking enzymes involved in cellular respiration. When these defensive molecules fail to prevent pathogens from infecting the plant, cell-mediated defences can involve self-destruction of infected or damaged cells (apoptosis), which helps to limit a pathogen's access to nutrients and, in turn, limit the pathogen's ability to spread to the rest of the plant. Many plants also produce a **hypersensitive response (HR)** when invaded by parasites, such as nematode larvae or bacteria. This response involves the rapid breakdown of cells around the parasite and the release of toxic substances.

There are four main components of plant **immune response** to pathogenic organisms:

1 basal resistance
2 gene-for-gene resistance
3 hypersensitive response (HR)
4 systemic acquired resistance (SAR).

Basal resistance

Plants are often able to detect pathogenic organisms through a variety of highly developed defence mechanisms. **Basal resistance** (innate immunity) is activated when plant cells detect elicitors or microbe-associated molecular patterns (MAMPs), proteins, fatty molecules and cell wall structures associated with fungal, bacterial and viral pathogens, along with other organisms that exhibit these molecular components in their cells. The activation of basal resistance results in the plant becoming largely impenetrable due to living plant cells becoming fortified against further pathogenic infection. Non-pathogens as well as pathogens are capable of triggering basal resistance in plants due to the widespread presence of these molecular components in their cells.

Gene-for-gene resistance

Gene-for-gene resistance was first proposed by Harold Flor in 1956 in his studies of rust disease in flax. Flor's gene-for-gene hypothesis states that for every gene in the plant that presents resistance, there is a corresponding gene in the pathogen that presents **avirulence** (lack of pathogenicity). To test this hypothesis, Flor bred flax rust and obtained several progeny (offspring) that he then used to infect flax plants. He determined the relationship between pathogen (flax rust) and host (flax plant) by noting whether the plants were immune, resistant, semi-resistant or susceptible to the rust pathogen. Flor bred both rust and flax creating several crosses to analyse subsequent generations. From his experiments he was able to establish that pathogenicity (the ability to cause disease) of the flax rust was inherited and concluded that resistance in plants and avirulence in pathogens were also inherited. Flor noted that even if the flax had the resistance gene (*R*) it would only be resistant to the pathogen if the invading pathogen had the avirulent gene (*Avr*).

The gene-for-gene hypothesis concludes that plants are able to produce proteins through specific plant disease resistance genes (*R* genes). Plants that produce the *R* gene will be resistant to the pathogen that produces the corresponding avirulence (*Avr*) gene. Pathogens exhibit avirulence genes (*Avr* genes) that are recognised by the plants *R* genes. These genes are pathogen-specific and without the specific *R* gene the plant will not be able to defend itself against the invader (Figure 11.1.7).

(a) Gene-for-gene hypothesis quadratic check

pathogen avirulence (*A*) or virulence (*a*) genes	plant resistance (*R*) or susceptibility (*r*) genes	
	R (resistant)	*r* (susceptible)
A (avirulent)	*AR*(−)	*Ar*(+)
a (virulent)	*aR*(+)	*ar*(+)

− = resistant
+ = susceptible

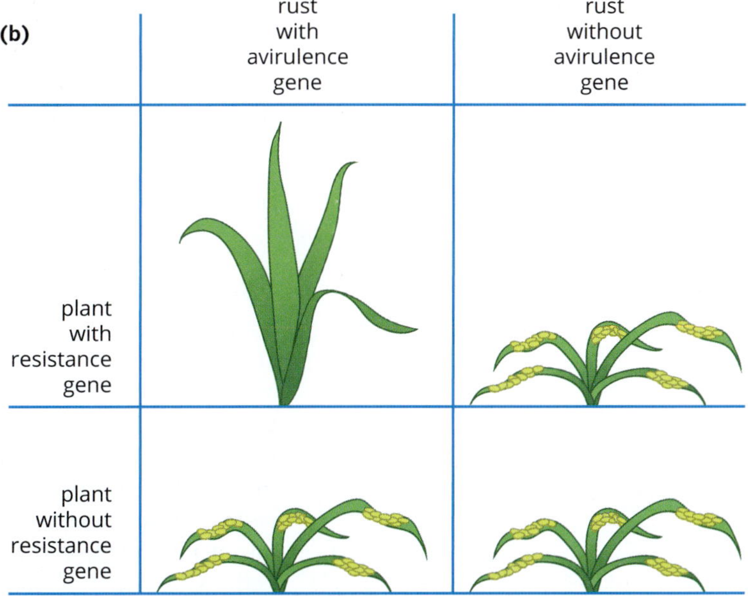

FIGURE 11.1.7 (a) The gene-for-gene hypothesis states that resistance in the host and virulence in the pathogen is determined by pairs of genes. In plants, the gene is referred to as the resistance (*R*) gene. In pathogens, the gene is called the avirulence (*Avr*) gene. (b) Harold Flor's experiments with flax and rust disease showed that for a plant to be unaffected by the pathogen (i.e. resistant), it must have the resistance (*R*) gene and the rust pathogen must have the avirulence (*Avr*) gene. In all other cases, the plant is susceptible to the pathogen.

FIGURE 11.1.8 Two sibling corn (maize, *Zea mays*) plants. The one on the left has a hypersensitive disease-resistant gene but the one on the right does not. This gene causes hypersensitive-response lesions to form spontaneously all over the plant, whether the pathogen is present or not.

Hypersensitive response

If basal resistance in plants is bypassed, the hypersensitive response (HR) is quickly activated to limit pathogenic access to the host plant (Figures 11.1.8 and 11.1.9). The hypersensitive response involves programmed cell death (apoptosis) by the host plant. This mechanism allows the plant to restrict the pathogen to the infected site and limit its access to the rest of the plant.

Pathogens produce molecules (elicitors) that are detected by the plant immune system. The plant cell will produce compounds that attach to bacterial pathogens and destroy their cell walls or increase **lignification** of plant cell walls to prevent further infection. Pathogens such as bacteria, fungi and nematodes induce this response in host plants.

Pathogens, which have been recognised as avirulent and therefore contain the *R-Avr* match will produce a stronger immune response in the host plant. The plant will localise the infection, producing small lesions to control the pathogen. Portions of the leaf may be destroyed and the plant will survive in most cases.

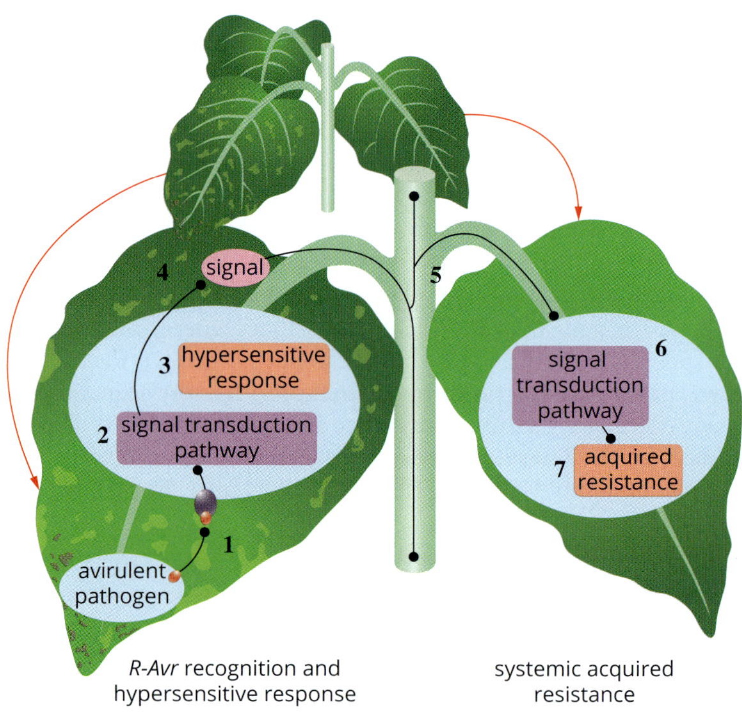

FIGURE 11.1.9 Summary of plant defences against an avirulent pathogen *(Avr)*

Systemic acquired resistance

Systemic acquired resistance (SAR) (Figure 11.1.9) is induced in a host plant after it has been exposed to molecules from virulent, avirulent or non-pathogenic microbes. SAR can also be initiated by salicylic acid where accumulation of salicylic acid stimulates defence mechanisms that usually prompts a localised hypersensitive response. Salicylic acid is a plant hormone that is proposed to be the first chemical in the initiation of pathogenesis related genes and plays a role in preventing or inhibiting pathogenic colonisation within a host plant. Accumulation of salicylic acid within the host plant is triggered by the presence of pathogens or microbe-associated molecular patterns (MAMPs). Salicylic acid is also thought to play a role during plant response to abiotic stresses (drought, extreme heat or cold and osmotic stress).

The hypersensitive response will further aid in the plant's defences in **signalling transduction pathways** (cellular response to chemicals or hormones). SAR is a non-specific mechanism of defence and will combat several types of pathogenic organisms. Once a plant activates SAR it is able to quickly and effectively mount an attack on subsequent exposures to the same invading pathogen.

BIOLOGY IN ACTION A CCT

A common plant could help fight deadly viruses

Dengue fever, zika and yellow fever are responsible for illness and death in many tropical and subtropical countries. These diseases are spread by the *Aedes aegypti* mosquito (Figure 11.1.10). A recent discovery by a Filipino teenager has found that a common San Francisco plant may be able to kill the larvae of the *Aedes aegypti* mosquito, potentially preventing the spread of these viruses.

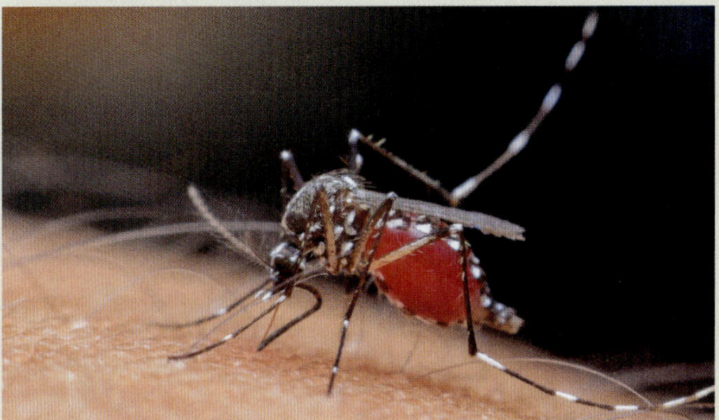

FIGURE 11.1.10 *Aedes aegypti* mosquito is a vector of many human viruses, including dengue fever virus, zika virus and yellow fever virus.

Jerouen Lumabao, a seventeen-year-old high school student from the Philippines, determined that phytochemicals (biological chemicals found in plants) from the San Fransisco plant, *Codiaeum variegatum*, (Figure 11.1.11) killed larvae of the *Aedes aegypti* mosquito.

Jerouen tested a variety of plants including, tawa-tawa (*Euphorbia hirta*), lemongrass (*Cymbopogon citratus*), and San Francisco plant (*Codiaeum variegatum*). Initially, he picked the leaves off each of his test plants and allowed them to dry for a week. He soaked the leaves in alcohol and allowed the alcohol to evaporate leaving behind the concentrated oily plant phytochemicals he required. Each extract was then placed into water containing mosquito eggs and larvae. Jerouen concluded that the tawa-tawa and the lemongrass caused the mosquito larvae to develop irregularly, preventing them from maturing into healthy adults. However, he found that the San Francisco plant killed all the mosquito larvae and eggs in his test.

FIGURE 11.1.11 The San Franscisco plant (*Codiaeum variegatum*) has been found to contain phytochemicals that kill mosquito larvae. These phytochemicals could be used to prevent the spread of the *Aedes aegypti* mosquito, a vector of deadly diseases such as dengue fever, zika virus and yellow fever.

Jerouen's experimental results could help scientists develop plant extracts to curtail potentially deadly diseases.

11.1 Review

SUMMARY

- Many plants are affected by disease, which may be a consequence of abiotic and biotic factors.
- Pathogenic organisms in plants can be bacteria, fungi and viruses.
- Common plant fungal diseases include wilting, mildews, smut, damping off, dieback, rusts and mosaics.
- Common plant bacterial diseases include blight, galls, lesions, bacterial wilt and soft rot.
- Vector organisms, such as some insects, transmit pathogenic organisms between plants.
- Plants have physical barriers involved in defence against pathogenic organisms. These include cell walls, waxy epidermal cuticles and bark which are composed of lignin, cutin and other macromolecules, which inhibit the entrance of many pathogens and protect the plant from a variety of biotic pests.
- Pathogenic organisms move through the plant either through the apoplastic or symplastic pathways.
- Plants produce and respond to many chemical messages to initiate their defence mechanisms. Several chemical defence molecules that enable them to fight off pathogens are lignin, cutin and salicylic acid.
- Plants have a variety of chemical defence molecules that they use in order to avoid being infiltrated by pests: saponin, caffeine, tannins and citronella.
- There are four main components of a plant's innate immune response to pathogens: basal resistance, gene-for-gene resistance, hypersensitive response (HR) and systemic acquired resistance (SAR). These responses enable plants to survive disease even after they have been infected.

KEY QUESTIONS

1. **a** List four common pathogens that infect plant hosts.
   b List some examples of disease symptoms in plants.
2. Distinguish between disease caused by viruses and those caused by bacteria.
3. Propose reasons for eliminating diseased plant parts in a plant population.
4. Describe how fungal pathogens overcome plant physical defence mechanisms to infect the plant.
5. Determine how you would distinguish between a healthy and a diseased plant.
6. Which of the following is a physical defence barrier in plants?
   A caffeine
   B broad leaves
   C tannin
   D cuticle

11.2 Animal responses to pathogens

Animals have developed a variety of ways in which to defend themselves against pathogens. Preventative strategies in animals, including physical, chemical and microbiological barriers, act to contain or inhibit pathogens from causing damage to their host (Figure 11.2.1). However, complex systems such as immune systems are also necessary to protect most complex animals against invaders. Once animal defences have been penetrated, the innate immune response is initiated. Vertebrates, including humans also have an adaptive immune response. You will learn more about the human immune systems in Chapter 12.

GO TO ➤ Section 12.1 page 422

In this section you will learn about the types of physical and chemical barriers animals have to protect themselves against foreign invaders. You will learn about the different types of chemicals involved in detecting and destroying pathogens in order to prevent infection and disease and how the body recognises and detects antigens in the body.

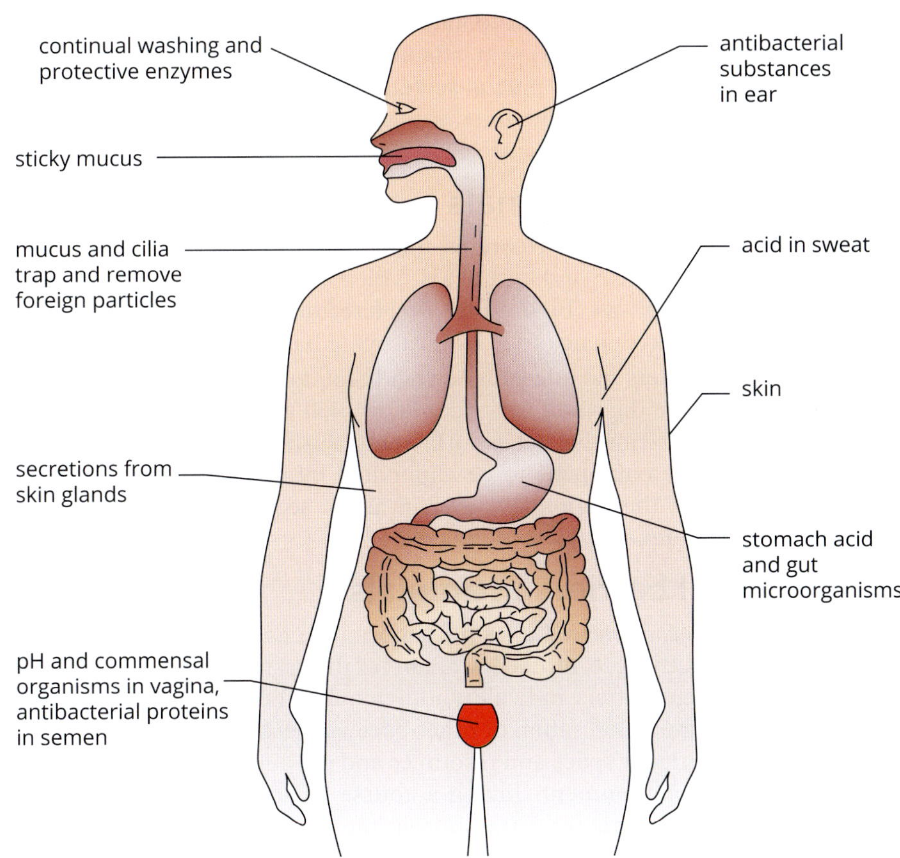

FIGURE 11.2.1 Some of the physical and chemical defence mechanisms that prevent foreign organisms from gaining access to the human body

RESPONDING TO PATHOGENS

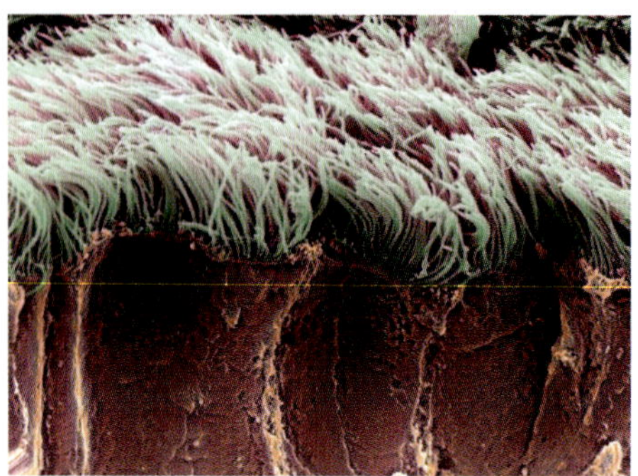

FIGURE 11.2.2 SEM of the mucous membrane (or bronchial epithelium) that lines the major airways of the lung. Mucus traps potential pathogens and foreign particles, and the rhythmic movement of hair-like cilia moves bacteria and other particles away from the lung and towards the throat.

Most animals are able to fight off different types of foreign invaders and infection from pathogens through a variety of physical barriers. These barriers are typically toughened outer layers of skin and play a major role in preventing microbial infiltration. It is only through a wound or injury that pathogenic organisms encounter the chemical barriers animals use to fight off potential infection that has been able to penetrate the outer layer of the skin. Various glands continuously produce antimicrobial chemicals, which are moved through the body by **cilia** (hair-like projections) and **peristalsis** (muscular contractions) in order to destroy the pathogen in less than optimal conditions, such as the acidic environment of the stomach. Intestinal **microflora** also play an important role in keeping pathogenic organisms from inhabiting an animal by outcompeting invading pathogens.

Physical barriers in animals

In vertebrates, **epithelial** cells create a physical barrier that prevents pathogens from entering the organism. Epithelial cells line the skin, as well as the respiratory, gastrointestinal and urogenital tracts. They are joined tightly by specialised membrane proteins, which form a continuous barrier against pathogens. In addition to toughened (keratinised) skin, adaptations that provide physical barriers to pathogens in animals include mucus-secreting membranes that trap invading organisms in mucus and membranes lined with cilia that sweep foreign bodies away (e.g. those that line the airways) (Figure 11.2.2).

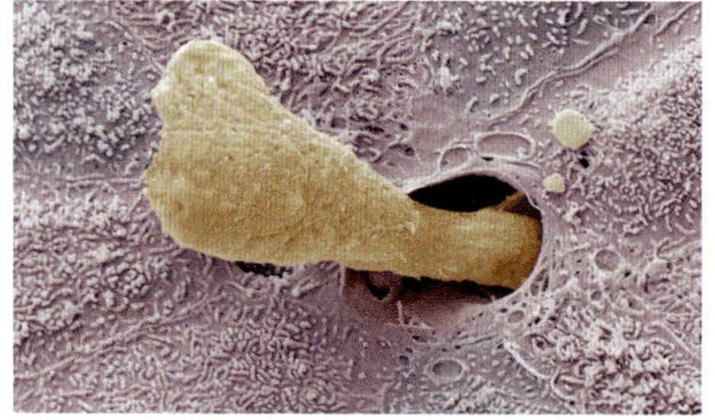

FIGURE 11.2.3 SEM of cerumen (ear wax) being secreted by a gland in the ear canal

Chemical barriers in animals

External chemical barriers in vertebrates include lysozyme enzymes and toxic metabolites, such as lactic acid and fatty acids, which are found in secretions such as tears, sweat, saliva and ear wax (Figure 11.2.3). Here, they have protective functions and provide a generalised defence (e.g. by destroying bacterial cell walls).

Other chemical barriers include stomach acid and digestive enzymes, which are primarily involved in the digestion of food, but also kill many pathogens. The fluid in the lungs contains proteins that act as surfactants. Surfactants coat the pathogens, making it easier for the pathogens to be eliminated by **macrophages**. In female mammals, the lining of the vagina is coated in acidic secretions that serve several functions, including defence against pathogens.

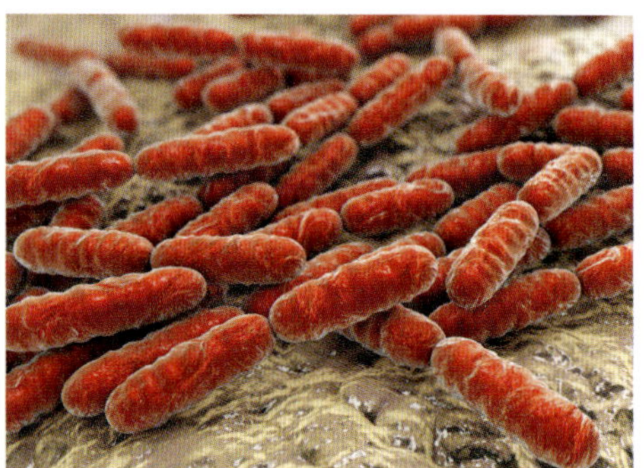

FIGURE 11.2.4 A 3D illustration of bacteria, *Lactobacillus* (lactic acid bacteria), which are part of normal human intestinal flora and are used as probiotics and in yoghurt production

Microbiological barriers in animals

Non-pathogenic bacteria, referred to as microflora (or normal flora), are found on the skin and in the mouth, nose, throat, lower part of the gastrointestinal tract (Figure 11.2.4) and the urogenital tract in healthy individuals. The presence of microflora prevents the growth and colonisation of other bacteria because microflora compete with pathogenic bacteria for space and resources and produce chemicals that reduce the pH of the micro-environment. Taking a course of antibiotics can disrupt the composition of the microflora of the intestine, as the antibiotics do not discriminate between beneficial bacteria and harmful bacteria. This can disturb the normal gut function and predispose a person to various infections until the microflora return to their pre-treatment levels. Although not a problem in healthy individuals, in people with weakened immune systems microflora can sometimes grow unchecked and cause disease. Table 11.2.1 outlines some common diseases that affect the physical and chemical barriers.

TABLE 11.2.1 Physical and chemical barriers of animals and some common diseases that affect them

Barrier	Non-specific mechanism	Description and location	Disease
skin	first line of defence physical	outer body covering dry outer layer difficult to penetrate unless wounded	ringworm leprosy impetigo
mucous membranes	first line of defence chemical	a viscous slimy secretion used predominantly for protection and lubrication contained within epithelial cells lining the openings of the body such as digestive tract, reproductive system, respiratory system and excretory systems	syphilis food poisoning cholera
cilia	first line of defence physical	slender hair-like structures that protrude from the surface of most mammalian cells found in nasal passages, ear canal, oesophagus, collecting duct of the kidney and other body structures	urinary tract infections sinus infections caused by bacterial pathogens
sebum	first line of defence chemical	light yellow oily secretion produced by sebaceous glands keeps the outer surface of the skin moisturised and protected from pathogens by making the skin slightly acidic and producing a non-favourable environment for pathogens	folliculitis acne
microflora	first line of defence competitive	beneficial microorganisms (bacteria) located on the skin or other internal structures such as the intestines and mouth	stomach ulcers caused by *Helicobacter pylori* bacteria
cerumen (ear wax)	first line of defence chemical	brown, grey or yellow waxy secretions made in the ear canal by ceruminous and sebaceous glands located on the outer part of the ear canal and used to trap dust and other particles	ear infections (particularly the Eustachian tube)
lacrimal secretions	first line of defence chemical	a watery fluid containing mucin, lipids, lysozyme, immunoglobulins and other substances to lubricate and protect the eye from foreign matter and pathogens produced by the tear ducts in the eyes	conjunctivitis trachoma
acidic secretions	first line of defence chemical	fluid secretions that have a lowered pH (low acidity) to prevent pathogenic entry into the body creating a hostile environment for foreign invaders. Produced by the vagina, stomach, ear canal, skin (sebaceous glands)	gonorrhoea herpes
lysozymes	first line of defence chemical	an enzyme that aids in the destruction of bacterial cell walls found in secretions such as saliva, mucus (nasal secretions) and tears	common cold measles
peristalsis	first line of defence physical	muscle movement that occurs in wave like motions to enable the movement of the oesophagus via muscular contraction this aids in the movement of secretions (mucus) or foreign invaders to the stomach in order to kill pathogenic organisms	esophagitis caused by bacterial or fungal infections

ANTIGENS

The human immune system is one of the most complex protectors against pathogenic invaders. All pathogens contain molecules or parts of molecules that initiate an immune response in the host. **Antigens** are unique molecules, or parts of molecules, that can often elicit an immune response and play a crucial role in immunity (Figure 11.2.5). Specialised cells of the immune system, called **B lymphocytes** and **T lymphocytes** play an important role in the recognition of antigens and the immune response. You will learn more about B and T lymphocytes in Chapter 12.

GO TO ➤ Section 12.2 page 432

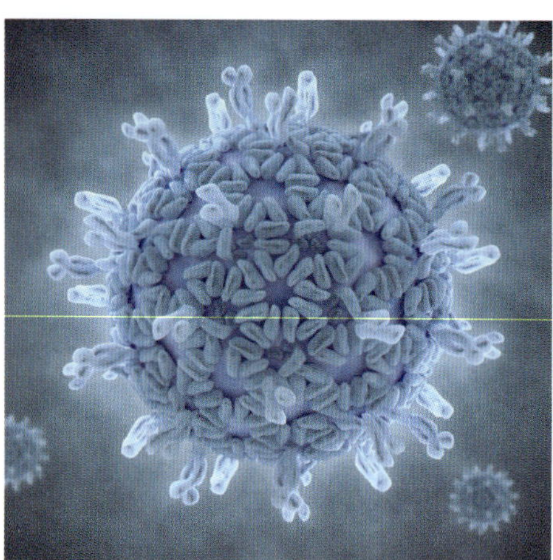

FIGURE 11.2.5 An artist's impression of rotavirus, a virus that is a common cause of gastroenteritis and diarrhoea in infants. Proteins on the surface of the virus act as antigens, which are recognised by the body's immune cells.

GO TO ➤ Section 12.1 page 422

The nature of antigens

Antigens are important because they allow the body to recognise potentially harmful pathogens and mount an immune response against them. Although many antigens trigger an immune response, some do not. Antigens that elicit an immune response are also known as **immunogens**; however, in the context of an immune response it is still common to simply refer to them as antigens.

There are two main types of antigens that are recognised by the immune system:

- **non-self antigens** (foreign or heteroantigens)
- **self-antigens** (autoantigens).

Non-self antigens are external to the body and can trigger an immune response. Examples of non-self antigens include components or molecules produced by viral pathogens or microorganisms (bacteria and protozoans), substances found in insect or animal venom, certain proteins found in foods, transplanted organs or donated blood from another individual.

Antigens, which are recognised as self (autoantigens) are produced internally. Typically, the immune system is able to differentiate self from non-self; however, in people with **autoimmune** disorders this recognition often does not occur and ordinary body molecules initiate an immune response. You will learn more about antigens and their role in the immune response in Chapter 12.

DEFENSIVE MOLECULES

There are several molecules that are produced in order to fight off specific pathogens. These are known as defensive molecules and include the following:

- **Complement proteins**—an array of more than 30 proteins that circulate in the blood and help to kill foreign cells.
- **Cytokines**—small signalling molecules of the immune system that coordinate many aspects of immune responses. Cytokines can be peptides, proteins or glycoproteins and are released by body cells in response to cell damage or the presence of pathogens.
  - **Interferons**—a class of cytokine that are produced by, and act on, a host cell infected by a virus. Interferons act in an autocrine manner, activating the infected cells to produce enzymes that breakdown viral RNA and proteins that block translation.
  - **Chemokines**—a type of cytokine that acts as a chemical attractant (or chemo-attractant). Chemokines are important for attracting **leukocytes** to sites of infection and **inflammation**.

You will learn about these defensive molecules in more detail in Chapter 12.

GO TO ➤ Section 12.1 page 422

INFLAMMATION

Inflammation is the accumulation of fluid, plasma proteins and leukocytes that occurs when tissue is damaged or infected and results in heat, pain, swelling, redness and loss of function. The interaction between leukocytes (especially **phagocytes**) and pathogens triggers the inflammatory response that results from the production, activation or release of peptides and proteins such as complement proteins and cytokines. Acute inflammation involves phagocytes and occurs soon after infection as part of the innate immune response, but inflammation can also involve lymphocytes and occur later as part of the adaptive immune response.

A number of steps are involved in the initiation of an inflammatory response to infection (Figure 11.2.6):

1 Bacteria or other pathogens breach the barriers that provide a first line of defence, (e.g. through an open cut or wound in the skin).
2 Injured cells release cytokines (chemokines) that attract **neutrophils**, and **mast cells** release **histamine**, which increases blood vessel dilation and permeability. The dilated, more permeable blood vessels allow leukocytes and fluid containing peptides and proteins, such as complement proteins, to enter the infected tissue. Platelets release clotting factors at the site of the wound.
3 Neutrophils migrate towards the cytokines and are activated causing the neutrophils to recruit macrophages and secrete factors, such as **defensins** and hydrogen peroxide, which degrade and kill pathogens.
4 Macrophages in turn become activated and secrete cytokines and, along with neutrophils, phagocytose pathogens and debris at the site of infection. This may lead to pus, which is fluid containing leukocytes, dead pathogens and cell debris.
5 The inflammatory response continues until the pathogen is eliminated and the wound has healed.

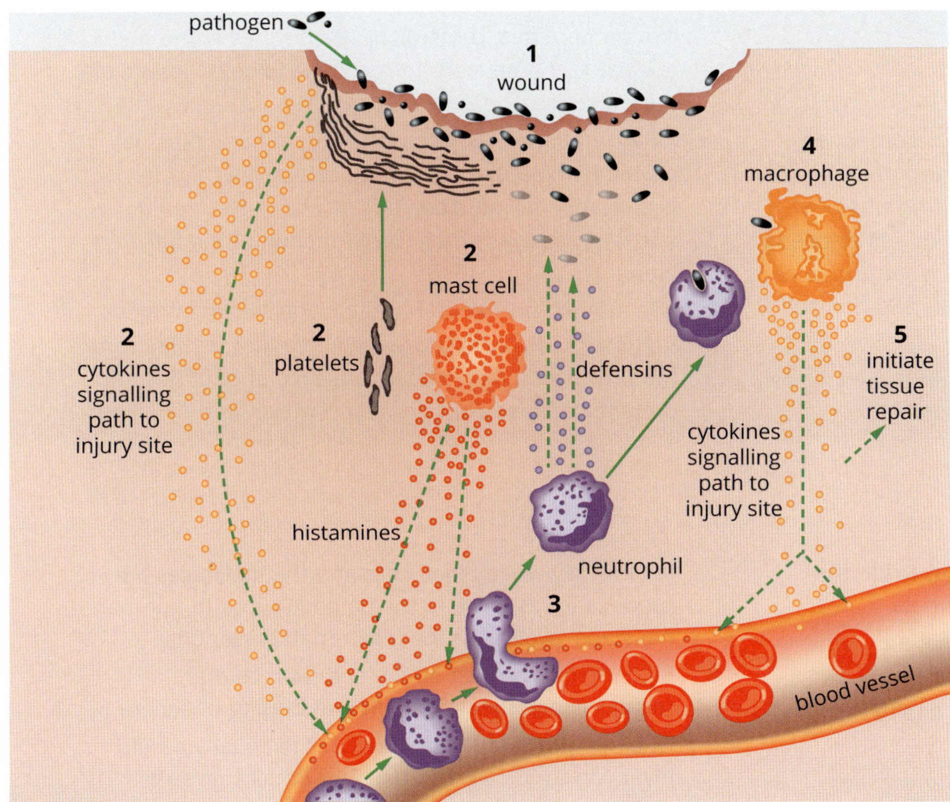

FIGURE 11.2.6 The process of inflammation in response to injury and infection

You will learn more about the cells involved in the inflammatory response in Chapter 12.

GO TO ➤ Section 12.1 page 422

Fever

A **fever** is an increase in body temperature that occurs as a consequence of inflammatory cytokines altering the body's set temperature (regulated by the hypothalamus in the brain). In humans, normal body temperature is around 37°C. Fever occurs when body temperature is above normal. Fever slows the replication of bacteria and viruses by shifting the temperature away from their optimal range, and so allows more time for other defences to intervene. Additionally, moderate increases in temperature increase the activity and proliferation of leukocytes, so fever also improves the immune response.

11.2 Review

SUMMARY

- Most animals are able to fight off different types of foreign invaders and infection from pathogens through a variety of physical barriers and chemical barriers. These include:
  - skin
  - mucous membranes
  - sebum
  - cilia
  - microflora
  - cerumen
  - lacrimal secretions
  - acidic secretions
  - lysozymes
  - peristalsis
- Antigens are protein-based molecules composed of one or more polypeptide chains, or parts of molecules, that can often induce an immune response. Antigens can be classified as self-antigens or non-self antigens, and an organism's immune cells can usually differentiate between self- and non-self antigens.
- Antigens can be recognised by T lymphocytes or by antibodies produced by B lymphocytes.
- Complement proteins and cytokines are defensive molecules involved in both the innate and adaptive immune responses. Complement proteins are a variety of more than 30 proteins that circulate in the blood and help to kill foreign cells. They are found in body fluids in an inactive form, and are activated as part of the non-specific immune response to certain antigens and carbohydrates on the surfaces of some bacteria and parasites.
- Cytokines are small signalling molecules such as interferons and interleukins, which help to coordinate many aspects of our immune responses. Cytokines can be peptides, proteins or glycoproteins and are released by body cells in response to cell damage or the presence of pathogens.
- Inflammation is the accumulation of fluid, plasma proteins and leukocytes that occurs when tissue is damaged or infected, and results in heat, pain, swelling, redness and loss of function of the affected area.
- The interaction between phagocytes and pathogens triggers the inflammatory response that results from the production, activation or release of peptides and proteins such as complement proteins and cytokines.

KEY QUESTIONS

1. List at least three physical barriers that aid in the prevention of pathogen entry into animals.
2. List at least three chemical barriers that aid in the prevention of pathogen entry into animals.
3. How are microflora necessary for the health of an animal?
4. Explain how mucus and cilia remove or prevent foreign invaders from entering the body.
5. Compare the roles of tears (lacrimal secretions), acidic secretions and sweat in preventing foreign substances from entering the human body.
6. Are pathogens sources of self-antigens or non-self antigens? Explain the difference between self- and non-self antigens in your answer.
7. Draw the process of inflammation that occurs when bacteria enter the skin through an open wound. Be sure to label key white blood cells and the molecules they produce in response to the pathogen, and to number the steps involved in the inflammatory response.
8. Mild infections such as the common cold are a regular experience of children attending childcare.

   One symptom of these infections is usually a mild fever (up to 39°C). The usual treatment given to children is a medication such as paracetamol, which reduces their temperature to normal.

   a Explain why reducing the body temperature of a patient with a mild fever may prolong the infection.

   b Patients who have a high fever (over 41°C) should always be treated to reduce their body temperature. Why would such high temperatures reduce the body's ability to fight off an infection?

Chapter review

11

KEY TERMS

adaptive immune response
antigen
apoplastic pathway
apoptosis
appressorium (pl. appressoria)
autoimmune disease
avirulence
bacteria
basal resistance
B lymphocyte (or B cell)
biotroph
canker
cell wall
chemokine
chlorosis
cilium (pl. cilia)
complement protein
cuticle
cutin
cytokine
defensin
defensive molecule
disease
elicitor
enzyme
epithelium (adj. epithelial)
fever
fungi
gall
gene-for-gene resistance
haustoria
histamine
hypersensitive response (HR)
hypha (pl. hyphae)
immune response
immunogen
infection
inflammation (or inflammatory response)
innate immune response
interferon
lesion
leukocyte
lignification
lignin
macrophage
mast cell
microbe-associated molecular pattern (MAMP)
microflora
necrosis
necrotroph
neutrophil
non-self antigen
pathogen
pattern recognition receptor (PRR)
peristalsis
phagocyte
phloem
photosynthesis (adj. photosynthetic)
plasmodesma (pl. plasmodesmata)
self-antigen
senescence
sieve tube
signalling transduction pathway
stoma (pl. stomata)
symplastic pathway
systemic acquired resistance (SAR)
T lymphocyte
transpiration
virus
wilt
xylem

REVIEW QUESTIONS

1. Distinguish between fungal and bacterial infections in plants, providing examples.
2. Formation of galls are usually a result of:
   A viruses and mycoplasmas
   B bacteria and parasites
   C fungi and bacteria
   D insect vectors and bacteria
3. Distinguish between the apoplastic and symplastic pathways for movement of pathogens through plants.
4. Compare biotrophic and necrotrophic organisms.
5. *Erwinia amylovora* is a bacterium that causes fire blight in the apple crops of New Zealand. Explain, with reference to the available treatment options, why Australian apple growers would be concerned about the import of New Zealand apples.
6. Describe the hypersensitive response (HR) in plants.
7. List the physical barriers used by plants in response to pathogens.
8. Determine the different ways by which plant pathogens gain entry to their host plant.
9. 'Many plant viruses go unnoticed.' Using your knowledge of virus replication and pathogenesis discuss this statement.
10. You have been given three plant samples (one plant shoot, new leaf and brown leaf) with the following characteristics: several unidentified waxy covered insects, the shoot exhibits stunted growth and sooty mould covers the brown leaf. Describe how you would identify the cause of the symptoms in each of your specimens.
11. Define the term vector and name three vectors that cause disease in plants.
12. Determine the effectiveness of enzymes in plant defence mechanisms.
13. Define chlorosis and name two causes of this symptom in plants.
14. Distinguish between a disease and a symptom in plants.
15. Compare the physical and chemical responses of an animal to pathogens.
16. Describe two different types of cells involved in pathogen destruction in animals.
17. What happens if the human immune response is directed against a self-antigen?
18. What components of the human immune system are responsible for antigen recognition?
19. List four defensive molecules in animals that play a role in responding to foreign invaders.
20. Analyse how fever and the inflammatory response combat pathogens in the body.
21. After completing the Biology Inquiry on page 404, reflect on the inquiry question: How does a plant or animal respond to infection? Using secondary sources assess the use of medications in reducing fever when responding to pathogens.

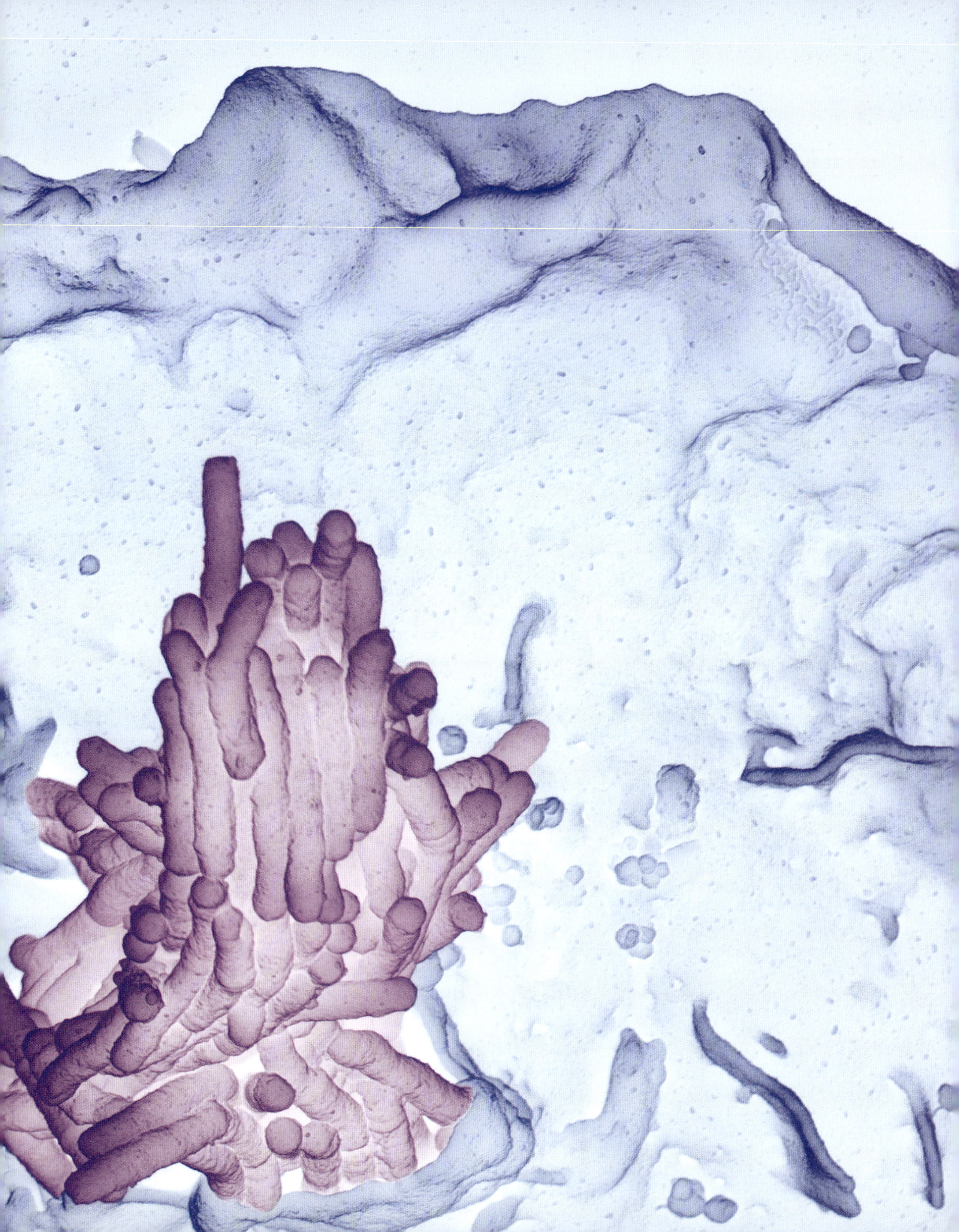

CHAPTER

12 Immunity

In this chapter you will learn about the innate and adaptive human immune systems. You will understand how these immune systems function, their components and the mechanisms they use to respond to pathogens. You will learn about the specific nature of lymphocytes and their role in immune response. You will also learn how the immune system responds after primary and secondary exposure to pathogens and develop an understanding of immune cell interaction and activation. You will examine the molecules necessary for immunological memory and how they continue to protect our bodies long after we are exposed to pathogens.

Content

INQUIRY QUESTION

How does the human immune system respond to exposure to a pathogen?

By the end of this chapter you will be able to:

- investigate and model the innate and adaptive immune systems in the human body (ACSBL119)
- explain how the immune system responds after primary exposure to a pathogen, including innate and acquired immunity

12.1 Innate immunity

BIOLOGY INQUIRY   

Human immune responses to pathogens

How does the human immune system respond to exposure to a pathogen?

COLLECT THIS...

- tablet or computer to access the internet
- large sheet of paper
- coloured pens, pencils or craft supplies

DO THIS...

1 Working in groups of two to four choose one of the following pathogens:
   - *Staphlococcus aureus*
   - human immunodeficiency virus (HIV)
   - *Salmonella*
   - *Candida albicans*
   - *Escherichia coli*
   - *Vibrio cholerae*
   - yellow fever virus

2 Use the internet to investigate how the body responds to your chosen pathogen.

3 Record information about the interaction of the body to the pathogen.

4 As a group create an infographic or pamphlet to explain how the immune system responds to your chosen pathogen. Make sure to include pictures with descriptions and explanations.

RECORD THIS...

Describe the immune response when exposed to your chosen pathogen.

Present your infographic or pamphlet to the class.

REFLECT ON THIS...

How does the human body respond to non-self antigens?

How can science, technology and medicine be used to support the immune system when it is exposed to unfamiliar pathogens?

Given the rise in antibiotic resistance, what are potential future directions of medicine for treating infectious disease?

The human body comes under attack from a barrage of foreign invaders and is continually exposed to infections. The body's exposure to **antigens** make it necessary for the body to be able to defend itself against foreign substances. The ability for the body to recognise these foreign substances and stimulate cells to destroy or remove them is termed **immunity**. Often when a person exhibits disease symptoms they are suffering from a disease caused by a **pathogen**. At times symptoms may be the last response to pathogen exposure. The three stages of being exposed to an infectious disease are:

1 entry of the pathogen (bypassing first line of defence and gaining entry into the body)

2 **incubation period** (no symptoms are exhibited by the **host**)

3 immune response activation (in response to antigen recognition).

Immune responses can be classified into two main categories:

- **innate immune responses**
- **adaptive immune responses** (specific).

In Chapter 11 you learnt about the physical, chemical and microbiological barriers of plants and animals that provide innate resistance to infection. In this section you will learn about the innate immune response to infection in humans that occurs when these barriers are breached (Figure 12.1.1).

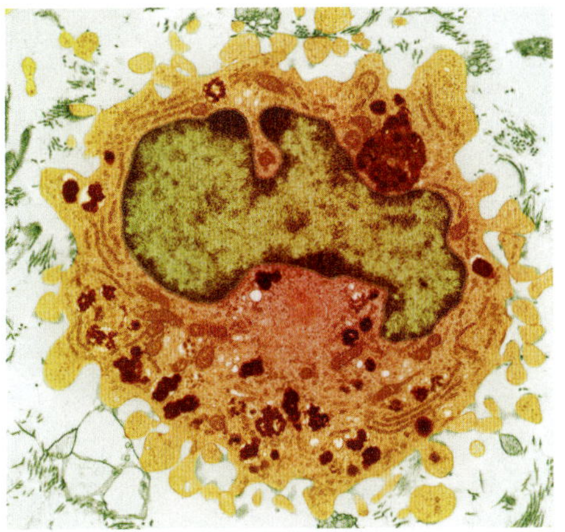

FIGURE 12.1.1 Coloured transmission electron micrograph (TEM) of a macrophage. Macrophages are cells of the innate immune response in vertebrates that recognise and engulf foreign material.

PRIMARY EXPOSURE TO PATHOGENS IN HUMANS

If pathogens manage to breach the barriers that act as a first line of defence, they are immediately met by attacking cells and molecules. Non-specific mechanisms are activated to prevent the pathogen from continuing to cause damage to the body. A series of events take place to prevent the pathogen from damaging the host including the inflammatory response and the action of **phagocytes** (Figure 12.1.2), and the production and mobilisation of antimicrobial proteins and chemicals.

The innate immune response is found in all organisms and its persistence over millions of years of evolution indicates its fundamental importance. Innate immune responses:

- are non-specific—they do not target a specific antigen
- are rapid—they occur within hours
- are present in all animals and plants
- are fixed responses—they do not adapt
- do not lead to an **immunological memory** of the pathogen that caused the infection.

Even when the innate immune response is unable to eliminate a pathogen, it remains critical for keeping infections under control until the adaptive immune response, which can take up to several days to develop, takes over.

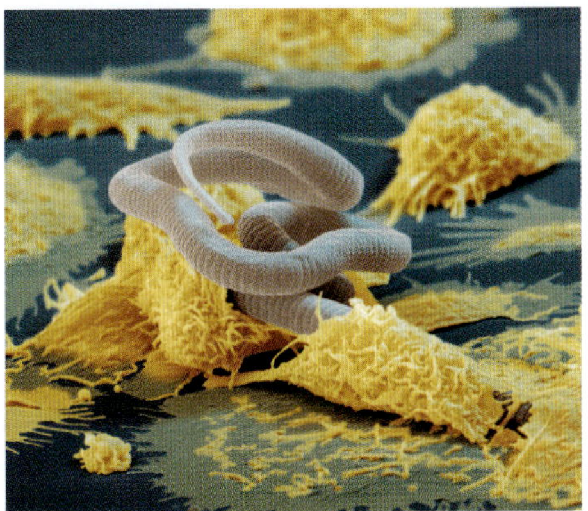

FIGURE 12.1.2 Macrophage (large white blood cell) attacking a parasitic larva. Macrophages play an important role in recognising and engulfing foreign substances in the body.

CELLS OF THE INNATE IMMUNE SYSTEM

White blood cells (or **leukocytes**) are immune cells that are present in blood and other tissues. Leukocytes have pattern recognition molecules, also known as **toll-like receptors (TLRs)** on their surface, which are able to recognise microbial molecules called **microbe-associated molecular patterns (MAMPs)**. There are different TLRs that recognise different PAMPs. For example, TLR-2 recognises the lipoproteins and peptidoglycan of gram-positive bacteria, TLR-4 recognises lipopolysaccharide of gram-negative bacteria, TLR-5 recognises bacterial flagellin, and TLR-7 and 8 recognise single-stranded viral RNA.

PAMPs are common to a range of pathogens, meaning that the innate immune response to them is not specific to a particular pathogen.

Phagocytes

Phagocytes are leukocytes that are able to engulf and break down pathogens in a process known as **phagocytosis**. Phagocytes include **neutrophils**, **macrophages**, **monocytes** and **dendritic cells**.

Phagocytosis

TLRs on a phagocyte interact with a microbe's PAMPs, causing **signal transduction** events to occur, which lead to the activation of the phagocyte. Once activated, the phagocyte engulfs the microbe, with the cell membrane forming a vacuole called a **phagosome** around it. Then a **lysosome** containing digestive enzymes (**lysozymes**) fuses with the phagosome, forming a **phagolysosome**, which breaks down the foreign material. The fragments can then be expelled from the cell by exocytosis.

> Phagocytosis is a type of endocytosis in which foreign particles are taken into the cell and enclosed in a section of cell membrane which is pinched off to form a vacuole.

> Exocytosis is the release of substances enclosed within a vesicle to the outside of a cell. It occurs by fusion of the vesicle with the plasma membrane.

> Signal transduction is the process of sending a signal into or out of a cell or changing the form of a signal.

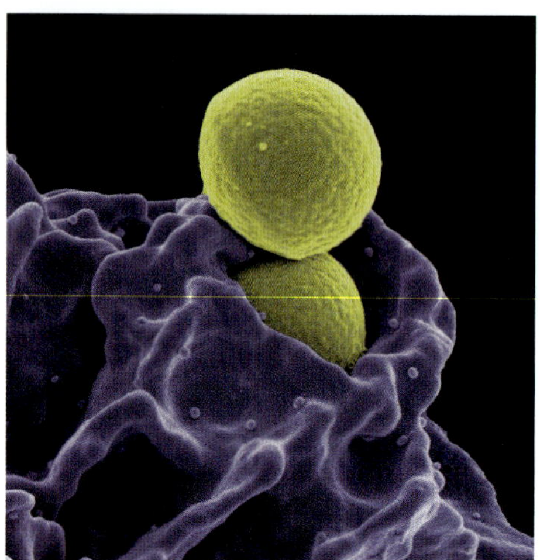

FIGURE 12.1.3 Phagocytes are a key part of the innate immune response. This neutrophil (purple) is engulfing *Staphylococcus aureus* bacteria (yellow), which it will then phagocytose.

Neutrophils

Neutrophils (Figure 12.1.3) are one type of phagocyte. Neutrophils are the most abundant white blood cells (about 50–70% of all white blood cells) involved in the innate immune response. Neutrophils are **granulocytes** and contain vesicles (compartments) within their cytoplasm filled with antimicrobial chemicals that are used to destroy pathogens. These antimicrobial proteins or peptides have a membrane affinity that enable them to penetrate bacterial membranes which results in bacterial **lysis** (rupture of cell membrane). All neutrophils are produced in the **bone marrow** and only live for a few hours in the body. Neutrophil **apoptosis** (programmed cell death) is an important aspect of phagocytosis and as such they are used as the frontline defence cells that die in order to control the infection (which may take days or weeks to fight off).

Macrophages

Macrophages are a type of phagocyte that differentiate from the precursor monocytes when the inflammatory response is activated (Figure 12.1.4). These cells are responsible for detecting, targeting and eliminating foreign substances or necrotic (dead) cells that have infiltrated or are present in the body. Macrophages work by ingesting the foreign or dead material and using enzymes to digest it. Macrophages are able to live for several months and destroy hundreds of pathogens before they die in an innate (non-specific) immune response (Figure 12.1.5). Macrophages act as **antigen-presenting cells (APCs)**. Macrophages present antigens on the cell surface by protein molecules called **major histocompatibility complex (MHC)** II and present them to **helper T lymphocytes** which will enable the adaptive immune response to be activated in a series of reactions that produce large numbers of T lymphocytes and B lymphocytes in a specific way (Figure 12.1.6).

Some phagocytes, namely macrophages and dendritic cells, also act as **antigen-presenting cells (APCs)**. When antigen-presenting cells phagocytose a pathogen, fragments of digested antigen are linked to MHC-II proteins and displayed (or presented) on the surface of the membrane (Figure 12.1.5).

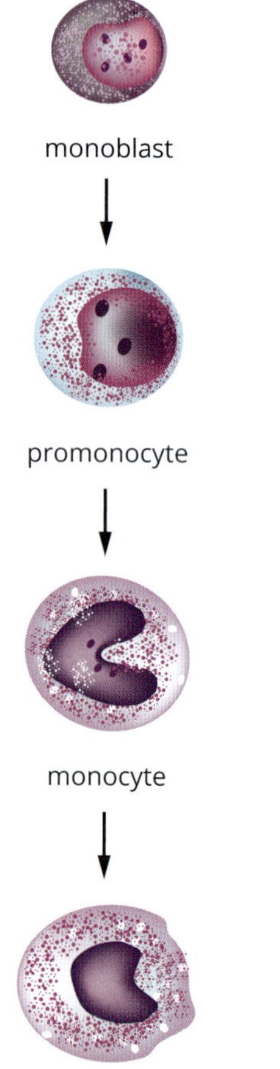

FIGURE 12.1.4 Development of monocytes and macrophages

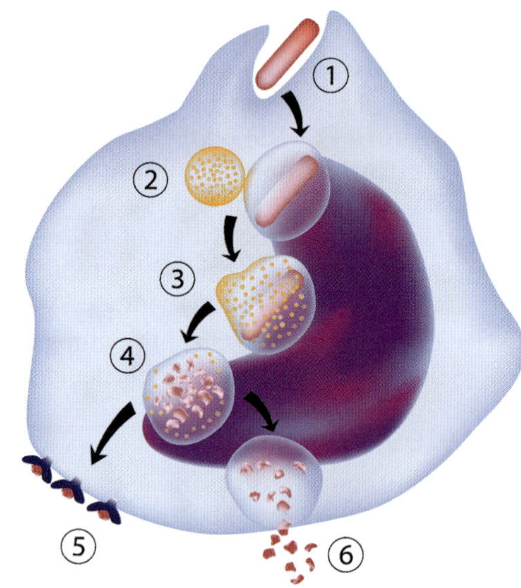

FIGURE 12.1.5 Phagocytosis and antigen presentation in an antigen-presenting cell (APC). APCs communicate with other immune cells by presenting antigens or fragments of antigens on the cell surface.

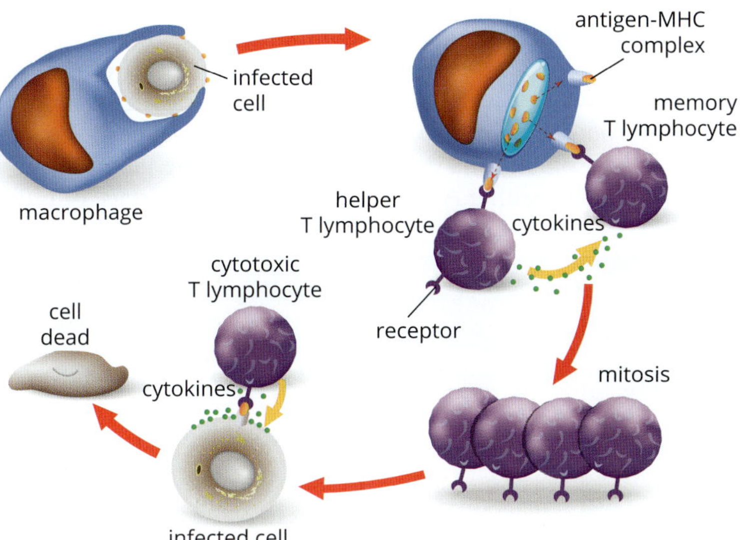

FIGURE 12.1.6 Macrophage presenting antigen to T lymphocytes

Antigen presentation links the innate and adaptive immune responses

GO TO ➤ Section 11.2 page 413

As you learnt in Chapter 11, antigens are molecules that **lymphocytes** recognise as self or non-self. Antigens that are recognised as non-self trigger an innate immune response.

Antigens have regions on their surface, called **epitopes**, that bind to corresponding receptor molecules (**paratopes**) of lymphocyte **antibodies** (Figure 12.1.7). Several different antibodies may attack antigens in different areas. Having a variety of antibodies that can detect different antigens on the same pathogen allows the immune system to recognise and destroy the pathogenic organism. As well as this, if an antigen alters its structure, and consequently the structure of the epitope, a specific antibody paratope will no longer recognise it for destruction. Multiple antibodies are necessary to efficiently combat pathogens.

An epitope is the specific part of an antigen which is recognised by the immune system's antibodies. The part of the antibody that binds to the epitope is called a paratope.

Once the lymphocytes have recognised and bound to the epitope they are stimulated to multiply and counteract the pathogen with an adaptive immune response, producing antibodies and activating **cytotoxic T lymphocytes** (also known as killer T cells). The number of cells activated and the amount of antibodies produced is dependent on the type of antigen detected, the type of host and the point of pathogenic entry into the host.

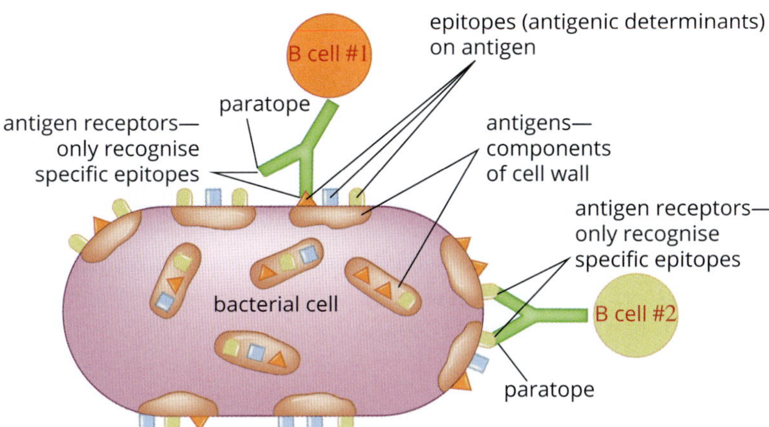

FIGURE 12.1.7 Epitopes are the specific regions on an antigen recognised by specific paratopes, the matching antibody receptors.

Structure of antigens

Most antigens are protein-based and can be composed of one or more polypeptide chains. However, antigens can also be composed of carbohydrates, lipids and even nucleic acids. For example, the complex carbohydrates of the human ABO blood group are antigens. It is the structure of the carbohydrate that makes the A antigen different from the B antigen. The presence or absence of A and B antigens on the surface of red blood cells determines whether the blood group is A, B or AB. Group O blood has neither A nor B antigens on the surface of red blood cells (Figure 12.1.8).

Blood type	Red blood cells	Antibodies present in plasma	Antigens present on cells
A		anti-B	A
B	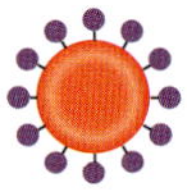	anti-A	B
AB	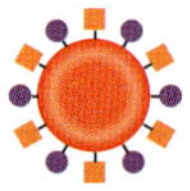	none	A and B
O		anti-A and anti-B	none

FIGURE 12.1.8 The A and B blood type antigens are carbohydrate molecules attached to proteins and lipids in the red blood cell membrane. If the blood type transfused into a patient is different from the patient's own blood type, an immune response will be elicited by the patient's immune system, which can lead to death.

There are many different receptors and they are specific for particular antigens. The major histocompatibility complex (MHC) proteins, are proteins on the surface of your body's cells that present **self-antigens** or **non-self antigens** to T lymphocytes. There are different classes of MHC proteins, which you will learn more about in the following sections. If self-tolerance breaks down and the immune system responds to self-antigens, it results in autoimmune diseases such as insulin dependent diabetes, rheumatoid arthritis or multiple sclerosis. As mentioned earlier, not all antigens (including not all non-self antigens) elicit an immune response. Antigens that elicit an immune response are called **immunogens**. In allergic reactions certain antigens elicit an allergic immune response. Antigens that trigger an allergic response are called **allergens**.

Responding to antigens

Antigen recognition is dependent on the detection of antigens by receptors as described below.

- The receptors on **B lymphocytes** are membrane-bound antibodies that recognise free antigens or antigens that are on the surface of a pathogen (Figure 12.1.9a).
- Antibodies can also be secreted by the B lymphocytes (Figure 12.1.9b).
- The receptors on **T lymphocytes** are different from the membrane-bound antibodies of B lymphocytes and recognise antigens presented by the organism's own cells (Figure 12.1.9c).

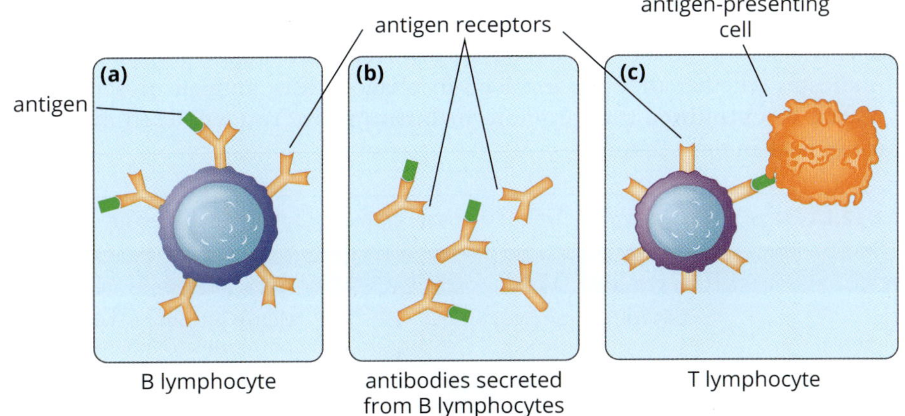

FIGURE 12.1.9 Antigen recognition by B and T lymphocytes. (a) Antigen receptors (antibodies) on the surface of B lymphocytes recognise antigens on the surface of a pathogen. (b) B lymphocytes also secrete antibodies. (c) Antigen-presenting cells present self- or non-self antigens to T lymphocytes.

There are different classes of MHC proteins, including MHC class I and class II, which are both involved in antigen presentation.

MHC class I (MHC-I) proteins are normally found on all nucleated cells, and present peptide antigens derived from the proteins of pathogens in the cytoplasm of non-phagocytic cells to cytotoxic T lymphocytes in the adaptive immune response. You will learn more about T lymphocytes and their role in the adaptive immune response in Section 12.2. MHC-I proteins are also important in the innate immune response of animals, as it allows **natural killer (NK) cells** to identify and destroy infected or damaged cells by releasing cytotoxic chemicals such as perforins (Figure 12.1.10).

Perforins are proteins released by killer cells in animals that cause lysis of target cells by forming pores in their membranes.

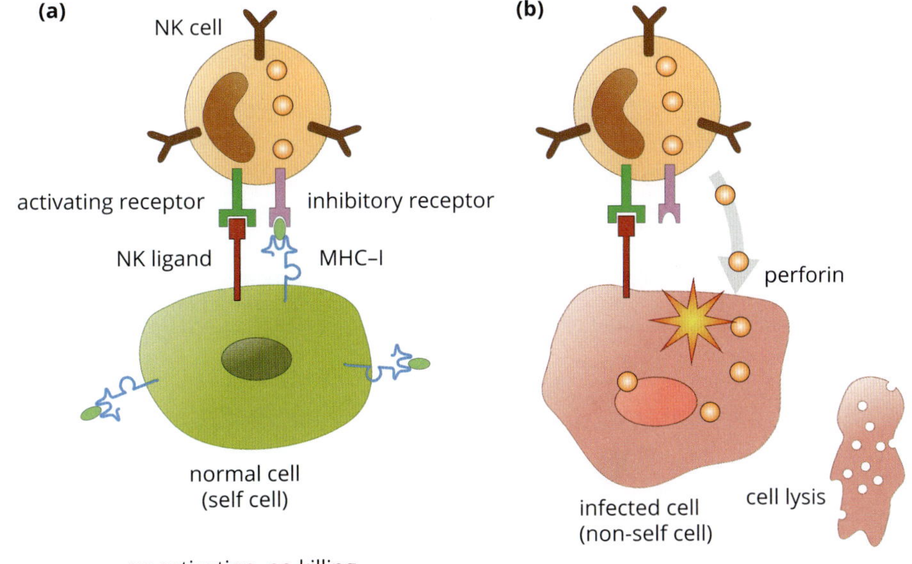

FIGURE 12.1.10 The action of natural killer (NK) cells. (a) The NK cell recognises a normal host cell by the presence of MHC-I and does not elicit an attack. (b) MHC-I is absent from the host cell's surface and the NK cell recognises that the cell is infected or damaged. The NK cell then elicits a response to destroy the infected cell.

MHC class II (MHC-II) proteins can be conditionally expressed on all cells, but are most commonly found on the surface of antigen-presenting cells such as dendritic cells, macrophages and B lymphocytes. This presentation of antigens activates the helper T lymphocytes of the adaptive immune response, linking the innate and adaptive immune responses (Figure 12.1.11).

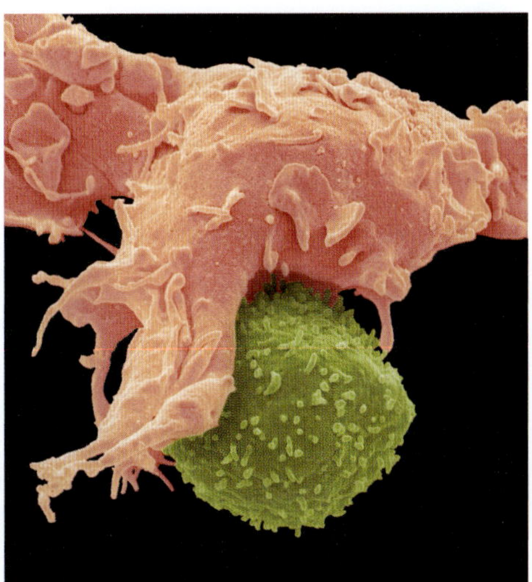

FIGURE 12.1.11 A coloured scanning electron micrograph (SEM) showing the interaction between a macrophage (pink) and a helper T lymphocyte (yellow).

Summary of innate immune cells

Table 12.1.1 shows some of the leukocytes involved in the innate immune responses and indicates whether they are involved in phagocytosis, antigen presentation, or the release of **cytokines** that promote **inflammation**. You will learn more about cytokines later in this section.

TABLE 12.1.1 Some of the leukocytes involved in innate immune responses and their function.

Cell type	Function	Cell type	Function
neutrophil (granulocyte)	• phagocytosis • release antimicrobial compounds, such as defensins and hydrogen peroxide, that disrupt bacterial and fungal membranes • release cytokines that attract other immune cells and cause inflammation	basophil (granulocyte)	• release histamine, which contributes to inflammation and therefore blood vessel dilation • have a limited role in phagocytosis
macrophage	• phagocytosis • antigen presentation • release of cytokines	eosinophil (granulocyte)	• antigen presentation • release cytokines and cytotoxic chemicals • have a limited role in phagocytosis • eosinophils are found in high numbers in parasitic infections
monocyte	• largest of the leukocytes • circulate in blood to site of infection where they differentiate into macrophages or dendritic cells	mast cell (granulocyte)	• play a key role in inflammation, and therefore blood vessel dilation, by releasing histamines • have a limited role in phagocytosis
dendritic cell	• phagocytosis • antigen presentation • have many grooves that increase their surface area and permit contact with a large number of nearby cells	natural killer (NK)	• recognise virus-infected and cancerous cells • release cytotoxic chemicals from granules, such as perforin, which punches holes in cell membranes, triggering apoptosis and cell death of virus-infected cells and abnormal cells • release cytokines to attract and activate cells of the adaptive immune system

DEFENSIVE MOLECULES OF THE INNATE IMMUNE SYSTEM

There are several proteins that are produced in order to fight off specific pathogens. You learnt about some of these defensive molecules in Chapter 11.

GO TO ➤ Section 11.2 page 413

Complement proteins

The **complement proteins** are an array of more than 30 proteins that circulate in the blood and help to kill foreign cells. They are found in body fluids in an inactive form, and are activated as part of the non-specific (innate) immune response to certain antigens and carbohydrates on the surfaces of some bacteria and parasites. Activation of complement proteins results in an enzyme-triggered reaction that leads to the lysis of the invading pathogens. For example, complement proteins destroy bacteria directly by punching holes in their cell walls, causing them to lyse. The release of the bacterial contents attracts phagocytes to the site of infection. Complement proteins activated by **antigen–antibody complexes** are also involved in specific (adaptive) immune responses.

Cytokines

Cytokines are small signalling molecules of the immune system and coordinate many aspects of our immune responses. Cytokines can be peptides, proteins or glycoproteins, and are released by body cells in response to cell damage or the presence of pathogens. Cytokines have a direct affect on the cells they target, particularly granulocytes, monocytes and macrophages.

There are many different cytokines and they trigger a variety of responses, both non-specific and specific. **Interferons** and **chemokines** are two different types of cytokines and they each have different functions. For example, cytokines can promote the proliferation of lymphocytes, induce inflammation and **fever**, promote antibody responses and activate macrophages. Cytokines that are secreted by leukocytes are called **interleukins** whereas those secreted by lymphocytes are called **lymphokines**. Once a cytokine has been released, it will bind to a receptor (protein molecule on a target cell) which will ultimately produce signals that will inhibit cell growth, differentiation or proliferation.

Interferons

Interferons are a class of cytokine that are produced by, and act on, a host cell infected by a virus. Interferons act in an autocrine manner, activating the infected cells to produce enzymes that breakdown viral RNA and proteins that block translation. This limits viral replication and release from the cell. Interferons also attract NK cells which release cytotoxic peptides to kill the virus-infected cell. Interferons are non-specific and will act against any virus. However, viruses vary widely in their susceptibility to interferons. Many viruses can evade interferon-induced defences and the more virulent viruses may be able to inhibit the production of interferon.

> Autocrine refers to a substance secreted by a cell that also has an effect on that cell.

> Although an important defence against viruses, interferons also play a smaller role in combating bacterial and parasitic infections. Interferons also regulate the immune system in a number of ways, such as enhancing T lymphocyte activity.

Chemokines

Chemokines are a type of cytokine and act as chemical attractants (or chemo-attractants). Chemokines are important for attracting leukocytes to sites of infection and inflammation.

INFLAMMATORY RESPONSE

As previously mentioned in Chapter 11, the inflammatory response is responsible for releasing several types of chemicals that enable the activation of phagocytes and other white blood cells to fight off foreign substances. In an attempt to fight off foreign invaders the body sends blood and fluid to the site of infection or injury and it becomes red, hot and swollen in the process. The site of injury releases chemokines that prompt **basophils** (a type of B lymphocyte) and **mast cells** (a type of B lymphocyte) to release **histamines**. The damaged tissue's cells release **prostaglandins** such as thromboxane, which is released from blood platelets

Prostaglandins are hormone-like substances that act as messengers to initiate or activate other processes.

to induce platelet clumping and blood vessel constriction, activating platelets to clot an area in order to contain the site of infection or damage. The circulating blood sends phagocytes and blood clotting factors such as **tissue factor (TF)**, **factor VII (FVII)**, **factor X (FX)** and **thrombin** to the area. Clotting factors prevent pathogens from further circulating in the body and allow the phagocytes to complete their work and engulf any remaining invaders. Macrophages will also engulf any dead neutrophils that have undergone apoptosis to form pus which is then excreted or broken down by the body.

Histamines and prostaglandins increase the permeability of capillaries and in doing so allow more blood flow to the site of damage or infection. This permeability allows phagocytes to enter tissues and begin phagocytosis. **Pyrogens** (substances that initiate or cause a fever) are another chemical that may be released by the body in response to infiltration by foreign substances. This chemical increases body temperature, and as most pathogens prefer the optimal body temperature of 37°C, an increase in body temperature decreases the host's suitability for the pathogen. You learnt about fever in Chapter 11.

GO TO ➤ Section 11.2 page 413

APOPTOSIS

Cells can be programmed to die in order to seal off a pathogen that can't be destroyed. Layers of white blood cells undergo apoptosis (programmed cell death) to form cysts and granulomas that isolate pathogens from causing further infection.

12.1 Review

SUMMARY

- Innate immune responses occur when barriers of the first line of defence are breached.
- Innate immune responses are:
  - non-specific—they do not target a specific antigen
  - rapid—they occur within hours
  - present in all animals and plants
  - fixed responses—they do not adapt.
  - not able to create an immunological memory.
- Leukocytes have pattern recognition molecules on their surface, which are able to recognise microbe-associated molecular patterns (MAMPs).
- Phagocytes are leukocytes that are able to engulf and break down microbes in a process known as phagocytosis.
- Some phagocytes also act as antigen-presenting cells.
- Defensive molecules include complement proteins and cytokines:
  - Activation of complement proteins results in an enzyme-triggered reaction that leads to the lysis of the invading pathogens.
  - Cytokines are small signalling molecules of the immune system and coordinate many aspects of our immune responses.
- Cytokines include interferon and chemokines:
  - Interferons are produced by virus-infected cells and inhibit viral replication by resulting in the transcription of antiviral genes and expression of antiviral proteins.
  - Chemokines attract white blood cells to the site of infection.
- Inflammation is the accumulation of fluid, plasma proteins and leukocytes that occurs when tissue is damaged or infected. It results in heat, pain, swelling, redness and loss of function.
- Fever is an increase in body temperature that results from the regulated body temperature set point in the hypothalamus of the brain being set to a higher level by inflammatory cytokines. It slows the replication of bacteria and viruses by shifting the temperature away from their optimal range, and improves the immune response by increasing the activity and proliferation of leukocytes.

KEY QUESTIONS

1 Innate immune responses are:
   A specific and delayed
   B non-specific and rapid
   C non-specific and delayed
   D specific and rapid

2 Outline the main components of the innate immune system.

3 Phagocytes are important cells of the innate immune system. What is the function of some phagocytes that links the innate and adaptive immune systems?

4 Distinguish between neutrophils and macrophages.

5 The diagram at right shows a natural killer (NK) cell attacking an infected cell. What is the missing inhibitory receptor that indicates this cell is an infected 'non-self' cell and not a healthy 'self' cell?

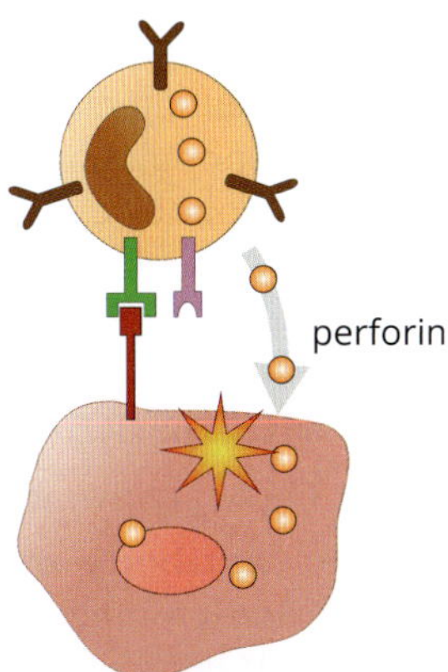

12.2 Adaptive immunity

In Section 12.1, you learnt that if the first line of defence is breached by a pathogen, that pathogen is met with a non-specific innate immune response. However, this innate immune response may not be successful in eliminating the invader. Vertebrates have an additional immune response—adaptive immunity. In this section, you will learn about the adaptive immune response.

Adaptive immunity, also known as acquired immunity (or specific immunity), is a rapid and efficient response to previously encountered antigens. Cells of the adaptive immune system (B and T lymphocytes) are specifically produced and cloned to combat antigens that they recognise and react accordingly. Due to this, an adaptive immune system has immunological memory and an ability to make antibodies for specific antigens. It provides long-lasting defence against recurrent infections and prevents the body from experiencing the same disease again. Large numbers of memory cells (clones) are able to be produced in response to secondary pathogen exposure allowing a faster and more effective way to fight off foreign substances.

THE NATURE OF THE ADAPTIVE IMMUNE RESPONSE

There are two distinguishing features of the adaptive immune response:

- **specificity**—the ability to recognise and respond exclusively to specific antigens (Figure 12.2.1). On recognising a specific foreign antigen on a pathogen, cells of the adaptive immune system trigger an array of defensive mechanisms that destroy the pathogen.
- immunological memory—the ability of cells of the adaptive immune system to 'remember' antigens after primary exposure, and to mount a larger and more rapid response when exposed to the same antigen again.

FIGURE 12.2.1 B lymphocytes produce antibodies for a specific antigen.

Lymphocytes are a type of white blood cell (or leukocyte) that are specialised for adaptive immune responses.

B lymphocytes develop and mature in the bone marrow and become activated in the secondary lymphoid tissues.

T lymphocytes develop in the bone marrow and mature in the thymus.

Lymphocytes—cells of the adaptive immune response

The cells that are crucial to the adaptive immune response are lymphocytes. Each lymphocyte has a different receptor for a particular antigen, and is able to proliferate, creating clones of the initial lymphocyte with the specific receptor for the antigen. This is called **clonal selection**.

Lymphocytes are classified as either B lymphocytes or T lymphocytes according to their interaction with the antigen and their response to it. The B and T lymphocyte sub-populations have distinct roles but are both key to the adaptive immune response. Lymphocytes travel through the **lymphatic system** and become activated when they encounter antigens specific to their receptors. You will learn more about the lymphatic system in Section 12.3.

Mechanisms of adaptive immune responses

There are two mechanisms of adaptive immunity (Figure 12.2.2):

- **antibody-mediated immunity** (also called humoral immunity), in which macromolecules, such as complement proteins, and antibodies produced by B lymphocytes, are secreted into the extracellular fluid

- **cell-mediated immunity**, which involves the action of T lymphocytes (which you will learn more about on page 437) and phagocytes.

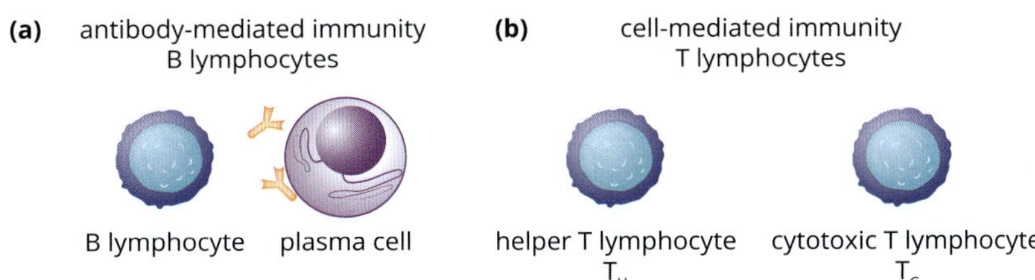

FIGURE 12.2.2 (a) B lymphocytes are involved in humoral or antibody-mediated immunity. They originate and mature in the bone marrow and become activated in the secondary lymphoid tissues. (b) T lymphocytes are involved in cell-mediated immunity. They originate in bone marrow and mature in the thymus. Except for plasma cells, the different types of lymphocytes look very similar under a microscope. The only way to know which is which is to identify their different surface proteins.

+ ADDITIONAL

Clonal selection theory

An almost infinite number of different antigens exist, and the immune system is able to produce lymphocytes specific to each antigen upon exposure. The clonal selection theory is a scientific theory that explains how lymphocytes are able to produce a large number of antibodies specific to an antigen.

When B and T lymphocytes form, each has a receptor that will react to a single antigen. The clonal selection theory states that a specific antigen will only activate a lymphocyte with a receptor that specifically recognises that antigen. Once activated, this lymphocyte will proliferate into a clone of millions of effector cells dedicated to eliminating the specific antigen that stimulated the immune response (Figure 12.2.3).

Sir Frank Macfarlane Burnet, one of Australia's most celebrated scientists, developed clonal selection theory in 1957. This theory is now a widely accepted view of how the immune system is able to determine self-antigens from foreign (non-self) antigens by selecting both B and T lymphocytes for cloning and subsequent destruction of the antigen present in the body.

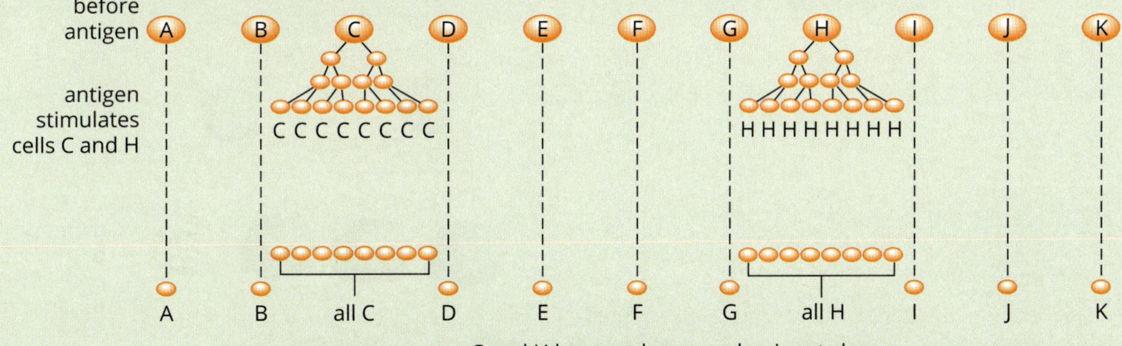

FIGURE 12.2.3 The clonal selection theory and expansion. Lymphocytes that encounter or interact with an antigen, such as lymphocytes C and H, begin to proliferate. This increases the number of lymphocytes with identical receptors, or clones, for the specific antigen that was first encountered.

In medieval times, the term 'humour' referred to body fluids.

ANTIBODY-MEDIATED IMMUNITY

Antibody-mediated immunity (also known as humoral immunity) involves B lymphocytes, which produce specific antibodies against foreign antigens and release them into the blood and **lymph**. Lymph is the fluid that circulates through the lymphatic system. You will learn more about the lymphatic system in Section 12.3.

B lymphocytes

B lymphocytes (or B cells) originate and mature in the bone marrow and become activated in the **secondary lymphoid tissues**. At any time there are billions of B lymphocytes circulating in the blood.

When a B lymphocyte meets and binds to a specific antigen, the B lymphocyte can be triggered (or activated) to differentiate and divide (or proliferate). Cytokines released by helper T lymphocytes are also important for helping to activate B lymphocytes. When B lymphocytes are activated, they divide and further differentiate into two types of daughter cells:

- **plasma cells**
- **memory B lymphocytes.**

Plasma cells

Activation of B lymphocytes leads to the production of plasma cells, which are essentially 'factories' specialising in antibody production (Figure 12.2.4). The antibodies produced are specific to the antigen that activated the B lymphocyte. Plasma cells can produce thousands of antibodies per second.

Memory B lymphocytes

Memory B lymphocytes remain in lymphoid tissues for long periods (even for the lifetime of the animal) and are responsible for the immunity that follows infection or vaccination. These cells can divide and give rise to the antibody-secreting plasma cells if secondary exposure to the antigen occurs (Figure 12.2.4).

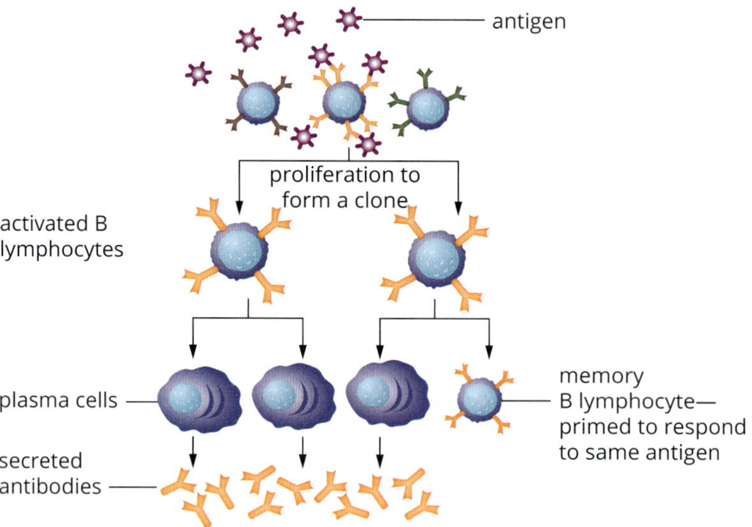

FIGURE 12.2.4 Many B lymphocytes differentiate into plasma cells, which produce and secrete antibodies for immune protection. Others become memory B lymphocytes and are retained in lymph nodes. Helper T lymphocytes are often involved in activating B lymphocytes to produce antibodies.

Activated B lymphocytes divide to form antibody-secreting plasma cells or memory B lymphocytes.

Antibodies

Antibodies, also known as **immunoglobulins (Ig)**, are produced by B plasma cells and released into the blood and lymph. Antibodies are proteins that bind to specific antigen molecules.

The basic unit of an antibody molecule is a Y-shaped protein, formed by four polypeptide chains: two long **heavy chains**, and two short **light chains** (Figure 12.2.5). The amino acid sequences that form the top of the 'arms' of the Y-shaped antibody are known as the **variable regions**. It is the variation of these variable regions that allows antibodies to bind to different antigens. The two variable regions are identical antigen-binding sites and attach to identical antigens. The single 'stem' of the Y-shaped antibody is a conserved sequence in all antibodies and is called the **constant region**. The constant region recruits other components of the immune system.

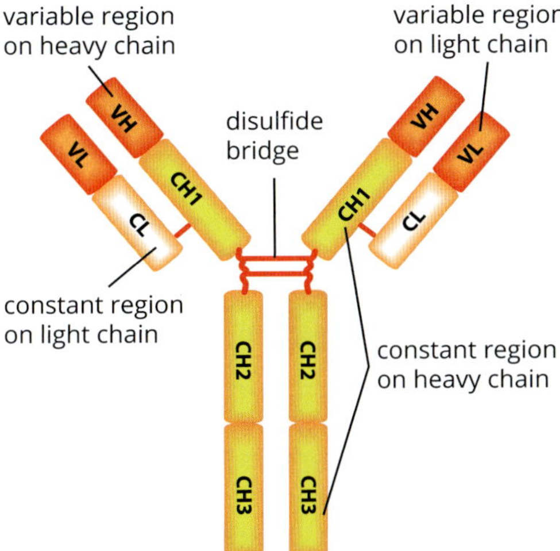

FIGURE 12.2.5 The structure of a basic unit of an antibody. Antibodies have two long heavy (H) chains and two short light (L) chains. Both heavy and light chains have a variable (V) and constant (C) region. Naturally-produced antibodies consist of two identical variable regions that are specific for a particular antigen. The constant region is capable of binding to and initiating other immune components, such as the complement proteins.

Antibody function

Antibodies do not directly destroy pathogens, but carry out several important mechanisms to interfere with the function of the pathogen (Figure 12.2.7):

- **neutralisation** of bacterial toxins—antibodies bind to bacterial toxins, blocking the action of the toxin
- neutralisation of pathogens—antibodies bind to antigens on the surface of the pathogen, which are required for entry into host cells, thus preventing pathogen invasion of host cells
- **agglutination**—antibodies bind to antigens on the surface of cells and form antigen–antibody complexes, which activate phagocytes and the complement cascade, leading to antigen/cell destruction
- **precipitation**—antibodies bind to soluble antigens, causing them to become insoluble and precipitate out of solution.

BIOFILE PSC

Agglutination of red blood cells

If a person is given a blood transfusion with the wrong blood type, antibodies will recognise the transfused blood cells as foreign and will bind to their antigens. This causes clumping (or agglutination) of red blood cells (Figure 12.2.6). Agglutination destroys the red blood cells, which transport oxygen throughout the body, and so can result in severe anaemia and even death.

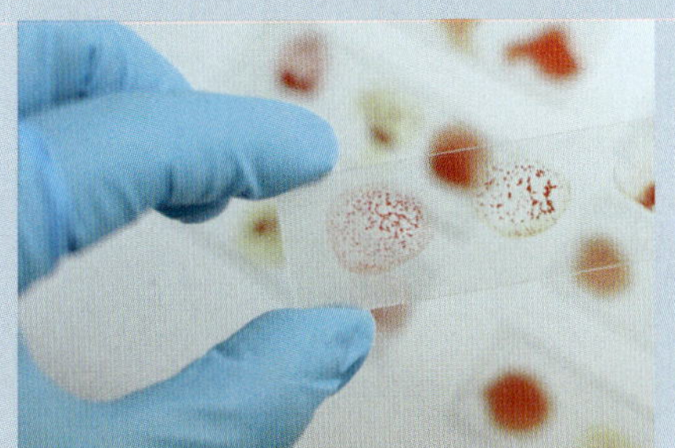

FIGURE 12.2.6 Agglutination test—red blood cells have clumped together (or agglutinated) in these drops of blood on a microscope slide

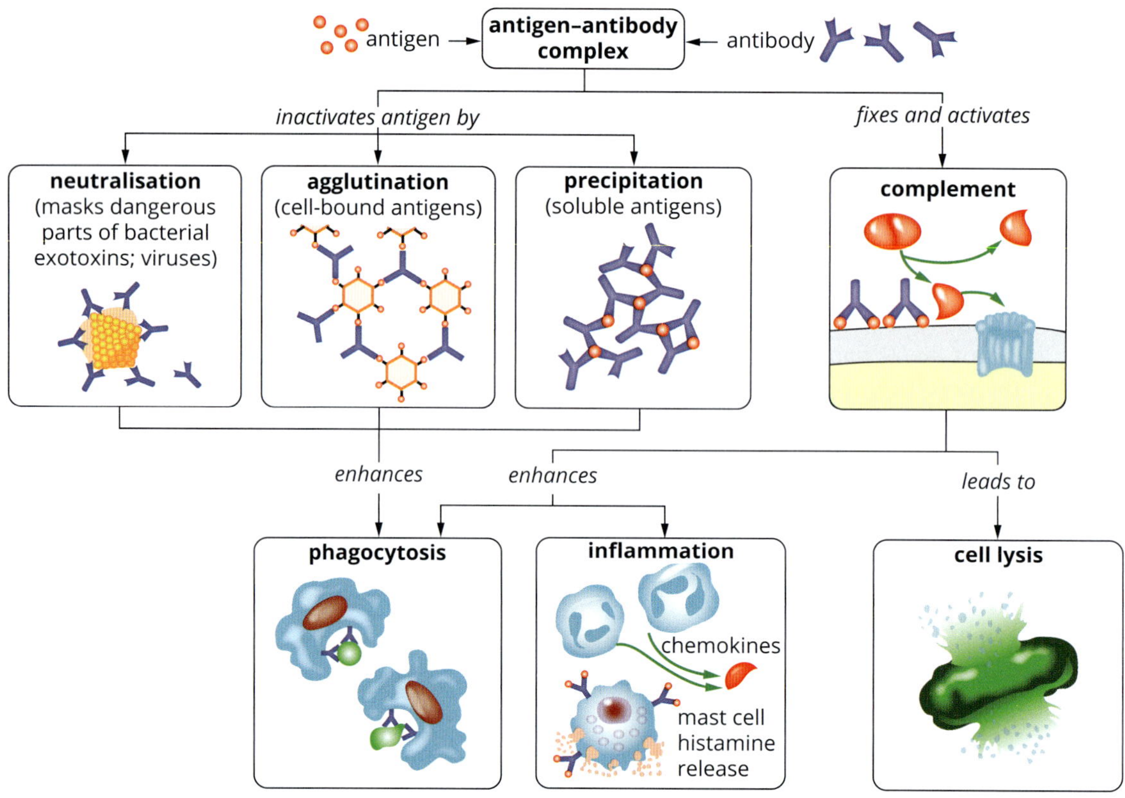

FIGURE 12.2.7 Antibodies function in a number of different ways to help eliminate pathogens.

Antibodies may act singly (monomers), in pairs (dimers) or in groups of five (pentamers). Mammals have five main classes of antibody molecules with different structures and functions (Table 12.2.1).

TABLE 12.2.1 Structure and function of mammalian antibodies (immunoglobulins)

Class	Half-life in serum	Presence	Functions	Structure
IgG	21 days	blood, lymph and extracellular fluid; most circulating antibodies (>80%); crosses placenta	agglutination, complement activation	**Antibody classification** IgG, IgE, IgD IgM: disulfide bond, joining chain IgA: joining chain, secretory protein
IgM	10 days	blood and lymph; produced early in infection response	agglutination, complement activation	
IgA	6 days	found in secretions such as tears, saliva and milk	mucosal immunity	
IgD	3 days	blood and lymph; mostly present on B lymphocyte surfaces; small amount in circulation; binds to basophils and mast cells	functions not well understood; possible role in regulating innate immune responses	
IgE	2 days	blood and lymph; attaches to mast cells	involved in allergic reactions	

CELL-MEDIATED IMMUNITY

Cell-mediated immunity is regulated by T lymphocytes, unlike antibody-mediated immunity which involves the B lymphocytes (Figure 12.2.8). The T lymphocyte response is mediated by the **T cell receptors (TCRs)**.

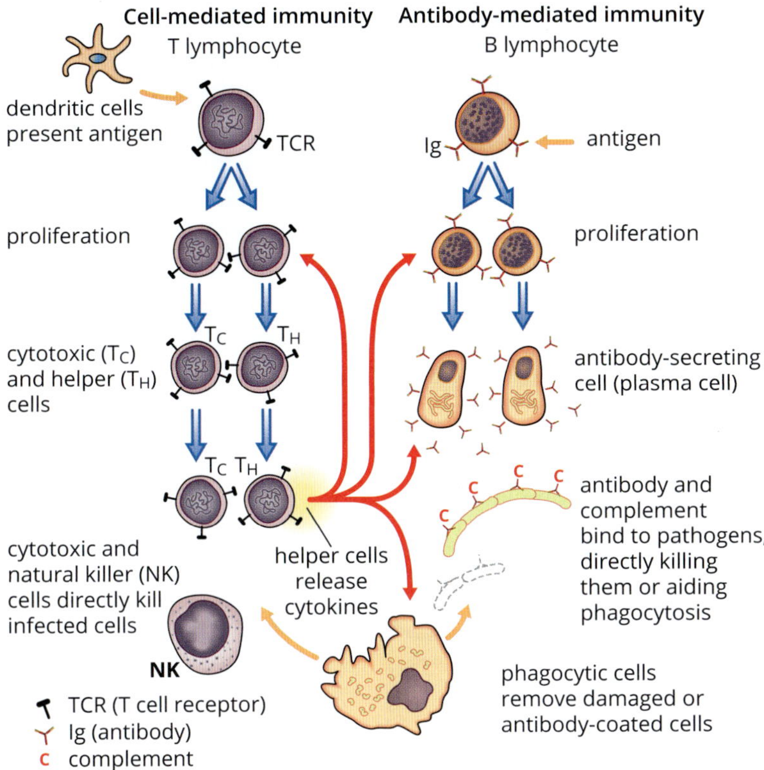

FIGURE 12.2.8 Summary of cell-mediated and antibody-mediated (humoral) immunity

T lymphocytes

T lymphocytes (or T cells) are made in the bone marrow and mature in the **thymus** gland which is located in the chest cavity. They multiply and are released by the thymus into the bloodstream and remain in the body long after it has been exposed to a pathogen. They are more concentrated in the **spleen**, **tonsils** and other parts of the lymphatic system.

Depending on their function, T lymphocytes are classified as helper, cytotoxic, suppressor or memory T lymphocytes.

Helper T lymphocytes

Helper T lymphocytes are antigen specific and do not directly kill pathogens, rather, as their name suggests, they 'help' with immune responses. These cells become activated in recognition to an antigen presented by a phagocyte (macrophage). This prompts the T lymphocytes to release cytokines that stimulate B lymphocytes to differentiate into plasma cells which multiply and secrete antibodies and memory B lymphocytes as a response. Helper T lymphocytes also stimulate the production of cytotoxic T lymphocytes (killer T cells) and cause macrophages to phagocytose more rapidly.

Cytotoxic T lymphocytes

Cytotoxic T lymphocytes (also known as killer T cells) recognise and kill foreign, infected or abnormal host cells by releasing toxic compounds. This includes virus-infected host cells, cancer cells and foreign cells such as those in transplanted tissue (Figure 12.2.9). They attach themselves to body cells that have been infected by pathogens. Once attached they destroy the body cell by deteriorating its cell membrane or injecting enzymes that will kill the cell and prevent the pathogen from infecting any further.

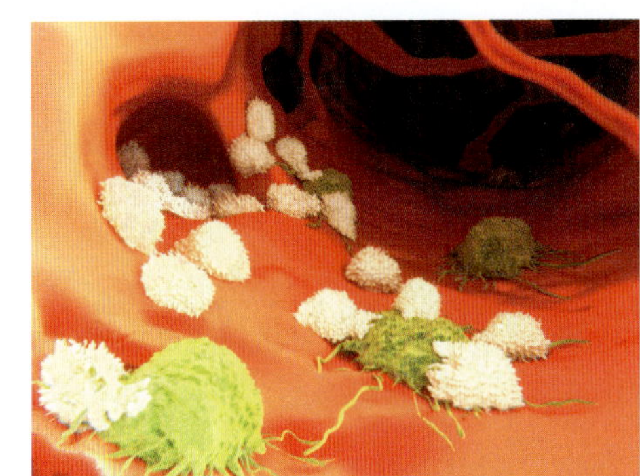

FIGURE 12.2.9 Digital illustration of cytotoxic (white) T lymphocytes attacking migrating cancer cells (yellow)

Suppressor T lymphocytes

Suppressor T lymphocytes are responsible for turning off the immune response once the antigen has been successfully contained, destroyed or removed.

Memory T lymphocytes

Memory T lymphocytes are produced after helper and cytotoxic T lymphocytes have been activated during an infection. Once activated, these lymphocytes differentiate into memory T lymphocytes that are antigen-specific. The memory T lymphocytes persist after the infection is resolved to ensure a prompt response should the same pathogen reinfect the organism.

T cell receptors

T cell receptors (TCRs) are central to the function of T lymphocytes in the adaptive immune response (remember that T lymphocytes are also known as T cells). TCRs are made up of two polypeptide chains. Like antibodies, TCRs have a variable and constant region (Figure 12.2.10). Unlike antibodies, which have two antigen-binding sites, TCRs have only one antigen-binding site.

TCRs do not bind to antigens on pathogens, as B lymphocyte receptors do; instead, they bind to fragments of antigens that are displayed or presented on the surface of antigen-presenting cells. Receptor binding triggers signal transduction in the T lymphocyte, resulting in proliferation, cytokine release and activation of cytotoxic function.

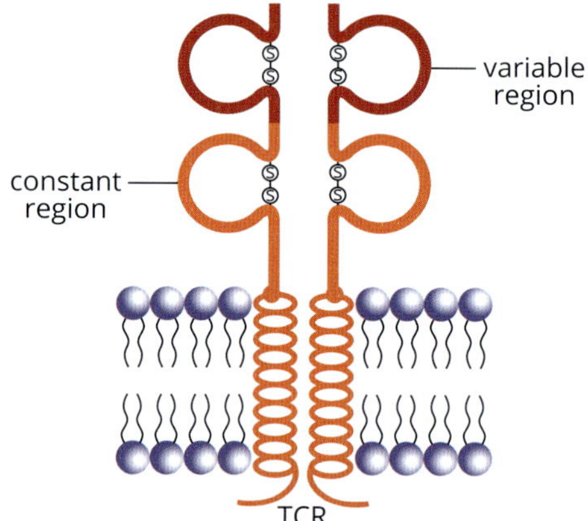

FIGURE 12.2.10 The structure of the T cell receptor (TCR), which is found on helper T and cytotoxic T lymphocytes and binds to fragments of antigen

ANTIGEN RECOGNITION BY T LYMPHOCYTES

T lymphocytes check the antigens of cells they come into contact with in the body, differentiating between cells that belong to the organism (self) and cells that are foreign (non-self). Remember that during their development, lymphocytes that react to self-antigens are normally destroyed. This inability of lymphocytes to respond to self-antigens is known as **self-tolerance**.

All nucleated cells have surface proteins that present peptide antigens of the proteins being synthesised in that cell. These antigens are presented to cytotoxic T lymphocytes by major histocompatibility complex I (MHC-I) molecules. Antigen-presenting cells are specialised for presenting antigens. When an antigen-presenting cell engulfs a pathogen, the antigens of the pathogen are broken into small peptides in the cell. These antigen fragments bind to MHC-II molecules inside the cell. The antigen–MHC-II complexes then move to the cell surface to present the antigens to helper T lymphocytes. The TCRs on the helper T lymphocytes recognise the antigen–MHC-II complex. Signal transduction in the T lymphocyte leads to activation of the cell, which then proliferates and releases cytokines (Figure 12.2.11).

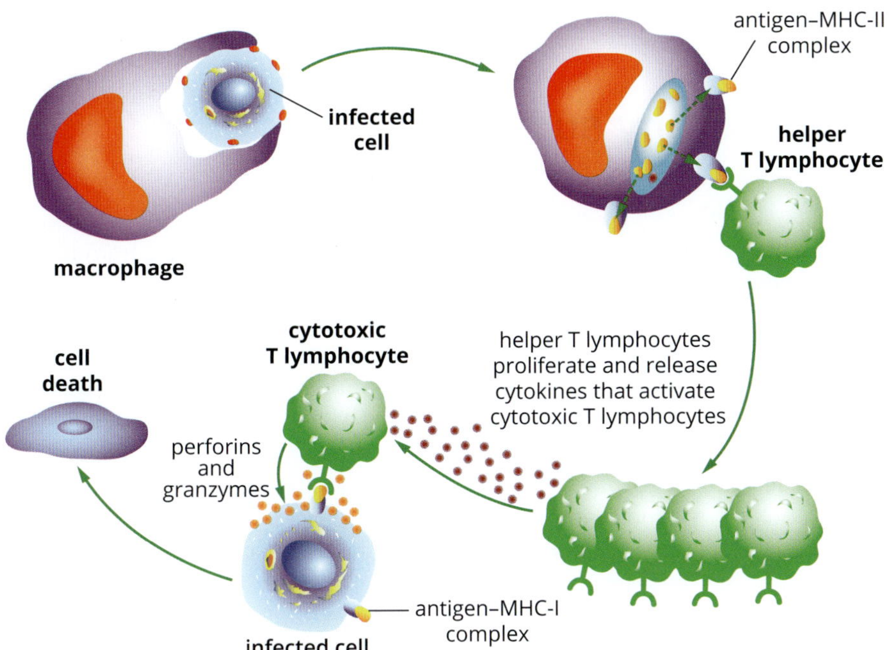

FIGURE 12.2.11 T lymphocytes are activated in cell-mediated immunity. Infected cells are detected and destroyed by cytotoxic T lymphocytes.

IMMUNOLOGICAL MEMORY

The response arising from the first encounter of a T or B lymphocyte with a specific antigen is known as the **primary immune response** (Figure 12.2.12). After the initial exposure, B and T lymphocytes form B and T memory lymphocytes. IgM antibodies are the predominant antibodies produced in a primary response.

The response arising from subsequent encounters with the same antigen is known as the **secondary immune response**. Lymphocyte proliferation and production of antibodies occurs much more quickly during the secondary immune response, because the existing memory lymphocytes, which were produced during the first encounter and which remain for months or years, allow faster proliferation of the required lymphocytes (those with the receptor specific to the antigen). IgG antibodies are the predominant antibodies produced in the secondary response.

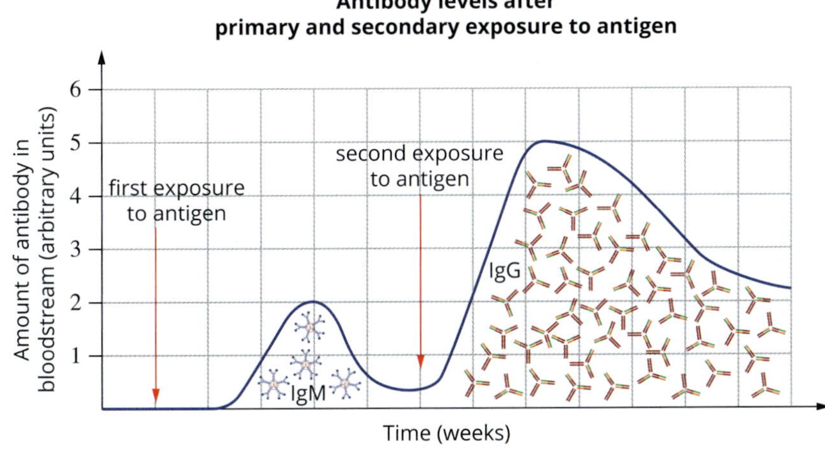

FIGURE 12.2.12 Primary and secondary immune responses after initial and secondary exposure to the same antigen

BIOLOGY IN ACTION CCT

Gut microbes boost immunotherapy success in fighting cancer

Immunotherapy is a rapidly growing field of medicine that has shown a lot of promise as a highly effective treatment for cancer. Immunotherapy works by using the patient's immune system to fight cancer cells. There are four main types of immunotherapy:

- monoclonal antibodies, which are synthetic antibodies designed to destroy, slow the growth, or directly deliver medicine (e.g. chemotherapy) to cancer cells
- immune checkpoint inhibitors, which prevent immune cells from being switched off by cancer cells, enabling the immune system to recognise and destroy cancer cells (Figure 12.2.13)
- cancer vaccines, which trigger the immune system to prevent cancer cell growth or destroy existing cancer cells
- non-specific immunotherapies, which boost the immune system in a non-targeted way to slow or halt cancer cell growth.

In clinical trials, melanoma (skin cancer) patients who received immunotherapy treatments had increased survival rates, with less toxicity and fewer negative side effects than patients who received chemotherapy or radiotherapy. Immunotherapy has also been successful in treating several other types of cancer, including lung, breast and colon. Although this form of treatment holds a lot of promise, the success rates with patients have been varied and there is still much research to be done.

Scientists have recently discovered an unexpected clue as to why some patients may respond well to immunotherapy while others have little to no success. Researchers from the University of Chicago found that introducing the bacteria species *Bifidobacterium* to the digestive systems of mice with melanoma markedly increased their anti-tumour T lymphocyte response. When mice with the bacterial strain were compared to mice without the bacteria but which were receiving immunotherapy via the drug anti-PD-L1 (an immune checkpoint inhibitor), the researchers found that tumour growth was slowed in both groups. By combining the bacterial treatment with immunotherapy, tumour control was dramatically improved. Another study by researchers at the Institut Gustave Roussy in Paris found that antibiotics reduced the effects of an immunotherapy drug. By replenishing gut microbes in antibiotic-treated and infection-free mice, the anti-cancer effects of the immunotherapy drug were restored.

Further investigation revealed that the *Bifidobacterium* triggered an immune response by interacting with dendritic cells in the intestinal tract. Dendritic cells are antigen-presenting cells that are responsible for detecting potential threats to the immune system and presenting them to the T lymphocytes, thereby triggering an immune response. A genome-wide scan of mice with *Bifidobacterium* also showed upregulation of several genes involved in anti-tumour responses. Both of these studies have demonstrated the important role that a healthy gut microbiome plays in the immune response and the significant implications for the treatment of cancer using immunotherapy.

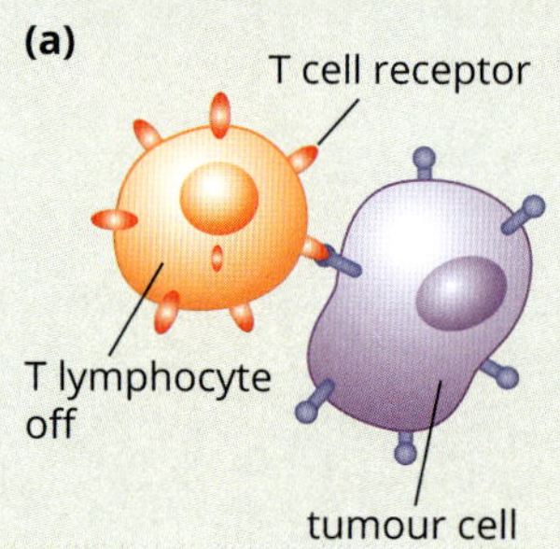

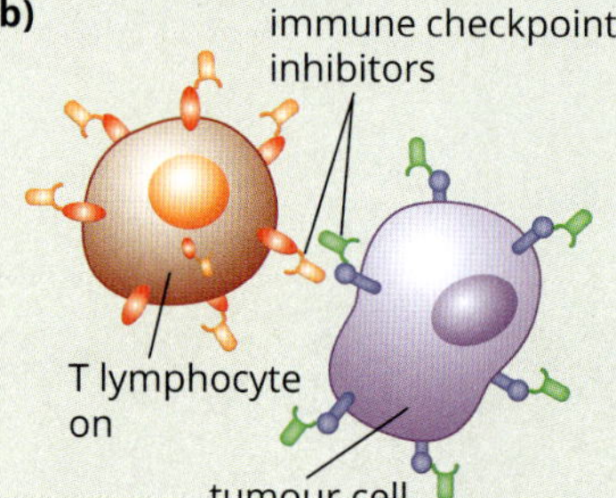

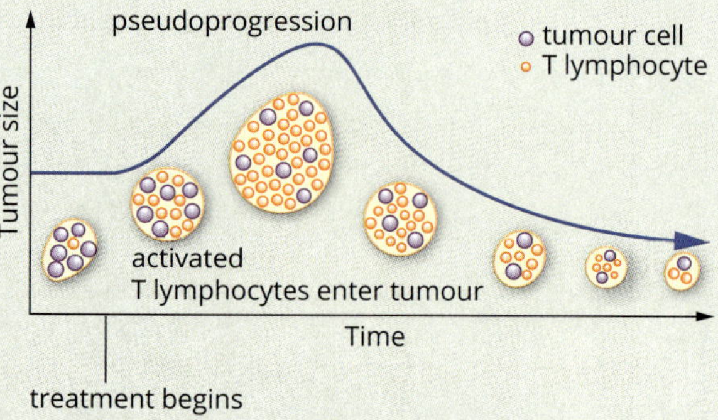

FIGURE 12.2.13 (a) Some tumour cells switch off T lymphocytes by binding to T cell receptors. (b) Immune checkpoint inhibitors, a type of immunotherapy drug, block inhibitory molecules on tumour cells from binding to T cell receptors, allowing T lymphocytes to remain active and infiltrate tumours, keeping them from growing. (c) Treatment with immune checkpoint inhibitors will briefly increase the size of a tumour (pseudoprogression) due to T lymphocyte infiltration, before its size is reduced due to tumour cell death.

12.2 Review

SUMMARY

- An adaptive immune response is one that is specific to a certain antigen.
- The adaptive immune response in vertebrates is classified as antibody-mediated (humoral) or cell-mediated.
- Antibody-mediated immunity involves B lymphocytes, which become activated and proliferate when stimulated by specific antigens or cytokines released by helper T lymphocytes. Activated B lymphocytes become plasma cells that produce antibodies and memory lymphocytes that remain in lymphoid tissues and provide immunological memory.
- Antibodies, also known as immunoglobulins, are proteins that bind to specific antigen molecules.
- Antibodies are Y-shaped proteins that have a constant 'tail' and variable 'arm' regions. The variable regions have antigen-binding sites and the constant region recruits components of the immune system.
- Cell-mediated immunity involves T lymphocytes as described below.
  - Cytotoxic T lymphocytes recognise and kill foreign, infected or abnormal host cells by releasing toxic compounds.
  - Helper T lymphocytes secrete cytokines that promote inflammation, and activate macrophages and B lymphocytes.
  - The major histocompatibility complex (MHC) is important in antigen presentation:
  - MHC-I is expressed on all nucleated cells and presents peptide antigens of proteins being produced within the cell to cytotoxic T lymphocytes
  - MHC-II is expressed on antigen-presenting cells and presents peptides of phagocytosed antigens to helper T lymphocytes.
- Antigen presentation is carried out by antigen-presenting cells (APCs), including dendritic cells, macrophages and B lymphocytes, and involves antigen fragments of a pathogen being presented on the MHC-II of the host cell. Helper T lymphocytes identify and bind to the complex and become activated to produce and release cytokines.
- Cytotoxic T lymphocytes kill infected cells, which are identifiable by pathogen antigens on MHC-I.
- Memory B and T lymphocytes persist after an infection to enable a larger and faster response upon reinfection with the same pathogen.
- The first infection with a pathogen produces a primary immune response, while reinfection with the same pathogen produces a secondary response due to the presence of memory cells from the primary response (known as immunological memory).

KEY QUESTIONS

1 What part of an antibody interacts with antigens on a pathogen?

A the constant region
B the disulfide bridge
C the constant region of the heavy chain
D the variable region

2 Which MHC class interacts with helper T lymphocytes, and which interacts with cytotoxic T lymphocytes?

3 Identify the different types of T lymphocytes and their functions.

4 Look at the graphs to the right and determine which label on the graphs (A, B, C, D or E) represents:

a the period when the virus has just entered the body and started to multiply (the incubation period)
b the day the patient became infected
c the period the patient felt most ill
d the normal body temperature of 37°C
e the period when the patient's antibodies destroy the virus.

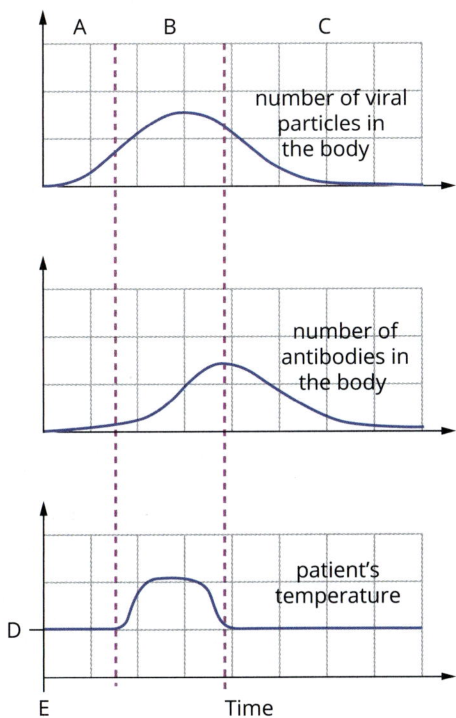

12.3 The lymphatic system

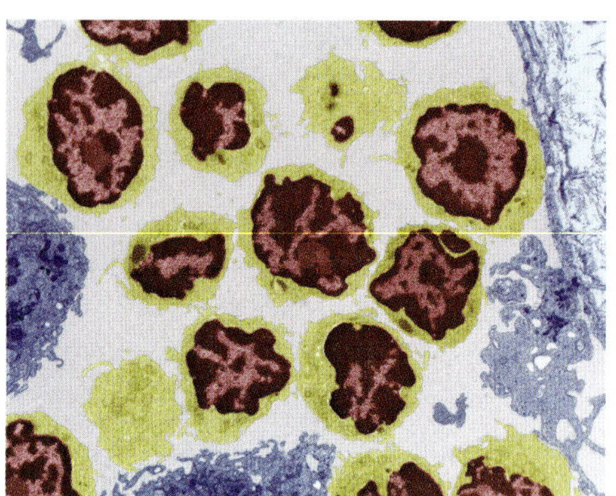

FIGURE 12.3.1 Coloured TEM of a section through a lymph node, showing a variety of lymphocytes (yellow). Lymph nodes are one of the structures of the lymphatic system.

The lymphatic system plays a key role in the adaptive immune responses of mammals. It transports immune cells, including antigen-presenting cells, throughout the body, and is where antigen recognition by lymphocytes occurs. In this section, you will learn about the lymphatic system and how its structures (Figure 12.3.1) are involved in adaptive immune responses.

THE ROLE OF THE LYMPHATIC SYSTEM

The lymphatic system is an arrangement of interconnected tissues and organs that aid in the removal of toxins, cellular wastes, excess fluids and pathogens. Its primary role is to carry white blood cells around the body in the lymph. Several lymphatic vessels (similar to veins and capillaries) are connected to **lymph nodes** which filter the lymph and remove unwanted material from the body.

Lymph nodes are small bean-shaped glands containing white blood cells. There are about 600 lymph nodes around the body which filter pathogens, cancer cells, wastes and other foreign substances that may harm the body.

The mammalian lymphatic system has several roles, including:

- returning fluid that seeps out of the blood vessels into tissues back to the circulatory system
- absorbing and transporting fatty acids and fats from the digestive system
- providing a place for lymphocytes to mature
- transporting lymphocytes and antigen-presenting cells to the lymph nodes, stimulating the adaptive immune response.

The lymphatic system is vital to the immune response. Invading pathogens are transported in the lymph to the lymph nodes, where bacteria, viruses and cancer cells are trapped and destroyed by phagocytes and lymphocytes. This is why your lymph nodes swell up when you have an infection.

THE STRUCTURE OF THE LYMPHATIC SYSTEM

The lymphatic system is made up of lymph, lymphatic vessels and primary and secondary lymphoid tissues (Figure 12.3.2).

When the fluid that surrounds the tissues (or interstitial fluid) is drained into the lymphatic vessels, it is considered lymph. Lymph contains immune cells such as lymphocytes and phagocytes.

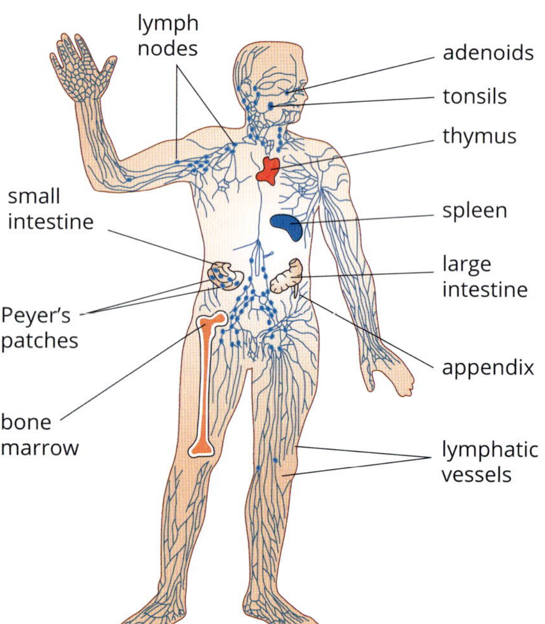

FIGURE 12.3.2 The human immune system is made up of primary lymphoid tissues (the thymus and bone marrow), which produce new lymphocytes, and the secondary lymphoid tissues (lymph nodes, the spleen, tonsils, adenoids, appendix, and Peyer's patches of the small intestine) in which immune responses occur.

The structure of the lymphatic system is similar to the venous part of the circulatory system. Fine lymphatic capillaries join to form increasingly larger vessels that eventually empty into the large veins near the heart (Figure 12.3.3a). Lymphatic capillaries are widespread, but they are absent from bones and the central nervous system (where excess tissue fluid drains into cerebrospinal fluid). Although blood and lymph capillaries are closed to each other, cells and fluid are able to pass between them through a process called extravasation (Figure 12.3.3b). Some of the larger lymph vessels can contract, but most lymph flow results from the external compression of lymph vessels by muscular activity, such as during movement and breathing. When vessels are compressed, the lymph fluid is forced in one direction because of numerous one-way valves, like those in veins, located along the vessels (Figure 12.3.3c). When a person is inactive (such as standing still or sitting) for a long time, the fluid drainage from tissues decreases and causes swelling. This is especially so in the legs, because fluid drainage must work against gravity.

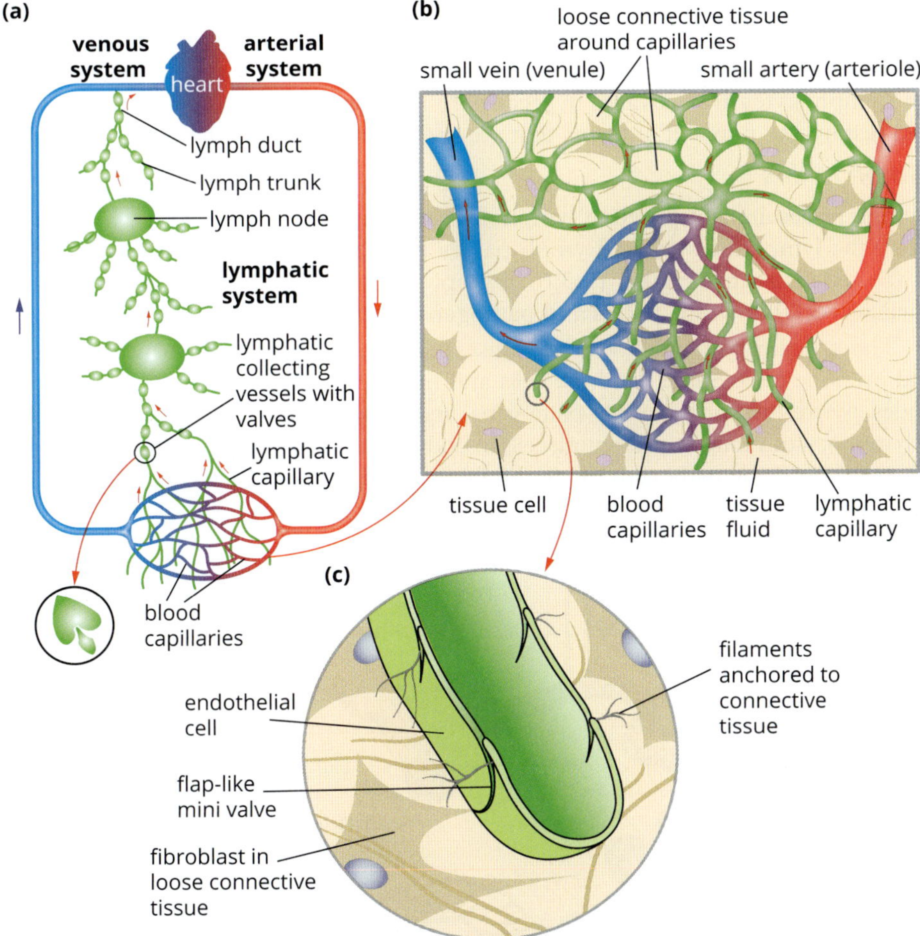

FIGURE 12.3.3 Lymphatic vessels weave through tissue cells and blood capillaries in loose connective tissues of the body. (a) Blood flows from the veins to the heart, then to the lungs to become oxygenated, then through the arteries to tissues. Lymph drains through a lymphatic duct called the thoracic duct, and into a vein called the left subclavian vein, as well as through the right lymphatic duct into the right subclavian vein and the right internal jugular vein. (b) Blood and lymph capillaries are closed to each other, but cells and fluids are able to pass from one vessel to another through a process called extravasation. (c) Lymphatic capillaries are closed-ended tubes in which adjacent endothelial cells overlap each other, forming flap-like mini valves.

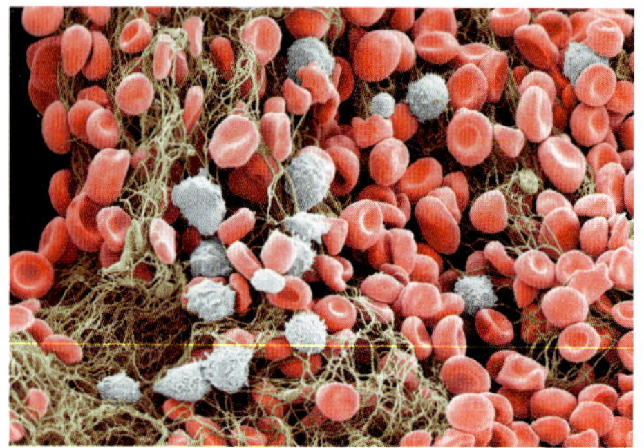

FIGURE 12.3.4 Coloured SEM of a fractured rib. Bone marrow lies between the spongy bone and contains stem cells that give rise to red blood cells (red) and white blood cells such as B and T lymphocytes (grey).

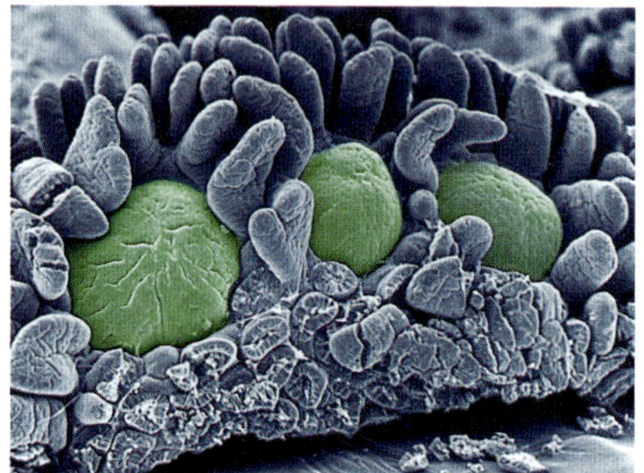

FIGURE 12.3.5 Coloured SEM of Peyer's patches (green) of the small intestine. Peyer's patches defend against infection by supplying lymphocytes to the local intestinal tissue, and are named after the Swiss anatomist Johann Conrad Peyer, who first described them in 1677.

Primary lymphoid tissues

The **primary lymphoid tissues** are the bone marrow and thymus.

Bone marrow contains stem cells from which B and T lymphocytes originate (Figure 12.3.4). B lymphocytes develop and mature in the bone marrow then enter the bloodstream and travel to the spleen and other secondary lymphoid tissues where they become activated after being exposed to antigen.

Immature T lymphocytes travel from the bone marrow to the thymus where they mature. The thymus is considered a primary lymphoid tissue because of its role in the maturation of T lymphocytes. The size of the thymus peaks at puberty then gradually shrinks each year as it becomes replaced by fat (or adipose) tissue. The shrinking of the thymus doesn't have an immediate disastrous effect on immunity, because of the already established pool of peripheral T lymphocytes, but it contributes to the higher risk of infection and cancer that comes with age.

Secondary lymphoid tissues

The secondary lymphoid tissues are the lymph nodes, spleen, tonsils, **adenoids**, **appendix** and **Peyer's patches** (Figure 12.3.5). It is in these organs and tissues that adaptive immune responses begin.

Lymphocytes are activated in secondary lymphoid tissues where they recognise and respond to non-self antigens that are specific to their receptors.

Lymph nodes

Lymph nodes are composed of lymphoid tissue, and are located at regular intervals along the lymphatic system. Lymph passes through lymph nodes on its way back to the bloodstream (Figure 12.3.6). Lymph nodes act as filters, trapping foreign particles, cellular waste, toxins and pathogens.

The structure of lymph nodes maximises the chance of encounters between antigens and immune cells. Some dendritic cells and macrophages are stationed in the lymph nodes, where they phagocytose pathogens, and present the foreign antigens to helper T lymphocytes. Antigen-presenting cells in body tissues also migrate to the lymph nodes after phagocytising pathogens to present foreign antigens to helper T lymphocytes.

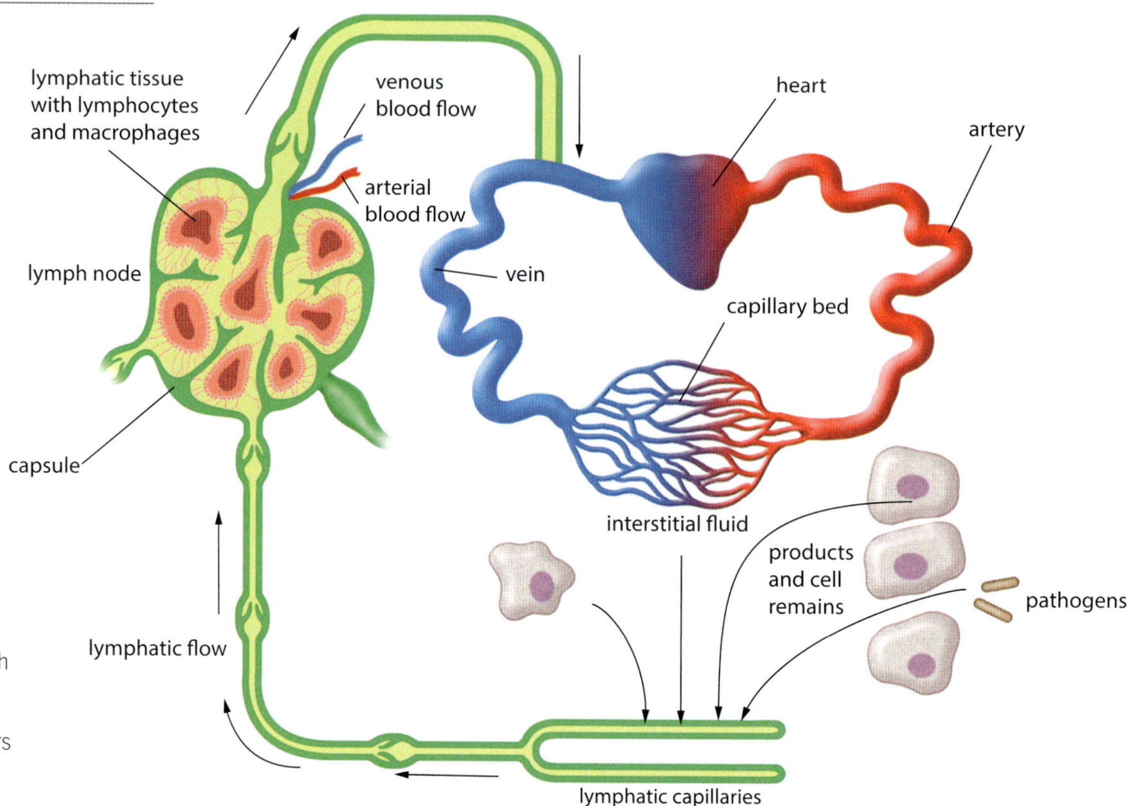

FIGURE 12.3.6 The flow of lymph through the lymphatic system is one-way due to the presence of valves. Lymph nodes act as filters and are important centres of immune cell activity.

B and T lymphocytes interact inside the lymph nodes. B lymphocytes that identify an antigen undergo clonal expansion and differentiation to plasma cells. Antibodies are released into the bloodstream to travel throughout the body. Cytotoxic T lymphocytes are activated, proliferate, and travel through the bloodstream to sites where they are needed.

The size of lymph nodes can expand markedly when cell proliferation is occurring in response to an infection. For example, during a respiratory tract infection, it is common for swollen lymph nodes to occur on the side of the neck (Figure 12.3.7).

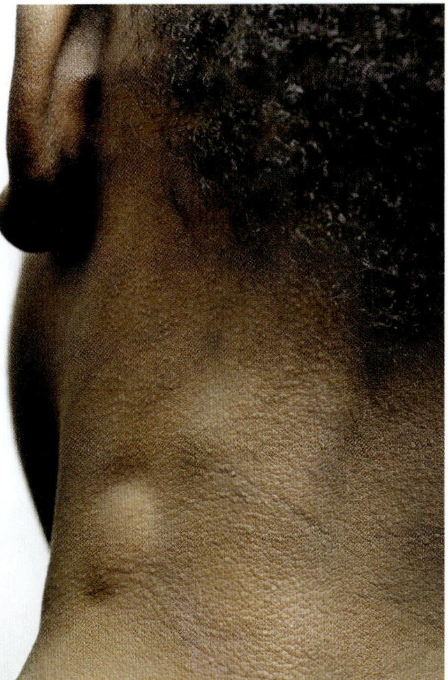

FIGURE 12.3.7 Swollen lymph nodes in a boy's neck. Inflammation of a lymph node (lymphadenitis) is usually a response to an infection. The increased lymph node activity produces more infection-fighting B and T lymphocytes.

Spleen

The spleen's primary function is to control the number of red blood cells in the body by destroying old and defective red blood cells. The spleen also stores up to a quarter of the body's lymphocytes and is a site of B lymphocyte activation.

BIOFILE CCT

Sentinel lymph nodes

Lymph nodes are filters for antigens and invading microbes. They can also trap abnormal cells such as cancer cells that have separated from a primary tumour and travelled in the lymph until reaching a lymph node.

A sentinel lymph node, which is the first node to which cancer cells are most likely to spread, may be removed and examined under the microscope. The presence of cancer cells in the lymph node indicates that a tumour is malignant.

Tattoos do not cause cancer, but tattoo ink migrates through to the lymph nodes and mimics the appearance of certain types of cancer (Figure 12.3.8) making proper diagnosis of some cancers more difficult.

FIGURE 12.3.8 Whether or not tattoos look good on the outside is open to interpretation, but on the inside tattoo ink migrates to your lymph nodes and can make them look cancerous.

12.3 Review

SUMMARY

- The lymphatic system produces lymphocytes and transports them, along with antigen-presenting cells, to the lymph nodes to stimulate adaptive immune responses.
- Primary lymphoid tissues (bone marrow and the thymus) are responsible for the production of B and T lymphocytes.
- T lymphocytes mature in the thymus.
- Secondary lymphoid tissues include lymph nodes, spleen, tonsils, adenoids, appendix and Peyer's patches of the small intestine.
- B lymphocytes develop and mature in the bone marrow and become activated in the secondary lymphoid tissues (e.g. spleen and lymph nodes)
- Secondary lymphoid tissues are the sites where lymphocytes identify and interact with antigen-presenting cells and are then activated to divide and differentiate.

KEY QUESTIONS

1. Which of the following are the primary lymphoid tissues?
   A bone marrow and lymph nodes
   B bone marrow and spleen
   C lymph nodes and spleen
   D bone marrow and thymus
2. Which of the following is where T lymphocytes mature?
   A bone marrow
   B the thymus
   C Peyer's patches
   D the spleen
3. Explain why the lymphatic system is an important part of the adaptive immune response.
4. Fill in the missing labels A–E for the following diagram of the lymphatic system.

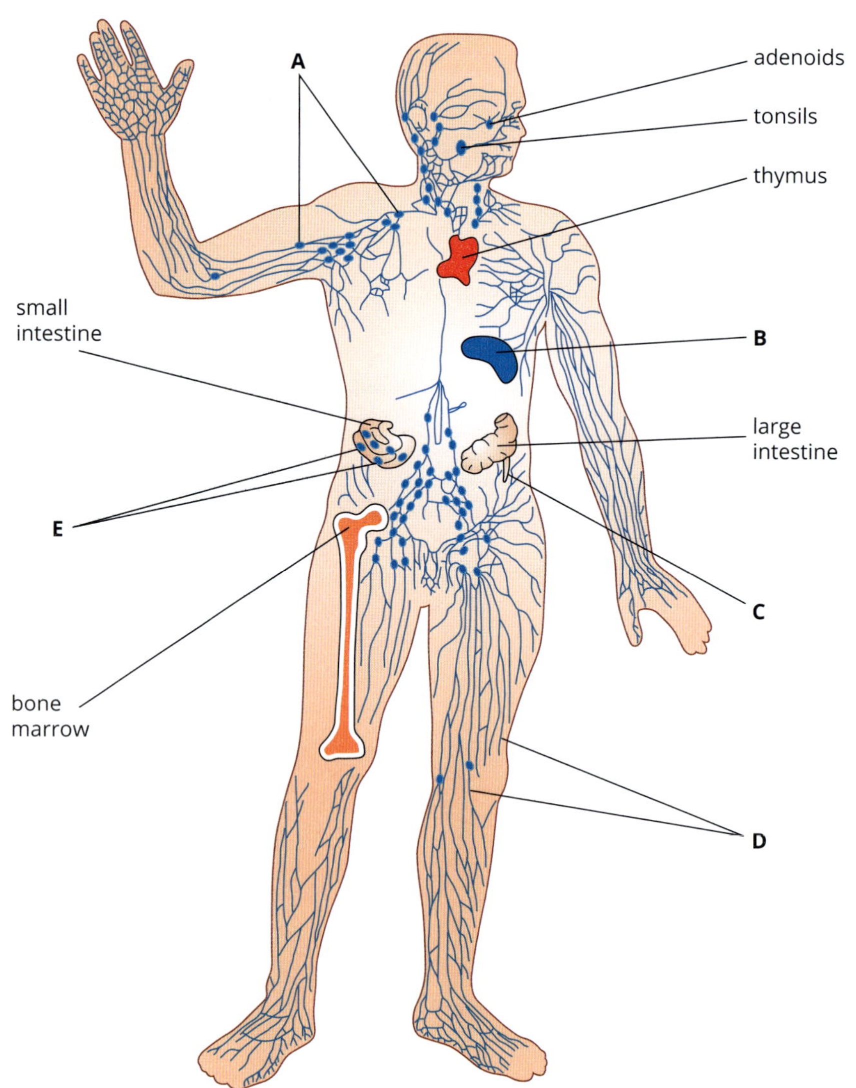

Chapter review

12

KEY TERMS

adaptive immune response
adenoid
agglutination
allergen
antibody
antibody-mediated immunity
antigen
antigen–antibody complex
antigen-presenting cell (APC)
apoptosis
appendix
B lymphocyte
basophil
bone marrow
cell-mediated immunity
chemokine
clonal selection
complement protein
constant region
cytokine
cytotoxic T lymphocyte
dendritic cell
epitope
factor VII (FVII)
factor X (FX)
fever
granulocyte
heavy chain
helper T lymphocyte
histamine
host
immunity
immunogen
immunoglobulin (Ig)
immunological memory
incubation period
inflammation
innate immune response
interferon
interleukin
leukocyte
light chain
lymph
lymph node
lymphatic system
lymphocyte
lymphokine
lysis
lysosome
lysozyme
macrophage
major histocompatibility complex (MHC)
mast cell
memory B lymphocyte
memory T lymphocyte
microbe-associated molecular pattern (MAMP)
monocyte
natural killer (NK) cell
neutralisation
neutrophil
non-self antigen
paratope
pathogen
Peyer's patch
phagocyte
phagocytosis
phagolysosome
phagosome
plasma cell
precipitation
primary immune response
primary lymphoid tissue
prostaglandin
pyrogen
secondary immune response
secondary lymphoid tissue
self-antigen
self-tolerance
signal transduction
specificity
spleen
suppressor T lymphocyte
T cell receptor (TCR)
T lymphocyte
thrombin
thymus
tissue factor (TF)
tonsil
variable region
white blood cell

REVIEW QUESTIONS

1 Define 'immunity'.

2 Describe the process of phagocytosis.

3 Distinguish between antibodies and antigens.

4 Describe the process of antigen presentation by a macrophage.

5 Working in the laboratory you have cut your skin with a sharp dissecting instrument. Describe the process by which the body will protect the site of injury/damage.

6 Compare histamines and antimicrobial proteins.

7 Explain how apoptosis provides non-specific immunity.

8 Why is it important for patients that have undergone transplant surgery to take immunosuppressant medication?

9 Identify five different chemicals involved in the innate immune response and outline their functions.

10 Contrast the innate and adaptive immune responses.

11 List the main cells involved in the innate and adaptive immune response.

12 Explain the specificity of the adaptive immune response.

13 Justify the need for both the innate and adaptive immune system in responding to pathogens.

14 Explain how the innate and adaptive immune system would respond to an initial exposure to a new strain of the influenza virus.

15 An important function of phagocytes is to destroy bacteria. This is done through the endocytosis of the bacteria, followed by its digestion by lysosomal enzymes. As with all cellular functions, this process is regulated by proteins. Chédiak–Higashi Syndrome is a rare inherited disorder of the immune system in which the proteins that regulate the joining of lysosomes with phagosomes are defective.

a What is the name of the structure formed when a phagosome fuses with a lysosome?

b The failure of lysosomal breakdown of engulfed bacteria will seriously undermine not only the innate immune response, but also the adaptive immune response. Explain why.

16 The first successful human-to-human blood transfusion is reported to have occurred in the 1800s. At that time a blood transfusion was a lottery that might help but could equally make you much worse. This was because the knowledge that humans have blood groups (ABO) was not discovered until 1901 and the idea that transfusions should be matched to the recipient's blood group was not suggested until 1907. It was not until 1939–40 that the rhesus protein (antigen D) was identified in humans.

a A patient has presented to the emergency department of the local hospital. The patient needs a blood transfusion. The plates below show the results of the test to determine the patient's blood group.

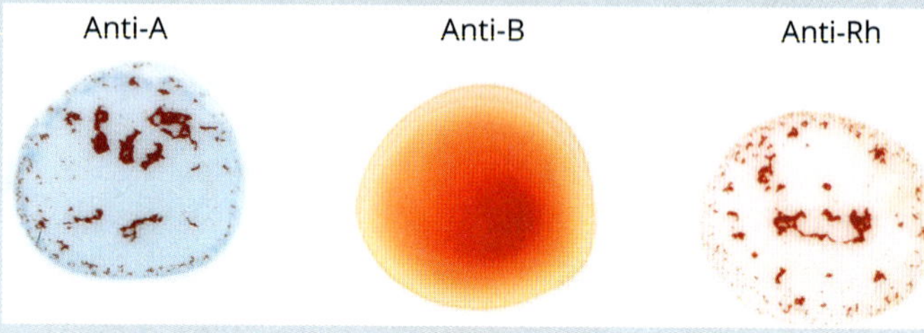

i What is the patient's blood group?
ii How do you know?
iii What blood groups could be used for the patient's transfusion?

b Occasionally, in hospitals when Rh negative blood is in short supply a man who is Rh negative is given blood that is Rh positive.
i This procedure can only be done once. Why?
ii This procedure would not be recommended for a young female patient. Why not?

17 Mammalian antibodies (or immunoglobulins) are generally grouped into five classes. Summarise the role of each class.

18 Outline the clonal selection theory.

19 Define immunological memory.

20 Explain how T lymphocytes recognise antigens.

21 Under what circumstances will a B lymphocyte be activated.

22 What structures comprise the lymphatic system?

23 What is the role of the lymphatic system?

24 Which of the following is not a role of lymphatic vessels?
A returning fluid that seeps out of the blood vessels into tissues back to the circulatory system
B absorbing and transporting fatty acids and fats from the digestive system
C providing a site for lymphocytes to mature
D transporting lymphocytes and antigen-presenting cells to the lymph nodes

25 One way in which the structure of lymph nodes enhances their efficiency is that the nodes have more vessels carrying fluid into the nodes than carrying fluid out.
a i What is the fluid travelling in lymph vessels called?
ii What effect would more vessels leading into the node than out of the node have on the rate of flow?
b One effect of inflammation is to make capillaries more permeable.
i How does the increased permeability of capillaries help the immune response?
ii What effect would the increased permeability of blood vessels have on flow through the lymphatic vessels?
c The lymphatic system is responsible for draining the fluid that leaks from blood vessels. How might the flow of fluid in the lymphatic system be affected by inflammation?
d Explain why it is beneficial to the immune process that lymphocytes accumulate in the lymph nodes.

26 After completing the Biology Inquiry on page 422, reflect on the inquiry question: How does the human immune system respond to exposure to a pathogen? Describe two key features that make the adaptive immune response different from the innate immune response.

CHAPTER 13

Prevention, treatment and control of infectious disease

This chapter examines the factors preventing and controlling the spread of infectious disease at local, regional and global scales. We will explore the range of treatment strategies to control disease and discuss ways of predicting and controlling epidemics and pandemics. You will also learn about contemporary applications of Indigenous Australian medicine and recognise the importance of protecting Indigenous intellectual property in relation to culturally significant biological materials and medicines.

Content

INQUIRY QUESTION

How can the spread of infectious diseases be controlled?

By the end of this chapter you will be able to:

- investigate and analyse the wide range of interrelated factors involved in limiting local, regional and global spread of a named infectious disease EU IU
- investigate procedures that can be employed to prevent the spread of disease, including but not limited to: (ACSBL124) CCT EU ICT IU
  - hygiene practices
  - quarantine
  - vaccination, including passive and active immunity (ACSBL100, ACSBL123) EU ICT
  - public health campaigns
  - use of pesticides
  - genetic engineering
- investigate and assess the effectiveness of pharmaceuticals as treatment strategies for the control of infectious disease, for example: CCT EU ICT IU
  - antivirals
  - antibiotics
- investigate and evaluate environmental management and quarantine methods used to control an epidemic or pandemic CCT IU
- interpret data relating to the incidence and prevalence of infectious disease in populations, for example: ICT N
  - mobility of individuals and the portion that are immune or immunised (ACSBL124, ACSBL125)
  - malaria or dengue fever in Southeast Asia A
- evaluate historical, culturally diverse and current strategies to predict and control the spread of disease (ACSBL125) AHC A CCT IU L
- investigate the contemporary application of Aboriginal protocols in the development of particular medicines and biological materials in Australia and how recognition and protection of Indigenous cultural and intellectual property is important, for example: AHC S
  - bush medicine
  - smokebush in Western Australia

13.1 Spread of infectious disease

BIOLOGY INQUIRY  

Host an infection party

How can the spread of infectious diseases be controlled?

COLLECT THIS...

- plastic cups (one per student)—the teacher will mark the cup with a number and add solution to it
- one pipette per student
- universal indicator
- whiteboard/large sheet of paper
- pen/pencil and paper
- teacher supplies—water (150 mL per student), 75 mL water + 75 mL vinegar solution, waterproof marker.

DO THIS...

1 Each student must dress in personal protective equipment before proceeding.
2 Each student must take a pre-prepared plastic cup with a number marked on it.
3 Students must use a pipette to move the solution to and from their cup with as many or as few students (party-goers!) as they choose.
4 Each student must record the number of the cups they pipetted solution from.
5 To check if they have been infected by the disease at the party, students must add one to two drops of universal indicator to their cup (if the solution turns a yellow/orange/red colour, they have been infected by the disease).
6 Using the whiteboard/sheet of paper, as a class, use logical working to determine which cup was the carrier of the disease.

RECORD THIS...

Describe how the transmission of infectious disease affects its incidence and prevalence in a population compared to non-infectious disease.

Present a logical diagram of how the infection spread from the host cup.

REFLECT ON THIS...

How was the infection transmitted at the party?

What human activities may spread infectious disease through a population?

How can we control the spread of infectious disease?

Infectious diseases continue to cause significant illness and loss of life in the human population. The bubonic plague, small pox, malaria and Ebola are some of the most devastating examples but there are many more. As world populations become larger and humans come into closer contact with other animals, new diseases have emerged. Diseases can be limited to a local area, such as an outbreak of gastroenteritis at a local school, or affect a larger region, such as the outbreak of influenza in NSW during the winter season. More recently, quick transport between countries and larger cities, which brings people into closer contact (Figure 13.1.1), has presented a global challenge for new and rapidly spreading diseases. Countries are now implementing strategies to prevent the worldwide spread of these diseases.

The study and surveillance of emerging and re-emerging diseases aims to find ways to predict, prevent and respond to outbreaks of disease. Many diseases of local and global importance are infectious diseases. Emerging infectious diseases may be defined as:

- new or previously unrecognised diseases
- diseases that have increased in incidence, **virulence** (the ability of a **pathogen** to cause disease) or geographic range over the past 20 years
- diseases that may increase in incidence in the near future.

FIGURE 13.1.1 As transport and larger cities bring people in closer contact with each other, diseases can become more prevalent and spread more rapidly.

Organisations that monitor global disease, such as the US Centers for Disease Control and Prevention (CDC) and the World Health Organization (WHO), generate priority lists of emerging diseases to direct global surveillance and data collection. They focus on early detection, containment (preventing the spread to other areas) and early treatment. As an island nation, Australia is highly susceptible to the introduction of new disease-causing agents. Australia has a strong emphasis on **biosecurity** to prevent, respond to and recover rapidly from pests and diseases. This helps secure our agriculture, biodiversity and human health.

Scientific knowledge has increased over the decades, allowing us to identify, diagnose and treat more diseases. Other factors, however, have led to new infectious diseases becoming a real challenge for governments and medical personnel around the world.

THE EMERGENCE AND SPREAD OF NEW DISEASES

GO TO ➤ Section 10.1 page 378

As discussed in Chapter 10, pathogens such as **viruses**, **bacteria** and **parasites** can all cause infectious diseases. New diseases can emerge when pathogens adapt to new environments or develop mutations in their genome. Mutations may occur when a pathogen moves from one host to another. The new mutation may increase the pathogen's ability to move into a wider range of host organisms, including humans (Table 13.1.1). When a disease passes from an animal to a human host, it is known as a **zoonotic** disease. Many emerging and re-emerging diseases are zoonotic infections. The World Health Organization and experts in the field expect that the next **pandemic** is likely to be a zoonotic disease.

TABLE 13.1.1 Examples of emerging infectious diseases and their mode of transmission to humans via hosts. Some infectious diseases, such as dengue fever, may increase in incidence due to climate change.

New or previously unrecognised diseases	HIV/AIDS—other primate → human severe acute respiratory syndrome (SARS)—bat → human Middle East respiratory syndrome (MERS)—camel → human Hendra virus—bat → horse → human Zika virus—mosquito → human Creutzfeldt-Jakob disease (vCJD) prion—bovine spongiform encephalopathy (BSE) (also known as mad cow disease—cattle → human
Increased in incidence, virulence or range over past 20 years	Ebola—bat → human dengue virus—mosquito → human West Nile virus cholera methicillin-resistant *Staphlococcus aureus* (MRSA) *Clostridium difficile*
May increase in the near future	influenza antibiotic-resistant bacteria cholera dengue virus prion diseases

There are many factors that interplay with and influence the emergence and spread of diseases. Some of these factors are:

- human migration
- human behaviour
- farming practices and food production.

Human migration

Over the centuries, there has been movement of populations due to factors such as war, or for socio-economic reasons. Migration facilitates the spread of pathogens to a new area. Pathogens that were contained in a small area may now be exposed to new hosts, who might not have the same **immunity** against them. Today, with people travelling in short periods of time, at large volumes and across large geographical ranges, emergent diseases have the potential to spread rapidly at a global scale (Figure 13.1.3). This was demonstrated by the 2009 Swine Flu outbreak, caused by the H1N1 virus, which killed over 18 000 people. This virus was a derivative of two strains of influenza virus. Although believed to have originated in a local geographic area in Mexico, the virus rapidly spread through the USA, with cases identified in most countries within months. The first case in Australia was a woman in her mid-30s who arrived in Brisbane after a flight from Los Angeles.

BIOFILE  EU  IU

Refugees and displaced persons

Conflict, natural disasters and political shifts can cause a large, and often quick, migration of people to new regions. Refugees and displaced persons often reside in high-density, temporary accommodation that may lack appropriate sanitation, medical care and infrastructure. It is in these environments that disease can quickly spread. In 1994 almost one million people fled from Rwanda to Zaire where they sheltered in refugee camps. Epidemics of cholera and dysentery soon broke out, killing 50 000 people. The recent displacement of Syrian refugees (Figure 13.1.2) has also seen an increase in disease, particularly Leishmaniasis, a disease historically known as 'Aleppo Boil'. Leishmaniasis is a parasitic infection spread by sandflies and causes skin ulcerations that can fatally spread to internal organs.

FIGURE 13.1.2 Leishmaniasis has become prevalent in many displaced Syrian refugees, many of whom reside in refugee camps such as the Atma Refugee Camp on the Turkish-Syrian border.

FIGURE 13.1.3 Methods of travel such as passenger jets have allowed for the fast and frequent transport of large volumes of people across geographic areas, facilitating the spread of disease.

BIOFILE 

Culture and the cycle of parasites

Human behaviour can facilitate the emergence of new or existing diseases, and often this behaviour is slow to change because of factors such as culture and socio-economic status (e.g. traditional practices in Southeast Asia continue to spread *Opisthorcis viverrini*, a parasite also known as liver fluke). This is a food-borne pathogen, contracted by eating fish from the waters of Laos and other Southeast Asian countries. Many traditional dishes of the region involve raw or undercooked fish (Figure 13.1.4), allowing the parasite to enter the human body and cause disease of the liver and bile duct. Further exacerbating the spread of disease is the low socio-economic status of many areas of Southeast Asia—open defecation occurs, polluting waterways and cycling the parasite back into the marine environment.

FIGURE 13.1.4 Sushi and other traditional Southeast Asian dishes that contain raw or undercooked fish facilitate the transmission of the liver fluke parasite.

Human behaviour

Human behaviour also has an impact on the emergence and re-emergence of diseases, increasing the spread of infectious disease, and may include:

- **hygiene** practice
- sanitation
- dietary habits
- human-to-human contact
- sexual activity
- medical procedures
- exposure to environmental agents of disease.

The H1N1 virus is highly contagious, and the 2009 pandemic was facilitated by the high rate of person-to-person transmission. The virus spread through saliva and mucus particles, which could also survive as airborne droplets (such as from coughing or sneezing), or on infected surfaces such as doorknobs or desks. Poor hygiene and increased human contact while infected with the virus promoted its rapid spread.

Farming practices and food production

Farming practices and food production have changed over the centuries to adapt to an increasing global population. A greater demand for meat has created pressures on the farming industry. Throughout the world, dense farming practices have developed and this has occurred in close proximity to human populations (Figure 13.1.5). This has allowed diseases in farm animals to transfer to humans more easily. These new diseases can have severe effects as the population has no previous resistance to them (e.g. the 2009 H1N1 swine flu virus has been traced to pig farms in a small area of central Mexico). The parent virus is believed to have existed for over 10 years in the pigs, before a strain developed with the capacity to infect humans. This new strain of influenza virus meant humans had little immunity to it. This virulence highlighted the need for worldwide surveillance of animal and human viruses on local, regional and global scales.

FIGURE 13.1.5 Increasing human populations worldwide has led to a high food demand, often resulting in high-density farming practices close to human settlements. This can facilitate the transfer of zoonotic diseases.

EMERGING DISEASES IN AGRICULTURE AND WILDLIFE

Wildlife and agricultural species are at risk of new diseases emerging from pathogens already in the environment and from introduced species. Australian biosecurity agencies and research organisations conduct intensive surveillance to identify potential pathogens, prevent their entry into the country and prevent them spreading throughout vulnerable habitats and farmlands. Some examples of emerging diseases of importance to agriculture and wildlife biodiversity are listed in Table 13.1.2.

TABLE 13.1.2 Examples of emerging diseases affecting Australian wildlife and agricultural plants and animals

New or previously unrecognised diseases	Increased in incidence, virulence or range over the last 20 years
• Hendra virus in horses • bat lyssavirus • devil facial tumour disease • bovine spongiform encephalopathy (BSE) in cattle • bee *Varoa* mite • amphibian chytrid fungus • sugar cane orange rust • Coral Sea fan fungus	• foot-and-mouth disease virus in cattle • blue tongue virus in sheep • wheat stem rust fungus • jarrah dieback *Phytophthora cinnamomi* • loggerhead turtle fungus

BIOLOGY IN ACTION CC

Bats, horses and Hendra

A new disease struck racehorses and their handlers on a Queensland property in 1994. Within just a few days, 14 horses and their owner had died. Government departments and the CSIRO Australian Animal Health Laboratory in Geelong moved into action to identify the mystery disease. The culprit was the Hendra virus (Figure 13.1.6).

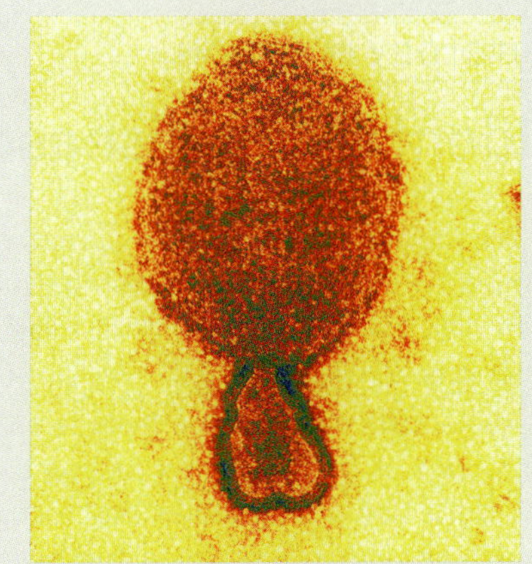

FIGURE 13.1.6 Coloured electron micrograph of the Hendra virus. Hendra virus is an RNA virus in the genus *Henipavirus*, which resides in bats.

The natural host of the Hendra virus is the Australian fruit bat, also called the flying fox. Exactly how the Hendra virus is transmitted from bats to horses is not fully understood. It is likely that horses are infected by ingesting or inhaling the virus in droplets of fluid secretions from bats, when horse properties overlap with bat roosts. The virus is transmitted from horse to horse through direct contact and infectious body fluids (Figure 13.1.7).

The virus can move from horse to human through contact with infected mucus and/or blood or other body fluids. Hendra virus does not appear to be transmitted from person to person. Horse owners and vets need to be aware of the signs of Hendra infection and seek treatment early. Signs include fever and increased heart rate and breathing difficulty. Prevention by immunisation is best.

CSIRO has developed a vaccine against Hendra virus, using a viral coat protein. Equivac HeV is for immunising horses. This is the most effective way to prevent the spread of Hendra virus and protect horses, horse owners and vets.

FIGURE 13.1.7 Risk factors for horses being infected by Hendra virus. (a) The concentration of virus in the environment is affected by the quantity of the virus shed by bats, as well as the survival of the virus out of the bat host. (b) Exposure of horses to the virus is affected by several factors. (c) The effectiveness of the horse's immune response determines whether exposure to the virus leads to infection.

Exotic species originate in another country. Native species have not coevolved with them and so often do not have any defence against exotic pathogenic species.

BIOFILE 

Thawing new viruses

With climate change occurring worldwide, and global warming melting ice shelves and permafrost soils that have been solid for thousands of years, viruses and bacteria contained in these land masses are beginning to emerge (Figure 13.1.8). Scientists recently discovered a new 'giant virus' exposed when a 30 000-year-old piece of Siberian permafrost thawed. The virus, named *Pithovirus sibericum* is the largest on record, measuring 1500 nanometres (ten times the size of HIV). While targeting amoebae rather than humans, the scientists were surprised to find that the virus was still infectious.

FIGURE 13.1.8 The melting of permafrost soils as a result of climate change is uncovering ancient bacteria and viruses.

BIOLOGY IN ACTION CCT

Identifying disease

To control a new or rapidly-spreading disease, the cause of the disease must be identified and a test for the infection developed that does not rely on diagnosing the symptoms, especially for diseases with long incubation periods.

Scientists can develop tests to identify the presence of the pathogen's proteins or DNA, typically testing tissue or blood samples. Gel electrophoresis and the use of antibodies that recognise these molecules can detect proteins, such as viral coat proteins and bacterial toxins. When testing for a pathogen's genetic material, PCR is commonly used to amplify the genetic material (DNA, or RNA in the case of RNA viruses). This is then used to identify genes for toxins, analyse DNA fingerprints of variable repeat sequences, analyse single nucleotide polymorphisms (SNPs) or undergo full genome sequencing. In order to determine the origin of the H1N1 Swine Flu virus (Figure 13.1.9), researchers collected 58 whole-genome sequences from Mexican pigs. They identified that all samples of the virus contained re-assortments of genetic material (an exchange of genetic segments that occurs when two different strains of a virus infect a single cell). The eight RNA strands of H1N1 include one strand from a human influenza strain, two strands from bird (or avian) influenza strains and five strands from pig (or swine) influenza strains.

The development of rapid molecular techniques, databases of protein and gene sequences and bioinformatics has made identifying pathogens much faster, enabling a more rapid response and a greater chance of control and containment of disease outbreaks.

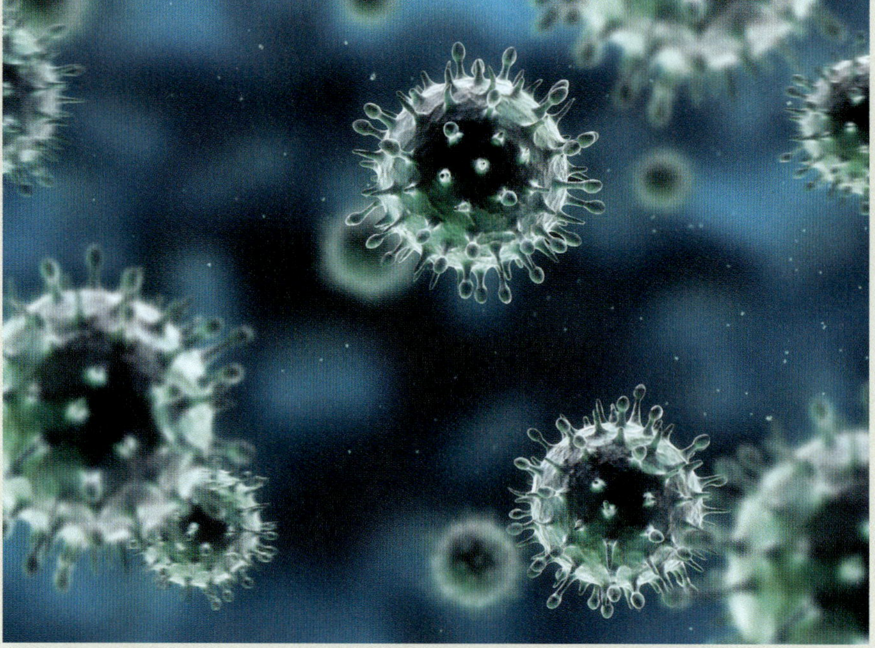

FIGURE 13.1.9 Digital illustration of the H1N1 strain of swine influenza virus

13.1 Review

SUMMARY

- The study and surveillance of newly emerging and re-emerging diseases aims to find ways to predict, prevent and respond to outbreaks of disease.
- Infectious diseases are caused by pathogens such as viruses, bacteria and parasites.
- Many of the emerging and re-emerging diseases are zoonotic infections.
- Today, with more people travelling across large geographical ranges in short periods of time, emergent diseases have the potential to spread rapidly around the globe.
- Human behaviours, such as hygiene practices, can have an impact on the emergence and re-emergence of disease.
- Changes in food and farming practices have led to higher density farming closer to human settlements, facilitating the transmission of new zoonotic diseases.
- Wildlife and agricultural species are at risk of new diseases emerging from pathogens already in the environment and from introduced species.
- To control a new or fast-spreading disease, the cause of the disease must be identified and a test for the infection developed that does not rely on diagnosing the symptoms.

KEY QUESTIONS

1. What is meant by the term 'emerging infectious disease'?
2. What is a zoonotic disease? Give an example of this type of disease.
3. How can an organism become pathogenic and cause disease?
4. Explain how human migration can influence the spread of diseases.
5. Identify three human behaviours that can facilitate the spread of infectious disease.
6. Explain what led to the 2009 outbreak of H1N1 swine flu in humans.

13.2 Preventing the spread of infectious disease

FIGURE 13.2.1 Sneezing (air droplets) is one way to spread infectious diseases.

Infectious diseases are transmitted between people through direct or indirect contact. Direct contact includes touching people, exchange of bodily fluids or spread of droplets (such as coughing or sneezing) (Figure 13.2.1). Indirect contact occurs when pathogens are transferred to people through contaminated objects, animal-to-person contact, or other mechanisms such as food or water.

Controlling the spread relies on preventing the pathogen coming into contact with more people or preventing infection in those people. Disease control can be aided with measures including improved hygiene practices, **quarantine** and **immunisation programs**. Such containment measures depend on the capacity of communities. Factors that can limit the effectiveness of the measures include poverty, high-density accommodation and access to quality medical services.

HYGIENE PRACTICES

Effective hygiene practices and an understanding of the transmission of infectious diseases play an important role in limiting the spread of pathogens.

Personal hygiene

FIGURE 13.2.2 Thorough and frequent hand washing can help limit the spread of pathogens.

Pathogens require the right conditions to continue to survive and proliferate—cleanliness and good personal hygiene can limit this growth and transmission. Personal hygiene is one of the easiest and most effective measures to limit the spread of infectious disease, with measures including:

- thorough and frequent hand washing (after using the toilet or encountering dirty objects, animals or someone who is sick) (Figure 13.2.2)
- thorough and frequent body and hair washing
- ensuring good dental hygiene (including brushing and flossing teeth)
- frequent washing of clothes, towels and bed linen
- cleaning houses and gardens thoroughly and frequently (including disinfecting high-risk areas such as bathrooms and kitchens)
- avoiding coughing or sneezing on others
- placing tissues and other personal products in the bin
- using protection during sex
- wearing personal protective equipment, such as gloves and face masks, when in contact with infected people or objects
- avoiding sharing food or items such as lipstick with others.

Government regulations

Government regulation of pubic services also plays a role in preventing the occurrence of infectious disease and controlling disease outbreaks. Public services that prevent the spread of infectious disease include collecting and disposing of sewage, household, medical and industrial waste, and providing clean water. Water supplies should be filtered, chlorinated and frequently tested to check for the presence of pathogens. In the event of emergency accommodation following a natural disaster or conflict event, the government should establish temporary infrastructure as soon as possible. This may include portable toilets, rubbish containers, water purification supplies and bottled water.

Safe food practices

Infection can easily spread when food is contaminated by pathogens. Food handling can facilitate the spread of viruses such as influenza, which may be infecting a population. Contaminated food sources can also spread bacteria such as *Salmonella*, which can cause gastroenteritis and other illnesses. Safe food practices to limit the spread of disease include:

- hand washing frequently before and during food preparation
- wearing food-handling gloves (Figure 13.2.3)
- cleaning food utensils and surfaces thoroughly
- storing food correctly (freezing or chilling at correct temperatures)
- avoiding sneezing or coughing in food preparation areas
- covering any wounds before handling food
- cooking foods thoroughly and at correct temperatures.

All Australian states and territories are governed by legislation for the correct preparation and handling of food. The *Food Act 2003* and the Food Regulation 2015 set out requirements for the safe service of food in commercial establishments in NSW.

FIGURE 13.2.3 The use of food-handling gloves can help maintain safe food practices.

QUARANTINE

Quarantine is a period of isolation used to prevent the spread of infectious disease. Quarantine measures may include the isolation or compulsory hospitalisation of infected people, closing schools or workplaces, and surveillance and isolation of people, animals and goods moving across borders.

To prevent the spread of infection, people may be isolated for the period of communicability. This might include confining people to their homes, isolating patients within a hospital or containing them at a specialist facility. Those who are potentially infected, but asymptomatic (not presenting any symptoms of the disease) may also be isolated. In day-to-day life, many people will voluntarily isolate themselves while sick (e.g. staying home from school or work with chicken pox to avoid the spread of the disease). However, enforced isolation and quarantine can also occur where a public health threat is identified. This often raises moral and ethical concerns regarding the freedom of the individual.

Border screening and quarantine has been used for centuries to limit the spread of infectious disease. In an increasingly globalised world, border controls are essential for preventing the spread of disease. Quarantine at airports, border stations and seaports include entry and exit screening, health-alert notifications, collection of passenger/cargo information, public health reporting, and physically examining sick people and animals (Figure 13.2.5). Inspecting goods in private or commercial travel is also a quarantine practice, with contaminated animals or goods inspected, disinfected, fumigated, isolated or destroyed. Australia has very strict quarantine laws, enforced by the Australian Quarantine and Inspection Service (AQIS), which prohibit the entry of people, animals or goods that are considered a risk, as well as enforced isolation. State borders in Australia also have quarantine, such as banning the movement of fruit across borders to protect against fruit flies, which are present in some states but not in others.

FIGURE 13.2.5 Quarantine officers and sniffer dogs help inspect and screen humans, animals and goods at airports and other borders.

BIOFILE EU

Typhoid Mary

Typhoid, a disease caused by the bacteria *Salmonella paratyphi*, is spread through the ingestion of faecally contaminated water and food. The disease is more prevalent in less developed countries that lack adequate sewage and waste disposal services, as well as clean water or safe food practices.

It is common for those in western cultures to call someone who is sick 'Typhoid Mary' (Figure 13.2.4). This was the name given to Mary Mallon who worked as a cook in households in the United States during the early 1900s. It is believed that she was a typhoid carrier as people would fall sick with the disease in households where she worked. However, by the time the disease had been traced to its source, Mary had already left the household and disappeared. Fifty-one cases in total were sourced to Mary, who was immune from the disease.

FIGURE 13.2.4 Mary 'Typhoid Mary' Mallon, depicted in a 1909 newspaper illustration. Mary was an asymptomatic carrier of the typhoid bacteria and infected 51 people during her time working as a household cook.

BIOFILE A IU

Foot-and-mouth disease in Southeast Asia

Foot-and-mouth disease is a highly contagious viral disease from genus *Aphthovirus*. It affects ungulate species, particularly domestic livestock such as cattle, sheep and pigs. The virus is endemic in many countries, including Southeast Asia where the virus can spread so rapidly that whole animal herds can be infected within two days. The virus is particularly difficult to control as it can survive for up to six months in food and animal products, on clothing and even hard surfaces such as vehicles and fences (Figure 13.2.6). The virus can also be transmitted in air droplets, urine, faeces, saliva, milk and semen. Symptoms include fever, drooling, lethargy and blisters on the mouth, tongue, snout, hooves and foot pads. The blisters can rupture, leading to open wounds and infection.

Southeast Asian farmers are hard-hit by the virus, with whole herds potentially infected and effects on agriculture including:

- decreased productivity
- treatment costs
- reduced value of livestock
- reduced income.

Australian quarantine classifies foot-and-mouth disease as a high biosecurity risk, with an uncontrolled outbreak potentially leading to tens of billions of dollars of loss in agriculture, exports and control methods. Border controls inspect food and animal products for the virus and ban live animal imports from Southeast Asia.

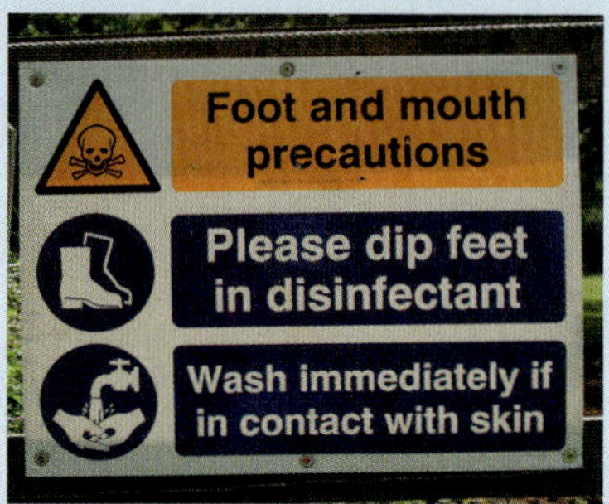

FIGURE 13.2.6 Foot-and-mouth disease can spread rapidly and remain on surfaces and in products for up to six months, making controlling its spread difficult.

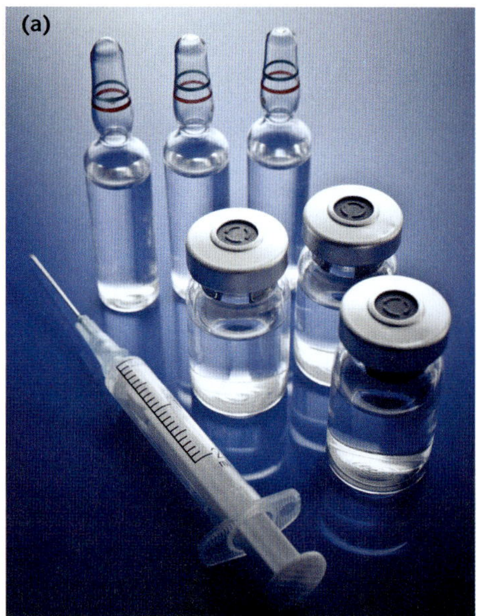

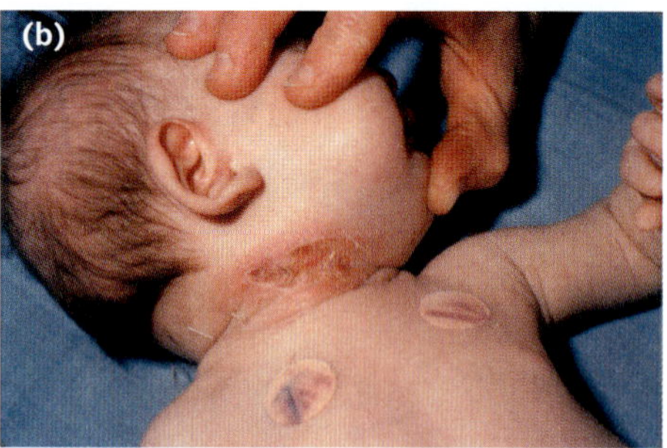

FIGURE 13.2.7 (a) Vaccines have helped protect populations from many infectious diseases such as (b) diphtheria, which previously caused large-scale illness and fatalities.

VACCINATION

Vaccines (Figure 13.2.7a) often take some time to develop but are one of the most effective measures in preventing future infections. Immunisation programs (also known as **vaccination** programs) aim to decrease the incidence of many diseases, with the purpose of ultimately eradicating them. Introducing immunisation programs has led to many countries dramatically reducing new cases of diseases that were common a century ago. Approximately 2.5 million lives are saved each year due to immunisation programs for potentially fatal diseases such as diphtheria.

Diphtheria is a contagious bacterial infection of *Corynebacterium diphtheriae* and *Corynebacterium ulcerans*. The infection is spread through airborne droplets or person-to-person contact and symptoms include membrane formation on the throat, tonsils and nose that make breathing and swallowing difficult. Toxins can also affect cardiac muscles, the nervous system, and cause large open wounds on the skin (Figure 13.2.7b). Diphtheria was a common cause of childhood deaths globally and outbreaks still occur in various countries. Diphtheria vaccination is part of the standard vaccination schedule in Australia—the disease has been almost eradicated from Australia due to its high vaccination rate, with the last **epidemic** occurring in 1921.

Vaccines can protect an immunised individual from infection, as well as unimmunised people in the community through herd immunity. You will learn about herd immunity in Section 13.4. Australians enjoy the benefits of vaccination not only because safe and effective vaccines have been developed, but also because everyone in Australia receives many vaccinations free of charge as part of the National Immunisation Program (Table 13.2.1). Since the introduction of vaccination for children in 1932, deaths from vaccine-preventable diseases have fallen by 99%, despite a threefold increase in the Australian population over that time.

TABLE 13.2.1 Australian child and school immunisation programs

Child immunisation programs	
Age	**Vaccine**
Birth	hepatitis B (hepB)
2 months	hepatitis B, diphtheria, tetanus, acellular pertussis (whooping cough), *Haemophilus influenzae* type B, inactivated poliomyelitis (polio), (hepB-dTP-a-Hib-IPV) pneumococcal conjugate (13vPVC) rotavirus
4 months	hepatitis B, diphtheria, tetanus, acellular pertussis (whooping cough), *Haemophilus influenzae* type B, inactivated poliomyelitis (polio), (hepB-dTP-a-Hib-IPV) pneumococcal conjugate (13vPVC) rotavirus
6 months	hepatitis B, diphtheria, tetanus, acellular pertussis (whooping cough), *Haemophilus influenzae* type B, inactivated poliomyelitis (polio), (hepB-dTP-a-Hib-IPV) pneumococcal conjugate (13vPVC) rotavirus
12 months	*Haemophilus influenzae* type B, meningococcal C (Hib-MenC) measles, mumps and rubella (MMR)
18 months	measles, mumps, rubella and varicella (chickenpox) (MMRV)
4 years	diphtheria, tetanus, acellular pertussis (whooping cough) and inactivated poliomyelitis (polio) (dTPa-IPV) measles, mumps and rubella (MMR) (to be given only if MMRV vaccine was not given at 18 months)
School immunisation programs	
Age	**Vaccine**
10 to 15 years	varicella (chickenpox) human papillomavirus (HPV) diphtheria, tetanus and acellular pertussis (whooping cough) (dTpa)

Immunity is active or passive depending on the origin of the **immune response**. It can develop naturally through exposure to a pathogen, or be induced artificially through purposeful introduction of antigens or antibodies into the body (such as vaccinations). Both active and passive immunity can arise naturally or artificially.

Passive immunity

Passive immunity is protection provided to an individual by the transfer of antibodies produced by another organism. This type of immunity is immediate, but will only protect the recipient for a limited time because it does not result in **immunological memory**, and the transferred antibodies degrade over time and are removed from the body.

Natural passive immunity

Natural passive immunity involves the passive transfer of antibodies from mother to fetus through the placenta prior to birth, and from mother to baby through breastfeeding. These maternal antibodies provide protection to the baby for weeks or months, while its own **immune system** is developing.

BIOFILE CC

Human papillomavirus vaccine

Human papillomavirus (or HPV) can result in the development of certain types of cancer, including cervical cancer (Figure 13.2.8). Professor Ian Frazer and Dr Jian Zhou from the University of Queensland developed a sub-unit vaccine for HPV. In 2007, Australia became the first country to roll out a national HPV vaccination program. The program originally only covered females, but was extended to cover males in 2013.

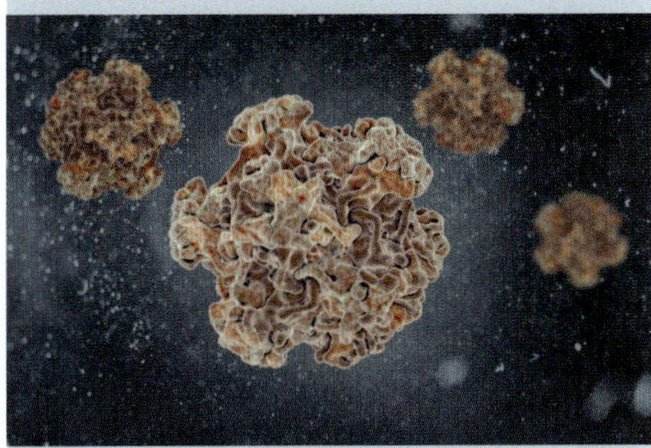

FIGURE 13.2.8 Digital illustration of human papillomavirus

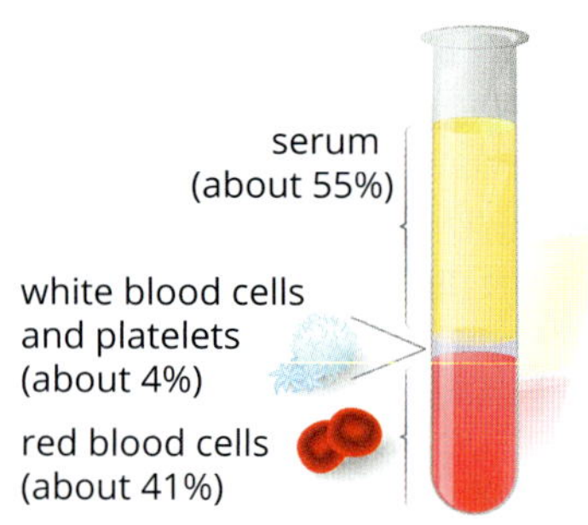

FIGURE 13.2.9 Serum is the fluid portion of blood that remains after blood cells and clotting factors (platelets) have been removed. Antiserum contains specific antibodies injected to treat or protect against disease.

Artificial passive immunity

Artificial passive immunity involves an individual receiving, usually by injection of **antiserum**, antibodies produced by another organism. Antiserum contains specific antibodies. **Serum** is the fluid portion of blood that remains after blood cells and material involved in blood clotting have been removed (Figure 13.2.9). When these transferred antibodies bind to the antigens on the pathogen or toxin, they form an antigen–antibody complex that inhibits the pathogen or toxin before it does much damage. However, introducing antibodies to contain the threat before the person's own **adaptive immune response** can be mobilised means the protection provided is only temporary, as no immunological memory is formed.

Artificial passive immunisation can be a useful way to treat an infection by a pathogen, or a bite or sting by a venomous animal, when death or injury is likely to occur before the primary immune response has developed. For example, administering tetanus antiserum to protect at-risk patients, such as those with a wound that may have been exposed to the bacteria *Clostridium tetani*.

Active immunity

Active immunity is protection provided by an individual's own adaptive immune response. This type of immunity takes time to develop, but the memory B and T **lymphocytes** that result can provide immunological memory that can last for many years, even a lifetime. You will learn more about the adaptive immune response in Chapter 12.

GO TO ➤ Section 12.2 page 432

The adaptive immune system is able to recognise and target specific pathogens. It does this using immunological memory—memory B and T lymphocytes that have previously been exposed to a pathogen recognise the pathogen's antigens and mount an immune response.

Natural active immunity

Natural active immunity develops from the adaptive immune response to a natural infection, and the immunological memory that results. This means that if exposed to the same antigen again in the future, the immune system will recognise it immediately and a secondary immune response will occur (Figure 13.2.10). Secondary immune responses are much faster and stronger than primary immune responses and are therefore more likely to minimise disease (e.g. if you have had chickenpox, you are unlikely to get it again because your immune system has developed immunological memory specific to the antigens of the varicella-zoster virus, the virus that causes chickenpox).

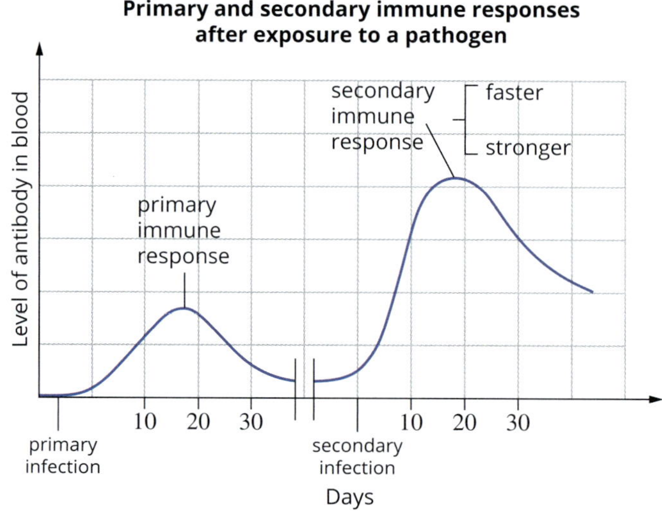

FIGURE 13.2.10 A subsequent infection with the same infectious agent will trigger a secondary immune response. The secondary immune response is faster and stronger than the primary immune response.

Artificial active immunity

Artificial active immunity results from administering antigens to induce an adaptive immune response. This is known as vaccination and the material used to induce artificial active immunity is called a vaccine. By administering a specific vaccine, usually made of altered, weakened or killed microorganisms such as bacteria or viruses, or inactivated forms of toxins or proteins, active immunity is induced.

As with natural active immunity, the primary response to vaccination takes time to develop. Booster vaccines are often needed to stimulate the stronger secondary immune response that provides longer-lasting immunity (Figure 13.2.11).

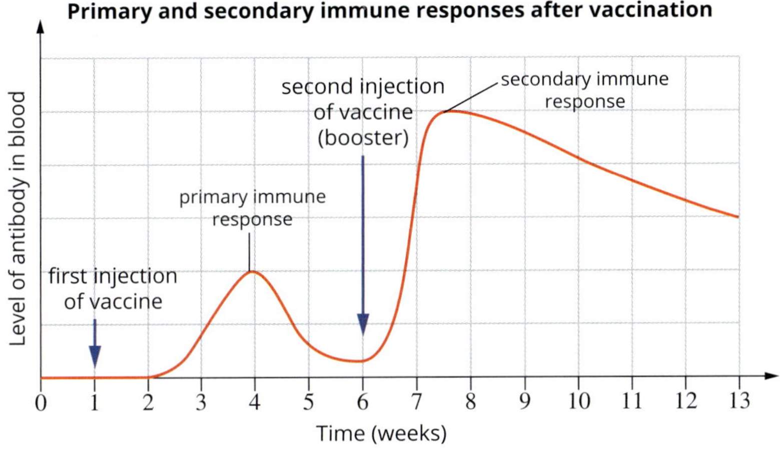

FIGURE 13.2.11 Vaccination induces an adaptive immune response that results in immunological memory. A booster vaccine is often needed to stimulate a stronger secondary response to provide sufficient, longer-lasting protection.

Vaccines need to be highly specific to initiate an adaptive immune response that results in immunological memory. Increased understanding of microbiology and immunology has led to the development of very safe vaccines that induce the desired immune response with minimal side effects.

Table 13.2.2 provides a summary of examples of the different types of immunity you have just learnt about. Table 13.2.3 provides a summary of the advantages and disadvantages of the types of vaccines available.

TABLE 13.2.2 Examples of types of immunity

	Active	Passive
Natural	the adaptive immune response of an individual • immunologial memory • long-term protection	the passive transfer of antibodies from mother to fetus through the placenta before birth, and from mother to baby through breastfeeding • no immunologial memory • short-term protection
Artificial	vaccination to stimulate an adaptive immune response • immunologial memory • long-term protection	administering antibodies that have been produced by another organism against a specific pathogen or toxin (e.g. antiserum) • no immunologial memory • short-term protection

BIOFILE CCT

The first vaccine

Edward Jenner (1740–1823) (Figure 13.2.12) was an English doctor who discovered a vaccine for smallpox. Jenner had heard of a milkmaid who could not catch smallpox because she had already had cowpox (a mild infection that is never fatal). To test this, in 1796, Jenner infected a young boy with cowpox in the hope of preventing infection with smallpox. He then infected the boy with smallpox, by injecting pus from a smallpox lesion under the boy's skin. The boy did not develop smallpox, providing evidence that inoculating a person with cowpox virus provided immunity against smallpox.

FIGURE 13.2.12 Edward Jenner, the English doctor who discovered the vaccine for smallpox in 1796

BIOLOGY IN ACTION 

The eradication of smallpox

Despite the development of the smallpox vaccine in 1796, smallpox persisted in Asia, Africa and South America for many years. In 1966, there was worldwide action to eradicate smallpox, and the last case of epidemic smallpox was registered in 1977 in Somalia, Africa. The only confirmed cases of smallpox after this time were in 1978. They involved a medical photographer called Janet Parker from the University of Birmingham, and her mother. Parker became infected after smallpox travelled through air ducting connecting her office and a laboratory. Although her mother survived, Parker did not.

On 8 May 1980, the World Health Organization announced smallpox had been eradicated. Since then, there have been no vaccinations against smallpox. It was Australian virologist, Frank Fenner (1914–2010) (Figure 13.2.13), who made that momentous announcement. Professor Fenner already had a distinguished career in microbiology, including work on the malaria parasite and the myxoma virus for biological control of rabbits. Then he was appointed Chairman of the Global Commission for the Certification of Smallpox Eradication. Wealthy nations had made great progress in eliminating smallpox from their own countries, but in 1959, when the World Health Assembly decided to eliminate the disease across the world, two million people per year were still dying from it. Fenner helped rule out the theory that monkey populations were acting as a reservoir for the disease, demonstrating that eradication was possible. Intensive intervention, with the cooperation of governments across the world, saw the 'ring vaccination' of populations around each outbreak. This technique led to the successful eradication. Factors that contributed to the eradication of smallpox included the lasting immunity achieved by those who recovered from infection, as well as the fact that the smallpox virus only infected humans and there were no natural reservoirs of infection.

Laboratories have greatly improved infection control standards since the time of Parker, and the smallpox virus is now stored safely in only two laboratories (one in Atlanta, United States and the other in Novosibirsk, Russia). The virus is stored in case there is a need to produce the vaccine again in the future.

FIGURE 13.2.13 Professor Frank Fenner (1914–2010) won numerous awards for his work as a microbiologist. He led the successful campaign to eradicate smallpox.

TABLE 13.2.3 Types of vaccines

Type of vaccine	Description	Advantages	Disadvantages	Examples
live attenuated vaccine	a living microbe that has been weakened in the laboratory, usually through repeated culturing	• single dose can provide long-lasting immunity because the vaccine induces a strong adaptive immune response, producing many types of antibodies directed against multiple antigens	• although safe for most people, it may cause disease in those with weakened immune systems • may cross the placenta of pregnant women, posing a risk to the developing fetus	• more commonly used for viruses than for bacteria • measles, mumps, rubella and polio vaccines
inactivated vaccine (or killed vaccine)	contain microbes that have been inactivated by heat, radiation or chemical means; contain a range of antigens	• stimulates the production of many different antibodies • can safely be used for people who have weakened immune systems • easier to store than live attenuated vaccines	• stimulates a weak immune response compared to live, attenuated vaccines, so they require booster doses to achieve and maintain long-term immunity	• most vaccines against bacteria • inactivated rabies and hepatitis A vaccines
sub-unit vaccine	contain only parts of microbes (e.g. specific proteins or toxins) that have been chemically extracted or genetically engineered using recombinant DNA technology	• can safely be used for people who have weakened immune systems • easier to store than live attenuated vaccines	• require booster doses to improve the strength of the immune response	• foot-and-mouth disease and diphtheria vaccines

The great vaccination debate

Most parents in Australia elect for their children to be immunised against potentially deadly infectious diseases such as pertussis and diphtheria, with 93% of Australian children fully immunised. However, some parents choose not to vaccinate their children because of health risks and philosophical opposition. The 'vaccine debate' has become prominent in the media in recent years and individuals are having to make a decision on whether or not to vaccinate.

Decision-making

Good decision-making is an essential life skill—many believe it is not the outcome that is important, but the process itself.

Elements of good decision-making

What do I need to know?

- Source useful and reliable information.
- Consult online scientific journal articles and government health websites for up-to-date information.

What am I really deciding?

- Use this information to create a focus question.
- Am I being logical?
- Use solid reasoning when making decisions.

What do I want?

- Ensure clear and personal values guide decision-making.

What can I do?

- Identify and consider options and alternatives.

PUBLIC HEALTH CAMPAIGNS

Governments, public health authorities, and non-government organisations (such as the Australian Cancer Council) organise and distribute **public health campaigns** to promote the adoption of healthy behaviours. When dealing with an outbreak of infectious disease, campaigns help inform people of the cause of the disease, how it is transmitted and the reasons for the measures employed to control the spread of the disease. Information may be disseminated through media outlets, community organisations, directly to citizens through postal services, or other organisations and institutions such as charity groups and schools. Often during a severe disease outbreak or epidemic, public anxiety rises and misinformation can be spread. Information in public health campaigns should be accurate, clear and as reassuring as possible.

Some health campaigns may be short-term to educate and inform the public about a sudden disease outbreak. The focus will be on disseminating information quickly to educate people on what measures to follow. In July 1998 the Greater Western Sydney water supply was infected by high concentrations of the parasites *Cryptosporidium* and *Giardia*. An immediate public health campaign was launched and included:

- educating the public about the parasites and possible causes of water supply contamination
- informing the public about the treatment of the water supply
- bans on drinking water from all sources

- information on which bottled water was safe to drink
- notifications of local public pool closures (Figure 13.2.14a)
- instruction on boiling water for safe use (Figure 13.2.14b)
- information on where to seek medical help if infected by the parasites.

Information was disseminated through the media, schools and community organisations, as well as anywhere contact with contaminated water was likely (such as public pools and water fountains).

FIGURE 13.2.14 The public health campaign for the short-term outbreak of parasites, *Cryptosporidium* and *Giardia*, in Greater Western Sydney's water supply included disseminating information to notify people about (a) pool closures and (b) safe procedures to treat water before using it.

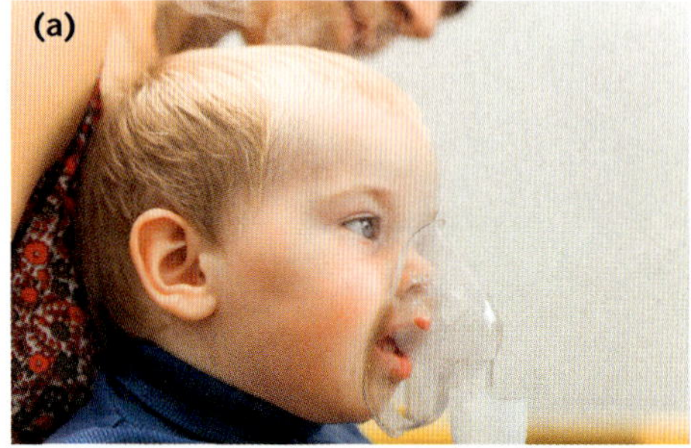

FIGURE 13.2.15 To protect the population against (a) whooping cough, NSW medical services promote (b) vaccination, isolation and hygiene practices.

Other health campaigns are long-term and aim to educate the public on positive health and lifestyle changes they can make to prevent disease, such as in the NSW Government's campaign for the prevention of whooping cough (pertussis), a serious respiratory infection that results in long and serious coughing episodes. It is especially serious for the vulnerable in the community such as babies (Figure 13.2.15a), the elderly and those with chronic illnesses. Newborns who are unable to successfully cough, may stop breathing and turn blue, leading to possible brain damage, pneumonia, vomiting and even death. For the elderly and those with chronic and complex health issues, complications can include rib fractures, vomiting, weight loss, low blood pressure and incontinence. While treatment includes **antibiotics** and isolation, the best prevention of this disease is vaccination. The NSW campaign information focuses on promoting:

- vaccination for:
  - pregnant mothers in their third trimester (this gives the baby short-term protection)
  - newborns and infants (first vaccine from two months of age)
  - children and teenagers (four years and 10 to 15 years) for booster vaccines
  - new parents or those in contact with newborns (under two months) (Figure 13.2.15b)
  - health care and child care workers
- isolation of those infected with whooping cough
- good hygiene practices.

Information is primarily disseminated through health care service providers (such as medical centres and hospitals), as well as in online media. You will learn about public health campaigns for non-infectious diseases in Chapter 17.

GO TO ➤ Section 17.1 page 582

USE OF PESTICIDES

Pesticides are chemicals often used to prevent the spread of infectious plant and animal diseases, as well as to eradicate insect **vectors** such as mosquitoes and midges to control malaria and other vector-borne diseases. Potato leaf roll infects Australian potato crops, and is caused by a small aphid. Farmers regularly spray potatoes with pesticides during growing seasons to eliminate this pest. The eradication of mosquito larvae and adults is common in most countries, including Australia, to prevent the spread of vector-borne diseases such as malaria. Pesticides may be used:

- as sprays or baits (Figure 13.2.16)
- in irrigation water
- as 'dips' for farm animals (i.e. a liquid bath containing insecticides and fungicides).

However, because of the overuse of many pesticides, genetic resistance has developed among many pests. It is estimated that 90% of sheep parasites are now resistant to farming pesticides. As a result, new pesticides constantly need to be developed. Integrated pest management is being promoted by the government to better deal with pests, combining physical, biological and chemical methods to prevent the spread of pests and disease.

FIGURE 13.2.16 Pesticides are sprayed on a large-scale over crops.

GENETIC ENGINEERING

Genetic engineering is used to modify the genetic structure of an organism using biotechnology. You learnt about biotechnology in Chapter 9. Genetic engineering helps to prevent the spread of disease by producing:

- plants and animals that are resistant to common pests and diseases
- vaccines for human and animal use
- vectors (such as mosquitos) with a diminished capacity to spread disease
- transgenic animals for harvesting biomedical products.

Currently, cows are being genetically engineered to produce large quantities of human antibodies to treat a wide range of infectious diseases (Figure 13.2.17). The idea came about during the 2014 West African Ebola outbreak. In the absence of an effective drug treatment, health care workers began injecting infected patients with plasma from individuals who had recovered from the disease in the hope that the donor plasma proteins would help remove the harmful pathogens. Those patients injected with the plasma had a 7% greater chance of survival.

+ ADDITIONAL

Health impacts of pesticide use

While pesticides are used, mostly successfully, to prevent the spread of disease in humans, animals and plants, health issues have arisen for both humans and the environment. The accumulation of pesticide in the food chain is one such problem, a term known as biomagnification. Biomagnification was first identified with the use of DDT, which was widely used to control mosquitoes and agricultural pests. DDT had an extremely long half-life, so did not break down easily. Instead, the pesticide amplified in concentration as it worked its way up the food chain. Effects included top-level carnivore deaths due to biomagnification of DDT in their tissues, and the thinning of predatory bird eggshells.

Human health is also impacted, with three million cases of pesticide poisoning reported each year, and hundreds of thousands of deaths. Young children are particularly susceptible to pesticide poisoning. Adverse health effects can include: asthma, allergies, hypersensitivity, cancer, hormone imbalances, reproductive problems, neurological effects including memory loss, coordination and mobility changes, visual impairment and speech impairment. Neurological effects are often very subtle and can remain undiagnosed.

GO TO ➤ Section 9.2 page 349

FIGURE 13.2.17 Plasma from transgenic cows may lead to future treatment of infectious diseases among humans.

Researchers have engineered cows by replacing part of their genome with a human artificial chromosome to generate human antibodies—vaccinating the cow with a target disease antigen leads to the production of antibodies. Plasma from the cow is then harvested and antibodies isolated and used to produce a therapeutic drug. Immunoglobulin therapy is promising in cows because they produce 150 to 600 mg of antibodies per month, much higher than humans can produce.

BIOLOGY IN ACTION  

Ebola vaccine to protect wildlife

While Ebola is commonly associated with mass fatalities among human populations, the disease also decimates populations of wildlife, including primates. An outbreak of Ebola in 2002–2003 wiped out over 5000 gorillas in the Republic of Congo (Figure 13.2.18). This is a serious concern given the threatened status of many primate species due to other threats such as habitat destruction. Genetic engineers in the UK have developed a vaccine for Ebola in primates, by inserting a protein from the surface of the virus into an existing live rabies vaccine used in fox populations. The vaccine has been tested both as an injection and given orally, with large amounts of antibodies produced in test subjects. The subjects also survived a test exposure of the virus. The oral vaccine is particularly promising as it can be added to baits for wild primates to find, rather than requiring the capture and sedation of the animals to inject the vaccine. However, ethical and environmental issues arise, with groups arguing that the baits may be taken by other animals, there may be knock-on effects for those who feed on primates, as well as the question of whether vaccinating wild animals interferes with their 'wild' status.

FIGURE 13.2.18 Genetic engineering holds the potential to develop a vaccine to protect primate species, such as the mountain gorilla (*Gorilla beringei beringei*) from the Ebola virus.

13.2 Review

SUMMARY

- Infectious diseases are transmitted between people through direct or indirect contact.
- Infection can be reduced if people understand how the disease is spread and are educated in simple hygiene and prevention methods.
- To prevent the spread of infection, people infected with a transmissible disease may be isolated in quarantine.
- Quarantine at airports, border stations and seaports includes entry and exit screening, health-alert notifications, collection of passenger/cargo information, public health reporting, as well as physically examining sick people and animals.
- Immunity can develop naturally or be induced artificially.
- Passive immunity involves the transfer of antibodies produced in another organism. It does not result in immunological memory and is temporary.
- Active immunity involves the individual's adaptive immune response. It results in immunological memory that can be long-lasting.
- When dealing with an outbreak of infectious disease, public health campaigns help inform people of the cause of the disease, the ways in which the disease is transmitted and the reasons for the measures used to control the spread of the disease.
- Pesticides are often used to prevent the spread of infectious plant and animal diseases, as well as to eradicate insect vectors.
- Genetic engineering is being used to research and develop better disease preventatives and treatments.

KEY QUESTIONS

1 Name six procedures that can be used to prevent the spread of infectious disease.

2 Identify two hygiene practices that can help control the spread of disease.

3 Explain the difference between active and passive immunity.

4 The following graph represents changes in antibody concentrations that occur during a primary and secondary adaptive immune response. Provide appropriate text for labels A, B, C and D.

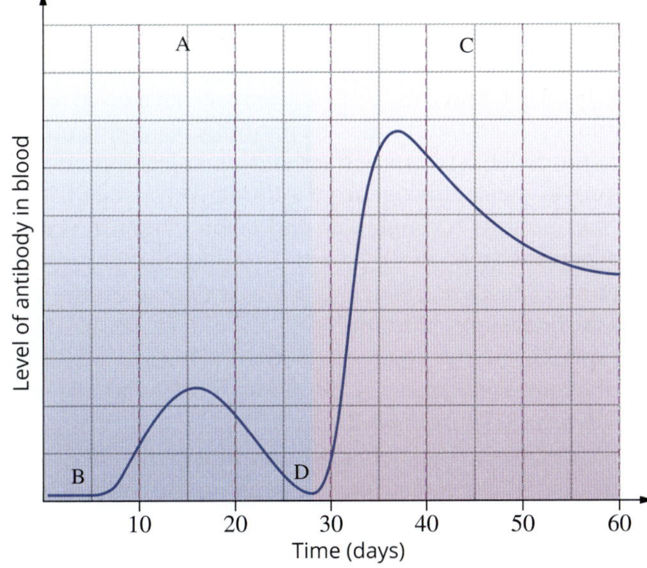

5 Vaccination is an example of:
A artificial passive immunity
B natural active immunity
C artificial active immunity
D natural passive immunity

6 Provide an example of a public health crisis in Australia caused by infectious disease and describe the public health campaign.

13.3 Treatment for infectious disease

Antibiotics treat bacterial infections. Antivirals treat viral infections.

The treatment for diseases will vary depending on their cause and effects (Table 13.3.1). In this section you will learn about some of the treatment options that are used to control infectious diseases, including pharmaceuticals such as **antivirals** and antibiotics.

TABLE 13.3.1 Types of diseases and their treatment

Type of disease	Treatment	Example
viral	antiviral drugs to minimise the viral load along with fluids and pain-relieving medication	• Ebola • human immunodeficiency virus (HIV)
bacterial	antibiotics to slow bacterial growth or kill bacteria	• whooping cough (pertussis) • meningitis
prion	currently there is no treatment available. Antibodies against prions are currently being tested as potential treatments for prion diseases	• variant Creutzfeldt-Jakob disease (vCJD)

BIOFILE CCT

Death by hospital

In the early 1800s, one of the greatest causes of death was admission to hospital. Doctors knew nothing of bacteria or viruses and did not know how infections occurred. A man called Joseph Lister argued that doctors were somehow transferring disease from one patient to another. He was not listened to by his colleagues, who thought his ideas were ridiculous.

Lister conducted trials in a hospital ward and showed that the washing of hands by doctors and nurses after seeing each patient significantly decreased the number of patients that became ill. He also suggested that operations should be performed in sterile environments and persuaded surgeons to operate under a mist of antiseptic to reduce the number of pathogens that might infect the wounds (Figure 13.3.1). Surgeons also used to wear their everyday clothes while operating, a practice that no longer occurs to limit the number of microbes entering the operating theatre.

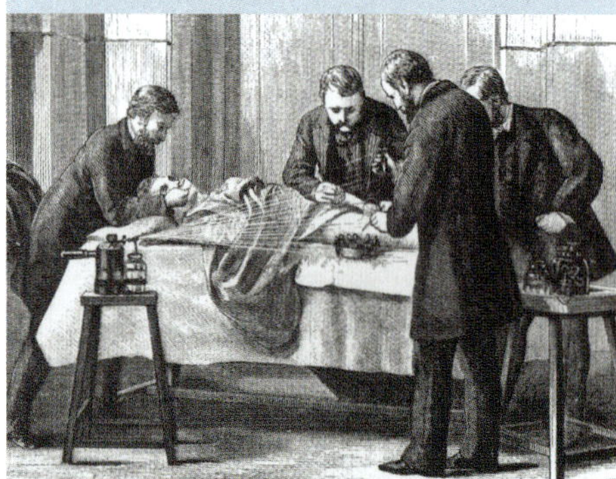

FIGURE 13.3.1 Joseph Lister using antiseptics during an operation

DISINFECTANTS AND ANTISEPTICS

One way of controlling pathogens and their spread is to reduce the number of pathogens in the outside environment. **Disinfectants** are used to kill pathogens on surfaces such as door handles and hospital equipment and include ethanol, chlorine and hydrogen peroxide. **Antiseptics**, such as ethanol, chlorhexidine, iodine and some detergents, are used to kill pathogens on the body. By reducing the number of pathogens, the chance that an infection will occur is reduced.

Disinfectants and antiseptics are non-specific antimicrobial agents, that is, they deactivate or destroy most biological agents (bacteria, viruses, fungi). Disinfectants and antiseptics have various modes of action to kill microorganisms, including denaturing proteins, disrupting cell membranes and dissolving lipids. Bacteria with waxy surfaces and microbes that form spores and cysts may be resistant to the effects of these compounds. Disease-causing prions are resistant to these chemicals and to strong acids, including stomach acid.

ANTIBIOTICS

Many bacteria are pathogens and can be treated using antibiotics. Penicillin, discovered in 1928, was the first commercial antibiotic. When it was first mass produced, it was often described as a 'magic bullet' because diseases that had once been fatal could now be easily cured. Since the 1930s, scientists have discovered and developed many other antibiotics. Many antibiotics are naturally occurring molecules produced by fungi or bacteria. Scientists can extract and chemically modify these natural molecules to improve their effectiveness. Alternatively, scientists have discovered new antibiotic compounds through trial and error, chemical screening, or by design to fit a particular pathogen.

To be successful, antibiotics ideally kill bacteria without damaging the cells of the organism being treated. Therefore, the antibiotic should target biochemical pathways and molecules specific to the microbe. Different classes of antibiotic target bacterial cell walls, ribosomes, enzymes for DNA and RNA synthesis, protein synthesis and metabolic pathways (Figure 13.3.2).

Some antibiotics slow bacterial growth—they are **bacteriostatic**. Others kill the bacteria—they are **bactericidal**. These effects may depend on the concentration of antibiotic used.

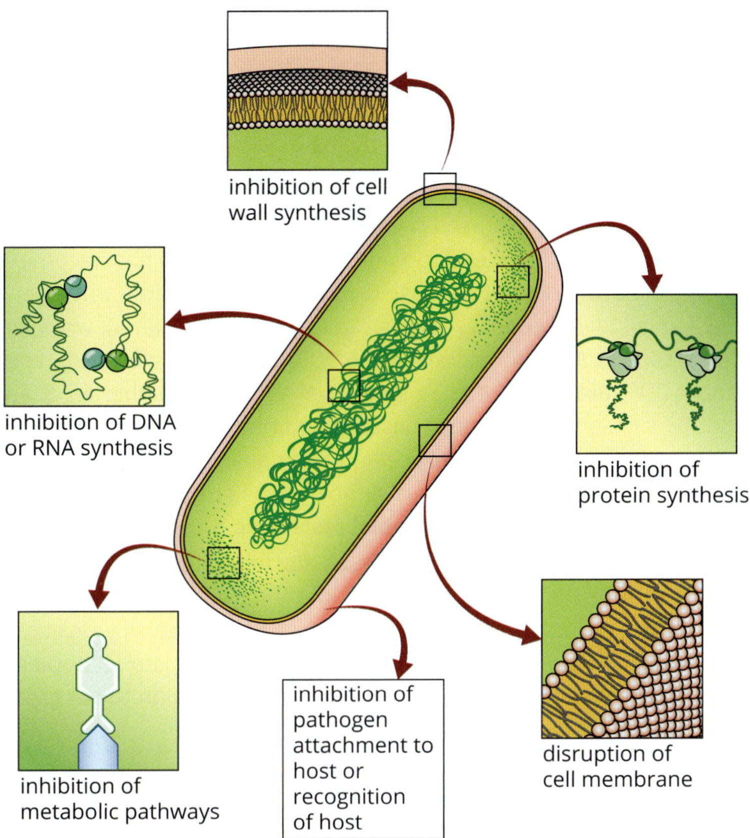

FIGURE 13.3.2 Mechanisms of antibiotic action

One group of antibiotics, including penicillin, inhibits the synthesis of a bacterial cell wall component called **peptidoglycan**, which is composed of amino acids and sugars linked into a mesh-like substance. These antibiotics prevent the bacteria from fully developing their cell walls, causing the cell walls to weaken and the bacteria to die. Bacteria with a thick layer of peptidoglycan in the cell wall are more susceptible to the effects of these antibiotics than bacteria with little peptidoglycan in their cell wall. Animals do not produce peptidoglycan and so these antibiotics have no effect on their cells.

Another group of antibiotics, called sulfonamide drugs, act as competitive inhibitors in the metabolic pathway for folic acid (folate) production. Folate is a B vitamin essential for DNA and RNA synthesis in all cells. Animal cells can absorb this vitamin, which is obtained in the diet, through their cell membranes, but the thick bacterial walls will not allow folic acid to cross. Without the ability to make their own folic acid, the bacteria die.

Other antibiotics are inhibitors of transcription, blocking mRNA synthesis (e.g. actinomycin). Protein synthesis may be blocked by antibiotics that interfere with the ribosomes (e.g. tetracycline), or that block transfer RNA (e.g. puromycin). Bacterial ribosomes differ from eukaryotic ribosomes, and the enzymes for RNA synthesis and DNA replication are specific in prokaryotes and eukaryotes. Most antibiotics inhibit only the bacterial molecules.

BIOFILE CCT

Adverse effects of antibiotics

Although antibiotics specifically target bacterial structures and proteins, some unfortunately cause damage to eukaryotic cells, particularly when used at high concentrations. Damage to the sensory hair cells in the ear, leading to deafness, and disruption of kidney tubules and tendons may be caused by some antibiotics. This may be due to cross-reaction of the drug with eukaryotic membranes and enzymes.

Recent studies have identified a potential new mechanism for some of the adverse effects of antibiotics—alteration of mitochondrial function. Mitochondria most likely evolved from bacteria by endosymbiosis, and indeed mitochondria share structural and molecular similarities with bacteria, including related ribosomes. Antibiotics that inhibit protein synthesis in bacteria also inhibit the mitochondrial ribosomes. Other antibiotic effects observed in mitochondria include decreased energy generation and excessive production of reactive oxygen species (free radicals), which damage proteins and lipids in the cell.

A positive side to these effects is the potential use of antibiotics to inhibit mitochondrial activity in cancer stem cells and thus slow their growth.

BIOFILE CCT

The last line of defence

Recently in mainland China, scientists have found bacteria in pigs and humans that are resistant to the 'last line of defence' antibiotic polymyxins. The resistance has come about because antibiotics are widely used in agriculture in China and a mutation occurring in a gene in plasmids of common bacteria has allowed these bacteria to survive when polymyxins are present. Unfortunately, these mutated plasmids can be passed between species of bacteria and this will increase the number of different species that will have this resistance.

BIOFILE CC

Australian Group on Antimicrobial Resistance (AGAR)

A group of medical scientists called the Australian Group on Antimicrobial Resistance (AGAR) tests and gathers information on the level of antibiotic resistance in bacteria causing important and life-threatening infections, particularly *Staphylococcus aureus* and *Enterococcus*. To increase global awareness of this problem, and to help educate the population on appropriate uses of antibiotics (Figure 13.3.4), the World Health Organization runs World Antibiotic Awareness Week each year.

FIGURE 13.3.4 The World Health Organization supports World Antibiotic Awareness Week.

REVISION

GO TO ➤ Year 11 Section 10.2

Drug resistance

There is an increasing concern that the antibiotics humans have relied on for so long to fight bacterial infections may be losing their effect. Bacteria that are exposed to antibiotics can develop a resistance to them. The development of drug resistance has become a huge problem, limiting the usefulness of all these drugs. There was a time when penicillin killed more than 97% of all *Staphylococcus aureus*, ('golden staph' bacteria), which can cause blood infection (sepsis). Now in Australia 80 to 90% of these bacteria are resistant to penicillin and about 20% are resistant to a higher-level antibiotic, methicillin. *S. aureus* normally resides on the skin, and in the nasal passage and mouth, but causes infection when it breaches the epithelial layer. The overuse and mismanagement of antimicrobial drugs in humans, along with the extensive use of antibiotics as growth-promoting agents in livestock is driving the development of **antimicrobial resistance**.

Bacteria can resist antibiotics in a variety of ways, as illustrated in Figure 13.3.3. For example, bacteria may reduce the intake of the drug into the cell, alter the target molecule to which the drug attaches, pump the drug out of the cell or enzymatically deactivate the drug. If these resistance properties are present in members of the bacterial population, the bacteria possessing them will survive the drug treatment and go on to become the dominant population.

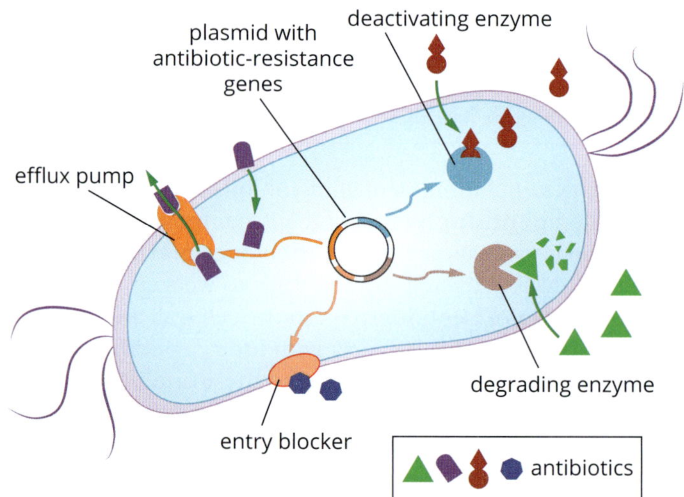

FIGURE 13.3.3 Mechanisms of bacterial resistance to antibiotics

Some bacteria in a population already carry resistance genes—they are not killed by the antibiotic and pass on the resistance genes. Alternatively, spontaneous mutations may occur in the bacterial DNA, making the bacteria resistant to the antibiotic.

These bacteria are then able to multiply rapidly without competition from sensitive strains, and produce cloned offspring that carry the resistance gene. Bacterial cells also exchange plasmid DNA between each other in a process known as horizontal gene transfer. Plasmids carrying a drug resistance gene can be acquired in this way. Transmitting resistant bacteria can be indirect, through food and water, or through direct contact with an infected person or animal. Methicillin-resistant *S. aureus* (MRSA) and vancomycin-resistant *Enterococcus* (VRE) are examples of bacteria that are now resistant to top-line antibiotics, making them almost impossible to treat. They are frequently acquired in hospitals.

Although scientists are working to develop new antibacterial drugs, the threat of antibiotic resistance is currently a major global health concern.

Resistance can also evolve to antiviral drugs, especially in patients with compromised immune systems. Once this happens, new forms of treatment may be necessary.

BIOFILE EU IU PSC

Antibiotic resistance sweeps the developing world

While access to medical treatment against infectious diseases is a positive step for developing countries where treatable illnesses often led to complications or even fatalities, the developing world has some of the highest statistics of antimicrobial resistance. For example, some strains of the bacteria, *Escherichia coli*, have shown high levels of resistance to most of the antibiotics currently in use.

The cause of these high rates of resistance are believed to include:

- medical personnel prescribing drugs inappropriately
- an oversupply of antimicrobial drugs available
- access to antibiotics available over the counter, without prescription (Figure 13.3.5)
- insufficient monitoring of prescription practices
- poor quality of drugs
- antimicrobial use in animal husbandry
- patients selecting to stop their treatment early, before returning to hospital with a more serious infection.

Combining these factors with environmental and human factors such as overcrowding, inadequate sanitation infrastructure, a lack of personal hygiene and ineffective hospital infection control practices leads to the perfect storm for building antibiotic resistance. This resistance is a major issue as the resistant pathogens can be spread on local, regional and global scales.

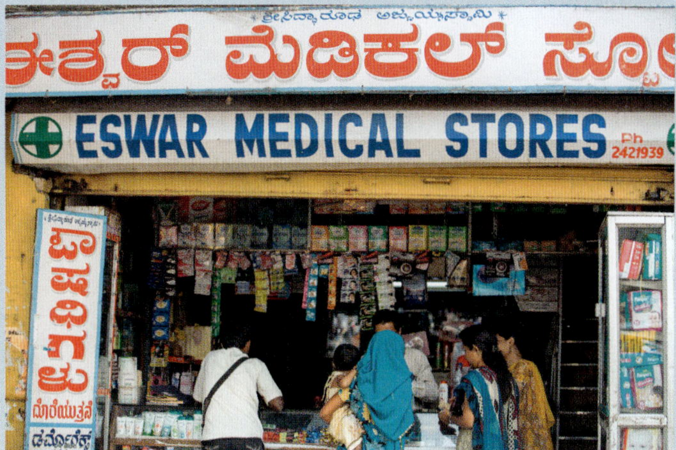

FIGURE 13.3.5 Many developing countries, including parts of India, allow the sale of antibiotics over the counter without a prescription.

ANTIVIRALS

Viruses are non-cellular pathogens that must be inside living cells to replicate (Figure 13.3.6). Viruses may be either RNA or DNA viruses. RNA viruses tend to have a higher rate of mutation than DNA viruses. Viruses can survive for a limited time outside their host cells. This extracellular form of a virus is called a **virion**. A virion consists of genetic material in the form of DNA or RNA, surrounded by a protective protein coat called a **capsid**.

Once the virion has penetrated the host cell, it disassembles, freeing its genetic material to transcribe and translate new viral proteins. Some viruses, called enveloped viruses, are surrounded by a lipid envelope that is picked up when they bud from the host cell in which they replicated. The envelope contains viral proteins that are used as recognition molecules for binding to, and entering, target host cells. Because replicating viruses are found inside cells, they are difficult to destroy without damaging the host cell. A virus uses the host ribosomes for protein synthesis and may use host cell enzymes for nucleic acid (DNA or RNA) replication.

Good targets for antiviral drugs are the capsid proteins, envelope proteins, and the DNA/RNA polymerase enzymes encoded by viral genes. Drugs that cross-react with host cell enzymes cause serious side effects. Antiviral drugs have only been in use since 1960, but a broad range has already been developed.

Antiviral drugs can work by several possible methods (Figure 13.3.6):

- preventing the virus from entering the cell by binding to receptors that allow the virus to enter
- inhibiting enzymes that catalyse reproduction of the virus genome
- blocking transcription and translation of viral proteins
- preventing the viruses from leaving the cell, and so preventing the infection of other cells.

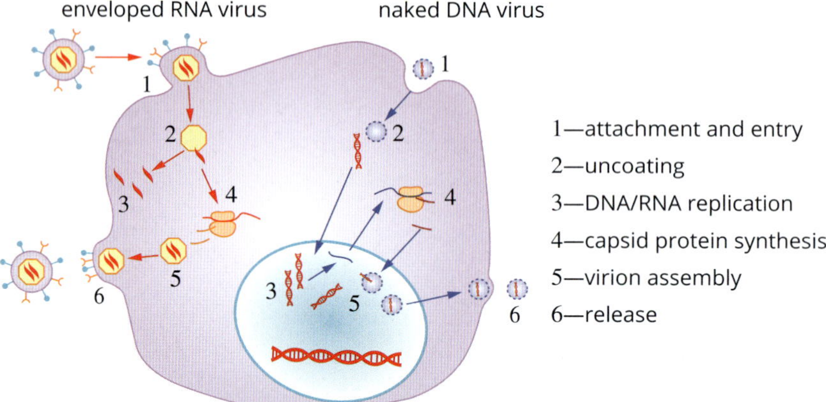

FIGURE 13.3.6 Targets for antiviral drug development. Any one of the key steps 1 to 6 in virus replication is a potential target for designing an antiviral drug. The illustration represents simplified life cycles for an enveloped RNA virus and a naked (non-enveloped) DNA virus.

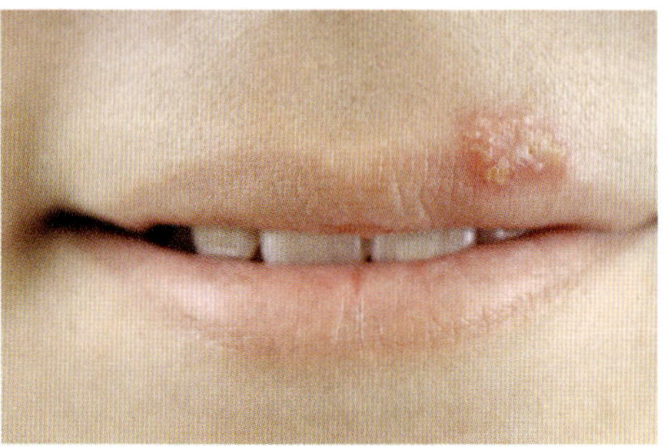

FIGURE 13.3.7 Cold sores caused by herpes simplex virus often appear around the mouth, nose and eyes. Aciclovir specifically targets herpes virus replication.

DNA viruses

DNA viruses have a genome of DNA. Some use the host DNA polymerase while others have a gene for DNA polymerase, which may be a target for antiviral drugs. DNA viruses include the herpes simplex virus that causes cold sores and varicella-zoster virus that causes chicken pox. The antiviral drug aciclovir lessens the symptoms by blocking the production of the herpes DNA polymerase and therefore blocking replication of the virus. Aciclovir comes as an ointment that is applied to cold sores (Figure 13.3.7). It is also used in tablet form for other herpes viruses, such as herpes zoster, which causes shingles.

RNA viruses

RNA viruses have a genome of RNA. They have their own RNA replication enzymes, which may be a target for antiviral drugs. Examples of RNA viruses are those that cause influenza, mumps, hepatitis C, dengue fever and Ebola. Experimental drugs targeting viral RNA, to block translation, are being trialled against dengue virus and hepatitis C virus.

Some RNA viruses use an enzyme called reverse transcriptase to produce DNA from their RNA. These viruses are called **retroviruses**, and drugs used to treat them are called antiretroviral drugs. They also have an enzyme called integrase, which inserts, or integrates, the retroviral DNA into a host chromosome where it stays permanently, making such viruses very difficult to treat. In addition to the antiviral targets mentioned above, antiretroviral drugs can work by specifically inhibiting the reverse transcriptase and integrase enzymes (Figure 13.3.8).

The human immunodeficiency virus (HIV) is an example of a retrovirus. Using a number of antiretroviral drugs together, often called a 'cocktail' of drugs, has been shown to be more effective against HIV than using a single antiretroviral drug. A combination of drugs, called combination antiretroviral therapy or highly active antiretroviral therapy, has improved the health and life expectancy of people with HIV/AIDS and reduced their chances of transmitting the virus to others.

The cocktail includes inhibitors of reverse transcriptase, DNA polymerase, and the enzyme that allows the virus to exit from the cell.

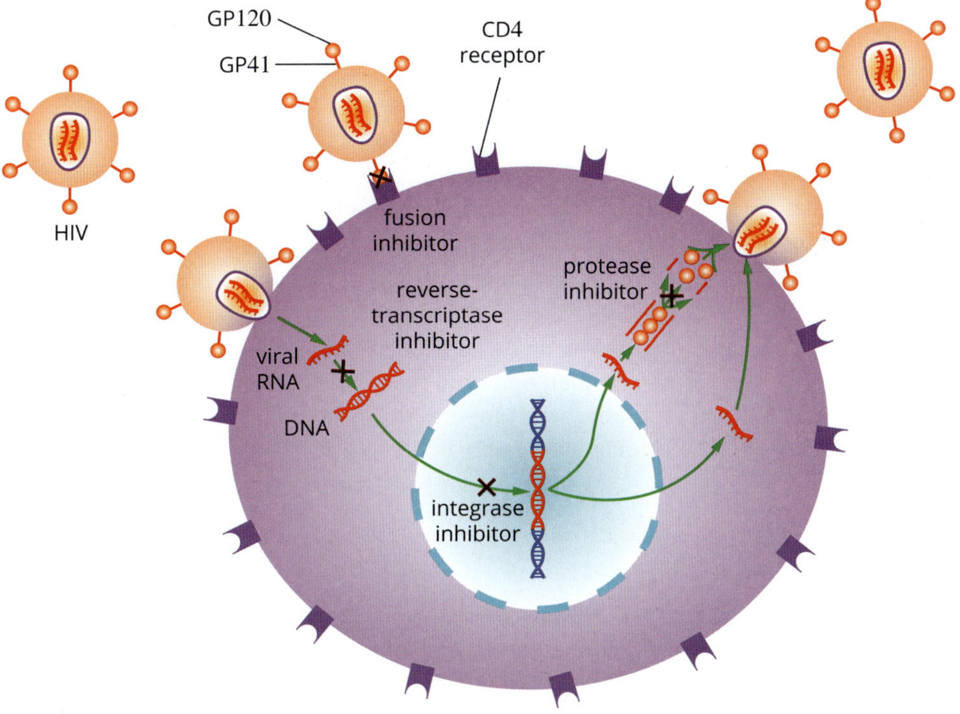

FIGURE 13.3.8 Targets for antiretroviral drug development, illustrating HIV

+ ADDITIONAL

Rational drug design

Rational drug design is a targeted approach to designing new drugs. It involves analysing the structure of a pathogen or disease-causing molecule and using this information to design a drug that will mimic or block the action of the disease-causing agent.

Designed drugs have complementary shapes and charges to the active sites of the pathogen or molecule they are targeting. Using methods such as X-ray crystallography and other sophisticated imaging techniques, scientists can identify the detailed structure of the active site or receptor-binding site on a molecule. It is then a case of finding an existing drug from databases or manufacturing a new drug that has a complementary shape to the active site or receptor site. Scientists can then test the interaction of the drug with the target molecule in the laboratory.

Figure 13.3.9 summarises the process of rational drug design. Examples of drugs developed using rational drug design are Relenza to treat influenza, Gleevec to treat myeloid leukaemia, and selective serotonin re-uptake inhibitors (SSRIs) to treat depression.

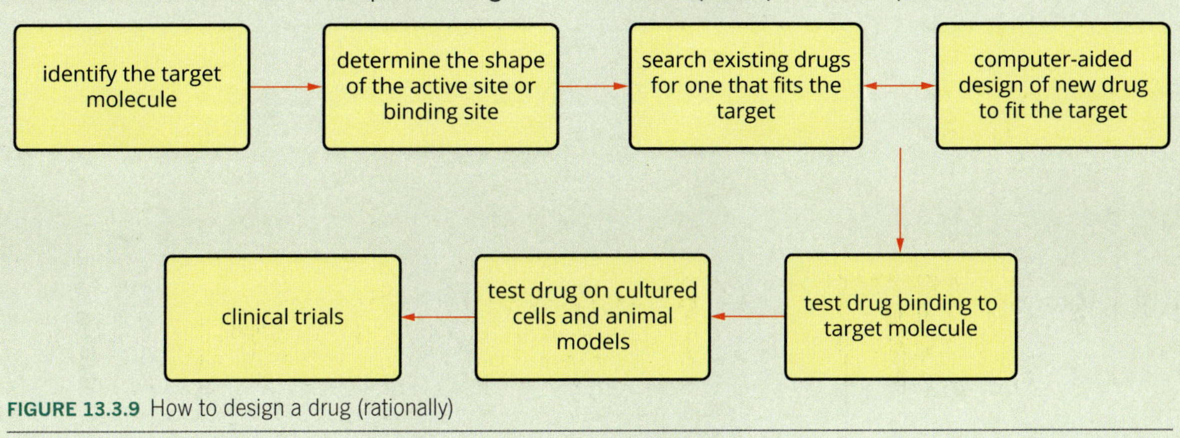

FIGURE 13.3.9 How to design a drug (rationally)

13.3 Review

SUMMARY

- Antiseptics and disinfectants control the number of pathogenic organisms or agents in the environment and on surfaces.
- Antibiotics are natural or synthetic molecules that slow bacterial growth (bacteriostatic) or kill bacteria (bactericidal).
- Antibiotics affect different aspects of the bacterium—cell wall, cell membrane, protein synthesis, DNA or RNA synthesis or metabolic pathways.
- Antiviral drugs target proteins for virus entry into cells, replication within cells or exit from cells.
- Antiretroviral drugs are specific for retroviruses, such as HIV—they inhibit specific enzymes such as reverse transcriptase.
- Bacteria, protists, fungi and viruses develop resistance to antimicrobial drugs.
- Drug resistance arises by mutation and transfer of these new gene variants.
- Rational drug design specifically develops a drug to be complementary to the shape and/or charge on a target molecule—the drug fits the active site or binding site to prevent its action.

KEY QUESTIONS

1 Explain how antiseptics and disinfectants differ from antibiotics.

2 List several ways in which antibiotics slow down or kill bacteria.

3 List some ways in which antiviral drugs may limit the spread of a virus.

4 Describe the cause of cold sores and how they may be treated.

5 HIV/AIDS is caused by a retrovirus.

  a How is this type of virus different from other viruses?

  b If you were designing a drug to treat HIV/AIDS, which molecule(s) would be a good target?

6 How does antibiotic resistance arise and spread through bacterial populations?

7 List some mechanisms in bacteria for antibiotic resistance.

13.4 Controlling epidemics and pandemics

The outbreak of an infectious disease on a local, regional or global scale can occur when a disease is identified in greater numbers in a population than is generally expected. The outbreak may last from days to years and affect many people, or just be a single case. For example, two people diagnosed in Sydney, NSW with tuberculosis may be called an outbreak, as the disease is very rare in NSW. However, a strain of influenza that affects 850 000 people in Sydney may be also described as a disease outbreak due to its higher than normal seasonal prevalence.

An epidemic occurs when an infectious disease becomes widespread within a local or regional level at a particular time. For example, the cholera epidemic that broke out in Haiti in 2010 claimed over 7000 lives—the epidemic is yet to be eradicated. In contrast, a pandemic is a global outbreak of an infectious disease. An example is the Spanish influenza outbreak that killed 40 to 50 million people worldwide in 1918.

While many consider epidemics and pandemics to be confined to history, along with the bubonic plague, the threat of modern-day epidemics and pandemics remains a reality. This was demonstrated by the 2014–2017 Ebola outbreak. With increasing globalisation and movement of people, animals and goods across borders, the risk of disease outbreaks continues to rise and requires monitoring and controls.

> An epidemic is the sudden increase in the number of cases of a disease above what is normally expected in that population in that area.

> A pandemic is an epidemic that has spread over several countries or continents, usually affecting a large number of people.

OUTBREAK CONTROL

Herd immunity

One of the most effective controls against the outbreak of infectious diseases at local, regional and global levels are immunisation programs. Immunisation is critical, not only for the person immunised, but also for the health of the wider community. For an immunisation program to be successful, enough people need to be vaccinated—a phenomenon called **herd immunity** (Figure 13.4.1). The more people who are vaccinated, the less chance there is of an infectious agent spreading throughout a population, because there will be fewer potential carriers. Herd immunity is essential for the protection of those who cannot be vaccinated or who have suppressed immune systems. This includes newborn babies, the elderly, people suffering from an immune disease and people taking immunosuppressant medication.

Breakdown of herd immunity—whooping cough

Immunological memory reduces over time, thereby reducing the herd immunity of immunised populations. An example of the breakdown in herd immunity is the recent spike in Australia in cases of whooping cough, a disease caused by the bacteria *Bordetella pertussis*. Although it only causes a persistent cough in adults, approximately one in 200 babies under the age of six months who become infected will die. Babies cannot be vaccinated until they are six weeks old and they are not fully protected by this vaccine until about six months of age.

One of the reasons for this breakdown in herd immunity is that not enough people get booster vaccinations. A public education campaign has been implemented to encourage adults to receive a booster vaccination to maintain herd immunity against whooping cough. New parents are offered the booster vaccination when their baby is born, and are encouraged to recommend the vaccination to family and friends who will be in close contact with their baby.

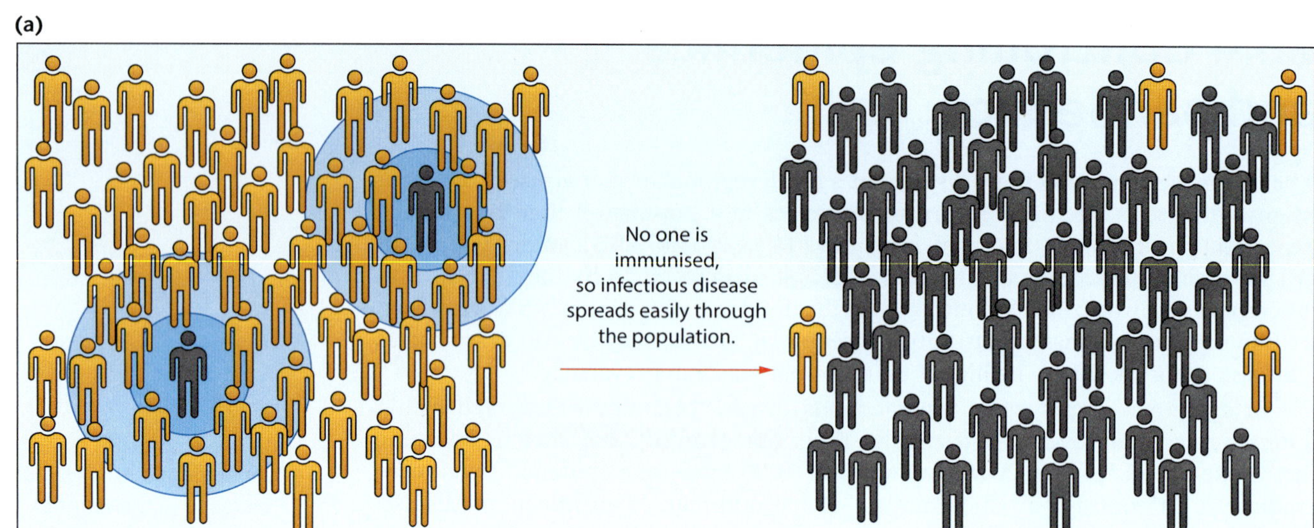

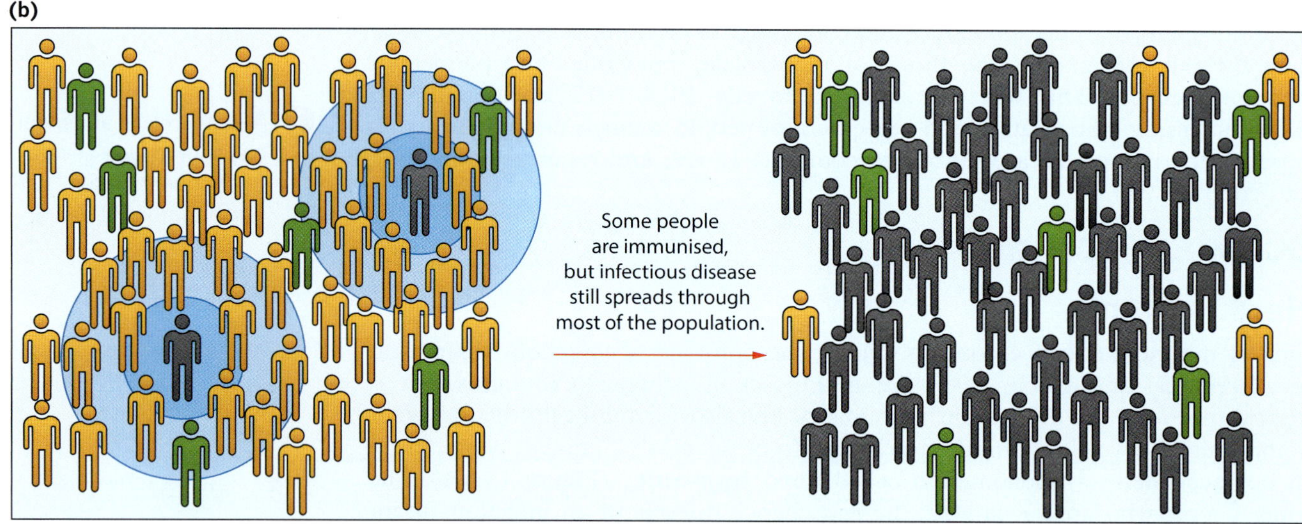

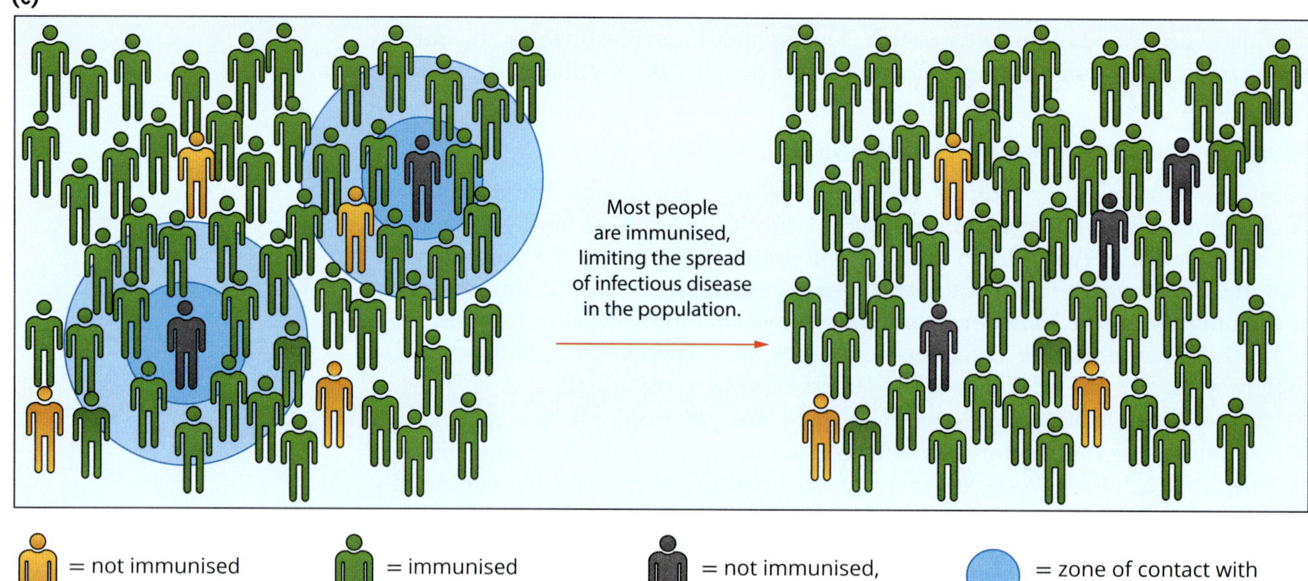

FIGURE 13.4.1 Three diagrams illustrating the effectiveness of herd immunity (a) with no immunisation in a community, infectious diseases spread easily; (b) with some community immunisation, infectious diseases spread less easily; (c) when most of the community is immunised, there are few carriers or infected people and minimal spread of infectious diseases. This is known as herd immunity.

Environmental management

Environmental management is imperative for effective disease control, with many environmental factors such as water supply, air quality, sanitation facilities and food sources influencing the spread or control of infectious disease (Figure 13.4.2). Effective environmental management can limit the spread of disease outbreaks and improve the health of those affected. Conversely, uncontrolled environmental factors can facilitate the spread of an outbreak.

Water supply

Contaminated water can facilitate the growth of water-borne diseases such as cholera, while unclean water can spread parasites such as *Cryptosporidium* sp., which lead to disease. Water supplies often become contaminated during emergency situations, such as during natural flooding events, or in less developed countries where water supply infrastructure is not sufficient. Further, diseases spread by vectors such as mosquitos can thrive in water supplies. To prevent or control an infectious disease outbreak, water supplies should be protected by:

- boiling water before drinking
- chlorinating water
- importing contained water from safe supplies
- sealing, containing, and/or draining water bodies to control the spread of disease.

Food sources

Similarly, food supplies can often become contaminated and act as an agent of transfer for infectious diseases. To prevent or control an infectious disease outbreak, food supplies should be protected by:

- limiting food preparation to those who are not infected by the contagion (and have not been in contact with those infected)
- following proper hygiene practices when handling food (e.g. hand washing)
- importing food from safe supplies
- storing food in sealed, temperature-controlled containers
- disposing of all affected food items and avoiding consumption if food is an agent of disease transfer (e.g. a batch of food contaminated by *Salmonella* sp.)
- avoiding sharing food with others.

FIGURE 13.4.2 Sanitation kits, sealed water and food are often distributed by relief organisations following a natural disaster to avoid the spread of disease.

Sanitation

Adequate sanitation facilities control the outbreak of disease by removing human waste from areas of habitation, as well as hygienically disposing of food and animal waste, along with other rubbish. In states of emergency, or in less developed countries, sanitation services and infrastructure are often not available. Adequate sanitation for the control of disease includes:

- sealed and functioning sewage systems
- safe disposal of animal waste, food scraps, and rubbish (e.g. removal of rubbish from towns to a sealed, secure rubbish tip)
- safe, secure disposal of hospital and health services waste.

Air quality

Poor air quality and ventilation can facilitate the transfer of infectious diseases that are spread by airborne droplets (e.g. Legionnaire's disease or influenza can be spread by sneezing or coughing). Polluted areas, such as cities with smog, can exacerbate the spread of disease, while contained areas such as buildings should have adequate ventilation to ensure the dispersement of airborne pathogens. Some infectious diseases are so highly contagious that hospitals will put patients in air-locked rooms with separate ventilation systems.

Human behaviour

Controlling the outbreak of an epidemic or pandemic becomes more complex in an increasingly globalised world, with human behaviour favouring increased travel, communication and trade. Controlling the spread of an epidemic or pandemic often requires the following:

- fast identification of the disease outbreak (by health workers and the general public)
- local, regional and global disease alerts (new technologies are helping facilitate fast communication, including social media and online news programs)
- travel bans to and from regions of high disease prevalence
- protective behaviour (e.g. using protective clothing and face masks, good hygiene practices, self-induced quarantine).

BIOLOGY IN ACTION  

Control of Ebola in West Africa

Usually, multiple factors contribute to an epidemic or pandemic. Existing pathogens may emerge in new locations and among populations that lack the ability to recognise and deal with it early enough to prevent it spreading.

In 2014, an epidemic of the Ebola virus occurred in Western Africa, affecting over 11 000 people in multiple countries. Ebola was first identified in 1976 and, although this was the largest and longest outbreak, Ebola had affected small numbers of people previously at least 26 times, mainly in equatorial African countries.

It is thought that Ebola entered the human population when local people ate bat meat. Fruit bats are one of the known natural reservoirs of Ebola as they can carry the virus without showing any symptoms of disease. The Ebola virus is spread through bodily fluids and has a relatively short incubation period of as little as two days.

The 2014 epidemic was an example of an old virus in a new context. Equatorial African countries with a history of local Ebola outbreaks had the experience to recognise it early, the laboratory facilities to identify it and the resources to contain an outbreak. West African countries, however, had none of this experience. Two significant factors that facilitated the spread of the disease included poor environmental management and no effective quarantine measures.

Environmental management

Fruit bats are believed to have been the original hosts of the virus, with land degradation and deforestation possibly causing a higher concentration of hosts carrying the virus to congregate in a smaller area. Agriculture now shares land with previously native forests, and bat species such as the straw-coloured fruit bat (*Eidolon helvum*) often feed on fruit crops, possibly leading to transmission of infection when humans eat the crops.

The virus, which was restricted to rural areas in previous outbreaks, reached urban centres where the disease spread rapidly in urban slums (Figure 13.4.3a).

Quarantine

West Africa has a highly mobile population, with people moving daily across borders to look for food, to travel to work, or to visit friends and relatives. However, while the population moved, health responders did not. As the situation in one country began to improve, patients seeking treatment would cross borders re-infecting a recovering population (Figure 13.4.3b). Some African cultures return people to their native village to be buried with their ancestors, further promoting the movement of the disease.

FIGURE 13.4.3 (a) Environmental management control was poor during the Ebola outbreak, with the outbreak escalating as it infected urban slums. (b) A sign warning people that the area is Ebola infected, Africa.

Quarantine

As discussed earlier in the chapter, quarantine separates and restricts the movement of people, animals and materials that may spread infectious disease. Quarantine stations globally have the power to detain, medically examine and conditionally release people and animals. Automatically quarantinable diseases include those with high transmissibility rates, and those capable of reaching epidemic or pandemic prevalence, such as tuberculosis, smallpox and cholera, among others. During epidemics, governments may restrict the movement of people, animals, plants and biological products into and out of the affected countries.

PATTERNS OF INFECTIOUS DISEASE

Epidemiology is the study of the patterns, causes and effects of health and disease in populations. It helps shape preventative health care, control measures and identify risk factors. The accuracy of the measures of the disease depends on accurate diagnosis and reporting of the disease.

Incidence

The incidence of disease is the rate of occurrence of new cases, which indicates the risk of people contracting the disease. Incidence is expressed as a fraction of the population within a period of time. For example, during the years 2000 to 2011 there were 990 new cases of measles in Australia. Therefore, the average annual rate was 0.4 per 100 000 population. The measure of incidence can also categorise subsets of the population, for example between the years 2000 and 2011, children 0–4 years had a measles incidence rate of 1.6 per 100 000 population.

Prevalence

The prevalence of disease measures the proportion of cases in the population at a given time, which indicates the spread of the disease. For example, on 1 January 2017, out of 10 000 people in town A, 40 have measles. The prevalence of measles in town A is calculated as 40/10 000 = 0.004 or 0.4%. You will learn about epidemiology and non-infectious diseases in Chapter 16.

GO TO ➤ Section 16.1 page 552

The mobility of populations and immunity

The increase of globalisation, as well as air travel facilitating the fast movement of people across geographic areas in shorter times, increases the risk of people's exposure to infectious diseases. Travellers (Figure 13.4.4) transmit and disseminate disease—travellers almost exclusively introduced HIV to most areas of the world. The prevalence of disease varies from place to place, and for this reason, many public health campaigns urge travellers to have vaccinations before travelling abroad, and to become informed about risk factors such as food and water quality.

FIGURE 13.4.4 The high frequency of modern travel facilitates the fast spread of disease and exposure of travellers to new diseases.

BIOFILE 

Dengue fever in Southeast Asia

Dengue fever is caused by the *Flavivirus*, which is transmitted by a vector (predominantly the *Aedes aegypti* (Figure 13.4.5) and *Aedes alpopictus* mosquitos). The infection manifests in flu-like symptoms with potentially fatal complications such as dengue haemorrhagic fever and dengue shock syndrome. There is no targeted treatment or vaccine for dengue fever, and so the only way to control the disease is by limiting the population of the *Aedes* mosquitos.

Prevalence of the disease in Southeast Asia increased following World War II when ecological and demographic changes led to the transport of *Aedes* to new regions, and an increase in the number of hosts. Further, during the war, containers of stored water and discarded junk were more common and provided an ideal breeding reservoir for the mosquito. Post-war, the rapid urbanisation of Southeast Asia meant that infrastructure such as sewerage systems were inadequate and left humans susceptible to infectious disease.

As Table 13.4.1 demonstrates, dengue fever epidemics are cyclical. During the years 2000 to 2003, the highest incidence of dengue fever occurred in 2001, and the highest prevalence occurred in Thailand.

FIGURE 13.4.5 The *Aedes aegypti* is a common vector of dengue virus

TABLE 13.4.1 Incidence of dengue fever in Southeast Asian countries, 2000 to 2003

Country	Cases per 100000 population			
	2000	2001	2002	2003
Cambodia	25	75	20	0
Malaysia	35	50	100	112
Thailand	35	150	125	70
Vietnam	35	50	50	35

Worked example 13.4.1

CALCULATING PREVALENCE—WHOOPING COUGH IN NSW

The prevalence of disease measures the proportion of cases in the population at a given time, indicating how widespread the disease is. From June 2010 to June 2011, there were 22000 cases of whooping cough (pertussis) in NSW. With a population during that time of approximately 7382200, we can calculate the prevalence of pertussis in the NSW population from 2010 to 2011.

Calculate the prevalence of pertussis in the NSW population from June 2010 to June 2011.

Thinking	Working
Identify the population size.	7382200
Identify the number of new cases.	22000
Calculate the prevalence rate.	22000 / 7382200 = 0.00298 0.00298 × 100 = 0.298%

The prevalence rate for pertussis in NSW between 2010 and 2011 was 0.298%.

Worked example: Try yourself 13.4.1

CALCULATING PREVALENCE—WHOOPING COUGH IN NSW

In 2015, NSW had a population of 7620000. During this year there were 12240 reported cases of pertussis in the state.

Calculate the prevalence of pertussis in the NSW population in 2015.

STRATEGIES TO PREDICT AND CONTROL THE SPREAD OF DISEASE

The prediction, monitoring and control of infectious disease dates back at least 2500 years, with management strategies and knowledge of disease control varying over time, and between cultures.

Prediction and monitoring

Epidemiology dates back as far as Hippocrates, with his suggestion that environmental and human factors influence the development of disease. In 1662, John Gruant published a quantitative analysis of patterns of disease among populations, and in 1800 William Farr added to the work by classifying disease and reporting to health authorities and the general public. In 1854, John Snow conducted a famous study to discover the cause of two outbreaks of cholera, and prevent their recurrence. By the twentieth century, biomedical and technological advancements meant that scientists could accurately test for the presence of disease using techniques such as serological and molecular testing.

Surveillance of infectious diseases involves detecting disease and notifying public health organisations at local, regional and global scales. Public health surveillance can vary between cultures, with less developed countries often not having public health bodies to monitor and control disease, or not having the resources for effective management. Further, identifying and treating diseases can vary from culture to culture. For example, some cultures believe that disease is caused wholly or in part by supernatural forces and so seek spiritual help rather than medical. Also, many family-oriented, or private cultures, will treat the ill without medical assistance. Both examples demonstrate reasons why disease incidence may not be reported.

Historical monitoring—the Indigenous Australian smallpox epidemic

In 1789, a smallpox epidemic devastated Indigenous Australian populations in NSW, particularly in the Sydney area. The epidemic soon spread to other states and territories. It is estimated that between 50 and 90% of individuals were killed in the affected areas. The fatalities had devastating effects on social structures within Indigenous clans, with Elders, pregnant women and young children most affected. This historic epidemic remains a unique example of the outbreak of an infectious disease. Historical diary entries provide evidence of the limited monitoring of the outbreak—conflicting theories about the outbreak's origin and control continue to be debated.

Outbreak

Theory 1—British 'accidental' introduction

British authorities excluded any person exhibiting smallpox symptoms from boarding vessels of the First Fleet, and harbour authorities at ports along the passage (such as Cape Town) would have also reported any cases of disease. However, the small pox virus can be retained in materials such as blankets, which may have been present in the fleet. This could have easily caused an epidemic if any material came into contact with local Indigenous peoples. It was noted, though, that none of the forty children born during the voyage or during settlement exhibited traces of the disease, which would have been likely if smallpox was present in any form.

Theory 2—British 'deliberate' introduction

The journal of marine captain Watkin Tench details the First Fleet carrying samples of smallpox, supposedly to mildly expose young children to promote immunity to the virus. Many authors and historians have claimed that smallpox was deliberately released by authorities, or rogue marines or convicts. It is claimed that the marine force was not strong enough to protect the settlement and supervise convicts, and that the strength of Indigenous Australian opposition was inhibiting the expansion of settlement to more fertile areas such as Parramatta. Indigenous Australian tradition supports this theory, with details of the outbreak told in traditional songs. The deliberate introduction of smallpox for military advantage was promoted in 1763 by the British forces who distributed contaminated smallpox blankets to Native Americans.

Theory 3—Macassan introduction

During the 1700s, Macassan people from Indonesia temporarily camped and fished along the north Australian coastline. It is believed that smallpox may have been introduced to local Indigenous Australian communities this way. Despite frequent smallpox epidemics in the East Indies, Dutch medical personnel only reported smallpox in Macassar in 1789, after the outbreak in Sydney. Further, there was no evidence of pock marks on Indigenous Australians along the traditional trade routes spreading from northern Australia. However, journal entries by early settlers in 1789 report finding, providing medical assistance to and quarantining Indigenous Australians affected by the virus in Sydney cove. Further, diarists noted that the Eora people of Sydney already had a name for the virus—'gal-gal-la'.

Control

While little is known from historic records, it is believed that little was done to provide medical aid to Indigenous Australians infected with smallpox, and the epidemic lasted until at least 1790. Indigenous individuals were not quarantined until 1837, when populations were placed on reserves with one of the purposes to stop the spread of infectious diseases to non-Indigenous persons. With up to 90% of infections fatal, it is believed that the virus 'ran its course' and stopped transmitting between people.

SKILLBUILDER CCT ICT

Global health maps

Global health maps present geographic location data on infectious diseases. National and international bodies use monitoring and surveillance to collect this data. Medical health personnel, public health regulators and private citizens can then use the data, which is presented in the health maps.

Map data

Global health maps can show:

- incidence and prevalence of disease according to geographic location
- changing patterns of disease (Figure 13.4.6)
- mortality rates according to geographic location
- designations of medical aid.

Sourcing maps

Public health and government organisations, such as the World Health Organization, have websites with interactive global health maps. These are updated with the latest data on current epidemics and pandemics.

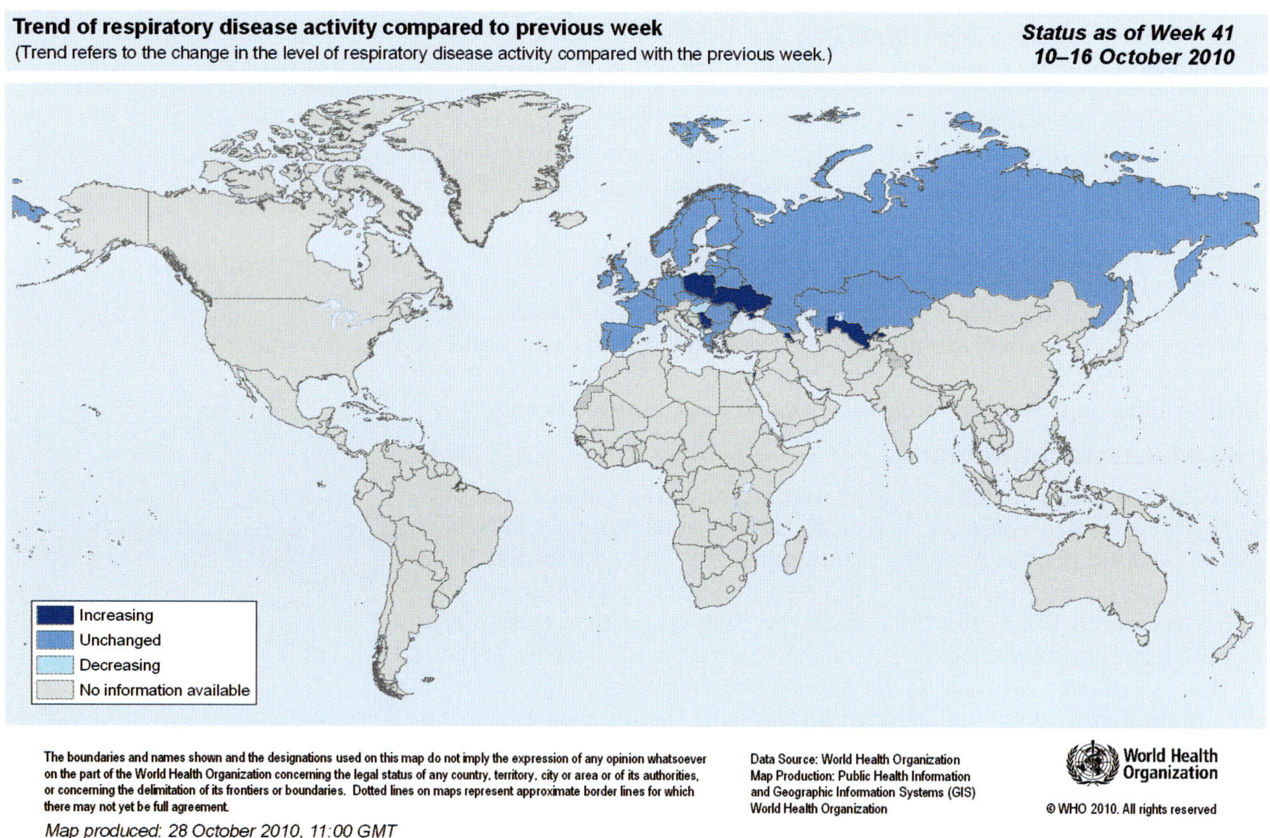

FIGURE 13.4.6 World Health Organization world map showing the trend in respiratory disease activity compared to the previous week.

BIOFILE CCT

The 'father of epidemiology'

When a cholera epidemic broke out in London in 1854, physician John Snow began an investigation. He hypothesised that water was the source of infection, and developed the 'spot map', or geographic distribution of cases, to map incidence of disease. By mapping the location of local water pumps on the spot map, he identified the relationship between prevalence of cholera and water supply. After successfully identifying the source of the outbreak, area officials removed the water pump and the epidemic ended. Snow successfully identified that water could act as a reservoir for disease and that epidemiological studies could inform public health strategy and action. A John Snow pub (Figure 13.4.7) exists in London, with a plaque commemorating Snow's work.

FIGURE 13.4.7 The John Snow pub in London commemorates his pioneering epidemiological work.

BIOLOGY IN ACTION  

Rabies in Southeast Asia

The rabies virus is transmitted to humans through animal bites, with Southeast Asian countries having some of the highest prevalence of the disease worldwide. It is estimated that 23 000 to 25 000 people die from the disease each year. While it is preventable with pre- and post-exposure vaccines, if the disease takes hold, fatality is almost inevitable.

Within Southeast Asia, the disease is perceived as rare and medical attention is not given a high priority. In addition, treatment is controlled by two government sectors, leading to a lack of agency and coordination. There is no national control program in most countries and canine populations continue to rise without an increase in canine vaccination.

However, two countries have proven how effective well-coordinated control programs can be. Sri Lanka and Thailand have recorded a sharp decrease in the number of rabies deaths, implementing mass dog vaccination programs (Figure 13.4.8) and improved access to human pre- and post-exposure vaccines. This proves how a deadly infectious disease can be prevented with the right prevention and control methods.

FIGURE 13.4.8 Rabies vaccination teams vaccinate local dogs in Bali.

13.4 Review

SUMMARY

- The outbreak of an infectious disease on a local, regional or global scale can occur when a disease is identified in greater numbers in a population than is generally expected.
- An epidemic is the rapid spread of a disease to a large number of people.
- When the spread of the disease reaches global proportions, it is known as a pandemic.
- Herd immunity is the result of large numbers of people being immune to a pathogen, reducing the chance of the pathogen successfully spreading through a population.
- Environmental management is imperative for effective disease control, with many environmental factors such as water supply, air quality, sanitation facilities and food sources influencing the spread or control of infectious disease.
- The study of the patterns, causes and effects of health and disease in populations is known as epidemiology. It helps shape preventative health care, control measures and identify risk factors.

KEY QUESTIONS

1. Distinguish between an epidemic and a pandemic.
2. Describe herd immunity.
3. Provide an example of an environmental factor that can be managed to control the spread of disease during an epidemic/pandemic.
4. Differentiate between the incidence and prevalence of a disease.
5. Describe the significance of John Snow's work to disease monitoring and control.
6. Provide an example of a program from outside Australia that has been used to eradicate disease.

13.5 Indigenous Australian medicines

Traditional medicine includes the knowledge, skills and practices of indigenous cultures to prevent, diagnose and treat physical and mental ailments. Knowledge of Australian bush medicine, in comparison, remains relatively limited. While Indigenous Australian culture dates back at least 65 000 years, traditional knowledge is not generally recorded in written form, but rather through practices such as art (Figure 13.5.1), singing, and dancing ceremonies. Further, little research has been conducted to understand the complexities of Indigenous Australian remedies and practices. In this section you will learn about the some of the contemporary applications of Indigenous Australian medicine and the importance of recognising and protecting Indigenous cultural and **intellectual property**.

FIGURE 13.5.1 While Indigenous Australian culture dates back at least 65,000 years, little is known about bush medicine, with knowledge passed to younger generations through oral tradition, art and ceremony. Rock paintings such as the above cave painting in Carnarvon National Park, is one way knowledge is recorded and shared.

INDIGENOUS REMEDY AND RITUAL

Traditional bush medicine indigenous to Australia encompasses the natural environment and spiritual world, and as such, healing practices often involve treating both physical and spiritual components. Treatment often requires both a spiritual 'doctor' and 'medicine man'. Indigenous Australian terms for these healers include *ngangkari*, *marrnggitj*, and *garraaji*. Healers diagnose, advise, reassure, and perform healing treatments. Before European settlement, all Indigenous Australian adults would have had general knowledge of bush medicine, while healers were considered to have access to spiritual powers and assistance. Indigenous Australians traditionally lived in isolated, dispersed populations and illness was often attributed to supernatural causes, with the protection of an individual's spirit of paramount importance to maintaining good health. As hunter-gatherers, traditional bush medicine varied with seasons, locations and available resources. The natural environment plays a large role in bush medicine, particularly in the use of plants (Figure 13.5.2), which are highly diverse, unique and harbour many pharmacological compounds.

FIGURE 13.5.2 Native plants have traditionally been used in Indigenous Australian food and medicine.

BUSH MEDICINE

Australian flora

Australian flora plays a significant role in Indigenous Australian bush medicine (Table 13.5.1). Plants were prepared and administered or used in healing ceremonies (e.g. smoking ceremonies).

TABLE 13.5.1 Australian flora species are used in many Indigenous Australian bush medicine applications.

Common name	Scientific name	Ailment being treated	Application
kangaroo apple	*Solanum laciniatum/Solanum aviculare*	swollen joints	poultice
goat's foot	*Ipomoea pes-caprae*	pain caused by marine stings	leaves crushed, heated and applied to skin
sticky hopbush	*Dodonaea viscosa*	ear ache	boiled and applied
digging stick tree	*Pemphis acidula*	toothache	tip of the stick burnt and applied to teeth
lemongrass	*Cymbopogon* sp.	fever diarrhoea ear ache	boiled, cooled, applied to skin liquefied and consumed direct contact with ear
snake vine	*Tinospora smilacina*	headaches and arthritis (acts as an anti-inflammatory)	crushed and applied
eucalyptus oil	*Eucalyptus* sp.	aches, pains, fevers and chills	leaf infusion

There are a variety of ways that plants can be used in traditional medicine, including:

- crushing, heating and applying to skin
- boiling and inhaling
- boiling for drinking
- saps applied to skin
- bark smoked and burned
- leaves burnt to a stick and applied to a wound
- mashed up and swallowed
- combined with animal fat for treatment
- chewed
- bruised and pounded to make a poultice.

Research has found that many native flora species contain antibacterial and anti-inflammatory compounds. Australia is also rich in endemic aromatic plants, which have naturally high levels of essential oils. Other medicinal properties of Australia's flora include:

- tannins—chemicals that act as astringents to contract tissues, often used to clean wounds
- mucilage—to soothe inflamed mucous membranes
- latex—containing enzymes that can digest proteins and are often used to treat ulcers and wounds
- alkaloids—highly therapeutic organic compounds, examples include morphine and caffeine.

FIGURE 13.5.3 Tea tree (*Melaleuca alternifolia*), native to NSW, is used in traditional Indigenous bush medicine. Tea tree is now commonly used in commercial products such as cosmetics and household cleaners.

Tea tree oil

The tea tree (*Melaleuca alternifolia*) (Figure 13.5.3) was commonly used by the Bundjalung Aboriginal people from the NSW coast. The tea tree leaves or bark are traditionally crushed and applied as a paste to treat wounds, or brewed and consumed to soothe sore throats. The antiseptic properties of tea tree oil have been scientifically proven, inhibiting the growth of many bacteria and fungi. Tea tree oil is today globally used as a household cleaner, to treat fungal infections of the feet and nails as well as to relieve skin complaints such as minor burns, stings and acne.

Kakadu plum

The Kakadu plum (*Terminalia ferdinandiana*), considered a gift of the Dreamtime, holds the richest source of vitamin C of any food source in the world. Kakadu plums are an Indigenous bush tucker staple—high in folate, vitamin E, iron and antioxidants. The fruit is also used in Indigenous Australian bush medicine as an antiseptic and natural healing agent. Research has discovered that the fruit contains phytochemicals that function as antibacterials, antivirals, antifungals, anti-inflammatories and protect against carcinogens and mutagens. Today, the Kakadu plum is used in global commercial production of cosmetics, vitamin supplements, pharmacological products and food.

FIGURE 13.5.4 Emu bush (*Eremophila* sp.) has been used traditionally in Indigenous bush medicine, and now in modern research, for its strong antibacterial properties.

Emu bush

Emu bush (*Eremophila* sp.) leaves (Figure 13.5.4) are traditionally used to cleanse surface wounds, gargled to treat ailments, or burnt to produce a wet steamy smoke for inhalation. The vapours are believed to inhibit bacterial and fungal pathogens, and to stimulate lactation in breastfeeding mothers. Modern scientific research has discovered the properties of the plant are equal in strength to antibiotics and antibacterial agents. Traditionally, the emu bush was placed on hot embers to produce smoke and sterilise tools in the purification processes of circumcision rituals. Researchers in South Australia are currently trialling the use of emu bush as a sterilisation medium for prosthetic implants.

BIOLOGY IN ACTION AHC

Maroon bush—powerful antimicrobial

The Australian native, *Scaevola spinescens*, commonly referred to as maroon bush, murin murin, prickly fan flower, or currant bush, is a medium-sized shrub with prickly stems, creamy white flowers and a small purple berry (Figure 13.5.5). It grows in most semi-arid parts of Australia and has traditionally been used in Indigenous Australian bush medicine. It is believed to be beneficial in treating heart disease, intestinal problems, urinary and kidney ailments and even cancer. Research has identified the maroon bush's strong antibacterial and antiviral properties. The plant's antibacterial action is so potent that researchers are investigating its possible use against bacteria and fungi that are resistant to existing antibiotics and antifungals, and the plant's potential as a broad-spectrum antiviral against RNA viruses.

FIGURE 13.5.5 The maroon bush's (*Scaevola spinescens*) highly potent antibacterial action is providing hope for future therapeutics, with many strains of bacteria and fungi resistant to existing antibiotics.

Australian fauna and other natural resources

Witchetty grub

The witchetty grub (*Endoxyla leucomochla*) (Figure 13.5.6) is a traditional food source, strengthening the body and promoting healing. It is also used to treat burns by crushing into a paste, applying to the affected area and covering with a bandage.

FIGURE 13.5.6 Witchetty grubs (*Endoxyla leucomochla*) are used in traditional Indigenous bush medicine to promote strength and healing.

BIOFILE AHC

Duboisia species

Walk into any Australian chemist store and you are likely to come across shelves of products containing alkaloids from the *Duboisia* species of plant, an Australian native. These alkaloids are used in antispasmodics, travel sickness tablets and stomach relaxants. However, in traditional Indigenous Australian culture, *Duboisia hopwoodii* was used for its alkaloid, nicotine. The plant material was made into a ball, and chewed like tobacco. It is called 'pituri', and is a highly valuable trading commodity in Indigenous communities (Figure 13.5.7).

FIGURE 13.5.7 (a) *Duboisia hopwoodii* was used in traditional Indigenous communities as a trading commodity, as well as for tobacco. (b) Pharmaceutical companies isolated the alkaloids, which are commonly used today in antispasmodic, stomach relaxant and anti-nausea medications.

Mud, sand, dirt and clay

Sediment is applied to wounds to heal the skin, provide a physical barrier and treat infection. Indigenous Australians also traditionally ingested small amounts of clay to deactivate gastrointestinal toxins caused by the presence of infection.

Desert mushrooms

Desert mushrooms (*Pycnoporus* sp.) are used to treat ailments of the mouth, such as dry lips and sores, as well as used as natural teething rings. The mushrooms are traditionally sucked on to deliver relief.

INTELLECTUAL PROPERTY AND CULTURAL PROTECTION

Indigenous Australian bush medicine holds a wealth of potentially therapeutic products and practices that could be used to improve all aspects of health care. This may be a way to preserve traditional Indigenous Australian bush medicine, however, it also raises the question of how to best protect Indigenous cultural heritage from commercialisation.

British settlers first documented Indigenous Australian bush medicines as they were forced to rely on the bush for remedies when supplies from Britain were scarce. Some early settlers did not consider Indigenous communities to have any sophisticated health or medical systems, while others were in awe of their ability to recover from serious illness and injury. Some historians have claimed that the entire European–Australian medical industry owes its existence to Indigenous Australian bush medicine practices. Many British settlers adopted bush medicine to treat their ailments, such as applying the old man's beard plant (*Clematis microphylla*) (Figure 13.5.8a) as a poultice to relieve joint pain. These remedies soon entered the general health knowledge of bushmen, drovers and others living and working on the land (Figure 13.5.8b). Soon the pharmacological potential of many Indigenous Australian plants was being studied and used, including the corkwood tree (*Duboisia myoporoides*), which was used to dilate pupils during ophthalmic surgery.

In Indigenous communities, knowledge of bush medicine is passed down through generations. Only those who have been properly initiated can practice these skills. However, since European colonisation, significant knowledge of bush medicine has been recorded by historians and scientists. Many pharmaceutical companies have seen this as a quick way to identify resources with potentially therapeutic properties.

FIGURE 13.5.8 (a) Many traditional Indigenous bush medicine remedies, such as using old man's beard (*Clematis microphylla*) to relieve joint pain, were adopted by (b) British settlers and their descendants who lived and worked on the land.

Intellectual property rights

FIGURE 13.5.9 Indigenous bush medicine knowledge is regarded as intellectual property, but is not currently adequately protected by legislation.

Intellectual property refers to ideas, inventions, images and other creations of the mind that can be used commercially. Knowledge of bush medicine is a type of intellectual property, and can also be used commercially (e.g. by pharmaceutical companies). Currently, Australian laws do not adequately protect Indigenous bush medicine intellectual property (Figure 13.5.9), and many Indigenous peoples believe that there is not adequate recognition or protection of their culture.

Many pharmaceutical companies have been given exclusive rights to certain native Australian plants, and have patented certain plant properties. As a result, by law, Indigenous people cannot use the plant species without approval from the drug companies.

However, the protection of Indigenous Australian intellectual property is a complex notion, as traditionally knowledge in Indigenous communities is not 'owned'. Knowledge can be held by a few people, or many people, some knowledge may be secret, or only known by a specific group. This concept is at odds with legal concepts of intellectual property.

Australia is moving towards protection against 'bio-piracy', which is the use of biological resources or traditional knowledge without consent, by signing the Nagoya Protocol, an official agreement to protect Australian biological resources, and those under custodianship of Indigenous peoples. Macquarie University recently partnered with the NSW Yaegl Local Aboriginal Land Council to record traditional medicinal knowledge and research the antibacterial and antifungal properties of native plants. A contract between both the university and council ensures joint ownership of any commercial outcomes of the studies.

Smokebush licencing

The Australian native smokebush (*Conospermum* sp.) (Figure 13.5.10), grown in coastal Western Australia, is used in traditional Indigenous Australian bush medicine because of its healing properties. From the 1960s to the 1980s, the USA's National Cancer Institute was given licensing rights to collect specimens of smokebush for research into its effectiveness in treating cancer and HIV. Research discovered that the smokebush plant contained Conocurovone, a property which has the potential to destroy low concentrations of HIV. The Institute subsequently gave the Victorian pharmaceutical company, Amrad, a global license to patent a product from Conocurovone. Should the product be commercially successful, the Western Australian government and Amrad would earn billions of dollars in profits and royalties. Meanwhile, Indigenous people, who originally discovered the healing properties of this plant, would receive no acknowledgment or financial return.

FIGURE 13.5.10 The smokebush (*Conospermum*), traditionally used in Indigenous Australian bush medicine, has been licensed by the Western Australian Government and a private pharmaceutical company after the discovery of potential HIV-killing properties.

13.5 Review

SUMMARY

- Traditional medicine includes the knowledge, skills and practices of indigenous cultures to prevent, diagnose and treat physical and mental ailments.
- While Indigenous Australian culture dates back at least 65 000 years, little is still known about traditional bush medicine.
- Traditional Indigenous Australian bush medicine encompasses the natural environment as well as the spiritual world.
- Australian flora plays a large role in bush medicine, along with fauna such as Witchetty grubs and the use of sediment and clay.
- Despite bush medicine knowledge being traditionally passed down through generations, historians and scientists that recorded this knowledge allowed pharmaceutical companies to use the healing properties of many native flora and fauna.
- Knowledge of bush medicine is a form of intellectual property but has been used commercially by pharmaceutical companies without adequate acknowledgement or consent.
- While Australia is looking towards better protection of Indigenous intellectual property, many Indigenous peoples believe that there is not adequate recognition or protection of Indigenous culture.

KEY QUESTIONS

1 Provide two reasons why knowledge of Indigenous Australian bush medicine remains limited.

2 Identify some traditional roles of Indigenous Australian healers in bush medicine.

3 Identify two ways plants can be administered to a patient in Indigenous Australian bush medicine.

4 Describe the traditional Indigenous Australian bush medicine and modern commercial uses of the tea tree (*Melaleuca alternifolia*) plant.

5 Explain why the maroon bush (*Scaevola spinescens*) is being researched in the scientific community.

6 **a** What is bio-piracy?
b How does bio-piracy relate to Indigenous Australian bush medicine?

7 Provide an example of a biological resource used in Indigenous Australian bush medicine that has been exploited by commercialism.

Chapter review

13

KEY TERMS

active immunity
adaptive immune response
antibiotic
antimicrobial resistance
antiseptic
antiserum
antiviral
bacteria
bactericidal
bacteriostatic
biosecurity
capsid
disinfectant
epidemic
epidemiology
herd immunity
hygiene
immune response
immune system
immunisation program
immunity
immunological memory
infectious disease
intellectual property
lymphocyte
pandemic
parasite
passive immunity
pathogen
peptidoglycan
public health campaign
quarantine
rational drug design
retrovirus
serum
vaccination
vaccine
vector
virion
virulence
virus
zoonotic

REVIEW QUESTIONS

1 Give an example of an emerging disease influenced by factors a–g.

a human demographics, demographic change and/or mobility

b human behaviour

c farming practices

d overuse of an antimicrobial agent

e poor sanitation

f limited social, transport or health infrastructure

g close association between wildlife and domestic animals

2 Q fever is an example of a zoonotic disease. It is caused by the bacteria called *Coxiella burnetii*. In most people it causes a flu-like illness, but in some individuals, it can cause liver or heart disease. One pathway for infection is shown below.

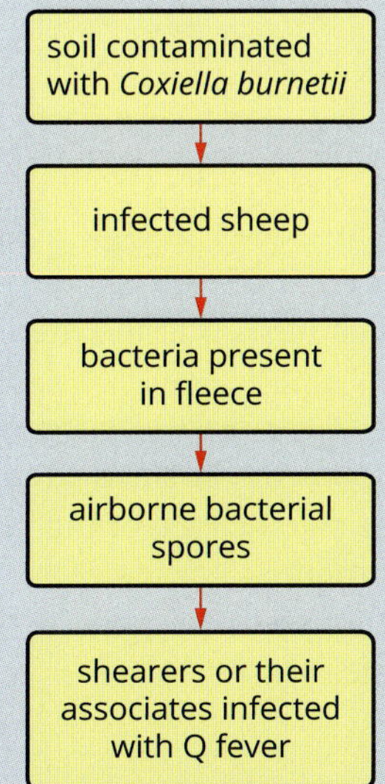

a What is a zoonotic disease?

b Identify three possible practices that could reduce the chance of infection with *C. burnetii*.

3 Pregnant women who are rhesus negative but who may be carrying a rhesus positive fetus will be given injections to protect the fetus from maternal antibodies. These injections make the mother immune to rhesus proteins. What type of immunity is this?

A artificial passive immunity

B artificial active immunity

C natural passive immunity

D natural active immunity

4 Breastfed babies tend to be healthier than bottle-fed babies. Give a reason why.

5 Chickenpox (varicella) is a disease caused by the herpes zoster virus. Vaccination is available and has been a part of the childhood immunisation schedule in Victoria since 2005. The vaccine is a live virus vaccine that is currently administered at 18 months of age.

a The vaccine contains a live virus. Explain why this virus does not cause the vaccine recipient to develop chickenpox.

b Before the vaccination program, it was common for the eldest child in a family to develop chickenpox in their kindergarten or first year of school. Following the eldest child's infection, the younger child would then develop the disease, but babies in the family rarely caught the disease and even if they did it was generally very mild. Explain this observation.

c Since introducing chickenpox vaccine to the immunisation schedule there has been an increase in the incidence of chickenpox among adults. Why?

6 On occasion the blood bank will advertise for people who have recently recovered from chickenpox to donate blood. After taking a blood donation from these individuals, the bank will then separate the blood using a centrifuge. The blood serum is collected and the blood cells may be injected back into the donor. The serum is then purified and some of the proteins are extracted.

a What are these proteins?
b These proteins are given to patients—which group of patients is most likely to need them?
c Why does extracting these proteins not expose the blood donors to the possibility of future bouts of chickenpox?
d Explain whether the injection of these proteins gives the recipients long-term immunity.

7 Diphtheria is a potentially fatal disease caused by either *Corynebacterium diphtheriae* or *Corynebacterium ulcerans*. Around 10% of individuals infected will die. For this reason, we have been vaccinating children against diphtheria since 1921. By 1929, contacts of people with diphtheria were being vaccinated. School-based vaccination programs began in 1932.

Diphtheria is still a major health issue in some countries. Fewer than 10 cases have occurred in Australia in the last 10 years, all of which have been linked to overseas travel. The last recorded death in Australia was in 2011.

The graph shows the incidence of diphtheria in Australia between 1917 and 1994.

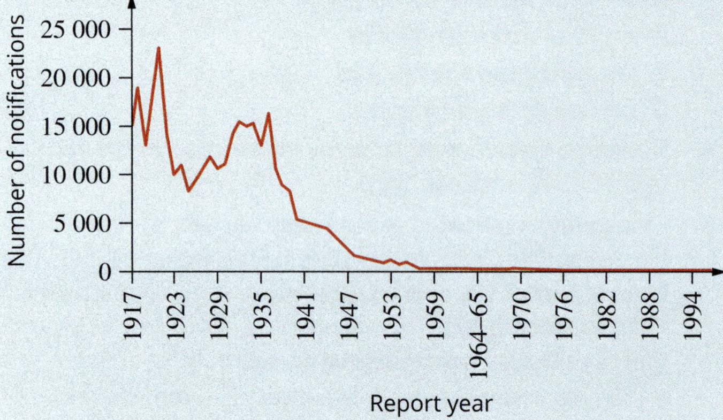

a Using the data in the graph evaluate the effectiveness of the vaccination program in Australia.
b Explain why, despite the high incidence of diphtheria in some overseas countries and the ease of international travel, Australia has had only isolated cases and no outbreaks of diphtheria.

8 In 1969, when astronauts returned from landing on the Moon, they were immediately ushered into quarantine for extensive testing before they could re-enter the community. Explain the reasoning behind this procedure.

9 The use of bactericidal hand-washes has become very common. These hand-washes are types of:
A antiseptics
B disinfectants
C antibiotics
D antivirals

10 Zika virus is a mosquito-transmitted virus that has been linked to an increase in birth defects in South America. It has been postulated that it is a cause of an increase in the number of babies born with microcephaly (a small skull and brain).

There is currently no vaccine or treatment for Zika virus, but considerable research efforts are being directed towards the development of both.

Preliminary research indicates that the virus uses a receptor on cell surfaces called AXL to attach to cells and gain entry. This receptor is one in a family that is very important in cell signalling.

a i Draw a labelled diagram showing a hypothetical AXL receptor with the virus attached to it.
ii What can you say about the shape of the protein the virus uses to attach to the cell and the shape of the AXL receptor?
b How might the spread of Zika virus be limited even in the absence of a vaccine or treatment?

11 A medical student studying the impact of antibiotics on pathogenic bacteria set up the following experiment:
- Three sterilised nutrient agar plates were prepared.
- Plate A remained sealed with nothing added to it.
- Plate B was exposed to bacterial spores, then sealed.
- Plate C was treated with three drops of a common antibiotic and then exposed to the bacteria in the same way as plate B, then it too was sealed.
- All three plates were incubated at 37°C for 24 hours and then examined for the growth of bacterial colonies.

The results are set out in the following diagram.

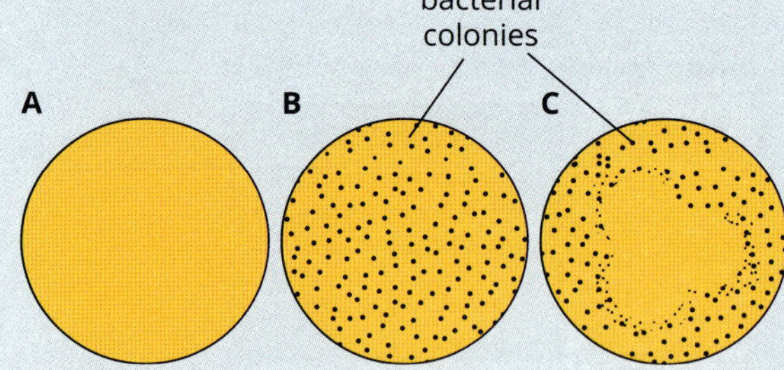

a Suggest a hypothesis that the medical student was testing.
b i State which agar plate was the negative control and which was the positive control.
ii Explain the significance of the positive control in this experiment.
c Explain whether the experimental results support the hypothesis.

12 Classify the following as prevention or cure.
- **a** placing an antiseptic on a fresh cut
- **b** taking a course of antibiotics for a bacterial chest infection
- **c** having a vaccination against cholera before going overseas
- **d** washing hands after going to the toilet
- **e** isolating an exotic bird in quarantine after it has been found in luggage by customs officials

13 How does herd immunity protect unvaccinated newborn babies?

14 Identify two human behaviours that can help control the spread of a disease outbreak.

15 Identify and describe two common medicinal properties of native Australia flora.

16 Choose three plants used in traditional Indigenous Australian bush medicine and complete the table below:

Common name	Scientific name	Ailment treated	Application

17 Explain how sediment is used in Indigenous Australian bush medicine.

18 Some historians have claimed that the entire European–Australian medical industry owes its existence to Indigenous Australian bush medicine practices. Justify this statement.

19 Explain how intellectual property rights can be at odds with Indigenous Australian culture.

20 After completing the Biology Inquiry on page 450, reflect on the inquiry question: How can the spread of infectious diseases be controlled? Discuss how government agencies and research organisations can help prevent the emergence of new diseases that could impact agriculture and wildlife biodiversity.

MODULE 7 • REVIEW

REVIEW QUESTIONS

Infectious disease

Multiple choice

1 Select the list that correctly identifies infectious diseases compared to non-infectious ones.

A Infectious—aneurysm, AIDS, anthrax, anaemia. Non-infectious—cancer, cholera, leprosy, rickets.
B Infectious—melanoma, mumps, meningitis, anorexia. Non-infectious—bulimia, tetanus, warts, hepatitis.
C Infectious—tinea, malaria, scabies, measles. Non-infectious—cancer, diabetes, scurvy, allergies.
D Infectious—asthma, asbestosis, chickenpox, herpes. Non-infectious—Ebola, cystic fibrosis, bronchitis, kuru.

2 Pathogens are organisms that can produce disease or illness in another organism. Select the list that names types of pathogens only:

A prions, bacteria, protozoa, fungi
B parasites, viruses, cancer, prokaryotes
C macroparasites, prions, protozoa, pertussis
D viroids, obesity, oomycetes, clones

3 Persistent infection by a virus is best described as:

A the protein synthesis mechanisms of an infected host cell producing many new virions
B virions dormant inside an infected cell until they break out and trigger symptoms of disease
C the release of new virions from the infected host cell to infect surrounding cells
D the infected host cell producing viral DNA from the virus RNA

4 A non-cellular pathogen that consists of protein particles and can replicate itself inside infected cells without DNA or RNA is called a:

A virus
B virion
C viroid
D prion

5 Koch's postulates provide a scientific method of:

A finding out how fast bacteria reproduce
B identifying the causative organism of infectious diseases
C identifying the causes of non-infectious diseases
D preventing the spread of diseases

6 Antibiotics:

A are produced by B lymphocytes
B trigger the immune response
C combat bacterial infections
D kill mainly viruses and bacteria

7 Which of the following diseases has been eliminated by a global vaccination program?

A smallpox
B polio
C measles
D diphtheria

8 The diagram below identifies some of the steps involved in one method of production of monoclonal antibodies.

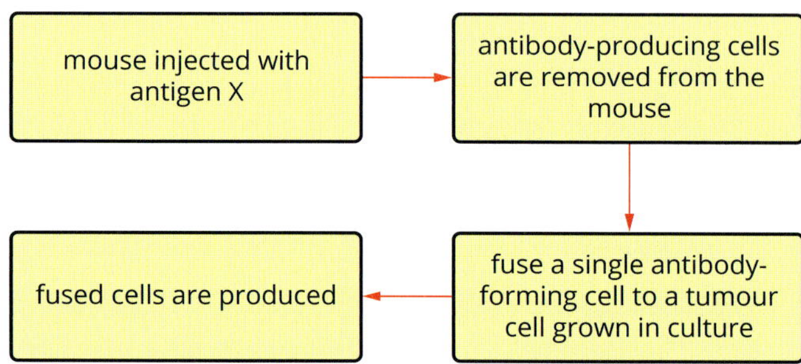

What is the resulting cell called?

A monoclonal antibody
B hybridoma
C antibody
D tissue culture

9 Media reports sometimes refer to large scale diseases as epidemics or pandemics. Identify the correct distinction between the two terms?

A the bubonic plague bacterium caused epidemics and pandemics throughout earlier, less hygienic times
B pandemic is a rapid and unusual spread of an infectious disease; epidemic is a pandemic that has become more global
C epidemic is a rapid and unusual spread of an infectious disease; pandemic is an epidemic that has become more global
D human movements around the globe tend to cause epidemics and pandemics, such as the SARS outbreak

10 Which one of the following statements is correct?

A antibodies are chemicals produced by B lymphocytes
B antibiotics are chemicals produced by T lymphocytes
C antigens initiate the second line of defence
D antibodies, antibiotics and antigens are different words with similar meanings

11 The innate immune system has responses that include:

A cilia, macrophages and B cells
B phagocytosis, inflammation response and macrophages
C macrophages, T cells and B cells
D inflammation response, immune response, antigens

12 Identify the statement that is not true for antibodies:
- A they are large protein molecules called immunoglobulins
- B they have two identical arms that bind to specific antigens
- C they have an arm that interacts with other immune response cells, like mast cells
- D they are produced by neutrophils

13 Which of the following is not true for complement proteins?
- A Complement proteins are part of the innate immune response.
- B Complement proteins attract phagocytes to the site of the infection.
- C Complement proteins include antibodies.
- D Complement proteins can be activated in the presence of antibody–antigen reactions.

14 Natural active immunity is achieved because of:
- A exposure to live or attenuated vaccines
- B infection by particular bacteria or virus
- C the administration of antibodies or antitoxin specific to a particular microorganism
- D adequate breast feeding in newborn infants

15 A pathogen shown below enters a human body and antibodies are produced against it.

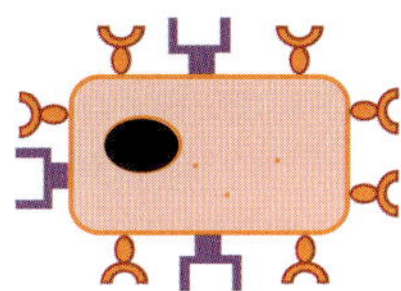

Which antibody would be made in response to this pathogen?

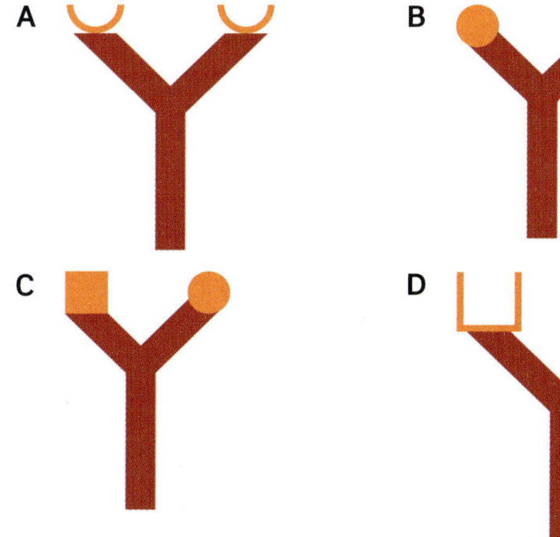

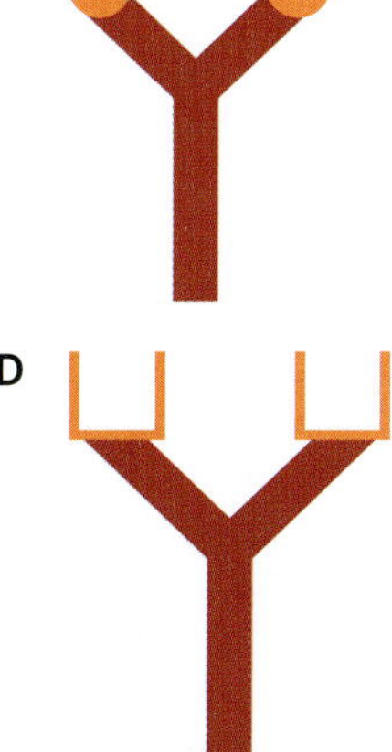

16 Innate immune responses are critical to maintaining the health of an individual because:
- A they are specific to the antigens on pathogenic organisms
- B the innate response produces antigens, which bind antibodies to mast cells
- C they provide immediate and continuous protection against foreign antibodies
- D none of the above

17 Antiviral medications provide treatments to reduce the seriousness of viral infections. Which of the following best describes how they do this?
- A They block the receptors used by the virus to attach to and enter the cells.
- B They block transcription in the infected cells.
- C They prevent the virus particles from leaving the cell.
- D all of the above

18 Different drugs are appropriate for different pathogens. Canestan fights fungi, penicillin fights bacteria, Relenza is effective against influenza virus. The plasmodium that causes malaria is a protozoan. Select the most effective treatment for malaria.
- A Canestan
- B penicillin
- C Relenza
- D none of the above

19 Which of the following involves gene delivery by nasal sprays as a treatment strategy?
- A removal of the genetic material from the adenovirus
- B using bacteria as vectors for the required gene
- C aerosolisation of plasmid DNA
- D all of the above

20 Rational design of a drug to combat a virus involves targeting a particular viral protein. Select the characteristics that would be most useful in designing a drug to combat the virus.
- A The protein is used by the virus in the early stages of its reproduction.
- B The protein is used by the virus to attach to its host cells.
- C The protein is used by the virus in the later stages of its reproduction.
- D The protein is common to several strains of the virus

Short answer

21 Use the list of pathogens supplied to complete a table with information that answers these points:

- **a** Is it a cellular or non-cellular type of pathogen?
- **b** How is this type of pathogen transmitted?
- **c** Does it infect animals, plants or both?
- **d** What are two examples of diseases caused by this type of pathogen?

The pathogens to be included in your table of information are—bacteria, fungi, oomycetes, protozoa, viruses, viroids, prions.

22 **a** Describe how Louis Pasteur tested his theory that microbial contamination can be carried in the air. Include diagrams to support your answer.

- **b** Outline the results of his experiments.
- **c** Was his initial hypothesis supported by his results?

23 Use at least one plant and one animal example to assess the impact that diseases have on agricultural production.

24 The image shows stages in the formation of a blood clot. Identify the type of immune response that includes blood clotting and outline how it helps the human body protect against infection.

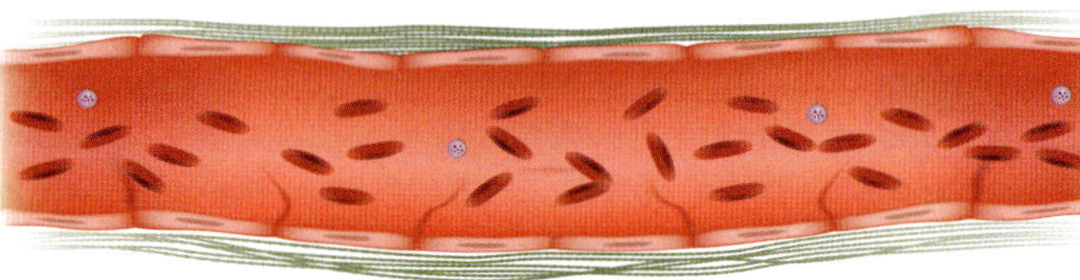

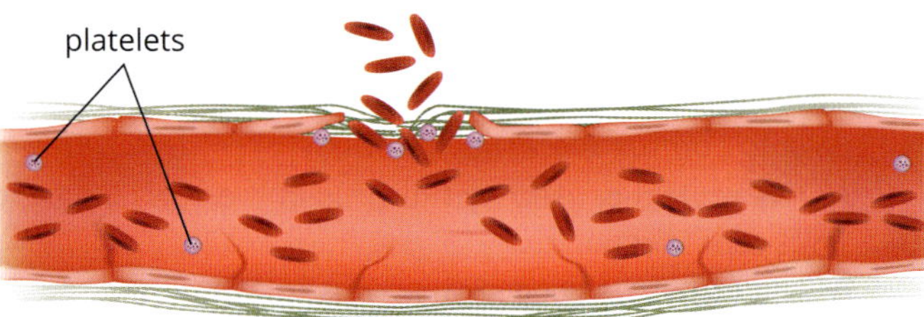

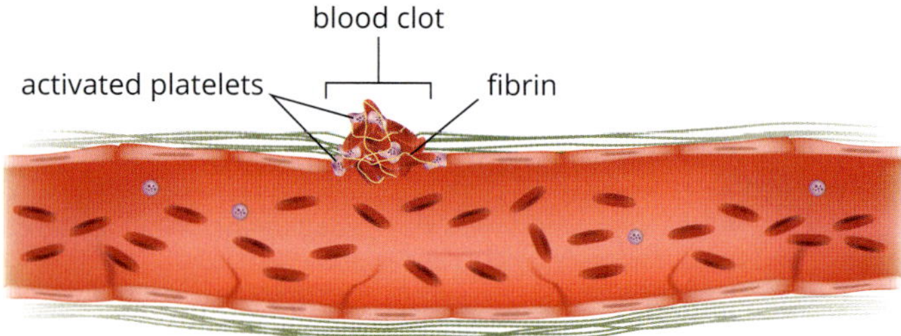

25 The lymphatic system has several important functions, some of which involve the immune system.

- **a** List the major functions of the lymphatic system.
- **b** How do lymph nodes assist the efficiency of the adaptive immune response?
- **c** How do primary and secondary lymphoid tissues differ?

26 **a** Clarify the difference between artificial passive immunity and natural passive immunity. Give examples to support your answer.

- **b** Predict if either of these types of immunity will deliver long-term protection.

27 Assess the importance of herd immunity for programs using vaccination to prevent the spread of infectious diseases.

28 Plants and animals respond to infection by pathogens in a variety of ways.

- **a** Discuss the response of a named Australian plant to a specific pathogen.
- **b** Suggest at least two ways to prevent the spread of the plant disease caused by this pathogen.
- **c** Compare animal responses to infection with the ways that plants respond.

29 The World Malaria Report 2016 from WHO contained the following data about the Southeast Asian region. Use this data to answer the following questions.

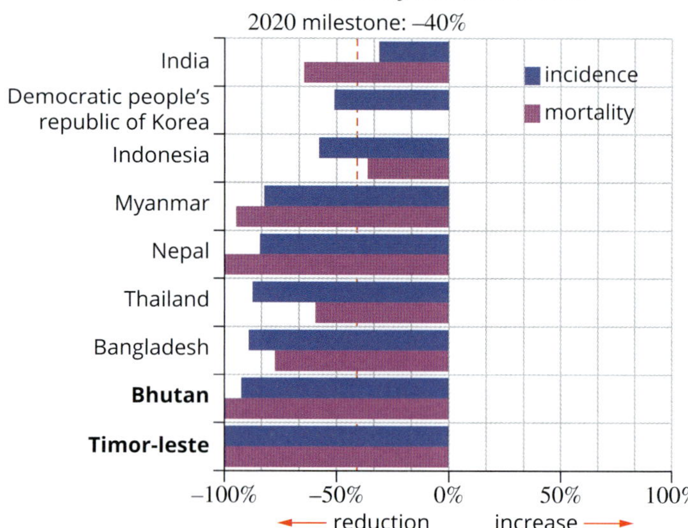

- **a** Which countries reported a decrease in (i) incidence and (ii) mortality for malaria?
- **b** Which countries have already reached the 2020 target of reducing the incidence of malaria by at least 40%?
- **c** Predict if this data gives cause for optimism in the battle against malaria.

30 There are claimed to be alternatives to some contemporary medications, as supplied by today's pharmacies, in the traditional medical practices of Australia's Indigenous people. Identify two named examples of effective bush medicines and discuss protocols for using the knowledge of these materials.

31 Students set up an experiment to test for the presence of microbes in food and water samples. They also wanted to test the effect of an antibiotic on any such microbes.

Nine nutrient agar plates were prepared with a suitable medium for growth of microbes and then sterilised.

Four plates were inoculated with different food samples and labelled F1 to F4. Each plate had three drops of a named antibiotic placed in the centre.

Another four plates were inoculated with water from different sources and labelled W1 to W4. Each plate had three drops of the same antibiotic placed in the centre.

One plate (C) was left sealed and labelled.

All plates were taped closed and incubated for 48 hours at 37°C.

The diagram illustrates some of the experimental results.

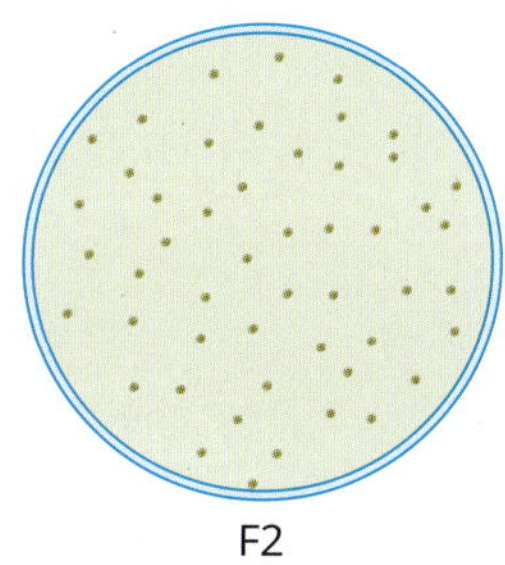
F2

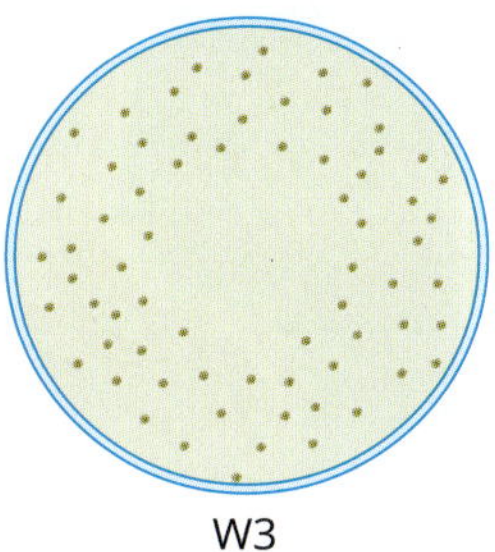
W3

a **i** Why was it important to sterilise the nutrient agar plates at the start of the experiment?
ii Account for the use of four plates for each set of samples.
iii Justify why the antibiotic was placed in the centre of each test plate.
vi Explain the purpose of plate C.

b Write a hypothesis that was being tested in this experiment.

c Assess if the hypothesis was supported by the results.

d Discuss how this investigation could be improved.

32 Tetanus is a serious, often fatal disease caused by the bacterium *Clostridium tetani*. The most serious of the symptoms are caused by the toxin produced by the bacteria. The toxin enters neurons in the central nervous system where it blocks the release of the neurotransmitters glycine and GABA. These neurotransmitters stimulate neural pathways that inhibit the contraction of muscles.

a Explain what effect the toxin will have on muscle behaviour.

b **i** A vaccination is available for tetanus. It involves injection of extremely minute amounts of the toxin. Explain how the injection of this minute amount of toxin gives protection from tetanus.
ii Does this produce active or passive immunity? Provide evidence to support your answer.
iii It is recommended that everyone has a booster vaccination for tetanus every 10 years. Why is this necessary?

c An individual who steps on a rusty nail is at significant risk of developing tetanus if they have never been vaccinated. Such individuals are given an injection of a preparation that has been created in horses. This preparation will protect the person against tetanus.
i What is the active constituent of the injection?
ii Explain why this is an example of passive immunity and why it would not give long-term protection.

d Predict if herd immunity would prevent outbreaks of tetanus.

33 Several cases of measles have been reported to Australian authorities in recent times. Measles is a preventable disease caused by a virus. In 2013 there were 96 000 deaths worldwide from measles. It is the disease with the highest mortality rate of all vaccine-preventable diseases. Most people in Australia are vaccinated against it in childhood.

The vaccination schedule requires children to be given one injection at 12 months and a second injection at 18 months. The antibody response to the vaccinations is shown in the graph below.

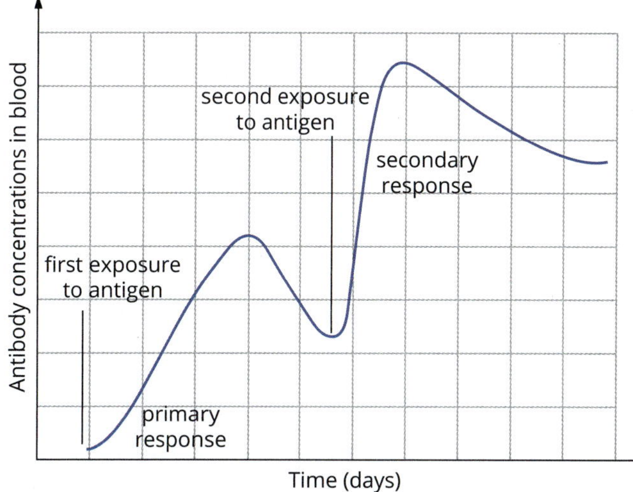

a Explain why the secondary response is so much greater than the primary response.

b T lymphocytes play a significant role in the adaptive immune response of the body against viruses such as the one that causes measles. Describe the role of T cells in immunity.

c In February 2016, Department of Health statistics showed that 93.58% of five-year-old children in Victoria were fully immunised against measles. In some areas the immunisation rate is as low as 73%. Why is this statistic of such concern?

d Do you think the statistics for other parts of Australia will be similar? Predict where there might be higher or lower statistics. Why are the statistics state-based?

e Conduct an investigation from secondary sources to see if your predictions are accurate.

34 Rational drug design is the new area of development in pharmaceuticals. It is based on learning all it is possible to know about the cause of a disease, whether its cause is a pathogen or a malfunction of a body system due to genetics or lifestyle. New and returning diseases are an important target for rational drug design. One disease that has been studied extensively is HIV. It has been discovered that about 1% of Caucasian people are highly resistant to infection by the virus. The resistance of these people to infection has provided an important line of research.

It was determined that HIV enters a cell by attaching to a receptor called CD4. This is an important receptor for many cell-signalling pathways and so is not a suitable target for an anti-HIV drug. However, further investigation showed that HIV also needs a second receptor (a co-receptor) that usually works with CD4. This receptor is called CCR5. In most people this is a transmembrane protein that assists HIV to enter the cell. The resistant individuals have a mutation in the gene for CCR5 that causes the protein to be much shorter so that it is totally enclosed within the cell and is thus not able to help HIV enter the cell. People heterozygous for the mutation are resistant to HIV and those homozygous for it are highly resistant. These people show no ill effects from having this mutation.

a i Describe a possible reason why this molecule would make an ideal target for a possible drug against HIV.

ii Describe the design process to develop such a drug.

b Are there likely to be any issues associated with trialling an anti-HIV drug?

35 Imagine you are a member of a group of scientists who have been working on finding a treatment for Ebola, a deadly viral disease. Your team has used rational drug design techniques to design a drug that will block the ability of Ebola to enter and infect cells. You are still in the initial stages of trialling and have decided that the initial tests will take place in mice. You have 200 genetically similar mice at your disposal.

a i Design an experiment to test the effectiveness of your drug in slowing the progress of the disease.

ii In your answer, make sure you name the dependent and independent variables.

b Identify what results would be required to indicate that trials should move on to the next stage with this drug.

c Outline the kinds of safety precautions that would be needed in performing your trials.

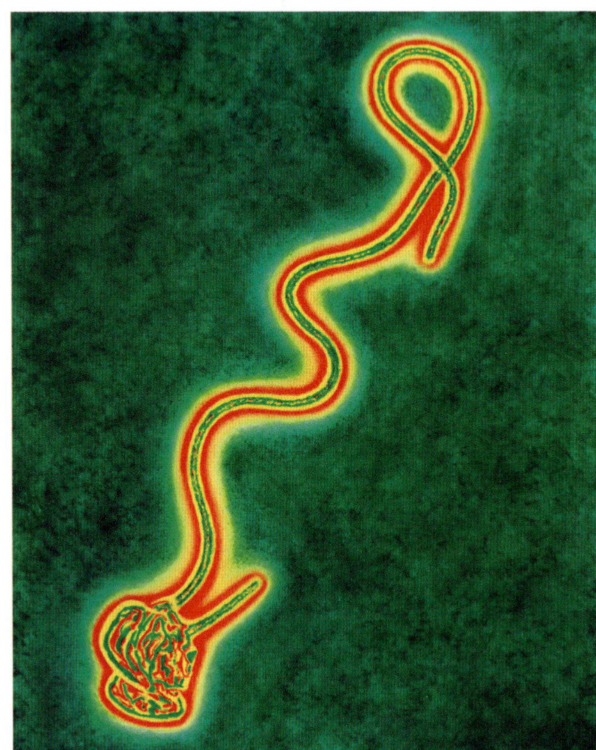

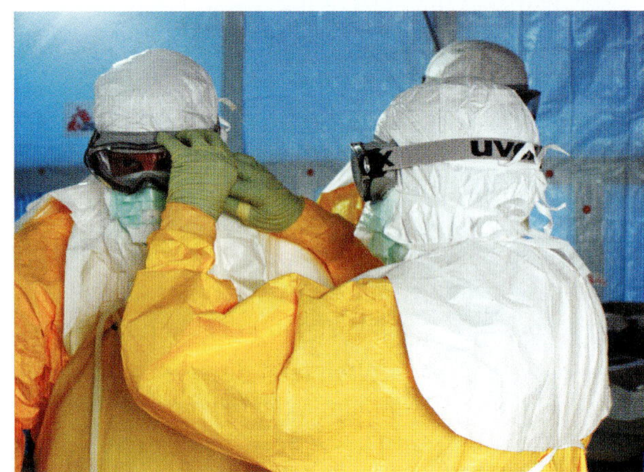

Non-infectious disease and disorders

You will engage with the study of non-infectious disease and disorders, including their causes and effects on human health. You will explore technologies and their uses in treating disease and disorders, as well as the epidemiology of non-infectious disease in populations.

This module examines the practical applications of STEM. It looks at the importance of understanding the multidisciplinary nature of science applications. It also examines physiology and engineered solutions to problems related to the management of human disorders.

Outcomes

By the end of this module you will be able to:

- analyse and evaluate primary and secondary data and information BIO12-5
- solve scientific problems using primary and secondary data, critical thinking skills and scientific processes BIO12-6
- communicate scientific understanding using suitable language and terminology for a specific audience or purpose BIO12-7
- explain non-infectious disease and disorders, as well as a range of technologies and methods used to assist, control, prevent and treat non-infectious disease BIO12-15

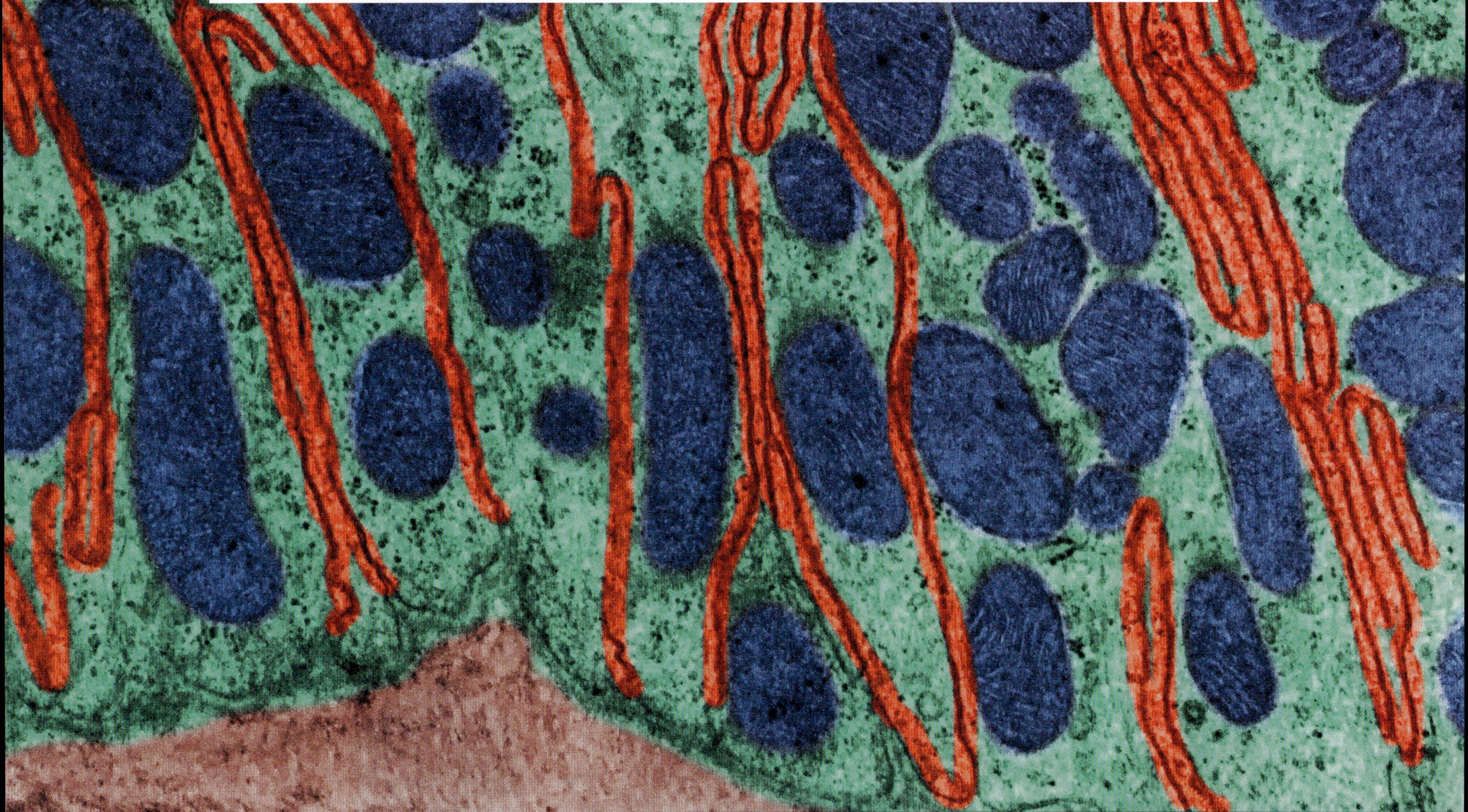

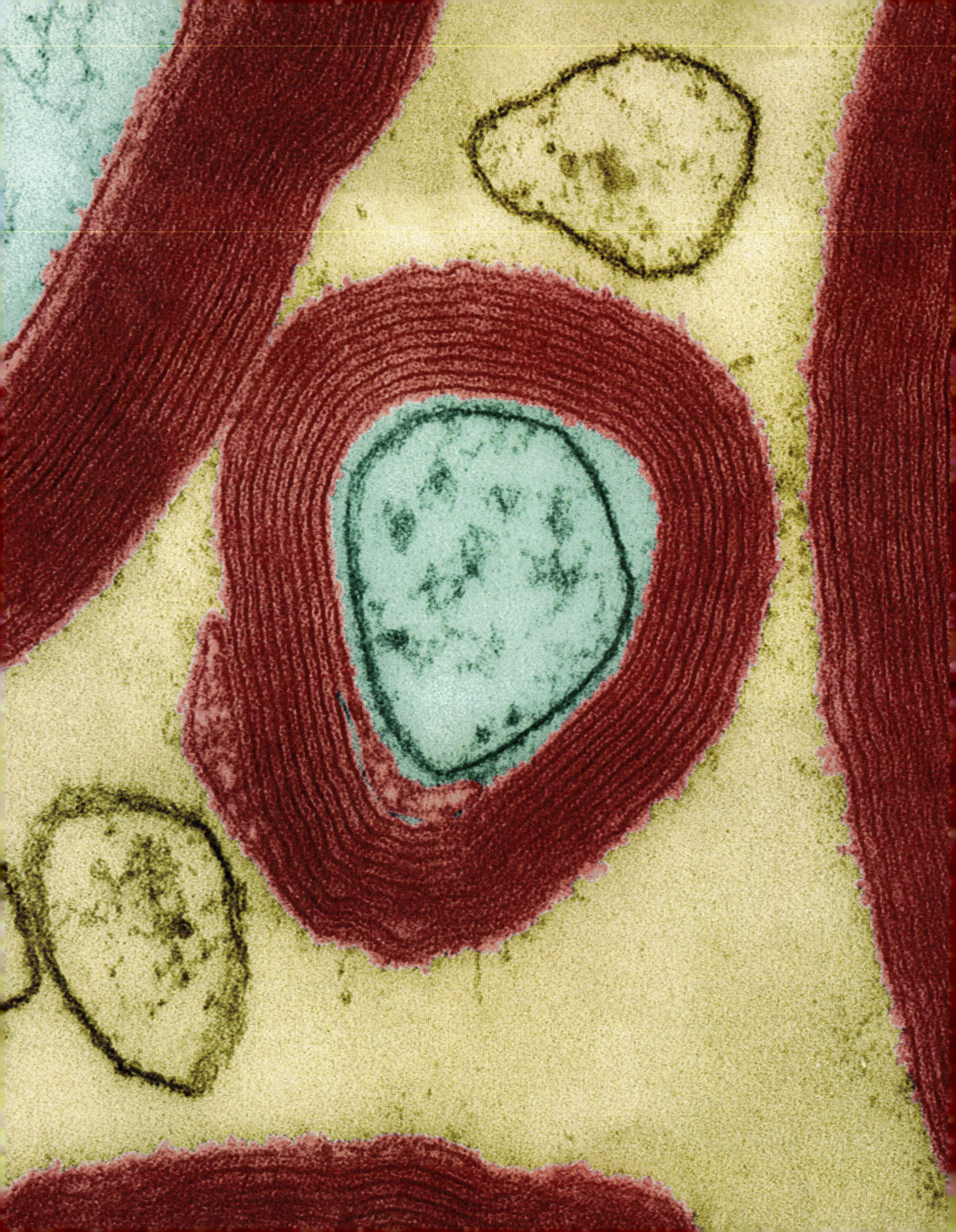

CHAPTER 14

Homeostasis

All organisms need to maintain varying degrees of internal stability despite environmental changes. Maintenance of this stability is called homeostasis, which organisms achieve through simple and complex mechanisms. Simplest organisms control internal chemistry only by excreting wastes—more complex organisms actively maintain concentrations of numerous substances, blood pressure and in some cases temperature. These mechanisms involve negative feedback loops regulated by hormones and the nervous system, in conjunction with chemical receptors.

By the end of this chapter you will be able to construct and interpret negative feedback loops that show homeostasis, including temperature and glucose and investigate the various mechanisms used by organisms to maintain their internal environment within tolerance, including hormones and neural pathways. You will also be able to investigate the trends and patterns in behavioural, structural and physiological adaptations for homeostasis in plants and endothermic animals.

Content

INQUIRY QUESTION

How is an organism's internal environment maintained in response to a changing external environment?

By the end of this chapter you will be able to:

- construct and interpret negative feedback loops that show homeostasis by using a range of sources, including but not limited to: (ACSBL101, ACSBL110, ACSBL111) CCT ICT
  - temperature (ACSBL098)
  - glucose
- investigate the various mechanisms used by organisms to maintain their internal environment within tolerance limits, including:
  - trends and patterns in behavioural, structural and physiological adaptations in endotherms that assist in maintaining homeostasis (ACSBL099, ACSBL114) ICT
  - internal coordination systems that allow homeostasis to be maintained, including hormones and neural pathways (ACSBL112, ACSBL113, ACSBL114)
  - mechanisms in plants that allow water balance to be maintained (ACSBL115) ICT

14.1 Negative feedback loops

BIOLOGY INQUIRY  

Volume and surface area

How is an organism's internal environment maintained in response to a changing external environment?

COLLECT THIS...

- electric kettle
- various saucepans, bake tins and cups
- kitchen thermometer
- paper and pen
- ruler or tape measure

DO THIS...

1 Working in groups of two to four, fill the electric kettle with water and bring to the boil.
2 Collect a few saucepans, frying pans, bake tins or trays and cups. They should be of different shapes, yet at least two should be the same shape but different sizes.
3 Make a prediction about which container will cool down fastest and which slowest.
4 When the water has boiled, fill each container to approximately 2 cm from the top. Note the time.
5 Every two minutes, record the temperature in each of the containers.
6 For each container, calculate the surface area, volume, and surface-area-to-volume ratio. Relate this data to your predictions.

RECORD THIS...

Describe the patterns of cooling you discovered.

Present your data in the form of a table and graph.

REFLECT ON THIS...

How is an organism's internal environment maintained in response to a changing external environment?

What is the relationship between volume and surface area?

What implications does this relationship have for an animal's proportions and thermoregulation?

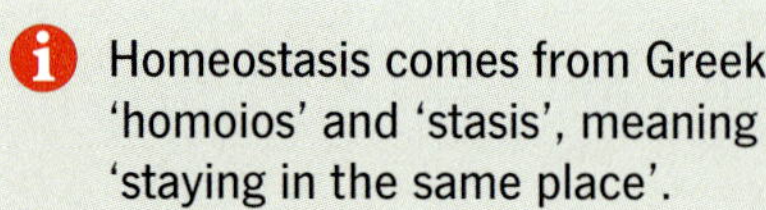

Homeostasis comes from Greek 'homoios' and 'stasis', meaning 'staying in the same place'.

When an organism is healthy and functioning well, its systems are in **homeostasis**. Breakdown of homeostatic mechanisms results in cellular damage, causing disease and eventually death. Homeostatic mechanisms can malfunction because of factors such as genetic disorders, ageing, poor nutrition, insufficient physical activity or exposure to harmful substances.

In this section, you will learn about two sets of **negative feedback loops** in the human body—thermoregulation (temperature control) and the control of blood glucose levels.

NEGATIVE FEEDBACK LOOPS IN HOMEOSTASIS

Negative feedback loops promote stability in an organism's internal environment. They maintain homeostasis by responding to changes in the body and adjusting the variables to their original or optimal state. For example, if the concentration of a substance in the blood is too high, a negative feedback loop will lower the concentration. If the concentration is too low, another negative feedback loop will increase the concentration. Most feedback loops in biological systems are negative.

The endrocrine system is the collection of glands that produce and secrete hormones.

Negative feedback loops are so named because the information produced by the feedback causes a reversal in the effect of the **stimulus**. Negative feedback loops maintain stability through the action of the **nervous system** or **hormones** (the **endocrine system**) or both, acting together . You will learn more about these systems in Section 14.2.

A negative feedback loop acts as described below (Figure 14.1.1).

- The system is in a stable state.
  1. A change (stimulus) occurs.
  2. The change is detected by an appropriate **receptor**.
  3. The receptor sends a signal to a control centre (**hypothalamus** or transmission molecules).
  4. The control centre sends a signal to an appropriate **effector** (a specific effector cell, tissue or organ).
  5. The effector responds to the signal.
- The original state is restored.

In the control centre of the brain (the hypothalamus) information from sensory receptors is received and compared with a set point (the optimal value for the functioning of that organism). This information is processed with other information about the state of the organism and an appropriate response is initiated.

Regulation therefore involves fluctuations around the set point. The size of the fluctuations depends on the:

- sensitivity of the receptor
- tolerance of the control centre to variation from the set point
- efficiency of the effector.

Some features of the internal environment, such as blood glucose levels, can vary considerably. Others, such as body temperature in mammals, are tightly controlled.

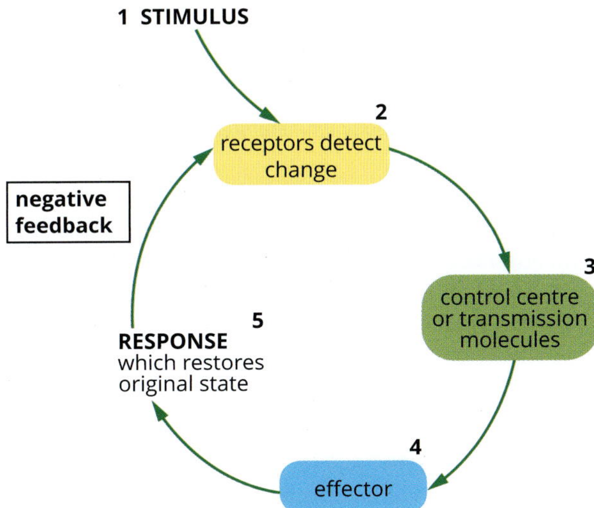

FIGURE 14.1.1 A negative feedback loop reduces the initial stimulus or disturbance, to restore homeostasis (stability) in the system.

THERMOREGULATION

Internal regulation of an animal's body temperature is called **thermoregulation**. In humans, thermoregulation involves a complex negative feedback pathway with several sensory inputs and many effector responses that act together to maintain a stable body temperature. If body temperature rises too far above or below the set point of 37°C, **thermoreceptors** in the hypothalamus detect the change. Then the hypothalamus initiates regulatory responses that can reduce heat loss, or initiate heat production or heat exchange (Figure 14.1.2).

The thermoregulatory response involves two separate negative feedback loops. If temperature rises above the set point, the hypothalamus triggers sweating and blood vessel dilation to promote cooling. However, if body temperature drops, the hypothalamus shuts down sweating, constricts blood vessels and (if necessary) initiates shivering and increases the metabolic rate.

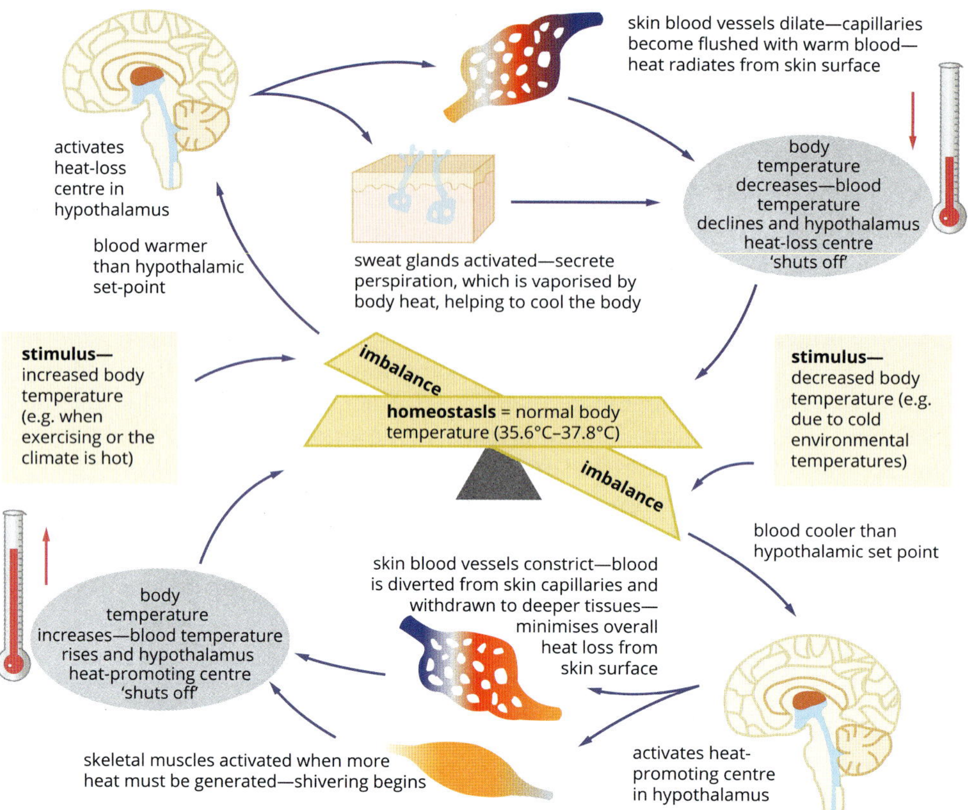

FIGURE 14.1.2 Thermoregulation in humans involves a range of regulatory mechanisms that function to maintain thermal homeostasis (normal body temperature) for optimal functioning of the organism.

Heat exchange

Organisms are constantly exchanging heat with their environment. This heat exchange occurs through the following four mechanisms (Figure 14.1.3).

- **Conduction** occurs when the temperature of the organism and the environment are different. Heat exchange is a result of direct contact (e.g. a lizard basking on a warm rock).
- **Convection** is the transmission of heat from a warmer region to a colder region, resulting from the movement of liquid or gas (e.g. heat moves from the inside of living organisms to the body surface by convection).
- **Radiation** is heat transfer via electromagnetic energy (infra red). This occurs without direct contact, regardless of temperature differences between the organism and their environment (e.g. heat radiating from dark coloured surfaces).
- **Evaporative cooling** occurs as energy is lost through a phase-state change from water to water vapour (e.g. sweating). This occurs most rapidly when the air is hot and dry.

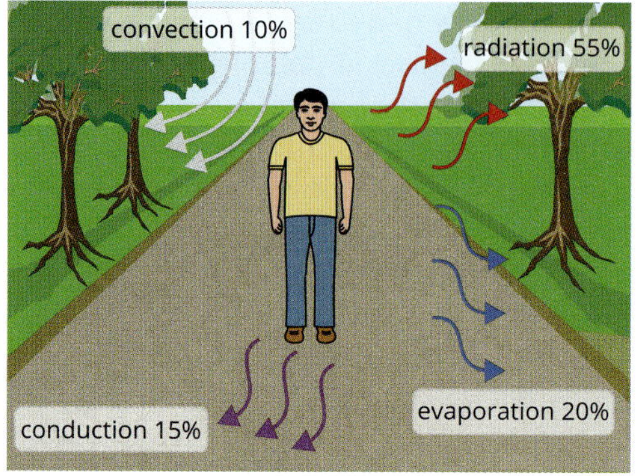

FIGURE 14.1.3 Methods of heat exchange with the environment and the average amount lost via each route in a human at rest

Detecting temperature change

Regulation of temperature in humans is an example of the way different sensory receptors work together to produce an integrated homeostatic response.

In **endotherms** (warm-blooded animals) a group of temperature-sensitive cells in the hypothalamus act as misalignment detectors, triggering homeostatic responses if blood temperature deviates from the optimal temperature range, or set point. Lowering or raising the temperature of the hypothalamus initiates regulatory changes in heat production or heat exchange.

Temperature receptors are also found in the skin. A decrease in environmental temperature detected by these receptors will initiate negative feedback responses that reduce heat loss and increase heat production, such as decreased blood flow to the skin and behavioural changes such as moving to a warmer or more sheltered environment. A similar negative feedback loop operates in the reverse direction if environmental temperature increases. These responses take place long before there has been any change in the internal temperature of the body.

If the arterial blood temperature falls despite the regulatory responses that have been initiated, or if it rises because the responses made have been too effective, these changes will be detected by the misalignment receptors in the brain, which will fine-tune the temperature-regulating mechanisms.

The value of the temperature receptors in the skin is to reduce fluctuations in arterial blood temperature, providing more precise control around the set point level than there would be if misalignment detectors (the brain's temperature receptors) alone were involved.

Responding to cold

In humans a number of nervous system and endocrine responses occur rapidly to reduce heat loss from the body and increase heat production when the body becomes too cold. Some of these responses are similar to the adaptations seen in non-human animals. You will learn more about these adaptations in Section 14.3.

In humans, various voluntary and involuntary responses are initiated in response to cold. Thermoreceptors in the hypothalamus detect a drop in body temperature, responding by stopping skin production of sweat and initiating other skin responses such as constricting blood vessels in order to restrict blood flow. Doing so reduces heat loss via evaporation and convection. The following steps are part of a negative feedback loop to reduce heat loss:

- **Vasoconstriction** (constriction of the blood vessels) near the body's surface reduces heat loss from the skin, as the amount of blood moving close to the exposed surface is reduced.
- **Piloerection** is the constriction of the piloerector muscles around hair follicles ('goose bumps'), which increases the insulating effect of the hairs (Figure 14.1.4). This response has a minimal effect in humans but in animals with thick fur, the layer of trapped air increases significantly and reduces heat loss from the body.
- Seeking shelter, (e.g. moving to a warmer place with less exposure to harsh winds and cold temperatures to conserve body heat).
- Changing body shape or decreasing surface area (e.g. curling up to make yourself small) reduces the area exposed to the cold and reduces the rate at which heat is lost from the skin.
- Putting on more clothes. Clothes trap a layer of air (which is the best insulator). Some fabrics are better than others at trapping air and increasing insulation.

Separately, the hypothalamus also initiates shivering and raises the metabolic rate to increase heat production. The following responses increase heat production in the human body:

- Voluntary movement—during physical effort, the amount of heat produced by the muscles is increased (Table 14.1.1).

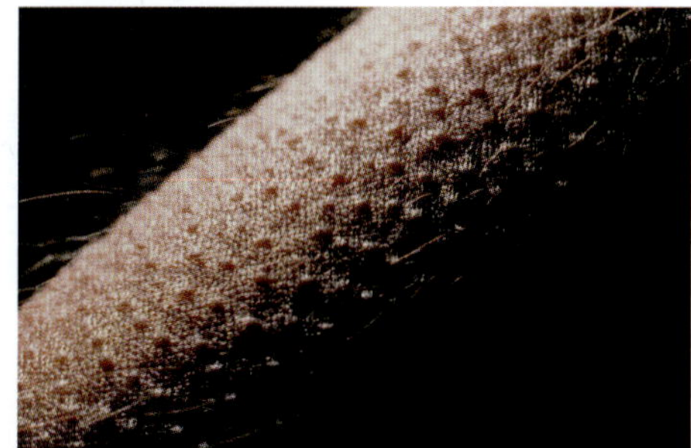

FIGURE 14.1.4 A close-up of a human forearm with goose bumps. The contraction of blood vessels and small muscles (arrectores pilorum) that are attached to the base of each hair follicle pull the hair into an upright position. In this position the skin resembles plucked goose skin.

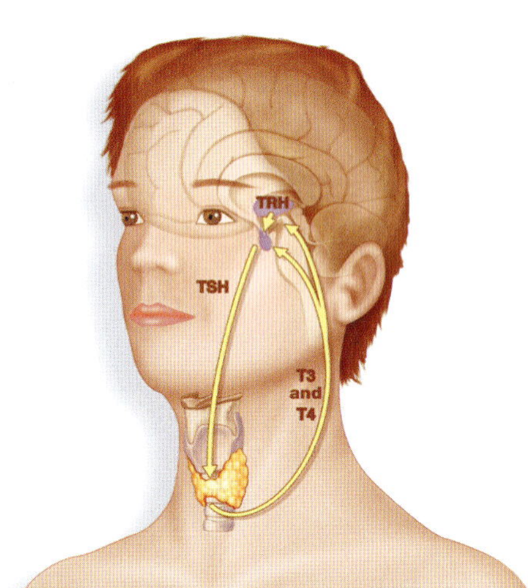

FIGURE 14.1.5 The thyroid is a gland that produces hormones that stimulate cellular respiration. The hypothalamus releases TRH, which stimulates the secretion of TSH by the pituitary gland, which in turn stimulates production of T3 and T4 hormones by the thyroid. The increase in cellular respiration creates thermal energy.

- **Shivering thermogenesis**—the production of metabolic heat is increased through shivering. This involuntary movement of the muscles generates especially large amounts of heat. Shivering thermogenesis is stimulated by **adrenaline**.
- **Non-shivering thermogenesis** in brown fat (brown adipose tissue or BAT)—increased cellular activity in BAT, which is a tissue specialised for heat production, causes the tissues to warm up. The heat produced is carried to other parts of the body in the blood. Brown fat contains a huge number of mitochondria (which give it its brown colour), fat-metabolising enzymes and an extensive vascular network. Brown fat is capable of high rates of aerobic metabolism—it uses a pathway that breaks down fats to produce large amounts of heat, but little ATP.
- Increasing **metabolism** (the rate of cellular respiration)—metabolic processes in the internal organs are the main source of heat production when the organism is at rest (Table 14.1.1). In humans, around 60% of the energy released during cellular respiration is transformed into thermal energy. In humans, the overall metabolic rate, and therefore the rate of heat production, is controlled by hormones.
- TRH (thyrotropin releasing hormone) secretion by the hypothalamus—TRH acts on the anterior pituitary to secrete TSH (thyroid stimulating hormone). As the name suggests, TSH acts on the thyroid gland to release thyroid hormones, tri-iodothyronine (T3) and thyroxine (T4) (Figure 14.1.5). T3 and T4 hormones regulate metabolic processes, increasing heat production and body temperature.

The amount of T3 and T4 in the bloodstream is regulated by the **pituitary gland** via a negative feedback loop. If there is too much or too little T3 or T4, the pituitary gland reduces or increases the amount of TSH it secretes. This mechanism allows a very delicate regulation of the level of thyroid hormones in the blood.

As a result of the above mechanisms, the body returns to its preset temperature following cooling. Thus homeostasis is maintained via negative feedback loops.

TABLE 14.1.1 Major sources of heat production in humans. Metabolic processes in the internal organs are the main source of heat production when a person is at rest. During physical activity the heat generated in the muscles increases.

Organs	Participation in heat production at rest (%)	Participation in heat production during physical effort (%)
brain	16	2
internal organs	56	22
skeletal muscles	15	73
other organs	13	3

Responding to heat

When thermoreceptors in the hypothalamus detect a temperature rise, the hypothalamus triggers a negative feedback pathway having a cooling effect. The pathway involves simultaneous initiation of sweating and other responses such as dilation of blood vessels. Both cause heat loss, thereby restoring the animal to the preset temperature.

Sweating

Evaporative cooling is a very effective way of losing heat energy from the body. A change of state from liquid to gas is an endothermic process. It requires an input of energy. In evaporative cooling of the skin, this energy comes from your body, in the form of heat energy.

Humans have two types of sweat glands—**eccrine glands** and **apocrine glands**. The function of apocrine glands is thought to be mainly scent or pheromone production while the eccrine sweat glands function to control body temperature. These glands are distributed over much of the body and release sweat onto the skin surface via pores when body temperature rises. These glands extend just below the surface of the dermis and secrete odourless sweat that is high in electrolytes and sodium. In humans, the rate of sweat secretion increases from almost zero in cold conditions to about 1.5 litres per hour in a hot environment. After about six weeks of acclimatisation to heat, this rate can increase to 4 litres per hour, enabling a much greater evaporative cooling capacity.

FIGURE 14.1.6 Long-distance runners spray water on their skin to take advantage of evaporative cooling. This enables their bodies to work at maximum efficiency, and helps to prevent them from becoming dangerously hot.

Other responses

Some other voluntary and involuntary ways that your body responds to heat are described below.

- Slowing the rate of cellular respiration in internal organs, which decreases heat generation thus decreasing body temperature.
- **Vasodilation** (dilation of the blood vessels in the skin). Dilation means more blood is sent to the extremities. Heat is lost to the environment by radiation and convection (especially if it is windy).
- Covering your body with water (Figure 14.1.6). Spraying water on your skin produces the same evaporative cooling effect as sweating.
- Swimming or bathing in cool water, which causes heat loss through conduction across the skin. Evaporative cooling also occurs when you are out of the water and your skin is still wet.
- Changing your body shape to increase its surface area (e.g. by standing with your arms and legs outstretched).
- Removing clothing, which reduces the insulating effect of clothing layers and allows heat to escape from the skin.
- Moving out of the sun into shade.
- Decreasing the amount of activity undertaken.

BIOFILE PSC

Rosy cheeks

Children often have rosy cheeks in cold weather (Figure 14.1.7). This is a result of increased blood flow to the cold tissues following vasodilation (opening of the blood vessels). Cold-induced vasodilation functions to warm parts of the body that have been exposed to the cold. By increasing the blood to the exposed area, the risk of injury from extreme cold is reduced. As vasodilation allows heat to escape the body, it can be maintained only for short periods in cold conditions. If the body does not warm up after a while, vasoconstriction occurs to minimise heat loss.

FIGURE 14.1.7 Rosy cheeks from exposure to cold are caused by the dilation of blood vessels (vasodilation). Vasodilation increases blood flow to exposed skin to counteract the cold.

CONTROL OF BLOOD GLUCOSE LEVEL

Blood glucose level (BGL), sometimes called blood sugar level, is the concentration of **glucose** in your blood. This level is constantly changing in your body and is tightly regulated by homeostatic negative feedback loops. Glucose is the main source of energy for your body's cells. Eating carbohydrates and doing physical activity will change blood glucose levels throughout the day. The more carbohydrates you eat, the more **insulin** you will produce. Glucose is stored in the body in the form of a complex carbohydrate called **glycogen**. When the body needs glucose, the glycogen is broken back down into usable glucose for cellular respiration, which is the energy-producing reaction in cells. If blood glucose level is not maintained within its optimal range, **hyperglycaemia** (BGL is too high) or **hypoglycaemia** (BGL is too low) can develop, leading to a range of long-term health problems, such as diabetes (Figure 14.1.8).

As with temperature, the body has two separate negative feedback loops working in combination to maintain stability. One loop raises glucose levels, the other lowers it (Figure 14.1.9).

Detecting blood glucose level

Cells in the pancreas detect changes in BGL. Blood glucose concentration is regulated so it remains within a range of about 3.5–8 mmol/L. A deviation from these levels in either direction will result in a response by clusters of specialised cells in the pancreas, called the islets of Langerhans. These cells detect blood glucose levels and release the hormones insulin and **glucagon** to maintain blood glucose levels within the normal range.

glucagon—raises blood glucose level
insulin—lowers blood glucose level

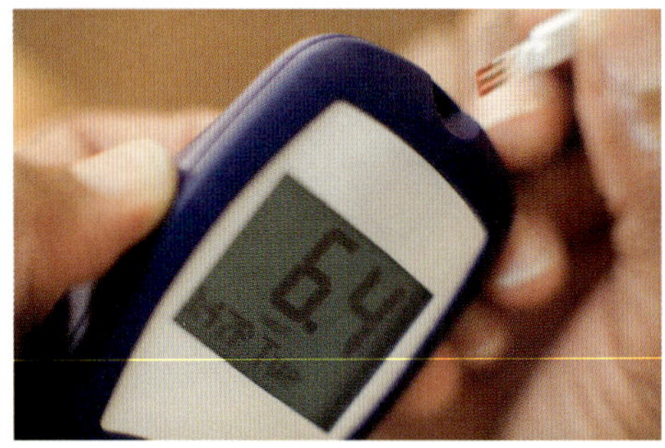

FIGURE 14.1.8 Blood glucose levels (BGL) can be tested by pricking the finger and placing a small drop of blood on a piece of absorbent material, which is tested by an electronic device. The person pictured here has a BGL of 6.4 mmol/L.

After an animal eats, its blood glucose rises. Sensor cells in the pancreas detect the rise. In response, insulin is released from the beta cells of the pancreas. Insulin in turn signals the liver to absorb glucose from the blood and convert it to glycogen, fats or fatty acids for storage in the liver and skeletal muscles. The overall effect of insulin is to lower BGL (Figure 14.1.9).

If an animal has not eaten and glucose levels decrease, different pancreatic sensor cells detect the fall. Glucagon is released from the alpha cells of the pancreas and stimulates the conversion of glycogen to glucose, which raises BGL (Figure 14.1.9). When glucose levels rise above about 5 mmol/L, the beta cells of the islets of Langerhans in the pancreas release insulin. When glucose levels fall below about 5 mmol/L, the alpha cells of the islets of Langerhans in the pancreas release glucagon. Adrenaline also raises BGL by its actions on fat cells and the liver.

There are also glucose-sensing nerve cells in the hypothalamus in the brain. These glucose-sensing nerve cells probably play a key role in food intake, thus helping to regulate blood glucose concentrations.

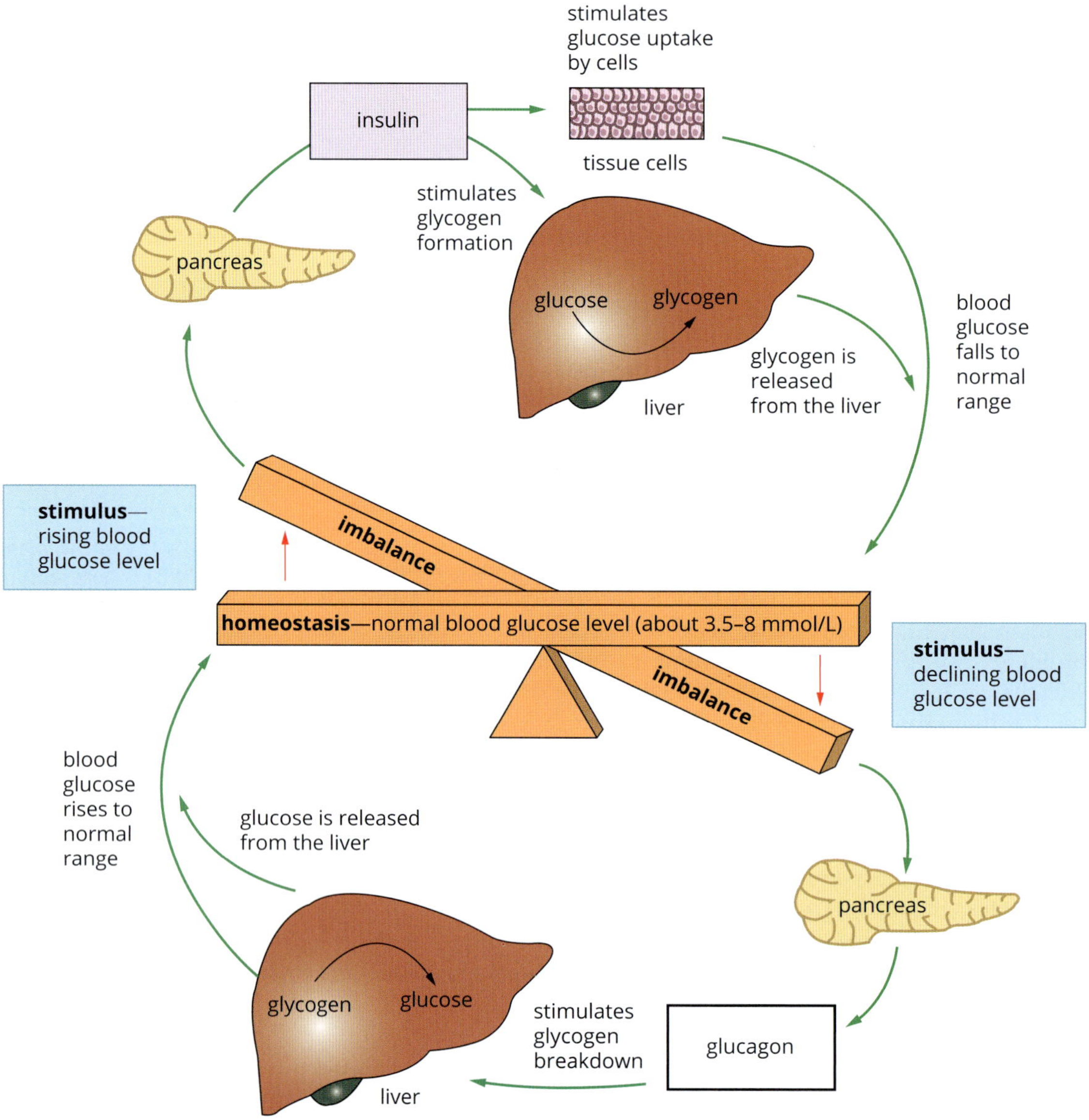

stimulates glucose uptake by cells

insulin

tissue cells

stimulates glycogen formation

pancreas

glucose

glycogen

blood glucose falls to normal range

glycogen is released from the liver

liver

stimulus—rising blood glucose level

imbalance

homeostasis—normal blood glucose level (about 3.5–8 mmol/L)

imbalance

stimulus—declining blood glucose level

blood glucose rises to normal range

glucose is released from the liver

pancreas

glycogen

glucose

stimulates glycogen breakdown

glucagon

liver

FIGURE 14.1.9 The regulation of blood glucose level (BGL) by insulin and glucagon. When BGL is too high, insulin is secreted by the beta cells in the pancreas. This stimulates glucose uptake and storage as glycogen in the liver, reducing BGL. When BGL is too low, glucagon is secreted by the alpha cells in the pancreas. This stimulates glycogen breakdown and the release of glucose from the liver, increasing BGL to within the normal range.

14.1 Review

SUMMARY

Homeostasis

- Homeostasis is the maintenance of variables in a biological system within certain limits.
- Malfunctions in homeostatic mechanisms lead to imbalances and a subsequent oversupply or undersupply of substances needed by cells, leading to disease.
- Homeostasis depends on negative feedback loops. These are stimulus–response mechanisms that respond to changes in the body by adjusting variables back to their original or optimal state, reversing the direction of the stimulus.
- Most feedback loops are negative.

Temperature control

- A change in the temperature of the hypothalamus initiates regulatory responses that can involve heat production or heat exchange.
- Temperature receptors are found in the skin and the hypothalamus.
- Heat is lost from the body by:
  - conduction
  - convection
  - radiation
  - evaporation.
- Responses to the cold involve reducing heat loss as well as producing heat.
- The responses to reduce heat loss are:
  - vasoconstriction
  - piloerection
  - behavioural changes such putting on more clothes or seeking shelter.
- The responses to generate heat are:
  - voluntary movement
  - shivering thermogenesis
  - non-shivering thermogenesis
  - increasing rate of cellular metabolism.
- The responses to heat are:
  - sweating or perspiring
  - covering your body with water
  - vasodilation
  - behavioural adaptations including removing clothing, seeking shade, increasing surface area and decreasing physical activity.

Control of blood glucose

- Blood glucose levels are detected by receptor cells in the pancreas and nerve cells in the hypothalamus. The normal concentration for blood glucose is 3.5–8 mmol/L.
- When glucose levels rise, insulin is released from the beta cells in the islets of Langerhans in the pancreas. Insulin causes a decrease in blood glucose levels by acting on a number of tissues to:
  - increase conversion of glucose to fat in fat cells
  - increase uptake of glucose in muscle and fat cells
  - increase conversion of glucose to the storage compound glycogen for storage in the liver.
- When glucose levels decrease, glucagon is released from alpha cells in the islets of Langerhans, which stimulates the conversion of glycogen to glucose.

KEY QUESTIONS

1. How do organisms exchange heat with their environment? Explain each of the four methods of heat exchange using examples.
2. What are three mechanisms humans use to produce heat?
3. What are three mechanisms humans use to lose heat?
4. How do receptors and effectors maintain homeostasis?
5. What is the main role of insulin, and how does this hormone function?
6. Describe the role of glucagon in glucose homeostasis.
7. Discuss how malfunctions in homeostatic mechanisms can lead to disease, using an example.

14.2 Internal coordination systems for homeostasis

Homeostasis involves several body systems working together. The key systems include the endocrine system and the nervous system. Communication between these systems involves signalling molecules, such as hormones and neurotransmitters, which pair with specific receptors on target cells to activate homeostatic effects.

In this section, you will learn about signalling molecules, hormones and glands, nerves and neurotransmitters, and osmoregulation.

SIGNALLING MOLECULES

Multicellular organisms survive, grow and reproduce because they can detect and respond to signals from their internal and external environments. Many of these responses involve negative feedback mechanisms through which homeostasis is maintained. To achieve these responses, communication between cells is essential and there are many types of molecules involved.

Cellular communication

Multicellular organisms have developed mechanisms to communicate between cells and convey messages about changes to internal and external environmental conditions. A change in conditions that elicits a response from a cell is referred to as a stimulus. Stimuli are varied in nature. Changes in pressure, light, temperature or chemical molecules are all examples of stimuli.

Hormones are a diverse group of compounds that act as intercellular messengers to regulate cell functions.

Not all cells are capable of detecting all stimuli, and not all cells are able to respond to all stimuli. Cells that are able to detect stimuli can pass this information to other cells by producing and releasing **signalling molecules**. Effector cells respond to signalling molecules. The signal may be passed to a number of different cells before reaching the ultimate effector cells, and different signalling molecules may be released at each step. Signalling molecules can trigger a response even at very low concentrations.

Once the signal reaches the effector cells, various cellular processes may be activated, depending on the nature of the original stimulus. These changes in cellular activities form the ultimate response.

Communication via signalling molecules can occur in nearby and distant environments.

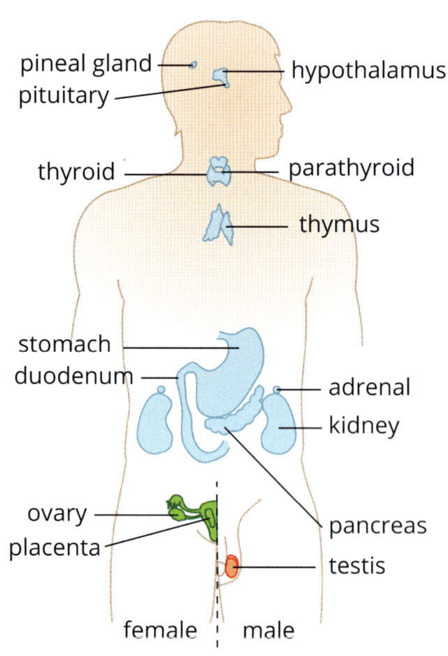

FIGURE 14.2.1 The major endocrine glands of the human body

THE ENDOCRINE SYSTEM

A hormone is a signalling molecule that helps regulate homeostasis in most animals and plants. Hormones, the glands that produce them, and other specialised tissues, are collectively called the endocrine system (Figure 14.2.1). The regulated functions of the endrocrine system include: rates of chemical reactions in cells (metabolism); transport of substances across cell membranes; secretion of other hormones; and the growth and reproduction of cells. Malfunctions in the endocrine system often lead to disruption in homeostatic mechanisms; which can have an adverse effect on the body.

Hormones are transported to where they are needed via the circulatory system (in animals) or by **diffusion** through the **extracellular fluid** (in both plants and animals). Thus hormonal effects are generally slower than nervous responses, longer in duration, and often affect cells that are widely distributed throughout the body.

Hormonal responses are slower and more indirect than nervous responses, but their effects last longer and can target cells that are widely distributed throughout the body.

While a few animal hormones affect most cells in the body, hormones generally affect specific organs, and often only one type of cell within that organ. Only those cells that possess specific receptors are capable of responding to the hormone. For example, a sudden shock causes the release of adrenaline from the adrenal gland. Only those cells that have adrenaline receptors in their cell membranes, such as muscle cells of the heart and blood vessels, can respond to the adrenaline circulating in the blood.

Hormones exert their effects either directly by passing through the cell membrane into the cell, or indirectly by interacting with a receptor on the surface of the cell.

Types of hormones

Hormones can be broadly grouped into three main classes (Figure 14.2.2).

- Lipid hormones are a class of hydrophobic signalling molecules derived from fatty acids (eicosanoids, e.g. prostaglandins) or cholesterol (steroids, e.g. testosterone, oestrogen and cortisol).
- Peptide hormones and protein hormones belong to a class of hydrophilic signalling molecules. An example of a peptide hormone is insulin and an example of a protein hormone is growth hormone.
- Amino acid-derived hormones are a class of small signalling molecules derived from the amino acids tyrosine and tryptophan. They can be further divided into catecholamines (e.g. adrenaline and dopamine) and thyroid hormones (e.g. thyroxine).

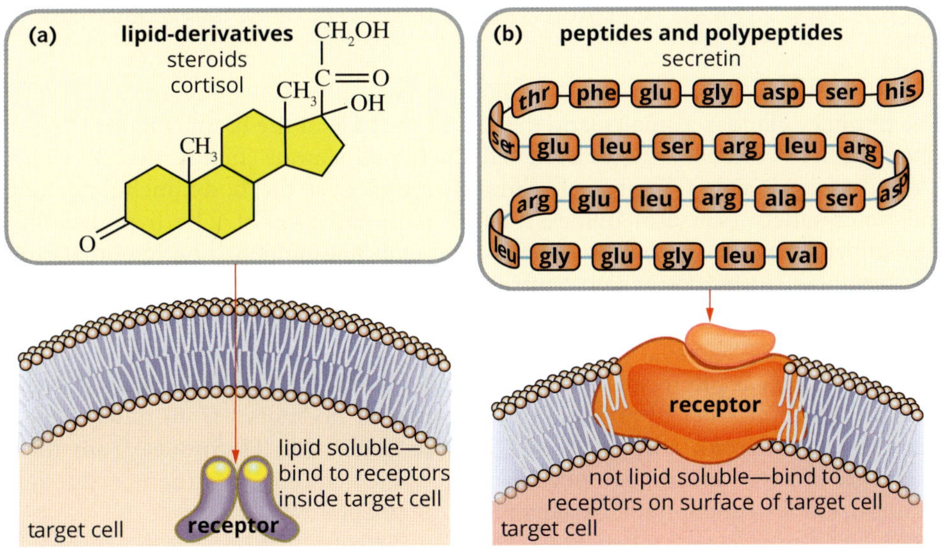

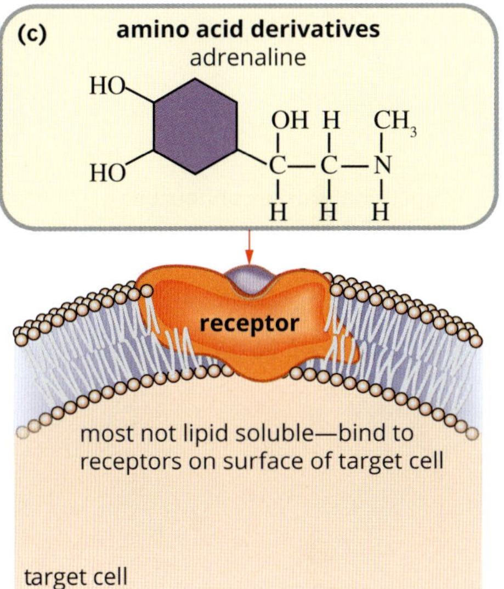

FIGURE 14.2.2 The three types of hormones: (a) lipid derivatives, (b) peptides and polypeptides and (c) amino acid derivatives. These hormones interact with the membrane of cells differently because of differences in their chemical structure.

A single hormone can trigger different responses in multiple target cells at the same time. For example, adrenaline targets cardiac muscle cells, vascular smooth muscle cells and the various glands and organs of the digestive system. An increase in adrenaline in the bloodstream will result in an increase in heart rate and blood pressure and will simultaneously decrease digestive functions, preparing for a 'fight or flight' response.

Endocrine glands

Animals usually have specialised cells for producing hormones. In more complex animals (vertebrates, insects, crustaceans and some molluscs), these cells are often clustered into discrete organs called endocrine glands (Figure 14.2.1). Initially, the term 'hormone' was restricted to the products of endocrine glands. However, it is now evident that hormones are secreted by a wide variety of tissues and that they are able to reach their site of action by simple diffusion. Mammals have many major hormones and many more molecules that act as minor hormones.

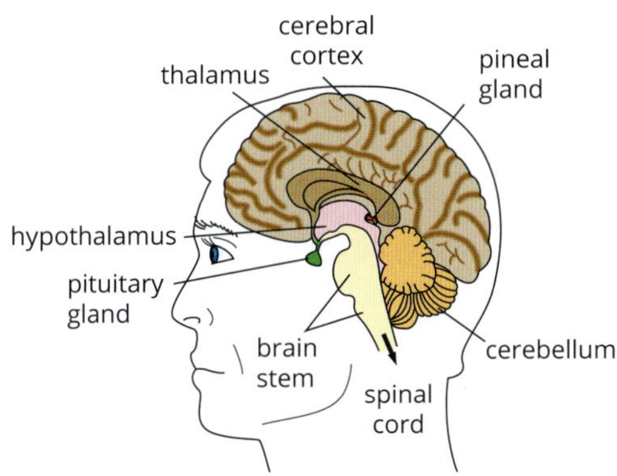

FIGURE 14.2.3 The human brain is subdivided into regions that are associated with different functions. The pituitary gland (green) lies immediately beneath the hypothalamus (pink).

BIOFILE DD PSC

Biorhythms and the pineal gland

All vertebrates have a **pineal gland** that detects and responds to light. In mammals the gland is deep within the brain (Figure 14.2.3) and does not sense light directly, but instead receives messages from the eyes about the brightness of light.

The pineal gland produces the hormone melatonin. When light levels are high, the gland stops producing the hormone, decreasing levels of melatonin during the day and increasing levels at night. This allows animals to sense the time and seasons internally (Figure 14.2.4). The pineal gland therefore functions like an internal 'biological clock'.

Travel from one time zone to another disrupts the internal rhythms, producing the symptoms of jet lag, which include disorientation, sleepiness or wakefulness. Similar problems are experienced by shift-workers. Melatonin or bright light can synchronise the internal clock, which has led to the use of melatonin as a 'jet-lag pill'. Alternatively, spending a day out in the bright sunlight will also help to reset your biological clock.

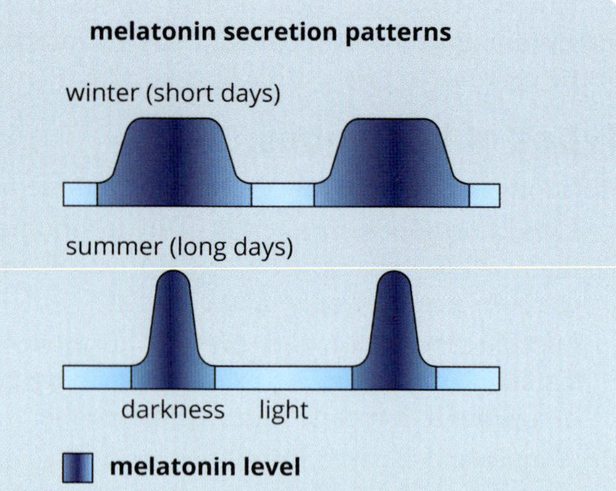

FIGURE 14.2.4 The pineal gland secretes the hormone melatonin in darkness. Because long and short nights produce different patterns of melatonin secretion, animals can tell what season it is.

The pituitary gland

The pituitary gland is a small gland that lies at the base of the brain (Figure 14.2.3). The pituitary is intimately associated with the hypothalamus (Figure 14.2.3), which is a collecting centre for internal stimuli from all over the body, including food and water satiety (sense of fullness), smell, pain and arousal. The hypothalamus determines whether optimal conditions are being maintained. The hypothalamus also produces hormones, which control the release of certain hormones from the pituitary gland. The combined functions of the hypothalamus and pituitary gland are vital for homeostasis.

The pituitary is often called the 'master gland' of the endocrine system. Several pituitary hormones act to regulate the secretion of hormones from other glands, including the thyroid, adrenal glands, ovaries and testes. Hormones from these glands are involved in a range of cellular processes including growth, reproduction, lactation, kidney function, fat tissue and skin pigmentation.

THE NERVOUS SYSTEM

Animals receive information about their environment and respond to external stimuli though the network of neural pathways that makes up the nervous system. The nervous system works closely with the endocrine system to respond to environmental changes and regulate the internal environment.

The rapid responses characteristic of most animals are brought about by the nervous system. Nervous systems typically involve a more direct pathway of communication between parts of the body than hormonal responses.

Neurons

The **neuron** (or nerve cell) is the functional unit of nervous systems. Neurons are specialised cells with structures that enable rapid transmission of information between cells.

All neurons consist of the same basic components—one or more dendrites, a cell body and an axon (Figure 14.2.5). Branching dendrites receive signals from other cells, and transmit these to the cell body. A single axon conducts a signal from the nerve cell body to nerve endings, which form synapses with other cells.

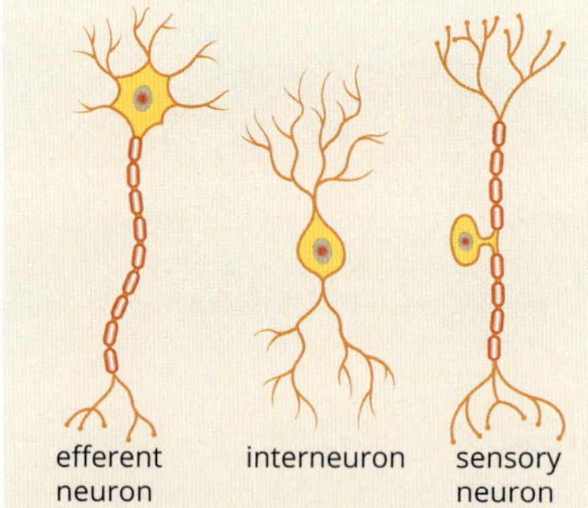

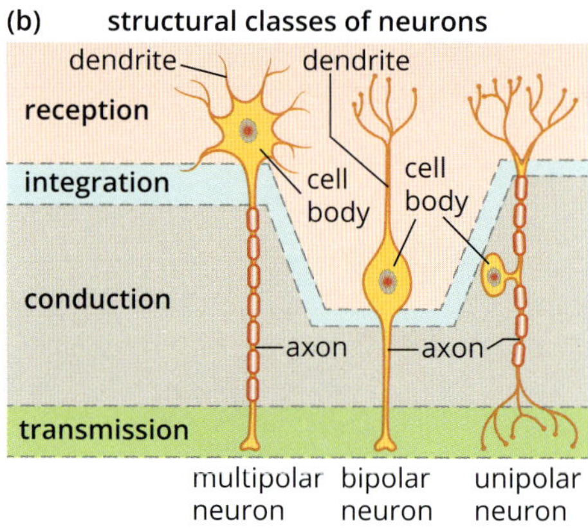

FIGURE 14.2.5 Different types of neurons can be divided into classes based on their (a) function or (b) structure.

Neurons can be grouped by function (Figure 14.2.5a) or structure (Figure 14.2.5b).

- Functional classification—sensory (or afferent) neurons transmit impulses towards the **central nervous system (CNS)**. Most sensory neurons are unipolar. Motor (or efferent) neurons transmit impulses away from the CNS. Motor neurons are multipolar. **Interneurons** (or association neurons) link motor and sensory neurons. Most interneurons are multipolar.
- Structural classification—multipolar neurons are the most common type of neuron in vertebrates. They have two or more dendrites that extend directly from the cell body and a single long axon (Figure 14.2.5b). Other types of neurons include bipolar neurons (single dendrite and two processes from the cell body) and unipolar neurons (have just one extension from the cell body, made up of an axon and dendrite so the cell body appears to sit in the middle of the axon).

The basic structure and function of neurons is very similar across all groups of the animal kingdom. In fact, much of what we know of the function of mammalian nerves comes from studies of frog neurons and giant neurons found in squids.

Neuron function

In nervous systems, nerve cells connect to form pathways between receptor cells and responsive organs, sometimes via a central region such as the brain or spinal cord. When sensory cells detect an environmental disturbance, a signal is generated that passes as an electrical message along two or more neurons to reach particular effector cells, such as muscle or gland. Chemical transmitters communicate from sensory cell to nerve cell, between nerve cells, and from nerve cell to the effector cell that produces the response.

Neural response pathways are highly specific. The 'wiring' of neuronal pathways is direct, and to respond, effector cells must possess specific receptors. Because of the directness of the pathway and the speed of conduction along nerve fibres, control by nerves is usually extremely rapid and short in duration.

Nervous responses require more energy than hormonal responses. Hormones are carried to their target cells in the circulatory system, while neurons use active transport to produce an electrical impulse or action potential. The passage of signals along nerve axons requires considerable energy to re-establish the balance of ions across the cell membrane after each signal. Therefore neurons need many mitochondria and a large supply of glucose.

> Nervous responses are faster and more direct than hormonal responses, but their effects are generally short in duration and targeted to a specific region.

Nervous system complexity

In simple animals, such as *Hydra* and sea stars, the nervous system is a nerve net that links sensory receptors in the skin to muscle cells (Figure 14.2.6).

More complex animals have nerve cells grouped together to form one or more structures similar to small 'brains' (known as ganglia), which can receive and coordinate information from sensory cells in all parts of the body in order to produce an appropriate response.

As animals evolved, locomotion became more efficient, with movement always in the direction of one end of the organism. It was therefore an advantage to have sense organs concentrated at the front (anterior) end, because that end is the first to meet the new environment, with the coordinating brain close to these sensory receptors. That is why most animals have heads. Bundles of nerve fibres pass from the anterior brain to innervate (form a functional connection with) muscle and sensory cells in more distant parts of the body (Figure 14.2.6).

The vertebrate nervous system

In vertebrates, coordination occurs largely in the brain and spinal cord, which together are known as the central nervous system (CNS) (Figure 14.2.7). The CNS has a major role in controlling the activity of the other organs. Information is relayed to and from the CNS by neurons lying outside the CNS, which make up the **peripheral nervous system (PNS)**. The PNS is made up of the system of nerves which branch throughout the body to and from the receptors and effectors. These are nerves that originate in the CNS and connect to all body parts, along with nerves that originate in the organs and connect to the CNS.

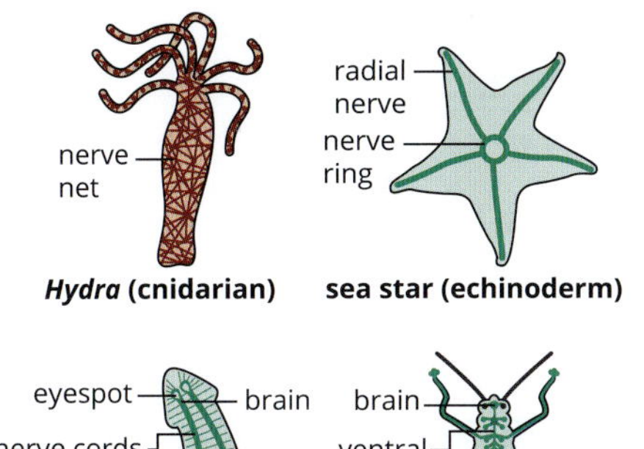

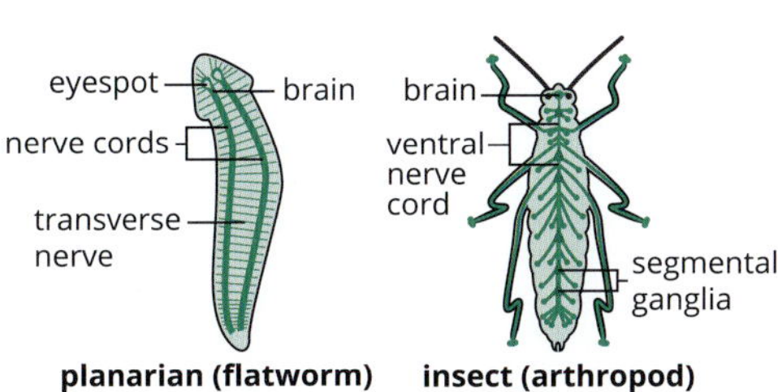

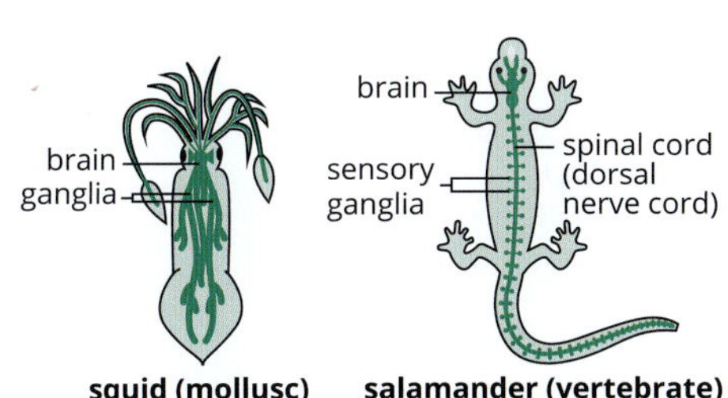

FIGURE 14.2.6 Nervous systems range from simple nerve nets in cnidarians (sea jellies and anemones) and sea stars, to complex networks of nerves connected to a brain in other invertebrates, and to a brain and spinal cord in vertebrates.

Neurotransmitters

Neurotransmitters are a group of hydrophilic signalling molecules secreted by neurons. Neurotransmitters are produced in the synaptic terminal of the neuron.

The mode of transmission of neurotransmitters is paracrine (cell-to-cell) signalling. Neurons secrete neurotransmitters into specialised junctions they form with local target cells, such as other neurons, gland cells or muscle cells. These specialised junctions are called **synapses**. When an electrical signal arrives at the synapse, it triggers the release of neurotransmitters into the intersynaptic space. The neurotransmitters will diffuse across the gap and bind to receptors on the surface of the **postsynaptic neuron**, which in turn cause the transmission of the nervous impulse from one neuron to another.

Neurotransmitters are involved in many cellular responses including movement, regulating hormone production and organ function. Examples of neurotransmitters include serotonin (which influences mood, sleep and appetite) and dopamine (which is involved in reward-motivated behaviour; dopamine acts as both a neurotransmitter and a hormone).

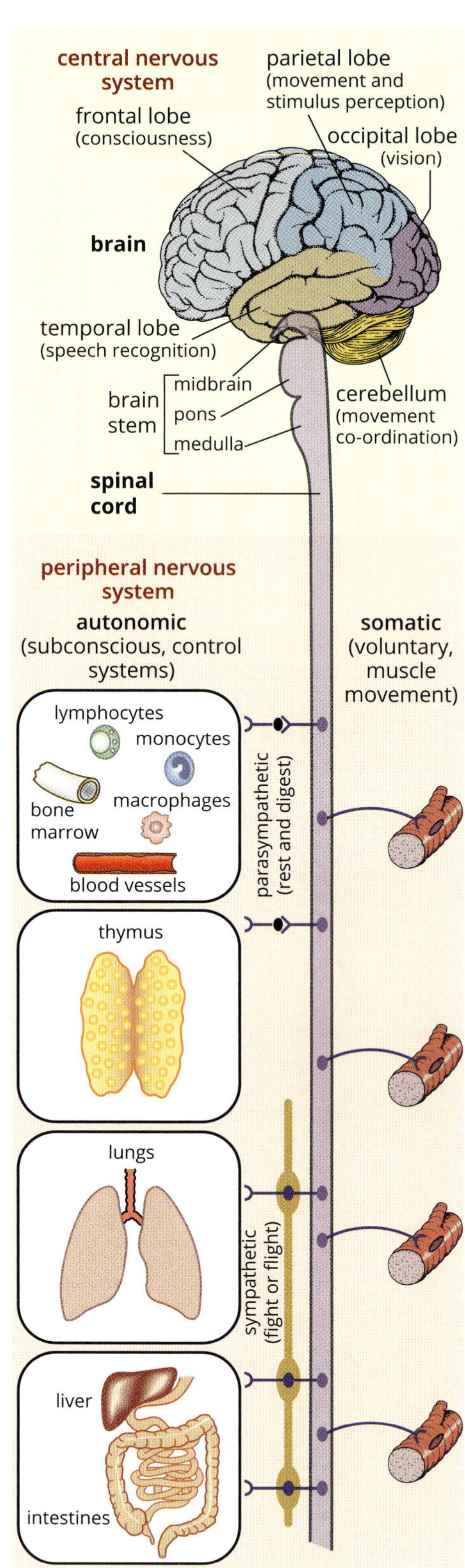

FIGURE 14.2.7 The vertebrate central nervous system (CNS) is made up of the brain and spinal cord. The peripheral nervous system (PNS) is a network of nerves that branch throughout the body and send signals to and from the CNS.

Processing stimuli

After a stimulus has been detected by a receptor, a message is sent along the axon of the receptor cell to the processing centre of the nervous system. In complex animals, this processing centre is the CNS (Figure 14.2.7). Different regions of the brain are associated with particular functions as described below.

- The cerebral cortex (cerebrum) has areas associated with motor activity, sensory input, speech, sight and hearing.
- The hypothalamus receives information relating to the wellbeing of the body, and functions in maintaining homeostasis.
- The cerebellum is involved in the coordination of muscular activity, including posture, balance and movement.
- The brainstem has centres associated with the control of the heart, blood vessels and lung ventilation.

As well as coordinating information from all parts of the body to produce appropriate responses, the brain stores information so that responses can take into account past experiences and learned information (memories).

The PNS includes sensory nerves, which carry information towards the CNS, and motor nerves, which carry signals to effector organs such as muscles and glands. The motor component of the PNS can also be subdivided into **somatic nervous system** (voluntary) and **autonomic nervous system** (involuntary). The voluntary nervous system involves functions over which you have voluntary control, such as movement of the body by skeletal muscles.

The autonomic nervous system includes nerves involved in unconscious responses, such as constriction of pupils, secretion from glands and heart rate changes. It conveys signals to smooth muscle (internal organs), heart muscle and glandular tissues, and regulates the activities of the digestive, cardiovascular, excretory, respiratory and endocrine systems.

The autonomic nervous system consists of two major subdivisions (sympathetic and parasympathctic) and a third smaller subdivision (enteric).

- The **sympathetic division** increases energy use and prepares the body for action in emergency situations by increasing the heart and metabolic rate (the so-called 'fight or flight' response).
- The **parasympathetic division** enhances activities that conserve energy, such as digestion and slowing the heart rate.
- The **enteric nervous system** is an extensive network of nerve cells (and reflexes) within the wall of the gut that coordinate digestive functions.

Having the correct balance of activity in different divisions of the nervous system along with their interactions with the endocrine system is essential to maintaining homeostasis.

Detecting external and internal stimuli

Animals have sensory receptors to detect aspects of their environment that may affect their ability to survive and reproduce. The types of sensory receptors present and their sensitivity differ substantially between animals, and are related to the way they have adapted to their environments. For example, a wombat has less visual acuity for distinguishing small objects than an eagle, dogs use chemical scents much more than humans, and some moths have **chemoreceptors** that can detect a single molecule of pheromone. Some animals respond to different parts of the energy spectrum. For example, snakes can detect infrared radiation, bees see ultraviolet light and platypuses can detect weak electric currents.

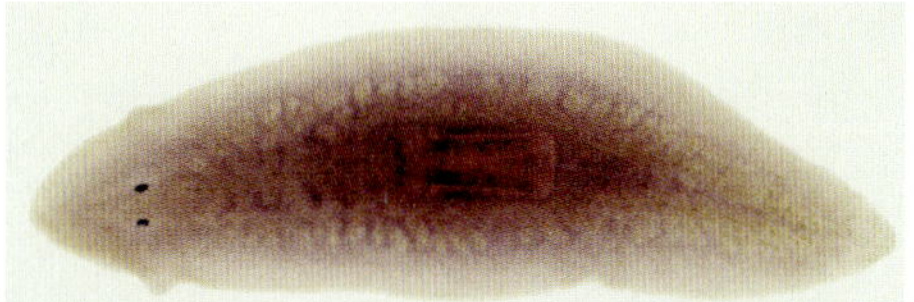

FIGURE 14.2.8 Planarians have simple light-sensitive receptors, called eye-spots, that help them detect and move away from light.

In humans, the five senses (vision, hearing, taste, smell and touch) are perceived through sense organs (eyes, ears, tongue, nose and skin) that collect and process sensory information. Receptors that detect external stimuli are known as **exteroreceptors**. These are usually located close to the surface of the body and detect stimuli such as pain and pressure. Some receptors detect internal states, such as blood pressure and blood chemistry (e.g. oxygen and carbon dioxide levels), and these are known as **interoreceptors** or visceral receptors. From a functional point of view, the types of sensory receptors involved in these senses can be classified as **photoreceptors** (vision) (Figure 14.2.8), chemoreceptors (taste, smell, communication) (Figure 14.2.9), **mechanoreceptors** (hearing, balance, pressure, touch) (Figure 14.2.10) and **thermoreceptors** (temperature).

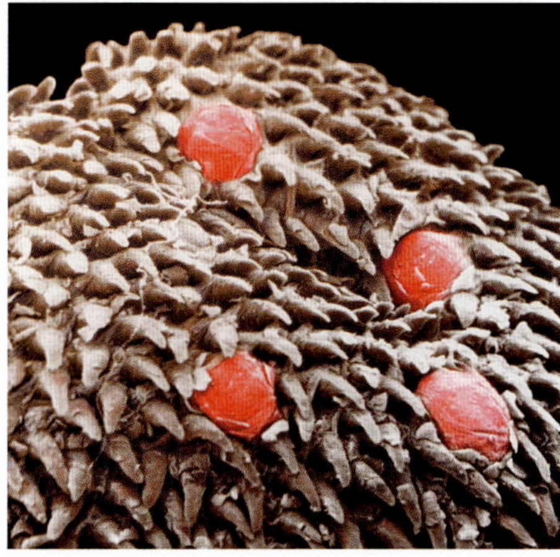

FIGURE 14.2.9 Coloured scanning electron microscope (SEM) image of the surface of the human tongue. The scale-like projections are filiform papillae, which sense pressure. The red areas are fungiform papillae, which contain the taste buds.

OSMOREGULATION

Some animals maintain water balance simply by living in environments where freshwater is freely available. Others can regulate the composition of their internal environment, allowing them to live in drier or saltier environments. The maintenance of water balance in an organism is known as **osmoregulation**.

Maintaining water balance is necessary to control salt concentrations. Salts form ions in solution and cells require the concentrations of ions to be held within narrow limits for biochemical processes to occur efficiently. Some ions (such as the hydrogen carbonate ion) are also important for regulating the pH of body fluids, which must be at a suitable pH for enzymes and other molecules to function efficiently. Maintaining the correct concentrations of ions is achieved by regulating both water and salt balance.

Water balance involves regulating the intake and loss of both water and salts. In organisms, net movement of water occurs as a result of **osmosis**, which is regulated by solute concentrations. Water moves across a semipermeable membrane from regions of lower solute concentration (higher concentration of free water molecules) to regions of higher solute concentration (lower concentration of free water molecules).

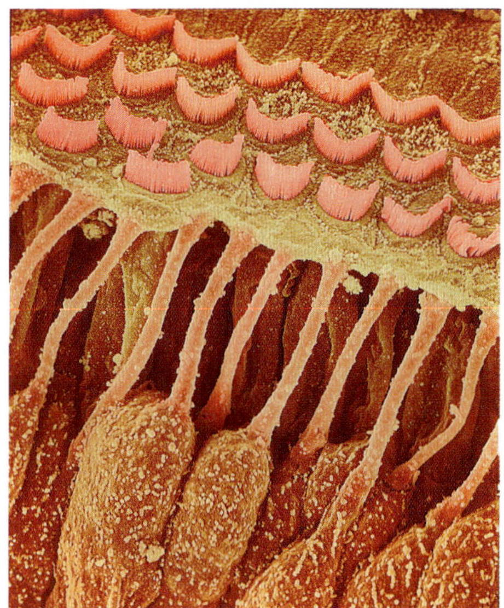

FIGURE 14.2.10 Coloured SEM of sensory hair cells in the cochlea. The inner ear converts sound waves into nerve impulses by stimulation of the stereocilia (the pink, crescent-shaped 'brushes') at the ends of the hair cells.

Water gain and loss

The total volume of fluid taken into the body depends on diet and activity levels, and typically varies from 2–6 L per day. The minimum water requirement for fluid replacement in a 70 kg person in a cool climate is about 3000 mL per day. Of this, about 400–600 mL is obtained by eating, and about 400 mL is produced by aerobic respiration. (This is called metabolic water because it is produced in cellular respiration.) The remainder of about 2000–2200 mL must be obtained by drinking.

For this person, water will be lost mainly in urine (500–1500 mL per day), evaporation from the respiratory system (400–800 mL per day), sweat (100–800 mL per day) and faeces (100–200 mL per day).

Salt gain and loss

Salt intake varies greatly depending on diet. The three major salt groups in the human diet are sodium salts, potassium salts and calcium salts.

Daily sodium salt intake (mostly as sodium chloride) ranges from about 1.0 to 10.0 g per day, mainly in bread, meat and processed cereal products. Highly processed foods usually contain more sodium salts than unprocessed foods.

Daily potassium salt intake (mainly potassium chloride and potassium citrate) ranges from about 2.0 to 4.0 g per day. Highly processed foods usually have a much lower potassium salt content that unprocessed foods.

Daily calcium salt intake (mainly in dairy foods and green vegetables) is up to about 2.4 g per day.

Salts are lost mainly in urine but also in sweat and faeces. The **nephrons** in the kidney filter excess salts from the blood and excretes them into the urinary system. However, most salts are reabsorbed into the blood plasma for recirculation to tissues.

Hormonal control of water balance

Osmolality is a measure of the concentration of particles (such as sodium and chloride ions) that affect osmosis.

Water and solute concentration are monitored by **osmoreceptors** in the hypothalamus and **baroreceptors** in the atria of the heart. Osmoreceptors are sensitive to blood solute concentrations, while baroreceptors detect changes in blood pressure, which is an indication of the volume of blood. Collectively, these receptors detect the solute concentration in blood and extracellular fluid. The unit of measurement used for these blood solute concentrations is **osmolality**, because they contribute to osmotic effects on cells.

Since cell membranes are permeable to water, the osmolality in the extracellular fluid is about the same as in the **intracellular fluid** (cytosol). Changes in the osmolality of the extracellular fluid will therefore affect cytosol concentrations, which can cause problems with cellular metabolic reactions. Compared to extracellular fluid, the cytosol of cells is high in potassium and magnesium and low in sodium and chloride ions.

Antidiuretic hormone (ADH), also called vasopressin, regulates water reabsorption. It is synthesised in the hypothalamus and transported to the posterior pituitary gland, where it is stored (Figure 14.2.11). When osmoreceptors in the hypothalamus detect an increase in the osmolality of the blood a signal is sent to the posterior pituitary gland and ADH is released.

ADH acts on the kidneys to increase the permeability to water of the distal tubules and collecting ducts. The collecting ducts run through the medulla of the kidney, which has high salt levels (and therefore a higher osmotic potential). This causes the absorption of water from the tubules back into the blood by osmosis, decreasing urine output—the urine becomes more concentrated and has a darker yellow colour. As the blood returns to a normal concentration, negative feedback stops the production of ADH (Figure 14.2.11).

Conversely, if the osmoreceptors detect a decrease in osmolality (e.g. if too much water has been taken into the body), ADH release will be stopped. This reduces the reabsorption of water and consequently increases urine volume—the urine becomes more dilute and has a paler yellow colour.

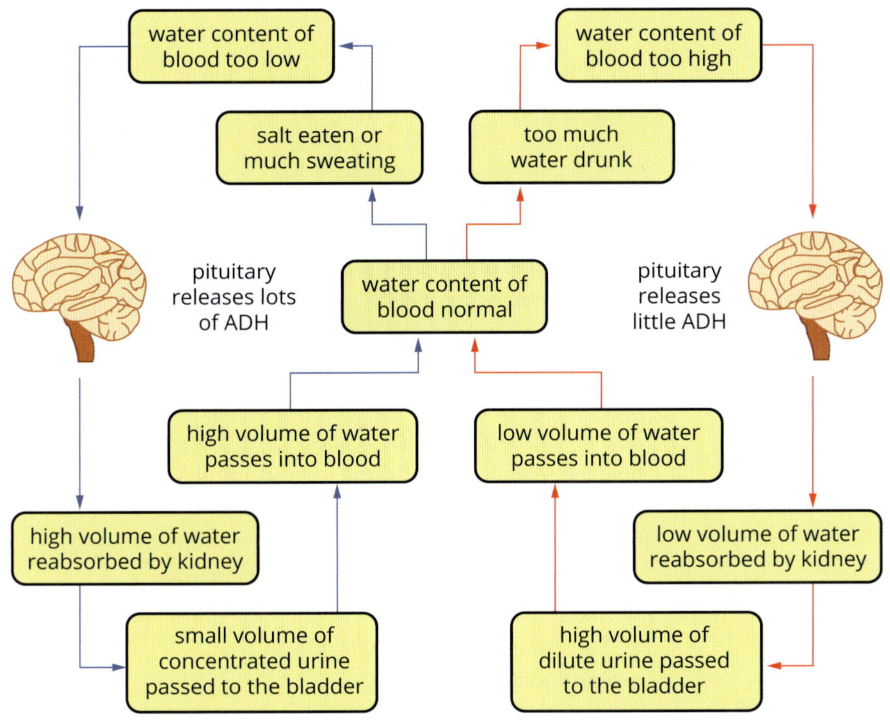

FIGURE 14.2.11 Hormonal control of water balance by antidiuretic hormone (ADH)

A number of substances such as nicotine, alcohol and narcotics can interrupt the feedback control of water balance in the body. This can also occur because of pain, stress or hypothermia (lowered body temperature).

Changes in blood osmolality or blood pressure also stimulate counteracting responses. Initially an enzyme called **renin** is secreted from the kidneys in response to these changes. Renin then triggers a series of reactions involving other hormones that results in the release of **aldosterone** from the adrenal glands situated above the kidney. Aldosterone simultaneously regulates sodium and potassium levels by increasing potassium excretion into the urine and causing sodium reabsorption into the blood. This causes more water to be drawn into the blood by osmosis, thus increasing blood volume and pressure (Figure 14.2.12).

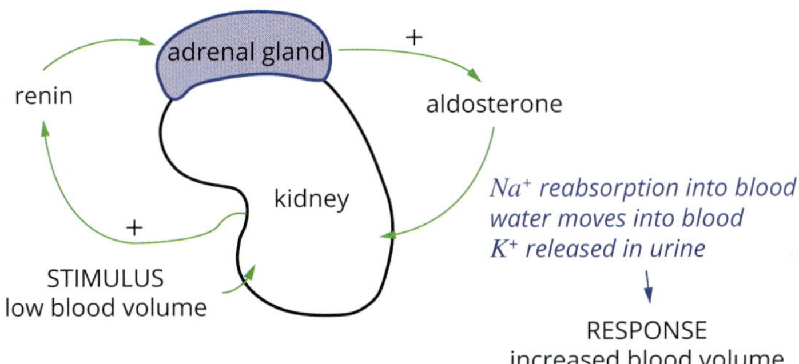

FIGURE 14.2.12 Hormonal control of sodium and potassium levels by renin and aldosterone

A lack of aldosterone can result in low sodium levels, high potassium levels and high acid levels in the blood. These are potentially dangerous conditions. People with an aldosterone deficiency suffer from Addison's disease and must take a synthesised hormone called fludrocortisone acetate.

You will learn more about kidney function in Chapter 18.

GO TO ➤ Section 18.3 page 629

14.2 Review

SUMMARY

Signalling molecules

- Signalling molecules transmit information about an internal or external stimulus to other cells, including, ultimately, the effector cells, which enact the response. These components are essential to feedback loops.
- Animal hormones are produced by organs and glands in animals.
- Neurotransmitters transmit signals to various types of cells that form synapses with neurons.

The endocrine system

- Hormones are signalling molecules that regulate the growth or activity of specific target cells, as a result of interaction with specific receptors.
- Hormones are generally slower and more indirect than nervous responses.
- Hormones can affect particular cells in widely separated tissues because they are effective at low concentrations and target cells have specific receptors.
- Complex animals have endocrine glands that typically release hormones directly into the circulatory system. In vertebrates the pituitary gland has a pivotal role in overall endocrine regulation. The pineal gland is involved in many biorhythms.

The nervous system

- The nervous system involves a more direct pathway of communication than hormonal responses. Control by nerves is generally rapid, short in duration and precisely located.
- Highly organised nervous systems have a central coordinating group of neurons (CNS—brain and spinal cord), and peripheral pathways of sensory and motor nerve fibres (PNS).
- The peripheral nervous system is composed of the somatic system and the autonomic (sympathetic, parasympathetic and enteric divisions) systems.
- Sensory receptors, often grouped into sense organs, monitor conditions in an animal's internal and external environment allowing a homeostatic response.
- The neuron is the functional unit of the nervous system and its basic structure and function are very similar across all groups of animals.

Water and salt balance

- Osmoreceptors in the hypothalamus and baroreceptors in the atria of the heart detect the osmolality of the blood.
- An increase in osmolality causes:
  - release of ADH from the pituitary
  - ADH acts on the kidney to increase water absorption back into the blood.
- As a result osmolality of the blood decreases and volume increases.
- A decrease in osmolality causes:
  - decrease in ADH levels
  - increase in urine volume.
- Low blood volume stimulates the secretion of aldosterone, which increases absorption of sodium into the blood and potassium excretion into the urine.

KEY QUESTIONS

1. What are 'signalling molecules'?
2. Describe briefly how signalling molecules are transmitted.
3. Describe the feature of a target cell that makes it receptive to a particular signalling molecule.
4. Hormones and neurotransmitters are both signalling molecules. Distinguish between the two.
5. Which type of nervous system response protects the body from further pain and injury?
6. Explain how hormones that circulate throughout the blood can act only on a specific type of target cell.
7. Name the types of neurons involved in a simple monosynaptic reflex and outline their function.

14.3 Adaptations for homeostasis

Adaptations are characteristics that increase an organism's likelihood of survival and reproduction in a particular environment. Adaptations are the result of the evolutionary process of natural selection—those organisms that are best suited to their environment survive and reproduce, passing on their advantageous adaptations to their offspring.

In this section you will learn about the different types of adaptations for homeostasis in plants and animals. Although they are presented below as primarily structural or physiological adaptations (and in animals, behavioural), all types function together to maintain homeostasis.

ANIMAL ADAPTATIONS

Animal adaptations for homeostasis include various behaviours for controlling temperature. The other adaptations animals rely on include anatomical structures and physiological mechanisms. Animals very rarely employ just one homeostatic system and most often use all three in combination.

Behavioural adaptations of animals

Most animal behavioural adaptations for homeostasis relate to thermoregulation.

Animals living in hot environments generally avoid the sun. This can mean simply finding shade and taking it easy or only going out at the day's beginning or end. For example, kangaroos are most active at dawn and dusk, while many birds are only active during cool mornings. Some animals are fully **nocturnal** or vary their active periods depending on the season.

To avoid hot environments, certain species construct underground burrows, which can be substantially cooler than the surface (Figure 14.3.1). An example is the white-throated wood rat (Figure 14.3.2), which makes a den within a clump of prickly pear cactus with tunnels underneath that lead to a subterranean nest. The underground environment is cool and stable. Many desert rodents spend the whole day in their burrows. Some burrowing animals also press their bellies to the cool soil to reduce their body temperature.

FIGURE 14.3.1 Coloured X-ray of a mole (Family Talpidae) in its burrow. The network of tunnels can be quite extensive.

FIGURE 14.3.2 The white-throated wood rat (*Neotoma albigula*). These animals dig tunnels, which are covered with an insulating layer of desert plant material.

Very few animals sweat like humans. One exception is the American hairless terrier, having sweat glands all over its body. A few other dogs sweat from pads on their paws. Sweating makes humans exceptionally tolerant of desert heat, although it comes at a tremendous water cost. When active on a hot day, humans may require 12 L of water per day. Yet various other forms of evaporative cooling are quite common, albeit less effective than sweating.

FIGURE 14.3.3 A Labrador retriever panting to reduce its body temperature. Panting is a form of evaporative cooling where water evaporates from the surface of the tongue and cools the blood.

Many animals, including kangaroos and lions, lick their forearm fur to moisten it. Kangaroo forearms have special capillaries for this purpose. Dogs and many other animals pant when hot (Figure 14.3.3). The rapid breathing over the tongue allows water to evaporate from the wet surface, cooling the blood, which circulates around the body. A dog's large nose also substantially cools its brain when exercising. Some large hairless animals (elephants, hippopotamuses and rhinoceroses) coat themselves with water and/or mud to reduce their body temperature via evaporative cooling.

Among birds, variations on the theme include the black vulture's habit of urinating on their own legs if daytime temperatures exceed 21°C. Many birds also engage in **gular fluttering**, rapidly vibrating the throat muscles, which is equivalent to panting. Some, such as flamingos, avoid the heat by standing on one leg, which reduces the area of exposed skin.

Termites build huge thermoregulating structures out of soil. These maintain very stable internal conditions despite high temperatures outside.

Other animal thermoregulation behaviours involve using their **structural adaptations**. Hot birds sometimes pat down their feathers to be thinner and less insulating, or stand up to expose their naked legs. Cold birds often fluff up their feathers for warmth. Many bird species spread their wings in a sunning posture, allowing them to capture the sun's warmth.

FIGURE 14.3.4 Emperor penguins (*Aptenodytes forsteri*) huddling in Antarctica

Lastly, many animals living in extremely cold conditions **huddle** together. For example, during winter, male emperor penguins stand together in groups of many thousands (Figure 14.3.4) for up to 115 days, while the females are away searching for food. Temperatures outside the huddle can reach −70°C and the winds 160 km/h. Huddling reduces body surface area in contact with the elements, while increasing the area in contact with other penguins. By huddling, the penguins reduce heat loss by 80%. The prolonged fasting can consume 50% of the birds' body weight over the season, and reduces core body temperature. Even so, the average temperature inside the huddle stays around 37°C. Individual penguins take turns being on the outside or inside of the huddle.

Musk oxen of northern Canada and Greenland congregate in similar huddles, although not so tightly packed as penguins. Huddling behaviour was probably common among ice-age animals.

Shivering is arguably a behaviour, but equally a physiological adaptation. When warm-blooded animals become too cold, their muscles reflexively start twitching uncontrollably to create warmth.

BIOFILE CCT

Camels—homeostatic marvels

Dromedaries (*Camelus dromedaries*) (Figure 14.3.5) combine numerous adaptations to a desert environment. Desert conditions push homeostatic mechanisms to the limit, creating huge physiological variations that would kill most other animals.

Most mammals can lose a maximum of 15% of body fluids before becoming seriously dehydrated. Dromedaries can lose up to 25% and are able to last a week without drinking, even during temperatures over 50°C. During this time they only lose 1.3 L of water per day, compared to 20–40 L per day for livestock. One reason is dromedaries' water-conserving nostrils, which collect water vapour and return it to the body. Camel kidneys are also very efficient, producing extremely concentrated, syrupy urine. When they do drink, a 600 kg dromedary can consume 200 L in three minutes.

Camel adaptations also permit them to cope with temperature extremes. A camel starts the day with a body temperature of 33.8°C and finishes at 41.6°C. The fat hump acts as insulation, while the total lack of fat on the rest of the body permits cooling. Finally, camels have a countercurrent blood flow (see Figure 14.3.13) in their heads, which protects the brain from overheating.

FIGURE 14.3.5 Dromedary (single-humped) camel (*Camelus dromedaries*)

Structural adaptations of animals

Structural adaptations are anatomical or morphological features that help organisms survive in a particular environment. Structural adaptations include specific internal and external homeostatic organs, insulating body coverings, plus an animal's overall shape and proportions.

Homeostatic organ—the liver

Hormones regulate the functions of the liver (Figure 14.3.6), which is a vital homeostatic organ. After a meal, blood returning to the body from the stomach and intestines has concentrations of sugars and amino acids too high for the maintenance of life processes. For the body to cope, blood from these organs goes to the liver for treatment before returning to the heart.

The liver removes most of the excess sugar molecules, converting them to glycogen, which is the molecule animals use for long-term carbohydrate storage. If incoming glucose exceeds the body's immediate needs, and if the liver is full of glycogen, the liver converts the excess glucose to fat. Fat is stored in **adipose tissue** (fat cells) at various locations around the body.

After this treatment blood returning to the body from the liver has only a slightly higher glucose concentration than normal. Liver-treated blood mixes in the vena cava (the main vein going to the heart) with glucose-depleted blood from the remainder of the body, thereby maintaining a normal concentration.

The process works in reverse when exercise depletes blood-glucose. Body capillaries give up glucose to muscle cells, so blood reaching the liver is glucose-depleted. The liver converts stored glycogen to glucose, which it adds to the blood.

Thus the liver maintains blood-glucose levels within tight tolerances despite large input variation.

The liver has an additional role in the homeostatic maintenance of amino acids. Most animals cannot store large amounts of amino acids or proteins. Any excess must be converted and excreted. In humans and most mammals, the liver removes the NH_2 (amidogen) group from amino acids and converts it to NH_3 (ammonia). Ammonia is combined with CO_2 to produce **urea** (CH_4N_2O), which is less toxic. In birds, the liver coverts the NH_3 to **uric acid** ($C_5H_4N_4O_3$), a white substance that is excreted.

In this case, the liver ensures wastes never reach a high concentration. The liver and kidneys intimately work together to maintain homeostasis in a variety of ways.

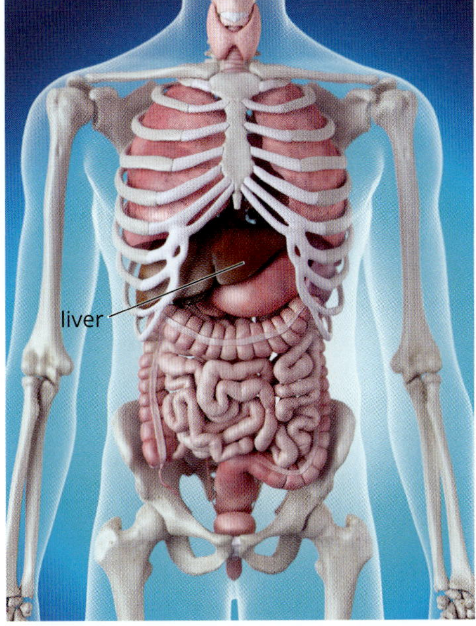

FIGURE 14.3.6 Position of the human liver, a vital homeostatic organ. The stomach is just below the liver and slightly to the right in this picture.

Shape and proportion

An organism's shape and **surface-area-to-volume ratio** (Figure 14.3.7) affect its ability to thermoregulate. Organisms able to generate internal heat do so throughout their entire volume. However, organisms can only lose heat through their surface. The lower the surface area, the less heat is lost. However, surface area and volume do not increase proportionally. Area is measured in square units, whereas volume is measured in cubic units. Thus cubic units increase at an exponential rate, whereas square units increase at a linear rate. That is why, when comparing two animals of the same shape, the smaller individual will lose heat more quickly—its surface area is greater in proportion to its volume.

To compensate for the heat loss, small warm-blooded animals have very rapid metabolism. By comparison, the metabolism of large warm-blooded animals is relatively slow.

A sphere is the most heat-conserving shape. It has the lowest possible surface area relative to its volume. Therefore animals adapted to cold environments tend to be ball-shaped, as do very small warm-blooded animals. In contrast, animals adapted to a hot climate, and all cold-blooded animals, tend to be long and thin. This shape increases the surface area in relation to volume. This permits heat loss in the case of warm-blooded animals and allows cold-blooded animals to gain and lose heat rapidly.

FIGURE 14.3.7 The surface-area-to-volume relationship can be represented by cubes. The largest cube has the smallest surface-area-to-volume ratio (SA:V), while the smallest cube has the largest SA:V. This relationship between surface area and volume explains why small animals lose proportionally more heat compared to large animals of the same shape.

BIOFILE CCT

Allen's rule

In 1877, American ornithologist Joel Allen proposed that that appendage (limbs, tail and ears) and body length depend on latitude and temperature. Warm-climate animals have long bodies and appendages, whereas those of cool-climate animals are short. This relationship is now known as Allen's rule.

Again, the reason concerns surface-area-to-volume ratio (Figure 14.3.7). Long appendages lose heat rapidly, which is suitable for warm climates (Figure 14.3.9a). Cold-hardy animals have short appendages, which conserve heat (Figure 14.3.9b). Allen's rule meshes very well with Bergmann's rule.

FIGURE 14.3.9 (a) The Ethiopian wolf (*Canis simensis*) lives in a hot climate and (b) the Arctic wolf, (*Canis lupus arctos*) lives in a cold climate. There are subtle differences in shape of these animals. The warm-adapted Ethiopian wolf's body, legs, tail, head and ears are all proportionally longer than its Arctic relative. This relationship between body appendage size and climate is consistent with Allen's rule.

BIOFILE CCT

Bergmann's rule

In 1847, German biologist Carl Bergmann described the principle that now bears his name. In species distributed across a temperature gradient, individuals living in cold climates will be generally larger than individuals living in warm climates. This is most obvious in a north-south direction, but can vary with altitude too. Fossil evidence also reveals species changing size over recent geological time as they adapt to changing climates.

Bergmann's rule is not a perfect relationship and a few exceptions are known. However, the rule is a robust generalisation in mammals and birds, especially for large species that cannot seek shelter from climatic conditions in burrows.

Some of the numerous examples include the white-tailed deer, which ranges across most North American latitudes, and koalas (*Phascolarctos cinereus*) in Australia. Koalas that live in the cooler climate of Victoria weigh almost twice as much as koalas living in tropical Queensland (Figure 14.3.8). The rule also applies to humans—people that inhabit regions close to the North Pole are generally larger than people that inhabit regions close to the equator.

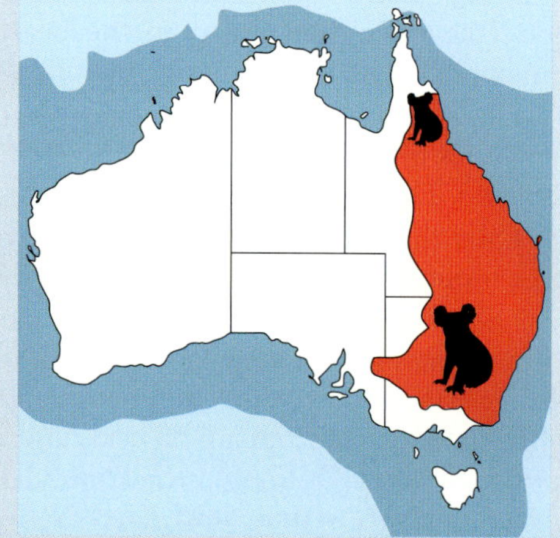

FIGURE 14.3.8 Koala (*Phascolarctos cinereus*) distribution and relative sizes. Koalas vary in size along a north-to-south gradient, in response to temperature; the cool-adapted southern variety weighs almost twice as much as the warm-adapted variety in Queensland. This relationship is consistent with Bergmann's rule.

Radiators

Desert animals often have disproportionately large body parts, usually ears. Examples include the greater bilby (*Macrotis lagotis*) found in arid areas in Australia (Figure 14.3.10a) and the fennec fox (*Vulpes zerda*) (Figure 14.3.10b) of the Sahara desert. These enlarged structures are essentially the opposites of countercurrent circulation (Figure 14.3.13). The ears are richly supplied with arteries that radiate heat to the air cooling the blood and the animal.

Body insulation

Metabolic warm-bloodedness has limits. In cold conditions, warm-blooded animals need help to maintain body temperature in the form of **insulation**. The first way is to trap air against their skin, inside fur or feathers.

Moving air can carry heat away via convection while still air is highly insulating. The more air fur or feathers can trap, the greater the insulation. Body coverings for the coldest conditions are especially thick and fluffy, often with extra microscopic hollow spaces for holding more air. Often animals fluff up their coverings to make them thicker and hold more air.

Many mammals and birds have several types of covering, including a waterproof outer layer and a fluffy inner layer for warmth. Some Arctic mammals produce and shed several coats of varying thicknesses throughout the year. Summer coats may be relatively thin, while winter coats are especially thick and warm (also often white for camouflage in snow). Some mammal furs, particularly that of the polar bear, reflect infrared radiation back to the animal's skin. Polar bear skin is dark, which absorbs heat.

Water conducts heat 27 times more effectively than air. Even so, many marine mammals—such as otters and fur seals—have thick fur to help them cope with icy water (Figure 14.3.11). However, water pressure forces air out of fur. At a certain depth, fur and feathers can no longer hold air, at which point these coverings become useless for insulation. Thus deep-diving mammals, including whales, seals, sea lions and penguins, have a thick layer of insulating fat underneath their skins, called blubber (Figure 14.3.12). Blubber is much less heat conductive than skin or muscle. In some whales, the blubber can be 60 cm thick. These animals thrive in polar waters that would be deadly for humans after just a few minutes.

Cold-blooded animals do not need insulation, nor do large warm-blooded animals living in warm climates. The latter usually have the opposite problem—needing to lose excess heat.

FIGURE 14.3.10 (a) The greater bilby (*Macrotis lagotis*) found in arid areas of Australia and (b) the fennec fox (*Vulpes zerda*) from the Sahara. Both species have exceptionally large ears, which help the animal radiate heat.

FIGURE 14.3.12 The beluga whale (*Delphinapterus leucas*) has particularly thick blubber suited to its Arctic habitat.

FIGURE 14.3.11 The sea otter (*Enhydra lutris*) has the thickest fur of any animal, with more than 155 000 hairs per cm^2.

Countercurrent circulation

In penguins and marine mammals, blubber or other insulation does not cover the whole body. A penguin's feet stick out from beneath its feathers. A whale's face, fins and tail flukes are quite lean. Animals can lose heat through these exposed body parts.

In such cases, these animals use a **countercurrent circulation** (also known as countercurrent heat exchange) (Figure 14.3.13). That is where internal arteries carrying warm blood are right next to surface veins carrying cold blood. The warm blood warms the cold blood, minimising heat loss. Countercurrent circulation can also occur in the opposite direction, cooling the blood of animals living in hot environments. You learnt about countercurrent circulation in Year 11.

GO TO ➤ Year 11 Section 8.2

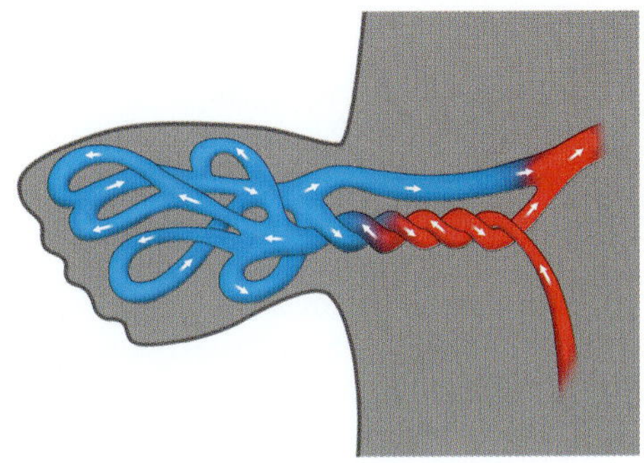

FIGURE 14.3.13 Countercurrent circulation in a seal flipper. This system is used to warm cold blood (blue) from the flipper before it returns back into the main body of the animal.

Physiological adaptations of animals

FIGURE 14.3.14 Thirsty impala (*Aepyceros melampus*) drinking at a river in Namibia, Africa

Animals have a wide variety of physiological adaptations to maintain homeostasis. These adaptations allow animals to live in some of the most extreme environments on Earth, from dry, hot deserts to below-freezing tundras.

Thirst

Thirst, the feeling of wanting to drink water, protects against dehydration (Figure 14.3.14). However, the process is much more complex than it seems. There are two kinds of thirst. The hypothalamus monitors both and controls separate homeostatic mechanisms for dealing with them. One set of receptors monitors blood concentration, the other monitors blood volume.

The body has two kinds of fluids—intracellular fluid and extracellular fluid. Homeostatic mechanisms maintain both at around the same concentration (called **isotonic**). If the fluid concentration becomes unbalanced, water will move from one to the other. If the intracellular fluid is concentrated relative to the extracellular fluid, water will flow into cells, potentially causing them to burst. When intracellular fluid is dilute relative to the extracellular fluid, water is drawn from the cells, causing them to dehydrate.

The first kind of thirst is caused by loss of fluid that does not lower extracellular fluid concentration. This form of thirst may result from blood loss, vomiting or diarrhoea. Low blood volume means that the heart cannot pump efficiently, so it works excessively and may fail. The kidneys detect the reduced blood volume and trigger the sensation of thirst.

The body has several mechanisms for detecting low blood volume, but the main one in this case is the renin-angiotensin system. Low blood volume reduces blood flow. When kidney cells detect the decreased blood flow, they secrete the enzyme called renin. Renin catalyses a protein called angiotensinogen to **angiotensin** I. A second enzyme converts angiotensin I to angiotensin II, which is the active form. Angiotensin II causes the posterior pituitary gland and the adrenal cortex to release sodium, increasing blood pressure, plus the numerous hormones that cause the kidneys to retain water. Angiotensin II, via the subfornical organ (in the hypothalamus), also increases salt appetite and stimulates drinking.

The second form of thirst is activated by increased concentration of extracellular fluid, which draws water from the body cells. The increase can be caused by a high-sodium intake, or reduced volume of body fluids (e.g. blood plasma) due to excessive loss of water (e.g. heavy perspiration). Osmoreceptors in the brain detect the low concentration of blood plasma and the presence of angiotensin II. The osmoreceptors activate the median preoptic nucleus (part of the hypothalamus), which is responsible for water seeking. Damage to this structure results in lack of desire to drink, even given very high salt concentrations in the extracellular fluids.

Water in the stomach is almost immediately absorbed into the blood; however, in humans it takes up to 45 minutes for the water to rehydrate the respective fluids. If humans or animals kept drinking for that entire time they would over-hydrate and die. So sensations involved with drinking temporarily quench thirst. A combination of mouth moistening/cooling and distension of the stomach/intestines tricks the brain into thinking that the body is sufficiently hydrated. The effect lasts long enough for the fluids to actually rehydrate.

> Birds and mammals are endotherms. This means that they can generate their own body heat and maintain a constant body temperature. Nearly all other animals are ectotherms (cold-blooded). The body temperature of endotherms fluctuates substantially depending on external conditions.

Metabolism

The great majority of animals do not maintain a constant body temperature. The ability to do so is a relatively recent evolutionary adaptation only seen in mammals and birds. These animals are called endothermic (warm-blooded) because they generate their own heat internally through metabolic processes.

Warm-bloodedness allows animals to exploit many ecological habitats that are unavailable to **ectotherms** (cold-blooded animals). Yet warm-bloodedness comes at a high energetic cost. Whereas cold-blooded animals can easily go many months or a year without food, warm-blooded animals must eat frequently to obtain the energy required to maintain a constant body temperature.

Size and shape determines whether animals are warm- or cold-blooded. Very large animals are nearly always warm-blooded. If they were ever to cool down, it would take too long for them to heat up again.

Hibernation and torpor

Some animals can be cold-blooded part of the time. Animals that do not migrate yet endure severe winters often sleep through the season (this is known as **hibernation**). Bears are the most famous example, although they are not true hibernators because they regularly wake-up. The group of true hibernators includes several Australian mammals, including four species of pygmy possum, numerous bat species and the short-beaked echidna. The eastern pygmy possum (Figure 14.3.15) has the longest hibernation of any mammal, which is up to one year.

FIGURE 14.3.15 Australia's eastern pygmy possum (*Cercartetus nanus*) a champion hibernator.

During hibernation the body temperature drops and metabolism slows. This conserves energy when the cost of thermoregulation is high and food is scarce. During such times, hibernating animals live off their fat reserves.

Animals with the highest metabolic rates generally eat the most energy-rich foods, meaning sugars. Examples include hummingbirds and some bats. However, the animals cannot sustain their high calorie intake when they are sleeping so they enter a state called **torpor**. It is very much like hibernation, although much briefer.

Torpor is common among small mammals and birds living in cold climates, including around 43% of Australian mammals. Torpor generally lasts overnight, but a few animals remain in that state for several days. At night, these animal's body temperature and metabolic rate drop sharply. Yet in the morning, when they wake up from their cold sleep, they warm up rapidly and resume their normal activities throughout the day. Animals weighing more than 10 kg do not experience torpor.

Cold-blooded animals often hibernate too (although in them it is called **brumation**, since the physiological mechanism is different) and can take it a step further, even tolerating freezing. Their intracellular fluid may freeze, but the high concentration of solutes inside cells lowers the freezing point. Thus the animal can remain alive even while partially frozen. **Aestivation** is a process similar to hibernation, but brought about due to dryness. During aestivation, some turtles and frogs bury themselves under mud for six or more months.

Vasodilation and vasoconstriction

Many animals combine structures that radiate heat (e.g. large ears, Figure 14.3.10) with vasoconstriction or vasodilation. Vasodilation means for blood vessels near the skin to open and let more blood through, allowing heat loss to the external environment to cool the body. Vasoconstriction is the opposite—it is narrowing of the blood vessels to reduce blood flow and heat loss, keeping the body warm. Efficient heat exchange with the environment allows an animal to maintain its core body temperature, protecting its vital organs. In doing so, the extremities are kept cooler than the core region (Figure 14.3.16).

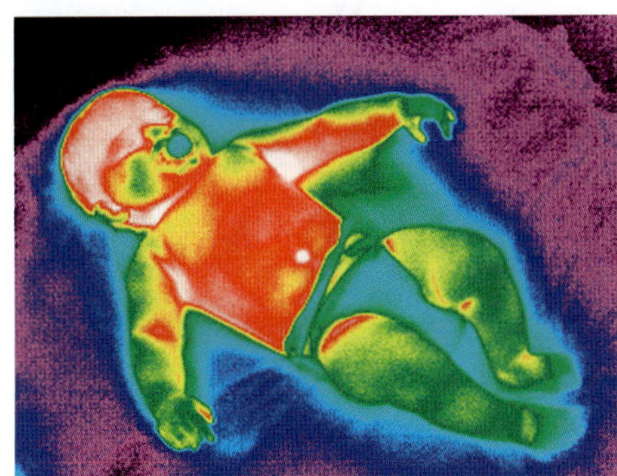

FIGURE 14.3.16 A thermogram of a baby. Different colours indicate varying temperatures throughout the body. The warmest areas are regions where vital organs are (e.g. brain and core region) and the coolest regions are the extremities (e.g. hands and legs).

Vasoconstriction and vasodilation are also methods by which the body regulates blood pressure.

ADAPTATIONS FOR WATER BALANCE IN PLANTS

Like all living things, plants have evolved homeostatic mechanisms. Plant homeostasis is simpler than animals—plants have fewer homeostatic systems and many are not as specialised. Although plants do use hormones, they have no nervous system to control them. For the same reason, plants do not exhibit any behavioural adaptations. However, they do exhibit movement adaptations in response to stimuli (e.g. phototropism). Movement in plants is not under the control of a nervous system, as in animals, but is a result of changes in hormones and water pressure (also called turgor pressure) in cells.

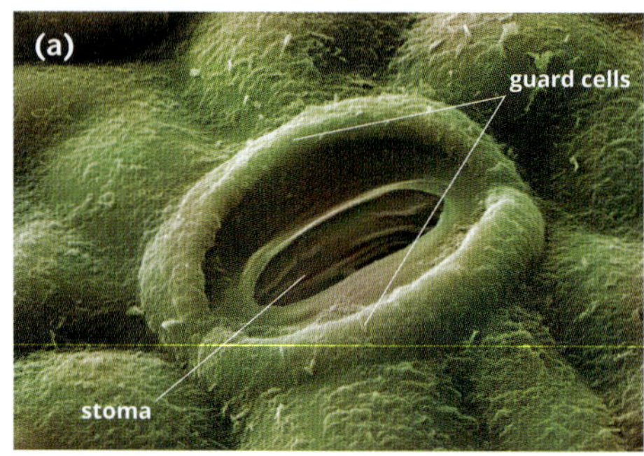

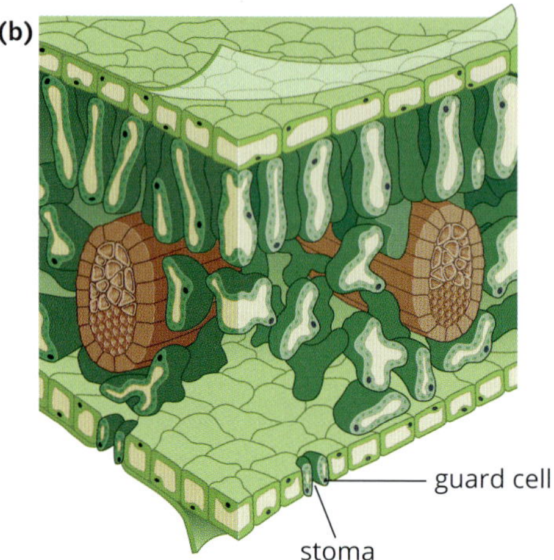

FIGURE 14.3.17 (a) False-colour SEM of an oak leaf stoma. (b) Diagram of a cross-section of a leaf showing the position of the stomata and guard cells on the underside

FIGURE 14.3.18 Resin oozing from tree bark

FIGURE 14.3.19 The Netherlands graveland water lily is a typical hydrophyte.

By far the main concern for plants is managing water balance. Plants must also manage removal of metabolic wastes, including oxygen and carbon dioxide. Additionally, plants need to control intracellular concentrations of major nutrient groups.

Plants have one main structural homeostatic adaptation—the **stomata** (singular stoma), which are small pores in leaves. Stomata have a nonspecific excretory function—to regulate gaseous wastes and water. Plants also have many other structural adaptations for water management that vary depending on habitat.

All other homeostatic processes in plants are physiological. Through such mechanisms, plants can control hormones and the concentration of salts and other electrolytes.

Structural adaptations of plants

Stomata are slit-like pores in leaves, generally on the underside (Figure 14.3.17). Depending on the plant species, stomata lengths range from 10–80 μm (micrometres) and can be up to 50 μm wide. Stomata function as valves, controlling the diffusion of gases into or out of the plant tissues. The gases include oxygen, carbon dioxide and water.

The opening and closing of stomata depends on a water feedback loop. When conditions are good for the plant, that is, when it has enough water and light, the two **guard cells** surrounding the stomata swell up (become **turgid**). The swelling bends the guard cells, which pulls the stomata open (Figure 14.3.17a). In that case, the plant can obtain as much carbon dioxide as it needs, and photosynthesis is not limited. However, if the water level drops, the guard cells shrink, which closes the stomata. This prevents evaporation, but also limits carbon dioxide diffusion and photosynthesis. The stomata remain closed until the plant receives more water.

Although stomata are the plant's excretory structures plants occasionally use other structures for excretion of non-gaseous metabolic wastes. In some cases, wastes (including tannins and essential oils) build up in the **vacuoles** of leaves. When the leaves eventually die and fall off, they take the wastes with them. This is called **abscission**. Similarly, wastes sometimes build up in tree bark, to be shed along with the bark. Many trees exude sticky gums or resins from their bark (Figure 14.3.18). Although humans sometimes find use for these materials, as in the case of latex, they are actually plant waste.

Habitat availability of water affects the position and number of stomata in plant species adapted to that habitat. The **hydrophytes**, **mesophytes** and **xerophytes** are groups of plants that have independently evolved similar traits to manage water conservation and loss in order to maintain homeostasis.

Hydrophytes

Hydrophytes are plants living full-time in freshwater, or very close to it (Figure 14.3.19). For such plants, water supply is never a problem so they have no adaptions for conserving it. Instead, their adaptations help to avoid tissue flooding. Thus hydrophytes have extraordinarily high numbers of stomata on the upper surface of leaves to maximise loss of water. Additionally, these plants have leaves with very large surface areas for maximum transpiration and photosynthesis.

Mesophytes

Mesophytes are land plants adapted to an adequate and regular supply of water. Mesophyte stomata are on the underside of leaves and at medium density. When water is available, as it is in these habitats most of the time, the stomata remain open—the plant can maximise photosynthesis and promote the evaporation of any excess water. Mesophytes can conserve water if they need to by closing the stomata, but otherwise they have no particular adaptations for water conservation. These plants are not especially drought-tolerant because they seldom experience serious water shortages.

Xerophytes

Xerophytes are adapted to dry conditions and have many adaptations for conserving water. In general, xerophytes have few stomata, which may also be protected within a localised depression and/or with stomatal hairs. These hairs help maintain a humid microclimate. Xerophyte leaves are often small and tough to limit the surface area in proportion to volume. These plants often have thick, waxy cuticles (outer layer), as in the case of eucalypts. *Eucalyptus* leaves also typically hang vertically to minimise exposure to sunlight and water loss via evaporation (Figure 14.3.20). Xerophytes also have extensive root systems to maximise water absorption from the soil.

FIGURE 14.3.20 *Eucalyptus* leaves, with a thick cuticle and vertical position minimise exposure to sunlight and therefore water loss.

Sometimes xerophytes have strap-like leaves, which roll up in self-defence when faced with water shortages. For example, marram grasses (*Ammophila* species) are xerophytes that grow in coastal soils. During hot and dry conditions, thin-walled bulliform (bubble-shaped) cells partially collapse, causing the leaves to roll inwards. Hairs on the inside of the leaf trap moisture. The humid microclimate reduces the concentration gradient between the outside and inside of the leaf, reducing evaporation.

A few desert plants (e.g. desert brittlebrush) and plants adapted to the arid conditions of high altitudes (e.g. edelweiss) have leaves coated with fine hairs (Figure 14.3.21). These hairs give the leaves a reflective whitish colour, which helps cool the surface and minimise evaporation. Other desert plants create microclimates, a small space with favourable temperatures, by grouping with other individuals of the same species (Figure 14.3.22).

FIGURE 14.3.21 Edelweiss cultivar (*Leontopodium alpinum*) has a fuzzy surface that prevents water loss.

Cacti and succulents take water-conserving adaptations to extremes. The leaves and stems of these plants often store water. In cacti, the leaves are reduced to zero. The stem is the photosynthesising part of the plant, and may be defended with sharp spines. In extreme cases, cacti stems may be spherical to reduce surface area and decrease light density (Figure 14.3.22).

FIGURE 14.3.22 The giant barrel cactus (*Echinocactus platyacanthus*). The species shows many adaptations to a dry environment, including complete loss of leaves, spines, a water-storing stem that can to swell and a round shape.

14.3 Review

SUMMARY

- Adaptations for homeostasis include structural and physiological types. Animals demonstrate behavioural adaptations as well.
- Structural adaptations of animals include:
  - homeostatic organs such as liver and kidney
  - body-proportions suited to climate
  - insulation
  - counter-current circulation.
- Structural adapations of plants include:
  - leaf shape, size and position to maximise water loss or conservation
  - stomata count and protection (e.g. hairs) appropriate to water availability
  - waxy cuticle.
- Physiological adaptations of animals include:
  - a complex, hormonally-controlled thirst response to dehydration
  - endothermic metabolism to maintain temperature, or if not possible, the ability to endure cold conditions via hibernation and torpor
  - temperature regulation via reflexive control of skin blood flow (vasoconstriction or vasodilation).
- Behavioural adaptations of animals include:
  - seeking sun and/or shade
  - avoiding daylight hours through nocturnal activity or burrowing
  - evaporative cooling via wetting the fur/skin, panting or gular fluttering
  - fluffing up or patting down fur or feathers to adjust the degree of insulation
  - thermoregulating structures made of soil, as in termites
  - huddling for warmth
  - shivering.

KEY QUESTIONS

1. Describe how adaptations are beneficial to the survival of an individual organism and to a species.
2. What are the three main types of adaptations in animals and plants?
3. List and describe two adaptations from each of the three main types described in Question 2, and state whether they are found in plants or animals.
4. What environments would each of the adaptations you mentioned in Question 3 be suited to? Explain your answers.
5. Imagine large and small individuals of the same species. Which would be more vulnerable to heat loss? Explain your answer.
6. How does evaporative cooling work? Give examples of evaporative cooling in animals.
7. What is the liver's role in homeostasis? Outline the mechanism for this control.
8. Give at least three examples of water-conserving adaptations in plants.

Chapter review

14

KEY TERMS

abscission
adaptation
adipose
adrenaline
aestivation
aldosterone
angiotensin
antidiuretic hormone (ADH)
apocrine gland
autonomic nervous system
baroreceptor
blood glucose level
brumation
central nervous system (CNS)
chemoreceptor
conduction
convection
countercurrent circulation
diffusion
eccrine gland
ectotherm
effector
endocrine system
endotherm
enteric nervous system
evaporative cooling
exteroreceptor
extracellular fluid
glucagon
glucose
glycogen
guard cell
gular fluttering
hibernation
homeostasis
hormone
huddle
hydrophyte
hyperglycaemia
hypoglycaemia
hypothalamus
insulation
insulin
interneuron
interoreceptor
isotonic
mechanoreceptor
mesophyte
metabolism
negative feedback loop
nephron
nervous system
neuron
neurotransmitter
nocturnal
non-shivering thermogenesis
osmolality
osmoreceptor
osmoregulation
osmosis
parasympathetic division
peripheral nervous system (PNS)
photoreceptor
piloerection
pineal gland
pituitary gland
postsynaptic neuron
radiation
receptor
renin
shivering thermogenesis
signalling molecule
somatic nervous system
stimulus
stomata
structural adaptation
surface-area-to-volume ratio
sympathetic division
synapse
thermoreceptor
thermoregulation
torpor
turgid
urea
uric acid
vacuole
vasoconstriction
vasodilation
xerophyte

REVIEW QUESTIONS

1 Why is it important for an organism to be able to regulate its internal environment?

2 Arrange the following terms from first to last in the order of their involvement in a physiological response—control centre, effector, receptor, response, stimulus.

3 a Explain the principle of negative feedback in homeostasis.

b Using a diagram, explain how a decrease in body temperature can be brought back up again. In your diagram, draw and label an arrow to show where negative feedback occurs.

4 Briefly explain how cells communicate with each other, with reference to the stimulus–response model.

5 Provide labels for boxes 1, 2, 3 and 4 in the following diagram to complete the steps of the feedback mechanisms that regulate glucose levels in the blood.

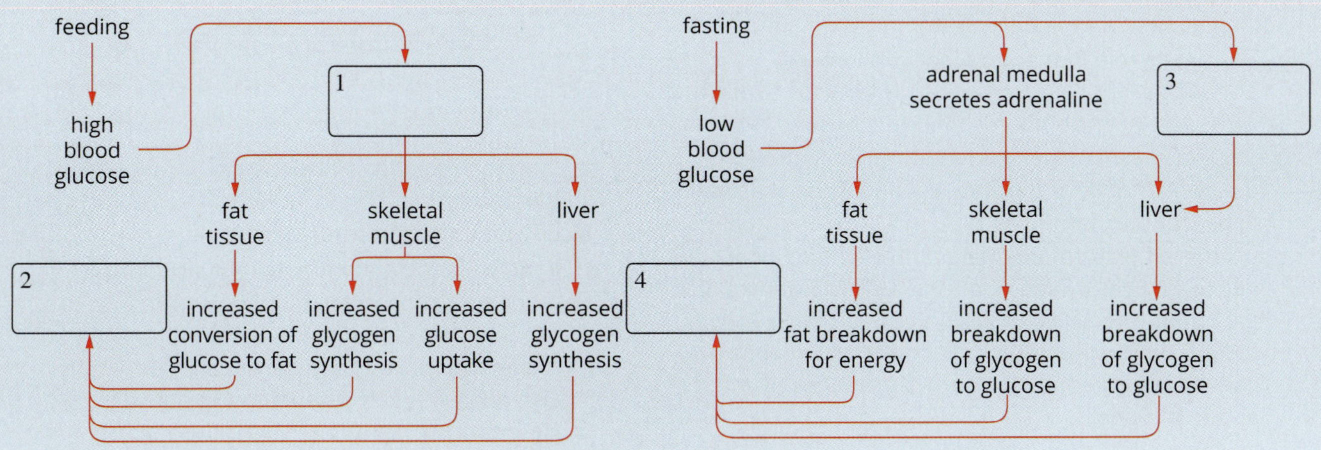

6 Blood glucose levels in normal and diabetic individuals after eating similar meals are shown in the following graph.

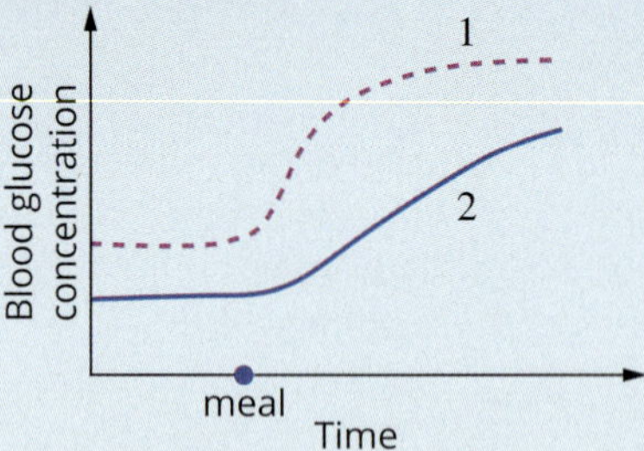

a Which line on the graph represents the diabetic individual? Explain your reasoning.

b i Name the hormone that is produced in insufficient amounts in a diabetic individual.

ii In which organ is this hormone produced in a normal individual?

c i Draw the negative feedback loop for the control of blood glucose levels in the body.

ii Use the example of blood glucose level to explain what is meant by a 'feedback mechanism'.

d Explain how the hormones glucagon and insulin work in the body to regulate blood glucose concentrations.

7 Complete the table below to summarise the five main types of signalling molecules in terms of their sources, modes of transmission and their chemical properties.

Signalling molecule	Source	Mode of transmission	Chemical properties
animal hormone			
plant hormone			
neurotransmitter			
cytokine			
pheromone			

8 Cortisol is an important human hormone. It plays a role in glucose regulation, immune system regulation and regulation of metabolic rate.

a Despite its role in many aspects of human physiology, not all cells respond to cortisol. Explain why some cells do not respond to cortisol.

b Receptors for cortisol are found in the cytosol of the cell. What does this indicate about the chemical nature of this hormone?

c Insulin is the hormone that stimulates the uptake of glucose by cells. Fat and muscle cells are generally particularly sensitive to insulin but cortisol is known to limit their response to this hormone. Propose how cortisol could reduce the normal response by fat and muscle cells.

9 Indicate which statements are true or false.

a One hormone can affect every cell.

b Hormones affect target cells.

c Target cells contain receptors.

d Receptors recognise hormones specific for them.

e Receptors recognise groups of hormones that are specific for them (e.g. peptide hormones).

f Receptors for steroid hormones are located in the cytoplasm and receptors for peptide hormones are located in the cell surface membrane.

10 Outline the difference in the source of plant and animal hormones.

11 Label the parts of the nervous systems of these organisms.

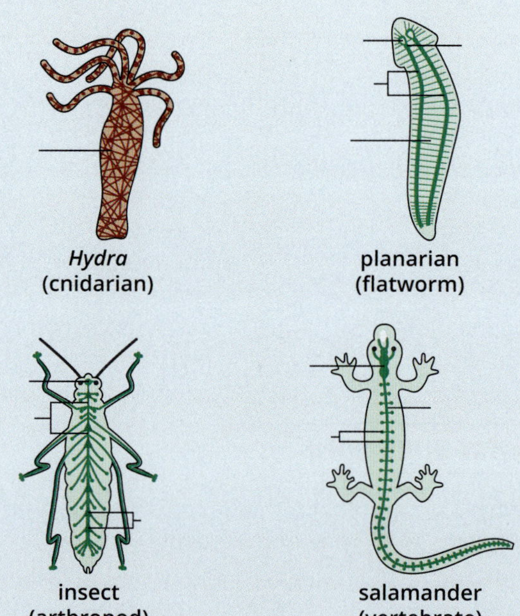

12 Describe the two main parts of the vertebrate nervous system, and state the primary function of each part.

13 Examine the following diagram of a neuron.

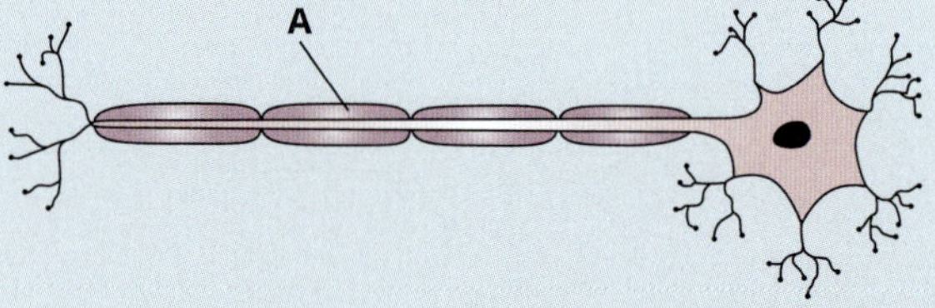

a In which direction will electrical impulses flow along the neuron?

b i Name the structure labelled A.

ii What is the main function of structure A?

c When an action potential passes into a nerve terminal, vesicles containing neurotransmitter molecules move to the nerve cell membrane and release their contents to the outside. Neurotransmitter molecules diffuse across the narrow gap and bind to specific receptors on the membrane of the subsequent neuron. Name the junction between a neuron and the cell it stimulates.

d Explain the similarity between neurotransmitters and hormones.

e Explain why nerve impulses can only be transmitted in one direction.

f Using a table, outline the differences in speed, nature of signal, method of transmission and cells affected between the nervous and endocrine systems.

14 Why is it beneficial for the neural circuit entering a body joint to contain interneurons?

15 Which section of the peripheral nervous system—somatic or autonomic—is involved in unconscious control of internal organs?

16 What are the primary functions of the kidneys? Which structures in the kidney are responsible for these functions?

17 a What does osmolality measure?

b Which two receptors detect changes in osmolality in the blood?

18 What is ADH and what is its role?

19 What change in the blood acts as a stimulus for ADH release?

20 Complete the table below by placing each of the following stimuli next to the receptor that it responds to. body position, temperature, internal stimulus, electrical current, touch and pressure, chemical stimulus, external stimulus, blood pressure

Receptor	Stimulus
baroreceptor	
chemoreceptor	
electroreceptor	
exteroreceptor	
interoreceptor	
mechanoreceptor	
proprioreceptor	
thermoreceptor	

21 Describe how negative feedback is involved in ADH action and water balance.

22 Define the term 'adaptation'. Describe how adaptations are beneficial for individuals.

23 What is the homeostatic purpose of the behavioural adaptations discussed in this chapter?

24 The ability of the wood frog to survive freezing conditions in Canada is best interpreted as a:

A structural adaptation

B behavioural adaptation

C physiological adaptation

D limiting factor in reproduction

25 Describe the relationship between surface-area-to-volume ratio and thermoregulation in animals.

26 What are vasoconstriction and vasodilation? Under what conditions do animals respond with each, and why?

27 Endothermic and exothermic animals regulate their body temperatures in different ways. Consider the following graph, which shows the body temperatures of a bobcat (pink line) and a snake (blue line) for different ambient environmental temperatures.

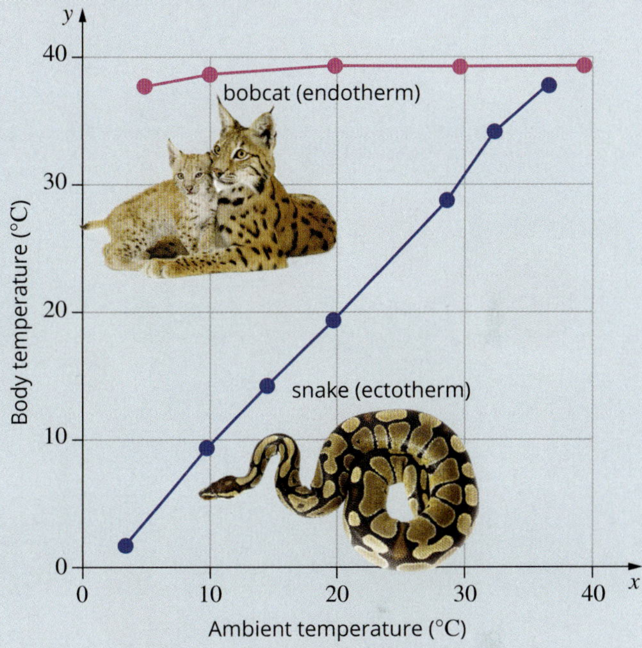

a When the ambient temperature is 30°C, what is the body temperature of the snake?

b Why does the body temperature of the snake continue to increase as the ambient temperature increases, but the body temperature of the bobcat does not?

28 What is the most important structure for osmoregulation in plants? Explain the differences in this structure between hydrophytes, mesophytes and xerophytes.

29 After completing the Biology Inquiry on page 502, reflect on the inquiry question: How is an organism's internal environment maintained in response to a changing external environment? Describe the relationship between surface-area-to-volume ratio and homeostasis. Provide at least two examples of adaptations that improve an organism's ability to maintain their internal environment in response to changes in the external environment.

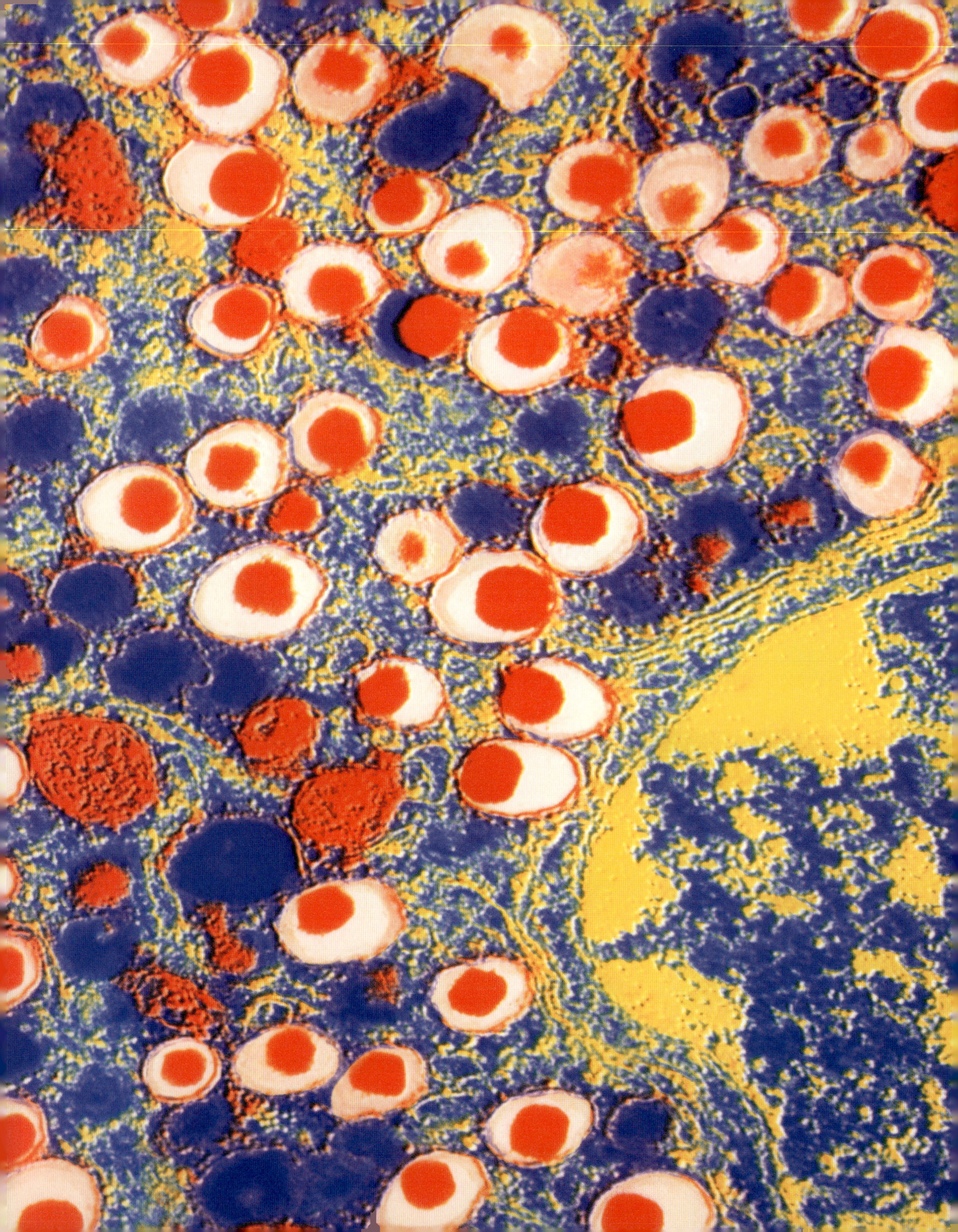

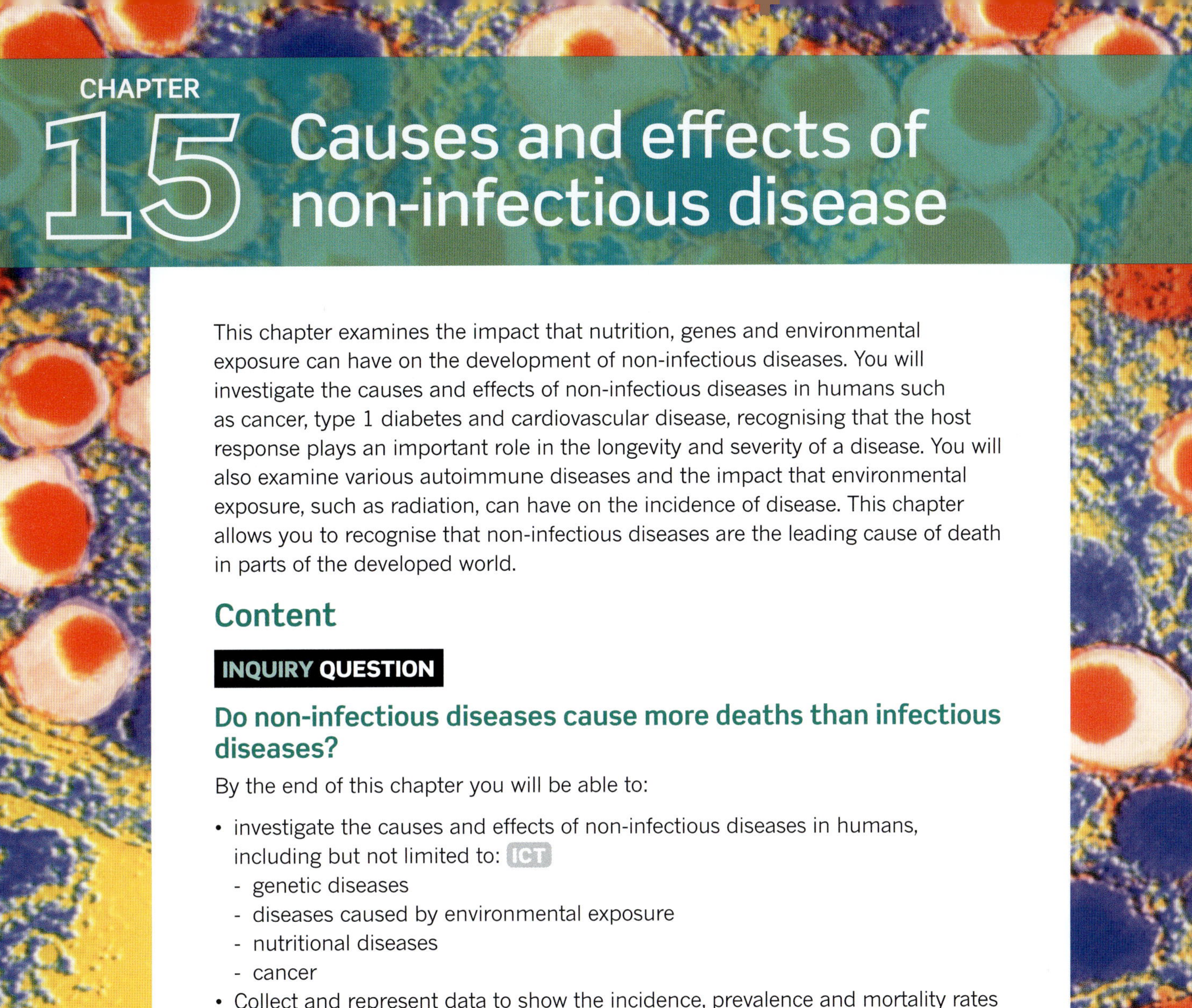

CHAPTER

15 Causes and effects of non-infectious disease

This chapter examines the impact that nutrition, genes and environmental exposure can have on the development of non-infectious diseases. You will investigate the causes and effects of non-infectious diseases in humans such as cancer, type 1 diabetes and cardiovascular disease, recognising that the host response plays an important role in the longevity and severity of a disease. You will also examine various autoimmune diseases and the impact that environmental exposure, such as radiation, can have on the incidence of disease. This chapter allows you to recognise that non-infectious diseases are the leading cause of death in parts of the developed world.

Content

INQUIRY QUESTION

Do non-infectious diseases cause more deaths than infectious diseases?

By the end of this chapter you will be able to:

- investigate the causes and effects of non-infectious diseases in humans, including but not limited to: ICT
  - genetic diseases
  - diseases caused by environmental exposure
  - nutritional diseases
  - cancer
- Collect and represent data to show the incidence, prevalence and mortality rates of non-infectious diseases, for example: ICT N CCT DD
  - nutritional diseases
  - diseases caused by environmental exposure

15.1 Non-infectious disease in humans—causes and effects

BIOLOGY INQUIRY  

Disease in the news

Do non-infectious diseases cause more deaths than infectious diseases?

COLLECT THIS...

- tablet or computer to access the internet
- pen
- notebook
- large sheet of paper
- coloured pencils

DO THIS...

1 Working in groups of two to four, search the internet using the term 'disease Australia'.
2 Narrow your search results to news articles only.
3 Record the number of articles that report on non-infectious disease and the number that report on infectious disease in the top 10 search results.
4 Discuss your observations with your group members. Consider the main themes of the articles, such as research, education campaigns, prevention, control or outbreaks. Note any trends you see.
5 Now choose a developing country and add it to the end of your search term. For example, 'disease Sudan'.
6 Record the number of articles that report on non-infectious disease and the number that report on infectious disease in the top 10 search results.
7 Discuss your observations with your group members, considering the main themes of the articles and any trends you see.
8 If you have time, search for two or three different developed and developing countries with the search term 'disease'.

RECORD THIS...

Describe the trends you observed in your news article survey.

Present your observations and ideas in a mind map or poster.

REFLECT ON THIS...

What factors influence the incidence, prevalence and mortality rates of non-infectious and infectious diseases in a community?

How do news articles reflect the current priorities for health and disease prevention in a society?

Do non-infectious diseases cause more deaths than infectious diseases?

A disease is any condition that impairs, or has the potential to impair, the normal functioning of the body.

GO TO ➤ Section 14.1 page 502

Just like infectious diseases, **non-infectious diseases** (also called non-communicable diseases) can be life threatening and very harmful to the host. Non-infectious diseases are not caused by a pathogen and therefore cannot be transmitted to another individual in the way infectious diseases can. Many non-infectious diseases are a result of genetic disorders, poor nutrition, insufficient physical activity or exposure to harmful substances.

You have seen in Chapter 14 that regulation of the internal environment, and thus the functionality and health of your body, is maintained by **homeostasis**. These mechanisms are extremely sensitive to changes in internal conditions. A malfunction in a homeostatic mechanism causes an imbalance and a subsequent oversupply or undersupply of substances to cells. Many diseases are associated with malfunctions in homeostatic mechanisms. Hormonal balance plays a key role in regulating the internal environment. The **endocrine system** makes, stores and releases **hormones**, which act as chemical messengers that trigger and direct functions in the body. The endocrine system controls vital functions in growth and development, **metabolism**, reproduction and tissue repair among many others. Malfunctions in the endocrine system often lead to a disruption in a homeostatic mechanism, which can have an adverse effect on the body.

GENETIC DISEASES

Many dysfunctions and diseases of the human body are the result of altered or incorrect expression of genes. Inheriting a genetic mutation can result in disruption to normal **gene expression** and protein production. The symptoms and severity of **genetic diseases** can vary greatly. Technological advances have made the sequencing of whole genomes possible. This has revealed disease-causing alleles (gene variants) and identified their patterns of inheritance. Fortunately, most genetic diseases are rare. This section will investigate genetic diseases and specific autoimmune diseases.

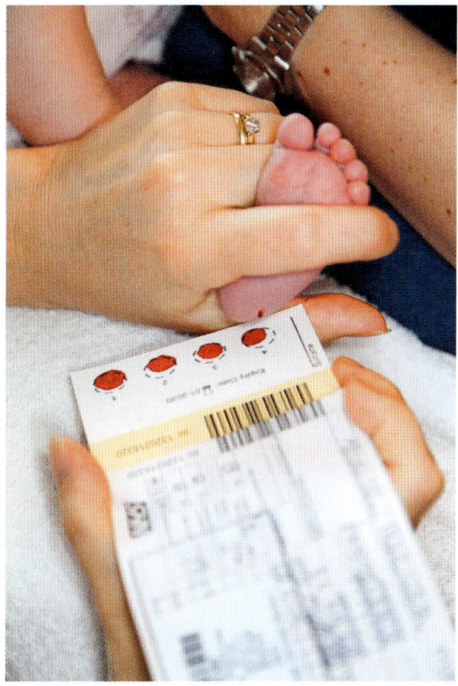

FIGURE 15.1.1 A Guthrie blood test, which screens for the rare genetic disease, phenylketonuria (PKU). PKU causes the amino acid phenylalanine to accumulate in the blood, which can cause severe brain damage. The baby's blood is smeared onto absorbent filter paper and cultured with bacteria that grow in the presence of phenylalanine. The amount of bacterial growth on the paper indicates how much phenylalanine is present in the baby's blood.

Phenylketonuria

Phenylketonuria (PKU) is a genetic disease caused by a mutation in the gene coding for the enzyme phenylalanine hydroxylase, which catalyses the breakdown of **phenylalanine** to tyrosine. This mutation means the body cannot convert phenylalanine to tyrosine, resulting in phenylalanine accumulating in the body. PKU is an autosomal recessive trait, which means that people with PKU have two mutated (recessive) alleles. Approximately 1 in 10 000 babies born in NSW are diagnosed with PKU every year. The consequences of PKU for an individual can include:

- mental retardation
- lack of skin and hair pigmentation
- stunted growth of the head
- seizures.

During pregnancy, a mother's phenylalanine hydroxylase enzyme helps remove excess phenylalanine created from an affected fetus. A few days after birth, breastfed babies that suffer from PKU will have a higher than desired level of the phenylalanine amino acid from consuming and digesting milk proteins. PKU needs to be diagnosed and treated in the first 24 hours after birth and can be detected using a simple Guthrie test (Figure 15.1.1). If diagnosed with PKU, patients need to maintain a diet low in phenylalanine by consuming low protein foods, such as fruits and vegetables; and avoid high protein foods, such as dairy, fish, red meat; and food containing the artificial sweetener aspartame, such as chewing gum and many soft drinks.

Albinism

A condition known as **albinism** is characterised by an absence of the pigment melanin from the skin, hair and eyes. Individuals with albinism appear pale, have pink skin, white hair and pale-coloured eyes. Their eyes may look red because light reflects off normal blood vessels in the back of the eye and the red blood vessels can be seen through the pale iris (Figure 15.1.2). Albinism also occurs in vertebrates other than humans, including birds, reptiles and other mammals (Figure 15.1.3).

Pigment production in vertebrates is controlled by genes and is therefore an inherited condition. Albinism is caused by mutations in the genes that produce the pigment melanin. In humans, mutations in the *TYRP1* gene cause oculocutaneous albinism type 3. The gene encodes the enzyme **tyrosinase**, which is involved in melanin production. Individuals that cannot produce the enzyme tyrosinase cannot produce the pigment melanin.

Babies with albinism are diagnosed at birth and have normal mental development. However, they do have a sensitivity to ultraviolet (UV) light because of the lack of pigment in their skin, making sun protection such as sunscreens and minimal exposure to UV light a priority. Other symptoms of albinism include nystagmus (rapid eye movement), minimised visual acuity and photophobia (discomfort in bright lights). Prescription glasses can assist with the visual problems caused by albinism.

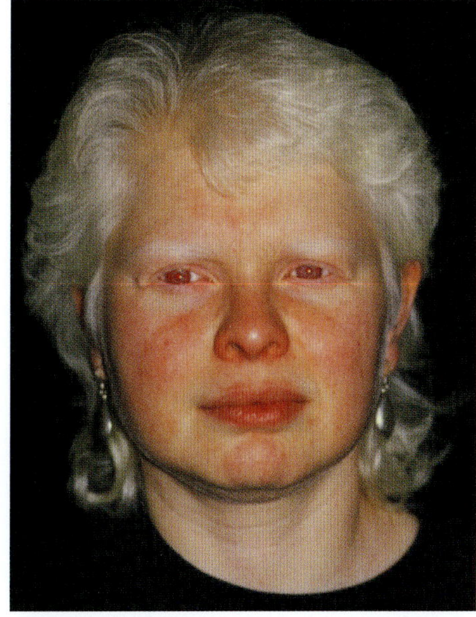

FIGURE 15.1.2 Albinism is a genetic disorder caused by mutations in the genes involved in producing the pigment melanin. A lack of melanin results in white hair, pale skin and pale eyes that can sometimes appear red.

BIOFILE CCT

Genetic markers

Each human genome is unique—even identical twins have slightly different genomes. These small differences are due to mutations and recombination of genetic material during meiosis and sexual reproduction. Although mutations are a source of healthy genetic variation, some can also disrupt normal functioning and cause genetic disease. Predisposition to a genetic disease can be detected through the presence of markers in an individual's DNA. Genetic markers can identify if someone has alleles that are associated with a genetic disease and if they are likely to develop disease symptoms during their lifetime.

Some of the most abundant genetic markers in our DNA are single nucleotide polymorphisms (SNPs). It is estimated that humans have approximately 10 million SNPs throughout our genome. SNPs can act as a type of 'genetic flag' for genetic disease alleles or other alleles. For an SNP to be useful as a genetic marker, it must be found in a large sample of the population. However, to map a genetic disease, multiple SNPs are usually examined. By comparing variations of genetic markers in many members of a population, scientists can identify patterns that link the markers to diseases and other unique differences in human physiology.

Genetic markers can have many functions such as:

- tracing inheritance of genetic diseases
- identifying specific genes that cause a disease
- determining the mode of inheritance for a genetic disease
- helping scientists develop appropriate disease treatments.

FIGURE 15.1.3 Bennett's wallaby (*Macropus rufogriseus rufogriseus*) albino mother carrying her non-albino young in her pouch

Cystic fibrosis

Cystic fibrosis is a genetic disorder that causes an accumulation of mucus in the lungs, resulting in breathing difficulties and frequent lung infections. Although respiratory problems are the most common symptoms, other organs can also be affected. Cystic fibrosis is one of the most common genetic disorders in people of European descent, with 1 in 2500 babies born with the condition every year in Australia. It is an autosomal recessive condition, which means that both male and females are equally affected and they have inherited two copies of the gene (one from each parent). Individuals with one copy of the cystic fibrosis gene are carriers and show no symptoms of the disease. In Australia, 1 in 25 people are carriers of the cystic fibrosis gene. If two carriers of the cystic fibrosis gene have a child, the probability that their child will have cystic fibrosis is one in four (or a 25% chance).

The gene associated with cystic fibrosis is known as the *CFTR* (cystic fibrosis transmembrane conductance regulator) gene and is on chromosome 7. A normal copy of the gene produces the CFTR protein, which forms a channel in the outer cell membrane. This channel is important for the secretion of sweat, mucus and digestive fluids. Mutations in the *CFTR* gene disrupt the production of the CFTR protein, causing sodium ions to move into cells and dry out the mucus layer. When this mucus becomes thick and sticky, cilia in the airways are not able to clear debris, leading to infections in the lungs. There are over 1500 known mutations in the *CFTR* gene that cause cystic fibrosis—different mutations cause varying degrees of severity of the disease.

Symptoms of cystic fibrosis

Most babies born with cystic fibrosis are diagnosed within the first two months with a routine heel prick blood test conducted during newborn screening programs. The symptoms of cystic fibrosis vary from person to person. Because the *CFTR* gene is involved in the secretion of sweat, mucus and digestive fluids, individuals with cystic fibrosis have abnormal mucus secretions in the lungs, pancreas and small intestine. Breathing difficulties and lung infections present the greatest threat to the lives of people with cystic fibrosis. People with cystic fibrosis have high salt levels in their sweat and frequently produce sticky mucus in the lungs, which blocks the airways and increases the chance of infection. The pancreas does not work efficiently in most people with cystic fibrosis and some individuals may also have liver problems.

Common symptoms of cystic fibrosis may include:

- coughing
- shortness of breath
- frequent respiratory infections
- excessive appetite or no appetite at all
- poor weight gain, small in stature
- late onset of puberty
- reflux
- constipation
- tires easily
- sinusitis.

Management of cystic fibrosis

In earlier years, babies born with cystic fibrosis rarely lived beyond childhood. Today, with a greater understanding of the condition and improvements in treatments, medication and research, the majority of babies born with cystic fibrosis in Australia are able to live well into adulthood.

For most suffers of cystic fibrosis, complications in the respiratory system are the most serious. For this reason, many methods of cystic fibrosis management involve breathing assistance. Daily physiotherapy treatment can clear the build-up of thick sticky mucus in the airways. Inhaled medications taken with a nebuliser are also common daily practice. Some cystic fibrosis patients use intermittent positive pressure breathing machines (IPPB). IPPB machines deliver a controlled pressure of gas to help ventilate or expand the lungs, increase tidal volume and ease breathing (Figure 15.1.4). Other management methods are short-lived and last from one to two hours.

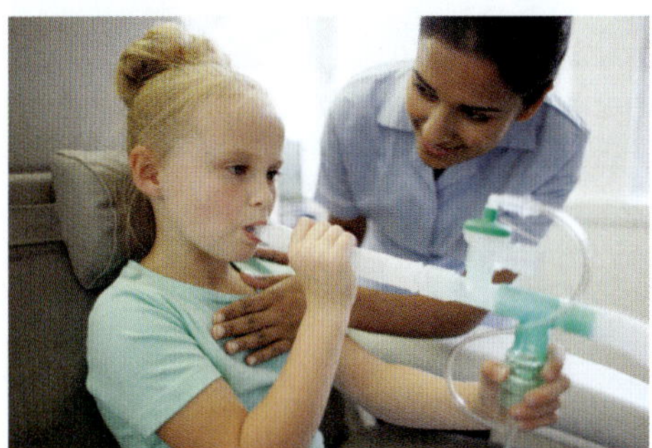

FIGURE 15.1.4 A physiotherapist using an intermittent positive pressure breathing (IPPB) machine to help clear the lungs of a patient with cystic fibrosis.

The lung-related symptoms can be so severe that lung transplants are performed. Lung transplants can greatly enhance the quality of life and extend life expectancy for people with cystic fibrosis. The cells of the transplanted lungs carry two normal copies of the *CFTR* gene but all other cells in the patient's body have two copies of the *CFTR* gene with mutations. For this reason, lung transplants do not alleviate non-respiratory symptoms of cystic fibrosis, such as digestive difficulties and male sterility. Transplant recipients also face problems associated with tissue rejection and therefore need ongoing treatment with **immunosuppressant** drugs.

Gene therapy may be a viable approach for treating lung disease caused by cystic fibrosis. Clinical trials to test cystic fibrosis gene therapies began in 1993. They aimed to identify the appropriate dosage and delivery method of the gene therapy in humans. During these trials, researchers used an adenovirus vector to deliver a normal full length *CFTR* gene into patients. They used the nasal passage as the initial site of delivery, as it was easy to access and measure gene activity from the patient's airways. During trials in 1993, a liposome vector was used and in 1998 an adeno-associated virus was trialled. Recent research has used pigs with cystic fibrosis to identify different virus-based vectors that can restore a working version of the CFTR protein that is faulty in the pigs' airways.

Gene therapy is the delivery of functional genes into an individual's cells or tissues to replace faulty genes.

DISEASES CAUSED BY ENVIRONMENTAL EXPOSURE

In some cases, scientists cannot use inheritance to trace the occurrence of a non-infectious disease. Some diseases are caused by mutations or reactions that are the result of environmental factors. For example, exposure to radiation or certain chemicals can damage DNA. Agents that cause damage to DNA are called **mutagens**. If the damage occurs to proto-oncogenes or tumour-suppressor genes, the control of cell division can be disrupted. Mutagens are covered in detail in Chapter 7.

GO TO ➤ Section 7.2 page 295

Other diseases caused by environmental exposure are due to overreaction of the **immune system** to **antigens** found in the environment, such as those in grass pollen, poison ivy and certain medications. This is known as a **hypersensitivity reaction**. Other types of hypersensitivity reactions involve an immune response against the body's own cells (**self-antigens** or auto-antigens), resulting in what is known as an **autoimmune disease** (see page 540).

Hypersensitivity reactions

GO TO ➤ Section 12.1 page 422

In Chapter 12 you learnt that the role of the immune system is to eliminate invading pathogens. However, when the immune system malfunctions it can react to antigens that are normally harmless, such as those found on **pollen** grains (Figure 15.1.5). When this happens, a hypersensitivity reaction can occur. Although often tolerated by the body, some hypersensitivity reactions can be severe and even fatal.

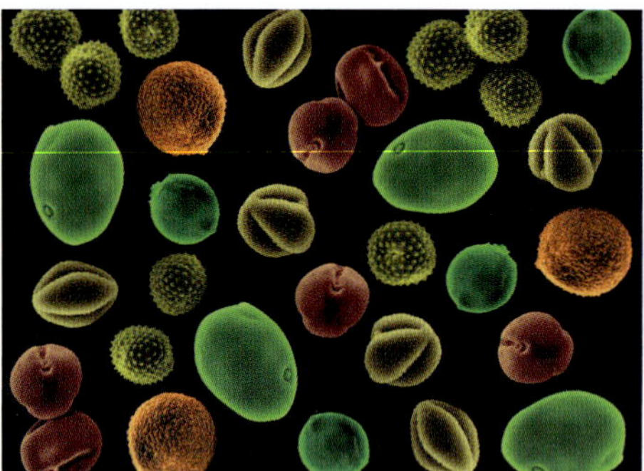

FIGURE 15.1.5 A digitally enhanced coloured scanning electron micrograph (SEM) of pollen grains. Pollen can cause an allergic immune response known as hay fever (or allergic rhinitis).

Hypersensitivity reactions have an underlying genetic component but also often have a strong environmental component as many of the antigens that trigger hypersensitivity reactions are found in the environment (e.g. pollen).

Hypersensitivity reactions occur when the immune system reacts to antigens that pose no threat to the body, causing cell damage and disease. They are classified into four types, depending on the antigens and immune mechanisms that result in the reaction:

- type I (or immediate) hypersensitivity
- type II (or **cytotoxic**) hypersensitivity
- type III (or immune complex) hypersensitivity
- type IV (or delayed-type) hypersensitivity.

Type I hypersensitivity is also known as **allergy**. Autoimmune diseases can result in type II, type III and type IV hypersensitivity reactions. However, autoimmune diseases are not the only causes of these types of hypersensitivity.

Immediate hypersensitivity (type I)—allergic reactions

Immediate hypersensitivity reactions (or allergic reactions) are due to a rapid and vigorous overreaction of the immune system to antigens that would otherwise be harmless. Antigens that result in type I reactions are called **allergens**. Typical allergenic substances include pollen, fur, house dust, latex and foods such as peanuts, lobster and monosodium glutamate (MSG). Depending on the individual and antigen, the hypersensitivity reaction can range from mild to life-threatening. Severe and potentially life-threatening allergic reactions are known as **anaphylaxis**.

> An allergen is a substance that can cause an allergic reaction. The body treats the allergen as foreign and produces a strong immune response to a substance that is generally harmless to the body.

An allergic reaction to pollen is called hay fever (or allergic rhinitis). It is triggered by pollen particles (Figure 15.1.5), which carry allergenic antigens on their surfaces. Grass and tree pollens are the most common cause of hay fever in Australia and New Zealand. Pollen sensitivity has a seasonal pattern as pollen is most abundant during spring and early summer.

Allergic reactions are mediated by a specific type of **antibody** called **immunoglobulin E (IgE)**, which is produced by plasma cells and travels in the bloodstream. When IgE comes into contact with **mast cells**, which are common in epithelial and mucosal tissues, the tail end of the IgE antibody binds to **receptors** on the cell surface. Upon subsequent exposure to the same allergen, the allergen binds to a pair of adjacent IgE molecules, bridging the gap between the two IgE molecules. This binding triggers a cascade of cellular signals that cause the mast cells to release **histamine** (and other mediators of inflammation) (Figure 15.1.6).

Histamine is a compound that is involved in immune response, acting to eliminate foreign particles from your body. When an allergen is detected, histamine is released from mast cells and basophils (types of white blood cells) and binds to histamine receptors on the surface of cells. Histamine causes inflammatory responses such as:

- blood vessel dilation and a subsequent decrease in blood pressure
- increased blood vessel permeability, which allows white blood cells to access the site of antigen contact and rid the body of the foreign particles
- increased blood vessel permeability, which allows fluid to leak from capillaries into tissues causing a runny nose and watery eyes
- contraction of smooth muscles lining the airways, which can make it more difficult to breathe
- sensory stimulation, which triggers sneezing to expel foreign antigens.

> An antihistamine is a drug that counteracts the effects of histamine by blocking histamine receptors and therefore suppressing some allergy symptoms.

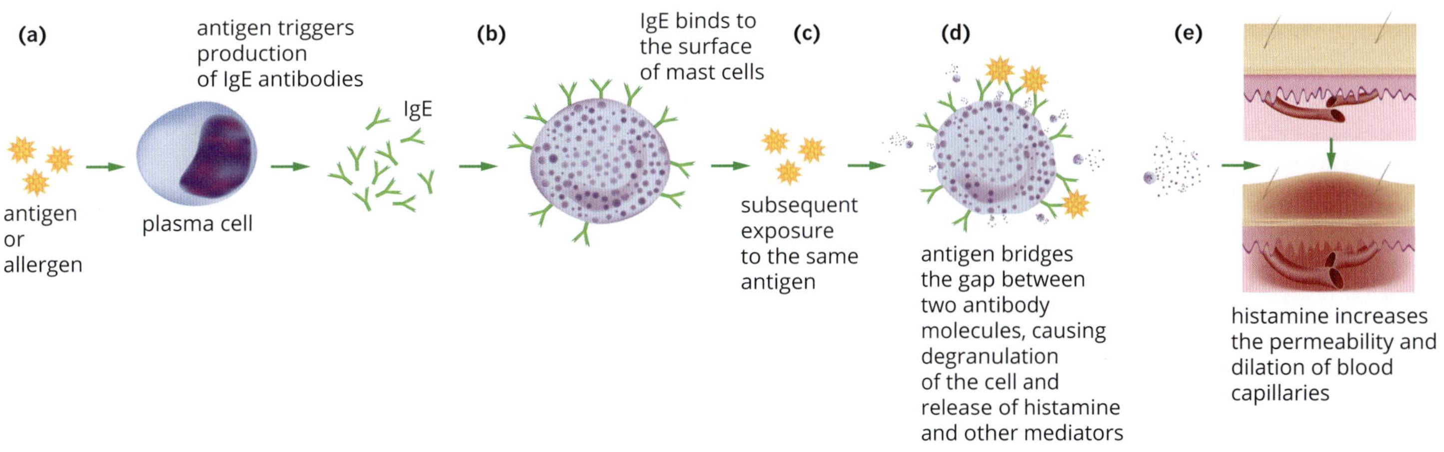

FIGURE 15.1.6 (a) Initial exposure to allergens (e.g. pollen) triggers plasma cells to produce IgE antibodies specific to the antigen. (b) The tail end of the IgE binds to receptors on the surface of mast cells. (c) Subsequent exposure to the same antigen, causes the antigen to bind to two adjacent IgE antibodies on a mast cell. This binding triggers a cascade of cellular signals that causes the mast cells to release histamine. (e) Histamine binds to receptors on various cells in the body, which produces the classic features of an allergic reaction.

BIOLOGY IN ACTION  

Treatment for allergic reactions

Medications

Antihistamines block the effects of histamine by binding to the same receptors as histamine, thereby preventing histamine from binding to them. Other medications used to treat allergies (e.g. cortisone) suppress the immune system more broadly and reduce the immune response in general.

If the allergic reaction is severe enough to cause anaphylaxis, an immediate intramuscular injection of adrenaline (or epinephrine) is needed. Adrenaline auto-injectors are commonly known and marketed as the EpiPen® (Figure 15.1.7). Adrenaline counters the actions of histamine by causing:

- blood vessels to constrict (decreasing swelling and increasing blood pressure)
- muscles in the airways to relax (so the airways open)
- the heart to beat faster, which increases the blood flow to the heart. (This prevents cardiovascular collapse, a condition that can result from decreased blood flow to the heart, which is caused by excessive blood vessel dilation following the release of too much histamine.)

FIGURE 15.1.7 This girl is demonstrating the use of an adrenaline auto-injector to prevent herself going into anaphylactic shock

Allergen immunotherapy

Allergen **immunotherapy** (or desensitisation) is used to treat hypersensitive reactions to allergens, such as bee sting toxin. Beginning with extremely small amounts, an allergen is injected multiple times, in increasing amounts over a period of months. This causes specific immunoglobulin G (IgG) antibodies to form against the allergen. IgG antibodies are a key component of the humoral immune response and are the main immunoglobulins in blood serum.

If IgG antibodies react with the allergen before it can bind to IgE antibodies to cause an allergic response, the allergic response is prevented. During treatment, the individual slowly becomes less and less sensitive (desensitised) to the allergen.

BIOFILE DD PSC

Thunderstorm asthma

Australia has the highest incidence of asthma in the world, with 9% of adults and 12% of children affected. Climate has a significant effect on the risk of an asthma attack among susceptible people. Thunderstorms are one event that is known to trigger attacks. Thunderstorm asthma occurs as a result of an uncommon combination of high pollen (usually during late spring and early summer) and a certain kind of thunderstorm. Grass pollen grains are swept up into the clouds as a storm forms and, when the pollen absorbs moisture, it bursts open and releases large amounts of much smaller allergen particles (Figure 15.1.8). These particles are then able to enter a person's airways and can be taken into the lungs. In some people this causes the lungs to become irritated, leading to swelling, narrowing of the airways and secretion of mucus, which makes it difficult to breathe and can cause an asthma attack. Not all thunderstorms result in thunderstorm asthma. Large thunderstorm asthma events have occurred in parts of Australia and around the world. On 21 November 2016, a cool change and thunderstorm swept across Melbourne, resulting in a thunderstorm asthma attack that hospitalised 8500 people and caused nine deaths. It is recommended that during a thunderstorm that might trigger an asthma attack, people avoid exposure to pollen particles by staying indoors.

FIGURE 15.1.8 Thunderstorms can cause asthma attacks when pollen grains absorb moisture and burst open, releasing smaller particles that are dispersed by high winds.

Cytotoxic hypersensitivity (type II)

Type II cytotoxic hypersensitivity reactions involve immunoglobulin M (IgM) and IgG antibodies directed against cell surface or extracellular matrix antigens, and can take hours to develop. This is in contrast to type I allergic reactions (discussed above), which usually occur within minutes of exposure to allergens.

An example of type II sensitivity is the antibody-mediated destruction of red blood cells (haemolytic anaemia) that occurs in the newborn when the mother produces antibodies directed against rhesus antigen on fetal red blood cells. Type II hypersensitivity reactions can also be a side effect of taking certain medications. For example, penicillin binds to red blood cells, and if anti-penicillin antibodies are present they bind to the drug and trigger the destruction of the red blood cells to which the penicillin is bound.

Immune complex hypersensitivity (type III)

Like type II reactions, type III immune complex hypersensitivity reactions also involve IgM and IgG. However, the antibodies in type III reactions are directed against soluble antigens, not antigens on the cell surface or extracellular matrix. When the antibodies bind to soluble antigens, they form antigen–antibody immune complexes that can be deposited in tissues, causing inflammation and damage. Type III reactions can also take hours or days to develop.

An example of type III hypersensitivity is serum sickness, which is an immune response to foreign antigens in medication or antiserum (e.g. snake venom antiserum). The most common cause of serum sickness is the antibiotic penicillin.

Delayed-type hypersensitivity (type IV)

Unlike type I, II and III hypersensitivity reactions, which are all mediated by antibodies, type IV hypersensitivity is mediated by helper T lymphocytes. These activate **macrophages** and **eosinophils** to produce inflammatory responses, and cytotoxic T lymphocytes, which directly attack and kill other cells. As the name suggests, delayed-type hypersensitivity reactions take days to develop.

An example of type IV hypersensitivity is the rash caused by contact with poison ivy (Figure 15.1.9), which has a lipid-soluble compound in its sap. This means it can cross the cell membrane. Once inside the cells, it causes new peptides to be produced. These peptides are then delivered to the cell surface, where they are recognised by cytotoxic T lymphocytes. The T-cell mediated immune response to the poison ivy causes a blistering, itchy rash and in severe cases, anaphylaxis.

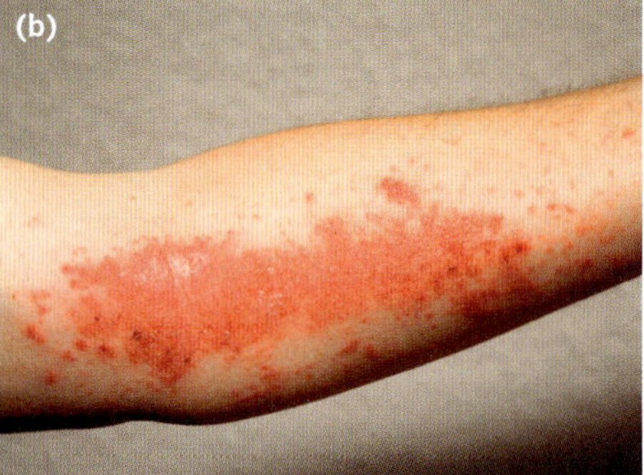

FIGURE 15.1.9 (a) Contact with sap from the poison ivy (*Toxicodendron radicans*) can cause a type IV hypersensitivity reaction, (b) which leads to itching, a rash and in severe cases, anaphylaxis.

AUTOIMMUNE DISEASES

Many diseases are a result of a combination of genetic and environmental factors. For autoimmune diseases, the causes are largely unknown. Research suggests that both genetic and environmental factors play a role in the development and severity of these diseases. Over 80 autoimmune diseases are currently known, including Crohn's disease, systemic lupus erythematosus, rheumatoid arthritis and type 1 diabetes.

All autoimmune diseases are caused by the body triggering an immune response against its own cells, leading to tissue and organ inflammation and damage. The symptoms of these are diseases are extremely variable between individuals, making diagnosis difficult. Researchers are working to understand the causes and risk factors, as well as the rise of autoimmune diseases in industrialised countries.

Autoimmune diseases result from an adaptive immune response directed against self-antigens.

Autoimmunity

When your immune system is working properly it is directed against foreign cells (**non-self antigens**), not against the body's own cells (self-antigens or autoantigens). This is known as **self-tolerance**. As you learnt in Chapter 12, normally T and B **lymphocytes** that are reactive against the body's own cells are destroyed. Autoimmune diseases result from a failure of self-tolerance, which leads to an **adaptive immune response** directed against specific self-antigens. An antibody that acts against a self-antigen is called an **autoantibody**.

An autoimmune disease occurs when a person's immune system mistakenly targets the body's own cells. Antibodies that attack the body's cells are called autoantibodies. Examples of autoimmune disorders are type 1 diabetes, Graves' disease and multiple sclerosis.

GO TO ➤ Section 12.1 page 422

When autoimmune diseases occur, the cytotoxic T lymphocytes of the adaptive immune response attack the tissues directly and B lymphocytes act indirectly by secreting antibodies. Mast cells are a type of white blood cell that play a key role in immune responses. They can be activated and release histamines, which results in inflammation around the affected tissues.

Autoimmune diseases can be organ-specific or generalised.

Organ-specific autoimmune diseases are localised to a particular part of the body. For example, multiple sclerosis only affects the brain and spinal cord.

Generalised autoimmune diseases occur widely throughout the body. For example, systemic lupus erythematosus affects various organs and tissues of the body, such as the joints, skin, kidneys and the brain.

Autoimmune haemolytic anaemia

Autoimmune haemolytic anaemia is an example of an autoimmune disease that results in a type II hypersensitivity reaction, because it involves autoantibodies directed against self-antigens on the surface of red blood cells (Figure 15.1.10).

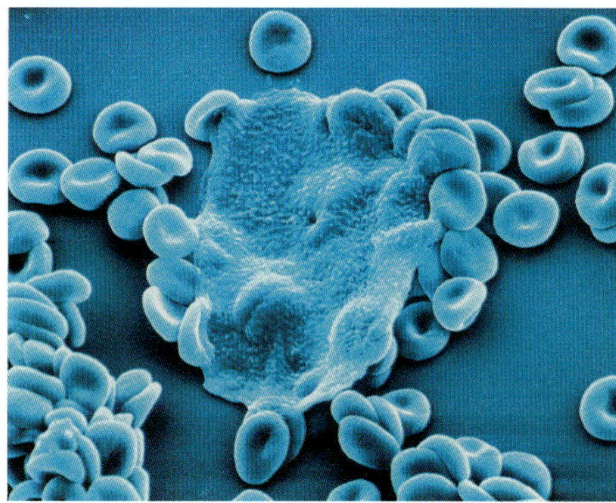

FIGURE 15.1.10 Coloured SEM of a macrophage (centre) engulfing multiple smaller red blood cells around it. Red blood cells recognised by autoantibodies are rapidly phagocytosed by macrophages.

Rheumatoid arthritis

Rheumatoid arthritis is an example of an autoimmune disease that results in a type III hypersensitivity reaction. It involves the deposition of antigen–antibody immune complexes in tissue, which results in inflammation and damage. Rheumatoid arthritis mainly affects the joints. Commonly affected joints are those of the knees and hands (Figure 15.1.11). Rheumatoid arthritis is also thought to result in type IV hypersensitivity reactions, in which T lymphocytes attack an unidentified antigen in the joints.

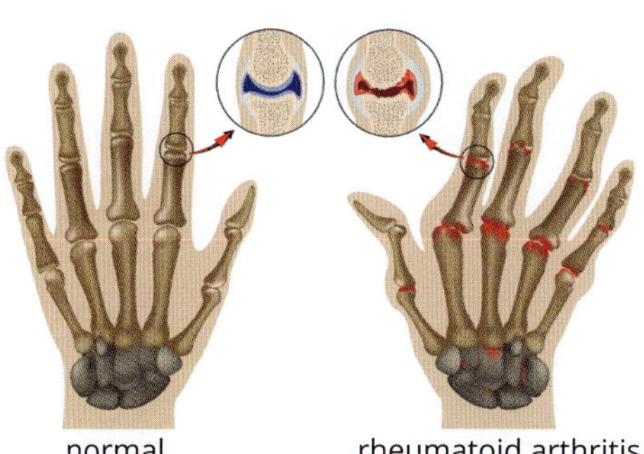

FIGURE 15.1.11 Rheumatoid arthritis causes inflammation in the joints of the hand, and over time can lead to disfigurement.

Type 1 diabetes

Type 1 diabetes is another example of an autoimmune disease. In this case T lymphocytes attack and destroy **beta cells** in the pancreas (Figure 15.1.12). Beta cells produce **insulin**, which regulates the levels of **glucose** in the blood. People with type 1 diabetes inject insulin to maintain glucose balance.

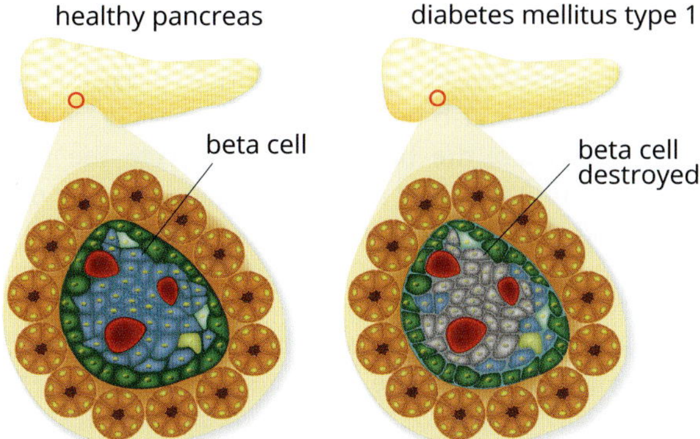

FIGURE 15.1.12 In type 1 diabetes, T lymphocytes attack and destroy beta cells of the islets of Langerhans in the pancreas.

There are two types of diabetes: type 1, which is genetic and may be caused by some types of viruses; and type 2, which is often late onset and related to lifestyle. Type 1 diabetes is a common non-infectious disease in the developing world. People with type 1 diabetes need to keep their blood glucose values within a certain range for normal bodily functioning. Those values are between 4.0 and 6.0 mmol/L (Figure 15.1.13).

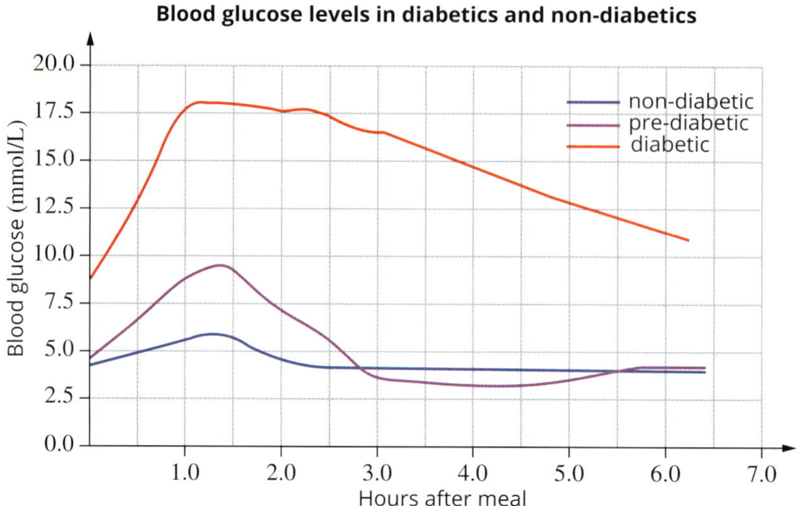

FIGURE 15.1.13 Graph indicating the blood glucose levels after a meal in pre-diabetic and diabetic patients compared with a non-diabetic control group

If a body's sensors detect blood glucose levels above or below these limits, the body releases the hormones insulin and **glucagon**. When blood glucose levels are too high, the body secretes insulin. Insulin stimulates glucose uptake into muscle and liver cells and converts it into the stored form of **glycogen** (Table 15.1.1). However, if the body detects lower than normal blood glucose levels, it produces the hormone glucagon. Glucagon stimulates the breakdown of glycogen into glucose in the liver, which releases glucose into the blood (Table 15.1.1).

TABLE 15.1.1 Changes in blood glucose levels—effects on the body and hormones released

Blood glucose level	Hormone secreted	Effect on blood glucose concentration in the body
higher than normal	Insulin is produced and secreted by the beta cells of the islets of Langerhans.	Insulin stimulates glucose uptake into muscle and liver cells. Glucose is then converted into glycogen and blood glucose levels fall.
lower than normal	Glucagon is produced and secreted by the alpha cells of the islets of Langerhans.	Glucagon stimulates the breakdown of glycogen into glucose in the liver. This releases glucose into the blood leading to raised blood glucose levels.

Cause of type 1 diabetes

Scientists are unsure what causes the beta cells in the islets of Langerhans in the pancreas to be destroyed in type 1 diabetes. There is some evidence for a link between the Coxsackie A and B4 viruses (which are common in children) and the onset of the autoimmune disease. Other childhood viruses, including enterovirus, mumps, polio and rubella, have also been suggested as triggers for type 1 diabetes. Without functioning beta cells, the body cannot secrete the insulin required to convert glucose to glycogen in the liver and to stimulate glucose uptake into muscle and fat. This causes blood glucose levels to increase to dangerously high levels.

Symptoms of type 1 diabetes

Insulin deficiency results in **hyperglycaemia** (high blood glucose levels) and accelerates the breakdown of fat for the body to use as energy. Symptoms of the disease include:

- glucose in the urine
- increased urine production

- excessive thirst
- excessive hunger
- ketoacidosis
- weight loss
- fatigue
- blurred vision
- irritability
- muscle cramps
- skin infections
- delayed wound healing
- tingling or numbness in the feet.

Longer-term consequences are kidney and eye disease. These symptoms occur because of the elevated levels of glucose in the blood. Glucose is excreted in the urine because the raised blood glucose levels exceed the filtration capacity of the kidneys (normally the kidneys prevent glucose from entering the urine). Glucose escaping into the nephron tubules draws in more water, by osmosis, increasing the volume of urine produced. As a result, more frequent urination leaves the body dehydrated and feeling thirsty. The presence of glucose in urine is a simple test for diabetes (Figure 15.1.14).

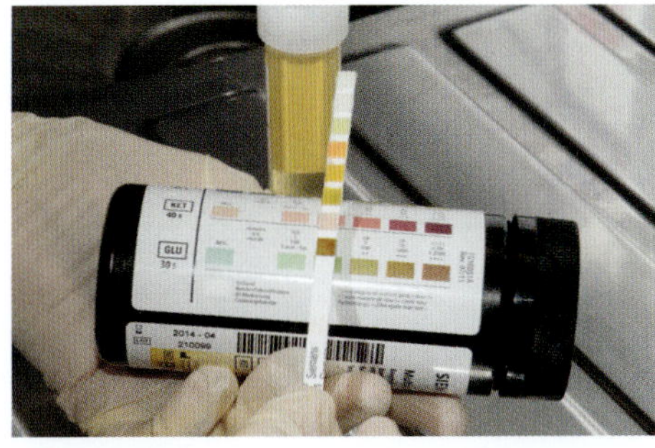

FIGURE 15.1.14 A dipstick test for glucose in urine. The pad is dipped in the urine, and the colour of the pad is checked against the chart. This gives an estimate of the glucose level in the urine. A high level is usually an indication of diabetes, although other conditions or the use of certain medications can cause high glucose levels.

Dehydration can lead to blurred vision as the lens loses moisture and the blood vessels are damaged. This can result in blindness if left untreated. The raised glucose levels in blood cause chemical reactions with molecules on the surface of neurons and cells lining the small blood vessels. The resulting damage to the body's nerves can lead to a loss of sensation in limbs, while damage to capillaries contributes to kidney malfunction and eye disease (diabetic retinopathy).

Management of type 1 diabetes

Type 1 diabetes can be managed throughout a person's life, allowing the patient to take responsibility for their health and wellbeing and live life to the fullest. As well as managing their diet, people with type 1 diabetes must receive insulin artificially.

Artificial insulin

This is usually administered by injection (Figure 15.1.15a). Patients monitor blood glucose levels by pricking a finger and testing a small drop of blood with a blood glucose meter or a chemical strip.

Alternatively, an electrode placed under the skin and connected to a continuous glucose monitoring device warns a person when their glucose level is reaching a high (or low) level. The monitor can be coupled to an electronic pump that delivers insulin when blood glucose levels reach a predetermined level (Figure 15.1.15b). An improved system called an 'artificial pancreas' uses a monitoring and feedback system to deliver insulin as the body requires it, in the same way that the pancreas produces insulin.

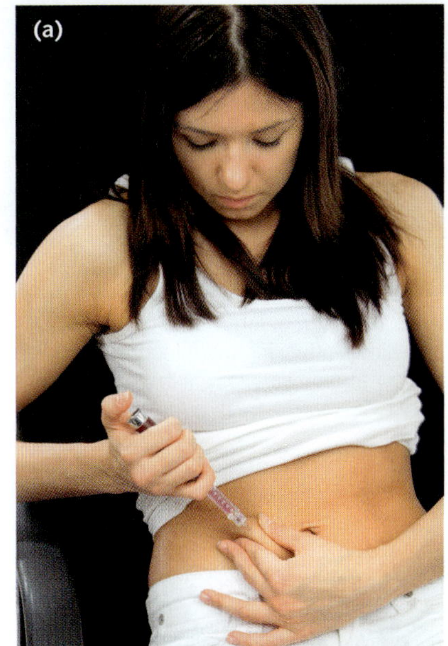

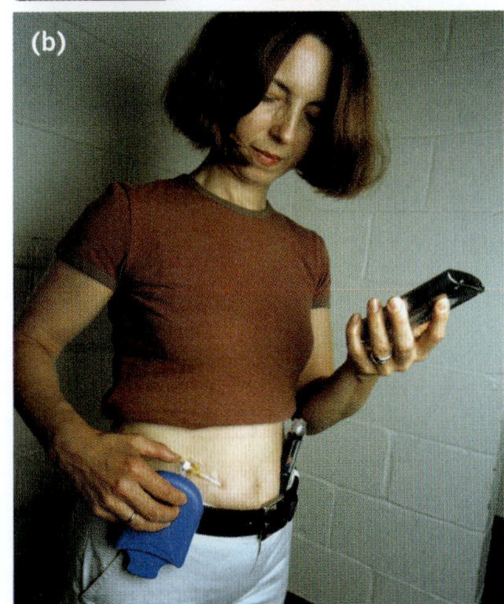

FIGURE 15.1.15 (a) A diabetic woman injecting insulin with an insulin pen, which delivers the correct dose of insulin. (b) A person wearing a continuous glucose monitoring device and insulin pump. The monitor measures blood glucose levels and sends a signal to the pump when insulin is needed.

Transplants

Pancreas transplants from deceased donors are usually given to patients with serious complications from diabetes. Human pancreas cells can also be transplanted into a patient's liver, where they begin to produce insulin. This process is called pancreatic islet transplantation. Although it is still in the experimental stage, it may become widely available in the next few years. Recipients of pancreas or pancreatic islet transplants must take immunosuppressant **drugs** for the rest of their lives to prevent their bodies from rejecting the transplanted organ. These drugs can have side effects such as high blood pressure, fatigue, and increased risk of bacterial and viral infections.

Gene therapy

Gene therapy, in which the gene that codes for insulin is inserted into the patient's cells, is a potential future treatment for diabetes. Trials in the USA have been successful in diabetic rats, targeting the liver because of the organ's regenerating ability. A major benefit of gene therapy is that patients would not require immunosuppressant drugs.

Multiple sclerosis

Neurons are the cells of the nervous system that relay signals along their axons. The central nervous system is made up of several different types of cell other than neurons. **Oligodendrocytes** are one of these other types of cells, and they produce a substance called myelin, which is composed mostly of lipids and some protein. Myelin forms an insulating sheath around the axons. Although some nerves in the peripheral nervous system also have a myelin sheath, **multiple sclerosis (MS)** generally only affects the myelin sheath in the central nervous system (Figure 15.1.16).

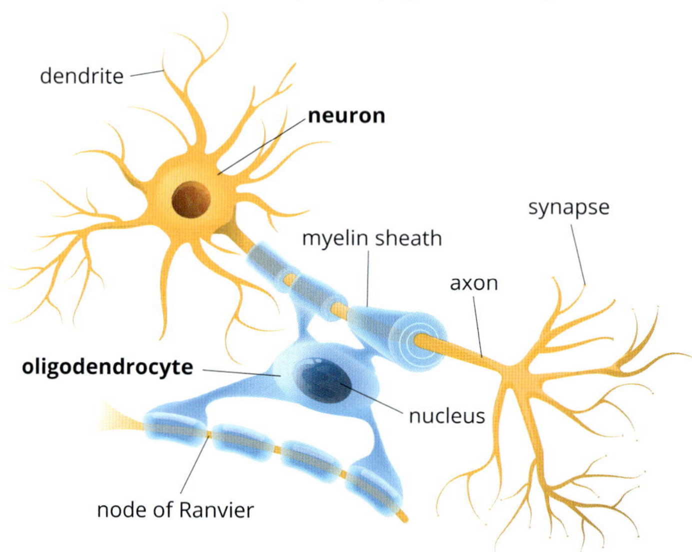

FIGURE 15.1.16 Healthy oligodendrocytes produce multiple myelin sheaths, which surround nerve axons.

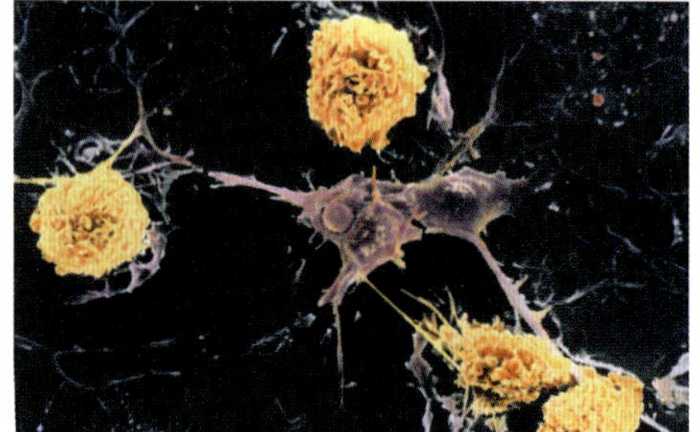

FIGURE 15.1.17 Coloured SEM of microglial cells (yellow), attacking oligodendrocytes (pink).

In MS, the nerves lose the myelin sheaths insulating them. This impairs the conduction of signals along the nerves, and eventually damages the nerve.

Both helper and cytotoxic T lymphocytes are involved in MS, and plasma cells produce antibodies that target proteins and lipids in the myelin sheath. Mitochondria in oligodendrocytes are damaged, which releases signalling molecules that induce **apoptosis** (programmed cell death) in these cells. Macrophages called microglia specific to the central nervous system are also involved in oligodendrocyte destruction (Figure 15.1.17). The causes of MS are still not fully understood, but it resembles a type IV hypersensitivity reaction because it is mediated by T lymphocytes and involves the activation of macrophages, which results in inflammation.

However it occurs, damage to and death of oligodendrocytes leads to destruction of the myelin sheaths (or **demyelination**). Once the myelin sheaths are damaged, damage also occurs to the nerve axons themselves (Figure 15.1.18).

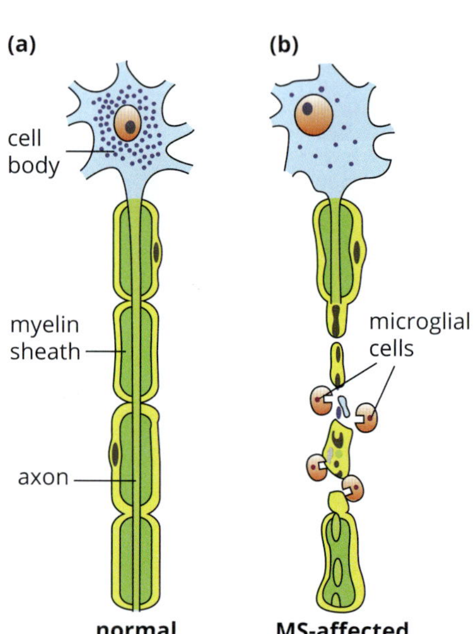

FIGURE 15.1.18 Comparison of the structure and function of myelin sheaths surrounding an axon in (a) a healthy neuron and (b) an MS-affected neuron

Symptoms vary from person to person and may include visual, motor and sensory problems, such as double vision, tiredness, numbness, muscle weakness, sensitivity to heat, and difficulty with balance and coordination. Symptoms can also include mental problems such as memory lapses, mood swings, depression and epilepsy.

There is no cure for MS. However, certain medications can manage symptoms and delay the progression of the disease. Standard treatments for MS include medications that reduce inflammation (steroids) and suppress the immune response (immunosuppressants).

NUTRITIONAL DISEASES

Many nutrients are essential for the health and survival of humans. The body is not able to synthesise these molecules and must obtain them from consuming a balanced diet. Essential nutrients are grouped into:

- **minerals**—such as calcium, magnesium
- **vitamins**—such as vitamins A, C, D and K

- amino acids—such as tryptophan and histidine
- fatty acids—such as omega-3 fatty acids.

Not all diets contain the required nutrients for a healthy body. Many nutritional diseases result from a lack of, imbalance or sometimes over-consumption of certain nutrients. This can lead to different types of malnutrition. For example, not having enough protein can lead to kwashiorkor, a disease that causes fluid retention, anorexia, ulcerating skin and an enlarged liver. A lack of other nutrients can lead to a number of different diseases such as scurvy, rickets and coronary heart disease.

A nutrient is a chemical substance that can be used by the human body.

Scurvy

Citrus fruits such as oranges and grapefruits are a good source of vitamin C (Figure 15.1.19). A lack of vitamin C in the diet can cause illness and a disease known as scurvy. **Scurvy** can lead to other health problems such as exhaustion, anaemia and swelling.

FIGURE 15.1.19 Citrus fruits and many vegetables are an importance source of vitamin C.

Scurvy was common among sailors in past centuries because of a lack of fresh fruits and vegetables in their diets. Many sailors on long voyages died as a result of scurvy. In countries where sufficient amounts of fruits and vegetables are not available, scurvy outbreaks still occur. Scurvy was common during the Irish potato famine in 1845 and in Afghanistan in 2002 following a period of war and drought. However, modern causes of scurvy are rare, especially in countries where vitamin C enriched breads and cereals are available.

Vitamin C is also known as L-ascorbic acid. Humans, and some other primates, cannot synthesise this compound. It is required to produce collagen, found in skin, connective tissues, blood vessels and tendons. Animals that have a mutation in the GLO gene, which codes for an enzyme needed in the synthesis of L-ascorbic acid, cannot synthesise their own vitamin C. The recommended daily intake (RDA) of vitamin C for humans is approximately 50 mg per day.

Rickets

Rickets is a disease associated with a lack of vitamin D in children. This condition in adults is known as **osteomalacia** and is a milder illness. Vitamin D is necessary to absorb calcium in the digestive system. Because calcium is required to build healthy bones, a lack of vitamin D can lead to softening of the bones. Children with rickets appear to have bowed legs (Figure 15.1.20).

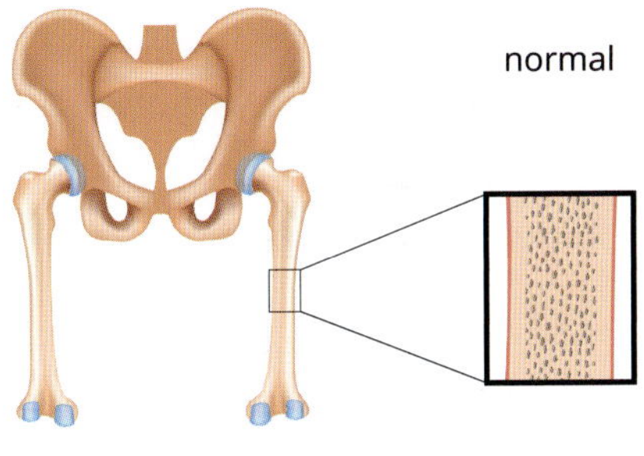

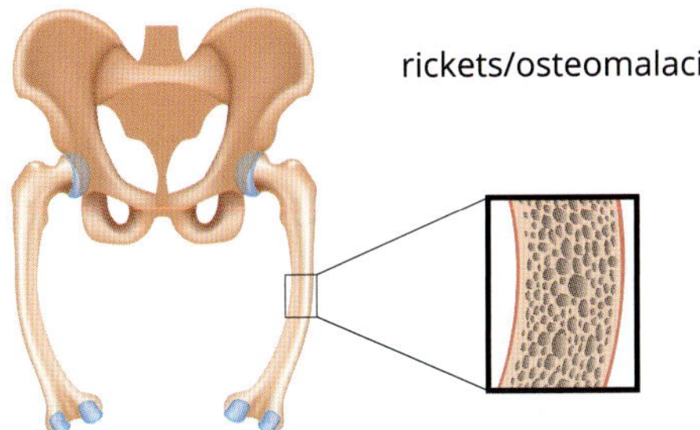

FIGURE 15.1.20 Rickets (known as osteomalacia in adults) results in a softening of the bones.

Vitamin D is essential for:

- maintaining calcium balance by acting to promote calcium absorption in the intestines
- maintaining calcium and phosphate levels for bone formation.

A diet high in vitamin D foods, including fish, avocados, egg yolks and dairy products such as cheese, milk and yogurt, will help prevent low vitamin D levels. Exposing human skin to low levels of UV light triggers the production of vitamin D in the skin. If skin is exposed to the open air and light for a short period of time each day, enough vitamin D will be supplied to your body. For this reason, vitamin D does not really fit the definition of a vitamin because it is one of few vitamins that can be made in the body. The World Health Organization recommends 5 to 15 mins of sun exposure of the hands, face and arms two to three times a week in the summer months. However, closer to the equator, where UV light levels are greater, even shorter periods of sun exposure are recommended.

In winter and in parts of the world at very northerly and southerly latitudes, UV light availability is limited and there is not enough sunlight to trigger the production of vitamin D in the skin. Because vitamin D is a fat-soluble vitamin, most people can store some vitamin D in the liver, which can be used until UV light is available again.

In countries such as Norway and Patagonia where light is limited for extended periods of the year, supplements are introduced into common foods like flour, cereals and milk. Other residents may take a vitamin D supplement to obtain the required daily intake.

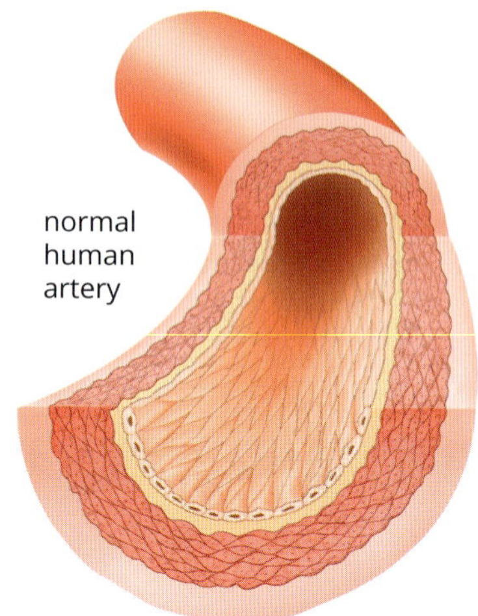

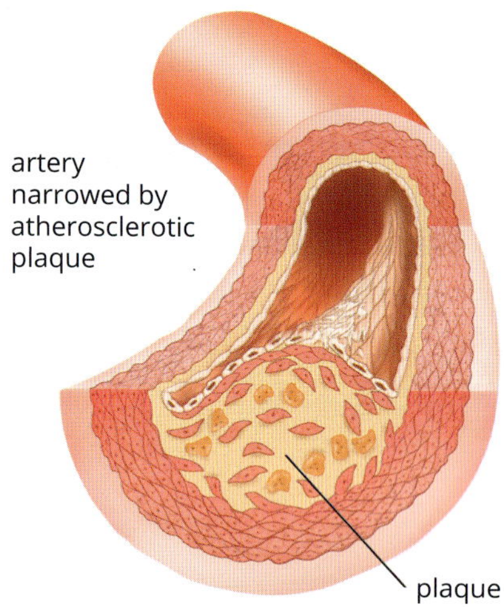

FIGURE 15.1.21 Cross-sections of (a) a normal artery and (b) an artery with atherosclerosis, showing the thickening of the arterial wall caused by a build-up of plaque

A lack of vitamin D for extended periods of time poses health issues. Research has shown that vitamin D is important to brain function and insufficient nutrient levels may contribute to mental illnesses such as depression. In breastfeeding mothers whose vitamin D intake is low, bone mineralisation of their infants can be affected.

Hypertension and coronary heart disease

Coronary heart disease involves the build-up of plaque inside the coronary arteries of the heart. These arteries supply blood that contains oxygen to the heart muscle itself.

With age, the arteries lose collagen and elastin filaments and gradually become less elastic and harden, referred to as **arteriosclerosis**. This puts stress on the heart because it has to pump harder to push the blood through the inflexible arteries. It can cause high blood pressure (also known as **hypertension**).

Over time and with a fatty diet, fatty substances, cholesterol and calcium can build up inside these hardened arteries, causing them to narrow (Figure 15.1.21). This specific type of arteriosclerosis is called **atherosclerosis**.

Both arteriosclerosis and atherosclerosis can affect arteries and arterioles in all parts of the body and restrict the flow of blood to tissues and organs. If atherosclerosis has developed, plaque can break away or blood clots can form around the plaque. Both these situations can cause a stroke or heart attack.

Atherosclerosis can affect the coronary blood vessels that supply blood to the heart muscle. A build-up of plaque restricts the supply of nutrients and oxygen to the heart tissue. If the coronary vessels become too narrow or completely blocked, a heart attack can result, possibly leading to the death of heart tissue.

Everyone will eventually develop some degree of arteriosclerosis, but what causes it to develop more rapidly in some individuals and progress to the more life-threatening atherosclerosis is not fully understood. What is known is that high blood pressure, along with high levels of cholesterol and triglycerides in the blood, increase the chance of developing atherosclerosis. Smoking, poor diet, lack of exercise and obesity are risk factors. Some medications are effective at lowering cholesterol levels in the human body.

Liver disease

People who drink alcohol excessively are prone to severe and often fatal liver disease. Medical evidence indicates that the addition of vitamins to alcoholic drinks, while good for nutrition, will not prevent chronic liver damage.

Alcohol is a toxic substance. The special enzymes that are needed to break it down are found in the liver. Because the biochemical pathways in the liver cells of a heavy drinker are involved with removing alcohol, the cells cannot carry out their normal levels of cellular respiration. Substances that should have been broken down for energy are converted to fats instead, and these fats accumulate in the liver.

For a while the situation is reversible, but then the cells filled with fat start to die, causing alcoholic hepatitis. This is followed by cirrhosis, which is the formation of scar tissue in the liver (Figure 15.1.22). Finally, death may occur when the liver is unable to carry out its normal functions.

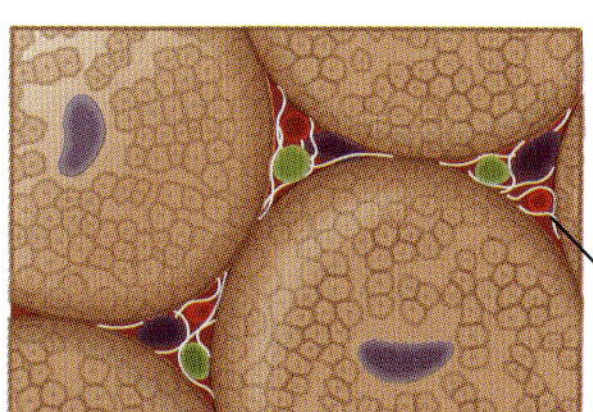

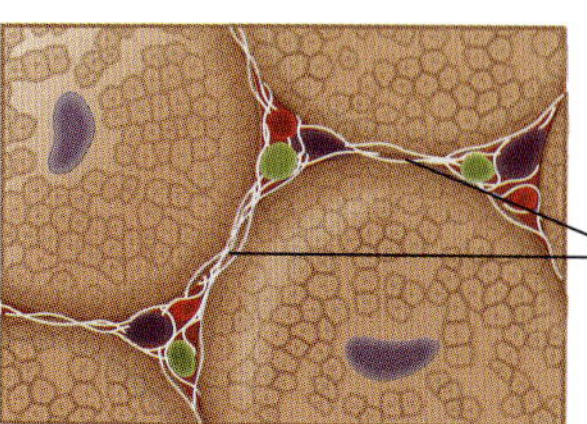

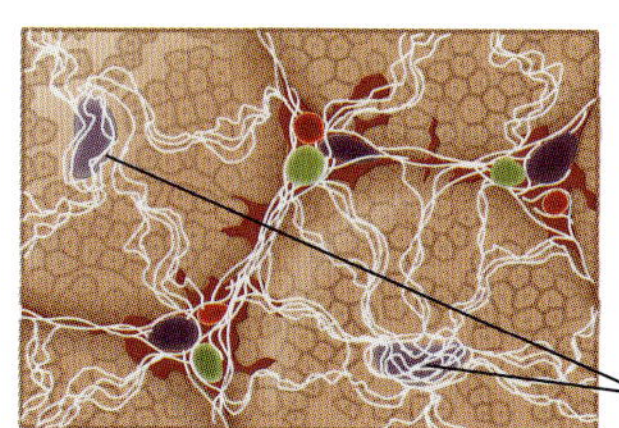

FIGURE 15.1.22 The stages of scar tissue formation (cirrhosis) in the liver. Tissue inflammation (stage 1), followed by cell death, leads to scar tissue build up (stage 2). Scar tissue replaces healthy tissue and eventually causes cirrhosis of the liver (stage 3). Excessive alcohol consumption can cause this form of liver damage.

CANCER

Cancers are a group of diseases that commonly involve unregulated and abnormal cell growth and division (Figure 15.1.23). Cancer can be caused by genetic mutations in the cells that either increase the rate of cell division and/or result in the suppression of apoptosis (programmed cell death). Either case can lead to the growth of tumours.

In Chapter 7 you learnt that cancer can be caused by mutagens called **carcinogens** (cancer-causing agents) and that there are three types of carcinogens—chemical, physical and biological. Fortunately, such mutations cannot pass from one generation to the next unless the mutation arises in the gametes (egg and sperm) of the parents. The treatment and management of cancer is covered in detail in Chapter 16.

GO TO ➤ Section 7.2 page 295
GO TO ➤ Section 16.2 page 564

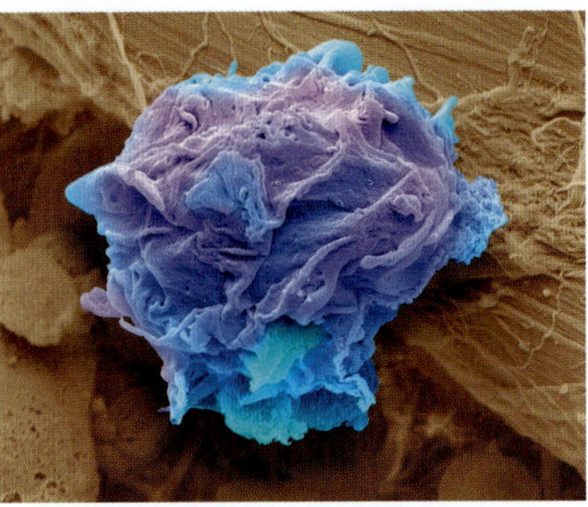

FIGURE 15.1.23 Coloured SEM of a lymphoma cell

INCIDENCE, PREVALENCE AND MORTALITY RATES OF NON-INFECTIOUS DISEASES

Chapter 16 discusses the importance of studying non-infectious diseases in a population. Carefully designed studies can help us recognise factors that may cause a non-infectious disease, compared with factors that are merely chance associations or related but that do not actually cause an effect. Scientists can learn a lot about a disease by studying where it happens, how often it happens, who is affected and how many people die as a result of such a disease.

Two common statistics in the study of disease are:

- the number of new cases diagnosed each year (incidence)
- the number of deaths (mortality).

You will learn more about the study of non-infectious disease in Chapter 16.

GO TO ➤ Section 16.3 page 571

SKILLBUILDER 

Interpreting scientific notation

Scientists often use graphs to display data in an easy-to-view format. Graphs commonly show quantitative data presented against the *x* and/or *y*-axis. Often the units of measurement used are taken from large sample sizes and written in a scientific form of notation.

For example, instead of writing 1 million as 1 000 000 or 1 billion as 1 000 000 000, scientists would write 10^6 for 1 million and 10^9 for 1 billion.

A number in scientific notation (also called standard form or power of 10 notation) is written as $a \times 10^n$ where:

a is a number equal to or greater than 1 and less than 10, that is $1 \le a < 10$

n is an integer (a positive or negative whole number)

n is the power that 10 is raised to and is called the index value.

To transform a very large or very small number into scientific notation:

1. Write the original number as a decimal number greater than or equal to 1 but less than 10.
2. Multiply the decimal number by the appropriate power of 10. The index value is determined by counting the number of places the decimal point needs to be moved to form the original number again.

   If the decimal point is moved n places to the right, n will be a positive number (e.g. 51 ≤ 5.1 < 101).

   If the decimal point is moved n places to the left, n will be a negative number (e.g. 0.51 ≤ 5.1 < 10–1).

   You will notice from these examples that when large numbers are written in scientific notation, the 10 has a positive index value. Very small numbers are written with 10 to a negative index.

15.1 Review

SUMMARY

- Many non-infectious diseases are associated with malfunctions in homeostatic mechanisms. The endocrine system is particularly important for maintaining homeostasis. Malfunctions in this system can affect the whole body.

Genetic diseases

- Inheriting a genetic disorder or a mutation in a gene can disrupt normal gene expression, leading to abnormal protein production.
- The gene (or genes) responsible for a disease can be determined by comparing the DNA sequence of a healthy individual with that of an individual suffering from the disease. Knowledge of the genes involved enables diseases to be detected early, and some may even be prevented.
- Cystic fibrosis (CF) is a genetic disease caused by mutations in the *CFTR* (cystic fibrosis transmembrane conductance regulator) gene and disruption to the production of the CFTR protein. Symptoms of cystic fibrosis vary from person to person but include abnormal mucus secretions in the lungs, pancreas and small intestine.
- Autoimmune diseases result from a failure of self-tolerance, leading to adaptive immune responses against self-antigens. They can be organ-specific or generalised. Autoimmune diseases can result in type II, type III and type IV hypersensitivity reactions.
- The cause of type 1 diabetes is the autoimmune destruction of the insulin-producing beta cells in the islets of Langerhans. People with type 1 diabetes do not produce enough insulin. Without treatment, blood glucose concentration can rise to dangerous levels.
- Multiple sclerosis (MS) is an organ-specific autoimmune disease that affects the central nervous system. Damage to the myelin sheath, which insulates nerve cell axons, impairs the conduction of nerve impulses.

Diseases caused by environmental exposure

- Environmental factors such as exposure to radiation or certain chemicals can damage DNA. Agents that damage DNA are called mutagens.
- There are three types of carcinogens (cancer-causing agents)—chemical, physical and biological.
- Hypersensitivity reactions occur when the immune system overreacts. They are classified into four types:
  - type I (immediate) hypersensitivity, also known as allergy
  - type II (cytotoxic) hypersensitivity
  - type III (immune complex) hypersensitivity
  - type IV (delayed-type) hypersensitivity.
- Antigens that trigger an allergic reaction are called allergens, and include pollens, dust, fur and foods.
- Allergic reactions involve the production of IgE, which attaches to the surface of mast cells. When an allergen cross-links two IgE molecules on a mast cell, it triggers a signal transduction cascade that releases histamine.
- Histamine is a compound that is involved in immune response.

Nutritional diseases

- Many nutritional diseases are caused by a lack of, imbalance or over-consumption of certain nutrients.
- Scurvy and rickets are examples of diseases caused by nutritional deficiencies (vitamin C and vitamin D respectively).
- A poor diet increases the chance of developing atherosclerosis and coronary heart disease. High blood pressure, along with high levels of cholesterol and triglycerides in the blood can result from poor nutrition.
- Malfunctions of the excretory system include polycystic kidney disease and glomerulonephritis and can be a result of drinking excessive amounts of alcohol.

Cancer

- Cancers are a group of diseases that commonly involve unregulated and abnormal cell growth leading to the growth of tumours.

KEY QUESTIONS

1. Why are all newborn babies tested for phenylketonuria (PKU) disease?
2. Outline both the short-term and long-term risks for someone with type 1 diabetes.
3. What are oligodendrocytes?
4. What is myelin composed of?
5. What does a nutritional disease result from?
6. Discuss the role of vitamin D in bone mineralisation.
7. Discuss if a poor diet is the only contributor to the development of coronary heart disease.
8. List the three types of carcinogens.

Chapter review

15

KEY TERMS

adaptive immune response
albinism
allergen
allergy (adj. allergic)
anaphylaxis
antibody
antigen
apoptosis
arteriosclerosis
atherosclerosis
autoantibody
autoimmune disease
beta cell
cancer
carcinogen
coronary heart disease
cystic fibrosis
cytotoxic
demyelination
endocrine system
eosinophil
gene expression
gene therapy
genetic disease
glucagon
glucose
glycogen
histamine
homeostasis
hormone
hyperglycaemia
hypersensitivity reaction
hypertension
immune system
immunoglobulin E (IgE)
immunosuppressant drug
immunotherapy
insulin
lymphocyte
macrophage
mast cell
metabolism
mineral
multiple sclerosis (MS)
mutagen
non-infectious disease
non-self antigen
oligodendrocyte
osteomalacia
phenylalanine
phenylketonuria (PKU)
pollen
receptor
rheumatoid arthritis
rickets
scurvy
self-antigen
self-tolerance
type 1 diabetes
vitamin

REVIEW QUESTIONS

1 Which foods can be eaten by people with phenylketonuria (PKU)?

2 Describe the cause of PKU.

3 A blood sample from someone who has not eaten for 24 hours would reveal high levels of which hormone? Discuss your response.

4 Why do people with diabetes sometimes need to have sweet drinks or food?

5 Why does type 1 diabetes cause high blood glucose levels?

6 Complete the following table, which outlines the main two hormones involved in blood glucose regulation.

Hormone	Site of production	Target organs	Main functions
insulin			
glucagon			

7 What happens when the beta cells of the pancreas release insulin into the blood?

8 Which of the following statements about genetic markers is correct?

A Genetic markers are mutations that cause disease
B Genetic markers are different lengths of DNA sequences
C Genetic markers are sequences that are polymorphic (variable)
D Genetic markers are only present inside liver cells

9 What type of hypersensitivity reaction are allergic reactions?

A immediate hypersensitivity (type I)
B cytotoxic hypersensitivity (type II)
C immune complex hypersensitivity (type III)
D delayed-type hypersensitivity (type IV)

10 Discuss the possible causes of type I diabetes and outline the host's response to this disease.

11 The table below shows the blood glucose levels of seven different patients. Identify the patients that are diabetic and justify your answer.

Patient number	1	2	3	4	5	6	7
Mean blood (mg/dL)	80	90	180	315	90	50	85
Glucose (mmol/L)	4.4	5	9	12	6	4.4	5.5

12 The graph below shows blood glucose levels measured in two patients after a meal. One patient has been diagnosed with diabetes and the other patient has not. Identify which patient, patient A or B is diabetic and provide reasons to support your answer.

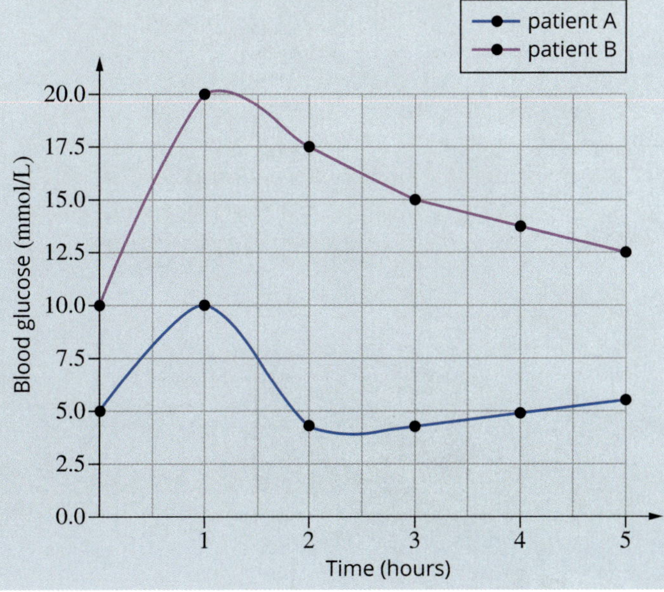

13 Explain the potential management options available for patients with type 1 diabetes.

14 Outline the difference between organ-specific and generalised autoimmune diseases. Provide an example of each.

15 Describe the cause of cystic fibrosis and the body's response to this condition, including potential symptoms.

16 What is coronary heart disease and what type of disease can it be classified as?

17 Name one non-infectious disease you have studied and summarise its:
- occurrence
- symptoms
- causes
- treatment/management.

18 Distinguish between non-infectious and infectious diseases.

19 What is the reason that vitamin D is not, strictly speaking, a vitamin?

20 Rickets, scurvy and liver disease can all be grouped under which type of non-infectious disease?

21 Using a table like the one below, classify the following carcinogens into the categories of biological, chemical or physical:
- sunburn
- alcoholism
- heavy metal poisoning
- nuclear radiation
- some viruses

Carcinogen type	Carcinogen
biological	
chemical	
physical	

22 Which list contains only non-infectious diseases?

A lung cancer, diabetes, multiple sclerosis, asthma
B cystic fibrosis, anaemia, heart disease, hepatitis B
C rickets, warts, scurvy, cancer
D lead poisoning, heart disease, diphtheria, thyroid cancer

23 After completing the Biology Inquiry on page 534, reflect on the inquiry question: Do non-infectious diseases cause more deaths than infectious diseases? After learning about the some of the causes of non-infectious and infectious diseases, explain why the incidence and mortality rates of these diseases varies throughout the world.

CHAPTER 16

Epidemiology

Studies in epidemiology aim to investigate the relationships between disease distribution, patterns and causes within a population. This chapter explores the usefulness of epidemiological studies in understanding the impacts that nutrition and environmental exposure can have on disease. You will analyse population data of certain non-infectious diseases such as cardiovascular disease and cancer to recognise a cause-and-effect relationship that exists with certain risk factors. You will learn about the treatment and management of cancer, recognising the potential for future research into gene transfer for non-infectious diseases. You will explore the importance of collecting population data. Epidemiology studies enable large quantities of data to be gathered and statistically analysed. The results can then be applied to health programs and health services. These studies also act to assess the value of treatments and preventative strategies in a population.

Content

INQUIRY QUESTION

Why are epidemiological studies used?

By the end of this chapter you will be able to:

- analyse patterns of non-infectious diseases in populations, their incidence and prevalence, including but not limited to: ICT IU L CC
  - nutritional diseases
  - diseases caused by environmental exposure
- investigate the treatment/management and possible future directions for further research of a non-infectious disease using an example from one of the non-infectious diseases categories listed above ICT IU L
- evaluate the method used in an example of an epidemiological study
- evaluate, using examples, the benefits of engaging in an epidemiological study

16.1 Non-infectious disease in populations

BIOLOGY INQUIRY  

The study of disease

Why are epidemiological studies used?

COLLECT THIS...

- tablet or computer to access the internet
- note paper
- coloured pens or pencils
- poster paper

DO THIS...

1. Access information about the top 10 diseases worldwide, using reliable sources such as government websites.
2. Select one disease to research.
3. Take 15 minutes to research the data available for your chosen disease, taking notes about key facts such as:
   - the leading causes of death due to infectious and non-infectious diseases
   - why it is important to collect data on mortality rates and how this data is used
   - other forms of data that might be collected to help understand the causes of disease worldwide.
4. Create a poster about the epidemiology of your chosen disease.

RECORD THIS...

Describe the leading causes of death worldwide.

Present your findings in a poster.

REFLECT ON THIS...

Why are epidemiological studies used?

Why don't all countries have the same ability to collect data on the causes of disease and mortality rates?

Why does the pattern of infectious and non-infectious disease prevalence vary across the world?

As you have learnt in Chapter 15, there are many different **non-infectious diseases** that affect various populations all around the world. **Epidemiology** is the study of the distribution, patterns and causes of diseases in a population. Diseases studied can include both infectious and non-infectious diseases. By analysing the spread, patterns and causes of diseases, predictions can be made and preventative measures undertaken. Epidemiological studies can predict where outbreaks of a disease may occur. It can help to determine measures to prevent the spread of infectious disease or reduce the **incidence** of non-infectious disease. Epidemiological data can also show the impact that **public health programs** have on the occurrence of a particular disease.

It is easier to establish a cause-and-effect relationship between pathogens and infectious disease than it is with the factors that cause non-infectious diseases. For example, in the Middle Ages, the spread of the bubonic plague (also called the black death) across Europe took years. By the 14th century an estimated 30 to 60% of the European population (25 million people) were killed as a result of the disease. The disease was estimated to progress at the rate of 15 km per day. Compare this to an infected person today. Someone can board a plane and fly to the other side of the world in 16 hours, potentially infecting all the other passengers on the flight and those at the new destination. It is much easier to track the path of an infectious disease (Figure 16.1.1) as non-infectious diseases are often slow to develop and difficult to diagnose. This makes the cause-and-effect relationship of a non-infectious disease a great challenge for epidemiologists to pin point.

> Epidemiologists are scientists who study diseases within populations of people. They analyse what causes a disease in order to treat existing diseases and prevent future outbreaks.

FIGURE 16.1.1 The path and spread of infectious diseases can be tracked between individuals. The cause and effect of non-infectious disease can be more difficult to monitor.

ANALYSING DATA—DISEASES CAUSED BY NUTRITIONAL IMBALANCES

As seen in Chapter 15, diseases resulting from a nutritional imbalance are varied. Some are the result of a lack of nutrients, other diseases are the result of an over-consumption of certain foods. Although each nutritional disease is specific, a common factor is the importance of a balanced and healthy diet.

BIOFILE IU

Culture and diet

Food and cultures are strongly linked to each other. By comparing what humans from different cultures eat, it is possible to draw conclusions that some diets are healthier than others. Some diets have been correlated with long life expectancy and evidence supports the finding that some foods offer better protection against certain types of cancer. Diets from Asian and Mediterranean countries, which include fresh fruit and vegetables and a low intake of red meat, have been celebrated for their health benefits (Figure 16.1.2). Such diets have been associated with lower levels of oxidised low-density lipoprotein (LDL) cholesterol, which is more likely to build up as deposits in the body and over time can lead to atherosclerosis and coronary heart disease. There has been some association found with consuming a Mediterranean diet and a reduced risk of cardiovascular disease and other non-infectious diseases such as cancer.

It is important to recognise that correlation does not always represent causation (i.e. a direct link) but refers to an association between the variables (e.g. diet and disease). Because it is difficult to rule out other contributing factors, such as genetic and environmental influences, and because studies cannot be directly preformed on humans, correlation rather than causation is the appropriate term.

FIGURE 16.1.2 Asian and Mediterranean diets have been linked to reduced risk of non-infectious diseases such as cardiovascular disease and cancer.

Malnutrition

A balanced diet is one that contains the required carbohydrates, fats, proteins, vitamins, minerals and water needed to maintain good health. Some diets have a deficiency, imbalance or excess of nutrients, which can lead to **malnutrition**. Remember that malnutrition is not just a lack of essential nutrients, it can also result from an excess of essential nutrients. There are many reasons that may cause a person to have an unbalanced diet including poverty, war, famine, drought, natural disaster, lack of accessible foods (Figure 16.1.3), lack of access to medical professionals or simply not understanding that what you eat is important to your overall health. Whatever the reason, a lack of essential nutrients may lead to disease, illness and potentially death. You learnt about nutritional disease in Chapter 15.

The term malnutrition covers two broad categories:

- **undernutrition**—includes stunting (low height for age), wasting (low weight for height), underweight (low weight for age) and micronutrient deficiencies (including a lack of vitamins and minerals)
- **overnutrition**—includes **obesity** and diet-related non-infectious diseases (e.g. heart disease, stroke, diabetes and cancer).

Malnutrition refers to deficiencies, excesses or imbalances in a person's intake of energy and/or nutrients.

FIGURE 16.1.3 Rice is a major food source for many people living in poverty. This reliance on one food and lack of accessibility to a balanced diet often leads to malnutrition.

The consequences of malnutrition

Malnutrition affects people from all around the world, in every society or culture, not just those living in low income countries. Even in nations with a high income, cases of malnutrition can exist, caused by an excessive intake of fats and sugars. A diet lacking in essential nutrients combined with a lack of exercise can lead to obesity.

It is estimated that around 1.9 billion adults worldwide are overweight, while 462 million are underweight. Many families cannot afford or do not have access to nutritious foods like fresh fruits, legumes, dairy products and meat. Poverty amplifies the risk of malnutrition. It is for this reason that people who live in poverty are more likely to be affected by forms of malnutrition. Malnutrition also increases the cost of health care, reduces productivity in a society and slows economic growth. All of which are factors that perpetuate the cycle of poverty and poor health.

Foods that are pre-packaged and high in sugar and fat are often readily available, cheaper than fresh foods and easy to prepare. Consumption of these foods leads to the rapid increase in children and adults who are overweight and obese in both low income as well as high income countries. It is possible for people within the same country, suburb or even family to be both overweight and micronutrient deficient. The consequences of malnutrition are quite drastic and if left untreated can lead to serious non-infectious diseases later in life.

Woman, infants and children are at the greatest risk of malnutrition. These groups require specific nutrients to ensure growth and development occur. Early in a child's life, a significant period of growth occurs and adequate nutrition ensures the best start to life possible and assists in preventing various non-infectious diseases such as kwashiorkor, a form of severe protein malnutrition. Kwashiorkor is characterised by an enlarged liver, oedema, ulcerating dermatoses and generally occurs in areas of the world where famine or poor food supply is common (Figure 16.1.4). Cases in high income countries are rare (Figure 16.1.5).

In 2015 there were 793 million undernourished people in the world, with **mortality** due to malnutrition accounting for 58% of total mortality in 2012.

Despite the world's farmers producing enough food to feed around 12 billion people (almost double the current world population), food distribution and lack of available trade to famine-affected countries means that malnutrition is still a significant problem in the 21st century. According to the World Health Organization (WHO), malnutrition is the biggest contributor to child mortality in the modern world.

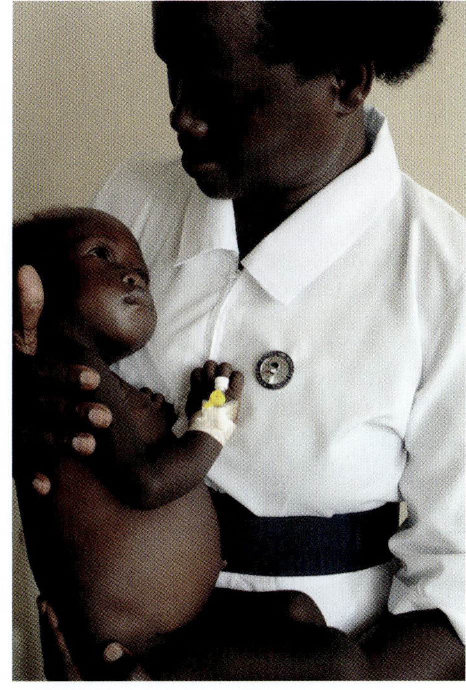

FIGURE 16.1.4 Nurse holding a malnourished baby in the nutrition ward at St. Mary's Hospital in Lacor, Gulu, Uganda. The baby's distended abdomen is a sign of lack of food and malnutrition, a condition known as kwashiorkor.

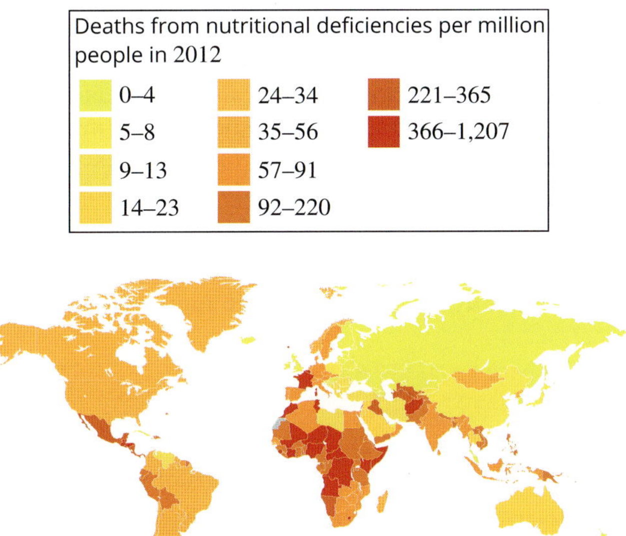

FIGURE 16.1.5 Global death rate from nutritional deficiencies per million people in 2012

Mortality is a measure of the number of deaths by place, time and cause.

BIOFILE IU

Decade of Action on Nutrition

In April 2016, the United Nations (UN) General Assembly drew a resolution outlining the UN Decade of Action on Nutrition from 2016 to 2025. This resolution has been made to ensure that all people, from all countries around the world have access to nutritious diets with the aim of eradicating malnutrition worldwide. The UN Decade of Action on Nutrition recognises the importance of local, national and global movements to end all forms of malnutrition and seeks to provide a framework that governments can use to assist in the implementation of policies and programs that create sustainable food systems and promote healthy dietary practices.

The UN Decade of Action on Nutrition:

- recognises the importance and urgency to address and end malnutrition worldwide
- acts to establish a focused period to track and achieve country specific commitments to health
- provides an opportunity to establish country specific national policy change
- provides global connectivity between countries to achieve a common goal
- puts in place a transparent mechanism for tracking progress and ensures a level of accountability for countries involved to act and work to achieve the goals set.

Obesity is defined as abnormal or excessive fat accumulation that may impair health.

Chronic diseases persist for a long time or constantly recurring.

Acute diseases are severe but of short duration.

Obesity

At the other end of the malnutrition scale, obesity is one of societies most concerning health problems. Obesity is a contributing factor to various non-infectious diseases, such as diabetes and stroke. With an estimated 41 million overweight children under five years of age in 2016, obesity is proving to be one of the world's most significant health problems.

Obesity is a complex condition that affects all age brackets and socioeconomic groups around the world. It has links with various social and psychological factors, which contributes to the complexity of the condition and the difficulty in slowing the ongoing rise in obesity rates worldwide. This rapid rise in rates of obesity across the world is known as the 'obesity epidemic'. For both males and females, obesity poses a major health risk for many non-infectious diseases, such as diabetes, cardiovascular disease and certain types of cancer. Health consequences of obesity range from increased risk of premature death to serious **chronic diseases** that reduce the overall quality of a person's life.

Causes of obesity

The main cause of obesity is an energy imbalance between calories consumed and calories expended. Changes in dietary and physical activity levels are often the result of environmental and societal changes. There are various lifestyle factors that can contribute to obesity including lack of access to health care, insufficient exercise, lack of nutrition education, marketing of unhealthy foods and excessive consumption of processed foods. Modern working conditions, which often involve sitting for most of the day or minimal physical movement, have also been linked to an increase in obesity.

The consequences of obesity

Having a raised level of body mass is a major **risk factor** for non-infectious diseases such as:

- cardiovascular disease
- stroke
- diabetes
- musculoskeletal disorders (such as osteoarthritis of the joints)
- some cancers (including breast, ovarian, prostate, liver, kidney and colon).

The risk of contracting these non-infectious diseases increases with increased **body mass index (BMI)**. Childhood obesity is also associated with a higher chance of obesity later in life and therefore a higher probability of contracting a non-infectious disease as a result in adulthood.

The **prevalence** of obese people in Australia is expected to increase (Figure 16.1.6). At least 80% of premature heart disease, stroke and type 2 diabetes and 40% of cancer could be prevented through a reduction in obesity levels within Australia. A healthy diet, regular physical activity and not smoking are all lifestyle factors that can be applied to assist in reducing a person's body mass.

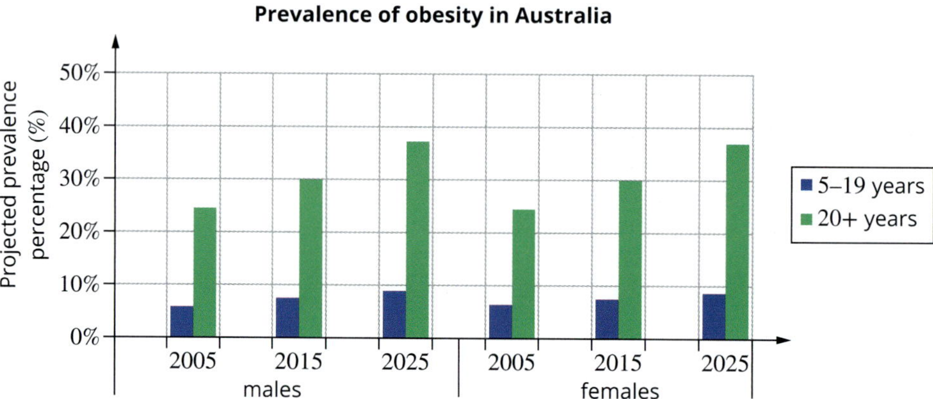

FIGURE 16.1.6 Projected prevalence expressed as a percentage (%) of obese males and females in Australia aged 5–19 years and 20+ years. The 2005 and 2015 values are actual data and 2025 are projections.

Body mass index (BMI)

Body mass index (BMI) is a measure of an individual's weight for height and is used to determine whether a patient is within the normal range of mass. BMI provides the most useful measure of obesity in a population as it is the same measure for both biological sexes and for all ages. It is quite a useful tool to estimate whether you are underweight or overweight, but must be used with care, as there are other factors to keep in mind, such as the proportion of muscular and adipose tissues. For children, age needs to be considered when referring to BMI levels.

For adults, WHO considers a BMI greater than or equal to 30 an indication of obesity.

Remember that BMI is a rather coarse measurement. Muscles in the body have a higher density than fat, so some professional athletes and individuals with large muscle mass have BMI values above 30 but are not considered obese.

Body mass index (BMI) is a simple index of weight-for-height that is commonly used to classify overweight and obesity in adults. It is defined as a person's weight in kilograms divided by the square of the person's height in metres (kg/m^2).

$$BMI = \frac{mass\ (kg)}{height\ (m)^2}$$

Worked example 16.1.1

CALCULATING BMI

Calculate the BMI of Alex who has a mass of 8000 g and a height of 191 cm

Thinking	Working
Write Alex's mass in kilograms	8000 g = 80 kg
Write Alex's height in metres	191 cm = 1.91 m
Substitute the mass and height values into the BMI equation: $BMI = \frac{mass\ (kg)}{height\ (m)^2}$	$BMI = \frac{mass\ (kg)}{height\ (m)^2}$ $BMI = \frac{80\ kg}{1.91\ m^2}$ $BMI = 21.9$

Worked example: Try yourself 16.1.1

CALCULATING BMI

Two friends want to compare their BMI levels. Calculate the BMI of Joel who has a mass of 7500 g and a height of 176 cm and the BMI of Omar who has a mass of 9500 g and a height of 171 cm.

Monitoring dietary intake

A balanced diet is an important lifestyle factor that can assist in reducing the incidence of many non-infectious diseases. A diet that contains all essential and non-essential nutrients in the right proportions is considered a balanced diet. The Nutrition Australia **Healthy Eating Pyramid** (Figure 16.1.7) has been a source of reference for many years, outlining the accepted intake of foods that you should favour such as fruit and vegetables in your diet. The scientific evidence for this food pyramid is based on epidemiological and statistical studies that look into the correlation between diet and the incidence of certain non-infectious diseases.

Keeping calorie intake to an acceptable level is also recommended for good health. The Recommended Dietary Allowances (RDA) can be used as a reference. This resource indicates accepted values for calorie intake per day. It is important to recognise that these recommended values should be guidelines only, as other lifestyle factors, such as exercise, alters the calorie intake required. All packaged foods now contain information on nutritional content. Many websites and apps are also available to assist people in tracking their daily calorie intake.

SKILLBUILDER 

International system of units for energy

In the International System of Units, energy is measured in joules (J). The kilojoule (kJ) is most often used for food-related quantities. Often the units provided on a food package are listed in kilojoules and can be converted into joules. Joules are the Standard International (SI) unit of energy.

The energy in joules (J) is equal to 1 kilojoule (kJ) multiplied by 1000:

$$\text{Energy (J)} = 1\,\text{kJ} \times 1000 = 1000\,\text{J}$$

So, 1 kilojoule (kJ) is equal to 1000 joules (J):

$$1\,\text{kJ} = 1000\,\text{J}$$

A widely used but older unit of energy is the kilocalorie (kcal) or calorie (Cal). 1 kcal = 4.184 kilojoules.

Energy in food can be measured in kilocalories (kcal) or kilojoules (kJ).

1 kilojoule (kJ) converted into kilocalorie equals = 0.24 kcal

$1\,\text{kJ} = 0.02\,\text{kcal}$

To convert a food with 2500 kJ of energy into kcal

$\text{Energy (kcal)} = 2500\,\text{kJ} \times 0.24\,\text{kcal} = 600\,\text{kcal}$

BIOLOGY IN ACTION  

The Healthy Eating Pyramid

Nutrition Australia first introduced the Healthy Eating Pyramid in 1980, based on a 'more to less' concept developed in Sweden in the 1970s. The Healthy Eating Pyramid (Figure 16.1.7) was designed as a visual model for people to use as an introduction to nutrition. It uses information based on the Australian Dietary Guidelines, to indicate the foods you should eat every day for good health and the foods that should only be eaten in moderation. The pyramid contains five core groups plus healthy fats.

The pyramid has undergone changes since its development and evolved with advancements in scientific evidence on nutritional requirements. It is a simple tool to encourage Australians to eat a healthy and varied diet and is often used in education for health workers and the general public.

The latest version of the Healthy Eating Pyramid was launched in 2015 to include information on limiting sugar and salt in your diet.

HEALTHY EATING PYRAMID
HEALTHY FATS
LIMIT SALT & ADDED SUGAR
MILK, YOGHURT, CHEESE & ALTERNATIVES
LEAN MEAT, POULTRY, FISH, EGGS, NUTS, SEEDS, LEGUMES
GRAINS
VEGETABLES & LEGUMES
FRUIT
ENJOY HERBS & SPICES
CHOOSE WATER
Enjoy a variety of food and be active every day!
Nutrition Australia
© Copyright The Australian Nutrition Foundation Inc. 3rd edition, 2015

FIGURE 16.1.7 The Nutrition Australia Healthy Eating Pyramid showing the recommended daily foods for good health.

BIOFILE DD

Leading cause of deaths in Australia

Coronary heart disease is the leading underlying cause of death in Australia, followed by dementia, Alzheimer's disease and cerebrovascular disease (which includes stroke). Lung cancer and chronic obstructive pulmonary disease make up the top five leading underlying causes of death in Australia for males and females of all ages combined.

Figure 16.1.8 shows the number of male and female deaths contributing to the top five causes of death in Australia in 2016. Males account for more deaths due to coronary heart disease, lung cancer and chronic obstructive pulmonary disease. Females account for the majority of deaths due to cerebrovascular disease, dementia and Alzheimer's disease. All diseases listed are non-infectious diseases.

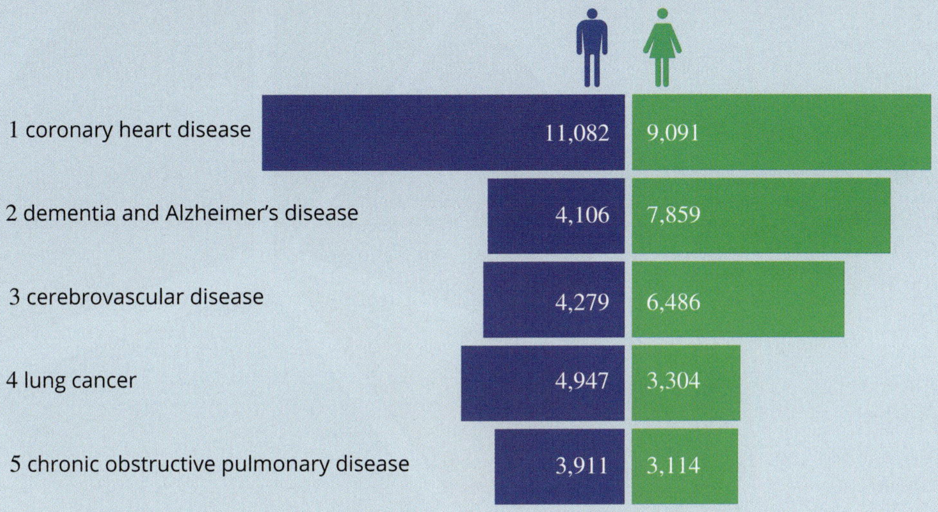

	Male	Female
1 coronary heart disease	11,082	9,091
2 dementia and Alzheimer's disease	4,106	7,859
3 cerebrovascular disease	4,279	6,486
4 lung cancer	4,947	3,304
5 chronic obstructive pulmonary disease	3,911	3,114

FIGURE 16.1.8 Leading causes of death in Australian males and females in 2016

ANALYSING PATTERNS—DISEASES CAUSED BY ENVIRONMENTAL EXPOSURE

Chapter 15 described the many diseases associated with exposure to environmental factors such as radiation or certain chemicals. Epidemiological studies into environmental exposure aim to identify if a disease is the result of exposure to an external factor. Identifying if a single factor alone is the cause of a disease is very difficult. Scientists often classify factors that can contribute to the cause of a disease as risk factors.

Environmental risk factors are modifiable and can be avoided in most cases. Having one or more risk factors does not mean that you will develop a disease as a result. Many people have at least one risk factor in their lives (such as exposure to **ultraviolet (UV) radiation** from the Sun) but never develop skin cancer, while others with skin cancer may have had no known risk factors or limited exposure to them. It is difficult for epidemiologists to pin point if a risk factor contributed to the development of a disease from a single case, so they often analyse large scale **population data** to recognise patterns and identify the potential for a cause and effect relationship.

A risk factor is any factor that is associated with increasing the chance of developing a disease.

Understanding data terminology

The following terms are used widely in epidemiology studies and are important to understand when analysing data.

- Incidence—the number of new cases of a disease diagnosed during a specific time period (usually one year).
- Mortality—the number of deaths occurring as a result of a disease during a specified time period (usually one year).
- Prevalence—the number of people alive with a prior diagnosis of a disease at a given time. It is distinct from incidence, which is the number of new cases of the disease diagnosed within a given period of time.
- Estimate—future estimates for incidence and mortality are a mathematical prediction of past trends. They assume that the most recent trends will continue into the future. Actual future incidence and mortality rates are likely to vary slightly from these estimates, due to the development of public health and prevention programs.

Risk factors of cancer

Cancer deaths are the second leading cause of death in Australia (Figure 16.1.9) and the incidence of cancer is increasing. Cancer is one of the world's most concerning diseases as it develops in a variety of different ways and is a very aggressive disease. You will learn more about cancer in Section 16.2 and develop an understanding around the treatment and management of this broad disease.

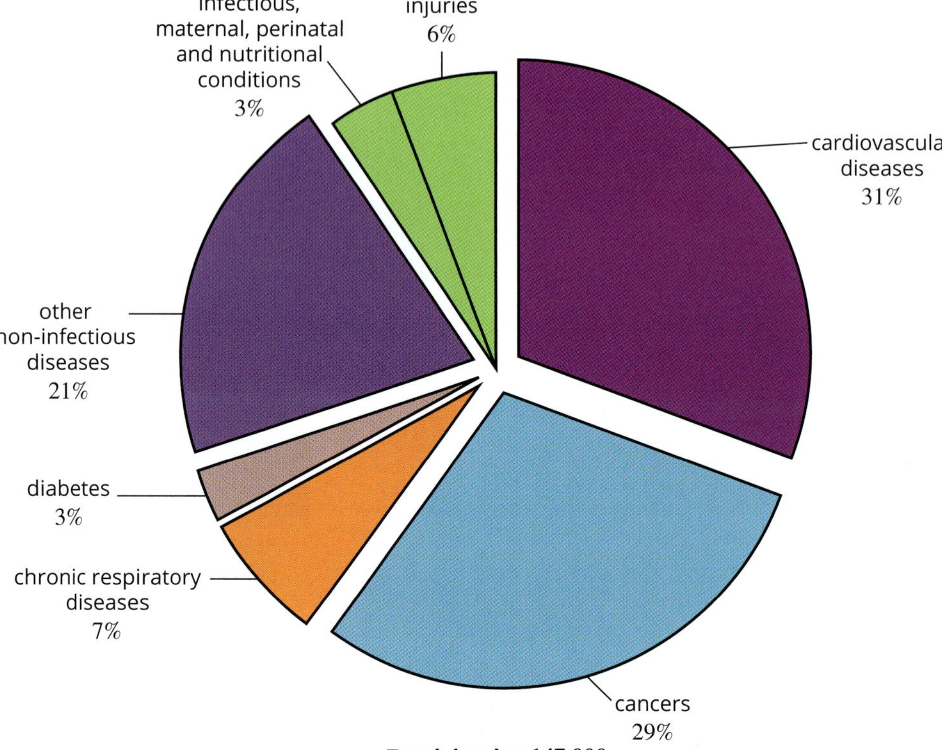

FIGURE 16.1.9 Proportional mortality rates for Australia in 2016 (total deaths, all ages, both sexes)

Risk factors of melanoma skin cancer

Melanoma skin cancer was the fourth most commonly diagnosed cancer in Australia in 2013. It is estimated that it will remain the fourth most commonly diagnosed cancer in 2017 (Table 16.1.1).

Some risk factors associated with developing skin cancer include:

- exposure to sunlight and other sources of UV radiation (e.g. tanning beds)
- a history of sunburns that caused blistering, especially in childhood
- having some large moles, many small moles, or moles that look different from normal moles
- a family history of unusual moles or melanoma
- exposure to radiation and some chemicals (e.g. solvents).

It does not mean that if you have been exposed to chemicals that you will at some stage in your life develop skin cancer, but it does mean that there has been a direct association found linking this factor to developing melanoma.

Risk factors of lung cancer

Like melanoma, lung cancer also has an association with exposure to environmental factors, acting as risk factors. Lung cancer occurs when abnormal cells in the lung grow in an uncontrolled way. It often spreads (metastasises) to other parts of the body before the cancer can be detected in the lungs. Lung cancer is the leading cause of cancer-related deaths in Australia.

Tobacco smoking is a major risk factor in developing lung cancer during a lifetime. In 2016 it was estimated that 14% of the Australian population smoked.

Risk factors that are associated with a higher risk of developing lung cancer include:

- smoking cigarettes, pipes or cigars currently or in the past—this is the greatest risk factor for lung cancer, and the risk is greatest for people who began smoking early in life, smoked for longer and smoked more often
- exposure to second hand smoke
- radiotherapy treatment to the chest
- exposure to asbestos fibres—this also increases the risk of developing **mesothelioma**, which starts in the lining surrounding the lungs (the pleura) and is not considered a type of lung cancer
- exposure to other workplace substances, including radioactive ores (e.g. uranium), chromium compounds, nickel, arsenic, soot, tar or diesel fumes
- exposure to air pollution.

Lung cancer was the fifth most commonly diagnosed cancer in Australia in 2013. It was estimated that it would remain the fifth most commonly diagnosed cancer in 2017 (Table 16.1.1).

TABLE 16.1.1 Estimated most common cancers diagnosed in 2017 from the Australian Institute of Health and Welfare

Cancer type	New cases 2017	New cancers 2017 (%)
breast (total population)	17730	13.2
breast (among females)	17586	28.4
colorectal (bowel)	16682	12.4
prostate (among males)	16665	23.1
melanoma	13941	10.4
lung	12434	9.3

BIOLOGY IN ACTION AHC DD

Lung cancer in our mob

Lung cancer is the most common cause of cancer death in Aboriginal and Torres Strait Islander people. Compared with non-Indigenous Australians, Aboriginal and Torres Strait Islander people are 70% more likely to die from lung cancer. Lung cancer in our mob is an online learning module for Aboriginal and Torres Strait Islander health workers, health practitioners and Aboriginal liaison officers involved in the care of Aboriginal and Torres Strait Islander people with lung cancer (Figure 16.1.10). It is also used by nurses and other allied health professionals, as well as students.

Aboriginal and Torres Strait Islander health workers play a vital role in improving cancer health outcomes in their communities. This online learning module provides information about lung cancer symptoms, the importance of early detection and treatment options. The online learning module can also be used as a reference tool by nurses and other allied health professionals. The resource includes a series of short videos showing a health worker supporting a patient to overcome some of the barriers they face when starting and completing lung cancer treatment. The module provides medical information and details on sources of support for patients and their families.

FIGURE 16.1.10 Lung cancer in our mob is a Cancer Australia online learning module for Aboriginal and Torres Strait Islander health workers involved in the care of Aboriginal and Torres Strait Islander people with lung cancer.

Lung cancer in our mob provides support by:

- raising awareness of the symptoms of lung cancer and the importance of early detection
- promoting a healthy lifestyle
- ensuring patient care is culturally appropriate
- providing support to the patient, their family and community
- explaining medical information in easy to understand language.

16.1 Review

SUMMARY

- Epidemiology is the study of the distribution, patterns and causes of diseases in a population.
- Diseases studied by epidemiologists can include both infectious and non-infectious diseases.
- By analysing the spread, patterns and causes of diseases, predictions can be made and preventative measures undertaken.
- Nutritional diseases can result from a lack of nutrients, or be the result of an over-consumption of certain foods. Although each nutritional disease is specific, a common factor is the importance of a balanced and healthy diet.
- Diets, which have a deficiency, imbalance or excess of nutrients can lead to malnutrition.
- Malnutrition is a factor associated with many nutritional diseases.
- The term malnutrition covers two broad categories:
  - undernutrition, which includes stunting, wasting, underweight and micronutrient deficiencies
  - overweight, which includes obesity and diet-related non-infectious diseases.
- Obesity is a factor associated with non-infectious diseases. Obesity is defined as abnormal or excessive fat accumulation that may impair health.
- Body mass index (BMI) is used to determine whether a person is within a normal range of mass for their height. Having a raised BMI is associated with nutritional diseases.
- Monitoring dietary intake can assist in preventing obesity.
- Environmental risk factors are modifiable and can be avoided in most cases. Environmental exposure to certain chemical, physical and biological factors are associated with certain non-infectious diseases.

KEY QUESTIONS

1. Outline the term malnutrition and discuss the difference between undernutrition and overnutrition.
2. Body mass index (BMI) is a useful tool in epidemiology studies.
   a. What is BMI?
   b. How is BMI calculated?
   c. Why is BMI a useful measure in population studies?
3. List some of the major non-infectious disease risk factors associated with having a raised body mass index (BMI).
4. Discuss if a poor diet is the only contributor to the development of coronary heart disease.
5. The graph on the right represents death rates for Australian males and females from 2000 to 2012.
   a. List the leading cause of deaths for males in the year 2000 and in the year 2012.
   b. List the leading cause of deaths for females in the year 2000 and in the year 2012.
   c. Describe the pattern of diabetes in males and females from 2000 to 2012.

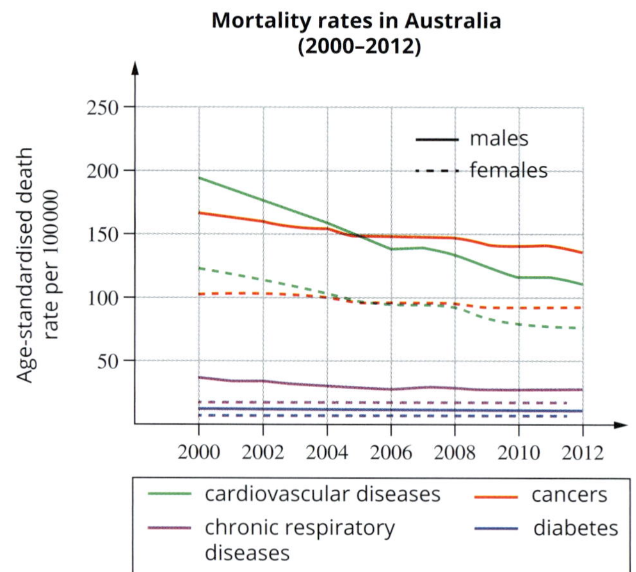

6 Using the information provided in the table below, create a graph showing the number of cancer deaths in Australia for 2017 against the six cancer types presented.

Cancer type	Number of deaths 2017
lung	9021
colorectal (bowel)	4114
prostate (among males)	3452
breast (total population)	3114
breast (among females)	3087
pancreatic	2915

7 Describe the difference between incidence, mortality and prevalence when discussing diseases.

8 The graph below shows the estimated age-specific incidence and mortality rates for melanoma skin cancer by biological sex in 2017.

Using the graph describe:

a Which biological sex has the highest mortality rate due to melanoma?

b What is the rate of incidence for melonama in females of the 80–84 age bracket?

c What is the rate of mortality for melonama in males in the 60–64 age bracket and the 80–84 age bracket?

d Discuss the correlation between age and rate of melanoma for males and females.

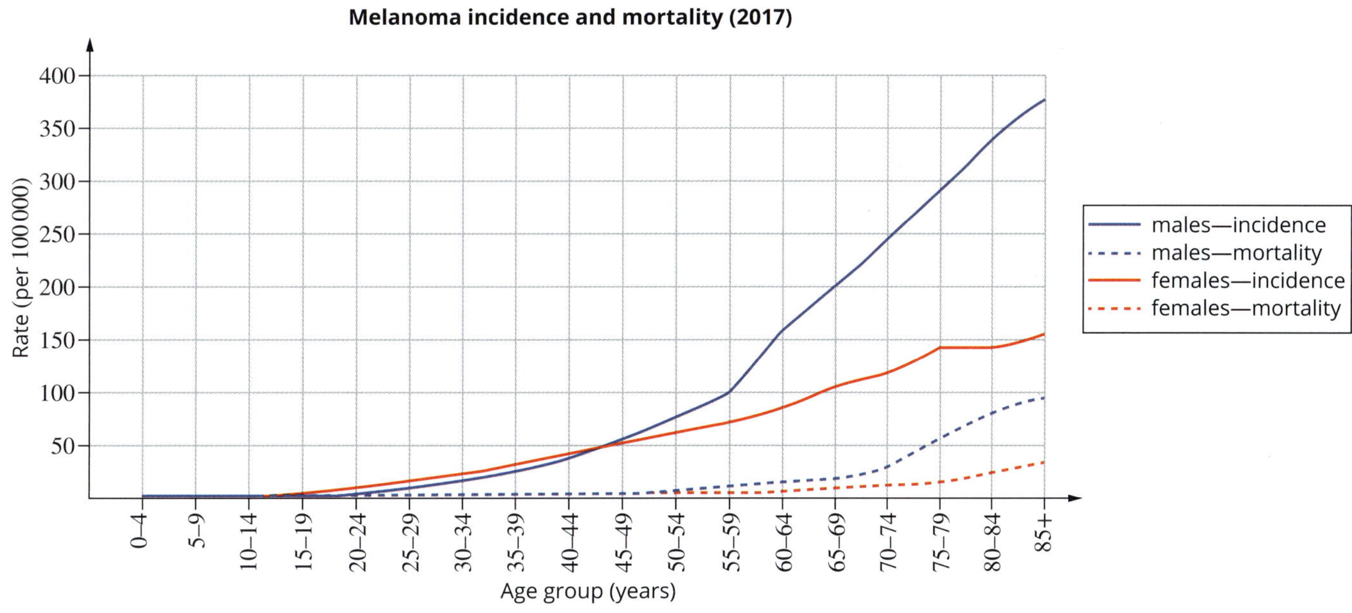

16.2 Treatment and management of non-infectious disease

It is important to recognise that just like infectious diseases, non-infectious diseases cause stress on the body. Many non-infectious diseases are life long and have no cure. The toll such a disease takes on its host can drastically limit quality of life and affect an individual's ability to overcome the disease once it is acquired. As you learnt in Chapter 15, many non-infectious diseases are the result of lifestyle factors, nutrition, genetic factors and exposure to environmental factors (e.g. carcinogens). Carcinogens were covered in detail in Chapter 7. In this section, you will learn about some of the management and treatment options available to people with non-infectious diseases.

GO TO ➤ Section 7.2 page 295

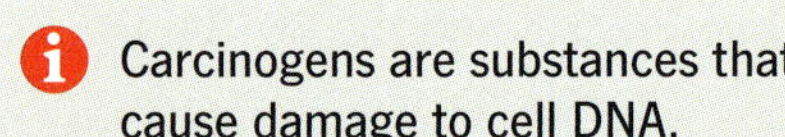

Carcinogens are substances that cause damage to cell DNA.

CANCER TREATMENT

Cancer is one of the most prevalent non-infectious diseases in society today. It is estimated that in 2017, 134 174 new cases of cancer will be diagnosed in Australia. **Immunotherapy** is a new frontier in the treatment of cancers. It enables more specific (or personalised) medicines than other types of treatments and improves outcomes while simultaneously decreasing side effects.

You learnt about the cause and effects of cancer in Chapter 15. In this section, you will learn about preventative measures for some cancers such as vaccines (Figure 16.2.1) and different types of immunotherapies for treating cancer, including monoclonal antibody therapy.

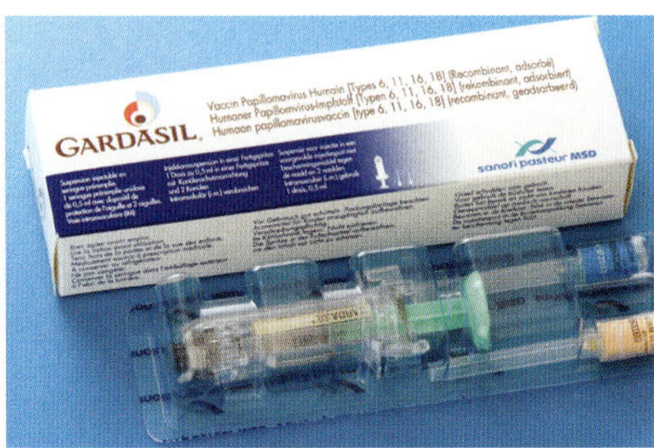

FIGURE 16.2.1 Gardasil® is the brand name for a vaccine that provides immunity against certain strains of human papillomavirus (HPV). The vaccine reduces the risk of developing certain types of cancers associated with HPV infection.

Cancer treatments over several decades have included chemotherapy, radiation therapy and surgery to remove **tumours**.

- **Chemotherapy** involves administering drugs that are **cytotoxic** to cells that multiply rapidly. Although the chemotherapy drugs available today are more specific than those of the past, they are not yet specific enough to avoid damage to healthy cells that also divide rapidly, such as bone marrow and hair follicle cells. This is why chemotherapy has negative side effects.
- **Radiation therapy** indiscriminately kills cells by damaging their DNA. In cancer treatment, radiotherapy is directed at the cancerous cells, but some damage to surrounding tissue is inevitable.
- **Surgery** is beneficial in removing solid tumours but it also has its disadvantages. Any surgery takes a toll on the body and often it is very difficult to ensure that all malignant cells are removed from the body.

The immune response to cancer

People who are older, who use **immunosuppressant** medications for a long period of time, or who have **immunodeficiency**, have weakened immune systems and are at increased risk of developing cancer. However, people do not need to be immunodeficient for their immune system to be ineffective against cancer.

Even people with a normal immune system are sometimes unable to control the growth of tumours. This is because tumour cells evade the immune response in a number of ways. They do this by expressing defective MHC-I molecules (so that cytotoxic T lymphocytes cannot detect that the cells are defective), by producing immunosuppressive **cytokines** or by releasing enzymes that suppress T lymphocyte responses.

Cytokines are a group of peptides and proteins released from cells that are important in cell signalling, particularly between cells of the immune system.

Cancer immunotherapy

Immunotherapy is any treatment that harnesses the immune system of the patient to fight diseases such as cancer (Figure 16.2.2). Immunotherapy can be non-specific or specific:

- Non-specific immunotherapies don't specifically target cancer cells but stimulate the immune system (e.g. by the injection of cytokines). Cytokines do not directly target cancer cells, but the stimulation of the immune system can result in a better immune response against cancer cells.

- Specific immunotherapies act on cancer cells by directly stimulating the adaptive immune response against them. Specific immunotherapies include cancer vaccines and monoclonal antibody therapy.

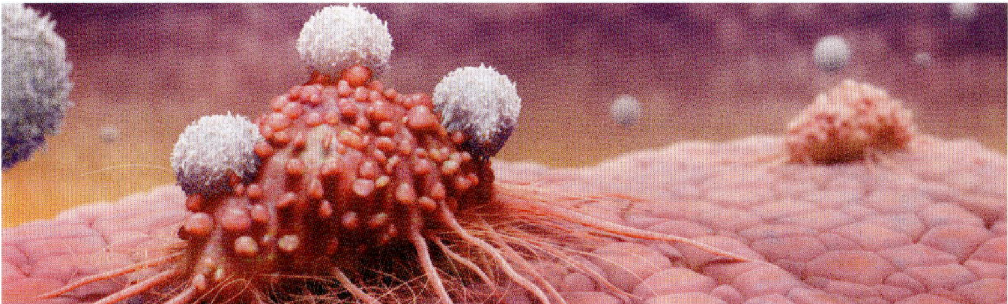

FIGURE 16.2.2 Digital representation of cytotoxic T lymphocytes (white) attacking a cancer cell (red)

Cancer vaccines

Cancer vaccines stimulate the immune system to attack cancer cells. Some cancer vaccines contain peptides or whole proteins of cancer cells and **adjuvants** (a substance which boosts the amount of antibodies) to help stimulate an immune response against them. Sometimes a patient's own immune cells are harvested, exposed to these antigens and then injected back into the body to produce an immune response. Cancer vaccines have few known side effects. They can be classified as preventive, therapeutic or personalised.

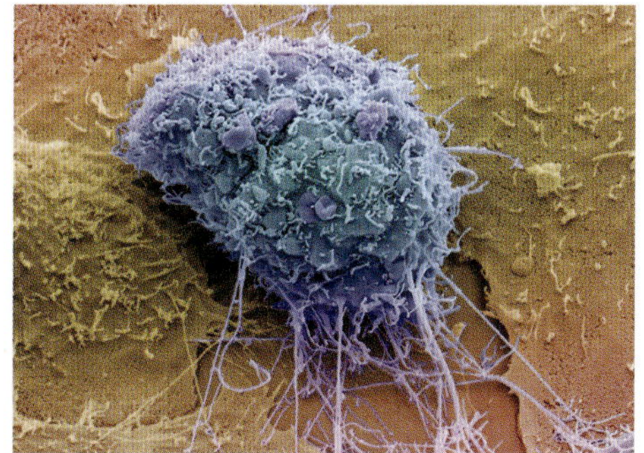

FIGURE 16.2.3 Coloured SEM of a cervical cancer cell

Preventive cancer vaccines

Preventative cancer vaccines are vaccines directed against viruses that cause cancer. Examples include vaccines for human papillomavirus (HPV), which causes cervical cancer (Figure 16.2.3), and hepatitis B virus (HBV), which causes cancer of the liver. These vaccines introduce specific viral antigens into the body, creating an adaptive immune response that will lead to immunological memory and a stronger and faster response towards the virus if the body is exposed to it.

An example of a preventative cancer vaccine is the HPV vaccine called Gardasil®. Professor Ian Frazer and Dr Jian Zhou from the University of Queensland developed Gardasil using a gene for the coat protein of HPV (Figure 16.2.4).

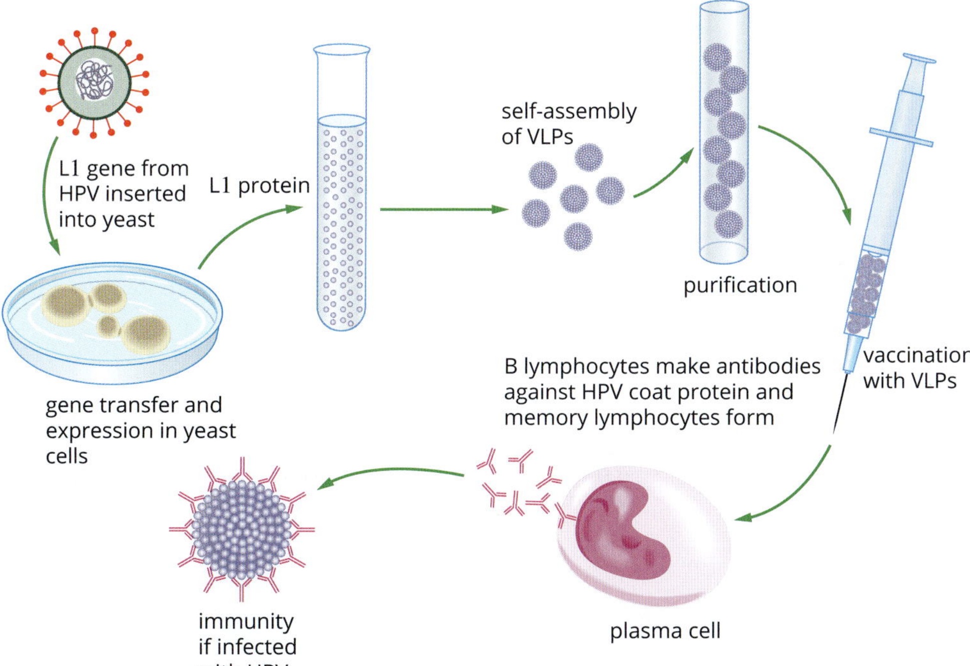

FIGURE 16.2.4 The gene for the HPV coat protein L1 was taken from the virus and inserted into yeast cells, which produce large amounts of the protein. This L1 protein 'self assembles' into particles that look like the virus and are therefore known as virus-like particles (VLPs). These VLPs do not contain any viral DNA, so they do not cause disease and are not infectious. The VLPs are the antigen used in the Gardasil vaccine.

Therapeutic cancer vaccines

Therapeutic cancer vaccines are given to people who already have cancer. These vaccines are made up of antigens for a specific type of cancer cell (usually proteins or parts of proteins). Often adjuvants are included to help boost the immune response, increasing its ability to identify and destroy cancer cells.

Personalised cancer vaccines

Personalised cancer vaccines are therapeutic vaccines developed for an individual patient. Some involve tumour cells that have been removed from the patient, altered in the laboratory to make them more obvious to the immune system, and then injected back into the patient.

One of the most effective personalised cancer vaccines involves the patient's own tumour and **dendritic cells**, which are antigen-presenting cells of the immune system. This therapy works by presenting the antigen isolated from the patient's tumour sample to the patient's dendritic cells. The activated dendritic cells are then injected back into the patient, where they present antigens to helper T lymphocytes, eliciting an adaptive immune response.

Antigen-presenting cells present foreign antigens attached to MHC-II molecules on their surface.

Monoclonal antibody therapy

Monoclonal antibodies (mAbs) are antibodies produced by a single clone of a B lymphocyte that is grown in culture to produce a large volume of the same clone. The mAbs produced by the clones are all identical and specific to the same antigen. One of the ways mAbs are used to treat cancer is by targeting specific antigens present on tumour cells (Figure 16.2.5). But they can also be used to target cells of the immune system and direct the immune response in a way that helps destroy tumour cells.

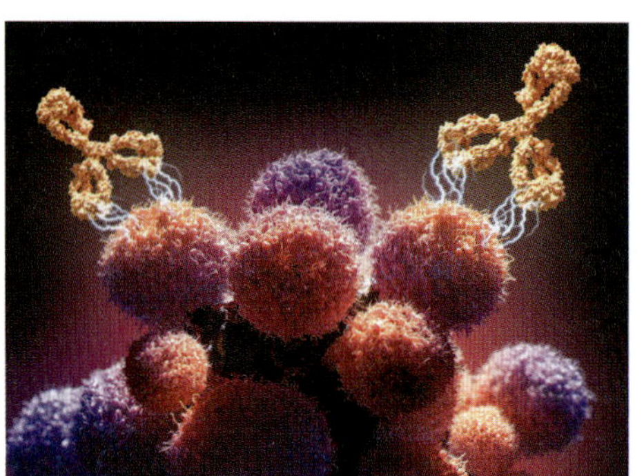

FIGURE 16.2.5 Digital representation of monoclonal antibodies binding to antigens on cancer cells

Production of monoclonal antibodies

Figure 16.2.6 illustrates the steps required to make monoclonal antibodies.

- First, mice are injected with a particular antigen, which in the case of cancer therapy is an antigen from a cancer cell.
- This induces the mice's B lymphocytes to produce antibody specific to the antigen. These B lymphocytes are then isolated from the spleens of the mice.
- In isolation the B lymphocytes only have a limited lifespan, so in order to produce the large quantity of antibodies needed, the isolated B lymphocytes are fused with **myeloma cells**, which are an **immortal cell line**.
- The fusion of the two cells results in a hybridised cell called a **hybridoma**.
- The hybridoma is more stable in tissue culture conditions and the cell secretes multiple copies of the specific antibody (the mAbs), which are then harvested.

Immortal cell lines can continually undergo division without mutation, which would normally occur as a cell ages, and can therefore be cultured for long periods.

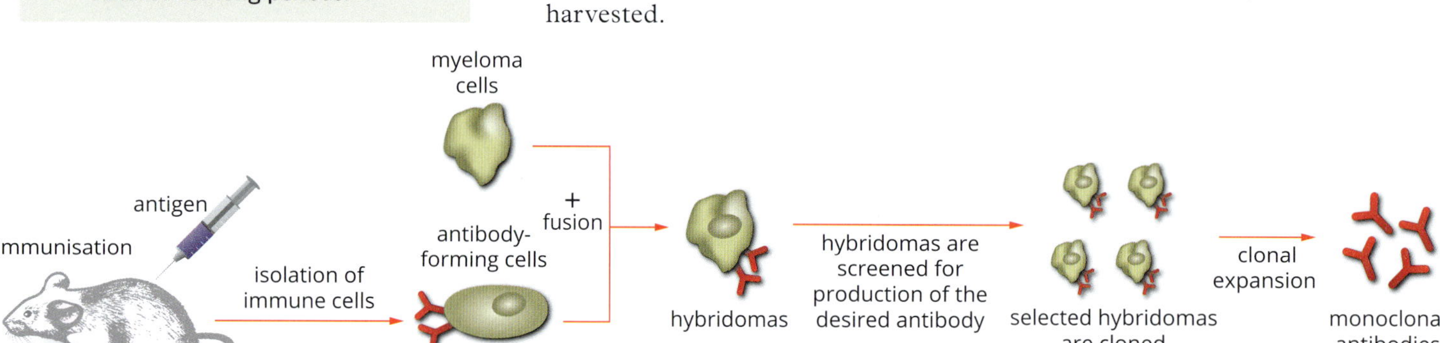

FIGURE 16.2.6 Production of monoclonal antibodies (mAbs). Mice are injected with an antigen. B lymphocytes sensitised to the antigen are then taken from the mice and fused with myeloma cells. The fused cells, called hybridomas, make antibodies to the antigen. These are grown in culture dishes to produce large quantities of monoclonal antibody specific to the antigen.

Chimeric, humanised and human monoclonal antibodies

The first mAbs produced were mouse mAbs made entirely by mouse B lymphocytes, and many still are today. Although these types of mAbs are initially effective when used in human therapy, an immune response is mounted against them once they are identified as foreign (mouse) proteins. Immunological memory is formed and the adaptive immune response recognises and destroys them faster when the same mAbs are subsequently used again.

To help prevent an immune response directed against them, researchers have been able to replace some components of mouse antibodies with human components using recombinant DNA techniques (Figure 16.2.7). Antibodies with a mixture of mouse and human components are known as chimeric mAbs. As mAbs contain more and more human components they are termed humanised mAbs (Figure 16.2.9a). Some mAbs are now fully human antibodies produced by transgenic mice. Although chimeric, humanised and human mAbs are all still produced by mice, these mAbs may be safer and potentially more effective than earlier mAbs. Antibodies that contain only human components are known as human monoclonal antibodies (Figure 16.2.9b).

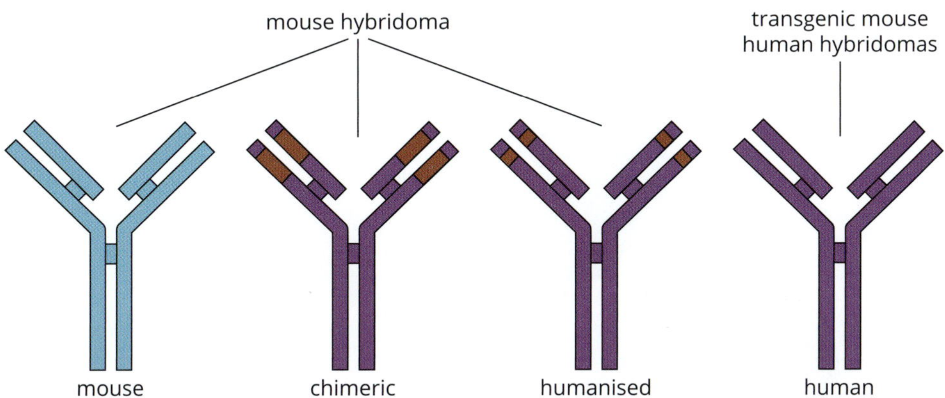

FIGURE 16.2.7 Monoclonal antibodies can be categorised into four types based on their protein composition: animal (most commonly mouse), chimeric (combination of mouse and human), humanised (mostly human) and human.

Conjugated monoclonal antibodies

Conjugated monoclonal antibodies are mAbs that have been attached to a chemotherapy drug, a toxin or a radioactive particle (Figure 16.2.9c). In this way, conjugated mAbs are used as carriers modified to deliver treatments directly to the specific cancer cells. For example, radioimmunotherapy can be used to treat pancreatic cancer. A radioactive isotope (lead-212) is combined with a specific antibody capable of targeting cancerous cells. This combination of antibody and lead-212 radioisotope is injected intravenously into the body. When it reaches the pancreas, it locks onto the cancerous cells' antigens and the lead-212 destroys the cells by irradiating them. This treatment limits the toxic effects on healthy cells to those located near the cancerous cells.

Transgenic mice have been genetically modified to contain genes from other species.

BIOFILE CCT

Immunotherapy for melanoma—pembrolizumab

Pembrolizumab (Keytruda®) is an immunotherapy medication approved for use in Australia to treat metastatic melanoma (Figure 16.2.8). Pembrolizumab is a humanised monoclonal antibody and works by binding to a receptor on T lymphocytes called programmed cell death 1 (or PD-1). PD-1 normally interacts with two ligands on antigen-presenting cells, called PD-L1 and PD-L2. This interaction between PD-1 and the PD-L1 and PD-L2 ligands inhibits T lymphocyte activation and cytokine production. (A ligand is a substance that binds specifically and reversibly to another substance, forming a complex.)

A range of tumour cells, including melanoma cells, have PD-L1 expressed on their surface, so binding of T lymphocyte PD-1 and tumour cell PD-L1 inhibits T lymphocyte responses against the tumour cells (PD-L2 is also expressed on a variety of tumour cells, but it has not been studied as extensively as PD-L1 and its impact on anti-tumour immunity is less clear). The binding of pembrolizumab to PD-1 receptors blocks the inhibition of T lymphocytes, allowing them to become activated and produce inflammatory cytokines. This improves the immune response against tumour cells, but it also results in autoimmune reactions in which healthy cells are damaged.

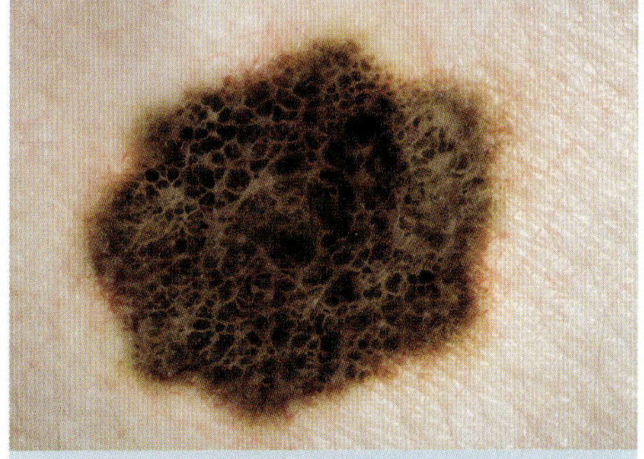

FIGURE 16.2.8 Photograph of melanoma

Bispecific monoclonal antibodies

Bispecific monoclonal antibodies are artificially produced using recombinant DNA technology and are used to target cancer cells and activate the immune system simultaneously (Figure 16.2.9d). This type of mAb is used to indirectly activate an adaptive immune response. Naturally produced mAbs have two binding sites, but each site binds to the same antigen. Bispecific mAbs can attach to two different antigens at the same time, because they are composed of parts from two different mAbs and have two different antigen-binding sites (Figure 16.2.10).

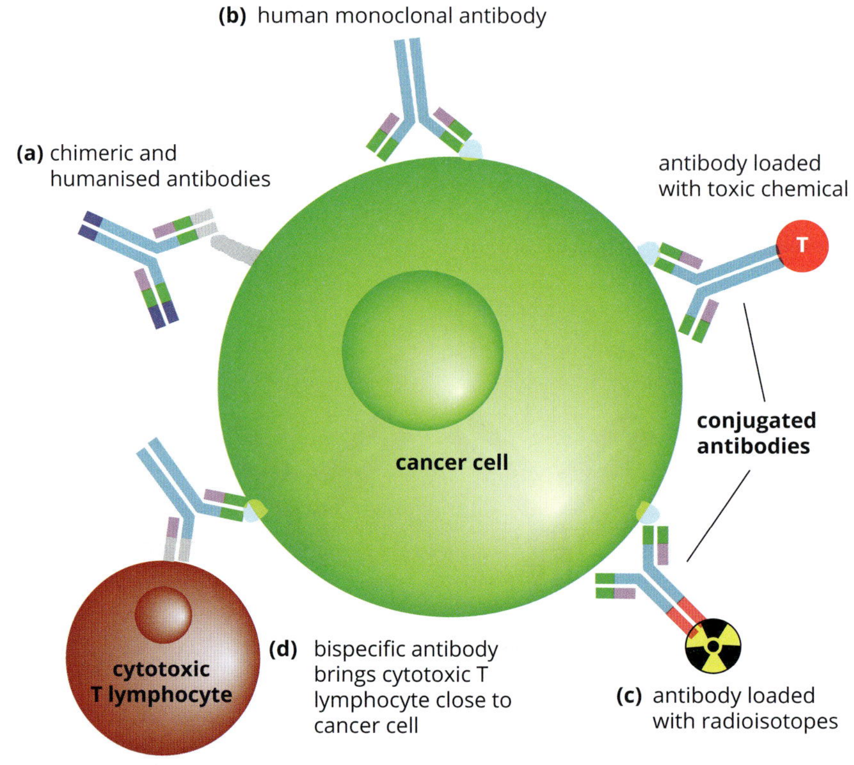

FIGURE 16.2.9 (a) Chimeric antibodies contain parts from mouse and human antibodies, and humanised antibodies contain mostly human parts, but are still produced in mice. (b) Human monoclonal antibodies contain only human components. (c) Conjugated antibodies are joined to toxins or radioisotopes to specifically target and kill cancer cells and spare normal body cells. (d) Bispecific antibodies are those engineered to bind two different antigens, in order to bring cytotoxic T lymphocytes close to tumour cells.

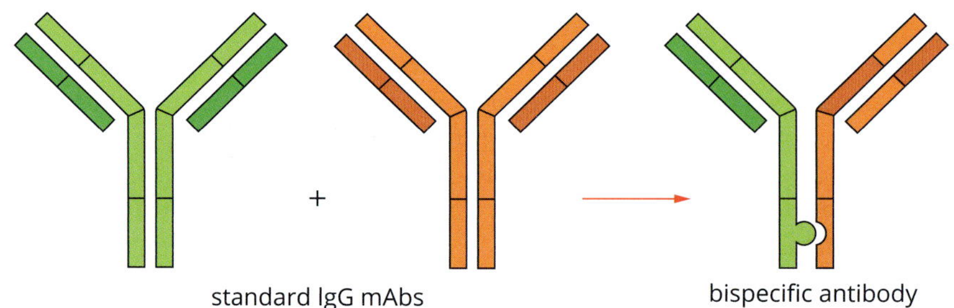

FIGURE 16.2.10 Diagram of a bispecific monoclonal antibody, in which parts of two different antibodies have been fused to form a hybrid that can bind to cancer cells and to T lymphocytes

An example of a bispecific mAb is Blincyto®, which is used to treat some types of acute lymphocytic leukaemia. One part of this mAb attaches to a protein on the surface of the leukaemia cells, while the other part attaches to a protein found on T lymphocytes of the immune system. By binding to both these proteins the bispecific mAbs are effectively 'identifying' the cancer cells as foreign and 'delivering' them to the immune system.

BIOLOGY IN ACTION  

A cancer vaccine for Tasmanian devils

Tasmanian devils (*Sacrophilus harrisii*) have been plagued for decades by a contagious facial cancer called devil facial tumour disease (DFTD) (Figure 16.2.11a). The cancer originated from a single cancerous Schwann cell in a single Tasmanian devil many years ago and spread from one Tasmanian devil to another by bites. DFTD has devastated the Tasmanian devil population and threatens the survival of the species.

Contagious cancer is very unusual and at first it was thought the reason DFTD spread so easily between Tasmanian devils was that they had weakened immune systems, or low genetic variation. Low genetic variation would make it easy for the cancer to spread, because it would mean the immune system of an individual Tasmanian devil wouldn't recognise the cancer cells from another Tasmanian devil as foreign. However, both these theories have been ruled out.

In fact, it appears the cancer cells evade the Tasmanian devil's immune system by destroying their major histocompatibility complex (MHC) molecules, which are vital for immune recognition. Without MHC molecules on the cancer cells, the Tasmanian devils' immune cells do not detect them.

Researchers from the University of Tasmania have shown that a vaccine using killed DFTD tumour cells and adjuvant is able to stimulate a protective adaptive immune response (Figure 16.2.11b). The vaccine is currently being trialled, and if successful, it may help bring the Tasmanian devil back from the brink of extinction.

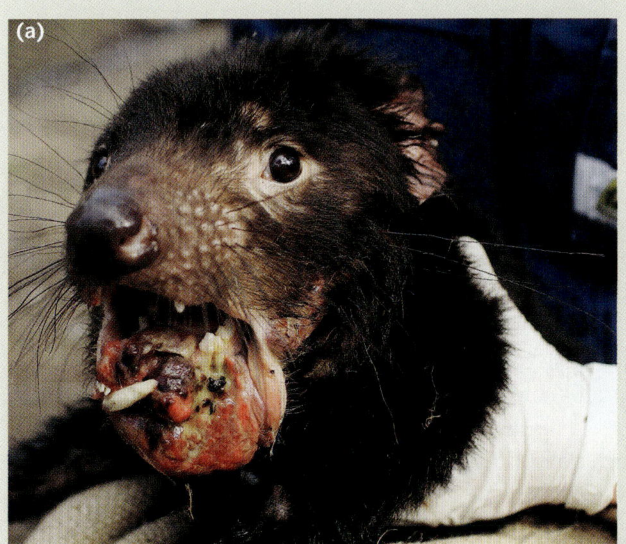

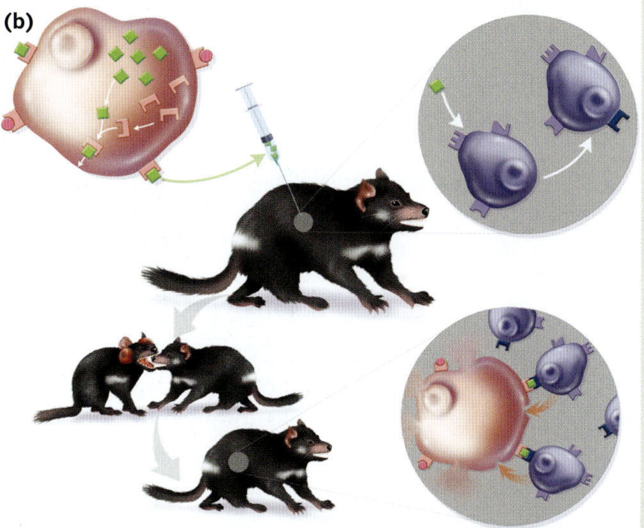

FIGURE 16.2.11 (a) A Tasmanian devil with the devil facial tumour disease (DFTD). (b) The cancer vaccine for devil facial tumour disease (DFTD) uses tumour-specific antigens from killed tumour cells to stimulate an adaptive immune response involving T lymphocytes.

16.2 Review

SUMMARY

- Cancer occurs when a single abnormal cell multiplies uncontrollably and spreads throughout the body.
- Cancer treatments over the last several decades have included chemotherapy, radiation therapy and surgery. These treatments come with significant side effects.
- Immunotherapy is any treatment that harnesses the immune system of the patient to fight diseases such as cancer.
- Immunotherapies can be non-specific, such as the injection of cytokines, or specific, such as cancer vaccines, personalised immunotherapy or monoclonal antibody therapy.
- Cancer vaccines are made using specific antigens from cancer cells or infectious agents that cause cancer. Like other vaccines, they are administered to a patient to stimulate an immune response that results in the production of an immunological memory.
- Cancer vaccines can be preventative, therapeutic or personalised.
- Preventative cancer vaccines are directed at viruses that cause cancer such as human papillomavirus (HPV), which causes cervical cancer. Antigens specific to the virus are injected into the body, triggering an adaptive immune response that will lead to immunological memory.
  - Therapeutic cancer vaccines are made up of antigens for a specific type of cancer. These antigens are injected into the body along with adjuvants that help boost the immune response.
  - Personalised cancer vaccines involve cells or substances from the patient, which are removed, altered in a laboratory and then reintroduced by injection. The most effective personalised immunotherapies involve the patient's own tumour and dendritic cells.
- Monoclonal antibody (mAb) therapy involves antibodies produced by a single clone of B lymphocytes that are replicated in culture. mAbs are all identical and specific to the same antigen. Targeting specific cells reduces harm to healthy cells, but identifying the specific antigen to create mAbs is a difficult task.
- The first monoclonal antibodies (mAbs) were mouse mAbs made entirely by mouse B lymphocytes. To avoid an immune response against mAbs, chimeric, humanised and human monoclonal antibodies can now be produced using transgenic mice:
  - chimeric mAbs are a mix of human and mouse components
  - humanised mAbs are also a mix but are mostly human
  - human mAbs are fully human.
- mAbs can be used as carriers of treatments (drugs, toxins and radioactive particles) for specific delivery to cancer cells. These types of mAbs are called conjugated mAbs.
- Bispecific mAbs are made up of two different mAbs and have two different binding sites: one is usually for a cancer cell and the other for an immune cell, such as a T lymphocyte. This enables an 'identify' and 'deliver' approach.

KEY QUESTIONS

1. What is the difference between non-specific and specific immunotherapies?
2. Explain the difference between a preventative cancer vaccine and a therapeutic cancer vaccine, including reference to the type of antigen in each vaccine.
3. What is the difference between chimeric, humanised and human monoclonal antibodies?
4. If a monoclonal antibody has a toxin or radioactive substance attached to it, what kind of monoclonal antibody is it?
   A bispecific
   B conjugated
   C chimeric
   D humanised
5. Bispecific antibodies are produced artificially. How are these antibodies different from those produced naturally by the immune system?
6. Describe how a tumour cell is able to evade the host immune response.

16.3 Epidemiological research

Epidemiological studies of non-infectious diseases involve the collection and careful statistical analysis of large quantities of data. In the past epidemiological studies have established links between diet and heart disease, smoking and lung cancer, as well as causes of other diseases. When large quantities of data are gathered and statistically analysed, information found can then be applied to health programs and health services. These studies also act to assess the value of treatments and preventative strategies in a population.

EPIDEMIOLOGICAL STUDIES

There are three main categories of epidemiological studies classified as analytical studies, descriptive studies and intervention studies.

- **Analytical studies** are planned investigations that test a specific hypothesis (e.g. comparing the lifestyles of people with a particular disease to determine if they have any risk factors in common).
- **Descriptive studies** show patterns in the way that a disease is distributed in a population (e.g. where a disease occurs in the world or which groups in a city suffer from a certain disease).
- **Intervention studies** act to measure the effectiveness and safety of certain interventions (e.g. the effectiveness of a new drug treatment in curing a disease).

Distribution, frequency and pattern of disease

Epidemiology is concerned with the geographic distribution, frequency and pattern of disease in a population.

- **Geographic distribution** refers to the relationship between the prevalence of a disease and the specific area and environmental conditions of the region (Figure 16.3.1).
- **Frequency** refers not only to the number of health events such as the number of cases of tuberculosis or diabetes in a population, but also to the relationship of that number to the size of the population.
- **Pattern** refers to the occurrence of health-related events by time, place and person. Time patterns may be annual, seasonal, weekly, daily, hourly, weekday versus weekend, or any other break down of time that may influence disease or injury occurrence. Place patterns include geographic variation, urban/rural differences and location of work sites or schools.

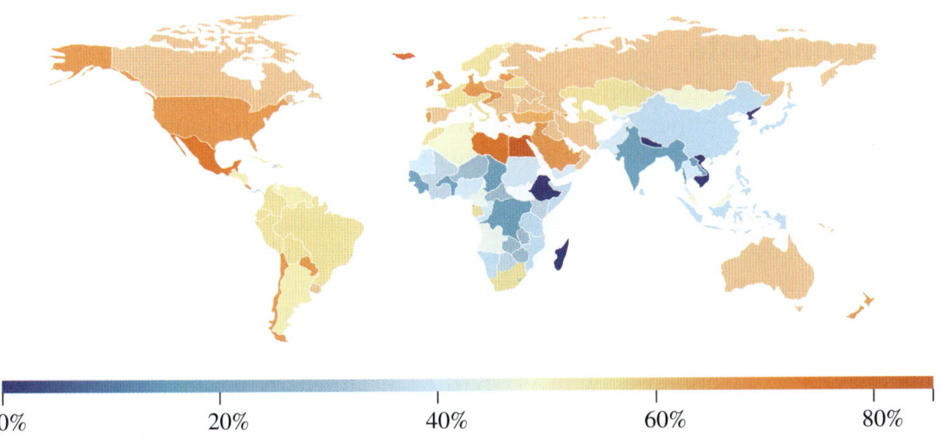

FIGURE 16.3.1 Map of obesity patterns around the world. The regions with the highest rates of obesity are indicated in orange and the regions with the lowest rates of obesity are indicated in blue.

Benefits of epidemiological studies

Epidemiological data aims to identify the nature and causes of diseases within a population. The results obtained are essential to decisions about public health programs and the risk factors contributing to the cause of a disease. The first epidemiological studies in the 19th century were essential towards the development of better sanitation, housing and nutrition. Today, the findings from epidemiological studies focus on drawing correlations between disease and demographic and economic factors such as age, ethnic group, socioeconomic status and education. Studies are able to identify and raise awareness about social injustices in populations and draw attention to groups of people who require greater assistance.

The importance of gathering data

Measuring how many people are affected by a disease each year, how many die or why they died is one of the most important means for assessing the effectiveness of a country's health care system.

Cause-of-death **statistics** help government health authorities determine the focus of their public health actions. For example, if a country in which deaths from heart disease and diabetes rise rapidly over a period of a few years, the statistics may act to promote programs to encourage lifestyle changes that help prevent these illnesses from continuing. If a country recognises that many children in their population are affected by a certain nutritional disease, this may prompt authorities to dedicate a larger proportion of the yearly budget to disease prevention or nutritional aid.

Data that is collected from epidemiological studies can be used to:

- compare changes over time
- estimate trends
- predict possible outbreaks and epidemics.

Statistical data collected about human disease can tell us about the incidence of disease in a population. There are various methods used to collect data for epidemiological studies. These methods are normally conducted in a series of stages as described below.

1. A diagnostic phase, in which the presence of the disease is confirmed.
2. A descriptive phase, which describes the populations at risk and the distribution of the disease within a population. This may then allow a hypothesis to be formed about the likely cause of a disease.
3. An investigative phase, which involves the implementation of a series of field studies.
4. An experimental phase, in which experiments are performed under controlled conditions to test the hypotheses in more detail.
5. An analytical phase, in which the results produced are analysed.
6. A decision-making phase, in which knowledge of the epidemiology of the disease is used to explore the various options available for managing health practices.

Collecting data

Epidemiological studies often involve the collection of large amounts of data from groups of people, rather than individuals. Collecting information from a large sample size helps to ensure that any relationship found between disease and an external factor represents a real association. The data can be used to compare changes over time, to estimate trends and even to predict possible outbreaks and epidemics. Cause-and-effect relationships are often difficult to set in stone, as there are many additional factors, which might also contribute to the cause of a disease such as socioeconomic status, diet and availability of food.

Data for epidemiological studies is often obtained through interviews and questionnaires. In countries with established health care systems, data can be collected quickly from doctor's clinics and health care providers. Data is often gathered through census documents, which people can complete online or send in the post. Epidemiological data collection can be difficult in developing countries due to limited health care services and challenges with communication, such as language barriers, illiteracy or limited communication technology.

EPIDEMIOLOGICAL PATTERNS

The study of epidemiological disease data allows scientists to identify patterns and trends in disease distribution and occurrence. Being able to recognise patterns in epidemiological data enables government funding and health care programs to be directed to specific target groups within a population.

Variations based on economic group

The collection of epidemiological disease data varies amongst countries, in many cases due to the economy and type of health care system. High income countries have systems in place for collecting information on causes of death. Many low and middle income countries do not have such systems and the numbers of deaths from specific causes have to be estimated from incomplete data (Figure 16.3.2). Improvements in producing high quality cause-of-death data are crucial for improving health and reducing preventable deaths in these countries.

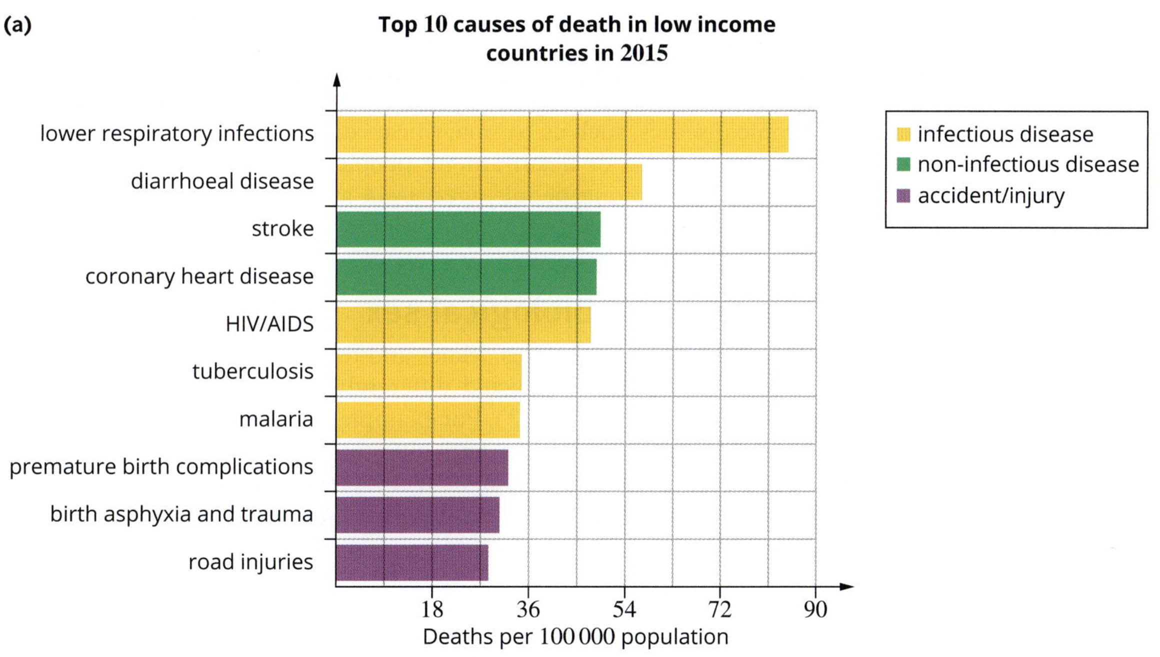

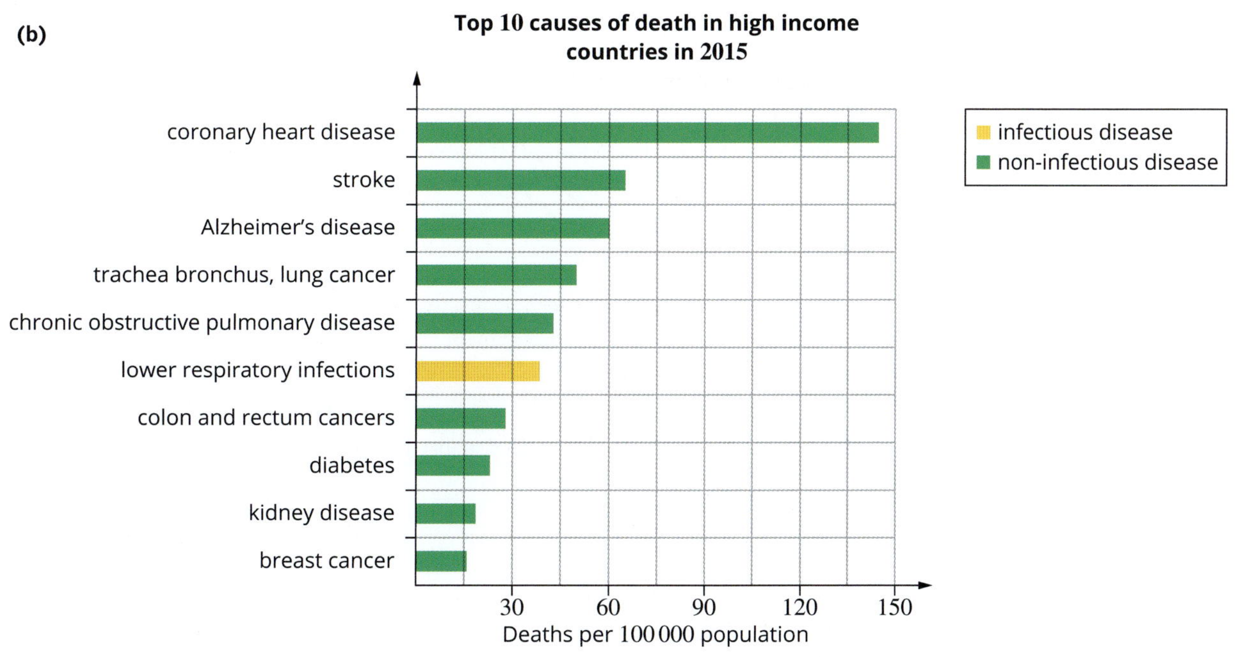

FIGURE 16.3.2 Graphs representing the top 10 causes of death in (a) low income countries in 2015 and (b) high income countries in 2015

More than half (52%) of all deaths in low-income countries in 2015 were caused by preventable conditions, which include non-infectious diseases, conditions arising during pregnancy and childbirth as well as nutritional deficiencies. By contrast, less than 7% of deaths in high income countries were due to such causes. In 2015, non-infectious diseases caused 70% of deaths globally, ranging from 37% in low income countries to 88% in high income countries. The higher rate of deaths due to non-infectious disease in high income countries has been linked to risk factors such as exposure to tobacco smoke, unhealthy foods and less active lifestyles.

The graphs in Figure 16.3.2 represent the top 10 causes of death in (a) low income countries in 2015 and (b) high income countries in 2015. It is evident from each graph that non-infectious diseases pose a much larger problem in high income countries.

Variations based on age group

As well as variations based on economy and health care, epidemiological data also indicates variations based on age group. It is important to gather data across the wide spectrum of a population, which means that all age groups from infancy to the elderly should be included in population studies.

The leading underlying causes of deaths are different at different ages. In Australia, chronic diseases, such as coronary heart disease and Alzheimer's disease, are a more common cause of death among people aged 45 and over, while the leading causes of death among people aged 1–44 are external causes such as car accidents. Among infants, certain conditions originating in the perinatal period and congenital conditions were responsible for most deaths.

Variations based on biological sex

Variations also exist between the sexes. Males and females have different biological responses to disease and can be affected differently or be more prone to certain non-infectious diseases (Figure 16.3.3). For example, males have a higher incidence of cancer than females in Australia and although there may be contributing environmental factors there are also genetic factors involved.

There are fields of medicine that study the biological differences between the human sexes and how the differences influence disease presentation, treatments and outcomes.

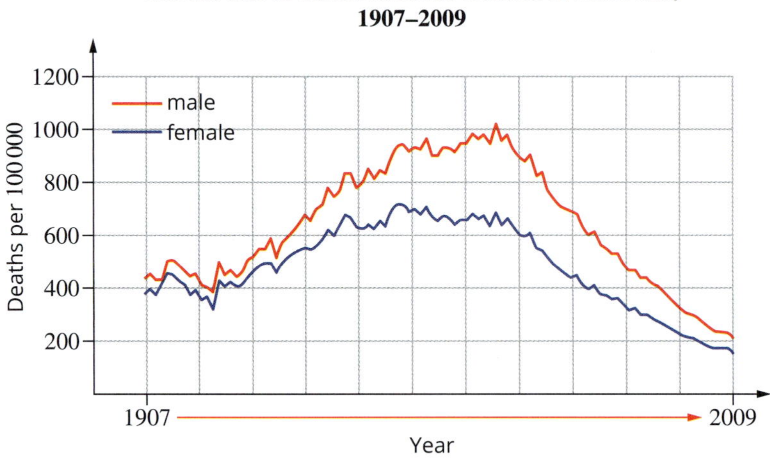

FIGURE 16.3.3 Deaths due to cardiovascular disease (per 100 000) in males (red line) and females (blue line) in Australia between 1907 and 2009

Factors that influence biological sex differentials in relation to the risk and presentation of disease include:

- biological (genetic, physiological, and hormonal) differences between males and females
- different behaviours and lifestyle choices of males and females
- differential access to social and health care
- cultural practices and religious beliefs which may determine gender roles in a population
- differences in access to education
- income difference.

BIOLOGY IN ACTION  PSC

Epidemiological studies of an association between cancer and physical activity

In Australia, being physically inactive ranks highly as a risk factor in contributing to the development of many non-infectious diseases. Being active also helps to maintain a healthy body weight. WHO found that there is convincing evidence that regular physical activity decreases the risk of weight gain and obesity while an inactive lifestyle increases the risk.

It is thought that being obese can increase the risk of developing cancers. In 2002, the International Agency for Research on Cancer (IARC) published a handbook on the evidence for body weight and physical activity in relation to cancer risk. Recent Australian data suggests that physical inactivity accounts for 5.6% of total cancers and 6.6% of all disease.

An epidemiological study in 2007 found that there is strong evidence for an association between leisure time, physical activity and post-menopausal breast cancer. However the evidence was much weaker for pre-menopausal breast cancer. While duration of physical activity (hours/week) was the primary measure in the review, some studies also reported on metabolic intensity, which refers to the rate at which an activity is performed or the magnitude of the effort required.

In 2003, a further review examining health behaviors in early adulthood found that there appears to be a weak link between physical activity in early life (i.e. adolescence) and risk of breast cancer (both pre- and post-menopausal). Yet, these results only reflect a particular life stage (i.e. adolescence) and it is likely that the accumulation of physical activity over many years is important for cancer protection.

Overall, epidemiological studies suggest that being physically active probably reduces the risk of developing breast cancer, especially in post-menopausal women. However, conflicting evidence is common due to other contributing factors, such as age, health status and diet, playing a role in the development of diseases such as cancer. Studies such as this provide great potential for further research into the risk factors of disease (Figure 16.3.4).

FIGURE 16.3.4 Research suggests that physical activity decreases the risk of developing breast cancer. The Mother's Day Classic national fun run raises money for breast cancer research and promotes exercise.

16.3 Review

SUMMARY

- Epidemiological studies involve the collection and careful statistical analysis of large quantities of data.
- There are three main categories of epidemiological studies classified as analytical studies, descriptive studies and intervention studies.
- Distribution patterns in epidemiology studies can be concerned with the frequency or the pattern of disease events.
- Data that is collected from epidemiological studies can be used to compare changes over time, estimate trends or predict possible outbreaks and epidemics.
- Collecting information from a large sample size helps to ensure that associations found between diseases and external factors are representative of real relationships.
- When large quantities of data are gathered and statistically analysed, information found can then be applied to health programs and health services.
- Epidemiological studies show variations in populations based on differences in income, age and biological sex.

KEY QUESTIONS

1 Identify the correct description of epidemiology.

A Epidemiology is the study of epidemics of infectious diseases in humans from developed countries.

B Epidemiology studies the patterns and causes of diseases in groups of people.

C Epidemiology is useful in the identifying causes of infectious diseases only.

D Epidemiology is the collection and statistical analysis of very small volumes of data on health and health services.

2 The data presented in the figure below outlines the leading causes of death in Australia in the year 2002.

a List the causes in order from the largest percentage of deaths to the smallest.

b Outline which causes can be classified as non-infectious diseases.

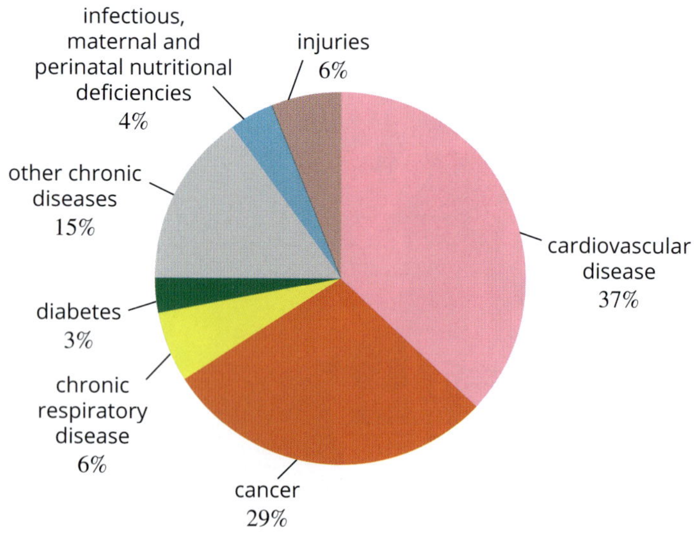

3 What is the difference between analytical studies in epidemiology and intervention studies?

4 In the study of epidemiology, the term 'distribution' refers to:

A who

B when

C where

D why

5 What is the difference between 'frequency' and 'pattern' in epidemiological studies?

6 During an epidemiological study, what does the descriptive phase of design aim to do?

7 Outline the six stages of collecting data for epidemiological studies.

8 Discuss the benefits of epidemiological studies of disease.

Chapter review

16

KEY TERMS

acute disease
adjuvant
analytical study
body mass index (BMI)
cancer vaccine
chemotherapy
chronic disease
cytokine
cytotoxic
dendritic cell
descriptive study
epidemiology
frequency
geographic distribution
Healthy Eating Pyramid
hybridoma
immortal cell line
immunodeficiency
immunosuppressant
immunotherapy
incidence
intervention study
malnutrition
melanoma
mesothelioma
monoclonal antibody (mAb)
mortality
myeloma cell
non-infectious disease
obesity
overnutrition
pattern
population data
prevalence
public health program
radiation therapy
risk factor
statistic
surgery
tumour
ultraviolet (UV) radiation
undernutrition

REVIEW QUESTIONS

1 a Which group of people are most at risk of malnutrition?
  b What factors increase the risk of malnutrition to this group of people?

2 The body mass index (BMI) of an athlete and a school teacher were measured. Both were Australian males, 33 years old who were non-smokers. Benjamin is a professional weight lifter who eats a healthy diet and trains regularly. Andy is a Year 12 school teacher who exercises on occasions and enjoys eating out with friends on the weekends. Benjamin is 180 cm tall and weighs 111 kg. Andy is 179 cm tall and weighs 78 kg.
  a Calculate the BMI for both Benjamin and Andy.
  b Classify both males as having normal or obese body mass.
  c Should the BMI results alone be used as a measure of health for these two males? Why or why not?

3 Study the graph below that illustrates the estimated incidence and mortality rate of all cancers in Australia in 2017.
  a Using the graph, compare the incidence and mortality rates of cancer.
  b Using the graph, compare the incidence and mortality rates of cancer between males and females.
  c Why is the incidence rate higher than the mortality rate in each category?

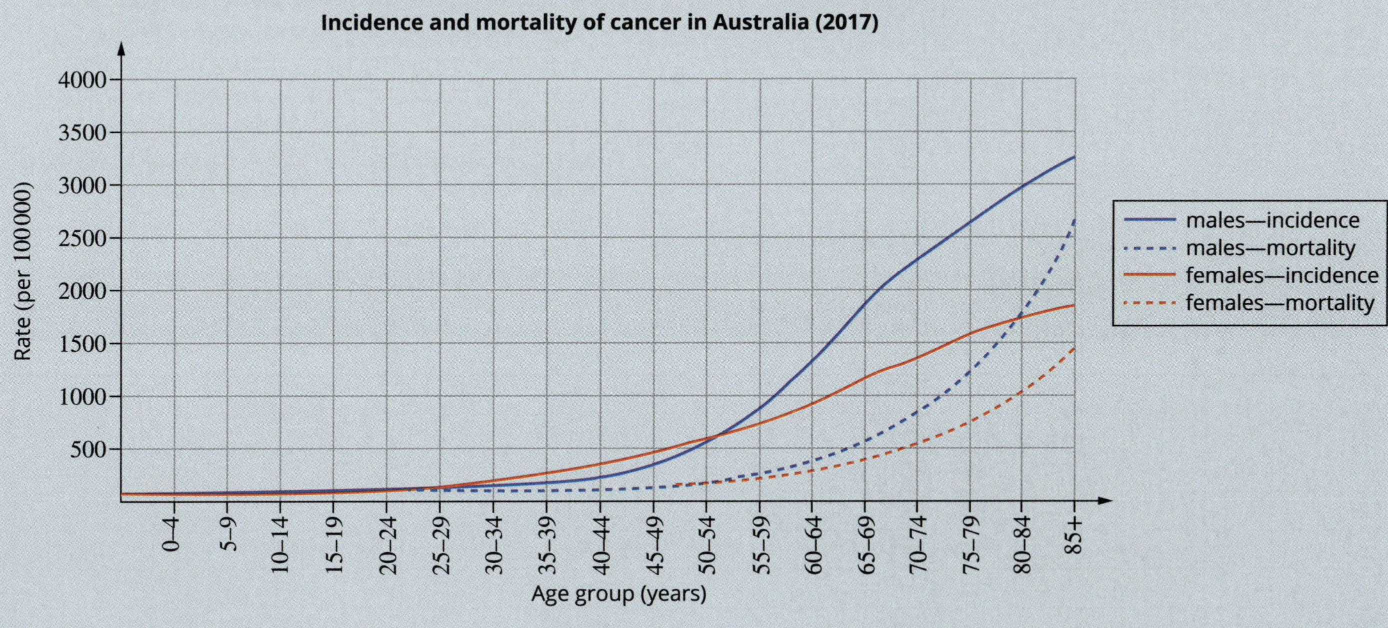

4 Lung cancer is associated with a number of risk factors.
  a Describe what lung cancer is and why it is the leading cause of cancer-related deaths in Australia compared to other forms of cancer.
  b Outline some of the risk factors associated with lung cancer.

5 Effective immunotherapy is one of the goals of modern research into the treatment of cancer.
  a Briefly describe the three traditional methods of combatting cancer.
  b Immunotherapy has advantages compared to traditional cancer treatments. Explain why.

6 The antibody shown to the right is:
  A chimeric
  B bispecific
  C conjugated
  D humanised

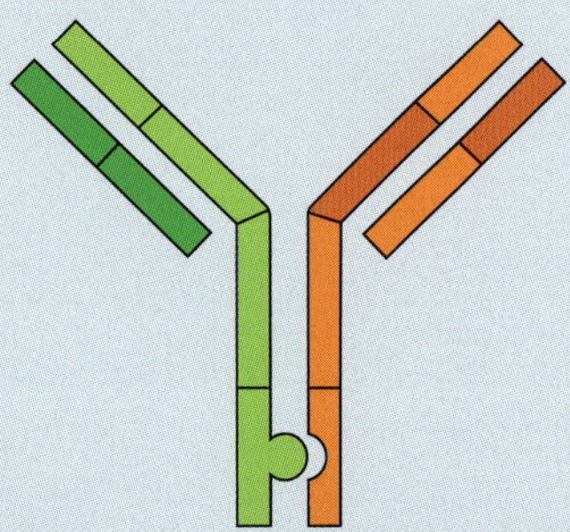

7 The first monoclonal antibodies produced for research were created in mice. These antibodies, however, were not successfully used in humans as they resulted in an immune response.
  a Explain why the mouse antibodies caused an immune response.
  b Suggest a possible solution to the problem of immune reaction to the antibodies.

8 Traditional cancer treatments have many side effects. In order to reduce these side effects researchers have sought ways to make anti-cancer drugs more specific to cancer cells. An important advance in this came about with the development of monoclonal antibodies. One group of researchers suspect that monoclonal antibodies can increase the efficacy and reduce the side effects of a commonly used anti-cancer drug by attaching the drug to a monoclonal antibody that has a binding site specific to an antigen expressed only on cancer cells.
  a The results of a preliminary trial comparing administration of the drug attached to the antibody with conventional administration are shown in the table below. In both cases there was one round of treatment, which consisted of administration of the drugs once per week for four successive weeks.
    i Complete the table by calculating the percentage change in the size of the tumour.
    Use the formula:
    $$\frac{\text{change in tumour size}}{\text{original tumour size}} \times 100 = \%\text{ change in size}$$
    ii Why is it necessary to calculate percentage change in tumour size before analysing the results of the trial?
    iii Do the results indicate that further trials of this approach should be undertaken? Explain your reasoning by referring to the data.
    iv Suggest a possible explanation for the results observed in patient 5.
  b The use of monoclonal antibodies to deliver cancer drugs has been shown to have fewer side effects than conventional methods of drug delivery and that much lower doses of the drug are needed to achieve the same level of response in the patients.
  Using your understanding of how monoclonal antibody therapy works, explain these two observations.

Tumour response to antibody-boosted or conventional drug treatment

Patient number	Antibody (A) or conventional (C) therapy	Tumour size at commencement of treatment	Tumour size after one round of treatment	Change in tumour size	Change in tumour size (%)
1	A	12.5 mm^3	9.6 mm^3		
2	A	23.9 mm^3	15.5 mm^3		
3	A	54.2 mm^3	26.8 mm^3		
4	A	46.8 mm^3	27.9 mm^3		
5	A	53.6 mm^3	56.4 mm^3		
6	C	54.8 mm^3	48.5 mm^3		
7	C	84.1 mm^3	66.9 mm^3		
8	C	36.9 mm^3	30.8 mm^3		
9	C	56.1 mm^3	49.1 mm^3		
10	C	38.9 mm^3	31.5 mm^3		

9 Cancer is characterised by cells that undergo uncontrolled replication, competing with body tissues for space and nutrients and disrupting cell function.

Leukaemia is a cancer of the white blood cells. In this condition the white blood cells of the bone marrow reproduce prolifically. One treatment for this cancer is radiation therapy to destroy a large percentage of the cancerous cells, followed by a bone marrow transplant. The stem cells of the donor marrow are critical to the patient's recovery potential.

a Suggest why radiation therapy is an important part of the treatment regime for leukaemia.

b What are stem cells? How are they important in such a transplant?

c Explain why it is important that the closest possible tissue match is made between a donor and the recipient for a bone marrow transplant to be successful.

10 The graphs below illustrate the estimated incident and mortality rates of (a) lung cancer (b) and melanoma cancer in Australia in 2017.

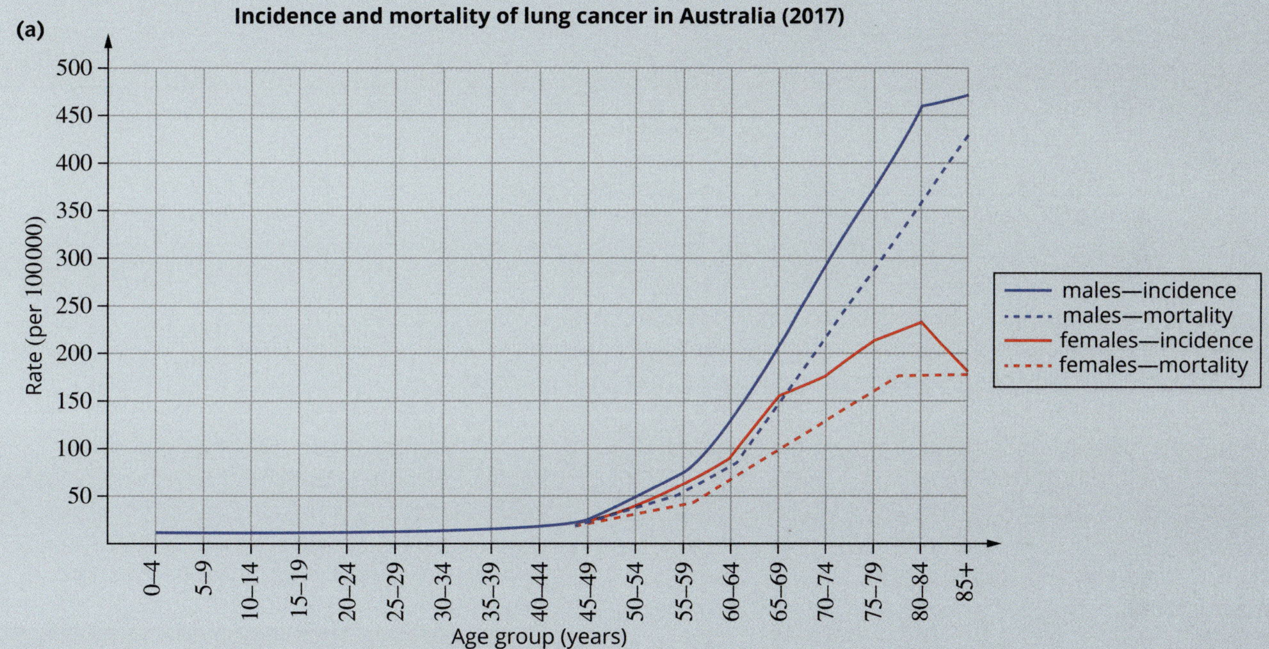

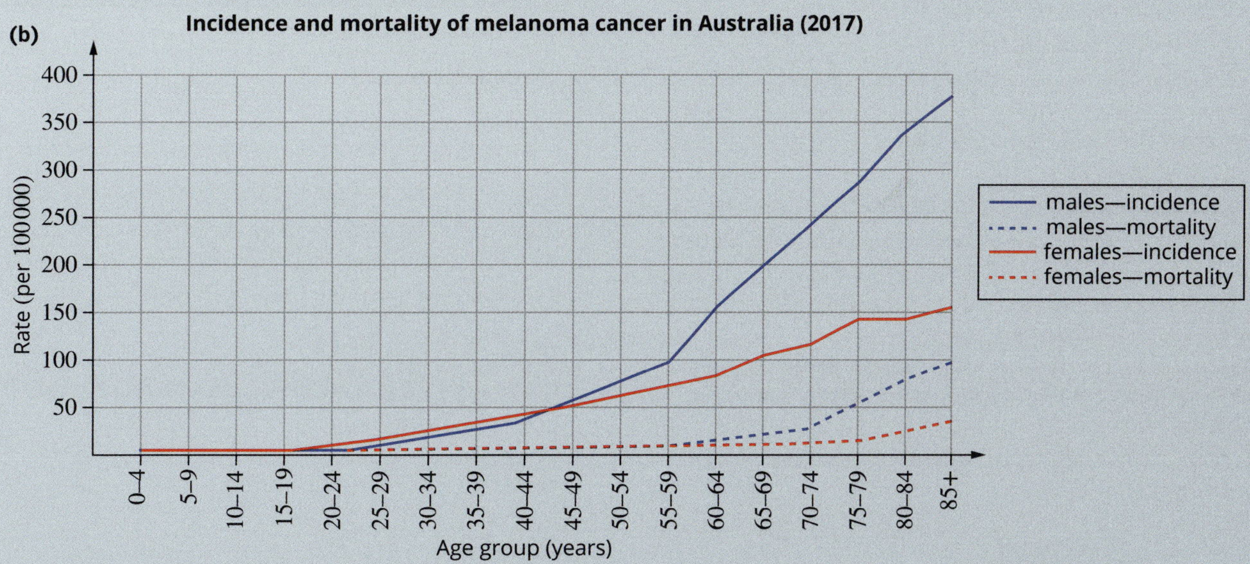

a Referring to graph (a) compare the incidence and mortality rates of lung cancer for both sexes.

b Referring to graphs (a) and (b) discuss why the incidence and mortality rates of lung cancer are quite similar, while those for melanoma cancer are quite different.

c Explain why mortality rates of both lung and melanoma cancer are highest in the 80+ age category.

11 The graph below indicates the trends in coronary heart disease deaths from 1985–2014.

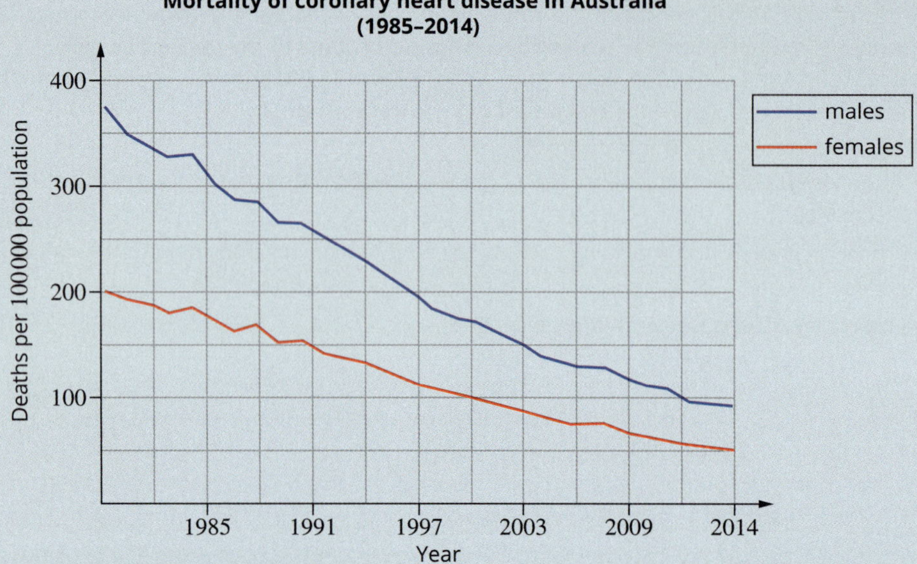

a What is coronary heart disease?
b Outline one major risk factor associated with developing coronary heart disease.
c Using the graph provided, discuss the trends in coronary heart disease from 1985 to 2014.
d Outline reasons why coronary heart disease deaths may have decreased over time.

---

12 What is the 'obesity epidemic' and why is it considered a major health problem?

13 Define the term 'obese'.

14 Which of the following are frequency (number) measures in epidemiological studies?
A birth rate
B time of year
C mortality rate
D geographic distribution

15 Outline the difference between incidence and prevalence.

16 Why are estimates useful in studies of non-infectious diseases? Are these estimates likely to be the same as incidence rates in the future?

17 Discuss how epidemiological data can be used.

18 Discuss how data from epidemiological studies differs between high and low income countries.

19 What is a common method for collecting data about disease in populations?

20 Match the following example studies with the three main categories of epidemiology research using the table to the right.
a Data indicating the location of tuberculosis deaths in an African population.
b A study into the association between drinks containing large quantities of sugar and diabetes in adolescents.
c Data on the effect that a public health campaign had on wearing sun smart hats in the school playground.

Type of epidemiology study	Example
analytical study	
descriptive study	
intervention study	

21 Outline at least three aims of epidemiological studies.

22 After completing the Biology Inquiry on page 552, reflect on the inquiry question: Why are epidemiological studies used? Describe the advantages of undertaking epidemiological studies for individuals and populations. Outline the methods that epidemiologists use to ensure that the data they collect is accurate, reliable and valid.

CHAPTER 17

Prevention of non-infectious disease

By the end of this chapter you will understand the impact that public health programs and campaigns can have on preventing disease in individuals and within a population. You will look at how educational programs can help prevent non-infectious diseases and how an individual's lifestyle choices can be influenced by such programs. You will also learn about the role of genetic engineering in preventing non-infectious diseases, with a focus on gene therapy.

Content

INQUIRY QUESTION

How can non-infectious diseases be prevented?

By the end of this chapter you will be able to:

- use secondary sources to evaluate the effectiveness of current disease-prevention methods and develop strategies for the prevention of a non-infectious disease, including but not limited to: CCT
  - educational programs and campaigns PSC
  - genetic engineering EU

17.1 Methods for preventing non-infectious disease

BIOLOGY INQUIRY  

Researching non-infectious disease

How can non-infectious diseases be prevented?

COLLECT THIS...

- tablet or computer to access the internet
- books and scientific journals

DO THIS...

1. Choose one non-infectious disease from the list below:
   - type 1 diabetes
   - stroke
   - coronary heart disease
   - cancer
   - severe combined immunodeficiency (SCID).
2. Research information about your selected disease using a reliable source such as a government website or scientific journal. You must read information carefully and be critical about the information that you use.
3. Analyse the information you find and investigate the following points:
   - **a** name of disease
   - **b** type of disease (genetic, environmental or nutritional)
   - **c** causes
   - **d** treatment
   - **e** prevention
4. Present information about your disease, referencing all sources of information used. When referencing use the APA style of referencing or a style indicated to you by your teacher.

RECORD THIS...

Describe how you found reliable sources of information for this task.

Present your findings about the disease as an illustrated pamphlet, using the dot points listed above as headings. You could also share your findings with the class as an oral presentation or multimedia presentation.

REFLECT ON THIS...

How can we prevent non-infectious diseases?

Why are non-infectious diseases so prevalent in developed countries?

What treatment types are available for non-infectious diseases in Australia?

Many **chronic** non-infectious diseases, such as heart disease, stroke, diabetes and some **cancers**, cause the largest number of deaths in developed countries and are largely preventable. Many **non-infectious diseases** have genetic components, but they often also have contributing lifestyle factors. For example, melanoma cancer has a genetic link, but people at high risk can limit their exposure to environmental factors such as ultraviolet (UV) radiation and have regular skin checks to allow early detection. By making healthy lifestyle choices, other types of cancer can also be prevented. For example, avoiding carcinogens, such as cigarette smoke, will prevent or reduce the likelihood of developing lung cancer.

Because non-infectious diseases cannot be passed from one person to another like infectious diseases can, we cannot use the same strategies of **prevention**, such as avoiding exposure to pathogens, washing our hands or keeping surfaces clean. As seen in Chapter 15, non-infectious diseases are the result of genetic conditions, the environment and lifestyle factors. Therefore, non-infectious disease prevention must address these factors. While it is difficult to change our genetic code, there are many ways to prevent some non-infectious diseases. In this section you will learn about some strategies to prevent non-infectious disease and improve the health of populations.

BIOLOGY IN ACTION   

Tanning beds and skin cancer

Some people think that tanning beds (or solariums) (Figure 17.1.1) are a safe alternative to sunbathing outside in the UV light. However, this is a myth—tanning beds also expose the skin to UV radiation. Any exposure to UV radiation increases the risk of developing skin cancer, whether the radiation comes from artificial light or the sun. The Cancer Council of Australia has warned that people who use solariums before the age of 35 have a 59% greater risk of developing melanoma compared to those who do not use solariums.

The International Agency for Research on Cancer re-classified solariums as a Group 1 carcinogen in 2009, placing them in the same risk group as tobacco and asbestos. Since 1 January 2015, it is now illegal to operate solariums commercially in Australia. This ban was the result of almost a decade of campaigning.

In 2007, a 26-year-old Melbourne woman named Clare Oliver told her story of battling end-stage melanoma. Clare had only weeks to live, yet decided to use her remaining time and energy to outline the dangers of solarium use to others. Clare's story received significant public and media attention and led to major changes to solarium regulation.

The Victorian Health Minister at the time, Daniel Andrews, regulated the solarium industry, with other states following this path of action. Clare's story had overwhelming public support and an advertisement against solarium use was aired. The message of the public campaign was 'no tan is worth dying for'. The advertisement was launched in 2008 and distributed to national television as a community service announcement.

FIGURE 17.1.1 Operating tanning beds (solariums) is now illegal in Australia.

PUBLIC HEALTH PROGRAMS

Messages about the importance of a healthy lifestyle are important to all members of a population. **Public health programs** aim to improve health and wellbeing as well as prevent disease in the wider community. They achieve this by promoting awareness about health issues and helping people make informed lifestyle choices to improve their overall health. Public health programs apply to a variety of settings such as schools, homes and workplaces. Programs may target small areas, such as a local neighbourhood, or be as far-reaching as an entire country or region of the world.

 Public health is defined as the science of protecting the safety and improving the health of communities through education, policy making and research for disease prevention.

Within a community there is a great diversity of health issues, age groups, gender identities, socioeconomic backgrounds, languages and religious and cultural beliefs. Public health programs therefore need to be diverse to reach the many different groups within a population.

Australia has many public health programs and campaigns to promote health issues and available programs. These help educate members of the public about the importance of current health issues and how best to prevent disease in the community. A lot of work goes into developing health programs and campaigns. They are based on research into preventing disease and use reliable, evidence-based information to educate people, promote healthy lifestyle choices and change policy and legislation for positive public health outcomes.

Public health programs employ a wide variety of methods to prevent non-infectious diseases. This includes health education, lifestyle advice, risk factor monitoring, taxes to discourage unhealthy lifestyle choices, and changes to policy and legislation to support healthy lifestyle choices. There is an emphasis on preventing disease rather than trying to cure it. Public health programs can also help to prevent infectious diseases by targeting strategies at the pathogen, the host and the environment. For example, public health campaigns to encourage hand washing and prevent the spread of infection.

Local and federal governments or non-government organisations, such as the Cancer Council and the Heart Foundation, develop and fund public health programs. Researchers play an essential role in improving our understanding of health by investigating diseases, their causes and methods for prevention, treatment and cure. Not-for-profit organisations, health professionals, educators and the media often promote public health programs and campaigns.

BIOFILE PSC

Slip! Slop! Slap!

Most Australians are now aware of the dangers of UV radiation and the consequences it can have for your health. Public health campaigns are run by state Cancer Councils every year to inform people about how to protect themselves from sun damage. These campaigns are credited with improving awareness of the early warning signs of skin cancer and helping to change public attitudes and behaviour towards sun exposure.

One of the most successful and well-known health campaigns in Australia's history—Slip! Slop! Slap!—was launched in 1981 to promote the importance of protecting your skin when in the sun. At the time of creation, melanoma rates were climbing and the link between UV radiation and skin cancer was becoming clearer. The Slip! Slop! Slap! health campaign featured Sid the Seagull in board shorts, T-shirt and hat singing a song encouraging people to 'slip' on sun-protective clothing, 'slop' on sunscreen and 'slap' on a broad-brimmed hat. In 2009, the SunSmart campaign added two more sun protection measures—'Seek' shade and 'Slide' on sunglasses (Figure 17.1.2). The SunSmart campaign is based on the importance of making responsible choices to protect yourself from the dangers of UV radiation while enjoying time outdoors.

FIGURE 17.1.2 The SunSmart 'Slip! Slop! Slap! Seek! Slide!' health campaign is one of the most recognisable public health programs in Australia's history.

Health education programs

Health education programs educate the public about health risk factors such as excessive sun exposure, poor nutrition and physical inactivity. They also promote healthy lifestyle choices, the importance of vaccination and health check-ups, as well as support for mental health and positive relationships. In Australia there are many different health education programs. For example, an education program that aims to reduce obesity in the population will provide information about healthy eating habits and promote the benefits of regular exercise.

BIOLOGY IN ACTION CC PSC S

Stephanie Alexander Kitchen Garden Foundation

The Stephanie Alexander Kitchen Garden Foundation promotes positive food habits to students around Australia. The Foundation provides information, training and support to schools and learning centres who are delivering pleasurable food education through hands-on kitchen garden programs.

FIGURE 17.1.3 The Stephanie Alexander Kitchen Garden Foundation promotes healthy eating for school students by engaging children in growing, harvesting, preparing and eating healthy fruits and vegetables.

Stephanie Alexander (AO) established the Stephanie Alexander Kitchen Garden Foundation as a means to promote the importance of learning about food, how it grows and the pleasures it can bring to developing a happy and healthy community.

Kitchen garden programs aim to engage students in the pleasures of fresh, seasonal and delicious food in a relaxed, happy and engaging environment. The program teaches children to grow, harvest, prepare and share their own fruit, vegetables and herbs (Figure 17.1.3), and with learning extending beyond the classroom walls students are able to engage with their families and the wider community. The program aims to develop positive eating habits in children for life.

Environmental health

Environmental health is associated with the public health aspects of environmental protection and involves research, environmental monitoring and education programs (e.g. on safe storage of chemical waste and water pollutants). Poor environmental health can lead to the development of non-infectious diseases. For example, poor government control of air pollution emissions results in reduced air quality and is a risk factor associated with developing lung cancer (Figure 17.1.4). Living in a building with inadequate lighting and ventilation can lead to reduced access to UV light and result in vitamin D deficiency.

Environmental and public health are improved by measures such as:

- pollution monitoring programs
- controls on waste dumping and pollution emissions
- clean water supplies
- domestic sanitation and sewerage
- garbage collection
- building designs that include adequate space, lighting and ventilation
- medical and hospital facilities
- quarantine controls.

Health education is any combination of learning experiences that help individuals and communities improve their health by increasing knowledge or influencing people's attitudes towards health and disease.

FIGURE 17.1.4 Air pollution is an example of an environmental health issue that requires monitoring to reduce the risk of developing non-infectious diseases such as lung cancer.

POPULATION SCREENING

Population screening involves testing or examining individuals for non-infectious diseases. Screening programs may be offered to groups based on biological sex, age or ethnic group, or to individuals known to be at risk of inheriting or developing a non-infectious disease based on family or medical history. Screening involves simple tests to look for particular changes or early signs of disease, before the disease has developed or symptoms arise. Screening tests are not 100% accurate because the human body changes over time. This is why regular screening throughout an individual's lifetime is recommended.

The screening tests aim to provide early diagnosis for individuals. Examples of active programs in Australia are:

- Cancer screening—detects early signs of cancer or indicators that the person is at risk of developing cancer in the future. In most cases, the early detection increases the chance of successful treatment.
- Breast, cervical, prostate and bowel cancer screening programs are well-known examples.
- Prenatal screening—blood tests and ultrasound of pregnant women to test for genetic conditions in the fetus, such as Down syndrome and neural tube defects.
- Newborn screening—blood tests to detect potentially serious medical conditions (e.g. phenylketonuria [PKU]), allowing babies to receive prompt treatment.

If an individual already has signs of a disease, or they are at high risk of developing a disease, it is recommended that they seek advice from their doctor and undergo **diagnostic tests**, rather than waiting to be tested through a screening program. Diagnostic tests aim to confirm if an individual has a particular disease and results are often received much faster than those of a screening test. They differ from screening tests in that they are usually conducted when a patient already has symptoms of a disease.

Population screening refers to a test that is offered to individuals in a target group to test for disease.

GENETIC ENGINEERING FOR DISEASE TREATMENT AND PREVENTION

In Chapters 8 and 9 you learnt about the applications of genetic technology in plants and animals. Its use in preventing infectious disease was examined in Chapter 13. Preventing non-infectious diseases using **genetic engineering** is also an important area of research. Genetically modifying cells for cancer **immunotherapy** and correcting defective genes via **gene therapy** is a promising area for treating and preventing life-threatening non-infectious diseases.

Genetic engineering is the alteration of DNA to correct genetic defects, manufacture proteins or make alterations to plant and animal phenotypes (characteristics).

If individuals could be born without life-threatening disease it would solve many problems related to treatment, reduced quality of life and health care costs. However, there are many ethical issues surrounding the use of genetic engineering for individuals, society and the environment.

BIOLOGY IN ACTION  

Cervical cancer screening

Cervical cancer is one of the most preventable cancers in the world. In Australia, it is estimated that in 2016, 903 women will be diagnosed with cervical cancer and 250 women will die from the disease. The cervix is the entrance to the uterus and cervical cancer is found in the cells of this region (Figure 17.1.5). Risk factors associated with cervical cancer include infection with human papillomavirus (HPV), increased age, smoking and lowered immunity.

Two types of cervical cancer exist:

1. Squamous cell carcinoma is the most common type of cervical cancer. It begins in the squamous cells that line the outer surface of the cervix.
2. Adenocarcinoma develops in the glandular cells often located higher in the cervix.

It is estimated that 99.7% of cervical cancers are caused by HPV infections. HPV is common in both females and males, with many people contracting it at some point in their lives. Most infections do not cause problems and go unnoticed, clearing up without treatment. There are over 100 different types of HPV. Genital HPV can be sexually transmitted and long-term infections (usually over ten years) are most commonly associated with cervical cancer. While infection with HPV can lead to cervical cancer, cervical cancer itself is not infectious and is considered a non-infectious disease.

Australia has a National Cervical Screening Program, which aims to detect early changes in the cervix and reduce illness and death associated with cervical cancer. It is currently recommended that all women in Australia aged between 18 and 69, who have ever been sexually active, have Pap smear tests every two years. This involves taking a small sample of cells from the cervix to be examined in a pathology laboratory.

In 2017, the national program changed to improve early detection and outcomes for women diagnosed with cervical cancer. Changes are the result of new evidence and better medical technology. The two-yearly Pap smear test, which detects abnormal cervical cells, was replaced with the more accurate five-yearly Cervical Screening Test, which includes a test for HPV infection. Because HPV can cause abnormal cell changes that can lead to cervical cancer, testing for HPV allows detection and monitoring of the infection before cancer cells develop. The procedure for collecting the sample for HPV testing is the same as that of a Pap smear.

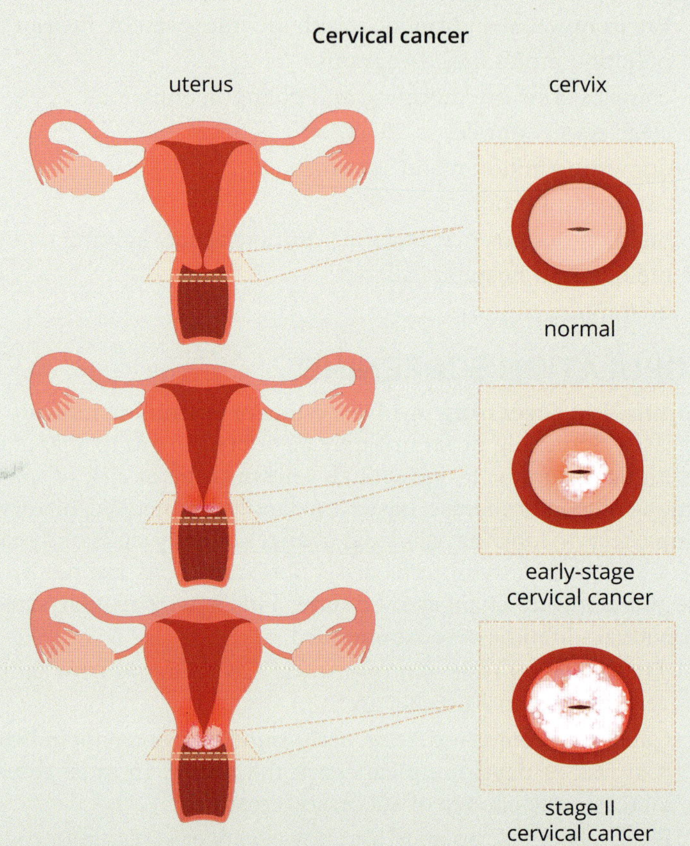

FIGURE 17.1.5 Changes in the cells of the cervix showing normal, early-stage and late stage cervical cancer

The screening program involves:

- encouraging all woman between the ages of 18 and 69 to enter the screening program
- replacing the Pap smear test with a more accurate Cervical Screening Test
- providing a system for notifying woman of the results of their cervical screen
- ensuring prompt follow-up of abnormal cervical screening results with patients
- increasing the time between tests from two years to five years.

Genetically modified cells

Scientists regularly produce **genetically modified cells** and cell lines to study normal and abnormal cellular processes and to expand the range of cell-based therapies. One area of application is cancer immunotherapy. In Chapter 16, you learnt about the use of **monoclonal antibodies (mAbs)** for cancer immunotherapy. Not only are transgenic mice being used to produce fully human monoclonal antibodies, but genetically modified lymphocytes are also being tested. Recall that T lymphocytes use their specific T cell receptor to recognise and attack cancer cells. Scientists at the Peter MacCallum Cancer Centre in Melbourne are investigating ways to boost the number of specific T cells in cancer patients. T cells are removed from the patient and genetically modified with genes coding for receptors that make the T cells better able to target the cancer cells. The genetically modified T cells are grown in vitro and then returned to the patient to attack the cancer cells (Figure 17.1.6).

GO TO ➤ Section 16.2 page 564

Transgenic refers to an organism whose genome has been altered by the transfer of a gene or genes from another species or breed.

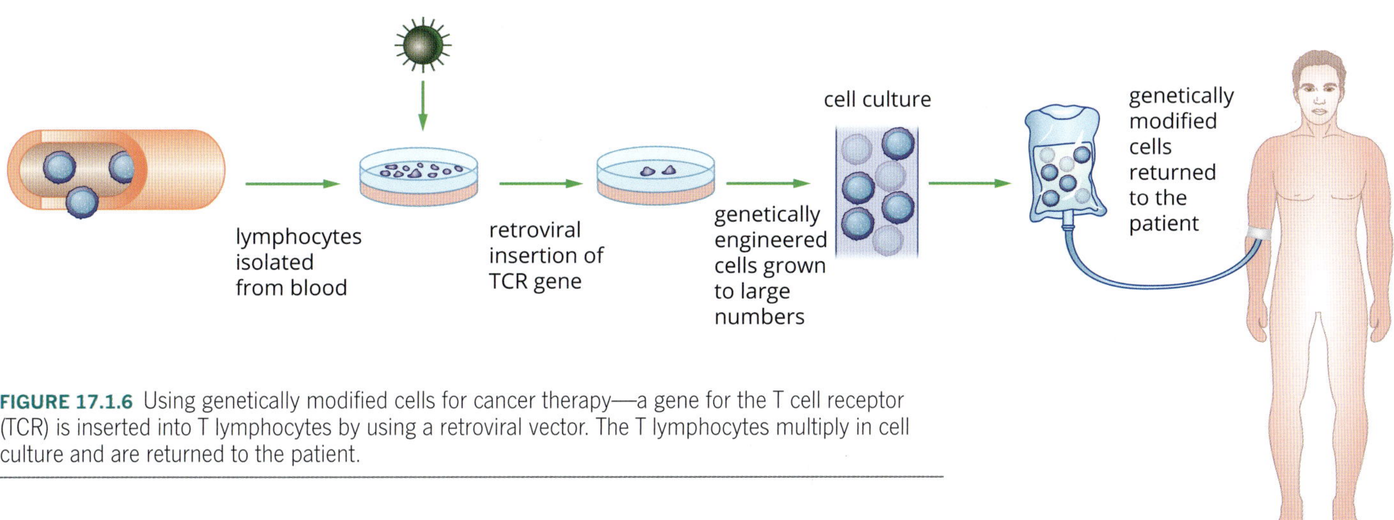

FIGURE 17.1.6 Using genetically modified cells for cancer therapy—a gene for the T cell receptor (TCR) is inserted into T lymphocytes by using a retroviral vector. The T lymphocytes multiply in cell culture and are returned to the patient.

BIOFILE EU

Insulin from animals

To treat type 1 diabetes, scientists used to extract insulin from the pancreas of other animals such as pigs and cattle (Figure 17.1.7). This was an expensive and time-consuming process that also involved the risk of an allergic reaction to the foreign molecule and potential for contracting diseases. Porcine (pig) and bovine (cattle) insulin are similar, but not identical, to human insulin. Their biological activity is not as effective as human insulin, so it is preferable to use the human hormone. Recombinant human insulin became available for treatment in the 1980s.

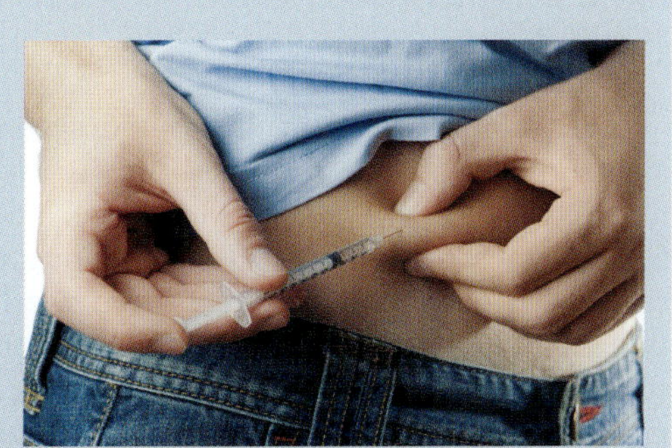

FIGURE 17.1.7 Woman giving herself an injection of insulin to treat diabetes

Gene therapy

Gene therapy is a rapidly growing area of science that involves the use of DNA or genetic material to treat diseases caused by missing or dysfunctional genes and proteins. Although it is an expanding area of research, many difficulties and challenges remain, including issues of safety, ethics and cost. Gene therapy has been a work in progress since the first trial in 1990. Delivering genes effectively and safely into eukaryotic cells is more difficult than transforming bacteria (prokaryotic cells). It is for this reason that proceeding with gene therapy is considered risky for most patients. Researchers continue to investigate and improve on the methods for gene delivery, including safer and more effective **vectors**, better targeting of cells, more control over where the gene is inserted and how it is regulated, and managing and minimising immune responses in patients. The methods that seem most effective are illustrated in Figure 17.1.8.

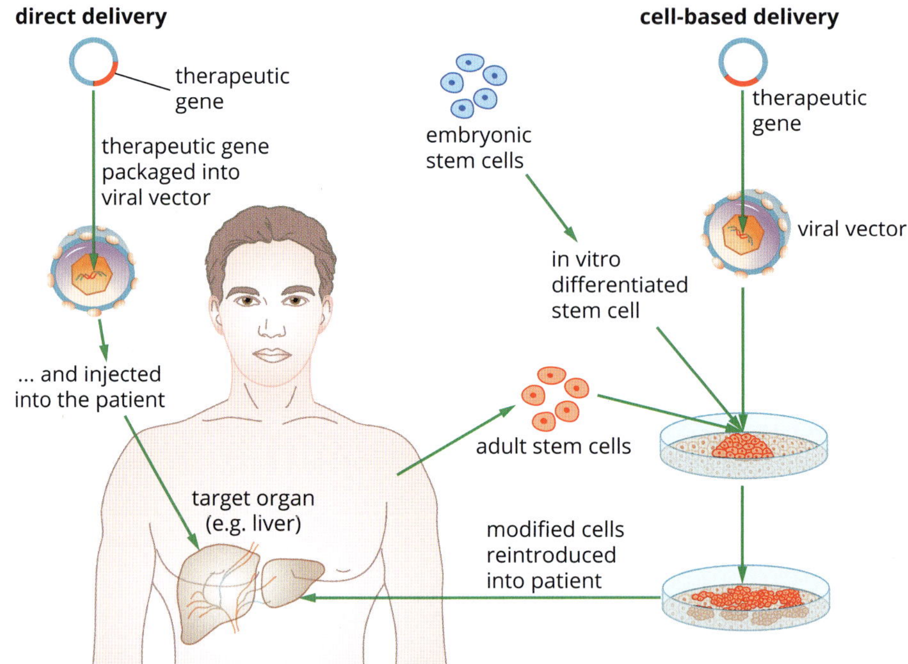

FIGURE 17.1.8 Recombinant DNA and cell biology methods are used to deliver genes to cells for gene therapy.

Only somatic cells (not egg and sperm cells) are used in gene therapy, such as blood cells and muscle cells. This is why treating a patient with gene therapy is targeted only to the treated individual. The altered protein cannot be passed onto future generations and is therefore not a cure to disease. Some methods of gene therapy involve removing cells from the body, such as bone marrow stem cells, inserting genes into these cells and then returning the cells to the patient. Other forms of gene therapy deliver a gene directly into the affected tissue in the body using a suitable vector. For example, scientists can insert a copy of the normal gene coding for the membrane protein that is faulty in cystic fibrosis into some of the patient's lung cells. The expression of the normal target gene allows a functional protein to be produced, relieving the symptoms of the disease.

Specialised vectors are used to deliver genes into eukaryotic cells for gene therapy.

Viral vectors

Viral vectors are often used in gene therapy, which involves transferring a therapeutic gene into a patient's cells. Viral vectors naturally insert genes into cells (Figure 17.1.9). The gene that is delivered by the viral vector produces proteins that will correct a genetic disorder or acquired disease. Adeno-associated virus (AAV) is one example of a viral vector. It is a small, non-pathogenic virus that can enter a wide variety of human cells and can insert DNA into a defined region of a chromosome. The virus is easily modified to carry a specific gene and usually does not induce an immune response.

The gene that is delivered by the viral vector will start to produce proteins that will correct a genetic disorder or acquired disease. Scientists have developed viral vectors that insert DNA into the patient's chromosomes. This is one of the most successful methods for introducing genetic material into patients' cells and tissues. Inserting it into the chromosome allows the therapeutic gene to be produced permanently. However, there is potential risk that the inserted viral gene may produce an unwanted mutation, making this procedure potentially dangerous.

A number of different viruses have been used for gene therapy. Some have a limited range of cells that they can infect, or can only infect actively dividing cells. Viruses that can be used for gene therapy must have the following features:

- the viral vector must have had its genes removed or altered so that the virus is not virulent
- the viral vector has cloning sites to allow a therapeutic gene to be inserted in the genome
- the virus can bind and enter the target cells.

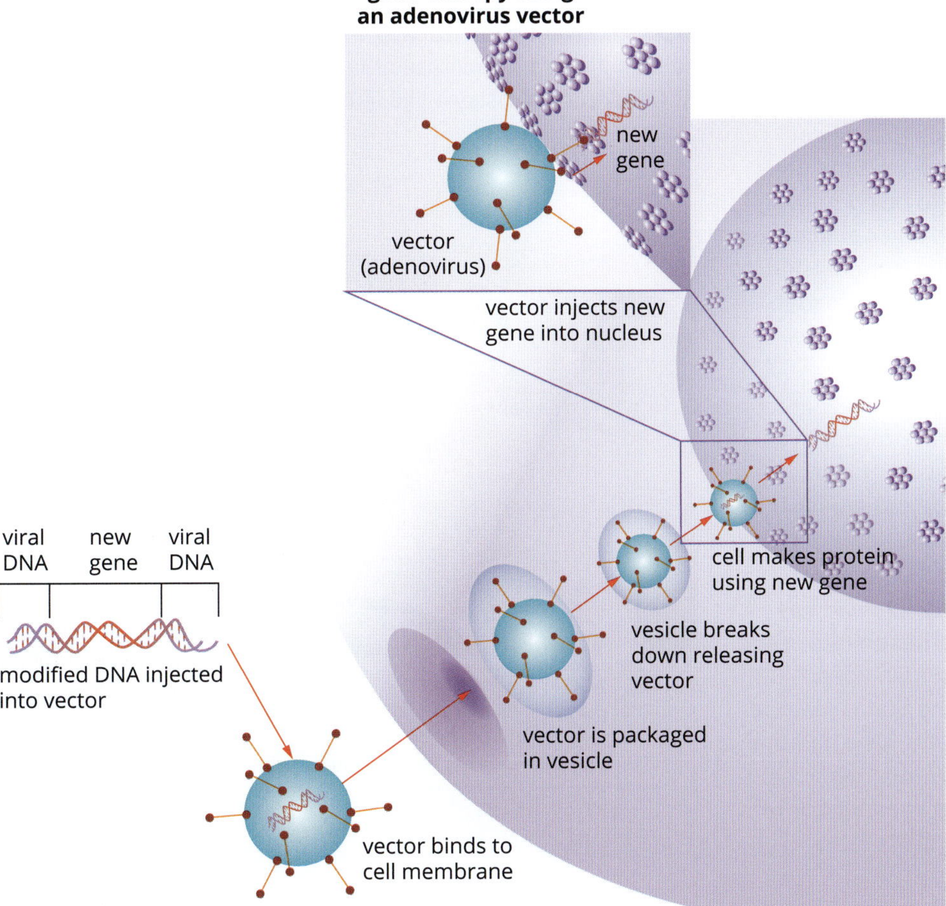

FIGURE 17.1.9 Illustration of a viral vector delivering recombinant DNA to a eukaryotic cell nucleus

BIOFILE DD EU

Gene therapy for haemophilia patients

Haemophilia is a disease in which the blood doesn't clot properly because of reduced levels of clotting factor proteins. People with severe haemophilia can lose large amounts of blood from a minor cut or injury that causes internal bleeding. If not managed quickly, internal bleeding can lead to pain, swelling and permanent damage, such as arthritis and chronic pain. Usually, people with haemophilia require regular injections to replace the missing clotting factor protein.

Researchers have shown promising results for haemophilia using gene therapy. An adeno-associated viral vector delivers a functional gene for the clotting protein, Factor IX, which can be delivered to the liver. Patients receiving the gene therapy had a significant increase in the level of clotting protein, reducing haemophilia symptoms from severe to mild.

Gene therapy has also shown promising results for some forms of cancer. Melanoma skin cancer has been successfully treated using a type of immunotherapy called T-VEC. A genetically modified form of the herpes virus is injected directly into melanoma tumours, where the virus replicates and ruptures the cancerous cells. The viral vector also stimulates the patient's immune system to kill cancer cells.

Liposomes

Liposomes are very small phospholipid vesicles that can diffuse across cell membranes or enter cells by endocytosis. DNA is inserted into the liposome and carried into the cell, where it is released. Liposomes can also be used to deliver drugs to cells. Liposomes may be safer gene delivery vehicles than viruses.

At present, different clinical trials are underway, where researchers are carefully testing treatments to ensure that any gene therapy conducted is both safe and effective.

Diseases that have been treated successfully with gene therapy include:

- degenerative blindness
- haemophilia
- beta-thalassemia
- lipoprotein lipase defects
- certain cancers, melanoma
- severe combined immunodeficiency.

WS 8.6

Severe combined immunodeficiency

Severe combined immunodeficiency (SCID) is a genetic disorder often caused by a deficiency in the enzyme adenosine deaminase. It leads to problems with the patient's immune system.

There are many different forms of SCID, many of which involve the destruction of T cells required for a functional immune response. SCID became known as the 'boy in a bubble' disease in the 1970s, because at the time the only way that patients could survive was to be completely isolated from the outside world and the danger of infection (Figure 17.1.10). Being completely isolated from outside air and contact meant that patients were not exposed to fungal, bacterial or viral infections, which would further compromise their immune system and likely result in death.

Developments in modern technology have meant that SCID patients can receive a hematopoietic stem cell transplant (these cells give rise to all blood cell types), if a matching donor is available.

Gene therapy is an option for those without a hematopoietic stem cell donor match available and has been successful for some forms of SCID. Treatments use a retrovirus viral vector that contains the human adenosine deaminase gene or restored a similar gene that is inactive. Overall, a high percentage of young patients have had their immune systems restored. However, early treatment using a gene therapy retro virus vector resulted in a number of treated children later developing leukaemia as a result of the therapy. Unfortunately for these patients, the retro virus inserted into their chromosomes promoted growth of T cells. Current gene therapy treatments for SCID now use self-inactivated retroviruses that decrease the likelihood of developing a disease like leukaemia from gene therapy treatment.

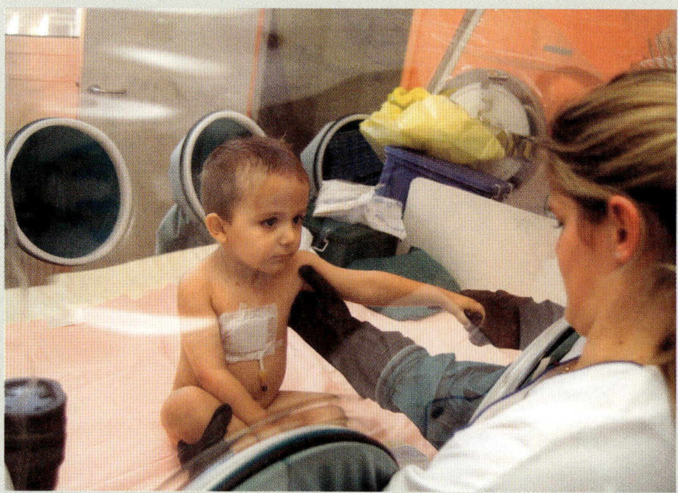

FIGURE 17.1.10 SCID is a genetic disorder that compromises the immune system, making patients vulnerable to infectious diseases. Patients need to live in a sterile environment to avoid exposure to pathogens such as viruses, bacteria and fungi.

17.1 Review

SUMMARY

- Many chronic non-infectious diseases are largely preventable.
- Non-infectious diseases often have contributing lifestyle factors, which if reduced can help prevent some non-infectious diseases from developing.
- Public health programs aim to prevent disease in the wider community by promoting health and wellbeing, as well as discouraging unhealthy lifestyle choices. They draw on a variety of methods including health education, lifestyle advice and monitoring risk factors.
- Successful public health programs can change the social environment, policy and legislation to support a healthier population.
- Public health programs are usually funded by the government and non-government organisations.
- Health care professionals, educators and the media promote public health programs to the community.
- Researchers play an essential role in improving our understanding of health by investigating diseases, their causes and methods for prevention, treatment and cure.
- Health education programs are designed to help individuals and communities improve their health by increasing understanding of disease, raising awareness about risk factors of disease and promoting healthy lifestyle choices.
- Environmental health strategies aim to improve the health of the environment for the safety and health of the public. Strategies include pollution monitoring, clean water supplies and building designs.
- Population screening tests are conducted for some diseases, such as cervical cancer, skin cancer and phenylketonuria (PKU), and aim to provide early diagnosis for individuals. Population screening tests are usually conducted before individuals experience any symptoms of the disease. These are different from diagnostic tests, which aim to determine if a disease is present in individuals experiencing symptoms.
- Genetic engineering for disease prevention can involve the use of genetically modified organisms—with modifications made to one or more genes—or genetically modified cells.
- Gene therapy research investigates methods for gene delivery, including using more effective vectors, better targeting of cells and greater regulation of an inserted gene.
- Gene therapy in eukaryotic cells can involve the use of specialised vital vectors or liposome vectors to deliver genes into a cell.
- Viral vectors are often used in gene therapy, which involves the transfer of a therapeutic gene into a patient's cells. The gene delivered by the viral vector will start to produce proteins that will correct a genetic disorder or acquired disease.

KEY QUESTIONS

1 Define the term 'public health program'.

2 Outline one major public health program that you are familiar with.
  a What are some of the strategies this program uses?
  b How long has the program been running?
  c Which organisation runs and funds the program?

3 What is population screening?

4 Describe one example of a population screening test in Australia.
  a What disease does the test screen for?
  b Who are the target group for this test?
  c What age is this test conducted at?

5 What is genetic engineering? Why is it an important area of research for non-infectious disease prevention?

6 List at least two benefits associated with using gene therapy in preventing non-infectious diseases.

7 Choose the correct statement about gene therapy:
  A Gene therapy uses egg and sperm cells.
  B Gene therapy uses both somatic cells and egg and sperm cells.
  C Gene therapy uses somatic cells.
  D Gene therapy uses muscle cells only.

8 Outline the features that a viral vector must have to be used in gene therapy.

Chapter review

17

KEY TERMS

cancer
chronic
diagnostic test
environmental health
genetic engineering
genetically modified cell
gene therapy
health education program
immunotherapy
liposome
monoclonal antibody (mAb)
non-infectious disease
population screening
prevention
public health program
vector
viral vector

REVIEW QUESTIONS

1 Distinguish between screening tests and diagnostic tests.

2 A large amount of statistical data is collected each year on the incidence of lung cancer in Australia.
  a Name a public health program aimed at preventing lung cancer.
  b Describe how public health programs aim to reduce the incidence of lung cancer.

3 Evaluate the importance of non-infectious disease prevention and research to our society. Use examples to support your ideas.

4 Use online sources to research an Australian program or organisation that has been developed to manage or prevent a non-infectious disease in humans.
  a What is the disease?
  b Why does this disease require 'managing'?
  c How can the disease be prevented?
  d Which group (e.g. age, biological sex, ethnic group) is this program aimed at?

5 Explain how public health programs can be beneficial to the health of the community. Use a specific example in your answer.

6 In what way are public health programs effective?
  A They provide expert medical advice.
  B They direct the public to change their lifestyle and eating habits.
  C They help treat diseases.
  D They inform the public about disease prevention and control.

7 Environmental health is an important area of environmental protection for public health. Discuss the types of activities that are involved in maintaining the environmental health of a community.

8 Describe the features of population screening and diagnostic testing by completing the table below.

	Population screening	Diagnostic testing
timing		
group(s) tested		
examples		

9 Cervical cancer can be detected through a screening test.
  a What is cervical cancer?
  b What are the two types of cervical cancer?
  c List some risk factors associated with cervical cancer.
  d What is HPV? And what is its link with cervical cancer?
  e What does the cervical cancer screening test detect?

10 What are the benefits of the National Cervical Screening Program?

11 What does 'genetic modification' of an organism mean? Include an example in your answer.

12 Give an example of how genetic modification may be used to prevent or treat non-infectious disease. List any potential benefits or disadvantages to the person, the environment and/or to society.

13 What is one important outcome of genetic engineering in preventing non-infectious disease in humans?
  A improved appearance
  B shorter life cycles
  C reduced cost
  D improved health and quality of life

14 Identify what you consider to be the positive and negative outcomes of using genetically modified cells for preventing non-infectious diseases. Draw a table like the one below in your notebook to complete this question.

	Positives	Negatives
social		
biological		
ethical		

15 From the cases described in this section, identify one example of a genetic modification that has been shown to improve human health.

16 Outline the two main methods used to deliver genes in human gene therapy.

17 Describe how genetically modified cells could be used in cancer therapy.

18 One of the results of genetic engineering is the production of a non-faulty protein. To do this, the gene that triggers production of the desired protein is inserted into a host organism. Define the term 'vector' in relation to genetic engineering.

19 Describe the difference between viral vectors and liposomes in gene therapy.

20 Why is the use of a viral vector in gene therapy not considered a risk?

21 After completing the Biology Inquiry on page 582, reflect on the inquiry question: How can non-infectious diseases be prevented? Describe the role that educational programs play in the prevention of non-infectious disease in your community.

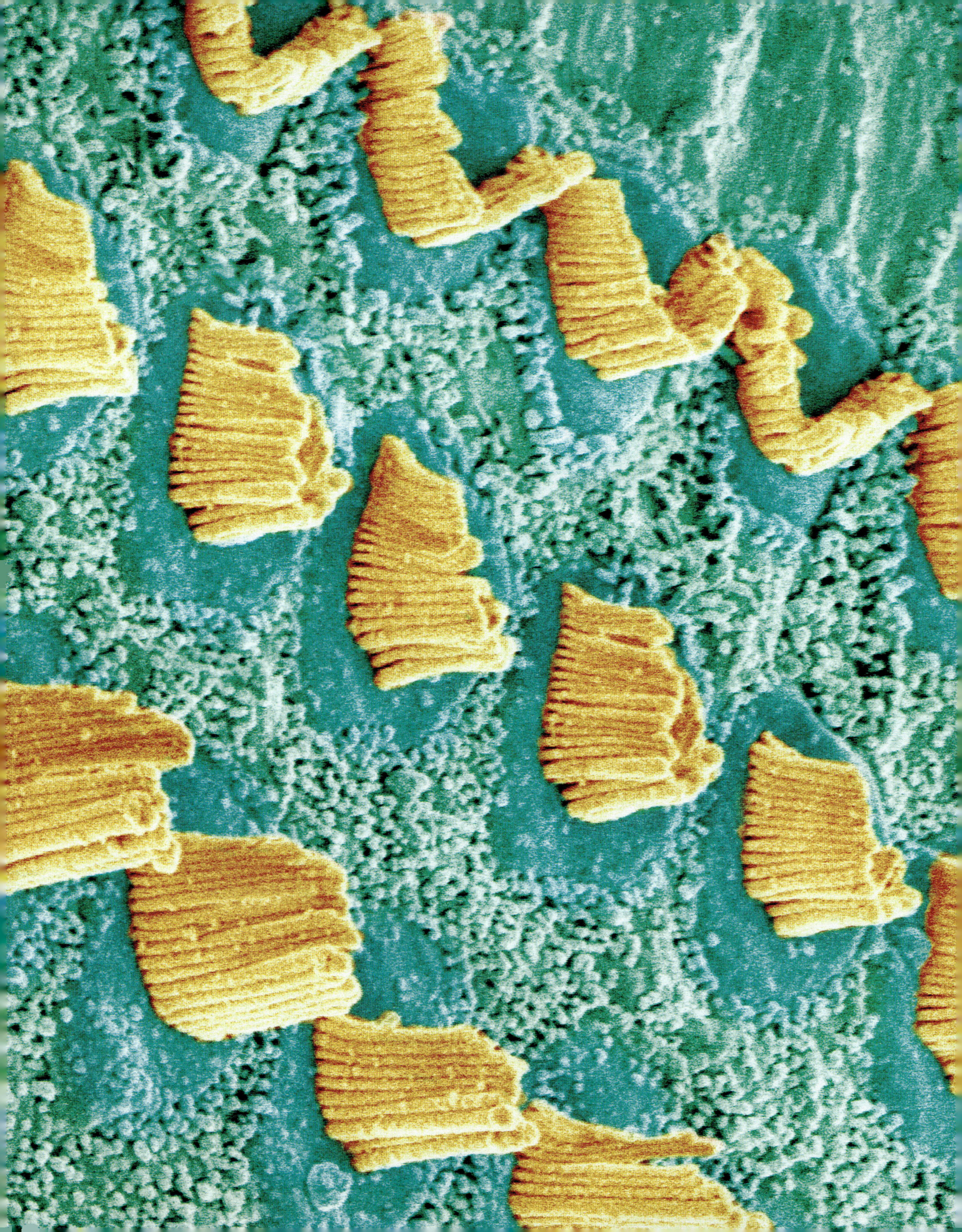

CHAPTER 18 Technologies and disorders

By the end of this chapter, you will have learnt about the normal structures and functions of the human ear for hearing, the eye for vision and the kidney for homeostatic balance in the body, including waste removal.

By understanding these organs you will appreciate the significance of disorders that cause loss of function in them. You will investigate a range of such disorders and their causes and study some technologies that have been developed to assist with the effects of these disorders. You will then be in a position to evaluate how effective a technology is for managing and assisting with a disorder that causes hearing loss, vision disorders or loss of kidney function.

Content

INQUIRY QUESTION

How can technologies be used to assist people who experience disorders? CCT

By the end of this chapter you will be able to:

- explain a range of causes of disorders by investigating the structures and functions of the relevant organs, for example:
  - hearing loss
  - visual disorders
  - loss of kidney function
- investigate technologies that are used to assist with the effects of a disorder, including but not limited to: (ACSBL100) ICT L
  - hearing loss—cochlear implants, bone conduction implants, hearing aids ICT PSC
  - visual disorders—spectacles, laser surgery ICT PSC
  - loss of kidney function—dialysis PSC
- evaluate the effectiveness of a technology that is used to manage and assist with the effects of a disorder (ACSBL100) EU L

18.1 Hearing loss

BIOLOGY INQUIRY    

Testing hearing and seeing abilities

How can technologies be used to assist people who experience disorders?

COLLECT THIS...

- tablet or computer to access the internet
- a set of tuning forks

DO THIS...

1. Enter search terms such as 'hearing test' and 'vision test' into an Internet browser.
2. Choose a website that you think will be a reliable source of information.
3. Take an online hearing or vision test.
4. Spend some time doing your own evaluation of at least three different online hearing or vision tests.
5. Mosquito sound—test the ability of people of various ages to hear the sound of a mosquito. You can search for and download an audio file of the sound.
6. Plan how you would conduct a valid scientific investigation of people's ability to hear the mosquito sound frequency. Try to collect enough data to generate a graph of results.
7. Use a set of tuning forks to listen to different sound frequencies (hertz, Hz). Complete a table listing the frequencies versus length of tines for the set of tuning forks.
8. Find some examples online of images that test for colour vision deficiencies and screen your classmates. Complete a table with the outcomes of the colour vision tests.

RECORD THIS...

Present the data you collected in steps 6–8 in a report that includes tables and graphs.

Describe your conclusions from the investigations for steps 6–8.

REFLECT ON THIS...

How can technologies be used to assist people who experience disorders?

Are your personal results from the hearing or vision test reliable?

Suggest some technological devices that might be used to assist people who have hearing loss in the high frequency range or colour vision deficiencies.

The human ear has two functions: hearing and balance. Damage to the specialised hearing structures may cause hearing loss and affect the balance functions of the inner ear. Having two outer ears in position at opposite sides of the head gives people the ability to accurately locate the direction a sound is coming from.

Depending on the cause, there are many technologies available to assist people with hearing loss. These solutions have increased and improved greatly with the arrival of the digital age and miniaturisation of equipment. You will first gain an understanding of the fascinating structure and function of the human ear. Next, a wide variety of causes of hearing loss are outlined. Then you will investigate some of the technological solutions for overcoming hearing loss, at least in part, because complete restoration of hearing is not yet possible.

BIOFILE CCT N

Properties of sound

Sound is produced by a vibrating object, for example the speaker of a sound system, a musical instrument or vocal cords in the human voice box. Vibrations require particles of a medium, be it gas, liquid or solid. Therefore sound cannot travel in a vacuum. Frequency and amplitude are two properties that determine the nature of a sound.

- Frequency is the property of a sound wave that determines pitch (a high or low pitched sound). The SI unit for frequency is hertz (Hz) and is defined as the number of wave cycles (vibrations) passing a point per second. For example, the musical note of middle C is 256 Hz meaning that 256 sound wave cycles would pass a point every second. To understand what a wave cycle is refer to the SkillBuilder on page 598.
- Amplitude is the property of sound that determines loudness. It is the maximum displacement of the vibrating particles and is measured as decibels (dB). The threshold of hearing is defined as 0 dB and equals the quietest sound that can be heard by the human ear. Humans detect sounds between 0 to 180 dB. Above 140 dB the loudness is painful and can cause damage. Deafness is defined as profound when sounds have to be >90 dB to be heard by the person.
- The frequency (pitch) of a sound is the same as the frequency of the vibration at the source producing the sound. The voice pitch produced by a typical adult male has a fundamental frequency of 85—180 Hz and that of a typical adult female is higher pitched, at 165—255 Hz.
- The hearing frequency range for humans is much wider than their speech frequency. A standard hearing range is regarded as 20–20 000 Hz but this is age dependent. Figure 18.1.1 compares hearing ranges of several species with the hearing range of humans.

Animal hearing frequency range

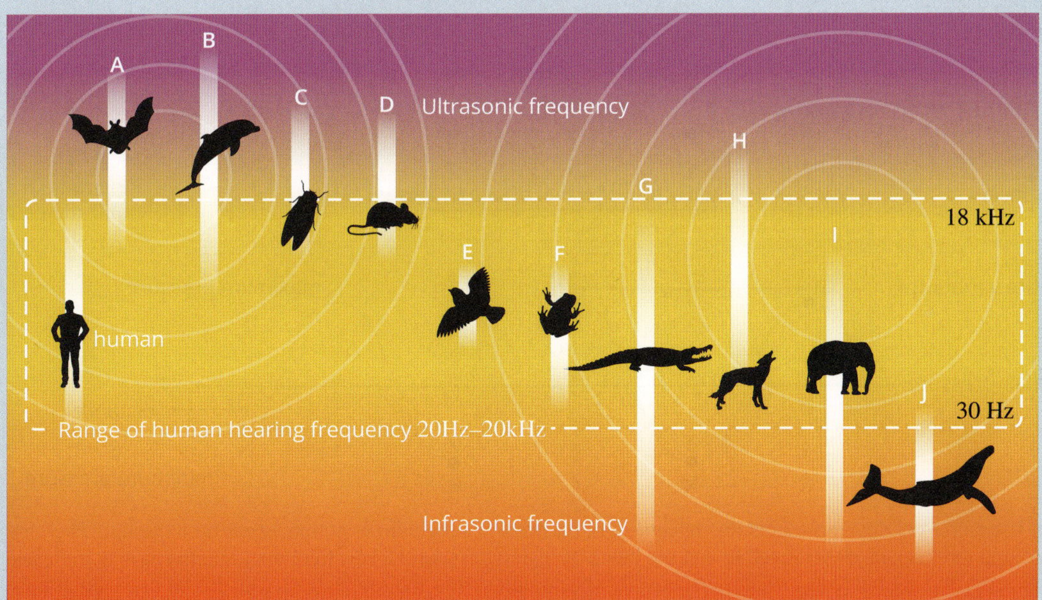

FIGURE 18.1.1 The ranges of frequencies heard by different animals compared to humans

SKILLBUILDER CCT N

Sound waves and hearing

Sound waves originate when a source causes vibration of particles in a medium like air, a fluid or a solid. The kinetic energy of these moving particles is passed from one to the next causing a compression wave to move forwards. It is the energy that travels forward not the particles. Individual particles vibrate back and forth around one spot rather than moving the full distance the sound travels. The vibration forms a sequence of compressions and rarefactions. A slinky spring is often used as a model for a sound wave. You can depict the vibration as dots (Figure 18.1.2a), and it is also useful to represent the sound wave as cycles (Figure 18.1.2b).

Frequency for a sound wave is the number of cycles passing a point per second and is measured in hertz (Hz). High frequency vibrations have shorter wavelengths and result in high pitch sounds; low frequency vibrations have longer wavelengths and are heard as low pitch sounds (Figure 18.1.3).

Under age 25, the normal hearing range is around 20–20 000 Hz. The ability to hear high pitch sounds decreases with age to the extent that many people over age 65 cannot hear above 5000 Hz. Inside the inner ear, where sounds are detected, the short sensory hairs that vibrate with high frequencies are the first to receive sound pressure waves. They become fatigued and damaged over time.

Loudness (volume) of a sound is determined by the amplitude of the compression wave. This is the maximum displacement of the particles as they vibrate back and forth (Figure 18.1.3).

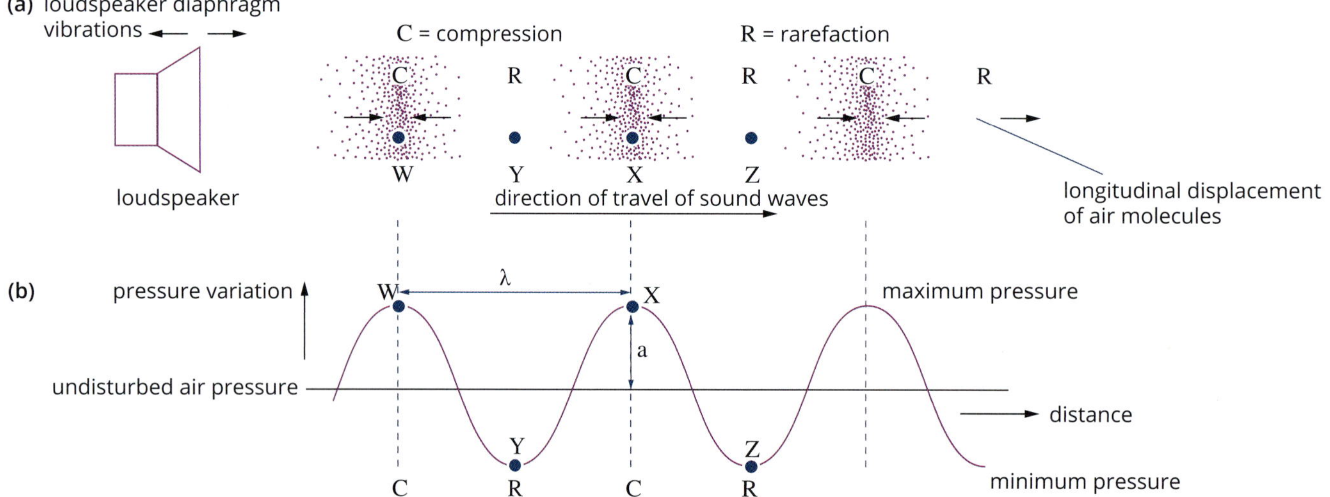

FIGURE 18.1.2 (a) The pattern of dots represents the movement of particles in a medium as sound waves pass through. High frequency sounds have compressions and rarefactions closer together and are heard as high pitch sounds. Low frequency sounds have compressions and rarefactions further apart and are heard as low pitch sounds. (b) The compressions and rarefactions can also be depicted as a cycle diagram.

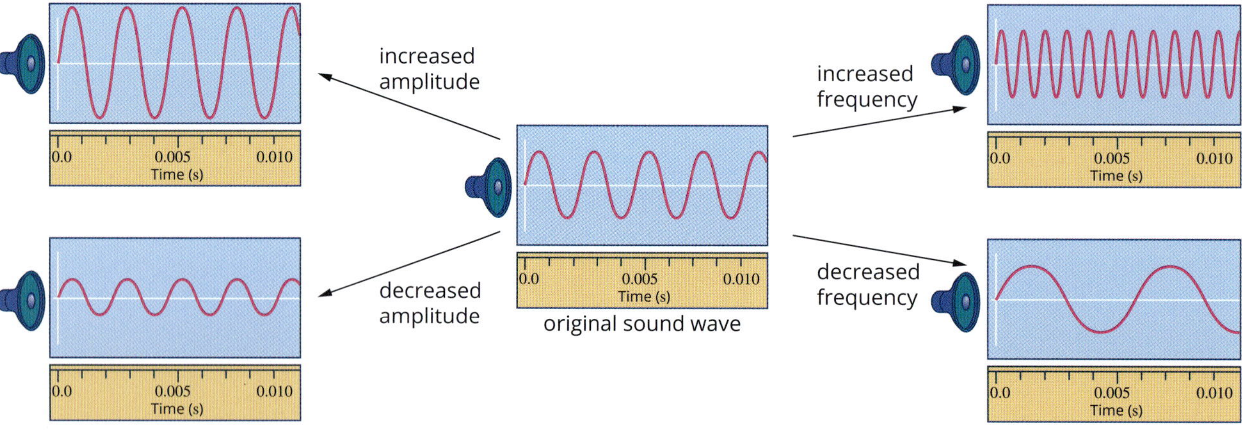

FIGURE 18.1.3 Increased amplitude means increased wave energy, which results in a louder sound. Higher frequency means that more wavelengths pass a point every second and is heard as a higher pitch.

STRUCTURE AND FUNCTION OF THE EAR

The main focus of this section is to study hearing and the causes of hearing loss. The hearing function of the ears plays a very important role for humans in awareness of their external environment and for communication purposes. As children we spend a lot of time learning how to interpret and form sounds, particularly those of language (Figure 18.1.4).

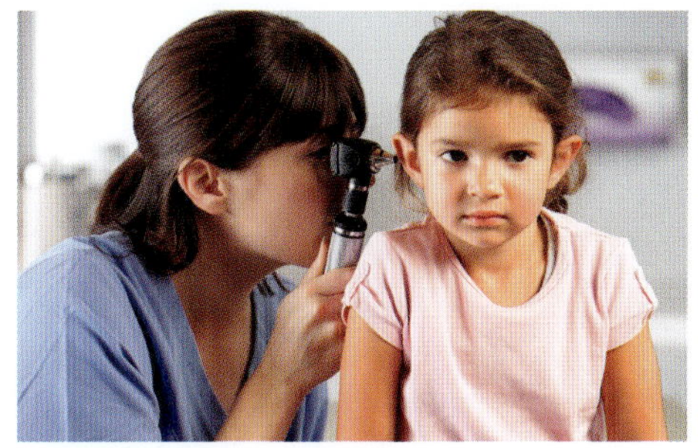

FIGURE 18.1.4 It is important to diagnose any hearing disorders early in childhood because much of our sensory input for early learning comes via hearing. In particular, speech development happens from six months of age and is dependent on hearing sounds produced by others.

Outer, middle and inner ear structure

The human ear has three sections that are physically divided from each other. They contain different structures and media (air, solid bone or fluid) through which sound is transmitted before it stimulates receptors to send messages to the brain.

Structures that enable hearing

Each ear has three sections called the outer (external), middle and inner ears (Figure 18.1.5), described below.

- **Outer ear**—the part visible on the side of the head and referred to by most people as the ear. There is actually a lot more of the ear structure set inside and protected by the skull (Figure 18.1.6). The visible part of the outer ear is called the **pinna** and it gathers in the sound waves. The **ear canal** and **tympanic membrane** (**eardrum**) are also part of the outer ear. These are all of the parts that are exposed to the outside atmosphere.

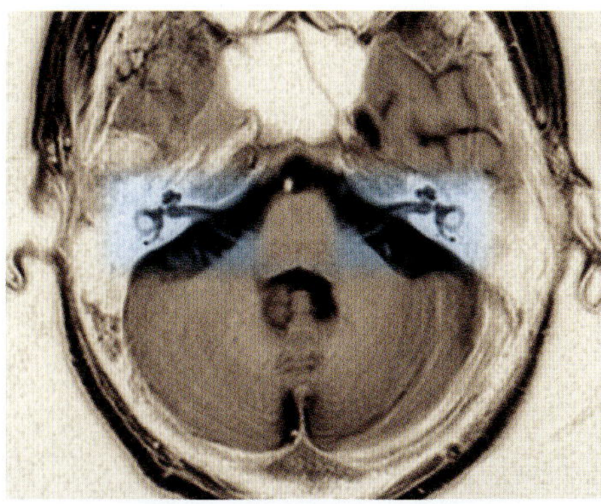

FIGURE 18.1.6 Coloured axial MRI scan through the head of a healthy 30-year-old, showing the anatomy of the left and right auditory systems. The structures of the inner ears (highlighted in blue, centre left and right) can be seen, including the cochlea and semicircular canals.

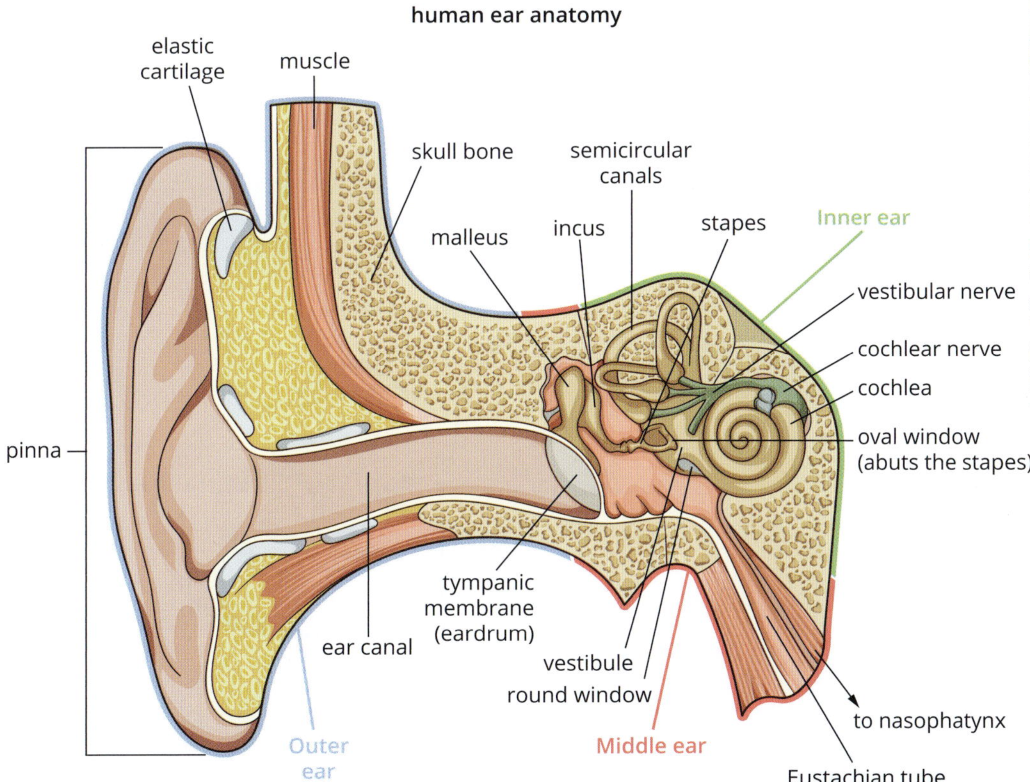

FIGURE 18.1.5 The structure of the human ear can be thought of as having three sections: outer, middle and inner ear.

- **Middle ear**—contains three tiny bones called **ossicles** and the opening to the **Eustachian tube** that connects to the back of the nose and throat (Figure 18.1.5 on page 599). The middle chamber is usually filled with air. Pressure equalisation takes place through the Eustachian tube when swallowing or coughing. The three ossicles are called **malleus**, **incus** and **stapes**, using Latin words that refer to their shape (Figures 18.1.7 and 18.1.8). Their common names are hammer, anvil and stirrup, again in reference to each shape. They are in end-to-end contact from the malleus at the tympanic membrane, to the incus in the middle, then the tiny stapes. The stapes has a base locked onto the matching shape of flexible membrane called the **oval window**—the interface with the canals of the inner ear (Figure 18.1.8). There is also a smaller, flexible **round window** close to the oval window that terminates the inner ear's cochlear canals.
- **Inner ear**—sometimes called the labyrinth, is protected by skull bone and contains the **cochlea** and the **semicircular canals** both of which are filled with fluid (Figures 18.1.5 on page 599 and Figure 18.1.8). Both structures contain many **mechanoreceptor** neurons that transmit nerve impulses to the brain for hearing or balance. The mechanoreceptors are sensory cells with hair-like projections called **stereocilia** (often just called hair cells). The oval and round windows are both small areas of flexible membrane that lie at the junction of the middle and inner ears. While they have similar structure their functions are different. The oval window transfers sound vibration from the solid stapes bone into pressure waves in the fluid filling the tapered canals of the cochlea. The round window is where these fluid vibrations terminate and have their energy absorbed so there is no reflection back into the canals to interfere with hearing (Figure 18.1.9). Otherwise it could sound like a very annoying repeating echo.

BIOFILE CCT

Smallest human bone

The bone that is essential for hearing in the human ear is the smallest in the body. The stapes bone has the appearance of a horse-riding stirrup and is only about 3.5 × 3 × 1.4 mm in size, no more than one grain of rice. It is also the lightest bone at 3–4 mg. It is this bone that knocks on the oval window, conducting the sound wave information into the cochlea for eventual transmission to the brain.

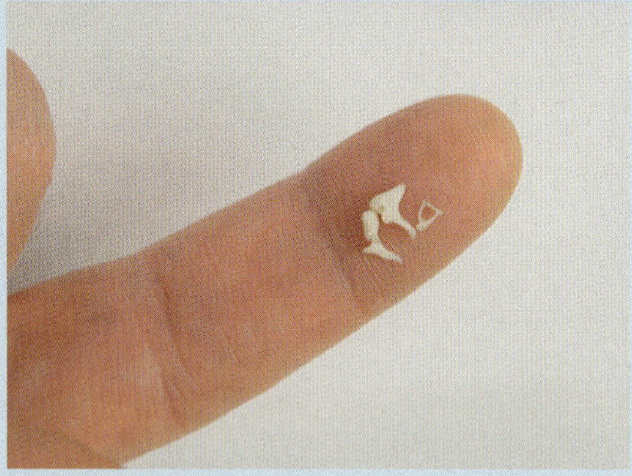

FIGURE 18.1.7 The three ear bones, malleus, incus and stapes, easily fit onto the tip of a finger.

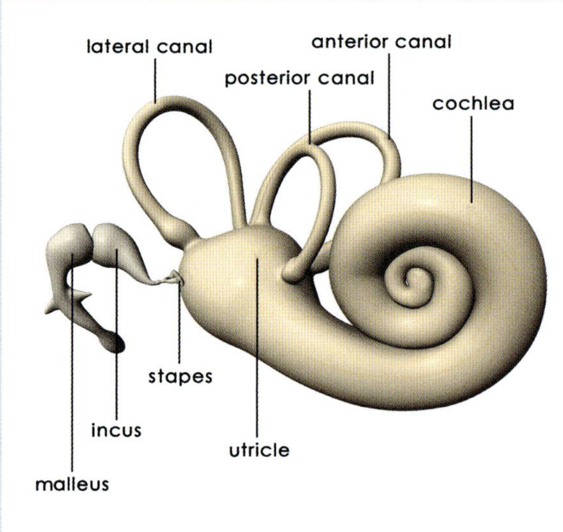

FIGURE 18.1.8 Three ossicle bones transmit sound vibrations into the cochlea. They are the smallest bones in the human body, with the tiniest being the stapes that interfaces with the oval window of the inner ear.

Mechanoreceptors

The nature of sound is that it travels as vibrations through air, water and solids, each of which is referred to as a medium. Animals detect sound using mechanoreceptors, which are sensory neurons (nerve cells) that can detect minute vibrations. Vibration-sensitive neurons may be attached to larger vibrating structures that select, filter and sometimes amplify the sound frequencies that are important to the animal. In human ears, vibrations of the air are amplified by the system of three ear bones (ossicles) that act as solid, mechanical levers in the middle ear. The vibrations are then transmitted to a fluid-filled canal (cochlea) in the inner ear that contains sensory hair cells with groups of brush-like projections on the ends called stereocilia (Figure 18.1.9). Vibrations entering the inner ear displace the fluid that surrounds the groups of stereocilia, causing them to bend and generating nerve impulses that travel to the brain along the **auditory nerve** (Figure 18.1.9). The inner ear can transmit information about the frequency (pitch) and volume of a sound to the auditory processing centres in the brain.

Sound detection

A **sound wave** starts as a form of kinetic (movement) energy. The vibrations are transmitted through a series of conduction media in the ear before it is detected in the brain and the person becomes aware of the sound.

- In the outer ear, the pinna acts like a collecting funnel to channel air particle vibrations (sound waves) along the ear canal. These cause the cone-shaped tympanic membrane to vibrate and conduct the sound energy further into the solid medium of the ossicle bones in the middle ear (Figure 18.1.10 on page 602).
- In the middle ear, the kinetic energy of moving air particles is converted to a mechanical vibration that moves through the solid medium of the three ossicles. As it travels along, the energy is amplified. First in line is the malleus with its handle connected to the vibrating tympanic membrane (ear drum), then incus, then stapes. The base of the stapes connects closely against the oval window at the junction of the middle and inner ear. When the amplified vibration reaches the stapes it pushes on the oval window and causes the fluid of the inner ear to move (Figure 18.1.10 on page 602).
- In the inner ear, sound energy continues as mechanical energy when the oval window bulges inwards and causes pressure waves in the fluid of the cochlear canals. The fluid movement bends the stereocilia of the hair cells. Since these are sensory neurons, the movement sets up a nerve impulse that travels to the brain along the auditory nerve. Once a neuron is stimulated, the mechanical energy converts to electrochemical energy for transmission via nerves to the brain (Figure 18.1.11 on page 602).

In an adult human ear, the cochlea is about 35 mm long and coils into two and three-quarter turns, resembling a snail shell.

Each ear has around 15 000 mechanoreceptor hair cells on the organ of Corti. Each sensory hair cell has a group of 50–200 hairs at one end, forming a kind of sensitive brush projecting into the fluid of the cochlear canals.

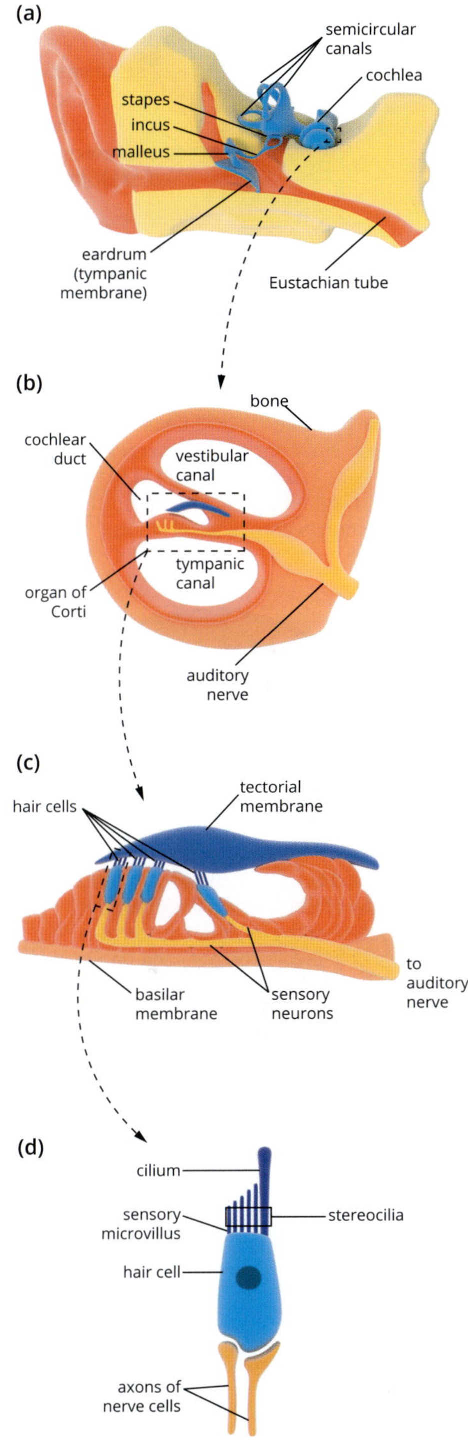

FIGURE 18.1.9 (a) The structure of the ear and cochlea (blue). (b) An enlarged cross-section through the cochlea shows the cochlear duct with the vestibular canal and tympanic canal either side. (c) The organ of Corti consists of the basilar membrane (orange), tectorial membrane (blue), hair cells (dark blue and light blue) and sensory neurons (yellow) leading to the auditory nerve (d) A single hair cell is at lower left, with its stereocilia (dark blue).

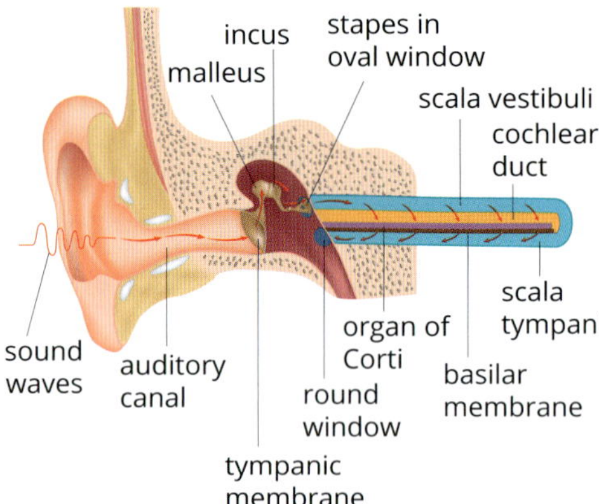

FIGURE 18.1.10 The conduction of sound energy is shown, from air vibrations in the outer ear to pressure waves in the fluid of the inner ear. In this diagram the cochlea is depicted as unwound from its normal spiral shape.

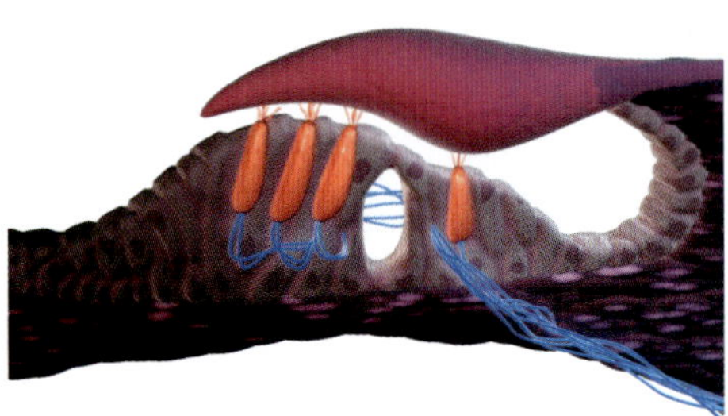

FIGURE 18.1.11 Artwork showing the structure of the organ of Corti, which lines the spiral of the cochlea in the inner ear. It contains four rows of hair cells (orange) and on top of these cells are stereocilia, which touch the tectorial membrane (purple, top), detecting tiny movements in the membrane. The hair cells translate these movements into electrical impulses, which travel down the nerve fibres (blue) to the brain, where they are deciphered as meaningful sounds.

Organ of Corti

Inside the cochlea of the inner ear is a sensory structure called the **organ of Corti**. The cochlea is a spiral of three internal canals running side by side and separated by membranes (Figure 18.1.9 on page 601 and Figure 18.1.10). The **basilar membrane** separates the second and third chambers and the organ of Corti is embedded along this basilar membrane. It is the organ of Corti that holds the mechanoreceptor hair cells organised into four rows (Figure 18.1.11). These tunnel-like chambers of the cochlea are filled with fluids called **endolymph** and **perilymph**, similar to spinal fluid.

When mechanical vibrations of sound energy are conducted across the oval window into the cochlear fluid, they act like ripples on a pond. The fluid movement is one-way along a canal, around the apex and back along another canal where it causes the round window to bulge out and absorb the energy, thus dampening the vibration (Figure 18.1.10). The longest stereocilia of each group of hairs on a hair cell reach across the fluid-filled chamber from the organ of Corti onto the overlying **tectorial membrane** (Figure 18.1.11). When the membrane flexes in response to fluid movement, hair cells at specific locations are stimulated and send a nervous impulse along cochlear nerves to the main auditory nerve. By registering which part of the organ of Corti is sending the nerve impulse, the brain can interpret the pitch of the sound being detected. Complex sounds usually contain several different frequencies and the brain learns from experience how to interpret the messages from multiple locations along the organ of Corti.

Frequency

The vibrations of different sound frequencies travel different distances along the canals. Individual hair cells respond to specific sound frequencies (pitches) so that, depending on the pitch of the sound, only certain hair cells are stimulated. The high frequency sounds (1500–20 000 Hz) are detected closest to the oval window, known as the base of the cochlea. Medium frequency sounds (600–1500 Hz) are registered in the middle region and low frequency (20–600 Hz) at the furthest part of the tapered cochlea, which is also the narrowest section at the apex of the spiral.

The hair fibres (stereocilia) are of different lengths to enable them to vibrate at different frequencies. The highest pitch sounds of 20 000 Hz vibrate the shortest hairs, closest to the oval window. Towards the apex of the cochlea, the longest hairs respond to the very lowest pitch sounds, even down to 20 Hz.

Amplitude

The loudness (amplitude) of a sound wave causes more or less vigorous pressure vibrations inside the cochlea. Hair cells are bent more for loud sounds and more nerve impulses are sent to the brain making it aware of the sound volume. Repeated stimulation from loud sound can cause the stereocilia hairs to stay bent and not recover to their normal upright position, or even for the cells to die, leading to permanent hearing loss for that section of the cochlea.

DISORDERS CAUSING HEARING LOSS

Hearing loss can be categorised by which part of the auditory system is damaged. The extent of a person's hearing loss may vary depending on the frequency (pitch) of the sounds being heard. This is particularly true for age-related or noise-induced hearing loss. There are four types of hearing loss: hearing loss due to auditory processing disorders, conductive hearing loss, sensorineural hearing loss (SNHL) and mixed hearing loss.

Auditory processing disorders

Some hearing problems are due not to damage or defects in ear structure, but to processing problems in the auditory areas of the brain. Lack of understanding of speech or being unable to detect the direction of a sound are two such disorders. Very little can be done to correct these apart from educational intervention. The causes may be congenital (present from birth; inherited condition) or from a physical injury to the brain or a medical condition such as a stroke.

Conductive hearing loss

Conductive hearing loss occurs when sound is not conducted efficiently through the outer and middle ear. Conductive hearing loss usually involves hearing reduced sound volume. This type of hearing loss can often be corrected medically or surgically. There are several causes of conductive hearing loss, which are described below.

- Malformation of outer ear, ear canal, or middle ear structures, which may be congenital.
- Fluid in the middle ear from upper respiratory tract infections (URTI), commonly called a head cold, is temporary and results from infection in the nose and throat travelling up to the middle ear via the Eustachian tube or sinuses (Figure 18.1.12).
- Ear infection in the outer ear canal such as swimmer's ear (Figure 18.1.13), or in the middle ear where it is called **otitis media**. An accumulation of fluid in the middle ear may interfere with the movement of the eardrum and ossicles.
- Allergies causing fluid build-up in the middle ear (Figure 18.1.12).
- Poor function of the Eustachian tube that is necessary to adjust air pressure in the middle ear, resulting in the feeling of blocked ears. Both scuba diving and airplane travel are activities that rely on the Eustachian tube to equalise air pressures.
- Perforated eardrum, which can result from external injury, for example a loud explosion close to the ear or damage from a sharp object pushed into the ear canal. The other cause is internal pressure from fluid build-up in the middle ear causing the eardrum to split. Infants are prone to this problem resulting from more frequent ear infections due to them having a shorter, straighter Eustachian tube that allows infection to spread up more easily from the throat.
- Benign tumours may grow as clusters of polyps in the middle ear or as a cholesteatoma tumour that arises from a perforation of the tympanic membrane with ingrowth of skin into the middle ear (Figure 18.1.14). Malignant cancer tumours are also possible such as squamous cell cancer of the outer ear or bone cancer in the skull near the ear.
- Hard plugs of excess dry earwax forming in the ear canal.
- Foreign body, having been pushed into the ear canal.
- Otosclerosis is a hereditary disorder in which a bony growth forms around an ossicle in the middle ear, preventing it from vibrating when stimulated by sound. The problem usually becomes apparent with hearing loss in early adulthood. It can often be managed with surgery or a hearing aid. Research suggests that the measles virus may contribute to the stapes being overgrown with bone and becoming rigid in people with a genetic predisposition to otosclerosis.

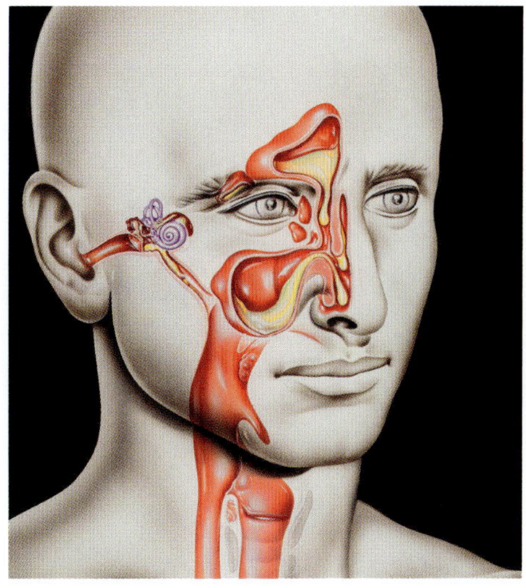

FIGURE 18.1.12 Illustration showing the effects of upper respiratory tract infections (URTI) or allergy problems: rhinitis (inflamed nasal membranes); inflammation of paranasal sinuses (sinusitis) and middle ear infection (otitis media). Mucus (yellow) is shown accumulating and flowing within the sinuses, in the middle ear and the Eustachian tube (connecting ear to throat).

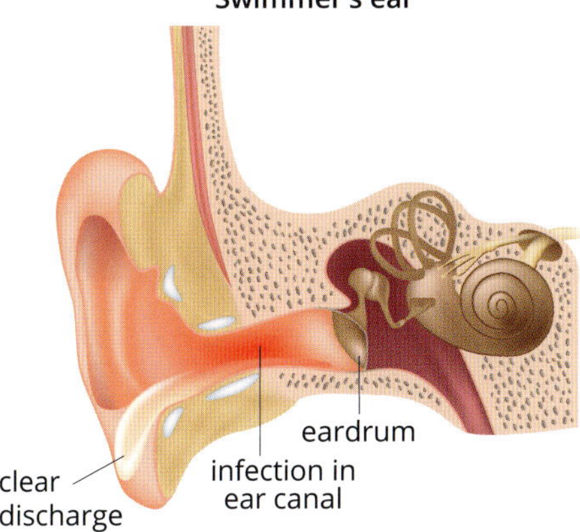

FIGURE 18.1.13 Swimmer's ear is an infection of the outer ear canal caused by dampness and spreading of pathogens in pool and spa water.

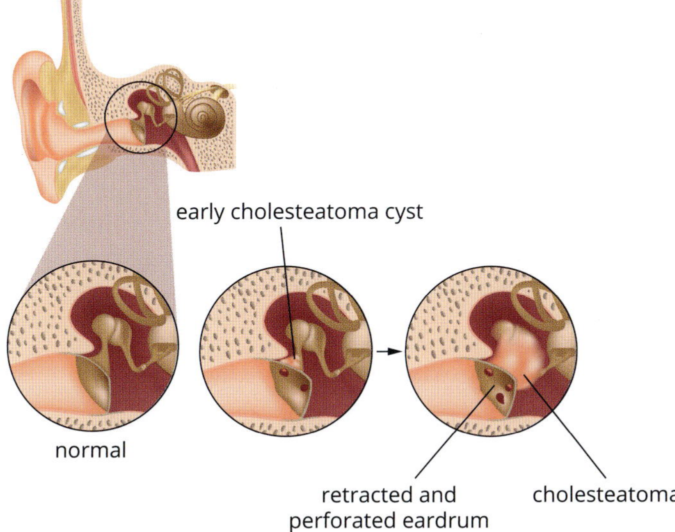

FIGURE 18.1.14 Development of a cholesteatoma. Growth of this tumour starts from perforation of the eardrum and subsequent ingrowth of benign tissue over the ossicles, inhibiting their vibration and causing hearing loss.

Sclerosis is a general term applied to hardening of body tissues such as multiple sclerosis that damages CNS neurons, arteriosclerosis which is hardening of the artery walls and otosclerosis which causes hearing loss when the ear bones are overgrown and hardened.

Sensorineural hearing loss

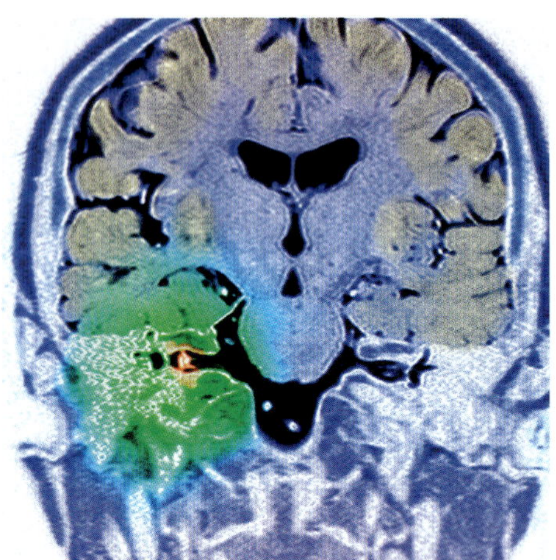

FIGURE 18.1.15 Coloured MRI scan taken from the front of a woman's head showing an acoustic neuroma (3–4 mm mass (red), visible at centre left in the inner ear). The brain is at upper centre and the right ear shaded green.

Sensorineural hearing loss (SNHL) occurs when there is damage to the cochlea of the inner ear or to the nerve pathways from the inner ear to the brain. Most of the time, the disorder cannot be medically or surgically corrected. This is relatively rare but is the most common type of permanent hearing loss and a **cochlear implant** may be the only solution. SNHL reduces the ability to hear faint sounds. Even when speech is loud enough to hear, it may still be unclear or muffled. Some possible causes of SNHL are described below.

- Exposure to damaging noise causes fatigue of the cochlear hair cells, even death. It has been estimated that 37% of hearing loss problems in Australia are due to preventable and repeated exposure to loud noise through work or leisure. Live music concerts and listening to music through headphones are known causes, as well as noise from airplanes and operating machinery like lawnmowers, chainsaws and power tools.
- Suffering a head injury and trauma.
- Benign tumours known as acoustic neuromas grow in the canal connecting the brain to the inner ear. Without treatment, important nerves can be affected, hearing loss may occur and the growing tumours eventually encroach on the brain (Figure 18.1.15).
- Ageing (presbycusis) because over time the hair cells may become enlarged, lose their resilience and their ability to respond to the fluid vibrations (Figure 18.1.16). Hair cells are not replaced if they die during an individual's lifetime and are some of the only cells that last for the whole life of the body.
- Malformation of the inner ear, which is usually a congenital problem that cannot be corrected.
- Some drugs are toxic to hearing. There are more than 200 known ototoxic medications (prescription and over-the-counter) on the market today, including medicines used to treat serious infections, cancer, and heart disease. In these cases, the risk of ear damage is known but no other alternatives are available for treating the life-threatening conditions. Drugs known to cause temporary damage include salicylate pain relievers (aspirin used for pain relief and some heart conditions), quinine (to treat malaria), and loop diuretics (to treat certain heart and kidney conditions; loop refers to the loop of Henle, part of the kidney structure). In some instances, exposure to loud noise while taking certain drugs will further increase their damaging effects.
- Hereditary hearing loss such as Ménière's disease, an inner ear problem that affects hearing and balance. Normally it occurs in only one ear at a time, but for many of those with the disorder it develops over time in the other ear. The disease usually occurs in people aged 40 to 60 although it is possible for anyone to have it. The cause of Ménière's disease is not known, except that it is related to the fluid of the inner ear. In people with Ménière's disease, too much of this fluid builds up creating pressure in the inner ear, affecting balance and hearing. Experts are still not sure why the fluid builds up. Some symptoms of Ménière's disease are hearing loss (temporary or permanent), **tinnitus** (a low roaring, ringing or buzzing in the ear in the absence of actual sound) and vertigo (the feeling that you or your surroundings are spinning).

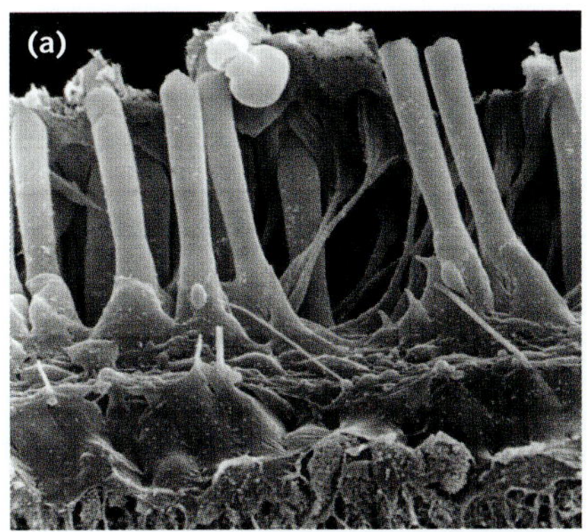

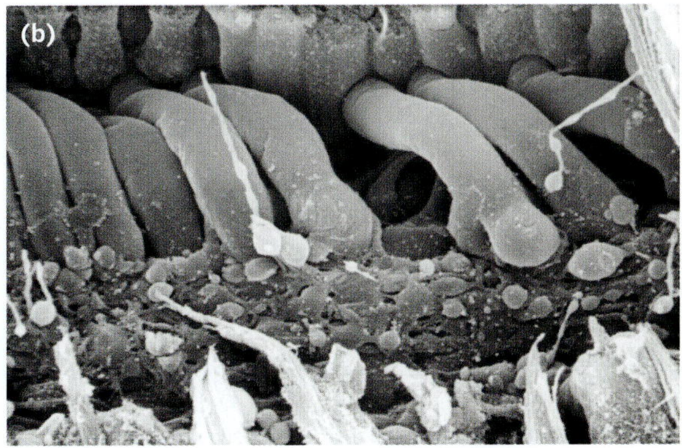

FIGURE 18.1.16 (a) Scanning electron micrograph (SEM) of normal hair cells within the inner ear. The stereocilia line the organ of Corti inside the cochlea. Sound waves entering the inner ear displace fluid around stereocilia, causing them to bend and trigger nerve impulses. This process transmits information about the loudness and pitch of a sound to the brain. (b) SEM of damaged hair cells within the inner ear. The age-related degeneration (presbycusis) of stereocilia has caused them to become swollen and collapsed, meaning they are unable to respond to pressure waves in the cochlear fluid, resulting in deafness.

Mixed hearing loss

Sometimes a conductive hearing loss occurs in combination with a SNHL. There may be damage in the outer or middle ear and in the inner ear (cochlea) or auditory nerve. When this occurs, the hearing loss is referred to as a mixed hearing loss. For example, otosclerosis may cause a SNHL (damaged sensory cells and/or nerve fibres of the inner ear), as well as a conductive hearing loss.

Tinnitus

Tinnitus occurs for around one in 10 of the population and may be permanent or temporary, such as the after-effects of listening to a prolonged period of very loud music. Tinnitus is when you hear a sound that isn't actually there. It is often described as ringing in the ears, but is commonly experienced as buzzing, roaring, clicking and hissing. It can vary from a low roar to a high squeal in one or both ears. Exposure to loud sound is one of the main risk factors for developing tinnitus—though getting older and even earwax can also contribute. It can also be triggered or made worse by emotional events such as depression or anxiety, accident and injury (not necessarily to the ear).

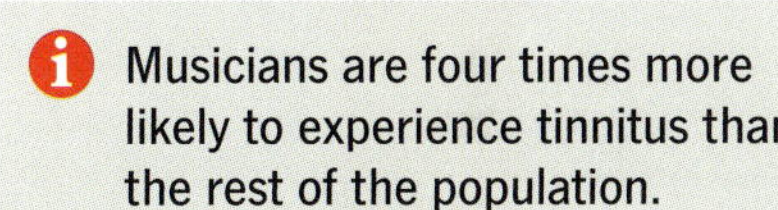

Musicians are four times more likely to experience tinnitus than the rest of the population.

Causes of tinnitus vary, can be in any part of the ear and may result from many of the hearing disorders already listed above. These causes include wax built up against the eardrum, otosclerosis when the stapes becomes fixed in position, Ménière's disease (fluid build-up), ototoxic medications and exposure to prolonged loud noise when the hair cells are damaged (Figure 18.1.16). Tinnitus is usually associated with some hearing loss and should be treated as a warning symptom.

TECHNOLOGIES TO ASSIST WITH HEARING LOSS

There are several technologies available to help with hearing loss, depending on the cause of the disorder. Table 18.1.1 provides a summary of the technologies based on the extent of loss for one ear. The figure shown in decibels (dB) is the cut off for the sound volume that can be heard. For example, profound loss means a sound can only be heard if it is above 90 dB (the volume of a jackhammer or a jet plane take-off at 300 m) and mild loss is for sounds quieter than 40 dB (whispers, rustling leaves, some bird calls). For loss in both ears, a combination of technologies may be used.

BIOFILE N PSC

Decibel scale and hearing loss

The decibel (dB) scale is a logarithmic scale not a linear scale (Figure 18.1.17). This means it works in powers of ten. The sound level of leaves rustling at 20 dB is not twice as loud but ten times louder than 10 dB, a sound level just audible to those with keen hearing such as hearing a gentle brushing of a fingertip over skin.

Any sound above 85 dB can cause hearing loss and the longer the exposure the more likelihood of damage. Normal speech is around 60 dB. Eight hours at 90 dB usually causes damage. Exposure to 140 dB causes pain and immediate damage.

When listening to music or podcasts through headphones, the volume should be kept low as the sound energy is transmitted directly through to the sensitive cochlea.

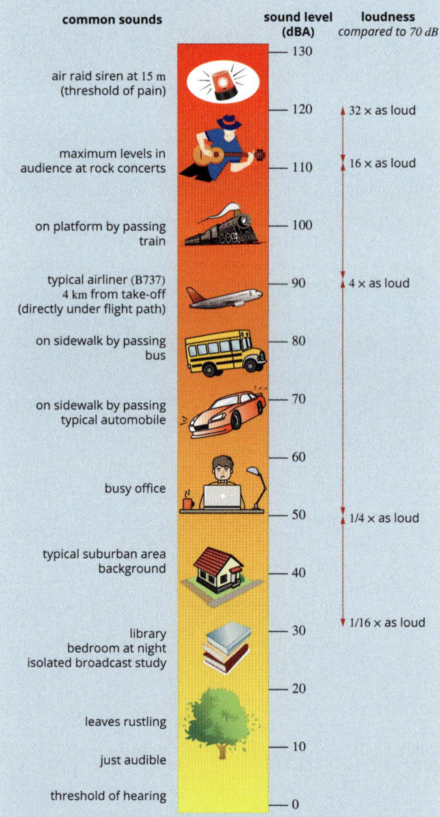

FIGURE 18.1.17 The loudness of some common sounds. The decibel scale that measures loudness is logarithmic, meaning that loudness increases dramatically as the decibels rise.

TABLE 18.1.1 Technologies to aid hearing disorders

Hearing disorder	Mild loss (<40 dB)	Moderate loss (40–60 dB)	Severe loss (60–90 dB)	Profound loss (>90 dB)
conductive hearing loss	usually not treated	bone conduction hearing devices and hearing aids	bone conduction implant	cochlear implant
sensorineural hearing loss	hearing aids, assistive listening devices and hearables	hearing aids and hybrid cochlear implant	hearing aids and cochlear implant	cochlear implant and auditory brainstem implant
mixed hearing loss	usually not treated	middle ear implant and hearing aids	middle ear implant and hearing aids	cochlear implant

Assistive listening devices

Most assistive listening devices are not wearable devices but are inbuilt into household devices such as alarms, phones or TVs to amplify the volume output or create a noticeable visual or vibration signal.

Hearables

Hearables, sometimes called smart headphones, are a relatively new technology available in the general marketplace with potential to assist people who have mild hearing loss. Existing devices such as hearing aids, headphones, earbuds and personal sound amplifiers are adding this new technology to transform them into hearables. A hearable is a wearable earpiece device that contains a microcomputer. It fits into the ear canal and uses wireless technology to connect with other devices such as a smartphone. The sound output of music, audio from a TV or computer, phone conversation and biometric monitoring (e.g. of the wearer's heart rate) is fed directly into the ear canal. The potential for those with hearing loss is that there is no confusion with extraneous sounds and the input sound can be amplified.

Hearing aids

Hearing aids are electronic amplification devices that assist with many types of hearing loss. They contain a microphone to detect sounds, a device to modulate the sound, an amplifier to increase the sound volume and a battery to power these functions. Different styles exist to suit individual needs. Some are worn behind the ear, some fit inside the ear canal and others have components in both positions. Hearing aids rely on amplifying the sound vibrations that pass into the ossicles, so these bones, the tympanic membrane and the cochlea must still have an adequate level of function.

While a hearing aid does not completely restore normal hearing, with perseverance by the wearer, it can vastly improve the impact of many hearing disorders. It is recommended that hearing aids are fitted as early as possible when there is hearing loss. The aim is to avoid a period of sensory deprivation in the brain that comes from a decrease in hearing. Otherwise, it may be that the auditory regions of the brain forget how to process the complexities of sound, especially speech, if left too long without stimulation.

The technology of hearing aids has advanced as new features have become available, such as improved batteries, miniaturisation of technology, connection with hearing loops in public venues and the use of wireless technology and microcomputers for hearables. Aids are now much more effective because they can be tailored to individual needs but this comes at an extra cost.

Conventional hearing aids

Conventional hearing aids (Figures 18.1.18 and 18.1.19a) may be uncomfortable to wear for some people and are sometimes quite visible, which is a disadvantage for people conscious of their appearance. Battery life may be an issue and replacement batteries can be expensive. Having background noise also amplified can be painful and confusing for many wearers. This is more of a problem with the cheaper quality hearing aids and in social situations. High quality, modern hearing aids are fitted with devices that can distinguish between background noise, such as traffic, and foreground noise, such as conversation. More modern hearing aids can be very small and discreet, some can be worn inside your ear but the type of aid required varies with the individual and expert help is needed for correct diagnosis and fitting.

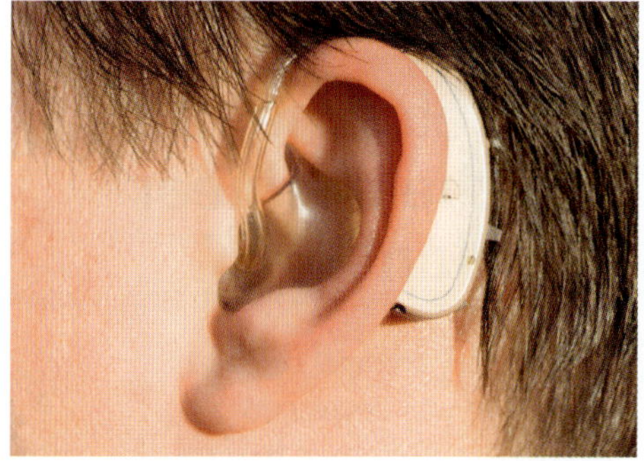

FIGURE 18.1.18 A conventional hearing aid worn behind the ear

Bone conducting hearing aids

Bone conduction hearing aids are a specialised type of hearing aid that relies on bone conduction of sound vibrations. This technology is suitable for people who get no benefit from conventional hearing aids because they lack any function in the middle ear. For this device, a hearing aid is worn behind the ear and coupled with an external vibrator on a tight headband or on the arm of a specially constructed pair of spectacles. Vibrations are created that move across the skull bone direct to the cochlea.

Bone conduction hearing aids are quite uncomfortable and not very efficient, so it is usually a temporary solution for people waiting for surgery to install a **bone conduction hearing implant**. The hearing aid has the benefit of not requiring surgery and can easily be removed or adjusted. Again, battery life and cost may be an issue for wearers.

Hearing implants

Bone conduction implant

Unlike the external bone conduction hearing aid mentioned in the hearing aid section, a bone conduction hearing implant requires specialised surgery to insert a component into the skull bone behind the ear (Figure 18.1.19b). It also has an external sound processor that detects sound waves and converts them into vibrations to be passed through the bone to the implant. From here the vibrations are moved into the inner ear where they are sensed naturally as sound by the cochlea. The outer and middle ears are bypassed completely for people with permanent hearing loss in these regions. However, the cochlea must be functioning normally.

(a) hearing aids in the canal

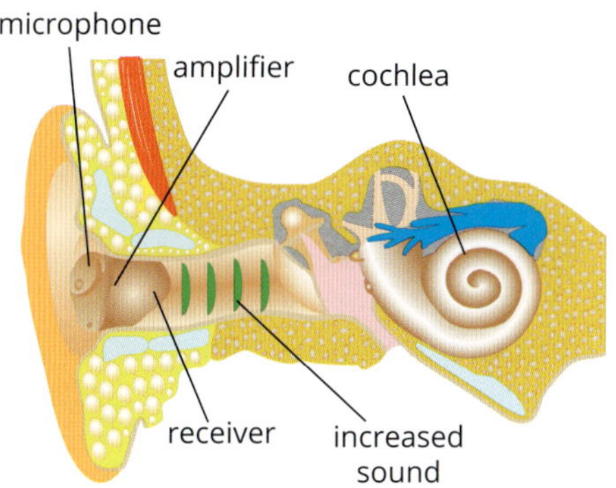

(b) implantable hearing aids

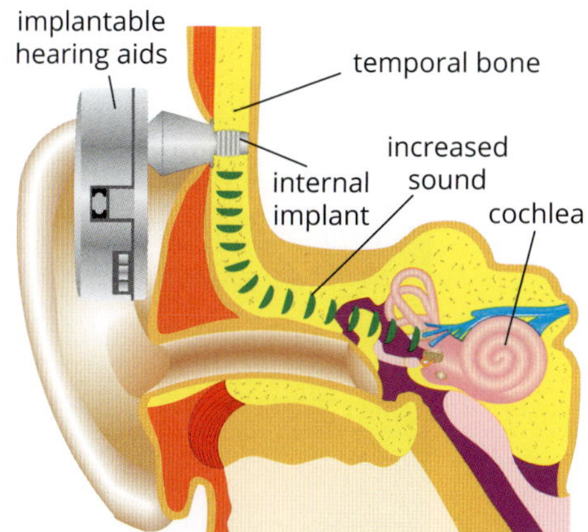

FIGURE 18.1.19 (a) A conventional hearing aid worn in the ear canal compared with (b) a bone conduction implant that has been surgically inserted

Cochlear implant

A cochlear implant (Figure 18.1.20 on page 608) is also known as the bionic ear. It is for people who have severe sensorineural hearing loss. They gain no benefit from hearing aid or bone conduction technologies because their cochlea does not respond to sound vibrations in the normal way.

Delicate surgery is required to insert an electrode array inside the cochlea and a receiver into a hole cut into the mastoid bone behind the ear. The device works by having an external microphone positioned behind the ear and a sound processor box clipped on close by. The microphone detects sound in the wearer's surroundings, a device amplifies it and the processor converts this sound energy into electrical impulses. A transmitter is positioned on the head, usually held in place by magnetic attraction to the receiver, which has been implanted just under the skin (Figure 18.1.20). The electrical signals are passed wirelessly into the internal receiver where they are relayed by a conducting wire to the line of electrodes in the cochlea. These electrodes directly stimulate several neurons in different locations on the organ of Corti, as if they were receiving sound of specific frequencies. The stimulation is not to the hair cells but direct to the sensory nerves.

BIOLOGY IN ACTION CC ICT

Development of the cochlear implant

The multi-channel cochlear implant was developed by Professor Graeme Clark, an Australian scientist who has since researched other bionic devices. His work on the bionic ear took 18 years from 1967, before it become a successful commercial reality in 1985. Prior to this, in 1978, his team made a world-first breakthrough when their patient was the first profoundly deaf person to hear speech with the aid of artificial sensory stimulation. Since then over 180000 cochlear implants have been completed (Figure 18.1.21). Graeme Clark tells the story that he was inspired in his early research by a seashell with the same spiral shape as a human cochlea. He threaded a long blade of grass around the shell and realised that perhaps this could be done with wires to reach the sensory nerves of the cochlea.

FIGURE 18.1.21 Close-up of electrode array of a cochlear implant. Multiple electrodes (silver) are seen on this thin wire that is surgically inserted deep into the cochlear of the inner ear, enabling sound stimuli to reach the brain. The electrodes are attached to a receiver implanted under the scalp. The patient wears an external transmitter, microphone with amplifier and sound processor box, all battery powered.

Cochlear implants do not restore a person's hearing completely and the quality of sound created by the devices is low compared to hearing sounds naturally. They do provide some hearing abilities for profoundly deaf people that allow their wearers to have a better understanding of speech and to detect sounds in their immediate environment. A professional educational therapy program is vital to assist implant patients with decoding the new sensory input they are receiving and how to interpret it.

A successful cochlear implant should enable the wearer to detect more sound frequencies than with a hearing aid but not as many as with normal hearing. Battery life is shorter for implants than hearing aids. The speech processor that converts sound into electrical signals uses more energy than the amplifier of a hearing aid. Also, the wireless feature of implants requires more energy input to push signals through the skin of the scalp and the electrode array in the cochlea uses energy to deliver signals of up to 50000 pulses per second. Button type batteries are used and various types are available including rechargeable ones that last six to eight hours per charge.

The cochlear implant is an expensive device that is limited for distance. The wearer must contend with issues of background noise. They usually still rely on lip reading to fully understand speech. With training, the implant makes lip reading easier. It is considered to give most benefit when implanted in young children before they reach five years, as they are more adaptable to learning how to decode the signals from a cochlear implant into meaningful sounds.

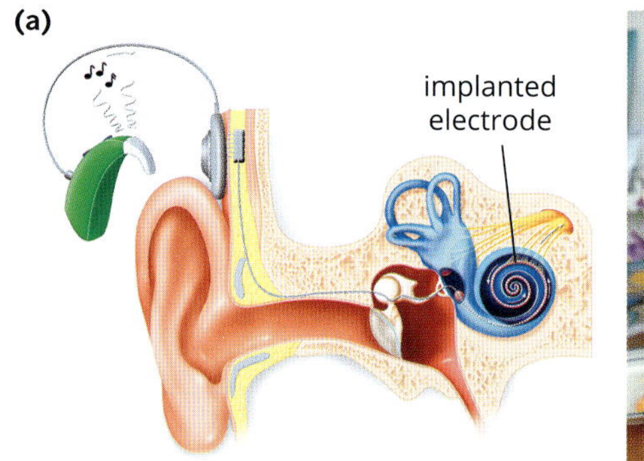

FIGURE 18.1.20 (a) Illustration of a cochlear implant. (b) Cochlear implant on the head of an eight-year-old boy whose profound deafness was caused by gentamicin, an ototoxic drug used to treat meningitis when he was aged one. The implant sound processor behind the ear passes signals to the transmitter (circular) attached magnetically to the scalp. The transmitter sends messages to an array of electrodes in the cochlea in the inner ear, which stimulates the auditory nerve at different frequencies. Nerve impulses from the cochlea are passed to the brain, allowing the boy to be aware of sound and improve his lip reading skills.

Hybrid cochlear implant

Hybrid cochlear implants are for people with severe, high frequency hearing loss, who still have the ability to hear lower frequencies (pitches). Age-related hearing loss (presbycusis) occurs in the higher frequencies first, so the elderly are one group of people who could benefit from the hybrid technology. It uses electro-stimulation technology, such as that applied in normal cochlear implants and the acoustic amplification technology used for hearing aids. Combining the two technologies allows the wearer to have sound sensations in the higher frequencies, while the hearing aid technology amplifies the sounds being heard in the lower frequencies. The hybrid implant is tailored to individual needs so that the implant technology takes over from the hearing aid at a frequency unique to each implantee. A more streamlined electrode of a slimmer shape causes less damage when inserted into the cochlea. The low frequency hair cells further from the oval window are left undamaged to keep their residual hearing ability.

Middle ear implant

For some people with severe mixed hearing loss, a **middle ear implant** may be the best available technology. If a traditional hearing aid is of no benefit for them, the middle ear implant is a device surgically attached onto the stapes bone. Like some other hearing technologies, a microphone and sound processor are worn behind the ear. These external devices collect and amplify the sound, then transmit it as electronic signals to the internal device on the stapes bone. Here the signals are changed back to mechanical vibrations to be passed across the oval window where they stimulate normal sound sensations to the hair cells.

Auditory brainstem implant

In the most extreme cases of auditory nerve damage or sensorineural hearing loss, the nerve pathways to the brain are damaged and cannot relay sound information. Hearing aids and implants cannot help these individuals. At present the only technology that may assist is an **auditory brainstem implant**. The device is similar to the cochlear implant but the electrodes are surgically placed on the surface of the brainstem rather than inserted into the cochlea. The best outcome is that individual sounds cannot be distinguished but the person can detect when a sound is present and this assists them with lip reading.

Fortunately there are only a very small group of people with this type of hearing disorder. One cause is when a certain kind of malignant tumour known as neurofibromatosis type 2 (NF2) occurs on the auditory nerve. In the process of removing a NF2 tumour, the auditory nerve is often severed resulting in the complete loss of hearing in the affected ear. If this damage happens, an auditory brainstem implant is the only viable hearing technology currently available.

Other treatments for hearing disorders

Although the following are not innovative digital technologies, they are worth noting because they are in common use to deal with some hearing disorders.

Otitis media and eardrum grommets

As already noted, otitis media is infection of the middle ear that allows fluid to build up and inhibit the hearing function of the ossicles. The Eustachian tube connecting throat, nose and middle ear becomes blocked and the pressure of fluid pushes painfully against the eardrum from inside, sometimes causing it to rupture. The condition is quite common in infants and can lead to some permanent hearing loss if scar tissue forms. A simple surgery has been developed to insert a **grommet** in the tympanic membrane to keep a small hole open that reduces pressure, allows pus to drain out and keeps the middle ear aerated (Figure 18.1.22). The grommet is removed when there is less risk of infection and the immune system has developed more. As the child grows older, the Eustachian tube lengthens and is positioned at more of an angle, lessening the risk of pathogens moving up to the middle ear.

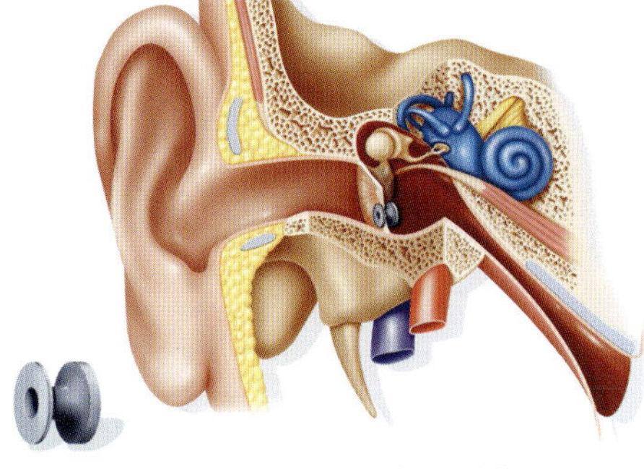

FIGURE 18.1.22 A treatment for middle ear infection (otitis media) is insertion of a tympanostomy tube (bottom left) in the eardrum (tympanic membrane). The tube, also called a grommet, is shown in position. It relieves the pressure and keeps the middle ear aerated.

Surgical solutions

Stapedectomy is a surgical procedure of the middle ear performed to improve hearing if the stapes footplate has become fixed in position. The rigidity results in conductive hearing loss. There are two major causes of stapes fixation. The first is otosclerosis—abnormal mineralisation bone growth into the ear from the skull. The second is a congenital malformation (birth defect) of the stapes.

The surgical solution is a **prosthesis**—either to replace the immobile stapes with a mobile, micro-prosthesis stapes (stapedectomy) or create a small hole in the fixed stapes footplate and insert a tiny, piston-like prosthesis (stapedotomy).

Like all surgery, there are risks. Of the patients who undergo the surgical procedure, 90% come out with significantly improved hearing and less than 1% will experience worse hearing function. Successful surgery usually provides an increase in hearing acuity of about 20 dB. That is the difference between having your hands over your ears or not. Most results from surgery report more success in the lower speech frequencies of 500–2000 Hz than in the high frequencies. However, the higher frequencies have usually been less impacted by otosclerosis in any case.

A prosthesis is an artificial body part, such as a replacement limb, hip joint, dentures or, in this case, an artificial stapes bone for the middle ear.

Ménière's disease may also have to be treated with surgery to reduce the amount of fluid and pressure in the inner ear. The outlook here is less optimistic. Currently there is no cure for Ménière's disease and the aim of surgery is to reduce balance problems and conserve hearing if possible. If non-surgical methods have failed to help with this distressing condition, and the patient is experiencing severe vertigo, then surgery is the last resort. The surgery may destroy the balance mechanism and nerves, which carries the risk of hearing loss. In Australia, there is a support organisation called the Whirled Foundation that exists to assist people who suffer dizziness, vertigo and loss of balance attacks, many of them from Ménière's disease.

BIOFILE CCT DD PSC

Welcome to my whirled

As already explained, the inner ear deals with both hearing and balance functions. The problem of deafness is widely recognised and has many support systems in place. However, there is little understanding in the general community about the devastating effects that loss of balance can have on an individual. Whirled Foundation in Australia exists to assist people who suffer dizziness, vertigo and loss of balance attacks (Figure 18.1.23). The word, vertigo, comes from Latin 'verter' which means 'to whirl'. Vestibular balance disorders include Ménière's disease that also causes hearing loss. Sufferers describe attacks as feeling like being rapidly whirled around out of control and being much more severe than just feeling dizzy. During an attack they may stagger and collapse, perhaps appear to onlookers as if they are drunk or suffering a migraine. They often have hearing loss and tinnitus.

FIGURE 18.1.23 An impression of the vertigo feeling, caused by inner ear balance problems

18.1 Review

SUMMARY

- The human ear is responsible for hearing and balance.
- Understanding properties of sound waves is important to understanding how the ears hear. All sound is made and carried by vibration of particles of a medium (gas, liquid or solid). The hearing process uses all three mediums.
- Volume (loudness) of a sound in decibels (dB) is determined by the amplitude of a sound wave, a measure of how far the medium's particles are displaced. Pitch of a sound is determined by frequency.
- The ear can be divided into three sections—outer, middle and inner ears. Each has unique structures and functions.
- The outer ear channels sound waves along the ear canal to the tympanic membrane (eardrum).
- The middle ear contains three small bones called ossicles that amplify and transmit vibrations from the eardrum to the inner ear.
- Inside the inner ear there is a fluid-filled coiled structure called the cochlea that contains the organ of Corti.
- The organ of Corti holds sensory hair cells (stereocilia) that respond to pressure waves in the cochlear fluid. Different locations distinguish frequencies.
- Stimulation of hair cells sends nerve impulses to auditory centres in the brain to tell the person that they are hearing sounds. The brain interprets the sounds into meaning, such as speech.
- Hearing loss is caused by a variety of disorders and can be classified by the part of the ear that is affected—conductive loss is from the outer and middle ears; sensorineural loss is from the inner ear. There is also mixed hearing loss when both types of loss are experienced and auditory processing loss when the hearing parts of the brain do not function properly to process sounds that are heard.
- Technologies to assist with hearing loss include a variety of hearing aids and implants. Each individual should have a technology tailored to their specific needs. None of the technologies completely restore normal hearing.
- Hearing aids are external and work by detecting and amplifying sound before transmitting it to be heard in the natural way. Cochlear and other implants require surgery and aim to give the person an awareness of sound for practical functioning and allow them to improve their lip reading skills for more reliable social interaction.
- Problems with the technologies include limited battery life and expense, issues with background noise, discomfort with wearing and visibility of the device and frustration at not hearing normal sounds in the same way as other people.
- Great advances have been possible with hearing loss technologies by using digital innovations and micro-computers. No doubt the progress will continue, especially as it is still not possible to completely restore hearing using current technologies.

KEY QUESTIONS

1 Outline the function of each of these structures and identify the location for each in the human ear.
- **a** pinna
- **b** tympanic membrane
- **c** Eustachian tube
- **d** semicircular canals
- **e** organ of Corti
- **f** round window

2 Describe the two functions of the ear ossicles.

3 Draw a flow chart for the sequence of events between a sound travelling in the air and a person's awareness of hearing that sound.

4 Explain how different sound frequencies are detected in the inner ear. Use a diagram to support your answer.

5 If the hair cells in the organ of Corti nearest the oval window were damaged, explain the impact of this damage on a person's hearing.

6 List four causes of hearing loss and identify which part of the human ear is affected for each.

7 Compare these three technologies that could be used to assist with hearing loss: hearing aid, bone conduction implant, cochlear implant.

8 Evaluate if surgery or a hearing aid or a cochlear implant will be more effective for a person with loss of hearing due to each of the following.
- **a** ageing
- **b** otosclerosis (bony growth around an ossicle)
- **c** use of ototoxic drugs in cancer chemotherapy

18.2 Visual disorders

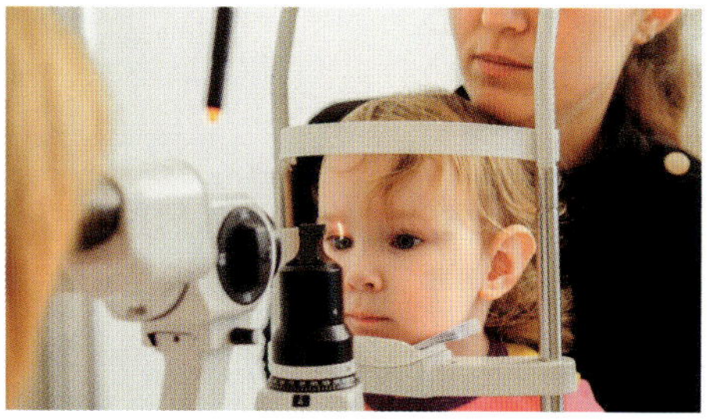

FIGURE 18.2.1 A child has her eyes checked by an ophthalmologist

Light is a very different form of communication wave to the sound waves discussed in Section 18.1. Visible light is a form of electromagnetic radiation (EMR) along with radio waves, microwaves, X-rays, infrared (heat) and ultra-violet (UV). These energy forms are transverse waves that travel without displacement of particles, which is why they are much faster than sound and can reach us from the Sun through the vacuum of outer space.

You should be aware that what humans detect, and call visible light, is only a small part of the whole electromagnetic (EM) spectrum. For most of us who see in colour, the light we detect ranges from 380–780 nm in wavelength. This allows us to see seven colours from red to violet. At night, our vision is more limited and does not include colour but does allow us to detect some movement in dim light.

Impaired vision can create barriers to learning and pose risks to personal safety so it is important to take advantage of any technology that prevents or reduces the effects of visual disorders (Figure 18.2.1).

In this section you will learn the basics of eye structure and function as well as how images are formed to be transmitted to, and interpreted by, the visual centres in our brain. This knowledge will support your investigation of visual disorders and how technologies may be used to offset the effects of such disorders.

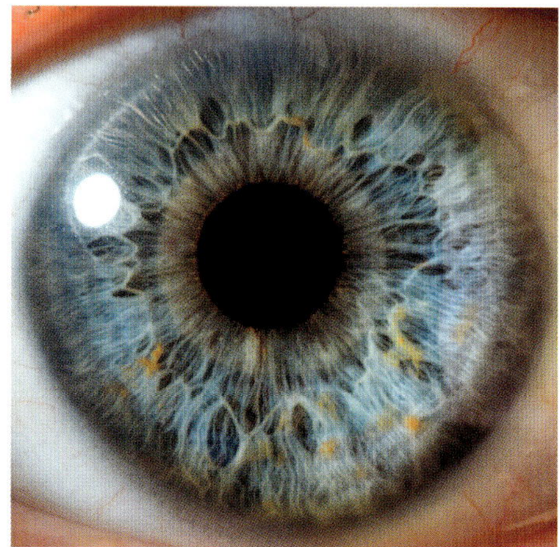

FIGURE 18.2.2 The iris is a pigmented muscle that responds to a light stimulus by dilating or constricting to regulate the amount of light entering the human eye. Sensory and motor neurons control this automatic response in a reflex arc to protect the eye.

STRUCTURE AND FUNCTION OF THE EYE

Sight is a very important sense for gathering information and as a means of communication for humans, who usually say they value it the most out of all their senses. Seeing relies on light entering the eye through the pupil, being detected by **photoreceptors** and then nerve impulses being sent to the visual centres of the brain. Finally these messages have to be interpreted in the brain as images.

How well we see is described as **visual acuity**. This is the resolving power of eyes to distinguish between point sources of light and form discrete, clear images. Thinking about the use of a microscope helps to understand properties of the eye such as resolving power. Simply put, if two dots are close together, are they seen as one or two? If two points are 100 μm (0.1 mm) apart, most human eyes can see them as two separate points. Any closer and they blur into one point. A light microscope improves on this with resolution of up to 0.2 μm and the best electron microscope at 0.0002 μm, which is around five million times better visual acuity than human eyes.

Also, like a microscope, the human eye has the ability to change focus and bring the image of an object being viewed into clear definition. However, the eye has a different way of achieving focus by using the flexibility of its lens, which is made of living cells. The microscope works by changing the position of its fixed shape lenses or the specimen to achieve focus.

Both the eye and the light microscope work best if the amount of light entering each lens can be adjusted (Figure 18.2.2).

Figure 18.2.3 shows the basic structure of a human eye in section from front of the face on the left. Table 18.2.1 outlines the various parts of the human eye structure and their functions that enable vision.

> Each human retina is estimated to have about three million cones (detect bright light and colour). In the fovea these are grouped at a density around 150 000/mm^2.

> The retina has three to four times more rods (dim light receptors) than cones, with 125–130 million in each eye. They are spread across the retina, except in the fovea, but are more numerous at the outer edges of the retina.

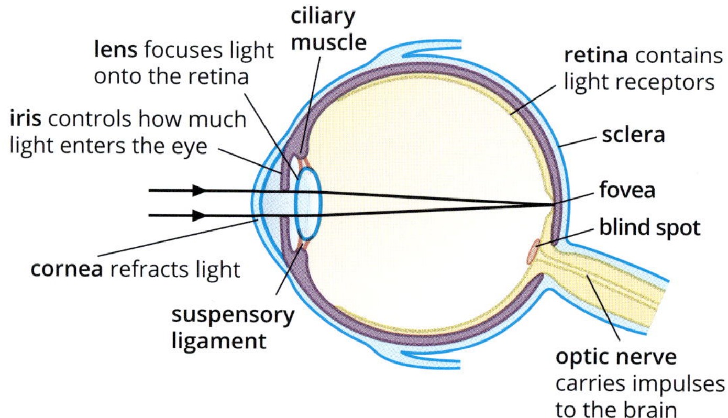

FIGURE 18.2.3 The structure of the human eye

TABLE 18.2.1 Structures of the eye and their functions listed in order from the outer eye surface to the back of the eyeball

Structure	Description	Function
conjunctiva	layer of clear epithelial cells across front surface of eye and continuing inside upper and lower eyelids	protects front of eye
cornea	transparent membrane over front of eyeball, made from collagen protein, covers iris and white of eye	protects eye, blocks rays shorter than 300 nm wavelength, refracts light rays to direct them through the pupil into the lens
sclera	white of the eye, continuous with the clear cornea, the tough layer encompasses the whole eyeball	protects eye, holds fluid and keeps eye in a spherical shape
choroid	black layer between sclera and retina with many blood vessels; at the front of the eye the choroid becomes the iris	provides blood supply for the retina, black pigment prevents false images by absorbing stray light, forms the iris
aqueous humour	clear thick liquid held between the cornea and lens	lubricates cornea and lens, helps hold eyeball shape, refracts light
iris	extension of choroid, coloured muscular ring around pupil	muscles contract or dilate to regulate size of pupil, colour pigment blocks excess light
pupil	opening at centre front of eye, circular in humans (can be other shapes, e.g. slit in some species)	changes size to control amount of light entering lens
lens	biconvex flexible disc made of cells with clear crystallin protein	blocks UV rays, changes shape to focus light rays onto retina
ciliary body	encircles the lens with suspensory ligaments and ciliary muscle (Figure 18.2.4)	holds lens in position and changes its thickness
vitreous humour	clear, jelly-like filling in back of eyeball between lens and retina	maintains spherical shape of eyeball, refracts light, keeps eye structures hydrated
retina	sensory layer inside back of eyeball, holds cone and rod photoreceptor cells and neurons (Figure 18.2.5)	converts light stimuli to electrochemical signals in nerves
cone cells	three types of conical-shaped photoreceptor neurons with photopsin pigments for detecting colours	respond to red, green, blue colours and bright light, used in daylight
rod cells	cylindrical shaped photoreceptors with rhodopsin pigment, denser towards edges of retina	respond to dim light, used for night vision, detect movement but not colours
fovea (part of macula)	shallow pit in central area of retina; highest density of cone cells (colour-detecting) and no rod cells	the position for clearest visual acuity, sharpest image and best colour detection
optic nerve	connection between eye and brain for visual messages, right optic nerve goes to left side of the brain and left optic nerve to right side of the brain	carries nerve signals about images seen from retina to visual cortex of brain on opposite side of the head
blind spot	small area of retina where optic nerve exits	no photoreceptors
eye muscles	four sets of muscles attached from eyeball to bone	holds eyeball in position; rotates it within eye socket of the skull

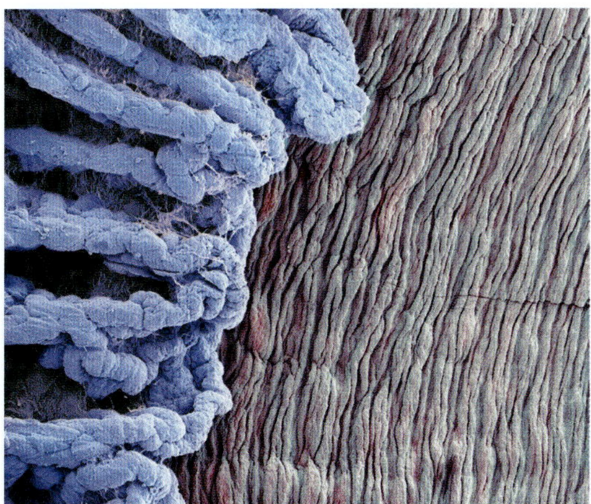

FIGURE 18.2.4 Coloured SEM showing part of the ciliary body (blue) and pigmented iris (right). The ciliary body is a ring-shaped structure that holds the lens in place and can alter the curvature of the lens to focus light on the retina.

FIGURE 18.2.5 Coloured SEM of a section through the retina of a human eye, showing a single cone cell (green, centre) and numerous rod cells (purple).

BIOFILE N PSC

20/20 vision

Having 20/20 vision is regarded as having perfect sight. The term persists from when the imperial system measured distance in feet and inches. The letter chart used for eye tests was placed 20 feet away. If the person could read the smallest row of letters at 20 feet their visual acuity was recorded as 20/20.

Today, an optometrist or other health practitioner places the letter chart approximately 6 m away (Figure 18.2.6). If all letters can be easily read, the score is 6/6 or normal vision. The first number is the test distance. The second number represents the size of the letter seen. For example, 6/12 means that at 6 m the smallest letters that can be read are the letters that can be read at 12 m with normal vision (6/6 vision) (Table 18.2.2). It is possible to have a 6/5 result meaning better than normal 6/6 vision.

For obvious personal and public safety reasons an eyesight test is required when applying for or renewing a driver's licence and to operate certain types of machinery.

TABLE 18.2 2 Examples of visual acuity levels

Acuity Level	Description
6/6	normal vision
6/12	reduced vision, Australian legal driving limit
6/18	low vision (World Health Organization definition)
Less than 6/60	legal blindness (eligible for assistance)

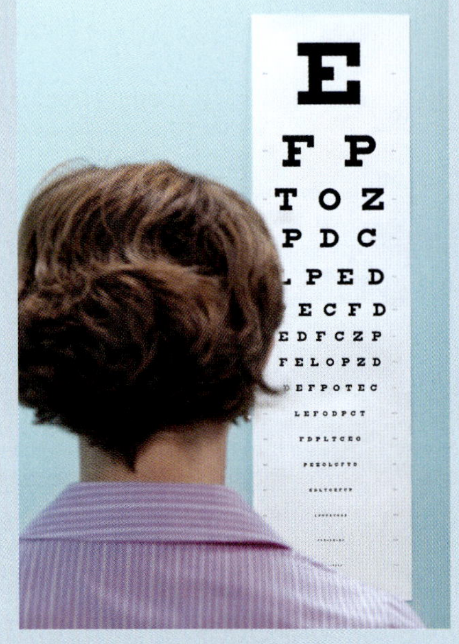

FIGURE 18.2.6 A standard eye test chart

Structure of the lens

Like many other parts of the human body, the living eye lens is a remarkable structure. By having the ability to change shape, the lens functions to alter the focal distance of the eye. This means it can usually form clear images on the **retina** of objects being viewed at various distances. If the lens cannot accommodate enough to create a sharply defined image, it can almost always be assisted by fairly simple, established technologies like spectacles or contact lenses.

The lens alone is responsible for about one-third of the light refraction and all of the adjustment needed to form a retinal image. The other two-thirds of refraction is achieved by the **cornea**, plus **aqueous humour** and **vitreous humour**.

There are three main parts to each lens (Figure 18.2.7):

- **lens capsule**—a transparent membrane that encapsulates the lens; made of collagen, an elastic protein that allows shape changes
- **lens epithelium**—transparent, cube-shaped cells that regulate nutrient, ion and water balance inside the lens; generates new lens fibres throughout life from the embryo stage; the lens grows to an elliptical shape until about age 20, then grows rounder

- **lens fibres**—long, thin transparent cells that form the bulk of the lens; each lens fibre cell is up to 12 mm long and only 4–7 μm wide (which is why they are referred to as fibres even though they are cells); arranged in concentric, tightly-packed layers (like the layers of an onion) with new layers being formed on the outside (Figure 18.2.8). A mature lens fibre cell has no organelles or nucleus, which aids transparency. These cells are filled with **crystallin proteins** that probably evolved from the chaperone family of protein molecules (Figure 18.2.9). You learnt about chaperone proteins in Chapter 4.

The density of the lens fibres and absence of internal light-scattering organelles maintains the refractive power and transparency of the lens. Once a new lens fibre cell nucleus breaks down, the crystallin protein has to be sustained for life. **Cataracts** usually develop with age as patches of crystallins start to denature. People with diabetes are also at greater risk of developing cataracts, as are people who are obese, smoke or have suffered eye injuries. If the opaque area causes significant vision problems, it has become quite common to have surgery and replace the natural lens with an artificial one, tailored to the individual's needs. Another consideration is that UV radiation, in the 300–400 nm range, is mostly blocked by the lens preventing harm to the retina. Intraocular lenses (IOL) are manufactured to also block UV rays. You will learn more about intraocular lenses later in the section.

Nutrients and wastes of the lens cells are exchanged via the aqueous humour. Compared to other cells in the body, cells of the lens have low energy needs because most of them are without organelles. Their main requirement is glucose and because they are without mitochondria, energy must be extracted from glucose by anaerobic respiration. Therefore, there is very little demand for the oxygen carried in the blood as required by other cells to carry out aerobic cellular respiration. You learnt about cellular respiration in Year 11.

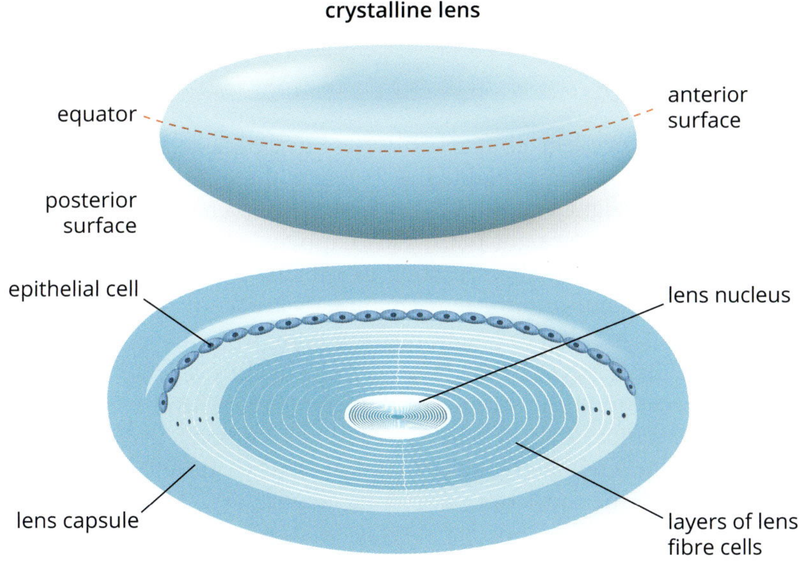

FIGURE 18.2.7 The internal structure of the eye lens

Accommodation by the lens

The lens of the eye adjusts to form a clearly focused image on the retina when an object is close or distant (Figure 8.2.10 on page 616). This ability, called **accommodation**, makes the vertebrate eye the incredible optical instrument that it is. No technologies have been invented with the property of flexibility like this for a single lens. Artificial devices use other solutions for getting an image into clear focus.

In the fovea, each cone cell connects to one neuron giving the greatest visual acuity. In other areas of the retina there may be several rods and cones connected to one neuron, so individual pinpoints of light cannot be distinguished as separate images, reducing the visual acuity.

The eye lens is sometimes called the crystalline lens because it is transparent and the bulk of its structure is crystallin protein molecules.

GO TO ➤ Section 4.3 page 181

GO TO ➤ Year 11 Section 3.3

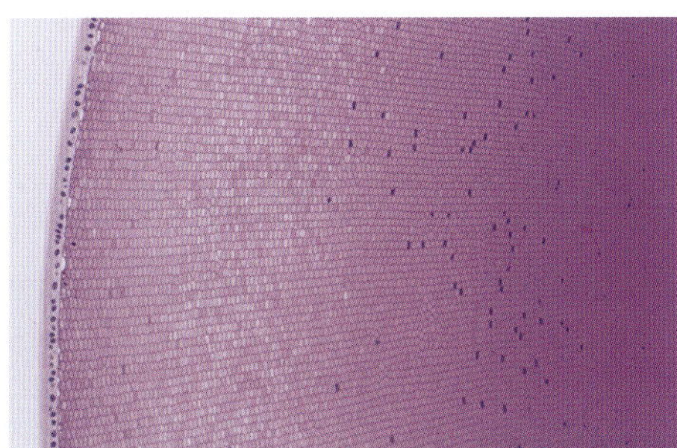

FIGURE 18.2.8 Light micrograph (LM) of a section through the lens of an eye, showing the arrangement of lens fibre cells. The tightly packed, orderly arrangement and orientation of the crystallin protein in the fibres contributes to the transparency of the lens. Some of the younger fibres still contain their nuclei (purple dots). These will be lost as the lens fibres mature.

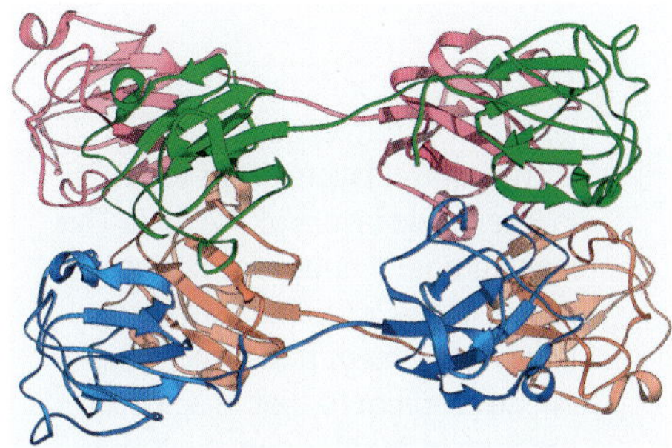

FIGURE 18.2.9 Molecular model of beta-crystallin, a protein found in the lens of the eye. The regular arrangement of the protein in the lens is thought to be responsible for its transparency.

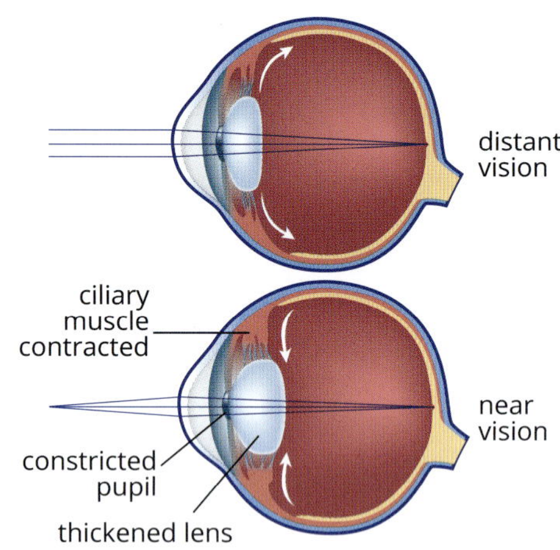

FIGURE 18.2.10 The lens changes shape by the action of the ring of ciliary muscle to accommodate distant and near vision.

Light rays emitted by, or reflected by, an object enter the eye and are refracted (bent) by a combination of cornea, aqueous and vitreous humour and the accommodating lens to reach a **focal point** on the retina. If the eye is working correctly, a clear, inverted image is formed from the combination of many focal points, such as the tree shown in Figure 18.2.11. This is a bit like having pixel resolution for a digital image such that it is seen as a single picture. Photoreceptors in the retina are stimulated and nerve messages sent to the brain for interpretation.

Images detected by the retina are upside down because light rays cross over before they reach the retina (Figure 18.2.11). From soon after birth, the brain has learnt to make automatic corrections of the inverted images to upright ones. The visual cortex of the brain will subconsciously turn the image upright and compare it with the memory bank of previous images.

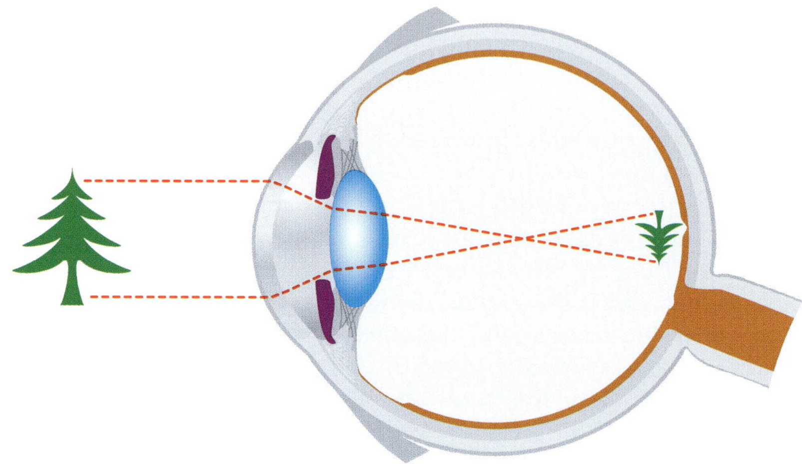

FIGURE 18.2.11 Images seen by the eye are inverted on the retina. The brain interprets them as being the right way up.

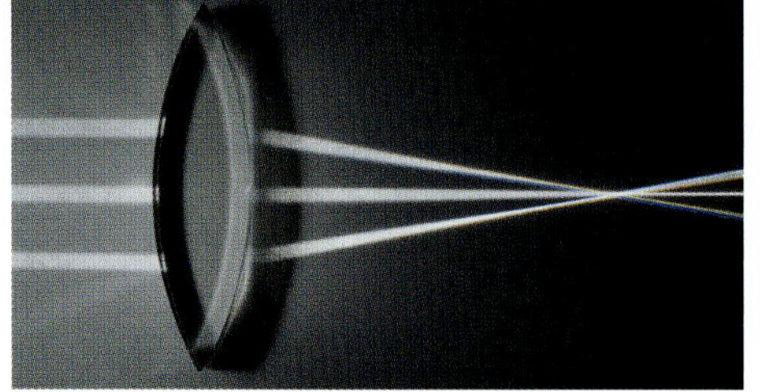

FIGURE 18.2.12 Refraction occurs when light passes into a different medium and changes its direction of travel.

When the natural eye lens changes thickness its power to refract light changes. **Refraction** of light is often explained as bending the light rays. Given that light always travels in a straight line, refraction is actually a change of direction of the light ray onto a different straight path. This happens whenever light enters a new medium and changes speed slightly, such as moving from air into the cornea or from the aqueous humour into the lens (Figure 18.2.12).

Experimental observation of a light beam shining from the side and at an angle into a clear prism like a biconvex lens (same shape as the eye lens) shows that there is a change of direction at the surface of the lens (Figure 18.2.12). A thick lens is able to refract rays more than a thin lens. The ciliary body (ring of muscle plus ligaments around the lens) alters the shape of the lens to make it thicker or thinner, which in turn adjusts the focus onto the retina (Figure 18.2.10).

When the ring of ciliary muscle relaxes, it flattens, tugs on the suspensory ligaments and pulls the lens into a thinner disc. This position works for viewing distant objects. For viewing objects that are close, the lens must be thicker to refract light rays more. To achieve this, the ring of ciliary muscle contracts back into a tighter, thicker bundle that moves closer to the lens. The suspensory ligaments therefore become looser, pulling less on the lens, which relaxes into a wider shape. Careful observation of the ciliary body in Figure 18.2.10 will make these changes easier to understand.

As people age, their lens becomes less flexible and does not accommodate into the thickest shape. This is called **presbyopia** and the outcome is that light rays are not refracted as much. Near vision, such as being able to focus on print in a book, deteriorates and older people frequently need to use spectacles (reading glasses) to aid their reading.

The refractive index of a human lens varies from approximately 1.406 in the central layers down to 1.386 in less dense layers of the lens. This index gradient enhances the optical power of the lens. The refractive index is a ratio of the speed of light in one medium to that in a second medium of greater density. It puts a numerical value on how much the light ray will be bent (refracted).

BIOFILE CCT L

Properties of light waves

Below are some useful terms to know when discussing light rays.

- Reflection and refraction—both refer to the way light behaves at a boundary between two media. With reflection, the light bounces back from the surface, like from a mirror surface. With refraction the light passes from one transparent medium into the next and is bent (changes its line of travel). Snell's law quantifies the angle of refraction by comparison with an imaginary normal line at 90° to the boundary—light passing into a denser medium (air to liquid) bends towards the normal line and vice versa.
- Convergent and divergent—light rays refract toward the normal and converge (come together) to a focal point or they diverge away from the normal and spread out, creating either a real or a virtual image (Figure 18.2.13).
- Focal length—the point where the converging light rays meet after refraction or reflection. Here an image will be clearly defined (in focus). In the eye it is measured from the centre of the lens to the retina.
- Convex and concave lenses: refers to the surface shape of a lens—convex curves out and concave curves in. The eye has a biconvex lens with outward curves on each side. The anterior (front) surface has a flatter curve than the posterior (inner) surface (Figure 18.2.7 on page 615).
- A convex lens—also known as a converging lens; a concave lens diverges light waves. The eye lens refracts light rays so that they converge.

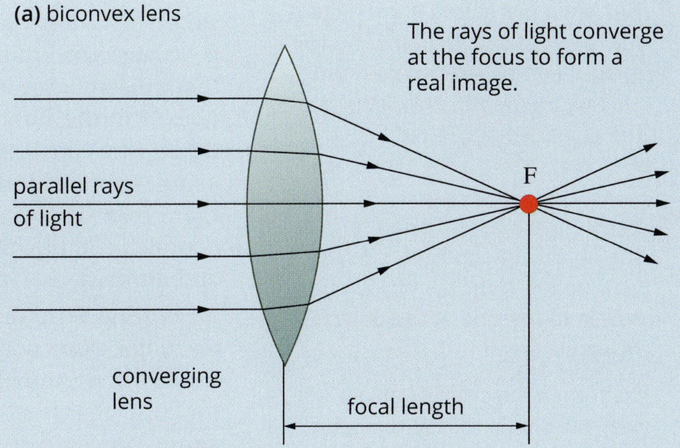

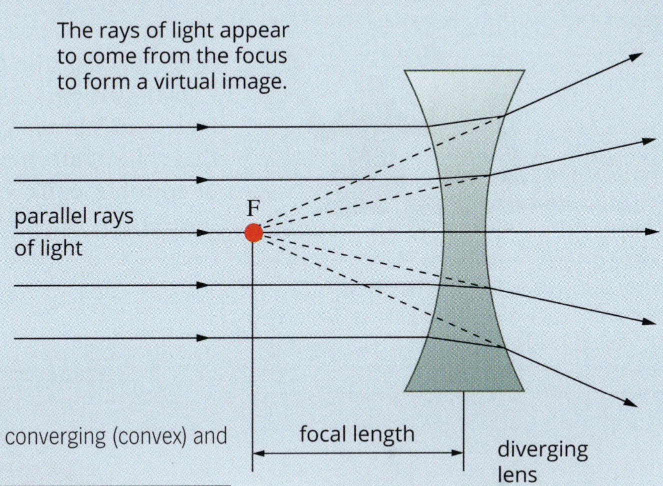

FIGURE 18.2.13 Light rays shining through (a) converging (convex) and (b) diverging (concave) lenses

Altering the size of the pupil

A coloured ring of two sets of muscles that form the iris changes the size of the pupil. The **dilator muscle** enlarges the pupil in dim light to gather as many light rays as possible and allow them to reach the edges of the retina where the most rod cells are present. The **sphincter muscle** makes the pupil smaller to restrict the light entering the eye when it is very bright. This is called the **light reflex** and it has two functions.

- Protecting the eye from bright light that may damage the retina. When a torch beam shines directly into the pupil, the light reflex rapidly closes the pupil to a pinprick size. This is one quick and easy way to check a person for concussion after a blow to the head.
- Constricting the pupil size directs the light to the fovea, the most sensitive area with the greatest density of cone cells. When a person changes from looking at a distant object to a close one, the pupil constricts to assist the lens to change focus. In this case the pupil change is known as the **accommodation reflex**. The word reflex is used to designate the nerve actions that happen automatically without conscious thought. There is still a slight time delay while the reflex action takes place.

Binocular vision enables depth perception

Having two eyes side by side means that the images formed in each eye and transmitted to the brain are slightly different. Because both human eyes are close together and face forward, the brain can combine the two similar viewed images into a single image based on a memory bank of experiences (Figure 18.2.14).

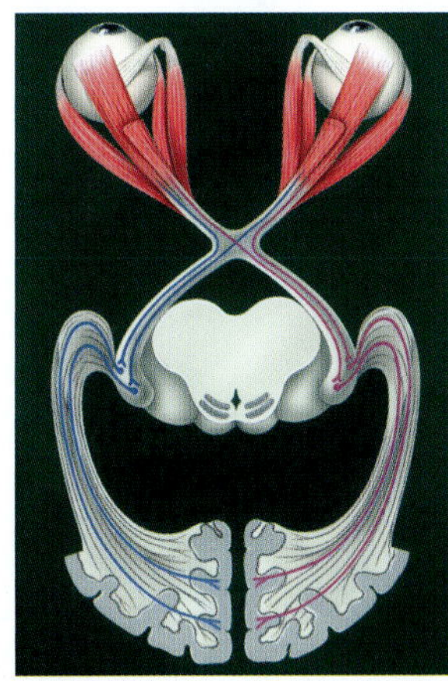

FIGURE 18.2.14 Visual pathways from eyes (top) to brain (bottom), seen from below. Light that falls on the right side of each retina (blue) coming from the left field of view passes as impulses to the right visual cortex of the brain via the optic nerves that cross at the chiasma. This is reversed for light that falls on the left side of each retina (pink). Nerve impulses are processed in the brain to create a single 3D image of what is seen by the eyes. The muscles that move the eyeball around can also be seen.

> Our brain has stored knowledge that allows us to combine the two images into one 3D image from judging the depth and distance of the object being viewed.

> People use averted vision to see objects in dim light, such as viewing the night sky by looking a little to the side of the object being observed.
>
> Light from the celestial object will then fall on the rods at the outer edges of the retina. It is also a safer way to view very bright objects.

This ability is called stereoscopic or **binocular vision**. The brain can interpret a single three dimensional (3D) image as being from a close or distant object. This is because the brain has stored memories of the sizes of objects and we understand that a distant object will appear small on the retina even if it is tall. We use clues from objects in the same field of view to judge relative sizes (Figure 18.2.15). We also detect movement using rod photoreceptors on the outer edges of the retina. So, by using slight head movements and our knowledge that close objects appear to move more than distant ones, we get extra clues as to distance. The brain processes the complex combination of all these messages and a judgement is made about distance and dimensions.

Beyond 60 m distance, **depth perception** is not very effective but it is certainly useful for closer activities like catching a ball or going down stairs. A person with sight in only one eye has more difficulty with estimating depth and distance, especially for close objects. Actions like catching a ball come automatically to adults who don't appreciate the complexity of the processes involved for the eyes and brain, but it is more obvious when watching a child learning the task for the first time.

FIGURE 18.2.15 On a salt flat in Bolivia, travellers create a visual illusion. Without familiar spatial clues from the immediate surroundings, the brain cannot interpret distance and depth correctly to determine the size of the visual images.

Colour vision and the retina

The retina has layers consisting of different types of cells (Figure 18.2.16). Closest to the back of the eyeball is the layer of photoreceptor cells, both rods and cones. Light must pass through the other layers first before it reaches the photoreceptor layer and stimulates these cells to send visual messages to the brain. The neurons that collect information from each photoreceptor are organised into two other layers of **bipolar cells** and **ganglion cells**. Bipolar cells relay messages and the nerve fibres from ganglion cells join up to form the optic nerve.

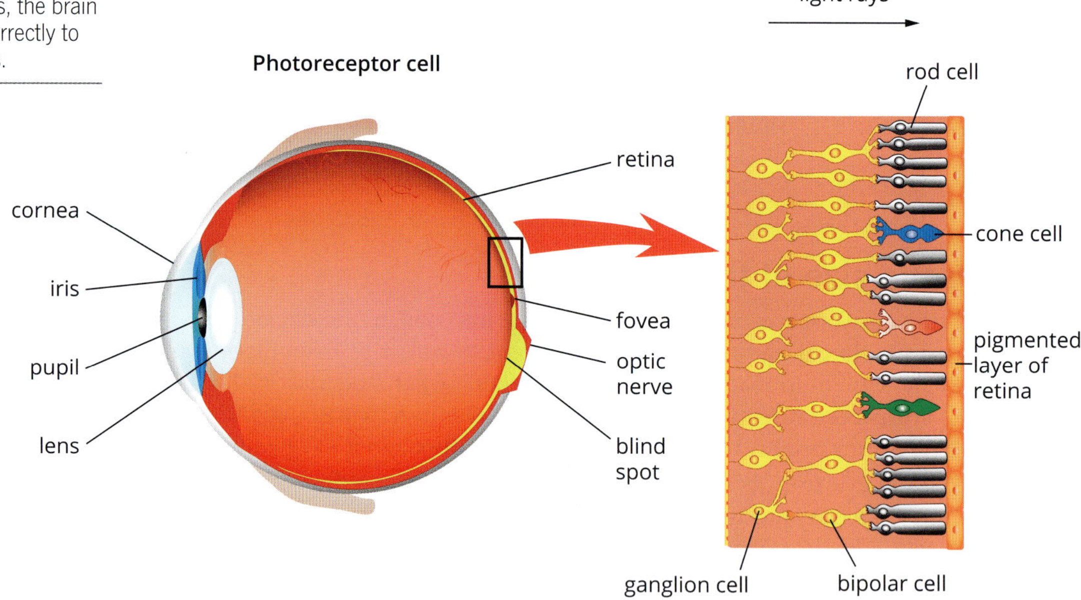

FIGURE 18.2.16 The structure of the retina: there are two types of photoreceptor cells in the layer at the back of the retina: rods for dim light, cones for bright light and colour vision. The other layers hold the nerve cells that collect messages and transmit them to the brain through the optic nerve. Each cone cell links to only one bipolar cell to increase visual acuity and only detects one colour—red, blue or green. Multiple rod cells link to a single bipolar cell for transmission of visual messages in dim light.

Photoreceptor cells contain light-sensitive pigments that convert stimulation by light into nerve messages which travel to the brain for interpretation. These visual pigments are made up of a retinal molecule joined to an **opsin** protein molecule. Retinal is derived from vitamin A and common to all photoreceptors.

There are different types of opsins:

- **rhodopsin**—found in rods and sensitive to shades of black, grey and white
- **photopsins**—the three types found in cones are:
  - blue-sensitive, around 430 nm wavelength
  - green-sensitive, around 550 nm wavelength
  - red-sensitive, around 600 nm wavelength.

Each cone contains only one of these three types of photopsins. For colours other than these three, a mix of cone stimulations by different wavelengths of light alerts the brain to hues between red, green and blue, depending on the degree that each type of opsin is stimulated (Figure 18.2.17). Equal stimulation of the three is recognised by the brain as white. No stimulation of any opsin is seen as black. Much of colour detection relies on interpretation by the brain based on previous experience and memory.

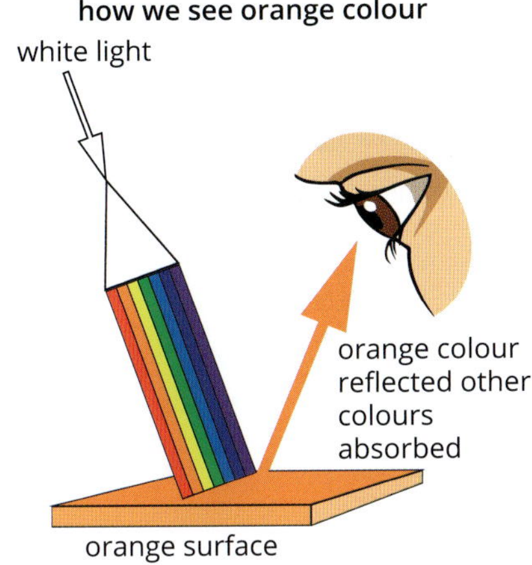

FIGURE 18.2.17 How we see orange colour—the orange wavelength of visible light is reflected to the eye. The object absorbs all other colours. The orange wavelength falls in the range detected by both the red and green cone cells, so the brain interprets that it must be a colour that falls between red and green on the spectrum.

When a photoreceptor in the retina receives light, the pigment absorbs the light energy and undergoes a structural change that breaks the bonds between the retinal and opsin molecules (Figure 18.2.19). This change triggers an electrochemical nerve impulse that passes to the bipolar cells then to the ganglion cells whose axon extensions group together to form the optic nerve. After the impulse has been sent to the brain, enzymes in the photoreceptors convert the molecules back into the receptive state again. There is a time lag for the chemical conversion process, which is why we need time for our eyes to adjust when moving from bright to dim light. **Colour blindness** can now be understood as being a disorder of one or more of the cone cells and their three types of colour-sensitive photopsins (Figure 18.2.20 on page 620).

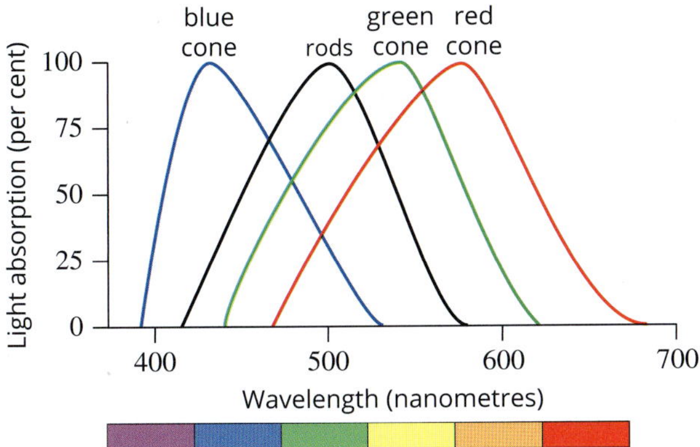

FIGURE 18.2.19 Light absorption by the four opsin pigments in the rods and cones of the human retina

BIOFILE CCT

Binocular and monocular vision

Binocular vision is the ability to combine two images, perceived through two forward-facing eyes, into one 3D image. Binocular vision, perception of movement and previous experience tells us if an approaching vehicle is close or far away, moving fast or slow—very useful when crossing a road or driving a car. From birth, a child starts learning how to interpret images but has to build considerable experience to be effective with depth perception. This can be observed by watching a baby learning to reach for an object or trying to catch a ball or an infant negotiating stairs, something that may be done with a backwards caterpillar crawl at first.

Most birds, lizards and some mammals have **monocular vision**—their eyes are on each side of their head. This gives them a greater field of view, which is useful for spotting predators or prey. However, they have poorer depth perception than humans. Many species have a combination of monocular and binocular vision (Figure 18.2.18). Humans use monocular vision in dim light when they are aware of movement out of the side of an eye.

FIGURE 18.2.18 Chameleons can swivel each eye independently, using a wide field of monocular vision to search for food. Once a prey has been located, they switch to binocular vision to pinpoint the location so they can catch it.

Because the rhodopsin in the rods is so sensitive to light, the visual pigment is in an excited state for most of the day. Overnight, enzymes restore the molecule to its resting state.

It is commonly said that eating carrots improves night vision. Whether this statement is evidence-based or not, it is true that carrots contain vitamin A (indicated by the orange colour) which is used in the body to make retinal. Retinal is part of rhodopsin, the light sensitive pigment in rods, cells that are most sensitive to dim light.

Rhodopsin is sometimes called visual purple because it is stimulated by blue-green light (around 500 nm) allowing us to see shades of grey in dim light.

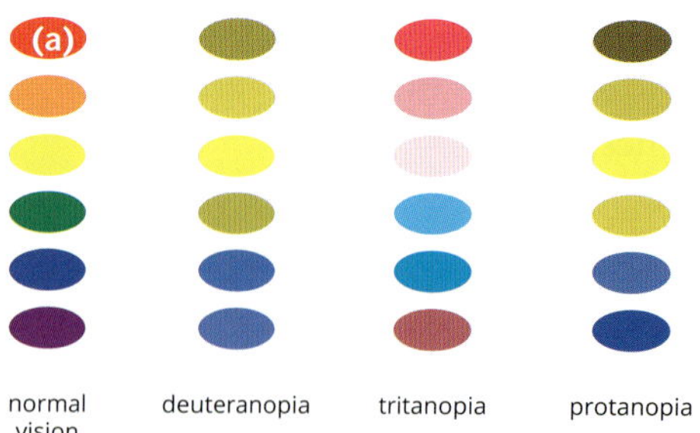

FIGURE 18.2.20 (a) Three of the types of colour blindness compared to the same colours in normal vision. (b) A sign viewed with normal colour vision. Someone with colour blindness would see the colours shown differently. (c) The same sign as it would be seen by a person with protanopia. Protanopes lack the long-wavelength cone cells in the eye's retina and are unable to distinguish between colours in the green/yellow/red section of the visible spectrum. (d) The same sign as it would be seen by a person with tritanopia. Tritanopes lack the short-wavelength cone cells in the eye's retina and are unable to distinguish between colours in the blue/yellow section of the spectrum.

REFRACTIVE ERRORS AND THEIR CAUSES

Many visual disorders are inherited and have a relatively minor impact on daily lives because there are convenient technologies to manage them. For example, being long- or short-sighted is often an inherited condition of the eye's structure that people are born with and have diagnosed early in childhood. Other visual disorders develop naturally with age. There are some that are more severe and alarming to the individual. We cannot restore natural sight but in most cases technologies can be used to assist with improved vision.

Focusing disorders of the lens

The lens and eyeball are subject to several disorders, known as **refractive errors**, which can prevent clear focus of an image on the retina (Figure 18.2.23). In many people, the lens and cornea do not properly focus light onto the retina and this causes blurry vision for either close or distant objects. Often the problem is due to the shape of the eyeball or the curve of the cornea and is an inherited disorder. It is worth remembering that one eye may have a different disorder to the other eye on the same person.

Myopia is a condition that causes distant objects to appear blurry and is commonly referred to as near-sightedness or short-sightedness (Figure 18.2.21). Myopia is caused when light focuses in front of the retina. An eyeball that is longer than normal or a cornea that is too curved are the usual causes. It can be corrected by placing a concave lens in front of the eye (either in the form of eyeglasses or contact lenses) that causes the light to diverge before it enters the eye. **Laser surgery** to reshape the cornea is another option.

Hyperopia occurs when light focuses behind the retina and causes near objects to appear blurry, a condition also called far-sightedness or long-sightedness (Figure 18.2.21). An eyeball that is shorter than normal, especially if it occurs in younger people, is a common cause. Hyperopia can also be due to a cornea that is too flat. It is corrected by placing a convex lens in front of the eye to cause the light to begin to converge before it enters the eye. Laser surgery to reshape the cornea is another option.

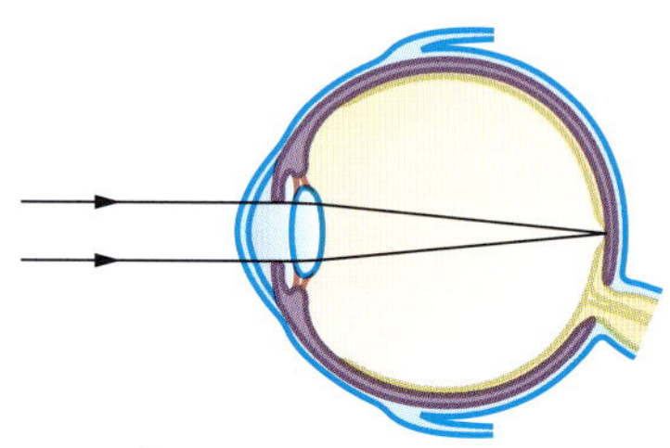

normal eye

long sight—shorter than normal eyeball or cornea curve is too flat, light rays do not converge enough and focus falls past the retina

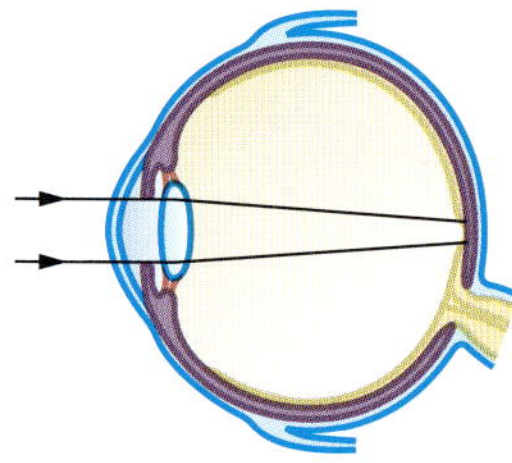

short sight—longer than normal eyeball or cornea too curved; light rays converge too soon and focus falls before the retina

FIGURE 18.2.21 The visual disorders of hyperopia (long sight) and myopia (short sight) are usually due to shorter or longer eyeballs. Alternatively, the cornea could be too curved or not curved enough. The lens cannot accommodate to set the focal point on the retina for a clear image.

Presbyopia occurs as people age and the lens begins to lose elasticity so that it no longer changes shape as easily to achieve the widest shape for maximum refraction. Like hyperopia, this causes difficulty in focusing on nearby objects and can also be corrected with a convex lens.

Astigmatism is an imperfection in the curvature of the cornea or the lens (Figure 18.2.22). Ideally the eye has a cornea and lens that are smooth and curved equally in all directions, helping to focus light rays sharply onto the retina at the back of your eye. However, if the cornea or lens isn't evenly curved, light rays aren't refracted properly creating distorted images like those in fun house mirrors. Most people are born with some degree of astigmatism but for many it doesn't cause significant visual problems. Children may not be aware they have astigmatism because their view of the world has always been this way. It can cause problems with learning in school and playing sport if left uncorrected. Fortunately the technology of making lenses, or the use of laser surgery, is sophisticated enough to correct most cases of astigmatism.

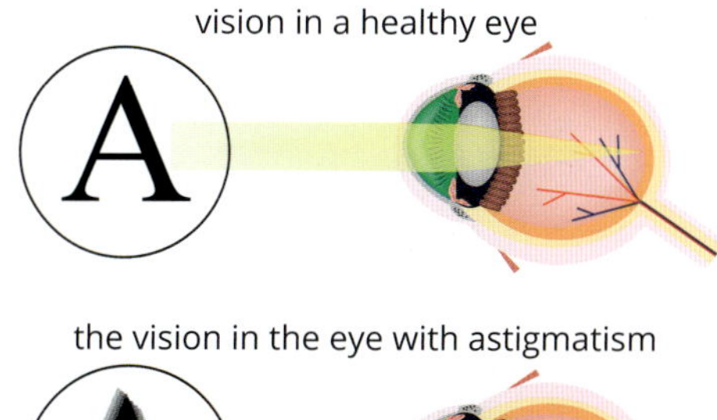

FIGURE 18.2.22 Impaired vision due to astigmatism

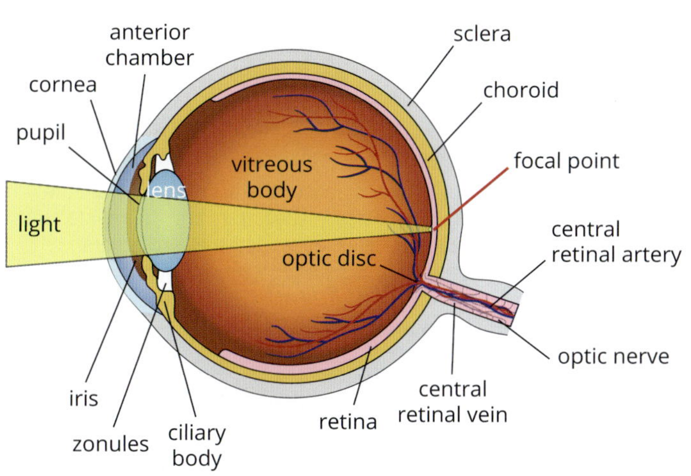

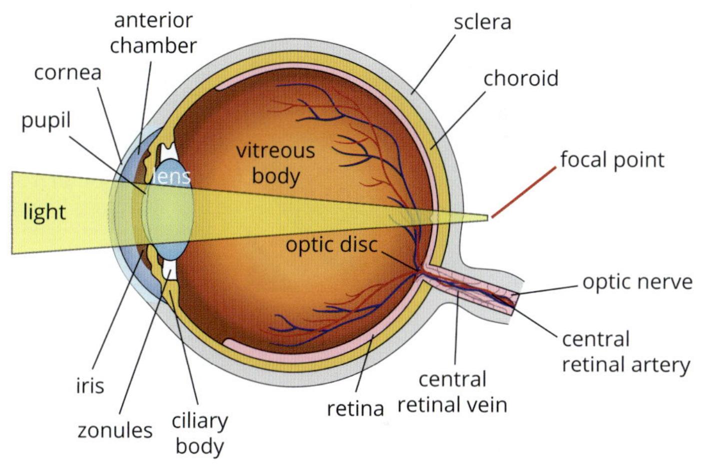

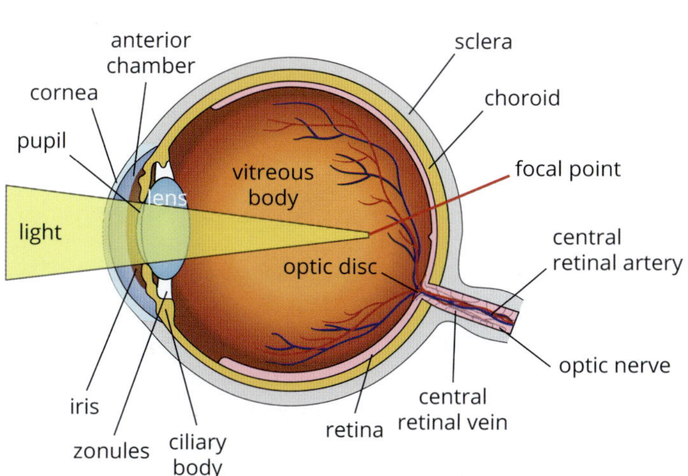

FIGURE 18.2.23 The refractive errors of hyperopia, myopia and astigmatism

TECHNOLOGIES TO ASSIST WITH VISION LOSS FROM REFRACTIVE DISORDERS

There is no complete cure for refractive errors yet, but there are ways to improve vision for many people with these visual disorders. These include:

- regular professional eye tests to detect refractive errors or other changes to the eye (Figure 18.2.23 on page 621)
- wearing **spectacles** (also known as eyeglasses)—a simple and safe way to correct vision problems when professionally prescribed; they are most useful in cases of hyperopia, myopia, presbyopia and astigmatism. Concave and convex lenses are shaped to correct individual needs (Figure 18.2.24a, b)
- wearing **contact lenses**—these fulfil the same function as spectacles but are worn directly on the surface of the cornea (Figures 18.2.24c and 18.2.25)

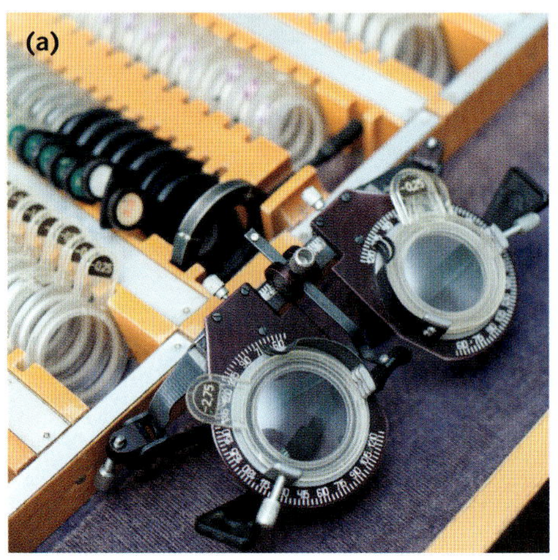

FIGURE 18.2.24 (a) A specialised optometry frame and a variety of lenses are used to prescribe the corrective lenses for refractive eye disorders. (b) Prescription lenses in spectacles bring print into focus. (c) Wearing prescribed contact lenses is an alternative to spectacles for refractive eye disorders.

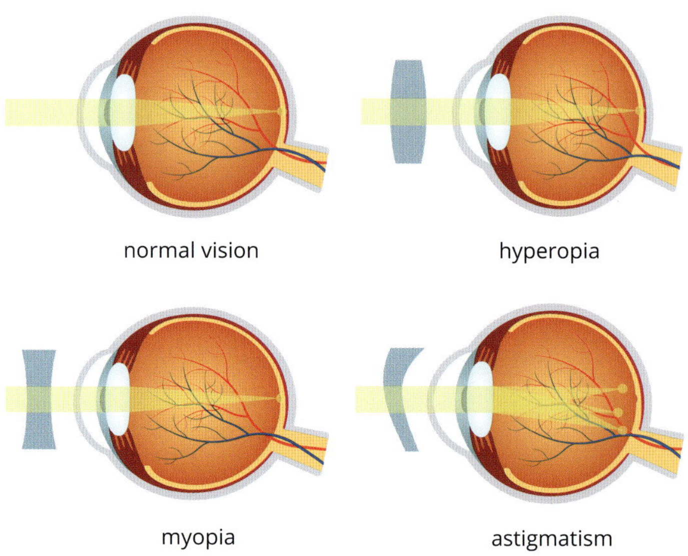

FIGURE 18.2.25 Correction of common visual disorders using prescription lenses.

- having laser surgery—using a laser beam to change the shape of your cornea (Figure 18.2.26). Laser techniques vary in the way they lift the outer corneal epithelium and replace it after the surgery. In all types of laser eye surgery, a computer-programmed excimer laser beam is used to remove microscopic amounts of tissue so that the cornea is reshaped with a new curve that will refract light more accurately into the pupil. The excimer laser is a cool laser that doesn't burn tissue but vaporises small amounts under precise control from the computerised technology.

Each eye can be treated in less than 10 minutes but the laser must have been carefully programmed for each individual patient. In myopia, the central peak of the cornea has to be flattened by the right amount. For hyperopia, the laser is applied to the edges of the cornea to increase the curvature of the cornea. Astigmatism is corrected by sculpting the cornea into a more even curve.

- having intraocular lens surgery to treat cataracts—the natural lens is replaced with a small plastic lens (intraocular lens).

Effectiveness of spectacles and laser surgery

Both methods of assisting visual disorders are very effective when prescribed and prepared by experts. Spectacles can be a nuisance for people playing some sports, in outdoor activities or certain occupations. In these cases, contact lenses are a commonly used alternative. Annual eye checks are important to determine if the lens prescription needs to be changed. This can be expensive but is worthwhile to regain full vision.

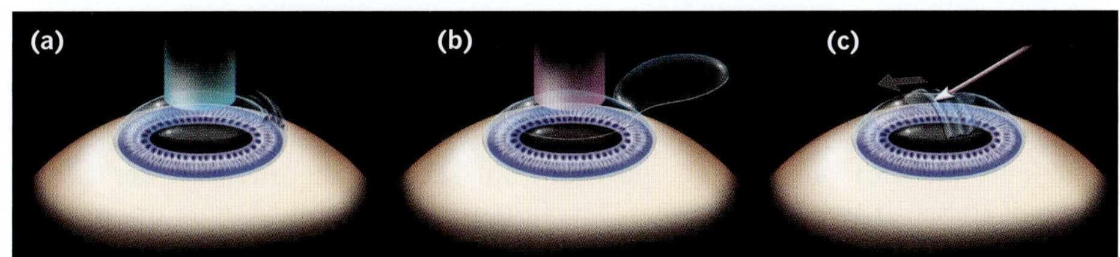

FIGURE 18.2.26 Lasik, laser-eye surgery, which treats myopia, astigmatism, hyperopia and presbyopia. (a) The first laser cuts a flap in the cornea. (b) The second laser remodels the cornea to correct the pathology. (c) The cornea flap is put back in place.

Laser surgery is expensive although within the range of affordability for most working people in western countries. It carries a small risk and should only be conducted by a highly skilled practitioner under sterile conditions and with high quality equipment (Figure 18.2.26). The technology has become highly accurate in recent years and the cost has reduced. It usually provides an effective and permanent solution to the visual disorder, at least until some of the age-related disorders start to develop.

OTHER VISUAL DISORDERS AND THEIR CAUSES

Not all visual disorders are caused by refractive errors. Many are caused by injury, ageing or the inheritance of a genetic disorder.

Cataracts

As described earlier, someone who develops cataracts will view things as being blurry or foggy and less colourful, like looking through a dusty or lightly-frosted window. In developed countries with good health care, most cataracts are age-related and emerge gradually, so early deterioration of vision may not be noticed. The crystallin protein starts to denature and is no longer completely transparent (Figure 18.2.27). Light passing through a cloudy part of the lens does not focus as clearly on the retina. A clear image may have a consistently blurred patch in the same place.

It is thought that increased exposure to UV light may hasten the development of cataracts.

Day surgery can replace the natural eye lens with an intraocular lens manufactured to suit each individual (Figure 18.2.28).

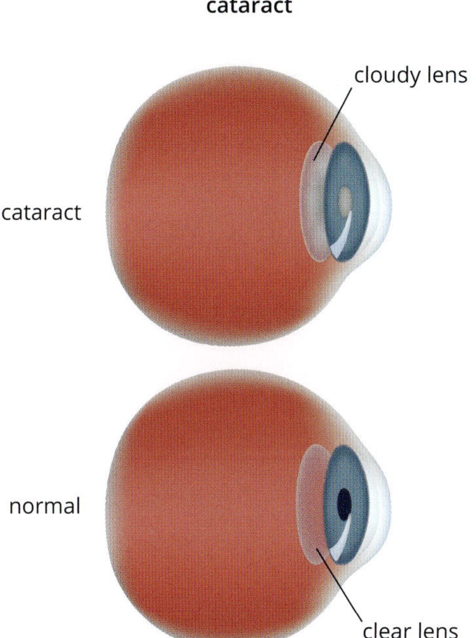

FIGURE 18.2.27 A cataract is a cloudy area on the lens where crystallin proteins have started to become less transparent.

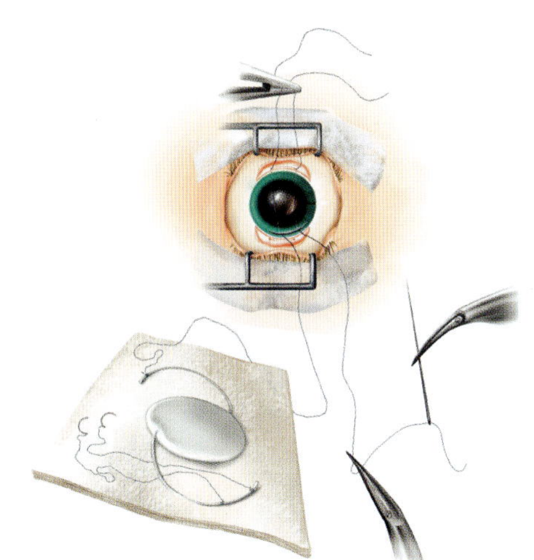

FIGURE 18.2.28 The technique for correcting a cataract by implanting an artificial lens, called an intraocular lens (IOL)

Glaucoma

Glaucoma is a major cause of blindness for people over the age of 60 or with a family history of the problem. It happens when fluid builds up in the eye, putting pressure on the optic nerve that may cause permanent disruption to transmission of visual messages to the brain (Figure 18.2.29 on page 624).

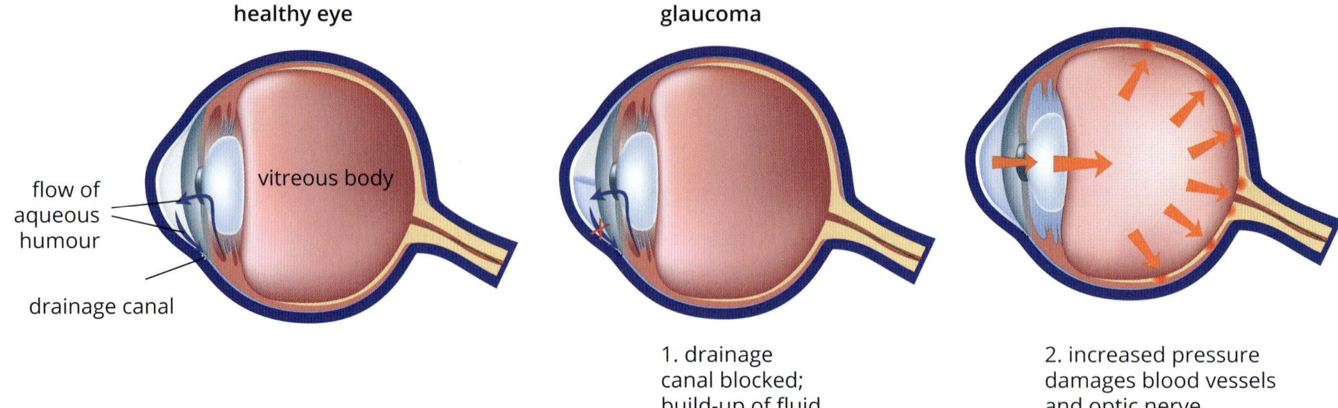

FIGURE 18.2.29 Glaucoma is a serious visual disorder caused by excess fluid inside the eye.

The eye continually refreshes the aqueous humour. When new fluid drains into the eye, the same amount should drain out keeping the internal pressure of the eyeball stable. If the drainage is not working properly, fluid builds up and the increased pressure starts to cause some nerve fibres to die. The result is minor blind spots in the vision at first, leading to increasing damage and permanent blindness if not treated. Glaucoma is serious and should be treated as early as possible. Treatments available include medications, laser surgery to help drain fluid from the eye or surgery to insert technology such as a drainage tube device.

Retinal tear or detachment

Partial or full detachment of the retina is another common age-related disorder of the eye. The most likely cause is shrinkage of the jelly-like ball of vitreous humour. When this happens it can pull on the retina enough to tear it. The person may report seeing flashes. If fluid passes through a retinal tear it can lift the retina off the back of the eye—much like damp wallpaper peeling off a wall. When the retina pulls away from the back of the eye like this, it is called a **retinal detachment** and vision becomes blurry. This is a very serious problem that almost always causes blindness unless treated with specialised surgery.

Some people have an increased risk for retinal detachment, including those who are near-sighted, have a family history of retinal detachment, have had cataract, glaucoma or other eye surgery or a serious eye injury.

Retinopathy

Retinopathy is damage to the retina, often caused by changes to the blood vessels in this region. The reasons for it are not well understood yet. There are various forms of retinopathy as described below:

- diabetic retinopathy—with diabetic patients, controlling blood sugar and blood pressure can slow or halt the progress of the disorder. There are some medications that can alleviate the damage (Figure 18.2.30 on page 625)
- retinopathy of prematurity (ROP)—a premature baby born at 30 weeks or less has a greater risk of retinopathy. With ROP, unwanted blood vessels grow on the baby's retina. ROP often goes away as an infant grows but if not outgrown, it must be treated
- hypertensive retinopathy—lowering the blood pressure with diet, exercise and medication can stop ongoing damage to the retina. Any existing damage is usually permanent

- central serous retinopathy—fluid accumulates in the membrane behind the retina, causing the layers to separate. The result is blurred vision or poor night vision. Most cases go away without treatment in three to four months. For persistent cases, laser treatment is often used and full vision can return within six months.

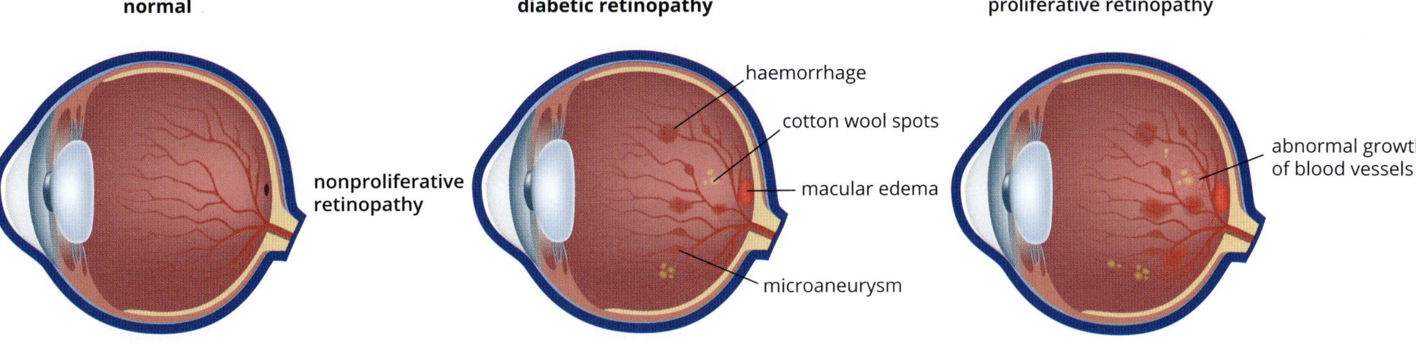

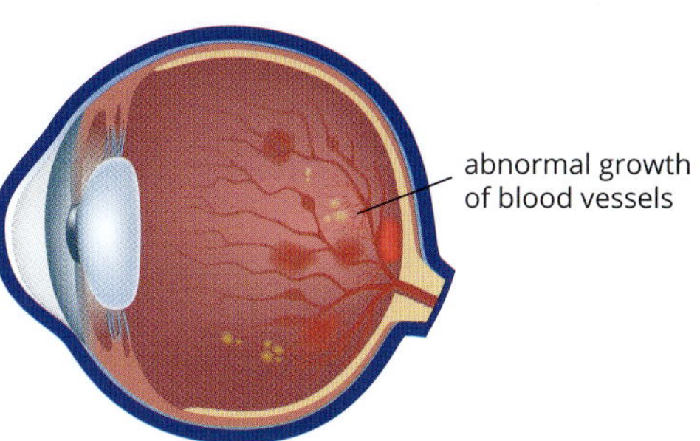

FIGURE 18.2.30 Retinopathy is a visual disorder of the retina that has several causes. Usually it is a problem with the blood vessels of this region in the eye.

Macular degeneration

Macular degeneration is yet another condition that is generally age-related and is most common in people over 50 years of age. There is vision loss due to a distortion or absence of the central field of retinal vision, an area known as the fovea or macula. The peripheral (side) vision is usually not affected and macular degeneration does not tend to cause complete blindness (Figure 18.2.31). Some tasks that require fine image details, like reading and craftwork, will be affected.

FIGURE 18.2.31 A scene as viewed by a patient with dry macular degeneration. This visual disorder is usually age-related and affects the macula (fovea) area of the retina where the clearest colour images are formed. It leads to slow loss of vision in the central field of view. While macular degeneration rarely causes a complete loss of vision it can cause enough vision loss for a person to be considered legally blind.

Two types of the disorder are recognised:

- dry macular degeneration—the most common type (about 90%) results in a gradual loss of central vision.
- wet macular degeneration—abnormal growth of blood vessels under the retina, which leads to fluid leakage and results in sudden loss of vision. This is a very frightening disorder.

Damage to the cornea

Keratopathy is the term used for damage to the cornea at the front of the eye. It can become swollen or thin, develop an irregular curve and lose light refractive ability. The causes of this problem include eye injury, herpes virus infection of the eye, severe bacterial infection or an inherited condition. A corneal transplant is possible to provide a new window to the world for the affected eye. Only the cornea is transplanted and, unlike organ transplants, there are few problems matching donor and recipient. In fact, corneal transplants are the oldest type and the most common of transplants. There are at least 2000 performed in Australia every year and they have about a 90% success rate. Corneas can also be grown from live cell cultures in a laboratory.

Colour blindness

GO TO ➤ Section 5.2 page 210

Colour blindness ranges from severe to mild when one or more of the three types of colour-detecting cone cells is absent, non-functional or detects a different colour than normal. Some people with mild colour disorders can see colours normally in good light but have difficulty in dim light. Others cannot distinguish certain colours in any light. The most severe form of colour blindness, in which everything is seen in shades of grey, is uncommon. Colour blindness usually affects both eyes equally and remains stable throughout life. It is usually genetically inherited and something that you have from birth although it can be acquired later in life in rare cases.

The symptoms include an inability to tell the difference between shades of the same or similar colours, particularly between red and green (protanopia) or blue and yellow (tritanopia) (Figure 18.2.20 on page 620). Except in the most severe form, colour blindness does not affect the visual acuity. Men are at much higher risk of being born with colour blindness than women because it is a sex-linked inheritance. It is more common among men of northern European descent and it is estimated that 10% of males have some form of colour vision disorder. You learnt about the inheritance pattern of colour blindness in Chapter 5.

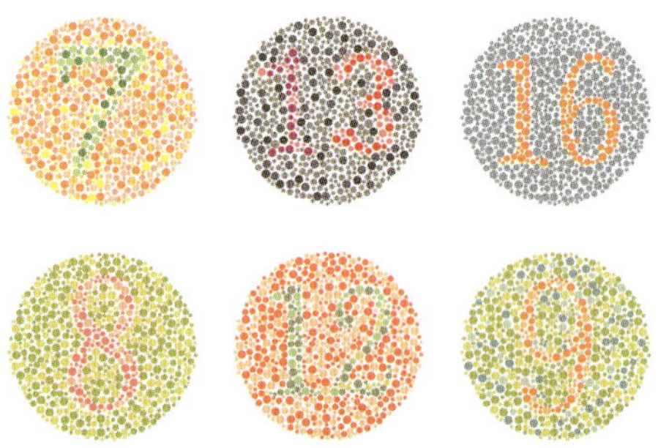

FIGURE 18.2.32 Ishihara test images for colour blindness. A person with colour deficiencies will not be able to see some, or all, of the following numbers amongst the dots: 7, 13, 16, 8, 12, 9. Young children can be asked to trace a coloured pathway through the dots or recognise a picture rather than numbers.

Diagnosis of a colour disorder is usually done with a set of Ishihara test images (Figure 18.2.32), especially if there is a family history of the problem. There is no cure or treatment for inherited colour blindness but mostly it does not cause significant disability. There are special contact lenses and glasses that may help people with colour disorders to tell the difference between similar colours.

BIOLOGY IN ACTION   

Fred Hollows and the intraocular lens

Fred Hollows (1929–1993; Figure 18.2.33) is well known for his dedicated humanitarian work in eye health care, both in Australia and developing countries. He personally pioneered affordable eye care in countries like Vietnam, Nepal, Burma, India, Sri Lanka, Bangladesh and Eritrea, as well as in impoverished Indigenous Australian communities. Today the Fred Hollows Foundation continues his work with a goal of ending avoidable blindness for Indigenous Australians and for people in 25 developing countries. In 2016 the foundation completed more than one million eye treatments worldwide and two million since their work began.

When his work as an ophthalmologist in remote Australian communities began, he found many people suffering from an eye disease called blinding trachoma. Trachoma starts as a bacterial infection of the conjunctiva (Figure 18.2.34). Repeated infections lead to scarring and inverted eyelids. When this happens, the eyelashes continually rub the cornea and it can lead to irreversible blindness from corneal damage. The condition is more likely to develop and spread in hot dusty areas where flies carry the infection and there is inadequate water for personal hygiene. A combination of these conditions, poverty and lack of access to health care and education all contribute to eye problems. Children and women are at highest risk. Fred implemented programs to treat the infection, improve sanitation and awareness. The same disorders either do not arise or are easily prevented in places where early diagnosis and intervention with antibiotics is available.

It is estimated that around the world at least 32 million people are blind and 90% of these live in developing countries. Here cataracts are a leading cause of blindness and not just in older people, as is the case in Australia. Fred's belief was that visual disorders, such as cataracts, are largely preventable or treatable. Access to the technologies at an affordable cost was the main barrier. He worked to establish intraocular lens factories in Eritrea and Nepal, and to train local people to do the cataract surgery. This initiative has reduced the cost of cataract surgery in these two countries to just under $25 per person. The foundation focuses on continuing and expanding his work.

Another breakthrough in affordable technology from the Fred Hollows Foundation is an inexpensive, pocket-sized ophthalmoscope, made for diagnosing eye diseases in the developing world. It costs around $10, is lightweight and solar powered. Early eye examination by local health care workers with the ophthalmoscope will detect the most common visual disorders in poor communities before they progress to blindness.

FIGURE 18.2.33 An Australian stamp issued in 1995 to celebrate Fred Hollows and his work in eye health care, including pioneering work for affordable intraocular lenses in developing countries.

FIGURE 18.2.34 A Ugandan man with a severe bacterial eye infection that has already damaged his right eye leading to trachoma. Blindness can result if the infection and resulting condition is not treated.

18.2 Review

SUMMARY

- The human eye has structures that can form a focused image on the retina where photoreceptor cells are stimulated to send messages to the brain.
- Formation of a clearly focused image on the retina depends on a flexible lens that accommodates to change its thickness for the appropriate refraction of light rays if they come from a distant or near object.
- Visual acuity is a measure of how clearly a point can be distinguished from another close point.
- Some knowledge of properties of light waves is important to understanding how the eyes function.
- Visible light is part of the electromagnetic (EM) spectrum with wavelengths between those of infrared and ultraviolet (UV) radiation. All EM waves are transverse and travel at the speed of light without displacement of particles.
- The properties of reflection and refraction for light waves are important in acquiring and forming images.
- The retina has two types of photoreceptor cells with opsin molecules that are sensitive to light:
  - rods that are more numerous towards the edges of the retina and allow some black and white vision in dim light; they contain a light-sensitive pigment called rhodopsin
  - cones that are more concentrated in the centre of the retina, are used for sharp images in bright light and to detect colours. Three types of light-sensitive pigments in different cones detect red, blue and green using photopsins.
- Within the visible light range of wavelengths there are seven colours that can be detected by the human eye.
- Colour is determined by the wavelength and frequency of the light energy, with red having the longest wavelength.
- The eye's natural lens is a living structure made of cells containing transparent proteins called crystallins.
- The iris controls the opening of the pupil to regulate the amount of light entering the eye.
- Humans have binocular vision that allows the brain to interpret distance and depth of objects being viewed.
- Visual loss is caused by a variety of disorders, many of them inherited and some that are acquired with age.
- Refractive error disorders of the lens and/or cornea have technologies to assist with vision loss, including spectacles, contact lenses or laser surgery. Each individual should have a technology prescribed to their specific needs.
- The technologies do not completely restore normal vision but they are very effective in supporting near normal ability.
- There are disorders that cause serious vision loss and do not yet have effective treatment, particularly in developing countries where health care is less accessible.
- Advances in cell culture research may offer a way forward for assisting with visual problems that are not assisted by current technologies.

KEY QUESTIONS

1 Match the structure to its function in the human eye.

Structure	Function
cornea	black layer surrounding eyeball (between retina and sclera) that absorbs stray light and holds a rich supply of blood vessels
iris	protects the eye surface and refracts light rays through the pupil
choroid	coloured ring of muscle that contracts or dilates to regulate size of pupil and control amount of light entering the eye
fovea	transmits nerve messages from eye to brain on opposite side of the body, that is, the left eye sends messages to right side of brain
optic nerve	central part of the retina with the highest density of cones for clearest colour vision in good light

2 Describe two functions of the retina.

3 Draw a flow chart for the sequence of events between a light ray travelling from text on a page and a person's awareness of seeing the written text.

4 Explain how the eye detects different colours.

5 Discuss the value of having binocular vision.

6 If the cornea is damaged, explain the impact of this damage on a person's vision.

7 List four causes of visual disorders. Identify which part of the human eye is affected for each and state if it can be corrected or assisted.

8 Compare the following two technologies that could be used to assist with some visual disorders: spectacles and laser surgery.

18.3 Loss of kidney function

The **kidneys** are vital organs of the human excretory system (Figure 18.3.1). As such, they are essential for removal of toxic wastes from the circulatory system and to maintaining homeostatic balance in the body. You learnt about homeostasis in Chapter 14.

GO TO ➤ Section 14.1 page 502

Processing of the blood by the kidneys must be continuous for good health. Any kidney disorders lead to a myriad of health problems and can eventually be fatal. The technology of **renal dialysis** maintains life in the case of complete **renal failure** (kidney failure) but is an expensive, imperfect measure and has a significant negative impact on quality of life. Even a kidney transplant is not regarded as a complete cure.

Renal is a descriptive word (adjective) for anything to do with the kidneys. For example, the renal artery is the vessel bringing blood to each kidney and renal failure is when the kidneys are not filtering the blood adequately.

STRUCTURE AND FUNCTION OF THE KIDNEY

The human kidneys are a pair of bean-shaped organs that regulate concentrations of soluble substances in the blood, maintain a healthy water-salt balance for the body and remove **nitrogenous waste** from the blood to be excreted in **urine** via the bladder. The two kidneys are at the back of the abdominal cavity on the right and left sides in the small of the back, below the rib cage (Figure 18.3.2). Blood flow to the kidneys is always kept high, because they are so important in maintaining the stability of the internal environment. Although kidneys are only about 1% of body tissue, they receive approximately 25% of the body's blood flow and filtering of the blood occurs on average 12 times per hour. Blood flow to the kidneys is directly from the liver and any discussion of kidney disorders should start with the liver because it is here that waste products are prepared before delivery to the kidneys.

The pair of kidneys has two main functions: elimination of nitrogenous wastes and osmoregulation (water-salt balance) for the body.

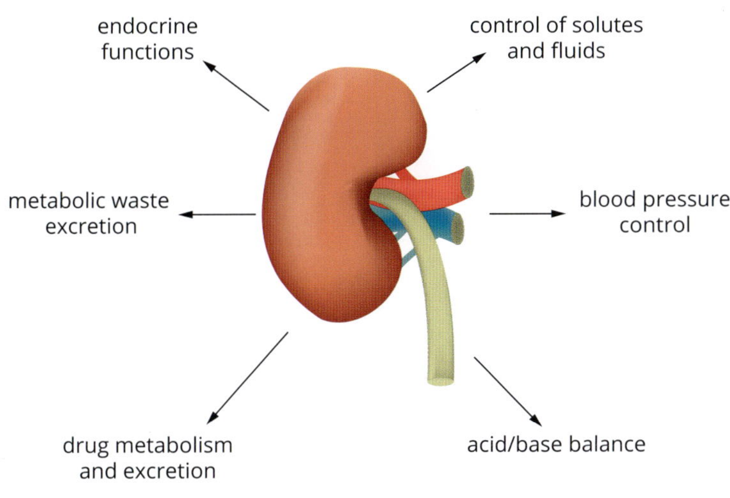

FIGURE 18.3.1 The functions of the kidney

+ ADDITIONAL

The liver prepares wastes

The liver in mammals performs many different functions and has a central role in the maintenance of a stable internal environment. As part of this role, it is responsible for preparing various substances for excretion. It detoxifies a variety of harmful chemicals such as alcohol and some drugs. It is also responsible for removal of amino acids (from cellular and dietary protein waste) and natural break down of nucleotides (from waste DNA and RNA) that releases **ammonia** (NH_3; toxic). In humans, the ammonia is converted largely into **urea** (CH_4N_2O; soluble and less toxic than ammonia). The waste products from the detoxification processes travel in the bloodstream to the kidneys for excretion in the urine.

To maintain good health it is vital for the liver and kidneys to function properly and in partnership.

> Kidneys function to excrete waste using the processes of filtration, reabsorption, secretion and elimination.

> The functional units within each kidney are microscopic nephrons, each with a filtering unit and long tube. There are around a million nephrons in each adult human kidney.

The kidney excretes wastes

The kidneys of all vertebrates function by filtration of blood; **reabsorption** of useful substances back into the blood, active secretion of extra wastes into the **filtrate** to form urine and then elimination of urine from the organism. Within the kidney, there are large numbers of microscopic functioning units called **nephrons** (Figures 18.3.2 and 18.3.3). Mammals have the most complex nephrons, with approximately one million nephrons in each kidney; all within a vital organ that is roughly fist-sized in an adult human. Each kidney has three regions, **cortex**, **medulla** and **pelvis**, as shown in Figure 18.3.2. The nephrons are tightly aligned side by side and even though they are microscopic, the parallel alignment gives a faintly striped appearance inside a dissected kidney.

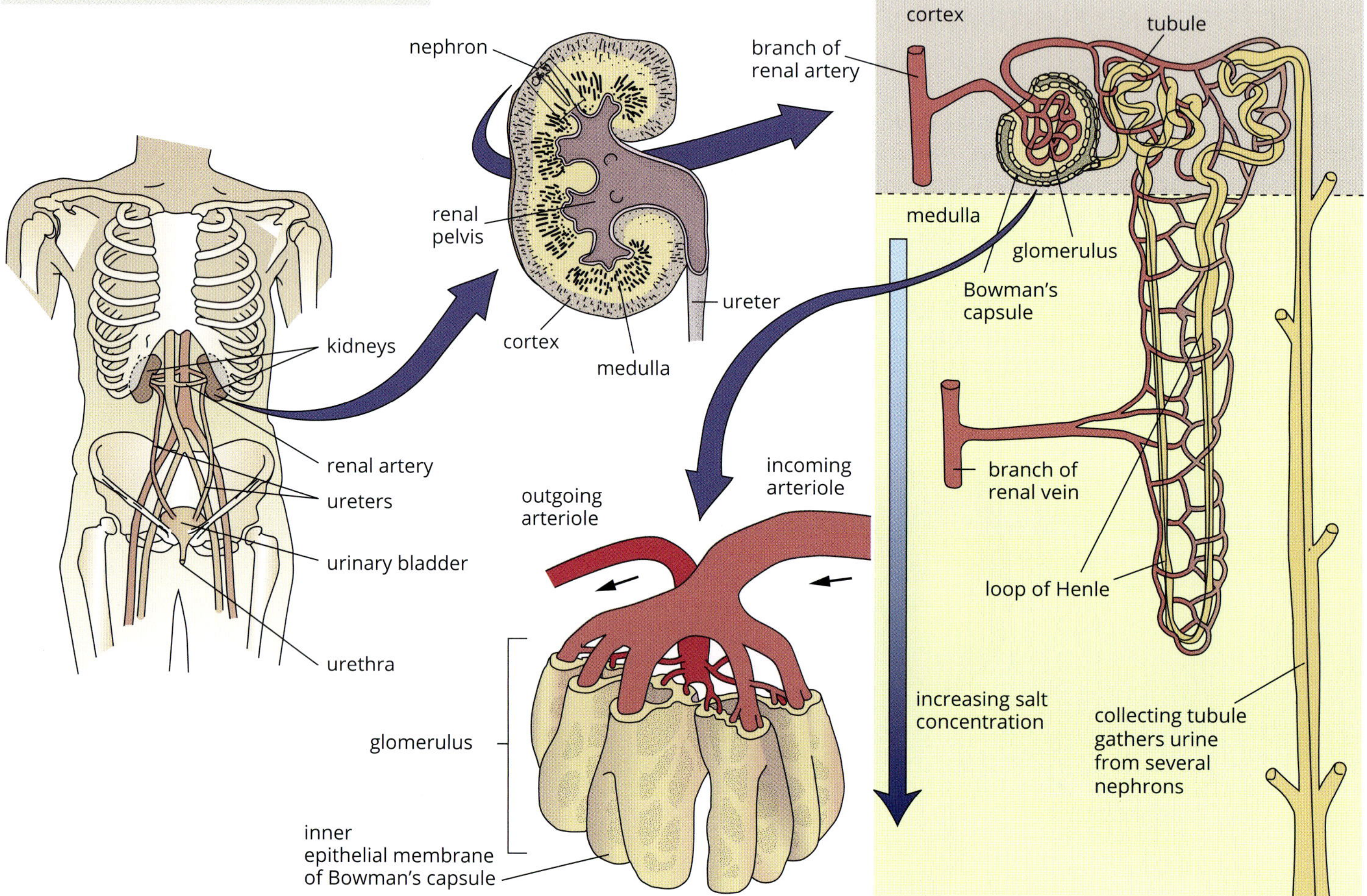

FIGURE 18.3.2 The human excretory system

Blood enters each kidney from the aorta through the **renal artery** and leaves through the **renal vein**. Blood vessels branch throughout the kidney in a complex fashion to individual nephrons (Figure 18.3.2). Urine is formed in the kidneys after blood filtration then drains via the **ureters** into the bladder for storage until it is an appropriate time for elimination through the urethra.

The excretory and regulatory processes in the mammalian kidney are carried out by the nephrons, which are the functional units of each kidney. A nephron is composed of a **Bowman's capsule** surrounding a **glomerulus** of clumped, narrow blood capillaries, and a nephron tubular structure consisting of the **proximal convoluted tubule**, **loop of Henle** and **distal convoluted tubule** that leads into a **collecting tubule** (Figures 18.3.2 and 18.3.3). The formation of urine involves passive filtration, selective reabsorption, active secretion and the passive removal of water. The functioning of nephrons regulates water and salt levels of the blood as they form urine.

During its routine circulation around the body, blood passes from liver to kidney and is filtered through the blood vessel walls to form a primary glomerular filtrate inside the nephron tubules. The filtrate has the same composition as blood plasma, except that large protein molecules, as well as blood cells, have been filtered out. Most of the useful substances in the primary filtrate are quickly reabsorbed as it passes along each nephron tubule. Some extra, unwanted substances may be secreted into the fluid in the tubule before it passes as urine from the kidney to the bladder (Figure 18.3.3).

The kidney processes regulate the concentration of different salts in the blood, including those salts that are responsible for maintaining the pH of body fluids within closely controlled limits (Figure 18.3.4). Mammals are able to conserve water by producing urine that is more concentrated than body fluids. The ability to produce concentrated urine requires energy and is related in some mammals to the degree of water stress experienced in their normal environments.

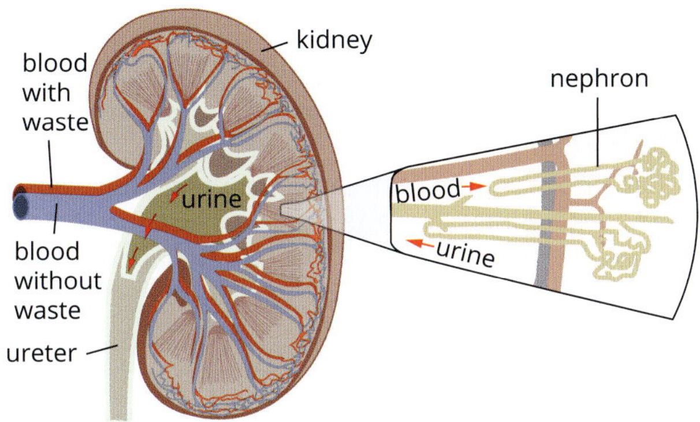

FIGURE 18.3.3 Approximately one million nephrons and a complex system of associated blood vessels are found in each kidney.

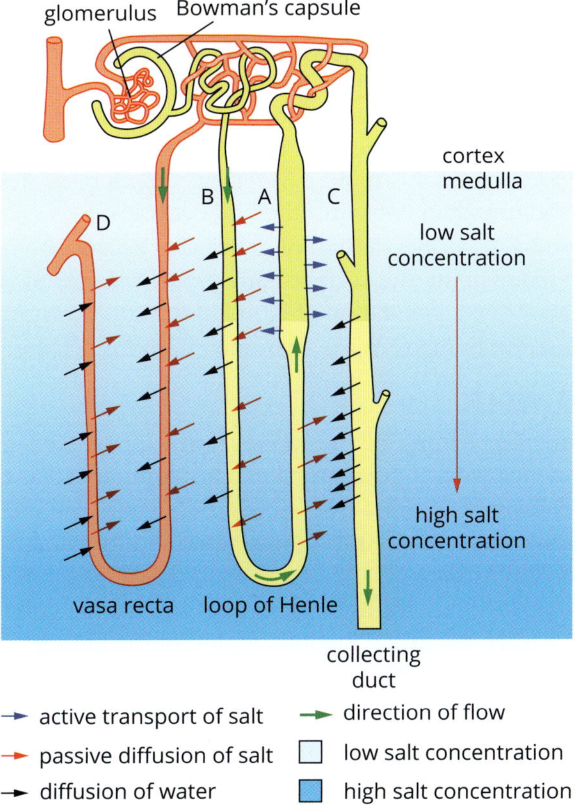

FIGURE 18.3.4 Specialised osmoregulation processes in the kidney's medulla region allow production of concentrated urine to conserve water. The vasa recta arrangement of blood vessels in the medulla contributes to the required concentration gradient.

BIOFILE CCT

Urine in space

Much of what we know about the composition of human urine comes from a NASA study that researched the extraction of drinking water from urine for use on the International Space Station (ISS).

Hauling tons of water to the ISS is inefficient and costly. In 2009, NASA astronauts began recycling urine using the Urine Processor Assembly, which was able to reclaim 75% of water from urine. In 2016, an astronaut installed an improved urine-recycling system aboard the orbiting lab by incorporating a chemical solution called Alternate Urine Pretreatment (AUP). A few millilitres of AUP are dispensed when the toilet is flushed on the space station, allowing for a greater percentage of reclaimed water afterwards (Figure 18.3.5). The treated urine is added to the ISS water reclamation system that also recycles air condensate and shower runoff. Around 93% of all water aboard is recycled in the NASA section, about 6000 extra litres each year. The claim is that the recycled water tastes like bottled water.

A difference between American and Russian astronauts on the ISS is that the Americans drink their filtered, recycled urine but the Russians don't. Occasionally, the NASA astronauts will even go over to the Russian side of the ISS and take the Russian supplies of urine to process it.

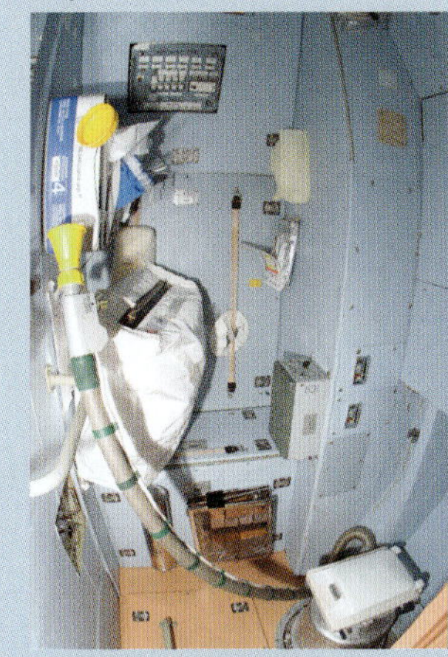

FIGURE 18.3.5 A toilet compartment aboard the ISS

BIOFILE CCT

Urea

Nitrogenous waste is excreted by animals in a form that is suitable for their environmental needs. All freshwater fish and most aquatic invertebrates excrete ammonia; birds, reptiles and insects excrete uric acid. Mammals and most marine fish excrete urea.

Urea (CH_4N_2O) is a larger molecule than ammonia (NH_3) with more nitrogen atoms and it also contains carbon and oxygen (Figure 18.3.6). It is much less toxic than ammonia and is highly soluble in water, but converting ammonia into urea requires energy. So although urea is less toxic, an animal spends more energy excreting urea instead of ammonia. Land vertebrates need to conserve water. Mammals excrete their nitrogenous waste largely in the form of urea, but their kidneys are capable of regulating and minimising water loss. This strategy is a successful adaptation to life on land.

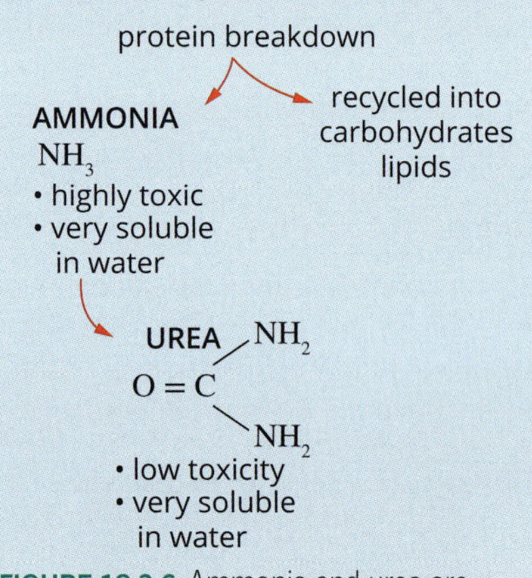

FIGURE 18.3.6 Ammonia and urea are two of the nitrogenous wastes produced from breakdown of proteins in animals. Freshwater fish excrete ammonia. Mammals convert ammonia to urea for excretion.

GO TO ➤ Section 14.2 page 510

Nitrogenous wastes

As noted earlier, a main function of the kidneys in vertebrates is to excrete nitrogenous wastes. Excretion usually involves the loss of water and is therefore closely linked to water balance in terrestrial animals including humans. The excretory system works closely with the circulatory system, filtering waste products from the blood stream and collecting them in urine. For heterotrophs, it is sometimes inevitable that toxic substances are absorbed from the food and liquids they ingest—these toxic substances must also be excreted.

Depending on the availability of water, the urine excreted by most animals will be in a liquid form but will vary in concentration. In humans, the waste that carries nitrogen atoms from the body is urea, dissolved in water to form liquid urine of a yellowish colour. First the liver has to process toxic ammonia into the less toxic urea and pass this form of nitrogenous waste to the kidneys for excretion.

Salt and water balance

Salts form ions in solution. The most common ions found dissolved in body fluids are sodium (Na^+), potassium (K^+), chloride (Cl^-), calcium (Ca^{2+}), magnesium (Mg^{2+}), nitrate (NO_3^-), sulfate (SO_4^{2-}) and phosphate (PO_4^{3-}). These all derive from our food intake. The concentrations of certain ions in cells must be held within narrow limits. Some of these ions are important for body functions, like conducting nerve impulses and for regulating the pH of body fluids, which must remain at a suitable level for enzymes and other molecules to function efficiently.

In regulating water and salt ions, the kidneys use a combination of processes. Some substances require active transport against a concentration gradient. Increased ion concentration allows water to follow via osmosis, which is passive. Generally in biological systems water moves passively along concentration gradients. However, the kidney is able to produce concentrated urine, thereby reducing water loss, via the counter-current arrangement of the loops of Henle and of blood vessels (**vasa recta**) in the medulla of the kidney (Figures 18.3.2 and 18.3.4). Careful study of Figure 18.3.4 will help with understanding the complexity of osmoregulation processes.

It might be expected that blood flowing through the kidney medulla would pick up and carry away any excess salt in the medulla and prevent the development of a high salt concentration. This does not occur because of the counter-current arrangement of blood vessels in the vasa recta (Figure 18.3.4). When blood is too concentrated, the antidiuretic hormone (ADH) acts to increase the reabsorption of water from the collecting duct back into the blood vessels.

Osmoregulation by the kidneys is complex and vital because many body functions are dependent on maintaining the correct homeostatic balance of water and salts in the body, not the least being circulation of blood of the correct consistency and pH.

Osmoregulation and homeostasis is examined in more detail in Chapter 14.

Nephrons at work

Nephrons, the functional units of the kidney were introduced in Figure 18.3.3. Their intricate and vital work to form urine can be summarised as occurring in four stages (Figures 18.3.4 and 18.3.7):

1 filtration—passive process, in the glomerulus and Bowman's capsule in the cortex region of the kidney
2 reabsorption—passive process in the proximal and distal nephron tubules in the cortex and loop of Henle in the medulla region of the kidney
3 secretion—active process in the proximal and distal nephron tubules in the cortex
4 excretion—removal of urine via the collecting tubule and pelvis region of the kidney into the ureter to the bladder; from the bladder to the outside via the urethra.

Each nephron is very closely associated with blood vessels, particularly in the region of the renal corpuscle made up of the glomerulus and Bowman's capsule, as well as the networks of capillaries wrapped around the remainder of the nephron tubule (Figure 18.3.2). This close association enables the exchange of substances between the blood and the fluid that becomes urine.

Useful substances are reclaimed into the blood, urea remains in the nephron tubule along with some water, salts and other wastes to be removed as urine.

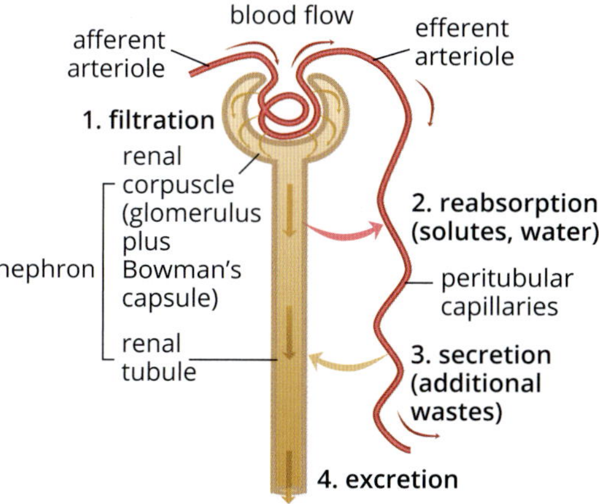

FIGURE 18.3.7 The basic steps in urine formation are filtration, reabsorption, secretion then excretion. The process starts with filtration from the glomerulus into Bowman's capsule, continues with reabsorption and secretion as the filtrate passes along the nephron tubule, then finishes with excretion of the urine end product from the body.

DISORDERS OF THE KIDNEY AND THEIR CAUSES

The kidneys are of critical importance for continual cleansing of the blood by removal of waste products and as part of the body's homeostatic mechanisms. Loss of kidney function can be due to multiple causes and it is unfortunate that many cases are not diagnosed until late in the disorder.

Symptoms of kidney problems

Kidney disorders vary from temporary infections or blockages to chronic kidney disease (CKD) to complete renal failure that occurs when the kidney function is less than 15%, a state called end stage renal disease (ESRD). At this point, dialysis is required because the kidneys are irreparably damaged and can no longer effectively remove waste products and maintain the fluid levels of the body. It is possible to have one damaged kidney and still survive well if the remaining kidney is healthy.

The causes of kidney disorders vary widely and include swelling of the nephrons (glomerulonephritis), diabetes, bladder valve failure, high blood pressure (hypertension), **polycystic kidney disease (PKD)**, build-up of toxic nitrogenous waste, kidney stones, liver problems, cancers, injuries, bladder infections, some medications and drug use. Physical damage from blows to the kidney area can also cause loss of kidney function. In some patients, the exact cause of their chronic kidney disease cannot be diagnosed because the kidneys are already too damaged by the time they seek treatment and there may be multiple factors at play.

Kidney disease is sometimes called a silent disease as there are often few or no symptoms until late in development of chronic kidney disease. There may be loss of up to 90% of kidney functionality before any symptoms of concern are noted. The symptoms develop slowly and are not specific to kidney disorders. Some signs and symptoms include: fatigue, high blood pressure, blood in the urine, foaming urine, pain in the lower back, nausea, loss of appetite, severe weight loss, abnormal heart rhythm, fluid in the lungs, insufficient urine production, frequent need to urinate during the night, itching, puffy eyes and ankle swelling due to water-salt imbalance.

Detection of kidney disorders

Loss of kidney filtration and less excretion of wastes results in higher levels of waste products like urea, creatinine and uric acid in the blood, plus fluid build up in tissues. Urine and blood tests are used to check for kidney problems. They measure the creatinine level and estimate the **glomerular filtration rate (GFR)**. Creatinine is a normal waste product continually created by muscular action and cleared by the kidney for elimination in the urine. Urea is the waste product that removes nitrogen waste via urine. Neither creatinine nor urea is directly toxic but they are useful substances for measuring kidney function. Testing needs to account for age, body size and biological sex as part of the calculation. As kidney function declines further, other waste products such as potassium, uric acid and phosphate accumulate. A high blood uric acid level may lead to gout (Figure 18.3.8) and retention of phosphate may lead to parathyroid overactivity and bone disease.

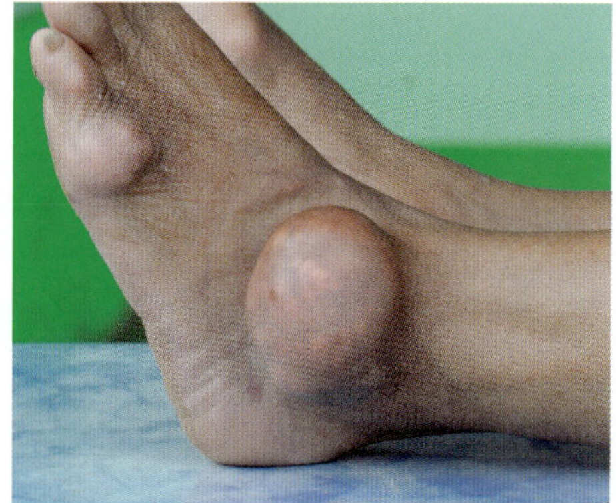

FIGURE 18.3.8 Swellings on the foot of a patient suffering from gout, a build up of nitrogenous waste

The normal level for creatinine in a blood sample is up to 110 μmol/L for adult males and 90 μmol/L for females. Because males have greater muscle mass they have higher blood serum creatinine levels than females. A rise in blood creatinine is observed only after significant damage of kidney nephrons, so it is not suited to detect early stage kidney disorders.

The measure of creatinine clearance over a 24-hour period is a more sensitive measure for earlier detection. It can be used to calculate glomerular filtration rate (GFR), which has a normal range of 90–120 mL/min. This is taken as 100% of kidney function. So, a GFR of 15 mL/min represents about 15% of kidney function and a critical stage has been reached where the need for dialysis is imminent. GFR varies noticeably with age, decreasing about 1% per year from the mid thirties. At age 70 years, GFR on average is often 60–70% of the normal.

The mole is the SI chemistry unit (mol) for the amount of a chemical substance. In the measurement of creatinine to test kidney function, a result of 100 μmol/L in blood serum is a normal result and is the equivalent of 1.13 mg/100 mL.

Causes of kidney problems

There are five known common causes of kidney disorders:

1. glomerulonephritis (inflammation of kidney filters)
2. diabetic nephropathy (kidney damage resulting from type 1 and type 2 diabetes)
3. reflux nephropathy (kidney damage caused by a leaking valve in the bladder allowing urine to flow back into the kidney)
4. hypertension (high blood pressure) leading to nephrosclerosis
5. polycystic kidney disease (cysts in both kidneys, not detectable until adolescence, and often heritable).

The number of Australians requiring treatment for kidney disorders is increasing. For example, from 1993 to 1997 there was a 26% increase over just four years, and the number of new cases of end stage renal disease more than doubled from 2000 to 2015. Renal clinics have expanded to cope and home dialysis has become more common. The Kidney Health Australia Foundation has an informative website, including videos that demonstrate dialysis treatments, and advice about choices of dialysis and kidney transplants.

Glomerulonephritis

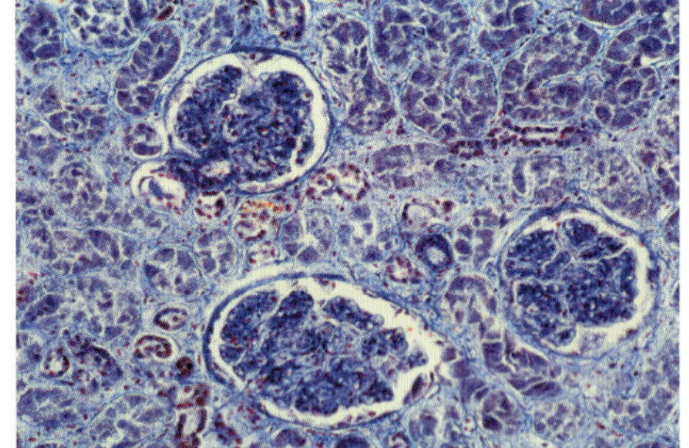

FIGURE 18.3.9 LM of three glomeruli (dark blue) in a human kidney affected by glomerulonephritis. In this condition there is a proliferation of enlarged epithelial cells that line the glomerular capillaries, and the presence of increased numbers of neutrophils (red/purple, blood cells). The glomeruli take on a dense appearance, as seen here.

Glomerulonephritis is the most common cause of kidney failure, often occurring after a bacterial infection. The inflamed and obstructed glomeruli (plural of glomerulus) cause red blood cells and proteins to leak into the urine and the patient retains fluid (Figure 18.3.9). Diabetic **nephropathy** has somewhat similar symptoms but from a different cause.

Diabetic nephropathy

Filtration from the glomerulus into Bowman's capsule of a nephron would normally leave blood cells and large protein molecules in the capillary. Both type 1 and type 2 diabetes can interfere with this filtering process and allow larger substances to pass through with the primary filtrate (Figure 18.3.10). A blood test called BUN will diagnose diabetic nephropathy (not to be confused with neuropathy) by measuring the amount of urea in a blood sample.

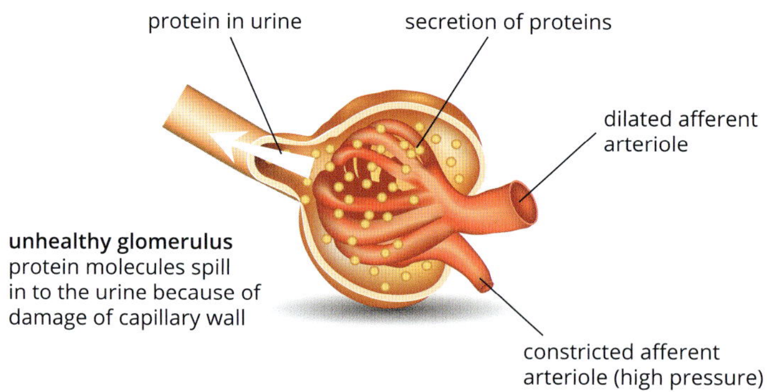

FIGURE 18.3.10 Nephropathy is a kidney malfunction often caused by type 1 and type 2 diabetes. The walls of the capillaries in the glomerulus are damaged and large protein molecules enter the nephron tubule. They are then excreted in the urine.

Reflux nephropathy

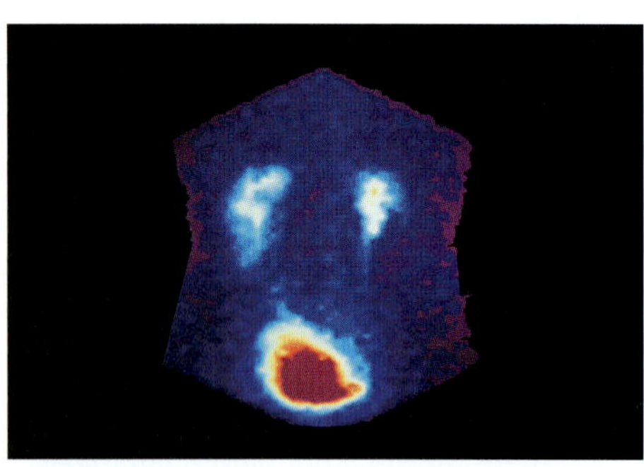

FIGURE 18.3.11 Coloured scintigram of the human abdomen, showing urinary reflux from the bladder to the kidneys. The partly emptied bladder is red; kidneys are at upper frame (faint white outlines). Urine is coloured white and has flowed back (refluxed) up to the kidneys.

Reflux nephropathy is due to a failure of the valves at the lower end of the ureters to keep urine in the bladder. The urine from a full bladder pushes back up the ureters into the kidneys. It is known as **urinary reflux** and may lead to kidney damage. Figure 18.3.11 is a scintigram, which is an image taken by a gamma camera scan. Gamma ray scanning uses a radioactive tracer, in this case technetium-99m, injected here into the urine in the bladder and detected as flashes of light by the external gamma camera. The image shows that white coloured urine has refluxed back up into the kidneys.

BIOFILE N

BUN: blood urea nitrogen

A blood urea nitrogen (BUN) test (Figure 18.3.12) is used to determine how well your kidneys are working. It does this by measuring the amount of urea nitrogen in a blood sample.

Normal BUN levels are in the following ranges:

- adult men: 8–20 mg/100 mL of blood
- adult women: 6–20 mg/100 mL of blood
- children: 5–18 mg/100 mL of blood.

Higher or lower levels indicate possible liver or kidney problems such as diabetic nephropathy, dietary protein too high or too low (malnutrition) or blood volume problems.

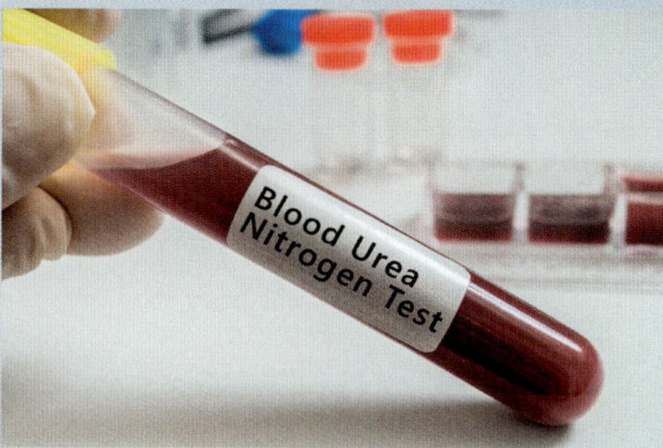

FIGURE 18.3.12 Test-tube with blood for blood urea nitrogen (BUN) test

Hypertension

The relationship between hypertension (increased blood pressure) and diseased kidneys works both ways. Kidney disorders are a common cause of secondary hypertension. On the other hand, high blood pressure can lead to renal problems (benign or malignant **nephrosclerosis**) and is an important factor in the progression of chronic kidney disease to end stage kidney failure. Nephrosclerosis is a progressive kidney disease where the small blood vessels harden (this is known as sclerosis). The kidneys are especially vulnerable to the changes to blood vessels that occur in hypertension. In chronic hypertension the arteriole walls are thickened. As a result, reduced renal blood flow causes chronic undersupply of blood and oxygen to the renal tissues and gradual atrophy of the nephron tubules. Figure 18.3.13 shows the extremity of the effect that hypertension can have on the kidney. Slow progressive loss of functioning nephrons may eventually lead to chronic renal failure and the shrunken 'end-stage' kidney seen here.

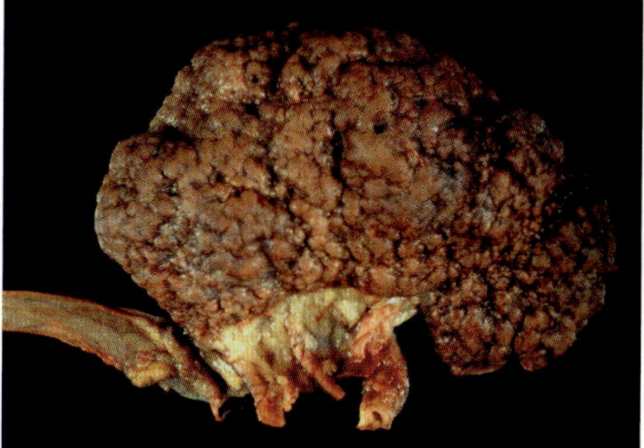

FIGURE 18.3.13 Hypertension (high blood pressure) has caused atrophy and nephrosclerosis of this kidney. The normal smooth and dark red surface appearance has been replaced by a noduled surface and dead areas at the pelvis of the kidney as the renal blood vessels hardened.

Polycystic kidney disease (PKD)

Polycystic kidney disease (PKD) causes cysts to grow inside the kidneys. These cysts make the kidneys much larger than normal and damage the renal tissue. The cysts are small fluid-filled sacs growing from the nephrons and there may be thousands of them (Figure 18.3.14 on page 636). PKD is a genetic condition, almost always inherited from one or both parents. Autosomal dominant PKD (ADPKD) is the most common type of PKD. About nine out of every 10 people with PKD have the autosomal dominant form and it is the most common inherited kidney disease. Because it is dominant, a child only needs to have one parent with ADPKD to give them a 50% chance of inheriting the disease. If both parents have ADPKD their child will be born with the disease. ADPKD causes cysts to form only in the kidneys and symptoms of the disease may not appear until a person is between 30 and 50 years old. At that age, people with ADPKD commonly have kidney pain and high blood pressure. Maintaining a healthy lifestyle and careful monitoring of the condition can delay chronic kidney disease.

Autosomal recessive PKD (ARPKD) is a less common and more severe form of PKD. It causes cysts to form in both the kidneys and the liver. Symptoms of the disease can begin even before birth and be life-threatening to the fetus.

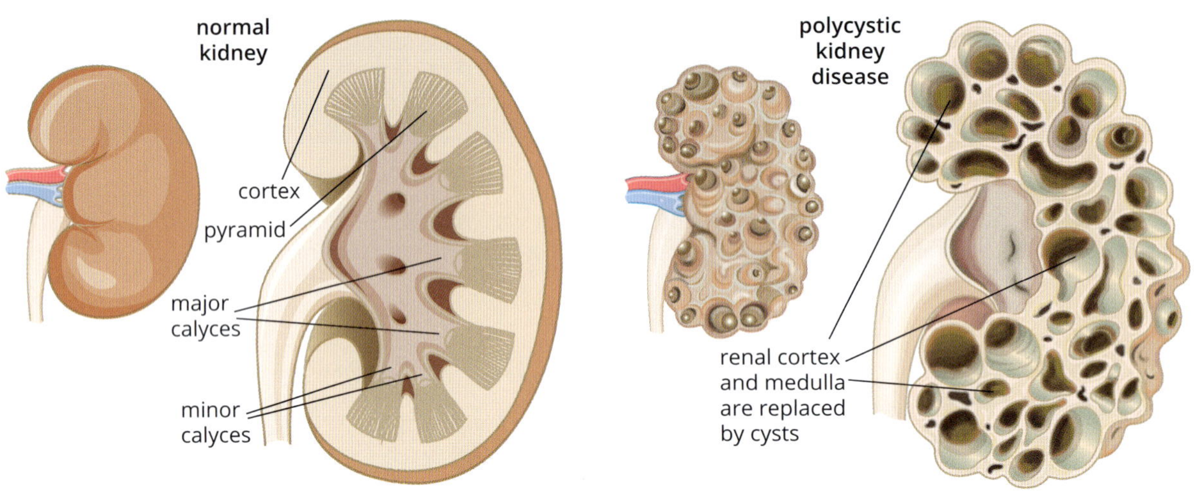

FIGURE 18.3.14 (a) A healthy kidney and (b) a kidney with polycystic kidney disease. Polycystic kidney disease is a genetic condition that causes kidney malfunction and may eventually lead to kidney failure.

Other kidney problems—kidney stones

FIGURE 18.3.15 Kidney stones passed after ultrasound treatment. The scale is in centimetres.

With the many waste products handled by the kidney, it is not surprising that from time to time, some of them crystallise into hard lumps of varying sizes and shapes, known as kidney stones (Figure 18.3.15). They can range from the size of a grain of sand to that of a golf ball. The larger ones cause blockages and a great deal of pain. Fortunately they can be dealt with fairly easily. Treatment may include medication to dissolve the stones, ultrasound therapy called lithotripsy to break the stone into smaller pieces that can pass out with the urine, endoscope removal or in a severe case, surgical removal.

There are four main types of kidney stones:

- the most common are formed from calcium combined with oxalate or phosphate
- struvite stones caused by a urine infection are often horn-shaped and large
- uric acid stones are often softer than other forms of kidney stone
- cystine stones are rare and hereditary and look more like crystals than stones.

Drinking good quantities of water each day will reduce the likelihood of formation of kidney stones.

Other kidney problems—liver disease

Remember that the liver is essential to healthy functioning of the excretory system because it removes toxins like alcohol from the blood and prepares urea to be sent to the kidney for elimination.

People who drink alcohol excessively are prone to severe and often fatal liver disease. Medical evidence indicates that the addition of vitamins to alcoholic drinks, while they may be good for nutrition, will not prevent chronic liver damage.

Alcohol is a toxic substance. The special enzymes that are needed to break it down are found in the liver. Because the biochemical pathways in the liver cells of a heavy drinker are involved with removing alcohol, the cells cannot carry out their normal levels of cellular respiration. Substances that should have been broken down for energy are converted to fats instead, and these fats accumulate in the liver.

> Hepatic is a descriptive word (adjective) for anything to do with the liver. For example, the hepatic artery is the vessel bringing blood to each kidney and hepatitis refers to liver diseases.

For a while the situation is reversible, but then the cells filled with fat start to die, causing alcoholic hepatitis. This is followed by cirrhosis, which is the formation of scar tissue in the liver (Figure 18.3.16). Finally, death may occur when the liver is unable to carry out its normal functions, including the initial preparation of waste products to be excreted by the kidney.

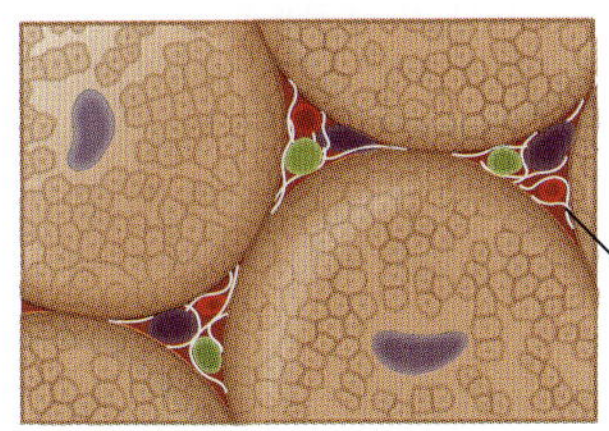

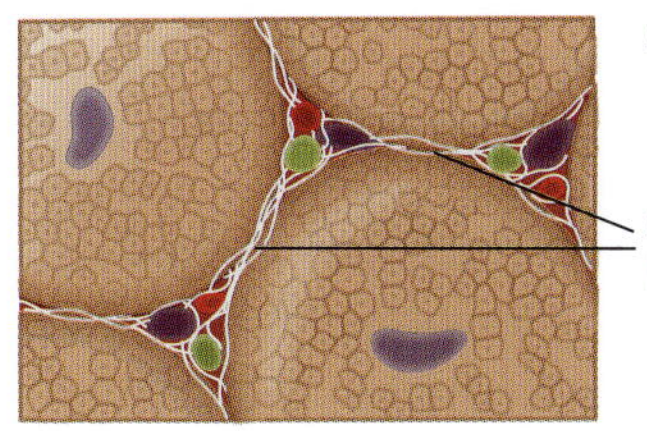

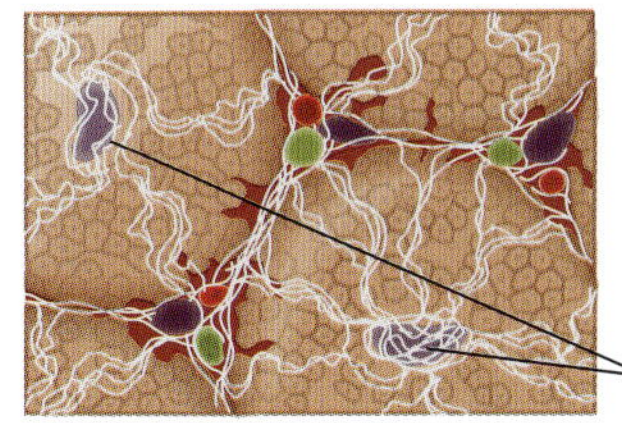

FIGURE 18.3.16 The stages of scar tissue formation (cirrhosis) in the liver. Tissue inflammation (stage 1), followed by cell death, leads to scar tissue build up (stage 2). Scar tissue replaces healthy tissue and eventually causes cirrhosis of the liver (stage 3). Excessive alcohol consumption can cause this form of liver damage.

BIOFILE CCT

Glomerulus

There can be as many as 50 capillaries with very thin walls within each glomerulus. Their narrow diameter, large surface area and close contact with the membranes of Bowman's capsule ensure a fast and voluminous filtering process. The blood vessel walls contain numerous pores (called fenestrae) 50–100 nm in diameter. These pores allow for the free filtration of fluid, plasma solutes and small protein molecules. However, they are not large enough that red blood cells can pass through. The outside surface of the glomerular capillaries is lined with folds called podocytes that control the filtration of protein molecules (Figure 18.3.17). These very complex structures and processes are vital to maintaining homeostasis and healthy functioning of the body. If the walls of the glomerular capillaries are damaged in a diabetic patient, this can allow larger protein molecules to pass into the urine, a condition called nephropathy.

(a) podocyte; macula densa; juxtaglomerular cells; renal nerve; afferent arteriole; proximal convoluted tubule

(b) macula densa; proximal convoluted tubule; glomerulus; distal convoluted tubule; basement membrane

FIGURE 18.3.17 (a) The structure of the glomerulus allows specialised cells to monitor the composition of the fluid in the distal convoluted tubule and adjust the glomerular filtration rate into the proximal tubule. (b) This micrograph shows a glomerulus and surrounding structures in a transverse section.

TECHNOLOGIES TO ASSIST WITH KIDNEY DISORDERS

Given the essential nature of kidney function, the only way to survive complete renal failure is to use the technology of dialysis or to receive the transplant of a healthy kidney.

Dialysis

If kidney function has decreased to 10–15%, dialysis treatment or a kidney transplant are the only alternatives to maintain life. Kidney transplants are now extremely successful and can offer the best solution to total kidney failure. However, there are always many people waiting for suitable donor kidneys. While they wait, these people depend on the technology of renal dialysis, which has been used since 1945 to keep patients with kidney failure alive.

Haemodialysis is the diffusion of small solute molecules through a partially permeable membrane in a dialysis machine (this is different from osmosis, which is the diffusion of water through a partially permeable membrane). Small molecules are free to diffuse through the dialysing membrane, but larger proteins and cells do not. This is similar to the filtration stage of normal kidney function but is a time consuming and expensive medical process carried out regularly with external equipment under sterile conditions (Figure 18.3.18). As such, the process of renal dialysis maintains life but imposes severe restraints for work and travel on the patient who has to spend four to five hours at least three times per week hooked up to the dialysis unit. People with kidney failure rely on a dialysis machine for life-saving treatment, but it does not provide a cure.

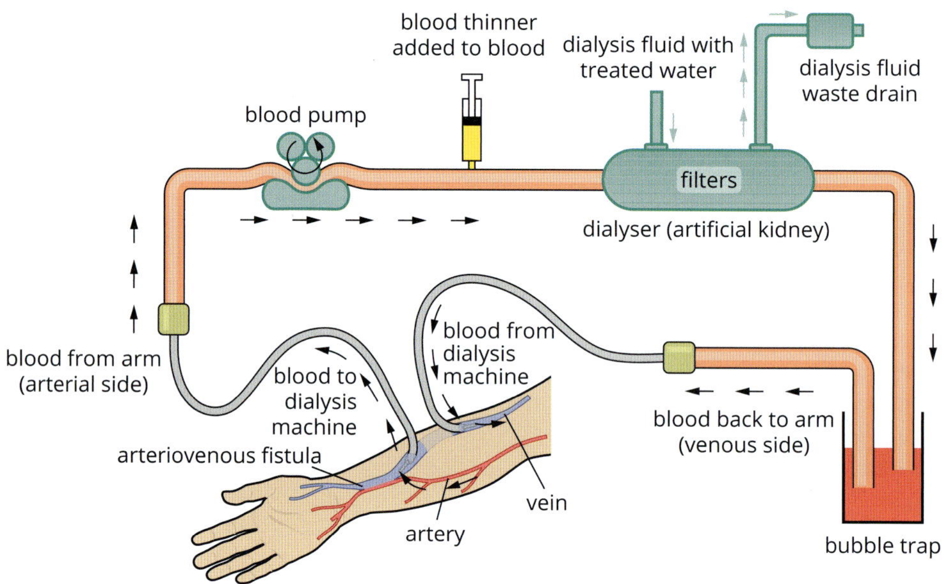

FIGURE 18.3.18 Blood flow through a renal dialysis machine, from arterial blood via a fistula (a connection between an artery and a vein) created from a vein in the arm. The blood passes through a pump, a filter filled with dialysis solution, a bubble trap and returns to the venous circulation in the same arm.

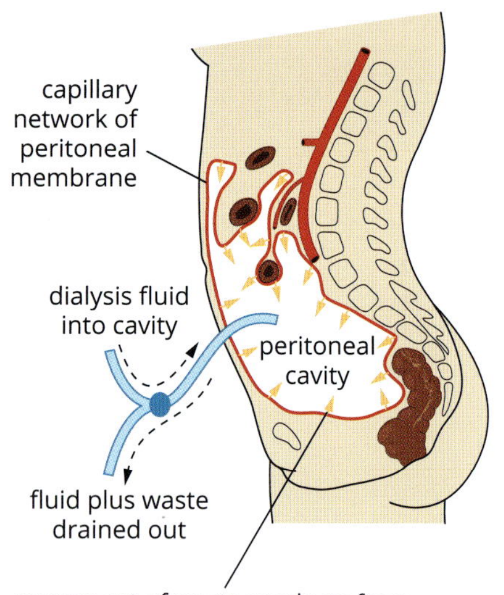

FIGURE 18.3.19 The technique of peritoneal dialysis, where dialysis fluid is placed in, and then removed from, the peritoneal cavity (abdomen) for the flushing away of waste materials.

Types of dialysis

There are two different types of dialysis: peritoneal dialysis and haemodialysis.

- **Peritoneal dialysis** is done at home after the initial set up (Figure 18.3.19). This dialysis treatment occurs inside the body and the patient can continue with a somewhat normal life while the dialysis is working. Basically, peritoneal dialysis is flushing of the whole abdominal cavity. A catheter tube of about 30 cm length and 0.5 cm width is inserted into the abdomen in a simple operation and remains there. This method is likely to be used for children with renal failure.
- Haemodialysis can be done either at home but requires specialised equipment, or travelling regularly to a dialysis clinic. During haemodialysis, a machine acting as an artificial kidney cleans the patient's blood (Figure 18.3.18). (See Biology in Action page 639).

Effectiveness of dialysis technology

Dialysis maintains the life of a person with kidney failure. However, it is not a cure or a trouble-free existence. Having a catheter for peritoneal dialysis increases risk of infection and it must be checked and cared for as part of a daily routine. With haemodialysis, a fistula has fewer complications than the other types of access, but care must be taken to prevent infections and blockages. A fistula may take a few months to mature before it can be used for dialysis (Figure 18.3.18). It can then last for many years. Some of the complications associated with the vascular access include blood clots, bleeding, aneurysms (weak spots that allow a bulge in the blood vessel wall) and infections.

Problems experienced with the dialysis treatment include nausea and fatigue, muscle cramps, loss of appetite and infection. The use of heparin to prevent clotting can damage blood cells. Diet must be regulated to limit protein, salts and fluid intake. And of course, the amount of time involved in dialysis treatment impacts heavily on a patient's freedom and ability to work (Figure 18.3.20).

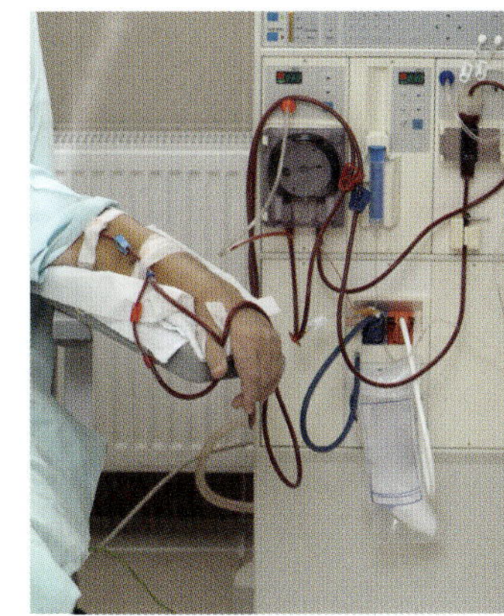

FIGURE 18.3.20 Haemodialysis treatment in a specialised renal unit

BIOLOGY IN ACTION DD  PSC

Dialysis

In the event of kidney failure or removal due to disease, homeostatic maintenance of blood chemistry breaks down. In that case, an artificial blood-filtration process is available.

The most common of the two main forms of dialysis, called haemodialysis, essentially involves diverting all of the body's blood through a filtering machine (Figure 18.3.21). The process removes salt, metabolic waste urea and excess water from the blood. Dialysis also maintains certain ions at safe concentrations.

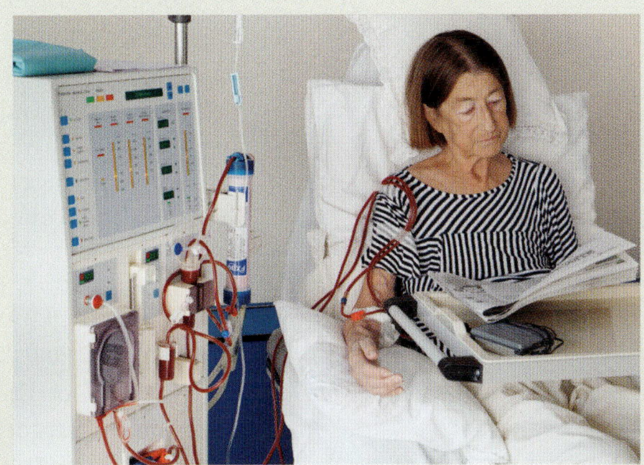

FIGURE 18.3.21 Patient undergoing haemodialysis

The treatment works via diffusion. Blood flows along one side of a semipermeable membrane, and on the other is a special fluid called dialysate. Since the kidneys are not working, the concentration of blood solutes (potassium, phosphorous and urea) is high. The concentrations of these substances are intentionally low in the dialysate. Therefore the solutes (being small molecules) can diffuse from the blood through the membrane into the dialysate. Constant removal of the dialysate maintains the concentration imbalance and the transfer of solutes to the dialysate. The dialysate is also roughly isotonic with healthy blood for potassium and calcium, so these substances are not filtered. The dialysate contains higher concentrations of bicarbonate than the blood, which diffuses into the blood and acts as a pH buffer to counteract the acidosis (acidic blood) that dialysis patients often suffer. Large molecules or cells are too large to pass through the membrane's micro-pores. This artificial filtration is similar to the natural process in the glomerulus.

Dutch physician Dr Willem Kolff developed dialysis during the 1940s. The procedure has been available as a routine hospital treatment since the 1960s. It is generally performed at a hospital, but can also be arranged in patient's homes.

The process generally takes about four hours, four times per week. Patients remain conscious throughout, albeit immobile in a seated position and may read or do light tasks. The time requirement is an inconvenience, but otherwise patients generally lead normal lives.

Kidney transplant

An organ transplant is a treatment for end stage kidney disease but is not regarded as a cure. A transplant potentially offers a longer, more active life, free from dialysis.

A deceased donor is a person who has died with their organs still relatively healthy and who has previously registered to be an organ and tissue donor or whose family has made the decision after the death. Relatively few of these situations occur but fortunately when they do, there are usually two kidneys available. A recipient receives only one kidney but it must be a suitable match. In Australia, the waiting time for a kidney transplant is over three years. Allocation of a donated kidney is decided by a set of national guidelines and is kept anonymous.

Kidney transplants from living donors now make up around three out of every ten kidney transplants in Australia each year. Living donors of a single kidney can be:

- related—a relative (parent, brother, sister, or adult children), related by blood to the recipient
- unrelated but known to the recipient—partner, non-blood relative or friend of the recipient
- non-directed kidney donation or altruistic—this is where someone anonymously donates a kidney to a recipient on the transplant waiting list. In this situation the living donor has no say in who receives their kidney.

The surgical procedure for a kidney transplant normally takes two to three hours and recovery should be quite rapid in a matter of days. The patient's natural kidneys are usually left in place unless a polycystic kidney has become too large. The new kidney is connected to the bladder for urine flow and carefully monitored until normal function commences.

Effectiveness of kidney transplants

Kidney transplants are very successful. In Australia, over 94% of transplants are working one year later.

Apart from the kidney failure, a person needs to be quite healthy to have a transplant. Surgery and the drugs used to prevent rejection of the transplanted organ place a strain on the body, making it less likely to succeed for an unhealthy person.

Medications that depress the immune system to prevent rejection of the new organ must be taken for the life of the transplanted kidney. They can make the person more prone to serious illnesses such as infections and cancers. Testing is carried out regularly, usually on a weekly basis.

So, while freedom from dialysis treatment allows the possibility of travel, a transplant recipient must be very careful to avoid travel-related infections. There may also be restrictions on vaccinations. For example, live vaccines like those used for yellow fever, typhoid, tuberculosis, measles, mumps and rubella are not safe for those with transplants. The anti-rejection medications may not combine well with some medications for malaria, altitude sickness and other common traveller's treatments, leading to transplant drug toxicity.

Generally, transplant recipients can lead a more normal life and have more effective function than dialysis patients.

BIOLOGY IN ACTION   

Organ and tissue donation

Organs that can be transplanted in Australia are: kidney, heart, lungs, liver, pancreas and intestines. Tissues that can be transplanted in Australia are: parts of the eye (sclera and cornea), bone, heart valves, skin, tendons and ligaments.

People can either become donors when they die (deceased donor) or they can donate a kidney or part of their liver while they are still alive (live donor). There are two steps to becoming an organ or tissue donor after death, outlined below.

1 Register on the Australian Organ Donor Register through Medicare. Even if you have ticked 'yes' to organ donation on your driver's licence, you should make sure you are on the register.

2 Discuss your decision with your family. Even if a deceased person had registered their wish on the Australian Organ Donor Register, the next of kin's consent will always be requested. For this reason, it is important to discuss your decision with your family.

The Australian Paired Kidney Exchange (AKX) Program is an initiative to increase the options for living kidney donation. The program is increasing live donor kidney transplants by identifying matches for patients who are eligible for a kidney transplant, and have a living donor who is willing but unable to donate to them, because of an incompatible blood type or tissue type (Figure 18.3.22). This option is known as paired kidney exchange or paired kidney donation. The program uses computer software to search the entire available database of registered recipient/donor pairs to look for combinations where the donor in an incompatible pair can be matched to a recipient in another pair. If the computer finds a compatible match, two or more simultaneous transplants can occur by exchanging donors.

In 2015, Australia's largest ever paired kidney exchange took place, which involved six hospitals across two states and saved seven lives. It started with a Victorian man making an altruistic kidney donation and the paired kidney exchanges grew to a chain of seven.

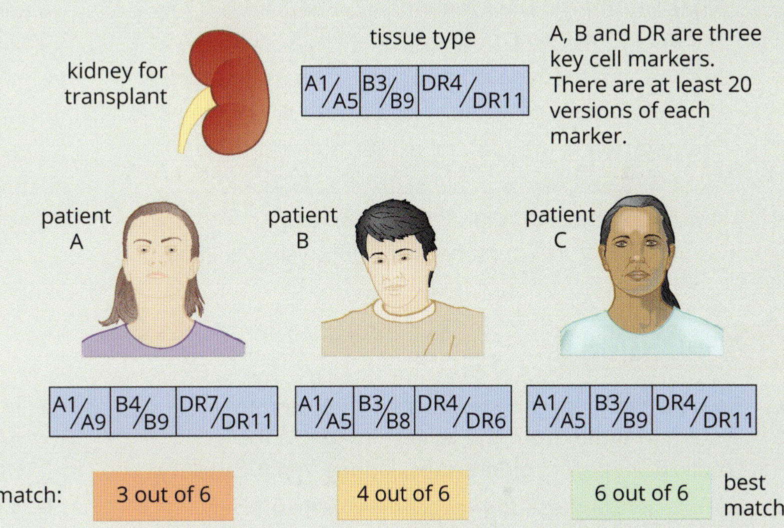

FIGURE 18.3.22 The complexity of finding suitable matches for kidney transplants. Diagram showing tissue type, cell markers and number of matches for three different patients

18.3 Review

SUMMARY

- In humans, removal of waste and toxic substances, control of pH, ion concentrations and water balance are carried out largely by the liver and a pair of kidneys.
- Proteins are broken down into carbohydrates or lipids to be used for energy. This creates nitrogenous wastes which must be removed from the cells before they become toxic.
- Ammonia is the first nitrogen product formed and the most toxic. In mammals it is converted by the liver into a less toxic substance called urea and passed to the kidney for elimination in urine.
- The environment of an animal is a major factor in determining the type of nitrogenous waste produced and the amount of water excreted with it.
- The nephron is the functional unit of the kidney.
- A mammalian nephron consists of a Bowman's capsule (surrounding a glomerulus) leading into a tubular structure (proximal convoluted tubule, loop of Henle and distal convoluted tubule) and then into the collecting tubule.
- The four main stages of urine formation are filtration, reabsorption, secretion and excretion. The first three stages all happen within the nephron.
- Two hormones, ADH and aldosterone, control water, salt and pH balance.
- There are a wide range of causes of kidney disorders, including high blood pressure, infections, injury, diabetes, cancer and polycystic kidney disease, an inherited condition.
- Symptoms of kidney disorders may be slow to appear and not distinctive.
- When kidneys fail, it is life threatening. Dialysis or a kidney transplant are the only technologies to sustain life. Neither is a cure and both have potential complications.

KEY QUESTIONS

1 Arrange the following structures of the excretory system in order, from the largest in size to the smallest: nephron, glomerulus, kidney, renal capillary.

2 **a** For each of these four processes, write a brief definition and state where they occur: reabsorption, excretion, filtration, secretion.
 b List the same four excretory processes in the sequence of their function, starting from Bowman's capsule.
 c Identify if each is an active or passive process.

3 **a** Identify which nitrogenous waste is accumulated and excreted by humans out of the following—nitrogen, ammonia, urea, uric acid.
 b Explain why this form is used in humans.

4 Name the two hormones responsible for regulating salt and water in the human kidney.

5 Describe two examples of a kidney disorder.

6 Discuss the technologies available to assist people with chronic kidney disease.

7 Evaluate the effectiveness of dialysis technologies.

Chapter review

18

KEY TERMS

accommodation
accommodation reflex
ammonia
aqueous humour
astigmatism
auditory brainstem implant
auditory nerve
basilar membrane
binocular vision
bipolar cell
bone conduction hearing aid
bone conduction hearing implant
Bowman's capsule
cataract
cochlea
cochlear implant
collecting tubule
conductive hearing loss
colour blindness
contact lens
conventional hearing aid
cortex
cornea
crystallin protein
depth perception
dilator muscle
distal convoluted tubule
ear canal
endolymph
Eustachian tube
filtrate
focal point
ganglion cell
glaucoma
glomerular filtration rate (GFR)
glomerulus
grommet
haemodialysis
hybrid cochlear implant
hyperopia
incus
inner ear
keratopathy
kidney
laser surgery
lens capsule
lens epithelium
lens fibre
light reflex
loop of Henle
macular degeneration
malleus
mechanoreceptor
medulla
middle ear
middle ear implant
monocular vision
myopia
nephron
nephropathy
nephrosclerosis
nitrogenous waste
opsin
organ of Corti
ossicle
otitis media
outer ear
oval window
pelvis
perilymph
peritoneal dialysis
photopsin
photoreceptor
pinna
polycystic kidney disease (PKD)
presbyopia
prosthesis
proximal convoluted tubule
reabsorption
refraction
refractive error
renal artery
renal dialysis
renal failure
renal vein
retina
retinal detachment
retinopathy
rhodopsin
round window
semicircular canal
sensorineural hearing loss (SNHL)
sound wave
spectacles
sphincter muscle
stapedectomy
stapes
stereocilia
tectorial membrane
tinnitus
tympanic membrane (eardrum)
urea
ureter
urinary reflux
urine
vasa recta
visual acuity
vitreous humour

REVIEW QUESTIONS

1 What are ossicles?
- **A** fluid-filled canals of the inner ear
- **B** a set of bones in the middle ear
- **C** solid calcium particles inside the canals of the inner ear
- **D** membranes that amplify sound before it passes to the auditory nerve

2 Which technology might be used to assist a person with complete hearing loss of the inner ear?
- **A** hearing aid
- **B** bone conduction implant
- **C** cochlear implant
- **D** amplifying device

3 State whether the following are true or false statements about the human ear.
- **a** The inner ear uses mechanoreceptors to detect sound.
- **b** The tympanic membrane transmits sound messages to the brain.
- **c** The middle ear is normally filled with fluid.
- **d** The inner ear has canals filled with fluid.

4 Identify at least six labels for this diagram of the human ear.

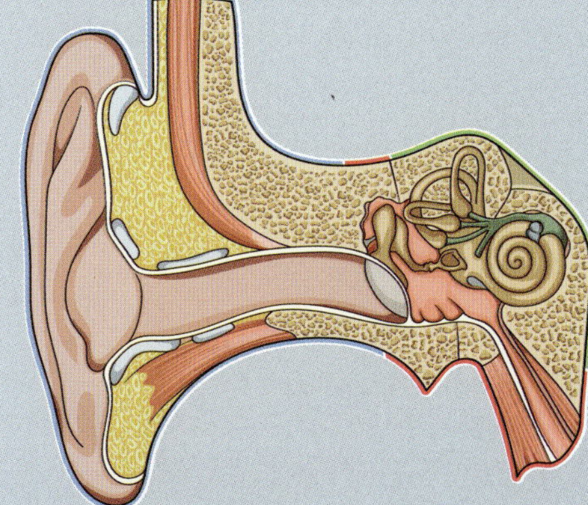

5 Distinguish between pitch and volume of a sound relating both to properties of sound waves.

6
- **a** Outline four examples to illustrate safe and unsafe levels of sound commonly present in everyday life.
- **b** Explain why very loud sound over a prolonged period can cause hearing loss.

7 Classify the following causes of hearing loss as 'conductive' or 'sensorineural' and list technologies that may assist with each type of hearing loss: Ménière's disease, otitis media, ageing (presbycusis), otosclerosis, cholesteatoma tumour, acoustic neuromas, ototoxic medication

Hearing disorder	Causes	Assistive technology
conductive hearing loss		
sensorineural hearing loss		

8 Outline the circumstances when a bone conduction implant would be the best technology to assist a person with hearing loss.

9 Which structure in the human eye focuses light to form an image?
- A cornea
- B vitreous humour
- C lens
- D all of the above

10 Which technology might be used to assist a person who is short sighted?
- A laser surgery on the retina
- B insertion of an IOL
- C spectacles with converging lenses
- D spectacles with diverging lenses

11 State whether the following are true or false statements about the human eye.
- a The eyeball is a hollow air-filled chamber.
- b The choroid helps to refract light.
- c The lens is a biconcave disc.
- d The iris is a ring of muscle.

12 Identify at least six labels for this diagram of the human eye.

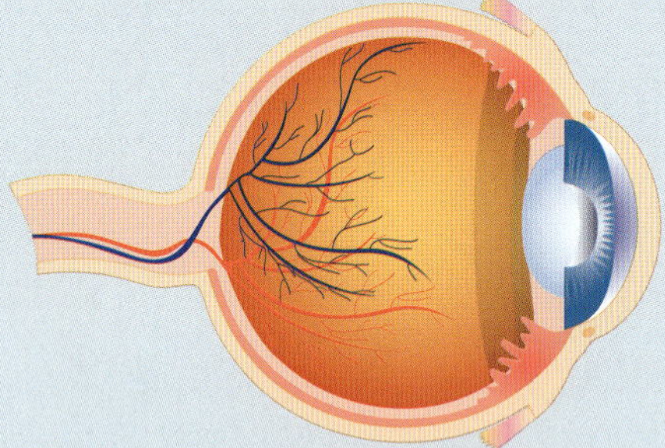

13 Apply your knowledge of the structure of the retina to suggest the safest way to look at a very bright light if protective equipment is not available.

14 Complete the table of visual disorders.

Visual disorder	Causes	Assistive technology
refractive errors		
cataracts		
glaucoma		
colour blindness		

15 a Compare the two shapes of lenses—diverging and converging.
b Explain why one lens would be used in spectacles to correct myopia and the other type would be used for hyperopia. Use a diagram to support your answer.

16 Evaluate the use of laser surgery to correct astigmatism in one eye.

17 Identify the statement that best describes a nephron.
- A microscopic filtering unit in the kidney
- B nerve cell
- C a sensory cell in the ear
- D a sensory cell in the eye

18 State whether the following are true or false statements about the human kidney.
- a The kidneys maintain water-salt balance in the body.
- b The kidneys excrete ammonia as a waste product.
- c People can have a normal life with 5% kidney function.
- d Good health requires the kidneys to operate continuously.

19 Identify at least four labels for this diagram of the human kidney.

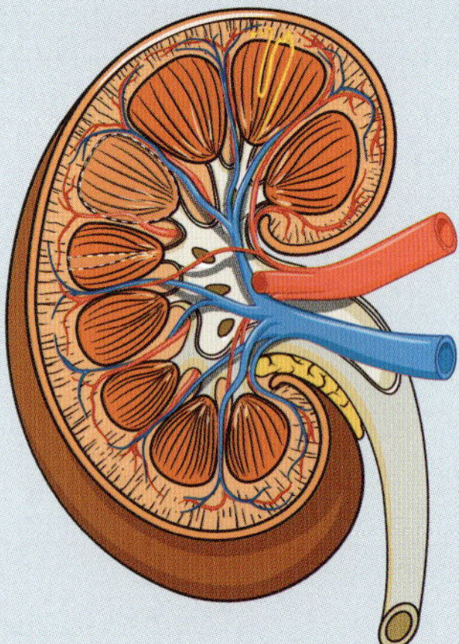

20 Draw a diagram of the kidney nephron with at least six labels for nephron structures.

21 Name the type of nitrogenous waste excreted by the human kidney and analyse the reasons why it is of this type.
22 Identify the kidney disorder shown and evaluate the impact it would have on the patient.

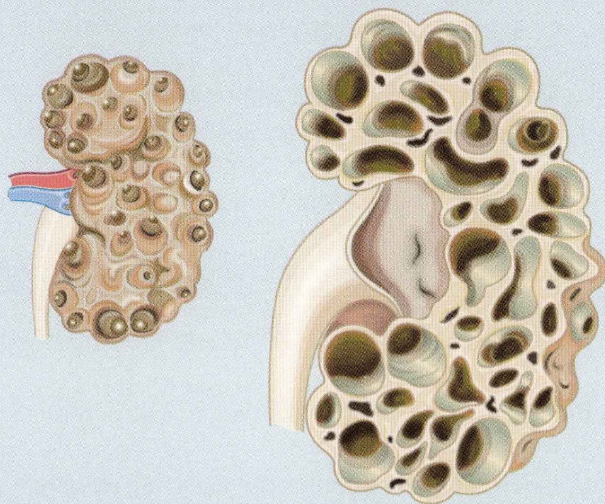

23 Discuss the effects of renal dialysis treatment on a person's lifestyle.
24 Summarise the only alternative to renal dialysis for a patient with complete kidney failure.
25 You have studied the effects of hearing, visual and kidney disorders and some of the technologies currently available to assist with these problems. Propose which is the most important of these problems for future development of improved or new technologies. Justify your choice and include examples.
26 After completing the Biology Inquiry on page 596, reflect on the inquiry question: How can technologies be used to assist people who experience disorders? Identify some causes of kidney disorders and a technology that would assist with renal failure.

MODULE 8 • REVIEW

REVIEW QUESTIONS

Non-infectious disease and disorders

Multiple choice

1 Identify the best description of homeostasis.
  - **A** Homeostasis is controlling pH and temperature within all the cells of the body.
  - **B** Homeostasis is keeping conditions within the body relatively stable and constant.
  - **C** Homeostasis is reducing metabolism when conditions are not ideal for functioning.
  - **D** Homeostasis is maintaining identical temperatures under different conditions.

2 Feedback mechanisms are important in the homeostatic control of the body's internal conditions. In a feedback system of control, which of the following is true?
  - **A** The stimulus alters the original response.
  - **B** The response alters the original stimulus.
  - **C** The response reduces the effect of the original stimulus.
  - **D** The response increases the effect of the original stimulus.

3 Which of the following describes paracrine signalling?
  - **A** The signalling molecule is carried to the target cell through the bloodstream.
  - **B** The signalling molecule acts on the same cell that secretes the molecule.
  - **C** The signalling molecule acts on target cells near the cell that secreted it.
  - **D** The signalling molecule acts on target cells distant from the cell that produced it.

4 A person is swimming in water that has a temperature of 20°C. What actions happen to regulate that person's body temperature?

	blood circulation to the skin	sweat glands	skeletal muscle
A	increased blood flow	increased secretion	decreased shivering
B	decreased blood flow	decreased secretion	increased shivering
C	decreased blood flow	increased secretion	increased shivering
D	increased blood flow	decreased secretion	increased shivering

5 Four adaptations of the Australian red kangaroo for regulating their body temperature are:
  - **i** a dense network of blood vessels close to the skin in the forelimbs
  - **ii** licking the forelimbs in hot weather
  - **iii** a powerful tail that acts as a counterbalance when hopping, allowing movement into and out of shade
  - **iv** the blood vessels in the forelimbs widen (dilate) in hot weather

  Which one of the following groups correctly classifies these four adaptations?
  - **A** i = physiological, ii = behavioural, iii = structural, iv = physiological
  - **B** i = structural, ii = behavioural, iii = structural, iv = physiological
  - **C** i = structural, ii = behavioural, iii = structural, iv = behavioural
  - **D** i = physiological, ii = behavioural, iii = physiological, iv = structural

6 Which feature is common to endotherms found in cold climates?
  - **A** a small surface-area-to-volume ratio
  - **B** a large surface-area-to-volume ratio
  - **C** large external organs such as ears
  - **D** a large blood supply

7 Plants growing in areas of water scarcity need adaptations to reduce water loss. Select the adaptation that conserves water:
  - **A** ability to close stomates during very hot conditions
  - **B** leaves with a thick, waxy cuticle
  - **C** hairs on leaves
  - **D** all of the above

8 Select the list that contains only non-infectious diseases:
  - **A** lung cancer, diabetes, Down syndrome, asthma
  - **B** cystic fibrosis, anaemia, heart disease, hepatitis B
  - **C** rickets, warts, scurvy, cancer
  - **D** lead poisoning, heart disease, diphtheria, cancer of the cervix

9 Identify the best summary description for control of the cell cycle.
   - **A** Tumour suppressor genes produce proteins that slow down cell growth and cell division.
   - **B** Damaged or mutated genes are inactivated by special genes at set points in the cycle.
   - **C** Proto-oncogenes and tumour suppressor genes repair any chromosomal damage throughout the cell cycle.
   - **D** A balance of stimulation and suppression of cell divisions results in control of the cell cycle.

10 Cancer refers to a group of diseases caused by some common malfunctions at the cellular level. Identify the most accurate description of the processes that result in cancer.
   - **A** very rapid cell division and growth
   - **B** uncontrolled meiosis and growth of abnormal cells
   - **C** production of mutant cells that do not differentiate normally
   - **D** activation of oncogenes and tumour growth

11 An expectant mother gave birth to a baby that was severely jaundiced and anaemic. The baby recovered after a blood transfusion, but concerned doctors tested the blood of both the baby and its mother. It was discovered that the baby and her older brother were positive to the rhesus factor while their mother was rhesus negative. The baby's haemolytic disease occurred because:
   - **A** The mother's antigens recognised the baby's antibodies as foreign and attacked them.
   - **B** The mother's antibodies to the rhesus factor developed in response to the blood of her first child, and then caused the immune response to the blood of her second child.
   - **C** The baby developed antibodies in response to the mother's rhesus negative blood and this caused the second baby's haemolytic disease.
   - **D** The rhesus antigen present on the mother's red blood cells caused an immune response to the baby's blood, which was free of the antigen.

12 Blood glucose levels rise after a meal is eaten, then fall again as the body uses the energy provided by the glucose, or stores the excess for later use. Diabetes is a condition that results in excess glucose in the blood. When compared to a 'normal' person, the blood glucose levels in a diabetic rise higher after eating a meal and fall at a slower rate. If the glucose levels do not fall, this would suggest the diabetic is suffering from:
   - **A** type 1 diabetes
   - **B** type 2 diabetes
   - **C** hyperglycaemia
   - **D** hypoglycaemia

13 Identify the most accurate description of epidemiology.
   - **A** Epidemiology is the collection and statistical analysis of large volumes of data on health and health services.
   - **B** Epidemiology is the study of epidemics of infectious diseases in humans.
   - **C** Epidemiology is useful for identifying the causes of infectious and non-infectious diseases.
   - **D** Epidemiology studies the patterns and causes of diseases in groups of people.

---

14 In 2006, residents of the mining town of Mount Isa in Queensland expressed concern at rising rates of childhood learning disabilities and the possible link with exposure to environmental lead. Queensland Health tested 400 children between September 2006 and December 2007 and the results are displayed in the graph.

Levels of lead in the blood greater than 10 micrograms (μg) per decilitre (dL) are considered potentially dangerous.

Note: To compare with international standards it was convenient to use the dL unit: 1 dL = 100 mL.

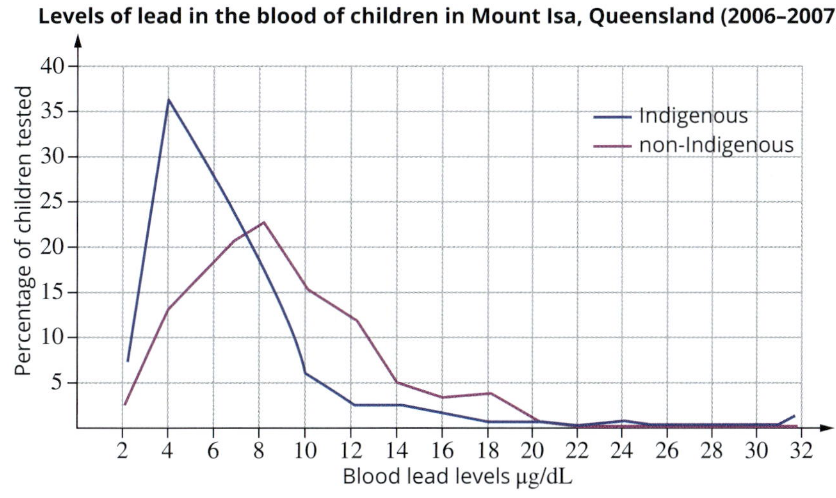

Select the most accurate interpretation of this graph.

A Fewer than half the children tested had dangerous levels of lead in their blood.

B There is a causal relationship between lead exposure and learning disabilities.

C 15% of Indigenous children and 5% of non-Indigenous children had levels >10 μg/dL.

D All Mount Isa children have some level of lead in their blood.

15 Identify the correct set of labels for structures of the human ear.

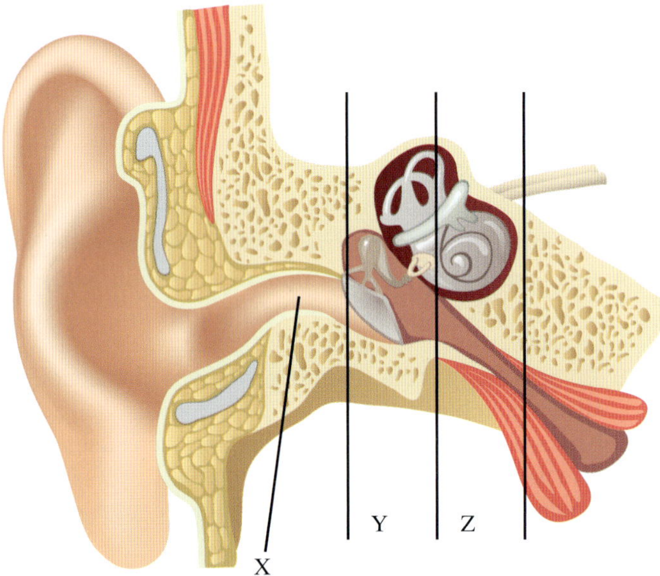

	label X	label Y	label Z
A	outer ear	semicircular canals	oval window
B	middle ear	auditory nerve	ossicles
C	tympanic membrane	middle ear	inner ear
D	outer ear	cochlea	round window

16 Hearing aids are particularly suitable for which of the following groups of people?

A people with sensorineural hearing loss

B all hearing-impaired individuals

C totally or profoundly hearing-impaired people

D people with poorly formed cochlea

17 Select the correct way to complete the sentence. Photoreceptors in the human eye:

A are located only in the fovea

B include rods and cones

C only work in dim light

D control the amount of light entering the eyes

18 Select the correct way to complete the sentence: In myopia, light entering the eye is focused:

A before the retina

B on the retina

C behind the retina

D before the lens

19 An athlete who has just completed a 20 km run will have lost an excessive volume of water through sweating. To regulate water and salt balance, which of the following occurs?

A Dilute urine will be produced.

B An increased volume of urine will be produced.

C ADH will be released into the bloodstream, increasing the reabsorption of water from the collecting ducts in the kidney.

D Vasopressin will be released into the bloodstream, decreasing the reabsorption of water from the collecting ducts in the kidney.

20 Dialysis treatment for chronic loss of kidney function requires:

A surgical removal of the diseased part of the kidney

B insertion of artificial replacement filters into the kidney

C a complete blood transfusion once a week

D lengthy treatment to cleanse the blood three to four times a week

Short answer

21 **a** Explain why homeostasis is important in organisms, using an example to support your answer.

b Describe the stages of homeostasis, again using an example to support your answer.

22 The following diagram shows how the body regulates the glucose concentration in blood.

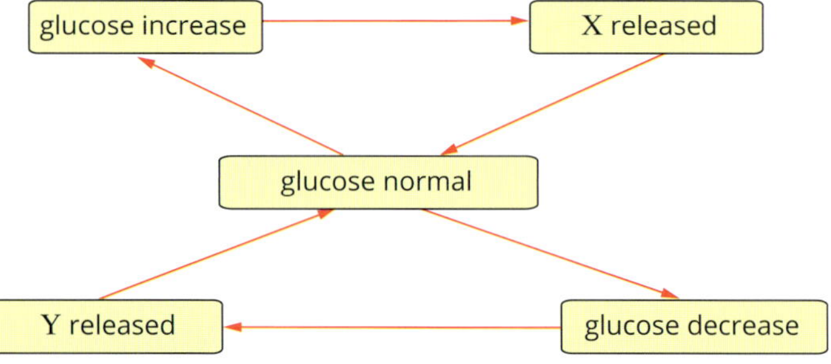

a Identify X and Y.

b Explain how the body regulates the glucose concentration in blood.

c Draw a circle on the diagram to show which process is stopped when a person suffers from type 1 diabetes.

d List the parts of the diagram that represent negative feedback.

23 Choose one plant adapted to arid conditions that you have studied in this course.

a Describe the features of the plant that make it suited to conditions of water scarcity.

b Explain how the features you have described assist the plant in maintaining water balance.

24 a Distinguish between infectious and non-infectious diseases.

b List three types of non-infectious disease and name one example of each.

25 Acetylcholine is produced by neurons and binds to receptors on the muscle cell membrane. This initiates a sequence of steps that result in muscle contraction. When acetylcholine receptors on muscle cells are blocked by an individual's own antibodies, it results in myasthenia gravis, which is characterised by weakness in the skeletal muscles.

a Which group of diseases does myasthenia gravis belong to?

b i Which group of signalling molecules does acetylcholine belong to?

ii How would the blocking of acetylcholine receptors on the muscles cause muscle weakness?

c Some patients with myasthenia gravis have an abnormally large thymus gland. What role does the thymus gland play in the adaptive immune response?

---

26 Hay fever is an allergic response in which the immune system overreacts to the presence of a previously encountered foreign antigen. The diagram below illustrates the key steps that occur in an allergic reaction.

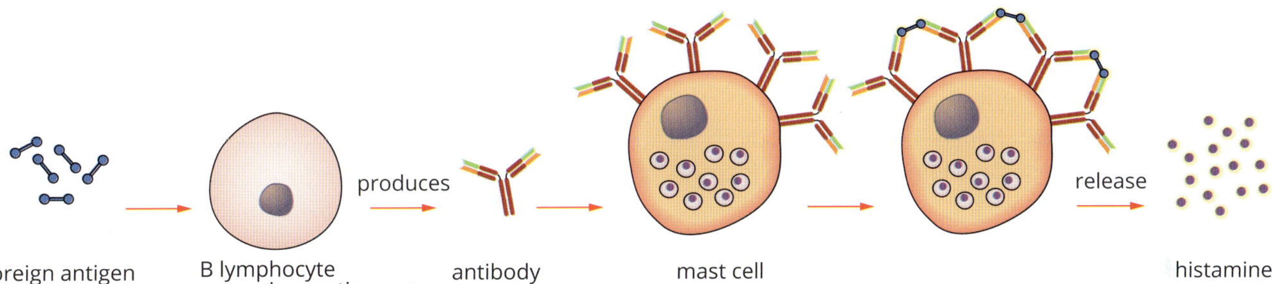

a Define the term 'antigen'.

b Identify the type of antibody involved in allergic reactions.

c Where in the body are mast cells located?

d Describe the event that causes mast cells to release histamine.

e Describe two symptoms resulting from the release of histamine.

27 Medical authorities spend large amounts of time and money on epidemiological studies.

a Explain the term epidemiology and outline its value.

b Describe how epidemiology has contributed to the effective treatment or management of a named non-infectious disease.

28 Explain how public education campaigns can benefit the health of the community at large, as well as individuals. Use at least two specific examples in your answer.

29 Account for the following:

a A colour-blind person with excellent visual acuity has difficulty finding a red object lost in the grass.

b The cause of colour blindness in humans.

30 As part of your study in this course you investigated the structure and function of the human kidney. Use your knowledge and the diagram to answer the following questions.

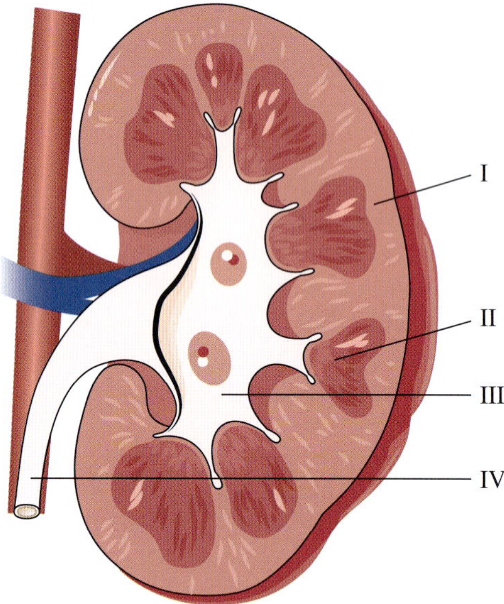

a Name structures I, II and III.

b Name the substance transported by structure IV.

c Describe the function of the kidney.

d Compare the composition of blood leaving the kidney with that of blood entering.

31 The diagram shows the changes in the core body temperature and skin surface temperature of a jogger. He jogged from $t = 0$ to $t = 60$ minutes. From $t = 60$ to $t = 90$ minutes, the jogger stopped and sat on a chair.

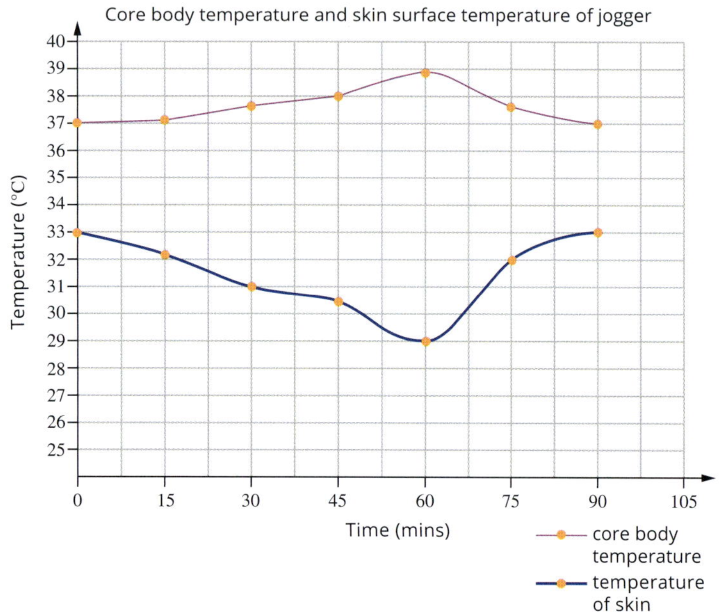

a Describe the changes in core body temperature and temperature of the skin from $t = 0$ to $t = 60$ minutes.

b Explain the patterns described in part a.

c Suggest why the temperature of the jogger's skin started to rise after $t = 60$ minutes.

d Outline the thermoregulation mechanisms in the human body that control temperature.

32 RAS is a family of proteins that are involved in cell signalling. When RAS proteins are switched on they promote cell division and growth. Mutations in the genes that encode the RAS proteins can result in these genes being permanently switched on. Normal RAS proteins are also important in the initiation of cell apoptosis (natural programmed cell death).

a Explain how mutations in *RAS* genes lead to cancer.

b Another cause of cancer in humans is the overproduction of certain proteins. One human breast cancer variant is caused by a mutation that results in the overproduction of the HER2 protein. This protein forms part of a receptor found on the surface of cell membranes. This receptor is involved in the signalling pathway that results in cell division.

Suggest how overproduction of HER2 leads to breast cancer.

c One important treatment for HER2-positive breast cancer is a drug called trastuzumab, marketed in Australia as Herceptin. This medication is a monoclonal antibody. It binds to the external domain of the HER2 receptor.

i What are monoclonal antibodies?

ii Herceptin is a humanised monoclonal antibody. What does this mean?

iii Why do scientists humanise monoclonal antibodies?

d A drug company has invented a new monoclonal antibody that it claims will be effective against HER2-positive breast cancers that have been shown to be resistant to current therapies. The company is ready to begin human trials of its proposed treatment.

i Describe an experiment to test the effectiveness of the new treatment.

ii What results would suggest that the drug company's claims are justified?

iii Outline some issues that the researchers could face in performing human trials?

iv Many human trials enlist only a few individuals. How does that affect the validity of the results?

33 Severe Combined Immune Deficiency (SCID) is an inherited disease. The most common form of the disease has been linked to mutations in the gene *IL2RG*, which is found on the X chromosome, so this form of the disease is called SCID-X1. Because of the mutation the body lacks the ability to make natural killer (NK) cells, T cells and B cells, effectively resulting in a total absence of the adaptive immune response and a depleted innate response to viruses. Typically, individuals die from viral infections within the first year of life. As girls need to inherit the trait from both parents and boys with SCID-X1 do not survive to maturity, this condition is only seen in boys.

Genetic engineering was used in 1999 for the first trials of a new treatment for SCID-X1. Bone marrow stem cells were collected from the patients and the active form of *IL2RG* was inserted into the nuclei of the cells using a viral vector.

a **i** What is a vector in this context?

ii Why is a virus used as a vector?

b By 2002, 20 boys had been treated and 18 of them actually developed the ability to make effective lymphocytes. Long-term monitoring of the patients showed that eventually five of the boys developed leukaemia and, of those, one eventually died. Further research into the causes of the leukaemia discovered that the cancer developed because the *IL2RG* gene inserted into the chromosomes by the virus was inserted in such a way as to activate a cancer-causing gene.

How might the insertion of a virus cause a cancer-causing gene to become activated?

c Gene therapy trials recommenced after the cause of the leukaemia was identified and the vector was subsequently modified to better regulate its insertion into the chromosomes. Even so, there are issues associated with the use of this therapy, especially because the recipients are all infants.

What ethical issues may arise in gene therapy of infants that do not arise when the patients are adults?

d How would you evaluate the effectiveness of this treatment and its future directions?

34 Sound is an important means of communication for humans. Hearing disorders affect personal safety, learning and social abilities.

a Draw a large, clearly labelled flow diagram to show the path of a sound wave through the ear. Use different colours to identify the different forms of energy involved as the sound is transmitted.

b If you damaged your hair cells in the organ of Corti near the oval window, explain the effect this would have on your hearing.

c Evaluate one hearing aid and one cochlear implant in terms of how they function, their value and their limitations.

35 Using diagrams, summarise the causes of two disorders of the human eye that cause vision problems and compare technologies that could assist.

Glossary

5′ cap (five prime cap) A special nucleotide that is added to the 5′ end of primary transcripts in eukaryotes. The process is known as mRNA capping and it functions to make stable, mature mRNA that is ready to undergo translation.

A

abscission The shedding of various parts of an organism, such as the shedding of leaves from a deciduous plant or the dropping of ripened fruit. From the Latin ab, away, and scindere, to cut (i.e. 'to cut away').

accommodation The ability of the eye to adjust its focal length to maintain clear images of objects at different distances.

accommodation reflex The automatic adjustment of the eye's focal length to maintain clear images of objects at different distances.

accuracy The ability to obtain an exact (or true) value.

acrosome Cap-like, membrane-bound structure at the tip of a sperm cell. The acrosome contains enzymes that break down the outer surface of the ovum, allowing fertilisation to take place.

active immunity Protection provided by the adaptive immune response. Natural active immunity develops after exposure to a pathogen while artificial active immunity develops after vaccination.

acute disease Condition that is severe but of short duration.

acute radiation syndrome (ARS) A severe condition caused by extremely high doses of ionising radiation.

adaptation (1) An inherited character that increases the likelihood of survival and reproduction of an organism or species. (2) The process by which a species becomes well suited to its lifestyle and environment.

adaptive immune response The antigen-specific immune response where an antigen is processed and recognised by the cells of the immune system; only present in vertebrates. Also known as acquired immunity.

adenine (A) Nitrogen-containing base (a purine) that occurs in nucleotides of DNA and RNA.

adenoid Mass of lymphatic tissue between the throat and nose. Adenoids are a type of secondary lymphoid tissue.

adipose Fat cells (also known as adipocytes).

adjuvant A substance added to vaccines which enhances the immune response, resulting in increased antibody production.

adrenaline A hormone that is secreted by the adrenal glands in response to stress. Adrenaline increases the heart rate, blood pressure and levels of glucose and lipids in the blood.

aestivation A long period of torpor in hot and dry conditions. See also *hibernation*.

agglutination The process in which antibodies bind to antigens on the surface of cells and form antigen–antibody complexes that clump together and activate phagocytes and the complement cascade, which leads to antigen/cell destruction.

aim A statement describing in detail what will be investigated. See also *purpose*.

albinism Inherited condition in which the skin, hair and eyes lack pigment.

aldosterone A hormone that is secreted by the adrenal glands and regulates the balance of water and electrolytes in the blood by stimulating the excreting of potassium into the urine and the reabsorption of sodium into the blood.

allele Different forms of a gene. Different alleles of a single gene will have differences in their genetic code, resulting in different forms of the phenotype. For example, the gene that codes for hairline shape in humans has two alleles—straight and widow's peak.

allele frequency The relative proportion of a particular allele in a gene pool. Typically presented as a decimal or percentage of the allele of that gene in the gene pool.

allergen An antigen that elicits an allergic response or reaction.

allergy (adj. allergic) The rapid and vigorous overreaction of the immune system to antigens called allergens. Allergic reactions involve the production of IgE by B lymphocytes and the release of histamine by mast cells.

alpha helix A coiled secondary protein structure within a polypeptide chain stabilised by hydrogen bonds between adjacent amino acids.

alternation of generation The alternation between haploid (n) and diploid ($2n$) life cycles in sexually reproducing organisms. In most animals, the diploid stage is the body of the animal and the haploid stage is the internal production of gametes (sperm and eggs). Plants alternate between a haploid gametophyte stage and a diploid sporophyte stage. See also *gametophyte* and *sporophyte*.

amine group A $-NH_2$ group. Amino acids contain an amine group, a carboxyl group and an R group.

amino acid An organic compound containing an amine group ($-NH_2$) at one end of the molecule and a carboxyl group ($-COOH$) at the other end. Amino acids are the monomer of polypeptides. Linked amino acids form the peptide chains in protein molecules.

ammonia A compound (formula NH_3) that is the first nitrogenous waste to be formed from the breakdown of proteins. Ammonia is highly toxic and it is excreted mainly by aquatic animals.

amniotic cavity The membranous fluid-filled sac surrounding the embryo, and later the fetus, of a mammal, bird or reptile. Also known as the amniotic cavity.

analytical study In relation to epidemiology, planned investigation that tests a specific hypothesis.

anaphase The stage of mitosis after metaphase and before telophase. During anaphase, chromosome pairs separate at the centromeres and move to the opposite poles of the cells.

anaphylaxis A severe allergic reaction that can be life-threatening.

ancient DNA DNA from very old fossil evidence. Special techniques must be used to extract and handle such old samples.

aneuploidy A condition in which an individual does not have the correct number of chromosomes. As a result of non-disjunction of homologous chromosomes during meiosis, individuals typically have an extra copy (trisomy) or only one copy of a particular chromosome. Triple X syndrome is an example of a human disorder caused by aneuploidy and is characterised by three copies of the X chromosome.

angiosperm A vascular plant that produces flowers and seeds enclosed within an ovary (fruit, grain or nut). Angiosperms include most trees, shrubs, herbs and grasses.

angiotensin A hormone involved in the thirst response.

animal husbandry The science of breeding and rearing animals, especially farm animals.

anonymous DNA A sequence of DNA whose location on a chromosome is not known, but which can be reliably analysed by sequencing or genotyping.

anther A structure containing chambers called pollen sacs in which pollen grains develop following meiosis. Anthers are located at the tips of stamens.

antibiotic (1) A medicine used to control bacterial infection, via restriction of bacterial reproduction. (2) A substance, produced by a microorganism or synthesised, that inhibits the growth of a type of bacteria.

antibiotic resistance The ability of a microbe to survive in the presence of an antimicrobial drug.

antibody Also known as immunoglobulins, antibodies are proteins produced by plasma cells that are highly selective for, and bind to, specific antigen molecules.

antibody-mediated immunity An immune response involving B lymphocytes that produce specific antibodies against foreign antigens. Also known as humoral immunity. Compare with *cell-mediated immunity*.

anticodon The three nucleotides on a transfer RNA (tRNA) molecule that join to the codons on mRNA by complementary base pairing during the process of translation.

antidiuretic hormone (ADH) A hormone that increases the permeability of the collecting duct of the kidney to water. This increases the amount of water reabsorbed, resulting in a smaller volume (and therefore more concentrated) urine. ADH is secreted by the pituitary gland. Also called vasopressin.

antigen A substance that stimulates antibody production and is capable of binding with an antibody produced by the immune system; antigens that induce an immune response are immunogens.

antigen presentation The presentation of antigens by antigen-presenting cells.

antigen-presenting cell (APC) A cell that uses MHC-II on its surface to present foreign antigens to helper T lymphocytes to elicit an adaptive immune response. Examples include dendritic cells and macrophages.

antigen–antibody complex A specific chemical interaction between an antibody (immunoglobulin) molecule and an antigen molecule.

antigenic variation The mechanism of changing surface antigens, usually to avoid detection or an immune attack. Employed by certain protozoans such as Plasmodium sp.

antimicrobial drug A compound that inhibits the growth of microorganisms, including bacteria, fungi, protists and viruses.

antiparallel Running in opposite directions; in relation to DNA, one strand runs in the 5′ to 3′ direction and the other in the 3′ to 5′ direction.

antiseptic A non-specific chemical that is used to kill disease-causing organisms on body surfaces.

antiserum A serum containing specific antibodies.

antiviral A drug that inhibits replication of a virus by blocking entry or exit from the cell or blocking viral replication enzymes.

apocrine gland A type of sweat gland that has a role in scent or pheromone production.

apoplastic pathway The space between the cell membrane and cell wall of plants, through which water and solutes move.

apoptosis Regulated and programmed cell death. The controlled destruction of the cell does not spill the contents and does not trigger an immune response.

appendix Small pouch-like tube that is located in lower right abdomen, connected to the colon.

appressorium (pl. appressoria) Specialised cell typical of many fungal plant pathogens that is used to infect host plants.

aqueous humour Clear fluid in the space between the cornea and the lens of the eyeball.

arteriosclerosis Thickening and hardening of the artery walls due to a build-up of fatty deposits (plaque).

artificial embryo twinning An artificial technique that splits an embryo to result in identical twins.

artificial insemination A reproductive technology that involves the deliberate introduction of male sperm into the female reproductive tract by a method other than sexual intercourse.

artificial pollination The process where humans manually facilitate the natural plant pollination process. The pollen is taken from the stamen (male part) of one flower and dusted over the stigma (female part) of another flower, resulting in fertilisation and the developments of seeds.

artificial selection The human-driven process where organisms with desirable traits are specifically selected for reproduction in order to pass on select traits to the next generation. Also known as selective breeding.

asexual reproduction Reproduction in which one parent gives rise to a new individual from its body cells. The resulting offspring are genetically identical to their parent.

assisted reproductive technology Technology that is used to achieve pregnancy. In vitro fertilisation (IVF) and surrogacy are examples of assisted reproductive technology.

astigmatism An eye condition caused by imperfection in the curvature of the cornea or lens. Astigmatism is a type of refractive error because the light entering the eye is not refracted equally in all directions, causing blurred or distorted vision.

artherosclerosis The narrowing of arteries due to the build up of fats, cholesterol and calcium.

auditory brainstem implant Surgically implanted device that connects to the brain stem to directly stimulate neurons and provide hearing sensations to people with hearing loss due to damage to the cochlear or auditory nerve. Auditory brainstem implants bypass the inner ear and auditory nerve and so are useful for people who do not benefit from hearing aids or cochlear implants. See also *auditory nerve*.

auditory nerve A cranial nerve that transmits information about sound and balance from the inner ear to the brain. Also known as the cochlear nerve or acoustic nerve.

autoantibody An antibody that acts against a self-antigen.

autoantigen An antigen that is of normal body make up to which the immune system produces antibodies. Also known as self-antigen.

autoimmune disease Disease in which there is a failure of tolerance and an adaptive immune response is directed against a self-antigen, causing T lymphocytes to attack tissues directly and B lymphocytes to produce antibodies against the self-antigen. It occurs as a result of an impaired ability of the immune system to recognise self. Autoimmune diseases can be organ-specific or generalised.

autonomic nervous system In vertebrates, the part of the nervous system that supplies nerves to the visceral organs and is under involuntary control.

autosome Chromosomes that are not sex chromosomes.

avirulence Lack of pathogenicity; unable to cause disease.

B

B lymphocyte (or B cell) Lymphocytes that when stimulated produce large quantities of antibodies specific to a particular antigen. They are responsible for the antibody-mediated immune response and include both memory and plasma cells.

Bacteria A domain of living things consisting of bacteria that live on or in animals, plants, soil or water, in environments of moderate conditions. See also *Archaea*, *prokaryote*.

bacterial transformation A type of horizontal gene transfer in which a bacterium incorporates DNA from another organism into its own DNA or takes up a plasmid containing foreign DNA. Used in biotechnology to incorporate recombinant DNA into bacteria. See also *recombinant DNA*.

bactericidal The action of any substance that kills bacteria.

bacteriostatic The action of any substance that slows the growth of bacteria.

bar graph A graph in which categorical data are represented by horizontal bars. Each bar represents one category of independent variable (such as a range of values, or a particular type of thing) and the length of the bar represents the value of the dependent value for that range or thing.

baroreceptor A receptor that detects blood pressure in vertebrates, sending the information to the brain to regulate blood pressure.

base Any of the four compounds adenine (A), thymine (T), guanine (G) and cytosine (C) present in the nucleotides of the nucleic acids DNA and RNA, forming the linking points between strands.

basal resistance A defence mechanism in plants which is activated when a pathogen is detected.

base analogue Chemical whose molecular structure is similar enough to DNA bases that it can be incorporated into sequence instead of the appropriate bases, rendering the DNA non-functional.

basilar membrane Structure that separates the second and third chambers within the cochlea. The organ of Corti rests on the basilar membrane.

basophil A type of white blood cell (leukocyte) that releases histamine during an allergic reaction.

beta cell Type of cell present in the islet of the Langerhans in the pancreas. Beta cells store and release insulin in response to changes in blood glucose concentration.

beta-pleated sheet A secondary protein structure stabilised by hydrogen bonds between different regions of a polypeptide chain that create pleat-like formations.

bias A form of systematic error resulting from the researcher's personal preferences or motivations.

binary fission A form of asexual reproduction in unicellular organisms, in which the parent cell divides into two approximately equal parts.

binocular vision Vision using two eyes positioned side by side (for example, at the front of the head). Two separate images are transmitted to the brain where they are combined into one 3D image. Many predatory animals have binocular vision (for example, humans and cats), which improves their perception of depth and distance. Contrast with *monocular vision*.

biochemical genetic testing Testing for the amount or activity of certain proteins, in order to detect any abnormality. An abnormality may be an indicator of a genetic disorder caused by changes in DNA.

biodiversity The variety of all life forms—the different plants, animals and micro-organisms, the genes they contain and the ecosystems in which they exist.

biofilm An aggregate group of bacterial cells connected to each other in a sticky extracellular matrix.

biofortification A process in which the nutritional properties of food crops are improved by biotechnology.

bioinformatics The use of mathematics, statistics and computer science to analyse and understand biological data.

biological mutagen A biological factor that can cause mutation (a change in DNA). Examples include viruses and transposable elements (TEs).

biomacromolecule A large molecule formed by joining together many monomers to form a chain. Examples include proteins and polysaccharides.

biomolecule A molecule involved in the maintenance or metabolism of living organisms.

biosecurity Measures to detect, respond rapidly to and recover from pests and diseases, including introduced species, to protect agricultural production and wildlife biodiversity.

biotechnology An interdisciplinary science that uses biological processes and organisms to create new products and develop new technologies.

biotroph A plant parasite that cannot survive in a dead host and therefore keeps it alive.

bipolar cell Cells between photoreceptors (rod and cone cells) and retinal ganglion cells in the retina of the eye. Bipolar cells relay messages from the photoreceptors to the retinal ganglion cells. See also *retinal ganglion cell.*

blastocyst The blastula stage in the development of a mammalian embryo. The blastocyst consists of an outer layer of cells that will develop into the placenta, an inner mass of cells that will develop into the embryo and a fluid-filled cavity.

blood glucose level The amount of glucose (sugar) in the blood. Glucose is used for cellular respiration and its uptake by cells is regulated by the hormone insulin.

body mass index (BMI) An index of weight-for-height that is commonly used to classify overweight and obesity in adults. It is defined as a person's weight in kilograms divided by the square of the body height in meters (kg/m^2).

bone conduction hearing aid External hearing device that creates vibrations that travel across the skull to the inner ear where they are perceived as sounds by the cochlea.

bone conduction hearing implant Surgically implanted hearing device with an external sound processor that detects sound waves and converts them into vibrations to be passed through the bone to the implant. Vibrations move from the implant to the inner ear where they are sensed as sound by the cochlea.

bone marrow Spongy tissue in the interior cavities of bones. Red bone marrow produces red blood cells, white blood cells and platelets, while yellow bone marrow consists of fat cells.

bottleneck effect The resulting impact when a large portion of a population is removed from the habitat by chance, typically as a result of a natural disaster. The effect of genetic drift is more significant on the smaller population, as the remaining gene pool has reduced diversity.

Bowman's capsule The region within the renal corpuscle of the kidney nephron into which filtered plasma flows from the glomerulus. See also *glomerulus* and *renal corpuscle.*

brumation A type of torpor undergone by many reptiles. It is similar to hibernation but differs in the metabolic processes involved. Brumation begins just before winter and lasts between one and eight months.

budding A form of asexual reproduction in which a new individual arises as an outgrowth or bud from the parent.

C

cancer A group of diseases characterised by uncontrolled cell division.

cancer vaccine Vaccine made up of cancer cells, parts of cells, or pure antigens that stimulate the immune system to prevent or fight cancer cells.

canker Abnormal woody growth caused by pathogenic organisms. A localised diseased or necrotic area on a plant part, especially on a trunk, branch, or twig of a woody plant, usually caused by fungi or bacteria.

capsid The protein coat of a virus. A coiled or polyhedral structure, composed of proteins, that encloses the nucleic acid of a virus.

carboxyl group A –COOH group containing a carbonyl and hydroxyl group. Amino acids contain a carboxyl group, an amine group and an R group.

carcinogen A substance that damages cell DNA. A carcinogen can be physical, chemical or biological.

carpel The female reproductive organ of a flower, which consists of an ovary, stigma and style. The female reproductive organ of a flower may also be called a pistil, which can consist of one or many carpels. See also *pistil.*

carrier (1) An organism infected by a pathogen and capable of transmitting the pathogen to another organism, usually without itself being affected by the pathogen. See also *channel protein.* (2) An individual that has an allele for a condition but does not express the condition because it is masked by a dominant phenotype. The carrier can pass the allele to its offspring, who will express the condition if they receive the same allele from the other parent.

cataract Condition in which the lens of the eye becomes cloudy due to the denaturation of crystallin proteins. Light passing through a cloudy part of the lens does not focus clearly on the retina causing vision impairment.

cell The smallest structural and functional unit in a living thing. All cells have a cell membrane and contain cytoplasm, organelles and genetic material (DNA). In plants, fungi and monerans, cells also have a cell wall.

cell culture Procedure in which cells are grown in a laboratory, under controlled conditions, outside an organism.

cell cycle The events in the life of a cell, from its formation by cell division through its growth and function until it divides again. It begins with the G1 stage and proceeds through S, G2, mitosis and finally cytokinesis. A G0 resting stage may also be entered during G1.

cell membrane A bilayer (double layer) of phospholipids that encloses the contents of a cell and controls the movement of substances into and out of the cell. Also called plasma membrane.

cell-mediated immunity An immune response that is mediated by T lymphocytes. Compare with *antibody-mediated immunity*.

cell replication The process by which a single cell divides into two or more daughter cells.

cellular pathogen Cellular organism that is a source of non-self antigens and causes disease. Cellular pathogens include bacteria, protozoa, oomycetes, fungi, several types of worms and arthropods.

cell wall An external structure that surrounds the cell membrane for structural support and protection. Composed of cellulose (in plants) or peptidoglycan (in bacteria).

central nervous system (CNS) The system of nerve tissues that controls most of the activities of an organism. In vertebrates it consists of the brain and spinal cord.

centriole A small cylindrical organelle consisting of a group of microtubules and occurring as a pair in the centrosome in the cells of animals and some other organisms. Centrioles are replicated in the S phase and the two pairs formed separate during mitosis and move towards the opposite ends (poles) of the cell.

centromere A part of the chromosome that attaches to spindle fibres during mitosis and where the two sister chromatids of a double-stranded chromosome are joined.

cervix The lower part of the uterus in the female reproductive system, separating the uterus from the vagina.

chaperonin (chaperone protein) A group of proteins that ensure the correct folding of newly synthesised proteins into their tertiary structures.

chemical code Symbol used to identify hazardous chemicals to ensure the correct storage, handling and disposal practices are adhered to when working with the chemical. HAZCHEM and GHS are examples of chemical coding systems used in Australia.

chemical mutagen A chemical factor that can cause mutation (a change in DNA). Examples include intercalating agents, base analogues and DNA reactive chemicals.

chemokine Cytokine that attracts white blood cells to the site of infection.

chemotaxis The movement of an organism or living cell in response to certain chemicals in the environment.

chemotherapy The use of one or more chemicals administered to a person for the treatment of cancer.

chiasma (pl. chiasmata) A point of crossing of strands of non-sister chromatids observed during the first division of meiosis.

chlorosis Loss of the normal green coloration of the leaves of plants.

chromatid One of two copies of a chromosome formed during the S stage of interphase. The two copies, called sister chromatids, are joined at a centromere.

chromosomal abnormality A change to the number of chromosomes, or the genetic composition of an individual or multiple chromosomes, which can lead to disease.

chromosomal mutation A type of mutation that typically involves entire genes or multiple neighbouring genes on a chromosome. There are five types of chromosomal mutations: duplication, chromosomal deletion, inversion, chromosomal insertion and translocation. Also known as block mutation.

chromosome A complex structure consisting of DNA strands coiled around histone proteins, carrying the hereditary information of the cell in the form of genes. All body cells in a particular species have the same number of chromosomes.

chronic disease Condition that persists for a long period of time or constantly recurs.

cilium (pl. cilia) A hairlike structure on the surface of some eukaryotic cells, consisting of a '9 + 2' arrangement of microtubules enclosed by an extension of the cell membrane. Cilia move with an oar-like motion and are usually shorter and more numerous than flagella.

cleavage The early division of a zygote into smaller cells by mitosis.

clonal selection The theory that in a group of lymphocytes, a specific antigen will activate only the lymphocyte that has a receptor that specifically recognises it. This lymphocyte will proliferate into clones of itself.

clone A biological entity (such as a gene, cell, tissue or organism) that is a genetically identical copy of another entity.

cloning (1) The process of replication that creates a new biological entity, such as a gene, cell, tissue or organism. (2) In animals, the creation of a new individual by transferring the nucleus of a somatic cell into an enucleated egg, which is then implanted for development. The resulting individual will be genetically identical to the parent that provided the nucleus.

clotting factor Protein that causes blood to clot, controlling bleeding. Examples are factor VII (FVII), factor X (FX) and tissue factor (TF).

co-dominant phenotype A phenotype in a heterozygote that results from the expression of both alleles. An example is the AB blood group in humans. Contrast with *dominant phenotype* and *recessive phenotype*.

cochlea Fluid-filled inner ear structure where vibrations are transformed into neural signals to be sent to the brain. The cochlea is a spiral-shaped tube that resembles a snail shell ('cochlea' comes from the Greek word 'kokhliās', meaning 'snail').

cochlear implant Surgically implanted hearing device for people with severe sensorineural hearing loss. Cochlear implants bypass the inner ear and send sound signals to the brain by stimulating the auditory nerve at different frequencies. Also known as the bionic ear.

coding strand The strand of DNA that has the same nucleotide base sequence as the mRNA strand produced by transcription (uracil in the mRNA in place of thymine in the DNA). Contrast with *non-coding strand*

codon Basic unit of the genetic code. A sequence of three nucleotides on mRNA that codes for a particular amino acid, or indicates the beginning or end of translation.

cofactor A chemical component such as a metal ion or coenzyme that is required for the proper function of proteins.

collecting tubule Long narrow tube in the kidneys that concentrates and transports urine from the nephrons to the renal pelvis and ureters.

colour blindness Vision disorder in which certain colours cannot be distinguished (for example, blue and purple). Colour blindness is an inherited, sex-linked disorder caused by a lack of functional colour-detecting cone cells in the eye.

complement protein Protein that is able to kill foreign cells by lysis. There are over 30 different complement proteins that are activated in response to antigen–antibody complexes, antigens and carbohydrates on the surfaces of some bacteria and parasites.

complementary base pair The pairing in DNA and RNA molecules of the nitrogenous bases between two strands. In DNA adenine always pairs with thymine and cytosine always pairs with guanine.

complete dominance In sexual reproduction, the expression of only one phenotype in all heterozygous individuals.

complete penetrance When all individuals with an affected genotype display the affected phenotype.

conception The time when fertilisation occurs in humans.

condensation polymerisation The reaction in which monomers are joined to create a polymer by the removal of water; a bond is formed between them.

conduction In relation to thermal conduction in biology, the transfer of heat via physical contact.

conductive hearing loss Hearing disorder in which sound is not conducted efficiently through the outer and middle ear, resulting in reduced sound volume.

conjugated protein A protein that contains a non-protein (prosthetic) group.

conjugation In eukaryotes, a form of sexual reproduction in which two single-celled organisms fuse together and exchange genetic material.

conservation The preservation and protection of biodiversity.

constant region The region of antibody molecules that remains the same and interacts with receptors on the body's cells.

contact lens A thin lens designed to be placed on the surface of the eye to correct refractive disorders. See also *refractive error*.

continuous variable A variable that can have any number value within a given range.

continuous variation Variation within a population that is smoothly graded.

convection Transfer of heat via currents in liquid or gas.

conventional hearing aid External electronic hearing device in which sounds are detected, amplified and transmitted to the ossicles in the middle ear.

cornea The transparent, outermost layer of the eye that covers the iris and white of the eye.

coronary heart disease Condition in which arteries are narrowed and blood flow to the heart is reduced due to the build-up of fat deposits (plaque) on artery walls. Coronary heart disease is a leading cause of heart attack.

corpus luteum The remains of a Graafian follicle after ovulation. The corpus luteum becomes an endocrine organ in the ovary, secreting oestrogen and progesterone during the latter part of the oestrous cycle in mammals.

cortex Outer layer of the kidney.

counter current circulation An animal adaptation that avoids heat loss via extremities by passing warm blood alongside cold blood.

cross The intentional breeding of two genetically different organisms which results in offspring that inherit genetic material from each parent.

cross-breeding The breeding of an individual through the hybridisation or mating of two different species, breeds of varieties of organism.

cross-pollination Transfer of pollen from the male reproductive organ of a flower to the female reproductive organ of a flower on another plant of the same species.

crossing over The exchange of chromosomal material between non-sister chromatids of a homologous chromosome pair during prophase I of meiosis.

cryopreservation The process of cooling biological samples to very low temperatures (usually −135°C) for long term storage and transport.

cryptic species Species that are morphologically very similar or indistinguishable but do not interbreed. Cryptic species are often classified as a single species based on morphology but DNA sequence data can reveal two or more distinct species.

crystallin protein Structural protein in the lens and cornea of the eye.

cuticle A protective waxy coating on the surface of plant organs (e.g. leaves).

cutin A waxy water-repellent substance in the cuticle of plants, consisting of highly polymerised esters of fatty acids.

cystic fibrosis Genetic disorder that causes the secretion of thick sticky mucus that accumulates in the lungs and leads to frequent infections and lung damage over time. The pancreas and digestive system of people with cystic fibrosis can also be affected.

cytogenetic testing A genetic testing method used to identify chromosomal abnormalities.

cytokine One of a group of peptides and proteins released from cells that are important in cell signalling, particularly between cells of the immune system.

cytokinesis The division of a cell following mitosis or meiosis, when the cytoplasm divides and the cell splits into two daughter cells.

cytosine (C) A nitrogen-containing base (a pyrimidine) that occurs in nucleotides of DNA and RNA.

cytotoxic A substance or process that is toxic to a cell and can cause death of that cell.

cytotoxic T lymphocyte T lymphocyte that is stimulated by cytokines to bind to antigen–MHC I complexes on infected host cells and release cytotoxic compounds that destroy the infected cells.

D

data The measurements or observations collected during an investigation.

daughter cell A new cell formed by cell replication.

decibel (dB) A unit of measurement that indicates how loud a sound is. The decibel scale is a logarithmic scale not a linear scale.

deep intronic mutation A mutation in an intron sequence that is more than 100 base pairs from either junction site.

de-extinction The concept of using new genetic technologies to bring extinct species back to life.

defensin Molecules active against bacteria, fungi and certain viruses.

defensive molecule Molecules that are produced to fight off pathogens. Examples include complement proteins and cytokines.

degenerate More than one codon may code for a particular amino acid.

deletion mutation When a point mutation: a type of mutation that involves the loss of one or two nucleotides in a sequence. This type of point mutation causes a frameshift mutation. When a block mutation: multiple genes are cut from a chromosome.

demyelination Damage to the myelin sheath that surrounds nerve cell axons, which limits or stops the ability of the nerve to transmit electrical impulses.

denature (n. denaturation) A change in the shape of a protein caused by alteration to the hydrogen bonds, disulfide bridges, hydrophobic interactions and van der Waals forces that create the tertiary structure of the protein. Misshapen proteins are biologically inactive and non-functional. If a protein becomes fully denatured, the reaction is non-reversible and the protein remains non-functional. See also *renature*.

dendritic cell A type of antigen-presenting cell.

deoxyribonucleic acid (DNA) See *DNA (deoxyribonucleic acid)*

deoxyribose The five-carbon sugar molecule found in DNA. Deoxyribose is derived from ribose but lacks an oxygen molecule; it has a hydrogen atom rather than a hydroxyl group.

dependent variable A variable that may change in response to a change in the independent variable and is measured or observed.

depth perception The visual ability to judge spatial relationships between objects in three dimensions (3D).

descriptive study In relation to epidemiology, a study that show patterns in disease distribution in a population.

diagnostic test A test used to identify an individual's specific condition, disease or illness.

dilator muscle A muscle in the eye that can enlarge the pupil, allowing more light to reach the retina.

dialysate Fluid that passes through a dialyser (machine that performs dialysis) during kidney dialysis.

diffusion The passive movement of a solute from a region of higher concentration to a region of lower concentration.

dihybrid cross A cross between pure lineages that exhibit two different phenotypes. An example is a cross between a pea with dominant phenotypes of yellow seeds and red flowers, and a pea with recessive phenotypes of green seeds and white flowers.

dinucleotide Two nucleotides joined together through a condensation polymerisation reaction that joins the phosphate of one nucleotide to the 3' end of the other nucleotide's sugar molecule. Water is removed in the process.

dipeptide Two amino acids joined by a peptide bond through a condensation reaction, which results in the production of a dipeptide and a molecule of water.

diploid A condition in which a cell has two (di-) sets of chromosomes, typically one set inherited from each parent in sexually reproducing organisms. The chromosomes with the same genes exist in homologous pairs. Denoted as $2n$.

discontinuous variation Variation within a population that is segmented in two or more groups.

discrete variable A variable that can have only certain values. For example, the number of individuals in a population can only be whole numbers.

disease Disorder in the structure and function of an organism.

disinfectant A substance that inhibits the growth of disease-causing microorganisms, or kills them. A disinfectant is generally toxic to human cells, but if diluted sufficiently can be used as an antiseptic.

distal convoluted tubule Structure of the nephron in the kidney, between the loop of Henle and the collecting tubule. The distal convoluted tubule plays an important role in regulating pH and levels of sodium, potassium and calcium.

DNA amplification The process of creating millions of identical copies of a DNA sample using the polymerase chain reaction (PCR).

DNA (deoxyribonucleic acid) A double stranded nucleic acid that contains the genetic code in its sequence of bases. DNA is found in all organisms and most viruses, in chromosomes, as well as in mitochondria and chloroplasts.

DNA polymerase An enzyme that catalyses the formation of polymers of DNA by linking nucleotides into a chain by complementary base pairing with a template strand.

DNA profile Genetic pattern that is generated by amplifying multiple genetic markers called short tandem repeats (STRs). DNA profiles are unique to individuals (except identical twins) and are used to establish identity. Also known as a DNA fingerprint.

DNA reactive chemical Any chemical that is known to react directly with DNA.

DNA sequencing The determination of the sequence of bases in a fragment of DNA. DNA sequencing can be used to determine relationships between individuals of a species and for determining the entire genome of an organism.

dominant phenotype A phenotype or trait that is expressed when only one allele is present (i.e. in a heterozygote), masking the recessive phenotype. Contrast with co-*dominant phenotype* and *recessive phenotype*.

double helix The structural shape of a DNA molecule, consisting of two linear lengths of nucleotides twisted spirally about each other and connected by phosphodiester bonds.

duplication mutation A block mutation that involves the repetition of a section of a chromosome, usually containing multiple genes. The duplication may involve thousands of repeats, significantly lengthening the chromosome.

E

ear canal Tube running from the outer ear to the middle ear. The ear canal is part of the outer ear and functions to channel sound waves from the external environment to the middle ear.

eccrine gland A type of sweat gland that has a role in controlling body temperature.

ecosystem service Processes carried out in ecosystems that contribute to the quality of life for humans. For example, oxygen production by plants, food production, climate control, crop pollination and cultural and recreational benefits.

ectoderm The outermost layer of the three primary germ layers in the early embryo.

ectoparasite Parasite that lives on the outside of its host.

ectotherm (adj. ectothermic) A cold-blooded animal.

effector cell A cell that responds to signalling molecules. For example, a cell in which a signal transduction pathway activates an enzyme and causes a metabolic change in the cell.

egg (ovum) The female reproductive cell in animals which, once fertilised by a sperm, will develop into an embryo. Also known as a gamete or ovum.

electromagnetic radiation (EMR) The energy that is emitted by radio waves, microwaves, infrared, visible light, ultraviolet light, X-rays and gamma rays.

elicitor Extrinsic or foreign molecules often associated with plant pests, diseases or synergistic organisms. Elicitor molecules can attach to special receptor proteins located on plant cell membranes.

embryo The stage in the development of a vertebrate, between the fertilisation of the ovum and the development of the characteristics of the adult organism (the fetus).

endocrine system The animal body system that is responsible for the production of hormones.

endoderm The innermost layer of the three primary germ layers in the early embryo.

endolymph Fluid within the chambers of the inner ear.

endometrium The inner lining of the uterus in mammals. It is the site of implantation of the embryo after fertilisation.

endoparasite A parasite that lives within its host.

endotherm (adj. endothermic) An animal that maintains a more or less constant body temperature, which is usually higher than the temperature of the surrounding environment.

enteric nervous system An extensive network of nerve cells (and reflexes) within the wall of the gut that coordinate digestive functions.

environmental health Branch of public health concerned with the effect of the natural and built environment on human health (for example, water or air quality).

enzyme A protein molecule that acts as a biological catalyst. Enzymes speed up rates of reactions that would otherwise take place much more slowly. Their action is often specific to only one type of reaction.

eosinophil A type of white blood cell that promotes inflammation. See also *granulocyte*.

epidemic The sudden increase in the number of cases of a disease above what is normally expected in that population in that area.

epidemiology Study of the distribution patterns and causes of disease.

epididymis Tightly coiled tube behind the testis that stores and transports sperm from the testis to the ductus deferens. It is present in the male reproductive systems of reptiles, birds and mammals, including humans. See also *ductus deferens*.

epigenetics The study of molecular events such as methylation that occur on DNA but do not alter the DNA sequence, but result in different phenotypes.

epithelium (adj. epithelial) A thin layer of tissue covering the external surfaces of a multicellular organism and also lining the inner surfaces of internal structures such as intestines and lungs.

epitope A part of an antigen to which an antibody attaches.

error The difference between the true value and the measured value.

eukaryote An organism whose cells contain a membrane-bound nucleus and other membrane-bound organelles. Protists, fungi, plants and animals are eukaryotes.

Eustachian tube Tube that links the upper portion of the pharynx to the middle ear. The Eustachian tube plays roles in pressure equalisation and mucus drainage.

evaporative cooling The process of cooling by evaporation. Some animals wet their skin or fur to achieve cooling by evaporation. See also *evaporation*.

ex situ Meaning 'out of place' or 'out of position'. Ex situ conservation refers to conservation of species outside their natural habitat.

exon The region of a gene that codes for a protein.

exteroreceptor Receptors that detect external stimuli.

extracellular fluid All fluid outside of cells.

F

F1 generation The generation consisting of offspring of a cross between members of the parental generation.

F2 generation The offspring of a cross between members of the F1 generation.

factor VII (FVII) A blood coagulation factor; one of the proteins that causes blood to clot.

factor X (FX) A blood coagulation factor; one of the proteins that causes blood to clot.

fallopian tube See *oviduct*.

fermentation Stage in the breakdown of glucose to yield energy for the production of ATP, which follows glycolysis when there is no oxygen present. Produces either lactic acid (in most animals) or alcohol in the form of ethanol (in most plants and microorganisms).

fertilisation Penetration of an egg by sperm and fusion of the egg and sperm nuclei. In animals, fertilisation may be external, occurring outside an animal's body, or internal, occurring in the reproductive tract of an animal.

fetus The stage in the development of a mammal, following the embryonic stage. The fetus has all of the major structures of the adult mammal. In humans this stage lasts from the eighth week of gestation until birth.

fever An increase in body temperature that results from the regulated body temperature set point in the hypothalamus of the brain being set to a higher level by inflammatory cytokines, to slow the replication of bacteria and improve the adaptive immune response.

fibrous protein A type of protein that forms long fibres and provides structural support to cells and tissues.

filament The stalk of a stamen that bears the anther.

filtrate A substance that has passed through a filter, such as the liquid that passes through the kidneys. Kidney filtrate is formed when fluid passes from a Bowman's capsule into a nephron.

fixed allele An allele whose frequency in a population is 1.0.

fluorescence in situ hybridisation (FISH) A method used in genetic testing to identify the specific location of a gene or allele on a chromosome, as well as to detect changes in chromosome structure such as deletion or duplication of certain sections. A fluorescent DNA probe binds to the target sequence on the chromosome.

focal point Point on the retina of the eye where light rays enter the eye and are refracted (bent) by a combination of cornea, aqueous and vitreous humour as well as the accommodating lens.

follicle A mature ovarian follicle that ruptures during ovulation to release the ovum. Also called Graafian follicle.

follicle stimulating hormone (FSH) Peptide hormone that promotes the development of ovarian follicles and the secretion of oestrogen. FSH is produced and secreted by the anterior pituitary gland.

founder effect Occurs when a small portion of a population disperses to a new location and becomes genetically isolated from the main population. The allele frequencies of the founding population are completely dependent on those of the specific individuals that were relocated and therefore may be significantly different from those of the original population.

fragmentation A form of asexual reproduction in which an organism breaks into two or more parts, each of which regenerates the missing pieces to form a complete new organism.

frameshift mutation A type of gene mutation that involves the insertion or deletion of one or two nucleotides, affecting every triplet (and codon) in that gene from the point of mutation.

frequency In relation to epidemiology, the number of health events in a population (for example, the number of cases of tuberculosis or diabetes). Frequency measures may also take into account the size of the population and/or the number of people within the population who can potentially be affected by the disease (for example, adult females are only considered in frequency measures of cervical cancer).

Fungi One of the five kingdoms of eukaryotic organisms, consisting of heterotrophs that are composed of hyphae and reproduce by spores. They include mushrooms, lichens (lichenised fungi), yeasts and moulds.

G

gall Abnormal outgrowth of plant tissue.

gametangia Cell or organ in which gametes are formed in ferns, algae or some plants.

gamete A haploid cell capable of fusion with another haploid cell to form a zygote. In vertebrates the gametes are sperm and egg cells.

gametophyte The sexual, gamete-forming, haploid stage in the life cycle of a plant. The gametophyte produces haploid gametes by mitosis. The haploid gametes fuse to form a diploid zygote, which develops into a sporophyte. In bryophytes (for example, mosses), the gametophyte stage is the prominent stage, while the sporophyte stage less conspicuous. See also *sporophyte* and *alternation of generations*.

ganglion cells A cluster of nerve cells. A layer of ganglion cells is found in the retina.

gastrula An early stage in the development of an embryo, consisting of three layers of cells: ectoderm (outer layer), mesoderm (middle layer) and endoderm (inner layer).

gastrulation A series of cell and tissue movements at the blastocyst stage of animal development, during which the embryo is reorganised to form a gastrula.

gel electrophoresis A technique used for separating fragments of DNA, or different proteins, based on their molecular weight (or length). Fragments migrate through a gel at rates that are dependent on their length and charge.

gemmule Cell produced asexually by sponges via budding, from which a new individual develops. Gemmules can remain dormant until conditions for growth are present.

gene A section of DNA that contains instructions for making a protein or RNA molecule. Particular genes have specific locations on chromosomes. Genes are copied and passed from one generation to the next during reproduction.

gene cloning The production of identical copies of a gene.

gene editing The modification of genes by removal, substitution or alteration by mutation, without necessarily introducing a foreign gene.

gene expression The process that leads to the transformation of the information stored in a gene into a functional gene product (usually a protein or RNA molecule).

gene flow The movement of alleles between individuals of different populations; includes the dispersal of pollen and seeds in plants.

gene-for-gene resistance The relationship between plant host resistance genes and pathogen virulence genes which determines if infection of the host is successful. At interacting loci, plants may have alleles for resistance or susceptibility, while pathogens may have alleles for virulence or avirulence.

gene linkage The tendency of genes that are located close together on a chromosome to be inherited together.

gene mapping The determination of the location of genes and the distance between them, on a chromosome.

gene pool All the alleles possessed by members of a population, which may potentially be passed to the next generation.

gene probe A section of DNA with a base sequence complementary to a particular gene, to which it base pairs. When labelled with a fluorescent dye or radioactive marker, it is used to find a particular DNA sequence within a DNA fragment.

gene regulation Processes that control gene expression, turning genes on or off.

gene therapy The replacement of faulty genes by genetic engineering techniques.

genetic code The linear sequence of three nucleotides in DNA or RNA that determines, or codes for, the sequence of amino acids in a protein.

genetic disease Disease caused by abnormalities in the genome. Genetic diseases are inherited (passed from parents to offspring) or caused by mutations to the DNA during gamete formation. Also known as a genetic disorder. See also *germline mutation* and *mutation*.

genetic drift Random changes to allele frequencies in a gene pool as the result of a chance event. This has a more significant impact on smaller populations, as the chance death of one individual could eliminate an allele from the gene pool.

genetic engineering The direct manipulation of an organism's genetic material.

genetic marker A segment of DNA that can be reliably analysed by sequencing or genotyping. When associated with a trait, genetic markers usually have a known location on a chromosome. When used for population analysis, the location of the genetic marker in the genome is not always known.

genetic screening Genetic testing of a large number of individuals to identify carriers for a particular disease or those at risk of developing the genetic disease.

genetic testing DNA analysis to determine the genetic status of an individual or embryo. Genetic testing is carried out to detect specific alleles, mutations, genotypes or karyotypes that are associated with heritable traits, diseases or predispositions to diseases, such as cystic fibrosis, Down syndrome or Turner syndrome.

genetic variation Variation in genes or alleles within a population or species.

genetically modified cell Cell that contains genetic material that has been directly manipulated. Genetically modified cells can be grown in vitro to study cellular processes or to produce cells for therapeutic purposes (for example, cancer immunotherapy).

genetically modified organism (GMO) An organism with a genetic modification (GM) made by transfer of specific genes from another organism (transgenic) or by gene editing techniques.

genome The complete set of genes or DNA in an organism.

genomics Field of science involving the sequencing and analysis of genomes.

genotype (adj. genotypic) (1) The total set of genes of an organism. (2) The combination of alleles for a trait carried by an individual.

genotypic ratio The number of times a genotype would be expected to appear in the offspring of a test cross.

geographic distribution (1) The geographic extent of a group of organisms. It is commonly applied to the extent of a population or species. (2) In epidemiology, the relationship between the prevalence of a disease and the area in which it occurs.

germ cell Any cell in an organism that gives rise to gametes.

germ layer The primary layer of cells that are formed during embryogenesis. Animals with bilateral symmetry have three layers: endoderm, mesoderm and ectoderm. Animals with radial symmetry have two layers: endoderm and ectoderm.

germ theory The scientific theory that all living matter comes from existing living matter.

germination The development and growth of a plant from a spore or seed, sometimes after a period of dormancy. Germination is usually triggered by environmental factors such as temperature or moisture.

germline mutation A mutation that can affect gamete formation and can therefore be inherited by offspring.

gestation The process and period of embryonic and fetal development inside the uterus of viviparous animal, from fertilisation to birth. Also known as pregnancy. During a single pregnancy, there may be one or multiple gestations (for example, twins or triplets). See also *viviparous*.

glaucoma Eye condition which causes damage to the optic nerve and vision loss. Glaucoma commonly results from fluid accumulation in the eye, which puts pressure on the optic nerve and disrupts the transmission of visual messages to the brain. If left untreated, glaucoma can lead to blindness.

globular protein Type of protein that is folded and coiled to form a compact spherical shape. It has a tertiary or quaternary structure specific to its function; for example, enzymes.

glomerular filtration rate (GFR) Measure of the flow rate of fluid filtered through the kidney. GFR is used to measure the level of kidney function.

glomerulus (pl. glomeruli) A cluster of capillaries in the renal corpuscle of the kidney nephron. Filtration occurs through the walls of the capillaries that form the glomerulus, across into the Bowman's capsule. See also *Bowman's capsule* and *renal corpuscle*.

glucagon A hormone produced in the pancreas that causes glycogen to be broken down in the liver, releasing glucose into the blood, thus opposing the effect of insulin.

glucose A simple sugar (formula $C_6H_{12}O_6$) that is an important energy source in living organisms and is a component of many carbohydrates. Glucose is the product of photosynthesis and the substrate for respiration.

glycogen The main carbohydrate storage molecule in animals. Glycogen is a complex carbohydrate consisting of glucose subunits. It is stored in the liver and muscle cells and easily converted to glucose.

gonad Reproductive organ in which gametes are formed; testes in male animals and ovaries in female animals.

granulocyte A type of white blood cell containing granules (sacs filled with enzymes that digest pathogenic microorganisms). Neutrophils, basophils and eosinophils are granulocytes.

grommet A surgically implanted tube that allows fluid to drain from the middle ear, preventing fluid build-up and infection.

guanine (G) A nitrogen-containing base (a purine) that occurs in nucleotides of DNA and RNA.

guard cell Specialised epidermal leaf cell, a pair of which surround stoma on the leaf surface. Guard cells open and close the stoma depending on water availability.

gular fluttering A cooling behaviour in birds involving flapping the throat.

gymnosperm Group of plants that produces seeds by cones rather than flowers. Gymnosperms include conifers, cycads and ginkgo.

H

haemodialysis An artificial blood filtration process duplicating the function of kidneys.

haploid Containing one set of chromosomes (half the normal number of chromosomes of a diploid cell). Denoted as *n*.

haplotype Haploid genotype; genetic variation (genes or alleles) present on one chromosome.

Hardy-Weinberg equilibrium A state in which the allele frequencies in a population will remain constant between generations, assuming no outside influences.

haustoria A slender projection from the root of a parasitic plant, such as a dodder, or from the hyphae of a parasitic fungus, enabling the parasite to penetrate the tissues of its host and absorb nutrients from it.

hCG (human chorionic gonadotropin) A hormone produced by the placenta during pregnancy.

health education program Education program designed to improve the health and wellbeing of a population.

healthy eating pyramid A set of recommended guidelines for a healthy, balanced diet.

heavy chain The polypeptide chain that forms the 'stem' of a Y-shaped antibody molecule.

helminth Parasitic worm that lives in or on a host. Examples are hookworms and threadworms.

helper T lymphocyte Helper T lymphocytes bind to antigen-MHC II complexes on antigen-presenting cells and activate B lymphocytes to secrete antibodies, macrophages to phagocytose and cytotoxic T cells to kill infected cells.

hemizygote (adj. hemizygous) A diploid cell or organism with only one copy of a particular chromosome. Human males are hemizygotes because they have one X chromosome rather than two.

herd immunity Phenomenon in which vaccination of a large proportion of a population provides protection from a pathogen to non-immune or non-vaccinated individuals.

hereditary Able to be passed from parent to offspring, or from one generation to the next.

heritable trait Traits passed down from parents to offspring via genetic processes.

hermaphrodite (1) In plants, having both stamens and carpels in the same flower. (2) In animals, producing both male and female gametes.

heterogametic Having different sex chromosomes; for example, human males with XY.

heterozygosity A measure of genetic variation (number of alleles) within a population. The higher the heterozygosity, the higher the variation in a population.

heterozygote (adj. heterozygous) A diploid individual with different alleles for a particular gene.

hibernation A long period of torpor during the colder months of the year. See also *aestivation*.

histamine An organic compound involved in inflammatory responses and allergic reactions, which causes surface blood vessels to dilate and become more permeable to immune cells and fluids. Common hay fever symptoms such as runny nose and eyes and sneezing are the result of histamine action and are aimed at flushing out allergens.

histone One of various alkaline proteins that arrange DNA into nucleosomes.

homeostasis (adj. homeostatic) The maintenance of a more or less stable internal environment, even when external conditions change.

homogametic Having two similar sex chromosomes; for example, human females with XX.

homologous chromosome Matching pairs of chromosomes in a diploid organism. Homologous chromosomes carry the same genes in the same loci.

homozygote (adj. homozygous) A diploid individual with two identical alleles at a particular genetic locus.

horizontal gene transfer Transfer of genes between cells, sometimes of different species, such as transfer of plasmids between bacteria. Contrast with *inheritance of genes from parent to daughter cell through cell division (vertical gene transfer)*.

hormone A molecule that regulates the growth or activity of those cells capable of responding to it (target cells). Hormones are produced by specialised cells within an organism.

host An organism that carries a parasite.

huddle An animal behaviour in which individuals gather in a group and press their bodies together to provide protection from the cold.

Human Genome Project A worldwide project started in 1990 to determine the human genome. The project was completed in 2003.

hybrid The result of mixing, through sexual reproduction, two individuals of different breeds, varieties, species or genera. This process is known as hybridisation.

hybrid cochlear implant Hearing device that combines cochlear implant technology and hearing aid technology. The electro-stimulation technology of cochlear implants is used to process high frequency sounds and the acoustic amplification technology of hearing aids is used to amplify low frequency sounds.

hybridisation See *hybrid*.

hybridoma The product of the fusion of an immortal cell line with a B lymphocyte to produce an immortal B lymphocyte; used in the production of antibodies.

hydrogen bond A weak chemical bond between an electronegative atom (for example, oxygen or nitrogen) and a hydrogen atom. Water molecules (H_2O) are held together by hydrogen bonds.

hydrophyte A plant that lives in water.

hygiene Practices that prevent the spread of disease and maintain cleanliness.

hyperglycaemia A higher than normal blood glucose level.

hyperopia Disorder of the eye in which light is focused behind, rather than on, the retina, causing near objects to appear blurry. Hyperopia is a refractive error cause by an eyeball that is shorter than normal or a cornea that is too flat. Also known as far-sightedness or long-sightedness. Contrast with *myopia*.

hypersensitive response (HR) A defence mechanism used by plants to prevent the spread of infection. The host plant induces programmed cell death (apoptosis) to restrict the pathogen to the infected site, limiting its access to healthy plant tissue.

hypersensitivity reaction Undesirable reaction of the normal immune system to antigens, including allergens (allergic reaction) and self-antigens (autoimmunity). There are four types of hypersensitivity reactions: type I, II, III and IV.

hypertension Condition in which blood pressure in the arteries is elevated increasing the risk of heart attack and stroke. Also known as high blood pressure.

hypha (pl. hyphae) Minute thread-like structures that make up the mycelium of a fungus.

hypoglycaemia A lower than normal blood glucose level.

hypothalamus In vertebrates, the base and part of the sides of the brain immediately below the thalamus. In mammals the hypothalamus directly or indirectly controls aspects of the internal environment, particularly through the secretion of various hormones such as anterior pituitary hormones.

hypothesis A tentative explanation that is based on observation and prior knowledge. A hypothesis must be testable and falsifiable (can be proven false).

I

immortal cell line A cell line that can continually undergo division without the mutations that would normally occur as a cell ages and can therefore be cultured for long periods.

immune response Activation of the body's immune system due to the presence of antigens from a foreign body or organism.

immune system System of specialised cells, tissues and organs that defends the body against agents of disease, such as pathogens, mutagens and cancerous cells.

immunisation program A series of vaccines recommended by the government with the aim of preventing and controlling the spread of infectious diseases.

immunity Ability of organisms to defend against infection and disease. See also *adaptive immunity* and *innate immunity*.

immuno-suppressant drug A drug that inhibits the immune response against foreign particles or tissues. Immuno-suppressant drugs are used to prevent the rejection of transplanted organs or tissues.

immunodeficiency An inadequate response by the immune system to the presence of antigens. Immunodeficiency diseases can be acquired (for example, AIDS) or congenital (for example, DiGeorge syndrome).

immunogen Antigens that elicit an immune response.

immunoglobulin (Ig) Alternate name for an antibody; a type of protein produced by B lymphocytes in an immune response to the presence of a particular antigen, to which the immunoglobulin binds.

immunoglobulin E (IgE) A type of antibody that mediates allergic reactions.

immunological memory The ability of lymphocytes of the adaptive immune system to 'remember' antigens after primary exposure and to mount a larger and more rapid response when exposed to the same antigen again.

immunotherapy Any treatment that harnesses the immune system of the patient to fight diseases; for example, monoclonal antibody therapy.

implantation Attachment and embedding of the blastocyst into the lining of the uterus. Implantation commences the development of the blastocyst into a fetus and occurs in all mammals except monotremes.

in situ Meaning 'in place' or 'in position'. Studies that are in situ are undertaken on biological structures in their natural position, such as cells functioning in a tissue or organ. In situ conservation refers to the conservation of species within their natural habitat.

in vitro Meaning 'in glass'. In vitro processes are undertaken outside a living organism, in a test tube or Petri dish (for example, cell or tissue cultures). Contrast to *'in vivo'*, a process taking place in a living organism.

in vitro fertilisation (IVF) Fertilisation of an ovum with sperm outside the body, in a laboratory dish. IVF is an assisted reproductive technology that is used in humans if normal fertilisation cannot occur or in agriculture for animal breeding purposes.

in vivo Meaning 'in the living'. Studies undertaken in living organisms.

inactivated vaccine Vaccines made from inactivated (killed) forms of pathogens. Inactivation destroys the pathogen's ability to replicate, but keeps it 'intact' so it can be recognised by the immune system.

inbreeding The loss of genetic variation within a species or population due to small population sizes and the breeding of closely related individuals.

incidence In epidemiology, the number of new cases of a disease in a population over a given time period. Incidence data can be used to determine the risk of contracting a disease in a given population. Contrast with *prevalence*.

incomplete dominance A form of inheritance in which neither the dominant or recessive phenotype is expressed completely. In heterozygotes, both alleles are partially expressed, producing an intermediate phenotype.

incomplete penetrance Penetrance is said to be incomplete when some individuals carrying the allele for a trait do not express the trait.

incubation period In disease, the period of time from exposure to a pathogen to the first appearance of symptoms.

incus One of three small bones in the middle ear that transmits vibrations. See also *malleus* and *stapes*.

independent variable The variable that is altered during an experiment to test its effect on another variable (the dependent variable). Also called experimental variable.

infectious agent A pathogen capable of infecting a host and causing disease.

infection (adj. infectious) The severe, uncontrolled growth of a foreign organism inside another.

infectious disease A medical condition or disease that is caused by infectious agents (pathogens). Infectious diseases are transmissable between hosts.

inflammation (or inflammatory response) A protective response triggered by damaged tissue or invading pathogens, that leads to increased blood flow and migration of white blood cells to the site of damage/infection. It results in heat, pain, swelling, redness and loss of function.

innate immune response Non-specific defense mechanisms (physical, chemical and microbiological barriers) in response to the presence of an antigen, a toxin, or other foreign substance capable of inducing an immune response.

inner ear The innermost part in the ear of vertebrates which plays important roles in sound detection and balance. The cochlea and vestibular system are found in the inner ear. See also *cochlea*.

inquiry question A question that defines the focus of an investigation.

insertion mutation When a point mutation: a type of mutation that involves the addition of one or two nucleotides into the existing sequence. When a block mutation: occurs as a result of one part of a chromosome breaking off and attaching to another chromosome.

insulation In relation to animals, a body covering that retains body heat (for example, fur, feathers or blubber).

insulin A peptide-based hormone produced in the pancreas by beta cells in the islets of Langerhans, which regulates the amount of glucose in the blood. Insulin is secreted in response to high glucose levels and acts by suppressing the breakdown of glycogen to glucose in the liver, stimulating the storage of glucose as glycogen in the liver and muscles and stimulating the formation of fat using glucose. A lack of insulin causes a form of diabetes.

intellectual property Ideas, inventions, images and other creations of the mind that may have commercial applications.

intercalating agent Chemical that inserts into the bonds between DNA nucleotides, altering the shape of the DNA and leading to subsequent errors in replication.

interferon A type of cytokine important in antiviral immunity. Interferons are produced by virus-infected cells to inhibit viral replication by resulting in the transcription of antiviral genes and the expression of antiviral proteins; have a lesser role in bacterial and parasitic immune responses. They regulate the immune response in a number of ways, such as enhancing T lymphocyte activity.

interleukin A type of cytokine produced by leukocytes for regulating immune responses. Cytokines are small signalling molecules of the immune system.

interneuron A neuron that transmits information from a sensory neuron to a motor neuron. Most of the nerve cells of the brain and spinal cord are interneurons.

interoreceptor A sensory receptor that detects internal stimuli.

interphase The phase in the cell cycle when the cell is not undergoing mitosis.

intervention study In relation to epidemiology, a study that aims to measure the effectiveness and safety of epidemiological interventions.

intracellular fluid The fluid inside cells.

interuterine insemination (IUI) Technique used to achieve pregnancy by depositing sperm into the uterus using a fine tube.

intron Section of DNA that does not code for proteins and is spliced during mRNA processing in eukaryotes.

inversion mutation A type of block mutation that involves the reversal of a sequence on a chromosome.

ionising radiation High energy radiation that can cause atoms and molecules to become ionised. This radiation can cause damage to DNA and is therefore a mutagen. Particle radiation and high-frequency electromagnetic radiation (EMR) are both forms of ionising radiation.

isotonic Two or more solutions having equal concentrations of solutes.

K

karyogamy The second stage of sexual reproduction in fungi, involving the fusion of two haploid nuclei to form a single diploid ($2n$) zygote nucleus.

karyotype A visual depiction of the number, size and shape of chromosomes in an individual. Karotyping allows large genetic changes, such as an extra copy of a chromosome or chromosomal mutations, to be identified.

keratopathy Damage to the cornea at the front of the eye. Keratopathy may be caused by injury, infection or an inherited disorder.

kidney Organ involved in filtering the blood and producing urine in complex animals.

Koch's postulates A set of four criteria formulated by German physician, Robert Koch, in 1884. The four postulates were used to establish whether a specific microorganism was the cause of a particular disease.

L

lactation Milk-production by mammals to feed their young.

laser eye surgery A type of surgery that uses a laser to cut tissue. Laser eye surgery can be used to correct refractive errors. See also *refractive error*.

lacZ gene A gene in the lac operon that codes for beta galactosidase; used in recombinant plasmids for detecting transformed bacteria.

Law of Independent Assortment The principle, first stated by Gregor Mendel, that inherited traits assort independently, so that the occurrence of a trait (such as brown eyes) is independent of the occurrence of any other trait (such as attached ear lobes). Because of gene linkage, the law applies only to alleles on different chromosomes. Also called Mendel's second law of inheritance. See also *gene linkage*.

Law of Segregation The principle, first stated by Gregor Mendel, that explains the inheritance of alleles. The law of segregation states that each gene has two alleles that separate (or segregate) during meiosis (haploid gamete formation) and randomly unite during fertilisation, resulting in offspring inheriting one allele from each parent. The phenotype of dominant alleles will mask the phenotype of recessive alleles, explaining why offspring can have different phenotypes to their parents. Also called Mendel's first law of inheritance.

lens capsule A transparent membrane made of collagen that encapsulates the lens of the eye.

lens epithelium A transparent, cube-shaped cell that regulates nutrient, ion and water balance inside the lens of the eye.

lens fibre Long, thin transparent cell in the lens of the eye. Lens fibres form tightly packed layers and are the dominant cell type in the lens.

lesion A region in an organ or tissue of a plant that has suffered damage through injury or disease.

leukocyte A white blood cell that counteracts antigens and other foreign substances that enter the body. Includes phagocytes and lymphocytes.

ligase An enzyme that joins together two molecules or fragments of molecules.

ligation The process of joining two fragments of DNA using a DNA ligase enzyme.

light chain The short polypeptide chains that form the 'arms' of a Y-shaped antibody molecule.

light reflex A reflex of the eye that changes the diameter of the pupil in response to light. The light reflex controls the amount of light entering the eye and functions to protect the eye and direct light to the fovea during accommodation (focusing on objects at different distances). See also *accommodation* and *iris sphincter muscle*.

lignification To turn into wood or become woody through the formation and deposit of lignin in cell walls.

lignin A complex organic compound deposited in the cell walls in the xylem vessels, tracheids and supporting tissue of vascular plants, making them rigid and woody. Lignin gives strength to the stem and other plant parts. It is not present in non-vascular plants such as mosses.

line graph A graph in which the relationship between the variables is represented by a straight line, curved line, or series of line segments. Line graphs are useful for representing continuous quantitative data.

line of best fit A straight line drawn between data points on a graph that shows the overall trend in the data and can be used to predict values between data points. See also *trend line*.

linear relationship A mathematical relationship in which variables are directly proportional to each other and produce a straight trend line when graphed.

linkage The tendency for two or more genes on the same chromosome to be inherited together because they are close together on the chromosome. Linked genes may be separated if crossing over occurs between them.

liposome Small phospholipid vesicle that can diffuse across cell membranes or enter cells by endocytosis. Liposomes are used as carriers for genes or drugs across the cell membrane into cells (for example, in gene therapy).

literature review Evaluation of or report on the current literature in a particular area of study.

locus (pl. loci) The site on a chromosome where a particular gene is located.

loop of Henle A U-shaped loop in a mammalian kidney between the proximal and distal convoluted ducts, dipping into the medulla. Its main function is to recover water and sodium chloride from urine, thus making the urine more concentrated and reducing the amount of water that needs to be taken in.

luteinising hormone (LH) Peptide hormone that stimulates the secretion of the male steroid hormone testosterone in the testes, promotes ovulation, maintains the corpus luteum and stimulates the secretion of progesterone. LH is produced by the anterior pituitary gland.

lymph The colourless fluid that circulates in the lymph system. It consists mainly of interstitial fluid (fluid forced from capillaries by blood pressure into the spaces between tissues) and contains lymphocytes, macrophages, proteins and fats. It has an important role in defending the body against harmful bacteria and other particles and also in the absorption and transport of fatty acids.

lymph node Gland of the lymphatic system and secondary lymphoid tissue. Lymph nodes contain lymphocytes (B and T cells), filter lymph fluid and fight infections.

lymphatic system The body system that transports immune cells including antigen-presenting cells throughout the body and is where antigen recognition by lymphocytes occurs; important for adaptive immune responses in mammals.

lymphocyte A type of leukocyte involved in adaptive immune responses; includes B and T lymphocytes.

lymphokine A type of cytokine secreted by lymphocytes. Cytokines are small signalling molecules of the immune system.

lysis The destruction of a cell, usually by rupturing the cell membrane.

lysosome An organelle vesicle containing digestive enzymes used in the digestion of waste and foreign material.

lysozyme An antibacterial enzyme present in body secretions such as saliva and tears. It disrupts the bacterial cell wall.

M

macronucleus Large nucleus in ciliates which contains many sets of chromosomes (polyploid). Contrast with *micronucleus*.

macrophage A type of large white blood cell that engulfs and digests foreign matter in the body, as well as damaged cells or the remnants of apoptosis, via phagocytosis.

macroscopic Structures that are visible to the naked eye, without the aid of a microscope. Contrast with *microscopic*.

macular degeneration Condition of the eye which causes blurred or loss of vision in the centre of the visual field. The vision loss is due to a distortion or absence of the central field of retinal vision, an area known as the fovea or macula.

major histocompatibility complex (MHC) A group of major histocompatibility complex proteins on the surface of cells, involved in antigen presentation to T cells. MHC proteins are also known as human leukocyte antigens.

malleus One of three small bones in the middle ear that transmits vibrations. See also *incus* and *stapes*.

malnutrition Condition caused by an imbalance in the dietary intake of nutrients, resulting in overnutrition or undernutrition.

marsupial A subclass of mammals characterised by a pouch for carrying the young, which are born immature and complete their development in the pouch. Contrast with *monotreme* and *placental*.

mast cell An immune cell containing granules of histamine. This cell mediates allergic responses by binding IgE–allergen complexes and releasing histamines during inflammation and allergic reactions.

mean The average value of a set of values, calculated by dividing the sum of the values by the number of values.

mechanoreceptor Receptors that detect hearing, balance, pressure and touch.

median The value in the middle of an ordered list of values.

medulla In the kidney, the innermost region where the renal pyramids and nephrons run through.

megaspore Large spore produced by seed-producing plants (gymnosperms and angiosperms). Contrast with *microspore*.

meiosis A division of a nucleus that results in one copy of each homologous chromosome and one sex chromosome in each daughter cell. Meiosis produces four genetically unique daughter cells, each with half the number of chromosomes of the parent cell.

meiosis I The first stage of meiosis in which homologous chromosomes are separated, producing two haploid daughter cells. Meiosis I is a reduction division because the chromosome number is reduced. See also *reduction division*.

meiosis II The second stage of meiosis in which sister chromatids are separated, producing four haploid daughter cells. Meiosis II is an equational division like mitosis because the chromosome number is not reduced.

meiospore Haploid spore produced sexually by meiosis. Contrast with *mitospore*.

melanoma A type of skin cancer formed from the pigment-producing cells, melanocytes. Melanomas usually occur on parts of the body that have been exposed to ultraviolet (UV) radiation.

memory B lymphocyte An antigen-specific B lymphocyte that remains in lymphoid tissues for long periods, divides and gives rise to plasma cells if secondary exposure to an antigen occurs. Memory B lymphocytes are responsible for the immunity that often follows infection or vaccination.

memory T lymphocyte Antigen-specific T lymphocyte that persists after an infection has been resolved, to ensure a prompt response should the same pathogen reinfect the organism.

menopause The natural cessation of ovulation and menstruation. Menopause occurs at around 50 years of age in humans.

menstruation Periodic shedding of the uterine lining occurring approximately two weeks after ovulation if fertilisation does not occur. The average menstrual cycle length is 28 days but individuals vary from 21 to 35 days and it may not be a regular pattern.

mesoderm The middle layer of the three primary germ layers in the early embryo. The mesoderm is present only in animals with bilateral symmetry; animals with radial symmetry have only two primary germ layers.

mesophyte A plant adapted to a medium water availability.

mesothelioma Type of cancer that develops in the thin tissue that covers the internal organs (mesothelial cells).

messenger RNA (mRNA) An RNA molecule that is transcribed from DNA in the nucleus, then passes into the cytoplasm and binds to a ribosome. At the ribosome, mRNA is translated into a polypeptide.

metabolism The total of the physical and chemical processes by which energy and matter are made available by an organism for its own use. Metabolism is controlled by enzymes.

metaphase Stage of cell division between prophase and anaphase in which chromosomes align at the equatorial plane of cell and spindle fibres attach to the centromeres of chromosomes.

metastasis The process by which cancer cells break away from the original (or primary) tumour, travel through the blood and lymph vessels and form secondary tumours at other locations.

microbe-associated molecular pattern (MAMP) Conserved molecules that are common to microbes. MAMPs are recognised by the pattern recognition receptors (PRRs) of the host's innate immune system. See also *pattern recognition receptor (PRR)*.

microflora Microorganisms that colonise particular sites (for example, the digestive system); normal microflora do not usually cause disease.

micronucleus Small nucleus in ciliates which contains a diploid set of chromosomes. Contrast with *macronucleus*.

microsatellite A short repeated sequence of nucleotides (for example, ATGATGATG) within the genome. The number of repeats within a microsatellite varies between individuals making them useful genetic markers in population studies or DNA profiling of individuals.

microspore Small spore produced by seed-producing plants (gymnosperms and angiosperms). Contrast with *megaspore*.

middle ear Region of the ear between the eardrum and inner ear. The middle ear contains three tiny bones called ossicles (malleus, incus and stapes) and the opening to the Eustachian tube that connects to the back of the nose and throat.

middle ear implant Hearing device that is surgically attached to the stapes bone. An external microphone and sound processor collect and amplify sound and transmit it as electronic signals to the internal device on the stapes bone. The signals are then changed back to mechanical vibrations that are passed across the oval window, stimulating normal sound sensations to the hair cells.

mineral Any naturally occurring inorganic substance. In nutrition, important minerals include elements such as magnesium, potassium, calcium, iron and sodium. Minerals in foods are essential for maintaining biological functions.

missense mutation A type of substitution mutation that results in a different amino acid in the sequence.

mistake An avoidable error.

mitochondrial Eve The most recent woman from whom all living humans are descended. She lived between 100 000 and 170 000 years ago.

mitosis A division of a nucleus that results in two cells that are genetically identical to the parent cell. Asexual reproduction and cell replication for growth occur by mitosis.

mitospore Haploid spore produced asexually by mitosis. Contrast with *meiospore*.

mode The value that appears most often in a data set.

model Representations of structures or processes, such as physical models or digital models, that are used to create and test theories and explain concepts.

model organism Organisms that are commonly used in scientific experiments such as mice, the fruit fly *Drosophila melanogaster*, the plant *Arabidopsis thaliana* and the bacterium *Escherichia coli*.

molecular genetic testing A genetic testing method to identify single genes or short lengths of DNA.

monoclonal antibody (mAb) Antibody produced by a single clone of B lymphocytes grown in culture. The antibodies produced by the clone are identical and specific to the same antigen.

monocular vision Vision in which eyes are positioned far apart (for example, on either side of the head) and see objects separately. Many prey animals have monocular vision (for example, horses, birds, lizards) which improves their field of view and allows them to quickly respond to movement but restricts their depth perception. Contrast with *binocular vision*.

monocyte Large white blood cell (leukocyte) that ingests foreign particles (phagocyte).

monogenic inheritance The inheritance of an observable trait that is determined by one gene.

monohybrid cross A cross between individuals that have different pairs of alleles of a particular gene. For example, one individual might have T and t alleles and the other might have T and T alleles. Monohybrid crosses are used to study the inheritance of one characteristic.

monomorphic A genetic marker that is not variable for a particular population.

monotreme The group of egg-laying mammals including platypus and echidna. Contrast with *marsupial* and *placental*.

mortality The death rate in a population, usually expressed as number of deaths per unit of population in a given time period. For example, the mortality rate in Australia in 2013 was 6.4 per 1000 population.

morula An embryo consisting of an unorganised mass of 16 cells, resulting from a series of divisions of the zygote.

mosaicism A genetic condition in which a single individual carries cells with different genotypes.

mRNA (messenger RNA) See *messenger RNA (mRNA)*.

mucous membrane Epithelial tissue that produces mucous lining in most body cavities and tubular organs such as the gut and respiratory system.

multiple sclerosis (MS) An autoimmune disease of the central nervous system, interfering with nerve impulses within the brain and spinal cord.

mutagen A physical, chemical or biological agent that can cause mutations in DNA.

mutant (1) A cell or organism carrying an altered (mutated) gene. (2) An individual with a phenotype that is different from the wild type.

mutation A permanent change in a genetic sequence, including changes to the nucleotide sequence of DNA or chromosomal arrangement. Mutations can have a beneficial effect, a harmful effect or no effect at all on the survival of the individual. Mutations may occur spontaneously or in response to exposure to radiation or harmful substances.

mutualism disruption The disturbance of a mutualistic relationship between two species. For example, the extinction of a species that is involved in a symbiotic relationship with another species would result in mutualistic disruption.

mycelium Mass of branching, threadlike hyphae in fungi. The hyphae are the vegetative feeding state of the fungus and absorb the food digested by secreted enzymes.

myeloma cell Myeloma is a cancer of B lymphocytes. Myeloma cells are B lymphocytes that proliferate uncontrollably.

myopia Disorder of the eye in which light is focused in front of, rather than on, the retina, causing distant objects to appear blurry. Myopia is a refractive error caused by an eyeball that is longer than normal or a cornea that is too curved. Also known as near-sightedness or short-sightedness. Contrast with *hyperopia*.

N

natural killer cell A lymphocyte that can bind to tumour cells and virus-infected cells and kill them.

natural selection The mechanism by which evolution occurs. Some individuals in a population have inherited characteristics that make them more likely to survive and reproduce than others in the population. These individuals then pass these characteristics on to their offspring. Over time this removes less suitable variations, so that evolutionary change gradually occurs.

necrosis The death of most or all of the cells in an organ or tissue due to disease.

necrotroph A parasitic organism that kills the living cells of its host and then feeds on the dead matter.

negative feedback loop A control system in which the response produced by a stimulus reduces the size of the original disturbance. This eventually leads to homeostasis.

neoplasm Abnormal tissue growth, caused by unusually rapid cell replication. Neoplasms may be malignant (cancerous) or benign (not cancerous).

nephron The functional unit of the kidney; consisting of a Bowman's capsule surrounding a glomerulus and a tubular region leading into a collecting duct. About one million nephrons are found in each human kidney.

nephropathy Kidney disease or damage.

nephrosclerosis Disease of the kidneys that results from the hardening (sclerosis) of the blood vessel walls in the kidney.

nervous system The network of nerve cells that transmits signals throughout the body in response to internal and external stimuli.

neuron A nerve cell, including its various processes and attachments. The neuron is the fundamental unit of the nervous system in animals.

neurotransmitter A group of signalling molecules produced by neurons and used to carry a signal across synapses between cells.

neutral marker A genetic marker that is not under selection in a population.

neutralisation The binding of neutralising antibodies to toxins or antigens on the surface of pathogens that inhibits their action or ability to enter cells.

neutrophil Type of white blood cell (leukocyte) that uses phagocytosis to engulf foreign particles. Neutrophils are the most abundant white blood cell.

next generation sequencing (NGS) DNA sequencing technology that allows the entire genome of an organism to be rapidly sequenced.

nitrogenous base Nitrogen-containing molecule that is the building block of the nucleic acids, DNA and RNA. Five nitrogenous bases make up DNA and RNA: adenine (A), cytosine (C), guanine (G), thymine (T) and uracil (U). Hydrogen bonds between complementary nitrogenous bases form the 'rungs' in the DNA double helix 'ladder'. Adenine pairs with thymine (or uracil in RNA) and cytosine pairs with guanine.

nitrogenous waste Waste products from the breakdown of proteins, including ammonia, urea and uric acid.

nocturnal An animal that is active during the night.

nominal variable A categorical variable in which there is no inherent order. Nominal variables can be counted but not ordered.

non-cellular pathogen A non-cellular, non-living agent that causes disease. Non-cellular pathogens include viruses, viroids and prions.

non-coding RNA A section of RNA that produces structural RNA that does not code for a protein. This includes transfer RNAs (tRNA) and ribosomal RNAs (rRNA).

non-coding strand A strand of DNA or RNA used as a template for building a complementary strand of a precise nucleotide sequence. Also known as template DNA. Contrast with *coding strand*.

non-disjunction The failure of homologous pairs of chromosomes to separate during metaphase I of meiosis. Non-disjunction results in aneuploidy because two of the gametes formed will have two copies of the chromosome, while the other two gametes will be missing that chromosome entirely.

non-homologous chromosomes Chromosomes that contain alleles for different types of genes.

non-infectious disease A medical condition or disease that is not caused by infectious agents (pathogens). Non-infectious diseases are non-transmissible.

non-ionising radiation Low frequency electromagnetic radiation (EMR) that does not emit enough energy to ionise atoms and molecules. This type of radiation cannot directly cause damage to DNA.

non-self antigen Antigen that does not belong to an organism's own cells. Also known as foreign antigen or heteroantigen.

non-shivering thermogenesis The production of body heat by an increase in the metabolic rate in brown fat. Brown fat is rich in mitochondria and is capable of high rates of aerobic metabolism.

nonsense mutation A type of substitution mutation that results in the creation of a stop codon within the coding sequence.

nuclear membrane The lipid bilayer surrounding the nucleus in eukaryotic cells.

nucleic acid A long-chain molecule formed from nucleotides. The nucleic acids DNA and RNA are the genetic material of all organisms. They determine the physical appearance of an organism and how it functions.

nucleoid The structure in prokaryotes that contains genetic material.

nucleosome A particle made up of histone proteins around which DNA is coiled. Nucleosomes occur in chromosomes.

nucleotide A molecule consisting of a 5-carbon sugar (ribose in RNA, or deoxyribose in DNA), a nitrogenous base (purine or pyrimidine) and a phosphate group. Nucleotides are the building blocks of nucleic acids such as DNA and RNA.

nucleus The organelle in eukaryotic cells that contains the genetic information (DNA) of the cell and regulates the synthesis of proteins, cell growth, development and reproduction.

O

obesity Abnormal or excessive fat accumulation that may impair health.

observation Using all your senses, as well as available instruments, to allow closer inspection of things that the human eye cannot see.

oestrogen Any of a group of steroid hormones, produced mainly in the ovaries, that initiate the development of secondary sex characteristics and control the ovarian cycle.

oligodendrocyte A cell in the central nervous system that produces myelin, which forms myelin sheaths around the axons of neurons.

oncogene Genes that induce uncontrolled cell division leading to the development of neoplasms.

oncogenic virus A virus that is capable of inducing a tumour or neoplasm.

oocyte A precursor egg cell in the ovary that undergoes meiosis, resulting in the formation of a single egg cell.

oomycetes Fungus-like pathogens of plants with branching hyphae (haustoria) that penetrate living cells and absorb nutrients, or release enzymes that digest cytoplasm into molecules that can be absorbed. Oomycetes were formally classified in the kingdom Fungi but are now classified in the kingdom Protista and the phylum Oomycota.

oospore The sexual spore of some species of algae and fungi, that develops following fertilisation of an oosphere (female reproductive cell of algae or fungi). Oospores have thick-walls and food reserves to ensure long-term survival.

opsin Light-sensitive proteins in the photoreceptor cells of the retina of the eye.

ordinal variable A categorical variable in which there is an inherent order. Ordinal variables can be counted as well as ordered.

organ of Corti A sensory structure in the cochlea of the inner ear. The organ of Corti is embedded along the basilar membrane and holds the mechanoreceptor hair cells organised into four rows. Movement of these hair cells is translated into electrical impulses that travel along nerves and to the brain for processing.

origin In prokaryotes, the point at which the chromosome is attached to the cell membrane.

osmolality The osmotic pressure of a liquid, measured in osmoles of solute per kilogram of water.

osmoreceptor A sensory receptor that detects changes in osmotic pressure in the internal environment. Most osmoreceptors are located in the hypothalamus.

osmoregulation The maintenance of water balance in an organism.

osmosis Passive diffusion of free water molecules across a semipermeable membrane from a more dilute solution to a more concentrated solution.

ossicle Small bone in the middle ear that amplifies and transmits vibrations from the eardrum to the inner ear. The three ossicles are called incus, malleus and stapes.

osteomalacia Softening of the bones in adults, typically through a deficiency in vitamin D or calcium.

otitis media Inflammatory disease of the middle ear.

outer ear The external part of the ear that consists of the auricle (visible part of the ear), ear canal and outer layer of the eardrum.

outlier A data point that lies outside the main group of data.

oval window One of two membrane-covered openings between the middle ear and the inner ear. The oval window connects the stapes to the upper part of the upper part of the cochlea. Also called fenestra vestibuli. See also *round window*.

ovary A female gonad into which precursor egg cells migrate and develop into ripe eggs. The two ovaries are important hormone-secreting organs during pregnancy.

overnutrition Conditions that includes obesity and diet-related non-infectious diseases (for example, heart disease, stroke, diabetes and cancer).

oviduct The tube through which an ovum (egg) passes from the ovary to the uterus after ovulation. Fertilisation usually occurs in the oviduct. Also called fallopian tube or uterine tube.

oviparous Reproductive mode in which offspring develop in eggs outside the mother's body. Examples of oviparous animals are fish, amphibians, reptiles, birds and monotremes.

ovulation The release of a ripe egg from its follicle in the ovary. Ovulation usually occurs spontaneously during each period of oestrus in response to a cyclic surge of luteinising hormone.

ovule The female reproductive structure in plants in which female gametes are produced and stored. The ovule develops into a seed when fertilised.

ovum (pl. ova) The female reproductive cell in animals which, once fertilised by a sperm, will develop into an embryo. Also known as a gamete or egg.

oxytocin Peptide hormone that plays a role in sexual reproduction, maternal-infant bonding, labour, birth and lactation in mammals. Oxytocin is released by the posterior pituitary gland.

P

pandemic An epidemic that has spread over several countries or continents, usually affecting a large number of people.

parasite (adj. parasitism) An organism that lives in or on another organism and benefits by feeding on nutrients.

parasympathetic division A division of the autonomic nervous system that is responsible for directing the unconscious actions of the body and inhibiting the effects of the sympathetic nervous system; for example, by dilating blood vessels and decreasing the heart rate.

paratope A component of an antibody, also known as an antigen binding site, that recognises and binds to an antigen.

parthenogenesis The development of an egg in the absence of fertilisation by sperm. Parthenogenesis is a normal part of the life cycle in some insects and crustaceans.

particle radiation Energy emitted by fast-moving subatomic particles.

passive immunity Immunity provided by the transfer of antibodies produced in another organism.

pasteurisation The process of heating that kills microorganisms in foods such as milk. Pasteurisation was developed by Louis Pasteur in the late 1800s.

patent Authority or licence applied to a new product or technology to legally allow the patent-holder to exclude others from making, using or selling the invention for a particular period of time.

pathogen An organism that can produce disease in another organism; includes many microorganisms and parasites.

pattern In relation to epidemiology, the occurrence of health-related events by time, place and person. Time patterns may be annual, seasonal, weekly, daily, hourly, weekday versus weekend, or any other unit of time that may influence disease or injury occurrence.

pattern recognition receptor (PRR) Receptors that recognise molecular patterns common to various microbes (i.e. microbe-associated molecular patters (MAMPs)), allowing early detection of infection and rapid activation of the host's immune cells. See also *microbe-associated molecular patter (MAMP)*.

PCR See *polymerase chain reaction.*

pedigree analysis The study of the pattern of inheritance in a group of related individuals. The pattern of inheritance may be determined by recording the presence or absence of a trait over generations.

peer-review Evaluation of professional work or research by scientists to check that the work meets scientific standards and is appropriate for publication.

pelvis (of the kidney) Upper section of the kidney, through which urine flows to the urinary bladder.

penis Male reproductive and excretory organ.

peptide A polypeptide that consists of fewer than 50 amino acids.

peptide hormone Hydrophilic signalling molecules that are peptides, such as insulin, or proteins, such as growth hormone and follicle stimulating hormone. Also known as protein hormones.

peptide bond Covalent chemical bond linking amino acids in a polypeptide chain.

peptidoglycan A component of bacterial cell walls, composed of sugars and amino acids. Gram-positive bacteria have a thick layer of peptidoglycan. gram-negative bacteria have a thin layer. Penicillin-type antibiotics inhibit its synthesis. Also called murein.

percentage uncertainty The difference between an estimated or measured value and a known value, expressed as a percentage.

perilymph Fluid within the cochlea and semicircular canals of the inner ear.

peripheral nervous system (PNS) Nerve pathways and neurons located outside the central nervous system of a vertebrate. The peripheral nervous system includes spinal and sensory nerves and nerves supplying the internal organs.

peristalsis Wave-like muscular contractions that move food through the digestive tract.

peritoneal dialysis Kidney dialysis treatment that occurs inside the body by flushing the abdominal cavity with dialysis solution.

personal protective equipment (PPE) Clothing and equipment worn to improve safety. For example, a laboratory coat, safety glasses and latex gloves should be worn when working in a laboratory, while sunscreen, a hat and sturdy shoes should be worn when doing fieldwork.

Peyer's patch Secondary lymphoid tissue located in the ileum region of the small intestine. Peyer's patches defend against infection by supplying lymphocytes to the local intestinal tissue.

phagocyte A type of cell capable of engulfing and absorbing pathogens or foreign particles. Phagocytes are important cells in the innate immune system.

phagocytosis The process by which a solid particle in the extracellular fluid is taken into a cell. The particle is enclosed by a section of cell membrane, which then pinches off to form a vacuole (phagosome) within the cell's cytoplasm. Phagocytosis is a type of endocytosis.

phagolysosome The fusion of a lysosome containing digestive enzymes and a phagosome (vacuole in a phagocyte) to break down foreign material, which is then expelled from the cell by exocytosis. See also *lysosome* and *phagosome*.

phagosome A vacuole in a cell's cytoplasm that contains a phagocytosed particle. See also *phagocytosis*.

pharmacogenomics Field of study that analyses the way in which the genome influences a person's response to drugs.

phenotype (adj. phenotypic) An observable trait; expression of a genotype in an individual for a particular trait. The dominance of the alleles and the environmental conditions influence the phenotype of an individual. For example, nutrient availability may influence the pigment synthesis in flower petals or hair follicles in animals.

phenotypic ratio The number of times a phenotype would be expected to appear in the offspring of a test cross.

phenylalanine An amino acid that is essential in the diet of vertebrates.

phenylketonuria (PKU) Genetic disorder in which the amino acid phenylalanine accumulates in the body, potentially leading to intellectual disability and seizures. PKU is caused by mutations in the PAH gene, which codes for the enzyme phenylalanine hydroxylase. This enzyme catalyses the reaction that converts phenylalanine to tyrosine.

phloem Plant tissue through which sugars and other organic compounds are distributed to different parts of a plant. In flowering plants, phloem consists of sieve tubes, companion cells and fibres.

phosphodiester bond The bond that joins nucleotides into a chain of DNA and RNA by linking the phosphate group of one nucleotide and the sugar of another.

photopsin Light-sensitive pigment in the cone cells of the eye. Three types of photopsin in different cones detect red, blue and green wavelengths.

photoreceptor Chemical receptors that are sensitive to light.

photosynthesis (adj. photosynthetic) The process by which plants, algae and some prokaryotes use sunlight, carbon dioxide and water to produce chemical energy (glucose) for biological functions. Oxygen is released as a by-product of photosynthesis.

physical mutagen A physical factor that can cause mutation (a change in DNA). Examples include particle radiation and electromagnetic radiation (EMR).

pie chart A circular diagram divided into sections, with each section representing the value of one set of data as a proportion of the total data set; useful for presenting qualitative and categorical data.

pili Small hair-like projections on the surface of prokaryotes. Pili are involved in the transfer of plasmid DNA between organisms by conjugation and can also help generate movement.

piloerection The erection of hair cells in response to cold, fright or shock. The sympathetic nervous system triggers this reflex.

pineal gland A hormone-secreting gland in the brain.

pinna Visible part of the outer ear, positioned on the side of the head in humans. The pinna funnels sound waves to be transmitted to the middle and inner ear. Also known as the auricle.

pistil The female reproductive organ of a flower, which consists of an ovary, stigma and style. The female reproductive organ of a flower may also be called a carpel. A single pistil can consist of one or many carpels. See also *carpel.*

pituitary gland A gland at the base of the brain having a key role in hormone regulation.

placenta Organ that develops in the uterus of placental mammals during pregnancy. The placenta provides nutrients and oxygen to the developing embryo and fetus until birth.

placental Group of mammals that provide nutrients and oxygen to developing young via a placenta. Placentals include all mammals except marsupials and monotremes. Contrast with *marsupial* and *monotreme*.

plagiarism Claiming that another person's work is your own.

planula A free swimming ciliated larvae that are produced by polyps.

plasma cell Activated B lymphocytes that produce large quantities of the same type of antibody.

plasmid Small, circular pieces of double-stranded DNA found in bacterial cells. Plasmids replicate independently of the bacteria's chromosomal DNA and are used in genetic engineering to produce recombinant DNA.

plasmodesma (pl. plasmodesmata) A microscopic channel that connects the cytoplasm of adjacent cells in plants and some algae.

plasmogamy First stage in sexual reproduction in fungi when two haploid ($1n$) cells fuse, leading to a stage where two haploid nuclei coexist in the cytoplasm of a single cell.

ploidy The number of full sets of chromosomes in an organism's karyoptye. Haploid (one set, or n) and diploid (two sets, or $2n$) are the most common ploidy states, but other states are possible. See also *polyploid*.

point mutation A type of gene mutation that typically only affects a single nucleotide. Types of point mutations include substitution and frameshift mutations.

polar body Small haploid cell formed during meiosis that undergoes apoptosis soon after formation.

pollen Granules containing the male gametes of flowers. Pollen is an allergen that causes hay fever.

pollen tube Male structure in seed plants that transports male gamete cells to an ovule.

pollination The transfer of pollen from one flower to another.

poly(A) tail A long tail of adenine (A) nucleotides (100–250) that is added to the end of mRNA during processing. Poly-(A) tails increase the stability of mRNA.

polycystic kidney disease (PKD) Inherited disorder in which cysts develop in the kidneys causing damage.

polygene A gene that acts together with one or more other genes to control a quantitative character, such as height in the human population.

polygenic inheritance The inheritance of an observable trait that is determined by many genes.

polymerase chain reaction (PCR) A laboratory technique used to amplify (make millions of copies of) a piece of DNA in a short period of time.

polymorphism (adj. polymorphic) (1) The occurrence of different forms within a population (also referred to as alternative phenotypes). (2) Genetic variation (i.e. multiple alleles) in a gene or genetic marker.

polynucleotide A polymer of nucleotides joined together through a condensation polymerisation reaction. Can refer to DNA or RNA.

polyp Small, soft-bodied organisms with hard limestone skeletons; the limestone skeletons form coral reefs.

polypeptide A long, chain-like molecule consisting of many amino acids linked together. Each amino acid loses a water molecule when it is linked, so a polypeptide is actually a chain of amino acid residues. The group of atoms (–NH–CO–) that links each amino acid to the next one is called a peptide bond.

polypeptide synthesis The process by which cells assemble new proteins. Protein synthesis occurs in two stages: (1) transcription, in which DNA information is copied into messenger RNA (mRNA) and (2) translation, in which proteins are synthesised from amino acids using the information in mRNA.

polyploid (n. polyploidy) Having more than one copy of the full complement of chromosomes. Common polyploid states are diploid (two copies), triploid (three copies) and tetraploid (four copies).

population A group of organisms of the same species that interact with each other.

population data Information gathered about a defined group (e.g. females under the age of 25). Population data can be used to recognise patterns and understand potential cause and effect relationships in epidemiological studies of disease.

population genetics A field of genetic study concerned with the genetic differences within and between populations.

population screening A procedure used in a population to identify the possible presence of an as-yet-undiagnosed disease in individuals without signs or symptoms.

positive feedback loop A control system in which the response produced by a stimulus increases the size of the original disturbance.

postsynaptic neuron A neuron positioned after the synapse which receives signals from the presynaptic neuron. See also *synapse*.

precipitation (in immunity) One of the mechanisms used by antibodies to interfere with the function of the pathogens by binding to soluble antigens, causing them to become insoluble and precipitate out of solution.

precision The ability to consistently obtain the same measurement.

pregnancy The process and period of embryonic and fetal development inside the uterus of viviparous animal, from fertilisation to birth. Also known as gestation. See also *viviparous*.

preimplantation genetic diagnosis (PGD) Genetic testing conducted before implantation of a fertilised egg, to determine whether there are any genetic abnormalities that might lead to a disorder in the resulting individual.

presbyopia Reduced ability to focus on near objects due to hardening of the lens of the eye. Presbyopia is a type of refractive error that naturally occurs with ageing.

prevalence In epidemiology, the proportion of a population affected by a disease at a given point in time. Prevalence data can be used to determine the geographic and demographic distribution of a disease. Contrast with *incidence*.

prevention Measures taken to keep something from occurring (e.g. vaccines to prevent the spread of infectious disease).

primary data Data created by a person directly involved in an investigation (also called first-hand data). Contrast with *secondary data*.

primary immune response The immune response to an antigen that has been encountered for the first time.

primary investigation An investigation that you conduct yourself. Contrast with *secondary-sourced investigation*.

primary lymphoid tissue The major structures of the lymphatic system: the bone marrow and the thymus.

primary source A source that includes first-hand (primary) information, such as the results of an original experiment. Contrast with *secondary source*.

primary structure The linear sequence of amino acids in the polypeptide chain of a protein.

primer A short, synthetic segment of DNA that is complementary to the sequences of bases at one end of a DNA region to be amplified. Primers specify the starting and finishing points for DNA replication. A primer is added to single-stranded DNA to start DNA synthesis during a polymerase chain reaction (PCR).

prion A small protein particle that, when its shape is altered due to mutation, causes protein aggregation and is toxic to neurons. Prions are the cause of spongiform encephalopathy diseases, BSE in cattle and CJD in humans.

procedure Method or process for conducting an activity, experiment or investigation.

processed data Data that has been mathematically manipulated.

progeny Offspring or descendants of an organism.

progesterone The hormone that regulates the oestrus and menstrual cycles and maintains a pregnancy. Progesterone is produced in the corpus luteum of the ovary and in the placenta.

prokaryote A single-celled organism with a simple cellular structure, lacking a nucleus and other membrane-bound organelles. Bacteria and Archaea are prokaryotes. for example, bacteria.

prolactin Peptide hormone that stimulates lactation (milk secretion) in mammals. Prolactin is secreted from the anterior pituitary gland.

promoter Upstream region of a gene (a specific DNA sequence) to which RNA polymerase attaches, initiating transcription.

prophase The first phase in mitosis in which chromosomes condense and become visible, centrioles move to opposite sides of the nucleus and form poles, the nuclear membrane breaks down and centrioles form spindle fibres between the two poles.

prostaglandin Lipid that is produced at the site of injury or infection to control the processes of inflammation, smooth muscle contractions, blood flow and the induction of labour.

prosthesis Artificial device that replaces a body part.

prosthetic group An non-protein compound that is involved in protein structure or function. A protein with a prosthetic group is known as a conjugated protein.

protein An organic compound consisting of one or more long chains of amino acids connected by peptide bonds and has a distinct and varied three-dimensional (3D) structure.

proteome The entire set of protein products of the genome.

proteomics The study of proteomes, including the structure, function and interactions of proteins.

proto-oncogene A normal cellular gene which could become a gene that triggers molecular events that lead to cancer.

Protozoa See *protozoan.*

protozoan A unicellular, eukaryotic organism that may have multiple stages in a complete life cycle and may replicate within the cells of its host. Although once grouped into the phylum Protozoa, protozoans are now considered to belong to all the main lineages of protists. Also called protozoon.

proximal convoluted tubule Structure of the nephron in the kidney, between the Bowman's capsule and the loop of Henle. The proximal tubule plays an important role in regulating pH and secreting organic acids. The proximal tubule is divided into the proximal convoluted tubule and the proximal straight tubule.

pseudo-exon A partial intron sequence that is incorrectly incorporated as an exon in the formation of mRNA, usually due to a mutation in the intron sequence.

public health Measures implemented to improve the safety and health of communities through education, policy making and research of disease and injury prevention.

public health campaign Media designed to promote public health awareness and reduce the incidence of disease.

purine A nitrogenous base that has a double ring structure, including the bases adenine (A) and guanine (G). Each purine base (A or G) pairs with a specific pyrimidine base (C, T or U). Purines are present in the nucleotides of DNA and RNA. Contrast with *pyrimidine.*

public health program A framework designed to reduce the incidence of disease by improving public health practices and health education.

purpose A statement describing in detail what will be investigated (see also *aim*).

pyrimidine A nitrogenous base that has a single ring structure, including the bases cytosine (C), thymine (T) and uracil (U). Each pyrimidine base (C, T or U) pairs with a specific purine base (A or G). Pyrimidines are present in the nucleotides of DNA (C and T) and RNA (C and U). Contrast with *purine.*

pyrogen A toxin, produced by bacteria, that causes a fever by acting directly on the brain.

Q

qualitative data Data that consists of categorical variables.

qualitative variable A variable that can be observed but not measured. Qualitative variables can be sorted into groups such as colour or shape. Nominal and ordinal variables are qualitative variables.

quantitative data Data that consists of numerical variables.

quantitative variable A variable that can be measured, such as temperature or height. Discrete and continuous variables are quantitative variables.

quarantine A period of isolation used to prevent the spread of infectious disease.

quaternary structure Two or more polypeptide chains joined as a single functional protein.

R

R group A side chain found on an amino acid.

radiation The transfer of heat via electromagnetic waves, specifically infrared.

radiation therapy Treatment of disease using ionising radiation, especially of cancer to destroy cancerous cells.

radioactive Emitting radiation (a form of energy from the nucleus of an unstable atom).

random assortment The process by which homologous chromosome pairs are separated and randomly assorted into haploid cells during meiosis. Random assortment creates genetic variation and is the reason that siblings (other than identical twins) are genetically different.

random coil A polymer conformation with the monomers orientated randomly. Adjacent monomers are bonded together.

random error Unpredictable variations that can occur with each measurement.

range The difference between the highest and lowest values.

rational drug design Directed chemical and computer-aided design of a drug based on knowledge of the shape and charge of the target molecule.

raw data The data recorded during an experiment.

reabsorption In the kidney, the process by which the primary kidney filtrate is taken back into the tissues via nephrons.

reactive oxygen species (ROS) Highly reactive molecules containing oxygen that can cause breakages and cross-links in DNA strands.

receptor A molecule in a cell membrane that binds and responds to specific molecules such as hormones and neurotransmitters, triggering a response.

recessive phenotype A trait (encoded by an allele or gene) whose expression or appearance is masked by a dominant trait. Recessive phenotypes are only observed in homozygous individuals (i.e. individuals with two copies of the recessive allele). Contrast with *co-dominant phenotype* and *dominant phenotype.*

reciprocal cross A breeding experiment to determine if traits are sex-linked (i.e. carried on the X or Y chromosome). A reciprocal cross involves two crosses in which the phenotypes of the sexes are reversed (for example, cross 1: a male with phenotype A is crossed with a female with phenotype B; cross 2: a male with phenotype B is crossed with a female with phenotype A).

recognition site Short sequence of DNA that restriction enzymes recognise.

recombinant DNA Technology DNA that has been genetically engineered by joining fragments of DNA from two or more different organisms.

recombinant gametes Gametes carrying a combination of alleles not observed in the parents, as a result of crossing over during meiosis.

recombinant plasmid A plasmid containing a foreign gene that has been inserted by the use of restriction enzymes and DNA ligase.

recombinant protein Protein encoded by recombinant DNA. See also *recombinant DNA.*

recombination In offspring, the formation of a new combination of alleles from the total alleles available from the parents.

reduction division A nuclear division that halves the number of chromosomes in the daughter cells (gametes) to half ($1n$) of that in somatic cells ($2n$). Reduction division occurs in meiosis. See also *meiosis* and *meiosis I.*

refraction Bending of light rays. Because light always travels in a straight line, refraction is actually a change of direction of the light ray onto a different straight path. This happens whenever light enters a new medium and changes speed slightly, such as moving from air into the cornea or from the aqueous humour into the lens.

refractive error Eye condition that prevents clear focus of an image on the retina, causing blurred vision for either close or distant objects. Refractive errors are due to imperfections in the curvature of the cornea or lens, or variations in the shape or length of the eyeball (i.e. longer or shorter than normal).

regeneration The process by which a detached part of an individual grows into another individual or body part. Regeneration is a form of asexual reproduction in Hydra, flatworms, sponges and sea anemones.

reliability The ability to consistently reproduce results.

renal artery A vessel that brings blood to each kidney. Contrast with *renal vein.*

renal dialysis Technology that assists people with impaired kidney function. Small solute molecules diffuse through a partially permeable membrane in a dialysis machine, providing similar functions to the filtration stage of normal kidney function.

renal failure Condition in which the kidneys do not filter the blood adequately due to kidney disorders.

renal vein A vessel through which blood leaves each kidney. Contrast with *renal artery.*

renature (n. renaturation) The ability of a protein that is partially denatured to fold again. See also *denature.*

renin An enzyme that is produced and stored in the kidneys. It plays a role in regulating blood pressure by catalysing the conversion of angiotensinogen to angiotensin I. This is then converted to angiotensin II, an effective vasoconstrictor.

reproduction Production of new individuals via sexual or asexual reproduction.

restriction enzyme A type of enzyme, also called an endonuclease, that occurs naturally in bacteria and can cut DNA at a particular site (a recognition site). Restriction enzymes are widely used in DNA analysis, genetic testing and genetic engineering.

restriction fragment length polymorphism (RFLP) A type of analysis that provides information about DNA sequences based on where restriction enzymes cut the DNA. Patterns of DNA fragments are generated to compare individuals.

retina Sensory layer inside back of eyeball which holds cone and rod photoreceptor cells and neurons. The retina converts light stimuli to electrochemical signals in nerves.

retinal detachment Condition in which the retina pulls away from the back of the eye.

retinopathy Damage to the retina, often caused by changes to the blood vessels in this region.

retrovirus An RNA virus that uses reverse transcriptase to copy its RNA genome into DNA for integration into the chromosome of a host.

reverse transcriptase A type of polymerase enzyme used by retroviruses to copy their RNA genome into DNA; used in genetic engineering to copy messenger RNA (mRNA) into complementary DNA (cDNA).

rheumatoid arthritis An autoimmune disease causing inflammation in the joints and resulting in painful deformity often in the fingers, wrists, feet and ankles.

rhizoid Root-like structure in mosses and liverworts.

rhodopsin Light-sensitive pigment in the rod cells of the eye. Rhodopsin is sensitive to shades of black, grey and white.

ribonucleic acid (RNA) See *RNA (ribonucleic acid)*.

ribosomal RNA (rRNA) The RNA part of a ribosome. It is synthesised in the nucleolus and is essential for protein synthesis.

ribosome A small non-membrane bound organelle, made of RNA and protein. Ribosomes are often attached to rough endoplasmic reticulum and are the site of translation in the process of protein synthesis.

rickets A nutritional disease of children caused by vitamin D deficiency, characterised by softening and distortion of the bones often resulting in bow legs.

risk factor Things that can contribute to the cause of a disease or condition.

RNA (ribonucleic acid) A nucleic acid that is a single strand made up of a sequence of ribose sugars and bases (adenine, cytosine, guanine and uracil) linked by phosphodiester bonds. There are three forms: messenger RNA (mRNA), ribosomal RNA (rRNA) and transfer RNA (tRNA).

RNA polymerase An enzyme that catalyses the synthesis of RNA, using an existing strand of DNA as a template.

RNA processing The removal of introns from the primary transcript produced in transcription. The exons are joined to form mRNA, ready for translation. This stage of gene expression only occurs in eukaryotes.

round window One of two membrane-covered openings between the middle ear and the inner ear. When vibrations travel through the inner ear, the membrane of the round window moves in the opposite direction to the membrane of the oval window. This relieves pressure in the inner ear and allows the cochlea to move, stimulating the hair cells on the basilar membrane. See also *oval window*.

rRNA See *ribosomal RNA*.

S

safety data sheet (SDS) A document that contains important information about the possible hazards in using a substance and how the substance should be handled and stored.

scatter plot A graph in which two variables are plotted as points. The x coordinate of a particular point is one measured value of the independent variable and the y coordinate is the corresponding measured value of the dependent variable.

scientific method The experimental approach to the study of science that involves formulating a hypothesis, designing and performing an experiment to test the hypothesis and analysing whether the results support or refute the hypothesis.

scurvy A nutritional disease caused by a deficiency in vitamin C.

secondary data Data generated from a synthesis, review or interpretation of primary sources of information (e.g. peer-reviewed scientific journal articles). Contrast with *primary data*.

secondary immune response The immune response to an antigen that has previously been encountered and which elicited a primary immune response. The process activates memory cells and so is faster and more effective that the primary response.

secondary lymphoid tissue The structures of the lymphatic system in which adaptive immune responses initiate: the lymph nodes, spleen, tonsils, adenoids and appendix.

secondary source A resource that interprets primary documents, written after the event by a person who was not a witness to the event. Contrast with *primary source*.

secondary structure The folding or coiling of the polypeptide chains in proteins due to hydrogen bonds. The main forms are the alpha helix structure, beta-pleated sheets and random coils.

secondary-sourced investigation An investigation that uses data obtained by someone else (secondary data). Contrast with *primary investigation*.

selection pressure An environmental factor that affects the survival and reproductive success of an individual based on their particular phenotype.

selective breeding The process by which animals or plants that express desired characteristics are bred together to produce offspring that will also show these characteristics. Humans selectively breed both plant and animal species, creating specific strains or breeds. Also known as artificial selection.

self-antigen An organism's own antigens, which are normally tolerated (do not elicit an immune response).

self-pollination Fertilisation of a flower's ovule by pollen from the same plant.

self-tolerance The inability of the adaptive immune system to respond to self-antigen.

semen A secretion of the testes that contains sperm. Also called seminal fluid.

semicircular canal Structure in the inner ear that is fluid-filled and lined with microscopic hairs (cilia). Both the semicircular canal and cochlea transmit nerve impulses to the brain for hearing or balance. See also *cochlea*.

seminiferous tubule Structure within in the testes where sperm is produced.

senescence The process of ageing in plants.

sensorineural hearing loss A reduction in hearing due to damage to the inner ear structures (cochlea or hearing nerve).

serum Fluid portion of the blood that remains after blood cells and material involved in blood clotting have been removed.

sex chromosome A chromosome that is involved in sex determination. In humans the sex chromosomes are the X and Y chromosomes. Also called an allosome.

sex-limited inheritance The inheritance of a trait that is expressed only in one sex, even though both sexes carry the gene. An example is haemophilia A, an X-linked trait that occurs almost exclusively in males (cases in females are extremely rare).

sex-linked inheritance Inheritance related to genes that occur on the sex-chromosomes (X and Y in humans). An example is red-green colour blindness, which is caused by a mutation in a gene on the X chromosome. See also *X-linked*, *Y-linked*.

sexual reproduction Reproduction involving the fusion of two gametes (egg and sperm), which are the haploid products of meiosis.

shivering A behaviour in warm-blooded animals in response to cold. When core body temperature drops, muscles reflexively start twitching (shivering) to generate warmth and maintain a stable internal temperature.

shivering thermogenesis The production of heat energy through the involuntary movement of muscles (shivering).

short tandem repeat (STR) A short repeated sequence of nucleotides (for example, ATGATGATG) within the genome. The number of repeats within an STR varies between individuals making them useful genetic markers in population studies or DNA profiling of individuals. Also known as microsatellites.

sieve tube Elongated living cell found in phloem tissue, through which fluids and dissolved sugars are transported throughout the plant.

signal transduction The process of transmitting a signal into or out of a cell, or changing the form of the signal. Examples are the production of second messengers inside a cell when signalling molecules bind to a cell surface receptor and the conversion of an electrical signal (action potential) to a chemical signal (neurotransmitters) in neural pathways.

signalling molecule A molecule, such as a neurotransmitter or hormone, that is involved in chemical communication between cells.

signalling transduction pathway Cellular response to chemicals or hormones (signals). The path by which a signal from outside the cell is translated to a change inside the cell.

significant figure The number of digits that contribute meaning to a measurement. All non-zero digits and zeros between non-zero digits are significant figures. Zeros that follow non-zero digits after a decimal place are also significant figures. For example, 5.10 has three significant figures and 5.1 has two significant figures.

silent mutation A type of substitution mutation that results in a different codon that codes for the same amino acid as the original sequence. This type of mutation has no effect on the individual.

single nucleotide polymorphism (SNP) A single base difference in DNA, used for genome comparison and studies of the association between genes.

small nuclear ribonucleoprotein particle (snRNP) Small complex of protein and non-coding RNA. snRNPs are involved in the formation of a spliceosome during transcription.

social inequality Disparity in social conditions; when social advantages and opportunities exist for some people in society but not others.

somatic cell Any cell of an organism except a cell that gives rise to gametes (eggs and sperm). Somatic cells in mammals are diploid ($2n$); that is, they contain a full set of paired chromosomes.

somatic cell nuclear transfer (SCNT) A cloning technique that involves transferring the nucleus from a somatic cell to an egg cell with the nucleus removed. The egg cell is then induced to divide as though under the natural fertilisation process. The resulting embryo is genetically identical to the donor somatic cell.

somatic mutation A mutation that occurs in somatic, or non-gamete, cells of an organism. These types of mutations may affect the individual, but cannot be passed on to offspring. Cancer is a form of somatic mutation.

somatic nervous system Component of the peripheral nervous system (PNS) that controls voluntary movement (e.g. movement of limbs).

sound wave Vibration of particles in a medium like air, a fluid or a solid. The kinetic energy of these moving particles is passed from one to the next causing a compression wave to move forwards. Sensory hairs in the inner ear transmit vibrations as sound information to the brain.

specialised cell Cell that carries out a specific role within an organism.

speciation The formation of new species following a lineage splitting event. Speciation may result from geographic, anatomical, physiological or behavioural barriers to breeding, leading to divergence over evolutionary time, or may be rapid as a result of adaptive radiation.

species (1) A group of organisms that interbreed in the wild (or could do so) and produce viable, fertile offspring, but cannot produce viable, fertile offspring if they interbreed with organisms outside the group. (2) A category or group in the binomial system.

specificity (1) Relating to enzymes that bind to only one substrate and catalyse only one type of reaction. (2) The ability to recognise and respond exclusively to specific antigens.

spectacles Eyeglasses that are worn to correct refractive errors. See also *refractive error*.

sperm The male gamete in animals, which can move by the motion of a flagellum. Also called spermatozoon.

spermatocyte Cells in the testes that divide by meiosis to produce four sperm cells.

sphincter muscle A muscle that controls the dilation of a sphincter (an orifice), such as the muscle that controls the dilation of the pupil.

spindle An arrangement of microtubules that binds to a centromere of a chromatid, enabling the chromosome to be divided equally between two daughter cells during mitosis and meiosis.

spleen Secondary lymphoid organ. The spleen controls the number of red blood cells in the body by destroying old and defective red blood cells, stores lymphocytes and is a site of B lymphocyte activation.

spliceosome Enzyme that removes the introns from the primary transcript to create mRNA during RNA processing (in eukaryotes).

splicing The removal of introns from the primary transcript produced in transcription. The exons are joined to form mRNA, ready for translation.

spontaneous mutation Any naturally occurring random change in DNA.

sporangium (pl. sporangia) The structure in a spore-bearing plant in which spores develop. Also called spore case, spore capsule.

spore A haploid cell that can develop into a new organism without sexual reproduction. In plants, algae and fungi, spores are the products of meiosis, but fungi and some algae can also produce spores by mitosis.

sporophyte The asexual, spore-forming, diploid stage in the life cycle of a plant. The spores are haploid, are produced by meiosis and develop into haploid gametophytes. In seed plants (gymnosperms and angiosperms), the sporophyte stage is the prominent stage (i.e. the plant structure), while the gametophyte stage is the less conspicuous pollen and embryo sac. See also *gametophyte* and *alternation of generations*.

stamen The male reproductive organ of a flower, composed of a filament and an anther. Stamens produce pollen, which is the male gamete.

stapedectomy Surgical procedure of the middle ear to improve hearing when the stapes footplate has become fixed in position. The immobile stapes is replaced with a mobile, micro-prosthesis stapes.

stapes One of three small bones in the middle ear that transmits vibrations. See also *incus* and *malleus*.

start codon Codon that indicates where translation should begin in messenger RNA (mRNA). The most common start codon is AUG.

statistic Numerical information obtained from an investigation.

stem cell A cell that can differentiate into a specialised cell.

stem cell therapy Medical treatment based on the use of stem cells.

stereocilia Hair-like projections on mechanoreceptors in the inner ear. Vibrations entering the inner ear displace the fluid that surrounds the groups of stereocilia, causing them to bend, generating nerve impulses that travel to the brain along the auditory nerve.

steroid hormone Lipid hormones based on cholesterol that are hydrophobic.

stigma The receptive surface for pollen at the tip of the style in a flower.

stimulus An environmental factor that an organism can detect and respond to.

stoma (pl. stomata) Pores in the leaf epidermis, bounded by specialised guard cells that open and close the pore. Stomata are the main routes through which gas exchange occurs in plants, and through which water loss is regulated.

stop codon Codon that indicates where translation should stop in messenger RNA (mRNA). The most common stop codons are UAG, UAA and UGA.

strobilation A method of asexual reproduction in multicellular organisms, in which the organism divides along its shortest axis and each part grows into a new individual. See also *longitudinal fission*, *transverse fission*.

structural adaptation Anatomical or morphological features that improve an organism's ability to cope with abiotic and biotic factors in their environment, increasing their chances of survival and reproduction.

structural gene A gene that codes for proteins and RNA molecules that are not involved in gene regulation (for example, enzymes).

style The organ in a flower that bears the stigma and through which, after pollination, pollen tubes grow towards the ovules.

substitution mutation A type of point mutation in which individual nucleotides, typically one or two, are replaced by different nucleotides. Substitution mutations can result in silent, missense or nonsense mutations.

sugar–phosphate backbone Structural component of DNA and RNA that consists of alternating sugars (deoxyribose in DNA and ribose in RNA) and phosphate groups.

supercoil Tightly wound stands of DNA that make up a chromosome.

suppressor T lymphocyte Cell that turns off the immune response when an antigen has been contained, destroyed or removed.

surface-area-to-volume ratio The relationship between the surface area and volume of a structure. As the size of a structure increases, its surface-area-to-volume ratio decreases.

surgery Medical treatment that involves the manipulation of body structures to treat injuries or disease.

sympathetic division A division of the autonomic nervous system that is responsible for directing the unconscious actions of the body and inhibiting the effects of the parasympathetic nervous system (for example, contracting blood vessels and increasing the heart rate).

symplastic pathway The route through the cytoplasm of plant cells in which water and solutes move.

synapse The point of communication between two cells, where at least one of the cells is a neuron. It includes the membrane of the presynaptic neuron, the synaptic gap and the membrane of the postsynaptic cell (which may be, for example, a neuron, muscle cell or gland cell).

synapsis The process of pairing of two homologous chromosomes during meiosis.

systematic acquired resistance (SAR) Non-specific defence mechanism in plants that is activated after the plant is exposed to molecules from a microbe. Once SAR is activated, the plant is able to mount a rapid response on subsequent exposures to the same microbe.

systematic error A consistent error that occurs every time you take a measurement.

Systemic Acquired Resistance (SAR) Response in plants that occurs following secondary exposure to a pathogen. SAR is similar to the innate immune response in animals. See also *innate immunity*.

T

T cell receptor (TCR) A molecule found on the surface of T lymphocytes that is responsible for recognising fragments of antigen as peptides bound to major histocompatibility complex (MHC) proteins. It is made up of two polypeptide chains that have a variable and a constant region and only one antigen-binding site.

T lymphocyte (or T cell) A type of lymphocyte that originates in the bone marrow and matures in the thymus and is responsible for cell-mediated immune responses. See *cytotoxic T lymphocyte* and *helper T lymphocyte*.

tectorial membrane An acellular membrane in the cochlea.

telomere An area of repetitive DNA at either end of the DNA molecule of a eukaryotic chromosome. Telomeres protect the end of the chromosome from deteriorating and from clumping together with other chromosomes.

telophase Final stage in mitosis, between anaphase and interphase. During telophase a nuclear membrane re-forms around the chromosomes at each pole, the spindle is dismantled and disappears and the chromosomes become longer and thinner.

tertiary structure The structure in proteins created by further folding as a result of bonds forming between the R groups of the amino acids, leading to greater stability than the folding in secondary structures.

test cross A type of backcross in which an individual with the dominant phenotype is crossed with an individual of the recessive phenotype for the character being studied. A testcross is used to identify whether the individual with the dominant trait is homozygous or heterozygous.

testis (pl. testes) The male gonad into which precursor sperm cells migrate and develop into sperm. The testes are important hormone-secreting organs.

testosterone A steroid hormone produced in male vertebrates. It is produced by the testes in male mammals and to a lesser extent by the ovaries in females and is responsible for the development of various sex characteristics.

therapeutic cloning The process of creating an embryo and harvesting stem cells for research and potential treatment for disease.

thermoreceptor A sensory receptor that detects and responds to temperature.

thermoregulation Process used by some animals to maintain their internal (core) temperature.

thrombin An enzyme in the blood plasma involved in the blood clotting process.

thymine (T) A nitrogen-containing base (a pyrimidine) that occurs in nucleotides of DNA, but not RNA.

thymus Primary lymphoid tissue where T lymphocytes mature.

tinnitus Condition in which a sound that isn't actually present is heard. It is often described as ringing in the ears, but is commonly experienced as buzzing, roaring, clicking and hissing. Tinnitus may be caused by wax build-up against the eardrum, otosclerosis, Ménière's disease, ototoxic medications and damage to inner ear hair cells from exposure to prolonged loud noise.

tissue culture A method of growing cells or tissues in an artificial medium containing essential nutrients, salts and growth factors.

tissue factor (TF) A glycoprotein involved in blood clotting.

tonsil Secondary lymphoid tissue located at the back of the throat.

torpor A state of inactivity or dormancy in animals, in which the body temperature is lower and the metabolism is slower than normal.

trait A particular characteristic or feature of an organism.

transcription Process by which a base sequence in DNA is used to produce a base sequence in RNA.

transcription factor Proteins that control gene expression at the transcription stage by binding to DNA sequences close to the promoter region of a gene or to the RNA polymerase to induce or repress the expression of specific genes.

transcriptome The full range of RNA molecules expressed by a genome.

transfer RNA (tRNA) An RNA molecule that brings a specific amino acid to a ribosome so it can be joined to other amino acids during translation.

transformation A type of horizontal gene transfer in which a bacterium incorporates DNA from another organism into its own DNA or takes up a plasmid containing foreign DNA. Used in biotechnology to incorporate recombinant DNA into bacteria. See also *recombinant DNA*.

transgene A gene transferred from one organism to another.

transgenic organism An organism that has had its genome artificially modified with genes from another species.

translation The process in which the base sequence of a mRNA molecule is used to produce the amino acid sequence of a polypeptide.

translocation mutation A type of block mutation that involves sections of two different chromosomes switching positions.

transmission (of disease) The spread of infectious disease agents, such as bacteria and viruses, from one individual to another.

transpiration The loss of water from the leaves of plants through stomata. Transpiration causes suction, which draws water up xylem vessels from the roots.

transposable element (TE or transposon) A sequence of DNA that can migrate to different locations in the genome, often interrupting functional genes. Also known as 'jumping genes'.

trend line A straight line drawn between data points on a graph that shows the overall trend in the data and can be used to predict values between data points. See also *line of best fit*.

triplet Sequence of three nucleotides in DNA that carries the genetic information for the sequence of amino acids in a protein. Each triplet usually codes for one amino acid.

trisomy An abnormality in which there are three copies of a particular chromosome in a cell. Down syndrome is characterised by trisomy 21; that is, three copies of chromosome 21.

true-breeding Producing only progeny with a particular characteristic or trait seen in the parent. Also called pure-breeding.

tumour An abnormal growth of cells resulting from uncontrolled cell division or failure of programmed cell death. It may be benign or malignant.

tumour-suppressor gene A gene whose protein product inhibits cell division, thus preventing uncontrolled cell growth that would result in a cancer.

twin study An experimental method used to investigate genetic and phenotypic factors using twins.

tympanic membrane (eardrum) A thin membrane that separates the outer ear and the middle ear, also known as the eardrum.

type 1 diabetes An autoimmune condition in which the immune system is activated to destroy the cells in the pancreas which produce insulin. Type 1 diabetes has a genetic basis, but appears to also require an environmental factor such as a viral infection in order to be expressed.

U

ultraviolet (UV) radiation Electromagnetic radiation with wavelengths shorter than visible light but longer than X-rays (between 10 nm to 400 nm). Prolonged exposure to UV radiation can cause sunburn and skin cancer.

umbilical cord Structure through which oxygen and nutrients are delivered from the placenta to an embryo or fetus in placental mammals. The umbilical cord contains two arteries and one vein and is attached to an opening in the abdomen of the fetus until it is cut soon after birth.

uncertainty The range of values within which the true value of a measured quantity probably occurs. Uncertainty is caused by random and systematic errors.

undernutrition Condition which includes stunting (low height for age), wasting (low weight for height), underweight (low weight for age) and micronutrient deficiencies (including a lack of vitamins and minerals).

uracil (U) A nitrogenous base (purine) found in the nucleotides of RNA. It forms a base pair with adenine.

urea A water-soluble molecule (CH_4N_2O) that is a major product of protein breakdown. It is excreted by many vertebrates, including mammals.

ureter The tube that carries urine from a kidney to the bladder for storage, before release via the urethra.

urethra The tube that carries urine from the bladder to the exterior for excretion.

uric acid A complex nitrogenous compound ($C_5H_4N_4O_3$) that is produced and excreted by birds and most land reptiles.

urinary reflux Condition in which the urine from a full bladder pushes back up the ureters into the kidneys, potentially leading to kidney damage.

urine Liquid by-product of metabolism in complex animals.

uterus The hollow, muscular organ of the female reproductive system, in which the fetus develops.

V

vaccination The technique of artificially inducing an adaptive immune response by administering (usually by injection) a vaccine usually made of altered, weakened or killed microorganisms, or inactivated forms of toxins or antigens.

vaccine A substance containing weakened or killed forms of a pathogen. Vaccines are administered (usually by injection) to stimulate the adaptive immune response and provide long term immunity against the pathogen.

vacuole Organelles that function as storage spaces within cells and contain metabolic by-products.

vagina Muscle-lined canal that extends from the cervix to an external opening in female mammals.

validity The extent to which an experiment or investigation accurately tests the stated hypothesis and purpose.

variable A factor or condition that is tested (i.e. the independent variable) or changed (i.e. the dependent variable) during your experiment.

variable region The region of an antibody molecule that varies between different antibodies and allows them to interact with different antigens.

vas deferens A muscular duct that moves sperm from the epididymis into the urethra before ejaculation. Also called ductus deferens. See also *epididymis*.

vasa recta Network of blood vessels in the medulla of the kidney.

vasoconstriction The narrowing of the internal diameter of a blood vessel.

vasodilation The widening of the internal diameter of a blood vessel.

vector (1) In infectious disease: object or organism that transfers a pathogen from one host to another. (2) In molecular biology: a vehicle used to transfer foreign DNA into a cell (for example, a plasmid, virus or liposome).

vegetative reproduction A form of asexual reproduction found in plants, in which a piece of a plant (usually a growing tip of a stem) is separated from the plant and grows into a new plant. It is a method used widely in horticulture to produce new plants.

viral vector Virus used as a vehicle to transfer genes (foreign DNA) into cells; used in gene therapy, vaccine production and research.

virion A complete mature virus particle that is metabolically inert and is in the transmission (infectious) phase.

viroid Infectious agents of plants that are a type of self-cleaving RNA enzyme (or ribozyme); composed of short, circular strands of RNA that lack a protein coat.

virulence (adj. virulent) The ability of a pathogen to cause disease.

virus An infectious agent composed of genetic material (DNA or RNA) enclosed in a protein coat and sometimes also a lipoprotein envelope; is only able to multiply in a host cell.

visual acuity The ability to see objects clearly. Visual acuity is determined by the resolving power of eyes to distinguish between point sources of light and form discrete, clear images.

vitamin Any organic compound required in small amounts for cell processes. In humans there are 13 such compounds, called vitamins A, B group (eight vitamins), C, D, E and K.

vitreous humour Clear, jelly-like filling in the back of the eyeball between the lens and the retina.

viviparous Reproductive mode in which offspring develop inside the mother's uterus and are released as live young. Viviparity occurs in placental and marsupial mammals and in some species of fish, reptiles and invertebrates. Viviparity also occurs in plants in which the seeds germinate while still attached to the parent plant (for example, in some mangroves).

W

white blood cell Cell in the blood involved in immune system processes. Also called leucocyte.

whole genome sequencing The process of determining the order of nucleotides in the complete DNA sequence of an organism.

whole organism cloning Process of producing an identical copy of a whole multicellular organism using somatic cell nuclear transfer. The cloned organism has genetically identical DNA to the original somatic cell.

wild type The phenotype most commonly observed in a natural population.

wilt To become limp through heat, loss of water, or disease.

X

xenotransplantation Organ, tissue or cell transplant from one species to another, such as from pig into human.

xerophyte A plant adapted to dry conditions.

X-linked Resulting from the inheritance of a gene on the X chromosome. An X-linked trait may be inherited from either parent, because both have an X chromosome.

xylem The tissue in vascular plants that transports water and nutrients upwards from the roots. It consists of hollow chains of dead cells.

Y

Y-linked Resulting from the inheritance of a gene on the Y chromosome. A Y-linked trait is passed from father to son. It is never observed in females because they do not have a Y chromosome.

Z

zona pellucida Protective glycoprotein layer surrounding an egg cell in mammals.

zoonotic Infectious disease transmitted from non-human animal to human.

zoospore A spore of certain algae, fungi and protozoans, capable of swimming by means of a flagellum.

zygote The diploid cell resulting from the fusion of an egg and sperm. It is the first stage of the development of a unique new organism, occurring after fertilsation but before cell differentiation.

Index

Attributions

Cover image: Dr Keith Wheeler/Science Photo Library.

The following abbreviations are used in this list: t = top, b = bottom, l = left, r = right, c = centre.

123RF: pp. 38c, 329l, alila, pp. 91l , 117bl, 163br, 251; alycan, p. 394b; Parawat Isarangura Na Ayudhaya, p. 70r; Marilyn Barbone, p. 111; Miroslav Beneda, p. 351r; blueringmedia, p. 88; Daniel Bueno, p. 208c; Steven Coling, p. 464tr; luk Cox, p. 416; Nuwat Chanthachanthuek, pp. 214c, 214l; dbvirago, p. 351l; designua, pp. 191bl, 191br, 435t, 436b, 541l, 544r; elec, p. 187t; farinosa, p. 317tr; gozzoli, p. 100r; Jeff Grabert, p. 101tl; hxdbzxy, p. 106bl; lculig, p. 181; iqoncept, p. 288; ivolodina, p. 507b; Stuart Jenner, p. 507t; kasto, p. 322l; Sebastian Kaulitzki, p. 521; Valerii Kirsanov, p. 101bl; Ewelina Kowalska, p. 626; Vit Krajicek, p. 183; mitriy Krasko, p. 317br; Anton Lebedev, p. 182l; Dmitry Lobanov, p. 543c; Thi Hong Hanh Mac, p. 347t; Tatiana Makotra, p. 612b; molekuul, pp. 182r, 290; napat, p. 163bl; Jarun Ontakrai, p. 435b; parilovv, p. 520t; Denis and Yulia Pogostins, p. 58cr; Alexander Pokusay, pp. 599bl, 643; rioblanco, p. 456t; p. 460; rob3000, p. 116r; Sakarin Sawasdinaka, p. 208t; Anatoly Shevkunov, p. 607b; Ekaterina Smirnova, p. 210; Nico Smit, p. 84r; Meletios Verras, p. 163t; Vasiliy Vishnevskiy, p. 531 (snake); Jovanche Vitanovski, p. 327b; wavebreakmediamicro, p. 325t; Wong Sze Yuen, p. 220b; p. 228; Zlikovec, p. 62bl.

AAP: Australian Associated Press Pty Ltd, p. 193; AP Photo/Aqua Bounty Farms, p. 329r; Alan Porritt, p. 462; Brian Walker/AP, p. 9b.

Age Fotostock: Mint Frans Lanting, p. 150t.

Ahmed, Javed, Rajashree Khalap, & Sumukha J. N.: Discovery credit, p. 149.

Alamy Stock Photo: ASK Images, p. 489t; Bill Bachman, p. 487b; Franco Banfi/Steve Bloom Images, p. 67b; Amelie-Benoist/BSIP SA, pp. 301tr, 328t; B. Boissonnet/BSIP SA, p. 464bl; BSIP SA, pp. 458b, 564, 569r; Volodymyr Burdiak, p. 265bl; Marc Bruxelle, p. 311; Pat Canova, p. 202; Nigel Cattlin, pp. 379cb, 379t, 383; cbstockfoto, p. 523tr; Chronicle, pp. 217, 317tl, 457b; David Colbran, p. 483t; David Coleman, p. 551; Ashley Cooper pics, p. 471; James D Coppinger/Dembinsky Photo Associates, p. 540r; Gary Corbett/ age fotostock, p. 308l; Adrian Davies, p. 381br; DMA, p. 627r; Kathy deWitt, p. 608r; Stephen Dorey – Bygone Images, p. 488br; Dorling Kindersley Ltd, p. 342; dpa picture alliance archive, p. 321b; Mark Dunn, p. 458t; DUVAL/BSIP SA, p. 363; Encyclopaedia Britannica Inc. 2012/Universal Images Group North America LLC, p. 144r; Paul Fearn, p. 268b; Jean-Paul Ferrero/AUSCAPE, p. 81bl; Fraud/ BSIP/BSIP SA, p. 623br; Jason Freeman, p. 65tr; Juan Gaertner/ Science Photo Library, p. 432; GARO/PHANIE, pp. 298t, 625b; Jeff Foott/Nature Picture Library, p. 523bcr; Jon Frank, p. 483b; David R. Frazier Photolibrary, Inc., p. 308c; David Gee 4, p. 535t; GL Archive, pp. 324bl, 398tr; Bob Gibbons, p. 76bl; John Glover, p. 308r; Granger, NYC./Granger Historical Picture Archive, p. 267b; The Granger Collection, p. 333b; Manfred Grebler, p. 237t; Mike Greenslade/Australia, p. 343r; Steve Gschmeissner/Science Photo Library, pp. 62br, 428tcl, 547; Steve Horrell/Science Photo Library, p. 64cbl; Chris Howes/Wild Places Photography, p. 523cl; Wayne Hutchinson, pp. 343c, 343l, 366b; IanDagnall Computing, p. 256; Kaiser/Agencja Fotograficzna Caro, p. 451t; Karen Kasmauski/RGB Ventures /SuperStock, p. 326l; Mark Kerrison/Alamy Live News, p. 222t; Stephanie Jackson – Photographsofaustralia.com, p. 485t; Jacopin/BSIP/BSIP SA, p. 608c; Christopher Jones, p. 62tr; Laguna Design/Science Photo Library, pp. 37,188b; Laurent/BSIP/BSIP SA, p. 590; frans lemmens, p. 526b; Patti McConville, p. 320b; Colin McPherson, p. 350t; Mediscan, p. 307b; Ben Molyneux, p. 101r; MNS Photo, p. 527t; The Natural History Museum, p. 270r; Nature and Science, p. 386(nermatodes); Andrey Nekrasov, p. 265t; North Wind Picture Archives, p. 316; NZ Collection, p. 335b; Gerry Pearce, pp. 382t, 525t; Pedro Perez, p. 584l; Pictorial Press Ltd, pp. 109tl, 130l, 392br, 468; Tracy Powell/Glasshouse Images, p. 207; Photo Researchers/Science History Images, pp. 130r, 631; Robert Pickett, p. 346l; PJF Military Collection, p. 337b; Norbert Probst/imageBROKER, p. 266; William Radcliffe/RGB Ventures/ SuperStock, p. 400b; Alessandra Sarti/imageBROKER, p. 486b; Alec Scaresbrook, p. 333r; C.Slawik/juniors@wildlife/Juniors Bildarchiv GmbH, p. 237b; Martin Shields, pp. 241t, 261b, 357Tom Soucek/ Design Pics Inc, p. 265br; Lee Karen Stow, p. 478t; Krystyna Szulecka Photography, p. 489b; Ann and Steve Toon, p. 350b; Top Photo Corporation, p. 346r; TopPhoto/Alamy Live News, p. 457bl; Travelscape Images, p. 485b; Marli Wakeling, p. 67c; Chris Warham, p. 477; BJ Warnick/Newscom, p. 396; Ken Welsh/Vantage/Design Pics Inc, p. 639b; Jim West, p. 99br; WILDLIFE GmbH, pp. 64ctl, 81br 103t; Ray Wilson, p. 488bc; Tami Kauakea Winston/Photo Resource Hawaii, p. 347b; Martin Zwick/Danita Delimont, p. 263.

Stephanie Alexander Kitchen Garden Foundation: p. 584r.

Auscape International Photo Library: D. Parer & E. Parer-Cook, Minden Pictures, p. 87b.

Australian Institute of Health and Welfare: p. 563; Based on 'Cardiovascular deaths: trends (deaths per 100,000)', Data and figures for Australia's health 2012—in brief., p. 574; Based on Cancer compendium: information and trends by cancer type Web report, 30 Mar 2017. Accessed 8 October 2017, p. 577; Based on Cancer compendium: information and trends by cancer type Web report, 30 Mar 2017. Accessed 8 October 2017, p. 579b; Based on Cancer compendium: information and trends by cancer type Web report, 30 Mar 2017. Accessed 8 October 2017, p. 579t; Based on Cardiovascular health compendium Web report, 30 Mar 2017. Accessed 8 October 2017, p. 580.

Banwell, Tom: p. 398cl.

Biodiversity Heritage Library: p. 205l.

Botanic Gardens & State Herbarium of South Australia/ saseedbank.com.au: p. 488t.

Cancer Australia: 'Lung Cancer in our Mob'. This information is reproduced with permission from Cancer Australia. canceraustralia.gov.au, p. 561; Frances Andrijich, p. 6l; Adapted from OpenStax, Biology. OpenStax CNX. 10 Apr 2018 http://cnx.org/contents/185cbf87-c72e-48f5-b51e-f14f21b5eabd@11.1., p. 76br.

Center for Disease Control: Weekly US Map: Influenza Summary Update, week ending January 06, 2018, www.cdc.gov/flu/weekly/usmap.htm, p. 387b; Weekly US Map: Influenza Summary Update, week ending December 02, 2017, www.cdc.gov/flu/weekly/usmap.htm, p. 387c; Weekly US Map: Influenza Summary Update, week ending October 07, 2017, www.cdc.gov/flu/weekly/usmap.htm, p. 387t.

Center for Food Safety: Based on data from Genetically Engineered Food Labeling Laws map, www.centerforfoodsafety.org/ge-map (accessed 19 June, 2018), p. 322.

Corbis Images: p. 321t.

CSIRO: Electron Microscopy Unit AAHL, p. 453t.

Department of Health, NSW: Adapted from "Diphtheria in Australia, recent trends and future prevention strategies" by H. Gidding et. al., Communicable Diseases Intelligence, 24(6), 2000. © Commonwealth of Australia. Reproduced with permission., p. 492.

DK Images: Kevin Jones/Dorling Kindersley, p. 66tr.

Fleagle, John: p. 270l.

Fotolia: Jose Gil, p. 214r; Silvy K, p. 77cl; Kjersti, p. 71br; Kokhanchikov, pp. 84l, 113l; Kurhan, p. 298c; powell83, p. 297c; Sondem, p. 297t; tadamichi, p. 31t; Romolo Tavani, p. 298b; juanjo tugores, p. 360t; Edward Westmacott, p. 64tl; WoGi, p. 50; wongtanarak, p. 99tr.

Getty Images: Gregory Adams/Moment Open, p. 260; AJ PHOTO/ Science Photo Library, p. 254b; Auscape/Universal Images Group, p. 82t; Auscape/UIG/Universal Images Group, p. 381t; Robyn Beck/ AFP, p. 261t; Bettmann, pp. 5l, 391; Biology Media, p. 282; Volker Brinkmann/Visuals Unlimited, Inc, p. 197t; BSIP, pp. 63tr, 309; Camazine/Sue Trainor – Scott, p. 253tl; Garry DeLong/Science Source, p. 79br; David Doubilet/National Geographic, p. 82b; Encyclopedia Britannica/UIG, p. 546; Don Farrall/Photographer's Choice RF, p. 616bl; Maciej Frolow, p. 38b; Marc Hoberman/Corbis Documentary, p. 147t; Jacopin/BSIP/Corbis Documentary, p. 191cl; Akio Kon/ Bloomberg, p. 553b; Laguna Design/Science Photo Library, pp. 185t, 615b; Mint Images – Art Wolfe, p. 280t; Mint Images – Frans Lanting, p. 520c; Omikron Omikron/Science Source, pp. 156–7; Ralph Orlowski, p. 395t; p. 428bcl; Adam Pretty, p. 569l; Ed Reschke, pp. 28, 119 (all); Science Picture Co/Collection Mix: Subjects, p. 289; Science Source, p. 515t; Science & Society Picture Library, p. 128; Gail Shumway, p. 280b; Adrian T Sumner/The Image Bank, p. 286–7; Tim Vernon/ Science Photo LIbrary, p. 565l; Dr. Ralph Slepecky/Visuals Unlimited, p. 59; ZEPHYR/Brand X Pictures/ Science Photo Library, p. 225.

Kauter, Greg: courtesy of Cotton Australia, p. 358b.

Harvard University Press: From Systematics and the Origin of Species, from the Viewpoint of a Zoologist by Ernst Mayr, 1945, p. 145.

Institute for Health Metrics and Evaluation: Institute for Health Metrics and Evaluation (IHME). Overweight and Obesity Viz. Seattle, WA: IHME, University of Washington, 2017. Available from vizhub. healthdata.org/obesity (accessed 11 May 2018), p. 571.

Library of Alexandria: Augustin Pyramus de Candolle (1778–1841), Quoted in Darwinism (1889): An Exposition of the Theory of Natural Selection with some of its Applications, by Alfred Russel Wallace, 1901., p. 145.

Lochman Transparencies: p. 487t.

Low, Tim: p. 104r.

McGraw-Hill Australia Pty Ltd: Bruce Knox, Pauline Ladiges, Barbara Evans, Biology 5th Edition © 2016 McGraw-Hill Education, Australia. Reproduced with reproduced with permission of McGraw-Hill Education., p. 433.

McKenzie, John/Phil Batterham: p. 205r; p. 252.

Meagher, David: p. 74c; p. 75tr.

Bridgette Meinhold: The impossible bottle trick with pine cones by American artist, Bridgette Meinhold, p. 78br; From the research laboratory of Jonathan T. Butcher, PhD, Meinig School of Biomedical Engineering, Cornell University, USA, p. 9t.

Miyako, Eijiro: p. 366t.

Museums Victoria: Based on data from Lockwood, C. The Human Story. Museums Victoria, 2008., p. 26t.

Sumukha J N: image credit, p. 149.

National Human Genome Research Institute: Adapted from National Human Genome Research Institute, p. 168; p. 173; Darryl Leja, p. 253bl.

National Institutes of Health: National Heart, Lung and Blood Institute, pp. 233t, 233b.

Nature Picture Library: Jose B. Ruiz, p. 145b.

Nature Publishing Group: Based on data published in Fisher, M. C., Henk, D. A., Briggs, C. J., Brownstein, J. S., Madoff, L. C., McCraw, S. L., & Gurr, S. J. (2012). Emerging fungal threats to animal, plant and ecosystem health. Nature, 484(7393), 10.1038/nature10947. http://doi.org/10.1038/nature10947, p. 395b.

Nobel Committee for Physiology or Medicine: © The Nobel Committee for Physiology or Medicine. Illustrator: Annika Röhl, p. 6r.

Nutrition Australia: The Healthy Eating Pyramid, 2015 © Copyright The Australian Nutrition Foundation Inc., p. 558.

Pearson Education, Inc.: MARIEB, ELAINE N.; HOEHN, KATJA N., HUMAN ANATOMY & PHYSIOLOGY, 9th Ed., ©2013. Reprinted and Electronically reproduced by permission of Pearson Education, Inc., New York., p. 436t; REECE, JANE B.; URRY, LISA A.; CAIN, MICHAEL L.; WASSERMAN, STEVEN A.; MINORSKY, PETER V.; JACKSON, ROBERT B., CAMPBELL BIOLOGY, 10th Ed., ©2014. Reprinted and Electronically reproduced by permission of Pearson Education, Inc., New York, NY., pp. 188t, 189t, 355.

RCSB PDB: H.M. Berman et al. (2000) Nucleic Acids Research, 28: 235–242; www.rcsb.org; Molecule of the Month feature by David S. Goodsell. Licensed under CC BY 4.0 License [creativecommons. org/licenses/by/4.0/], p. 190.

Science Photo Library: pp. 19, 117tr, 303, 382b, 400t, 420–1, 444c, 445r, 449, 515b, 532–3, 537, 601; AMI Images, pp. 61br, 408b; AMI IMAGES/Dartmouth college – Louisa Howard, p. 389, 399b; Auscape Ferrero, p. 192; Peter Balint-Kurti/US Department of Agriculture, p. 410; Asklepios Medical Atlas, p. 444b; Barrington-Brown, Gonville and Caius College, p. 131l; John Bavosi, p. 603t; Juergen Berger, p. 399t; George Bernard/SPL, p. 64bl; Bernard Benoit, p. 81c; Thierry Berrod, Mona Lisa Production, p. 268t; Bildagentur-Online/Tschanz, p. 398bl; Biozentrum, University of Basel, pp. 284–5, 370–4; Ian Boddy, p. 539b; Dr Klaus Boller, p. 581; Ron Bonser, p. 407; Dr Tony Brain, p. 381c; Dr Goran Bredberg, pp. 605l, 605r; Massimo Brega, The Lighthouse, p. 359l; BSIP, Wladimir Bulgar, pp. 356, 372; John Bravos, p. 617; John Burbidge, p. 398tl; Dr Jeremy Burgess, pp. 122t, 262, 197b; Jose Calvo, pp. 499, 646–51; CDC, pp. 8, 384, 388r, 498b; Professor J.C. Chermann, p. 300t; Martyn F. Chillmaid, p. 222; Clouds Hill Imaging Ltd, p. 380bl; Michael Clutson, p. 201; CNRI, pp. 293l, 296; Herve Conge, p. 117br; Maurizio de Angelis, p. 566; Dept. of Clinical Cytogenetics, Addenbrookes Hospital, pp. 124t, 292r, 373; Dept. of Nuclear Medicine, Charing Cross Hospital, p. 634bl; Thomas Deerinck, p. 123; Martin Dohrn, p. 505; Laurent Douek/Look at Sciences, pp. 620bl, 620br, 620tr; A. Dowsett, Health Protection Agency, p. 388l, 498t; John Durham, p. 5c; Gunilla Elam, pp. 142, 602b; EM Unit, VLA, pp. 376–7; Eye of Science, pp. 71t, 75bl, 423b; Mauro Fermariello, pp. 253br, 555r; Simon Fraser, p. 604; Astrid & Hanns-Frieder, p. 634tl; James Gathany/CDC, p. 267r; Bob Gibbons, p. 65tl; Pascal Goetgheluck, pp. 272l, 301bl, 359r; Steve Gschmeissner, pp. 54–5, 114–5, 122b, 146, 277–83, 301tl, 414c, 414t, 423t, 427r, 428tcr, 428tl, 442, 444t, 565r, 613r, 613l; Prof. J.J. Hauw, p. 184r; Jan Hinsch, p. 61bcr; Ian Hooton, p. 508; Alex Hyde, p. 58c; Innerspace Imaging, p. 615tr; Jacopin, p. 506; Jacopin/ BSIP, pp. 609, 623tl; Cavallini James, p. 137b; Mikkel Juul Jensen, pp. 271, 523br; Manfred Kage, pp. 121t, 541tr; Kallista Images/ Custom Medical Stock Photo, pp. 176t, 189bl; Nancy Kedersha, p. 428br; Dr. Richard Kessel & Dr. Gene Shih, Visuals Unlimited, p. 61cl; James King-Holmes, pp. 253tr, 267cl, 543b, 608l; Dennis Kunkel Microscopy, pp. 61bl, 500–01, 538; Laguna Design, pp. 185br, 195, 358tl; Patrick Landmann, p. 267cr; Francis Leroy, Biocosmos, p. 281; Claus Lunau, p. 94tl; Dr. P. Marazzi, pp. 295, 365, 535b, 543t; Arno Massee, p. 241b; Dr Charles Mazel/Visuals Unlimited., INC, p. 360b; Tony McConnell, pp. 299b, 525b; Medical RF.Com, p. 220t; Peter Menzel, p. 250; Dr. Brad Mogen/Visuals Unlimited, p. 358tc;

Cordelia Molloy, p. 208b; Professor P.M. Motta, G. Macchiarelli, S.A Nottola, pp. 56–7; Dr Gopal Murti, pp. 159, 341, 403; NAID/National Institute of Health, p. 424t; Natural History Museum, London, p. 145t; Susumu Nishinaga, pp. 79tr, 515c, 594–5; Matthew Oldfield, p. 66bl; Joe Antonio Penas, p. 15; Alfred Pasieka, pp. 58l, 189br; Gerry Pearce, p. 83c; Phillipe Plailly, p. 358tr; P. Plailly/E. Daynes, p. 272r; Pr Philippe VAGO, ISM, pp. 292l, 294, 313b, 313t; Philippe Psaila, pp. 100tl, 249; Power and Syred, pp. 62tl, 74l, 118r, 526t; J.C. Revy, ISM, pp. 267l, 448; Alexis Rosenfeld, p. 87t; Rosenfeld Images Ltd., pp. 65b, 106br; SCI-COMM Studios, p. 7t; Science Pictures Limited, pp. 58r, 90; Science VU/B. John, Visuals Unlimited, p. 124b; Solvin Zankl/Visuals Unlimited, Inc., p. 204b; St Bartholomew Hospital, p. 428bcr; Sriram Subramaniam/National Institutes of Health, p. 428bl; Sinclair Stammers, p. 73t; Prof. Stanley N. Cohen, p. 300b; Dr Linda Stannard, UCT, pp. 375, 494–8; Claudia Stocker, p. 148t; Svalbard Global Seed Vault, The Lighthouse, p. 150b; TEK Image, pp. 2–3, 247; Wim Van Egmond, pp. 61bc, 73b, 112tr; Arie van 'T Riet, p. 519r; Veisland/Science Photo Library, p. 600l; Dr. E. Walker, p. 635r; Dr Keith Wheeler, p. i; .Dr John Zajicek, p. 544l; PR Michel Zanca, ISM, p. 305; Zephyr, pp. 254t, 301bc, 301br, 599br.

Shutterstock: p. 17; 18percentgrey, p. 327t; 269185, p. 336t; 270058, p. 43; 3000ad, p. 453c; Matusciac AJCespedes, p. 380c; AlessandroZocc, p. 76c; Alexander_P, p. 644r; Alexandru, p. 361t; Alila Medical Media, pp. 83b, 154, 177, 424br, 496l, 539t, 541br, 545b, 602t, 603bl, 603br, 616tl, 623tr, 624, 625t, 633t; All-stock-photos, p. 144tl; almaje, p. 545t; Protasov AN, p. 347tc; Marochkina Anastasiia, p. 586; ancroft, p. 345; Gudkov Andrey, p. 68; Ilya Andriyanov, p. 456b; Medvedev Andrey, p. 105br; Romanova Anna, p. 97; APCat, p. 144cl; ArCaLu, p. 522t; Anest, p. 76t; Artwork studio BKK, p. 597; Asholove, p. 411l; Sarine Arslanian, p. 102r; Australis Photography, p. 567; Awe Inspiring Images, p. 131r; Kitch Bain, p. 264; SK Bakker, p. 333c; marilyn barbone, p. 175; bart78, p. 94bl; Yuriy Bartenev, p. 144bl; Mrs_Bazilio, p. 621b; belizar, p. 536; Uwe Bergwitz, p. 83t; Zsolt Biczo, p. 91r; Blamb, p. 638l, 638r; Bos11, p. 147b; BlueRingMedia, pp. 78, 112l; BMCL, p. 87c; Bork, p. 456t; branislavpudar, p. 99l; Steve Byland, p. 519l; leonello calvetti, p. 336b; Jose Luis Calvo, p. 637; John Carnemolla, p. 105tr; carroteater, p. 379b; catolla, p. 5r; Chaiyabutra, p. 527b; Pavel Chagochkin, p. 327c; chiqui, p. 118l; Katarina Christenson, p. 386(lerps); corners74, p. 104bl; Tony Craddock, p. 74r; andras_csontos, p. 607t; Jim Cumming, p. 522b; Damerau, p. 248; DashaR, p. 610; julie deshaies, p. 102tl; Designua, pp. 75c, 129, 141r, 155, 158, 184l, 299t, 406, 425, 439, 615l, 618r; Dimarion, p. 137t; Mary Dimitropoulou, p. 526c; Phill Doherty, p. 378; ellepigrafica, pp. 427l, 431; Erni, p. 204tr; extender_01, p. 191cr; EcoPrint, p. 72l; Ezume Images, p. 386(slugs); Festa, p. 553t; Svetlana Foote, p. 335t; Martin Fowler, p. 66tl; paula french, p. 524; Lukiyanova Natalia frenta, p. 116l; Peter Hermes Furian, p. 616cr; Juan Gaertner, pp. 80t, 437; Gentoo Multimedia Limited, p. 71bl; gillmar, p. 100bl; Oleg Golovnev, p. 392bl; Dragana Gordic, p. 488bl; Grandpa, p. 109r; GTS Productions, p. 575; Guitar photographer, p. 452b; Jubal Harshaw, p. 75br; Damian Herde, p. 380br; igorstevanovic, p. 325b; iofoto, p. 445l; Eric Isselee, pp. 401, 531(bobcat); Tania Izquierdo, p. 369; Ammit Jack, p. 347bc; jsp, p. 317bl; jamestorm, p. 540l; joloei, p. 633b; joshya, p. 629; Jun MT, p. 392l; Raisa Kanareva, p. 361b; Kateryna Kon, p. 414b; Matej Kastelic, p. 60r; Sebastian Kaulitzki, p. 454; khlungcenter, p. 480; Kichigin, pp. 451b, 63tl; Kjuuurs, p. 81bc; Kletr, p. 397; Georgios Kollidas, p. 461; Levent Konuk, p. 472; Kzenon, p. 307t; D. Kucharski, K. Kucharska, p. 381bl; KuLouKu, p. 60l; Laborant, p. 520b; l i g h t p o e t, p. 178l; neil langan, p. 104tl; Johan Larson, p. 109bl; Henrik Larsson, p. 386(borers); Karin Hildebrand Lau, p. 464tc; Lebendkulturen.de, p. 58cl, 67t; Lightspring, p. 328bl; Gang Liu, p. 344r; Evan Lorne, p. 636; magnetix, pp. 187b, 64br; Kazakov Maksim, p. 64cr; margouillat photo, p. 297b; mangostock, p. 470; Marten_House, p. 178r; Matteo photos, p. 81t; Brian Maudsley, p. 77tr; Riccardo Mayer, p. 554; mejnak, p. 334br; Melory, p. 237c; Millenius, p. 394t; misolasop, p. 452t; molekuul_be, p. 185bl, 198; Monkey Business Images, pp. 320t, 324t, 599t; Christian Mueller, pp. 450, 479; Naeblys, p. 160b; Olga Nayashkova, p. 318b; nd3000, p. 583; Neokryuger, p. 622b; natthawut ngoensanthia, p. 64tr; niderlander, p. 326r; Gerasymovych Oleksandr, p. 148b; Vasik Olga, p. 82c; Maciej Olszewski, p. 80b; Sari ONeal, p. 112br; OnTheCoastPhotography, p. 79l; Jarun Ontakrai, p. 51; Martin Pateman, p. 333l; Heiti Paves, p. 324br; petarg, p. 141l; Anders Peter Photography, p. 204tl; Photodynamic., p. 466b; Mr. Aukid Phumsirichat, p. 386 (aphids); Picsfive, p. 639t; piotr.ma, p. 63b; polarman, p. 337t; Rhoeo, p. 151t; picturepartners, pp. 328br, 486t; RAJ CREATIONZS, p. 349; Alexander Raths, pp. 458c, 587; Valentina Razumova, p. 94br; Ian D M Robertson, p. 151b; Leena Robinson, p. 540c; Rogatnykh, p. 105bl; Armin Rose, p. 334bl; RossHelen, p. 318t; Federico Rostagno, p. 465; RTimages, p. 614; saravutpics, p. 315; Vladimir Sazonov, p. 622tr; science photo, p. 176b; sciencepics, p. 600r; sergey23, p. 334t; SL-Photography, p. 618l; snapgalleria, p. 644l; Heather Lucia Snow, p. 77cr; June Marie Sobrito, p. 622tc; stoonn, p. 103b; songwut tanoi, p. 105bc; Studio 72, p. 612t; Suwin, p. 89; Swapan Photography, p. 113r; tristan tan, p. 627l; Tefi, pp. 96, 160t, 634r; testing, p. 585; Rattiya Thongdumhyu, p. 61tcr; toeytoey, p. 428tr; Timonina, pp. 174, 424bl, 620tl, 621t; topimages, p. 411r; udaix, p. 121b; UMB-O, p. 408t; udaix, p. 619l; unoL, p. 241c; Sergey Uryadnikov, p. 478b; Jiri Vaclavek, p. 93b; VanderWolf Images, p. 466t; vchal, p. 635l; Vecton, p. 120; VectorPot, p. 93t; venars.original, p. 527c; Vilor, p. 380t; vitstudio, p. 459; vladsilver, p. 72r; Jana Vodickova, p. 619r; wavebreakmedia, p. 304; WilleeCole, p. 105cr; Wire_man, p. 61tr; withGod, p. 622tl; Vladimir Wrangel, p. 523tcr; yuris, pp. 102bl, 293r; Igor Zakowski, p. 344l; Serg Zastavkin, p. 106t; Jun Zhang, p. 72c.

Sinauer Associates: Based on data from Cooper, G. M. The Cell: A Molecular Approach, 2nd edition. Sunderland (MA): Sinauer Associates; 2000., p. 25c.

U.S. Food and Drug Administration: Based on Guidance for Industry: Clinical Studies Section of Labeling for Human Prescription Drug and Biological Products – Content and Format, January 23, 2006 (accessed 30 April 2018), www.fda.gov/RegulatoryInformation/Guidances/ucm127509.htm#f1, p. 32b.

Vigan, Tyler: Spurious Correlations, tylervigan.com. Data sources: National Science Foundation and Dept. of Energy, p. 31b.

World Health Organization: Reprinted from Projected changes in malaria incidence rates, by country, 2000–2015. Global Malaria Programme © WHO 2015, p. 482; World Malaria Report 2016, Annex 2 – F. Regional profile: South-East Asia Region, published 13 December 2016., p. 496r; Based on Mortality and Burden of Disease estimates for WHO member states in 2002, published December 2004, p. 555l; Based on Global Health Observatory (GHO) data, Australia: country profiles. WHO Chronic Diseases and Health Promotion database, p. 560; Based on Global Health Observatory (GHO) data, Noncommunicable diseases country profile – Australia, p. 562; Based on Fact sheet: The top 10 causes of death, WHO © 2015, p. 573b; Based on Fact sheet: The top 10 causes of death, WHO © 2015, p. 573t; Based on Global Health Observatory (GHO) data, Australia: country profiles. WHO Chronic Diseases and Health Promotion database, p. 576.

Wikimedia Commons: Alexbateman, p. 7b; Thomas Splettstoesser, p. 473.